D0792998

A History of Mathematics

An Introduction

A History of Mathematics

An Introduction

Third Edition

Pearson Modern Classic

Victor J. Katz
University of the District of Columbia

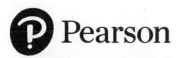 Pearson

Editor in Chief: Deirdre Lynch
Senior Acquisitions Editor:William Hoffman
Executive Project Manager: Christine O'Brien
Project Editor: Elizabeth Bernardi
Associate Editor: Caroline Celano
Senior Managing Editor: Karen Wernholm
Senior Production Supervisor: Tracy Patruno
Marketing Manager: Katie Winter
Marketing Assistant: Jon Connelly
Senior Prepress Supervisor: Caroline Fell
Manufacturing Manager: Evelyn Beaton
Production Coordination, Composition, and Illustrations:Windfall Software, using ZzTeX
Senior Designer: Barbara T. Atkinson
Text and Cover Design: Leslie Haimes
Cover photo: Tycho Brahe and Others with Astronomical Instruments, 1587, "Le Quadran Mural"
1663. Blaeu, Joan (1596–1673 Dutch). Newberry Library, Chicago, Illinois, USA © Newberry
Library/SuperStock.

Copyright © 2018, 2009 by Pearson Education, Inc. All Rights Reserved. Printed in the United States of America.
This publication is protected by copyright, and permission should be obtained from the publisher prior to any
prohibited reproduction, storage in a retrieval system, or transmission in any form or by any means, electronic,
mechanical, photocopying, recording, or otherwise. For information regarding permissions, request forms and
the appropriate contacts within the Pearson Education Global Rights & Permissions department, please visit
www.pearsoned.com/permissions/.

PEARSON is an exclusive trademark owned by Pearson Education, Inc. or its affiliates in the U.S. and/or other
countries.

Unless otherwise indicated herein, any third-party trademarks that may appear in this work are the property
of their respective owners and any references to third-party trademarks, logos or other trade dress are for
demonstrative or descriptive purposes only. Such references are not intended to imply any sponsorship,
endorsement, authorization, or promotion of Pearson's products by the owners of such marks, or any
relationship between the owner and Pearson Education, Inc. or its affiliates, authors, licensees or distributors.

Library of Congress Cataloging-in-Publication Data

Names: Katz, Victor J.
Title: A history of mathematics : an introduction / Victor J. Katz,
  University of the District of Columbia.
Description: Third edition [2018 reissue]. | New York, NY : Pearson, [2018] |
  Series: Pearson modern classic | Originally published in 2009, reissued as
  part of Pearson's modern classic series. | Includes bibliographical
  references and index.
Identifiers: LCCN 2016055054| ISBN 9780134689524 (pbk.) | ISBN 0134689526
  (pbk.)
Subjects: LCSH: Mathematics--History.
Classification: LCC QA21 .K33 2018 | DDC 510.9--dc23
LC record available at https://lccn.loc.gov/2016055054

 **Pearson**

ISBN 13: 978-0-13-468952-4
ISBN 10:   0-13-468952-6

Contents

PART TWO   *Medieval Mathematics*

PART THREE   *Early Modern Mathematics*

Preface

In *A Call For Change: Recommendations for the Mathematical Preparation of Teachers of Mathematics*, the Mathematical Association of America's (MAA) Committee on the Mathematical Education of Teachers recommends that all prospective teachers of mathematics in schools

> ... develop an appreciation of the contributions made by various cultures to the growth and development of mathematical ideas; investigate the contributions made by individuals, both female and male, and from a variety of cultures, in the development of ancient, modern, and current mathematical topics; [and] gain an understanding of the historical development of major school mathematics concepts.

According to the MAA, knowledge of the history of mathematics shows students that mathematics is an important human endeavor. Mathematics was not discovered in the polished form of our textbooks, but was often developed in an intuitive and experimental fashion in order to solve problems. The actual development of mathematical ideas can be effectively used in exciting and motivating students today.

This textbook grew out of the conviction that both prospective school teachers of mathematics and prospective college teachers of mathematics need a background in history to teach the subject more effectively. It is therefore designed for junior or senior mathematics majors who intend to teach in college or high school, and it concentrates on the history of those topics typically covered in an undergraduate curriculum or in elementary or high school. Because the history of any given mathematical topic often provides excellent ideas for teaching the topic, there is sufficient detail in each explanation of a new concept for the future (or present) teacher of mathematics to develop a classroom lesson or series of lessons based on history. In fact, many of the problems ask readers to develop a particular lesson. My hope is that students and prospective teachers will gain from this book a knowledge of how we got here from there, a knowledge that will provide a deeper understanding of many of the important concepts of mathematics.

DISTINGUISHING FEATURES

FLEXIBLE ORGANIZATION

Although the text's chief organization is by chronological period, the material is organized topically within each period. By consulting the detailed subsection headings, the reader can choose to follow a particular theme throughout history. For example, to study equation solving one could consider ancient Egyptian and Babylonian methods, the geometrical solution methods of the Greeks, the numerical methods of the Chinese, the Islamic solution methods for cubic equations by use of conic sections, the Italian discovery of an algorithmic solution of cubic and quartic equations, the work of Lagrange in developing criteria for methods of solution of higher degree polynomial equations, the work of Gauss in solving cyclotomic equations, and the work of Galois in using permutations to formulate what is today called Galois theory.

FOCUS ON TEXTBOOKS

It is one thing to do mathematical research and discover new theorems and techniques. It is quite another to elucidate these in such a way that others can learn them. Thus, in many chapters there is a discussion of one or more important texts of the time. These are the works from which students learned the important ideas of the great mathematicians. Today's students will see how certain topics were treated and will be able to compare these treatments to those in current texts and see the kinds of problems students of years ago were expected to solve.

APPLICATIONS OF MATHEMATICS

Two chapters, one for the Greek period and one for the Renaissance, are devoted entirely to mathematical methods, the ways in which mathematics was used to solve problems in other areas of study. A major part of both chapters deals with astronomy since in ancient times astronomers and mathematicians were usually the same people. To understand a substantial part of Greek mathematics, it is crucial also to understand the Greek model of the heavens and how mathematics was used in applying this model to give predictions. Similarly, I discuss the Copernicus-Kepler model of the heavens and consider how mathematicians of the Renaissance applied mathematics to its study. I also look at the applications of mathematics to geography during these two time periods.

NON-WESTERN MATHEMATICS

A special effort has been made to consider mathematics developed in parts of the world other than Europe. Thus, there is substantial material on mathematics in China, India, and the Islamic world. In addition, Chapter 11 discusses the mathematics of various other societies around the world. Readers will see how certain mathematical ideas have occurred in many places, although not perhaps in the context of what we in the West call "mathematics."

TOPICAL EXERCISES

Each chapter contains many exercises, organized in order of the chapter's topics. Some exercises are simple computational ones, while others help to fill gaps in the mathematical arguments presented in the text. **For Discussion** exercises are open-ended questions, which may involve some research to find answers. Many of these ask students to think about how they would use historical material in the classroom. Even if readers do not attempt many of the exercises, they should at least read them to gain a fuller understanding of the material of the chapter. (Answers to the odd numbered computational problems as well as some odd numbered "proof" problems are included at the end of the book.)

FOCUS ESSAYS

Biographies   For easy reference, many biographies of the mathematicians whose work is discussed are in separate boxes. Although women have for various reasons not participated in large numbers in mathematical research, biographies of several important women mathematicians are included, women who succeeded, usually against heavy odds, in contributing to the mathematical enterprise.

Special Topics   Sidebars on special topics also appear throughout the book. These include such items as a treatment of the question of the Egyptian influence on Greek mathematics, a discussion of the idea of a function in the work of Ptolemy, a comparison of various notions of continuity, and several containing important definitions collected together for easy reference.

ADDITIONAL PEDAGOGY

At the start of each chapter is a relevant quotation and a description of an important mathematical "event." Each chapter also contains an annotated list of references to both primary and secondary sources from which students can obtain more information. Given that a major audience for this text is prospective teachers of secondary or college-level mathematics, I have provided an appendix giving suggestions for using the text material in teaching mathematics. It contains a detailed list to correlate the history of various topics in the secondary and college curriculum to sections in the text; there are suggestions for organizing some of this material for classroom use; and there is a detailed time line that helps to relate the mathematical discoveries to other events happening in the world. On the back inside cover there is a chronological listing of most of the mathematicians discussed in the book. Finally, given that students may have difficulty pronouncing the names of some mathematicians, the index has a special feature: a phonetic pronunciation guide.

PREREQUISITES

A knowledge of calculus is sufficient to understand the first 16 chapters of the text. The mathematical prerequisites for later chapters are somewhat more demanding, but the various section titles indicate clearly what kind of mathematical knowledge is required. For example, a full understanding of chapters 19 and 21 will require that students have studied abstract algebra.

COURSE FLEXIBILITY

The text contains more material than can be included in a typical one-semester course in the history of mathematics. In fact, it includes adequate material for a full year course, the first half being devoted to the period through the invention of calculus in the late seventeenth century and the second half covering the mathematics of the eighteenth, nineteenth, and twentieth centuries. However, for those instructors who have only one semester, there are several ways to use this book. First, one could cover most of the first twelve chapters and simply conclude with calculus. Second, one could choose to follow one or two particular themes through history. (The table in the appendix will direct one to the appropriate sections to include when dealing with a particular theme.) Among the themes that could be followed are equation solving; ideas of calculus; concepts of geometry; trigonometry and its applications to astronomy and surveying; combinatorics, probability, and statistics; and modern algebra and number theory. For a thematic approach, I would suggest making every effort to include material on mathematics in the twentieth century, to help students realize that new mathematics is continually being discovered. Finally, one could combine the two approaches and cover ancient times chronologically, and then pick a theme for the modern era.

NEW FOR THIS EDITION

The generally friendly reception of this text's first two editions encouraged me to maintain the basic organization and content. Nevertheless, I have attempted to make a number of improvements, both in clarity and in content, based on comments from many users of those editions as well as new discoveries in the history of mathematics that have appeared in the recent literature. To make the book somewhat easier to use, I have reorganized some material into shorter chapters. There are minor changes in virtually every section, but the major changes from the second edition include: new material about Archimedes discovered in analyzing the palimpsest of the *Method;* a new section on Ptolemy's *Geography;* more material in the Chinese, Indian, and Islamic chapters based on my work on the new *Sourcebook* dealing with the mathematics of these civilizations, as well as the ancient Egyptian and Babylonian ones; new material on statistics in the nineteenth and twentieth centuries; and a description of the eighteenth-century translation into the differential calculus of some of Newton's work in the *Principia.* The text concludes with a brief description of the solution to the first Clay Institute problem, the Poincarè conjecture. I have attempted to correct all factual errors from the earlier editions without introducing new ones, yet would appreciate notes from anyone who discovers any remaining errors. New problems appear in every chapter, some of them easier ones, and references to the literature have been updated wherever possible. Also, a few new stamps were added as illustrations. One should note, however, that any portraits on these stamps—or indeed elsewhere—purporting to represent mathematicians before the sixteenth century are fictitious. There are no known representations of any of these people that have credible evidence of being authentic.

ACKNOWLEDGMENTS

Like any book, this one could not have been written without the help of many people. The following people have read large sections of the book at my request and have offered many valuable suggestions: Marcia Ascher (Ithaca College); J. Lennart Berggren (Simon Fraser University); Robert Kreiser (A.A.U.P.); Robert Rosenfeld (Nassau Community College); John Milcetich (University of the District of Columbia); Eleanor Robson (Cambridge University); and Kim Plofker (Brown University). In addition, many people made detailed suggestions for the second and third editions. Although I have not followed every suggestion, I sincerely appreciate the thought they gave toward improving the book. These people include Ivor Grattan-Guinness, Richard Askey, William Anglin, Claudia Zaslavsky, Rebekka Struik, William Ramaley, Joseph Albree, Calvin Jongsma, David Fowler, John Stillwell, Christian Thybo, Jim Tattersall, Judith Grabiner, Tony Gardiner, Ubi D'Ambrosio, Dirk Struik, and David Rowe. My heartfelt thanks to all of them.

The many reviewers of sections of the manuscript for each of the editions have also provided great help with their detailed critiques and have made this a much better book than it otherwise could have been. For the first edition, they were Duane Blumberg (University of Southwestern Louisiana); Walter Czarnec (Framington State University); Joseph Dauben (Lehman College–CUNY); Harvey Davis (Michigan State University); Joy Easton (West Virginia University); Carl FitzGerald (University of California, San Diego); Basil Gordon (University of California, Los Angeles); Mary Gray (American University); Branko Grunbaum (University of Washington); William Hintzman (San Diego State University); Barnabas Hughes (California State University, Northridge); Israel Kleiner (York University); David E. Kullman (Miami University); Robert L. Hall (University of Wisconsin, Milwaukee); Richard Marshall (Eastern Michigan University); Jerold Mathews (Iowa State University); Willard Parker (Kansas State University); Clinton M. Petty (University of Missouri, Columbia); Howard Prouse (Mankato State University); Helmut Rohrl (University of California, San Diego); David Wilson (University of Florida); and Frederick Wright (University of North Carolina at Chapel Hill).

For the second edition, the reviewers were Salvatore Anastasio (State University of New York, New Paltz); Bruce Crauder (Oklahoma State University); Walter Czarnec (Framingham State College); William England (Mississippi State University); David Jabon (Eastern Washington University); Charles Jones (Ball State University); Michael Lacey (Indiana University); Harold Martin (Northern Michigan University); James Murdock (Iowa State University); Ken Shaw (Florida State University); Svere Smalo (University of California, Santa Barbara); Domina Eberle Spencer (University of Connecticut); and Jimmy Woods (North Georgia College).

For the third edition, the reviewers were Edward Boamah (Blackburn College); Douglas Cashing (St. Bonaventure University); Morley Davidson (Kent State University); Martin J. Erickson (Truman State University); Jian-Guo Liu (University of Maryland); Warren William McGovern (Bowling Green State University); Daniel E. Otero (Xavier University); Talmage James Reid (University of Mississippi); Angelo Segalla (California State University, Long Beach); Lawrence Shirley (Towson University); Agnes Tuska (California State University at Fresno); Jeffrey X. Watt (Indiana University–Purdue University Indianapolis).

I have also benefited greatly from conversations with many historians of mathematics at various forums, including numerous history sessions at the annual joint meetings of the Mathematical Association of America and the American Mathematical Society. Among those who have helped at various stages (and who have not been mentioned earlier) are V. Frederick Rickey, United States Military Academy; Florence Fasanelli, AAAS; Israel Kleiner, York University; Abe Shenitzer, York University; Frank Swetz, Pennsylvania State University, and Janet Beery, University of Redlands. In addition, I want to thank Karen Dee Michalowicz, of the Langley School, who showed me how to reach current and prospective high school teachers, and whose untimely death in 2006 was such a tragedy. In addition, I learned a lot from all the people who attended various sessions of the Institute in the History of Mathematics and Its Use in Teaching, as well as members of the 2007 PREP workshop on Asian mathematics. My students in History of Mathematics classes (and others) at the University of the District of Columbia have also helped me clarify many of my ideas. Naturally, I welcome any additional comments and correspondence from students and colleagues elsewhere in an effort to continue to improve this book.

My former editors at Harper Collins, Steve Quigley, Don Gecewicz, and George Duda, who helped form the first edition, and Jennifer Albanese, who was the editor for the second edition, were very helpful. And I want to particularly thank my new editor, Bill Hoffman, for all his suggestions and his support during the creation of both the brief edition and this new third edition. Elizabeth Bernardi at Pearson Addison-Wesley has worked hard to keep me on deadline, and Jean-Marie Magnier has caught several errors in the answers to problems, for which I thank her. The production staff of Paul C. Anagnostopoulos, Jennifer McClain, Laurel Muller, Yonie Overton, and Joe Snowden have cheerfully and efficiently handled their tasks to make this book a reality.

Lastly, I want to thank my wife Phyllis for all her love and support over the years, during the very many hours of working on this book and, of course, during the other hours as well.

Victor J. Katz
Silver Spring, MD
May 2008

Egypt and Mesopotamia

Accurate reckoning. The entrance into the knowledge of all existing things and all obscure secrets.

—Introduction to *Rhind Mathematical Papyrus*[1]

Mesopotamia: In a scribal school in Larsa some 3800 years ago, a teacher is trying to develop mathematics problems to assign to his students so they can practice the ideas just introduced on the relationship among the sides of a right triangle. The teacher not only wants the computations to be difficult enough to show him who really understands the material but also wants the answers to come out as whole numbers so the students will not be frustrated. After playing for several hours with the few triples (a, b, c) of numbers he knows that satisfy $a^2 + b^2 = c^2$, a new idea occurs to him. With a few deft strokes of his stylus, he quickly does some calculations on a moist clay tablet and convinces himself that he has discovered how to generate as many of these triples as necessary. After organizing his thoughts a bit longer, he takes a fresh tablet and carefully records a table listing not only 15 such triples but also a brief indication of some of the preliminary calculations. He does not, however, record the details of his new method. Those will be saved for his lecture to his colleagues. They will then be forced to acknowledge his abilities, and his reputation as one of the best teachers of mathematics will spread throughout the entire kingdom. ◾

The opening quotation from one of the few documentary sources on Egyptian mathematics and the fictional story of the Mesopotamian scribe illustrate some of the difficulties in giving an accurate picture of ancient mathematics. Mathematics certainly existed in virtually every ancient civilization of which there are records. But in every one of these civilizations, mathematics was in the domain of specially trained priests and scribes, government officials whose job it was to develop and use mathematics for the benefit of that government in such areas as tax collection, measurement, building, trade, calendar making, and ritual practices. Yet, even though the origins of many mathematical concepts stem from their usefulness in these contexts, mathematicians always exercised their curiosity by extending these ideas far beyond the limits of practical necessity. Nevertheless, because mathematics was a tool of power, its methods were passed on only to the privileged few, often through an oral tradition. Hence, the written records are generally sparse and seldom provide much detail.

In recent years, however, a great deal of scholarly effort has gone into reconstructing the mathematics of ancient civilizations from whatever clues can be found. Naturally, all scholars do not agree on every point, but there is enough agreement so that a reasonable picture can be presented of the mathematical knowledge of the ancient civilizations in Mesopotamia and Egypt. We begin our discussion of the mathematics of each of these civilizations with a brief survey of the underlying civilization and a description of the sources from which our knowledge of the mathematics is derived.

 1.1

EGYPT

Agriculture emerged in the Nile Valley in Egypt close to 7000 years ago, but the first dynasty to rule both Upper Egypt (the river valley) and Lower Egypt (the delta) dates from about 3100 BCE. The legacy of the first pharaohs included an elite of officials and priests, a luxurious court, and for the kings themselves, a role as intermediary between mortals and gods. This role fostered the development of Egypt's monumental architecture, including the pyramids, built as royal tombs, and the great temples at Luxor and Karnak. Writing began in Egypt at about this time, and much of the earliest writing concerned accounting, primarily of various types of goods. There were several different systems of measuring, depending on the particular goods being measured. But since there were only a limited number of signs, the same signs meant different things in connection with different measuring systems. From the beginning of Egyptian writing, there were two styles, the hieroglyphic writing for monumental inscriptions and the hieratic, or cursive, writing, done with a brush and ink on papyrus. Greek domination of Egypt in the centuries surrounding the beginning of our era was responsible for the disappearance of both of these native Egyptian writing forms. Fortunately, Jean Champollion (1790–1832) was able to begin the process of understanding Egyptian writing early in the nineteenth century through the help of a multilingual inscription—the Rosetta stone—in hieroglyphics and Greek as well as the later demotic writing, a form of the hieratic writing of the papyri (Fig. 1.1).

FIGURE 1.1

Jean Champollion and a piece of the Rosetta stone

It was the scribes who fostered the development of the mathematical techniques. These government officials were crucial to ensuring the collection and distribution of goods, thus helping to provide the material basis for the pharaohs' rule (Fig. 1.2). Thus, evidence for the techniques comes from the education and daily work of the scribes, particularly as related in

FIGURE 1.2

Amenhotep, an Egyptian high official and scribe (fifteenth century BCE)

two papyri containing collections of mathematical problems with their solutions, the *Rhind Mathematical Papyrus*, named for the Scotsman A. H. Rhind (1833–1863) who purchased it at Luxor in 1858, and the *Moscow Mathematical Papyrus*, purchased in 1893 by V. S. Golenishchev (d. 1947) who later sold it to the Moscow Museum of Fine Arts. The former papyrus was copied about 1650 BCE by the scribe A'h-mose from an original about 200 years older and is approximately 18 feet long and 13 inches high. The latter papyrus dates from roughly the same period and is over 15 feet long, but only some 3 inches high. Unfortunately, although a good many papyri have survived the ages due to the generally dry Egyptian climate, it is the case that papyrus is very fragile. Thus, besides the two papyri mentioned, only a few short fragments of other original Egyptian mathematical papyri are still extant.

These two mathematical texts inform us first of all about the types of problems that needed to be solved. The majority of problems were concerned with topics involving the administration of the state. That scribes were occupied with such tasks is shown by illustrations found on the walls of private tombs. Very often, in tombs of high officials, scribes are depicted working together, probably in accounting for cattle or produce. Similarly, there exist three-dimensional models representing such scenes as the filling of granaries, and these scenes always include a scribe to record quantities. Thus, it is clear that Egyptian mathematics was developed and practiced in this practical context.

One other area in which mathematics played an important role was architecture. Numerous remains of buildings demonstrate that mathematical techniques were used both in their design and construction. Unfortunately, there are few detailed accounts of exactly how the mathematics was used in building, so we can only speculate about many of the details. We deal with a few of these ideas below.

1.1.1 Number Systems and Computations

The Egyptians developed two different number systems, one for each of their two writing styles. In the hieroglyphic system, each of the first several powers of 10 was represented by a different symbol, beginning with the familiar vertical stroke for 1. Thus, 10 was represented by ∩, 100 by ⟨symbol⟩, 1000 by ⟨symbol⟩, and 10,000 by ⟨symbol⟩ (Fig. 1.3). Arbitrary whole numbers were then represented by appropriate repetitions of the symbols. For example, to represent 12,643 the Egyptians would write

$$\text{⟨hieroglyphic numerals⟩}.$$

(Note that the usual practice was to put the smaller digits on the left.)

FIGURE 1.3

Egyptian numerals on the Naqada tablets (c. 3000 BCE)

The hieratic system, in contrast to the hieroglyphic, is an example of a ciphered system. Here each number from 1 to 9 had a specific symbol, as did each multiple of 10 from 10 to 90 and each multiple of 100 from 100 to 900, and so on. A given number, for example, 37, was written by putting the symbol for 7 next to that for 30. Since the symbol for 7 was 𝓁 and that for 30 was 𝟃, 37 was written 𝓁𝟃. Again, since 3 was written as 𝖑𝖑𝖑, 40 as ⟂, and 200 as ⤴, the symbol for 243 was 𝖑𝖑𝖑 ⟂ ⤴. Although a zero symbol is not necessary in a ciphered system, the Egyptians did have such a symbol. This symbol does not occur in the mathematical papyri, however, but in papyri dealing with architecture, where it is used to denote the bottom leveling line in the construction of a pyramid, and accounting, where it is used in balance sheets to indicate that the disbursements and income are equal.[2]

Once there is a system of writing numbers, it is only natural that a civilization devise algorithms for computation with these numbers. For example, in Egyptian hieroglyphics, addition and subtraction are quite simple: combine the units, then the tens, then the hundreds, and so on. Whenever a group of ten of one type of symbol appears, replace it by one of the next. Hence, to add 783 and 275,

put ⫿⫿⫿∩∩∩∩∩∩∩∩999 999   and   ⫿⫿∩∩∩∩∩99 99   together to get   ⫿⫿⫿∩∩∩∩∩∩∩∩∩∩99999 9999 .

Since there are fifteen ∩'s, replace ten of them by one 9. This then gives ten of the latter. Replace these by one 𝑥. The final answer is

⫿⫿⫿∩∩∩𝑥 ,

or 1058. Subtraction is done similarly. Whenever "borrowing" is needed, one of the symbols would be converted to ten of the next lower symbol. Such a simple algorithm for addition and subtraction is not possible in the hieratic system. Probably, the scribes simply memorized basic addition tables.

The Egyptian algorithm for multiplication was based on a continual doubling process. To multiply two numbers a and b, the scribe would first write down the pair 1, b. He would then double each number in the pair repeatedly, until the next doubling would cause the first element of the pair to exceed a. Then, having determined the powers of 2 that add to a, the scribe would add the corresponding multiples of b to get his answer. For example, to multiply 12 by 13, the scribe would set down the following lines:

$$\begin{array}{ll} `1 & 12 \\ 2 & 24 \\ `4 & 48 \\ `8 & 96 \end{array}$$

At this point he would stop because the next doubling would give him 16 in the first column, which is larger than 13. He would then check off those multipliers that added to 13, namely, 1, 4, and 8, and add the corresponding numbers in the other column. The result would be written as follows: Totals   13   156.

There is no record of how the scribe did the doubling. The answers are simply written down. Perhaps the scribe had memorized an extensive two times table. In fact, there is some evidence that doubling was a standard method of computation in areas of Africa to the south of Egypt and that therefore the Egyptian scribes learned from their southern colleagues. In addition, the scribes were somehow aware that every positive integer could be uniquely expressed as the sum of powers of two. That fact provides the justification for the procedure. How was it discovered? The best guess is that it was discovered by experimentation and then passed down as tradition.

Because division is the inverse of multiplication, a problem such as $156 \div 12$ would be stated as, "multiply 12 so as to get 156." The scribe would then write down the same lines as above. This time, however, he would check off the lines having the numbers in the right-hand column that sum to 156; here that would be 12, 48, and 96. Then the sum of the corresponding numbers on the left, namely, 1, 4, and 8, would give the answer 13. Of course, division does not always "come out even." When it did not, the Egyptians resorted to fractions.

The Egyptians only dealt with unit fractions or "parts" (fractions with numerator 1), with the single exception of 2/3, perhaps because these fractions are the most "natural." The fraction $1/n$ (the nth part) is in general represented in hieroglyphics by the symbol for the integer n with the symbol ⌒ above. In the hieratic a dot is used instead. So 1/7 is denoted in the former system by 𓏢𓏮 and in the latter by 𝓁̇. The single exception, 2/3, had a special symbol: 𓂝 in hieroglyphic and ⟓ in hieratic. Two other fractions, 1/2 and 1/4, also had special symbols: ⊂ and ✗, respectively. In what follows, however, the notation $\overline{n}$ will be used to represent $1/n$ and $\overline{3}$ to represent 2/3.

Because fractions show up as the result of divisions that do not come out evenly, surely there is a need to be able to deal with fractions other than unit fractions. It was in this connection that the most intricate of the Egyptian arithmetical techniques developed, the representation of any fraction in terms of unit fractions. The Egyptians did not view the question this way, however. Whenever we would use a nonunit fraction, they simply wrote a sum of unit fractions. For example, problem 3 of the *Rhind Mathematical Papyrus* asks how to divide 6 loaves among 10 men. The answer is given that each man gets $\overline{2}\,\overline{10}$ loaves (that is, 1/2 + 1/10). The scribe checks this by multiplying this value by 10. We may regard the scribe's answer as more cumbersome than our answer of 3/5, but in some sense the actual division is easier to accomplish this way. We divide five of the loaves in half, the sixth one in tenths, and then give each man one half plus one tenth. It is then clear to all that every man has the same portion of bread. Cumbersome or not, this Egyptian method of unit fractions was used throughout the Mediterranean basin for over 2000 years.

In multiplying whole numbers, the important step is the doubling step. So too in multiplying fractions; the scribe had to be able to express the double of any unit fraction. For example, in the problem above, the check of the solution is written as follows:

$$
\begin{array}{cl}
1 & \overline{2}\,\overline{10} \\
\backslash 2 & 1\,\overline{5} \\
4 & 2\,\overline{3}\,\overline{15} \\
\backslash 8 & 4\,\overline{\overline{3}}\,\overline{10}\,\overline{30} \\
10 & 6
\end{array}
$$

How are these doubles formed? To double $\overline{2}\,\overline{10}$ is easy; because each denominator is even, each is merely halved. In the next line, however, $\overline{5}$ must be doubled. It was here that the scribe had to use a table to get the answer $\overline{3}\,\overline{15}$ (that is, $2 \cdot 1/5 = 1/3 + 1/15$). In fact, the first section of the *Rhind Papyrus* is a table of the division of 2 by every odd integer from 3 to 101 (Fig. 1.4), and the Egyptian scribes realized that the result of multiplying $\overline{n}$ by 2 is the same as that of dividing 2 by n. It is not known how the division table was constructed, but there are several scholarly accounts giving hypotheses for the scribes' methods. In any case, the solution of problem 3 depends on using that table twice, first as already indicated and second in the next step, where the double of $\overline{15}$ is given as $\overline{10}\,\overline{30}$ (or $2 \cdot 1/15 = 1/10 + 1/30$). The final step in this problem involves the addition of $1\,\overline{5}$ to $4\,\overline{3}\,\overline{10}\,\overline{30}$, and here the scribe just gave the answer. Again, the conjecture is that for such addition problems an extensive table existed. The *Egyptian Mathematical Leather Roll*, which dates from about 1600 BCE, contains a short version of such an addition table.[3] There are also extant several other tables for dealing with unit fractions and a multiplication table for the special fraction 2/3. It thus appears that the arithmetic algorithms used by the Egyptian scribes involved extensive knowledge of

FIGURE 1.4

Transcription and hiero-glyphic translation of $2 \div 3$, $2 \div 5$, and $2 \div 7$ from the *Rhind Mathematical Papyrus* (Reston, VA: National Council of Teachers of Mathematics, 1967, Arnold B. Chace, ed.)

2 DIVIDED BY 3, 5, AND 7

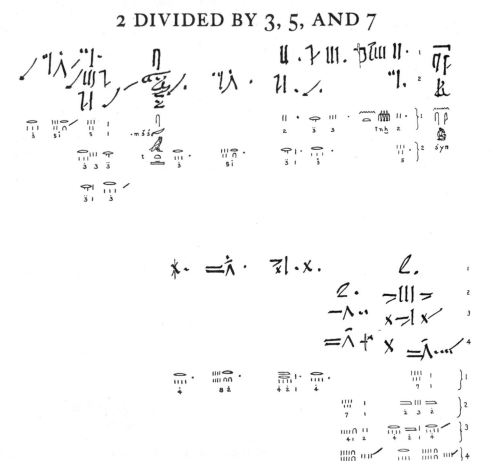

basic tables for addition, subtraction, and doubling and then a definite procedure for reducing multiplication and division problems into steps, each of which could be done using the tables.

Besides the basic procedures of doubling, the Egyptian scribes used other techniques in performing arithmetic calculations. For example, they could find halves of numbers as well as multiply by 10; they could figure out what fractions had to be added to a given mixed number to get the next whole number; and they could determine by what fraction a given whole number needs to be multiplied to give a given fraction. These procedures are illustrated in problem 69 of the *Rhind Papyrus*, which includes the division of 80 by $3\,\overline{2}$ and its subsequent check:

1	$3\,\overline{2}$	'1	$22\,\overline{\overline{3}}\,\overline{7}\,\overline{21}$
10	35	'2	$45\,\overline{3}\,\overline{4}\,\overline{14}\,\overline{28}\,\overline{42}$
20	70'	'$\overline{2}$	$11\,\overline{3}\,\overline{14}\,\overline{42}$
2	7'	$3\,\overline{2}$	80
$\overline{\overline{3}}$	$2\,\overline{3}'$		
$\overline{21}$	$\overline{6}'$		
$\overline{7}$	$\overline{2}'$		
$22\,\overline{\overline{3}}\,\overline{7}\,\overline{21}$	80		

In the second line, the scribe took advantage of the decimal nature of his notation to give immediately the product of $3\,\overline{2}$ by 10. In the fifth line, he used the 2/3 multiplication table mentioned earlier. The scribe then realized that since the numbers in the second column of the third through the fifth lines added to $79\,\overline{3}$, he needed to add $\overline{2}$ and $\overline{6}$ in that column to get 80. Thus, because $6 \times 3\,\overline{2} = 21$ and $2 \times 3\,\overline{2} = 7$, it follows that $\overline{21} \times 3\,\overline{2} = \overline{6}$ and $\overline{7} \times 3\,\overline{2} = \overline{2}$, as indicated in the sixth and seventh lines. The check shows several uses of the table of division by 2 as well as great facility in addition.

1.1.2 Linear Equations and Proportional Reasoning

The mathematical problems the scribes could solve, as illustrated in the *Rhind* and *Moscow Papyri*, deal with what we today call linear equations, proportions, and geometry. For example, the Egyptian papyri present two different procedures for dealing with linear equations.

First, problem 19 of the *Moscow Papyrus* used our normal technique to find the number such that if it is taken 1 1/2 times and then 4 is added, the sum is 10. In modern notation, the equation is simply $(1\,1/2)x + 4 = 10$. The scribe proceeded as follows: "Calculate the excess of this 10 over 4. The result is 6. You operate on 1 1/2 to find 1. The result is 2/3. You take 2/3 of this 6. The result is 4. Behold, 4 says it. You will find that this is correct."[4] Namely, after subtracting 4, the scribe noted that the reciprocal of 1 1/2 is 2/3 and then multiplies 6 by this quantity. Similarly, problem 35 of the *Rhind Papyrus* asked to find the size of a scoop that requires 3 1/3 trips to fill a 1 hekat measure. The scribe solved the equation, which would today be written as $(3\,1/3)x = 1$ by dividing 1 by 3 1/3. He wrote the answer as $\overline{5}\,\overline{10}$ and proceeded to prove that the result is correct.

The Egyptians' more common technique of solving a linear equation, however, was what is usually called the method of **false position**, the method of assuming a convenient but probably incorrect answer and then adjusting it by using proportionality. For example, problem 26 of the *Rhind Papyrus* asked to find a quantity such that when it is added to 1/4 of itself the result is 15. The scribe's solution was as follows: "Assume [the answer is] 4. Then $1\,\overline{4}$ of 4 is 5. . . . Multiply 5 so as to get 15. The answer is 3. Multiply 3 by 4. The answer is 12."[5] In modern notation, the problem is to solve $x + (1/4)x = 15$. The first guess is 4, because 1/4 of 4 is an integer. But then the scribe noted that $4 + 1/4 \cdot 4 = 5$. To find the correct answer, he therefore multiplied 4 by the ratio of 15 to 5, namely, 3. The *Rhind Papyrus* has several similar problems, all solved using false position. The step-by-step procedure of the scribe can therefore be considered as an algorithm for the solution of a linear equation of this type. There is, however, no discussion of how the algorithm was discovered or why it works. But it is evident that the Egyptian scribes understood the basic idea of proportionality of two quantities.

This understanding is further exemplified in the solution of more explicit proportion problems. For example, problem 75 asked for the number of loaves of *pesu* 30 that can be made from the same amount of flour as 155 loaves of *pesu* 20. (*Pesu* is the Egyptian measure for the inverse "strength" of bread and can be expressed as *pesu* = [number of loaves]/[number of hekats of grain], where a hekat is a dry measure approximately equal to 1/8 bushel.) The problem was thus to solve the proportion $x/30 = 155/20$. The scribe accomplished this by dividing 155 by 20 and multiplying the result by 30 to get 232 1/2. Similar problems occur elsewhere in the *Rhind Papyrus* and in the *Moscow Papyrus*.

On the other hand, the method of false position is also used in the only quadratic equation extant in the Egyptian papyri. On the *Berlin Papyrus*, a small fragment dating from approximately the same time as the other papyri, is a problem asking to divide a square area of 100 square cubits into two other squares, where the ratio of the sides of the two squares is 1 to 3/4. The scribe began by assuming that in fact the sides of the two needed squares are 1 and 3/4, then calculated the sum of the areas of these two squares to be $1^2 + (3/4)^2 = 1\,9/16$. But the desired sum of the areas is 100. The scribe realized that he could not compare areas directly but must compare their sides. So he took the square root of $1\,9/16$, namely, $1\frac{1}{4}$, and compared this to the square root of 100, namely, 10. Since 10 is 8 times as large as $1\frac{1}{4}$, the scribe concluded that the sides of the two other squares must be 8 times the original guesses, namely, 8 and 6 cubits, respectively.

There are numerous more complicated problems in the extant papyri. For example, problem 64 of the *Rhind Papyrus* reads as follows: "If it is said to thee, divide 10 hekats of barley among 10 men so that the difference of each man and his neighbor in hekats of barley is 1/8, what is each man's share?"[6] It is understood in this problem, as in similar problems elsewhere in the papyrus, that the shares are to be in arithmetic progression. The average share is 1 hekat. The largest share could be found by adding 1/8 to this average share half the number of times as there are differences. However, since there is an odd number (9) of differences, the scribe instead added half of the common difference (1/16) a total of 9 times to get $1\frac{9}{16}$ ($1\,\overline{2}\,\overline{16}$) as the largest share. He finished the problem by subtracting 1/8 from this value 9 times to get each share.

A final problem, problem 23 of the *Moscow Papyrus*, is what we often think of today as a "work" problem: "Regarding the work of a shoemaker, if he is cutting out only, he can do 10 pairs of sandals per day; but if he is decorating, he can do 5 per day. As for the number he can both cut and decorate in a day, what will that be?"[7] Here the scribe noted that the shoemaker cuts 10 pairs of sandals in one day and decorates 10 pairs of sandals in two days, so that it takes three days for him to both cut and decorate 10 pairs. The scribe then divided 10 by 3 to find that the shoemaker can cut and decorate 3 1/3 pairs in one day.

1.1.3 Geometry

As to geometry, the Egyptian scribes certainly knew how to calculate the areas of rectangles, triangles, and trapezoids by our normal methods. It is their calculation of the area of a circle, however, that is particularly interesting. Problem 50 of the *Rhind Papyrus* reads, "Example of a round field of diameter 9. What is the area? Take away 1/9 of the diameter; the remainder is 8. Multiply 8 times 8; it makes 64. Therefore, the area is 64."[8] In other words, the Egyptian scribe was using a procedure described by the formula $A = (d - d/9)^2 = [(8/9)d]^2$. A comparison with the formula $A = (\pi/4)d^2$ shows that the Egyptian value for the constant π in the case of area was $256/81 = 3.16049\ldots$. Where did the Egyptians get this value, and why was the answer expressed as the square of $(8/9)d$ rather than in modern terms as a multiple (here 64/81) of the square of the diameter?

FIGURE 1.5

Octagon inscribed in a square of side 9, from problem 48 of the *Rhind Mathematical Papyrus*

A hint is given by problem 48 of the same papyrus, in which is shown the figure of an octagon inscribed in a square of side 9 (Fig. 1.5). There is no statement of the problem, however, only a bare computation of $8 \times 8 = 64$ and $9 \times 9 = 81$. If the scribe had inscribed a circle in the same square, he would have seen that its area was approximately that of the octagon. What is the size of the octagon? It depends on how one interprets the diagram in the papyrus. If one believes the octagon to be formed by cutting off four corner triangles each having area 4 1/2, then the area of the octagon is 7/9 that of the square, namely, 63. The scribe therefore might have simply taken the area of the circle as $A = (7/9)d^2 [= (63/81)d^2]$. But since he wanted to find a square whose area was equal to the given circle, he may have approximated 63/81 by $(8/9)^2$, thus giving the area of the circle in the form $[(8/9)d]^2$ indicated in problem 50. On the other hand, in the diagram, the octagon does not look symmetric. So perhaps the octagon was formed by cutting off from the square of side 9 two diagonally opposite corner triangles each equal to 4 1/2 and two other corner triangles each equal to 4. This octagon then has area 64, as explicitly written on the papyrus, and thus this may be the square that the scribe wanted, which was equal in area to a circle.

It should be noted that problem 50 is not an isolated problem of finding the area of a circle. In fact, there are several problems in the *Rhind Papyrus* where the scribe used the rule $V = Bh$ to calculate the volume of a cylinder where B, the area of the base, is calculated by this circle rule. The scribes also knew how to calculate the volume of a rectangular box, given its length, width, and height.

Because one of the prominent forms of building in Egypt was the pyramid, one might expect to find a formula for its volume. Unfortunately, such a formula does not appear in any extant document. The *Rhind Papyrus* does have several problems dealing with the *seked* (slope) of a pyramid; this is measured as so many horizontal units to one vertical unit rise. The workers building the pyramids, or at least their foremen, had to be aware of this value as they built. Since the *seked* is in effect the cotangent of the angle of slope of the pyramid's faces, one can easily calculate the angles given the values appearing in the problems. It is not surprising that these calculated angles closely approximate the actual angles used in the construction of the three major pyramids at Giza.

The *Moscow Papyrus*, however, does have a fascinating formula related to pyramids, namely, the formula for the volume of a truncated pyramid (problem 14): "If someone says to you: a truncated pyramid of 6 for the height by 4 on the base by 2 on the top, you are to square this 4; the result is 16. You are to double 4; the result is 8. You are to square this 2; the result is 4. You are to add the 16 and the 8 and the 4; the result is 28. You are to take 1/3 of 6; the result is 2. You are to take 28 two times; the result is 56. Behold, the volume is 56. You will find that this is correct."[9] If this algorithm is translated into a modern formula, with the length of the lower base denoted by a, that of the upper base by b, and the height by h, it gives the correct result $V = \frac{h}{3}(a^2 + ab + b^2)$. Although no papyrus gives the formula $V = \frac{1}{3}a^2h$ for a completed pyramid of square base a and height h, it is a simple matter to derive it from the given formula by simply putting $b = 0$. We therefore assume that the Egyptians were aware of this result. On the other hand, it takes a higher level of algebraic skill to derive the volume formula for the truncated pyramid from that for the complete pyramid. Still, although many ingenious suggestions involving dissection have been given, no one knows for sure how the Egyptians found their algorithm.

No one knows either how the Egyptians found their procedure for determining the surface area of a hemisphere. But they succeeded in problem 10 of the *Moscow Papyrus*: "A basket with a mouth opening of 4 1/2 in good condition, oh let me know its surface area. First, calculate 1/9 of 9, since the basket is 1/2 of an egg-shell. The result is 1. Calculate the remainder as 8. Calculate 1/9 of 8. The result is 2/3 1/6 1/18 [that is, 8/9]. Calculate the remainder from these 8 after taking away those [8/9]. The result is 7 1/9. Reckon with 7 1/9 four and one-half times. The result is 32. Behold, this is its area. You will find that it is correct."[10] Evidently, the scribe calculated the surface area S of this basket of diameter $d = 4$ 1/2 by first taking 8/9 of $2d$, then taking 8/9 of the result, and finally multiplying by d. As a modern formula, this result would be $S = 2(\frac{8}{9}d)^2$, or, since the area A of the circular opening of this hemispherical basket is given by $A = (\frac{8}{9}d)^2$, we could rewrite this result as $S = 2A$, the correct answer. (It should be noted that there is not universal agreement that this calculation gives the area of a hemisphere. Some suggest that it gives the surface area of a half-cylinder.)

1.2   MESOPOTAMIA

The Mesopotamian civilization is perhaps a bit older than the Egyptian, having developed in the Tigris and Euphrates River valley beginning sometime in the fifth millennium BCE. Many different governments ruled this region over the centuries. Initially, there were many small city-states, but then the area was unified under a dynasty from Akkad, which lasted from approximately 2350 to 2150 BCE. Shortly thereafter, the Third Dynasty of Ur rapidly expanded until it controlled most of southern Mesopotamia. This dynasty produced a very centralized bureaucratic state. In particular, it created a large system of scribal schools to train members of the bureaucracy. Although the Ur Dynasty collapsed around 2000 BCE, the small city-states that succeeded it still demanded numerate scribes. By 1700 BCE, Hammurapi, the

FIGURE 1.6

Hammurapi on a stamp of Iraq

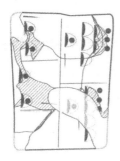

FIGURE 1.7

Tablet from Uruk, c. 3200 BCE, with number signs

FIGURE 1.8

Babylonian clay tablet on a stamp of Austria

ruler of Babylon, one of these city-states, had expanded his rule to much of Mesopotamia and instituted a legal system to help regulate his empire (Fig. 1.6).

Writing began in Mesopotamia, quite possibly in the southern city of Uruk, at about the same time as in Egypt, namely, at the end of the fourth millennium BCE. In fact, writing began there also with the needs of accountancy, of the necessity of recording and managing labor and the flow of goods. The temple, the home of the city's patron god or goddess, came to own large tracts of farming land and vast herds of sheep and goats. The scribes of the temple managed these assets to provide for the well-being of the god(dess) and his or her followers. Thus, in the temple of goddess Inana in Uruk, the scribes represented numbers on small clay slabs, using various pictograms to represent the objects that were being counted or measured. For example, five ovoids might represent five jars of oil. Or, as in the earliest known piece of school mathematics yet discovered, the scribe who wrote tablet W 19408,76[11] used three different number signs to represent lengths as he calculated the area of a field (Fig. 1.7). Small circles represented 10 rods; a large D-shaped impression represented a unit of 60 rods, whereas a small circle within a large D represented $60 \times 10 = 600$ rods. On this tablet, there are two other signs, a horizontal line representing width and a vertical line representing length. The two widths of the quadrilateral field were each $2 \times 600 = 1200$, while the two lengths were $600 + 5 \times 60 + 3 \times 10 = 930$ and $600 + 4 \times 60 + 3 \times 10 = 870$. The approximate area could then be found by a standard ancient method of multiplying the average width by the average length; that is, $A = ((w_1 + w_2)/2)((l_1 + l_2)/2)$. In this case, the answer was $1200 \times 900 = 1,080,000$. But since in the then current measurement system 1 square rod was equal to 1 *sar*, while 1800 *sar* were equal to 1 *bur*, the result here was 600 *bur*, a conspicuously "round" number, typical of answers in school tablets.

On this particular tablet, as in other situations where quantities were measured, there were several different units of measure and different symbols for each type of unit. Here, the largest unit was equal to 60 of the smallest unit. This was typical in the units for many different types of objects, and at some time, the system of recording numbers developed to the point where the digit for 1 represented 60 as well. We do not know why the Mesopotamians decided to have one large unit represent 60 small units and then adapt this method for their numeration system. One plausible conjecture is that 60 is evenly divisible by many small integers. Therefore, fractional values of the "large" unit could easily be expressed as integral values of the "small." But eventually, they did develop a sexagesimal (base-60) place value system, which in the third millennium BCE became the standard system used throughout Mesopotamia. By that time, too, writing began to be used in a wide variety of contexts, all achieved by using a stylus on a moist clay tablet (Fig. 1.8). Thousands of these tablets have been excavated during the past 150 years. It was Henry Rawlinson (1810–1895) who, by the mid-1850s, was first able to translate this cuneiform writing by comparing the Persian and Mesopotamian cuneiform inscriptions of King Darius I of Persia (sixth century BCE) on a rock face at Behistun (in modern Iran) describing a military victory.

A large number of these tablets are mathematical in nature, containing mathematical problems and solutions or mathematical tables. Several hundreds of these have been copied, translated, and explained. These tablets, generally rectangular but occasionally round, usually fit comfortably into one's hand and are an inch or so in thickness. Some, however, are as small as a postage stamp while others are as large as an encyclopedia volume. We are fortunate that these tablets are virtually indestructible, because they are our only source for Mesopotamian mathematics. The written tradition that they represent died out under Greek domination in the last centuries BCE and was totally lost until the nineteenth century. The great majority of the excavated tablets date from the time of Hammurapi, while small collections date from the earliest beginnings of Mesopotamian civilization, from the centuries surrounding 1000 BCE, and from the Seleucid period around 300 BCE. Our discussion in this section, however, will generally deal with the mathematics of the "Old Babylonian" period (the time of Hammurapi), but, as is standard in the history of mathematics, we shall use the adjective "Babylonian" to refer to the civilization and culture of Mesopotamia, even though Babylon itself was the major city of the area for only a limited time.

1.2.1 Methods of Computation

The Babylonians at various times used different systems of numbers, but the standardized system that the scribes generally used for calculations in the "Old Babylonian" period was a base-60 place value system together with a grouping system based on 10 to represent numbers up to 59. Thus, a vertical stylus stroke on a clay tablet Υ represented 1 and a tilted stroke $\triangleleft$ represented 10. By grouping they would, for example, represent 37 by

$$\triangleleft\triangleleft\triangleleft\triangleleft \begin{matrix} \Upsilon\Upsilon\Upsilon \\ \Upsilon\Upsilon\Upsilon \\ \Upsilon \end{matrix} \ .$$

For numbers greater than 59, the Babylonians used a place value system; that is, the powers of 60, the base of this system, are represented by "places" rather than symbols, while the digit in each place represents the number of each power to be counted. Hence, $3 \times 60^2 + 42 \times 60 + 9$ (or 13,329) was represented by the Babylonians as

$$\Upsilon\Upsilon\Upsilon \qquad \begin{matrix} \triangleleft\triangleleft \\ \triangleleft\triangleleft \end{matrix}\Upsilon\Upsilon \qquad \begin{matrix} \Upsilon\Upsilon\Upsilon \\ \Upsilon\Upsilon\Upsilon \\ \Upsilon\Upsilon\Upsilon \end{matrix} \ .$$

(This will be written from now on as 3,42,09 rather than with the Babylonian strokes.) The Old Babylonians did not use a symbol for 0, but often left an internal space if a given number was missing a particular power. There would not be a space at the end of a number, making it difficult to distinguish $3 \times 60 + 42$ (3,42) from $3 \times 60^2 + 42 \times 60$ (3,42,00). Sometimes, however, they would give an indication of the absolute size of a number by writing an appropriate word, typically a metrological one, after the numeral. Thus, "3 42 sixty" would represent 3,42, while "3 42 thirty-six hundred" would mean 3,42,00. On the other hand, the Babylonians never used a symbol to represent zero in the context of "nothingness," as in our $42 - 42 = 0$.

That the Babylonians used tables in the process of performing arithmetic computations is proved by extensive direct evidence. Many of the preserved tablets are in fact multiplication tables. No addition tables have turned up, however. Because over 200 Babylonian table texts have been analyzed, it may be assumed that these did not exist and that the scribes knew their addition procedures well enough so they could write down the answers when needed. On the other hand, there are many examples of "scratch tablets," on which a scribe has performed various calculations in the process of solving a problem. In any case, since the Babylonian number system was a place value system, the actual algorithms for addition and subtraction, including carrying and borrowing, may well have been similar to modern ones. For example, to add 23,37 (= 1417) to 41,32 (= 2492), one first adds 37 and 32 to get 1,09 (= 69). One writes down 09 and carries 1 to the next column. Then $23 + 41 + 1 = 1, 05$ (= 65), and the final result is 1,05,09 (= 3909).

Because the place value system was based on 60, the multiplication tables were extensive. Any given one listed the multiples of a particular number, say, 9, from 1×9 to 20×9 and then gave 30×9, 40×9, and 50×9 (Fig. 1.9). If one needed the product 34×9, one simply added the two results $30 \times 9 = 4, 30$ (= 270) and $4 \times 9 = 36$ to get 5,06 (= 306). For multiplication of two- or three-digit sexagesimal numbers, one needed to use several such tables. The exact algorithm the Babylonians used for such multiplications—where the partial products are written and how the final result is obtained—is not known, but it may well have been similar to our own.

One might think that for a complete system of tables, the Babylonians would have one for each integer from 2 to 59. Such was not the case, however. In fact, although there are no tables

FIGURE 1.9

A Babylonian multiplication table for 9 (Department of Archaeology, University of Pennsylvania)

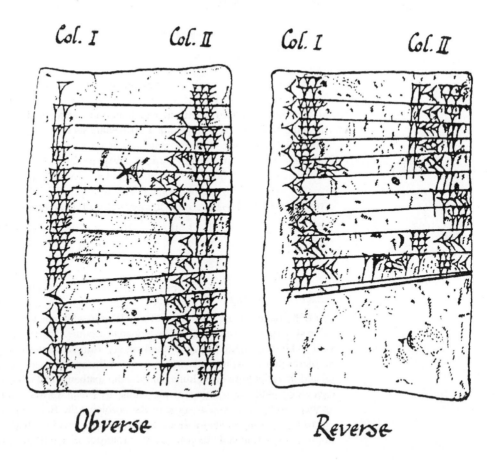

Col. I   Col. II   Col. I   Col. II

Obverse   Reverse

for 11, 13, 17, for example, there are tables for 1,15, 3,45, and 44,26,40. We do not know precisely why the Babylonians made these choices; we do know, however, that, with the single exception of 7, all multiplication tables so far found are for **regular** sexagesimal numbers, that is, numbers whose reciprocal is a terminating sexagesimal fraction. The Babylonians treated all fractions as sexagesimal fractions, analogous to our use of decimal fractions. Namely, the first place after the "sexagesimal point" (which we denote by ";") represents 60ths, the next place 3600ths, and so on. Thus, the reciprocal of 48 is the sexagesimal fraction $0;1,15$, which represents $1/60 + 15/60^2$, while the reciprocal of 1,21 (= 81) is $0;0,44,26,40$, or $44/60^2 + 26/60^3 + 40/60^4$. Because the Babylonians did not indicate an initial 0 or the sexagesimal point, this last number would just be written as 44,26,40. As noted, there exist multiplication tables for this regular number. In such a table there is no indication of the absolute size of the number, nor is one necessary. When the Babylonians used the table, of course, they realized that, as in today's decimal calculations, the eventual placement of the sexagesimal point depended on the absolute size of the numbers involved, and this placement was then done by context.

Besides multiplication tables, there are also extensive tables of reciprocals, one of which is in part reproduced here. A table of reciprocals is a list of pairs of numbers whose product is 1 (where the 1 can represent any power of 60). Like the multiplication tables, these tables only contained regular sexagesimal numbers.

2	30	16	3, 45	48	1, 15
3	20	25	2, 24	1, 04	56, 15
10	6	40	1, 30	1, 21	44, 26, 40

The reciprocal tables were used in conjunction with the multiplication tables to do division. Thus, the multiplication table for 1,30 (= 90) served not only to give multiples of that number but also, since 40 is the reciprocal of 1,30, to do divisions by 40. In other words, the Babylonians considered the problem $50 \div 40$ to be equivalent to $50 \times 1/40$, or in sexagesimal notation, to $50 \times 0;1,30$. The multiplication table for 1,30, part of which appears here, then gives 1,15 (or 1,15,00) as the product. The appropriate placement of the sexagesimal point gives $1;15 (= 1\ 1/4)$ as the correct answer to the division problem.

1	1,30	10	15	30	45
2	3	11	16,30	40	1
3	4,30	12	18	50	1,15

1.2.2 Geometry

The Babylonians had a wide range of problems to which they applied their sexagesimal place value system. For example, they developed procedures for determining areas and volumes of various kinds of figures. They worked out algorithms to determine square roots. They solved problems that we would interpret in terms of linear and quadratic equations, problems often related to agriculture or building. In fact, the mathematical tablets themselves are generally concerned with the solution of problems, to which various mathematical techniques are applied. So we will look at some of the problems the Babylonians solved and try to figure out what lies behind their methods. In particular, we will see that the reasons behind many of the Babylonian procedures come from a tradition different from the accountancy traditions with which Babylonian mathematics began. This second tradition was the "cut-and-paste" geometry of the surveyors, who had to measure fields and lay out public works projects. As we will see, these manipulations of squares and rectangles not only developed into procedures for determining square roots and finding Pythagorean triples, but they also developed into what we can think of as "algebra."

As we work through the Babylonian problems, we must keep in mind that, like the Egyptians, the scribes did not have any symbolism for operations or unknowns. Thus, solutions are presented with purely verbal techniques. We must also remember that the Babylonians often thought about problems in ways different from the ways we do. Thus, even though their methods are usually correct, they may seem strange to us.

As one example of the scribes' different methods, we consider their procedures for determining lengths and areas. In general, in place of our formulas for calculating such quantities, they presented coefficient lists, lists of constants that embody mathematical relationships between certain aspects of various geometrical figures. Thus, the number 0;52,30 (= 7/8) as the coefficient for the height of a triangle means that the altitude of an equilateral triangle is 7/8 of the base, while the number 0;26,15 (= 7/16) as the coefficient for area means that the area of an equilateral triangle is 7/16 times the square of a side. (Note, of course, that these results are only approximately correct, in that they both approximate $\sqrt{3}$ by 7/4.) In each case, the idea is that the "defining component" for the triangle is the side.

We too would use the length of a side as the defining component for an equilateral triangle. But for a circle, we generally use the radius r as that component and therefore give formulas for the circumference and area in terms of r. The Babylonians, on the other hand, took the circumference as the defining component of a circle. Thus, they gave two coefficients for the circle: 0;20 (= 1/3) for the diameter and 0;05 (= 1/12) for the area. The first coefficient means that the diameter is one-third of the circumference, while the second means that the area is one-twelfth of the square of the circumference. For example, on the tablet YBC 7302, there is a circle with the numbers 3 and 9 written on the outside and the number 45 written on the inside (Fig. 1.10). The interpretation of this is that the circle has circumference 3 and that the area is found by dividing $9 = 3^2$ by 12 to get 0;45 (= 3/4). Another tablet, Haddad 104, illustrates that circle calculations virtually always use the circumference. On this tablet, there is a problem asking to find the area of the cross section of a log of diameter 1;40 (= $1\frac{2}{3}$). Rather than determine the radius, the scribe first multiplies by 3 to find that the circumference

FIGURE 1.10

Tablet YBC 7302 illustrating measurements on a circle

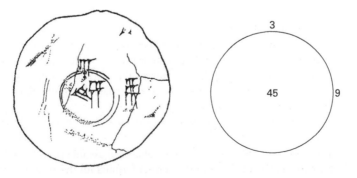

FIGURE 1.11

Babylonian barge and bull's-eye

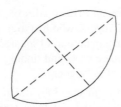

is equal to 5, then squares 5 and multiplies by 1/12 to get the area 2;05 ($= 2\frac{1}{12}$). Note further, of course, that the Babylonian value for what we denote as π, the ratio of circumference to diameter, is 3; this value produces the value $4\pi = 12$ as the constant by which to divide the square of the circumference to give the area.

There are also Babylonian coefficients for other figures bounded by circular arcs. For example, the Babylonians calculated areas of two different double bows: the "barge," made up of two quarter-circle arcs, and the "bull's-eye," composed of two third-circle arcs (Fig. 1.11). In analogy with the circle, the defining component of these figures was the arc making up one side. The coefficient of the area of the barge is 0;13,20 ($= 2/9$), while that of the bull's-eye is 0;16,52,30 ($= 9/32$). Thus, the areas of these two figures are calculated as $(2/9)a^2$ and $(9/32)a^2$, respectively, where in each case a is the length of that arc. These results are accurate under the assumptions that the area of the circle is $C^2/12$ and that $\sqrt{3} = 7/4$. Similarly, the coefficient of the area of the concave square (Fig. 1.12) is 0;26,40 ($= 4/9$), where the defining component is one of the four quarter-circle arcs forming the boundary of the region.[12] Clearly, the use of these coefficients shows that the scribes recognized that lengths of particular lines in given figures were proportional to the length of the defining component, while the area was proportional to the square of that component.

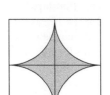

FIGURE 1.12

Babylonian concave square

The Babylonians also dealt with volumes of solids. They realized that the volume V of a rectangular block is $V = \ell w h$, and they also knew how to calculate the volume of prisms given the area of the base. But just like in Egypt, there is no document that explicitly gives the volume of a pyramid, even though the Babylonians certainly built pyramidal structures. Nevertheless, on tablet BM 96954, there are several problems involving a grain pile in the shape of a rectangular pyramid with an elongated apex, like a pitched roof (Fig. 1.13). The method of solution corresponds to the modern formula

$$V = \frac{hw}{3}\left(\ell + \frac{t}{2}\right),$$

where ℓ is the length of the solid, w the width, h the height, and t the length of the apex. Although no derivation of this correct formula is given on the tablet, we can derive it by breaking up the solid into a triangular prism with half a rectangular pyramid on each side. Then the volume would be the sum of the volumes of these solids (Fig. 1.14). Thus, $V =$ volume of triangular prism + volume of rectangular pyramid, or

$$V = \frac{hwt}{2} + \frac{hw(\ell - t)}{3} = \frac{hw\ell}{3} + \frac{hwt}{6} = \frac{hw}{3}\left(\ell + \frac{t}{2}\right),$$

as desired.[13] It therefore seems reasonable to assume from the result discussed here that the Babylonians were aware of the correct formula for the volume of a pyramid.

FIGURE 1.13

Babylonian grain pile

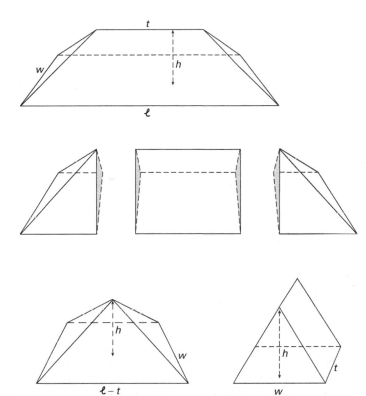

FIGURE 1.14

Dissection of grain pile

   That assumption is even more convincing because there is a tablet giving a correct formula for the volume of a truncated pyramid with square base a^2, square top b^2, and height h in the form $V = [(\frac{a+b}{2})^2 + \frac{1}{3}(\frac{a-b}{2})^2]h$. The complete pyramid formula, of course, follows from this by putting $b = 0$. On the other hand, there are tablets where this volume is calculated by the rule $V = \frac{1}{2}(a^2 + b^2)h$, a simple but incorrect generalization of the rule for the area of the trapezoid. It is well to remember, however, that although this formula is incorrect, the calculated answers would not be very different from the correct ones. It is difficult to see how anyone would realize that the answers were wrong in any case, because there was no accurate method for measuring the volume empirically. However, because the problems in which these formulas occurred were practical ones, often related to the number of workmen needed to build a particular structure, the slight inaccuracy produced by using this rule would have little effect on the final answer.

1.2.3 Square Roots and the Pythagorean Theorem

We next consider another type of Babylonian algorithm, the square root algorithm. Usually, when square roots are needed in solving problems, the problems are arranged so that the square root is one that is listed in a table of square roots, of which many exist, and is a rational number. But there are cases where an irrational square root is needed, in particular,

$\sqrt{2}$. When this particular value occurs, the result is generally written as 1;25 (= $1\frac{5}{12}$). There is, however, an interesting tablet, YBC 7289, on which is drawn a square with side indicated as 30 and two numbers, 1:24,51,10 and 42;25,35, written on the diagonal (Fig. 1.15). The product of 30 by 1;24,51,10 is precisely 42;25,35. It is then a reasonable assumption that the last number represents the length of the diagonal and that the other number represents $\sqrt{2}$.

FIGURE 1.15

Tablet YBC 7289 with the square root of 2

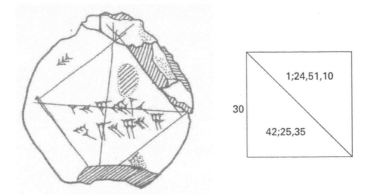

Whether $\sqrt{2}$ is given as 1;25 or as 1;24,51,10, there is no record as to how the value was calculated. But because the scribes were surely aware that the square of neither of these was exactly 2, or that these values were not exactly the length of the side of a square of area 2, they must have known that these values were approximations. How were they determined? One possible method, a method for which there is some textual evidence, begins with the algebraic identity $(x + y)^2 = x^2 + 2xy + y^2$, whose validity was probably discovered by the Babylonians from its geometric equivalent. Now given a square of area N for which one wants the side $\sqrt{N}$, the first step would be to choose a regular value a close to, but less than, the desired result. Setting $b = N - a^2$, the next step is to find c so that $2ac + c^2$ is as close as possible to b (Fig. 1.16). If a^2 is "close enough" to N, then c^2 will be small in relation to $2ac$, so c can be chosen to equal $(1/2)b(1/a)$, that is, $\sqrt{N} = \sqrt{a^2 + b} \approx a + (1/2)b(1/a)$. (In keeping with Babylonian methods, the value for c has been written as a product rather than a quotient, and, since one of the factors is the reciprocal of a, we see why a must be regular.) A similar argument shows that $\sqrt{a^2 - b} \approx a - (1/2)b(1/a)$. In the particular case of $\sqrt{2}$, one begins with $a = 1;20$ (= 4/3). Then $a^2 = 1;46,40$, $b = 0;13,20$, and $1/a = 0;45$, so $\sqrt{2} = \sqrt{1;46,40 + 0;13,20} \approx 1;20 + (0;30)(0;13,20)(0;45) = 1;20 + 0;05 = 1;25$ (or 17/12).

To calculate the better approximation 1;24,51,10, one would have to repeat this procedure, with $a = 1;25$. Unfortunately, 1;25 is not a regular sexagesimal number. The scribes could, however, have found an approximation to the reciprocal, say, 0;42,21,10, and then calculated

$$\sqrt{2} = \sqrt{1;25^2 - 0;00,25} \approx 1;25 - 0;30 \times 0;00,25 \times 0;42,21,10 = 1;24,51,10,35,25.$$

Because the approximation formula leads to a slight overestimate of the true value, the scribes would have truncated this answer to the desired 1;24,51,10. There is, however, no direct

FIGURE 1.16

Geometric version of $\sqrt{N} = \sqrt{a^2 + b} \approx a + \frac{1}{2} \cdot b \cdot \frac{1}{a}$

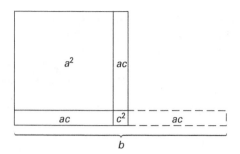

evidence of this calculation nor even any evidence for the use of more than one step of this approximation procedure.

One of the Babylonian square root problems was connected to the relation between the side of a square and its diagonal. That relation is a special case of the result known as the **Pythagorean Theorem**: In any right triangle, the sum of the areas of the squares on the legs equals the area of the square on the hypotenuse. This theorem, named after the sixth-century BCE Greek philosopher-mathematician, is arguably the most important elementary theorem in mathematics, since its consequences and generalizations have wide-ranging application. Nevertheless, it is one of the earliest theorems known to ancient civilizations. In fact, there is evidence that it was known at least 1000 years before Pythagoras.

In particular, there is substantial evidence of interest in Pythagorean triples, triples of integers (a, b, c) such that $a^2 + b^2 = c^2$, in the Babylonian tablet Plimpton 322 (Fig. 1.17).[14] The extant piece of the tablet consists of four columns of numbers. Other columns were probably broken off on the left. The numbers on the tablet are shown in Table 1.1, reproduced in modern decimal notation with the few corrections that recent editors have made and with one extra column, y (not on the tablet), added on the right. It was a major piece of mathematical detective work for modern scholars, first, to decide that this was a mathematical work rather than a list of orders from a pottery business and, second, to find a reasonable mathematical explanation. But find one they did. The columns headed x and d (whose headings in the original can be translated as "square-side of the short side" and "square-side of the diagonal") contain in each row two of the three numbers of a Pythagorean triple. It is easy enough to subtract the square of column x from the square of column d. In each case a perfect square results, whose square root is indicated in the added column, y. Finally, the first column on the left represents the quotient $(\frac{d}{y})^2$.

How and why were these triples derived? One cannot find Pythagorean triples of this size by trial and error. There have been many suggestions over the years as to how the scribe found these as well as to the purpose of the tablet. If one considers this question as purely a mathematical one, there are many methods that would work to generate the table. But since this tablet was written at a particular time and place, probably in Larsa around 1800 BCE, an understanding of its construction and meaning must come from an understanding of the context of the time and how mathematical tablets were generally written. In particular, it is important to note that the first column in a Babylonian table is virtually always written in numerical order (either ascending or descending), while subsequent columns depend on those to their left. Unfortunately, in this instance it is believed that the initial columns

FIGURE 1.17

Plimpton 322 (*Source:* George
Arthur Plimpton Collection,
Rare Book and Manuscript
Library, Columbia University)

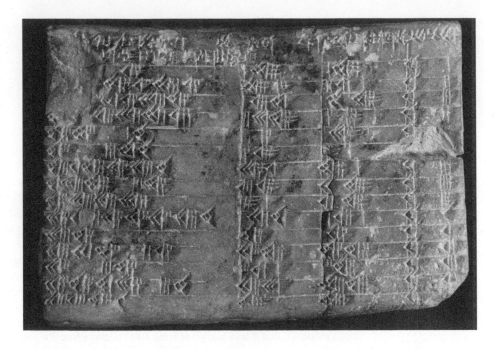

TABLE 1.1   *Numbers on the Babylonian tablet Plimpton 322, reproduced in modern decimal notation. (The column to the right, labeled y, does not appear on the tablet.)*

$\left(\frac{d}{y}\right)^2$	x	d	#	y
1.9834028	119	169	1	120
1.9491586	3367	4825	2	3456
1.9188021	4601	6649	3	4800
1.8862479	12,709	18,541	4	13,500
1.8150077	65	97	5	72
1.7851929	319	481	6	360
1.7199837	2291	3541	7	2700
1.6845877	799	1249	8	960
1.6426694	481	769	9	600
1.5861226	4961	8161	10	6480
1.5625	45	75	11	60
1.4894168	1679	2929	12	2400
1.4500174	161	289	13	240
1.4302388	1771	3229	14	2700
1.3871605	28	53	15	45

on the left are missing. However, some clues as to the meaning of the table reside in the words at the top of the column we have labeled $(\frac{d}{y})^2$. Deciphering the words was difficult because some of the cuneiform wedges were damaged, but it appears that the heading means "the holding-square of the diagonal from which 1 is torn out so that the short side comes up." The "1" in that heading indicates that the author is dealing with reciprocal pairs, very common in Babylonian tables. To relate reciprocals to Pythagorean triples, we note that to find integer solutions to the equation $x^2 + y^2 = d^2$, one can divide by y and first find solutions to $(\frac{x}{y})^2 + 1 = (\frac{d}{y})^2$ or, setting $u = \frac{x}{y}$ and $v = \frac{d}{y}$, to $u^2 + 1 = v^2$. This latter equation is equivalent to $(v + u)(v - u) = 1$. That is, we can think of $v + u$ and $v - u$ as the sides of a rectangle whose area is 1 (Fig. 1.18). Now split off from this rectangle one with sides u and $v - u$ and move it to the bottom left after a rotation of 90°. The resulting figure is an L-shaped figure, usually called a gnomon, with long sides both equal to v, a figure that is the difference $v^2 - u^2 = 1$ of two squares. Note that the larger square is the square on the diagonal of the right triangle with sides $(u, 1, v)$. The area of that square, $v^2 = (d/y)^2$, is the entry in the leftmost column on the extant tablet, and furthermore, that square has a gnomon of area 1 torn out so that the remaining square is the square on the short side of the right triangle, as the column heading actually says.

FIGURE 1.18

A rectangle of area 1 turned into the difference of two squares

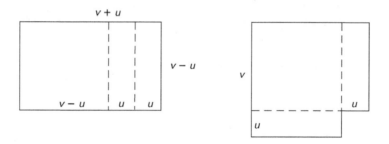

To calculate the entries on the tablet, it is possible that the author began with a value for what we have called $v + u$. Next, he found its reciprocal $v - u$ in a table and solved for $u = \frac{1}{2}[(v + u) - (v - u)]$. The first column in the table is then the value $1 + u^2$. He could then find v by taking the square root of $1 + u^2$. Since $(u, 1, v)$ satisfies the Pythagorean identity, the author could find a corresponding integral Pythagorean triple by multiplying each of these values by a suitable number y, one chosen to eliminate "fractional" values. For example, if $v + u = 2;15 (= 2\frac{1}{4})$, the reciprocal $v - u$ is $0;26,40 (= 4/9)$. We then find $u = 0;54,10 = 65/72$. We would find v by taking half the sum of $v + u$ and $v - u$, but our scribe found v as $\sqrt{1 + u^2} = \sqrt{1;48,54,01,40} = 1;20,50$, or $\sqrt{1 + u^2} = \sqrt{1.8150077} = 1\frac{25}{72}$. Multiplying the values for u, v, and 1 by $1,12 = 72$ gives the values 65 and 97 for x and d, respectively, shown in line 5 of the table, as well as the value 72 for y. Conversely, the value of $v + u$ for line 1 of the table can be found by adding $169/120 (= 1;24,30)$ and $119/120 (= 0;59,30)$ to get $288/120 (= 2;24)$.

Why were the particular Pythagorean triples on this tablet chosen? Again, we cannot know the answer definitively. But if we calculate the values of $v + u$ for every line of the tablet, we notice that they form a decreasing sequence of regular sexagesimal numbers of no more than four places from 2;24 to 1;48. Not all such numbers are included—there are five missing—but it is possible that the scribe may have decided that the table was long enough without them. He may also have begun with numbers larger than 2;24 or continued with numbers smaller than 1;48 on tablets that have not yet been unearthed. In any case, it is likely that this column of values for $v + u$, in descending numerical order, was one of the missing columns on the original tablet. And our author, quite probably a teacher, had thus worked out a list of integral Pythagorean triples, triples that could be used in constructing problems for students for which he would know that the solution would be possible in integers or finite sexagesimal fractions.

Whether or not the method presented above was the one the Babylonian scribe used to write Plimpton 322, the fact remains that the scribes were well aware of the Pythagorean relationship. And although this particular table offers no indication of a geometrical relationship except for the headings of the columns, there are problems in Old Babylonian tablets making explicit geometrical use of the Pythagorean Theorem. For example, in a problem from tablet BM 85196, a beam of length 30 stands against a wall. The upper end has slipped down a distance 6. How far did the lower end move? Namely, $d = 30$ and $y = 24$ are given, and x is to be found. The scribe calculated x using the theorem: $x = \sqrt{30^2 - 24^2} = \sqrt{324} = 18$. Another slightly more complicated example comes from tablet TMS 1 found at Susa in modern Iran. The problem is to calculate the radius of a circle circumscribed about an isosceles triangle with altitude 40 and base 60. Applying the Pythagorean theorem to the right triangle ABC (Fig. 1.19), whose hypotenuse is the desired radius, gives the relationship $r^2 = 30^2 + (40 - r)^2$. This could be easily transformed into $(1, 20)(r - 20) = 15,00$ and then, by multiplying by the reciprocal $0;0,45$ of $1,20$, into $r - 20 = (0;0,45)(15,00) = 11;15$, from which the scribe found that $r = 31;15$.

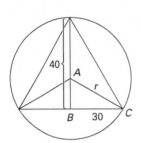

FIGURE 1.19

Circumscribing a circle about an isosceles triangle

1.2.4   Solving Equations

The previous problem involved what we would call the solution of an equation. Such problems were very frequent on the Babylonian tablets. Linear equations of the form $ax = b$ are generally solved by multiplying each side by the reciprocal of a. (Such equations often occur, as in the previous example, in the process of solving a complex problem.) In more complicated situations, such as systems of two linear equations, the Babylonians, like the Egyptians, used the method of false position.

Here is an example from the Old Babylonian text VAT 8389: One of two fields yields 2/3 *sila* per *sar*, the second yields 1/2 *sila* per *sar*, where *sila* and *sar* are measures for capacity and area, respectively. The yield of the first field was 500 *sila* more than that of the second; the areas of the two fields were together 1800 *sar*. How large is each field? It is easy enough to translate the problem into a system of two equations with x and y representing the unknown areas:

$$\frac{2}{3}x - \frac{1}{2}y = 500$$
$$x + y = 1800$$

A modern solution might be to solve the second equation for x and substitute the result in the first. But the Babylonian scribe here made the initial assumption that x and y were both equal to 900. He then calculated that $(2/3) \cdot 900 - (1/2) \cdot 900 = 150$. The difference between the desired 500 and the calculated 150 is 350. To adjust the answers, the scribe presumably realized that every unit increase in the value of x and consequent unit decrease in the value of y gave an increase in the "function" $(2/3)x - (1/2)y$ of $2/3 + 1/2 = 7/6$. He therefore needed only to solve the equation $(7/6)s = 350$ to get the necessary increase $s = 300$. Adding 300 to 900 gave him 1200 for x while subtracting gave him 600 for y, the correct answers.

Presumably, the Babylonians also solved complex single linear equations by false position, although the few such problems available do not reveal their method. For example, here is a problem from tablet YBC 4652: "I found a stone, but did not weigh it; after I added one-seventh and then one-eleventh [of the total], it weighed 1 *mina* [= 60 *gin*]. What was the original weight of the stone?"[15] We can translate this into the modern equation $(x + x/7) + 1/11(x + x/7) = 60$. On the tablet, the scribe just presented the answer, here $x = 48\frac{1}{8}$. If he had solved the problem by false position, the scribe would first have guessed that $y = x + x/7 = 11$. Since then $y + (1/11)y = 12$ instead of 60, the guess must be increased by the factor $60/12 = 5$ to the value 55. Then, to solve $x + x/7 = 55$, the scribe could have guessed $x = 7$. This value would produce $7 + 7/7 = 8$ instead of 55. So the last step would be to multiply the guess of 7 by the factor $55/8$ to get $385/8 = 48\frac{1}{8}$, the correct answer.

While tablets containing explicit linear problems are limited, there are very many Babylonian tablets whose problems can be translated into quadratic equations. In fact, many Old Babylonian tablets contain extensive lists of quadratic problems. And in solving these problems, the scribes made full use of the "cut-and-paste" geometry developed by the surveyors. In particular, they applied this to various standard problems such as finding the length and width of a rectangle, given the semiperimeter and the area. For example, consider the problem $x + y = 6\frac{1}{2}$, $xy = 7\frac{1}{2}$ from tablet YBC 4663. The scribe first halved $6\frac{1}{2}$ to get $3\frac{1}{4}$. Next he squared $3\frac{1}{4}$, getting $10\frac{9}{16}$. From this is subtracted $7\frac{1}{2}$, leaving $3\frac{1}{16}$, and then the square root is extracted to get $1\frac{3}{4}$. The length is thus $3\frac{1}{4} + 1\frac{3}{4} = 5$, while the width is given as $3\frac{1}{4} - 1\frac{3}{4} = 1\frac{1}{2}$. A close reading of the wording of the tablets indicates that the scribe had in mind a geometric procedure (Fig. 1.20), where for the sake of generality the sides have been labeled in accordance with the generic system $x + y = b$, $xy = c$. The scribe began by halving the sum b and then constructing the square on it. Since $b/2 = x - \frac{x-y}{2} = y + \frac{x-y}{2}$, the square on $b/2$ exceeds the original rectangle of area c by the square on $\frac{x-y}{2}$; that is,

$$\left(\frac{x+y}{2}\right)^2 = xy + \left(\frac{x-y}{2}\right)^2.$$

The figure then shows that if one adds the side of this square, namely,

$$\sqrt{(b/2)^2 - c},$$

to $b/2$, one finds the length x, while if one subtracts it from $b/2$, one gets the width y. The algorithm is therefore expressible in the form

$$x = \frac{b}{2} + \sqrt{(b/2)^2 - c} \qquad y = \frac{b}{2} - \sqrt{(b/2)^2 - c}.$$

FIGURE 1.20

Geometric procedure for
solving the system $x + y = b$, $xy = c$

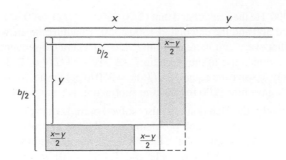

Geometry is also at the base of the Babylonian solution of what we would consider a single quadratic equation. Several such problems are given on tablet BM 13901, including the following, where the translation shows the geometric flavor of the problem: "I summed the area and two-thirds of my square-side and it was 0;35. You put down 1, the projection. Two-thirds of 1, the projection, is 0;40. You combined its half, 0;20 and 0;20. You add 0;06,40 to 0;35 and 0;41,40 squares 0;50. You take away 0;20 that you combined from the middle of 0;50 and the square-side is 0;30."[16] In modern terms, the equation to be solved is $x^2 + (2/3)x = 7/12$. At first glance, it would appear that the statement of the problem is not a geometric one, since we are asked to add a multiple of a side to an area. But the word "projection" indicates that this two-thirds multiple of a side is to be considered as two-thirds of the rectangle of length 1 and unknown side x. For the solution, the scribe took half of 2/3 and squared it ("combine its half, 0;20 and 0;20"), then took the result 1/9 (or 0;06,40) and added it to 7/12 (0;35) to get 25/36 (0;41,40). The scribe then noted that 5/6 (0;50) is the square root of 25/36 ("0;41,40 squares 0;50"). He then subtracted the 1/3 from 5/6 to get the result 1/2 ("the square-side is 0;30"). The Babylonian rule exemplified by this problem is easily translated into a modern formula for solving $x^2 + bx = c$, namely,

$$x = \sqrt{(b/2)^2 + c} - b/2,$$

recognizable as a version of the quadratic formula. Figure 1.21 shows the geometric meaning of the procedure in the generic case, where we start with a square of side x adjoined by a rectangle of width x and length b. The procedure then amounts to cutting half of the rectangle off from one side of the square and moving it to the bottom. Adding a square of side $b/2$ "completes the square." It is then evident that the unknown length x is equal to the difference between the side of the new square and $b/2$, exactly as the formula implies.

For the analogous problem $x^2 - bx = c$, the Babylonian geometric procedure is equivalent to the formula $x = \sqrt{(b/2)^2 + c} + b/2$. This is illustrated by another problem from BM 13901, which we would translate as $x^2 - x = 870$: "I took away my square-side from inside the area and it was 14,30. You put down 1, the projection. You break off half of 1. You combine 0;30 and 0;30. You add 0;15 to 14,30. 14,30;15 squares 29;30. You add 0;30 which you combined to 29;30 so that the square-side is 30."[17]

One should, however, keep in mind that the "quadratic formula" did not mean the same thing to the Babylonian scribes as it means to us. First, the scribes gave different procedures

FIGURE 1.21

Geometric version of the quadratic formula for solving $x^2 + bx = c$

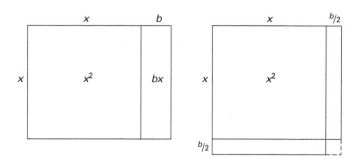

for solving the two types $x^2 + bx = c$ and $x^2 - bx = c$ because the two problems were different; they had different geometric meanings. To a modern mathematician, on the other hand, these problems are the same because the coefficient of x can be taken as positive or negative. Second, the modern quadratic formula in these two cases gives a positive and a negative solution to each equation. The negative solution, however, makes no geometrical sense and was completely ignored by the Babylonians.

In both of these quadratic equation problems, the coefficient of the x^2 term is 1. How did the Babylonians treat the quadratic equation $ax^2 \pm bx = c$ when $a \neq 1$? Again, there are problems on BM 13901 showing that the scribes scaled up the unknown to reduce the problem to the case $a = 1$. For example, problem 7 can be translated into the modern equation $11x^2 + 7x = 6\frac{1}{4}$. The scribe multiplied by 11 to turn the equation into a quadratic equation in $11x$: $(11x)^2 + 7 \cdot 11x = 68\frac{3}{4}$. He then solved

$$11x = \sqrt{\left(\frac{7}{2}\right)^2 + 68\frac{3}{4}} - \frac{7}{2} = \sqrt{81} - \frac{7}{2} = 9 - 3\frac{1}{2} = 5\frac{1}{2}.$$

To find x, the scribe would normally multiply by the reciprocal of 11, but in this case, he noted that the reciprocal of 11 "cannot be solved." Nevertheless, he realized, probably because the problem was manufactured to give a simple answer, that the unknown side x is equal to 1/2.

This idea of "scaling," combined with the geometrical coefficients discussed earlier, enabled the scribes to solve quadratic-type equations not directly involving squares. For example, consider the problem from TMS 20: The sum of the area and side of the convex square is 11/18. Find the side. We will translate this into the equation $A + s = 11/18$, where s is the quarter-circle arc forming one of the sides of the figure whose area is A. To solve this, the scribe used the coefficient 4/9 of the convex square as his scaling factor. Thus, he turned the equation into $(4/9)A + (4/9)s = 22/81$. But we know that the area A of the convex square is equal to $(4/9)s^2$. It follows that this equation can be rewritten as a quadratic equation for $(4/9)s$:

$$\left(\frac{4}{9}s\right)^2 + \frac{4}{9}s = \frac{22}{81}.$$

The scribe then solved this in the normal way to get $(4/9)s = 2/9$. He concluded by multiplying by the reciprocal 9/4 to find the answer $s = 1/2$.

Although the methods described above are the standard methods for solving quadratic equations, the scribes occasionally used other methods in particular situations. For example, in problem 23 of BM 13901, we are told that the sum of four sides and the (square) surface is 25/36. Although this problem is of the type $x^2 + bx = c$, in this case the b is four, the number of sides of the square, which is more "natural" than the coefficients we saw earlier. Modern scholars believe that this problem is an example of an original problem coming directly from the surveyors, a problem that then turns up in much later manifestations of this early tradition both in Islamic mathematics and in medieval European mathematics. The scribe's method here depends directly on the "four." In the first step of the solution, he took 1/4 of the 25/36 to get 25/144. To this he added 1, giving 169/144. The square root of this value is 13/12. Subtracting the 1 gives 1/12. Thus, the length of the side is twice that value, namely, 1/6. This new procedure is best illustrated by another diagram (Fig. 1.22). What the scribe intended is that the four "sides" are really projections of the actual sides of the square into rectangles of length 1. Taking 1/4 of the entire sum means that we are only considering the shaded gnomon, which is one-fourth of the original figure. When we add a square of side 1 to that figure, we get a square whose side we can then find. Subtracting the 1 from the side then gives us half of the original side of the square.

FIGURE 1.22

The sum of four sides and the square surface

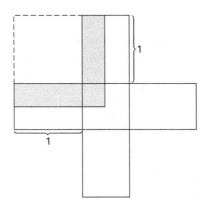

Other problems on BM 13901 deal with various situations involving squares and sides, with each of the solution procedures having a geometric interpretation. As a final example, we consider the problem $x^2 + y^2 = 13/36$, $x - y = 1/6$. The solution to this system, which we generalize into the system $x^2 + y^2 = c$, $x - y = b$, was found by a procedure describable by the modern formula

$$x = \sqrt{\frac{c}{2} - \left(\frac{b}{2}\right)^2} + \frac{b}{2} \qquad y = \sqrt{\frac{c}{2} - \left(\frac{b}{2}\right)^2} - \frac{b}{2}.$$

It appears that the Babylonians developed the solution by using the geometric idea expressed in Figure 1.23. This figure shows that

$$x^2 + y^2 = 2\left(\frac{x+y}{2}\right)^2 + 2\left(\frac{x-y}{2}\right)^2.$$

FIGURE 1.23

Geometric procedure for solving the system $x - y = b$, $x^2 + y^2 = c$

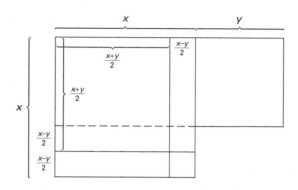

It follows that

$$c = 2 \left(\frac{x+y}{2} \right)^2 + 2 \left(\frac{b}{2} \right)^2$$

and therefore that

$$\frac{x+y}{2} = \sqrt{ \frac{c}{2} - \left(\frac{b}{2} \right)^2 }.$$

Because

$$x = \frac{x+y}{2} + \frac{x-y}{2} \quad \text{and} \quad y = \frac{x+y}{2} - \frac{x-y}{2},$$

the result follows.

 1.3

CONCLUSION

The extant papyri and tablets containing Egyptian and Babylonian mathematics were generally teaching documents, used to transmit knowledge from one scribe to another. Their function was to provide trainee scribes with a set of example-types, problems whose solutions could be applied in other situations. Learning mathematics for these trainees was learning how to select and perhaps modify an appropriate algorithm, and then mastering the arithmetic techniques necessary to carry out the algorithm to solve a new problem. The reasoning behind the algorithms was evidently transmitted orally, so that mathematicians today are forced to speculate as to the origins.

We note that although the long lists of quadratic problems on some of the Babylonian tablets were given as "real-world" problems, the problems are in fact just as contrived as the ones found in most current algebra texts. That the authors knew they were contrived is shown by the fact that, typically, all problems of a given set have the same answer. But since often the problems grew in complexity, it appears that the tablets were used to develop techniques of solution. One can speculate, therefore, that the study of mathematical problem solving, especially problems involving quadratic equations, was a method for training the minds of future leaders of the country. In other words, it was not really that important to solve quadratic equations—there were few real situations that required them. What was important was for the students to develop skills in solving problems in general, skills that could be used in dealing with the everyday problems that a nation's leaders need to solve. These skills included not only following well-established procedures—algorithms—but also knowing how and when to modify the methods and how to reduce more complicated problems to ones already solved. Today's students are often told that mathematics is studied to "train the mind." It seems that teachers have been telling their students the same thing for the past 4000 years.

EXERCISES

1. Represent 375 and 4856 in Egyptian hieroglyphics and Babylonian cuneiform.

2. Use Egyptian techniques to multiply 34 by 18 and to divide 93 by 5.

3. Use Egyptian techniques to multiply $2\,\overline{14}$ by $1\,\overline{2}\,\overline{4}$. (This is problem 9 of the *Rhind Mathematical Papyrus*.)

4. Use Egyptian techniques to multiply $\overline{28}$ by $1\,\overline{2}\,\overline{4}$. (This is problem 14 of the *Rhind Mathematical Papyrus*.)

5. Show that the solution to the problem of dividing 7 loaves among 10 men is that each man gets $\overline{\overline{3}}\,\overline{30}$. (This is problem 4 of the *Rhind Mathematical Papyrus*.)

6. Use Egyptian techniques to divide 100 by $7\,\overline{2}\,\overline{4}\,\overline{8}$. Show that the answer is $12\,\overline{\overline{3}}\,\overline{42}\,\overline{126}$. (This is problem 70 of the *Rhind Mathematical Papyrus*.)

7. Multiply $7\,\overline{2}\,\overline{4}\,\overline{8}$ by $12\,\overline{\overline{3}}$ using the Egyptian multiplication technique. Note that it is necessary to multiply each term of the multiplicand by $\overline{\overline{3}}$ separately.

8. A part of the *Rhind Mathematical Papyrus* table of division by 2 follows: $2 \div 11 = \overline{6}\,\overline{66}$, $2 \div 13 = \overline{8}\,\overline{52}\,\overline{104}$, $2 \div 23 = \overline{12}\,\overline{276}$. The calculation of $2 \div 13$ is given as follows:

1	13
$\overline{2}$	$6\,\overline{2}$
$\overline{4}$	$3\,\overline{4}$
$\overline{8}$	$1\,\overline{2}\,\overline{8}$
$\overline{52}$	$\overline{4}$
$\overline{104}$	$\overline{8}$
$\overline{8}\,\overline{52}\,\overline{104}$	$1\,\overline{2}\,\overline{4}\,\overline{8}\,\overline{8}$
	2

Perform similar calculations for the divisions of 2 by 11 and 23 to check the results.

9. Solve by the method of false position: A quantity and its 1/7 added together become 19. What is the quantity? (problem 24 of the *Rhind Mathematical Papyrus*)

10. Solve by the method of false position: A quantity and its 2/3 are added together and from the sum 1/3 of the sum is subtracted, and 10 remains. What is the quantity? (problem 28 of the *Rhind Mathematical Papyrus*)

11. A quantity, its 1/3, and its 1/4, added together, become 2. What is the quantity? (problem 32 of the *Rhind Mathematical Papyrus*)

12. Calculate a quantity such that if it is taken two times along with the quantity itself, the sum comes to 9. (problem 25 of the *Moscow Mathematical Papyrus*)

13. Problem 72 of the *Rhind Mathematical Papyrus* reads "100 loaves of *pesu* 10 are exchanged for loaves of *pesu* 45. How many of these loaves are there?" The solution is given as, "Find the excess of 45 over 10. It is 35. Divide this 35 by 10. You get $3\,\overline{2}$. Multiply $3\,\overline{2}$ by 100. Result: 350. Add 100 to this 350. You get 450. Say then that the exchange is 100 loaves of *pesu* 10 for 450 loaves of *pesu* 45."[18] Translate this solution into modern terminology. How does this solution demonstrate proportionality?

14. Solve problem 11 of the *Moscow Mathematical Papyrus*: The work of a man in logs; the amount of his work is 100 logs of 5 handbreadths diameter; but he has brought them in logs of 4 handbreadths diameter. How many logs of 4 handbreadths diameter are there?

15. Various conjectures have been made for the derivation of the Egyptian formula $A = (\frac{8}{9}d)^2$ for the area A of a circle of diameter d. One of these uses circular counters, known to have been used in ancient Egypt. Show by experiment using pennies, for example, whose diameter can be taken as 1, that a circle of diameter 9 can essentially be filled by 64 circles of diameter 1. (Begin with one penny in the center; surround it with a circle of six pennies, and so on.) Use the obvious fact that 64 circles of diameter 1 also fill a square

of side 8 to show how the Egyptians may have derived their formula.[19]

16. Some scholars have conjectured that the area calculated in problem 10 of the *Moscow Mathematical Papyrus* is that of a semicylinder rather than a hemisphere. Show that the calculation in that problem does give the correct surface area of a semicylinder of diameter and height both equal to $4\frac{1}{2}$.

17. Convert the fractions 7/5, 13/15, 11/24, and 33/50 to sexagesimal notation.

18. Convert the sexagesimal fractions 0;22,30, 0;08,06, 0;04,10, and 0;05,33,20 to ordinary fractions in lowest terms.

19. Find the reciprocals in base 60 of 18, 32, 54, and 64 (–1,04). (Do not worry about initial zeros, since the product of a number with its reciprocal can be any power of 60.) What is the condition on the integer n that ensures it is a regular sexagesimal, that is, that its reciprocal is a finite sexagesimal fraction?

20. In the Babylonian system, multiply 25 by 1,04 and 18 by 1,21. Divide 50 by 18 and 1,21 by 32 (using reciprocals). Use our standard multiplication algorithm modified for base 60.

21. Show that the area of the Babylonian "barge" is given by $A = (2/9)a^2$, where a is the length of the arc (one-quarter of the circumference). Also show that the length of the long transversal of the barge is $(17/18)a$ and the length of the short transversal is $(7/18)a$. (Use the Babylonian values of $C^2/12$ for the area of a circle and 17/12 for $\sqrt{2}$.)

22. Show that the area of the Babylonian "bull's-eye" is given by $A = (9/32)a^2$, where a is the length of the arc (one-third of the circumference). Also show that the length of the long transversal of the bull's-eye is $(7/8)a$, whereas the length of the short transversal is $(1/2)a$. (Use the Babylonian values of $C^2/12$ for the area of a circle and 7/4 for $\sqrt{3}$.)

23. For the concave square, the coefficient of the diagonal (the line from one vertex to the opposite vertex) is given as $1;20(= 1\frac{1}{3})$, while the coefficient of the tranversal (the line from the midpoint of one arc to the midpoint of the opposite arc) is given as $0;33,20(= 5/9)$. Show that both of these values are correct, given the normal Babylonian approximations.

24. Convert the Babylonian approximation 1;24,51,10 to $\sqrt{2}$ to decimals and determine the accuracy of the approximation.

25. Use the assumed Babylonian square root algorithm of the text to show that $\sqrt{3} \approx 1;45$ by beginning with the value 2. Find a three-sexagesimal-place approximation to the reciprocal of 1;45 and use it to calculate a three-sexagesimal-place approximation to $\sqrt{3}$.

26. Show that taking $v + u = 1;48 (= 1\frac{4}{5})$ leads to line 15 of Plimpton 322 and that taking $v + u = 2;05 (= 2\frac{1}{12})$ leads to line 9. Find the values for $v + u$ that lead to lines 6 and 13 of that tablet.

27. The scribe of Plimpton 322 did not use the value $v + u = 2;18,14,24$, with its associated reciprocal $v - u = 0;26,02,30$, in his work on the tablet. Find the smallest Pythagorean triple associated with those values.

28. Solve the problem from the Old Babylonian tablet BM 13901: The sum of the areas of two squares is 1525. The side of the second square is 2/3 that of the first plus 5. Find the sides of each square.

29. Solve the Babylonian problem taken from a tablet found at Susa: Let the width of a rectangle measure a quarter less than the length. Let 40 be the length of the diagonal. What are the length and width? Use false position, beginning with the assumption that 1 (or 60) is the length of the rectangle.

30. Solve the following problem from VAT 8391: One of two fields yields 2/3 *sila* per *sar*, the second yields 1/2 *sila* per *sar*. The sum of the yields of the two fields is 1100 *sila*; the difference of the areas of the two fields is 600 *sar*. How large is each field?

31. Solve the following problem from YBC 4652: I found a stone, but did not weigh it; after I subtracted one-seventh and then one-thirteenth [of the difference], it weighed 1 *mina* [= 60 *gin*]. What was the original weight of the stone?

32. Solve the following problem from YBC 4652: I found a stone, but did not weigh it; after I subtracted one-seventh, added one-eleventh [of the difference], and then subtracted one-thirteenth [of the previous total], it weighed 1 *mina* [= 60 *gin*]. What was the stone's weight?

33. Give a geometric argument to justify the Babylonian "quadratic formula" that solves the equation $x^2 - ax = b$.

34. Solve the following problem from tablet YBC 6967: A number exceeds its reciprocal by 7. Find the number and the reciprocal. (In this case, that two numbers are "reciprocals" means that their product is 60.)

35. Solve the following Babylonian problem about a concave square: The sum of the area, the arc, and the diagonal is $1;16,40(= 1\frac{5}{18})$. Find the length of the arc. (Recall that the coefficient of the area is 4/9 and the coefficient of the diagonal is 1 1/3—see Exercise 23.)

36. Solve the following problem from BM 13901: I added one-third of the square-side to two-thirds of the area of the square, and the result was 0;20 (= 1/3). Find the square-side.

37. Solve the following Babylonian problem from tablet IM 55357: Given the right triangle ABC with sides 0;45 and 1 and hypotenuse 1;15, as in Figure 1.24, suppose AD is perpendicular to BC, DE is perpendicular to AC, and EF is perpendicular to BC. Suppose further that the area of triangle ABD is 0;08,06, that of triangle ADE is 0;05,11,02,24, that of triangle DEF is 0;03,19,03,56,09,36, and that of EFC is 0;05,53,53,39,50,24. What are the lengths of AD, DE, EF, BD, DF, and FC?

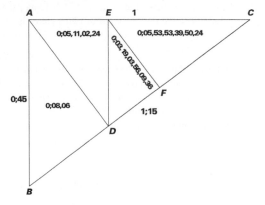

FIGURE 1.24

Tablet IM 55357 with a problem on triangles

38. Given a circle of circumference 60 and a chord of length 12, what is the perpendicular distance from the chord to the circumference? (This problem is from tablet BM 85194.)

39. Solve the following problem from tablet AO 8862: Length and width. I combined length and width and then I built an area. I turned around. I added half of the length and a third of the width to the middle of my area so that it was 15. I returned. I summed the length and width and it was 7. What are the length and width?

40. Construct two or three real-life division problems where giving the answer using just unit fractions, rather than other common fractions, makes sense.

41. Devise a lesson to teach ideas of proportionality by using the Egyptian method of false position.

42. Devise a lesson on place value using the Babylonian system and, in particular, using the multiplication table by 9 given in the text.

43. Devise a lesson teaching the quadratic formula using geometric arguments similar to the (assumed) Babylonian ones.

REFERENCES AND NOTES

The best source on Egyptian mathematics is Richard J. Gillings, *Mathematics in the Time of the Pharaohs* (Cambridge: MIT Press, 1972). See also Gillings, "The Mathematics of Ancient Egypt," *Dictionary of Scientific Biography* (New York: Scribners, 1978), vol. 15, 681–705. A more recent survey is James Ritter, "Egyptian Mathematics," in H. Selin, ed., *Mathematics across Cultures: The History of Non-Western Mathematics* (Dordrecht: Kluwer Academic Publishers, 2000), pp. 115–136. Finally, there is a new book in preparation on Egyptian mathematics by Annette Imhausen, soon to be published by Princeton University Press, which will probably supersede some of the earlier works. The *Rhind Mathematical Papyrus* is available in an edition by Arnold B. Chace (Reston, VA: National Council of Teachers of Mathematics, 1967). This work is an abridgement of the original publication by the Mathematical Association of America in 1927 and 1929. Another English version of that papyrus as well as the *Moscow Mathematical Papyrus* and other Egyptian mathematical fragments are available in Marshall Clagett, *Ancient Egyptian Science: A Source Book. Volume Three: Ancient Egyptian Mathematics* (Philadelphia: American Philosophical Society, 1999). New translations of some of these sources as well as a substantial number of Mesopotamian sources are in Victor Katz, ed., *The Mathematics of Egypt, Mesopotamia, China, India, and Islam: A Sourcebook* (Princeton: Princeton University Press, 2007).

The standard accounts of Babylonian mathematics are Otto Neugebauer, *The Exact Sciences in Antiquity* (Princeton: Princeton University Press, 1951; New York: Dover, 1969) and B. L. Van der Waerden, *Science Awakening I* (New York: Oxford University Press, 1961). More recent surveys are by Jens Høyrup, "Mathematics, Algebra, and Geometry," in *The Anchor Bible Dictionary*, David N. Freedman, ed. (New York: Doubleday, 1992), vol. IV, pp. 601–612, and by Jöran Friberg, "Mathematik," *Reallexikon der Assyriologie* 7 (1987–1990), pp. 531–585 (in English). Also, there will soon be a new book on Mesopotamian mathematics by Eleanor Robson, to be published by Princeton University Press, which will deal more with contextual issues than the earlier works cited here. It will undoubtedly be very much worth reading. Translations and analyses of the Babylonian tablets themselves are found principally in

Otto Neugebauer, *Mathematische Keilschrift-Texte* (New York: Springer, 1973, reprint of 1935 original), Otto Neugebauer and Abraham Sachs, *Mathematical Cuneiform Texts* (New Haven, CT: American Oriental Society, 1945), and Evert Bruins and M. Rutten, *Textes Mathématiques de Suse* (Paris: Paul Geuthner, 1961). Two technical works analyzing the Babylonian tablets are Jens Høyrup, *Lengths, Widths, Surfaces: A Portrait of Old Babylonian Algebra and Its Kin* (New York: Springer, 2002), and Eleanor Robson, *Mesopotamian Mathematics, 2100–1600 BC: Technical Constants in Bureaucracy and Education* (Oxford: Clarendon Press, 1999). More general surveys of Mesopotamian mathematics in the context of Mesopotamian society include chapter 3 of Jens Høyrup, *In Measure, Number, and Weight: Studies in Mathematics and Culture* (Albany: State University of New York Press, 1994), Eleanor Robson, "Mesopotamian Mathematics: Some Historical Background," in Victor Katz, ed., *Using History to Teach Mathematics* (Washington: MAA, 2000), pp. 149–158, and Eleanor Robson, "The Uses of Mathematics in Ancient Iraq, 6000–600 BC," in Selin, *Mathematics across Cultures*, pp. 93–113. Finally, the early Babylonian tokens are discussed in great detail and with many illustrations in Denise Schmandt-Besserat, *Before Writing: From Counting to Cuneiform* (Austin: University of Texas Press, 1992).

A book that discusses the mathematics of Egypt and Babylonia, along with that of other ancient societies comparatively, and also deals with the questions of transmission and a possible single origin of mathematics is B. L. Van der Waerden, *Geometry and Algebra in Ancient Civilizations* (New York: Springer, 1983).

1. Chace, *Rhind Mathematical Papyrus*, p. 27.

2. For the use of the zero in Egyptian architecture, see Dieter Arnold, *Building in Egypt* (New York: Oxford University Press, 1991), p. 17, and George Reisner, *Mycerinus: The Temples of the Third Pyramid at Giza* (Cambridge: Harvard University Press, 1931), pp. 76–77. For the use of the zero in Egyptian accounting, see Alexander Scharff, "Ein Rechnungsbuch des Königlichen Hofes aus der 13. Dynastie," *Zeitschrift für Ägyptische Sprache und Altertumskunde* 57 (1922), 58–59.

3. See Clagett, *Ancient Egyptian Science*, vol. 3, pp. 255–260, and Katz, *Sourcebook*, p. 21.

4. Clagett, *Ancient Egyptian Science*, vol. 3, p. 224.

5. Chace, *Rhind Mathematical Papyrus*, p. 69.

6. Gillings, *Mathematics in the Time of the Pharaohs*, p. 173.

7. Katz, *Sourcebook*, p. 39.

8. Gillings, *Mathematics in the Time of the Pharaohs*, p. 139. For further analysis, see Hermann Engels, "Quadrature of the Circle in Ancient Egypt," *Historia Mathematica* 4 (1977), 137–140.

9. Gillings, *Mathematics in the Time of the Pharaohs*, p. 188. See also Katz, *Sourcebook*, pp. 33–34, for further discussion.

10. Clagett, *Ancient Egyptian Science*, vol. 3, p. 218. See also Katz, *Sourcebook*, pp. 31–33, for a detailed discussion of this problem and its interpretation.

11. Mesopotamian tablets are generally, but not always, named by the library where they now reside. Thus, in the tablets mentioned in this chapter, we find W (Iraq Museum, Baghdad, Uruk excavations), YBC (Yale Babylonian Collection), BM (British Museum), Plimpton (Plimpton Collection at Columbia University), VAT (Vorderasiatischer Museum, Berlin), AO (Departement des Antiquités Orientales, Louvre, Paris), IM (Iraq Museum, Baghdad), and TMS (which stands for the book where these were first published: E. M. Bruins and M. Rutter, *Textes Mathématique de Suse* (Paris: Paul Geuthner, 1961)).

12. Robson, *Mesopotamian Mathematics*, pp. 50–54. This book has extensive discussions regarding geometrical coefficients.

13. Ibid., pp. 118–122. Besides the discussion here, there are translations of the tablets in which this and related grain piles appear on pp. 219–231.

14. This analysis of Plimpton 322 is based on Eleanor Robson's analysis as found in "Words and Pictures: New Light on Plimpton 322," *American Mathematical Monthly* 109 (2002), 105–120. A more technical discussion of the issues involved is in Eleanor Robson, "Neither Sherlock Holmes nor Babylon: A Reassessment of Plimpton 322," *Historia Mathematica* 28 (2001), 167–206. Both of these papers criticize an earlier assessment of Plimpton 322: R. Creighton Buck, "Sherlock Holmes in Babylon," *American Mathematical Monthly* 87 (1980), 335–345.

15. Neugebauer and Sachs, *Mathematical Cuneiform Texts*, p. 101.

16. See Katz, *Sourcebook*, p. 104. Also, see Jens Høyrup, "The Old Babylonian Square Texts—BM 13901 and YBC 4714," in Jens Høyrup and Peter Damerow, eds., *Changing Views on Ancient Near Eastern Mathematics* (Berlin: Dietrich Reimer Verlag, 2001). Both books have a complete translation of BM 13901, along with analyses of the various problems.

17. Ibid.

18. Chace, *Rhind Mathematical Papyrus*, p. 57.

19. This suggestion comes from Paulus Gerdes, "Three Alternate Methods of Obtaining the Ancient Egyptian Formula for the Area of a Circle," *Historia Mathematica* 12 (1985), 261–267. Two other possibilities are also presented in that article.

The Beginnings of Mathematics in Greece

Thales was the first to go to Egypt and bring back to Greece this study [geometry]; he himself discovered many propositions, and disclosed the underlying principles of many others to his successors, in some cases his method being more general, in others more empirical.

—Proclus's *Summary* (c. 450 CE) of Eudemus's *History* (c. 320 BCE)[1]

A report from a visit to Egypt with Plato by Simmias of Thebes in 379 BCE (from a dramatization by Plutarch of Chaeronea (first–second century CE)): "On our return from Egypt a party of Delians met us . . . and requested Plato, as a geometer, to solve a problem set them by the god in a strange oracle. The oracle was to this effect: the present troubles of the Delians and the rest of the Greeks would be at an end when they had doubled the altar at Delos. As they not only were unable to penetrate its meaning, but failed absurdly in constructing the altar . . . , they called on Plato for help in their difficulty. Plato . . . replied that the god was ridiculing the Greeks for their neglect of education, deriding, as it were, our ignorance and bidding us engage in no perfunctory study of geometry; for no ordinary or near-sighted intelligence, but one well versed in the subject, was required to find two mean proportionals, that being the only way in which a body cubical in shape can be doubled with a similar increment in all dimensions. This would be done for them by Eudoxus of Cnidus . . . ; they were not, however, to suppose that it was this the god desired, but rather that he was ordering the entire Greek nation to give up war and its miseries and cultivate the Muses, and by calming their passions through the practice of discussion and study of mathematics, so to live with one another that their relationships should be not injurious, but profitable."[2]

As the quotation and the (probably) fictional account indicate, a new attitude toward mathematics appeared in Greece sometime before the fourth century BCE. It was no longer sufficient merely to calculate numerical answers to problems. One now had to prove that the results were correct. To double a cube, that is, to find a new cube whose volume was twice that of the original one, is equivalent to determining the cube root of 2, and that was not a difficult problem numerically. The oracle, however, was not concerned with numerical calculation, but with geometric construction. That in turn depended on geometric proof by some logical argument, the earliest manifestation of such in Greece being attributed to Thales.

This change in the nature of mathematics, beginning around 600 BCE, was related to the great differences between the emerging Greek civilization and those of Egypt and Babylonia, from whom the Greeks learned. The physical nature of Greece with its many mountains and islands is such that large-scale agriculture was not possible. Perhaps because of this, Greece did not develop a central government. The basic political organization was the *polis,* or city-state. The governments of the city-state were of every possible variety but in general controlled populations of only a few thousand. Whether the governments were democratic or monarchical, they were not arbitrary. Each government was ruled by law and therefore encouraged its citizens to be able to argue and debate. It was perhaps out of this characteristic that there developed the necessity for proof in mathematics, that is, for argument aimed at convincing others of a particular truth.

Because virtually every city-state had access to the sea, there was constant trade, both in Greece itself and with other civilizations. As a result, the Greeks were exposed to many different peoples and, in fact, themselves settled in areas all around the eastern Mediterranean. In addition, a rising standard of living helped to attract able people from other parts of the world. Hence, the Greeks were able to study differing answers to fundamental questions about the world. They began to create their own answers. In many areas of thought, they learned not to accept what had been handed down from ancient times. Instead, they began to ask, and to try to answer, "Why?" Greek thinkers eventually came to the realization that the world around them was knowable, that they could discover its characteristics by rational inquiry. Hence, they were anxious to discover and expound theories in such fields as mathematics, physics, biology, medicine, and politics. And although Western civilization owes a great debt to Greek society in literature, art, and architecture, it is to Greek mathematics that we owe the idea of mathematical proof, an idea at the basis of modern mathematics and, by extension, at the foundation of our modern technological civilization.

This chapter discusses the Greek numerical system and then considers the contributions of the earliest Greek mathematicians beginning in the sixth century BCE. It then deals with the beginnings of the Greek approach to geometric problem solving and concludes with the work of Plato and Aristotle in the fourth century BCE on the nature of mathematics and the idea of logical reasoning.

2.1   THE EARLIEST GREEK MATHEMATICS

Unlike the situation with Egyptian and Babylonian mathematics, there are virtually no extant texts of Greek mathematics that were actually written in the first millennium BCE. What we have today are copies of copies of copies, where the actual written documents date from not much earlier than 1000 CE. And even then, the earliest complete texts (of which these are copies) are not from earlier than about 300 BCE. So to tell the story of early Greek mathematics, we are forced to rely on works that were originally written much later than the actual occurrences. Thus, given that these works do not always agree with each other, there is a considerable amount of controversy about some of the early developments. We will try to present the story as coherently as possible, but will note many areas in which scholarly opinion varies.

2.1.1 Greek Numbers

From what fragments exist from ancient times, and even from some of the copies, we do know that the Greeks represented numbers in a ciphered system using their alphabet, from as far back as the sixth century BCE. The representation was as shown in Table 2.1, where the letters ς (digamma) for 6, φ (koppa) for 90, and λ (sampi) for 900 are letters that by this time were no longer in use. Hence, 754 was written $\psi\nu\delta$ and 293 was written $\sigma\varphi\gamma$. To represent thousands, a mark was made to the left of the letters α through θ; for example, $'\theta$ represented 9000. Larger numbers still were written using the letter M to represent myriads (10,000), with the number of myriads written above: $M^\delta = 40,000$, $M^{'\zeta\rho o\epsilon} = 71,750,000$.

TABLE 2.1   *Representation of a number system used by the Greeks as early as the sixth century BCE.*

Letter	Value	Letter	Value	Letter	Value
α	1	ι	10	ρ	100
β	2	κ	20	σ	200
γ	3	λ	30	τ	300
δ	4	μ	40	υ	400
ϵ	5	ν	50	ϕ	500
ς	6	ξ	60	χ	600
ζ	7	o	70	ψ	700
η	8	π	80	ω	800
θ	9	φ	90	λ	900

Among the earliest extant inscriptions in this alphabetic cipher were numbers inscribed on the walls of the tunnel on the island of Samos constructed by Eupalinus around 550 BCE to bring water from a spring outside the capital city through a mountain to a point inside the city walls. Modern archaeological excavations of the tunnel have revealed that it was dug by two teams that met in the middle (Fig. 2.1). There are no records as to how the construction crews managed to keep digging in the correct direction, but there have been many theories as to how this was done. The latest archaeological evidence leads to the conclusion that the builders used the simplest possible mathematical techniques, such as lining up flags to make sure that the diggers kept digging in the right direction. And evidently the numbers on the walls, 10, 20, 30, . . . , 200 (from the south entrance) and 10, 20, 30, . . . , 300 (from the north entrance) were written to keep tabs on the distances dug. Although most of the tunnel is

FIGURE 2.1

Water tunnel on the island of Samos

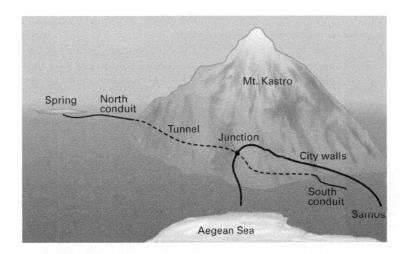

straight, there is one clear jog in the tunnel, probably necessitated by difficult soil conditions. Somehow, Eupalinus managed to figure out at that point how to get the digging back to the correct direction.

The numbers in the Eupalinus tunnel are integers. But Greek merchants and accountants, for example, needed fractions as well. Generally, in this early period, the Greeks used the Egyptian system of "parts." There was a special symbol $\angle$, which represented a half; $\overset{\smile}{\beta}$ represented two-thirds. For the rest, the system was standard: $\acute{\gamma}$ represented one-third, $\acute{\delta}$ one-fourth, and so on. More complicated fractions than simple parts are expressed as the sum of an integer and different simple parts. For example, the fraction we represent as 12/17 might be represented as $\angle \acute{\iota}\beta\acute{\iota}\zeta\acute{\lambda}\delta\acute{\nu}\alpha\acute{\xi}\eta$, which in modern notation would be $\frac{1}{2} + \frac{1}{12} + \frac{1}{17} + \frac{1}{34} + \frac{1}{51} + \frac{1}{68}$. We do not know if there was any systematic method for figuring out which unit fractions to use, for there are many possible ways to represent 12/17, or as the Greeks would say, the "seventeenth part of twelve." In addition, there is clearly the possibility of confusion between the representations of, for example, $\frac{1}{20} + \frac{1}{5}$ and $\frac{1}{25}$. But all those who needed to calculate evidently had methods of determining how they would use this system and how to avoid confusion.[3]

Fortunately for us, most of the early Greek mathematics we will discuss involves little calculation. As Aristotle wrote in his *Metaphysics*,

> At first, he who invented any art whatever that went beyond the common perceptions of man was naturally admired by men, not only because there was something useful in the inventions, but because he was thought wise and superior to the rest. But as more arts were invented, and some were directed to the necessities of life, others to recreation, the inventors of the latter were naturally always regarded as wiser than the inventors of the former, because their branches of knowledge did not aim at utility. Hence when all such inventions were already established, the sciences which do not aim at giving pleasure or at the necessities of life were discovered, and first in the places where men first began to have leisure. This is why the mathematical arts were founded in Egypt; for there the priestly caste was allowed to be at leisure.[4]

Although Aristotle referred only to Egypt, he certainly believed that in Greece as well mathematics was the province of a leisured class, people who did not deal with such mundane matters as measurement or accountancy problems. Thus, in Greece as in Egypt and Mesopotamia, mathematics of the type we will discuss in this chapter and the next was the province of a very limited group of people, virtually all of whom were part of the ruling groups. As we will see, this theoretical mathematics was to be a central part of the education of the rulers of the state.

2.1.2   Thales

FIGURE 2.2

Thales on a Greek stamp

The most complete reference to the earliest Greek mathematics is in the commentary to Book I of Euclid's *Elements* written in the fifth century CE by Proclus, some 800 to 1000 years after the fact. This account of the early history of Greek mathematics is generally thought to be a summary of a formal history written by Eudemus of Rhodes in about 320 BCE, the original of which is lost. In any case, the earliest Greek mathematician mentioned is Thales (c. 624–547 BCE), from Miletus in Asia Minor (Fig. 2.2). There are many stories recorded about him, most written down several hundred years after his death. These include his prediction of a solar eclipse in 585 BCE and his application of the angle-side-angle criterion of triangle congruence to the problem of measuring the distance to a ship at sea. He is said to have impressed Egyptian officials by determining the height of a pyramid by comparing the length of its shadow to that of the length of the shadow of a stick of known height. Thales is also credited with discovering the theorems that the base angles of an isosceles triangle are equal and that vertical angles are equal and with proving that the diameter of a circle divides the circle into two equal parts. Although exactly how Thales "proved" any of these results is not known, it does seem clear that he advanced some logical arguments.

Aristotle related the story that Thales was once reproved for wasting his time on idle pursuits. Therefore, noticing from certain signs that a bumper crop of olives was likely in a particular year, he quietly cornered the market on oil presses. When the large crop in fact was harvested, the olive growers all had to come to him for presses. He thus demonstrated that a philosopher or a mathematician could in fact make money if he thought it worthwhile. Whether this or any of the other stories are literally true is not known. In any case, the Greeks of the fourth century BCE and later credited Thales with beginning the Greek mathematical tradition. In fact, he is generally credited with beginning the entire Greek scientific enterprise, including recognizing that material phenomena are governed by discoverable laws.

2.1.3   Pythagoras and His School

FIGURE 2.3

Pythagoras on a Greek coin

There are also extensive but unreliable stories about Pythagoras (c. 572–497 BCE), including that he spent much time not only in Egypt, where Thales was said to have visited, but also in Babylonia (Fig. 2.3). Around 530 BCE, after having been forced to leave his native Samos, he settled in Crotona, a Greek town in southern Italy. There he gathered around him a group of disciples, later known as the Pythagoreans, in what was considered both a religious order and a philosophical school. From the surviving biographies, all written centuries after his death, we can infer that Pythagoras was probably more of a mystic than a rational thinker, but one who commanded great respect from his followers. Since there are no extant works ascribed to Pythagoras or the Pythagoreans, the mathematical doctrines of his school can only be surmised from the works of later writers, including the "neo-Pythagoreans."

One important such mathematical doctrine was that "number was the substance of all things," that numbers, that is, positive integers, formed the basic organizing principle of the universe. What the Pythagoreans meant by this was not only that all known objects have a number, or can be ordered and counted, but also that numbers are at the basis of all physical phenomena. For example, a constellation in the heavens could be characterized by both the number of stars that compose it and its geometrical form, which itself could be thought of as represented by a number. The motions of the planets could be expressed in terms of ratios of numbers. Musical harmonies depend on numerical ratios: two plucked strings with ratio of length $2 : 1$ give an octave, with ratio $3 : 2$ give a fifth, and with ratio $4 : 3$ give a fourth. Out of these intervals an entire musical scale can be created. Finally, the fact that triangles whose sides are in the ratio of $3 : 4 : 5$ are right-angled established a connection of number with angle. Given the Pythagoreans' interest in number as a fundamental principle of the cosmos, it is only natural that they studied the properties of positive integers, what we would call the elements of the theory of numbers.

The starting point of this theory was the dichotomy between the odd and the even. The Pythagoreans probably represented numbers by dots or, more concretely, by pebbles. Hence, an even number would be represented by a row of pebbles that could be divided into two equal parts. An odd number could not be so divided because there would always be a single pebble left over. It was easy enough using pebbles to verify some simple theorems. For example, the sum of any collection of even numbers is even, while the sum of an even collection of odd numbers is even and that of an odd collection is odd (Fig. 2.4).

FIGURE 2.4

(a) The sum of even numbers is even. (b) An even sum of odd numbers is even. (c) An odd sum of odd numbers is odd.

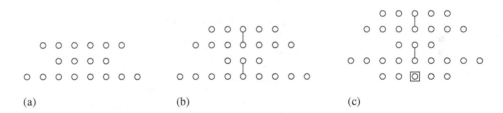

(a)   (b)   (c)

Among other simple corollaries of the basic results above were the theorems that the square of an even number is even, while the square of an odd number is odd. Squares themselves could also be represented using dots, providing simple examples of "figurate" numbers. If one represents a given square in this way, for example, the square of 4, it is easy to see that the next higher square can be formed by adding a row of dots around two sides of the original figure. There are $2 \cdot 4 + 1 = 9$ of these additional dots. The Pythagoreans generalized this observation to show that one can form squares by adding the successive odd numbers to 1. For example, $1 + 3 = 2^2$, $1 + 3 + 5 = 3^2$, and $1 + 3 + 5 + 7 = 4^2$. The added odd numbers were in the L shape generally called a gnomon (Fig. 2.5). Other examples of figurate numbers include the triangular numbers, also shown in Figure 2.5, produced by successive additions of the natural numbers themselves. Similarly, oblong numbers, numbers of the form $n(n + 1)$, are produced by beginning with 2 and adding the successive even numbers (Fig. 2.6). The first

FIGURE 2.5

Square and triangular numbers

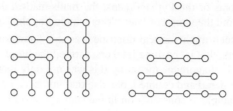

FIGURE 2.6

Oblong numbers

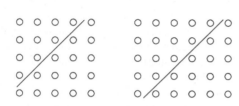

FIGURE 2.7

Two theorems on triangular numbers

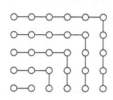

FIGURE 2.8

An odd square that is the difference of two squares

four of these are 2, 6, 12, and 20, that is, 1×2, 2×3, 3×4, and 4×5. Figure 2.7 provides easy demonstrations of the results that any oblong number is the double of a triangular number and that any square number is the sum of two consecutive triangular numbers.

Another number theoretical problem of particular interest to the Pythagoreans was the construction of Pythagorean triples. There is evidence that they saw that for an odd number n, the triple $(n, \frac{n^2-1}{2}, \frac{n^2+1}{2})$ is a Pythagorean triple, while if m is even, $(m, (\frac{m}{2})^2 - 1, (\frac{m}{2})^2 + 1)$ is such a triple. An explanation of how the Pythagoreans may have demonstrated the first of these results from their dot configurations begins with the remark that any odd number is the difference of two consecutive squares. Hence, if the odd number is itself a square, then three square numbers have been found such that the sum of two equals the third (Fig. 2.8). To find the sides of these squares, the Pythagorean triple itself, note that the side of the gnomon is given since it is the square of an odd number. The side of the smaller square is found by subtracting 1 from the gnomon and halving the remainder. The side of the larger square is one more than that of the smaller. A similar proof can be given for the second result. Although there is no explicit testimony to additional results involving Pythagorean triples, it seems probable that the Pythagoreans considered the odd and even properties of these triples. For example, it is not difficult to prove that in a Pythagorean triple, if one of the terms is odd, then two of them must be odd and one even.

The geometric theorem out of which the study of Pythagorean triples grew, namely, that in any right triangle the square on the hypotenuse is equal to the sum of the squares on the legs, has long been attributed to Pythagoras himself, but there is no direct evidence of this. The theorem was known in other cultures long before Pythagoras lived. Nevertheless, it was the knowledge of this theorem by the fifth century BCE that led to the first discovery of what is today called an irrational number.

For the early Greeks, number always was connected with things counted. Because counting requires that the individual units must remain the same, the units themselves can never be divided or joined to other units. In particular, throughout formal Greek mathematics, a number meant a "multitude composed of units," that is, a counting number. Furthermore, since the unit 1 was not a multitude composed of units, it was not considered a number in the same sense as the other positive integers. Even Aristotle noted that two was the smallest "number."

Because the Pythagoreans considered number as the basis of the universe, everything could be counted, including lengths. In order to count a length, of course, one needed a measure. The Pythagoreans thus assumed that one could always find an appropriate measure. Once such a measure was found in a particular problem, it became the unit and thus could not be divided. In particular, the Pythagoreans assumed that one could find a measure by which both the side and diagonal of a square could be counted. In other words, there should exist a length such that the side and diagonal were integral multiples of it. Unfortunately, this turned out not to be true. The side and diagonal of a square are **incommensurable**; there is no common measure. Whatever unit of measure is chosen such that an exact number will fit the length of one of these lines, the other line will require some number plus a portion of the unit, and one cannot divide the unit. (In modern terms, this result is equivalent to the statement that the square root of two is irrational.) We do not know who discovered this result, but scholars believe that the discovery took place in approximately 430 BCE. And although it is frequently stated that this discovery precipitated a crisis in Greek mathematics, the only reliable evidence shows that the discovery simply opened up the possibility of some new mathematical theories. In fact, Aristotle wrote in his *Metaphysics*,

> For all men begin, as we said, by wondering that things are as they are, as they do about self-moving marionettes, or about the solstices or the incommensurability of the diagonal of a square with the side; for it seems wonderful to all who have not yet seen the reason, that there is a thing which cannot be measured even by the smallest unit. But we must end in the contrary and, according to the proverb, the better state, as is the case in these instances too when men learn the cause; for there is nothing which would surprise a geometer so much as if the diagonal turned out to be commensurable.[5]

In other words, Aristotle seems to say that although the incommensurability is initially surprising, once one finds the reason—and clearly Greek thinkers did so—it then becomes very unsurprising.

So what is the "cause" of the incommensurability and how did a Greek thinker discover it? The only hint is in another work of Aristotle, who notes that if the side and diagonal are assumed commensurable, then one may deduce that odd numbers equal even numbers. One possibility as to the form of the discovery is the following: Assume that the side BD and diagonal DH in Figure 2.9 are commensurable, that is, that each is represented by the number of times it is measured by their common measure. It may be assumed that at least one of these numbers is odd, for otherwise there would be a larger common measure. Then the squares $DBHI$ and $AGFE$ on the side and diagonal, respectively, represent square numbers. The latter square is clearly double the former, so it represents an even square number. Therefore, its side $AG = DH$ also represents an even number and the square $AGFE$ is a multiple of four. Since $DBHI$ is half of $AGFE$, it must be a multiple of two; that is, it represents an even

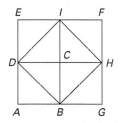

FIGURE 2.9

The incommensurability of the side and idagonal of a square (first possibility)

square. Hence, its side BD must also be even. But this contradicts the original assumption, that one of DH, BD, must be odd. Therefore, the two lines are incommensurable.

It must be realized that such a proof presupposes that by this time the notion of proof was ingrained into the Greek conception of mathematics. Although there is no evidence that the Greeks of the fifth century BCE possessed the entire mechanism of an axiomatic system and had explicitly recognized that certain statements need to be accepted without proof, they certainly had decided that some form of logical argument was necessary for determining the truth of a particular result. Furthermore, this entire notion of incommensurability represents a break from the Babylonian and Egyptian concepts of calculation with numbers. There is naturally no question that one can assign a numerical value to the length of the diagonal of a square of side one unit, as the Babylonians did, but the notion that no "exact" value can be found is first formally recognized in Greek mathematics.

Although the Greeks could not "measure" the diagonal of a square, that line, as a geometric object, was still significant. Plato, in his dialogue *Meno*, had Socrates question a slave boy about finding a square whose area is double that of square of side two feet. The boy first suggests that each side should be doubled. Socrates pointed out that this would give a square of area sixteen. The boy's second guess, that the new side should be three feet, is also evidently incorrect. So Socrates then led him to figure out that if one draws a diagonal of the original square and then constructs a square on that diagonal, the new square is exactly double the old one. But Socrates' proof of this is simply by a dissection argument (Fig. 2.10). There is no mention of the length of this diagonal at all.[6]

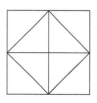

FIGURE 2.10

Dissection argument for determining the diagonal of a square

2.1.4 Squaring the Circle and Doubling the Cube

The idea of proof and the change from numerical calculation are further exemplified in the mid-fifth century attempts to solve two geometric problems, problems that were to occupy Greek mathematicians for centuries: the squaring of the circle (already attempted in Egypt) and the duplication of the cube (as noted in the oracle). The multitude of attacks on these particular problems and the slightly later one of trisecting an arbitrary angle serve to remind us that a central goal of Greek mathematics was geometrical problem solving, and that, to a large extent, the great body of theorems found in the major extant works of Greek mathematics served as logical underpinnings for these solutions. Interestingly, that these problems apparently could not be solved via the original tools of straightedge and compass was known to enough of the Greek public that Aristophanes could refer to "squaring the circle" as something absurd in his play *The Birds*, first performed in 414 BCE.

Hippocrates of Chios (mid-fifth century BCE) (no connection to the famous physician) was among the first to attack the cube and circle problems. As to the first of these, Hippocrates perhaps realized that the problem was analogous to the simpler problem of doubling a square of side a. That problem could be solved by constructing a mean proportional b between a and $2a$, a length b such that $a : b = b : 2a$, for then $b^2 = 2a^2$. From the fragmentary records of Hippocrates' work, it is evident that he was familiar with performing such constructions. In any case, ancient accounts record that Hippocrates was the first to come up with the idea of reducing the problem of doubling the cube of side a to the problem of finding two mean proportionals b, c, between a and $2a$. For if $a : b = b : c = c : 2a$, then

$$a^3 : b^3 = (a : b)^3 = (a : b)(b : c)(c : 2a) = a : 2a = 1 : 2$$

and $b^3 = 2a^3$. Hippocrates was not, however, able to construct the two mean proportionals using the geometric tools at his disposal. It was left to some of his successors to find this construction.

Hippocrates similarly made progress in the squaring of the circle, essentially by showing that certain lunes (figures bounded by arcs of two circles) could be "squared," that is, that their areas could be shown equal to certain regions bounded by straight lines. To do this, he first had to show that the areas of circles are to one another as the squares on their diameters, a fact evidently known to the Babylonian scribes. How he accomplished this is not known. In any case, he could now square the lune on a quadrant of a circle.

Suppose that semicircle ABC is circumscribed about the isosceles right triangle ABC and that around the base AC an arc ADC of a circle is drawn so that segment ADC is similar to segments AB and BC; that is, the arcs of each are the same fraction of a circle, in this case, one-quarter (Fig. 2.11). It follows from the result on areas of circles that similar segments are also to one another as the squares on their chords. Therefore, segment ADC is equal to the sum of segments AB and BC. If we add to each of these areas the part of the triangle outside arc ADC, it follows that the lune $ABCD$ is equal to the triangle ABC.

Although Hippocrates gave constructions for squaring other lunes or combinations of lunes, he was unable to actually square a circle. Nevertheless, it is apparent that his attempts on the squaring problem and the doubling problem were based on a large collection of geometric theorems, theorems that he organized into the first recorded book on the elements of geometry.

FIGURE 2.11

Hippocrates' lune on an isosceles right triangle

 2.2

THE TIME OF PLATO

The time of Plato (429–347 BCE) (Fig. 2.12) saw significant efforts made toward solving the problems of doubling the cube and squaring the circle and toward dealing with incommensurability and its impact on the theory of proportion. These advances were achieved partly because Plato's Academy, founded in Athens around 385 BCE, drew together scholars from all over the Greek world. These scholars conducted seminars in mathematics and philosophy with small groups of advanced students and also conducted research in mathematics, among other fields. There is an unverifiable story, dating from some 700 years after the school's founding, that over the entrance to the Academy was inscribed the Greek phrase ΑΓ ΕΩΜΕΤΡΗΤΟΣ ΜΗΔΕΙΣ ΕΙΣΙΤΩ, meaning roughly, "Let no one ignorant of geometry enter here." A student "ignorant of geometry" would also be ignorant of logic and hence unable to understand philosophy.

The mathematical syllabus inaugurated by Plato for students at the Academy is described by him in his most famous work, *The Republic*, in which he discussed the education that should be received by the philosopher-kings, the ideal rulers of a state. The mathematical part of this education was to consist of five subjects: arithmetic (that is, the theory of numbers), plane geometry, solid geometry, astronomy, and harmonics (music). The leaders of the state are "to practice calculation, not like merchants or shopkeepers for purposes of buying and selling, but with a view to war and to help in the conversion of the soul itself from the world of becoming to truth and reality. . . . It will further our intentions if it is pursued for the sake of knowledge and not for commercial ends. . . . It has a great power of leading the mind upwards and forcing it to reason about pure numbers, refusing to discuss collections of

FIGURE 2.12

Plato and Aristotle: A detail of Raphael's painting *The School of Athens* material things which can be seen and touched."[7] In other words, arithmetic is to be studied for the training of the mind (and incidentally for its military usefulness). The arithmetic of which Plato writes includes not only the Pythagorean number theory already discussed but also additional material that is included in Books VII–IX of Euclid's *Elements* and will be considered later.

Again, a limited amount of plane geometry is necessary for practical purposes, particularly in war, when a general must be able to lay out a camp or extend army lines. But even though mathematicians talk of operations in plane geometry such as squaring or adding, the object of geometry, according to Plato, is not to *do* something but to gain knowledge, "knowledge, moreover, of what eternally exists, not of anything that comes to be this or that at some time and ceases to be."[8] So, as in arithmetic, the study of geometry—and for Plato this means theoretical, not practical, geometry—is for "drawing the soul towards truth." It is importanat to mention here that Plato distinguished carefully between, for example, the real geometric circles drawn by people and the essential or ideal circle, held in the mind, which is the true object of geometric study. In practice, one cannot draw a circle and its tangent with only one point in common, although this is the nature of the mathematical circle and the mathematical tangent.

The next subject of mathematical study should be solid geometry. Plato complained in the *Republic* that this subject has not been sufficiently investigated. This is because "no state thinks [it] worth encouraging" and because "students are not likely to make discoveries without a director, who is hard to find."[9] Nevertheless, Plato felt that new discoveries would be made in this field, and, in fact, much was done between the dramatic date of the dialogue (about 400 BCE) and the time of Euclid, some of which is included in Books XI–XIII of the *Elements*.

In any case, a decent knowledge of solid geometry was necessary for the next study, that of astronomy, or, as Plato puts it, "solid bodies in circular motion." Again, in this field Plato distinguished between the stars as material objects with motions showing accidental irregularities and variations and the ideal abstract relations of their paths and velocities expressed in numbers and perfect figures such as the circle. It is this mathematical study of ideal bodies that is the true aim of astronomical study. Thus, this study should take place by means of problems and without attempting to actually follow every movement in the heavens.

Similarly, a distinction is made in the final subject, of harmonics, between material sounds and their abstraction. The Pythagoreans had discovered the harmonies that occur when strings are plucked together with lengths in the ratios of certain small positive integers. But in encouraging his philosopher-kings in the study of harmonics, Plato meant for them to go beyond the actual musical study, using real strings and real sounds, to the abstract level of "inquiring which numbers are inherently consonant and which are not, and for what reasons."[10] That is, they should study the mathematics of harmony, just as they should study the mathematics of astronomy, and should not be overly concerned with real stringed instruments or real stars. It turns out that a principal part of the mathematics necessary in both studies is the theory of ratio and proportion, the subject matter of Euclid's *Elements*, Book V.

Although it is not known whether the entire syllabus discussed by Plato was in fact taught at the Academy, it is certain that Plato brought in the best mathematicians of his day to teach and do research, including Theaetetus (c. 417–369 BCE) and Eudoxus (c. 408–355 BCE), who we will discuss later. The most famous person associated with the Academy, however, was Aristotle.

ARISTOTLE

Aristotle (384–322 BCE) (Fig. 2.13) studied at Plato's Academy in Athens from the time he was 18 until Plato's death in 347. Shortly thereafter, he was invited to the court of Philip II of Macedon to undertake the education of Philip's son Alexander, who soon after his own accession to the throne in 335 began his successful conquest of the Mediterranean world (Fig. 2.14). Meanwhile, Aristotle returned to Athens where he founded his own school, the Lyceum, and spent the rest of his days writing, lecturing, and holding discussions with his advanced students. Although Aristotle wrote on many subjects, including politics, ethics, epistemology, physics, and biology, his strongest influence as far as mathematics was concerned was in the area of logic.

FIGURE 2.13

Bust of Aristotle

2.3.1 Logic

Although there is only fragmentary evidence of logical argument in mathematical works before the time of Euclid, some appearing in the work of Hippocrates already mentioned, it is apparent that from at least the sixth century BCE, the Greeks were developing the notions of logical reasoning. The active political life of the city-states encouraged the development of argumentation and techniques of persuasion. And there are many examples from philosophical works, especially those of Parmenides (late sixth century BCE) and his disciple Zeno of Elea (fifth century BCE), that demonstrate various detailed techniques of argument. In particular, there are examples of such techniques as *reductio ad absurdum*, in which one assumes that a proposition to be proved is false and then derives a contradiction, and *modus tollens*, in which one shows first that if A is true, then B follows, shows next that B is not true, and concludes finally that A is not true. It was Aristotle, however, who took the ideas developed over the centuries and first codified the principles of logical argument.

FIGURE 2.14

Painting of Alexander on horseback

Aristotle believed that logical arguments should be built out of **syllogisms**, where "a syllogism is discourse in which, certain things being stated, something other than what is stated follows of necessity from their being so."[11] In other words, a syllogism consists of certain statements that are taken as true and certain other statements that are then necessarily true. For example, the argument "if all monkeys are primates, and all primates are mammals, then it follows that all monkeys are mammals," exemplifies one type of syllogism, whereas the argument "if all Catholics are Christians and no Christians are Moslem, then it follows that no Catholic is Moslem," exemplifies a second type.

After clarifying the principles of dealing with syllogisms, Aristotle noted that syllogistic reasoning enables one to use "old knowledge" to impart new. If one accepts the premises of a syllogism as true, then one must also accept the conclusion. One cannot, however, obtain every piece of knowledge as the conclusion of a syllogism. One has to begin somewhere with truths that are accepted without argument. Aristotle distinguished between the basic truths that are peculiar to each particular science and the ones that are common to all. The former are often called **postulates**, while the latter are known as **axioms**. As an example of a common truth, he gave the axiom "take equals from equals and equals remain." His examples of peculiar truths for geometry are "the definitions of line and straight." By these he presumably meant that one postulates the existence of straight lines. Only for the most basic ideas did Aristotle permit the postulation of the object defined. In general, however, whenever one defines an object, one must in fact prove its existence. "For example, arithmetic assumes the meaning of odd and even, square and cube, geometry that of incommensurable, . . . , whereas the existence of these attributes is demonstrated by means of the axioms and from previous conclusions as premises."[12] Aristotle also listed certain basic principles of argument, principles that earlier thinkers had used intuitively. One such principle is that a given assertion cannot be both true and false. A second principle is that an assertion must be either true or false; there is no other possibility.

For Aristotle, logical argument according to his methods is the only certain way of attaining scientific knowledge. There may be other ways of gaining knowledge, but demonstration via a series of syllogisms is the one way by which one can be sure of the results. Because one cannot prove everything, however, one must always be careful that the premises, or axioms, are true and well known As Aristotle wrote, "syllogism there may indeed be without these conditions, but such syllogism, not being productive of scientific knowledge, will not be demonstration."[13] In other words, one can choose any axioms one wants and draw conclusions from them, but if one wants to attain knowledge, one must start with "true" axioms. The question then becomes, how can one be sure that one's axioms are true? Aristotle answered that these primary premises are learned by induction, by drawing conclusions from our own sense perception of numerous examples. This question of the "truth" of the basic axioms has been discussed by mathematicians and philosophers ever since Aristotle's time. On the other hand, Aristotle's rules of attaining knowledge by beginning with axioms and using demonstrations to gain new results has become the model for mathematicians to the present day.

Although Aristotle emphasized the use of syllogisms as the building blocks of logical arguments, Greek mathematicians apparently never used them. They used other forms, as have most mathematicians down to the present. Why Aristotle therefore insisted on syllogisms is not clear. The basic forms of argument actually used in mathematical proof were analyzed in some detail in the third century BCE by the Stoics, of whom the most prominent was Chrysippus (280–206 BCE). This form of logic is based on **propositions**, statements that can be either true or false, rather than on the Aristotelian syllogisms. The basic rules of inference dealt with by Chrysippus, with their traditional names, are the following, where p, q, and r stand for propositions:

(1) *Modus ponens*
If p, then q.
p.
Therefore, q.

(2) *Modus tollens*
If p, then q.
Not q.
Therefore, not p.

(3) *Hypothetical syllogism*
If p, then q.
If q, then r.
Therefore, if p, then r.

(4) *Alternative syllogism*
p or q.
Not p.
Therefore, q.

For example, from the statements "if it is daytime, then it is light" and "it is daytime," one can conclude by *modus ponens* that "it is light." From "if it is daytime, then it is light" and "it is not light," one concludes by *modus tollens* that "it is not daytime." Adding to the first hypothesis the statement "if it is light, then I can see well," one concludes by the hypothetical syllogism that "if it is daytime, then I can see well." Finally, from "either it is daytime or it is nighttime" and "it is not daytime," the rule of the alternative syllogism allows us to conclude that "it is nighttime."

2.3.2 Number versus Magnitude

Another of Aristotle's contributions was the introduction into mathematics of the distinction between number and magnitude. The Pythagoreans had insisted that all was number, but Aristotle rejected that idea. Although he placed number and magnitude in a single category, "quantity," he divided this category into two classes, the discrete (number) and the continuous (magnitude). As examples of the latter, he cited lines, surfaces, bodies, and time. The primary distinction between these two classes is that a magnitude is "that which is divisible into divisibles that are infinitely divisible,"[14] while the basis of number is the indivisible unit. Thus, magnitudes cannot be composed of indivisible elements, whereas numbers inevitably are.

Aristotle further clarified this idea in his definition of "in succession" and "continuous." Things are **in succession** if there is nothing of their own kind intermediate between them. For example, the numbers 3 and 4 are in succession. Things are **continuous** when they touch and when "the touching limits of each become one and the same."[15] Line segments are therefore continuous if they share an endpoint. Points cannot make up a line, because they would have to be in contact and share a limit. Since points have no parts, this is impossible. It is also impossible for points on a line to be in succession, that is, for there to be a "next point." For between two points on a line is a line segment, and one can always find a point on that segment.

Today, a line segment is considered to be composed of an infinite collection of points, but to Aristotle this would make no sense. He did not conceive of a completed or actual infinity. Although he used the term "infinity," he only considered it as potential. For example, one can bisect a continuous magnitude as often as one wishes, and one can count these bisections. But in neither case does one ever come to an end. Furthermore, mathematicians really do not need infinite quantities such as infinite straight lines. They only need to postulate the existence of, for example, arbitrarily long straight lines.

2.3.3 Zeno's Paradoxes

One of the reasons Aristotle had such an extended discussion of the notions of infinity, indivisibles, continuity, and discreteness was that he wanted to refute the famous paradoxes of Zeno. Zeno stated these paradoxes, perhaps in an attempt to show that the then current notions of motion were not sufficiently clear, but also to show that any way of dividing space or time must lead to problems. The first paradox, the *Dichotomy*, "asserts the non-existence of motion on the ground that that which is in locomotion must arrive at the half-way stage before it arrives at the goal."[16] (Of course, it must then cover the half of the half before it reaches the middle, etc.) The basic contention here is that an object cannot cover a finite distance by moving during an infinite sequence of time intervals. The second paradox, the *Achilles*, asserts a similar point: "In a race, the quickest runner can never overtake the slowest, since the pursuer must first reach the point whence the pursued started, so that the slower must always hold a lead."[17] Aristotle, in refuting the paradoxes, concedes that time, like distance, is infinitely divisible. But he is not bothered by an object covering an infinity of intervals in a finite amount of time. For "while a thing in a finite time cannot come in contact with things quantitatively infinite, it can come in contact with things infinite in respect to divisibility, for in this sense time itself is also infinite."[18] In fact, given the motion in either of these paradoxes, one can calculate when one will reach the goal or when the fastest runner will overtake the slowest.

Zeno's third and fourth paradoxes show what happens when one asserts that a continuous magnitude is composed of indivisible elements. The *Arrow* states that "if everything when it occupies an equal space is at rest, and if that which is in locomotion is always occupying such a space at any moment, the flying arrow is therefore motionless."[19] In other words, if there are such things as indivisible instants, the arrow cannot move during that instant. Since if, in addition, time is composed of nothing but instants, then the moving arrow is always at rest. Aristotle refutes this paradox by noting that not only are there no such things as indivisible instants, but motion itself can only be defined in a period of time. A modern refutation, on the other hand, would deny the first premise because motion is now defined by a limit argument.

The paradox of the *Stadium* supposes that there are three sets of identical objects: the A's at rest, the B's moving to the right past the A's, and the C's moving to the left with equal velocity. Suppose the B's have moved one place to the right and the C's one place to the left, so that B_1, which was originally under A_4, is now under A_5, while C_1, originally under A_5, is now under A_4 (Fig. 2.15). Zeno supposes that the objects are indivisible elements of space and that they move to their new positions in an indivisible unit of time. But since there must have been a moment at which B_1 was directly over C_1, there are two possibilities. Either the two objects did not cross, and so there was no motion at all, or in the indivisible instant, each object had occupied two separate positions, so that the instant was in fact not indivisible. Aristotle believed that he had refuted this paradox because he had already denied the original assumption—that time is composed of indivisible instants.

FIGURE 2.15

Zeno's parardox of the *Stadium*

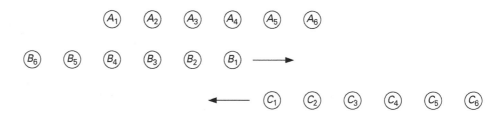

Interestingly, the four paradoxes exhaust the four possibilities of divisibility/indivisibility of space and time. That is, in the *Arrow* both space and time are assumed infinitely divisible, in the *Stadium* both are assumed ultimately indivisible, in the *Dichotomy* space is assumed divisible and time indivisible, and in the *Achilles* the reverse is assumed. So Zeno has shown each of the four possibilities leads to a contradiction.

Controversy regarding these paradoxes has lasted throughout history. The ideas contained in Zeno's statements and Aristotle's attempts at refutation have been extremely fruitful in forcing mathematicians to the present day to think carefully about their assumptions in dealing with the concepts of the infinite or the infinitely small. And in Greek times they were probably a significant factor in the development of the distinction between continuous magnitude and discrete number so important to Aristotle and ultimately to Euclid.

EXERCISES

1. Represent 125, 62, 4821, and 23,855 in the Greek alphabetic notation.

2. Represent 8/9 as a sum of distinct unit fractions. Express the result in the Greek notation. Note that the answer to this problem is not unique.

3. Represent 200/9 as the sum of an integer and distinct unit fractions. Express the result in Greek notation.

4. There are extant Greek land surveys that give measurements of fields and then find the area so the land can be assessed for tax purposes. In general, areas of quadrilateral fields were approximated by multiplying together the averages of the two pairs of opposite sides. In one document, one pair of sides is given as $a = 1/4 + 1/8 + 1/16 + 1/32$ and $c = 1/8 + 1/16$, where the lengths are in fractions of a *schonion*, a measure of approximately 150 feet. The second pair of sides is given as $b = 1/2 + 1/4 + 1/8$ and $d = 1$. Find the average of a and c, the average of b and d, and multiply them together to show that the area of the field is approximately $1/4 + 1/16$ square *schonion*. Note that the taxman has rounded up the exact answer (presumably to collect more taxes).

5. Thales is said to have invented a method of finding distances of ships from shore by use of the angle-side-angle theorem. Here is a possible method: Suppose A is a point on shore and S is a ship (Fig. 2.16). Measure the distance AC along a perpendicular to AC and bisect it at B. Draw CE at right angles to AC and pick point E on it in a straight line with B and S. Show that $\triangle EBC \cong \triangle SBA$ and therefore that $SA = EC$.

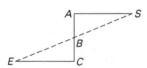

FIGURE 2.16

One method Thales could have used to determine the distance to a ship at sea

6. A second possibility for Thales' method is the following: Suppose Thales was atop a tower on the shore with an instrument made of a straight stick and a crosspiece AC that could be rotated to any desired angle and then would remain where it was put (Fig. 2.17). One rotates AC until one sights the ship S, then turns and sights an object T on shore without moving the crosspiece. Show that $\triangle AET \cong \triangle AES$ and therefore that $SE = ET$.

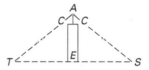

FIGURE 2.17

Second method Thales could have used to determine the distance to a ship at sea

7. Suppose Thales found that at the time a stick of length 6 feet cast a shadow of 9 feet, there was a length of 342 feet from the edge of the pyramid's side to the tip of its shadow. Suppose further that the length of a side of the pyramid was 756 feet. Find the height of the pyramid. (Assuming that the pyramid is laid out so the sides are due north-south and due east-west, this method requires that the sun be exactly in the south when the measurement is taken. When does this occur?[20])

8. Show that the nth triangular number is represented algebraically as $T_n = \frac{n(n+1)}{2}$ and therefore that an oblong number is double a triangular number.

9. Show algebraically that any square number is the sum of two consecutive triangular numbers.

10. Show using dots that eight times any triangular number plus 1 makes a square. Conversely, show that any odd square diminished by 1 becomes eight times a triangular number. Show these results algebraically as well.

11. Show that in a Pythagorean triple, if one of the terms is odd, then two of them must be odd and one even.

12. Construct five Pythagorean triples using the formula $(n, \frac{n^2-1}{2}, \frac{n^2+1}{2})$, where n is odd. Construct five different ones using the formula $(m, (\frac{m}{2})^2 - 1, (\frac{m}{2})^2 + 1)$, where m is even.

13. Show that if a right triangle has one leg of length 1 and a hypotenuse of length 2, then the second leg is incommensurable with the first leg. (In modern terms, this is equivalent to showing that $\sqrt{3}$ is irrational.) Use an argument similar to the proposed Pythagorean argument that the diagonal of a unit square is incommensurable with the side.

14. Show that the areas of similar segments of circles are proportional to the squares on their chords. Assume the result that the areas of circles are proportional to the squares on their diameters.

15. Here is another lune that was "squared" by Hippocrates: Construct a trapezoid $BACD$ such that $BA = AC = CD$ and the square on BD is triple the square on each of the other sides (Fig. 2.18). Then circumscribe a circle around the trapezoid and describe on side BD a circular arc similar to those on the other three sides, that is, an arc whose ratio to side BD is equal to that of the arc on BA to the side BA. Show that the segment on BD is equal to the sum of the segments on BA, AC, and CD. Conclude that the lune bounded by the arcs $BACD$ and BED is equal to the original trapezoid. (Note that you should first prove that the given trapezoid can be constructed and that it can be circumscribed by a circle.)

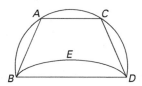

FIGURE 2.18

Hippocrates' lune with outer arc greater than a semicircle

16. Read the entire passage from Plato's *Meno* referred to in the text and write a short essay discussing Socrates' method of convincing the slave boy that he knows how to construct a square double a given square. Consider both the "Socratic method" that Socrates uses as well as the mathematics. (It may be a good idea to do this as a "play" with different students playing the various roles.[21])

17. Consider the quotation from Plato's *Republic:* "It will further our intentions if it [calculation] is pursued for the sake of knowledge and not for commercial ends." Discuss the relevance of this statement to current discussions on the purposes for studying mathematics in school.

18. Give two further examples of each of the two types of syllogisms mentioned in the text.

19. Make up a purposely incorrect syllogism that is related to the correct models in the text. Discuss why its conclusion may be false.

20. Give an example of each of the four rules of inference discussed in the text.

21. In Zeno's *Achilles* paradox, assume the quick runner Achilles is racing against a tortoise. Assume further that the tortoise has a 500-yard head start but that Achilles' speed is fifty times that of the tortoise. Finally, assume that the tortoise moves 1 yard in 5 seconds. Determine the time t it will take until Achilles overtakes the tortoise and the distance d he will have traveled. Note that Achilles must first travel 500 yards to reach the point where the tortoise started. This will take 50 seconds. But in that time the tortoise will move 10 yards farther. Continue this analysis by writing down the sequence of distances that Achilles must travel to reach the point where the tortoise had already been. Show that the sum of this infinite sequence of distances is equal to the distance d calculated first.

REFERENCES AND NOTES

A good source of basic information on Greek civilization is H. D. F. Kitto, *The Greeks* (London: Penguin, 1951). Two excellent general works on early Greek science are by G. E. R. Lloyd, *Early Greek Science: Thales to Aristotle* (New York: Norton, 1970) and *Magic, Reason and Experience* (Cambridge: Cambridge University Press, 1979). The latter work, in particular, deals with the beginnings of logical reasoning in Greece and the emergence of the idea of mathematical proof. The standard reference on Greek mathematics is Thomas Heath, *A History of Greek Mathematics* (New York: Dover, 1981, reprinted from the 1921 original). However, many of Heath's conclusions have been challenged in more recent works. The two best reevaluations of some central parts of the story of Greek mathematics are Wilbur Knorr, *The Ancient Tradition of Geometric Problems* (Boston:

Birkhäuser, 1986), which argues that geometric problem solving was the motivating factor for much of Greek mathematics, and David Fowler's *The Mathematics of Plato's Academy: A New Reconstruction* (Oxford: Clarendon Press, 1987; 2nd edition, 1999), which claims that the idea of *anthyphairesis* (reciprocal subtraction) provides much of the impetus for the Greek development of the ideas of ratio and proportion. A newer work, Serafina Cuomo's *Ancient Mathematics* (London: Routledge, 2001), provides an excellent survey of Greek mathematics, while claiming that many of Heath's (and others') conclusions are based on very flimsy evidence. The emergence of the deductive method in Greek mathematics is discussed in Reviel Netz, *The Shaping of Deduction in Greek Mathematics: A Study in Cognitive History* (Cambridge: Cambridge University Press, 1999). An earlier, but still useful, work on the same topic is I. Mueller, *Philosophy and Deductive Structure in Euclid's Elements* (Cambridge: MIT Press, 1981). Many of the available fragments from the earliest Greek mathematics are collected in Ivor Thomas, *Selections Illustrating the History of Greek Mathematics* (Cambridge: Harvard University Press, 1941).

1. From Proclus's *Summary*, translated in Thomas, *Selections*, I, p. 147.

2. *Plutarch's Moralia*, translated by Phillip H. De Lang and Benedict Einarson (Cambridge: Harvard University Press, 1959), VII, pp. 397–399.

3. See Fowler, *Mathematics of Plato's Academy*, chapter 7, for more on Greek numbers and fractions.

4. Aristotle, *Metaphysics*, 981^b, 14–24. The translations here and below are in the *Great Books* edition (Chicago: Encyclopedia Britannica, 1952), but the references here and to the works of Plato are to lines in the standard Greek text and can be checked in any modern translation.

5. Ibid., 983^a, 14–20.

6. The passage about Socrates and the slave boy is found in Plato, *Meno*, 82^b–85^b.

7. Plato, *Republic* VII, 525. The translation used is that of Frances Cornford.

8. Ibid., VII, 527.

9. Ibid., VII, 528.

10. Ibid., VII, 531.

11. Aristotle, *Prior Analytics* I, 1, 24^b, 19.

12. Aristotle, *Posterior Analytics* I, 10, 76^a, 40–76^b, 10.

13. Ibid., I, 2, 71^b, 23.

14. Aristotle, *Physics* VI, 1, 231^b, 15.

15. Ibid., V, 3, 227^a, 12.

16. Ibid., VI, 9, 239^b, 11. For more on Zeno's paradoxes, see F. Cajori, "History of Zeno's Arguments on Motion," *American Mathematical Monthly* 22 (1915), 1–6, 39–47, 77–82, 109–115, 145–149, 179–186, 215–220, 253–258, 292–297, and H. D. P. Lee, *Zeno of Elea* (Cambridge: Cambridge University Press, 1936).

17. Ibid., VI, 9, 239^b, 15.

18. Ibid., VI, 2, 233^a, 26–29.

19. Ibid., VI, 9, 239^b, 6.

20. For some speculation on how Thales might have accomplished his task, see Lothar Redlin, Ngo Viet, and Saleem Watson, "Thales' Shadow," *Mathematics Magazine* 73 (2000), 347–353.

21. For the text of a large portion of the *Meno* with discussions on how to use this in class, see Victor J. Katz and Karen Michalowicz, eds., *Historical Modules for the Teaching and Learning of Mathematics*, CD (Washington, DC: Mathematical Association of America, 2005), Geometry Module.

Euclid

Not much younger than these [Hermotimus of Colophon and Philippus of Mende, students of Plato] is Euclid, who put together the Elements, collecting many of Eudoxus's theorems, perfecting many of Theaetetus's, and also bringing to irrefragable demonstration the things which were only somewhat loosely proved by his predecessors. This man lived in the time of the first Ptolemy.

—Proclus's *Summary* (c. 450 CE) of
Eudemus's *History* (c. 320 BCE)[1]

Two legends about Euclid: Ptolemy is said to have asked him if there was any shorter way to geometry than through the *Elements*, and he replied that there was "no royal road to geometry." And, according to Stobaeus (fifth century CE), a student, after learning the first theorem, asked Euclid, "What shall I get by learning these things?" Euclid then asked his slave to give the student a coin, "since he must make gain out of what he learns."[2]

FIGURE 3.1

Euclid (detail from Raphael's painting *The School of Athens*). Note that there is no evidence of Euclid's actual appearance.

Since the first Ptolemy, Ptolemy I Soter, the Macedonian general of Alexander the Great who became ruler of Egypt after the death of Alexander in 323 BCE and lived until 283 BCE, it is generally assumed from the quotation from Proclus that Euclid flourished around 300 BCE (Fig. 3.1). But besides this date, written down some 750 years later, there is nothing at all known about the life of the author of the *Elements*. Nevertheless, most historians believe that Euclid was one of the first scholars active at the Museum and Library at Alexandria, founded by Ptolemy I and his successor, Ptolemy II Philadelphus. "Museum" here means a "Temple of the Muses," that is, a location where scholars meet and discuss philosophical and literary ideas. The Museum was to be, in effect, a government research establishment. The Fellows of the Museum received stipends and free board and were exempt from taxation. In this way the rulers of Egypt hoped that men of eminence would be attracted there from the entire Greek world. In fact, the Museum and Library soon became a focal point of the highest developments in Greek scholarship, both in the humanities and the sciences. The Fellows were initially appointed to carry on research, but since younger students gathered there as well, the Fellows soon turned to teaching. The aim of the Library was to collect the entire body of Greek literature in the best available copies and to organize it systematically. Ship captains who sailed from Alexandria were instructed to bring back scrolls from every port they touched until their return. The story is told that Ptolemy III, who reigned from 247–221 BCE, borrowed the authorized texts of the playwrights Aeschylus, Sophocles, and Euripides from Athens against a large deposit. But rather than return the originals, he returned only copies. He was quite willing to forfeit the deposit. The Library ultimately contained over 500,000 volumes in every field of knowledge. Although parts of the library were destroyed in various wars, some of it remained intact until the fourth century CE.

This chapter will be devoted primarily to a study of Euclid's most important work, the *Elements*, but we will also consider Euclid's *Data*.

 3.1 INTRODUCTION TO THE *ELEMENTS*

The *Elements* of Euclid is the most important mathematical text of Greek times and probably of all time. It has appeared in more editions than any work other than the *Bible*. It has been translated into countless languages and has been continuously in print in one country or another nearly since the beginning of printing. Yet to the modern reader the work is incredibly dull. There are no examples; there is no motivation; there are no witty remarks; there is no calculation. There are simply definitions, axioms, theorems, and proofs. Nevertheless, the book has been intensively studied. Biographies of many famous mathematicians indicate that Euclid's work provided their initial introduction into mathematics, that it in fact excited them and motivated them to become mathematicians. It provided them with a model of how "pure mathematics" should be written, with well-thought-out axioms, precise definitions, carefully stated theorems, and logically coherent proofs. Although there were earlier versions of *Elements* before that of Euclid, his is the only one to survive, perhaps because it was the first one written after both the foundations of proportion theory and the theory of irrationals had been developed and the careful distinctions always to be made between number and magnitude had been propounded by Aristotle. It was therefore both "complete" and well organized. Since the mathematical community as a whole was of limited size, once Euclid's work was recognized for its general excellence, there was no reason to keep another inferior work in circulation.

Euclid wrote his text about 2300 years ago. There are, however, no copies of the work dating from that time. The earliest extant fragments include some potsherds discovered in Egypt dating from about 225 BCE, on which are written what appear to be notes on two propositions from Book XIII, and pieces of papyrus containing parts of Book II dating from about 100 BCE. Copies of the work were, however, made regularly from the time of Euclid. Various editors made emendations, added comments, or put in new lemmas. In particular, Theon of Alexandria (fourth century CE) was responsible for one important new edition. Most of the extant manuscripts of Euclid's *Elements* are copies of this edition. The earliest such copy now in existence is in the Bodleian Library at Oxford University and dates from 888. There is, however, one manuscript in the Vatican Library, dating from the tenth century, which is not a copy of Theon's edition but of an earlier version. It was from a detailed comparison of this manuscript with several old manuscript copies of Theon's version that the Danish scholar J. L. Heiberg compiled a definitive Greek version in the 1880s, as close to what he believed the Greek original was as possible. The extracts to be discussed here are all adapted from Thomas Heath's 1908 English translation of Heiberg's Greek. (It should be noted that some modern scholars believe that one can get closer to Euclid's original by taking more account of medieval Arab translations than Heiberg was able to do.)

Euclid's *Elements* is a work in thirteen books. The first six books form a relatively complete treatment of two-dimensional geometric magnitudes while Books VII–IX deal with the theory of numbers, in keeping with Aristotle's instructions to separate the study of magnitude and number. In fact, Euclid included two entirely separate treatments of proportion theory—in Book V for magnitudes and in Book VII for numbers. Book X then provides the link between the two concepts, because it is here that Euclid introduced the notions of commensurability and incommensurability and showed that, with regard to proportions, commensurable magnitudes may be treated as if they were numbers. The book continues by presenting a classification of some incommensurable magnitudes. Euclid next dealt in Book XI with three-dimensional geometric objects and in Book XII with the method of exhaustion applied both to two- and three-dimensional objects. Finally, in Book XIII he constructed the five regular polyhedra and classified some of the lines involved according to his scheme of Book X.

It is useful to note that much of the ancient mathematics discussed in Chapter 1 is included in one form or another in Euclid's masterwork, with the exception of actual methods of arithmetic computation. The methodology, however, is entirely different. Namely, mathematics in earlier cultures always involves numbers and measurement. Numerical algorithms for solving various problems are prominent. The mathematics of Euclid, however, is completely nonarithmetical. There are no numbers used in the entire work aside from a few small positive integers. There is also no measurement. Various geometrical objects are compared, but not by use of numerical measures. There are no cubits or acres or degrees. The only measurement standard—for angles—is the right angle. Nevertheless, the question must be asked as to how much influence the mathematical cultures of Egypt and Mesopotamia had on Euclidean mathematics. In this chapter we discuss certain pieces of evidence in this regard, but a complete answer to this question cannot yet be given.

SIDEBAR 3.1 *Euclid's Postulates and Common Notions*

Postulates

1. To draw a straight line from any point to any point.
2. To produce a finite straight line continuously in a straight line.
3. To describe a circle with any center and distance.
4. That all right angles are equal to one another.
5. That, if a straight line intersecting two straight lines make the interior angles on the same side less than two right angles, the two straight lines, if produced indefinitely, meet on that side on which the angles are less than two right angles.

Common Notions (Axioms)

1. Things which are equal to the same thing are also equal to one another.
2. If equals are added to equals, the wholes are equal.
3. If equals are subtracted from equals, the remainders are equal.
4. Things which coincide with one another are equal to one another.
5. The whole is greater than the part.

 3.2   BOOK I AND THE PYTHAGOREAN THEOREM

As Aristotle suggested, a scientific work needs to begin with definitions and axioms. Euclid therefore prefaced several of the thirteen books with definitions of the mathematical objects discussed, most of which are relatively standard. He also prefaced Book I with ten axioms; five of them are geometrical postulates and five are more general truths about mathematics called "common notions." Euclid then proceeded to prove one result after another, each one based on the previous results and/or the axioms. If one reads Book I from the beginning, one never has any idea what will come next. It is only when one gets to the end of the book, where Euclid proved the Pythagorean Theorem, that one realizes that Book I's basic purpose is to lead to the proof of that result. Thus, in order to understand the reasons for various theorems, we begin our discussion of Book I with the Pythagorean Theorem and work backwards. This also enables us to see why certain unproved results must be assumed, namely, the axioms. Sidebar 3.1 does, however, list all of Euclid's axioms (called "postulates" and "common notions") and Sidebar 3.2 has selected definitions.

As we discuss the various propositions, the reader should keep in mind a few important issues. First, although Euclid has modeled the overall structure of the *Elements* using some of Aristotle's ideas, he did not use syllogisms in his proofs. His proofs were written out in natural language and generally used the notions of propositional logic. In fact, one can find examples of all four of the basic rules of inference among Euclid's proofs. Next, Euclid always assumed that if he proved a result for a particular configuration representing the hypotheses of the theorem and illustrated in a diagram, he had proved the result generally. For example, as we will see, he proved the Pythagorean Theorem by drawing some lines and marking some points on a particular right triangle, then arguing to his result on that triangle, and then concluding that the result is true for any right triangle. Of course, when mathematicians today use that strategy, they base it on explicit ideas of mathematical logic. Euclid, in contrast, never discussed his philosophy of proof; he just went ahead and proved

---

SIDEBAR 3.2   *Selected Definitions from Euclid's* Elements, *Book I*

1. A **point** is that which has no part.
2. A **line** is breadthless length.
3. The extremities of a line are points.
4. A **straight line** is a line which lies evenly with the points on itself.
5. A **surface** is that which has length and breadth only.
6. The extremities of a surface are lines.
7. A **plane surface** is a surface which lies evenly with the straight lines on itself.
8. A **plane angle** is the inclination to one another of two lines in a plane which meet one another and do not lie in a straight line.
9. And when the lines containing the angle are straight, the angle is called **rectilinear**.
10. When a straight line meeting another straight line makes the adjacent angles equal to one another, each of the equal angles is **right**, and the first straight line is called a **perpendicular** to the second line.
15. A **circle** is a plane figure contained by one line such that all the straight lines meeting it from one point among those lying within the figure are equal to one another.
16. And the point is called the **center** of the circle.
17. A **diameter** of the circle is any straight line drawn through the center and terminated in both directions by the circumference of the circle, and such a straight line also bisects the circle.
18. A **semicircle** is the figure contained by the diameter and the circumference cut off by it. And the center of the semicircle is the same as that of the circle.
23. **Parallel** straight lines are straight lines which, being in the same plane and being produced indefinitely in both directions, do not meet one another in either direction.

---

things. Of course, occasionally, he seems to have depended on the diagram more than modern mathematicians would allow. These so-called gaps in Euclid's logic were discussed extensively in the nineteenth century, so we will refer to them briefly when they occur here.

We are now ready to state Euclid's version of the Pythagorean Theorem:

PROPOSITION I–47   *In right-angled triangles the square on the hypotenuse is equal to the sum of the squares on the legs.*

Euclid proved the result for triangle ABC by first constructing the line AL parallel to BD meeting the base DE of the square on the hypotenuse at L and then showing that rectangle BL is equal to the square on AB and rectangle CL is equal to the square on AC (Fig. 3.2). To accomplish the first equality, Euclid connected AD and CF to produce triangles ADB and CBF. He then showed that these two triangles are equal to each other, that rectangle BL is double triangle ABD, and that the square on AB is double triangle CBF. His first equality then follows. The second one is proved similarly, while the sum of the two equalities proves the theorem, given common notion 2, that equals added to equals are equal.

We need to understand here what Euclid meant when he claimed that two plane figures are equal. Evidently, he meant that the figures have "equal area," but he nowhere defined this notion, nor did he calculate any areas. His alternative was generally to decompose the regions involved and to show that individual pieces are, in fact, identical. This process is justified by common notion 4, that things that coincide are equal. We will look at this in more detail later. But first, let us see what results we need to make Euclid's proof of I–47 work. First, of course, to make any sense of the theorem at all, we need to know how to construct a square on a given

FIGURE 3.2

The Pythagorean Theorem in
Euclid's *Elements*

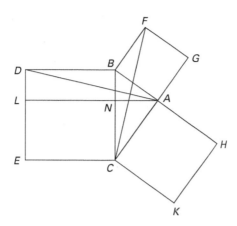

straight line segment. After all, the theorem states a relationship between certain squares. We are therefore led to

PROPOSITION I–46   *On a given straight line to describe a square.*

There are many ways to accomplish this construction, so Euclid had to make a choice. He began by constructing a perpendicular AC to the given line AB and determining a point D so that $AD = AB$. He then constructed a line through D parallel to AB and a line through B parallel to AD, the two lines meeting at point E. His claim now is that quadrilateral $ADEB$ is the desired square (Fig. 3.3). (Note that to get this far we need to be able to construct lines perpendicular and parallel to given lines—these constructions are given in Propositions I–11 and I–31, respectively—as well as cut off on one line segment a line segment equal to another one (Proposition I–3).) To prove that his construction is correct, Euclid began by noting that quadrilateral $ADEB$ has two pairs of parallel sides, so it is a parallelogram. And by Proposition I–34, the opposite sides are equal. It follows that all four sides of $ADEB$ are equal. To show that it is a square, it remains to show that all the angles are right angles. But line AD crosses the two parallel lines AB, DE. So by Proposition I–29, the two interior angles on the same side, namely, angles BAD and ADE, are equal to two right angles. But since we already know that angle BAD is a right angle, so is angle ADE. And since opposite angles in parallelograms are equal according to I–34, all four angles are right, and $ADEB$ is a square.

So although the actual construction of a square is fairly obvious, the proof that the construction is correct appears to require many other propositions. Before looking at some of those propositions, let us return to the main theorem and see what else we need.

The first result is the one that allows Euclid to conclude that triangles ADE and CBF are equal. That follows by the familiar side-angle-side theorem (SAS), proved by Euclid as

PROPOSITION I–4   *If two triangles have two sides equal to two sides respectively, and have the angles contained by the equal sides also equal, then the two triangles are congruent.*

The word "congruent" is used here as a modern shorthand for Euclid's conclusion that each part of one triangle is equal to the corresponding part of the other. Euclid proved this theorem

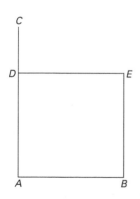

FIGURE 3.3

Elements, Proposition I–46

by superposition. Namely, he imagined the first triangle being moved from its original position and placed on the second triangle with one side placed on the corresponding equal side and the angles also matching. Euclid here tacitly assumed that such a motion is always possible without deformation. Rather than supply such a postulate, nineteenth-century mathematicians tended to assume this theorem itself as a postulate.

Euclid also needed the result that a rectangle is double a triangle with the same base and height. This follows from

PROPOSITION I–41   *If a parallelogram has the same base with a triangle and is in the same parallels, the parallelogram is double the triangle.*

Since "in the same parallels" means from a modern point of view that the two figures have the same height, it would seem that this proposition follows from the formulas for the areas of a triangle and a parallelogram, namely, $A = \frac{1}{2}bh$ and $A = bh$. But, as noted earlier, Euclid did not use formulas to deal with equal area; he used decomposition. So here he showed that the parallelogram can be divided into two triangles, each equal to the given one. In Figure 3.4, the given parallelogram is $ABCD$ and the given triangle is BCE. Euclid drew AC, the diagonal of the parallelogram, then noted that triangle ABC is equal to triangle BCE because they have the same base and are in the same parallels (Proposition I–37). But now parallelogram $ABCD$ is double triangle ABC (by Proposition I–34) and therefore is double triangle BCE.

FIGURE 3.4

Elements, Proposition I–41

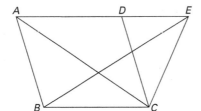

Recall that the construction of a square required the construction of both a perpendicular to a given line and a parallel to a given line. The first of these constructions (Proposition I–11) begins with the drawing of the equilateral triangle DFE in which the midpoint C of DE is the point at which the perpendicular is drawn (Fig. 3.5). The construction of an equilateral triangle is accomplished in Proposition I–1, in which Euclid drew circles of radius DE centered on each of the points D and E and then found F as the intersection of the two circles. This construction in turn requires the use of a compass and a straightedge. Namely, Euclid needed to postulate that a circle can be drawn with a given center and radius and that a line can be drawn connecting two points. These postulates are postulate 3 and postulate 1, respectively. But even with these two postulates, modern commentators have noted that there is a logical gap in this proof. How did Euclid know that the two circles drawn from the endpoints of DE actually intersect? It seems obvious in the diagram, but some postulate of continuity is necessary. This was supplied in the nineteenth century and will be discussed later. But once the triangle is constructed, the line from the vertex F to the midpoint C of the base is the desired perpendicular. To prove this, Euclid noted that the two triangles DCF and ECF are congruent by side-side-side (SSS), a result proved as Proposition I–8, by superposition, like

FIGURE 3.5

Elements, Proposition I–11

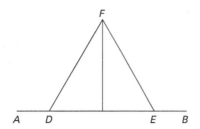

SAS. Since the sum of the equal angles DCF and ECF is two right angles, each of the angles DCF and ECF is right.

To construct a line through a given point A parallel to a given line BC (Proposition I–31), Euclid took an arbitrary point D on BC and connected AD (Fig. 3.6). By Proposition I–23, he then constructed the angle DAE equal to the angle ADC and extended AE into the straight line AF. That one can extend a straight line in a straight line is the substance of another construction postulate, postulate 2. To prove that EF is now parallel to BC, Euclid noted that the alternate interior angles DAE and ADC are equal. By Proposition I–27, the two lines are parallel.

FIGURE 3.6

Elements, Proposition I–31

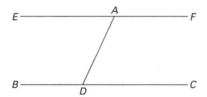

Let us now consider

PROPOSITION I–27   *If a straight line falling on two straight lines makes the alternate angles equal to one another, then the straight lines are parallel to one another.*

Here Euclid argued by *reductio ad absurdum*, a version of *modus tollens*. Namely, he assumed that even though the alternate angles AEF, EFD, formed by line EF falling on lines AB and CD are equal, the lines themselves are not parallel (Fig. 3.7). Therefore, they must meet at point G. It follows that in triangle EFG, the exterior angle AEF equals the interior angle EFD. But this contradicts Proposition I–16, so the original assumption must be false and AB is parallel to CD.

FIGURE 3.7

Elements, Proposition I–27

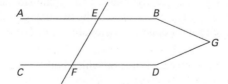

This then takes us back to

PROPOSITION I–16   *In any triangle, if one of the sides is produced, the exterior angle is greater than either of the interior and opposite angles.*

Suppose side BC of triangle ABC is produced to D (Fig. 3.8). Bisect AC at E and join BE. Euclid then claimed that BE may be extended to F so that $EF = BE$. Unfortunately, there is no postulate allowing him to extend a line to any arbitrary length. Of course, if that assumption is granted, then the proof is straightforward. One connects FC and shows that the triangles ABE and CFE are congruent. Thus, $\angle BAE = \angle ECF$. But $\angle ECF$ is part of the exterior angle ACD; thus, the latter angle is greater than $\angle BAE$. This last statement also requires a postulate, that the whole is always greater than the part (common notion 5).

FIGURE 3.8

Elements, Proposition I–16

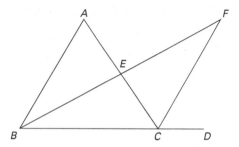

An immediate corollary is Proposition I–17, that two angles of any triangle are always less than two right angles. As will be discussed later, this proposition, based on the faulty proof of Proposition I–16, was important in the developments leading to the discovery of non-Euclidean geometry.

We could continue by analyzing the proof of I–23, which was used in I–31. This would force us to analyze most of the earlier results in Book I as well. So we will leave some of those results for the exercises and conclude this section by considering just two more important propositions that have already been quoted several times. First, we look at

PROPOSITION I–34   *In parallelograms the opposite sides and angles are equal to one another and the diameter bisects the areas.*

Note that in the proofs of Propositions I–46 and I–41, we have used all three conclusions of this proposition. To prove it, one thinks of the diagonal as first cutting one pair of parallel sides and then cutting the other. In each case, Proposition I–29 implies that the alternate interior angles are equal. It then follows (by angle-side-angle) that the two triangles into which the diagonal cuts the parallelogram are congruent. (The angle-side-angle triangle congruence theorem is Proposition I–26.) The congruence of the two triangles then implies that each pair of opposite sides and each pair of opposite angles are equal. The third part of the proposition follows immediately.

The final proposition we consider is one on which both I–34 and I–46 depend:

PROPOSITION I–29 *A straight line falling on parallel straight lines makes the alternate angles equal to one another, the exterior angle equal to the interior and opposite angle, and the interior angles on the same side equal to two right angles.*

It is easy enough to see that any two of the statements are simple consequences of the third. So we need to decide which one to prove. From hints in various Greek texts, we know that before Euclid, the situation regarding this theorem was very unclear. How do you prove one of these results? What must you assume? It is in his answer to these questions that Euclid showed his genius. He had already proved the converse of this theorem in Propositions I–27 and I–28. Evidently, however, he saw no way of proving any of the statements in this proposition directly. We can imagine that he struggled with this, but he eventually realized that he would have to take one of these results—or its equivalent—as a postulate. And so he decided, for reasons we cannot guess, to take the contrapositive of the third statement in the proposition as a postulate. Thus, at the beginning of Book I, he placed

POSTULATE 5 *If a straight line intersecting two straight lines make the interior angles on the same side less than two right angles, the two straight lines, if produced indefinitely, meet on that side on which the angles are less than two right angles.*

Given this postulate, the proof of Proposition I–29 is straightforward by a *reductio* argument: Assume that angle AGH is greater than angle GHD (Fig. 3.9). Then the sum of angles AGH and BGH is greater than the sum of angles GHD and BGH. The first sum equals two right angles (by Proposition I–13), so the second one is less than two right angles. Then by the postulate, the lines AB and CD must meet. But this contradicts the hypothesis that those lines are parallel.

FIGURE 3.9

Elements, Proposition I–29

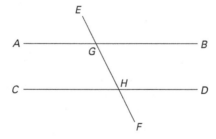

Thus, we see that the Pythagorean Theorem, the culminating theorem of Book I, besides requiring very many of the earlier results in Book I (including all three triangle congruence theorems), rests on the critical parallel postulate. The parallel postulate, alone among Euclid's postulates, has caused immense controversy over the years, because many people felt it was not self-evident. And for Euclid, as for Aristotle, a postulate should be "self-evident." Thus, almost from the time the *Elements* appeared, people have attempted to prove this result as a theorem, using as a basis just the other axioms and postulates. Many people thought they had accomplished this task, but a close examination of every such proof always reveals either an error or, more likely, another assumption—one that perhaps is more self-evident than Euclid's postulate but nevertheless cannot be proved from the other nine axioms. Probably the most familiar "other assumption" is what is now known as

PLAYFAIR'S AXIOM   *Through a given point outside a given line, exactly one line may be constructed parallel to the given line.*

We leave it as an exercise that this result is entirely equivalent to Euclid's postulate, at least under the assumption that lines of arbitrary length may be drawn and therefore that Proposition I–16 is true.

 3.3

BOOK II AND GEOMETRIC ALGEBRA

Book I of the *Elements*, with its familiar geometric results, was a major component of the Greek mathematician's "toolbox," a set of results that were frequently used in any advanced geometric argument. Book II, on the other hand, is quite different. It deals with the relationships between various rectangles and squares and has no obvious goal. In fact, the propositions in Book II are only infrequently used elsewhere in the *Elements*. Thus, the purpose of Book II has been the subject of much debate among students of Greek mathematics. One interpretation, dating from the late nineteenth century but still common today, is that this book, together with a few propositions in Books I and VI, can best be interpreted as "geometric algebra," the representation of algebraic concepts through geometric figures. In other words, the squares of side length a can be thought of as geometric representations of a^2; rectangles with sides of length a and b can be interpreted as the products ab; and relationships among such objects can be interpreted as equations. Of course, one of the issues in this debate is what one means by the term "algebra." If we think of algebra as meaning the finding of unknown quantities, given certain relationships between those and known quantities, regardless of how these quantities are expressed, then there is certainly algebra in Book II, as well as elsewhere in the *Elements*. It is also easy enough to apply some of Euclid's theorems to the solution of quadratic equations—and this was, in fact, done by medieval Islamic mathematicians. But the majority of scholars today believe that Euclid himself really intended in Book II only to display a relatively coherent body of geometric knowledge that could be used in the proof of further geometric theorems, if not in the *Elements* themselves, then in more advanced Greek mathematics such as the study of conic sections. We shall look at some of the arguments about geometric algebra in what follows.[3]

Euclid began Book II with a definition: *Any rectangle is said to be* **contained** *by the two straight lines forming the right angle.* This definition shows Euclid's geometric usage. The statement does not mean that the area of a rectangle is the product of the length by the width. Euclid never multiplied two lengths together, because he had no way of defining such a process for arbitrary lengths. At various places, he multiplied lengths by numbers (that is, positive integers), but otherwise he only wrote of rectangles contained by two lines. One question then is whether one can interpret Euclid's "rectangle" as meaning a "product."

As an example of Euclid's use of this definition, consider

PROPOSITION II–1   *If there are two straight lines, and one of them is cut into any number of segments whatever, the rectangle contained by the two straight lines is equal to the sum of the rectangles contained by the uncut straight line and each of the segments.*

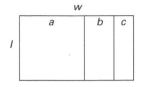

FIGURE 3.10

Elements, Proposition II–1:
$l(a + b + c) = la + lb + lc$

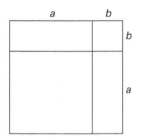

FIGURE 3.11

Elements, Proposition II–4:
$(a + b)^2 = a^2 + b^2 + 2ab$

We can interpet this algebraically as stating that given a length l and a width w cut into several segments, say, $w = a + b + c$, the area of the rectangle determined by those lines, namely, lw, equals the sum of the areas of the rectangles determined by the length and the segments of the width, namely, $la + lb + lc$. In other words, this theorem states the familiar distributive law: $l(a + b + c) = la + lb + lc$. But let us look more closely at Euclid's proof. Two lines A and BC are given, and the second is divided into three segments by the points D and E (Fig. 3.10). (Euclid had no way of representing "any number" of segments, so he used "three" as what we may call his **generalizable example**.) He then drew BG perpendicular to BC and of length equal to that of A and completed the rectangles $BDKG$, $DELK$, and $ECHL$. Since rectangle $BCHG$ is "the rectangle contained by A and BC," while $BDKG$, $DELK$, and $ECHL$ are the "rectangles contained by A and each of the segments," Euclid could conclude from the diagram that the result was true. At first glance, the proposition seems almost a tautology. But what Euclid seems to be doing here, as well as later in this book, is proving a result about "invisible" figures, that is, the figures stated in the theorem with respect just to the initial two lines and the segments, by using "visible" figures, the actual rectangles drawn. Euclid clearly believed that the "visible" result in the diagram was a correct basis for the proof of the "invisible" result of the proposition.[4] Another example of this process is in

PROPOSITION II–4 *If a straight line is cut at random, the square on the whole is equal to the squares on the segments and twice the rectangle contained by the segments.*

Algebraically, this proposition is simply the rule for squaring a binomial, $(a + b)^2 = a^2 + b^2 + 2ab$, the basis for the square root algorithms discussed in Chapter 1 (Fig. 3.11). Euclid's proof is quite complex, since he needed to prove that the various figures in the diagram are in fact squares and rectangles. But again, he needed to reduce the invisible statement to a visible diagram.

The next two propositions were interpreted in the ninth century CE as geometric justifications of the standard algebraic solutions of quadratic equations.

PROPOSITION II–5 *If a straight line is cut into equal and unequal segments, the rectangle contained by the unequal segments of the whole together with the square on the straight line between the points of section is equal to the square on the half.*

PROPOSITION II–6 *If a straight line is bisected and a straight line is added to it, the rectangle contained by the whole with the added straight line and the added straight line together with the square on the half is equal to the square on the straight line made up of the half and the added straight line.*

Figure 3.12 should help clarify these propositions. If AB is labeled in each diagram as b, AC and BC as $b/2$, and DB as x, Proposition II–5 translates into $(b - x)x + (b/2 - x)^2 = (b/2)^2$, while Proposition II–6 gives $(b + x)x + (b/2)^2 = (b/2 + x)^2$. The quadratic equation $bx - x^2 = c$ [or $(b - x)x = c$] can be solved using the first equality by writing $(b/2 - x)^2 = (b/2)^2 - c$ and then getting

$$x = \frac{b}{2} - \sqrt{\left(\frac{b}{2}\right)^2 - c}.$$

FIGURE 3.12

Elements, Propositions II–5
and II–6

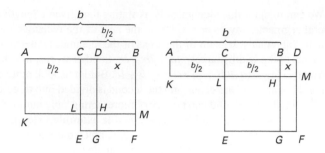

Similarly, the equation $bx + x^2 = c$ (or $(b + x)x = c$) can be solved from the second equality by using an analogous formula. Alternatively, one can label AD as y and DB as x in each diagram and translate the first result into the standard Babylonian system $x + y = b$, $xy = c$, and the second into the system $y - x = b$, $yx = c$. In any case, note that Figure 3.12 is essentially the same as Figure 1.20, the figure representing the Babylonian scribes' probable method for solving the first of these systems.

Euclid, of course, did not do any of the translations indicated. He just used the constructions in Figure 3.12 to prove the equalities of the appropriate squares and rectangles. He did not indicate anywhere that these propositions are of use in solving what we call quadratic equations.

What did these theorems then mean for Euclid? We can see how Proposition II–6 is used in the proof of Proposition II–11, and Proposition II–5 in the proof of Proposition II–14.

PROPOSITION II–11   *To cut a given straight line so that the rectangle contained by the whole and one of the segments is equal to the square on the remaining segment.*

The goal of this proposition is to find a point H on the line so that the rectangle contained by AB and HB equals the square on AH (Fig. 3.13). This is an algebraic problem, in terms of the definition given earlier, since it asks to find an unknown quantity given its relationship to certain known quantities. To translate this problem into modern notation, let the line AB be a and let AH be x. Then $HB = a - x$, and the problem amounts to solving the equation

$$a(a - x) = x^2 \quad \text{or} \quad x^2 + ax = a^2.$$

The Babylonian solution is

$$x = \sqrt{\left(\frac{a}{2}\right)^2 + a^2} - \frac{a}{2}.$$

Euclid's proof seemingly amounts to precisely this formula. To get the square root of the sum of two squares, the obvious method is to use the hypotenuse of a right triangle whose sides are the given roots, in this case, a and $a/2$. So Euclid drew the square on AB and then bisected AC at E. It follows that EB is the desired hypotenuse. To subtract $a/2$ from this length, he drew EF equal to EB and subtracted off AE to get AF; this is the needed value x. Since he wanted the length marked off on AB, he simply chose H so that $AH = AF$. To prove that this choice of H is correct, Euclid then appealed to Proposition II–6: The line AC has been bisected and a straight line AF added to it. Therefore, the rectangle on FC and AF

FIGURE 3.13

Elements, Proposition II–11

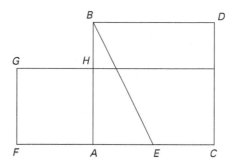

plus the square on AE equals the square on FE. But the square on FE equals the square on EB, which in turn is the sum of the squares on AE and AB. It follows that the rectangle on FC and AF (equal to the rectangle on FC and FG) equals the square on AB. By subtraction of the common rectangle AK, we get that the square on AH equals the rectangle on HB and AB, as desired.

Euclid has thus solved what we would call a quadratic equation, albeit in geometric dress, in the same manner as the Babylonians. Interestingly enough, he solved the same problem again in the *Elements* as Proposition VI–30. There he wanted to cut a given straight line in "extreme and mean ratio," that is, given a line AB to find a point H such that $AB : AH = AH : HB$. Naturally, this translates algebraically into the same equation as given above. The ratio $a : x$ from that equation, namely, $(\sqrt{5} + 1) : 2$, is generally known as the **golden ratio**. Much has been written about its importance from Greek times to today.[5]

Before considering an example of the use of Proposition II–5, a slight digression back to Book I is necessary.

PROPOSITION I–44 *To a given straight line to apply, in a given rectilinear angle, a parallelogram equal to a given triangle.*

The aim of the construction is to find a parallelogram of given area with one angle given and one side equal to a given line segment. That is, the parallelogram is to be "applied" to the given line segment. This notion of the "application" of areas is, according to some sources, due to the Pythagoreans. That this too can be interpreted algebraically is easily seen if the given angle is a right angle. If the area of the triangle is taken to be c^2 and the given line segment to have length a, the goal of the problem is to find a line segment of length x such that the rectangle with length a and width x has area c^2, that is, to solve the equation $ax = c^2$. Given that Euclid did not deal with "division" of magnitudes, a solution for him amounted to finding the fourth proportional in the proportion $a : c = c : x$. But since he could not use the theory of proportions in Book I, he was forced to use a more complicated method involving areas.

From a geometrical point of view, this construction enables one to compare the sizes of two rectangles. For if rectangle A is applied to one of the sides of rectangle B, then the new rectangle C, equal to A, will share a side with B. Thus, the ratio of the areas of $C = A$ to B will be equal to the ratios of the nonshared sides. Such comparisons, making use of this proposition, are found in the works of Archimedes and Apollonius.

In Proposition I–45, Euclid demonstrated how to construct a rectangle equal to any given rectilinear figure, by simply dividing the figure into triangles and using the result of I–44, among others. This proposition is then used in the first step of the solution of

PROPOSITION II–14   *To construct a square equal to a given rectilinear figure.*

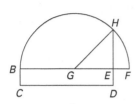

FIGURE 3.14

Elements, Proposition II–14

We can think of this construction as an algebraic problem, since we are asked to find an unknown side of a square meeting certain conditions. In modern notation, we are asked to solve the equation $x^2 = cd$, where c, d are the lengths of the sides of the rectangle constructed, using I–45, equal to the given figure (Fig. 3.14). Placing the sides of the rectangle BE, EF, in a straight line and bisecting BF at G, Euclid constructed the semicircle BHF of radius GF, where H is the intersection of that semicircle with the perpendicular to BF at E. Then, since the straight line BF has been cut into equal segments at G and into unequal segments at E, Proposition II–5 shows that the rectangle contained by BE and EF together with the square on EG is equal to the square on GF. But since $GF = GH$ and the square on GH equals the sum of the squares on GE and EH, it follows that the square on EH satisfies the condition of the problem. Like II–11, Euclid solved this problem a second time using proportions as Proposition VI–13, the construction of a mean proportional between two line segments.

Additionally, in Book VI, Euclid expanded the notion of "application of areas" to applications that are "deficient" or "exceeding." The importance of these notions will be apparent in the discussion of conic sections later. For now, however, we note that in the following two propositions, Euclid solved two types of quadratic equations geometrically.

PROPOSITION VI–28   *To a given straight line to apply a parallelogram equal to a given rectilinear figure and deficient by a parallelogram similar to a given one; thus the given rectilinear figure must not be greater than the parallelogram described on the half of the straight line and similar to the defect.*

PROPOSITION VI–29   *To a given straight line to apply a parallelogram equal to a given rectilinear figure and exceeding by a parallelogram figure similar to a given one.*

In the first case, Euclid proposed to construct a parallelogram of given area whose base is less than the given line segment AB. The parallelogram on the deficiency, the line segment SB, is to be similar to a given one. In the second case, the constructed parallelogram of given area has base greater than the given line segment AB, while the parallelogram on the excess, the line segment BS, is again to be similar to a given one (Fig. 3.15). To simplify matters, and to show why we can think of Euclid's constructions as solving quadratic equations, we will assume that the given parallelogram in each case is a square. This implies that the constructed parallelograms are rectangles.

FIGURE 3.15

Elements, Propositions VI–28 and VI–29

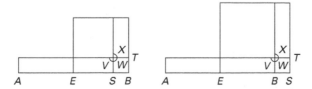

Designate AB in both cases by b, and the area of the given rectilinear figure by c. The problems reduce to finding a point S on AB (Proposition VI–28) or on AB extended (Proposition VI–29) so that $x = BS$ satisfies $x(b - x) = c$ in the first case and $x(b + x) = c$ in the second. That is, it is necessary to solve the quadratic equations $bx - x^2 = c$ and $bx + x^2 = c$, respectively. In each case, Euclid found the midpoint E of AB and constructed the square on BE, whose area is $(b/2)^2$. In the first case, S was chosen so that ES is the side of a square whose area is $(b/2)^2 - c$. That is why the condition is stated in the proposition that in effect c cannot be greater than $(b/2)^2$. This choice for ES implies that

$$x = BS = BE - ES = \frac{b}{2} - \sqrt{\left(\frac{b}{2}\right)^2 - c}.$$

In the second case, S was chosen so that ES is the side of a square whose area is $(b/2)^2 + c$. Then

$$x = BS = ES - BE = \sqrt{\left(\frac{b}{2}\right)^2 + c} - \frac{b}{2}.$$

In both cases, Euclid proved that his choice was correct by showing that the desired rectangle equals the gnomon XWV and that the gnomon is in turn equal to the given area c. Algebraically, that amounts in the first case to showing that

$$x(b - x) = \left(\frac{b}{2}\right)^2 - \left[\left(\frac{b}{2}\right)^2 - c\right] = c$$

and in the second that

$$x(b + x) = \left[\left(\frac{b}{2}\right)^2 + c\right] - \left(\frac{b}{2}\right)^2 = c.$$

There has long been a debate over whether the geometric algebra in Euclid stems from a deliberate transformation of the Babylonian quasi-algebraic results into formal geometry. Euclid's solution of several construction problems mirrors the Babylonian solutions of similar problems. One can then argue that the Greek adaptation into their geometric viewpoint, given the necessity of proof, was related to the discovery that not every line segment could be represented by a "number." One can further argue that, once one has translated the material into geometry, one might just as well state and prove certain results for parallelograms as for rectangles, since little extra effort is required. A further argument supporting the transmission and translation is that the original Babylonian methodology itself was couched in a "naive" geometric form, a form well suited to a translation into the more sophisticated Greek geometry.

Was there any opportunity for direct cultural contact between Babylonian mathematical scribes and Greek mathematicians? It used to be argued that this was virtually impossible, because there was no record of Babylonian mathematics at all during the sixth to the fourth centuries BCE, when this contact would have had to take place, and because those in the aristocracy to which the Greek mathematicians belonged would be disdainful of the activities of the scribes, who in Old Babylonian times were not themselves part of the elite. However, recent discoveries have indicated that mathematical activity did continue in the mid-first millennium BCE. Furthermore, by this time, the Mesopotamian languages were often being written in ink on papyrus using a new alphabet. Cuneiform writing on clay tablets was then restricted to important documents that needed to be preserved, and those who could perform this service were now members of the elite, experts in traditional wisdom who were central to the functioning of the state. Besides, from the sixth century BCE on, Mesopotamia was a province of the Persian empire, with whom the Greeks did maintain contact.

On the other hand, despite the possibilities for contact and the logic in the argument of how Babylonian mathematics could have been "translated" into Greek geometry, there is no direct evidence of any transmission of Babylonian mathematics to Greece during or before the fourth century BCE. One could then argue that although the Greeks did employ what we think of as algebraic procedures, their mathematical thought was so geometrical that all such procedures were automatically expressed that way. The Greeks of the period up to 300 BCE had no algebraic notation and therefore no way of manipulating expressions that stood for magnitudes, except by thinking of them in geometric terms. In fact, Greek mathematicians became very proficient in manipulating geometric entities. And finally, we note that there was no way the Greeks could express, other than geometrically, irrational solutions of quadratic equations.

A clear answer to the question of whether Babylonian algebra was transmitted in some form to Greece by the fourth century BCE cannot yet be given. Hopefully, further research in the original sources will enable us to find an answer in the future.

 3.4   CIRCLES AND THE PENTAGON CONSTRUCTION

Books I and II dealt with properties of rectilinear figures, that is, figures bounded by straight line segments. In Book III, Euclid turned to the properties of the most fundamental curved figure, the circle. The Greeks were greatly impressed with the symmetry of the circle, the fact that no matter how you turned it, it always appeared the same. They thought of it as the most perfect of plane figures. Similarly, they felt the three-dimensional analogue of the circle, the sphere, was the most perfect of solid figures. These philosophical ideas provided the basis for the Greek ideas on astronomy, which will be discussed in Chapter 5. Many of the theorems in Books III and IV dated from the earliest period of Greek mathematics. As such, they became part of the Greek mathematician's toolbox for solving other problems. As we saw, Hippocrates used results on circles in his quadrature of lunes.

If there is any organizing principle of Book III, it is to provide for the construction, in Book IV, of polygons, both inscribed in and circumscribed about circles. In particular, most of the propositions from the last half of Book III are used in the most difficult construction of Book IV, the construction of the regular pentagon. The constructions of the triangle, square, and hexagon are relatively intuitive and are probably the work of the Pythagoreans. On the other hand, the construction of the pentagon involves more advanced concepts, including the division of a line segment into extreme and mean ratio, and is therefore probably a later development, perhaps due to Theaetetus in the early fourth century BCE. This construction in turn is used in Euclid's construction of some of the regular solids in Book XIII.

SIDEBAR 3.3 *Selected Definitions from Euclid's* Elements, *Book III*

2. A straight line is said to **touch a circle** which, meeting the circle and being produced, does not cut the circle.

6. A **segment of a circle** is the figure contained by a straight line and a circumference of a circle.

8. An **angle in a segment** is the angle which, when a point is taken on the circumference of the segment and straight lines are joined from it to the extremities of the straight line which is the **base of the segment**, is contained by the straight lines so joined.

After presenting a few relevant definitions (Sidebar 3.3), Euclid began Book III with some elementary constructions and propositions, including the very useful result that diameters bisect chords to which they are perpendicular. He then showed how to construct a tangent to a circle:

PROPOSITION III–16 *The straight line drawn at right angles to the diameter of a circle from its extremity will fall outside the circle, and into the space between the straight line and the circumference another straight line cannot be interposed.*

This proposition asserts that the line perpendicular to the diameter at its extremity is what is today called a **tangent**. Euclid only remarked in a corollary that it "touches" the circle, as in definition 2. But the statement that no straight line can be interposed between the curve and the line ultimately became part of the definition of a tangent before the introduction of calculus. Euclid's proof of this result, as to be expected, was by a *reductio* argument.

Propositions III–18 and III–19 give partial converses to Proposition III–16. The former shows that the line from the center of a circle that meets a tangent is perpendicular to the tangent; the latter demonstrates that a perpendicular from the point of contact of a tangent goes through the center of the circle. Propositions III–20 and III–21 also give familiar results, respectively, that the angle at the center is double the angle at the circumference, if both angles cut off the same arc, and that angles in the same segment are equal. The proofs of both are clear from Figure 3.16 as is the proof of Proposition III–22, that the opposite angles of quadrilaterals inscribed in a circle are equal to two right angles.

FIGURE 3.16

Elements, Propositions III–20, III–21, and III–22

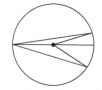

Proposition III–31 asserts that the angle in a semicircle is a right angle. One could conclude this immediately from Proposition III–20, if one is prepared to consider a straight angle as an angle. Then the angle in a semicircle is half of the straight angle of the diameter, which is in turn equal to two right angles. Euclid, however, did not consider a straight angle as an angle, so he gave a different proof.

Proposition III–32 is more complicated, but necessary for the pentagon construction.

PROPOSITION III–32   *If a straight line is tangent to a circle, and from the point of tangency there is drawn a straight line cutting the circle, the angles which that line makes with the tangent will be equal to the angles in the alternate segments of the circle.*

In other words, this proposition asserts that one of the angles formed by the tangent EF and the secant BD, say, angle DBF, is equal to any angle in the "alternate" segment BD of the circle, such as angle DAB (Fig. 3.17). Similarly, the other angle made by the tangent, angle DBE, is equal to any angle in the remaining segment, such as angle DCB. (We can say "any angle" in the segment, since by Proposition III–21, any two angles in the same segment are equal to one another.) To prove this result, we draw a perpendicular AB to the tangent at the point B of tangency. Since a perpendicular to a tangent passes through the center of the circle (Proposition III–19), the angle ADB, being an angle in a semicircle, is a right angle (Proposition III–31). Therefore, angles DAB and ABD sum to a right angle. But angles DBF and ABD also sum to a right angle. It follows that angle DAB equals angle DBF, as claimed. The equality of the other two angles can then be easily established.

FIGURE 3.17

Elements, Proposition III–32

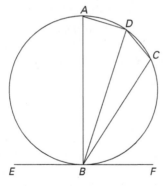

Proposition III–36 is also necessary for the pentagon construction, but because it is closely related to Proposition III–35, and because Propositions II–5 and II–6 make another appearance in these propositions, we first move to

PROPOSITION III–35   *If in a circle two straight lines cut one another, then the rectangle contained by the segments of the one equals the rectangle contained by the segments of the other.*

We note that the rectangles of the proposition are "invisible"; they will only make their appearance through Proposition II–5. For the proof, Euclid first noted that if the two lines meet at the center of the circle, then the result is obvious. Thus, we will assume that the lines AC and BD meet at a point E different from the center F (Fig. 3.18). Draw FG and FH from F perpendicular to AC and DB, and then join FB, FC, and FE. We know that G is then the midpoint of AC. Thus, we can apply II–5 to the line AC and conclude that the rectangle contained by AE and EC together with the square on EG equals the square on GC. By adding the square on GF to both sides and applying the Pythagorean Theorem, we conclude that the rectangle contained by AE and EC plus the square on FE equals the

FIGURE 3.18

Elements, Proposition III–35

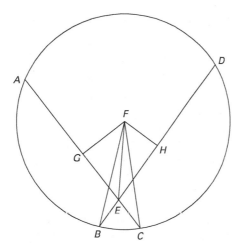

square on FC, which in turn equals the square on FB. By the same argument, the rectangle contained by DE and EB plus the square on FE equals the square on FB. It follows that the rectangle contained by DE and EB equals the rectangle contained by AE and EC, as claimed.

The next proposition deals with two lines cutting the circle that meet outside it:

PROPOSITION III–36   *If from a point outside a circle we draw a tangent and a secant to the circle, then the rectangle contained by the whole secant and that segment which is outside the circle equals the square on the tangent.*

The statement may remind the reader of Proposition II–6. And in fact that proposition is used in the proof. We will just consider the easier case here, where the secant line $DCFA$ goes through the center F (Fig. 3.19). Join FB to form the right triangle FBD. Proposition II–6 now asserts that the rectangle contained by AD and CD, together with the square on FC, equals the square on FD. But $FC = FB$, and the sum of the squares on FB and BD equals the square on FD. Therefore, the rectangle contained by AD and CD equals the square on DB, as claimed. The case where the secant line does not pass through the center is slightly trickier.

FIGURE 3.19

Elements, Proposition III–36

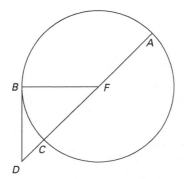

Proposition III–37 is a converse of III–36, asserting that if two straight lines are drawn to a circle from a point outside, one a secant and one touching the circle, and if the relationship between the rectangle and square of that proposition holds, then the second line is a tangent. The proof involves actually drawing a tangent and then showing, using Proposition III–36, that the given line equals the tangent.

The treatment of the pentagon begins in Book IV after Euclid first showed the simpler techniques of inscribing triangles and squares in circles, inscribing circles in triangles and squares, circumscribing triangles and squares about circles, and circumscribing circles about triangles and squares. Euclid then divided his construction of a regular pentagon into two steps, the first being the construction of an isosceles triangle with each of the base angles double the vertex (IV–10), and the second being the actual inscribing of the pentagon in the circle (IV–11). As usual, Euclid did not show how he arrived at the construction, but a close reading of it can well give a clue to his analysis of the problem. We will therefore assume the construction made and try to see where that assumption leads.

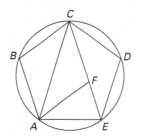

FIGURE 3.20

Construction of a regular pentagon

So suppose $ABCDE$ is a regular pentagon inscribed in a circle (Fig. 3.20). Draw the diagonals AC and CE. Since angles CEA and CAE each subtend an arc double that subtended by angle ACE, it follows that triangle ACE is an isosceles triangle with base angles double those of the vertex. We have therefore reduced the pentagon construction to the construction of that triangle. Assume then that ACE is such an isosceles triangle and let AF bisect angle A. It follows that triangles AFE and CEA are similar, so $EF : AF = EA : CE$. But triangles AFE and AFC are both isosceles, so $EA = AF = FC$. Therefore, $EF : FC = FC : CE$, or, in modern terminology, $FC^2 = EF \cdot CE$. The construction is therefore reduced to finding a point F on a given line segment CE such that the square on CF is equal to the rectangle contained by EF and CE. But this is precisely the construction of Proposition II–11. Once F is found, the isosceles triangle with base angles double the vertex angle can be constructed by drawing a circle centered on C with radius CE and another circle centered on E with radius CF. The intersection A of the two circles is the third vertex of the desired triangle.

Euclid performs this construction in Proposition IV–10 (Fig. 3.21), but could not use similarity arguments in his proof of its validity. He therefore used alternatives. The goal is to show that $\alpha = 2\delta$. If it is shown that $\beta = \delta$, then $\beta + \gamma = \delta + \gamma = \epsilon$. Also, since $\alpha = \beta + \gamma$,

FIGURE 3.21

Elements, Proposition IV–10

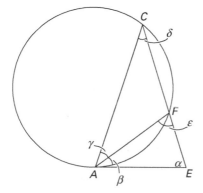

then $\epsilon = \alpha$. But then $AE = AF$, and since by construction $AE = FC$, it follows that triangle AFC is isosceles and that $\delta = \gamma$. Finally, $\alpha = \beta + \gamma = \delta + \delta = 2\delta$, as desired. To show that $\beta = \delta$, circumscribe a circle around triangle AFC. Since the rectangle contained by CE, FE, equals the square on FC, it follows that this rectangle also equals the square on AE. Proposition III–37 then asserts that under these conditions on the lines AE and CE, AE is tangent to the circle. Proposition III–32 then allowed Euclid to conclude that $\beta = \delta$ as desired, completing the proof of the construction.

Given the isosceles triangle with base angles double the vertex angle, the inscribing of the regular pentagon in a circle is now straightforward. Euclid first inscribed the isosceles triangle ACE in the circle. Next, he bisected the angles at A and E. The intersection of these bisectors with the circle are points D and B, respectively. Then A, B, C, D, E are the vertices of a regular pentagon.

Euclid completed Book IV with the construction of a regular hexagon and a regular 15-gon in a circle, but did not mention the construction of other regular polygons. Presumably, he was aware that the construction of a polygon of $2^n k$ sides ($k = 3, \ 4, \ 5$) was easy, beginning with the constructions already made, and even that, in analogy with his 15-gon construction, it was straightforward to construct a polygon of kl sides (k, l relatively prime) if one can construct one of k sides as well as one of l sides. Whether he was aware of a construction for the heptagon, however, is not known. In any case, that construction, the first record of which is in the work of Archimedes, would for Euclid be part of advanced mathematics, rather than part of the "elements," because it requires tools other than a straightedge and compass.

3.5   RATIO AND PROPORTION

The regular pentagon is part of the pentagram, evidently one of the symbols used by the Pythagoreans. Thus, it is believed that the Pythagoreans worked out a construction of the pentagon, although more likely their construction used similarity rather than the method described above. It is therefore plausible that the property of the pentagram in reproducing itself when one connects the diagonals of the inner pentagon (Fig. 3.22) could well have been an alternative path to the discovery of incommensurability, rather than the one described earlier. To explain this, we need to move to Book VII, the first of the three books of number theory in the *Elements*.

Book VII, like all the number theory books, deals with what we call the positive integers in contrast to the geometrical magnitudes of the earlier books. And the first item of business for Euclid here is the familiar process for finding the greatest common divisor of two numbers. This algorithm, usually called the **Euclidean algorithm** although certainly known long before Euclid, is presented in Propositions VII-1 and VII-2. Given two numbers, a, b, with $a > b$, one subtracts b from a as many times as possible; if there is a remainder, c, which of course must be less than b, one then subtracts c from b as many times as possible. Continuing in this manner, one eventually comes either to a number m, which "measures" (divides) the one before (Proposition VII–2), or to the unit (1) (Proposition VII–1). In the first case, Euclid proved that m is the greatest common measure (divisor) of a and b. In the second case, he showed that a and b are prime to one another. For example, given the two numbers 18 and 80, first subtract 18 from 80. One can do this four times, with remainder 8. Next subtract 8

FIGURE 3.22

Diagonals of inner pentagon of a pentagram

BIOGRAPHY

Theaetetus (417–369 BCE)

Because Plato dedicated a dialogue to him, something is known about Theaetetus's life. He was born near Athens into a wealthy family and was educated there. A meeting with Theodorus of Cyrene before he was 20 excited him about studying mathematics. Theodorus showed him the demonstration that not only was the square root of 2 incommensurable with 1 but so too were the square roots of the other nonsquare integers up to 17. Theaetetus then began research on this issue of incommensurability, both in Heraclea (on the Black Sea) and after 375 BCE in Athens at the Academy. In 369 BCE, he was drafted into the army during a war, was wounded in battle at Corinth, and soon after died of dysentery.

from 18; this can be done twice with remainder 2. Finally, one can subtract 2 exactly four times from 8. It then follows that 2 is the greatest common divisor of 18 and 80. In addition, this calculation shows that one can express the ratio of 80 to 18 in the form (4,2,4), in the sense that the algorithm applied to any other pair a, b, such that $a : b = 80 : 18$, will also give (4,2,4). As another example, take the pair 7 and 32. One can subtract 7 four times from 32 with remainder 4. One can then subtract 4 once from 7 with remainder 3. Finally, one can subtract 3 once from 4 with remainder 1. Thus, 7 and 32 are prime to one another and their ratio can be expressed in the form (4,1,1). (The notation (a,b,c) for ratio is, of course, a modern one.)

It was probably Theaetetus (417–369 BCE) who investigated the possibility of applying the Euclidean algorithm to magnitudes. The results appear as Propositions 2 and 3 of Book X, where we learn how to determine whether two magnitudes A and B have a common measure (are commensurable) or do not (are incommensurable). The procedure, called *anthyphairesis* (reciprocal subtraction), is basically the same as for numbers.[6] Thus, supposing that $A > B$, one first subtracts B from A as many times as possible, say, n_0, getting a remainder b that is less than B. One next subtracts b from B as many times as possible, say, n_1, getting a remainder b_1 less than b. Euclid showed in Proposition X–2 that if this process never ends, then the original two magnitudes are incommensurable. If, on the other hand, one of the magnitudes of this sequence measures the previous one, then that magnitude is the greatest common measure of the original two (Proposition X–3). A natural question here is how one can tell whether or not the process ends. In general, that is difficult. But in certain cases, one observes a repeated pattern in the remainders, which shows that the process cannot end.

For example, let us consider the case of the diagonal and side of the regular pentagon (Fig. 3.23). By the properties of the pentagon, we know that $CG = KG$. Therefore, we can subtract the side $CG = KG$ once from the diagonal GD, leaving remainder KD. We now must subtract KD from the side CG. But $CG = HD$, so KD can be subtracted once from $CG = HD$ with remainder KH. Note that KH is the side of another regular pentagon, whose diagonal is $KM = KD$. Therefore, at the next stage one is again subtracting a side from a diagonal of a pentagon. Since one can continue getting new smaller and smaller pentagons by connecting diagonals of previous ones, it is clear that the process never ends in this case.

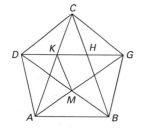

FIGURE 3.23

Incommensurability of diagonal and side of a regular pentagon

BIOGRAPHY

Eudoxus (408–355 BCE)

Eudoxus studied medicine in his youth in Cnidus, an island off the coast of Asia Minor. On a visit to Athens, he was attracted to the lectures at the Academy in philosophy and mathematics and began the study of these subjects. Later, he visited Egypt and was able to make numerous astronomical observations and study the Egyptian calendar. Returning to his home, he opened a school and conducted his own research. Although he returned at least one other time to Athens, this time with his own students, he spent most of the remainder of his life in Cnidus. He is famous not only for his work in geometry but also for his application of spherical geometry to astronomy.

Thus, the diagonal and side of a regular pentagon are incommensurable. In fact, the ratio of the diagonal to the side may be written as $(1, 1, 1, \dots)$.

Given now the existence of incommensurable magnitudes, the Greeks realized that they had to figure out a method of dealing with the ratios of such magnitudes. When they believed that any pair of quantities was commensurable, it was easy enough to see when two such pairs were proportional, or had the same ratio. Euclid in fact defined this concept in Book VII, when he was dealing with numbers: *Four numbers are* **proportional** *when the first is the same multiple, or the same part, or the same parts, of the second that the third is of the fourth.*

As an example, $3 : 4 = 6 : 8$, because 3 is 3 "fourth" parts of 4 while at the same time 6 is 3 "fourth" parts of 8. But for general magnitudes, one cannot use this definition. The side of a pentagon cannot be expressed either as a multiple or as a part or as parts of the diagonal.

Thus, using the anthyphairesis procedure, Theaetetus gave a new definition of "same ratio," which applied to all magnitudes. Suppose there are two pairs of magnitudes A, B, and C, D. Applying this procedure to each pair gives two sequences of equalities:

$$A = n_0 B + b \; (b < B) \qquad C = m_0 D + d \; (d < D)$$
$$B = n_1 b + b_1 \; (b_1 < b) \qquad D = m_1 d + d_1 \; (d_1 < d)$$
$$b = n_2 b_1 + b_2 \; (b_2 < b_1) \qquad d = m_2 d_1 + d_2 \; (d_2 < d_1)$$
$$\vdots \qquad\qquad\qquad \vdots$$

If the two sequences of numbers $(n_0, n_1, n_2, \dots)$, $(m_0, m_1, m_2, \dots)$, are equal term by term and both end at, say, $n_k = m_k$, then one can check that the ratios $A : B$ and $C : D$ are both equal to the same ratio of integers. Hence, Theaetetus could give the general definition that $A : B = C : D$ if the (possibly never ending) sequences $(n_0, n_1, n_2, \dots)$, $(m_0, m_1, m_2, \dots)$, are equal term by term. Although in general it may be difficult to decide whether two ratios are equal, we have seen that there are interesting cases in which the sequence $n_0, n_1, n_2, \dots$, is relatively simple to determine. In any case, Aristotle noted that this anthyphairesis definition of equal ratio was the one in use in his time.

Unfortunately, it turned out that Theaetetus's definition was very awkward to use in practice, so the mathematicians continued to search for a better one. It is not known what inspired Eudoxus (408–355 BCE) to give his new definition of same ratio, but a reasonable guess can be made.[7] Theaetetus's definition shows, for example, that if $A : B = C : D$, then

SIDEBAR 3.4 *Selected Definitions from Euclid's* Elements, *Book V*

1. A magnitude is a **part** of a magnitude, the less of the greater, when it measures (divides) the greater.
2. The greater is a **multiple** of the less when it is measured by the less.
3. A **ratio** is a sort of relation in respect to quantity between two magnitudes of the same kind.
4. Magnitudes are said to **have a ratio** to one another which are capable, when multiplied, of exceeding one another.
6. Let magnitudes which have the same ratio be called **proportional**.

7. When, of the equimultiples, the multiple of the first magnitude exceeds the multiple of the second, but the multiple of the third does not exceed the multiple of the fourth, then the first is said to **have a greater ratio** to the second than the third has to the fourth.
9. When three magnitudes are proportional, the first is said to have to the third the **duplicate ratio** of that which it has to the second.
10. When four magnitudes are continuously proportional, the first is said to have to the fourth the **triplicate ratio** of that which it has to the second.

$A > n_0 B$ while $C > n_0 D$ (since $m_0 = n_0$). Since $n_1 A = n_1 n_0 B + n_1 b = (n_1 n_0 + 1)B - b_1$, also $n_1 A < (n_1 n_0 + 1)B$ and similarly $n_1 C < (n_1 n_0 + 1)D$. A comparison of further multiples of A and B and corresponding multiples of C and D shows that for various pairs r, s, of numbers, $rA > sB$ whenever $rC > sD$ and $rA < sB$ whenever $rC < sD$. Thus, Eudoxus took for his definition of **same ratio** the one now included as definition 5 of Book V (see Sidebar 3.4 for other definitions from Book V):

> 5. Magnitudes are said to be **in the same ratio** (alternatively, **proportional**), the first to the second and the third to the fourth, when, if any equal multiples whatever are taken of the first and third, and any equal multiples whatever of the second and fourth, the former multiples alike exceed, are alike equal to, or alike fall short of, the latter multiples respectively taken in corresponding order.

Translated into algebraic symbolism, this definition says that $a : b = c : d$ if, given any positive integers m, n, whenever $ma > nb$, also $mc > nd$, whenever $ma = nb$, also $mc = nd$, and whenever $ma < nb$, also $mc < nd$. In modern terms, this is equivalent to noting that for every fraction $\frac{n}{m}$, the quotients $\frac{a}{b}$ and $\frac{c}{d}$ are alike greater than, equal to, or less than that fraction.

Of course, before one can define "same ratio," a definition of **ratio** itself is in order. This is given in definitions 3 and 4. Note that Euclid was quite clear that a ratio can only exist between magnitudes of the same kind, that is, lines, surfaces, solids, and so on. In addition, there must be a multiple of each that is greater than the other. So, for example, because no multiple of the angle between the circumference of a circle and a tangent line can exceed a given rectilinear angle, there can be no ratio between these two angles.

Definition 9 is Euclid's version of what is today called the square of a ratio, or, equivalently, the ratio of the squares: If $a : b = b : c$, then $a : c$ is the **duplicate** of the ratio $a : b$. A modern form would be $a : c = (a : b)(b : c) = (a : b)(a : b) = (a : b)^2 = a^2 : b^2$, or, in fractions, $\frac{a}{c} = (\frac{a}{b})^2 = \frac{a^2}{b^2}$. Euclid, however, did not multiply ratios, much less fractions, just as he did not multiply magnitudes. He only multiplied magnitudes by numbers. Similarly, he never divided magnitudes. One cannot interpret Euclid's ratio $a : b$ as a fraction corresponding to a particular point on a number line to which can be applied the standard arithmetical operations. On the other hand, Euclid did use the equivalence between the duplicate ratio of two quantities and the ratio of their squares in the cases where it made sense to speak of the "square" of a quantity (see Proposition VI–20).

The first proposition of Book V asserts, in modern symbols, that if $ma_1, ma_2, \ldots, ma_n$ are equal multiples of $a_1, a_2, \ldots, a_n$, then $ma_1 + ma_2 + \cdots + ma_n = m(a_1 + a_2 + \cdots + a_n)$. Similarly, Proposition V–2 asserts in effect that $ma + na = (m + n)a$, while the next result can be translated as $m(na) = (mn)a$. In other words, these first propositions of Book V give versions of the modern distributive and associative laws.

Proposition V–4 is the first in which the definition of same ratio is invoked. The result states that if $a : b = c : d$, then $ma : nb = mc : nd$, where m, n are arbitrary numbers. To show that equality, Euclid needed to show that if $p(ma)$, $p(mc)$, are equal multiples of ma, mc, and $q(nb)$, $q(nd)$, are equal multiples of nb, nd, then according as $p(ma) >=< q(nb)$, so is $p(mc) >=< q(nd)$. But since $a : b = c : d$, the associative law and the definition of same ratio for the original magnitudes allowed Euclid to conclude the equality of the ratios for the multiples.

The next two propositions repeat the first two with addition being replaced by subtraction. Proposition V–7 shows that if $a = b$, then $a : c = b : c$ and $c : a = c : b$, while Proposition V–8 asserts that if $a > b$, then $a : c > b : c$ and $c : b > c : a$. The proof of the first part of the latter shows Euclid's use of definitions 4 and 7. Since $a > b$, there is an integral multiple, say, m, of $a - b$ that exceeds c (by definition 4). Let q be the first multiple of c that equals or exceeds mb. Then $qc \geq mb > (q - 1)c$. Since $m(a - b) = ma - mb > c$, it follows that $ma > mb + c > qc$. Because also $mb \leq qc$, definition 7 implies that $a : c > b : c$. A similar argument gives the second conclusion.

Among other results of Book V are Proposition V–11, which asserts the transitive law, if $a : b = c : d$ and $c : d = e : f$, then $a : b = e : f$, and Proposition V–16, which states that if $a : b = c : d$, then $a : c = b : d$. The remaining results give other properties of magnitudes in proportion, in particular results dealing with adding or subtracting quantities to the antecedents or consequents in various proportions.

Although Book V gives numerous properties of magnitudes in proportion, the main application of this theory for Euclid was in the treatment of similarity in Book VI. The results of this book then became another major component of the Greek mathematician's toolbox. The book begins with the definition of similarity:

> **Similar rectilinear figures** are such as have their angles respectively equal and the sides about the equal angles proportional.

Recall that the foundation of the idea of similarity, the notion of same ratio (or proportionality), was originally based on the idea that all quantities could be thought of as numbers. So once the basis for the idea of proportionality was destroyed, the foundation for these results no longer existed. That is not to say that mathematicians ceased to use them. Intuitively, they knew that the concept of equal ratio made perfectly good sense, even if they could not provide a formal definition. In Greek times as also in modern times, mathematicians often ignored foundational questions and proceeded to discover new results. The working mathematician knew that eventually the foundation would be strengthened. Once this occurred, the actual similarity results could be organized into a logically acceptable treatise. It is not known who provided this final organization. What is probably true is that there was actually very little to

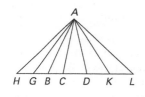

FIGURE 3.24

Elements, Proposition VI–1

redo except for the proof of the first proposition of the book. That is the only one that depends directly on Eudoxus's definition.

PROPOSITION VI–1   *Triangles and parallelograms which have the same height are to one another as their bases.*

Given triangles ABC, ACD, with the same height, Euclid needed to show that as BC is to CD, so is the triangle ABC to the triangle ACD. Proceeding as required by Eudoxus's definition, he extended the base BD to both right and left so that he could take arbitrary multiples of both BC and CD along that line (Fig. 3.24). As earlier, since he could not take an "arbitrary multiple," Euclid used a "generalizable example." So working with two line segments on each side, Euclid noted that because triangles with equal heights and equal bases are equal, whatever multiple the base HC is of the base BC, the triangle AHC is the same multiple of triangle ABC. The same holds for triangle ALC with respect to triangle ACD. Since again triangles AHC and ALC have the same heights, the former is greater than, equal to, or less than the latter precisely when HC is greater than, equal to, or less than CL. Equal multiples having been taken of base BC and triangle ABC, and other equal multiples of base CD and triangle ACD, and the results compared as required by Eudoxus's definition, it follows that $BC : CD = ABC : ACD$ as desired. The result for parallelograms is immediate, because each parallelogram is double the corresponding triangle.

After showing in Proposition VI–2 that a line parallel to one of the sides of a triangle cuts the other two sides proportionally and conversely, and in the following proposition that the bisector of an angle of a triangle cuts the opposite side into segments in the same ratio as that of the remaining sides and conversely, Euclid next gave various conditions under which two triangles are similar. Because the definition of similarity requires both that corresponding angles are equal and that corresponding sides are proportional, Euclid showed that one or the other of these two conditions is sufficient. He also stated the conditions under which the equality of only one pair of angles and the proportionality of two pairs of sides guarantees similarity. Proposition VI–8 then shows that the perpendicular to the hypotenuse from the right angle of a right triangle divides the triangle into two triangles, each similar to the original one.

Among the useful constructions of Book VI are the finding of proportionals. Given line segments a, b, c, Euclid showed how to determine x satisfying $a : b = b : x$ (Proposition VI–11), $a : b = c : x$ (Proposition VI–12), and $a : x = x : b$ (Proposition VI–13). This last result is equivalent to finding a square root, that is, to solving $x^2 = ab$, and is therefore nearly identical to the result of Proposition II–14. In fact, the constructions in the proof are the same; the only difference is that here Euclid used similarity to prove the result, while earlier he used II–5.

Proposition VI–16 is in essence the familiar one that in a proportion the product of the means is equal to the product of the extremes. But since Euclid never multiplied magnitudes, he could not have stated this result in terms of Book V. In the geometry of Book VI, however, he has the equivalent of multiplication, for line segments only:

PROPOSITION VI–16   *If four straight lines are proportional, the rectangle contained by the extremes is equal to the rectangle contained by the means; and if the rectangle contained by the extremes is equal to the rectangle contained by the means, the four straight lines will be proportional.*

Proposition VI–19 is of fundamental importance later. It also illustrates Euclid's notion of duplicate ratio:

PROPOSITION VI–19   *Similar triangles are to one another in the duplicate ratio of the corresponding sides.*

A modern statement of this result would replace "in the duplicate ratio" by "as the square of the ratio." But Euclid did not multiply either magnitudes or ratios. Ratios are not quantities; they are not to be considered as numbers in any sense of the word. Hence, for this particular proposition, Euclid needed to construct a point G on BC so that $BC : EF = EF : BG$ (Fig. 3.25). The ratio $BC : BG$ is then the duplicate of the ratio $BC : EF$ of the corresponding sides. To prove the result, he showed that the triangles ABG, DEF, are equal. Because triangle ABC is to triangle ABG as BC is to BG, the conclusion follows immediately. Proposition VI–20 extends this result to similar polygons. In particular, the duplicate ratio of two line segments is equal to the ratio of the squares on the segments.

FIGURE 3.25

Elements, Proposition VI–19

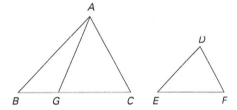

Two parallelograms, of course, can be equiangular without being similar. Euclid was also able to deal with the ratio of such figures, but only by using a concept not formally defined:

PROPOSITION VI–23   *Equiangular parallelograms have to one another the ratio compounded of the ratios of the sides.*

The proof shows what Euclid means by the term "compounded," at least in the context of ratios of line segments. If the two ratios are $a : b$ and $c : d$, one first constructs a segment e such that $c : d = b : e$. The ratio **compounded** of $a : b$ and $c : d$ is then the ratio $a : e$. In modern terms, the fraction $\frac{a}{e}$ is simply the product of the fractions $\frac{a}{b}$ and $\frac{c}{d} = \frac{b}{e}$. Interestingly enough, although Euclid never considered compounding again, this notion became quite important in later Greek times as well as in the medieval period.

 3.6   NUMBER THEORY

Book VII of the *Elements* is the first of three dealing with the elementary theory of numbers. There is no mention of the first six books in Books VII, VIII, and IX; these three books form an entirely independent unit. Only in later books is there some connection made between the three arithmetic books and the earlier geometric ones. The new start that Euclid made in Book VII is evidence of his desire to stick with Aristotle's clear separation between magnitude and number. The first six books dealt with magnitudes, in particular lengths and areas. The fifth book dealt with the general theory of magnitudes in proportion. But in Books VII–IX

SIDEBAR 3.5   *Selected Definitions from Euclid's* Elements, *Book VII*

1. A **unit** is that by virtue of which each of the things that exist is called one.
2. A **number** is a multitude composed of units.
3. A number is a **part** of a number, the less of the greater, when it measures the greater;
4. but **parts** when it does not measure it.
5. The greater number is a **multiple** of the less when it is measured by the less.
11. A **prime number** is that which is measured by the unit alone.

12. Numbers **prime to one another** are those which are measured by the unit alone as a common measure.
15. A number is said to **multiply** a number when that which is multiplied is added to itself as many times as there are units in the other, and thus some number is produced.
20. Numbers are **proportional** when the first is the same multiple, or the same part, or the same parts, of the second that the third is of the fourth.

Euclid dealt only with numbers. He did not consider these as types of magnitudes, but as entirely separate entities. Therefore, although there are many results in Book VII that appear to be merely special cases of results in Book V, for Euclid they are quite different. One should not be misled by the line segments Euclid used in these books to represent numbers. He did not use the fact of the representation in his proofs. Perhaps this representation was the only one that occurred to him.

It is reasonably certain that many of the propositions in the arithmetic books date back to the Pythagoreans. But from the use of Book VII in Book X, it appears that the details of the compilation of that book are due to the same mathematician who is responsible for Book X, namely, Theaetetus. That is, Theaetetus took the loosely structured number theory of the Pythagoreans and made it rigorous by introducing precise definitions and detailed proofs. It is these that Euclid included in his version of the material.

Book VII, like most of Euclid's books, begins with definitions (Sidebar 3.5). The first definition is, like the beginning definitions of Book I, mathematically useless in modern terms. For Euclid, however, the definition appears as the mathematical abstraction of the concept of "thing." What is more interesting is the second definition, that a number is a **multitude** of units. Since "multitude" means plurality, and the unit is not a plurality, it appears that for Euclid, as for the Pythagoreans earlier, 1 is not a number.

Definitions 3 and 5 are virtual word-for-word repetitions of definitions 1 and 2 in Book V, while definition 4 would make no sense in the context of arbitrary magnitudes. Definitions 11 and 12 are essentially modern definitions of prime and relatively prime, with the note that for Euclid a number does not measure itself. Definition 15 is somewhat curious in that this is the only arithmetic operation defined by Euclid. He assumes that addition and subtraction are known. Note that there is no analogue of this definition in Book V.

Recall that the first two propositions of Book VII deal with the Euclidean algorithm. Several of the next propositions are direct analogues of propositions in Book V. For example, Euclid proved in Propositions VII–5 and VII–6 what amounts to the distributive law $\frac{m}{n}(b + d) = \frac{m}{n}b + \frac{m}{n}d$. He had proved this for magnitudes as Proposition V–1, except that there the result dealt with (integral) multiples rather than the parts—here represented as fractions—of Book VII. Even the proofs of these results are virtually identical. That Euclid did not simply quote results from Book V is evidence that for Euclid number was not a type of magnitude.

Propositions VII–11 through VII–22 include various standard results on numbers in proportion, several of which Euclid proved for magnitudes in Book V. Most are used again in the following two books. In particular, Proposition VII–16 proves the commutativity of multiplication, a nontrivial result given Euclid's definition of multiplication. Proposition VII–19 gives the usual test for proportionality, that $a : b = c : d$ if and only if $ad = bc$. Recall that Euclid had already proved an analogue for line segments (Proposition VI–16). The proof here, however, is quite different. Given that $a : b = c : d$, it follows that $ac : ad = c : d = a : b$. Also $a : b = ac : bc$. Therefore, $ac : ad = ac : bc$. Hence, $ad = bc$. The converse is proved similarly. Proposition VII–20 shows that if a, b, are the smallest numbers in the ratio $a : b$, then a and b each divide c, d, the same number of times, where $c : d = a : b$. It then follows that relatively prime numbers are the least of those in the same ratio and conversely.

Propositions VII–23 through VII–32 deal further with primes and numbers relatively prime to one another. In particular, they present Euclid's theory of divisibility and give, together with Proposition IX–14, a version of the fundamental theorem of arithmetic—that every number can be uniquely expressed as a product of prime numbers.

PROPOSITION VII–31 *Any composite number is measured by some prime number.*

PROPOSITION VII–32 *Any number either is prime or is measured by some prime number.*

The latter proposition is a direct consequence of the former. That one in turn is proved by a technique Euclid used often in the arithmetic books, the least number principle. He began with a composite number a, which is therefore measured (divided) by another number b. If b were prime, the result would follow. If not, then b is in turn measured by c, which will then measure a, and c is in turn either prime or composite. As Euclid then said, "if the investigation is continued in this way, some prime number will be found which will measure the number before it, which will also measure a. For, if it is not found, an infinite series of numbers will measure the number a, each of which is less than the other; which is impossible in numbers." One can again note the distinction between number and magnitude. Any decreasing sequence of numbers has a least element, but the same is not true for magnitudes.

Although Euclid did not do so, it is straightforward to demonstrate from VII–32 that any number can be expressed as the product of prime numbers. To prove that this expression is unique, we need

PROPOSITION VII–30 *If a prime number measures the product of two numbers, it will measure one of them.*

Suppose the prime number p divides ab and p does not divide a. Then $ab = sp$, or $p : a = b : s$. But since p and a are relatively prime, they are the least numbers in that ratio. It follows that b is a multiple of p, or that p divides b. Euclid used this proposition to prove the uniqueness of any prime decomposition in

PROPOSITION IX–14 *If a number is the least of those that are measured by certain prime numbers, then no other prime number will measure it.*

Book VIII primarily deals with numbers in continued proportion, that is, with sequences $a_1, a_2, \ldots, a_n$, such that $a_1 : a_2 = a_2 : a_3 = \cdots$. In modern terms, such a sequence is called a geometric progression. It is generally thought today that much of the material in this book is due to Archytas (fifth century BCE), the person from whom Plato received his mathematical training. In particular, Proposition VIII–8 is a generalization of a result due to Archytas and coming out of his interest in music. The original result is that there is no mean proportional between two numbers whose ratio in lowest terms is equal to $(n + 1) : n$. Recall that the ratio of two strings whose sound is an octave apart is $2 : 1$. This ratio is the compound of $4 : 3$ and $3 : 2$, so the octave is composed of a fifth and a fourth. Archytas's result then states that the octave cannot be divided into two equal musical intervals. Of course, in this case, the result is equivalent to the incommensurability of $\sqrt{2}$ with 1. But the result also shows that one cannot divide a whole tone, whose ratio of lengths is $9 : 8$, into two equal intervals.

PROPOSITION VIII–8   *If between two numbers there are numbers in continued proportion with them, then, however many numbers are between them in continued proportion, so many will also be in continued proportion between numbers which are in the same ratio as the original numbers.*

Euclid concerned himself in several other propositions of Book VIII with determining the conditions for inserting mean proportional numbers between given numbers of various types. Proposition VIII–11 in particular is the analogue for numbers of a special case of VI–20. Namely, Euclid showed that between two square numbers there is one mean proportional and that the square has to the square the duplicate ratio of that which the side has to the side. Similarly, in Proposition VIII–12, Euclid showed that between two cube numbers there are two mean proportionals and the cube has to the cube the triplicate ratio of that which the side has to the side. This is, of course, the analogue in numbers of Hippocrates' reduction of the problem of doubling the cube to that of finding two mean proportionals.

The final book on number theory is Book IX. Proposition IX–20 shows that there are infinitely many prime numbers:

PROPOSITION IX–20   *Prime numbers are more than any assigned multitude of prime numbers.*

As in earlier proofs, Euclid used the method of generalizable example. He picked just three primes, A, B, C, and showed that one can always find an additional one. To do this, consider the number $N = ABC + 1$. If N is prime, a prime other than those given has been found. If N is composite, then it is divisible by a prime p. Euclid showed that p is distinct from the given primes A, B, C, because none of these divides N. It follows again that a new prime p has been found. Euclid presumably assumed that his readers were convinced that a similar proof will work, no matter how many primes are originally picked.

Propositions IX–21 through IX–34 form a nearly independent unit of very elementary results about even and odd numbers. They probably represent a remnant of the earliest Pythagorean mathematical work. This section includes such results as the sum of even numbers is even, an even sum of odd numbers is even, and an odd sum of odd numbers is odd. These elementary results are followed by two of the most significant results of the entire number theory section of the *Elements*.

PROPOSITION IX–35   *If as many numbers as we please are in continued proportion, and there is subtracted from the second and the last numbers equal to the first, then, as the excess of the second is to the first, so will the excess of the last be to all those before it.*

In effect, this result determines the sum of a geometric progression. Represent the sequence of numbers in "continued proportion" by $a, ar, ar^2, ar^3, \ldots, ar^n$, and the sum of "all those before [the last]" by S_n (since there are n terms before ar^n). Euclid's result states that

$$(ar^n - a) : S_n = (ar - a) : a.$$

The modern form for this sum is

$$S_n = \frac{a(r^n - 1)}{r - 1}.$$

The final proposition of Book IX, Proposition IX–36, shows how to find perfect numbers, those that are equal to the sum of all their factors. The result states that if the sum of any number of terms of the sequence $1, 2, 2^2, \ldots, 2^n$ is prime, then the product of that sum and 2^n is perfect. For example, $1 + 2 + 2^2 = 7$ is prime; therefore, $7 \times 4 = 28$ is perfect. And, in fact, $28 = 1 + 2 + 4 + 7 + 14$. Other perfect numbers known to the Greeks were 6, corresponding to $1 + 2$; 496, corresponding to $1 + 2 + 4 + 8 + 16$; and 8128, corresponding to $1 + 2 + 4 + 8 + 16 + 32 + 64$. Although several other perfect numbers have been found by using Euclid's criterion, it is still not known whether there are any perfect numbers that do not meet it. Leonhard Euler proved that any even perfect number meets Euclid's criterion, but it is not known whether there are any odd perfect numbers. It is curious, perhaps, that Euclid devoted the culminating theorem of the number theory books to the study of a class of numbers only four of which were known. Nevertheless, the theory of perfect numbers has always proved a fascinating one for mathematicians.

 3.7   IRRATIONAL MAGNITUDES

Many historians consider Book X the most important of the *Elements*. It is the longest of the thirteen books and probably the best organized. The purpose of Book X is evidently the classification of certain incommensurable magnitudes. One of the motivations for the book was the desire to characterize the edge lengths of the regular polyhedra, whose construction in Book XIII forms a fitting climax to the *Elements*. Euclid needed a nonnumerical way of comparing the edges of the icosahedron and the dodecahedron to the diameter of the sphere in which they were inscribed. In a manner familiar in modern mathematics, this simple question was to lead to the elaborate classification scheme of Book X, far past its direct answer. Much of this book is attributed to Theaetetus, since he is credited with some of the polyhedral constructions of Book XIII and since it was in Plato's dialogue bearing his name that the question of determining which numbers have square roots incommensurable with the unit was brought up. It is the answer to that question, given early in Book X, that then leads to the general classification.

The introductory definitions give Euclid's understanding of the basic terms "incommensurable" and "irrational" (Sidebar 3.6). The first two definitions are relatively straightforward. The third one, on the other hand, needs some comment. First of all, it includes a theorem, which is proved subsequently in Book X. But secondly, note that Euclid's use of the term "rational" is different from the modern usage. For example, if the assigned straight line has

---

SIDEBAR 3.6   *Selected Definitions from Euclid's* Elements, *Book X*

1. Those magnitudes are said to be **commensurable** which are measured by the same measure, and those **incommensurable** which cannot have any common measure.
2. Straight lines are **commensurable in square** when the squares on them are measured by the same area, and **incommensurable in square** when the squares on them cannot possibly have any area as a common measure.

3. With these hypotheses, it is proved that there exist straight lines infinite in multitude which are commensurable and incommensurable respectively, some in length only, and others in square also, with an assigned straight line. Let then the assigned straight line be called **rational**, and those straight lines which are commensurable with it, whether in length and in square or in square only, **rational**, but those which are incommensurable with it **irrational**.

---

length 1, then not only are lines of length $\frac{a}{b}$ called rational, but also lines of length $\sqrt{\frac{a}{b}}$ (where a and b are positive integers).

The first proposition of Book X is fundamental, not only in that book but also in Book XII.

PROPOSITION X–1   *Two unequal magnitudes being given, if from the greater there is subtracted a magnitude greater than its half, and from that which is left a magnitude greater than its half, and if this process is repeated continually, there will be left some magnitude less than the lesser of the given magnitudes.*

The result depends on definition 4 of Book V, the criterion that two given magnitudes have a ratio. That definition requires that some multiple n of the lesser magnitude exceeds the greater. Then n subtractions of magnitudes greater than half of what is left at any stage gives the desired result.

Propositions X–2 and X–3 are the results on anthyphairesis discussed earlier. But since Euclid used the same procedure for magnitudes as he did for numbers in Book VII, he could now connect these two distinct concepts. Namely, Euclid showed in Propositions X–5 and X–6 that magnitudes are commensurable precisely when their ratio is that of a number to a number. So even though number and magnitude are distinct notions, one can now apply the machinery of numerical proportion theory to commensurable magnitudes. The more complicated Eudoxian definition is then only necessary for incommensurable magnitudes.

Proposition X–9 is the result attributed to Theaetetus that provides the generalization of the Pythagorean discovery of the incommensurability of the diagonal of a square with its side, or, in modern terms, of the irrationality of $\sqrt{2}$. Namely, Euclid showed here in effect that the square root of every nonsquare integer is incommensurable with the unit. In Euclid's terminology, the theorem states that two sides of squares are commensurable in length if and only if the squares have the ratio of a square number to a square number. The more interesting part is the "only if" part. Suppose the two sides a, b, are commensurable in length. Then $a : b = c : d$ where c, d, are numbers. Hence, the duplicates of each ratio are equal. But Euclid already showed (VI–20) that the square on a is to the square on b in the duplicate ratio of a to b as well as (VIII–11) that c^2 is to d^2 in the duplicate ratio of c to d. The result then follows.

After some further preliminaries on criteria for incommensurability, Euclid proceeded to the major task of Book X, the classification of certain irrational lengths, lengths that are neither commensurable with a fixed unit length nor commensurable in square with it. The entire classification is too long to discuss here, so only a few of the definitions, those that are of use in Book XIII, will be mentioned to provide some of the flavor of this section. It is significant to note that although each of these irrational lengths can be expressed today as a solution of a polynomial equation, Euclid did not use any algebraic machinery. Everything is done geometrically. Nevertheless, for ease of understanding, numerical examples of each definition are presented.

A **medial** straight line is one that is the side of a square equal to the rectangle contained by two rational straight lines commensurable in square only. For example, because the lengths $1, \sqrt{5}$, are commensurable in square only, and because the rectangle contained by these two lengths has an area equal to $\sqrt{5}$, the length equal to $\sqrt[4]{5}$ is medial. A **binomial** straight line is the sum of two rational straight lines commensurable in square only. So the length $1 + \sqrt{5}$ is a binomial. Similarly, the difference of two rational straight lines commensurable in square only is called an **apotome**. The length $\sqrt{5} - 1$ provides a simple example. A final, more complicated example is given by Euclid's definition of a **minor** straight line. Such a line is the difference $x - y$ between two straight lines such that x, y, are incommensurable in square, such that $x^2 + y^2$ is rational, and such that xy is a medial area, that is, equal to the square on a medial straight line. For example, if $x = \sqrt{5 + 2\sqrt{5}}$ and $y = \sqrt{5 - 2\sqrt{5}}$, then $x - y$ is a minor.

 ## 3.8   SOLID GEOMETRY AND THE METHOD OF EXHAUSTION

Book XI of the *Elements* is the first of three books dealing with solid geometry. This book contains the three-dimensional analogues of many of the two-dimensional results of Books I and VI. The introductory definitions include such notions as pyramids, prisms, and cones (Sidebar 3.7). The only definition that is somewhat unusual is that of a sphere, which is defined not by analogy to the definition of a circle but in terms of the rotation of a semicircle about its diameter. Presumably, Euclid used this definition because he did not intend to discuss the properties of a sphere as he had discussed the properties of a circle in Book III. The elementary properties of the sphere were in fact known in Euclid's time and dealt with in other texts, including one due to Euclid himself. In the *Elements*, however, Euclid considered spheres only in Book XII, where he dealt with the volume, and in Book XIII, where he constructed the regular polyhedra and showed how they fit into the sphere. His constructions in Book XIII, in fact, show how these polyhedra are inscribed in a sphere by rotating a semicircle around them, as in his definition.

The propositions of Book XI include some constructions analogous to those of Book I. For example, Proposition XI–11 shows how to draw a straight line perpendicular to a given plane from a point outside it, whereas Proposition XI–12 shows how to draw such a line from a point in the plane. There is also a series of theorems on parallelepipeds. In particular, by analogy with Proposition I–36, Euclid showed that parallelepipeds on equal bases and with the same height are equal (Proposition XI–31), and then, in analogy with VI–1, that parallelepipeds of the same height are to one another as their bases (Proposition XI–32). Also, in analogy

SIDEBAR 3.7   *Selected Definitions from Euclid's* Elements, *Book XI*

12. A **pyramid** is a solid figure, contained by planes, which is constructed from one plane to one point.

13. A **prism** is a solid figure contained by planes two of which, namely those which are opposite, are equal, similar and parallel, while the rest are parallelograms.

14. When, the diameter of a semicircle remaining fixed, the semicircle is carried round and restored again to the same position from which it began to be moved, the figure so comprehended is a **sphere**.

18. When, one leg of a right triangle remaining fixed, the triangle is carried around and restored again to the same position from which it began to be moved, the figure so comprehended is a **cone**. And if the fixed leg is equal to the other leg, the cone will be **right-angled**; if less, **obtuse-angled**; and if greater, **acute-angled**.

with VI–19 and VI–20, he showed in Proposition XI–33 that similar parallelepipeds are to one another in the triplicate ratio of their sides. Hence, the volumes of two similar parallelepipeds are in the ratio of the cubes of any pair of corresponding sides. And in Proposition XI–34, in partial analogy with VI–14 and VI–16, he demonstrated that in equal parallelepipeds, the bases are reciprocally proportional to the heights and conversely. As before, Euclid computed no volumes. Nevertheless, one can easily derive from these theorems the basic results on volumes of parallelepipeds. The "formulas" for volumes of other solids are included in Book XII.

The central feature of Book XII, which distinguishes it from the other books of the *Elements*, is the use of a limiting process, generally known as the method of exhaustion. This process, developed by Eudoxus, is used to deal with the area of a circle as well as the volumes of pyramids, cones, and spheres. "Formulas" giving some of these areas and volumes were known much earlier, but for the Greeks a proof was necessary, and Eudoxus's method provided a proof. What it did not provide was a way of discovering the formulas to begin with.

The main results of Book XII are the following:

PROPOSITION XII–2   *Circles are to one another as the squares on the diameters.*

PROPOSITION XII–7 (COROLLARY)   *Any pyramid is a third part of the prism which has the same base with it and equal height.*

PROPOSITION XII–10   *Any cone is a third part of the cylinder which has the same base with it and equal height.*

PROPOSITION XII–18   *Spheres are to one another in the triplicate ratio of their respective diameters.*

The first of these results is Euclid's version of the ancient result on the area of a circle, a version already known to Hippocrates 150 years earlier. In modern terms, it states that the area of a circle is proportional to the square on the diameter. It does not state what the constant of proportionality is, but the proof does provide a method for approximating this. Proposition XII–1, that similar polygons inscribed in circles are to one another as the squares on the diameters, serves as a lemma to this proof. This result in turn is a generalization of the result of VI–20 that similar polygons are to one another in the duplicate ratio of the corresponding sides. It is not difficult to show first of all that one can take any corresponding lines in place of the "corresponding sides," even the diameter of the circle, and secondly that one can replace "duplicate ratio" by "squares."

The main idea of the proof of XII–2 is to "exhaust" the area of a particular circle by inscribing in it polygons of increasingly many sides. In particular, Euclid showed that one can inscribe in the given circle a polygon whose area differs from that of the circle by less than any given area. His proof of the theorem began by assuming that the result is not true. That is, if the two circles C_1, C_2, have areas A_1, A_2, respectively, and diameters d_1, d_2, he assumed that $A_1 : A_2 \neq d_1^2 : d_2^2$. Therefore, there is some area S, either greater or less than A_2, such that $d_1^2 : d_2^2 = A_1 : S$. (Note that Euclid has never proved the existence of a fourth proportional to three arbitrary magnitudes, but only to three lengths. This is therefore another unproved result in Euclid. Its truth needs to come from some kind of continuity argument, but perhaps Euclid ignored it because he did not require the actual construction of such a magnitude.)

Suppose first that $S < A_2$ (Fig. 3.26). Then beginning with an inscribed square and continually bisecting the subtended arcs, inscribe in C_2 a polygon P_2 such that $A_2 > P_2 > S$. In other words, P_2 is to differ from A_2 by less than the difference between A_2 and S. This construction is possible by Proposition X–1, since at each bisection one is increasing the area of the polygon by more than half of the difference between the circle and the polygon. Next inscribe a polygon P_1 in C_1 similar to P_2. By Proposition XII–1, $d_1^2 : d_2^2 = P_1 : P_2$. By assumption, this ratio is also equal to $A_1 : S$. Therefore, $P_1 : A_1 = P_2 : S$. But clearly, $A_1 > P_1$. It follows that $S > P_2$, contradicting the assumption that $S < P_2$. Therefore, S cannot be less than A_2. Euclid proved that S also is not greater than A_2 by reducing it to the case already dealt with. It then follows that the ratio of the circles must be equal to the ratio of the squares on the diameters, as asserted.

FIGURE 3.26

Elements, Proposition XII–2, the method of exhaustion

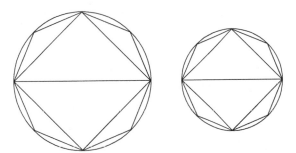

It is virtually certain that the theorem giving the volume of the pyramid was known to both the Egyptians and the Babylonians (Sidebar 3.8). Archimedes, however, wrote that although Eudoxus was the first to prove that theorem, the result was first discovered by Democritus (fifth century BCE) (Fig. 3.27). Unfortunately, we have no record of how the Egyptians, the Babylonians, or Democritus may have made their discovery. For the latter, we do have a hint in a report given by Chrysippus, in which Democritus discussed the problem of slicing a cone into "indivisible" sections by planes parallel to the base. He wondered whether these indivisible circles would be unequal or equal: "If they are unequal, they will make the cone

FIGURE 3.27

Democritus on a Greek stamp

SIDEBAR 3.8   *What Did the Greeks Learn from the Egyptians?*

Did the Greeks learn any mathematics from the Egyptians, or was their idea of mathematics so different from that of their predecessors that we may as well assume that they started from scratch? This question has been posed over the years, but because there is no extant documentation of transmission from Egypt to Greece before the third century BCE, we cannot give a definitive answer. Nevertheless, there are certainly hints.

The Greeks in general stated that they had learned from Egypt. The stories the Greeks told about many of their mathematicians, including Pythagoras, Thales, and Eudoxus, note that they studied in Egypt. And many Greek documents say that geometry was first invented by the Egyptians and then passed on to the Greeks. But what is meant here by geometry? It clearly cannot mean an axiomatic treatment such as we find in Euclid's *Elements*. What it could mean, however, is the results themselves. After all, one does not *discover* results by the axiomatic method. One discovers them by experiment, by trial and error, by induction; only after the discovery is made does one worry about actually proving that what one has proposed is correct. So it seems clear that what the Greek writers meant about the Egyptians inventing geometry was the results, not the method of proof. It also seems clear that the idea of proof from a system of axioms is original to the Greeks.

What geometric results could the Greeks have learned? One answer seems to be most of the formulas concerned with the measurement of geometric objects, such as the volume of a pyramid, the area of a circle, and the area of a hemisphere. They could also have learned the basic principles of similarity, since Egyptian sources reveal highly developed proportional

thinking connected with the use of scale models. And we are certain that the Greeks learned the use of unit fractions from the Egyptians, although these did not appear in formal Greek mathematics.

Just as in the case of the Babylonians, there is no documentary evidence of direct Egyptian influence on Greek mathematics, but the circumstantial evidence is relatively strong. And as in the case of Babylonian influence, we will have to await further research to answer the question.

There has been much recent historical controversy over the relationship of Greek civilization to Egyptian civilization and, in particular, of the relationship of Greek mathematics to Egyptian mathematics. The opening shot in this battle was the publication of Martin Bernal's *Black Athena: The Afroasiatic Roots of Classical Civilization* (New Brunswick: Rutgers University Press, 1987). This work asserted that classical Greek civilization has deep roots in Afroasiatic cultures, but that these influences have been systematically ignored or denied since the eighteenth century, chiefly for racist reasons. Bernal did not write much about science in this work, but summarized his views on the contributions of Egyptian science to Greek science in "Animadversions on the Origins of Western Science," *Isis* 83 (1992), 596–607. This article was answered by Robert Palter in his "*Black Athena*, Afro-Centrism, and the History of Science," *History of Science* 31 (1993), 227–287. Bernal responded in "Response to Robert Palter," *History of Science* 32 (1994), 445–464; and Palter answered Bernal in the same issue on pages 464–468. The last word on this issue has not yet been uttered.

---

irregular, as having many indentations, like steps, and unevennesses; but if they are equal, the sections will be equal, and the cone will appear to have the property of the cylinder, and to be made up of equal, not unequal, circles, which is very absurd."[8]

Although we do not know what Democritus's final conclusion was, he evidently did think that the cone and, analogously, the pyramid were "made up" of indivisibles. If so, he could have derived Euclid's Proposition XII–5, that pyramids of the same height and with triangular bases are to one another as their bases. For if one imagines the two pyramids cut respectively by planes parallel to and at equal distances from the bases, then the corresponding sections of the two pyramids would be in the ratio of the bases. Since Democritus conceived of each pyramid as being "made up" of these infinitely many indivisible sections, the pyramids themselves would be in this same ratio. He could then have completed the demonstration of the volume formula by noting, as in XII–7, that a prism with a triangular base can be divided into three pyramids, all of equal height and equal bases.

Euclid, of course, proved XII–5 as well as XII–10 and XII–18 by using *reductio* arguments. Assuming the falsity of the given assertion, he proceeded to construct inside the given solid other solids, whose properties are already known, such that the difference between the given solid and the constructed one is less than a given "small" solid, the "error" defined by the false assumption. That is, he exhausted the solid. The known properties of the constructed figure then led him to a contradiction as in the proof of XII–2. But the quotation from Democritus shows us that from the earliest period of Greek mathematics there were attempts to discover certain results by the use of infinitesimals, even though, as we have seen, Aristotle banned such notions from formal Greek mathematics.

The final book of the *Elements*, Book XIII, is devoted to the construction of the five regular polyhedra and their "comprehension" in a sphere (Fig. 3.28). This book is the three-dimensional analogue to Book IV. The study of the five regular polyhedra—the cube, tetrahedron, octahedron, dodecahedron, and icosahedron—and the proof that these are the only regular polyhedra are due to Theaetetus. The first three solids were known in pre-Greek times, and there is archaeological evidence of bronze dodecahedra dating back perhaps to the seventh century BCE. The icosahedron, however, was evidently first studied by Theaetetus. It was also he who recognized that these five were the only regular polyhedra, and that in fact the properties of the regular polyhedra were something to study.

FIGURE 3.28

The five regular polyhedra

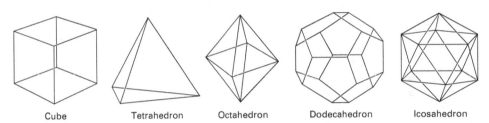

Cube   Tetrahedron   Octahedron   Dodecahedron   Icosahedron

Euclid proceeded systematically in Book XIII to construct each of the polyhedra, to demonstrate that each may be comprehended (inscribed) in a sphere, and to compare the edge length of the polyhedron with the diameter of the sphere. For the tetrahedron, Euclid showed that the square on the diameter is $1\frac{1}{2}$ times the square on the edge. In the cube the square on the diameter is triple the square on the edge, whereas in the octahedron the square on the diameter is double that on the edge. The other two cases are somewhat trickier. Euclid proved that the edge of the dodecahedron is an apotome equal in length to the greater segment of the edge of the inscribed cube when that edge is cut in extreme and mean ratio. Thus, if the diameter of the sphere is 1, then the edge of the cube is $c = \frac{\sqrt{3}}{3}$. Therefore, the edge length of the dodecahedron is the positive root of $x^2 + cx - c^2 = 0$ or $\frac{c}{2}(\sqrt{5} - 1) = \frac{1}{6}(\sqrt{15} - \sqrt{3})$. Because both $\sqrt{15}$ and $\sqrt{3}$ are rational by Euclid's definition, and because they are commensurable in square only, the edge length is in fact an apotome.

For the icosahedron, Euclid proved that the side is a minor straight line. In this case, the square on the diameter of the sphere is five times the square on the radius r of the circle circumscribing the five upper triangles of the icosahedron. The bases of these five triangles form a regular pentagon, each edge of which is an edge of the icosahedron. The side of a pentagon inscribed in a circle of radius r is equal to

$$\frac{r}{2}\sqrt{5 + 2\sqrt{5}} - \frac{r}{2}\sqrt{5 - 2\sqrt{5}} = \frac{r}{2}\sqrt{10 - 2\sqrt{5}}.$$

If the diameter of the sphere is 1, then $r = \frac{\sqrt{5}}{5}$, a rational value, and the edge length of the icosahedron is indeed a minor straight line. In particular, this edge length is

$$\frac{\sqrt{5}}{10}\sqrt{10 - 2\sqrt{5}} = \frac{1}{10}\sqrt{50 - 10\sqrt{5}}.$$

In a fitting conclusion to Book XIII and the *Elements*, Euclid constructed the edges of the five regular solids in one plane figure, thereby comparing them to each other and the diameter of the given sphere. He then demonstrated that there are no regular polyhedra other than these five.

 3.9   EUCLID'S *DATA*

Euclid wrote several mathematics books more advanced than the *Elements*. The most important of the ones that have survived is the *Data*. This was in effect a supplement to Books I–VI of the *Elements*. Each proposition of the *Data* takes certain parts of a geometric configuration as given, or known, and shows that therefore certain other parts are determined. ("Data" means "given" in Latin.) Generally, in his proofs, Euclid showed that these other parts were determined by showing exactly how to determine them. Thus, the *Data* in essence transformed the synthetic purity of the *Elements* into a manual appropriate to one of the goals of Greek mathematics, the solution of new problems.

As one example, consider

PROPOSITION 39   *If each of the sides of a triangle is given in magnitude, the triangle is given in form.*

In other words, this proposition claims that if the lengths of the three sides of a triangle are known, then the triangle itself is determined, that is, not only are the sides known but also the angles. In the demonstration, Euclid carefully constructed a triangle with sides equal to those of the given triangle. He then used parts of the "toolbox," in this case Proposition I–8 and definition 1 of Book VI of the *Elements*, to conclude that the constructed triangle was "equal and similar" to the given triangle. This means, then, that the original triangle was "given in form."

We can certainly consider several of the propositions of the *Data* as examples of geometric algebra, in that Euclid showed how to find unknown lengths, given certain known ones. For example, here are two propositions closely related to *Elements* VI–29.

PROPOSITION 84   *If two straight lines contain a given area in a given angle, and one of them is greater than the other by a given straight line, each of them will be given, too.*

If, as in the discussion of VI–29, it is assumed that the given angle is a right angle—and the diagram in the medieval manuscripts that survive shows such an angle—the problem is related to one of the standard Babylonian problems: Find x, y, if the product and difference are given. That is, solve the system

$$xy = c, \qquad x - y = b.$$

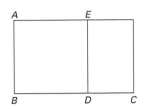

FIGURE 3.29

Data, Proposition 84

Euclid began by setting up the rectangle contained by the two straight lines AB, BC (Fig. 3.29). He then chose point D on BC so that $BD = AB$. Thus, $DC = b$ was the given straight line. He now had a given area, the rectangle ($= c$) applied to a given line b, exceeding by a square figure. He could then apply Proposition 59:

PROPOSITION 59   *If a given area be applied to a given straight line, exceeding by a figure given in form, the length and width of the excess are given.*

It is here that Euclid really solved the problem of Proposition 84, using a diagram similar to that of *Elements* VI–29 (Fig. 3.30). As there, he bisected the line $DE = b$ at Z, constructed the square on $ZE = b/2$, noted that the sum of that square and the original area (the rectangle $AB = c$) is equal to the square on $ZB = y + b/2$ (or $x - b/2$), and thereby showed how either of those quantities can be determined as the side of that square. Algebraically, this amounts to the standard Babylonian formula

$$y = \sqrt{\left(\frac{b}{2}\right)^2 + c} - \frac{b}{2}$$

$$x = \sqrt{\left(\frac{b}{2}\right)^2 + c} + \frac{b}{2}.$$

As before, Euclid dealt only with geometric figures and never actually wrote out a rule like the above. Nevertheless, given that the problem is in fact to find two lengths satisfying certain conditions, even its formulation is nearly identical to the Babylonian formulation. On the other hand, as in VI–29, the statement of the result enables one to deal with parallelograms as well as the rectangles discussed by the Babylonians. Euclid treated other similar geometric algebra

FIGURE 3.30

Data, Proposition 59

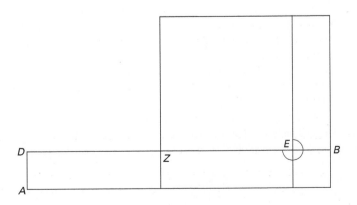

problems in the *Data*. Thus, in Propositions 85 and 58 he solved the geometric equivalent of the system

$$xy = c, \quad x + y = b,$$

while in Proposition 86 he solved the system

$$xy = a, \quad \frac{y^2 - b}{x^2} = \alpha.$$

Most probably, in this latter problem, Euclid was showing that if two hyperbolas each have their axes as the asymptotes of the other, then their points of intersection are determined.[9]

That Euclid would present a problem useful in the study of conic sections is not surprising, given that he is credited with a book on the subject. And, as we noted earlier, many of the propositions in Book II have application to that subject as well. Besides his work in conics, Euclid is also credited with works in such fields as spherical geometry, optics, and music. Thus, whoever Euclid was, it appears from the texts attributed to him that he saw himself as a compiler of the Greek mathematical tradition to his time. Certainly, this would be appropriate if he was the first mathematician called to the Museum at Alexandria. It would therefore have been his aim to demonstrate to his students not only the basic results known to that time but also some of the methods by which new problems could be approached. The two mathematicians in the third century BCE who most advanced the field of mathematics, Archimedes and Apollonius, probably received their earliest mathematical training from the students of Euclid, training that in fact enabled them to solve many problems left unsolved by Euclid and his predecessors.

EXERCISES

1. Prove Proposition I–5, that the base angles of an isosceles triangle are equal to one another.

2. Find a construction to bisect a given angle and prove that it is correct (Proposition I–9).

3. Prove Proposition I–15, that if two straight lines cut one another, they make the vertical angles equal to one another.

4. Construct a triangle out of three given straight lines and prove that your construction is correct. Note that it is necessary that two of the straight lines taken together in any manner should be greater than the remaining one (Proposition I–22).

5. On a given straight line at a point on it, construct an angle equal to a given angle and prove that your construction is correct (Proposition I–23).

6. Prove Proposition I–32, that the three interior angles of any triangle are equal to two right angles. Show that the proof depends on I–29 and therefore on postulate 5.

7. Solve the (modified) problem of Proposition I–44, to apply to a given straight line AB a rectangle equal to a given rectangle c. Use Figure 3.31, where $BEFG$ is the given rectangle, D is the intersection of the extension of the diagonal HB and the extension of the line FE, and $ABML$ is the rectangle to be constructed.

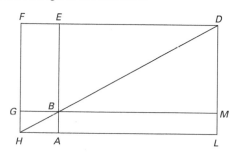

FIGURE 3.31

Elements, Proposition I–44

8. Give a proof of I–47 using similarity. Use the same diagram as in the text (Fig. 3.2) and begin by noting that triangles ABN, ACN, and ABC are all similar.

9. Show that Playfair's Axiom—through a given point outside a given line, exactly one line may be constructed parallel to the given line—is equivalent to Euclid's postulate 5, under the assumption that lines of arbitrary length may be drawn and therefore that Proposition I–16 is true.

10. Draw a geometric diagram that proves the truth of Proposition II–8: If a straight line is cut at random, four times the rectangle contained by the whole and one of the segments together with the square on the remaining segment is equal to the square on the whole and the former segment taken together. Then translate this result into algebraic notation and verify it algebraically.

11. Show that Proposition II–13 is equivalent to the law of cosines for an acute-angled triangle: In acute-angled triangles, the square on the side opposite the acute angle is less than the sum of the squares on the other two sides by twice the rectangle contained by one of the sides about the acute angle, namely, that on which the perpendicular falls, and the line segment between the angle and the perpendicular.

12. Prove Proposition III–3, that if a diameter of a circle bisects a chord, then it is perpendicular to the chord. And if a diameter is perpendicular to a chord, then it bisects the chord.

13. Provide the details of the proof of Proposition III–20: In a circle, the angle at the center is double the angle at the circumference, when the angles cut off the same arc.

14. Prove Proposition III–31, that the angle in a semicircle is a right angle.

15. Find a construction for circumscribing a circle about an arbitrary triangle.

16. Find a construction for inscribing a regular hexagon in a circle.

17. Given that a pentagon and an equilateral triangle can be inscribed in a circle, show how to inscribe a regular 15-gon in a circle.

18. Prove that the last nonzero remainder in the Euclidean algorithm applied to the numbers a, b, is in fact the greatest common divisor of a and b.

19. Use the Euclidean algorithm to find the greatest common divisor of 963 and 657; of 2689 and 4001.

20. Use Theaetetus's definition of equal ratio to show that $46 : 6 = 23 : 3$. Show that each can be represented by the sequence $(7, 1, 2)$.

21. Use Theaetetus's definition of equal ratio to show that $33 : 12 = 11 : 4$ and that each can be represented by the sequence $(2, 1, 3)$.

22. Suppose that a line of length 1 is divided in extreme and mean ratio, that is, that the line is divided at x so that $\frac{1}{x} = \frac{x}{x-1}$. Show by the method of the Euclidean algorithm that 1 and x are incommensurable. In fact, show that $1 : x$ can be expressed using Theaetetus's definition as $(1, 1, 1, \dots)$.

23. Show that the side and diagonal of a square are incommensurable by using the method of anthyphairesis. Show that the ratio $d : s$ can be expressed using Theaetetus's definition as $(1, 2, 2, 2, \dots)$. *Hint:* Draw the diagonal of the square; then cut off on it the side and draw a square on the remaining segment.

24. Prove the second half of Proposition V–8: If $a > b$, then $c : b > c : a$.

25. Prove Proposition V–12 both by using Eudoxus's definition and by modern methods: If any number of magnitudes are proportional, as one of the antecedents is to one of the consequents, so will all of the antecedents be to all of the consequents. (In algebraic notation, this says that if $a_1 : b_1 = a_2 : b_2 = \dots = a_n : b_n$, then $(a_1 + a_2 + \dots + a_n) : (b_1 + b_2 + \dots + b_n) = a_1 : b_1$.)

26. Use Eudoxus's definition to prove Proposition V–16: If $a : b = c : d$, then $a : c = b : d$.

27. Construct geometrically the solution of $8 : 4 = 6 : x$.

28. Solve geometrically the equation $\frac{9}{x} = \frac{x}{5}$ by beginning with a semicircle of diameter $9 + 5 = 14$.

29. Prove Proposition VI–14, that in equal and equiangular parallelograms, the sides about the equal angles are reciprocally proportional and conversely.

30. Prove Proposition VIII–8 and Archytas's special case that there is no mean proportional between $n + 1$ and n.

31. Find the one mean proportional between two squares guaranteed by Proposition VIII–11.

32. Find the two mean proportionals between two cubes guaranteed by Proposition VIII–12.

33. Prove Proposition VIII–14: If a^2 measures b^2, then a measures b and conversely.

34. Use Proposition VII–30 to prove the uniqueness (up to order) of the prime decomposition of any positive integer. (This is essentially Proposition IX–14.)

35. Give a modern proof of the result that there are infinitely many prime numbers. Compare your proof to Euclid's and comment on the differences.

36. Use Euclid's criterion in Proposition IX–36 to find the next perfect number after 8128.

37. Prove XIII–9: If the side of the hexagon and the side of the decagon inscribed in the same circle are placed together in a single straight line, then the meeting point divides the entire line segment in extreme and mean ratio, with the greater segment being the side of the hexagon. In Figure 3.32, BC is the side of a decagon and CD the side of a hexagon inscribed in the same circle. Show that $\triangle EBD$ is similar to $\triangle EBC$.

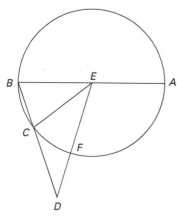

FIGURE 3.32

Elements, Proposition XIII–9

38. Prove XIII–10: If an equilateral pentagon, hexagon, and decagon are each inscribed in a given circle, then the square on the side of the pentagon equals the sum of the squares on the sides of the hexagon and the decagon. Do this by using the numerical values of the sides of the given polygons inscribed in a circle of radius 1.

39. Solve geometrically the system $x - y = 7$, $xy = 18$, using the propositions from the *Data*.

40. Solve the equations of Proposition 86 of the *Data* algebraically. Show that the two hyperbolas defined by the equations each have their axes as the asymptotes of the other.

41. Discuss the advantages and disadvantages of a geometric approach relative to a purely algebraic approach in the teaching of the quadratic equation in school.

42. Prepare a lesson proving a number of simple algebraic identities geometrically. (For example, prove $(a + b)^2 = a^2 + 2ab + b^2$ and $(a + b)(a - b) = a^2 - b^2$.)

43. Discuss whether Euclid's *Elements* fits Plato's dictums that the study of geometry is for "drawing the soul toward truth" and that it is to gain knowledge "of what eternally exists."

44. Should one base the study of geometry in high school on Euclid's *Elements* as was done for many years? Discuss the pros and cons of Euclid versus a "modern" approach.[10]

45. Read the Declaration of Independence. Note that Jefferson writes, "we hold these truths to be self-evident, . . . " and then gives a list of what could be called axioms. Comment on whether Jefferson modeled the argument in the Declaration after a Euclidean proof.

REFERENCES AND NOTES

Most of the books on Greek mathematics referred to in Chapter 2 are relevant to the study of Euclid. Other useful works include F. Lasserre, *The Birth of Mathematics in the Age of Plato* (Larchmont, NY: American Research Council, 1964), J. Klein, *Greek Mathematical Thought and the Origin of Algebra* (Cambridge: MIT Press, 1968), and Asger Aaboe, *Episodes from the Early History of Mathematics* (Washington, DC: MAA, 1964). I have also used material from the doctoral dissertation of Charles Jones, *On the Concept of One as a Number* (University of Toronto, 1979) in organizing some of the material in this chapter.

The standard modern English version of Euclid's *Elements* is the three-volume set edited by Thomas Heath and still in print from Dover Publications, New York. This edition contains Heath's extensive notes, as well as a long introduction. If you just want the text of Euclid in a convenient one-volume edition, such

a version was prepared by Dana Densmore and published in 2002 by Green Lion Press, Santa Fe. A theorem-by-theorem analysis of the *Elements*, which is very helpful to have alongside the text of Euclid, is Benno Artmann, *Euclid: The Creation of Mathematics* (New York: Springer, 1999). The entire text of the *Elements* is also available online at http://aleph0.clarku.edu/~djoyce/java/elements/elements.html. This website of David Joyce also includes interactive diagrams in the proofs and some analysis of the interrelationships among the various theorems.

The most extensive modern study of Euclid's *Data* is available in Christian Marinus Taisbak, *Euclid's Data or the Importance of Being Given* (Copenhagen: Museum Tusculanum Press, 2003). There are also two earlier translations. One is by Shuntaro Ito, *The Medieval Latin Translation of the Data of Euclid* (Boston: Birkhäuser, 1980). The second is by George L.

McDowell and Merle A. Sokolik, *The Data of Euclid* (Baltimore: Union Square Press, 1993).

1. From Proclus's *Summary*, translated in Thomas, *Selections*, I, p. 147.

2. See Heath, *History of Greek Mathematics*, pp. 354, 357.

3. The debate over geometric algebra was renewed with a vengeance in an article by Sabetai Unguru entitled "On the Need to Rewrite the History of Greek Mathematics," *Archive for History of Exact Sciences* 15 (1975), 67–114. He was answered by several other historians over the next two years. The most important responses were by B. L. Van der Waerden, "Defence of a Shocking Point of View," *Archive for History of Exact Sciences* 15 (1976), 199–210, and Hans Freudenthal, "What Is Algebra and What Has It Been in History?" *Archive for History of Exact Sciences* 16 (1977), 189–200. A reply to these was offered by Unguru and David Rowe, "Does the Quadratic Equation Have Greek Roots? A Study of Geometric Algebra, Application of Areas, and Related Problems," *Libertas Mathematica* 1 (1981) and 2 (1982). These articles are recommended as examples of the strong feelings historical controversy can bring out. Over the last decade, however, it appears that Unguru's views have, in general, won out, and most historians do not interpret any part of the *Elements* or other ancient Greek works using algebra.

4. See Ken Saito, "Book II of Euclid's *Elements* in the Light of the Theory of Conic Sections," *Historia Scientiarum* 28, 31–60, for more details on the interpretation of theorems of Book II.

5. For more details on division in extreme and mean ratio, see Roger Herz-Fischler, *A Mathematical History of Division in Extreme and Mean Ratio* (Waterloo, Ont.: Wilfrid Laurier Univ. Press, 1987).

6. D. H. Fowler, *The Mathematics of Plato's Academy*, p. 225. This work contains a detailed study of anthyphairesis and its possible implications in the development of Greek mathematics.

7. The discussion of the origins of Eudoxus's theory of proportion is adapted from the treatment in Knorr, *The Evolution of the Euclidean Elements*.

8. Thomas, *Selections*, I, p. 229.

9. See Taisbak, *Euclid's Data*, chapter 13, for more details on Proposition 86 of the *Data*. See also Saito, "Book II of Euclid's *Elements*," *Historia Scientiarum* 28, 31–60.

10. See Paul Daus, "Why and How We Should Correct the Mistakes in Euclid," *Mathematics Teacher* 53 (1960), 576–581.

4

Archimedes and Apollonius

The third book [of Conics] contains many incredible theorems of use for the construction of solid loci and for limits of possibility of which the greatest part and the most beautiful are new. And when we had grasped these, we knew that the three-line and four-line locus had not been constructed by Euclid, but only a chance part of it and that not very happily. For it was not possible for this construction to be completed without the additional things found by us.

—Preface to Book I of Apollonius's *Conics*[1]

Here is a story told by Vitruvius: "It is no surprise that Hiero [the king of Syracuse in the third century BCE], after he had obtained immense kingly power in Syracuse, decided, because of the favorable turn of events, to dedicate a votive crown of gold to the immortal gods in a certain shrine. He contracted for the craftsman's wages, and he [himself] weighed out the gold precisely for the contractor. This contractor completed the work with great skill and on schedule; it was approved by the king, and the contractor seemed to have used up the furnished supply of gold. Later, charges were leveled that in the making of the crown a certain amount of gold had been removed and replaced by an equal amount of silver. Hiero, outraged that he should have been shown so little respect, and not knowing by what method he might expose the theft, requested that Archimedes take the matter under consideration on his behalf. Now Archimedes, once he had charge of this matter, chanced to go to the baths, and there, as he stepped into the tub, he noticed that however much he immersed his body in it, that much water spilled over the sides of the tub. When the reason for this occurence came clear to him, he did not hesitate, but in a transport of joy he leapt out of the tub, and as he rushed home naked, he let one and all know that he had truly found what he had been looking for—because as he ran he shouted over and over in Greek: 'I found it! I found it! [Eureka! Eureka!]'"[2]

Greek mathematics in the third and early second centuries BCE was dominated by two major figures, Archimedes of Syracuse (c. 287–212 BCE) and Apollonius of Perga (c. 250–175 BCE), each heir to a different aspect of fourth-century Greek mathematics. The former took over the "limit" methods of Eudoxus and succeeded not only in applying them to determine areas and volumes of new figures, but also in developing new techniques that enabled the results to be discovered in the first place. Archimedes, unlike his predecessors, was neither reluctant to share his methods of discovery nor afraid of performing numerical calculations and exhibiting numerical results. And also, unlike Euclid, he did not write systematic treatises on a major subject, but instead what may be considered research monographs, treatises concentrating on the solution of a particular set of problems. These treatises were often sent originally as letters to mathematicians Archimedes knew, so many of them include prefaces describing the circumstances and purposes of their writing. Furthermore, several of the treatises presented mathematical models of certain aspects of what we would call theoretical physics and applied his physical principles to the invention of various mechanical devices.

Apollonius, on the other hand, was instrumental in extending the domain of analysis to new and more difficult geometric construction problems. As a foundation for these new approaches, he created his magnum opus, the *Conics*, a work in eight books developing synthetically the important properties of this class of curves, properties that were central in developing new solutions to such problems as the duplication of the cube and the trisection of the angle.

As is the case for Euclid, there are no surviving manuscripts of the works of either Archimedes or Apollonius dating from anywhere near their time of composition. For Archimedes, we know that an edition of some of his works with extensive commentaries was prepared by Eutocius early in the sixth century somewhere near Byzantium. This edition was the basis for some part of the three collections of Archimedes' works, written on parchment, that were available in Byzantium in the tenth or eleventh century. Only one of these is still extant and will be discussed in some detail below. The second oldest extant Archimedes manuscript is a 1260 Latin translation by Moerbeke, probably made from both of the two now missing Byzantine copies, but such a literal translation that from it we can practically re-create the Greek text. There are also several fifteenth- and sixteenth-century Greek copies of the missing Byzantine versions. Heiberg collated these manuscripts in the late nineteenth century and produced the now standard Greek text of Archimedes in 1880–81, with a revised version in 1910–15. Similarly, Eutocius prepared an edition of the first four books of Apollonius's *Conics* of which the Greek manuscripts available in tenth-century Byzantium were copies. The earliest surviving Greek manuscript was copied there in the twelfth or thirteenth century. But there are two older Arabic manuscripts of seven books of the *Conics*, one written in Egypt in the early eleventh century and now in Istanbul, and one written in Maragha toward the end of that century and now in Oxford. Again, Heiberg produced a definitive Greek edition of Books I–IV in 1891–93, while a definitive Arabic edition of Books V–VII was only produced in 1990 by Toomer.

This chapter surveys the extant works of both of these mathematicians, as well as the work of certain others who considered similar problems.

ARCHIMEDES AND PHYSICS

Archimedes was the first mathematician to derive quantitative results from the creation of mathematical models of physical problems on earth. In particular, Archimedes is responsible for the first proof of the law of the lever (Fig. 4.1) and its application to finding centers of gravity, as well as the first proof of the basic principle of hydrostatics and some of its important applications.

FIGURE 4.1

Archimedes and the law of the lever

4.1.1 The Law of the Lever

Everyone is familiar with the principle of the lever from having played on seesaws as children. Equal weights at equal distances from the fulcrum of the lever balance, and a lighter child can balance a heavier one by being farther away. The ancients were aware of this principle as well. The law even appears in writing in a work on mechanics attributed to Aristotle: "Since the greater radius is moved more quickly than the less by an equal weight, and there are three elements in the lever, the fulcrum . . . and two weights, that which moves and that which is moved, therefore the ratio of the weight moved to the moving weight is the inverse ratio of their distances from the fulcrum."[3]

As far as is known, no one before Archimedes had created a mathematical model of the lever by which one could derive a mathematical proof of the law of the lever. In general, a difficulty in attempting to apply mathematics to physical problems is that the physical situation is often quite complicated. Therefore, the situation needs to be idealized. One ignores those aspects that appear less important and concentrates on only the essential variables of the physical problem. This idealization is referred to today as the creation of a mathematical model. The lever is a case in point. To deal with it as it actually occurs, one would need to consider not only the weights applied to the two ends and their distances from the fulcrum, but also the weight and composition of the lever itself. It may be heavier at one end than the other. Its thickness may vary. It may bend slightly—or even break—when certain weights are applied at certain points. In addition, the fulcrum is also a physical object of a certain size. The lever may slip somewhat along the fulcrum, so it may not be clear from what point the distance of the weights should be measured. To include all of these factors in a mathematical analysis of the lever would make the mathematics extremely difficult. Archimedes therefore simplified the physical situation. He assumed that the lever itself was rigid, but weightless, and that the fulcrum and the weights were mathematical points. He was then able to develop the mathematical principles of the lever.

Archimedes dealt with these principles at the beginning of his treatise *Planes in Equilibrium*. Being well trained in Greek geometry, he began by stating seven postulates he would assume, four of which are reproduced here.

1. Equal weights at equal distances are in equilibrium, and equal weights at unequal distances are not in equilibrium but incline toward the weight that is at the greater distance.
2. If, when weights at certain distances are in equilibrium, something is added to one of the weights, they are not in equilibrium but incline toward the weight to which the addition was made.

BIOGRAPHY

Archimedes (287–212 BCE)

More biographical information about Archimedes survives than about any other Greek mathematician. Much is found in Plutarch's biography of the Roman general Marcellus, who captured Syracuse, the major city of Sicily, after a siege in 212 BCE during the Second Punic War. Other Greek and Roman historians also discuss aspects of Archimedes' life.

Archimedes was the son of the astronomer Phidias and perhaps a relative of King Hiero II of Syracuse, under whose rule from 270 to 216 BCE the city greatly flourished. It is also probable that Archimedes spent time in his youth in Alexandria, for he is credited with the invention there of what is known as the Archimedean screw, a machine for raising water used for irrigation (Fig. 4.2). Moreover, the prefaces of many of his works are addressed to scholars at Alexandria, including one of the chief librarians, Eratosthenes. Most of his life, however, was spent in his native Syracuse, where he was repeatedly called upon to use his mathematical talents to solve various practical problems for Hiero and his successor. Many stories are recorded about his intense dedication to his work. Plutarch, in *The Lives of the Noble Grecians and Romans* (*Great Books*, 14, Dryden translation), wrote that on many occasions his concen-tration on mathematics "made him forget his food and neglect his person, to that degree that when he was carried by absolute violence to bathe or have his body anointed, he used to trace geometrical figures in the ashes of the fire, and diagrams in the oil on his body, being in a state of entire preoccupation, and in the truest sense, divine possession with his love and delight in science" (p. 254). And it was this dedication that ultimately cost him his life.

His genius as a military engineer kept the Roman army under Marcellus at bay for months during the siege of Syracuse. Finally, however, probably through treachery, the Romans were able to enter the city. Marcellus gave explicit orders that Archimedes be spared, but Plutarch relates that, "as fate would have it, he was intent on working out some problem with a diagram and, having fixed his mind and his eyes alike on his investigation, he never noticed the incursion of the Romans nor the capture of the city. And when a soldier came up to him suddenly and bade him follow to Marcellus, he refused to do so until he had worked out his problem to a demonstration; whereat the soldier was so enraged that he drew his sword and slew him" (*Lives*, p. 252).

FIGURE 4.2

Archimedes and the Archime-dean screw

3. Similarly, if anything is taken away from one of the weights, they are not in equilibrium but incline toward the weight from which nothing was taken.

6. If magnitudes at certain distances are in equilibrium, other magnitudes equal to them will also be in equilibrium at the same distances.

These postulates come from basic experience with levers. The first postulate, in fact, is an example of what is usually called the **Principle of Insufficient Reason**. That is, one assumes that equal weights at equal distances balance because there is no reason to make any other assumption. The lever cannot incline to the right, for example, since what is the right side from one viewpoint is the left side from another. The second and third postulates are equally obvious. The sixth appears to be virtually meaningless. In Archimedes' use of it, however, it appears that the second clause means "other equal magnitudes, the centers of gravity of which lie at the same distances from the fulcrum, will also be in equilibrium." That is, the influence of a magnitude on the lever depends solely on its weight and the position of its center of gravity.

Although Archimedes used the term "center of gravity" in many of the book's propositions, he never gave a definition. Presumably, he felt that the concept was so well known to his readers that a definition was unnecessary. There are, however, later Greek texts that do give a definition, perhaps the one that was even used in Archimedes' time: "We say that the center of gravity of any body is a point within that body which is such that, if the body be conceived to be suspended from that point, the weight carried thereby remains at rest and preserves the original position."[4] But it was also clear to Archimedes, and this is what he expressed in postulate 6, that the downward tendency of gravitation may be thought of as being concentrated in that one point. Note that in neither the postulates nor the theorems is there any mention of the lever itself. It is just there. Its weight does not enter into the calculations. Archimedes in effect assumed that the lever is weightless and rigid. Its only motion is inclination to one side or the other.

The first two in Archimedes' sequence of propositions leading to the law of the lever are very easy:

PROPOSITION 1   *Weights which balance at equal distances are equal.*

PROPOSITION 2   *Unequal weights at equal distances will not balance but will incline toward the greater weight.*

The proof of the first result is by *reductio ad absurdum*. For if the weights are not equal, take away from the greater the difference between the two. By postulate 3, the remainders will not balance. This contradicts postulate 1, since now we have equal weights at equal distances. Our original assumption must then be false. To prove Proposition 2, again take away from the greater weight the difference between the two. By postulate 1, the remainders will balance. So if this difference is added back, the lever will incline toward the greater by postulate 2.

PROPOSITION 3   *Suppose A and B are unequal weights with $A > B$ which balance at point C (Fig. 4.3). Let $AC = a$, $BC = b$. Then $a < b$. Conversely, if the weights balance and $a < b$, then $A > B$.*

FIGURE 4.3

Planes in Equilibrium,
Proposition 3

The proof is again by contradiction. Suppose $a \not< b$. Subtract from A the difference $A - B$. By postulate 3, the lever will incline toward B. But if $a = b$, the equal remainders will balance, and if $a > b$, the lever will incline toward A by postulate 1. These two contradictions imply that $a < b$. The proof of the converse is equally simple.

In Propositions 4 and 5, Archimedes showed that the center of gravity of a system of two (and three) equally spaced equal weights is at the geometric center of the system. These results are extended in the corollaries to any system of equally spaced weights provided that those at equal distance from the center are equal. The law of the lever itself is stated in Propositions 6 and 7:

PROPOSITION 6, 7   *Two magnitudes, whether commensurable [Proposition 6] or incommensurable [Proposition 7], balance at distances inversely proportional to the magnitudes.*

First assume that the magnitudes A, B, are commensurable; that is, $A : B = r : s$, where r, s, are numbers. Archimedes' claim is that if A is placed at E and B at D, and if C is taken on DE with $DC : CE = r : s$, then C is the center of gravity of the two magnitudes A, B (Fig. 4.4). To prove the result, assume that units have been chosen so that $DC = r$ and $CE = s$. Choose H on DE so that $HE = r$ and extend the line past E to L so that EL also equals r. Also extend the line in the opposite direction to K, making $DK = HD = s$. Then C is the midpoint of LK. Now break A into $2r$ equal parts and B into $2s$ equal parts. Space the first set equally along LH and the second along HK. Since $A : B = r : s = 2r : 2s$, it follows that each part of A is equal to each part of B. From the corollary mentioned above, the center of gravity of the parts of A will be at the midpoint E of HL, while the center of gravity of the parts of B will be at the midpoint D of KH. By postulate 6, nothing is changed if A itself is considered situated at E and B at D. On the other hand, the total system consists of $2r + 2s$ equal parts equally spaced along the line KL. Hence, the center of gravity of the system is at the midpoint C of that line. Therefore, weight A placed at E and weight B placed at D balance about the point C.

FIGURE 4.4

Planes in Equilibrium,
Proposition 6

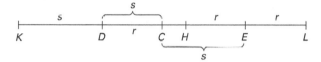

Archimedes concluded the proof in the incommensurable case by a *reductio* argument using the fact that if two magnitudes are incommensurable, one can subtract from the first an amount smaller than any given quantity such that the remainder is commensurable with the second. Interestingly enough, Archimedes made no use here of the Eudoxian proportion theory for incommensurables of *Elements*, Book V, nor even of Theaetetus's earlier version based on the Euclidean algorithm. He instead made use essentially of a continuity argument. But even so, his proof is somewhat flawed.

Nevertheless, Archimedes used the law of the lever in the remainder of the treatise to find the centers of gravity of various geometrical figures. He proved that the center of gravity of a parallelogram is at the intersection of its diagonals, of a triangle at the intersection of two medians, and of a parabolic segment at a point on the diameter three-fifths of the distance from the vertex to the base.

4.1.2 Applications to Engineering

Not only are there geometric consequences of the law of the lever, but there are also physical consequences. In particular, given any two weights A and B and any lever, there is always a point C at which the weights balance. If A is much heavier than B, they will balance when A is sufficiently close to C and B is sufficiently far away. But then any additional weight added to B will incline the lever in that direction and will cause weight A to be lifted. Archimedes therefore was able to boast that "any weight might be moved and . . . if there were another earth, by going into it he could move this one."[5] When King Hiero heard of this boast, he asked Archimedes to demonstrate his principles in actual experiment. Archimedes complied, but instead of using a lever, he probably made use of some kind of pulley or tackle system, which also provided a great mechanical advantage. Plutarch wrote that "he fixed accordingly upon a ship of burden out of the king's arsenal, which could not be drawn out of the dock without great labor and many men; and loading her with many passengers and a full freight, sitting himself the while far off, with no great endeavor, but only holding the head of the pulley in his hand and drawing the cords by degrees, he drew the ship in a straight line, as smoothly and evenly as if she had been in the sea."[6] Other sources give a variant of Plutarch's story, to the effect that Archimedes was responsible for the construction of a magnificent ship, named the *Syracusa*, and singlehandedly launched this 4200-ton luxury vessel.

Archimedes enjoyed the greatest fame in antiquity, however, for his design of various engines of war. These engines enabled Syracuse to hold off the Roman siege for many months. Archimedes devised various missile launchers as well as huge cranes by which he was able to lift Roman ships out of the water and dash them against the rocks or simply dump out the crew. In fact, he was so successful that any time the Romans saw a little rope or piece of wood come out from the walls of the city, they fled in panic.

Plutarch related that Archimedes was not particularly happy as an engineer: "He would not deign to leave behind him any commentary or writing on such subjects; but, repudiating as sordid and ignoble the whole trade of engineering, and every sort of art that lends itself to mere use and profit, he placed his whole affection and ambition in those purer speculations where there can be no reference to the vulgar needs of life."[7] In fact, however, there is evidence that Archimedes did write on certain mechanical subjects, including a book *On Sphere Making* in which he described his planetarium, a mechanical model of the motions of the heavenly bodies, and another one on water clocks.

The incident of the gold crown and the bath led Archimedes to the study of an entirely new subject, that of hydrostatics, in which he discovered its basic law, that a solid heavier than a fluid will, when weighed in the fluid, be lighter than its true weight by the weight of the fluid displaced. It is, however, not entirely clear how Archimedes' noticing the water being displaced in his bath led him to the concept of weight being lessened. Perhaps he also noticed that his body felt lighter in the water.

As in his study of levers, Archimedes began the mathematical development of hydrostatics, in his treatise *On Floating Bodies*, by giving a simplifying postulate. He was then able to show, among other results, that the surface of any fluid at rest is the surface of a sphere whose center is the same as that of the earth. He could then deal with solids floating or sinking in fluids by assuming that the fluid was part of a sphere. Archimedes was able to solve the crown problem by using the basic law, proved as Proposition 7. One way by which he could have applied the law is suggested by Heath, based on a description in a Latin poem of the fifth century CE.[8] Suppose the crown is of weight W, composed of unknown weights w_1 and w_2 of gold and silver, respectively. To determine the ratio of gold to silver in the crown, first weigh it in water and let F be the loss of weight. This amount can be determined by weighing the water displaced. Next take a weight W of pure gold and let F_1 be its weight loss in water. It follows that the weight of water displaced by a weight w_1 of gold is $\frac{w_1}{W} F_1$. Similarly, if the weight of water displaced by a weight W of pure silver is F_2, the weight of water displaced by a weight w_2 of silver is $\frac{w_2}{W} F_2$. Therefore, $\frac{w_1}{W} F_1 + \frac{w_2}{W} F_2 = F$. Thus, the ratio of gold to silver is given by

$$\frac{w_1}{w_2} = \frac{F - F_2}{F_1 - F}.$$

Vitruvius himself provided a somewhat different suggestion for solving the wreath problem, more clearly based on the story of the bath, but not on the basic law of hydrostatics. He also recorded that Archimedes indeed found that the goldsmith had cheated the king. What happened to the smith, however, is not mentioned.

4.2 ARCHIMEDES AND NUMERICAL CALCULATIONS

The brief treatise, *Measurement of the Circle*, contains numerical results, unlike anything found in Euclid's work. Its first proposition, in addition, gives Archimedes' answer to the question of squaring the circle, by showing that the area of a circle of given radius can be found once the circumference is known.

PROPOSITION 1   *The area A of any circle is equal to the area of a right triangle in which one of the legs is equal to the radius and the other to the circumference.*

Archimedes gave a rigorous proof, using a Eudoxian exhaustion argument. Namely, if K is the area of the given triangle, Archimedes first supposed that $A > K$. By inscribing in the circle regular polygons of successively more sides, he eventually determined a polygon of area P such that $A - P < A - K$. Thus, $P > K$. Now the perpendicular from the center of the circle to the midpoint of a side of the polygon is less than the radius, while the perimeter of the polygon is less than the circumference. It follows that $P < K$, a contradiction. Similarly, the assumption that $A < K$ leads to another contradiction and the result is proved.

The third proposition of this treatise complements the first by giving a numerical approximation to the length of the circumference:

PROPOSITION 3   *The ratio of the circumference of any circle to its diameter is less than $3\frac{1}{7}$ but greater than $3\frac{10}{71}$.*

Archimedes' proof of this statement provided algorithms for determining the perimeter of certain regular polygons circumscribed about and inscribed in a circle. Namely, Archimedes began with regular hexagons, the ratios of whose perimeters to the diameter of the circle are known from elementary geometry. He then in effect used the following lemmas (here given in modern notation) to calculate, in turn, the ratios to the diameter of the perimeters of regular polygons with 12, 24, 48, and 96 sides, respectively.

LEMMA 1   *Suppose OA is the radius of a circle and CA is tangent to the circle at A. Let DO bisect $\angle COA$ and intersect the tangent at D. Then $DA/OA = CA/(CO + OA)$ and $DO^2 = OA^2 + DA^2$ (Fig. 4.5).*

FIGURE 4.5

Lemmas 1 and 2 to *Measurement of the Circle*

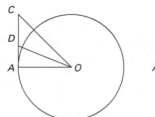
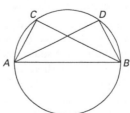

LEMMA 2   *Let AB be the diameter of a circle and ACB a right triangle inscribed in the semicircle. Let AD bisect $\angle CAB$ and meet the circle at D. Connect DB. Then $AB^2/BD^2 = 1 + (AB + AC)^2/BC^2$ and $AD^2 = AB^2 - BD^2$.*

Archimedes used the first lemma repeatedly to develop a recursive algorithm for determining the desired ratio using circumscribed polygons. He began by assuming that $\angle COA$ is one-third of a right angle (30°), so CA is half of one side of a circumscribed regular hexagon. Therefore, CA and CO are known. Since $\angle DOA = 15°$, it follows that DA is half of one side of a regular 12-gon. DA and DO are then calculated by use of the lemma. Next, $\angle DOA$ is bisected to get an angle of $7\frac{1}{2}°$. The piece of the tangent subtending that angle is then half of one side of a regular 24-gon. Its length can be calculated as well. If r is the radius of the circle, t_i half of one side of a regular 3×2^i–gon ($i \geq 1$), and u_i the length of the line from the center of the circle to a vertex of that polygon, the lemma can be translated into the recursive formulas

$$t_{i+1} = \frac{rt_i}{u_i + r}, \qquad u_{i+1} = \sqrt{r^2 + t_{i+1}^2}.$$

The ratio of the perimeter of the ith circumscribed polygon to the diameter of the circle is then $6(2^i t_i) : 2r = 3(2^i t_i) : r$.

Archimedes developed a similar algorithm for inscribed polygons by use of the second lemma, and in both cases provided explicit numerical results at each stage. For example, in his calculations involving hexagons in both the circumscribed and inscribed cases, he needed to evaluate the ratio $\sqrt{3} : 1$. What he wrote indicates that he knew that this ratio is greater than $265 : 163$ and less than $1351 : 780$. Although it is not known exactly how Archimedes found these results, it is certain that he, like many great mathematicians of later times, was a superb calculator. After four steps of both algorithms, in fact, he concluded that the ratio of the perimeter of the circumscribed 96-sided polygon to the diameter is less than

$$14{,}688 : 4673\frac{1}{2} = 3 + \frac{667\frac{1}{2}}{4673\frac{1}{2}} < 3\frac{1}{7}$$

while the ratio of the inscribed 96-sided polygon to the diameter is greater than

$$6336 : 2077\frac{1}{4} > 3\frac{10}{71},$$

thus proving the theorem.

Archimedes' proof is the first recorded method for actually computing π. Once the method was known, it was merely a matter of patience to calculate π to as great a degree of accuracy as desired. Archimedes does not tell us why he stopped at 96-sided polygons. But his value of $3\frac{1}{7}$ has become a standard approximation for π to the present day.

It was Nicomedes (late third century BCE), a successor of Archimedes, who used an entirely new method to determine the length of the circumference of a circle and, therefore, by Proposition 1 above, to square the circle. Namely, he used the **quadratrix**, a curve probably introduced a century earlier, defined via a combination of two motions: In the square $ABCD$, imagine that the ray AB rotates uniformly around A from its beginning position to the ending position on AD, while at the same time the line BC moves parallel to itself from BC to AD (Fig. 4.6). The quadratrix BZK is then the curve traced out by the moving intersection point. It follows from this definition that a point Z on the quadratrix satisfies the proportion $ZL : BA = $ arc $DG :$ arc BD, or $ZL :$ arc $DG = AB :$ arc BD. In modern notation, if the polar equation of the curve is given by $\rho = \rho(\theta)$, ρ satisfies the equation

$$\frac{\rho(\theta)\sin\theta}{a\theta} = \frac{a}{\frac{1}{2}\pi a},$$

where a is the length of a side of the square.

FIGURE 4.6

The quadratrix

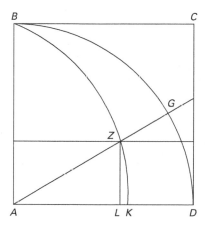

If we take the limit of the left side of the equation as θ approaches 0, we get the result

$$\frac{\rho(0)}{a} = \frac{a}{\frac{1}{2}\pi a}.$$

Naturally, the Greeks did not present such a limit argument, but the result, in the form $AK : AB = AB :$ arc BD, was proved, probably by Nicomedes, through a double *reductio* argument. It then follows that arc BD, a quarter of the circumference of the circle, is a third proportional to the known lines AK and AB and thus can be constructed by Euclidean means. (It should be noted that even in ancient times this construction was criticized, because the actual position of the terminal point K is not determined by the definition of the curve. It can only be approximated.)

4.3   ARCHIMEDES AND GEOMETRY

What distinguishes Archimedes' work in geometry from that of Euclid is that Archimedes often presented his method of discovery of the theorem and/or his analysis of the situation before presenting a rigorous synthetic proof. The methods of discovery of several of his results are collected in a treatise called *The Method*, which was unexpectedly discovered in 1899 in a Greek monastery library in Constantinople. The manuscript, containing several other works of Archimedes as well, is the oldest extant manuscript of Archimedes. It dates from the tenth century, but the writing was partially washed out in the thirteenth century and the parchment reused for a religious work. (Parchment was a very valuable commodity in the middle ages; a reused parchment is called a **palimpsest**.) Fortunately, the old writing is in large part still readable. Heiberg deciphered much of it in 1906 and soon after published the Greek text.

Interestingly, the original palimpsest disappeared during the First World War, only to reappear in an auction in 1998. Evidently, it had been owned by a French family for many years, who finally decided to sell. Despite some legal challenges to the sale, the manuscript with Archimedes' *The Method* was sold for about $2 million to an anonymous buyer, who then contracted with the Walters Art Gallery in Baltimore to preserve it and restore it where possible. At this writing, it is still at the Gallery, but scholars have been permitted to inspect it using modern techniques. Although it seems that Heiberg's original reading of the manuscript is relatively accurate, there have been a few new discoveries from the manuscript in the past several years, including many of the original diagrams that Heath had been unable to see. Some of these discoveries are noted below.

4.3.1 Archimedes' Method of Discovery

In the introductory letter to *The Method*, written to Eratosthenes, the chief librarian at the Library in Alexandria, Archimedes described his purpose in writing it:

> Since, as I said, I know that you are diligent, an excellent teacher of philosophy, and greatly interested in any mathematical investigations that may come your way, I thought it might be appropriate to write down and set forth for you in this same book a certain special method, by means of which you will be enabled to recognize certain mathematical questions with the aid of mechanics. I am convinced that this is no less useful for finding the proofs of these same theorems. For some things, which first became clear to me by the mechanical method, were afterwards proved geometrically, because their investigation by the said method does not furnish an actual demonstration. But it is of course easier, when we have previously acquired, by the method, some knowledge of the questions, to supply the proof than it is to find it without any previous knowledge. . . . I now wish to describe the method in writing, partly because I have already spoken about it before, . . . partly because I am convinced that it will prove very useful for mathematics; in fact, I presume there will be some among the present as well as future generations who by means of the method here explained will be enabled to find other theorems which have not yet fallen to our share.[9]

The Method contains Archimedes' method of discovery by mechanics of many important results on areas and volumes, most of which are rigorously proved elsewhere. The essential features of *The Method* are, first, the assumption that figures are "composed" of their indivisible cross sections and, second, the balancing of cross sections of a given figure against corresponding cross sections of a known figure, using the law of the lever. Archimedes knew that this method did not give a rigorous proof, because neither mechanical principles nor "indivisible" cross sections could appear in a formal mathematical argument. Therefore, as he noted in his preface, those proofs would have to come later.

The first proposition of *The Method*, that a segment of a parabola is 4/3 of the triangle inscribed in it, is presented here in detail as a typical example of that work. By a segment ABC of a parabola, Archimedes meant the region bounded by the curve and a line AC, where B is the point at which the line segment through the midpoint D of AC drawn parallel to the axis of the parabola meets the curve (Fig. 4.7). The point B is called the **vertex** of the parabolic segment. The vertex is also that point of the curve whose perpendicular distance to AC is the greatest. Now given the parabolic segment ABC with vertex B, draw a tangent at C meeting the axis produced at E and a line through A parallel to the axis meeting the tangent line at F. Produce CB to meet AF in K and extend it to H so that $CK = KH$. Archimedes now considered CH as a lever with midpoint K. The idea of his demonstration is to show that triangle CFA placed where it is in the figure balances the segment ABC placed at H. He did this, line by line, by beginning with an arbitrary line segment MO of triangle CFA parallel to ED and showing that it balances the line PO of segment ABC placed at H. To show the balancing, two properties of the parabola are needed, first that $EB = BD$, and second that $MO : PO = CA : AO$. (It is evident that Archimedes was quite familiar with the elementary properties of parabolas.) From $EB = BD$, it follows that $FK = KA$ and $MN = NO$, and from the proportion and the fact that CK bisects AF, it follows from *Elements* VI–2 that $MO : PO = CA : AD = CK : KN = HK : KN$. If a line TG equal to PO is placed with its center at H, this latter proportion becomes $MO : TG = HK : KN$. Therefore, since N is the center of gravity of MO, by the law of the lever, MO and TG will be in equilibrium about K.

FIGURE 4.7

Balancing a parabolic segment

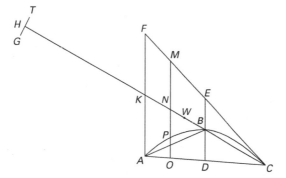

Archimedes continued, "since the triangle CFA is made up of all the parallel lines like MO, and the segment CBA is made up of all the straight lines like PO within the curve, it follows that the triangle, placed where it is in the figure, is in equilibrium about K with the segment CBA placed with its center of gravity at H."[10] Because nothing is changed by considering the triangle as located at its center of gravity, the point W on CK two-thirds of the way from C to K, Archimedes derived the proportion $\triangle ACF$: segment ABC = $HK : KW = 3 : 1$. Therefore, segment $ABC = (1/3)\triangle ACF$. But $\triangle ACF = 4\triangle ABC$. Hence, segment $ABC = (4/3)\triangle ABC$ as asserted. Archimedes concluded this demonstration with a warning: "Now the fact here stated is not actually demonstrated by the argument used; but that argument has given a sort of indication that the conclusion is true. Seeing then that the theorem is not demonstrated, but at the same time suspecting that the conclusion is true, we shall have recourse to the geometrical demonstration which I myself discovered and have already published."[11]

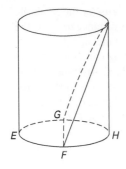

FIGURE 4.8

A plane cutting off a segment
of a cylinder

The Method contains several other similar proofs by the use of indivisibles and balancing, where the theorems are rigorously proved elsewhere, but the final propositions in the extant part of the work deal with a theorem that only appears in this work:

PROPOSITION 14   *If a cylinder is inscribed in a rectangular parallelepiped with square base, and if a plane is drawn through the center of the circle at the base of the cylinder and through one side of the square forming the top of the parallelepiped, then the segment of the cylinder cut off by this plane has a volume equal to one-sixth of the entire parallelepiped (Fig. 4.8).*[12]

Archimedes first gave a mechanical proof of this proposition similar to that of the other propositions in the work, but then gave a proof only using indivisibles and followed with a rigorous geometric proof using the method of exhaustion. We will look at the proof using indivisibles, because it brings out some other facets of Archimedes' mathematics. In the diagram accompanying the proof, we have the square base $ABCD$ of the parallelepiped, the circle $EFHG$ inscribed in the square forming the base of the cylinder, and a parabola EFH cutting the circle at H and E with axis KF (Fig. 4.9). (If we think of F as the origin of a coordinate system, then the parabola, drawn in the diagram with straight lines, has equation $x^2 = \ell y$, where ℓ is the length of FK, usually called the "parameter" of the parabola. Similarly, the circle has equation $x^2 + y^2 - 2\ell y = 0$. Of course, Archimedes himself had nothing about coordinates, only the geometric description of the curves.) Archimedes then noted that the plane drawn through the center of the circle and one side of the top cuts off a prism that is one-fourth of the entire parallelepiped.

FIGURE 4.9

Finding the volume of the
cylindrical segment

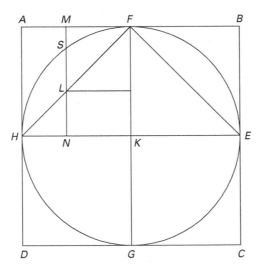

Archimedes' goal was to compare the desired segment with the prism cut off by the plane. To do this, he compared the "indivisible" right triangles in each that were above an arbitrary line MN drawn in the rectangle $ABEH$ parallel to KF, where MN intersects the circle at S and the parabola at L. The triangle in the prism has one side MN, a second side the line drawn through M perpendicular to AB, and hypotenuse a line in the cutting plane itself. The triangle in the segment has one side NS, a second perpendicular side in the surface of the cylinder, and hypotenuse also in the cutting plane. Because the two triangles are similar, their areas are in the duplicate ratio of MN to NS, that is, in modern terms, in the ratio of MN^2 to NS^2. But by the defining property of the parabola, we know that $KN^2 = MN \cdot ML$. Also,

$$MN \cdot NL + MN \cdot ML = MN^2 = KS^2 = KN^2 + NS^2.$$

By subtraction, we then get that $MN \cdot NL = NS^2$, or that $MN : NS = NS : NL$. It follows that $MN^2 : NS^2 = MN : NL$, or

triangle in prism : triangle in cylinder segment =

line segment in rectangle : line segment in parabolic segment. (4.1)

Given that the ratio holds for any line segment MN in rectangle $ABEH$ parallel to KF, the aim now is somehow to "add up" all the segments and all the triangles and compare the ratio of the "sums." What Archimedes did was use a special case of a lemma from another treatise, *Conoids and Spheroids*, which he recalled in the introduction to *The Method*. In modern terms, the special case states that if there are four finite sets $A = \{a_i\}$, $B = \{b_i\}$, $C = \{c_i\}$, $D = \{d_i\}$, each with k elements, such that $a_1 = a_2 = \cdots = a_k$, $b_1 = b_2 = \cdots = b_k$, and $a_i : c_i = b_i : d_i$ for all i, then $\sum a_i : \sum c_i = \sum b_i : \sum d_i$. Archimedes applied the result in a new way, however, to the four (infinite) sets of triangles and line segments that appear in the basic proportionality result (4.1). Namely, let A be the set of triangles in the prism, B the set of line segments in the rectangle, C the set of triangles in the segment of the cylinder, and D the set of line segments in the parabolic segment. Although these sets are infinite, they are, as Archimedes wrote, "equal in multitude," evidently because there is an obvious one-to-one correspondence among the elements of each of the four sets. In addition, since the elements in set A are equal to one another as are the elements in set B, and since the basic proportion (4.1) fits the other requirement of the lemma, Archimedes evidently believed that the conclusion would be true, even though the sets are infinite rather than finite. As we have noted earlier, Aristotle and Greek mathematicians in general did not deal with actual infinities, but only with potential infinities. Yet here Archimedes has violated this injunction. Not only that, but given that he noted that the sets involved were "equal in multitude," it is possible that he even conceived of infinite sets that would not be "equal in multitude" with these particular sets.

So, given the lemma, Archimedes could now "add up" his infinite sets and get a conclusion. He noted that since the rectangle $ABEH$ is "filled" by the lines drawn parallel to KF, the rectangle itself is "all the lines." Similarly, the prism is composed of "all the triangles" in it; the segment of the cylinder is composed of "all the triangles" in it; and the parabolic segment is composed of "all the lines" in it. It then followed that the prism is to the segment of the cylinder as the rectangle is to the parabolic segment. The remainder of the argument is straightforward: By the first theorem in *The Method*, the rectangle is 3/2 the parabolic segment. Therefore, the prism is 3/2 the segment of the cylinder. And since the prism is one-fourth of the entire rectangular parallelepiped, that figure is six times the segment of the cylinder, as claimed.

Interestingly, although Archimedes used indivisibles throughout *The Method*, he did not explain how they are to be used, even heuristically. This leads us to believe that his contemporaries, and especially the mathematicians in Alexandria with whom he corresponded, understood the use of indivisibles and, perhaps, used them in similar arguments even though they knew that these arguments did not form a rigorous geometrical proof.[13]

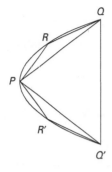

FIGURE 4.10

Area of a parabolic segment by summation of a geometric series

4.3.2 Sums of Series

The geometrical proof of the result on the segment of a parabola that Archimedes considered valid occurs in his treatise *Quadrature of the Parabola* and is based on Eudoxus's method of exhaustion. The idea as before is to construct rectilinear figures inside the parabolic segment whose total area differs from that of the segment by less than any given value. The figures Archimedes used for this purpose are triangles. Thus, in each of the two parabolic segments PRQ, $PR'Q'$, left by the original triangle PQQ', he constructed a triangle PRQ, $PR'Q'$; in each of the four segments left by these triangles, he constructed new triangles, and so on (Fig. 4.10).

Archimedes next calculated that the total area of the triangles constructed at each stage is one-fourth of the area of the triangles constructed in the previous stage. The more steps taken, the more closely the sum of the areas approaches the area of the parabolic segment. Therefore, to complete the proof, Archimedes in effect needed to find the sum of the geometric series $a + \frac{1}{4}a + (\frac{1}{4})^2 a + \cdots + (\frac{1}{4})^n a + \cdots$, where a is the area of $\triangle PQQ'$. Archimedes did not use Euclid's formula for the sum of a geometric progression from *Elements* IX–35, but instead gave that sum in the form

$$a + \frac{1}{4}a + \left(\frac{1}{4}\right)^2 a + \cdots + \left(\frac{1}{4}\right)^n a + \frac{1}{3}\left(\frac{1}{4}\right)^n a = \frac{4}{3}a.$$

He completed the argument through a double *reductio ad absurdum*, which began with the assumption that $K = \frac{4}{3}a$ is not equal to the area B of the segment. If K is less than this area, then triangles can be inscribed as above so that $B - T < B - K$, where T is the total area of the inscribed triangles. But then $T > K$. This is impossible because the summation formula shows that $T < \frac{4}{3}a = K$. On the other hand, if $K > B$, n is determined so that $(\frac{1}{4})^n a < K - B$. Because also $K - T = \frac{1}{3}(\frac{1}{4})^n a < (\frac{1}{4})^n a$, it follows that $B < T$, which is again impossible. Hence, $K = B$.

The important lemma to this proof shows how to find the sum of a geometric series. Archimedes' demonstration of this result was given for a series of five numbers, because, like Euclid, he had no notation to express a series with arbitrarily many numbers. But since his method generalizes easily, we will here use modern notation with n denoting an arbitrary positive integer. Archimedes began by noting that $(\frac{1}{4})^n a + \frac{1}{3}(\frac{1}{4})^n a = \frac{1}{3}(\frac{1}{4})^{n-1}a$. Then he calculated

$$a + \frac{1}{4}a + \left(\frac{1}{4}\right)^2 a + \cdots + \left(\frac{1}{4}\right)^n a + \frac{1}{3}\left[\frac{1}{4}a + \left(\frac{1}{4}\right)^2 a + \cdots + \left(\frac{1}{4}\right)^n a\right]$$

$$= a + \left(\frac{1}{4}a + \frac{1}{3}\cdot\frac{1}{4}a\right) + \left[\left(\frac{1}{4}\right)^2 a + \frac{1}{3}\left(\frac{1}{4}\right)^2 a\right] + \cdots + \left[\left(\frac{1}{4}\right)^n a + \frac{1}{3}\left(\frac{1}{4}\right)^n a\right]$$

$$= a + \frac{1}{3}a + \frac{1}{3}\cdot\frac{1}{4}a + \cdots + \frac{1}{3}\left(\frac{1}{4}\right)^{n-1} a$$

$$= a + \frac{1}{3}a + \frac{1}{3}\left[\frac{1}{4}a + \cdots + \left(\frac{1}{4}\right)^{n-1} a\right].$$

Subtracting equals and rearranging gives the desired result:

$$a + \frac{1}{4}a + \left(\frac{1}{4}\right)^2 a + \cdots + \left(\frac{1}{4}\right)^n a + \frac{1}{3}\left(\frac{1}{4}\right)^n a = \frac{4}{3}a.$$

Another formula for a sum led to another area result in *On Spirals*, a result again proved by Eudoxian methods. In Proposition 10 of that book, Archimedes demonstrated a formula for determining the sum of the first n integral squares,

$$(n+1)n^2 + (1 + 2 + \cdots + n) = 3(1^2 + 2^2 + \cdots + n^2),$$

as a corollary to which he showed that

$$3(1^2 + 2^2 + \cdots + (n-1)^2) < n^3 < 3(1^2 + 2^2 + \cdots + n^2).$$

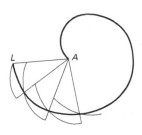

FIGURE 4.11

Area of the Archimedean spiral

Archimedes needed the last inequality to determine the area bounded by one turn of the "Archimedean spiral," the curve given in modern polar coordinates by the equation $r = a\theta$. In Proposition 24 of *On Spirals*, he demonstrated that the area R bounded by one complete circuit of that curve and the radius line AL to its endpoint equals one-third of the area C of the circle with that line as radius. Archimedes first noted that one can inscribe and circumscribe figures about the region R whose areas differ by less than any assigned area ϵ (Fig. 4.11). By continued bisection (according to *Elements* X–1), one can determine an integer n such that the circular sector with radius AL and angle $(360/n)°$ has area less than ϵ. Then, inscribing a circular arc in and circumscribing a circular arc about the part of the spiral included in each of the n sectors with this angle, one notes that the difference between the complete circumscribed figure and the complete inscribed figure is equal to the area of the sector chosen initially and thus is less than ϵ.

The proof of the area result by a double *reductio* argument is now straightforward. For suppose that $R \neq \frac{1}{3}C$. Then either $R < \frac{1}{3}C$ or $R > \frac{1}{3}C$. In the first case, circumscribe a figure F about R as described above so that $F - R < \frac{1}{3}C - R$. Therefore, $F < \frac{1}{3}C$. From the defining equation of the curve, it follows that the radii of the sectors making up F are in arithmetic progression, which can be considered as $1, 2, \ldots, n$. Because $n \cdot n^2 < 3(1^2 + 2^2 + \cdots + n^2)$ and because the areas of the sectors (and the circle itself) are proportional to the squares on their radii, it follows that $C < 3F$ or $\frac{1}{3}C < F$, a contradiction. A similar argument using an inscribed figure shows that $R > \frac{1}{3}C$ also leads to a contradiction, and the proposition is proved.

4.3.3 Analysis

Our final examples of Archimedes' work show again his concern that his readers learn not only the solution to a geometric problem but also how the solution was found. Namely, he often used the method of analysis; he assumed the problem was solved and then deduced consequences until he reached a result or construction already known. Then, assuming that each step was reversible, he could provide a synthetic proof by working backwards. Consider Proposition 3 of *On the Sphere and the Cylinder II*:

PROBLEM   *To cut a given sphere by a plane so that the surfaces of the segments may have to one another a given ratio.*

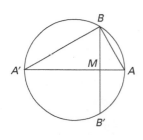

FIGURE 4.12

On the Sphere and Cylinder II, Proposition 3

Archimedes assumed that the plane BB' cuts the sphere so that the surface of BAB' is to the surface of $BA'B'$ as H is to K (Fig. 4.12). He had already proved in Proposition 42 of *On the Sphere and the Cylinder I* that the areas of such segments equal the area of the circles on the radii AB, $A'B$. Hence, he concluded that $AB^2 : A'B^2 = H : K$ and therefore that $AM : A'M = H : K$ (since the areas of the similar triangles ABM, $A'BM$ are both as the squares on corresponding sides and as the two bases with a common altitude). But the dividing of a line segment in a given ratio is a known procedure. Archimedes could therefore solve the original problem by beginning with that step and proceeding in reverse. Namely, given diameter AA', he chose M so that $AM : MA' = H : K$. The results already quoted then show that $AM : MA' = AB^2 : A'B^2 = $ (circle with radius AB) : (circle with radius $A'B$) = (surface of segment BAB') : (surface of segment $BA'B'$). The problem is solved.

Archimedes presented the analysis of a more complex problem in Proposition 4 of the same book, where he proposed to cut a given sphere by a plane so that the volumes of the segments are in a given ratio.[14] In this case, his analysis reduced the problem to a special case of the general problem: To cut a given straight line AB at a point E such that $AE : AG = \Delta : BE^2$, where AG is a given line segment (here drawn perpendicular to AB) and Δ is a given area (Fig. 4.13). Assuming the construction completed, we draw GE, continue it to Z, draw GH parallel and equal to AB, and complete rectangle $GTZH$. Further, we draw KEL parallel to ZH and continue GH to M such that the rectangle on GH and HM is equal to the given area Δ. Now,

$$\Delta : BE^2 = AE : AG = GH : HZ \text{ (by similarity)} = GH^2 : GH \cdot HZ,$$

where the multiplication in the final consequent is modern shorthand for "rectangle." Since $BE^2 = KZ^2$, we have $GH^2 : GH \cdot HZ = \Delta : KZ^2$, or $GH \cdot HZ : KZ^2 = GH^2 : \Delta$. But $\Delta = GH \cdot HM$. Therefore,

$$GH^2 : \Delta = GH^2 : GH \cdot HM = GH : HM = GH \cdot HZ : HM \cdot HZ.$$

It follows that $HM \cdot HZ = KZ^2$. But this relationship, as before, defines a parabola. So Archimedes noted that the parabola through H with axis HZ and parameter HM passes through K. (As above, if we use a coordinate system with origin H, and set $HM = \ell$, then the parabola has equation $x^2 = \ell y$.) Furthermore, rectangles $GTKL$ and $AGHB$ are equal, or

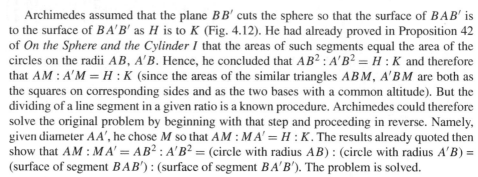

FIGURE 4.13

*On the Sphere and Cylinder
II*, Proposition 4

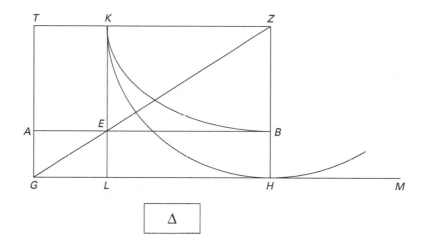

$TK \cdot KL = AB \cdot BH$. For Archimedes, such an equality showed that the hyperbola through B with asymptotes TG and GH passed through K. (If $AB = a$ and $AG - d$, this hyperbola has equation $(a - x)y = ad$.) Archimedes thus concluded the analysis by noting that K was on the intersection of the hyperbola and parabola. In the synthesis of the problem, Archimedes could then construct the parabola and hyperbola, determine K as their intersection, and then find E as in the diagram to solve the original problem.

The synthesis, however, is not complete unless we know in advance that the two curves will intersect. Archimedes considered this issue in detail. First, he noted that the original proportion implies that the desired solid with base the square on BE and height AE is equal to the given solid with base the area Δ and height AG. But because AE and BE are segments of line AB, the volume of the desired solid cannot be arbitrarily large. In fact, as Archimedes showed, there is a maximum volume that occurs when BE is twice AE. Thus, the problem cannot be solved unless the given solid is not larger than this maximum. Furthermore, Archimedes also showed that if this condition is met, there is one solution if the given solid equals the maximum and two if it is smaller than the maximum. Note that if we set $BE = x$, so $AE = a - x$, and set $\Delta = c^2$, then the relationship between the solids can be transformed into the modern cubic equation $x^2(a - x) = c^2 d$. Archimedes had therefore shown that the maximum of $x^2(a - x)$ occurs when $x = \frac{2}{3}a$, and therefore that the equation can be solved if $c^2 d \leq \frac{4}{9}a^2 \cdot \frac{1}{3}a = \frac{4}{27}a^3$. Finally, Archimedes found that the original sphere problem reduced to a special case of the general problem, where Δ is the square on 2/3 of the line AB, and AG is less than 1/3 of line AB. In that case, it is clear that the inequality condition is met, so the problem is solvable. And Archimedes could also easily determine which of the two solutions to the general problem actually split the diameter of the sphere so that the segments were in the given ratio.

It is often stated in connection with this problem that "Archimedes solved a cubic equation." Certainly, in terms of the definition of "algebra" given earlier, this problem is an example of an algebraic problem. But it also seems that Archimedes did not set out to "solve an equation," but instead to construct the solution to an interesting geometric problem. All of the expressions used in the solution are geometric, including the appearance of conic sections, which earlier mathematicians, including Euclid, had defined geometrically. So whether we can include the solution of a cubic equation as one of Archimedes' many accomplishments is, at least, debatable. What is not debatable is that his mathematical genius was far-reaching. Among numerous other results from the 14 extant treatises, Archimedes proved that the volume of a sphere is four times that of the cone with base equal to a great circle of the sphere and height equal to its radius, that the volume of a segment of a paraboloid of revolution is 3/2 that of the cone with the same base and axis, and that the surface of a sphere is four times the greatest circle in it. There is also evidence that Archimedes dealt with combinatorial problems in the *Stomachion*, of which only a small part is extant, and wrote more extensive works on balances and centers of gravity, on semiregular polyhedra, on optics, and on astronomy, which have completely disappeared. In fact, the Roman historian Livy refers to Archimedes as "an unrivalled observer of the heavens and the stars."[15]

Archimedes was buried near one of the gates of Syracuse. He had requested that his tomb include a cylinder circumscribing a sphere together with an inscription of what he evidently thought one of his most important theorems, that a cylinder whose base is a great circle in the sphere with height equal to the diameter is 3/2 of the sphere in volume and also has surface area 3/2 of the surface area of the sphere (Fig. 4.14). The tomb was found neglected by Cicero when he served as an official in Sicily about 75 BCE and was restored. Unfortunately, however, it no longer exists.

FIGURE 4.14

Stamp of Archimedes from San Marino that shows a cylinder with base equal to a great circle on a sphere

 4.4

CONIC SECTIONS BEFORE APOLLONIUS

We have already seen that Archimedes was quite familiar with the properties of the conic sections, and evidently expected his readers to be so as well. There are various indications that conics were being studied in detail a century earlier than Archimedes, but the exact origins of the theory are somewhat hazy. One possibility is that the origin may be connected to the problem of doubling the cube. Recall that Hippocrates in the fifth century BCE reduced the problem of constructing a cube double the volume of a given cube of side a to the finding of two mean proportionals x, y, between the lengths a and $2a$, that is, of determining x, y, such that $a : x = x : y = y : 2a$. In modern terms, this is equivalent to solving simultaneously any two of the three equations $x^2 = ay$, $y^2 = 2ax$, and $xy = 2a^2$, equations that represent parabolas in the first two instances and a hyperbola in the third.

It was Menaechmus (fourth century BCE) who first constructed curves that satisfy these algebraic properties and thus showed that the point of intersection of these curves would give the desired two means and solve the problem of doubling the cube. It is not known how he produced these curves, but a pointwise construction was certainly possible using Euclidean methods. To construct the points of a curve satisfying $y^2 = 2ax$, one could just apply repeatedly the method of *Elements* VI–13 (Fig. 4.15). First, put segments of length $2a$ and x together into a single line. Then, draw a semicircle having that line as diameter, and erect a perpendicular at the join of the two segments. This perpendicular has length y satisfying the equation. If this is done for various lengths x and the endpoints of the perpendiculars are

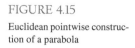

FIGURE 4.15

Euclidean pointwise construction of a parabola

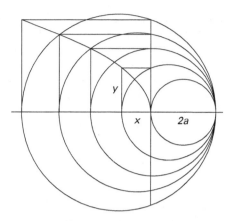

connected, the desired curve is drawn.[16] We note that although each point of this curve has been constructed using Euclidean tools, the completed curve is not a proper construction in Euclid's sense. In any case, it does appear that the conic sections were introduced as tools for the solution of certain geometric problems.

There can be only speculation as to how the Greeks realized that curves useful in solving the cube doubling problem could be generated as sections of a cone. Someone, perhaps Menaechmus himself, may have noticed that the circle diagram above could be thought of as a diagram of level curves of a certain cone, hence that the curve could be generated by a section of such a cone. Another possibility is that these curves appeared as the path of the moving shadow of the gnomon on a sundial as the sun traveled through its circular daily path, which in turn was one base of a double cone whose vertex was the tip of the gnomon. In this suggestion, the plane in which the shadow falls would be the cutting plane. It might further have been noted that the apparent shape of a circle viewed from a point outside its plane was an ellipse, and this shape comes from a plane cutting the cone of vision. In any case, by the end of the fourth century, there were in existence two extensive treatises on the properties of the curves obtained as sections of cones, one by Aristaeus (fourth century BCE) and one by Euclid. Although neither is still available, a good deal about their contents can be inferred from Archimedes' extensive references to basic theorems on conic sections.

Recall that Euclid (in Book XI of the *Elements*) defined a cone as a solid generated by rotating a right triangle about one of its legs. He then classified the cones in terms of their vertex angles as right angled, acute angled, or obtuse angled. A section of such a cone can be formed by cutting the cone by a plane at right angles to the generating line, the hypotenuse of the given right triangle. The "section of a right-angled cone" is today called a parabola, the "section of an acute-angled cone" an ellipse, and the "section of an obtuse-angled cone" a hyperbola. The names in quotation marks are those generally used by Archimedes and his predecessors.

BIOGRAPHY

Apollonius (250–175 BCE)

Apollonius was born in Perga, a town in southern Asia Minor, but few details are known about his life. Most of the reliable information comes from the prefaces to the various books of his magnum opus, the *Conics* (Fig. 4.16). These indicate that he went to Alexandria as a youth to study with successors of Euclid and probably remained there for most of his life, studying, teaching, and writing. He became famous in ancient times first for his work on astronomy, but later for his mathematical work, most of which is known today only by titles and summaries in works of later authors. Fortunately, seven of the eight books of the *Conics* do survive, and these represent in some sense the culmination of Greek mathematics. It is difficult for us today to comprehend how Apollonius could discover and prove the hundreds of beautiful and difficult theorems without modern algebraic symbolism. Nevertheless, he did so, and there is no record of any later Greek mathematical work that approaches the complexity or intricacy of the *Conics*.

FIGURE 4.16

Title page of the first Latin printed edition of Apollonius's *Conics*, 1566 (*Source*: Smithsonian Institution Libraries, Photo No. 86-4346)

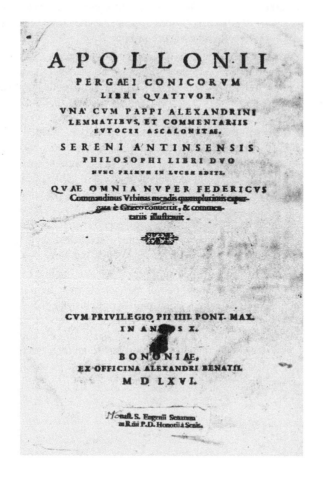

THE CONICS OF APOLLONIUS

Apollonius, in his *Conics*, decided to define the conic sections slightly differently. He decided that it was not necessary to restrict oneself to a cutting plane perpendicular to a generator, nor even to a right circular cone, to determine the curves. In fact, he generalized the notion of a cone as follows:

> If from a point a straight line is joined to the circumference of a circle which is not in the same plane as the point, and the line extended in both directions, and if, with the point remaining fixed, the straight line is rotated about the circumference of the circle . . . , then the generated surface composed of the two surfaces lying vertically opposite one another . . . [is] a **conic surface**. The fixed point [is] the **vertex** and the straight line drawn from the vertex to the center of the circle [is] the **axis**. . . . The circle [is] the **base** of the cone.[17]

For Apollonius, a conic surface was what is today called a double oblique cone. In general, its axis is not perpendicular to the base circle, but in what follows, for simplicity, we will take the axis perpendicular to the base.

To define the three curves, Apollonius first cut the cone by a plane through the axis. The intersection of this plane with the base circle is a diameter BC of that circle. The resulting triangle ABC is called the axial triangle. The parabola, ellipse, and hyperbola are then defined as sections of this cone by certain planes that cut the plane of the base circle in the straight line ST perpendicular to BC or BC produced (Figs. 4.17, 4.18, and 4.19). The straight line EG is the intersection of the cutting plane with the axial triangle. If EG is parallel to one side of the axial triangle, the section is a parabola. If EG intersects both sides of the axial triangle, the section is an ellipse. Finally, if EG intersects one side of the axial triangle and the other side produced beyond A, the section is a hyperbola. In this situation, there are two branches of the curve, unlike in the earlier obtuse-angled cone.

For each case, Apollonius derived the "symptom" of the curve, the characteristic relation between the ordinate and abscissa of an arbitrary point on the curve. Apollonius, of course, presented his results in geometric language. Nevertheless, it is not difficult to "translate" his words into modern algebraic language. We will, in general, do so, but please keep in mind that there is no evidence that Apollonius himself used algebra of any kind in his work.

Apollonius began by picking an arbitrary point L on the section and passing a plane through L parallel to the base circle. The section of the cone produced by the plane is a circle with diameter PR. Let M be the intersection of this plane with the line EG. Then LM is perpendicular to PR, and therefore $LM^2 = PM \cdot MR$.

If EG is parallel to AC, a side of the axial triangle, Apollonius found the standard symptom of a parabola, the relation between EM and LM, the abscissa and ordinate, respectively, of the point L on the curve (Proposition I–11). To do this (Fig. 4.17), he drew EH perpendicular to EK such that

$$\frac{EH}{EA} = \frac{BC^2}{BA \cdot AC}.$$

The right-hand side of this equation can be written as the product of BC/BA and BC/AC.

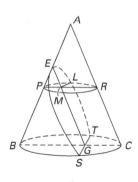

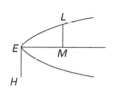

FIGURE 4.17

Derivation of the symptoms of a parabola

But by similarity,

$$\frac{BC}{BA} = \frac{PR}{PA} = \frac{PM}{EP} = \frac{MR}{EA} \quad \text{and} \quad \frac{BC}{AC} = \frac{PR}{AR} = \frac{PM}{EM}.$$

It follows that

$$\frac{EH}{EA} = \frac{MR \cdot PM}{EA \cdot EM}.$$

But also

$$\frac{EH}{EA} = \frac{EH \cdot EM}{EA \cdot EM}.$$

Therefore, $MR \cdot PM = EH \cdot EM$ and $LM^2 = EH \cdot EM$. If we set $LM = y$, $EM = x$, and $EH = p$, we have derived the standard equation of the parabola: $y^2 = px$. The name "parabola" comes from the Greek word *paraboli* (applied), because the square on the ordinate y is equal to the rectangle *applied* to the abscissa x. The constant p, which depends only on the cutting plane that determines the curve, is called the **parameter** of the parabola.

In the other two cases, let D be the intersection of EG with the second side of the axial triangle (ellipse) or with the second side produced (hyperbola) (Figs. 4.18, 4.19). Apollonius

FIGURE 4.18

Derivation of the symptoms of an ellipse

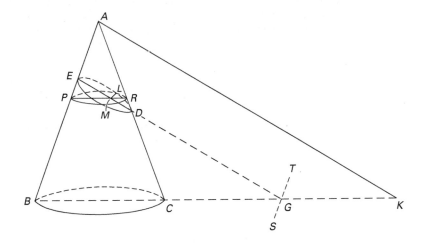

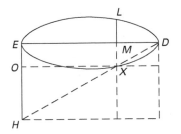

FIGURE 4.19

Derivation of the symptoms
of a hyperbola

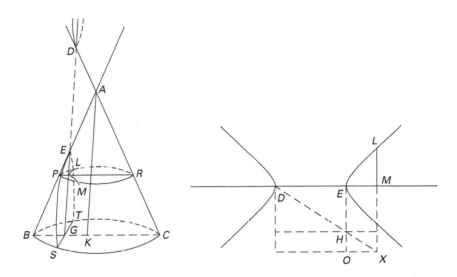

proved in these cases (Propositions I–12, I–13) that the square on LM is equal to a rectangle applied to a line EH with width equal to EM and exceeding (*yperboli*) or deficient (*ellipsis*) by a rectangle similar to the one contained by DE and EH, thus indicating the reason for the curves' names. He first chose EH, drawn perpendicular to DE, so that

$$\frac{DE}{EH} = \frac{AK^2}{BK \cdot KC}.$$

As before, the right side of this equation can be written as a product: $(AK/BK) \cdot (AK/KC)$, where AK is parallel to DE. By similarity,

$$\frac{AK}{BK} = \frac{EG}{BG} = \frac{EM}{MP} \quad \text{and} \quad \frac{AK}{KC} = \frac{DG}{GC} = \frac{DM}{MR}.$$

Therefore,

$$\frac{DE}{EH} = \frac{EM \cdot DM}{MP \cdot MR}.$$

But also

$$\frac{DE}{EH} = \frac{DM}{MX} = \frac{DM}{EO} = \frac{EM \cdot DM}{EM \cdot EO}.$$

It follows that $MP \cdot MR = EM \cdot EO$ and therefore that $LM^2 = EM \cdot EO$. In the case of the hyperbola, $EO = EH + HO$, while for the ellipse, $EO = EH - HO$. In either case, because the rectangle contained by EM ($= OX$) and HO is similar to the one contained by DE and EH, Apollonius has proved his result. In modern terms, because $EM/HO = DE/EH$, we have $HO = EM \cdot EH/DE$ and therefore, setting $LM = y$, $EM = x$, $EH = p$, and $DE = 2a$, Apollonius's symptoms become the modern equations

$$y^2 = x\left(p + \frac{p}{2a}x\right) \quad \text{and} \quad y^2 = x\left(p - \frac{p}{2a}x\right)$$

for the hyperbola and ellipse, respectively. As before, the parameter p depends only on the cutting plane determining the curve.

After giving the symptoms in this form, Apollonius, in Proposition I–21, gave a new version of the symptoms for both ellipse and hyperbola. In terms of Figures 4.18 and 4.19, he showed that $LM^2 : EM \cdot MD = EH : DE$. Turning this proportion into an algebraic equation gives us the equation for both the ellipse and hyperbola in the form $y^2 = \frac{p}{2a}x_1x_2$, where x_1 and x_2 are distances of the point M from the two ends, E and D, of the axis of the curve. We note, in the case of the ellipse, that if the point (x, y) is the endpoint of the minor axis (of length $2b$), then this equation shows that $b^2 = \frac{p}{2a}a^2$ or $b^2 = \frac{pa}{2}$. We thus have the basic relationship between the parameter and the lengths of the two axes. For the hyperbola, as we will see below, the same equation holds, with b being the perpendicular distance from a vertex to an asymptote.

Having derived the symptoms of the curves from their definitions as sections of a cone, Apollonius showed conversely in the final propositions of Book I that given a vertex (or a pair of vertices) at the end(s) of a given line (line segment) and a parameter, a cone and a cutting plane can be found whose section is a parabola (ellipse or hyperbola) with the given vertex (vertices), axis, and parameter. Henceforth, as we have already seen in some of Archimedes' work, an ancient or medieval author could assert the "construction" of a conic section with given vertices, axes, and parameter in the same manner as the construction of a circle with given center and radius. New construction postulates had thus been added to the basic ones of Euclid's *Elements*.

In deriving the properties of the conics, Apollonius generally used the symptoms of the curves, rather than the original definition, just as in modern practice these properties are derived from the equation. Although Apollonius always used geometric language, much of his work, including the symptoms themselves, can be easily translated into modern algebraic notation. Therefore, in our brief survey of highlights of the *Conics*, algebra will be used to simplify some of the statements and proofs.

4.5.1 Asymptotes

In Book II, Apollonius dealt with the asymptotes to a hyperbola. These are constructed in Proposition II–1 (Fig. 4.20). Drawing a tangent to the vertex A of the hyperbola and laying off on this tangent two segments AL, AL' (in opposite directions from the vertex), such that $AL^2 = AL'^2 = \frac{pa}{2}(= b^2)$, Apollonius showed that the lines CL, CL', drawn to L, L', from the center of the hyperbola do not meet either branch of the curve. (The word *asymptotos* in Greek means "not capable of meeting.") Furthermore, in Proposition II–14, Apollonius showed that the distance between the curve and these asymptotes, if both are extended indefinitely, becomes less than any given distance.

In Proposition II–4, Apollonius demonstrated how to construct a hyperbola given a point on the hyperbola and its asymptotes, thus providing a further construction postulate. In Proposition II–8, he then established the fact that segments cut off by a secant of a hyperbola between the hyperbola and the two asymptotes are equal. Then in Propositions II–10 and II–12, he showed that the symptom of a hyperbola can be expressed in terms of its asymptotes instead of its parameter and axis. Note that since $AL = b$, $AC = a$, and we are taking A as the origin of our coordinate system, the equation of the asymptotes can be expressed in modern

FIGURE 4.20

Constructing asymptotes to a
hyperbola

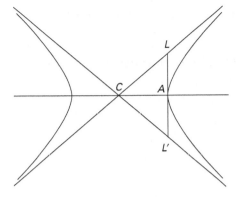

notation as $y = \pm \frac{b}{a}(x + a)$. Now take Q, q, on the hyperbola such that Qq is perpendicular
to the diameter (Fig. 4.21). If R, r, are the intersection points of Qq with the two asymptotes,
and if we write $Q = (x, y)$, then, since $b^2 = \frac{pa}{2}$,

$$QR \cdot qr = \left(\frac{b}{a}(x + a) - y\right)\left(\frac{b}{a}(x + a) + y\right) = \frac{b^2}{a^2}(x + a)^2 - y^2$$

$$= \frac{b^2 x^2}{a^2} + \frac{2b^2 ax}{a^2} + b^2 - px - \frac{p}{2a}x^2$$

$$= \left(\frac{b^2}{a^2} - \frac{p}{2a}\right)x^2 + \left(\frac{2b^2}{a} - p\right)x + b^2 = b^2.$$

Similarly, $qr \cdot qR = b^2$. If one draws from Q, q, a pair of parallel lines to each of the
asymptotes, intersecting one at H, h, respectively, and the second at K, k, one sees that
$RQ : Rq = HQ : hq$ and $qr : Qr = qk : QK$. It is then immediate that $HQ : hq = qk : QK$
or that $HQ \cdot QK = hq \cdot qk$. In other words, the product of the lengths of the two lines drawn

FIGURE 4.21

The symptom of a hyperbola
using asymptotes

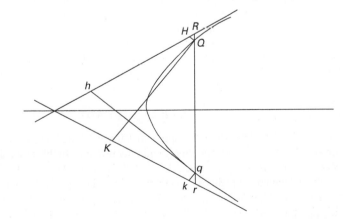

from any point of the hyperbola in given directions to the asymptotes is a constant. In modern notation, this result shows that a hyperbola can be defined by the equation $xy = k$.

4.5.2   Tangents, Minimum Lines, and Similarity

Although we do not know exactly what motivated Apollonius's choice of particular topics in his *Conics*, it does seem that among his motivations was the desire to generalize various theorems on circles, including those proved in *Elements*, Book III. For example, in Book I he discussed the problem of drawing tangents to the conic sections. For Apollonius, as for Euclid earlier, a "tangent" was a line that touches a curve but does not cut it. Recall that to draw a tangent to a circle only required drawing a line perpendicular to a radius. For a parabola, the situation was slightly more complicated:

PROPOSITION I-33   *Let C be a point on the parabola CET with CD perpendicular to the diameter EB. If the diameter is extended to A with AE = ED, then line AC will be tangent to the parabola at C (Fig. 4.22).*

FIGURE 4.22

Conics, Proposition I–33

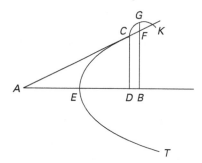

Set $DC = y$, $DE = x$, and $AE = t$. The theorem says that if $t = x$, then line AC is tangent to the curve at C. In other words, the tangent can be found by simply extending the diameter past E a distance equal to x and connecting the point so determined with C. To prove the result, Apollonius used a *reductio* argument and assumed that the line through A and C does cut the curve again, say, at K. Then the line segment from C to K lies within the parabola. Pick F on that segment and drop a perpendicular from F to the axis, meeting the axis at B and the curve at G. Therefore, $BG^2 : CD^2 > BF^2 : CD^2 = AB^2 : AD^2$. Also, since G and C lie on the curve, the symptom shows that $BG^2 = p \cdot EB$ and $CD^2 = p \cdot ED$, so $BG^2 : CD^2 = BE : DE$. Therefore, $BE : DE > AB^2 : AD^2$. So also $4BE \cdot EA : 4DE \cdot EA > AB^2 : AD^2$, and therefore

$$4BE \cdot EA : AB^2 > 4DE \cdot EA : AD^2.$$

Now note that *Elements* II–5 implies that for any lengths a, b, we have $ab \leq ((a + b)/2)^2$ or $4ab \leq (a + b)^2$, with equality if and only if $a = b$. In this case, since $AE = DE$, we have $4DE \cdot EA = AD^2$, while since $AE < BE$, we have $4BE \cdot EA < AB^2$. Thus, the left side of the displayed inequality is less than 1, while the right side equals 1, a contradiction.

Drawing a tangent to an ellipse or hyperbola is more involved, but the proof is similar to the previous one.

PROPOSITION I-34 *Let C be a point on an ellipse or hyperbola, CB the perpendicular from that point to the diameter. Let G and H be the intersections of the diameter with the curve, and choose A on the diameter or the diameter extended so that $AH : AG = BH : BG$. Then AC will be tangent to the curve at C.*

This result can be stated algebraically by letting $AG = t$ and $BG = x$. In the case of the ellipse, $BH = 2a - x$ and $AH = 2a + t$, while in the case of the hyperbola, $BH = 2a + x$ and $AH = 2a - t$ (Fig. 4.23). Therefore, for the ellipse $(2a + t)/t = (2a - x)/x$ and for the hyperbola $(2a - t)/t = (2a + x)/x$. Solving these for t gives $t = ax/(a - x)$ for the ellipse and $t = ax/(a + x)$ for the hyperbola. The tangent line can now be constructed. Apollonius completed his treatment of the tangents by proving the converses of these results as Propositions I–35 and I–36.

FIGURE 4.23

Conics, Proposition I–34

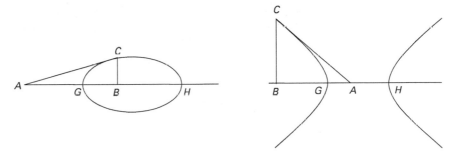

In Book III, Apollonius took up other properties of tangents to a conic section, for example, this one related to *Elements* III–36.

PROPOSITION III-16 *If two tangents to a conic section (parabola, ellipse, or one branch of a hyperbola) meet, and suppose a line is drawn across the section parallel to one tangent and meeting the other one, then as the squares on the tangents are to each other, so is the rectangle contained by the straight lines between the curve and the tangent to the square cut off at the point of contact (Fig. 4.24).*

FIGURE 4.24

Conics, Proposition III–16

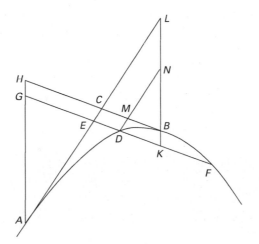

In other words, if the tangents to the section at A and B meet at C, and if FDE is parallel to CB and meets AC at E, then $BC^2 : AC^2 = FE \cdot ED : EA^2$. In the case of a circle, of course, the two tangents from a point are equal and, according to *Elements* III–36, $EA^2 = FE \cdot ED$.

Book IV generally deals with the number of ways conics can intersect. Thus, Apollonius showed, in contrast to *Elements* III–10, which proves that two circles can intersect in only two points, that two conic sections can intersect in at most four points. And, in contrast to *Elements* III–13, which states that two circles can be tangent only at one point, Apollonius showed that two conic sections can be tangent at most at two points.

In Book V, Apollonius considered minimal lines to conics from points on the axes. These lines turn out to be normals to the conic. Only the parabola is considered here, where we have combined three propositions dealing with that curve. (There are analogous propositions for the other conics.)

PROPOSITIONS V-8, V-13, V-27 *In a parabola with vertex A and symptom $y^2 = px$, let G be a point on the axis such that $AG > \frac{p}{2}$. Let N be taken between A and G so that $NG = \frac{p}{2}$. Then, if NP is drawn perpendicular to the axis meeting the curve in P, PG is the minimum straight line from G to the curve. Conversely, if PG is the minimum straight line from G to the curve, and PN is drawn perpendicular to the axis, $NG = \frac{p}{2}$. Finally, PG is perpendicular to the tangent at P (Fig. 4.25).*[18]

FIGURE 4.25

Conics, Propositions V–8,
V–13, and V–27

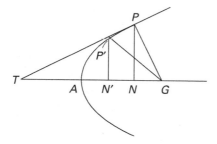

For the proof, suppose P' is another point on the parabola with abscissa AN'. By the defining property of the parabola, we have $P'N'^2 = p \cdot AN' = 2NG \cdot AN'$. Also $N'G^2 = NN'^2 + NG^2 \pm 2NG \cdot NN'$ (with the sign depending on the position of N'). Adding these two equations together and using the Pythagorean Theorem gives $P'G^2 = 2NG \cdot AN + NN'^2 + NG^2 = PN^2 + NG^2 + NN'^2 = PG^2 + NN'^2$. Thus, PG is the minimum straight line from G to the curve. The converse is proved by a *reductio* argument. Finally, to show that PG is perpendicular to the tangent TP, note that $AT = AN$. Therefore, $NG : p = \frac{1}{2} = AN : NT$, so $TN \cdot NG = p \cdot AN = PN^2$. Because the angle at N is a right angle, so is the angle TPG as desired.

In Book VI, Apollonius took up the subject of equality and similarity of conic sections and their segments. In particular, two conic sections AEH and aeh with axes AH and ah are defined to be **similar** if whenever points $B, D, F, \ldots$ are taken on AH and corresponding points $b, d, f, \ldots$ are taken on ah such that $AB : ab = AD : ad = AF : af = \cdots$, then the

FIGURE 4.26

Similarity of conic sections

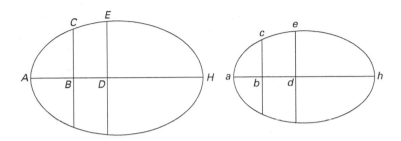

ratios of the corresponding ordinates to the abscissas are equal, that is, $CB : AB = cb : ab$, $ED : AD = ed : ad$, $GF : AF = gf : af$, ... (Fig. 4.26).

Among the theorems Apollonius proved using this definition are that all parabolas are similar to one another (IV–11) and that hyperbolas and ellipses are similar precisely when the figures that define them, that is, the rectangles contained by the parameter and the axis, are similar (IV–12). In algebraic notation, this latter result says that two ellipses (hyperbolas) with parameters p_1, p_2, and major axes $2a_1$, $2a_2$, are similar if and only if $p_1 : a_1 = p_2 : a_2$.

In Book VII, the final extant book of the *Conics*, Apollonius studied conjugate diameters of conic sections. We consider two examples of results from Book VII in the exercises to this chapter.

4.5.3 Foci

Although the notion of the foci of a conic section is an important modern concept, it was evidently not central to Apollonius's thinking. He dealt briefly with this topic in a series of propositions in Book III, but did not feel it necessary even to give these points a special name. Thus, as part of Proposition III–45, we are told to apply a rectangle equal to one-fourth of the rectangle on the parameter N and the major axis AB to the axis AB of an ellipse (on each side) that is deficient by a square figure. This application results in two points F and G on the axis "produced by the application." It is these points that were first named **foci** by Johannes Kepler in 1604. According to our algebraic translation of *Elements* VI–28 on applications with deficiencies, these points are the solutions of the equation $x(2a - x) = \frac{1}{4}(2ap)$, where $2a$ is the length of the axis AB, and p is the parameter of the ellipse. Because the solutions to this equation are $x = a \pm \sqrt{a^2 - ap/2} = a \pm \sqrt{a^2 - b^2}$, where $2b$ is the length of the minor axis, Apollonius has shown that the two foci are at a distance $c = \sqrt{a^2 - b^2}$ from the center of the ellipse. Given this definition, Apollonius then presented a series of propositions leading to the well-known results that the lines from the two foci to any point on the ellipse make equal angles with the tangent to the ellipse at that point (Proposition III–48) and that the sum of these lines equals the major axis (Proposition III–52). This latter proposition is the standard textbook definition of an ellipse today.

Although Apollonius presented similar results for the hyperbola, he did not deal at all with the focal properties of the parabola, perhaps because he had discussed these in an earlier work now lost. In any case, the analogous property for a parabola, that any line from the focus to a point on the parabola makes an angle with the tangent at that point equal to the one made by a line parallel to the axis, was probably first proved by Diocles (early second century BCE), a contemporary of Apollonius, in a treatise *On Burning Mirrors*, perhaps written a few years before the *Conics*. It is in fact that property of the parabola that gives this treatise its name. The problem is to find a mirror surface such that when it is placed facing the sun, the rays reflected from it meet at a point and thus cause burning. Diocles showed that this would be true for a paraboloid of revolution. There are stories told about Archimedes and others that such a mirror was used to set enemy ships on fire. However, there is no reliable evidence for the veracity of these stories.

To complete this topic of foci, then, we consider Diocles' proof of the focal property of the parabola from Proposition 1 of his treatise.[19] Given a parabola LBM with axis BW, lay off BE along the axis equal to half the parameter and bisect BE at D (Fig. 4.27). It is this point D, whose distance from the vertex is $p/4$, that is today called the **focus** of the parabola. Pick an arbitrary point K on the parabola, draw a tangent line AKC through K meeting the axis extended at A, draw KS parallel to the axis, and connect DK. The proposition then asserts that $\angle AKD = \angle SKC$.

FIGURE 4.27

Diocles' *On Burning Mirrors*

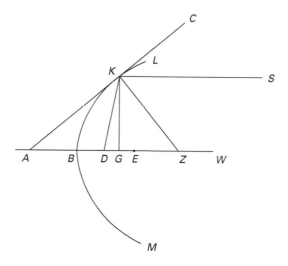

To prove this, first drop a perpendicular from K to the axis, meeting it at G. By *Conics* I–33, $AB = BG$. Next draw a line perpendicular to AK from K that meets the axis at Z. Because $KG^2 = AG \cdot GZ$ and also $KG^2 = p \cdot BG$, it follows that $GZ = p/2$. Then $GZ = BE$, so $GB = EZ$, $AB = EZ$, and finally $AD = DZ$. Because triangle AKZ is a right triangle whose hypotenuse is bisected at D, we have $AD = DK = DZ$. Therefore, $\angle DZK = \angle DKZ$. Since KS is parallel to AZ, it also follows that $\angle ZKS = \angle DKZ$. Subtracting these equal angles from the right angles ZKC and ZKA, we obtain the desired result. Diocles concluded the proposition by showing how to construct the burning mirror by rotating LBM about the axis AZ and covering the resulting surface with brass.

In Propositions 4 and 5 of his brief treatise, Diocles showed how to construct a parabola with given focal length. His construction in effect uses the focus-directrix property of a parabola, that the points of the parabola are equally distant from the focus and a given straight line called the **directrix**. There is no earlier reference in antiquity to this particular property of the parabola, although it was discussed by the fourth-century commentator Pappus. In fact, Pappus also noted that an ellipse is determined as the locus of a point moving so that the ratio of its distances from a fixed point (the focus) and a fixed line (the directrix) is a constant less than 1, while a hyperbola is described if this constant ratio is greater than 1. These latter properties were probably also discovered around the time of Diocles and Apollonius.

4.5.4 Problem Solving Using Conic Sections

One of Apollonius's aims in the *Conics* was to develop the properties of the conic sections necessary for the application of these curves to the solution of geometric problems. We therefore conclude this chapter with three examples of how the conics were so used in Greek times.

We first consider the angle trisection problem. Let angle ABC be the angle to be trisected (Fig. 4.28). Draw AC perpendicular to BC and complete the rectangle $ADBC$. Extend DA to the point E, which has the property that if BE meets AC in F, then the segment FE is equal to twice AB. It then follows that $\angle FBC = \frac{1}{3}\angle ABC$. For if FE is bisected at G, then $FG = GE = AG = AB$. Therefore, $\angle ABG = \angle AGB = 2\angle AEG = 2\angle FBC$ and the trisection is demonstrated. To complete the proof, however, it is necessary to show how to construct BE satisfying the given condition. Again an analysis will help. Assuming $FE = 2AB$, draw CH and EH parallel to FE and AC, respectively. It follows that H lies on the circle of center $= C$ and radius $FE(= 2AB)$. Moreover, since $DE : DB = BC : CF$, or $DE : AC = DA : EH$, we have $DA \cdot AC = DE \cdot EH$, so H also lies on the hyperbola with asymptotes DB, DE, and passing through C. Therefore, if one constructs the hyperbola and the circle and drops a perpendicular from the intersection point H to DA extended, the foot E of the perpendicular is the point needed to complete the solution.

FIGURE 4.28

Angle trisection by way of conic sections

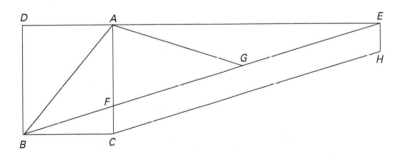

Cube duplication constructions virtually all begin with Hippocrates' reduction of the problem to the construction of two mean proportionals between given lines AB and AC. One of them from the time of Apollonius begins with the completion of the rectangle on these two lines, the drawing of the diagonal AD, and the construction of the circle with diameter AD passing through B (Fig. 4.29). Now, let F be the intersection of the circle with a hyperbola through D with asymptotes AB and AC. Extend line DF to meet AB

produced in E and AC produced in G. By *Conics* II–8, $FE = DG$ and therefore $DE = FG$. Furthermore, since F, D, C, A, and B all lie on the same circle, *Elements* III–36 implies that $GA \cdot GC = GF \cdot GD$ and $EA \cdot EB = ED \cdot EF$. Therefore, $GA \cdot GC = EA \cdot EB$ or $GA : EA = BEB : GC$. By similarity, $GA : EA = DB : BE = AC : BE$, and also $GA : EA = GC : DC = GC : AB$. It follows that $AC : BE = BE : GC = GC : AB$, so that BE, GC, are the two desired mean proportionals.

FIGURE 4.29

Doubling the cube using conic sections

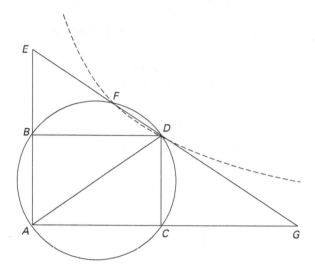

A final problem to be dealt with here, a problem that had reverberations down to the seventeenth century, is the three- and four-line locus problem. The problem in its most elementary form can be stated as follows: Given three fixed straight lines, to find the locus of a point moving so that the square of its distance to one line is in a constant ratio to the product of its distance to the other two lines. (Here distance is to be measured at a fixed angle to each line.) If one takes the special case where two of the lines are parallel and the third perpendicular to the first two, it is easy to see analytically that the given locus is a conic section. Recall that one version of the equation of the ellipse and hyperbola was $y^2 = \frac{p}{2a}x_1 x_2$, where y is the distance of a given point from the diameter of the conic, and x_1, x_2, are the distances of the abscissa of that point from the endpoints of the diameter. If tangents are drawn to the conic at those two endpoints, the curve then provides a solution to the three-line locus problem with respect to the diameter and the two tangents.

The problem for the Greek mathematicians was to generalize this solution, that is, to show that the locus was a conic whatever the position of the three lines. Apollonius wrote (see the chapter opening quotation) that the three-line locus problem had been only partially worked out by Euclid, but that his new results in Book III would enable the problem to be completely solved. The text of Book III does not mention the problem as such, but in fact one can derive the result that a conic has the property of the three-line locus relative to two tangents to the curve from a given point and the secant joining the two points of tangency from theorems in that book. Other theorems there enable one to show that a conic also solves the four-line locus problem, to find the locus of a point such that the product of its distances to one pair of lines is in a constant ratio to the product of its distances to the other pair. In later Greek times, an attempt was made, without great success, to find the locus with regard to greater numbers of lines. It was this problem that Descartes and Fermat both demonstrated they could solve through their new method of analytic geometry in the seventeenth century, a method whose germ came from a careful reading of Apollonius's work. Descartes in fact was able to derive the equations of curves that satisfied analogous conditions for various numbers of lines and to classify the solutions. As should be evident from our description of many of the Greek problems in modern notation, the Greek tradition of geometric problem solving, which was carried on in the Islamic world long after its demise in the Hellenic world, ultimately led to new advances in mathematical technique, advances that finally reduced much of this kind of Greek mathematics to mere textbook exercises.

EXERCISES

1. Find where to place the fulcrum in a lever of length 10 m so that a weight of 14 kg at one end will balance a weight of 10 kg at the other.

2. If a weight of 8 kg is placed 10 m from the fulcrum of a lever and a weight of 12 kg is placed 8 m from the fulcrum in the opposite direction, toward which weight will the lever incline?

3. An alternative method by which Archimedes could have solved the crown problem is given by Vitruvius in *On Architecture*. Assume as in the text that the crown is of weight W, composed of weights w_1 and w_2 of gold and silver, respectively. Assume that the crown displaces a certain quantity of fluid, V. Furthermore, suppose that a weight W of gold displaces a volume V_1 of fluid, while a weight W of silver displaces a volume V_2 of fluid. Show that $V = \frac{w_1}{W}V_1 + \frac{w_2}{W}V_2$ and therefore that $\frac{w_1}{w_2} = \frac{V_2 - V_1}{V - V_1}$.

4. Prove the two lemmas (see Section 4.2) that Archimedes used to derive his algorithms for calculating π.

5. Use a calculator (or program a computer) to calculate π by iterating the algorithm of Archimedes given by Lemma 1. How many iterations are necessary to get five-decimal-place accuracy?

6. Translate Lemma 2 into a recursive algorithm for calculating π. Iterate this algorithm to calculate π to five-decimal-place accuracy. How many iterations are necessary?

7. Given the parabolic segment with MO parallel to the axis of the segment and MC tangent to the parabola, show analytically that $MO : OP = CA : AO$ (see Fig. 4.7).

8. The proof of Proposition 2 of Archimedes' *The Method* is outlined here:

PROPOSITION 2   *Any sphere is (in respect of solid content) four times the cone with base equal to a great circle of the sphere and height equal to its radius.*

Let $ABCD$ be a great circle of a sphere with perpendicular diameters AC, BD. Describe a cone with vertex A and axis AC and extend its surface to the circle with diameter EF. On the latter circle erect a cylinder with height and axis AC. Finally, extend AC to H such that $HA = AC$. Certain pieces of the figures described are to be balanced using CH as the lever (Fig. 4.30).

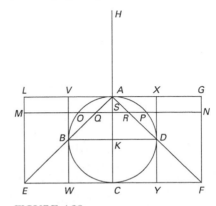

FIGURE 4.30

Archimedes' *The Method*, Proposition 2

Let MN be an arbitrary line in the plane of the circle $ABCD$ and parallel to BD with its various intersections marked as in the diagram. Through MN, draw a plane at right angles to AC. This plane cuts the cylinder in a circle with diameter MN, the sphere in a circle with diameter OP, and the cone in a circle with diameter QR.

a. Show that $MS \cdot SQ = OS^2 + SQ^2$.

b. Show that $HA : AS = MS : SQ$. Then, multiplying both parts of the last ratio by MS, show that $HA : AS = MS^2 : (OS^2 + SQ^2) = MN^2 : (OP^2 + QR^2)$. Show that this last ratio equals that of the circle with diameter MN to the sum of the circle with diameter OP and that with diameter QR.

c. Conclude that the circle in the cylinder, placed where it is, is in equilibrium about A with the circle in the sphere together with the circle in the cone, if both the latter circles are placed with their centers of gravity at H.

d. Archimedes concluded from the above that the cylinder, placed where it is, is in equilibrium about A with the sphere and the cone together, when both are placed with their center of gravity at H. Show therefore that $HA : AK = $ (cylinder) : (sphere + cone AEF).

e. From the fact that the cylinder is three times the cone AEF and the cone AEF is eight times the cone ABD, conclude that the sphere is equal to four times the cone ABD.

9. Derive, by the general technique of *The Method*, Proposition 4, that the volume of a segment of a paraboloid of revolution cut off by a plane at right angles to the axis is 3/2 the volume of a cone that has the same base and the same height. Begin with triangle ABC inscribed in the segment $BOAPC$ of a parabola (Fig. 4.31) with both inscribed

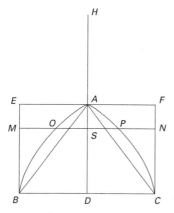

FIGURE 4.31

Archimedes' *The Method*, Proposition 4

in rectangle $EFCB$. Rotate the entire figure around the axis AD to get a cone inside of a paraboloid of revolution that is in turn inside of a cylinder. Extend DA to H so that $AD = AH$; draw MN parallel to BC; and imagine the plane through MN producing sections of the cone, the paraboloid, and the cylinder. Finally, imagine that HD is a lever with midpoint A and use Archimedes' balancing techniques to show that the circle in the cylinder of radius MS, placed where it is, balances the circle in the paraboloid of radius OS, if the latter is placed at H. Use the result that the volume of a cone is 1/3 that of the inscribing cylinder to conclude the proof of the theorem.

10. Use calculus to prove Archimedes' result from *The Method* that the volume of the segment of the cylinder described in the text is 1/6 the volume of the rectangular parallelepiped circumscribing the cylinder.

11. Use calculus to prove Archimedes' result that the area of a parabolic segment is four-thirds of the area of the inscribed triangle.

12. Show analytically that the vertex of a parabolic segment (see the definition in Section 4.3.1) is that point on the curve whose perpendicular distance to the base of the segment is greatest.

13. Use calculus to prove Archimedes' result that a cylinder whose base is a great circle in the sphere and whose height is equal to the diameter of the sphere has volume 3/2 that of the sphere and also has surface area 3/2 the surface area of the sphere.

14. Use calculus to prove Archimedes' result that the area bounded by one complete turn of the spiral given in polar coordinates by $r = a\theta$ is one-third of the area of the circle with radius $2\pi a$.

15. Consider Proposition 1 of *On the Sphere and Cylinder II*: Given a cylinder, to find a sphere equal to the cylinder. Provide the analysis of this problem. That is, assume that V is the given cylinder and that a new cylinder P has been constructed of volume $\frac{3}{2}V$. Assume further that another cylinder Q has been constructed equal to P but with height equal to its diameter. The sphere whose diameter equals the height of Q would then solve the problem, because the volume of the sphere is $\frac{2}{3}$ that of the cylinder. So given the cylinder P of given diameter and height, determine how to construct a cylinder Q of the same volume but whose height and diameter are equal.

16. There is a story about Archimedes that he used a "burning mirror" in the shape of a paraboloid of revolution to set fire to enemy ships in the harbor. What would be the equation of the parabola that one would rotate to form the appropriate

paraboloid if it were to be designed to set fire to a ship 100 m from the mirror? How large would the burning mirror need to be? What is the likelihood that this story is true?

17. Show that in the curve $y^2 = px$, the value p represents the length of the **latus rectum**, the straight line through the focus perpendicular to the axis.

18. Rewrite the equations $y^2 = x(p + \frac{p}{2a}x)$ and $y^2 = x(p - \frac{p}{2a}x)$ for the hyperbola and ellipse, respectively, in the current standard forms for those equations. What point is the center of the curve? Show in the case of the ellipse, where $2b$ is the length of the minor axis, that $b^2 = pa/2$.

19. Use calculus to prove *Conics* I–33.

20. Use calculus to prove *Conics* I–34.

21. Demonstrate analytically Apollonius's result from Book IV that two conic sections can intersect in at most four points.

22. Demonstrate analytically Apollonius's result from Book IV that two conic sections can be tangent at no more than two points.

23. Apollonius stated and proved many of the properties of conics in a more general form than we have given. Namely, instead of restricting himself to the principal diameters of the conic, such as the major and minor axes of the ellipse, he dealt with any pair of conjugate diameters. For an ellipse, given the tangent at any point, the parallel to this tangent passing through the center of the ellipse is **conjugate** to the straight line passing through the point of tangency and the center (Fig. 4.32). (A similar definition can be given for a hyperbola, but we restrict this problem to the case of the ellipse.)

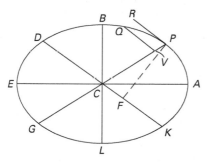

FIGURE 4.32

Conjugate diameters in an ellipse

a. Show that if DK is conjugate to PG, then PG is also conjugate to DK.

b. Assume the equation of the ellipse is given by $b^2x^2 + a^2y^2 = a^2b^2$ in rectangular coordinates x, y. Denote angle PCA by θ and angle DCA by α. Given that di-

ameter DK is parallel to the tangent to the ellipse at $P = (x_0, y_0)$, show that $\tan \theta = y_0/x_0$ and that $\tan \alpha = -b^2x_0/a^2y_0$.

c. Transform the equation of the ellipse to new oblique coordinates x', y', based on the conjugate diameters PG and DK. Show that the transformation is given by

$$x = x' \cos \theta + y' \cos \alpha$$
$$y = x' \sin \theta + y' \sin \alpha$$

and that the new equation for the ellipse is $Ax'^2 + Cy'^2 = a^2b^2$, where $A = b^2 \cos^2 \theta + a^2 \sin^2 \theta$ and $C = b^2 \cos^2 \alpha + a^2 \sin^2 \alpha$.

d. Let $Q = (x', y')$ be a point on the ellipse, with coordinates relative to the conjugate diameters PG and DK, and let QV be drawn to diameter PG with QV parallel to DK. Then set $QV = y'$, $PV = x'_1$, $GV = x'_2$, $PC = a'$, and $DC = b'$ and show that the equation of the ellipse can be written in the form

$$y^2 = \frac{b'^2}{a'^2}x'_1x'_2,$$

thus generalizing the version of Apollonius's Proposition I–21 given in the text.

e. Show that the parallelogram constructed on any pair of conjugate diameters (using the angle at which they meet) is equal to the rectangle constructed on the principal axes (Proposition VII–31). Namely, show that the parallelogram whose sides are PC and CD is equal to the rectangle whose sides are AC and BC. In other words, if PF is drawn perpendicular to DK, show that $PF \times CD = AC \times BC = ab$.

24. Prove analytically Proposition VII–12, that in any ellipse the sum of the squares on any two of its conjugate diameters is equal to the sum of the squares on its two axes. (In Figure 4.32, this means that $PG^2 + DK^2 = AE^2 + BL^2$.)

25. Use Proposition II–8 to show that the two line segments of a tangent to a hyperbola between the point of tangency and the asymptotes are equal. Then show, without calculus, that the slope of the tangent line to the curve $y = 1/x$ at $(x_0, 1/x_0)$ equals $-1/x_0^2$.

26. Given an ellipse with diameter $AA' = 2a$, center C, and symptom $y^2 = x(p - \frac{p}{2a}x)$, let G be any point on AA' such that $AG > \frac{p}{2}$ (Fig. 4.33). Choose N on AG so that $NG : CN = p : 2a$. Prove analytically that if NP is drawn perpendicular to the axis and meets the curve at P, then PG is the minimum straight line from G to the curve. Also show that PG is perpendicular to the tangent at P (Propositions V–10, V–28).

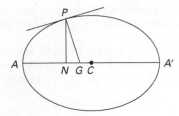

FIGURE 4.33

Finding the perpendicular to an ellipse

27. Prove that all parabolas are similar (Proposition VI–11), according to Apollonius's definition.

28. Use Apollonius's Proposition VI–12 to show that ellipses are similar if and only if the ratios of the major axes to the minor axes are equal. State and prove an analogous proposition for hyperbolas.

29. Prove Proposition III–45 for an ellipse, namely, if AC and BD are tangent to the ellipse at the two ends of the major axis, and if CD is tangent to the ellipse at E, and if one connects C, D, to the two foci F, G, respectively, then angles CFD and DGC are right angles (Fig. 4.34). (See section 4.5.3 for the definition of foci.)

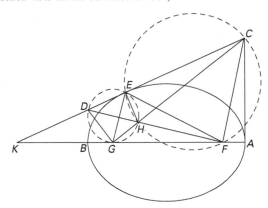

FIGURE 4.34

Conics, Propositions III–45, III–46, III–47, III–48

30. Prove Proposition III–46: Under the same assumptions as in III–45, angle ACF = angle DCG and angle CDF = angle BDG (Fig. 4.34).

31. Prove Proposition III–47: Under the same assumptions as in III–45 and III–46, let H be the intersection point of GC and FD. Join HE. Prove that EH is perpendicular to CD (Fig. 4.34).

32. Prove Proposition III–48: Under the same assumptions as in the three previous propositions, connect EF and

EG. Show that angle CEF = angle GED, that is, that the lines from an arbitrary point on the ellipse to the two foci make equal angles with the tangent at that point (Fig. 4.34).

33. Give the definition of foci of a hyperbola analogous to the definition for an ellipse presented in the text, and state the theorems analogous to III–48 and III–52 in this case.

34. Give a proof using calculus that the line from the focus to a point on a parabola makes an angle with the tangent at that point equal to that made by a line parallel to the axis.

35. Show analytically that the solution to the three-line locus problem is a conic section in the case where two of the lines are parallel and the third is perpendicular to the other two. Characterize the curve in reference to the distance between the two parallel lines and the given ratio.

36. Show analytically that the solution to the general three-line locus problem is always a conic section.

37. Fill in the details of the following solution to the angle trisection problem (given in Pappus but probably dating from much earlier).[20] Let the given angle AOG be placed at the center of the circle, cutting off the arc AG on the circumference (Fig. 4.35). To trisect this angle, it is sufficient to trisect arc AG, that is, to find a point B on the circle such that arc BG is one-half of arc AB. Using the method of analysis, suppose that this has been done. Then $\angle BGA = 2\angle BAG$. Draw GD to bisect $\angle BGA$ and draw DE, BZ, perpendicular to AG. Use *Elements* VI–3 and similarity to show that $BG : EZ = AG : AE = 2 : 1$. Use the focus-directrix property to conclude that B lies on a particular hyperbola, and then complete the synthesis.

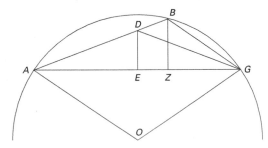

FIGURE 4.35

Angle trisection by way of conic sections, second method

38. Design a lesson for a precalculus course that will demonstrate the formula for the sum of a geometric series as in Archimedes' work.

39. Discuss whether one can adapt Archimedes' procedure for determining the area of a parabolic segment and/or the area bounded by one turn of the spiral to introduce a precalculus class (or even a calculus class) to the calculation of areas bounded by curves.

40. Design lessons for a precalculus course deriving the equations of the conic sections from their definitions as sections of a cone, as in the work of Apollonius. How does this method compare to the use of the standard modern textbook definitions?

41. Design a series of lessons for a precalculus course that will demonstrate the basic tangent and focal properties of the conic sections.

42. How is Apollonius's treatment of the conic sections using symptoms similar to a modern analytic geometry treatment of the same subject? Can one consider Apollonius as an inventor of analytic geometry?

43. Can one consider Archimedes as an inventor of the integral calculus?

REFERENCES AND NOTES

Many of the books on Greek mathematics referred to in the Chapter 2 references have sections on Archimedes, Apollonius, and Ptolemy. In particular, Thomas Heath's *A History of Greek Mathematics* and Wilbur Knorr's *The Ancient Tradition of Geometric Problems* are good sources of further reading on the material of this chapter as is B. L. Van der Waerden's *Science Awakening*, mentioned in the Chapter 1 references. Selections from the works of these three mathematicians as well as others discussed in this chapter can be found in Ivor Thomas, *Selections Illustrating the History of Greek Mathematics*, also mentioned in the Chapter 2 references.

The first complete English translation of the extant works of Archimedes is now appearing: Reviel Netz, *The Works of Archimedes: Translation and Commentary*, (Cambridge: Cambridge University Press, 2004–). The earlier English versions cannot be considered "translations" but are detailed summaries edited for modern readers. These two versions are Thomas Heath, *The Works of Archimedes* (New York: Dover, 1953) and E. J. Dijksterhuis, *Archimedes* (Princeton: Princeton University Press, 1987). This second edition of Dijksterhuis's work has a bibliographic essay by Wilbur Knorr, which gives further information regarding research on the work of Archimedes up to 1987. A detailed analysis of the newly rediscovered manuscript of *The Method* can be found in *The Archimedes Palimpsest*, published by Christie's, New York, in 1998, prior to the auction. A complete explication of that work is in Reviel Netz and William Noel, *The Archimedes Codex* (London: Weidenfeld & Nicolson, 2007).

An English translation of the first three books of Apollonius's *Conics* is available as *Apollonius of Perga Conics, Books I–III*, R. Catesby Taliaferro, trans. (Santa Fe: Green Lion Press, 2000). Book IV is published separately: *Apollonius of Perga Conics, Book IV*, Michael N. Fried, trans. (Santa Fe: Green Lion Press,

2002). Thomas Heath's older work, *Apollonius of Perga* (Cambridge: W. Heffer and Sons, 1961), contains all seven extant books of the *Conics*. But since Heath modifies the order and often combines several theorems, this cannot be considered a translation. Books V–VII are available in English translation as *Apollonius Conics Books V to VII: The Arabic Translation of the Lost Greek Original in the Version of the Banū Mūsā*, Gerald J. Toomer, trans. (New York: Springer-Verlag, 1990). A new analysis of Apollonius's work is Michael Fried and Sabetai Unguru, *Apollonius of Perga's Conics: Text, Context, Subtext* (Leiden, Netherlands: Brill Academic Publishers, 2001). Diocles' work is available in Gerald Toomer, *Diocles on Burning Mirrors* (New York: Springer, 1976). This book provides a complete translation as well as a discussion of its importance.

1. Taliaferro, *Apollonius of Perga*, 1.

2. Vitruvius, *Ten Books on Architecture*, Ingrid D. Rowland, trans. (Cambridge: Cambridge University Press, 1999), 108.

3. A discussion of the pseudo-Aristotelian *Mechanica* can be found in Thomas Heath, *A History of Greek Mathematics*, I, p. 344–346, and a translation of sections of this work is found in Thomas, *Selections*, I, p. 431.

4. Dijksterhuis, *Archimedes*, p. 299. The reference is to a translation of Pappus, *Collectio* VIII, 5, 11.

5. Plutarch, *The Lives of the Noble Grecians and Romans* (Dryden translation), in the *Great Books*, 14, p. 252. This reference and the succeeding ones are taken from the section on Marcellus.

6. Plutarch, *Lives*, p. 253.

7. Ibid.

8. The discussion of the crown problem is from Heath, *The Works of Archimedes*, pp. 259–260. Heath's introduction

provides insight into the various mathematical techniques of Archimedes.

9. Heath includes as an appendix to the previous work a translation of *The Method of Archimedes*. The present quotation from the introduction is found on p. 13. A valuable discussion of this work is also found in Asger Aaboe, *Episodes from the Early History of Mathematics* (Washington, DC: MAA, 1964). A brief account is in S. H. Gould, "The Method of Archimedes," *American Mathematical Monthly* 62 (1955), 473–476.

10. Ibid., p. 17.

11. Ibid.

12. This discussion of Proposition 14 of *The Method* is adapted from "A New Reading of *Method* Proposition 14: Preliminary Evidence from the Archimedes Palimpsest (Parts 1 and 2)," SCIAMVS, 2 (2001), 9–29, and 3 (2002), 109–125.

13. A discussion of the use of indivisibles in Greek mathematics is found in Wilbur Knorr, "The Method of Indivisibles in Ancient Geometry," in Ronald Calinger, ed., *Vita Mathematica* (Washington, DC: MAA, 1996), pp. 67–86, although Knorr was not aware of the new readings of *The Method* found in the recovered palimpsest and discussed in the reference in note 12.

14. A very detailed discussion of the history of Archimedes' problem on dividing a sphere can be found in Reviel Netz, *The Transformation of Mathematics in the Early Mediterranean World: From Problems to Equations* (Cambridge: Cambridge University Press, 2004).

15. Livy, *History of Rome* (Cambridge: Harvard University Press, 1934), XXIV, sec. 34.

16. This discussion and Figure 4.15 are adapted from Wilbur Knorr, *The Ancient Tradition of Geometric Problems*. Knorr has an extensive discussion of the contributions of Apollonius to Greek geometric problem solving.

17. The quotations from the first three books of Apollonius's *Conics* are taken from the R. Catesby Taliaferro translation.

18. These theorems can be found in the Toomer translation of Apollonius's Books V–VII.

19. The discussion of Diocles' work is adapted from Gerald Toomer, *Diocles on Burning Mirrors*.

20. Knorr, *The Ancient Tradition*, p. 128.

5

Mathematical Methods
in Hellenistic Times

*Plato ... set the mathematicians the
following problem: What circular motions,
uniform and perfectly regular, are to be
admitted as hypotheses so that it might be
possible to save the appearances presented
by the planets?*

—Simplicius's *Commentary on
Aristotle's* On the Heavens[1]

Egypt (c. 150 CE): Hiring now. Calculators wanted to perform extensive but routine calculations to create tables necessary for major work on astronomy. Must be able to follow detailed instructions with great accuracy. Compensation: Room and board plus the gratitude of the thousands of people who will use these tables for the next 1200 years. Contact: Claudius Ptolemy at the Observatory. (A classified advertisement in an Alexandrian newspaper)

Although such an advertisement did not actually appear, Claudius Ptolemy did write a major work answering Plato's challenge, a work studied, commented upon, extensively criticized, yet never replaced for 1400 years, a work in which Ptolemy not only used earlier ideas from plane and spherical geometry but also devised new ways to perform the extensive numerical calculations necessary to make his book a useful one. Ptolemy's text, and other ancient astronomical works from Babylonia and Egypt, were heavily used in astrology. Nevertheless, the evidence from all of these civilizations indicates that the primary reason for the study of astronomy was the solving of problems connected with the calendar, problems such as the determination of the seasons, the prediction of eclipses, and the establishment of the beginning of the lunar month.

In the process of using mathematics to study astronomy, the Greeks created plane and spherical trigonometry and also developed a mathematical model of the universe, a model they modified many times during the five centuries between the times of Plato and Ptolemy. Among the major contributors to the development of mathematical astronomy whose ideas will be discussed in this chapter are Eudoxus in the fourth century BCE, Apollonius late in the third century BCE, Hipparchus in the second century BCE, Menelaus around 100 CE, and finally Ptolemy. The chapter then concludes with a survey of other work in "practical mathematics" developed in the Greco-Roman world, mathematics applicable to problems on earth rather than the heavens, including work by Roman surveyors and architects, several practical works by Heron, and the *Geography* of Ptolemy.

5.1   ASTRONOMY BEFORE PTOLEMY

What did ancient peoples know about the heavens? The most important heavenly bodies were the sun and the moon. It was obvious that both rose in the east and set in the west, but the actual movements of each were considerably more subtle. For example, in the northern hemisphere, the sun rises at exactly the east point on the spring equinox, well north of east through the summer, due east again at the autumn equinox, and south of east during the winter. It was observed everywhere that this sun cycle repeated itself at intervals. Wherever there are records of the calculation, the length of this interval, the year, is specified to be about 365 days.

If one wants to identify the important days in this yearly calendar, one needs to be able to observe the sun's position. It was in part to do this that the great stone temple at Stonehenge in England was constructed beginning in the third millennium BCE (Fig. 5.1). Many similar but smaller such structures were built elsewhere in England and other parts of northern Europe. Although the reasons for the construction of these structures are not entirely clear, most scholars believe that among these reasons was the determination of the farthest north and farthest south sunrise and sunset positions.[2] For example, the passage grave constructed at Newgrange in County Meath, Ireland, about 3200 BCE is aligned so that on the three or four days surrounding the winter solstice—and just on those days—the rays of the rising sun shine through a slit in the roof and illuminate the rear of the structure (Fig. 5.2). In other constructions, an alignment between stones or between a stone and a prominent natural landmark on the horizon marks precisely the directions of the solstice sunrise or sunset.

FIGURE 5.1

British stamp of Stonehenge indicating its use as an astronomical observatory

FIGURE 5.2

Irish stamp of passage grave at Newgrange illuminated on the winter solstice

In theory, one can construct a calendar based on the sunrise positions of the sun. But in most civilizations of which records exist, it was the motions of the moon that determined the important intervals within the year, the months. The moon, like the sun, rises at varying positions on the eastern horizon. Patient observations over a period of many years evidently enabled the builders of Stonehenge to mark the most northerly and southerly positions of moonrise. They also may have noted the existence of an 18.6-year cycle of the moonrise positions that could have been used to help predict lunar eclipses. Eclipses, both lunar and solar, were of great significance to ancient peoples. The ability to predict such striking phenomena and by appropriate ritual to cause the heavenly body to reappear after being "consumed" was an important function of the priestly classes.

The most prominent feature of the moon's appearance in the sky is not its position of rising, however, but its phases. All early civilizations noted the times it took for the moon to change from tiny crescent to full moon to invisibility and back to tiny crescent again, and such observations may well have been the basis of the earliest numerical markings yet found. The Egyptians and Babylonians both used the phases of the moon to establish the months of their years, but in different ways. It was easy enough to determine that the time from the appearance of the moon's crescent in the western sky just after sunset through all the phases to the next appearance of the crescent was about 29 1/2 days. Unfortunately, there is no integral multiple of 29 1/2 that equals 365, the number of days in the solar year, so there was no simple way of constructing a calendar incorporating both the moon's phases and the sun's control of the seasons. The Egyptians from a fairly early period simplified matters entirely. They employed a 12-month calendar of 30 days each with an additional 5 days tacked on at the end to give the 365-day year. By necessity, this calendar ignored the moon's cycles. In addition, since the year is in fact 365 1/4 days long, eventually even the yearly calendar was out of step with the seasons. In other words, as the Egyptian priests were well aware, the beginning of the year would in 1460 years (4 × 365) make a complete cycle through the seasons. Thus, for various religious purposes the priests did keep track of the actual lunar months. They also discovered that the annual Nile flood, that most important agricultural event that brought rich silt to the fields, always began just after the bright star Sirius first appeared in the eastern sky shortly before dawn after a period of invisibility. They were thus able to make the accurate predictions that helped to justify their power.

The calendrical situation in Mesopotamia was different. The priests there wanted to accommodate the calendar to both the sun and the moon so that given agricultural events would always occur in the same month. Hence, the months generally alternated in length between 29 and 30 days, a new month always starting with the first appearance of the crescent moon in the evening. Because 12 of these months equal 354 days, they decided to add an extra month every several years. In earliest times, this was done by decree whenever it was believed necessary, but in the middle of the eighth century BCE, the Babylonians codified the calendar into a system of 7 leap years every 19 years, each leap year consisting of 13 months. The lengths of the months were occasionally adjusted too so that in each 19-year cycle of 235 months there were 6940 days. In fact, the Babylonians were aware that the mean value for the length of the moon's cycle was equal to about 29.53 days, which is in turn equal to 6940/235. The current Jewish calendar preserves the essence of the Babylonian calendar, with some minor modifications to keep it in agreement with Jewish law.

Besides keeping track of the calendar, the Babylonians were able to make relatively accurate predictions of the recurrence of various celestial phenomena, from such simple ones as the time of sunrise and sunset to such complicated ones as the times of lunar eclipses. But they never apparently applied more than arithmetic and simple algebra to this study, nor did they develop a model to connect the various celestial phenomena. The initial creation of such a model was a product of fourth-century BCE Greece, the time of Plato's Academy.

The basic model developed at that time contained two concentric spheres, the sphere of the earth and the sphere of the stars (or the celestial sphere). The immediate evidence of our senses indicates that the earth is flat, but more sophisticated observations, including the facts that the hull of a ship sailing away disappears before the top of the mast and that the shadow of the earth on the moon during a lunar eclipse has a circular edge, convinced the Greeks of the earth's sphericity. Their sense of esthetics—that a sphere was the most perfect solid shape—added to this conviction. That the shape of the heavens should mirror the shape of the earth was also only natural.

The evidence of the senses, and some logical argument as well, further convinced the Greeks that the earth was stationary in the middle of the celestial sphere. The second part of this conclusion came from the general symmetry of the major celestial phenomena, while the first part came from the lack of any sensation of motion of the earth. The Greeks also noted that if the earth rotated on its axis once a day, its motion would of necessity be so swift that "objects not actually standing on the earth would appear to have the same motion, opposite to that of the earth; neither clouds nor other flying or thrown objects would ever be seen moving toward the east, since the earth's motion toward the east would always outrun and overtake them, so that all other objects would seem to move in the direction of the west and the rear."[3] With the earth considered immovable, the observed daily motion in the sky must be due to the rotation of the celestial sphere, to which were firmly attached the so-called **fixed stars**, grouped into patterns called **constellations**. These never change their positions with respect to each other and form the fixed background for the **wandering stars** or planets (Sidebar 5.1).

The seven wanderers—the sun, the moon, Mercury, Venus, Mars, Jupiter, and Saturn— were more loosely attached to the celestial sphere. That they were attached was obvious; in

---

SIDEBAR 5.1   *Precursors of Copernicus*

Some ancient astronomers asserted a theory contrary to the immovable, central earth theory discussed in the text. Heraclides of Pontus (c. 388–310 BCE) is credited with having the earth's rotation account for the daily motion of the heavens, while Aristarchus of Samos (c. 310–230 BCE), as reported by Archimedes, hypothesized "that the fixed stars and, the sun remain unmoved [and] that the earth revolves about the sun in the circumference of a circle, the sun lying in the middle of the orbit."[4] The chief objection to Aristarchus' theory was that it implied that the appearance of the fixed stars would change

as one viewed them from different parts of the earth's orbit. Aristarchus met this objection by further assuming that the distance to the fixed stars was so enormous that this effect would be unnoticeable. Other astronomers at the time could not bring themselves to believe that these huge distances were possible. In addition, certain thinkers charged Aristarchus with impiety for having "set in motion the hearth of the universe"[5] in order to save the appearances. Conflicts between science and religion evidently date back to ancient times.

general they participated in the daily east-to-west rotation of the celestial sphere. But they also had their own motion, usually in the opposite direction (west to east) at much slower speeds. It is these motions that the Greek astronomers (and indeed all earlier astronomers) attempted to make sense of. The Greeks were limited in their attempts at solution, however, by an overriding philosophical consideration. Namely, since the universe outside the earth was thought to be unchanging and perfect, according to Aristotle, the only movements in the heavens were the"natural" movements of these perfect bodies. Because the bodies were spherical, the natural movements were circular. Thus, the astronomers and mathematicians (usually the same people) attempted to solve Plato's problem quoted at the opening of the chapter—that is, to develop a model that would explain the phenomena in the heavens ("save the appearances")—through a combination of geometrical constructs using circular and uniform motion. It was not the business of the astronomer-mathematicians to decide if or how such motions were physically possible, for celestial physics as we know it was never a topic of study in ancient Greece. But they did in fact succeed in finding several different systems that met Plato's challenge.

Because the basic Greek model of the heavens consisted of spheres, the first element of the study of celestial motion was the study of the properties of the sphere. Recall that Euclid's *Elements* contained virtually nothing about these properties. There were, however, other texts written in the fourth century BCE on the general subject of spherics, including ones by Autolycus of Pitane (c. 300 BCE) and by Euclid himself, which did cover the basics, mostly in the context of results immediately useful in astronomy. These books contained such definitions as that of a **great circle** (a section of a sphere by a plane through its center) and its **poles** (the extremities of the diameter of the sphere perpendicular to this plane). The texts also included three important theorems that prove very useful in what follows:

1. Any two points on the sphere that are not diametrically opposite determine a unique great circle.
2. Any great circle through the poles of a second great circle is perpendicular to the original one, and, in this case, the second circle also contains the poles of the first.
3. Any two great circles bisect one another.

There are several great circles on the celestial sphere that are important for astronomy. For example, the sun's path in its west-to-east movement through the stars is a great circle. This great circle, called the **ecliptic**, passes through the 12 constellations of the zodiac (Fig. 5.3). (These constellations were first mentioned in Babylonian astronomy and appear in Greek sources as early as 300 BCE.) The diameter of the earth through the North and South poles, extended to the heavens, is the axis around which the daily rotation of the celestial sphere takes place. The great circle corresponding to the poles of that axis is called the **celestial equator**. The equator and ecliptic intersect at two diametrically opposite points, the **vernal** and **autumnal equinoxes**, for on those dates the sun is located on those intersections (Fig. 5.4). The points on the ecliptic at the maximal distance north and south of the equator are the **summer** and **winter solstices**, respectively.

Since the Greeks knew that the earth was so small that it could in effect be considered as a point with respect to the sphere of the stars, they assumed that the horizon plane passed through the center of the celestial sphere and hence that the horizon itself was also a great circle. The horizon intersects the equator at the east and west points. Finally, the **local meridian** is the great circle that passes through the north and south points of the horizon and

FIGURE 5.3

Mosaic of the zodiac on an
Israeli souvenir sheet

FIGURE 5.4

Ecliptic and equator

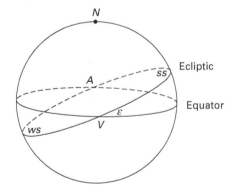

the point directly overhead, the **local zenith**. Because the meridian circle is perpendicular to both the horizon and the celestial equator, it also passes through the North and South poles of the latter. The angle ϵ between the equator and the ecliptic can be determined by taking half the distance (in degrees) between the noon altitudes of the sun at the summer and winter solstices. This value was measured to be $24°$ by the time of Euclid and was taken to be $23°51'20''$ by Ptolemy. (In fact, this value is slowly decreasing and is now about $23\frac{1}{2}°$.) The angle between the horizon and the equator is $90° - \phi$, where ϕ is the geographical latitude

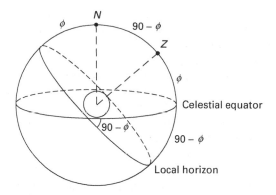

FIGURE 5.5

Horizon and equator

of the observer (Fig. 5.5). The measure of the arc between the north celestial pole and the horizon is also given by ϕ.

5.1.1 Eudoxus and Spheres

FIGURE 5.6

Spheres of Eudoxus (with earth at center) on lower left side of Liberian stamp

Eudoxus, famous for his work on ratios and the method of exhaustion, was the person largely responsible for turning astronomy into a mathematical science. He was probably the inventor of the two-sphere model as well as of the modifications necessary to account for the various motions of the sun, moon, and planets, nevertheless keeping to Plato's dictum to use only circular motion. In his scheme, each of the heavenly bodies was placed on the inner sphere of a set of two or more interconnected spheres, all centered on the earth, whose simultaneous rotation about different axes produced the observed motion (Fig. 5.6). For example, the sun requires two spheres to account for its two basic motions. The outer sphere represents the sphere of the stars; it rotates westward about its axis once in a day. The inner sphere, which contains the sun, is attached to the outer sphere so that its axis is inclined at angle ϵ to the axis of the outer sphere. If this sphere now rotates slowly eastward so that it makes a complete revolution in one year, the combination of the two motions will produce the apparent motion of the sun (Fig. 5.7). In the case of the moon, three spheres are necessary. The outer sphere again

FIGURE 5.7

Eudoxus's spheres for the sun

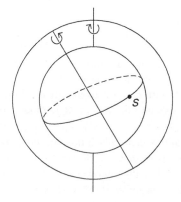

rotates westward about its axis once a day. The innermost sphere makes a complete eastward revolution in $27\frac{1}{3}$ days, the time it takes the moon to make one complete journey through the ecliptic. But since the moon deviates up to 5° from the ecliptic circle during its journey, Eudoxus postulated an intermediate sphere, inclined at angle ϵ to the outer sphere and 5° to the inner sphere, whose slow westward motion produces, at least qualitatively, the moon's north and south deviation. For the even more complicated motion of the planets, including not only their general eastward movement but also their occasional retrograde (westward) motion, Eudoxus required four spheres.[6]

In all probability, Eudoxus regarded these spheres only as a computational device rather than as objects having physical existence. And although numerical parameters could be found to permit the system to represent the various motions of the heavenly bodies, the system could not account for all of the observed phenomena. For example, the four-sphere theory of the planets did not predict the obvious changes in their brightness during their motion. Nevertheless, Aristotle took a modified version of Eudoxus's system of spheres as a physical reality, incorporating this system into his detailed cosmology. As such, it became part of the general conception of the heavens in Western civilization through the sixteenth century.

5.1.2   Apollonius: Eccenters and Epicycles

About 150 years after Eudoxus, Apollonius attempted a new answer to Plato's challenge. It had long been known that the velocity of the sun around the ecliptic was not constant. The Babylonians had already discovered this in connection with their attempts to determine, for example, the time of first visibility of the moon each month. The Greeks discovered this by determining that the seasons of the year were not equal in length; for example, the time from the vernal equinox to the summer solstice is two days longer than the time from the summer solstice to the autumnal equinox. Therefore, a simple model of the sun revolving in a circle centered on the earth at constant speed, even if the sun were attached to one of Eudoxus's spheres, could not account for this phenomenon. Because nonuniform motion would not satisfy Plato's rules, Apollonius or one of his predecessors proposed the following solution: Place the center of the sun's orbit at a point (called the **eccenter**) displaced away from the earth. Then if the sun moves uniformly around the new circle (called the **deferent circle**), an observer on earth will see more than a quarter of the circle against the spring quadrant (the upper right) than against the summer quadrant (the upper left) (Fig. 5.8a). The distance ED, or better, the ratio of ED to DS, is known as the **eccentricity** of the deferent. If line ED is extended to the deferent circle, the intersection point closest to the earth is called the **perigee** of the deferent, while the one farthest from the earth is called the **apogee**. Assuming that one can determine the correct parameters in this model (the length and direction of ED) so that the seasonal lengths come out right, the question in using the model is where the sun will be seen on a particular day. To answer this question, one needs to find angle DES. This requires solving triangle DES, which in turn requires trigonometry. In fact, it was the necessity for introducing numerical parameters into these geometric models that led to the invention of trigonometry.

Apollonius also noticed that one can replace this eccentric model by another geometric model, the epicyclic one. That is, instead of considering the sun as traveling on the eccentric circle, it may be imagined as traveling on a small circle, the **epicycle**, whose center travels on the original earth-centered circle (Fig. 5.8b). If the epicycle rotates once clockwise in the same

FIGURE 5.8

(a) Apollonius's eccenter model for the sun (b) Apollonius's epicycle for the sun

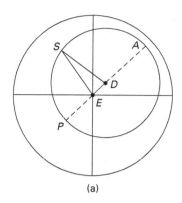

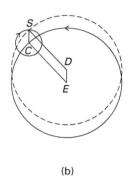

(a)   (b)

time as its center rotates around the earth—that is, if the two motions always keep $DECS$ a parallelogram—the actual path of the sun will be the same as it was using the deferent circle.

It then turns out that if one combines epicycles and eccentric circles, one can produce the more complicated motions of the planets. In fact, Apollonius initiated the study of this model. The planet P travels uniformly counterclockwise on an epicycle with center C. This latter point travels in the same direction on a deferent circle with center D at a distance DE from the earth (Fig. 5.9a). If the speeds along these circles are set appropriately, the planet as seen from the earth will in general travel eastward along the ecliptic, but during certain periods will travel in the opposite direction (when the planet is on the inner part of the epicycle) (Fig. 5.9b). To use this model, it is again necessary to find the various parameters involved, such as the lengths PC and ED and their relative directions. Once these are established for a given planet, however, the position of the planet at any time can be found by solving certain triangles.

FIGURE 5.9

(a) Motion of planet on deferent circle (b) Explanation of retrograde motion

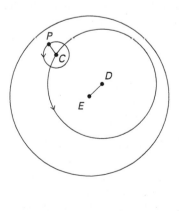

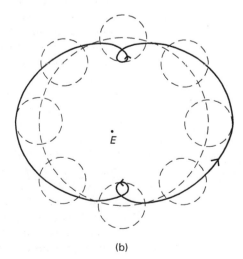

(a)   (b)

5.1.3 Hipparchus and the Beginning of Trigonometry

Apollonius himself did not possess the trigonometric machinery necessary to complete the solution of these problems. It was Hipparchus of Bithynia (190–120 BCE) who systematically carried out numerous observations of planetary positions, introduced a coordinate system for the stellar sphere, and began the tabulation of trigonometric ratios necessary to enable one to easily solve right triangles and successfully attack Apollonius's questions (Fig. 5.10).

To deal quantitatively with the positions of the stars and planets, one needs both a unit of measure for arcs and angles as well as a method of specifying where a particular body is located on the celestial sphere—that is, a system of coordinates. Euclid's unit of angle measure was simply a right angle. Other angles were referred to as parts or multiples of this angle. The Babylonians, however, sometime before 300 BCE, initiated the division of the circumference of the circle into 360 parts, called degrees, and within the next two centuries this measure, along with the sexagesimal division of degrees into minutes and seconds, was adopted in the Greek world. Hipparchus was one of the first to make use of this measure, although he also used arcs of $\frac{1}{24}$ of a circle and $\frac{1}{48}$ of a circle, so-called "steps" and "half-steps," in some of his work. Why the Babylonians divided the circle into 360 parts is not known. Perhaps it was because 360 is easily divisible by many small integers or because it is the closest "round" number to the number of days in the year. The latter reason gives us the convenient approximation that the sun travels 1° along the ecliptic each day.

It was also the Babylonians who first introduced coordinates into the sky. The system they used, later taken over by Ptolemy, is known as the ecliptic system. Positions of stars are measured both along and perpendicular to the ecliptic. The coordinate along the ecliptic (measured in degrees counterclockwise from the vernal point as seen from the North Pole) is called the **longitude** λ; the perpendicular coordinate, measured in degrees north or south of the ecliptic, is called the **latitude** β (Fig. 5.11a). This coordinate system is particularly useful when dealing with the sun, moon, and planets. The sun, since it travels along the ecliptic, always has latitude 0°. Its longitude increases daily by approximately 1° from 0° at the vernal equinox to 90° at the summer solstice, 180° at the autumnal equinox, and 270° at the winter solstice. Often, however, in both the Babylonian sources and the later Greek ones, longitudes were counted using the zodiacal signs. Namely, the ecliptic was divided into twelve intervals of 30° each, named by the zodiacal constellations. For example, Aries included longitudes from 0° to 30° and Taurus from 30° to 60°. Thus, if one noted that the sun had longitude Taurus 5°, one meant it had ecliptic longitude 35°.

FIGURE 5.10

Hipparchus on a Greek stamp

FIGURE 5.11

(a) Ecliptic coordinate system on the celestial sphere
(b) Equatorial coordinate system on the celestial sphere

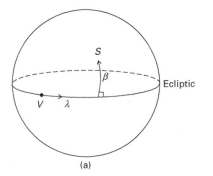

(a)

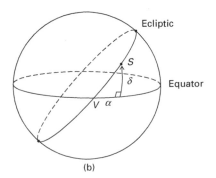

(b)

In place of this ecliptic coordinate system, Hipparchus used a system based on the celestial equator. The coordinate along the equator, also measured counterclockwise from the vernal point, is called the **right ascension** α. The perpendicular coordinate, measured north and south from the equator, is called the **declination** δ (Fig. 5.11b). Hipparchus drew up a catalogue of fixed stars in which he described some of their positions in terms of this coordinate system.

To be able to relate the coordinates of a point in one coordinate system to its coordinates in another—and this is necessary to solve astronomical problems—one needs spherical trigonometry. But before this could be developed, it was necessary to understand plane trigonometry. Hipparchus was evidently the first to attempt the detailed tabulation of lengths that would enable plane triangles to be solved. Although there are no explicit documents giving Hipparchus's table or his method, enough has been pieced together from various sources to give us a reasonable picture of his work.

The basic element in Hipparchus's (and also, later, in Ptolemy's) trigonometry was the chord subtending a given arc (or central angle) in a circle of fixed radius. Namely, both men gave a table listing α and chord(α) for various values of the arc α. Note that chord(α), henceforth abbreviated crd(α), is simply a length (Fig. 5.12). If the radius of the circle is denoted by R, then the chord is related to the sine by the equations

$$\frac{1}{2}\,\mathrm{crd}(\alpha)/R = \sin\frac{\alpha}{2} \quad \text{or} \quad \mathrm{crd}(\alpha) = 2R\sin\frac{\alpha}{2}.$$

FIGURE 5.12

Chord(α) and chord($180 - \alpha$)

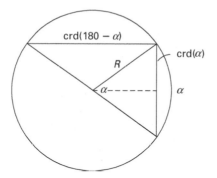

Because the angle or arc was to be measured in degrees and minutes, Hipparchus decided to use the same measure for the radius of the circle. Knowing that the circumference equaled $2\pi R$, and taking for π the sexagesimal approximation 3;8,30 (which is close to the mean between the two Archimedean values of $3\frac{10}{71}$ and $3\frac{1}{7}$), he calculated the radius R as $\frac{60 \cdot 360}{2\pi} = \frac{6,0,0}{6;17} = 57,18 = 3438'$ to the nearest integer. In a circle of this radius, the measure of an angle (defined as length cut off on the circumference divided by the radius) equals its radian measure.

To calculate a table of chords, Hipparchus began with a 60° angle. In this case, the chord equals the radius, or crd($60°$) = $3438' = 57,18$. For a 90° angle, the chord is equal to $R\sqrt{2} = 4862' = 81,2$. (Note that the mixed decimal and sexagesimal notation used here is common in both Greek and modern angle measure.) To calculate chords of other angles, Hipparchus used two geometric results. First, it is clear from Figure 5.12 that

$$\text{crd}(180 - \alpha) = \sqrt{(2R)^2 - \text{crd}^2(\alpha)}.$$

Because $\text{crd}(180 - \alpha) = 2R \cos \frac{\alpha}{2}$, this result is equivalent to $\sin^2 \alpha + \cos^2 \alpha = 1$. Second, Hipparchus calculated $\text{crd}(\frac{\alpha}{2})$ from a version of the half-angle formula. (It is conjectured that he used the method given later by Ptolemy.) Suppose $\alpha = \angle BOC$ is bisected by OD (Fig. 5.13). To express $\text{crd}(\frac{\alpha}{2}) = DC$ in terms of $\text{crd}(\alpha) = BC$, choose E on AC so that $AE = AB$. Then $\triangle ABD$ is congruent to $\triangle AED$ and $BD = DE$. Since $BD = DC$, also $DC = DE$. If DF is drawn perpendicular to EC, then $CF = \frac{1}{2}CE = \frac{1}{2}(AC - AE) = \frac{1}{2}(AC - AB) = \frac{1}{2}(2R - \text{crd}(180 - \alpha))$. But also, triangles ACD and DCF are similar, so $AC : CD = CD : CF$. Therefore,

$$\text{crd}^2 \left(\frac{\alpha}{2} \right) = CD^2 = AC \cdot CF = R(2R - \text{crd}(180 - \alpha)).$$

Putting this into modern notation gives

$$\left(2R \sin \frac{\alpha}{4} \right)^2 = R \left(2R - 2R \cos \frac{\alpha}{2} \right),$$

or, replacing α by 2α,

$$\sin^2 \frac{\alpha}{2} = \frac{1 - \cos \alpha}{2},$$

the standard half-angle formula.

FIGURE 5.13

Hipparchus-Ptolemy half-angle formula

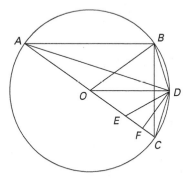

Hipparchus could now easily calculate the chord for every angle from $7\frac{1}{2}°$ to $180°$ in "half-steps" of $7\frac{1}{2}°$. For example, by applying the formula three times to $\text{crd}(60°)$, one finds $\text{crd}(7\frac{1}{2}°)$. By complements, one then finds $\text{crd}(172\frac{1}{2}°)$. This limited table enabled Hipparchus to make some progress in solving triangles and applying the results toward completing the models of the heavens. Because the actual works of Hipparchus are lost, however, it is necessary to turn to the most influential astronomical work of antiquity, the *Almagest* of Claudius Ptolemy.

FIGURE 5.14

Ptolemy (with crown and globe) (detail from Raphael's painting *The School of Athens*). The crown represents Raphael's mistaken assumption that Ptolemy was related to the rulers of Egypt

FIGURE 5.15

Woodcut from early printing of a summary of the *Almagest* (1496) (*Source*: Smithsonian Institution Libraries, Photo No. 76-14409)

PTOLEMY AND THE *ALMAGEST*

Nothing is known of the personal life of Claudius Ptolemy (c. 100–178 CE) other than that he made numerous observations of the heavens from locations near Alexandria and wrote several important books (Fig. 5.14). He is most famous today, however, for the *Mathematiki Syntaxis (Mathematical Collection)*, a work in 13 books that contained a complete mathematical description of the Greek model of the universe with parameters for the various motions of the sun, moon, and planets. The book was the culmination of Greek astronomy. Like Euclid's *Elements*, it replaced all earlier works on its subject. It was the most influential astronomical work from the time it was written until the sixteenth century, being copied and commented on countless times. More than any other book it gave impetus to the notion that one could create a mathematical model, that is, a quantitative description of natural phenomena that would yield reliable predictions. Virtually all subsequent astronomical works, both in the Islamic world and in the West, up to and including the work of Copernicus, were based on Ptolemy's masterpiece. Many centuries after it was written, it became known as the *megisti syntaxis* (the greatest collection), to distinguish it from lesser astronomical works. Islamic scientists then began to call the book *al-magisti*, and ever since it has been known as the *Almagest* (Fig. 5.15).

5.2.1 Chord Tables

Ptolemy began the *Almagest* with a basic introduction to the Greek concept of the cosmos, followed by strictly mathematical material detailing the plane and spherical trigonometry necessary for the computation of the planetary positions. The first order of business for Ptolemy was the construction of a table of chords more complete than that of Hipparchus. To construct this table of chords of all arcs from $\frac{1}{2}^\circ$ to 180° in intervals of $\frac{1}{2}^\circ$, as well as find a scheme for interpolating between the computed values, he needed somewhat more geometry than Hipparchus. Also, instead of taking $R = 57,18$, a rather difficult value to compute with, he took $R = 60$, a unit in the sexagesimal system in which all of Ptolemy's computations were made.

Ptolemy's first calculation established the chord of 36°, namely, the length of a side of a regular decagon inscribed in a circle. In Figure 5.16, ADC is the diameter of the circle with center D, BD is perpendicular to ADC, E bisects DC, and F is chosen so that $EF = EB$. By *Elements* II–6, we have $CF \cdot FD + ED^2 = BE^2$. Therefore, $CF \cdot FD = BE^2 - ED^2 = BD^2 = CE^2$, and the line CF has been divided at D in extreme and mean ratio. Recall from *Elements* XIII–9 that if the side of a hexagon and decagon inscribed in the same circle are placed together in a straight line, then the meeting point divides the entire line segment in extreme and mean ratio. Because CD, the radius, equals the side of a hexagon inscribed in the circle, Ptolemy had shown that DF is the side of a decagon; that is, $DF = \mathrm{crd}(36^\circ)$. To calculate its length, he noted that

$$DF = EF - ED = EB - ED = \sqrt{BD^2 + ED^2} - ED = \sqrt{3600 + 900} - 30 = 37;4,55.$$

FIGURE 5.16

Ptolemy's calculation of crd(36°)

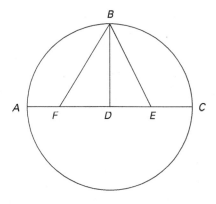

Ptolemy next noted that since the square on the side of a regular pentagon ($= \mathrm{crd}(72^\circ)$) equals the sum of the squares on the side of a regular decagon and the side of a regular hexagon (*Elements* XIII–10), it followed that

$$\mathrm{crd}(72^\circ) = \sqrt{R^2 + \mathrm{crd}^2(36^\circ)} = 70;32,3,$$

with, of course, $\mathrm{crd}(60^\circ) = 60$. Furthermore, $\mathrm{crd}(90^\circ) = \sqrt{2R^2} = \sqrt{7200} = 84;51,10$, and $\mathrm{crd}(120^\circ) = \sqrt{3R^2} = 103;55,23$. Finally, because $\mathrm{crd}^2(180 - \alpha) = (2R)^2 - \mathrm{crd}^2\,\alpha$, Ptolemy could also calculate the chord of the supplement to any arc whose chord was known. For example, $\mathrm{crd}(144°) = 114;7,37$. He was therefore well started on a chord table simply from propositions of Euclidean geometry and the ability to calculate square roots.

Ptolemy, like Archimedes four centuries earlier, never mentioned how he calculated these square roots, but merely presented the results. A commentary on Ptolemy's work by Theon in the late fourth century gave a method Ptolemy could well have used: "If we seek the square root of any number, we take first the side of the nearest square number, double it, divide the product into the remainder reduced to minutes, and subtract the square of the quotient; proceeding in this way, we reduce the remainder to seconds, divide it by twice the quotient in degrees and minutes, and we shall have the required approximation to the side of the square area."[7]

We give an example of Theon's method by calculating $\sqrt{7200}$. Note first that $84^2 = 7056$ and $85^2 = 7225$, so the answer must be of the form $84;x, y$. Since $7200 - 84^2 = 144$, we divide $144 \cdot 60$ ("the remainder reduced to minutes") by $2 \cdot 84$ and get 51 as the nearest integer. Therefore, the answer is now known to be of the form $84;51, y$. Finally, $7200 - (84;51)^2 = 0;28,39$, which, converted to seconds, is 1719. Dividing this by $2 \cdot 84;51 = 169;42$ gives 10 to the nearest integer. The desired square root approximation is thus $84;51,10$, as noted. The relative complexity of this operation, and the fact that Ptolemy simply stated the results of large numbers of such calculations, leads us to believe that Ptolemy must have had the assistance of numerous "calculators" who performed these tedious but necessary calculations. In particular, these calculators were necessary to help Ptolemy complete his chord table, using the basic values above, the half-angle formula due to Hipparchus, and a new theorem from which certain sum and difference formulas could be derived:

PTOLEMY'S THEOREM *Given any quadrilateral inscribed in a circle, the product of the diagonals equals the sum of the products of the opposite sides.*

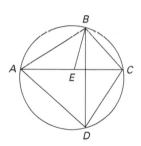

FIGURE 5.17

Ptolemy's Theorem

To prove that $AC \cdot BD = AB \cdot CD + AD \cdot BC$ in quadrilateral $ABCD$, choose E on AC so that $\angle ABE = \angle DBC$ (Fig. 5.17). Then $\angle ABD = \angle EBC$. Also $\angle BDA = \angle BCA$ since they both subtend the same arc. Therefore, $\triangle ABD$ is similar to $\triangle EBC$. Hence, $BD : AD = BC : EC$ or $AD \cdot BC = BD \cdot EC$. Similarly, since $\angle BAC = \angle BDC$, $\triangle ABE$ is similar to $\triangle DBC$. Hence, $AB : AE = BD : CD$ or $AB \cdot CD = BD \cdot AE$. Adding equals to equals gives $AB \cdot CD + AD \cdot BC = BD \cdot AE + BD \cdot EC = BD(AE + EC) = BD \cdot AC$, and the theorem is proved.

To derive a formula for the chord of a difference of two arcs α, β, Ptolemy used the theorem with $AC = \mathrm{crd}\,\alpha$ and $AB = \mathrm{crd}\,\beta$ given. Applying the result to quadrilateral $ABCD$ gives $AB \cdot CD + AD \cdot BC = AC \cdot BD$ (Fig. 5.18). Because $BC = \mathrm{crd}(\alpha - \beta)$,

$$120\,\mathrm{crd}(\alpha - \beta) = \mathrm{crd}\,\alpha \cdot \mathrm{crd}(180 - \beta) - \mathrm{crd}\,\beta \cdot \mathrm{crd}(180 - \alpha).$$

This is easily translated into the modern difference formula for the sine:

$$\sin(\alpha - \beta) = \sin\alpha\cos\beta - \cos\alpha\sin\beta.$$

A similar argument shows that

$$120\,\mathrm{crd}(180 - (\alpha + \beta)) = \mathrm{crd}(180 - \alpha)\,\mathrm{crd}(180 - \beta) - \mathrm{crd}\,\beta \cdot \mathrm{crd}\,\alpha,$$

FIGURE 5.18

The difference formula for chords

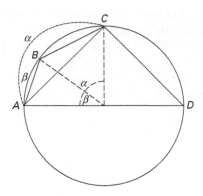

a formula equivalent to the sum formula for the cosine:

$$\cos(\alpha + \beta) = \cos \alpha \cos \beta - \sin \alpha \sin \beta.$$

Using the difference formula and the half-angle formula, Ptolemy then calculated $\mathrm{crd}(12°) = \mathrm{crd}(72° - 60°)$, $\mathrm{crd}(6°) = \mathrm{crd}(\frac{1}{2} \cdot 12°)$, $\mathrm{crd}(3°)$, $\mathrm{crd}(1\frac{1}{2}°)$, and $\mathrm{crd}(\frac{3}{4}°)$. His last two results are $\mathrm{crd}(1\frac{1}{2}°) = 1;34,15$, and $\mathrm{crd}(\frac{3}{4}°) = 0;47,8$. He could have used the addition formula to build up the table in intervals of $1\frac{1}{2}°$ or even $\frac{3}{4}°$. Because, however, he wanted his table to be in intervals of $\frac{1}{2}°$, and since, "if a chord such as the chord of $1\frac{1}{2}°$ is given, the chord corresponding to an arc which is one-third of the previous one cannot be found by geometrical methods (if this were possible we should immediately have the chord of $\frac{1}{2}°$)," Ptolemy could only find $\mathrm{crd}(1°)$ and $\mathrm{crd}(\frac{1}{2}°)$ by a procedure that, although "it cannot in general exactly determine the size [of chords], in the case of such very small quantities can determine them with a negligibly small error."[8] In other words, Ptolemy was convinced, although he offered no proof, that Euclidean tools ("geometrical methods") are not sufficient to determine $\mathrm{crd}(\frac{1}{2}°)$, or, in general, to trisect an angle. An alternative method was therefore necessary.

This alternative, an approximation procedure, is based on the lemma that if $\alpha < \beta$, then $\mathrm{crd}\,\beta : \mathrm{crd}\,\alpha < \beta : \alpha$, or, in modern notation, that $\frac{\sin x}{x}$ increases as x approaches 0. Applying this lemma first to $\alpha = \frac{3}{4}°$ and $\beta = 1°$, Ptolemy found $\mathrm{crd}(1°) < \frac{4}{3}\mathrm{crd}(\frac{3}{4}°) = \frac{4}{3}(0;47,8) = 1;2,50,40$. Applying it next to $\alpha = 1°$ and $\beta = 1\frac{1}{2}°$, he found $\mathrm{crd}(1°) > \frac{2}{3}\mathrm{crd}(1\frac{1}{2}°) = \frac{2}{3}(1;34,15) = 1;2,50$. Since all calculated values were rounded off to two sexagesimal places, it appears to that number of places that $\mathrm{crd}(1°) = 1;2,50$, and therefore $\mathrm{crd}(\frac{1}{2}°) = 0;31,25$. The addition formula now enabled Ptolemy to build up his table in steps of $\frac{1}{2}°$ from $\mathrm{crd}(\frac{1}{2}°)$ to $\mathrm{crd}(180°)$. To aid in interpolation for calculating chords of any number of minutes, he appended a third column to his table containing one-thirtieth of the increase from $\mathrm{crd}\,\alpha$ to $\mathrm{crd}(\alpha + \frac{1}{2}°)$. A small portion of the table, whose accuracy is roughly equivalent to that of a modern five-decimal-place table, is illustrated in Table 5.1.

TABLE 5.1   *A portion of Ptolemy's chord table.*

Arcs	Chords	Sixtieths	Arcs	Chords	Sixtieths
$\frac{1}{2}$	0;31,25	0;1,2,50	6	6;16,49	0;1,2,44
1	1;2,50	0;1,2,50	47	47;51,0	0;0,57,34
$1\frac{1}{2}$	1;34,15	0;1,2,50	49	49;45,48	0;0,57,7
2	2;5,40	0;1,2,50	72	70;32,3	0;0,50,45
$2\frac{1}{2}$	2;37,4	0;1,2,48	80	77;8,5	0;0,48,3
3	3;8,28	0;1,2,48	108	97;4,56	0;0,36,50
4	4;11,16	0;1,2,47	120	103;55,23	0;0,31,18
$4\frac{1}{2}$	4;42,40	0;1,2,47	133	110;2,50	0;0,24,56

5.2.2 Solving Plane Triangles

Given his chord table, Ptolemy could now solve plane triangles. Although he never stated a systematic procedure for doing so, he did seem to apply fixed rules. One difference to keep in mind when comparing Ptolemy's method to a modern one is that Ptolemy's table contains lengths of chords when the radius is 60 rather than ratios. Therefore, he always had to adjust his tabular values in a given problem to the actual length of the radius. We consider here three examples of his procedures.

First, to calculate the length CF of the noon shadow of a pole CE of length 60 at Rhodes (latitude 36°) at the vernal equinox, Ptolemy began by noting that at that time the sun is 36° below the zenith (that is, $\angle AEB = 36°$) (Fig. 5.19). Ptolemy considered CF as the chord of the circle circumscribing triangle ECF. Because the angle at the center is double the angle at the circumference, $CF = \text{crd}(72°) = 70;32,3$. Then $CE = \text{crd}(180° - 72°) = \text{crd}(108°) = 97;4,56$. Since Ptolemy wanted the shadow when $CE = 60$, he reduced this calculated value by the ratio $\frac{60}{97;4,56}$. Thus, the desired shadow is $\frac{60}{97;4,56} \cdot (70;32,3) = 43;36$. This calculation

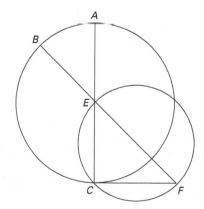

FIGURE 5.19

Calculating the length of a shadow

for finding the leg a of a right triangle, given α and b, can be rewritten as

$$a = b \cdot \frac{\text{crd}(2\alpha)}{\text{crd}(180 - 2\alpha)} = b \cdot \frac{2R \sin \alpha}{2R \cos \alpha} = b \tan \alpha$$

in agreement with modern procedure. It is Ptolemy's lack of a tangent function and his need to use actual chords in circles that forced him to calculate the chords of double both the given angle and its complement as well as their quotient.

A second example shows how Ptolemy calculated the parameters for the eccentric model of the sun.[9] The calculation amounts to solving the right triangle LDE, where D represents the center of the sun's orbit and E represents the earth (Fig. 5.20; compare Figure 5.8a). Divide the ecliptic into four quadrants by perpendicular lines through E and similarly divide the eccentric circle. To find LD and LE, one must first calculate the arcs $\theta = \frac{1}{2}\widehat{VV'}$ and $\tau = \frac{1}{2}\widehat{WW'}$ using the known inequalities of the seasons. Given that the spring path of the sun is 94.5 days while that of the summer is 92.5 days, and supposing that v is the mean daily angular velocity of the sun, the diagram shows that $90 + \theta + \tau = 94.5v$ for the spring while $90 + \theta - \tau = 92.5v$ for the summer. Because v equals the length of the year (observed to be 365;14,48 days) divided by 360°, or 0°59′8″ per day, it follows that $90° + \theta + \tau = 93°9′$ while $90° + \theta - \tau = 91°11′$. A simple calculation then shows that $\theta = 2°10′$ and $\tau = 0°59′$.

FIGURE 5.20

Calculating the parameters in
the eccentric model of the sun

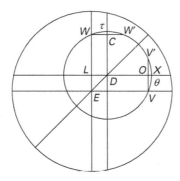

The sides of the triangle DLE can now be determined under the assumption that the radius DX of the deferent is 60. Since DX bisects arc VV', it is evident that $LE = OV = \frac{1}{2}VV' = \frac{1}{2}$ crd $2\theta = \frac{1}{2}$ crd($4°20′$) = 2;16. Similarly, $DL = \frac{1}{2}$ crd $2\tau = \frac{1}{2}$ crd($1°58′$) = 1;2. By the Pythagorean Theorem, $DE^2 = LE^2 + DL^2 = 6;12,20$, and $DE = 2;29,30$, or, approximately, $2;30 = 2\frac{1}{2}$. In modern terminology, Ptolemy has simply calculated $LE = OV = R \sin \theta$ and $DL = C\widehat{W} = R \sin \tau$. The necessity of calculating half the chord of double the angle so often led later astronomers to tabulate this quantity, the modern sine function.

To complete the solution of the triangle, Ptolemy calculated $\angle LED$ by circumscribing a circle around $\triangle LDE$. Since $LD = 1;2$ when $DE = 2;29,30$, it would be 49;46 if DE were 120. Using the table of chords in reverse, Ptolemy read off that the corresponding arc is about 49°, hence $\angle LED$ is half of that, or 24°30′. Then $\angle LDE = 65°30′$ and the triangle is solved. Again, in modern terminology, Ptolemy first calculated $120a/c = 2R \sin \alpha$ or $\sin \alpha = a/c$ and then used the inverse sine relation to determine α.

A final example is provided by Ptolemy's solution of an oblique triangle. The problem here is to find the direction $\angle DES$ of the sun, from the eccentric model, given that $DE = 2;30$ if DS is arbitrarily picked to be 60 (Fig. 5.21). For a given day, the angle PDS is known from the speed of the sun in its orbit and hence the angle EDS is known. Ptolemy made the calculation where $\angle PDS = 30°$ and $\angle EDS = 150°$. Ptolemy first constructed the perpendicular EK to SD extended. Considering as before the circle about triangle DKE, he concluded that arc $DK = 120°$. From the table he noted that if the radius were 60 (or $DE = 120$), then DK would be $\mathrm{crd}(120°) = 103;55$. Since, however, $DE = 2;30$, by proportionality $DK = 2;10$. Then $SK = SD + DK = 62;10$. Since $\angle KDE = 30°$, also $EK = \frac{1}{2}DE = 1;15$. Applying the Pythagorean Theorem to $\triangle SKE$ gives $SE = 62;11$. Next, consider the circle circumscribing $\triangle SKE$. Because $KE = 1;15$ when $SE = 62;11$, it would be $2;25$ if SE were 120. The chord table is now used in reverse to find that $2;25$ corresponds to an arc of $2°18'$. It follows that $\angle KSE = 1°9'$ and therefore that $\angle DES$ is $180° - 150° - 1°9' = 28°51'$.

FIGURE 5.21

Finding the position of the sun

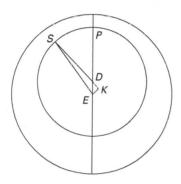

Ptolemy's procedure can be translated as follows. Given $\triangle ABC$ with a, b, and $\gamma > 90°$ known, drop AD perpendicular to BC extended (Fig. 5.22). If $AD = h$ and $CD = p$, then $p = \frac{\mathrm{crd}(2\gamma - 180) \cdot b}{2R}$ and $h = \frac{\mathrm{crd}(360 - 2\gamma) \cdot b}{2R}$. It follows that

$$c^2 = h^2 + (a + p)^2$$

$$= a^2 + \left(\frac{\mathrm{crd}^2(360 - 2\gamma)}{4R^2} + \frac{\mathrm{crd}^2(2\gamma - 180)}{4R^2} \right) b^2 + \frac{2ab\,\mathrm{crd}(2\gamma - 180)}{2R}$$

$$= a^2 + b^2 + 2ab\frac{\mathrm{crd}(2\gamma - 180)}{2R}$$

or

$$c^2 = a^2 + b^2 - 2ab \cos \gamma,$$

precisely the law of cosines for the case where two sides and the included angle are known. To find the angles, Ptolemy then noted that $\mathrm{crd}(2\beta) = h \cdot 2R/c$ and found β from the table. This translates as $\sin \beta = h/c = (b \sin \gamma)/c$. Hence, Ptolemy has also used the equivalent of the law of sines.

It should be noted that in giving the preceding example Ptolemy explicitly provided an algorithm for calculating c and β given values of a, b, and γ. In fact, such algorithms are

FIGURE 5.22

Ptolemy's law of cosines

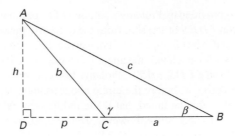

common in the *Almagest*. These algorithms of plane trigonometry can therefore be translated into modern formulas without doing injustice to Ptolemy's own procedure.

5.2.3   Solving Spherical Triangles

Ptolemy dealt even more extensively with algorithms for solving spherical triangles. Although spherical geometry had been studied as early as 300 BCE, the earliest work on spherical trigonometry appears to be the *Spherica* of Menelaus (c. 100 CE). A major result of that work, today known as Menelaus's theorem, gives the relationships among the arcs of great circles in the configuration on a spherical surface, illustrated in Figure 5.23. Two arcs AB, AC, are cut by two other arcs BE, CD, which intersect at F. With the arcs labeled as in the figure, and further with $AB = m$, $AC = n$, $CD = s$, and $BE = r$, Menelaus's theorem, written using sines rather than chords, states that

$$\frac{\sin(n_2)}{\sin(n_1)} = \frac{\sin(s_2)}{\sin(s_1)} \cdot \frac{\sin(m_2)}{\sin(m)} \tag{5.1}$$

and

$$\frac{\sin(n)}{\sin(n_1)} = \frac{\sin(s)}{\sin(s_1)} \cdot \frac{\sin(r_2)}{\sin(r)}. \tag{5.2}$$

Menelaus proved these results (and the same proof also appears in the *Almagest*) by first proving them for a similar plane configuration and then projecting the spherical diagram onto a plane.[10] Ptolemy then used Menelaus's theorem to solve spherical right triangles, triangles composed of arcs of great circles where two of the arcs meet in a right angle. Given such a triangle with the right angle at C, and the sides opposite angles C, B, A, labeled c, b, and

FIGURE 5.23

A Menelaus configuration

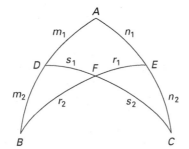

FIGURE 5.24

Ptolemy's double Menelaus configuration

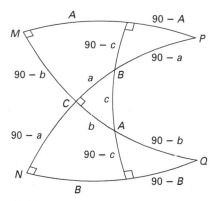

a, respectively (Fig. 5.24), Ptolemy constructed a Menelaus configuration containing it. For example, if ABC is the right triangle, construct the great circles PM, QN, which have A, B, respectively, as poles, and extend each side of the triangle to meet both of those great circles. There are then two Menelaus configurations, one with vertex at M, the other with vertex at N. Since the length of an arc on a great circle subtended by an angle at a pole of that circle is equal to the degree measure of the angle, and since P and Q are poles of QM, PN, respectively, the two equations can be simplified considerably to get results relating the angles and sides of the given triangle.

First, if one uses the configuration with vertex M, Equation 5.1 becomes

$$\frac{\sin(90 - A)}{\sin A} = \frac{\sin(90 - a)}{\sin a} \cdot \frac{\sin b}{\sin 90} \quad \text{or} \quad \tan A = \frac{\tan a}{\sin b}. \qquad (5.3)$$

Equation 5.2 becomes

$$\frac{\sin 90}{\sin A} = \frac{\sin 90}{\sin a} \cdot \frac{\sin c}{\sin 90} \quad \text{or} \quad \sin A = \frac{\sin a}{\sin c}. \qquad (5.4)$$

Second, if one uses the configuration with vertex N, Equation 5.1 becomes

$$\frac{\sin a}{\sin(90 - a)} = \frac{\sin c}{\sin(90 - c)} \cdot \frac{\sin(90 - B)}{\sin 90} \quad \text{or} \quad \cos B = \frac{\tan a}{\tan c}, \qquad (5.5)$$

while Equation 5.2 becomes

$$\frac{\sin 90}{\sin(90 - a)} = \frac{\sin 90}{\sin(90 - c)} \cdot \frac{\sin(90 - b)}{\sin 90} \quad \text{or} \quad \cos c = \cos a \cdot \cos b. \qquad (5.6)$$

Ptolemy's first application of these results was to find the declination δ and right ascension α of the sun, given its longitude λ (Fig. 5.25). Here, VA is the equator, VB the ecliptic, and V the vernal point. The angle ϵ between the equator and ecliptic, according to Ptolemy, is $23°51'20''$. Suppose the sun is at H, a point with longitude λ. To determine $HC = \delta$ and $VC = \alpha$, the right triangle VHC must be solved. From Equation 5.4, $\sin \epsilon = \sin \delta / \sin \lambda$ or $\sin \delta = \sin \epsilon \sin \lambda$. Ptolemy performed this calculation with both $\lambda = 30°$ and $\lambda = 60°$ to get in the first case, $\delta = 11°40'$, and in the second, $\delta = 20°30'9''$. Having thus demonstrated the algorithm, he presumably set his calculators to work to produce a table for δ, given

FIGURE 5.25

Method for determining the declination and right ascension of the sun, given its longitude

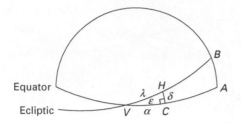

each integral value of λ from 1° to 90°. Similarly, from Equation 5.5, $\cos \epsilon = \tan \alpha / \tan \lambda$ or $\tan \alpha = \cos \epsilon \tan \lambda$. Again, Ptolemy calculated the value of α corresponding to $\lambda = 30°$ to be 27°50′, while that corresponding to $\lambda = 60°$ to be 57°44′. He then listed the values of α corresponding to other values of λ. Note further that, by symmetry, $\alpha(\lambda + 180) = \alpha(\lambda) + 180°$ and $\delta(\lambda + 180) = -\delta(\lambda)$.

Many of the other problems solved by Ptolemy are closely related to the determination of the **rising time** of an arc of the ecliptic. Namely, at a given geographical latitude, Ptolemy wanted to determine the arc of the celestial equator, which crosses the horizon at the same time as a given arc of the ecliptic. Since it is sufficient to determine this for arcs one endpoint of which is the vernal point, it is only necessary to determine the length VE of the equator, which crosses the horizon simultaneously with the given arc VH of the ecliptic (Fig. 5.26). This arc length is called the *rising time* because time is measured by the uniform motion of the equator around its axis. One complete revolution takes 24 hours, so 15° along the equator corresponds to 1 hour, and 1° corresponds to 4 minutes. In any case, to solve Ptolemy's problem, it suffices to solve the triangle HCE for $EC = \sigma(\lambda, \phi)$ and then subtract that value from $VC = \alpha(\lambda)$ already determined. For example, suppose that the latitude $\phi = 36°$ and that $\lambda = 30°$. By the calculation above, $\delta = 11°40′$. Equation 5.3 then gives $\sin \sigma = \tan \delta / \tan(90 - \phi) = \tan \delta \tan \phi$ and, therefore, $\sigma = 8°38′$. Since $\alpha = 27°50′$, the rising time $VE = 27°50′ - 8°38′ = 19°12′$. Ptolemy (or his staff) calculated the rising time $\rho(\lambda, \phi)$ for values of λ in 10° intervals from 10° to 360° at eleven different latitudes ϕ and presented the results in an extensive table.

This table can now be used to calculate the length of daylight $L(\lambda, \phi)$ at any date at any given latitude. If the sun is at longitude λ, the point at longitude $\lambda + 180$ is rising when the sun is setting. Hence, one simply needs to subtract the rising time of λ from that of $\lambda + 180$.

FIGURE 5.26

Calculating the rising time

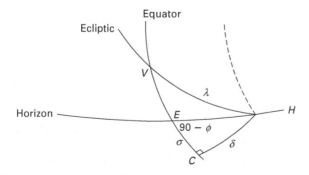

We can simplify matters somewhat by noting that, since $\sigma(\lambda + 180, \phi) = -\sigma(\lambda, \phi)$, we have

$$L(\lambda, \phi) = \rho(\lambda + 180, \phi) - \rho(\lambda)$$
$$= \alpha(\lambda + 180) - \sigma(\lambda + 180, \phi) - \alpha(\lambda) + \sigma(\lambda, \phi)$$
$$= 180° + 2\sigma(\lambda, \phi).$$

For example, when $\phi = 36°$ and $\lambda = 30°$, $L(30, 36) = 180° + 2\sigma(30, 36) = 180° + 17°16' = 197°16'$, which corresponds to approximately 13 hours, 9 minutes.

By use of Figure 5.26, we can also calculate the position of the sun when it rises, that is, the length of arc $EH = \beta$. To determine this at latitude 36° when $\lambda = 30°$, one uses Equation 5.4 to get

$$\sin \beta = \frac{\sin \delta}{\sin(90 - \phi)} = \frac{\sin 11°40'}{\sin 54°} = 0.25$$

and $\beta = 14°28'30''$. Therefore, on the day when $\lambda = 30°$, the sun will rise at 5:25 a.m. local time at a point $14°28'30''$ north of the east point on the horizon.

As a final calculation, we determine the distance of the sun from the zenith at noon. The sun on any given day is always at a distance δ from the equator. Hence at noon, when it crosses the meridian, it is (assuming $\delta > 0$) between the North Pole N and the intersection T of the meridian with the equator at a distance δ from that intersection (Fig. 5.27). Because arc $NT = 90°$ and arc $NY = \phi$, it follows that arc $SZ = 90° - (90° - \phi) - \delta = \phi - \delta$. Note that if $\phi - \delta > 0$, or $\phi > \delta$, the sun will be in the south at noon and hence shadows will point north. Because the maximum value of δ is $23°51'20''$, this will always be the case for latitudes greater than that value. On the other hand, when $\phi = \delta$, the sun is directly overhead at noon. The dates on which that occurs and also the dates when the sun is in the north at noon can easily be calculated for a given latitude. In any case, given the angular distance of the sun from the zenith, Ptolemy was able to calculate shadow lengths as previously described. He presented his results in a long table. For 39 different parallels of latitude, he gave the length of the longest day as well as the shadow lengths of a pole of length 60 at noon on the summer solstice, the equinoxes, and the winter solstice.

FIGURE 5.27

Calculation of the distance of the sun from the zenith

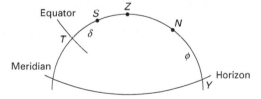

The examples above deal only with the sun and are taken from the first three books of the *Almagest*. In the remainder of the work, Ptolemy discussed the moon and the planets. For each heavenly body, he gave first a brief qualitative sketch of the phenomena to be explained, then an account of the postulated geometrical model, combined epicycles and eccenters, and finally a detailed deduction of the parameters of the model from certain observations that he had personally made or of which he had records. He generally concluded by showing that his model with the calculated parameters in fact predicted a new planetary position, which was verified by observation. Ptolemy is thus the first mathematical scientist of whom there is documented evidence of the use of mathematical models in actually "doing" science. He began with a model and then used observations to improve it to the point that it predicted observed phenomena to within the limits of his observational accuracy.

Ptolemy was proud of his accomplishments in "saving the appearances," that is, in showing that for all seven of the wandering heavenly bodies "their apparent anomalies can be represented by uniform circular motions, since these are proper to the nature of divine beings. . . . Then it is right that we should think success in such a purpose a great thing and truly the proper end of the mathematical part of theoretical philosophy. But, on many grounds, we must think that it is difficult and that there is good reason why no one before us has yet succeeded in it."[11] Ptolemy, however, overcame the difficulties and gave to posterity a masterful mathematical work that did predict the celestial phenomena, a work not superseded for 1400 years (Sidebar 5.2).

SIDEBAR 5.2   *Ptolemy and the Idea of a Function*

As a mathematical work, Ptolemy's *Almagest* raises the interesting question of whether one can see in it the germ of the modern idea of a function. First, there are many examples of tables displaying a functional relationship between sets of quantities. The Babylonians much earlier had compiled tables for square roots and reciprocals, for example, as well as astronomical ones giving the predicted time of various celestial phenomena. In general, however, they were only interested in discrete values. Ptolemy took the enormous step of providing a basis for the computational treatment of continuous phenomena by not only presenting tables but also by showing how to interpolate to provide functional values for any given value of the "independent variable." Thus, the chord is expressed as a function $\text{crd}(\alpha)$ of the arc, the declination of the sun as a function $\delta(\lambda)$ of the longitude, and the rising time $\rho(\lambda, \phi)$ as a function of the two variables representing the length of arc λ along the ecliptic and the geographical latitude ϕ. Ptolemy often used his tables in reverse as well, finding, for example, the arc from the chord, and thereby using what we would call the inverse function.

Second, however, given that Ptolemy's general aim was to predict planetary positions, in many places he wrote down an explicit algorithm describing how to do this for a particular time. For example, to calculate the sun's position at any given time, Ptolemy described the various steps required: first calculate the time t from epoch (the starting point for all calculations—February 26, 747 BCE) to the desired time; next

obtain the mean motion $\mu(t)$ from the "mean motion" table; add $\mu(t)$ to $265°15'$ and subtract multiples of $360°$ to get a value $\bar{\lambda}$ less than $360°$; enter $\bar{\lambda}$ in the table of the sun's anomaly (an entry of which was calculated in the example of Ptolemy's solution of an oblique triangle) to get $\theta(\bar{\lambda})$; and then add $\theta(\bar{\lambda})$ to $\bar{\lambda}$ and $65°30'$ to get the final result. In modern symbols, we can write this result as $p(t) = \theta(\bar{\lambda}(t)) + \bar{\lambda}(t) + 65°30'$ (mod $360°$), where $\bar{\lambda}(t) \equiv \mu(t) + 265°15'$ (mod $360°$) and where θ, μ, are themselves defined by tables derived from functional procedures. Although Ptolemy did not use modern symbolism, it is clear that he was well aware of the modern idea of a functional relationship. In many of his procedures, he even used appropriate symmetries to simplify his calculations.

Ptolemy did not, however, discuss the general notion of function. In fact, he apparently took the procedures for dealing with functions for granted. One concludes that such methods may well have been familiar to his readers and must have been used, at least by astronomers, before his time. Nevertheless, there is no evidence that any Greek mathematician wrote on the subject of functions, perhaps because there were no good theoretical methods of dealing with functions or their properties. There were no relevant postulates. It is, however, important to realize that behind the "geometrical facade of official Greek mathematics"[12] there existed areas of practical mathematics, the mathematics necessary to solve problems, both in the heavens and on earth.

5.3   PRACTICAL MATHEMATICS

By Ptolemy's time, the entire eastern Mediterranean—and much else besides—was part of the Roman Empire. The administrative center of the empire was, of course, in Rome, and the official language was Latin. Nevertheless, in much of the empire there were local rulers, and the Romans left the local language and culture intact. In particular, in the eastern Mediterranean, including what is now Egypt, Israel, Syria, mainland Greece, and Turkey, the prevalent "international" language remained Greek. And Alexandria itself remained an intellectual center, where Ptolemy, among others, found it conducive to work.

5.3.1 Roman Mathematics

Was there "Roman mathematics," or was all the mathematics accomplished under the aegis of the Roman Empire part of "Greek mathematics"? The great orator Cicero admitted that the Romans were not interested in mathematics: "The Greeks held the geometer in the highest honor; accordingly, nothing made more brilliant progress among them than mathematics. But we have established as the limit of this art its usefulness in measuring and counting."[13] But Cicero himself, as a magistrate and landowner, was certainly numerate enough to understand accounts and detect frauds. So, although it is certainly true that there was no Roman Euclid or Archimedes, in fact the Romans did have somewhat more to do with mathematics than "measuring and counting."

One person whose writings (in Latin) display a solid knowledge of mathematics is Vitruvius (first century BCE). In his famous work, *On Architecture*, he wrote that architects needed to have a comprehensive liberal education, including topics from draftsmanship to astronomy. In particular, he noted: "Geometry, in turn, offers many aids to architecture, and first among them, it hands down the technique of compass and rule, which enables the on-site layout of the plan as well as the placement of set-squares, levels, and lines. Likewise, through knowledge of optics, windows are properly designed so as to face particular regions of heaven. Through arithmetic the expenses of buildings are totaled up, and the principles of measurement are developed, the difficult issues of symmetry are resolved by geometric principles and methods."[14] But although Vitruvius recommended such knowledge for architects, *On Architecture* itself contains only a little mathematics. For example, Vitruvius showed how to determine true north. One draws a circle on a flat space on the ground and places a sundial gnomon in the center, long enough so that its shadow sometimes falls outside the circle. One then marks where the moving shadow crosses the circle both in the morning and in the afternoon. If one draws a straight line connecting the two points and then constructs the perpendicular bisector of the line, that bisector will point due north and south (Fig. 5.28). Vitruvius also discussed the problem from Plato's *Meno* of constructing a square that is double a given square and also showed that, according to the Pythagorean Theorem, one can make a set square out of rules of lengths 3, 4, and 5. But there was nothing in *On Architecture* more advanced mathematically than this.

The Roman Empire was famous for its surveyors. They laid out roads and aqueducts throughout a huge territory, many of which still survive. But an inspection of the extant surveying manuals shows that the Roman surveyors used only very elementary mathematics. Lucius Columella, a Roman gentleman farmer in the first half of the first century CE, wrote that one who deals with fields needs to be able to work out areas. So he gave basic formulas

FIGURE 5.28

Determining true north

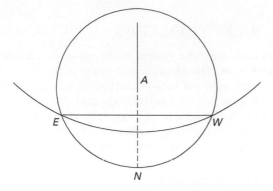

for the areas of squares, rectangles, triangles, circles, and so on, including the use of $(\frac{1}{3} + \frac{1}{10})s^2$ for the area of an equilateral triangle of side s, $\frac{22}{7}$ as an approximation for π, and $A = \frac{1}{2}(h + b)h + \frac{1}{14}(\frac{b}{2})^2$ for the area of a circle segment of base b and height h.

A manual by Marcus Junius Nipsius displayed a method for measuring the width of a river by using congruent triangles (Fig. 5.29). The distance BC is to be found. The point A is sighted in line with BC, and line AD is drawn at right angles to AC and bisected at G. Line DH is drawn perpendicular to AD to the point H from which G and C are sighted in a straight line. Then BC is equal to $DH - AB$. This is obviously quite an elementary method, but the records do not show the use of more sophisticated mathematics in surveying.

FIGURE 5.29

Finding the distance across a river

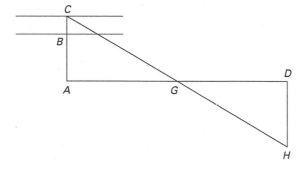

This is particularly surprising, since Greek mathematicians had developed better methods of indirect measurement. As we have seen, Hipparchus's and Ptolemy's trigonometry enabled the Greeks to "measure" triangles in the heavens as well as those on the earth related to occurrences in the heavens. And it would appear that these same methods would enable one to solve ordinary triangles on earth in order to make indirect measurements of distance and height. It would seem natural that, at least after the time of Hipparchus, the Greeks and Romans would use trigonometrical methods, that is, methods involving the table of chords. But the available historical evidence gives us no reason to believe that they did so.

Of course, before the time of Hipparchus one would only expect methods of indirect measurement coming directly from the notion of similarity. And this is exactly what is found in Euclid's *Optics*. This treatise is basically a work on the geometrical principles of vision, based on the assumption that light rays travel in straight lines. But Euclid does include several results on indirect measurement. Thus, Proposition 18 asks "to find the magnitude of a given height, the sun being visible."[15] In other words, with the sun at Γ, Euclid wanted to determine the height of a tower AB whose shadow has length $B\Delta$ (Fig. 5.30). Placing an object of known height EZ in such a way that its shadow also has tip Δ and therefore length $E\Delta$, Euclid concluded from the similarity of triangles $AB\Delta$ and $ZE\Delta$ that the height AB was determined.

FIGURE 5.30

Calculating heights using the sun, from Euclid's *Optics*

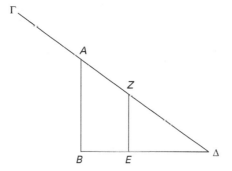

5.3.2 The Work of Heron

Some 350 years after Euclid, Heron of Alexandria (first century CE) wrote a detailed work on indirect measurement, his *Dioptra*. (The dioptra is a sighting instrument.) Heron too used similar triangles even though it appears from another of his books that he was familiar with a table of chords. Thus, Heron showed how to determine the distance from the observer (at A) to an inaccessible point B by first choosing Γ so that ΓAB is a straight line, then constructing the perpendicular ΓE to ΓAB, and finally sighting B from E, thereby establishing a point Δ on BE such that $A\Delta$ is also perpendicular to $BA\Gamma$ (Fig. 5.31). Since triangles $AB\Delta$ and ΓBE are similar, $\Gamma E : A\Delta = \Gamma B : BA$. The first ratio is known, because each length can be measured. Therefore, the second ratio is known. But $\Gamma B : BA = (\Gamma A + AB) : BA = \Gamma A : BA + 1$, and since ΓA is known, BA can be determined.

FIGURE 5.31

Calculating distance, from Heron's *Dioptra*

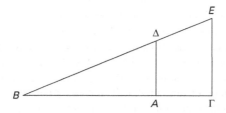

Heron used analogous methods to determine such quantities as the distance between two inaccessible points, the height of a tower (without using shadows), and the depth of a valley. He also showed how to use similar triangles to determine the direction to dig from each end in order to construct a straight tunnel through a mountain. (As we noted earlier, the tunnel on Samos was probably not constructed this way.)

Heron's many works include other significant ideas in applied mathematics. His *Catoptrica* contains an interesting proof that, for light rays impinging on a mirror, the angle of incidence equals the angle of reflection. Although the result was known earlier, Heron based his proof on the hypothesis that "Nature does nothing in vain,"[16] that is, that the path of the light ray from object C via the mirror to the eye D must be the shortest possible. Suppose A is the point on the mirror GE, which makes $\angle CAE = \angle DAG$ (Fig. 5.32). Extend DA to meet CE extended at F. It follows easily that $\triangle AEF$ is congruent to $\triangle AEC$ and therefore that the light path $DA + AC$ is equal to the straight line DAF. Now suppose B is any other point on the mirror. Connect BF, BD, and BC. Since $BF = BC$, we have $DB + BC = DB + BF > DAF$. Therefore, any other proposed light ray path is longer than the one making the angle of incidence equal the angle of reflection.

FIGURE 5.32

The angle of incidence equals
the angle of reflection

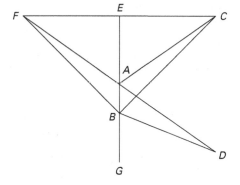

In Heron's *Mechanics*, there appears what is today called the parallelogram of velocities, although this idea too had appeared earlier in the work on mechanics attributed to Aristotle. Namely, suppose a point moves with uniform velocity along a straight line AB from A to B, while at the same time the line AB moves with uniform velocity parallel to itself, ending on the line $\Gamma\Delta$ (Fig. 5.33). Suppose EZ is any intermediate position of line AB and that G is the position of the moving point on it. Then $AE : A\Gamma = EG : EZ$ (by definition of the motion), so $AE : EG = A\Gamma : EZ = A\Gamma : \Gamma\Delta$ and G therefore lies on the diagonal $A\Delta$. In other words, the diagonal is the actual path of the moving point. In modern terms, the "velocity vector" $\vec{A\Delta}$ is the vector sum of the "velocity vectors" $\vec{AB}$ and $\vec{A\Gamma}$.

Naturally, the Greeks did not themselves consider "velocity vectors." Velocity was not considered as an independent quantity capable of being measured. There was no such concept as "miles per hour." Recall that according to *Elements* V, definition 3, ratios can be taken only between magnitudes of the same kind. One could not, therefore, consider the ratio of a distance to a time. One could only compare distances or compare times. Thus, an early definition of velocity by Autolycus states, "a point is said to be moved with equal movement

FIGURE 5.33

Parallelogram of velocities, from Heron's *Mechanics*

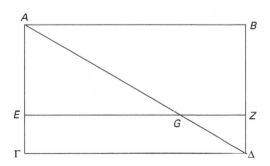

when it traverses equal and similar quantities in equal time. When any point on an arc of a circle or on a straight line traverses two lines with equal motion, the proportion of the time in which it traverses one of the two lines to the time in which it traverses the other is as the proportion of one of the two lines to the other."[17] In modern terms, Autolycus had stated that the velocity of a point is uniform when it covers equal distances in equal times, and further that if the point covers distance s_1 in time t_1 and distance s_2 in time t_2, then $s_1 : s_2 = t_1 : t_2$. It is from this definition that the initial proportions in the previous paragraph as well as those in the discussion of the quadratrix in Chapter 4 stem. Archimedes in fact discussed this matter in great detail at the beginning of his treatise *On Spirals*, because the spiral itself is defined as the locus of a point moving with uniform velocity along a line segment at the same time as the line segment revolves uniformly about one of its endpoints.

The Greeks certainly observed that falling bodies did not move with uniform velocity. Thus, they were aware of the notion of acceleration. One of the few extant explicit comments on accelerated motion, however, is from a sixth-century CE commentary on the lost treatise *On Motion* by the physicist Strato (third century BCE). Strato asserted first of all that a falling body "completes the last stage of its motion in the shortest time" and further that it traverses "each successive space more swiftly."[18] In other words, accelerated motion implies that successive equal distances are covered in shorter and shorter times and therefore with increasing velocities. It is not clear from the brief fragment, however, whether Strato meant to imply that the velocity of a falling body was proportional to distance fallen. A third-century CE commentator on Aristotle did claim, however, that "bodies move downward more swiftly in proportion to their distance from above."[19]

Although the Greeks were familiar with the basic notions of kinematics, there is no evidence that they performed numerical calculations using them, as was done in the field of astronomy. On the other hand, the *Metrica* of Heron is an example of a handbook of practical mensuration, a book that enabled its readers to learn how to measure areas and volumes of various types of figures. Here, Heron showed how to arrive at numerical answers, even where "irrational" quantities were involved. Heron sometimes gave proofs, but always his aim was to calculate, even though he often quoted the work of men such as Archimedes and Eudoxus in justifying his rules.

Book I of the *Metrica* gave procedures for calculating areas of plane figures and surface areas of solids. After the easy cases of the rectangle and the right and isosceles triangles, Heron dealt with finding the area of a scalene triangle whose sides are given. He presented

FIGURE 5.34

Area of a triangle from Heron

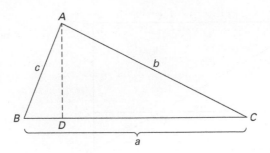

two methods. The first method is based on *Elements* II–12, II–13: Given a triangle ABC, drop the perpendicular AD to BC (or BC extended) and use the quoted theorems to show that $c^2 = a^2 + b^2 \mp 2a \cdot CD$ (Fig. 5.34). It follows that CD is known, hence that $AD = h$ is known. The area is then $\frac{1}{2}ah$.

The second method is known today as Heron's formula. Namely, if $s = \frac{1}{2}(a + b + c)$, then the area equals $\sqrt{s(s - a)(s - b)(s - c)}$. As Heron stated it, "let the sides of the triangle be 7, 8, and 9. Add together 7, 8, and 9; the result is 24. Take half of this, which gives 12. Take away 7; the remainder is 5. Again, from 12 take away 8; the remainder is 4. And again 9; the remainder is 3. Multiply 12 by 5; the result is 60. Multiply this by 4; the result is 240. Multiply this by 3; the result is 720. Take the square root of this and it will be the area of the triangle."[20]

Heron gave here a correct geometrical proof of this area result. The formula and proof, probably due originally to Archimedes, are unusual in Greek times in that they involve the product of four lengths, a completely "ungeometrical" concept. Heron made no special note of this seeming aberration, so presumably it was already present in his source. Although in the *Elements* only two or three lengths could be multiplied to give a rectangle or a rectangular parallelepiped, the practical requirements of such aspects of Greek mathematics as are discussed in this chapter led certain mathematicians to consider lengths as "numbers" and, as such, to multiply them. Naturally, this new concept violated Aristotle's basic philosophical tenets as to how mathematics should be understood. It does show again, however, that there was much going on in Greek mathematics behind its "geometrical facade."

Heron continued in this passage to show how to calculate the necessary square roots:

> Since 720 has not a rational square root, we shall make a close approximation to the root in this manner. Since the square nearest to 720 is 729, having a root 27, divide 27 into 720; the result is $26\frac{2}{3}$; add 27; the result is $53\frac{2}{3}$. Take half of this; the result is $26\frac{5}{6}$. Therefore, the square root of 720 will be very nearly $26\frac{5}{6}$. For $26\frac{5}{6}$ multiplied by itself gives $720\frac{1}{36}$; so that the difference is $\frac{1}{36}$. If we wish to make the difference less than $\frac{1}{36}$, instead of 729 we shall take the number now found, $720\frac{1}{36}$, and by the same method we shall find an approximation differing by much less than $\frac{1}{36}$.[21]

This square root algorithm is another piece of practical mathematics that is, interestingly enough, quite different from Theon's description of Ptolemy's algorithm. Perhaps Heron's method was the procedure when calculating in base ten, while Ptolemy's was the method in astronomical sexagesimal calculation. It is also quite possible that one or both of these algorithms were used by the Babylonians.

The *Metrica* also contains formulas for the area A_n of a regular polygon of n sides of length a, where n ranges from 3 to 12. For example, Heron showed that $A_3 \approx \frac{13}{30}a^2$ (the same result as Columella), $A_5 \approx \frac{5}{3}a^2$, and $A_7 \approx \frac{43}{12}a^2$. In each case he used approximations to the various square roots that appeared in the geometrical derivations. It was in his derivation of the formula for the regular 9-gon that Heron appealed to a "table of chords" in which he found that the chord of a central angle of $40°$ is equal to one-third of the diameter of the circle. Therefore, $AC^2 = 9AB^2$, $BC^2 = 8AB^2$ (Fig. 5.35), and $A_9 = 9\triangle ABO = \frac{9}{2}\triangle ABC = \frac{9}{4}BC \cdot AB = \frac{9}{4}\sqrt{8}a^2 \approx \frac{9}{4} \cdot \frac{17}{6}a^2 = \frac{51}{8}a^2$.

FIGURE 5.35

Calculating the area of a regular 9-gon

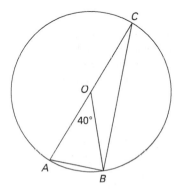

To find the area of a circle, Heron used Archimedes' value $\frac{22}{7}$ for π, thus giving the area of the circle as $\frac{11}{14}d^2$, where d is the diameter. He then quoted "the ancients" on the formula for the area of a segment of a circle, $A = \frac{1}{2}(b + h)h$, where b is the base and h is the height of the segment. A more accurate value, he said, is given by adding the extra term $\frac{1}{14}(\frac{b}{2})^2$. This new formula is certainly accurate for the semicircle, given that $\pi = \frac{22}{7}$, but is only approximate for other segments. Heron even noted that it is only "reasonably" accurate when $b \leq 3h$.

In the preface to Book II of the *Metrica*, Heron noted that the volume of a rectangular solid is the product of numbers representing the measure of its length, width, and depth, because the solid can be divided into that many unit cubes. But then he stated a more general result: If in a solid figure, all of the sections parallel to the base are equal, while the centers of the sections are on a straight line through the center of the base, either perpendicular or oblique to the base, then the volume is equal to the product of the area of the base and the perpendicular height of the figure. Heron did not justify the rule, nor could his initial explanation apply, because the solid cannot in general be divided into an integral number of unit cubes. Presumably, Heron understood that the justification could be given via an argument with indivisibles. If one takes a rectangular solid whose base equals the base of the given solid and whose height equals its height, then since each solid is "made up" of its parallel "indivisible" sections and since the sections of one solid are equal to the sections of the other, it follows that the volumes of the two solids are equal. Thus, because the volume of the rectangular solid is the product of its base and height, the same is true for the given solid. As we have noted in previous chapters, arguments by indivisibles seem to have been present in Greek mathematics for centuries, although they were never given any formal status and hence never "published."

In the remainder of Book II, Heron gave formulas for calculating the volumes of many other solid figures; for some he just quoted earlier results and for others he gave elementary arguments, though not formal proofs. Among other results here, he gave the formulas for the volume of a torus ($2\pi^2 ca^2$), where a is the radius of a circular section and c is the distance of the center of the section from the center of the torus, and for the volume of a regular octahedron ($\frac{1}{3}\sqrt{2}a^3$), where a is the edge length.

5.3.3   The *Geography* of Ptolemy

Another significant "applied mathematics" work from Alexandria during Roman times was the *Geography* of Ptolemy, a work that Ptolemy himself considered as significant as his *Almagest*. Certainly, with the Romans having conquered much of the known world, they needed maps in order to understand their domains and to know where their enemies lived. Although some mapping was accomplished by earlier mathematicians, including Eratosthenes (285–194 BCE; one of the first librarians at the Alexandria Library), Hipparchus, and Marinus of Tyre (c. 100 CE), it was Ptolemy who compiled all the information known about the position of places on the earth, combined this with some basic principles for representing the spherical earth on flat paper, and put together a work that, like his *Almagest*, became the standard reference in its field for close to a millennium and a half.

Of course, one of the issues Ptolemy had to deal with in mapping the earth was its actual size. That it was a sphere was not in question. But how large a sphere? Eratosthenes was the first to actually attempt to measure it. He noted that at noon on the summer solstice the sun was directly overhead at Syene, a place on the Tropic of Cancer, while at the same time at Alexandria, approximately 5000 stades due north, the sun was at $7\frac{1}{5}^{\circ}$ from the zenith. Given that the rays from the sun to the earth are all parallel, he concluded that $\angle SOA = 7\frac{1}{5}^{\circ}$ (Fig. 5.36). He therefore concluded that the total circumference of the earth was 250,000 stades, though it appears that at some point he modified this figure to 252,000, probably so that he could give the round figure of 700 stades per degree on the circumference. To determine the accuracy of Eratosthenes' calculation, we need to know how long a stade is in modern measures. There has been a good deal of scholarly disagreement on this point, but the general consensus today is that Eratosthenes' stade was approximately 185 meters, making the earth's circumference 46,620 km, approximately 16.5% higher than the actual value.[22]

FIGURE 5.36

Eratosthenes' determination of the size of the earth

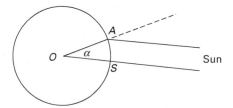

Interestingly, although it seems that Ptolemy was aware of Eratosthenes' value, in the *Geography* he used a much smaller value for the size of the earth, namely, 180,000 stades. This value, equivalent to 500 stades per degree, is approximately 17% too small. But this is the

FIGURE 5.37

Ptolemy's world map in the 1552 Basel edition of his *Geography* (*Source*: Smithsonian Insitution Libraries, Photo No. 90-15779)

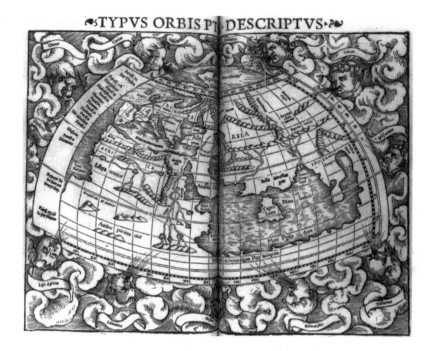

value that was passed down through the centuries and was the starting point for Columbus's own calculation of the size of the earth in the fifteenth century.

In his actual mapping, Ptolemy limited himself to what he called the *oikoumenē*, or the inhabited world. This was the region about which Ptolemy knew something, from travelers' stories and reports and from documents of earlier geographers. The northern boundary of Ptolemy's world is the parallel of Thule, near today's Shetland Islands at 63° north, while the southern boundary is at 16°25' south of the equator (Fig. 5.37). In east-west extent, Ptolemy's *oikoumenē* stretched close to 180°. The westernmost part of the inhabited world was the Islands of the Blest, identified with the Canary Islands off the coast of Spain; the meridian through those islands was Ptolemy's prime meridian, that is, the meridian from which degrees east were measured. The eastern boundary of the *oikoumenē* was somewhere on the east coast of China. A glance at a modern map shows that the actual longitudinal distance between those two points is about 135° rather than Ptolemy's 180°, thus decreasing the westward distance from Spain to China. Again, Columbus took his geographical knowledge from Ptolemy in making his case with the Spanish monarchs that he could reach China by sailing west.

But even though Ptolemy did not get the size or extent of the Eurasian continent correct, he did work out two different ways to map his *oikoumenē* on flat paper, beginning with a grid of the parallels (latitude lines) and the meridians (longitude lines). Marinus had used straight lines to represent both of these. Noting that the length of a degree along the parallel through Rhodes (latitude 36°) was in the ratio of 4 : 5 to the length of a degree along the equator (because, in modern terms, the length of the parallel through Rhodes is cos 36° times the length of the equator and cos 36° ≈ 0.8), Marinus simply spaced his parallels so that the distance between two parallels separated by n degrees was 5/4 of the distance between two meridians separated by n degrees. As Ptolemy noted, this caused distortions both to the north and to the south of the 36th parallel.

Ptolemy's first projection used straight lines for the meridians and circular arcs for the parallels. But we will look more closely at this second projection, which uses circular arcs for both sets of lines, as in Figure 5.37. About this projection, Ptolemy wrote, "we could make the map of the *oikoumenē* on the [planar] surface still more similar and similarly proportioned [to the globe] if we took the meridian lines, too, in the likeness of the meridian lines on the globe, on the hypothesis that the globe is so placed that the axis of the visual rays passed through both the intersection nearer the eye of the meridian that bisects the longitudinal dimension of the known world and the parallel that bisects its latitudinal dimension, and also the globe's center."[23] In other words, we are to imagine that we are viewing the globe from a point above the intersection of the central meridian (90° east of the Islands of the Blest, or near the east coast of the Arabian Peninsula) and the central parallel, taken to be at latitude 23°50′, the Tropic of Cancer.

To determine the center of the circular arcs of the parallels, Ptolemy described a circle $J\Gamma QV$ representing half of the globe with intersecting diameters $V\Theta\Gamma$ and $J\Theta Q$ (Fig. 5.38), and assumed that the four radii were all 90 units (representing a quarter of a circle). Since $V\Theta\Gamma$ represents the parallel at latitude $23\frac{5}{6}^{\circ}$, he placed H, a point on the equator, $23\frac{5}{6}$ units below Θ, connected VH, bisected VH at Δ, and chose the center L of his circles to be the intersection of the perpendicular bisector of VH and the extension of QJ. (Since the equator is to be a circular arc through H, V, and Γ, these two lines will both be perpendicular to that circle and thus will determine its center.) It is now straightforward to determine the length LH. We have $V\Theta = 90$, $\Theta H = 23\frac{5}{6}$, so $VH = 93\frac{1}{10}$. Because triangles $V\Theta H$ and $L\Delta H$ are similar, we know that $LH : \Delta H = VH : \Theta H$. Since we know three of these lengths, we can determine LH to be $181\frac{5}{6}$.

FIGURE 5.38

Beginning of the construction of the grid for a map projection, from Ptolemy's *Geography*

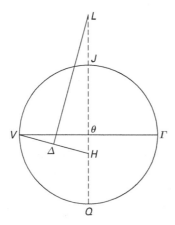

Now, Ptolemy could construct the grid for the projection. He set up a rectangle $ABDG$ with AB being twice AG, with $AE = EB$, and with EZ at right angles to AEB (Fig. 5.39). Because AB represents 180°, he considered EZ to have a length of 90 units. Since the southernmost parallel is $16\frac{5}{12}^{\circ}$ south of the equator, he put H at $16\frac{5}{12}$ units above Z. He then put Θ at $23\frac{5}{6}$ units farther north, and finally K, representing the latitude of Thule, at 63 units north of H (or $39\frac{1}{6}$ units north of Θ). He now placed L on the extension of EZ so that HL has length $181\frac{5}{6}$ units, as determined earlier. The arcs of circles centered at L with radii extending to Z, Θ, and K represent the parallels at $16\frac{5}{12}^{\circ}$ south latitude, $23\frac{5}{6}^{\circ}$ north latitude, and 63° north latitude, respectively.

FIGURE 5.39

Grid construction continued,
from Ptolemy's *Geography*

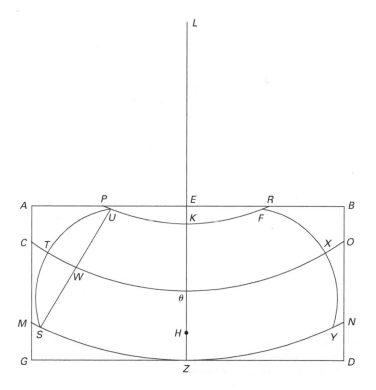

To draw the meridians, Ptolemy noted that a degree along the parallel of latitude 63° was in the ratio of approximately $2\frac{1}{4} : 5$ to a degree along the equator (because, in modern terms, $\cos 63^{\circ} \approx 0.45$). Similarly, the ratio for the parallel of latitude $23\frac{5}{6}^{\circ}$ was $4\frac{7}{12} : 5$ and the ratio for the parallel of latitude $16\frac{5}{12}^{\circ}$ was $4\frac{5}{6} : 5$. Since a chord of a five-degree arc is approximately equal to the arc itself, he then marked 18 points separated by $2\frac{1}{4}$ units in each direction from K along the arc PKR to represent five-degree intervals. The points 90° to the west and east of K are labeled U and F. Similarly, he could mark 18 points in each direction from the central line along each of the other two circular arcs, ending at T and X for the arc through Θ and at S and Y for the arc through Z. The meridians were then drawn by drawing circles through each set of three points at a given distance in degrees from the central line KZ. In particular, the arcs STU and YXF represented the bounding meridians of the *oikoumenē*. The remaining parallels were then filled in as well by circular arcs centered on L to get a complete grid upon which the map of the world could be drawn. As Ptolemy noted, this map preserved the correct ratio of latitudinal to longitudinal dimension along the three selected parallels and, at least "roughly" along the other parallels.

To complete his work, Ptolemy compiled a catalogue of about 8000 localities with their latitudes and longitudes, thus enabling actual maps to be produced, not only the map of the *oikoumenē*, but also 26 regional maps. Although much better map projections were to be discovered during the Renaissance, it was the editions of Ptolemy's maps, that began to reappear in Europe in the fourteenth century, that gave Europeans their first picture of the entire known world.

EXERCISES

1. Calculate crd(30°), crd(15°), and crd($7\frac{1}{2}$°) using the half-angle formula of Hipparchus, beginning with the fact that crd(60°) = R = 60.

2. Calculate crd(120°), crd(150°), crd(165°), and crd($172\frac{1}{2}$°) using Hipparchus's formula for crd(180° − α).

3. Use Theon's method to calculate $\sqrt{4500}$ to two sexagesimal places. The answer is 67;4,55.

4. Prove the sum formula,

$$120\,\text{crd}(180 - (\alpha + \beta))$$
$$= \text{crd}(180 - \alpha)\,\text{crd}(180 - \beta) - \text{crd}\,\alpha\,\text{crd}\,\beta,$$

   using Ptolemy's theorem on quadrilaterals inscribed in a circle.

5. Use Ptolemy's difference formula to calculate crd(12°) and then apply the half-angle formula to calculate crd(6°), crd(3°), crd($1\frac{1}{2}$°), and crd($\frac{3}{4}$°). Compare your results to Ptolemy's.

6. Compare the derivation of the half-angle formula of Hipparchus to the method used by Archimedes in Lemma 2 in *Measurement of a Circle*.

7. Prove that crd β : crd α < β : α or, equivalently, that $\frac{\sin \beta}{\sin \alpha} < \frac{\beta}{\alpha}$ for $0 < \alpha < \beta$.

8. Calculate, using Ptolemy's methods, the length of a noon shadow of a pole of length 60 at the vernal equinox at a place of latitude 40°.

9. Explain why the angle ϵ between the equator and the ecliptic can be determined by taking half the angular distance between the noon altitudes of the sun at the summer and winter solstices. (See Fig. 5.40.)

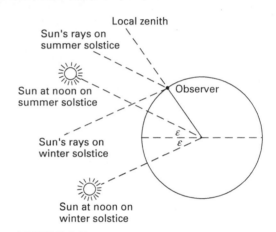

FIGURE 5.40

Calculating the inclination of the ecliptic

10. Calculate the shadow lengths at the summer and winter solstices of a pole of length 60 for latitude 36°. (Use the result of Exercise 9.)

11. Calculate the declination and right ascension of the sun when it is at longitude 90° (summer solstice) and longitude 45°. By symmetry, find the declination at longitudes 270° and 315°.

12. Calculate the rising times $\rho(\lambda, \phi)$ for $\phi = 45°$ and $\lambda = 60°$ and 90°.

13. Calculate the length of daylight on a day when $\lambda = 60°$ at latitude 36°. Calculate the local time of sunrise and sunset.

14. Suppose that the maximum length of day at a particular location is known to be 15 hours. Calculate the latitude of

that location and the position of the sun at sunrise on the summer and winter solstices.

15. The formula $\sin \sigma = \tan \delta \tan \phi$ only makes sense if the right-hand side is less than or equal to 1. Since the maximum value of δ is $23\frac{1}{2}^\circ$, show that the right-hand side will be greater than 1 whenever $\phi > 66\frac{1}{2}^\circ$. Interpret the formula in this case in terms of the length of daylight.

16. Calculate the angular distance of the sun from the zenith at latitude 45° when $\lambda = 45^\circ$ and 90°.

17. At approximately what dates is the sun directly overhead at noon at a place whose geographical latitude is 20°?

18. Calculate the sun's maximal northerly sunrise point for latitude 36°.

19. At approximately what date does the "midnight sun" begin at latitude 75°?

20. Compare the formula $A = (\frac{1}{3} + \frac{1}{10})s^2$ for the area of an equilateral triangle of side s, used by a Roman surveyor, with the exact formula. What approximation has the surveyor used for $\sqrt{3}$?

21. Show how to calculate the distance between two inaccessible points A, B, by the use of similar triangles. (Assume, for example, that the two points are on the bank of a river opposite your position.)

22. Calculate the area of a triangle with sides of lengths 4, 7, and 10 using both of Heron's methods.

23. Derive a formula for the area A_5 of a regular pentagon with side a (using plane geometry). Discuss the differences between Heron's formula $A_5 = \frac{5}{3}a^2$ and your formula.

24. Heron derived his formula for the area A_7 of a regular heptagon of side a, $A_7 = \frac{43}{12}a^2$, by assuming that $a = \frac{7}{8}r$, where r is the radius of the circumscribed circle. Use this approximation to derive Heron's result. What square root approximation is necessary here?

25. Derive $\frac{17}{6}$ as an approximation to $\sqrt{8}$ to complete the proof of Heron's formula for A_9.

26. Derive Heron's formula for the volume $\frac{1}{3}\sqrt{2}a^3$ of a regular octahedron of edge length a.

27. Check Eratosthenes' calculations on a modern map. That is, find the actual distance between Alexandria and Syene as well as the distance in degrees. (Note that Syene is not exactly on the same meridian as Alexandria.) If there were 5000 stades between Alexandria and Syene, what would be the length of a stade?

28. Show that the total length of the parallel at latitude α equals $\cos \alpha$ multiplied by the total length of the equator.

29. Confirm Ptolemy's results that the ratio of a degree along the parallel at latitude $23\frac{5}{6}^\circ$ to that of a degree along the equator is approximately $4\frac{7}{12} : 5$ and at latitude $16\frac{5}{12}^\circ$ is approximately $4\frac{5}{6} : 5$.

30. Outline a trigonometry course following Ptolemy's order of presentation. That is, begin with our modern definition of a sine and then derive the major formulas as tools for producing a sine table. Discuss the advantages and disadvantages of this approach compared to the standard textbook approach today.

31. Ptolemy must have been aware of a method of trisecting angles by the use of conic sections. Such a method would have enabled him to construct the chord of $\frac{1}{2}^\circ$ given that he knew the chord of $1\frac{1}{2}^\circ$. Why would Ptolemy not have considered this to be a construction by "geometrical methods"? Can one use such a construction to calculate the chord of $\frac{1}{2}^\circ$ numerically?

32. Discuss the potential for including some spherical trigonometry in courses on trigonometry, following the general lines of Ptolemy's approach.

33. Outline a lesson using the basic formulas of spherical trigonometry to calculate some simple astronomical phenomena.

34. What observations would have convinced the Greeks that the radius of the celestial sphere was so large that the earth could in effect be considered a point with respect to that sphere?

35. List evidence that convinces you that the earth (a) rotates on its axis once a day and (b) revolves around the sun once a year. Would this evidence have convinced the Greeks? How would you refute the reasons Ptolemy gives for the earth's immovability?

36. Look up in an astronomy work the "equation of time," and discuss why the times of sunrise and sunset calculated via the methods in the text are likely to be incorrect by several minutes.

37. "Quadratic equations were totally useless in solving problems necessary to the running of the Roman Empire." Give arguments for and against.

38. The Roman Empire in fact survived for several hundred years without apparently encouraging original mathematical research. Why do we generally believe today the opposite, that one of the factors on which the survival of the United States as a great power depends is the encouragement of original mathematical research?

REFERENCES AND NOTES

Thomas Heath's *A History of Greek Mathematics* and B. L. Van der Waerden's *Science Awakening*, referred to earlier, have sections on the material of this chapter. Selections from the works of Ptolemy and Heron can be found in Ivor Thomas, *Selections Illustrating the History of Greek Mathematics*. The standard reference on the subject of the applications of mathematics to astronomy from Babylonian times to the sixth century CE, however, is Otto Neugebauer, *A History of Ancient Mathematical Astronomy* (New York: Springer, 1975). This work provides a detailed study of the mathematical techniques used by Ptolemy and other astronomers as they worked out their versions of the system of the heavens. A more popular book is James Evans, *The History and Practice of Ancient Astronomy* (New York: Oxford University Press, 1998). The best available English translation of Ptolemy's *Almagest* is by Gerald T. Toomer: *Ptolemy's Almagest* (New York: Springer, 1984). An earlier translation by R. Catesby Taliaferro is available in the *Britannica Great Books*. The theoretical chapters of Ptolemy's *Geography* have recently been translated and annotated: J. Lennart Berggren and Alexander Jones, *Ptolemy's Geography* (Princeton: Princeton University Press, 2000). Among many books that deal with map projections, including those of Ptolemy, is John P. Snyder, *Flattening the Earth: Two Thousand Years of Map Projections* (Chicago: University of Chicago Press, 1993).

1. Simplicius's *Commentary* on Aristotle's *On the Heavens*, quoted in Pierre Duhem, *To Save the Phenomena* (Chicago: University of Chicago Press, 1969), p. 5. Duhem's work provides a detailed look at how the Greeks attempted to "save the appearances."

2. For Stonehenge astronomy, see Euan W. MacKie, *Science and Society in Prehistoric Britain* (London: Paul Elek, 1977). Also see Gerald Hawkins, *Stonehenge Decoded* (New York: Doubleday, 1965), and Fred Hoyle, *On Stonehenge* (San Francisco: Freeman, 1977).

3. Toomer, *Ptolemy's Almagest*, p. 45.

4. Thomas Heath, *The Works of Archimedes* (New York: Dover, 1953), p. 222.

5. Plutarch's *On the Face of the Moon*, in Thomas, *Selections*, II, p. 5.

6. This discussion is adapted from Thomas Kuhn, *The Copernican Revolution* (Cambridge: Harvard University Press, 1957), p. 58. This work provides excellent background reading for the nature of astronomy in Greek times. A

more recent work, especially designed for undergraduates, is Michael J. Crowe, *Theories of the World from Antiquity to the Copernican Revolution* (New York: Dover Publications, 1990). A very detailed mathematical description of the motion of the planetary spheres, with several diagrams, is found in Neugebauer, *Ancient Mathematical Astronomy*, pp. 677–685.

7. Theon's *Commentary on Ptolemy's Syntaxis*, quoted in Thomas, *Selections*, I, p. 61.

8. Toomer, *Ptolemy's Almagest*, p. 54.

9. This discussion is adapted from that of J. L. Berggren in "Mathematical Methods in Ancient Science: Astronomy," in I. Grattan-Guinness, ed., *History in Mathematics Education*, (Paris: Belin, 1987).

10. For a proof of Menelaus's theorem, see Toomer, *Ptolemy's Almagest*, pp. 64–69 or Neugebauer, *Ancient Mathematical Astronomy*, pp. 27–28.

11. Toomer, *Ptolemy's Almagest*, p. 420.

12. Olaf Pedersen, *A Survey of the Almagest* (Odense: University Press, 1974), p. 93. This work is an excellent companion to a translation of Ptolemy's work. It provides background and commentary on all of Ptolemy's mathematical and astronomical material.

13. Cicero, *Tusculan Disputations* (Cambridge: Harvard University Press, 1927), I, p. 2.

14. Vitruvius, *Ten Books on Architecture*, Ingrid Rowland and Thomas Howe, eds. (Cambridge: Cambridge University Press, 1999), p. 22.

15. Euclid, *L'Optique et la Catoptrique*, translated into French by Paul Ver Eecke (Paris: Descleé de Brouwer, 1938), p. 13.

16. Thomas, *Selections*, II, p. 497.

17. Quoted in Marshall Clagett, *The Science of Mechanics in the Middle Ages* (Madison: University of Wisconsin Press, 1961), p. 165. Although this work is chiefly concerned with medieval mechanics, there are summaries of Greek work at the beginning of most chapters.

18. Simplicius's *Commentary on Aristotle's Physics*, in Morris Cohen and I. E. Drabkin, *A Source Book in Greek Science* (Cambridge: Harvard University Press, 1948), p. 211. This book is an excellent source of original materials in Greek mathematics, astronomy, physics, and the other sciences.

19. Quoted in Marshall Clagett, *Greek Science in Antiquity* (New York: Collier, 1963), p. 92. Clagett provides here a succinct treatment of various aspects of Greek science from its beginnings through its effect on Latin science up through the early middle ages.

20. Thomas, *Selections*, II, p. 471.

21. Ibid.

22. For details on the length of the stade, see the recent article by Newlyn Walkup, "Eratosthenes and the Mystery of the Stades," *Convergence: Where Mathematics, History and Teaching Interact* (http://convergence.mathdl.org), 2006.

23. Berggren and Jones, *Ptolemy's Geography*, p. 88.

CHAPTER

6

The Final Chapters of Greek Mathematics

This tomb holds Diophantus . . . [and] tells scientifically the measure of his life. God granted him to be a boy for the sixth part of his life, and adding a twelfth part to this, He clothed his cheeks with down. He lit him the light of wedlock after a seventh part, and five years after his marriage He granted him a son. Alas! late-born wretched child; after attaining the measure of half his father's life, chill Fate took him. After consoling his grief by this science of numbers for four years, He ended his life.

—Epigram 126 of Book XIV of the *Greek Anthology* (c. 500 CE)[1]

March, 415 CE, Alexandria: "A rumor was spread among the Christians that [Hypatia], the daughter of Theon, was the only obstacle to the reconciliation of the prefect [Orestes] and the archbishop [Cyril]. On a fatal day in the holy season of Lent, Hypatia was torn from her chariot, stripped naked, dragged to the church, and inhumanly butchered by the hands of Peter the reader and a troop of savage and merciless fanatics. . . . The murder of Hypatia has imprinted an indelible stain on the character and religion of Cyril of Alexandria."[2]

Alexandria remained an important Greek mathematical center, even under the rule of Rome, beginning in 31 BCE. In Chapter 5, we discussed the work of several prominent "applied" mathematicians who flourished under Roman rule in Egypt. But there were other mathematicians in the first centuries of the common era whose "pure" mathematical works also had influence stretching into the Renaissance. This chapter deals with four of them.

We first discuss the works of Nicomachus of Gerasa, a Greek town in Judaea. He wrote in the late first century an *Introduction to Arithmetic*, based on his understanding of Pythagorean number philosophy. Besides Books VII–IX of Euclid's *Elements*, this is the only extant number theory work from Greek antiquity. However, there was another important work entitled *Arithmetica*, written by Diophantus of Alexandria in the mid-third century, which was destined to be of far more importance than Nicomachus's book. Despite its title, this was a work in algebra, consisting mostly of an organized collection of problems translatable into what are today called indeterminate equations, all to be solved in rational numbers. Like Heron's *Metrica*, the style of the *Arithmetica* is that of an Egyptian or Babylonian problem text rather than a classic Greek geometrical work. The third mathematician to be considered is also from Alexandria, the geometer Pappus of the early fourth century. He is best known not for his original work, but for his commentaries on various aspects of Greek mathematics and in particular for his discussion of the Greek method of geometric analysis. The chapter concludes with a brief discussion of the work of Hypatia, the first woman mathematician of whom any details are known. It was her death at the hands of an enraged mob that marked the effective end of the Greek mathematical tradition in Alexandria.

 6.1

NICOMACHUS AND ELEMENTARY NUMBER THEORY

Almost nothing is known about the life of Nicomachus, but since his work is suffused with Pythagorean ideas, it is likely that he studied in Alexandria, the center of mathematical activity and of neo-Pythagorean philosophy. Two of his works survive, the *Introduction to Arithmetic* and the *Introduction to Harmonics*. From other sources it appears that he also wrote introductions to geometry and astronomy, thereby completing a series on Plato's basic curriculum, the so-called quadrivium.

Nicomachus's *Introduction to Arithmetic* was probably one of several works written over the years to explain Pythagorean number philosophy, but it is the only one still extant. Since no text exists from the time of Pythagoras, it is the source of some of the ideas about Pythagorean number theory already discussed in Chapter 2. Because the work was written some 600 years after Pythagoras, however, it must be considered in the context of its time and compared with the only other treatise on number theory available, Books VII–IX of Euclid's *Elements*.

Nicomachus began this brief work, written in two books, with a philosophical introduction. Like Euclid, he followed the Aristotelian separation of the continuous "magnitude" from the discontinuous "multitude." Like Aristotle, he noted that the latter is infinite by increasing indefinitely, while the former is infinite by division. Continuing the distinction in terms of the four elements of the quadrivium, he distinguished arithmetic and music, which deal with the discrete (the former absolutely, the latter relatively), from geometry and astronomy, which deal with the continuous (the former at rest and the latter in motion). Of these four subjects, the one that must be learned first is arithmetic, "not solely because . . . it existed before all the others in the mind of the creating God like some universal and exemplary plan, relying upon which as a design and archetypal example the creator of the universe sets in order his material creations and makes them attain to their proper ends, but also because it is naturally prior in birth inasmuch as it abolishes other sciences with itself, but is not abolished together with them."[3] In other words, arithmetic is necessary for each of the other three subjects.

Most of Book I of Nicomachus's *Arithmetic* is devoted to the classification of integers and their relations. For example, the author divided the even integers into three classes, the even times even (those that are powers of two), the even times odd (those that are doubles of odd numbers), and the odd times even (all the others). The odd numbers are divided into the primes and the composites. Nicomachus took what appears to us as an inordinate amount of space discussing these classes and showing how the various members are formed. But it must be remembered that he was writing an introduction for beginners, not a text for mathematicians.

Nicomachus discussed the Euclidean algorithm of repeated subtraction to find the greatest common measure of two numbers and to determine if two numbers are relatively prime. He also dealt with the perfect numbers, giving the Euclidean construction (*Elements* IX–36) and, unlike Euclid, actually calculating the first four: 6, 28, 496, and 8128. However, also unlike Euclid, Nicomachus presented no proofs. He just gave examples.

The final six chapters of the first book are devoted to an elaborate tenfold classification scheme for naming ratios of unequal numbers, a scheme that probably had its origin in early music theory. The scheme was in common use in medieval and Renaissance arithmetics and is sometimes found in early printed editions of Euclid's *Elements*. Among the classes in this scheme of naming the ratio $A : B$, which reduces to lowest terms as $a : b$, are multiple, when $a = nb$; superparticular, when $a = b + 1$; and superpartient, when $a = b + k(1 < k < b)$.

It is Book II of Nicomachus that is, however, of most interest to us, since there he discussed plane and solid numbers, again in great detail but without proofs. This material is not mentioned at all by Euclid. Nicomachus not only dealt with triangular and square numbers (see Chapter 2) but also considered pentagonal, hexagonal, and heptagonal numbers and showed how to extend this series indefinitely. For example, the pentagonal numbers are the numbers 1, 5, 12, 22, 35, 51, . . . (although Nicomachus noted here that 1 is only the side of a "potential" pentagon). Each of these numbers can be exhibited, using the dot notation of Chapter 2, as a pentagon with equal sides (Fig. 6.1). Beginning with 5, each is formed from the previous one in the sequence by adding the next number in the related sequence 4, 7, 10, ... So $5 = 1 + 4$, $12 = 5 + 7$, $22 = 12 + 10$, and so on. This is in perfect analogy to the series of triangular numbers 1, 3, 6, 10, . . . , each of which comes from the previous one by adding numbers of the sequence 2, 3, 4, . . . , and the series of squares 1, 4, 9, 16, . . . , each of which results from the previous one by adding numbers of the sequence 3, 5, 7, ... Nicomachus continued this analogy and displayed the first 10 numbers of each of the polygonal classes mentioned.

Nicomachus further explored the solid numbers. A pyramidal number, on a given polygonal base of side n, is formed by adding together the first n polygonal numbers of that shape. For example, the pyramidal numbers with triangular base are 1, $1 + 3 = 4$, $1 + 3 + 6 = 10$, $1 + 3 + 6 + 10 = 20$, . . . , while those with square base are 1, $1 + 4 = 5$, $1 + 4 + 9 = 14$, $1 + 4 + 9 + 16 = 30$, ... One can similarly construct pyramidal numbers on any polygonal base.

FIGURE 6.1

Pentagonal numbers

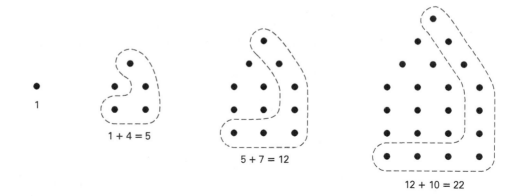

1

1 + 4 = 5

5 + 7 = 12

12 + 10 = 22

Another form of solid number is the cubic number. Nicomachus noted, again without proof, that the cubes are formed from odd numbers, not even. Thus, the first (potential) cube, 1, equals the first odd number, the second cube, 8, equals the sum of the next two odd numbers, the third cube, 27, equals the sum of the next three odd numbers, and so on. Thus, the cubes are closely related to the squares, which are also formed by adding odd numbers. And, Nicomachus concluded, these two facts show that the odd numbers, not the even, are the cause of "sameness."

The final topic of the treatise is proportion. Nicomachus, referring to pre-Euclidean terminology, used the word "proportion" in a different sense from Euclid's definition 2 of *Elements*, Book VII. For Euclid, three numbers are in proportion if the first is the same multiple (or part or parts) of the second that the second is of the third. Nicomachus noted that "the ancients" considered not only this type (the type he calls geometric), but also two others, the arithmetic and the harmonic. For Nicomachus, an **arithmetic** proportion of three terms is a series in which each consecutive pair of terms differs by the same quantity. For example, 3, 7, 11, are in arithmetic proportion. Among the properties of such a proportion are that the product of the extremes is smaller than the square of the mean by the square of the difference. In a **geometric** proportion, "the only one in the strict sense of the word to be called a proportion,"[4] the greatest term is to the next greatest as that one is to the next. For example, 3, 9, 27, are in geometric proportion. Among the properties of such a proportion is that the product of the extremes equals the square of the mean. Nicomachus quoted two results of Euclid in this regard, namely, that only one mean term lies between two squares while two lie between two cubes.

The third type of proportion among three terms, the **harmonic**, is that in which the greatest term is to the smallest as the difference between the greatest and mean terms is to the difference between the mean and the smallest terms. For example, 3, 4, 6, are in harmonic proportion because $6 : 3 = (6 - 4) : (4 - 3)$. Among the properties of this proportion is that when the extremes are added together and multiplied by the mean, the result is twice the product of the extremes. Nicomachus gave as a possible reason for the term "harmonic" that 6, 4, 3, come from the most elementary harmonies. The ratio $6 : 4 = 3 : 2$ gives the musical fifth; the ratio $4 : 3$ gives the fourth, and the ratio $6 : 3 = (4 : 3)(3 : 2) = 2 : 1$ gives the octave. Today, it is more common to use the names "arithmetic," "geometric," and "harmonic" for means rather than for proportions. Thus, 7 is the arithmetic mean of 3 and 11, 9, is the geometric mean of 3 and 27, and 4 is the harmonic mean of 3 and 6.

The *Introduction to Arithmetic* was obviously just that, a basic introduction to elementary ideas about the positive integers. Although it has some points in common with Euclid's *Elements*, it was written at a much lower level. There are no proofs at all, just a large number of examples. The book was therefore suitable for use by beginners in schools. It was in fact used extensively during ancient times, was translated into Arabic in the ninth century, and was used, in a Latin paraphrase by Boethius (c. 480–524) throughout the early Middle Ages in Europe. For these reasons, copies still exist. That it was so popular and that no more advanced work on the subject, including Euclid's *Elements*, was studied during much of the period in Europe, shows the level to which mathematical study there fell from its Greek heights. These elementary number properties were for many centuries the summit of the arithmetic curriculum.

6.2   DIOPHANTUS AND GREEK ALGEBRA

Little is known about Diophantus's life, other than what is found in the epigram at the beginning of the chapter, except that he lived in Alexandria. It is through his major work, the *Arithmetica*, that his influence has reached modern times. Diophantus wrote in his introduction that the *Arithmetica* is divided into thirteen books. Only six have survived in Greek. Four others were recently discovered in an Arabic version. From internal references it appears that these form the fourth through seventh books of the complete work, while the final three Greek books come later.[5] We will refer to the Greek books as I–VI and the Arabic ones as A, B, C, D. The style of the Arabic books is somewhat different from that of the Greek in that each step in the solution of a problem is explained more fully. It is quite possible, therefore, that the Arabic work is a translation not of Diophantus's original, but of a commentary on the *Arithmetica*, written by Hypatia around 400 CE.

Before dealing with the problems of the *Arithmetica*, it is worthwhile to discuss Diophantus's major advance in the solution of equations, his introduction of symbolism. The Egyptians and Babylonians wrote out equations and solutions in words. Diophantus, on the other hand, introduced symbolic abbreviations for the various terms involved in equations (Sidebar 6.1). And in a clear break with traditional Greek usage, he dealt with powers higher than the third.

Note that all of Diophantus's symbols are abbreviations, including the final two: ς is a contraction of the first two letters of $\alpha\rho\iota\theta\mu o\varsigma$ (*arithmos,* or number), while $\overset{\circ}{M}$ stands for $\mu o\nu\alpha\varsigma$ (*monas,* or unit). Thus, the manuscripts contain expressions such as $\triangle^{\Upsilon}\gamma\varsigma\iota\beta\overset{\circ}{M}\theta$, which stands for 3 squares, 12 numbers, and 9 units, or, as we will write it, $3x^2 + 12x + 9$. (Recall that the Greeks used an alphabetic cipher for representing numbers in which, for example, $\gamma = 3$, $\iota\beta = 12$, and $\theta = 9$.) Diophantus further used the symbols above with the mark χ to designate reciprocals. For example, $\triangle^{\Upsilon}x$ represented $\frac{1}{x^2}$. In addition, the symbol Λ, perhaps coming from an abbreviation for $\lambda\epsilon\iota\psi\iota\varsigma$ (*lepsis,* or wanting, or negation), is used for "minus," as in $K^{\Upsilon}\alpha\varsigma\gamma\Lambda\triangle^{\Upsilon}\gamma\overset{\circ}{M}\alpha$ for $x^3 - 3x^2 + 3x - 1$. (Negative terms are always collected, so a single Λ suffices for all terms following it.) In the discussion of Diophantus's problems, however, we use modern notation.

SIDEBAR 6.1 *Diophantus's Terms and Symbolism*

"All numbers are made up of some multitude of units. . . . Among them are—

squares, which are formed when any number is multiplied by itself; the number itself is called the *side of the square*;

cubes, which are formed when squares are multiplied by their sides;

square-squares, which are formed when squares are multiplied by themselves;

square-cubes, which are formed when squares are multiplied by the cubes formed from the same side;

cube-cubes, which are formed when cubes are multiplied by themselves;

and it is from the addition, subtraction, or multiplication of these numbers, or from the ratio which they bear one to another or to their own sides, that most arithmetical problems are formed; you will be able to solve them if you follow the method shown below.

"Now each of these numbers, which have been given abbreviated names, is recognized as an element in arithmetical science; the *square* [of the unknown quantity] is called *dynamis* and its sign is Δ with the index Υ, that is, Δ^{Υ}; the cube is called *kubos* and has for its sign K with the index Υ, that is, K^{Υ}; the square multiplied by itself is called *dynamo-dynamis* and its sign is two *deltas* with the index Υ, that is, $\Delta^{\Upsilon}\Delta$; the square multiplied by the cube formed from the same root is called *dynamo-kubos* and its sign is ΔK with the index Υ, that is, ΔK^{Υ}; the cube multiplied by itself is called *kubo-kubos* and its sign is two *kappas* with the index Υ, $K^{\Upsilon}K$.

"The number which has none of these characteristics, but merely has in it an undetermined multitude of units, is called *arithmos*, and its sign is ς. There is also another sign denoting the invariable element in determinate numbers, the unit, and its sign is M with the index O, that is, $\overset{\circ}{M}$." (From Thomas, *Selections*, II, pp. 519–523.)

Diophantus was also aware of the rules for multiplying with the minus: "A minus multiplied by a minus makes a plus, a minus multiplied by a plus makes a minus."[6] Of course, Diophantus was not here dealing with negative numbers, which did not exist for him. He was simply stating the rules necessary for multiplying algebraic expressions involving subtractions. But he did not explicitly state the rules for adding and subtracting with positive and negative terms, simply assuming they were known. Near the conclusion of his introduction, he stated the basic rules for solving equations:

> If a problem leads to an equation in which certain terms are equal to terms of the same species but with different coefficients, it will be necessary to subtract like from like on both sides, until one term is found equal to one term. If by chance there are on either side or on both sides any negative terms, it will be necessary to add the negative terms on both sides, until the terms on both sides are positive, and then again to subtract like from like until one term only is left on each side. This should be the object aimed at in framing the hypotheses of propositions, that is to say, to reduce the equations, if possible, until one term is left equal to one term; but I will show you later how, in the case also where two terms are left equal to one term, such a problem is solved.[7]

In other words, Diophantus's general method of solving equations was designed to lead to an equation of the form $ax^n = bx^m$, where, in the first three books at least, m and n are no greater than 2. On the other hand, he did know how to solve quadratic equations, for example, of the form $ax^2 + c = bx$.

6.2.1 Linear and Quadratic Equations

Most of Diophantus's problems are indeterminate; that is, they can be written as a set of k equations in more than k unknowns. Often there are infinitely many solutions. For these problems, Diophantus generally gave only one solution explicitly, but one can easily extend the method to give other solutions. For determinate problems, once certain quantities are made explicit, there is only one solution. Examples of both of these types are described in what follows.[8]

PROBLEM I–1 *To divide a given number into two having a given difference.*

Diophantus presented the solution for the case where the given number is 100 and the given difference is 40. If x is the smaller of the two numbers of the solution, then $2x + 40 = 100$, so $x = 30$, and the required numbers are 30 and 70. This problem is determinate, once the "given" numbers are specified, but Diophantus's method works for any pair. If a is the given number and $b < a$ the given difference, then the equation would be $2x + b = a$, and the required numbers would be $\frac{1}{2}(a - b)$ and $\frac{1}{2}(a + b)$.

PROBLEM I–5 *To divide a given number into two numbers such that given fractions (not the same) of each number when added together produce a given number.*

In modern notation, we are given a, b, r, s ($r < s$) and asked to find u, v, such that $u + v = a$, $\frac{1}{r}u + \frac{1}{s}v = b$. (Diophantus here, and usually, took his fractions to be unit fractions.) Diophantus noted that for this problem to be solvable, it is necessary that $\frac{1}{s}a < b < \frac{1}{r}a$. He then presented the solution in the case where $a = 100$, $b = 30$, $r = 3$, and $s = 5$: Let the second part (of 100) be $5x$. Therefore, the first part is $3(30 - x)$. Hence, $90 + 2x = 100$ and $x = 5$. The required parts are then 75 and 25.

Like Problem I–1, once the "given" numbers are specified, this problem is determinate, and the method works for any choice of the "givens" meeting the required condition. In the present case, Diophantus took 1/5 of the second part for his unknown. This allowed him to avoid fractions in the rest of his calculation because 1/3 of the first part must then equal $30 - x$ and the first part must be $3(30 - x)$. The remainder of the solution is clear. To check the generality, let sx represent the second part of a and $r(b - x)$ the first. The equation becomes $sx + r(b - x) = a$ or $br + (s - r)x = a$. Then $x = \frac{a - br}{s - r}$ is a perfectly general solution. Since x must be positive, $a - br > 0$ or $b < \frac{1}{r}a$, the first half of Diophantus's necessary condition. The second half, that $\frac{1}{s}a < b$, or $a < sb$, comes from the necessity that $sx < a$ or $s(\frac{a - br}{s - r}) < a$. In this particular problem, as in most of the problems in Book I, the given values are picked to ensure that the answers are integers. But in the other books, the only general condition on solutions is that they be positive rational numbers. Evidently, Diophantus began with integers merely to make these introductory problems easier. In what follows, then, the word "number" should always be interpreted as "rational number."

PROBLEM I–28 *To find two numbers such that their sum and the sum of their squares are given numbers.*

It is a necessary condition that double the sum of the squares exceeds the square of the sum by a square number. In the problem presented, the given sum is 20 and the sum of the squares is 208.

This problem is of the general form $x + y = a$, $x^2 + y^2 = b$, a type solved by the Babylonians. Three other Babylonian types appear in I–27, I–29, and I–30; namely, $x + y = a$, $xy = b$; $x + y = a$, $x^2 - y^2 = b$; and $x - y = a$, $xy = b$, respectively. As we have seen, results giving methods of solutions of these problems are also found in Euclid, Book II. Diophantus's solution to the present problem, although presented strictly algebraically, uses the same basic procedure as the Babylonians. Namely, he took as his "unknown" z half the difference between the two desired numbers. Therefore, since 10 is half the sum of the two numbers, the two numbers themselves are $x = 10 + z$ and $y = 10 - z$. The Babylonian result tells us that the sum of the squares, here 208, is twice the sum of the squares on half the sum and half the difference. In this case, then, we get $200 + 2z^2 = 208$. It follows that $z = 2$ and the required two numbers are 12 and 8. Diophantus's method, applicable to any system of the given form, can be translated into the modern formula

$$x = \frac{a}{2} + \frac{\sqrt{2b - a^2}}{2}, \qquad y = \frac{a}{2} - \frac{\sqrt{2b - a^2}}{2}.$$

His condition is then necessary to ensure that the solution is rational. Interestingly, the answers to problems I–27, I–29, and I–30 are also 12 and 8, reminding us of the common Babylonian practice of having the same answers to a series of related problems.

Did Diophantus have access to the Babylonian material? Or did he learn his methods from a careful study of Euclid's *Elements* or *Data*? These questions cannot be answered. It is, however, apparent that there is no geometric methodology in Diophantus's procedures. Perhaps by this time the Babylonian algebraic methods, stripped of their geometric origins, were known in the Greek world.

PROBLEM II–8 *To divide a given square number into two squares.*

Here we quote Diophantus exactly:

> Let it be required to divide 16 into two squares. And let the first square $= x^2$; then the other will be $16 - x^2$; it shall be required therefore to make $16 - x^2 =$ a square. I take a square of the form $(ax - 4)^2$, a being any integer and 4 the root of 16; for example, let the side be $2x - 4$, and the square itself $4x^2 + 16 - 16x$. Then $4x^2 + 16 - 16x = 16 - x^2$. Add to both sides the negative terms and take like from like. Then $5x^2 = 16x$, and $x = \frac{16}{5}$. One number will therefore be $\frac{256}{25}$, the other $\frac{144}{25}$, and their sum is $\frac{400}{25}$ or 216, and each is a square[9] [Fig. 6.2].

This is an example of an indeterminate problem. It translates into one equation in two unknowns, $x^2 + y^2 = 16$. This problem also demonstrates one of Diophantus's most common methods. In many problems from Book II onward, Diophantus required a solution, expressed in the form of a quadratic polynomial, which must be a square. To ensure a rational solution, he chose his square in the form $(ax \pm b)^2$, with a and b selected so that either the quadratic term or the constant term is eliminated from the equation. In this case, where the quadratic polynomial is $16 - x^2$, he used $b = 4$ and the negative sign, so the constant term is eliminated and the resulting solution is positive. The rest of the solution is then obvious. The method can be used to generate as many solutions as desired to $x^2 + y^2 = 16$, or, in general, to $x^2 + y^2 = b^2$. Take any value for a and set $y = ax - b$. Then $b^2 - x^2 = a^2x^2 - 2abx + b^2$ or $2abx = (a^2 + 1)x^2$, so $x = \frac{2ab}{a^2+1}$.

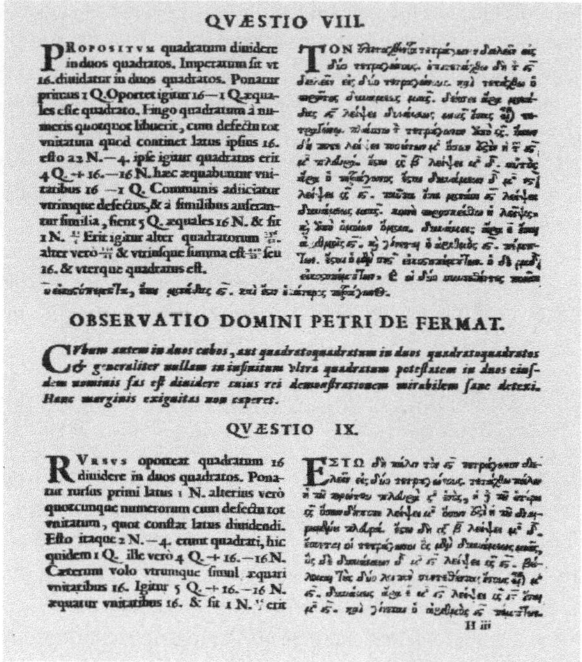

QVÆSTIO VIII.

PROPOSITVM quadratum diuidere in duos quadratos. Imperatum sit vt 16. diuidatur in duos quadratos. Ponatur primus 1 Q. Oportet igitur 16 − 1 Q. æquales esse quadrato. Fingo quadratum à numeris quotquot libuerit, cum defectu tot vnitatum quod continet latus ipsius 16. esto 2 N. − 4. ipse igitur quadratus erit 4 Q. + 16. − 16 N. hæc æquabuntur vnitatibus 16 − 1 Q. Communis adiiciatur vtrimque defectus, & à similibus auferantur similia, fient 5 Q. æquales 16 N. & fit 1 N. 16/5 Erit igitur alter quadratorum 256/25. alter verò 144/25 & vtriusque summa est 400/25 seu 16. & vterque quadratus est.

OBSERVATIO DOMINI PETRI DE FERMAT.

Cubum autem in duos cubos, aut quadratoquadratum in duos quadratoquadratos & generaliter nullam in infinitum vltra quadratum potestatem in duos eiusdem nominis fas est diuidere cuius rei demonstrationem mirabilem sane detexi. Hanc marginis exiguitas non caperet.

QVÆSTIO IX.

RVRSVS oporteat quadratum 16 diuidere in duos quadratos. Ponatur rursus primi latus 1 N. alterius verò quotcunque numerorum cum defectu tot vnitatum, quot constat latus diuidendi. Esto itaque 2 N. − 4. erunt quadrati, hic quidem 1 Q. ille verò 4 Q. + 16. − 16 N. Cæterum volo vtrumque simul æquari vnitatibus 16. Igitur 5 Q. + 16. − 16 N. æquatur vnitatibus 16. & fit 1 N. 16/5 erit

H iii

FIGURE 6.2
Part of page 61 from the 1670 edition of the *Arithmetica* of Diophantus. This page contains Problem II–8 and the note of Fermat in which he states the impossibility of dividing a cube into a sum of two cubes or, in general, any nth power ($n > 2$) into a sum of two nth powers. (*Source*: Smithsonian Institution Libraries, Photo No. 92-337)

As another example where Diophantus needed a square, consider

PROBLEM II–19 *To find three squares such that the difference between the greatest and the middle has a given ratio to the difference between the middle and the least.*

Diophantus assumed that the given ratio is 3 : 1. If the least square is x^2, then he took $(x + 1)^2 = x^2 + 2x + 1$ as the middle square. Because the difference between these two squares is $2x + 1$, the largest square must be $x^2 + 2x + 1 + 3(2x + 1) = x^2 + 8x + 4$. To make that quantity a square, Diophantus set it equal to $(x + 3)^2$, in this case choosing the coefficient of x so that the x^2 terms cancel. Then $8x + 4 = 6x + 9$, so $x = 2\frac{1}{2}$ and the desired squares are $6\frac{1}{4}$, $12\frac{1}{4}$, $30\frac{1}{4}$. One notices, however, that given his initial choice of $(x + 1)^2$ as the middle square, 3 is the only integer b Diophantus could use in $(x + b)^2$ that would give him a solution. Of course, with other values of the initial ratio, there would be more possibilities as there would with a different choice for the second square. In any case, in this problem as in all of Diophantus's problems, only one solution is required.

Problem II–11 introduces another general method, that of the double equation.

PROBLEM II–11 *To add the same (required) number to two given numbers so as to make each of them a square.*

Diophantus took the given numbers as 2 and 3. If his required number is x, he needed both $x + 2$ and $x + 3$ to be squares. He therefore had to solve $x + 3 = u^2$, $x + 2 = v^2$, for x, u, v. Again, this is an indeterminate problem. Diophantus described his method as follows: "Take the difference between the two expressions and resolve it into factors. Then take either (a) the square of half the difference between these factors and equate it to the lesser expression or (b) the square of half the sum and equate it to the greater."[10]

Since the difference between the expressions is $u^2 - v^2$ and this factors as $(u + v)(u - v)$, the difference of the two factors is $2v$ while the sum is $2u$. What Diophantus did not mention explicitly is that the initial factoring must be carefully chosen so that the solution for x is a positive rational number. In the present case, the difference between the two expressions is 1. Diophantus factored that as $4 \times 1/4$. Thus, $u + v = 4$ and $u - v = 1/4$, so $2v = 15/4$, $x + 2 = v^2 = 225/64$, and $x = 97/64$. Note, for example, that the factorization $2 \times 1/2$ would not give a positive solution, nor would the factorization $3 \times 1/3$. The factorization $1 = a \cdot 1/a$ needs to be chosen so that $\left[\frac{1}{2}\left(a - \frac{1}{a}\right)\right]^2 > 2$.

6.2.2 Higher-Degree Equations

Because the problems in Book A involve cubes and even higher powers, Diophantus began with a new introduction in which he described the rules for multiplying such powers. For example, since x^2, x^3, x^4, x^5, and x^6 are represented by Δ^{Υ}, K^{Υ}, $\Delta^{\Upsilon}\Delta$, ΔK^{Υ}, and $K^{\Upsilon}K$, respectively, Diophantus wrote, for example, that ΔK^{Υ} multiplied by ς equals K^{Υ} multiplied by itself, equals Δ^{Υ} multiplied by $\Delta^{\Upsilon}\Delta$, and all equal $K^{\Upsilon}K$. Similarly, if $K^{\Upsilon}K$ is divided by $\Delta^{\Upsilon}\Delta$, the result is Δ^{Υ}. Thus, although Diophantus's results are equivalent to our laws of exponents, his notation did not allow him to express it in our familiar way of "add the exponents" when you multiply powers and "subtract the exponents" when you divide.

Diophantus did, however, explain that, as before, his equations end up with a term in one power equaling a term in another, that is, $ax^n = bx^m$ $(n < m)$, where now m may be any number up to 6. To solve, one must use the rules to divide both sides by the lesser power and end up with one "species" equal to a number, that is, in our notation, $a = bx^{m-n}$. The latter equation is easily solved. Speaking to the reader, he concluded, "when you are acquainted with what I have presented, you will be able to find the answer to many problems which I have not presented, since I shall have shown to you the procedure for solving a great many problems and shall have explained to you an example of each of their types."[11]

As an example of Diophantus's use of higher powers of x, consider

PROBLEM A–25 *To find two numbers, one a square and the other a cube, such that the sum of their squares is a square.*

The goal is to find x, y, and z such that $(x^2)^2 + (y^3)^2 = z^2$. Thus, this is an indeterminate problem with one equation in three unknowns. Diophantus set x equal to $2y$ (the 2 is arbitrary) and performed the exponentiation to conclude that $16y^4 + y^6$ must be a square, which he took to be the square of ky^2. So $16y^4 + y^6 = k^2y^4$, $y^6 = (k^2 - 16)y^4$, and $y^2 = k^2 - 16$. It follows that $k^2 - 16$ must be a square. Diophantus chose the easiest value, namely, $k^2 = 25$, so $y = 3$. Therefore, the desired numbers are $y^3 = 27$ and $(2y)^2 = 36$. This solution is easily generalized. Take $x = ay$ for any positive a. Then k and y must be found so that $k^2 - a^4 = y^2$ or so that $k^2 - y^2 = a^4$. Diophantus had, however, already demonstrated in Problem II–10 that one can always find two squares whose difference is given.

Problem B–7 shows that Diophantus knew the expansion of $(x + y)^3$. As he put it, "whenever we wish to form a cube from some side made up of the sum of, say, two different terms—so that a multitude of terms does not make us commit a mistake—we have to take the cubes of the two different terms, and add to them three times the results of the multiplication of the square of each term by the other."[12]

PROBLEM B–7   *To find two numbers such that their sum and the sum of their cubes are equal to two given numbers.*

The problem asks to solve $x + y = a$, $x^3 + y^3 = b$. This system of two equations in two unknowns is determinate. It is a generalization of the "Babylonian" problem I–28, $x + y = a$, $x^2 + y^2 = b$, and Diophantus's method of solution generalized his method there. Letting $a = 20$ and $b = 2240$, he began as before by letting the two numbers be $10 + z$ and $10 - z$. The second equation then becomes $(10 + z)^3 + (10 - z)^3 = 2240$ or, using the expansion already discussed, $2000 + 60z^2 = 2240$, or $60z^2 = 240$, $z^2 = 4$, and $z = 2$. Diophantus gave, of course, a condition for a rational solution, namely, that $(4b - a^3)/3a$ is a square (equivalent to the more natural condition that $[b - 2(\frac{a}{2})^3]/3a$ is a square). It is interesting that the answers here are the same as in I–28, namely, 12 and 8.

When reading through the *Arithmetica*, one never quite knows what to expect next. There are a great variety of problems. Often there are several similar problems grouped together, one involving a subtraction where the previous one involved an addition, for example. But then one wonders why other similar ones were not included. For example, the first four problems of Book A ask for (1) two cubes whose sum is a square, (2) two cubes whose difference is a square, (3) two squares whose sum is a cube, and (4) two squares whose difference is a cube. What is missing from this list is, first, to find two squares whose sum is a square—but that had been solved in II–8—and second, to find two cubes whose sum is a cube. This latter problem is impossible to solve, and there are records stating this impossibility dating back to the tenth century. Probably Diophantus was also aware of the impossibility. At the very least, he must have tried the problem and failed to solve it. But he did not mention anything about it in his work. A similar problem with fourth powers occurs as V–29: to find three fourth powers whose sum is a square. Although Diophantus solved that problem, he did not mention the impossibility of finding two fourth powers whose sum is a square. Again, one assumes that he tried the latter problem and failed to solve it.

In his discussion of Problem D–11, he finally addressed an impossibility. After solving that problem, to divide a given square into two parts such that the addition of one part to the square gives a square and the subtraction of the other part from the square also gives a square, he continued, "since it is not possible to find a square number such that, dividing it into two parts and increasing it by each of the parts, we obtain in both cases a square, we shall now present something which is possible."[13]

PROBLEM D–12   *To divide a given square into two parts such that when we subtract each from the given square, the remainder is (in both cases) a square.*

Why is the quoted case impossible? To solve $x^2 = a + b$, $x^2 + a = c^2$, $x^2 + b = d^2$ would imply that

$$3x^2 = c^2 + d^2 \quad \text{or} \quad 3 = \left(\frac{c}{x}\right)^2 + \left(\frac{d}{x}\right)^2.$$

It is, in fact, impossible to decompose 3 into two rational squares. One can show this easily by congruence arguments modulo 4. Diophantus himself did not give a proof, nor later, when he stated in VI–14 that 15 is not the sum of two squares, did he tell why. The solution of D–12, however, is very easy.

6.2.3   The Method of False Position

In Book IV, Diophantus began use of a new technique, a technique reminiscent of the Egyptian "false position." Among many problems he solved using this technique, the following one will be important in our later discussion of elliptic curves.

PROBLEM IV–24   *To divide a given number into two parts such that their product is a cube minus its side.*

If a is the given number, the problem is to find x and y such that $y(a - y) = x^3 - x$. This is an indeterminate problem. As usual, Diophantus began by choosing a particular value for a, here $a = 6$. So $6y - y^2$ must equal a cube minus its side. He chose the side x to be of the form $x = my - 1$. The question is, What value should he choose for m? Diophantus picked $m = 2$ and calculated: $6y - y^2 = (2y - 1)^3 - (2y - 1)$, or $6y - y^2 = 8y^3 - 12y^2 + 4y$. We note immediately that the "1" in $x = my - 1$ was chosen so that there would be no constant term in this equation. Nevertheless, this is still an equation with three separate species, not the type Diophantus could solve most easily. So he noted that if the coefficients of y on each side were the same, then the solution would be simple. Now the "6" on the left is the "given number," so that cannot be changed. But the "4" on the right comes from the calculation $3 \cdot 2 - 2$, which in turn depends on the choice $m = 2$ in $x = my - 1$. Therefore, Diophantus needed to find m so that $3 \cdot m - m = 6$. Therefore, $m = 3$. We can then begin again: $x = 3y - 1$ and $6y - y^2 = (3y - 1)^3 - (3y - 1)$, or $6y - y^2 = 27y^3 - 27y^2 + 6y$. Therefore, $27y^3 = 26y^2$ and $y = \frac{26}{27}$. The two parts of 6, therefore, are $\frac{26}{27}$ and $\frac{136}{27}$, while the product of those two numbers is $(\frac{17}{9})^3 - \frac{17}{9}$. The general solution to this problem, for arbitrary a, is then given by

$$y = \frac{6a^2 - 8}{a^3}, \; x = \frac{3a^2 - 4}{a^2} - 1.$$

In Problem IV–31, Diophantus found again that his original assumption did not work. But here the problem is that a mixed quadratic equation, the first one to appear in the *Arithmetica*, fails to have a rational solution.

PROBLEM IV–31   *To divide unity into two parts so that, if given numbers are added to them respectively, the product of the two sums is a square.*

Diophantus set the given numbers at 3, 5, and the parts of unity as x, $1 - x$. Therefore, $(x + 3)(6 - x) = 18 + 3x - x^2$ must be a square. Since neither of his usual techniques for determining a square will work here (neither 18 nor -1 are squares), he tried $(2x)^2 = 4x^2$ as the desired square. But the resulting quadratic equation, $18 + 3x = 5x^2$ "does not give a rational result." He needed to replace $4x^2$ by a square of the form $(mx)^2$, which does give a rational solution. Thus, since $5 = 2^2 + 1$, he noted that the quadratic equation will be solvable if $(m^2 + 1) \cdot 18 + (3/2)^2$ is a square. This implies that $72m^2 + 81$ is a square, say, $(8m + 9)^2$. (Here, his usual technique succeeds.) Then $m = 18$ and, returning to the beginning, he set $18 + 3x - x^2 = 324x^2$. He then simply presented the solution: $x = 78/325 = 6/25$, and the desired numbers are 6/25, 19/25.

Although Diophantus did not give details in IV–31 on the solution of the quadratic, he did give them in Problem IV–39. His words in that problem are easily translated into the formula

$$x = \frac{\frac{b}{2} + \sqrt{ac + \left(\frac{b}{2}\right)^2}}{a}$$

for solving the equation $c + bx = ax^2$. This formula translates correctly the Babylonian procedure, which began by multiplying the equation through by a and solving for ax. Diophantus was sufficiently familiar with this formula and its variants that he used it in various later problems not only to solve quadratic equations but also to solve quadratic inequalities.

PROBLEM V–10 *To divide unity into two parts such that, if we add different given numbers to each, the results will be squares.*

In this problem the manuscripts have, for one of only two times in the entire work, a diagram (Fig. 6.3). Diophantus assumed that the two given numbers are 2 and 6. He represented them, as well as 1, by setting $DA = 2$, $AB = 1$, and $BE = 6$. The point G is chosen so that $DG \,(= AG + DA)$ and $GE \,(= BG + BE)$ are both squares. Since $DE = 9$, the problem is reduced to dividing 9 into two squares such that one of them lies between 2 and 3. If that square is x^2, the other is $9 - x^2$. Unlike the situation in previous problems, Diophantus could not simply put $9 - x^2$ equal to $(3 - mx)^2$ with an arbitrary m, for he needed x^2 to satisfy the inequality condition. So he set it equal to $(3 - mx)^2$ without specifying m. Then

$$x = \frac{6m}{m^2 + 1}.$$

Rather than substitute the expression for x into $2 < x^2 < 3$ and attempt to solve a fourth-degree inequality, he picked two squares close to 2 and 3, respectively, namely, $289/144 = (17/12)^2$ and $361/144 = (19/12)^2$, and substituted the expression into the inequality $17/12 < x < 19/12$. Therefore,

$$\frac{17}{12} < \frac{6m}{m^2 + 1} < \frac{19}{12}.$$

The left inequality becomes $72m > 17m^2 + 17$. Although the corresponding quadratic equation has no rational solution, Diophantus nevertheless used the quadratic formula and showed that since $\sqrt{(72/2)^2 - 17^2} = \sqrt{1007}$ is between 31 and 32, the number m must be chosen so that $m \leq 67/17$. The right inequality similarly shows that $m \geq 66/19$. Diophantus therefore picked the simplest m between these two limits, namely, 3 1/2. So

$$9 - x^2 = \left(3 - 3\frac{1}{2}x\right)^2 \quad \text{and} \quad x = \frac{84}{53}.$$

Then $x^2 = 7056/2809$ and the desired segments of 1 are 1438/2809 and 1371/2809.

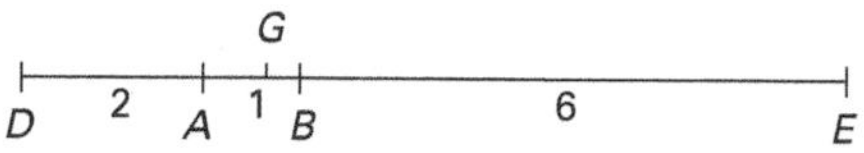

FIGURE 6.3

Diophantus's *Arithmetica*, Problem V–10

Diophantus's work, the only example of a genuinely algebraic work surviving from ancient Greece, was highly influential. Not only was it commented on in late antiquity, but it was also studied by Islamic authors. Many of its problems were taken over by Rafael Bombelli and published in his *Algebra* of 1572, while the initial printed Greek edition of Bachet, published in 1621, was carefully studied by Pierre Fermat and led him to numerous general results in number theory, about which Diophantus himself only hinted. Perhaps more important, however, is the fact that this work, as a work of algebra, was in effect a treatise on the analysis of problems. Namely, the solution of each problem began with the assumption that the answer x, for example, had been found. The consequences of this fact were then followed to the point where a numerical value of x could be determined by solving a simple equation. The synthesis, which in this case is the proof that the answer satisfies the desired conditions, was never given by Diophantus because it only amounted to an arithmetic computation. Thus, Diophantus's work is at the opposite end of the spectrum from the purely synthetic work of Euclid.

 6.3   PAPPUS AND ANALYSIS

Although analysis and synthesis had been used by all of the major Greek mathematicians, there was no systematic study of the methodology published, as far as is known, until the work of Pappus, who lived in Alexandria early in the fourth century (Sidebar 6.2). Pappus was one of the last mathematicians in the Greek tradition. He was familiar with the major and minor works of the men already discussed, and even extended some of their work in certain ways. He is best known for his *Collection*, a group of eight separate works on various topics in mathematics, probably put together shortly after his death by an editor attempting to preserve Pappus's papers. The books of the collection vary greatly in quality, but most of the material consists of surveys of certain mathematical topics collected from the works of his predecessors.

The preface to Book 3 provides an interesting sidelight to the work. Pappus addressed the preface to Pandrosian, a woman teacher of geometry. He complained that "some persons professing to have learned mathematics from you lately gave me a wrong enunciation of problems."[14] By that Pappus meant that these people attempted to solve problems by methods that could not work, for example, to solve the problem of the two mean proportionals using only circles and straight lines. There is no indication of how Pappus knew that such a construction was impossible. From his remark, however, we learn that women were involved in mathematics in Alexandria.[15]

Book 5, the most polished book of the *Collection*, deals with isoperimetric figures, figures of different shape but with the same perimeter. Pappus's introduction provided a counterpoint to the pure mathematics of the text as he wrote of the intelligence of bees:

> [The bees], believing themselves, no doubt, to be entrusted with the task of bringing from the gods to the more cultured part of mankind a share of ambrosia in this form, . . . do not think it proper to pour it carelessly into earth or wood or any other unseemly and irregular material, but, collecting the fairest parts of the sweetest flowers growing on the earth, from them they prepare for the reception of the honey the vessels called honeycombs, [with cells] all equal, similar and adjacent, and hexagonal in form.

SIDEBAR 6.2   *Who Were the Alexandrian Mathematicians?*

Raphael's painting *The School of Athens* depicts Ptolemy as a prince with Italian features, while the most common "portrait" of Hypatia, attributed to an artist named Gasparo, shows her as Italian as well. There is nothing surprising in this; artists usually use their contemporaries as models for figures from long ago. But the more serious question is to what extent the Alexandrian mathematicians of the period from the first to the fifth centuries CE were Greek. Certainly, all of them wrote in Greek and were part of the Greek intellectual community of Alexandria. And most modern studies of Hellenistic Egypt conclude that the Greek community and the native Egyptian community coexisted, with little mutual influence. So do we then conclude that Ptolemy and Diophantus, Pappus and Hypatia were ethnically Greek, that their ancestors had come from Greece at some point in the past and had remained effectively isolated from the Egyptians for many centuries?

The question is, of course, not possible to answer definitively. But the research in papyri dating from the early centuries of the common era also demonstrates that there was significant intermarriage between the Greek and Egyptian communities, chiefly by Greek men taking Egyptian wives. And it is known, for example, that Greek marriage contracts increasingly resembled Egyptian ones. In addition, even from the founding of Alexandria, small numbers of Egyptians were admitted to the privileged classes in the city to fulfill numerous civic roles. Of course, it was essential in this case for the Egyptians to become "Hellenized," to adopt Greek habits and the Greek language. Given that the Alexandrian mathematicians mentioned above were active several hundred years after the founding of the city, however, it would seem at least equally possible that they were ethnically Egyptian as that they remained ethnically Greek. In any case, it is unreasonable for us today to portray these mathematicians with pure European features when we have no physical descriptions of them whatsoever.

FIGURE 6.4

Honeycomb in hexagons on Luxembourg stamp

That they have contrived this in accordance with a certain geometrical forethought we may thus infer. They would necessarily think that the figures must all be adjacent one to another and have their sides common, in order that nothing else might fall into the interstices and so defile their work. Now there are only three rectilineal figures which would satisfy the condition, I mean regular figures which are equilateral and equiangular, inasmuch as irregular figures would be displeasing to the bees. . . . [These being] the triangle, the square and the hexagon, the bees in their wisdom chose for their work that which has the most angles, perceiving that it would hold more honey than either of the two others [Fig. 6.4].

Bees, then, know just this fact which is useful to them, that the hexagon is greater than the square and the triangle and will hold more honey for the same expenditure of material in constructing each. But we, claiming a greater share in wisdom than the bees, will investigate a somewhat wider problem, namely that, of all equilateral and equiangular plane figures having an equal perimeter, that which has the greater number of angles is always greater, and the greatest of them all is the circle having its perimeter equal to them.[16]

The most influential book of Pappus's *Collection*, however, is Book 7, *On the Domain of Analysis*, which contains the most explicit discussion from Greek times of the method of analysis, the methodology Greek mathematicians used to solve problems. The central ideas are spelled out in the introduction to this book:

That which is called the Domain of Analysis . . . is, taken as a whole, a special resource . . . for those who want to acquire a power in geometry that is capable of solving problems set to them; and it is useful for this alone. It was written by three men, Euclid the writer of the *Elements*, Apollonius of Perga, and Aristaeus the elder, and proceeds by analysis and synthesis.

Now analysis is the path from what one is seeking, as if it were established, by way of its consequences, to something that is established by synthesis. . . . There are two kinds of analysis; one of them seeks after truth and is called "theorematic," while the other tries to find what was demanded, and is called "problematic." In the case of the theorematic kind, we assume what is sought as a fact and true, then advance through its consequences, as if they are true facts according to the hypothesis, to something established; if this thing that has been established is a truth, then that which was sought will also be true, and its proof the reverse of the analysis; but if we should meet with something established to be false, then the thing that was sought too will be false. In the case of the problematic kind, we assume the proposition as something we know, then proceed through its consequences, as if true, to something established; if the established thing is possible and obtainable, which is what mathematicians call "given," the required thing will also be possible, and again the proof will be the reverse of the analysis; but should we meet with something established to be impossible, then the problem too will be impossible.[17]

According to Pappus, then, to solve a problem or prove a theorem by analysis, begin by assuming what is required, then consider the consequences flowing from it until a result is reached that is known to be true or "given." That is, begin by assuming that which is required, p, for example, and then prove that p implies q_1, q_1 implies q_2, . . . , q_n implies q, where q is something known to be true. To give the formal synthetic proof of the theorem, or solve the problem, reverse the process beginning with q implies q_n. This method of reversal has always been a controversial point; after all, not all theorems have valid converses. In fact, however, most important theorems from Euclid and Apollonius do have at least partial converses. Thus, the method does often provide the desired proof or solution, or at least demonstrates, when there are only partial converses, the conditions under which a problem can be solved.

There are few examples in the extant literature of theorematic analysis, because Euclid, for example, never shared his method of discovery of his proofs. But some of the manuscripts of *Elements*, Book XIII, contain, evidently as an interpolation made in the early years of the common era, an analysis of each of the first five propositions. Consider

PROPOSITION XIII–1   *If a straight line is cut in extreme and mean ratio, the square on the sum of the greater segment and half of the whole is five times the square on the half.*

Let AB be divided in extreme and mean ratio at C, AC being the greater segment, and let $AD = \frac{1}{2}AB$ (Fig. 6.5). To perform the analysis, assume the truth of the conclusion, namely, $CD^2 = 5AD^2$, and determine its consequences. Since also $CD^2 = AC^2 + AD^2 + 2AC \cdot AD$, therefore, $AC^2 + 2AC \cdot AD = 4AD^2$. But $AB \cdot AC = 2AC \cdot AD$ and, since $AB : AC = AC : BC$, also $AC^2 = AB \cdot BC$. Therefore, $AB \cdot BC + AB \cdot AC = 4AD^2$, or $AB^2 = 4AD^2$, or, finally, $AB = 2AD$, a result known to be true. The synthesis can then proceed by reversing each step: Since $AB = 2AD$, we have $AB^2 = 4AD^2$. Since also $AB^2 = AB \cdot AC + AB \cdot BC$, it follows that $4AD^2 = 2AD \cdot AC + AC^2$. Adding to each side the square on AD gives the result $CD^2 = 5AD^2$.

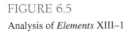

FIGURE 6.5

Analysis of *Elements* XIII–1

More important for Greek mathematics than theorematic analysis is the problematic analysis. We have already discussed several examples of this type of analysis, including the problems of angle trisection and cube duplication and Archimedes' problems on the division of a sphere by a plane. And although Euclid did not present the analysis as such, one can carry out the procedure in solving *Elements* VI–28, the geometric algebra problem leading to the solution of the quadratic equation $x^2 + c = bx$. The analysis there shows that an additional condition is required for the solution, namely, that $c \leq (\frac{b}{2})^2$.

Pappus's Book 7, then, is a companion to the Domain of Analysis, which itself consists of several geometric treatises, all written many centuries before Pappus. These works, Apollonius's *Conics* and six other books (all but one lost), Euclid's *Data* and two other lost works, and single works (both lost) by Aristaeus and Eratosthenes, even though the last-named author is not mentioned in Pappus's introduction, provided the Greek mathematician with the tools necessary to solve problems by analysis. For example, to deal with problems that result in conic sections, one needs to be familiar with Apollonius's work. To deal with problems solvable by "Euclidean" methods, the material in the *Data* is essential.

Pappus's work does not include the Domain of Analysis itself. It is designed only to be read along with these treatises. Therefore, it includes a general introduction to most of the individual books along with a large collection of lemmas that are intended to help the reader work through the actual texts. Pappus evidently decided that the texts themselves were too difficult for most readers of his day to understand as they stood. The teaching tradition had been weakened through the centuries, and there were few, like Pappus, who could appreciate these several-hundred-year-old works. Pappus's goal was to increase the numbers who could understand the mathematics in these classical works by helping his readers through the steps where the authors wrote "clearly . . . !" He also included various supplementary results as well as additional cases and alternative proofs.

Among these additional remarks is the generalization of the three- and four-line locus problems discussed by Apollonius. Pappus noted that in that problem itself the locus is a conic section. But, he says, if there are more than four lines, the loci are as yet unknown; that is, "their origins and properties are not yet known." He was disappointed that no one had given the construction of these curves that satisfy the five- and six-line locus. The problem in these cases is, given five (six) straight lines, to find the locus of a point such that the rectangular parallelepiped contained by the lines drawn at given angles to three of these lines has a given ratio to the rectangular parallelepiped contained by the remaining two lines and some given line (remaining three lines). Pappus noted that one can even generalize the problem further to more than six lines, but in that case, "one can no longer say 'the ratio is given between some figure contained by four of them to some figure contained by the remainder' since no figure can be contained in more than three dimensions." Nevertheless, according to Pappus, one can express this ratio of products by compounding the ratios that individual lines have to one another, so that one can in fact consider the problem for any number of lines. But, Pappus complained, "[geometers] have by no means solved [the multi-line locus problem] to the extent that the curve can be recognized. . . . The men who study these matters are not of the same quality as the ancients and the best writers. Seeing that all geometers are occupied with the first principles of mathematics . . . and being ashamed to pursue such topics myself, I have proved propositions of much greater importance and utility."[18]

Pappus concluded Book 7 by stating one of the "important" results he had proved, that "the ratio of solids of complete revolution is compounded of that of the revolved figures and that of the straight lines similarly drawn to the axes from the center of gravity in them."[19] The modern version of this theorem is that the volume of a solid formed by revolving a region Ω around an axis not intersecting Ω is the product of the area of Ω and the circumference of the circle traversed by the center of gravity of Ω. Unfortunately, there is no record of Pappus's proof. There is some indication that it is in one of the books of the *Collection* now lost.

Much of the explicit analysis in Greek mathematics has to do with material we generally think of as algebraic. The examples from *Elements* XIII–1 and VI–28 are clearly such. The examples using the conic sections are ones that today would be solved using analytic geometry, a familiar application of algebra. It is somewhat surprising, then, that Pappus does not mention the strictly algebraic *Arithmetica* of Diophantus as a prime example of analysis, because, in effect, every problem in Diophantus's work is solved according to Pappus's model. Perhaps Pappus did not include this work because it was not on the level of the classic geometric works. In any case, it was the algebraic analysis of Diophantus and the "quasi-algebraic" analysis of many of the other mentioned works, rather than the pure geometric analysis, that provided the major impetus for sixteenth- and seventeenth-century European mathematicians to expand on the notion of algebra and develop it into a major tool to solve even purely geometric problems.[20]

 ## 6.4   HYPATIA AND THE END OF GREEK MATHEMATICS

Pappus's aim of reviving Greek mathematics was unsuccessful, probably in part because the increasingly confused political and religious situation affected the stability of the Alexandrian Museum and Library. In his time, Christianity was changing from a persecuted sect into the official religion of the Roman Empire. In 313 the emperor Galerius issued an edict of toleration in the Eastern Empire, and two years later the same was done in the West by Constantine. The latter in fact converted to Christianity before his death in 337. Within 60 years, Christianity became the state religion of the empire and the ancient worship of the Roman gods was banned. Of course, the banning of paganism did not cause everyone to adopt Christianity. In fact, in the late fourth and early fifth centuries, Hypatia (c. 355–415), the daughter of Theon of Alexandria, was a respected and eminent teacher in that city, not only of mathematics but also of some of the philosophic doctrines dating back to Plato's Academy. And although she maintained her non-Christian religious beliefs, she enjoyed intellectual independence and even had eminent Christians among her students, including Synesius of Cyrene (in present-day Libya), who later became a bishop.

Although there is some evidence of earlier women being involved in Greek mathematics, it is only about Hypatia that the evidence is substantial enough to give some indication of her mathematical accomplishments. Hypatia was given a very thorough education in mathematics and philosophy by her father. Although the only surviving documents with a clear reference to Hypatia are Synesius's letters to her requesting scientific advice, recent detailed textual studies of Greek, Arabic, and medieval Latin manuscripts lead to the conclusion that she was responsible for many mathematical works. These include several parts of her father's commentary on Ptolemy's *Almagest*, the edition of Archimedes' *Measurement of the Circle*

SIDEBAR 6.3   *The Decline of Greek Mathematics*

Why did Greek mathematics decline so dramatically from its height in the fourth and third centuries BCE? Among the several answers to this question, the most important is the change in the sociopolitical scene in the region surrounding the eastern Mediterranean.

A consideration of mathematical development in the various ancient societies already studied shows that mathematical creativity requires some sparks of intellectual curiosity, whether or not these are stimulated by practical concerns. But this spark of curiosity needs a climate of government encouragement for its flames to spread. The Babylonians used their most advanced techniques, not for everyday purposes, but for solving intellectually challenging problems. The government encouraged the use of these mathematical problems to help train the minds of its future leaders. In Greek civilization, the intellectual curiosity ran even deeper. In the Greek homeland, the sociopolitical system provided philosophy and mathematics with encouragement. The Ptolemies continued this encouragement in Egypt after 300 BCE.

But even in Greek society, the actual number of those who understood theoretical mathematics was small. There were never many who could afford to spend their lives as mathematicians or astronomers and persuade the rulers to provide them with stipends. The best of the mathematicians wrote works that were discussed and commented on in the various mathematical schools, but not everything could be learned from the texts. An oral teaching tradition was necessary to keep mathematics progressing because, in general, one could not master Euclid's *Elements* or Apollonius's *Conics* on one's own. A break of a generation in this tradition thus meant that the entire process of mathematical research would be severely damaged.

One factor certainly weakening the teaching tradition, if not breaking it entirely, was the political strife around the eastern Mediterranean in the years surrounding the beginning of the common era. More important, because the Roman imperial government evidently decided that mathematical research was not an important national interest, it did not support it. There was little encouragement of mathematical studies in Rome. Few Greek scholars were imported to teach mathematics to the children of the elite. Soon, no one in Rome could even understand, let alone extend, the works of Euclid or Apollonius.

The Greek tradition did continue for several centuries, however, under the Roman governors of Egypt, particularly because the Alexandrian Museum and Library remained in existence. Anyone interested could continue to study and interpret the ancient texts. With fewer and fewer teachers, however, less and less new work was accomplished. The virtual destruction of the great library by the late fourth century finally severed the tenuous links with the past. Although there continued to be some limited mathematical activity for a while in Athens and elsewhere—wherever copies of the classic works could be found—by the end of the fifth century, there were too few people devoting their energies to mathematics to continue the tradition, and Greek mathematics ceased to be.

---

from which most later Arabic and Latin translations stem, a work on areas and volumes reworking Archimedean material, and a text on isoperimetric figures related to Pappus's Book 5.[21] She was also responsible for commentaries on Apollonius's *Conics* and, as noted earlier, on Diophantus's *Arithmetica*.

Unfortunately, although Hypatia had many influential friends in Alexandria, including the Roman prefect Orestes, they were primarily from the upper classes. The populace at large in general supported the patriarch Cyril in his struggle with Orestes for control of the city. So when Cyril spread rumors that the famous woman philosopher in reality practiced sorcery as part of her philosophical, mathematical, and astronomical work, a group emerged that was willing to eliminate this "satanic" figure. Hypatia's life was thus cut short as already described. Her death effectively ended the Greek mathematical tradition of Alexandria (Sidebar 6.3).

EXERCISES

1. Devise a formula for the nth pentagonal number and for the nth hexagonal number.

2. Derive an algebraic formula for the pyramidal numbers with triangular base and one for the pyramidal numbers with square base.

3. Show that in a harmonic proportion the sum of the extremes multiplied by the mean is twice the product of the extremes.

4. Nicomachus defined a **subcontrary** proportion, which occurs when in three terms the greatest is to the smallest as the difference of the smaller terms is to the difference of the greater. Show that 3, 5, 6, are in the subcontrary proportion. Find two other sets of three terms that are in subcontrary proportion.

5. Nicomachus claims that if three terms are in subcontrary proportion, then the product of the greater and mean terms is twice the product of the mean and smaller; for, he notes, 6 times 5 is twice 5 times 3. Show that Nicomachus is incorrect in general.

6. Nicomachus defined a "fifth proportion" to exist whenever among three terms the middle term is to the lesser as their difference is to the difference between the greater and the mean. Show that 2, 4, 5, are in fifth proportion. Find two more triples in this proportion.

7. Determine Diophantus's age at his death from his epigram at the opening of the chapter.

8. Solve Diophantus's Problem I–27 by the method of I–28: To find two numbers such that their sum and product are given. Diophantus gives the sum as 20 and the product as 96.

9. Solve Diophantus's Problem II–10: To find two square numbers having a given difference. Diophantus puts the given difference as 60. Also, give a general rule for solving this problem given any difference.

10. Generalize Diophantus's solution to II–19 by choosing an arbitrary ratio $n : 1$ and the value $(x + m)^2$ for the second square.

11. Solve Diophantus's Problem II–13 by the method of the double equation: From the same (required) number to subtract two given numbers so as to make both remainders square. (Take 6, 7, for the given numbers. Then solve $x - 6 = u^2$, $x - 7 = v^2$.)

12. Solve Diophantus's Problem B–8: To find two numbers such that their difference and the difference of their cubes are equal to two given numbers. (Write the equations as $x - y = a$, $x^3 - y^3 = b$. Diophantus takes $a = 10$, $b = $

2120.) Derive necessary conditions on a and b that ensure a rational solution.

13. Solve Diophantus's Problem B–9: To divide a given number into two parts such that the sum of their cubes is a given multiple of the square of their difference. (The equations become $x + y = a$, $x^3 + y^3 = b(x - y)^2$. Diophantus takes $a = 20$ and $b = 140$ and notes that the necessary condition for a solution is that $a^3(b - \frac{3}{4}a)$ is a square.)

14. Solve Diophantus's Problem D–12: To divide a given square into two parts such that when we subtract each from the given square, the remainder (in both cases) is a square. Note that the solution follows immediately from II–8.

15. Solve Diophantus's Problem IV–9: To add the same number to a cube and its side and make the second sum the cube of the first. (The equation is $x + y = (x^3 + y)^3$. Diophantus begins by assuming that $x = 2z$ and $y = 27z^3 - 2z$.)

16. Solve Diophantus's Problem V–10 for the two given numbers 3, 9.

17. Book VI of the *Arithmetica* deals with Pythagorean triples. For example, solve Problem VI–16: To find a right triangle with integral sides such that the length of the bisector of an acute angle is also an integer. (*Hint*: Use *Elements* VI–3, that the bisector of an angle of a triangle cuts the opposite side into segments in the same ratio as that of the remaining sides.)

18. Carry out the analysis of *Elements* VI–28: To a given straight line to apply a parallelogram equal to a given rectilinear figure and deficient by a parallelogram similar to a given one. Just consider the case where the parallelograms are all rectangles. Begin with the assumption that such a rectangle has been constructed and derive the condition that "the given rectilinear figure must not be greater than the rectangle described on the half of the straight line and similar to the defect."

19. Provide the analysis for *Elements* XIII–4: If a straight line is cut in extreme and mean ratio, the sum of the squares on the whole and on the lesser segment is triple the square on the greater segment.

20. Write an equation for the locus described by the problem of five lines. Assume for simplicity that all the lines are either parallel or perpendicular to one of them and that all the given angles are right.

21. Show that a regular hexagon of given perimeter has a greater area than a square of the same perimeter.

22. Find the volume of a torus by applying Pappus's theorem. Assume that the torus is formed by revolving the disk of radius r around an axis whose distance from the center of the disk is $R > r$.

23. Solve Epigram 116: Mother, why do you pursue me with blows on account of the walnuts? Pretty girls divided them all among themselves. For Melission took two-sevenths of them from me, and Titane took the twelfth. Playful Astyoche and Philinna have the sixth and third. Thetis seized and carried off twenty, and Thisbe twelve, and look there at Glauce smiling sweetly with eleven in her hand. This one nut is all that is left to me. How many nuts were there originally?[22]

24. Solve Epigram 130: Of the four spouts, one filled the whole tank in a day, the second in two days, the third in three days, and the fourth in four days. What time will all four take to fill it?

25. Solve Epigram 145: A. Give me ten coins and I have three times as many as you. B. And if I get the same from you, I have five times as much as you? How many coins does each have?

26. Devise a lesson teaching the method of problematic analysis. Use problems from ancient times and more recent problems.

27. Why were there so few women involved in mathematics in Greek times?

REFERENCES AND NOTES

Thomas Heath's *A History of Greek Mathematics*, cited in Chapter 2, and B. L. Van der Waerden's *Science Awakening*, referred to in Chapter 1, have sections on the material discussed in this chapter. A translation of Nicomachus's major work is found in M. L. D'Ooge, F. E. Robbins, and L. C. Karpinski, *Nicomachus of Gerasa: Introduction to Arithmetic* (New York: Macmillan, 1926). This translation can also be found in the *Great Books*, vol. 11. The six books of Diophantus still extant in Greek are found in Thomas L. Heath, *Diophantus of Alexandria: A Study in the History of Greek Algebra* (New York: Dover, 1964). Heath does not, however, translate Diophantus literally; he generally just outlines Diophantus's arguments. More literal translations of certain of the problems are found in Thomas, *Selections Illustrating the History of Greek Mathematics*. A translation of and commentary on the four newly discovered books of Diophantus's *Arithmetica* is J. Sesiano, *Books IV to VII of Diophantos's Arithmetica in the Arabic Translation of Qusṭā ibn Lūqā* (New York: Springer, 1982). A brief survey of Diophantus's work is in J. D. Swift, "Diophantus of Alexandria," *American Mathematical Monthly* 63 (1956), 163–170. The entire extant text of Pappus's *Collection* is translated into French in Paul Ver Eecke, *Pappus d'Alexandrie, La Collection Mathematique* (Paris: Desclée, De Brouwer et Cie., 1933). A recent English translation of Book 7, with commentary, is provided by Alexander Jones, *Pappus of Alexandria: Book 7 of the Collection* (New York: Springer, 1986). There is a recent biography of Hypatia: Maria Dzielska, *Hypatia of Alexandria*, translated by F. Lyra (Cambridge: Harvard University Press, 1995). Although the book has little discussion of her mathematics, that gap is filled by Michael A. B. Deakin in "Hypatia and Her Mathematics," *American Mathe-matical Monthly* 101 (1994), 234–243, who himself wrote a biography: *Hypatia of Alexandria: Mathematician and Martyr* (Amherst, NY: Prometheus Books, 2007).

1. W. R. Paton, trans., *The Greek Anthology* (Cambridge: Harvard University Press, 1979), Volume V, pp. 93–94 (Book XIV, Epigram 126).

2. Edward Gibbon, *The Decline and Fall of the Roman Empire* (Chicago: Encyclopedia Britannica, 1952) (*Great Books* edition), chapter 47, p. 139.

3. Nicomachus, *Introduction to Arithmetic*, I, IV, 2.

4. Nicomachus, *Introduction to Arithmetic*, II, XXIV, 1.

5. Details of this argument are presented in J. Sesiano, *Books IV to VII of Diophantos' Arithmetica*, pp. 71–75.

6. Thomas, *Selections*, p. 525.

7. Thomas L. Heath, *Diophantus of Alexandria*, pp. 130–131.

8. The problems from Books I–VI are adapted from Heath, *Diophantus*, while those from Books A–D are taken from Sesiano, *Books IV to VII of Diophantos' Arithmetica*.

9. Thomas, *Selections*, II, p. 553.

10. Heath, *Diophantus*, p. 146.

11. Sesiano, *Books IV to VII of Diophantos' Arithmetica*, p. 87.

12. Ibid., p. 130.

13. Ibid., p. 165.

14. Thomas, *Selections*, II, p. 567.

15. The arguments for concluding that Pandrosian is a woman are given in Jones, *Pappus of Alexandria*.

16. Thomas, *Selections,* II, pp. 589–593.

17. This translation is adapted from one in Michael Mahoney, "Another Look at Greek Geometrical Analysis," *Archive for History of Exact Sciences* 5 (1968), 318–348, and from Jones, *Pappus of Alexandria*. Mahoney's conclusions about Greek analysis are disputed in some respects in J. Hintikka and U. Remes, *The Method of Analysis: Its Geometrical Origin and Its General Significance* (Boston: Reidel, 1974).

18. Jones, *Pappus of Alexandria*, pp. 120–122, and Thomas, *Selections*, II, p. 601.

19. Jones, *Pappus of Alexandria*, p. 122.

20. An extensive discussion of the algebraic analysis of Diophantus and its effects on the development of algebra is found in J. Klein, *Greek Mathematical Thought and the Origin of Algebra* (Cambridge: MIT Press, 1968).

21. Details on the attribution of various mathematical works to Hypatia are found in Wilbur Knorr, *Textual Studies in Ancient and Medieval Geometry* (Boston: Birkhäuser, 1989), as well as in Michael Deakin, "Hypatia and Her Mathematics."

22. This problem and the next two are taken from Paton, *The Greek Anthology*.

CHAPTER

7

Ancient and Medieval China

Now the science of mathematics is considered very important. This book . . . therefore will be of great benefit to the people of the world. The knowledge for investigation, the development of intellectual power, the way of controlling the kingdom and of ruling even the whole world, can be obtained by those who are able to make good use of the book. Ought not those who have great desire to be learned take this with them and study it with great care?

—Introduction to *Precious Mirror of the Four Elements* by Zhu Shijie, 1303[1]

A report to the Throne by the Astronomical Observer, Wang Sibian, early in the seventh century noted that the 10 computational canons were riddled with mistakes and contradictions. Consequently, Li Chunfeng, together with Liang Shu, an Erudite of Mathematics, and Wang Zhenru, an Instructor from the National University, were ordered by imperial decree to annotate these works and remove the contradictions. Once their task was completed, the Emperor Gaozu ordered that these books be used at the National University.[2]

In the first six chapters, we discussed the mathematics of Greece as well as the mathematics of two civilizations known to have influenced Greek mathematics, Mesopotamia and Egypt. But mathematics was done in other parts of the world, even in ancient times. In this chapter, we look at some mathematical ideas from ancient and medieval China, some of which may have, through paths so far undiscovered, reached Europe.

 7.1   INTRODUCTION TO MATHEMATICS IN CHINA

Although there are legends that date Chinese civilization back 5000 or more years, the earliest solid evidence of such a civilization is provided by the excavations at Anyang, near the Huang River, which are dated to about 1600 BCE. It is to the society centered there, the Shang dynasty, that the "oracle bones" belong, curious pieces of bone inscribed with very ancient writing, which were used for divination by the priests of the period. The bones are the source of our knowledge of early Chinese number systems. Around the beginning of the first millennium BCE, the Shang were replaced by the Zhou dynasty, which in turn dissolved into numerous warring feudal states. In the sixth century BCE, there was a great period of intellectual flowering during which the most famous philosopher was Confucius. Academies of scholars were founded in several of the states. Other feudal lords hired individual scholars to advise them in a time of technological growth caused by the development of iron.

The feudal period ended as the weaker states were gradually absorbed by the stronger, until ultimately China was unified under the emperor Qin Shi Huangdi in 221 BCE. Under his leadership, China was transformed into a highly centralized bureaucratic state. He enforced a severe legal code, levied taxes evenly, and demanded the standardization of weights, measures, money, and especially the written script. Legend holds that this emperor ordered the burning of all books from earlier periods to suppress dissent, but there is some reason to doubt that this was actually carried out. The emperor died in 210 BCE, and his dynasty was soon overthrown and replaced by that of the Han, which was to last about 400 years.

At some time early in this dynasty, an official was buried in a tomb near Zhangjiashan in Hubei Province with several of his books. The tomb was opened in early 1984, and among the books was discovered a mathematics text written on 200 bamboo strips. This work, called the *Suan shu shu* (*Book of Numbers and Computation*), is the earliest extant text of Chinese mathematics. Like many later works, it consists of problems and their solutions, a few of which we will consider below. There were two other works we know of compiled during the Han dynasty, which may well have played a part in the education of the civil service at the time. These are the *Zhoubi suanjing* (*Arithmetical Classic of the Gnomon and the Circular Paths of Heaven*) and the *Jiuzhang suanshu* (*Nine Chapters on the Mathematical Art*). The first of these has come down to us with the commentaries of Zhao Shuang (third century CE), Zhen Luang (sixth century CE), and Li Chunfeng (seventh century CE). The latter work, which became central to Chinese mathematical practice over the centuries, has survived in the edition of Liu Hui (third century CE), who commented on it extensively and even added a tenth chapter, now known as the *Sea Island Mathematical Manual*. Li Chunfeng also made extensive comments, thus adding to our knowledge of the development of Chinese mathematics between the time the book was originally written and his time. We consider the commentaries of both of these men in what follows.

The Han dynasty in China disintegrated early in the third century CE, and China broke up into several warring kingdoms. The period of disunity lasted until 581, when the Sui dynasty was established, followed 37 years later by the Tang dynasty, which was to last nearly 300 years. Although another brief period of disunity followed, much of China was again united under the Song dynasty (960–1279), a dynasty itself overthrown by the Mongols under Ghenghis Khan. This dynasty was replaced by a native Chinese dynasty, the Ming, a hundred years later.

Despite the numerous wars and dynastic conflicts, a true Chinese culture was developing throughout most of east Asia, with a common language and common values. The system of imperial examinations for entrance into the civil service, instituted during the Han dynasty, lasted—with various short periods of disruption—into the twentieth century. Although the examination was chiefly based on Chinese literary classics, the demands of the empire for administrative services, including surveying, taxation, and calendar making, required that many civil servants be competent in certain areas of mathematics. Thus, in the Tang dynasty, as noted in the chapter opening, Li Chunfeng led the effort to collect and annotate what became known as the *Ten Mathematical Classics*. These included the *Arithmetical Classic of the Gnomon*, the *Nine Chapters*, Liu Hui's *Sea Island Mathematical Manual*, the *Mathematical Classic of Master Sun* (fourth century CE), and the *Mathematical Classic of Zhang Qiujuan* (late fifth century CE), among others. An incomplete version of this set exists from the Song dynasty, printed in 1213, and a more complete version from the Ming dynasty, printed in 1403–1407. In general, these mathematical texts studied by candidates for the civil service were collections of problems with methods of solution. New methods were rarely introduced. The examination system often required recitation of relevant passages from the mathematics texts, as well as the solving of problems in the same manner as described in these texts. Thus, even though the Chinese imperial government encouraged the study of applicable mathematics, as indicated in the opening quotation, there was no particular incentive for mathematical creativity.

Nevertheless, creative mathematicians did appear in China, mathematicians who applied their talents not only to improving old methods of solution to practical problems but also to extending these methods far beyond the requirements of practical necessity. We look at developments in four major areas: numerical calculations, geometry, equation solving, and the solution of linear congruences. In particular, new discoveries in the latter two areas were being made into the thirteenth century, especially by Qin Jiushao, Li Ye, Yang Hui, and Zhi Shijie.

7.2 CALCULATIONS

From earliest recorded times, the Chinese used a base-10 system of numbers. But the forms of the numbers and the mode of representation changed over the years.

7.2.1 Number Symbols and Fractions

The Chinese of the Shang dynasty used a multiplicative system of writing numbers, based on powers of 10. That is, they developed symbols for the numbers 1 through 9 as well as for each of the powers of 10. Then, for example, the number 659 would be written using the symbol for 6 (↑↑) attached to that for 100 (◎), then the 5 (✕) attached to the symbol for 10 (/), and finally the symbol Ƽ for 9: ◎ ✕ Ƽ. There are records dating from the fourth century BCE of a physical system of representing numbers by counting rods, small bamboo rods about 10 cm long. These were manipulated on a counting board in which rods were arranged in vertical columns standing for the various powers of 10. There were two possible arrangements of the rods to represent integers less than 10:

1	2	3	4	5	6	7	8	9
I	II	III	IIII	IIIII	⊤	⊤⊤	⊤⊤⊤	⊤⊤⊤⊤
—	=	≡	≣	≣̄	⊥	⊥̄	⊥̿	⊥̿̄

To represent numbers greater than 10, the rods were set up in columns with the rightmost column holding the units, the next the tens, the next the hundreds, and so on. A blank column in a given arrangement represented a zero. To help one read the numbers easily, the two arrangements of rods were alternated. The vertical arrangement was used in the units column, the hundreds column, the ten thousands column, and so on, while the horizontal arrangement was used in the other columns. Thus, 1156 was represented by — I ≡ ⊤ and 6083 by ⊥ ≣ III. These representations also occur in written records of counting-board computations. There is some evidence that a dot was used in this situation to represent an empty column (intermediate zero) as early as the eighth century CE, but it was not until the twelfth century that we have unambiguous evidence of the use of a small circle to represent zero in these situations. Thus, it is only by that time that we can say that Chinese number notation was in the form of a decimal place value system. Our earliest records of fractions in China are of common fractions, designated by symbols representing the words *fen zhi*. For example, 2/3 would be written 3 *fen zhi* 2 and could be translated as "2 parts from a whole broken into 3 equal parts." By medieval times, however, the Chinese were also using decimal fractions in many contexts.

Negative numbers, which were in use from at least the beginning of our era, were represented on the counting board by using some feature to distinguish "negative" rods from "positive." One way was to use red rods for positive numbers and black ones for negative numbers. A negative number was represented in written records by an oblique bar drawn across one of the digits in the rod numeral notation.

Rules for calculating with fractions appear near the beginning of the *Suan shu shu*. For example, the rule for reducing fractions to lowest terms is given as follows:

> Take the numerator and subtract it (successively) from the denominator; also take the denominator and subtract it (successively) from the numerator; (when) the amounts of the numerator and denominator are equal, this will simplify it. Another rule for simplifying fractions says: If it can be halved, halve it; if it can be (successively) divided by a certain number, divide by it. Yet another rule says: Using the numerator of the fraction, subtract it from the denominator; using the remainder as denominator, subtract it (successively) from the numerator; use what is equal to (both) numerator and denominator as the divisor; then it is possible to divide both the numerator and denominator by this number. If it is not possible to subtract but it can be halved, halve the denominator and also halve the numerator. 162/2016, simplified, is 9/112.[3]

In the example, we note that 162 can be subtracted 12 times from 2016, with a remainder of 72. Then 72 can be subtracted twice from 162 with a remainder of 18. Since 72 is now a multiple of 18, 18 is the number (the greatest common divisor) by which we divide both 2016 and 162 to reduce the fraction to 9/112. Note that this is the identical process to the Euclidean algorithm.

The rule for addition of fractions reads as follows:

> If the denominators are of the same kind, add the numerators together; if the denominators are not of the same kind, but some can be doubled to make the denominators equal, then double them; if some can be tripled, then triple (them); . . . likewise, the numerator should be doubled, so double it; when multiplied by 3, 4, or 5 times like the denominators, and if the denominators are the same amount, then add the numerators together. If the denominators are still not of the same kind, then mutually multiply all of the denominators together as the divisor, and after cross-multiplying the numerators with the denominators, add them together as the dividend; and then divide.[4]

The basic idea, illustrated in several problems, is to use as a common divisor the product of the original divisors. Thus, the sum of 2/5, 3/6, 8/10, 7/12, and 2/3 is given as 2 57/60. Rules are also given for the other arithmetic operations on fractions. As an example, the quotient of 7 + 1/2 + 1/3 by 5 is calculated as 1 17/30.

With the basic methods set, the *Suan shu shu* applied the methods to solving many interesting problems. Among these is the one called "Woman Weaving":

> There is a woman in the neighborhood who is displeased with herself, but happy that every day she doubles her weaving. In five days she weaves five *chi* (= 50 *cun*). How much does she weave in the first day, and how much in every day thereafter? The answer: the first day she weaves 1 38/62 *cun*; then 3 and 14/62 *cun*; then 6 and 28/62 *cun*; then 12 56/62 *cun*; then 25 and 50/62 *cun*. The method says: Put down the values 2, 4, 8, 16, 32; add these together as the divisor; taking the 5 *chi*, multiply this by each of them (2, 4, 8, 16, 32) as the dividend; dividing the dividend by the divisor gives the amount of *chi*. If the amount in *chi* is not even, multiply by 10 and express the remainder in *cun*. If the amount in *cun* is not even, use the remainder to determine the fractional amount left over.[5]

7.2.2 Roots

Another type of calculation, discussed in detail in chapter 4 of the *Nine Chapters*, is the determination of square and cube roots. The square root algorithm is based on the algebraic formula $(x + y)^2 = x^2 + 2xy + y^2$, but most probably the author had in mind a diagram like Figure 7.1. We illustrate this algorithm by problem 12 of that chapter, where we are

FIGURE 7.1

Algorithm for determination
of the square root

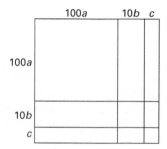

asked to determine the side of a square of area 55,225. The idea is to find digits a, b, c, so that the answer can be written as $100a + 10b + c$. First, find the largest digit a so that $(100a)^2 < 55,225$. In this case, $a = 2$. The difference between the large square (55,225) and the square on $100a$ (40,000) is the large gnomon in the figure. If the outer thin gnomon is neglected, it is clear that b must satisfy $55,225 - 40,000 > 2(100a)(10b)$ or $15,225 > 4000b$. So certainly $b < 4$. To check that $b = 3$ is correct, that is, that with the square on $10b$ included, the area of the large gnomon is still less than 15,225, it is necessary to check that $2(100a)(10b) + (10b)^2 < 15,225$. Because this is in fact true, the same procedure can be repeated to find c: $55,225 - 40,000 - 30(2 \times 200 + 30) > 2 \times 230c$ or $2325 > 460c$. Evidently, $c < 6$. An easy check shows that $c = 5$ gives the correct square root: $\sqrt{55,225} = 235$.

The Chinese algorithm for calculating square roots is similar to one that was taught in schools in recent years. This method gives a series of answers, in this case, 200, 230, 235, each a better approximation to the true result than the one before. Although it appears clear to a modern reader that, if the answer is not a whole number, the procedure could continue indefinitely using decimal fractions, the Chinese author used common fractions as a remainder in the cases where there was no integral square root.

The same chapter 4 also presents a cube root algorithm, essentially based on the binomial expansion $(r + s)^3 = r^3 + 3r^2s + 3rs^2 + s^3$, probably thought of geometrically as in Figure 7.2. For example, we find the cube root of 1,860,867. We begin by noting that the solution is a three-digit number starting with 1. In other words, the closest integer solution can be written as $x = 100 + 10b + c$. Ignoring temporarily the c, we need to find the largest b so that $(100 + 10b)^3 = 100^3 + 3 \cdot 100^2 \cdot 10b + 3 \cdot 100 \cdot (10b)^2 + (10b)^3 \leq 1,860,867$, or so that $3 \cdot 100^2 \cdot 10b + 3 \cdot 100 \cdot 100b^2 + 1000b^3 = b(300,000 + 30,000b + 1000b^2) \leq 860,867$. By trying in turn $b = 1, 2, 3, \ldots$, one discovers that $b = 2$ is the largest value satisfying the inequality. Since $2(300,000 + 60,000 \cdot 2 + 1000 \cdot 2^2) = 728,000$, one next subtracts this number from 860,867 and derives a similar inequality for c: $c(3 \cdot 120^2 + 3 \cdot 120c + c^2) \leq 132,867$. In this case, it turns out that $c = 3$ satisfies this as an equality, so the cube root is $x = 123$.

FIGURE 7.2

Diagram for cube root algorithm

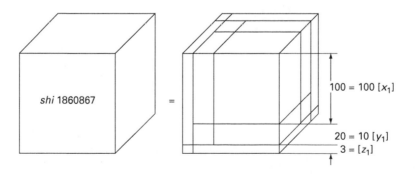

shi 1860867   =

100 = 100 [x_1]

20 = 10 [y_1]

3 = [z_1]

Note that in both of these algorithms, the solution of a quadratic or cubic equation (or, at least, an inequality) is part of the process. The Chinese ultimately developed these ideas into a detailed procedure for solving polynomial equations of any degree, a procedure to be discussed below.

7.3

GEOMETRY

Chinese geometry was generally practical, but in certain cases Chinese mathematicians developed important theoretical principles to enable difficult problems to be solved.

7.3.1 Areas and Volumes

The Chinese developed numerous formulas for calculating the areas and volumes of geometrical figures. Many of them are standard formulas, such as those for the areas of rectangles and triangles or for the volume of parallelepipeds. The *Nine Chapters* also gives the correct formula for the volume of a pyramid. Here, however, we will consider the formulas for the area of a circle and volume of a sphere.

For the area of a circle, the Chinese presented several versions. For example, consider problem 32 from the first chapter of the *Nine Chapters*:[6]

> There is a round field whose circumference is 181 yards and whose diameter is 60 1/3 yards. What is the area of the field? Answer: 2730 1/12 square yards.

The first thing to notice is that the stated diameter of the field is 1/3 of the circumference. In other words, at the time the *Nine Chapters* was written, the number used for the ratio of circumference to diameter of a circle was always taken as 3, the same value used by the Babylonians. Secondly, the Chinese scribe stated not one but four separate formulas by which the calculation of area could be made:

> 1. The rule is: Half of the circumference and half of the diameter are multiplied together to give the area.
> 2. Another rule is: The circumference and the diameter are multiplied together, then the result is divided by 4.
> 3. Another rule is: The diameter is multiplied by itself. Multiply the result by 3 and then divide by 4.
> 4. Another rule is: The circumference is multiplied by itself. Then divide the result by 12.

Of course, given that π is taken to be 3, all of the formulas are equivalent. We also note that it is the fourth rule that is the same as the usual Babylonian rule, but, like the Babylonians, the author of the *Nine Chapters* does not tell why these formulas work.

On the other hand, Liu Hui, in his own commentary, noted that the value "3" for the ratio of circumference to diameter must be incorrect. He did it in the context of the area situation, where the Chinese formula for the area of a circle of radius 1 is 3, but where he could easily calculate that the area of a regular dodecagon inscribed in that circle is also 3. Thus, he concluded, the area of the circle must be larger. In fact, Liu then proceeded to approximate this area by an argument involving the construction of inscribed polygons with more and more sides, an argument that reminds us of Archimedes' own determination of π by using perimeters of polygons. As he wrote, "the larger the number of sides, the smaller the difference between the area of the circle and that of its inscribed polygons. Dividing again and again until it cannot be divided further yields a regular polygon coinciding with the circle, with no portion whatever left out." That is, although he did not use a formal *reductio ad absurdum* argument as in the Eudoxian method of exhaustion, he assumed that, eventually, the polygons will in fact "exhaust" the circle.

We can describe Lui's argument by looking at an inscribed regular n-gon in a circle of radius r. Let c_n be the length of the side of the inscribed n-gon, a_n the length of the perpendicular from the center of the circle to the side, and S_n the total area of the n-gon (Fig. 7.3). We start with $c_6 = r$. In general,

$$a_n = \sqrt{r^2 - \left(\frac{c_n}{2}\right)^2} \quad \text{and} \quad c_{2n} = \sqrt{\left(\frac{c_n}{2}\right)^2 + (r - a_n)^2}.$$

Then

$$S_{2n} = 2n\frac{1}{2}\frac{c_n}{2}r = \frac{1}{2}nrc_n.$$

FIGURE 7.3

Inscribed regular n-gon in a circle of radius r

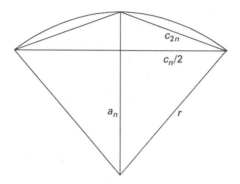

Liu calculated S_{2n} for $n = 96$ in the case of $r = 10$ to be $314\frac{64}{625}$, equivalent to a value for π of 3.141024, and then noted that it is "convenient" to take 3.14 as an approximation to π and neglect the fractional part. Two centuries later, however, Zu Chongzhi (c. 429–500) decided to carry out the calculations further. He found by use of S_{24576} that a better approximation to π was 3.1415926.

Chapter 4 of the *Nine Chapters* gave a rule for determining the diameter d of a sphere of given volume V, which is equivalent to proposing a formula for the volume of a sphere: "Lay down the given number (V). Multiply it by 16; divide it by 9; extract the cube root of the result." In other words, the rule is that $d = \sqrt[3]{(16/9)V}$, equivalent to the volume formula $V = \frac{9}{16}d^3$, or $V = \frac{9}{2}r^3$, where r is the radius. Even taking the usual approximation that $\pi = 3$, this result is incorrect—and Liu Hui described in his commentary how he knew that.

Consider a cylinder inscribed in a cube of side d and consider the cross section of this figure by a plane perpendicular to the axis of the cylinder (Fig. 7.4). The plane cuts the cylinder in a circle of diameter d and the cube in a square of side d. The ratio of the areas of these two plane figures is $\pi : 4$. Since this is true for each cross section, the ratio of the volumes must be the same, so the volume of the cylinder is $(\pi/4)d^3$. (This principle, similar to Archimedes' procedure in the *Method*, is what is now known as Cavalieri's principle.) Now let us consider a sphere inscribed in the cylinder. If the ratio of the volume of the sphere to that of the cylinder were also $\pi : 4$, then the volume of the sphere would be $(\pi^2/16)d^3$, which with π taken equal to 3, is exactly the value given in the *Nine Chapters*. But Liu knew that this was incorrect,

FIGURE 7.4

Cross section of a cylinder
inscribed in a cube

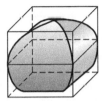

FIGURE 7.5

Intersection of two cylinders
inscribed in the same cube

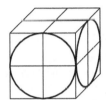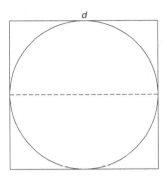

that in fact the ratio of the volume of the sphere to that of the cylinder was not $\pi : 4$. His argument was as follows: Inscribe a second cylinder in the cube, whose axis is perpendicular to that of the first cylinder, and consider the intersection of the two cylinders (Fig. 7.5). He called this intersection the "double box-lid." Since the sphere is contained in each cylinder, it is contained in their intersection. Now any cross section of the box-lid perpendicular to its axis is a square, so the ratio of the volume of the sphere to that of the box-lid is $\pi : 4$. But the box-lid is smaller than the cylinder, so the ratio of the volume of the sphere to the cylinder must be less than $\pi : 4$. So to find a correct formula for the volume of a sphere, it is necessary to find the volume of the box-lid.

Liu Hui could not find this volume; as he wrote, "Let us leave the problem to whomever can tell the truth." That person was Zu Geng (fifth–sixth century), the son of Zu Chongzhi. He formalized Cavalieri's principle as follows: "If the corresponding section areas of two solids are equal everywhere, then their volumes cannot be unequal." In the case of the double box-lid, his argument went like this. Consider 1/8 of the box-lid and inscribe it in a cube of side $r = d/2$ (Fig. 7.6). If we pass a plane through the box-lid at height h, the cross section is a square of side s, where $s^2 = r^2 - h^2$. Therefore, since the plane intersects the circumscribing cube in a square of area r^2, the difference between the two cross sections is h^2. But we know that if we take an inverted pyramid of height r and square base of side r and pass through it a plane at height h (from the vertex), the cross section is also a square of area h^2. It follows that the volume of that part of the inscribing cube outside of the box-lid is equal to the volume of the pyramid, namely, $(1/3)r^3$. Subtracting this from the volume of the cube itself, we find that the volume of 1/8 of the box-lid is $(2/3)r^3$ and therefore the volume of the entire box-lid is $(16/3)r^3$. But the ratio of the volume of the sphere to that of the box-lid is $\pi : 4$. Thus, the volume of the sphere is $(4/3)\pi r^3$.

FIGURE 7.6

One-eighth of the box-lid
inscribed in a cube of side r

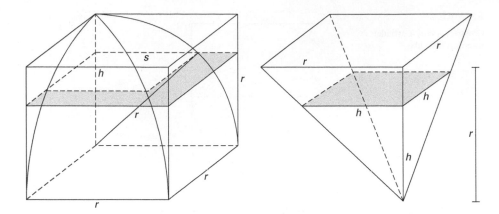

7.3.2   The Pythagorean Theorem and Surveying

The *Nine Chapters* and other ancient Chinese documents assume known the Pythagorean Theorem. And both Zhao Shuang's commentary on the *Arithmetical Classic of the Gnomon* and Liu Hui's commentary on chapter 9 of the *Nine Chapters* contain an argument for the theorem. Both of these arguments describe a diagram, but the original ones are lost. We have reproduced diagrams that later commentators believed to be close to those of the original authors.

Zhao Shuang's argument is as follows (where Figure 7.7 is what is believed to be "the hypotenuse diagram"):

> The base and altitude are each multiplied by themselves. Add to make the hypotenuse area. Take the square root, and this is the hypotenuse. In accordance with the hypotenuse diagram, you may further multiply the base and altitude together to make two of the red areas. Double this to make four of the red areas. Multiply the difference of the base and the altitude by itself to make the central yellow area. If one [such] difference area is added [to the four red areas], the hypotenuse area is completed.[7]

In essence, Zhao seems to be arguing that $c^2 = a^2 + b^2 = (a - b)^2 + 2ab$. Liu's argument, at the beginning of chapter 9, is slightly different (and refers to a diagram probably similar to Figure 7.8):

> The shorter side [of the perpendicular sides] is called the *gou*, and the longer side the *gu*. The side opposite to the right angle is called the hypotenuse. The *gou* is shorter than the *gu*. The *gu* is shorter than the hypotenuse. They apply in various problems. . . . Hence I mention them here so as to show the reader their origin. Let the square on the *gou* be red in color, the square on the *gu* be blue. Let the deficit and excess parts be mutually substituted into corresponding positions, the other parts remain unchanged. They are combined to form the square on the hypotenuse. Extract the square root to obtain the hypotenuse.

Are the arguments given here proofs? To meet modern standards, it would be necessary to show that all figures that appear to be squares are in fact squares and that all the pairs of regions assumed to be equal are in fact equal. To the Chinese, however, and probably to most students today, this was obvious. The Chinese had no notion of an axiomatic system from which theorems could be derived. Here "proof" means simply a convincing argument.

FIGURE 7.7

Zhao's hypotenuse diagram

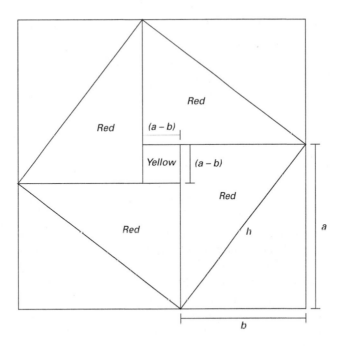

FIGURE 7.8

Possible diagram representing
Liu's argument

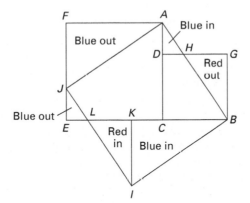

In fact, the Greek word "theorem" is derived from *theorein*, "to look at." If one looks at the diagrams, one sees the theorem at once.

Assuming knowledge of the Pythagorean Theorem, chapter 9 of the *Nine Chapters* contains many problems involving right triangles. Thus, problem 6 concerns a square pond with side 10 feet, with a reed growing in the center whose top is 1 foot out of the water. If the reed is pulled to the shore, the top just reaches the shore. The problem is to find the depth of the water and the length of the reed. In Figure 7.9, $y = 5$ and $x + a = d$, where, in this case, $a = 1$. A modern solution might begin by setting $d^2 = x^2 + y^2$ and substituting for d. A brief algebraic

FIGURE 7.9

Problem 6 of chapter 9 of the *Nine Chapters*

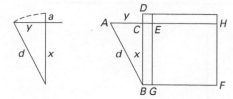

calculation gives $x = \frac{y^2 - a^2}{2a}$. With the given numerical values, $x = 12$ and therefore $d = 13$. The Chinese rule states: "Multiply half of the side of the pond by itself; decrease this by the product of the length of the reed above the water with itself; divide the difference by twice the length of the reed above the water. This gives the depth. Add this to the length of the reed above the water. This gives the length of the reed." A translation of this rule into a formula gives the same $x = \frac{y^2 - a^2}{2a}$ already derived. It is not clear, however, whether the Chinese author found the solution algebraically as above or by the equivalent geometric method illustrated, where $y^2 = AC^2 = AB^2 - BC^2 = BD^2 - EG^2 = DE^2 + 2 \times CE \times BC = a^2 + 2ax$. But what is certain is that the author was fluent in the use of the Pythagorean Theorem.

Most of the final problems in chapter 9 of the *Nine Chapters* deal with surveying questions. When writing his commentary on the *Nine Chapters*, Liu Hui (third century CE) decided to add an addendum on more complicated problems of that type. This addendum ultimately became a separate mathematical work, the *Haidao suanjing* (*Sea Island Mathematical Manual*).

In the continuing tradition of problem texts, the *Sea Island Mathematical Manual* was simply a collection of nine problems with solutions, derivations, illustrations, and commentary. Unfortunately, all that remains today are the problems themselves with the computational directions for finding the solutions. No reasons are given why these particular computations are to be performed, so the following discussion presents some possible methods by which Liu Hui worked out his rules.

The first of the nine problems, and the one for which the text is named, shows how to find the distance and height of a sea island. The others demonstrate how to determine such items as the height of a tree, the depth of a valley, and the width of a river. The sea island problem reads, "for the purpose of looking at a sea island, erect two poles of the same height, 5 feet, the distance between the front and rear pole being 1000 feet. Assume that the rear pole is aligned with the front pole. Move away 123 feet from the front pole and observe the peak of the island from ground level. Move backward 127 feet from the rear pole and observe the peak of the island from ground level again; the tip of the back pole also coincides with the peak. What is the height of the island and how far is it from the front pole?"[8]

Liu Hui's answer is that the height of the island is 1255 feet, while its distance from the pole is 30,750 feet. He also presents the rule for the solution (Fig. 7.10):

Multiply the distance between poles by the height of the pole, giving the *shi*. Take the difference in distances from the points of observations as the *fa* to divide the *shi*. Add what is thus obtained to the height of the pole; the result is the height of the island. [Thus, the height h is given by the formula $h = a + \frac{ab}{c-d}$, where a is the height of the pole, b the distance between the poles, and c and d the respective distances from the poles to the observation points.] To find the distance of the island from the front pole, multiply the distance of the backward movement from the front pole

FIGURE 7.10

Problem 1 of the *Sea Island Mathematical Manual*

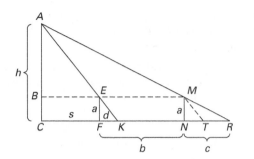

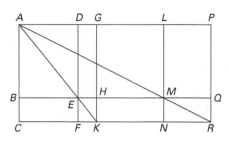

by the distance between the poles, giving the *shi*. Take the difference in distance at the points of observation as the *fa* to divide the *shi*. The result is the distance of the island from the pole. [The distance s is given by $s = \frac{bd}{c-d}$.]

Liu Hui called his method the **method of double differences**, because two differences are used in the solution procedure. A modern derivation of the method would use similar triangles: Construct MT parallel to EK. Then $\triangle AEM$ is similar to $\triangle MTR$ and $\triangle ABM$ is similar to $\triangle MNR$. Therefore, $ME : TR = AM : MR = AB : MN$, so

$$AB = \frac{ME \cdot MN}{TR} = \frac{FN \cdot EF}{TR}$$

and the height $h \ (= AB + BC)$ of the island is

$$h = \frac{FN \cdot EF}{TR} + EF = \frac{ab}{c-d} + a,$$

as noted above. A similar argument gives Liu Hui's result for the distance s of the island.

However, there are other ways of deriving Liu Hui's formula. In the mid-thirteenth century, Yang Hui commented on this particular problem and gave a justification using only congruent triangles and area relationships, a justification more in keeping with what is known about early Chinese mathematical techniques. Since triangles APR and ACR are congruent, as are triangles ALM and ABM, trapezoid $LPRM$ has the same area as trapezoid $BMRC$. Subtracting off the congruent triangles MQR and MNR shows that rectangles $LPQM$ and $BMNC$ are also equal in area. By a similar argument, rectangle $DGHE$ equals rectangle $BECF$. It follows that rectangle $EMNF$ ($=$ rectangle $BMNC -$ rectangle $BECF$) $=$ rectangle $LPQM -$ rectangle $DGHE$. Writing each of the areas of the rectangles as products gives

$$FN \cdot EF = PQ \cdot QM - GH \cdot HE$$
$$= PQ \cdot RN - PQ \cdot FK = PQ(RN - FK) = AB(RN - FK).$$

Therefore, $AB = \frac{FN \cdot EF}{RN - FK}$ and the height $h = AC$ is given by

$$h = AC = AB + BC = \frac{FN \cdot EF}{RN - FK} + EF$$

as desired. The distance $s = CF$ can then be determined by beginning with the equality of the areas of rectangles $DGHE$ and $BCFE$, that is, with $CF \cdot BC = DE \cdot EH$, and replacing $DE = AB$ by the value already found.

In problem 4, Liu Hui calculated the depth of a valley from two observations made along the valley wall. Figure 7.11 illustrates the situation, where x is the desired depth and the measurements are in feet. A modern solution would again use similar triangles. Namely, $6/8.5 = (y + 30)/z$ and $6/9.1 = y/z$. It follows that $6z = 8.5(y + 30) = 9.1y$. So $0.6y = 8.5(30)$ and $y = 8.5(30)/0.6 = 425$. Liu Hui gave precisely this calculation and then noted that the valley depth is 6 feet less than this value, or 419 feet. But again, it is more likely that Liu Hui used an area manipulation similar to that of problem 1 to justify his solution method.

FIGURE 7.11

Problem 4 of the *Sea Island Mathematical Manual*

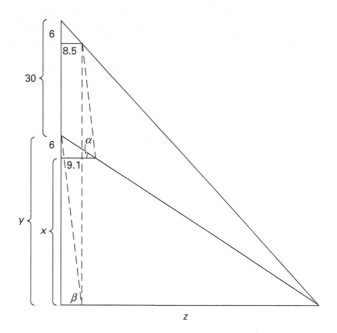

Calculations using similar triangles may often be thought of as "trigonometry" calculations. One can thus consider the instructions in problem 4 as instructions for finding y by multiplying 8.5 by the tangent $(30/0.6)$ of angle α (or angle β) in Figure 7.11. Other problems in the *Haidao suanjing* similarly involve multiplying lengths by tangents of angles. However, because neither Lui Hui nor his later commentators mention angles as such, it would be difficult to characterize the method of this text as trigonometric.

In the eighth century, however, Chinese astronomers did use genuine trigonometric methods involving tables of tangents calculated for various angles. The Chinese emperors, like rulers elsewhere, had always been interested in problems of the calendar, that is, in predicting various celestial events such as eclipses. Unfortunately, Chinese astronomers were not very successful in predicting eclipses, because they did not fully understand the motions of the sun and moon. Indian astronomers, because of Greek influence in the creation of a geometrical model, were more successful. Thus, in the eighth century, when Buddhism was strong both in India and China and there were many reciprocal visits of Buddhist monks, the Chinese emperors of the Tang dynasty brought in Indian scholars as well to provide a new expertise. These scholars, led by Chutan Hsita (early eighth century), prepared an astronomical work in Chinese in 718, the *Chiu-chih li* (*Nine Planets* [*sun, moon, five ordinary planets and two invisible ones*]), based on Indian sources. In particular, this work contained a description of the construction of a sine table in steps of $3°45'$ using a circle radius of $3438'$. (More details will be given in Chapter 8.)

FIGURE 7.12

Yi Xing on a Chinese stamp

In 724, the State Astronomical Bureau of the Tang dynasty began an extensive program of field research to determine the length of the shadows cast by a standard gnomon (of length 8 feet) at latitudes ranging from 29° to 52° along the same meridian (114° E), at the summer and winter solstices and at the equinoxes. These observations were then analyzed by the chief astronomer, Yi Xing (683–727), himself a Buddhist monk (Fig. 7.12). Yi Xing's goal was to use these and other observations, as well as various interpolation techniques, to calculate the length of such shadows, the duration of daylight and night, and the occurrence of eclipses, whatever the position of the observer. (Yi Xing was not aware of the sphericity of the earth and therefore could not make use of the classic Greek model.) Among the tables Yi Xing produced for these purposes in his *Ta yen li* was a shadow table based on the sun's zenith distance α rather than on the latitude and date. Yi Xing's table gave the length of a shadow of a gnomon of 8 feet for each integral value of the zenith angle α from 1 to 79. In modern terms, this is a table of the function $s(\alpha) = 8 \tan \alpha$ and is the earliest recorded version of a tangent table.[9]

It is not known how Yi Xing calculated the table, but a detailed comparison of Yi Xing's work with the standard Indian astronomical works and with the sine table in the *Chiu-chih li* leads one to the tentative conclusion that he interpolated in the sine table and used the resulting values to calculate shadow lengths by the formula $s(\alpha) = 8 \frac{\sin \alpha}{\sin(90-\alpha)}$. In any case, although the *Ta yen li* and even the *Chiu-chih li* were preserved in Chinese compendia, Yi Xing's tangent table ideas were not continued in his own country. Trigonometric methods do not appear again in China until after general contact with the West was opened in the seventeenth century. On the other hand, the next appearance of a shadow (tangent) table is in Islamic sources in the ninth century. Whether transmittal of this idea occurred across central Asia during that century is not known.

7.4 SOLVING EQUATIONS

The Chinese used two basic algorithms to solve systems of linear equations. For equations of higher degree, they developed various procedures for solving them numerically.

7.4.1 Systems of Linear Equations

The *Nine Chapters* contained both algorithms for solving systems. The first method, used chiefly for solving problems we would translate into systems of two equations in two unknowns, is called the method of surplus and deficiency and is found in chapter 7. The methodology, today called the method of "double false position," begins with the "guess-ing" of possible solutions and concludes by adjusting the guess to get the correct solution. Its use showed that the Chinese understood the concept of a linear relationship.

Consider problem 17: "The price of 1 acre of good land is 300 pieces of gold; the price of 7 acres of bad land is 500. One has purchased altogether 100 acres; the price was 10,000. How much good land was bought and how much bad?" A modern translation of this problem would be as a system of two equations in two unknowns:

$$x + y = 100$$

$$300x + \frac{500}{7}y = 10{,}000$$

The Chinese rule for the solution states: "Suppose there are 20 acres of good land and 80 of bad. Then the surplus is $1714\frac{2}{7}$. If there are 10 acres of good land and 90 of bad, the deficiency is $571\frac{3}{7}$." The solution procedure, as explained by the Chinese author, is then to multiply 20 by $571\frac{3}{7}$, 10 by $1714\frac{2}{7}$, add the products, and finally divide this sum by the sum of $1714\frac{2}{7}$ and $571\frac{3}{7}$. The result, $12\frac{1}{2}$ acres, is the amount of good land. The amount of bad land, $87\frac{1}{2}$ acres, is then easily found.

The author did not explain how he arrived at his algorithm, an algorithm that was to turn up in the Islamic world and then in western Europe over a thousand years later. We can express the algorithm by the formula

$$x = \frac{b_1 x_2 + b_2 x_1}{b_1 + b_2},$$

where b_1 is the surplus determined by the guess x_1 and b_2 is the deficiency determined by the guess x_2. One conjecture as to how this algorithm was found begins by noting that the change from the correct but unknown x to the guessed value 20 involves a change in the value of the "function" $300x + (500/7)y$ of $1714\frac{2}{7}$, while a change from 10 to x involves a change in the function value of $571\frac{3}{7}$. Since linearity implies that the ratios of each pair of changes are equal, we derive the proportion

$$\frac{20 - x}{1714\frac{2}{7}} = \frac{x - 10}{571\frac{3}{7}},$$

or, in the general case,

$$\frac{x_1 - x}{b_1} = \frac{x - x_2}{b_2}.$$

The desired solution for x then follows.

Each of the 20 problems in chapter 7 is solved by one or another modification of this algorithm of "surplus and deficiency." For example, two different guesses may both give a surplus or both give a deficiency. In every case, the author gave an explanation of the appropriate calculation. It is certainly possible using modern symbolism to write each of these problems in the same form and give a single (algebraic) solution. But for the Chinese author, there were several different types of problems, each requiring its own solution procedure. Interestingly, the scribes did not hesitate to present problems with unwieldy solutions, perhaps because they wanted to convince their students that a thorough mastery of the methods would enable even difficult problems to be solved.

Chapter 8 of the *Nine Chapters* describes a second method of solving systems of linear equations, again by presenting various examples with slightly different twists. In this case, however, the modern methods are no simpler. In fact, the Chinese solution procedure is virtually identical to the method of Gaussian elimination and is presented in matrix form on a counting board. As an example, here is problem 1 of that chapter. "There are three classes of grain, of which three bundles of the first class, two of the second, and one of the third make 39 measures. Two of the first, three of the second, and one of the third make 34 measures. And one of the first, two of the second, and three of the third make 26 measures. How many measures of grain are contained in one bundle of each class?" The problem can be translated into modern terms as the system

$$3x + 2y + z = 39$$
$$2x + 3y + z = 34$$
$$x + 2y + 3z = 26.$$

The algorithm for the solution is then stated: "Arrange the 3, 2, and 1 bundles of the three classes and the 39 measures of their grains at the right. Arrange other conditions at the middle and at the left." This arrangement is presented in the diagram below:

$$
\begin{array}{ccc}
1 & 2 & 3 \\
2 & 3 & 2 \\
3 & 1 & 1 \\
26 & 34 & 39
\end{array}
$$

The text continues: "With the first class on the right column multiply currently the middle column and directly leave out." This means to multiply the middle column by 3 (the first class on the right) and then subtract off a multiple (in this case, 2) of the right-hand column so that the first number in the middle column becomes 0. The same operation is then performed with respect to the left column. The results are presented as follows:

$$
\begin{array}{ccc}
1 & 0 & 3 \\
2 & 5 & 2 \\
3 & 1 & 1 \\
26 & 24 & 39
\end{array}
\qquad
\begin{array}{ccc}
0 & 0 & 3 \\
4 & 5 & 2 \\
8 & 1 & 1 \\
39 & 24 & 39
\end{array}
$$

"Then with what remains of the second class in the middle column, directly leave out." That is, perform the same operations using the middle column and the left column. The result is given below:

$$
\begin{array}{ccc}
0 & 0 & 3 \\
0 & 5 & 2 \\
36 & 1 & 1 \\
99 & 24 & 39
\end{array}
$$

Because this diagram is equivalent to the triangular system

$$3x + 2y + z = 39$$
$$5y + z = 24$$
$$36z = 99,$$

the author explained how to solve that system by what is today called "back substitution," beginning with $z = 99/36 = 2\frac{3}{4}$.

Although the original author did not explain why this algorithm worked or how it was derived, Liu Hui did give a justification in his commentary: "If the rates in one column are subtracted from those in another, this does not affect the proportions of the remainders." In other words, Liu was essentially justifying the procedure by quoting the "axiom" that when one subtracts equals from equals, the remainders are equal.

Given this procedure of subtracting columns, one might wonder what happened when such a matrix manipulation led to a negative quantity in one of the boxes. A glance at problem 3 of the same chapter shows that this was not a limitation. The method was carried through perfectly correctly for the system

$$\begin{aligned} 2x &+ y & &= 1 \\ & 3y &+ z &= 1 \\ x & &+ 4z &= 1, \end{aligned}$$

a system in which negative quantities appear in the process of completing the algorithm. In fact, the author gave here the rules for adding and subtracting with positive and negative quantities: "For subtraction—with the same signs, take away one from the other; with different signs, add one to the other; positive taken from nothing makes negative, negative from nothing makes positive. For addition—with different signs, subtract one from the other; with the same signs, add one to the other; positive and nothing makes positive; negative and nothing makes negative." Thus, interestingly, rules for dealing with negative numbers arose in China not in the context of solving equations that have no positive solution, but as an intermediate step in the use of a known algorithm designed to solve a problem that does have positive solutions.

As an example with a different difficulty, consider finally problem 13, a system of five equations in six unknowns:

$$\begin{aligned} 2x &+ y & & & & &= s \\ & 3y &+ z & & & &= s \\ & & 4z &+ u & & &= s \\ & & & 5u &+ v &= s \\ x & & & &+ 6v &= s. \end{aligned}$$

The matrix method leads ultimately to the equation $v = 76s/721$. If $s = 721$, then $v = 76$. This is the single answer given. Unfortunately, it is not known if the Chinese considered other possibilities for s or considered the implications of an infinite number of solutions. In general, the Chinese only considered problems with an equal number of equations and unknowns. And there are no records of any discussion of why that situation produces a unique solution or what happens in other situations.

7.4.2 Qin Jiushao and Polynomial Equations

Recall that the *Nine Chapters* contained at least some indications of the solution of quadratic and cubic equations in the description of the procedure for finding square and cube roots. Other polynomial equations appeared elsewhere in China through the centuries. For example, in *The Mathematical Classic of Zhang Quijian*, there appeared the following: Given a segment of a circle with chord $68\frac{3}{5}$ and area $514\frac{32}{45}$, find the height. The solution is given as $12\frac{2}{3}$, but the description of the method is missing from the manuscript. Presumably, the author used the formula $A = \frac{1}{2}h(h + c)$ and converted it into a quadratic equation for h. In this case, after clearing fractions, the equation becomes $45h^2 + 3087h = 46{,}324$. Cubic equations occurred in a work by Wang Xiaotong (early seventh century), but again no method of solution is given other than a cryptic reference to solve according to the rule of cube root extraction. Evidently, then, a method existed for solving such equations during the first millennium of the common era.

It was in the mid-eleventh century that Jia Xian in a work now lost, both generalized the square and cube root procedures of the *Nine Chapters* to higher roots by using the array of numbers known today as the Pascal triangle and extended and improved the method into one usable for solving polynomial equations of any degree. Jia Xian's methods are discussed in a work of Yang Hui written about 1261.

Jia's basic idea stemmed from the original square and cube root algorithms, which made use of the binomial expansions in degrees 2 and 3. He realized that this solution process could be generalized to nth-order roots for $n > 3$ by determining the binomial expansion $(r + s)^n$. In fact, as Yang Hui reports, not only did he write out the Pascal triangle of binomial coefficients through the sixth row (Fig. 7.13), but he also developed the usual method of generating the triangle: "Add the numbers in the two places above in order to find the number in the place below."[10] Yang Hui further explained how Jia used the binomial coefficients to find higher-order roots by a method analogous to that just described.

FIGURE 7.13

Yang Hui's diagram of the Pascal triangle (*Source: "The Chinese Connection between the Pascal Triangle and the Solution of Numerical Equations of Any Degree"* by Lam Lay-Yong, *Historia Mathematica* Vol. 7, No. 4, November 1980. Copyright © 1980 by Academic Press, Inc. Reprinted by permission of Academic Press, Inc., and Dr. Lam Lay-Yong.)

BIOGRAPHY

Qin Jiushao (1202–1261)

Qin Jiushao was probably born in Sichuan during the time when the Mongols under Ghenghis Khan were completing their conquest of North China. The Song dynasty's capital at this time was at Hangzhou, and it was there that Qin studied at the Board of Astronomy, the agency responsible for calendrical computations. Subsequently, Qin wrote, "I was instructed in mathematics by a recluse scholar. At the time of troubles with the barbarians [the mid-1230s], I spent some years at the distant frontier; without care for my safety among the arrows and stone missiles, I endured danger and unhappiness for ten years." To console himself, he then spent time thinking about mathematics. "I made inquiries among well-versed and capable [persons] and investigated mysterious and vague matters. . . . As for the details [of the mathematical problems], I set them out in the form of problems and answers meant for practical use. . . . I selected eighty-one problems and divided them into nine classes; I drew up their methods and their solutions and elucidated them by means of diagrams."[11] The "diagrams" of his *Mathematical Treatise in Nine Sections* are of the positions of the rods on the counting board as solutions to the various problems are described.

Qin served the government later in several offices, but since he "was extravagant and boastful [and] obsessed with his own advancement," he was often relieved of his duties because of corruption. Nevertheless, he became rich. On a magnificently situated plot of land that he obtained by trickery, he had an enormous house constructed, in the back of which was a "series of rooms for lodging beautiful female musicians and singers."[12] In fact, he developed an impressive reputation in love affairs.

Evidently, Jia went even further. He saw that his method could be used to solve arbitrary polynomial equations, especially since these appeared as part of the root extraction process, but that it would be simpler on the counting board to generate the various multiples by binomial coefficients step-by-step rather than from the triangle itself.

The first detailed account of Jia's method for solving equations, probably somewhat improved, appears in Qin Jiushao's *Shushu jiuzhang* (*Mathematical Treatise in Nine Sections*) of 1247. We consider his method in the context of a particular equation, $-x^4 + 763{,}200x^2 - 40{,}642{,}560{,}000 = 0$, where the equation comes from a geometrical problem of finding the area of a pointed field (see Exercise 20). The initial steps in solving such an equation are the same as those in the solution of the pure equation, $x^n = b$, namely, first, determine the number of decimal digits in the answer and, second, guess the appropriate first digit. In the case at hand, the answer is found, by experience or by trial and error, to be a three-digit number beginning with 8. Qin's approach, like that of the old cube root algorithm, was, in effect, to set $x = 800 + y$, substitute this value into the equation, and then derive a new equation in y whose solution would be only a two-digit number. One can then guess the first digit of y and repeat the process. Given the decimal nature of the Chinese number system, the Chinese could repeat this algorithm as often as desired to approximate the answer to any desired level of accuracy. Qin in fact did give answers to some problems to one or two decimal places, but in other cases where the solution is not a whole number, he stated the remainder as a fraction.

The Chinese did not, of course, use modern algebra techniques to "substitute" $x = 800 + y$ into the original equation as William Horner did in his essentially similar method of 1819. The problem was set up on a counting board with each row standing for a particular power of the unknown (Fig. 7.14). For reasons of space, however, we will write the coefficients horizontally. Thus, for the problem at hand, the opening configuration is

$$-1 \qquad 0 \qquad 763{,}200 \qquad 0 \qquad -40{,}642{,}560{,}000.$$

Given that the initial approximation to the root was 800, Qin described what is now called the repeated (synthetic) division of the original polynomial by $x - 800 \ (= y)$. The first step gives the following:

800	1	0	763200	0	40642560000
		-800	-640000	98560000	78848000000
	-1	-800	123200	98560000	38205440000

Qin's description of the counting-board process tells exactly what numbers to multiply and add (or subtract) to give the arrangement on the third line. For example, the -1 is multiplied by 800 and the result added to the 0. That result (-800) is then multiplied by 800 and the product subtracted from the 763,200. In algebraic symbolism, this first step shows that the original polynomial has been replaced by

$$(x - 800)(-x^3 - 800x^2 + 123200x + 98560000) + 38205440000$$

$$= y(-x^3 - 800x^2 + 123200x - 98560000) + 38205440000.$$

Qin repeated the procedure three more times, dividing each quotient polynomial by the same $y = x - 800$. The result is finally that

$$0 = -x^4 + 763200x^2 - 40642560000$$

$$= y\{y[y(-y - 3200) - 3076800] - 826880000\} + 38205440000$$

or

$$-y^4 - 3200y^3 - 3076800y^2 - 826880000y + 38205440000 = 0.$$

Of course, Qin only has numbers on the counting board. His diagrams (one for each step) are here combined into a single large diagram:

FIGURE 7.14

Initial counting-board configuration for solution of $-x^4 + 763{,}200x^2 - 40{,}642{,}560{,}000 = 0$

800	−1	0	763200	0	−40642560000
		−800	−640000	98560000	78848000000
800	−1	−800	123200	98560000	38205440000
		−800	−1280000	−925440000	
800	−1	−1600	−1156800	−826880000	
		−800	1920000		
800	−1	−2400	−3076800		
		−800			
800	−1	−3200			
	−1				

40	−1	−3200	−3076800	−82688000	38205440000
		−40	−129600	−128256000	−38205440000
	−1	−3240	−3206400	−955136000	0

The third line from the bottom contains the coefficients of Qin's equation for y along with his guess of 4 as the first digit of the two-digit answer. (This came simply from dividing 38205440000 by 826880000.) In the example, as is normally the case in our texts today, the answer "comes out even." The equation for y is exactly divisible by $y - 40$. The solution to the original equation is then $x = 840$.

To see the relationship of Qin's description to Jia's method by the Pascal triangle and how the binomial coefficients are generated step-by-step, consider how the equation $x^3 = 1,860,867$ would be solved using Qin's procedure. The layout of the figures in this case is as follows:

100	1	0	0	−1860867
		100	10000	1000000
100	1	100	10000	−860867
		100	20000	
100	1	200	30000	
		100		
100	1	300		
	1			

20	1	300	30000	−860867
		20	6400	728000
20	1	320	36400	−132867
		20	6800	
20	1	340	43200	
		20		
20	1	360		
	1			

3	1	360	43200	−132867
		3	1089	132867
	1	363	44289	0

One can easily see the binomial coefficients in this table. For example, the ninth line implies that the equation for the second-decimal digit b is $(10b)^3 + 3 \cdot 100 \cdot (10b)^2 + 3 \cdot 100^2 \cdot 10b + (100^3 - 1860867) = 0$, exactly as specified by Jia.

Qin himself gave no theoretical justification of his procedure, nor did he mention the Pascal triangle. But since he solved 26 different equations in the *Shushu jiuzhang* by the method and since several of his contemporaries solved similar equations by the same method, it is evident that he and the Chinese mathematical community in general were in possession of a correct algorithm for solving these problems. This algorithm, since it was rediscovered in Europe more than five centuries after Qin's time, deserves a few additional comments.

First, the texts only briefly state how the guessed values for the digits of the root are found. In some cases, it is clear that the solver simply made a trial division of the constant term by the coefficient of the first power of the unknown, as is generally done in the square root algorithm itself. Sometimes several trials are indicated and the author picks one that works. But in general, one can only surmise that the Chinese mathematicians possessed extensive tables of powers, which could be used to make the various guesses. Second, there is no mention in the texts of multiple roots. Qin's fourth-degree equation above, in fact, has another positive root, 240, as well as two negative ones. The root 240 could easily have been found by the same method, provided one had guessed 2 for the initial digit. But in this case, the geometric problem from which the equation was derived had only one solution, 840, and Qin did not deal with equations in the abstract. Third, operations with negative numbers were performed as easily as those with positives. Recall that the Chinese used different-colored counting rods to represent the two types of numbers and had long before discovered the correct arithmetic algorithms for computations. On the other hand, negative roots do not appear, again because the problems from which the equations arise have positive solutions. Fourth, because they could deal with negative numbers, the Chinese generally represented equations in a form equivalent to $f(x) = 0$. This represents a basic difference in approach compared to the ancient Babylonian method or to the medieval Islamic one. Finally, it appears that the Chinese method of solving quadratic equations is completely different from that of the Babylonians. The latter essentially developed a formula that could only be applied to such equations. The Chinese developed a numerical algorithm that they ultimately generalized to equations of any degree.

7.4.3 The Work of Li Ye, Yang Hui, and Zhu Shijie

Qin Jiushao had three contemporaries who also made significant contributions to the mathematics of solving equations, Li Ye (1192–1279), Yang Hui (second half of thirteenth century), and Zhu Shijie (late thirteenth century). But probably due to the war between the Mongols and the two Chinese dynasties of the Jin and the Southern Song, which lasted most of the century, there is doubt that any of these mathematicians had much influence on the others.

Li Ye wrote two major mathematical works, the *Ceyuan haijing* (*Sea Mirror of Circle Measurements*) in 1248 and the *Yigu yanduan* (*Old Mathematics in Expanded Sections*) in 1259, as well as numerous works in other fields. The *Ceyuan haijing* dealt with the properties of circles inscribed in right triangles but was chiefly concerned with the setting up and solution

---

BIOGRAPHY

Li Ye (1192–1279)

Li Ye was born into a bureaucratic family in Zhending in Hebei Province north of the Yellow River. In 1230 he passed the civil service examination and took a government post in the northern kingdom of Jin. But his district, and the entire Jin kingdom, fell to the Mongols within a few years, so

Li gave up hope of an official career and devoted the rest of his life to scholarship. After Kublai Khan ascended the throne in 1260, Li was asked to serve in the Mongol government, and did so briefly. He retired for good in 1266 and returned to seclusion in the Mt. Fenglong district of his birth.

---

of algebraic equations for dealing with these properties. The *Yigu yanduan* similarly dealt with geometric problems on squares, circles, rectangles, and trapezoids, but again its main object was the teaching of methods for setting up the appropriate equations, invariably quadratic, for solving the problem.

We give one example of Li Ye's methods from his *Yigu yanduan*:[13]

> Problem 8: There is a circular pond inside a square field and the area outside the pond is 3300 square feet. The sum of the perimeters of the square and the circle is 300 feet. Find the two perimeters.

Li's discussion was virtually identical to what one would find in a modern text. He set x to be the diameter of the circle and $3x$ ($\pi = 3$) to be the circumference. Then $300 - 3x$ is the perimeter of the square. Squaring that value, he found $90,000 - 1800x + 9x^2$ as the area of 16 square fields. Also, because $\frac{3x^2}{4}$ is the area of one circular pond, $12x^2$ is the area of 16 circular ponds. The difference of the two expressions, namely, $90,000 - 1800x - 3x^2$, is equal to 16 portions of the area outside the pond, or $16 \times 3300 = 52,800$. The desired equation is then $37,200 - 1800x - 3x^2 = 0$. In contrast to the work of Qin, Li Ye merely asserted that 20 was the root, or the diameter, and therefore that 60 was the circumference of the circle and 240 that of the square.

It is interesting that Li Ye nearly always followed his algebraic derivation with a geometric derivation (Fig. 7.15). Here the side of the large square is 300, the sum of the given perimeters. The shaded areas represent 16×3300. Since $300x$ is the area of each long strip, x^2 the area of each small square, and $12x^2$ the total area of the 16 circular ponds, he derived the equation $300^2 - 16 \times 3300 = 6 \times 300x - 9x^2 + 12x^2 = 1800x + 3x^2$, or $37,200 = 1800x + 3x^2$ as before. (Note that the diagram indicates the three small squares at the bottom right.)

The text thus provides more evidence for the development of Chinese mathematics. Not only did the solution method originally have a geometric basis, but the very setting up of the problems did as well. Because the numerical results were recorded and calculated on the counting board, the Chinese scholars ultimately recognized patterns on that board and developed them into numerical algorithms. At the same time, they probably began to abstract the geometrical concept of, for instance, square, into simply a position on the counting board and then into the algebraic idea of the square of an unknown numerical quantity. Once the notion of squares of an unknown became abstract, there was no barrier to considering equations of higher degree. Qin Jiushao's equations were based on real and even geometric

FIGURE 7.15

Problem 8 from Li Ye's *Yigu yanduan*

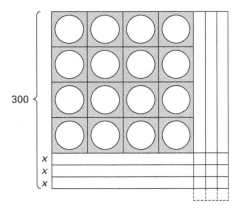

problems, but he had no hesitation about using powers of the unknown, which had no geometric meaning whatsoever.

About Yang Hui, whose reports on the work of Jia Xian were discussed earlier, little is known other than that he lived under the Song dynasty in the south of China. Two major works of his are still extant, the *Xiangjie jiushang suanfa* (*A Detailed Analysis of the Arithmetical Rules in the Nine Sections*) of 1261 and the collection known as *Yang Hui suanfa* (*Yang Hui's Methods of Computation*) of 1275. The latter work, like the work of Li Ye, contains material on quadratic equations. In contrast to Li's work, however, Yang Hui gave a detailed account of his methods. In general, Yang used the same method as Qin, but he also gave alternate methods more reminiscent of the Chinese method of square root extraction described earlier, namely, the explicit use of double the first approximation in deriving the second equation. In addition, Yang presented geometric diagrams consisting of squares and rectangles illustrating the various numerical methods used.

Little is known as well about the life of the last important thirteenth-century Chinese mathematician, Zhu Shijie. He was probably born near present-day Beijing, but spent most of his life as a wandering teacher, that is, as a professional mathematics educator. He wrote two major works, the *Suanxue Qimeng* (*Introduction to Mathematical Studies*) in 1299 and the *Sijuan yujian* (*Precious Mirror of the Four Elements*) in 1303. The first book was elementary, probably intended for beginners or for reference in the Office of Mathematics. In general, problems and methods are repeated, or only slightly modified, from the *Nine Chapters*.

In the *Precious Mirror*, however, we find an important new technique, Zhu's adaptation of Qin's method of solving polynomial equations into a procedure for solving systems of equations in several unknowns. In fact, he was able to work with up to four unknowns, by associating regions of the counting board to each possible combination of powers of one or two of the unknowns (Fig. 7.16). The coefficient of a given combination was then placed in the region associated to that term. For example, the expression $x^2 + y^2 + z^2 + u^2 + 2xy + 2xz + 2xu + 2yz + 2yu + 2zu$ would have been displayed as in Figure 7.17. Zhu then was able to manipulate the coefficients of his equations by manipulating the counting rods in such a way that the system was reduced to a single equation in one unknown. That equation

FIGURE 7.16

Zhu's counting-board representation in four unknowns

$\vdots$	$\vdots$	$\vdots$	$\vdots$	$\vdots$	$\vdots$	$\vdots$
.... y^3u^2	y^2u^2	yu^2	u^2	zu^2	z^2u^2	z^3u^2 ....
.... y^3u	y^2u	yu	u	zu	z^2u	z^3u ....
				$\boxed{yz}$		
.... y^3	y^2	y	z	z^2	z^3	....
		$\boxed{xu}$				
.... xy^3	xy^2	xy	x	xz	xz^2	xz^3 ....
.... x^2y^3	x^2y^2	x^2y	x^2	x^2z	x^2z^2	x^2z^3 ....
.... x^3y^3	x^3y^2	x^3y	x^3	x^3z	x^3z	x^3z^3 ....
$\vdots$	$\vdots$	$\vdots$	$\vdots$	$\vdots$	$\vdots$	$\vdots$

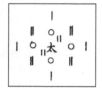

FIGURE 7.17

Representation of $x^2 + y^2 + z^2 + u^2 + 2xy + 2xz + 2xu + 2yz + 2yu + 2zu$ on Zhu's counting board

could then be solved by standard procedures. We illustrate Zhu's procedure by considering problem 2 from the *Precious Mirror*:

> Subtract from the square of the altitude of a right triangle the difference of the hypotenuse and the difference of the altitude and base to equal the product of the altitude and base. It is also given that the square of the base added to the sum of the hypotenuse and the difference of the altitude and base equals the product of the base and hypotenuse. Find the altitude. Answer: 4 *bu*.[14]

The problem concerns a right triangle; if the base is a, altitude b, and hypotenuse c, then the given data produce these equations:

$$b^2 - [c - (b - a)] = ba \quad \text{and} \quad a^2 + c + b - a = ac.$$

In addition, we have the Pythagorean Theorem equation:

$$a^2 + b^2 = c^2.$$

Zhu's first step was to set $x = b$ and $y = a + c$ and then to manipulate the three given equations into the following two:

$$x^3 + 2yx^2 + 2xy - xy^2 - 2y^2 = 0 \tag{7.1}$$

$$x^3 + 2yx - xy^2 + 2y^2 = 0 \tag{7.2}$$

Zhu next proceeded to eliminate the y^2 terms. Thus, he subtracted Equation 7.2 from Equation 7.1 and simplified to get

$$x^2 - 2y = 0. \tag{7.3}$$

Then he multiplied Equation 7.3 by x and substituted $2yx$ for x^3 in Equation 7.1. This simplified to

$$2x^2 + 4x - xy - 2y = 0. \tag{7.4}$$

Finally, he proceeded to eliminate y between Equations 7.3 and 7.4 by first rewriting the two in the form $A_1y + A_2 = 0$ and $B_1y + B_2 = 0$, where A_i and B_i do not contain y, then multiplying the first equation by B_1, the second equation by A_1, and subtracting. What remains is a polynomial without y: $A_2B_1 - A_1B_2 = 0$. Specifically, Equation 7.3 becomes $(-2)y + x^2 = 0$ and Equation 7.4 becomes $(x + 2)y - 2x^2 - 4x = 0$. Then the equation $A_2B_1 - A_1B_2 = 0$ is $(x^3 + 2x^2) - (4x^2 + 8x) = 0$, which simplifies to $x^2 - 2x - 8 = 0$. Zhu could then solve the quadratic equation to get $x = b = 4$, the desired answer.

In more complicated problems, Zhu applied this elimination technique over and over, sometimes to eliminate the square of an unknown before using it again to eliminate the unknown itself. But he was always able eventually to reduce the given system of equations to a single equation in one unknown, which could then be solved. Unfortunately, his description of the method was very cryptic and, in his discussions of several problems, he only wrote out a few of the many auxiliary equations he needed to complete his task. Thus, in Figure 7.18, which reproduces a page of Zhu's book, Equation 7.1 appears in columns g and h near the top; Equation 7.2 appears in the same columns in the middle; A_1B_2 is near the bottom of column h; A_2B_1 is near the top of column i; and the final quadratic equation appears near the bottom of column i.

FIGURE 7.18

Problem 2 from Zhu Shijie's *Precious Mirror*

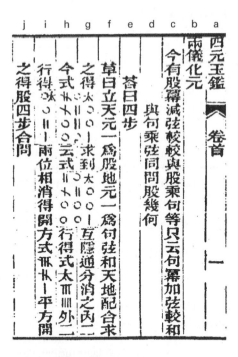

Interestingly, although Zhu, Qin, and others exploited the counting board to the fullest, its very use provided limits. Equations remained numerical, and there could be no development of any theory of equations as was to take place several centuries later in the West. Furthermore, the political changes in China associated with the Mongol and Ming dynasties resulted in a decline in mathematical activity, so that soon even these thirteenth-century works were no longer studied.

INDETERMINATE ANALYSIS

Calendrical problems apparently led the Chinese mathematicians to the question of solving systems of indeterminate linear equations. For example, the Chinese assumed that at a certain point in time, the *Shang yuan*, there occurred simultaneously the beginning of the 60-day cycle used in Chinese dating, the winter solstice, and the new moon. If in a certain other year, the winter solstice occurred r days into a 60-day cycle and s days after the new moon, then that year was N years after *Shang yuan*, where N satisfied the simultaneous congruences

$$aN \equiv r \pmod{60} \quad \text{and} \quad aN \equiv s \pmod{b},$$

where a is the number of days in the year and b is the number of days from new moon to new moon. In the extant records of ancient calendars, however, there is no indication as to how the Chinese astronomers solved such problems.

7.5.1 The Chinese Remainder Problem

Simpler versions of congruence problems occur in various mathematical works. In fact, probably the most famous mathematical technique coming from China is the technique long known as the Chinese remainder theorem. This result was so named after a description of some congruence problems appeared in one of the first reports in the West on Chinese mathematics, articles by Alexander Wylie published in 1852 in the *North China Herald*, which were soon translated into both German and French and republished in European journals. The earliest example in Chinese mathematics of this procedure for solving systems of linear congruences is in the *Sunzi suanjing* (*Mathematical Classic of Master Sun*), a work probably written late in the third century: "We have things of which we do not know the number; if we count them by threes, the remainder is 2; if we count them by fives, the remainder is 3; if we count them by sevens, the remainder is 2. How many things are there?" In modern notation, the problem is to find N, which simultaneously satisfies

$$N = 3x + 2 \qquad N = 5y + 3 \qquad N = 7z + 2$$

for integral values x, y, z, or, what amounts to the same thing, which satisfies the congruences

$$N \equiv 2 \pmod 3 \qquad N \equiv 3 \pmod 5 \qquad N \equiv 2 \pmod 7.$$

Sun Zi gave the answer, 23, as well as his method of solution: "If you count by threes and have the remainder 2, put 140. If you count by fives and have the remainder 3, put 63. If you count by sevens and have the remainder 2, put 30. Add these numbers and you get 233. From this subtract 210 and you get 23." Sun Zi explains further: "For each unity as remainder when counting by threes, put 70. For each unity as remainder when counting by fives, put 21. For each unity as remainder when counting by sevens, put 15. If the sum is 106 or more, subtract 105 from this and you get the result."[15]

In modern notation, Sun Zi apparently noted that

$$70 \equiv 1 \pmod 3 \equiv 0 \pmod 5 \equiv 0 \pmod 7,$$
$$21 \equiv 1 \pmod 5 \equiv 0 \pmod 3 \equiv 0 \pmod 7,$$

and

$$15 \equiv 1 \ (\mathrm{mod}\ 7) \equiv 0 \ (\mathrm{mod}\ 3) \equiv 0 \ (\mathrm{mod}\ 5).$$

Hence, $2 \times 70 + 3 \times 21 + 2 \times 15 = 233$ satisfies the desired congruences. Since any multiple of 105 is divisible by 3, 5, and 7, one subtracts off 105 twice to get the smallest positive value. Because this problem is the only one of its type presented by Sun Zi, it is not known whether he had developed a general method of finding integers congruent to 1 modulo m_i but congruent to 0 modulo m_j, $j \neq i$, for given integers $m_1, m_2, m_3, \ldots, m_k$, the most difficult part of the complete solution. The numbers in this particular problem are easy enough to find by inspection, but note for future reference that $70 = \frac{3 \times 5 \times 7}{3} \times 2$, $21 = \frac{3 \times 5 \times 7}{5} \times 1$, and $15 = \frac{3 \times 5 \times 7}{7} \times 1$.

Perhaps two centuries after Sun Zi, *Zhang Quijian's Mathematical Manual* contained the initial appearance of the problem of the "hundred fowls," famous because it also occurs in various guises in mathematics texts in India, the Islamic world, and Europe. Zhang's original problem was as follows: "A rooster is worth 5 coins, a hen 3 coins, and 3 chicks 1 coin. With 100 coins we buy 100 of the fowls. How many roosters, hens, and chicks are there?"[16] In modern notation, with x the number of roosters, y the number of hens, and z the number of chicks, the problem translates into two equations in three unknowns:

$$5x + 3y + \frac{1}{3}z = 100$$
$$x + y + z = 100$$

Zhang gave three answers: 4 roosters, 18 hens, 78 chicks; 8 roosters, 11 hens, 81 chicks; and 12 roosters, 4 hens, 84 chicks; but he only hinted at a method: "Increase the roosters every time by 4, decrease the hens every time by 7, and increase the chicks every time by 3." Namely, he noted that changing the values this way preserves both the cost and the number of fowls. It is possible to solve this problem by a modification of the "Gaussian elimination" method known from the *Jiuzhang suanshu* and get as a general solution $x = -100 + 4t$, $y = 200 - 7t$, $z = 3t$, from which Zhang's description follows. In fact, Zhang's answers are the only ones in which all three values are positive. It is not known, however, if Zhang used this method or some other one.

Several Chinese authors over the next centuries commented on this hundred fowls problem, but none succeeded in giving a reasonable explanation of the method or a way of generalizing it to other problems. No explanation of Sun Zi's remainder problem appeared either, although there is a record of a calendrical computation by Yi Xing in the early eighth century that used indeterminate analysis to relate several astronomical cycles by solving the simultaneous congruences

$$N \equiv 0 \ (\mathrm{mod}\ 1{,}110{,}343 \times 60),$$
$$N \equiv 44{,}820 \ (\mathrm{mod}\ 60 \times 3040),$$
$$N \equiv 49{,}107 \ (\mathrm{mod}\ 89{,}773).$$

The answer is given as $N = 96{,}961{,}740 \times 1{,}110{,}343$.

7.5.2   Qin Jiushao and the *Ta-Yen* Rule

It was Qin Jiushao who first published a general method for solving systems of linear congruences in his *Mathematical Treatise in Nine Sections*. Qin there described what he called the *ta-yen* rule for solving simultaneous linear congruences, congruences that in modern notation are written $N \equiv r_i \pmod{m_i}$ for $i = 1, 2, \ldots, n$. In fact, ten of the problems in the *Mathematical Treatise* are remainder problems of this type. In particular, we will follow Qin's method to solve Problem I, 5. In this problem, the m_i are relatively prime in pairs, although Qin dealt with the more general case in other problems.

> There are three farmers of the highest class. As for the rice they got by cultivating their fields, when making use of full *dou*, the amounts are the same. All of them go to different places to sell it. After selling his rice on the official market of his own prefecture, *A* is left with 3 *dou* and 2 *sheng*. After selling his rice to the villagers of Anji, *B* is left with 7 *dou*. After selling his rice to a middleman from Pingjiang, *C* is left with 3 *dou*. How much rice did each farmer have initially and how much did each one sell? Note: The *hu* [a dry measure] of the local office for *A* is worth 83 *sheng*, that of Anji is worth 110 *sheng*, and that of Pingjiang is worth 135 *sheng*. [Note: 1 *dou* = 10 *sheng*.] Answer: Total amount of rice: 7380 *dou* to be divided among the three men, or 2460 *dou* each; amount of rice sold by *A*, 296 *hu*; by *B*, 223 *hu*; by *C*, 182 *hu*. [17]

This problem results in the following congruence:

$$N \equiv 32 \pmod{83} \quad N \equiv 70 \pmod{110} \quad N \equiv 3 \pmod{27}.$$

The first step is to determine M, the product of the moduli. In this case, $M = 83 \times 110 \times 27 = 246{,}510$. Since any two solutions to the system will be congruent modulo M, once Qin found one solution, he generally found the smallest positive solution by subtracting off sufficient copies of this value.

For the second step, Qin divided M by each of the moduli m_i in turn to get values we will designate by M_i. Here $M_1 = M \div m_1 = 246{,}510 \div 83 = 2970$, $M_2 = 246{,}510 \div 110 = 2241$, and $M_3 = 246{,}510 \div 27 = 9130$. Each M_i satisfies $M_i \equiv 0 \pmod{m_j}$ for $j \neq i$.

In the third step, Qin subtracted from each of the M_i as many copies of the corresponding m_i as possible; that is, he found the remainders of M_i modulo m_i. These remainders, labeled P_i, are $P_1 = 2970 - 35 \times 83 = 65$, $P_2 = 2241 - 20 \times 110 = 41$, and $P_3 = 9130 - 338 \times 27 = 4$. Of course, $P_i \equiv M_i \pmod{m_i}$ for each i, so P_i and m_i are relatively prime.

It is finally time to solve congruences, in particular, the congruences $P_i x_i \equiv 1 \pmod{m_i}$. Once this is done, one answer to the problem is easily seen to be

$$N = \sum_{i=1}^{n} r_i M_i x_i,$$

in analogy with the solution to Sun Zi's problem. Because each m_i divides M, any multiple of M can be subtracted from N to get other solutions.

To solve $P_i x_i \equiv 1 \pmod{m_i}$ with P_i and m_i relatively prime, Qin used what he called the "technique of finding one," essentially the Euclidean algorithm. Qin described it using diagrams of the counting board. We can demonstrate the technique by solving $P_1 x_1 \equiv 1 \pmod{m_1}$, that is, $65 x_1 \equiv 1 \pmod{83}$. Qin began by placing 65 in the upper right of a counting board with four squares, 83 in the lower right, 1 at the upper left, and nothing in the lower left. As he wrote, "first divide right bottom by right top, multiply the quotient obtained by the top left and [add it to] the bottom left [at the same time replacing the bottom right by the remainder of the division]. And then use the right column top and bottom; using the smaller to divide the greater, dividing alternately, immediately multiply by the quotient obtained [and add it] successively . . . into the left column top or bottom until finally the top right is just 1, then stop. Then take the top left result [as the solution]."[18] The diagrams in Figure 7.19 represent the following computations:

$$83 = 1 \cdot 65 + 18 \qquad 1 \cdot 1 + 0 = 1$$
$$65 = 3 \cdot 18 + 11 \qquad 3 \cdot 1 + 1 = 4$$
$$18 = 1 \cdot 11 + 7 \qquad 1 \cdot 4 + 1 = 5$$
$$11 = 1 \cdot 7 + 4 \qquad 1 \cdot 5 + 4 = 9$$
$$7 = 1 \cdot 4 + 3 \qquad 1 \cdot 9 + 5 = 14$$
$$4 = 1 \cdot 3 + 1 \qquad 1 \cdot 14 + 9 = 23$$

FIGURE 7.19

Counting-board diagrams for solving $65x \equiv 1 \pmod{83}$ by the method of Qin Jiushao

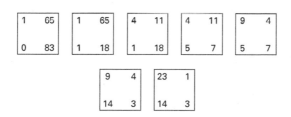

The last numbers in the second column can be thought of as representing the absolute values of the successive coefficients of 65 obtained by substitution. Namely, begin with $18 = 83 - 1 \cdot 65$ and substitute this into $11 = 65 - 3 \cdot 18$ to get $11 = 65 - 3 \cdot (83 - 1 \cdot 65) = 4 \cdot 65 - 3 \cdot 83$, where the 4 is the result of the second calculation in the second column. Similarly, $7 = 18 - 1 \cdot 11 = (83 - 1 \cdot 65) - 1 \cdot (4 \cdot 65 - 3 \cdot 83) = 4 \cdot 83 - 5 \cdot 65$. The final result is that $1 = 23 \cdot 65 - 18 \cdot 83$, and $x_1 = 23$ is a solution to the congruence. (Qin always adjusted matters so that the final coefficient is positive.)

To complete the original problem, we note that $x_2 = 51$ and $x_3 = 7$. It follows that

$$N = \sum_{i=1}^{3} r_i M_i x_i = 32 \cdot 2970 \cdot 23 + 70 \cdot 2241 \cdot 51 + 3 \cdot 9130 \cdot 7$$
$$= 2{,}185{,}920 + 8{,}000{,}370 + 191{,}730 = 10{,}378{,}020.$$

We then determine the smallest positive solution by subtracting off $42M = 42 \cdot 246{,}510 = 10{,}353{,}420$ to get our final answer, $N = 24{,}600$ *sheng*. The rest of the problem is then easily solved.

7.6 TRANSMISSION TO AND FROM CHINA

Not much is known about the possible transmission of mathematical ideas between China and other cultures before the sixteenth century. All that is known is that there are certain similarities in techniques in the mathematics of China, India, Europe, and the Islamic world.

For example, the Chinese essentially used a decimal place value system on their counting board and even represented an empty place by a dot by the seventh century. But whether the Chinese system influenced the Indian development of our modern decimal place value system is not known. Similarly, Indian mathematicians used a technique involving the Euclidean algorithm to solve simultaneous congruences, while Islamic mathematicians used a technique related to Horner's method to solve polynomial equations numerically. Similarly, Europeans eventually discovered a method of solving the Chinese remainder problem fully equivalent to Qin's method, although it took many years to prove that this method worked in the case where the moduli are not relatively prime in pairs. However, in all these cases, there are sufficient differences in detail to rule out direct copying from one civilization to the other. Whether the ideas traveled, however, is much more difficult to answer.

At the end of the sixteenth century, the Jesuit priest Mateo Ricci (1552–1610) came to China (Fig. 7.20). Ricci and one of his Chinese students, Xu Guangqi (1562–1633), translated the first six books of Euclid's *Elements* into Chinese in 1607. And although it took many years for the Chinese to understand that the form and content of Euclidean geometry were inseparable (to Western minds, at least), nevertheless from this time period forward, Western mathematics began to enter China and the indigenous mathematics began to disappear.

FIGURE 7.20

Matteo Ricci on a stamp from Taiwan

EXERCISES

1. The basic Chinese symbols for numbers from the Shang period are

   | 1 | 2 | 3 | 4 | 5 | 6 | 7 | 8 | 9 | 10 | 100 | 1000 |

   There were compound symbols for 20, 30, 40 (namely, ∪ ⋃ ⋓), but in general notation followed the plan indicated in the text. Hence, 88 is)( )(and 162 is ⊘↑=. Write the Chinese form of 56, 554, 63, and 3282.

2. Use the Chinese square root algorithm to find the square root of 142,884.

3. Use the Chinese cube root algorithm to find the cube root of 12,812,904.

4. Solve explicitly the "Woman Weaving" problem of the *Suan shu shu*, using the method described there. Is there a modern method that is easier?

5. Solve problem 3 of chapter 3 of the *Nine Chapters*: Three people, who have 560, 350, and 180 coins, respectively, are required to pay a total tax of 100 coins in proportion to their wealth. How much does each pay?

6. Solve problem 26 of chapter 6 of the *Nine Chapters*: There is a reservoir with five channels bringing in water. If only the first channel is open, the reservoir can be filled in 1/3 of a day. The second channel by itself will fill the reservoir in 1 day, the third channel in 2 1/2 days, the fourth one in 3 days, and the fifth one in 5 days. If all the channels are open together, how long will it take to fill the reservoir? (This problem is the earliest known one of this type. Similar problems appear in later Greek, Indian, and Western mathematics texts.)

7. Solve problem 28 of chapter 6 of the *Nine Chapters*: A man is carrying rice on a journey. He passes through three customs stations. At the first, he gives up 1/3 of his rice, at the second 1/5 of what was left, and at the third, 1/7 of what remains. After passing through all three customs stations, he has left 5 pounds of rice. How much did he have when he started? (Versions of this problem occur in later sources in various civilizations.)

8. Perform the calculations in Liu Hui's algorithm for determining π to find S_{2n} for $n = 6, 12, 24, 48,$ and 96.

9. Use calculus to confirm that the volume of the box-lid, the intersection of two perpendicular cylinders of radius r, is $\frac{16}{3}r^3$.

10. Turn Zhao Shuang's argument into a modern proof of the Pythagorean Theorem.

11. Turn Liu Hui's argument into a modern proof of the Pythagorean Theorem.

12. Solve problem 1 of chapter 7 of the *Nine Chapters* using the method of surplus and deficiency: Several people purchase

in common one item. If each person paid 8 coins, the surplus is 3; if each paid 7, the deficiency is 4. How many people were there and what is the price of the item?

13. Solve problem 8 of chapter 9 of the *Nine Chapters*: The height of a wall is 10 *ch'ih*. A pole of unknown length leans against the wall so that its top is even with the top of the wall. If the bottom of the pole is moved 1 *ch'ih* farther from the wall, the pole will fall to the ground. What is the length of the pole?

14. Show that the diameter D of the largest circle that can be inscribed in a right triangle with legs a and b and hypotenuse c is given by $D = 2ab/(a + b + c)$. (This is a generalization of problem 16 of chapter 9 of the *Nine Chapters*, which uses the specific 8-15-17 triangle.)

15. In the same situation as Exercise 14, show that D may be expressed as $D = a - (c - b)$ or as $D = \sqrt{2(c - a)(c - b)}$.

16. Solve problem 20 of chapter 9 of the *Nine Chapters*: A square walled city of unknown dimensions has four gates, one at the center of each side. A tree stands 20 *pu* from the north gate. One must walk 14 *pu* southward from the south gate and then turn west and walk 1775 *pu* before one can see the tree. What are the dimensions of the city?

17. Solve problem 24 of chapter 9 of the *Nine Chapters*. (This is an example of the type of elementary surveying problem that stimulated Liu Hui to write his *Sea Island Mathematical Manual*.) A deep well 5 feet in diameter is of unknown depth (to the water level). If a 5-foot post is erected at the edge of the well, the line of sight from the top of the post to the edge of the water surface below will pass through a point 0.4 feet from the lip of the well below the post. What is the depth of the well?

18. Solve problem 3 of the *Sea Island Mathematical Manual*: To measure the size of a square walled city $ABCD$, we erect two poles 10 feet apart at F and E (Fig. 7.21). By moving northward 5 feet from E to G and sighting on D, the line of observation intersects the line EF at a point H such that $HE = 3\frac{93}{120}$ feet. Moving to point K such that $KE = 13\frac{1}{3}$ feet, the line of sight to D passes through F. Find DC and EC. (Liu Hui gets $DC = 943\frac{3}{4}$ feet while $EC = 1245$ feet.)

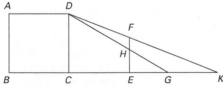

FIGURE 7.21

Problem 3 of the *Sea Island Mathematical Manual*

19. Find the solution to problem 3 of chapter 8 of the *Nine Chapters* using the Chinese method: None of the yields of 2 bundles of the best grain, 3 bundles of ordinary grain, and 4 bundles of the worst grain are sufficient to make a whole measure. If we add to the good grain 1 bundle of the ordinary, to the ordinary 1 bundle of the worst, and to the worst 1 bundle of the best, then each yield is exactly one measure. How many measures does 1 bundle of each of the three types of grain contain? Show that the solution according to the Chinese method involves the use of negative numbers.

20. The numerical equation from Qin Jiushao's *Shushu jiuzhang* analyzed in Section 7.4.2 came from the geometrical problem of finding the area of a pointed field. If the sides and one diagonal are labeled as in Figure 7.22, show that the area of the lower triangle is given by $B = (c/2)\sqrt{b^2 - (c/2)^2}$ and that of the upper triangle by $A = (c/2)\sqrt{a^2 - (c/2)^2}$. Then the area x of the entire field is given by $x = A + B$. Show that x satisfies the fourth-degree polynomial equation

$$-x^4 + 2(A^2 + B^2)x^2 - (A^2 - B^2)^2 = 0.$$

If $a = 39$, $b = 20$, and $c = 30$, show that this equation becomes the one solved by Qin in the text.

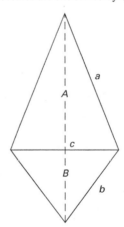

FIGURE 7.22

A pointed field from the *Shushu jiuzhang*

21. Solve the equation $16x^2 + 192x - 1863.2 = 0$ numerically using Qin Jiushao's procedure. This equation is taken from his text.

22. Use Qin's method to solve the pure cubic equation $x^3 = 12,812,904$. Compare this method with the old cube root algorithm discussed in the text. In each case, show where

the third-order coefficients of the Pascal triangle 3 3 1 appear in the solution procedures.

23. Solve the pure fourth-degree equation $y^4 = 279,841$ using Qin's procedure. Show how the fourth-order coefficients of the Pascal triangle 4 6 4 1 appear in the solution procedure.

24. Provide the details for the first step of Zhu Shijie's solution to problem 2 of his *Precious Mirror*. That is, let a be the base, b the altitude, and c the hypotenuse of a right triangle, and assume

$$b^2 - [c - (b - a)] = ba \quad \text{and} \quad a^2 + c + b - a = ac.$$

Then set $x = b$ and $y = a + c$. Show that the two given equations along with the Pythagorean Theorem imply that the following two equations hold:

$$x^3 + 2yx^2 + 2xy - xy^2 - 2y^2 = 0 \quad \text{and}$$
$$x^3 + 2yx - xy^2 + 2y^2 = 0.$$

25. Solve Problem I, 1, from the *Shushu jiuzhang*, which is equivalent to $N \equiv 0 \pmod 3$, $N \equiv 1 \pmod 4$.

26. Solve Problem I, 4, from the *Shushu jiuzhang*, which is equivalent to $N \equiv 0 \pmod{11}$, $N \equiv 0 \pmod 5$, $N \equiv 4 \pmod 9$, $N \equiv 6 \pmod 8$, $N \equiv 0 \pmod 7$.

27. Devise a lesson teaching the Pythagorean Theorem using material from Chinese sources.

28. Explain why Qin's method of solving the congruence $P_i x_i \equiv 1 \pmod{m_i}$ works.

29. Imbed your explanation in the previous problem into a lesson teaching the Chinese remainder theorem in a number theory course, based on the methods of Qin Jiushao.

30. Devise a lesson teaching a method of solving simultaneous linear equations using Chinese sources. In the lesson, you should explain why the method works.

31. Liu Hui's method for finding the height of a distant object was used in many cultures around the world up until the seventeenth century. Curiously, this method was even used in cultures that understood methods of solving triangles using trigonometry. Discuss why this method would continue to be used, even in those circumstances.

REFERENCES AND NOTES

There are two recent surveys of Chinese mathematics available in English: Li Yan and Du Shiran, *Chinese Mathematics—A Concise History*, translated by John N. Crossley and Anthony W. C. Lun (Oxford: Clarendon Press, 1987), and Jean-Claude Martzloff, *A History of Chinese Mathematics*, translated by Stephen S. Wilson (Springer: Berlin, 1997). Two older surveys, which are still useful, are J. Needham, *Science and Civilization in China* (Cambridge: Cambridge University Press, 1959), vol. 3, and Yoshio Mikami, *The Development of Mathematics in China and Japan* (New York: Chelsea, 1974). Briefer surveys include Frank Swetz, "The Evolution of Mathematics in Ancient China," *Mathematics Magazine* 52 (1979), 10–19, Philip D. Straffin, Jr., "Liu Hui and the First Golden Age of Chinese Mathematics," *Mathematics Magazine* 71 (1998), 163–181, and Man-Keung Siu, "An Excursion in Ancient Chinese Mathematics," in Victor J. Katz, ed., *Using History to Teach Mathematics: An International Perspective* (Washington, DC: MAA, 2000).

Two English translations of the *Suan shu shu* are available: Christopher Cullen, *The Suan shu shu, "Writings on Reckoning"* (Cambridge: Needham Research Institute, 2004), and Joseph W. Dauben, "*Suan shu shu* (A Book on Numbers and Computations): English Translation with Commentary," *Archive for History of Exact Sciences* (2007). The *Zhou bi suan jing* is available in Christopher Cullen, *Astronomy and Mathematics in Ancient China: The Zhou bi suan jing* (Cambridge: Cambridge University Press, 1995). An English translation of chapter 9 of the *Jiuzhang suanshu* with commentary has been published by Frank Swetz and T. I. Kao as *Was Pythagoras Chinese?* (Reston, VA: N.C.T.M., 1977). The entire work has now been translated into English, with much commentary, by Shen Kangshen, John N. Crossley, and Anthony W.-C. Lun, entitled *The Nine Chapters on the Mathematical Art: Companion and Commentary* (Oxford: Oxford University Press, 1999). The *Sea Island Mathematical Manual* is also available in English in Frank J. Swetz, *The Sea Island Mathematical Manual: Surveying and Mathematics in Ancient China* (University Park, PA: Pennsylvania State University Press, 1992). A detailed work on aspects of Chinese mathematics in the thirteenth century and its relationship to mathematics at other times and in other countries, which also includes translations of much of Qin Jiushao's *Mathematical Treatise in Nine Sections,* is Ulrich Libbrecht, *Chinese Mathematics in the Thirteenth Century: The Shu-shu chiu-chang of Ch'in Chiu-shao* (Cambridge: MIT Press, 1973). Excerpts from all of these translations as well as from many others are in Victor Katz, ed., *The Mathematics of Egypt, Mesopotamia, China, India, and Islam: A Sourcebook* (Princeton: Princeton University Press, 2007).

Finally, there is a guide to the literature on Chinese mathematics in Frank Swetz and Ang Tian Se, "A Brief Chronological and Bibliographic Guide to the History of Chinese Mathematics," *Historia Mathematica* 11 (1984), 39–56.

1. E. L. Konantz, "The Precious Mirror of the Four Elements," *China Journal of Arts and Science* 2 (1924), 304–310. This quotation is from the introduction to Zhu Shijie's work by Chien Chiu Shimoju.

2. Martzloff, *Chinese Mathematics*, p. 123.

3. Katz, *Sourcebook*, p. 206.

4. Ibid., pp. 206–207.

5. Ibid., p. 208.

6. All quotations from the *Nine Chapters* are taken from Shen, Crossley, and Lun, *The Nine Chapters on the Mathematical Art*.

7. Cullen, *Astronomy and Mathematics*, p. 208.

8. All quotations from the *Sea Island Mathematical Manual* are taken from the section on that book in Shen, Crossley, and Lun, *The Nine Chapters on the Mathematical Art*.

9. For more information on trigonometrical work in China, see Yabuuti Kiyosi, "Researches on the *Chiu-chih li*—Indian Astronomy under the T'ang Dynasty," *Acta Asiatica* 36 (1979), 7–48, and Christopher Cullen, "An Eighth Century Chinese Table of Tangents," *Chinese Science* 5 (1982), 1–33. The first article contains an English translation and commentary on the *Chiu-chih li*, while details of how and why Yi Xing developed his table of tangents are found in the second article.

10. Lam Lay Yong, "On the Existing Fragments of Yang Hui's Hsiang Chieh Suan Fa," *Archive for History of Exact Sciences* 5 (1969), 82–86. For more details on the use of the Pascal triangle in solving equations, see Lam Lay Yong, "The Chinese Connection between the Pascal Triangle and the Solution of Numerical Equations of Any Degree," *Historia Mathematica* 7 (1980), 407–424.

11. Libbrecht, *Chinese Mathematics*, p. 62.

12. Ibid., p. 31.

13. Lam Lay Yong and Ang Tian Se, "Li Ye and His Yi Gu Yan Duan," *Archive for History of Exact Sciences* 29 (1984), 237–266.

14. Katz, *Sourcebook*, p. 347.

15. Libbrecht, *Chinese Mathematics*, p. 269.

16. Ibid., p. 277.

17. Ibid., p. 399.

18. Li Yan and Du Shiran, *Chinese Mathematics*, p. 163.

Ancient and Medieval India

In all those transactions which relate to worldly, Vedic, or ... religious affairs, calculation is of use. In the science of love, in the science of wealth, in music and in the drama, in the art of cooking, and similarly in medicine and in things like the knowledge of architecture; in prosody, in poetics and poetry, in logic and grammar and such other things, ... the science of computation is held in high esteem. In relation to movements of the sun and other heavenly bodies, in connection with eclipses and the conjunction of planets ... it is utilized. The number, the diameter and the perimeter of islands, oceans and mountains, the extensive dimensions of the rows of habitations and halls belonging to the inhabitants of the world, ... all of these are made out by means of computation.

—Mahāvīra's *Gaṇitāsarasaṅgraha*[1]

This story, probably a myth, appears in the work of a Persian commentator on the work of the Indian mathematician Bhāskara (1114–1185). It seems that astrologers predicted that his daughter Līlāvatī would not wed. But her father, being an expert astronomer and astrologer himself, divined the one lucky moment for her marriage. The time was kept by a water clock, but shortly before the exact hour, while Līlāvatī was looking into the clock, a pearl from her headdress accidentally dropped into the clock unnoticed and stopped the flow of water. By the time it was discovered, the designated moment had passed. To console his daughter, Bhāskara named the chapter on arithmetic of his major work, the *Siddhāntaśiromaṇi*, after her.

8.1

INTRODUCTION TO MATHEMATICS IN INDIA

A civilization called the Harappan arose in India on the banks of the Indus River in the third millennium BCE, but there is no direct evidence of its mathematics. The earliest Indian civilization for which there is such evidence was formed along the Ganges River by Aryan tribes migrating from the Asian Steppes late in the second millennium BCE. From about the eighth century BCE, there were monarchical states in the area that had to deal with such activities as fortifications, administrative centralization, and large-scale irrigation works. These states had a highly stratified social system headed by the king and the priests (brahmins). The literature of the brahmins was oral for many generations, expressed in lengthy verses called Vedas. Although these verses probably achieved their current form by 600 BCE, there are no written records dating back beyond the current era.

Some of the material from the Vedic era describes the intricate sacrificial system of the priests, the bearers of the religious traditions that grew into Hinduism. It is in these works, the *Śulbasūtras*, that we find mathematical ideas. Curiously, however, although this mathematics deals with the theoretical requirements for building altars out of bricks, as far as is known the early Vedic civilization did not have a tradition of brick technology, while the Harappan culture did. Thus, there is a possibility that the mathematics in the *Sulbasutras* was created in the Harappan period, although the mechanism of its transmission to the later period is currently unknown. In any case, it is the *Śulbasūtras* that are the sources for our knowledge of ancient Indian mathematics.

In 327 BCE, Alexander the Great crossed the Hindu Kush mountains into northeastern India and, during the following two years, conquered the small Indian kingdoms of the area. Greek influence began to spread into India. Alexander came with scientists and historians in his entourage, not just as a conqueror interested in plunder but on a mission to "civilize" the East. Naturally, the Indians believed they were already "civilized." Each people considered the other "barbarians." Alexander's grand designs ended with his premature death in 323 BCE. His Indian provinces were soon reconquered by Chandragupta Maurya, who had earlier become king of Magadha, the major north Indian kingdom of the time. Chandragupta established friendly relations with Seleucus, Alexander's successor in western Asia, and through this relationship there was evidently some interchange of ideas. Shortly after Chandragupta's death, Ashoka succeeded to the throne. He proceeded to conquer most of India but then converted to Buddhism and sent missionaries both east and west to convert the neighboring kingdoms. Ashoka left records of his reign in edicts carved on pillars throughout his kingdom. These pillars contain some of the earliest written evidence of Indian numerals.

During the first century CE, northern India was conquered by Kushan invaders. The Kushan empire soon became the center of a flourishing trade between the Roman world and the East. Early in the fourth century, northern India was again united under a native dynasty, that of the Guptas. Under their rule, which only lasted about a century and a half, India reached a high point of culture with the flowering of art and medicine and the opening of universities. It was also during this period that Indian colonists spread Hindu culture to various areas of southeast Asia, including Burma, Malaya, and Indochina. The earliest identifiable Indian mathematician, Āryabhaṭa (b. 476), wrote his chief work, the *Āryabhaṭīya*, near the Gupta capital of Pāṭalipura (modern Patna) on the Ganges in Bihar in northern India. This work, although concentrating mainly on astronomy in its 123 verses in four chapters, contained a wide range of mathematical topics, as we will see below.

A northern Indian kingdom was revived in 606 by Harsha, a remarkably tolerant and just ruler. Two prominent mathematicians flourished during his reign, Bhāskara I (so called to distinguish him from a later mathematician with the same name) and Brahmagupta. The first probably came from what is now Maharashtra or Gujurat, while the second lived in what is now Bhinmal in Rajasthan, the capital of the kingdom of the Guyaras, part of Harsha's empire. It is not known whether these two mathematicians knew each other, but certainly by this time there was cultural unity in the Indian subcontinent, primarily based on the use of Sanskrit as a common learned language, so that one could speak of Indian astronomy and mathematics.

After Harsha's death in 647, his empire collapsed and northern India broke up into many small states, as was already true in the south. It was in one of these kingdoms that the ninth-century mathematician Mahāvīra composed the earliest Sanskrit textbook entirely devoted to mathematics, rather than having mathematics as an adjunct to astronomy. There were other mathematics texts written over the next few centuries, but the most influential of all were two works by Bhāskara II from the twelfth century, the *Līlāvatī* and the *Bījagaṇita*, on arithmetic and algebra, respectively. Bhāskara lived in Ujjain in what is now Madhya Pradesh and probably served the royal court of the small kingdom based there. Soon after he died, however, northern India was conquered by a Moslem army under Mohammed Ghori, and in 1206 the Moslem Sultanate of Delhi was established, an empire that was to last over 300 years. The sultanate even succeeded in conquering parts of the Hindu kingdoms in the south of India, kingdoms that had generally been independent even of the earlier native kingdoms of the north. But it was in the Vijayanagara empire in southern India, specifically in modern Kerala, that the mathematical "school" of Mādhava became established. From the fourteenth to the sixteenth centuries, there was a sequence of transmissions from teacher to pupil in this region, which resulted in the writing of proofs of many results that had been handed down in India for centuries, as well as the development of infinite series, particularly those related to the trigonometric functions.

Through the various invasions and new kingdoms, it does appear that the study of astronomy was always encouraged. Whoever ruled the country seemed to need astronomers to help with calendrical questions and, of course, to give astrological advice. Thus, much of Indian mathematics is recorded in astronomical works. Nevertheless, here, as elsewhere, creative mathematicians went beyond the strict requirements of practical problem solving to develop new areas of mathematics that they found of interest. We consider in this chapter the Indian number systems and methods of calculation, then the geometry of the *Śulbasūtras* and later, next the algebraic methods developed in the medieval period to solve equations (including the so-called Pell equation), next the beginning of combinatorics, and then the development of trigonometry and associated techniques. We conclude with a study of the development of power series in south India during the fourteenth and fifteenth centuries.

8.2

CALCULATIONS

The Indians, like the Chinese, used a base-10 system from as far back as records are available. But the decimal place value system itself first appeared in the middle of the first millennium CE.

8.2.1 The Decimal Place Value System

Our modern decimal place value system is usually referred to as the Hindu-Arabic system because of its supposed origins in India and its transmission to the West via the Arabs. However, the actual origins of the important components of this system, the digits 1 through 9 themselves, the notion of place value, and the use of 0, are to some extent lost to the historical record. We present here a summary of the most recent scholarship on the beginnings and development of these three ideas.

Symbols for the first nine numbers of our number system have their origins in the Brahmi system of writing in India, which dates back to at least the time of King Ashoka (mid-third century BCE). The numbers appear in various decrees of the king inscribed on pillars throughout India. There is a fairly continuous record of the development of these forms. Probably in the eighth century, these digits were picked up by the Moslems during the time of the Islamic incursions into northern India and their conquest of much of the Mediterranean world. These digits then appear a century later in Spain and still later in Italy and the rest of Europe (Fig. 8.1).

More important than the form of the number symbols themselves, however, is the notion of place value, and here the evidence is somewhat weaker. The Babylonians had a place value system, but it was based on 60. Although this system was adopted by the Greeks for astronomical purposes, it did not have much influence on the writing of numbers in other situations. The Chinese from earliest times had a multiplicative system with base 10. This probably was derived from the Chinese counting board, which itself contained columns each representing a different power of 10. In India, although there were number symbols to represent the numbers 1 through 9, there were also symbols to represent 10 through 90. Larger numbers were represented by combining the symbol for 100 or 1000 with a symbol for one of the first 9 numbers, except that 200 and 300 were written by adding one or two horizontal lines to the symbol for 100. Āryabhaṭa in fact lists names for the various powers of 10 in his *Āryabhaṭīya* : "*dasa* [ten], *sata* [hundred], *sahasra* [thousand], *ayuta* [ten thousand], *niyuta* [hundred thousand]. . . . "[2]

Around the year 600, the Indians evidently dropped the symbols for numbers higher than 9 and began to use their symbols for 1 through 9 in our familiar place value arrangement. The earliest dated reference to this use, however, does not come from India itself. In a fragment of a work of Severus Sebokht, a Syrian priest, dated 662, is the remark that the Hindus have a valuable method of calculation "done by means of nine signs."[3] Severus only wrote about nine signs; there is no mention of a sign for zero. However, in the Bakhshālī manuscript, a mathematical manuscript in fairly poor condition discovered in 1881 in the village of Bakhshālī in northwestern India, the numbers are written using the place value

FIGURE 8.1

Development of our modern numerals (*Source*: From *Number Words and Number Symbols, A Cultural History of Numbers*, by Karl Menninger. Translated by Paul Broneer from the revised German edition. English translation copyright © 1969 by Massachusetts Institute of Technology. Reprinted by permission.)

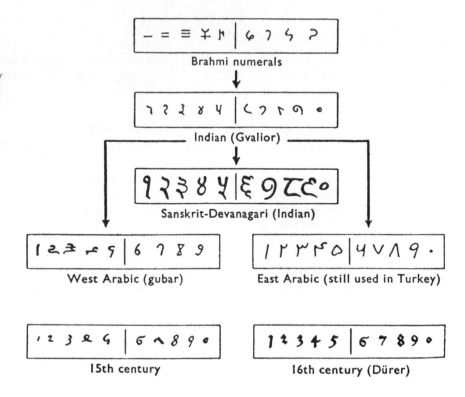

system and with a dot to represent zero. The best evidence we have is that this manuscript also dates from the seventh century. Perhaps Severus did not consider the dot as a "sign." In other Indian works from the same period, numbers were generally written in a quasi–place value system to accommodate the poetic nature of the documents. For example, in the work of Mahāvīra, certain words stand for numbers: moon for 1, eye for 2, fire for 3, and sky for 0. Then the word fire-sky-moon-eye would stand for 2103 and moon-eye-sky-fire for 3021. Note that the place value begins on the left with the units.

Curiously, the earliest dated inscriptions using the decimal place value system including the zero were found in Cambodia. The earliest one appeared in 683, where the 605th year of the Saka era there was represented by three digits with a dot in the middle and the 608th year by three digits with our modern zero in the middle. The dot as symbol for 0 as part of a decimal place value system also appeared in the *Chiu-chih li*, the Chinese astronomical work of 718 compiled by Indian scholars in the employ of the Chinese emperor. Although the actual symbols for the other Indian digits are not known, the author did give details of how the place value system works: "Using the [Indian] numerals, multiplication and division are carried out. Each numeral is written in one stroke. When a number is counted to ten, it is advanced into the higher place. In each vacant place a dot is always put. Thus the numeral is always denoted in each place. Accordingly there can be no error in determining the place. With the numerals, calculation is easy. . . . [4]

The question remains then as to why the Indians at the beginning of the seventh century dropped their earlier system and introduced the place value system including a symbol for zero. We cannot answer that definitively. It has been suggested, however, that the true origins of the system in India come from the Chinese counting board. The counting board was a portable object. Certainly, Chinese traders who visited India carried these along. In fact, since southeast Asia is the border between Hindu culture and Chinese influence, it may have well been in that area where the interchange took place. What may have happened is that the Indians were impressed with the idea of using only nine symbols. But they naturally took for their symbols the ones they had already been using. They then improved the Chinese system of counting rods by using exactly the same symbols for each place value rather than alternating two types of symbols in the various places. And because they needed to be able to write numbers in some form, rather than just have them on the counting board, they were forced to use a symbol, the dot and later the circle, to represent the blank column of the counting board.[5] If this theory is correct, it is somewhat ironic that Indian scientists then returned the favor and brought this new system back to China early in the eighth century.

In any case, we can certainly put a fully developed decimal place value system for integers in India by the eighth century, even though the earliest definitively dated decimal place value inscription there dates to 870. Well before then, though, this system had been transmitted not only to China but also west to Baghdad, the center of the developing Islamic culture. It is important to note, however, that although decimal fractions were used in China, again as places on the counting board, in India itself there is no early evidence of these. It was the Moslems who completed the Indian written decimal place value system by introducing these decimal fractions.

8.2.2 Arithmetic Algorithms

Even before the decimal place value system was fully developed, the Indians were adept at calculations. For example, in the second chapter of his *Āryabhaṭīya*, Āryabhaṭa presented the methods of calculating square and cube roots. We look at the latter, in the rather cryptic words necessitated partly by the limitations of Sanskrit verse:

STANZA II, 5   *One should divide the second noncube [place] by three times the square of the root of the cube. The square [of the quotient] multiplied by three and the former [quantity] should be subtracted from the first [noncube place] and the cube from the cube [place].*[6]

Counting from right to left, the first, fourth, and so on places of the given number are named the cube place; the second, fifth, and so on are called the first noncube place; and the third, sixth, and so on are called the second noncube place. We illustrate the procedure by calculating the cube root of 12,977,875. The first step is to note that the largest cube less than the 12 in the millions place is $8 = 2^3$. We subtract that from the 12, leaving 49 in the second noncube place. We now divide this by $3 \times 4 = 12$. The quotient is 4, but it turns out that this is too large, so we take 3, and subtract $3 \times 2^2 = 36$ from 49, leaving 137 in the first noncube place. We next multiply 3^2 by 3 and by 2, giving 54, and subtract this from 137, leaving 837 in the next cube place. We then continue, as shown here:

	1	2	9	7	7	8	7	5	)2		First digit $2 \approx \sqrt[3]{12}$
		8									2^3
1	2		4	9					)3		$12 = 3 \times 2^2$
			3	6							3 approximates $49 \div 12$ (4 is too large)
			1	3	7						$36 = 3 \times 2^2 \times 3$
				5	4						$54 = 3 \times 2 \times 3^2$
				8	3	7					
					2	7					3^3
1	5	8	7		8	1	0	8		)5	$1587 = 3 \times 23^2$
					7	9	3	5			5 approximates $8108 \div 1587$
						1	7	3	7		$7935 = 3 \times 23^2 \times 5$
						1	7	2	5		$1725 = 3 \times 23 \times 5^2$
							1	2	5		
							1	2	5		5^3

Of course, the basis for this algorithm is the expansion of $(a + b)^3$. In this case, for example, $23^3 = (20 + 3)^3 = 20^3 + 3 \times 20^2 \times 3 + 3 \times 20 \times 3^2 + 3^3$.

Brahmagupta gave many details of arithmetic calculation in his major work, the *Brāhma-sphuṭasiddhānta* (*Correct Astronomical System of Brahma*). Not only did he present the standard arithmetical rules for calculating with fractions, but in chapter 18 he gave the rules for operations on positive and negative numbers, as well as zero:

> The sum of two positives is positive, of two negatives negative; of a positive and a negative the sum is their difference; if they are equal it is zero. The sum of a negative and zero is negative, that of a positive and zero positive, and that of two zeros, zero. If a smaller positive is to be subtracted from a larger positive, the result is positive; if a smaller negative from a larger negative, the result is negative; if a larger negative or positive is to be subtracted from a smaller negative or positive, the sign of their difference is reversed—negative becomes positive and positive negative. A negative minus zero is negative, a positive minus zero positive; zero minus zero is zero. When a positive is to be subtracted from a negative or a negative from a positive, then it is to be added.
>
> The product of a negative and a positive is negative, of two negatives positive, and of positives positive; the product of zero and a negative, of zero and a positive, or of two zeros is zero. A positive divided by a positive or a negative divided by a negative is positive; a zero divided by a zero is zero; a positive divided by a negative is negative; a negative divided by a positive is also negative. A negative or a positive divided by zero has that zero as its divisor, or zero divided by a negative or a positive has that negative or positive as its divisor.[7]

These final rules for operating with zero sound strange to us. But Bhāskara II, after repeating essentially the same rules in his own work, justified them using the concept of infinity: "In this quantity also, which has zero as its divisor, there is no change even when many quantities have entered into it or come out of it, just as at the time of destruction and of creation, when throngs of creatures enter into and come out of him, there is no change in the infinite and unchanging one [i.e., Viṣṇu]."[8] Nevertheless, he could still set the problem:

"There is an unknown number whose multiplier is 0. Its own half is added. Its multiplier is 3; its divisor 0. The given number is 63."[9] Evidently, he was thinking of the equation

$$\frac{3\left(0x + \frac{1}{2}0x\right)}{0} = 63,$$

which, by factoring out the zeros in the numerator and "canceling," becomes $3x + \frac{3}{2}x = 63$, an equation whose solution is 14.

8.3   GEOMETRY

Many important geometric ideas were expressed in the *Śulbasūtras* as part of their treatment of the construction of altars. But since these literary pieces were not designed to teach mathematics as such, there are no derivations, just assertions. On the other hand, later commentators sometimes did give demonstrations. We will look at several results from the *Baudhāyana Śulbasūtra*, which probably dates to around 600 BCE. The first is the Pythagorean Theorem:

> The areas of the squares produced separately by the length and the breadth of a rectangle together equal the area of the square produced by the diagonal. This is observed in rectangles having sides 3 and 4, 12 and 5, 15 and 8, 7 and 24, 12 and 35, 15 and 36.[10]

A proof of this result was given in the *Yuktibhāsā*, written by Jyesthadeva (1530–1610) in the mid-sixteenth century. The idea is to put two right triangles together, then draw the square on each of the two sides and on the hypotenuse (Fig. 8.2). If one cuts along each of the two lines indicated, then rotates each of the triangles outside the large square, the two pieces together will fill up the square on the hypotenuse. Again, as in the Chinese proof, there is no principle of beginning with axioms. One just studies the diagram, rotates the pieces, and understands that the theorem is true. This procedure could be thought of as an "empirical" proof.

FIGURE 8.2

Proof of Pythagorean Theorem, as given in the *Yuktibhāsā*

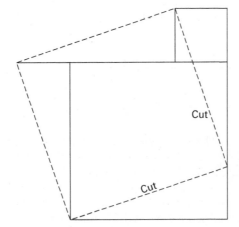

The Pythagorean Theorem was then used implicitly to justify each of the following constructions:

> If it is desired to remove a square from another, a rectangular part is cut off from the larger square with the side of the smaller one to be removed; the longer side of the cut-off rectangular part is placed across so as to touch the opposite side; by this contact the side is cut off. With the cut-off part the difference of the two squares is obtained (Fig. 8.3).[11]

FIGURE 8.3

Procedure for determining a square equal to the difference of two squares, from the *Baudhāyana Śulbasūtra*

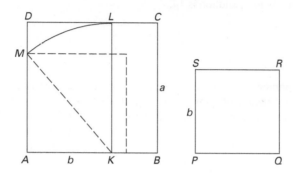

> If it is desired to transform a rectangle into a square, its breadth is taken as the side of a square and this square on the breadth is cut off from the rectangle. The remainder of the rectangle is divided into two equal parts and placed on two sides (one part on each). The empty space in the corner is filled up with a square piece. The removal of it has been stated [in the previous construction] (Fig. 8.4).

FIGURE 8.4

Procedure for transforming a rectangle into a square, from the *Baudhāyana Śulbasūtra*

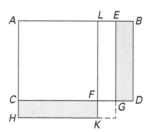

Note that this second construction uses the "completing the square" technique that we have seen in Babylonian mathematics. It is quite different, however, from Euclid's construction of the same problem found in *Elements* II–14. Later in the *Śulbasūtra* are two results involving circles:

> If it is desired to transform a square into a circle, a cord of length half the diagonal of the square is stretched from the center to the east, a part of it lying outside the eastern side of the square. With one-third of the part lying outside added to the remainder of the half diagonal, the requisite circle is drawn (Fig. 8.5).

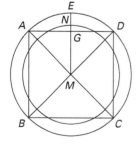

FIGURE 8.5

Indian procedure for "circling" the square

To transform a circle into a square, the diameter is divided into eight parts; one such part, after being divided into twenty-nine parts, is reduced by twenty-eight of them and further by the sixth of the part left less the eighth of the sixth part. [The remainder is then the side of the required square.]

In the first of these constructions, MN is the radius r of the desired circle. It is straightforward to show that if the side of the original square is s, then $r = (\frac{2+\sqrt{2}}{6})s$. This implies a value for π of 3.088311755. In the second construction, the author means for us to take the side of the required square equal to

$$\frac{7}{8} + \frac{1}{8 \times 29} - \frac{1}{8 \times 29 \times 6} + \frac{1}{8 \times 29 \times 6 \times 8}$$

of the diameter of the circle. This is equivalent to taking 3.088326491 for π. The Indian authors did not mention in either case that these constructions were approximations. What seems remarkable is that the two constructions imply values for π that are equal to four decimal places. Yet there is no indication whether one of these constructions was derived from the other. On the other hand, a further result in the *Śulbasūtra* indicates an easier but approximate construction:

Divide the diameter into fifteen parts and reduce it by two of them. This gives the approximate side of the desired square.

In other words, the side was given here as 13/15 of the diameter. It is easy to see that this results in a value for π of $4(13/15)^2 = 3.00444444$.

Āryabhaṭa too presented some geometric results:

STANZA II, 16 *The upright side is the distance between the tips of the two shadows multiplied by a shadow divided by the decrease. That upright side multiplied by the gnomon, divided by its shadow, becomes the base.*[12]

This stanza gives a method for finding the height of a pole with a light at the top by measuring various shadows. In Figure 8.6, we have two gnomons of length g. The lengths s_1 and s_2 of the shadows of the two gnomons cast by the light at height h are known, as well as the distance d between the shadow ends. The lengths h of the base and u of the upright side

FIGURE 8.6

Finding heights by use of shadows

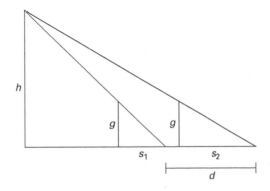

are to be found. (The terms "base" and "upright side" are somewhat counterintuitive here.) Bhāskara I, an early commentator on the *Āryabhaṭīya*, presented an example:

> The shadows of two equal gnomons [of height 12 *aṅgulas*] are observed to be respectively 10 and 16 *aṅgulas* and the distance between the tips of the shadows is seen as 30. . . . Procedure: The distance between the tips of the shadows is 30; it is multiplied by the first shadow, 300; the difference of the lengths of the shadows is 6; what has been obtained with this is the upright side, 50. Precisely this upright side is multiplied by the height of the gnomon; what has been obtained is 600, which when divided by the first gnomon's shadow is the base, 60.

The procedures of the stanza can be translated into the formulas

$$u = \frac{ds_1}{s_2 - s_1} \quad \text{and} \quad h = \frac{ug}{s_1}.$$

Note that this problem is very similar both in form and solution method to problem 1 in the Chinese *Sea Island Mathematical Manual*.

STANZA II, 17 . . . *In a circle, the product of both arrows is the square of the half-chord, certainly, for two bow fields.*[13]

Here, the "arrows" are the two segments s_1, s_2, of the diameter of a circle intersected at right angles by a chord of length $2h$, dividing the circle into two "bows" (Fig. 8.7). Thus, the result is that $h^2 = s_1 s_2$. Although this result was essentially proved by Euclid, there is no indication of how Āryabhaṭa discovered it. Bhāskara gave the following problem as an example:

> A hawk was resting upon a wall whose height was 12 *hastas*. The departed rat was seen by that hawk at a distance of 24 *hastas* from the foot of the wall; and the hawk was seen by the rat. There, because of his fear, the rat ran with increasing speed toward his own residence, which was in the wall. On the way, the rat was killed by the hawk moving along the hypotenuse. In this case I wish to know what is the distance not attained by the rat, and what is the distance crossed by the hawk. . . . Procedure: The square of the height of the hawk is 144; when that is divided by the size of the rat's roaming ground, 24, the quotient is 6. The rat's roaming ground, when increased by this difference, is 30, and when decreased is 18. Their halves in due order, the path of the hawk and the distance to the rat's residence: 15, 9.

FIGURE 8.7

Perpendicular chord and diameter

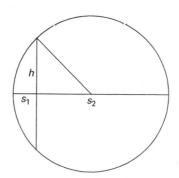

In this case, since $s_2 = 24$ and $h = 12$, Bhāskara calculated that $s_1 = 6$. Thus, the diameter of the circle was 30. The rat was killed at the center of the circle, a distance of 15 from the hawk's initial perch and 9 from the base of the wall.

Many other geometric formulas, some exact, some stated as exact but only approximate, and some stated explicitly as approximate, occur in various Indian mathematical texts. But we will conclude this section with two remarkable results of Brahmagupta dealing with cyclic quadrilaterals (quadrilaterals inscribed in circles), given in chapter 12 of the *Brāhmasphuṭasiddhānta*. The first is the following:

> The accurate area [of a cyclic quadrilateral] is the square root of the product of the halves of the sums of the sides diminished by each side of the quadrilateral.[14]

This result says that if $s = \frac{1}{2}(a + b + c + d)$, where a, b, c, d, are the sides of the quadrilateral in cyclic order (Fig. 8.8), then the area S is given by $S = \sqrt{(s - a)(s - b)(s - c)(s - d)}$. Heron's formula is a special case of this result, but how Brahmagupta discovered his formula, or whether he was aware of Heron's result, is unknown. A complete proof first appeared in the *Yuktibhāṣā*, based on a second result of Brahmagupta:

> One should multiply the sum of the products of the arms adjacent to the diagonals, after it has been mutually divided on either side, by the sum of the two products of the arms and the counter-arms. In an unequal cyclic quadrilateral, the two square roots are the two diagonals.

FIGURE 8.8

Area of a cyclic quadrilateral, from the *Brāhmasphuṭasiddhānta*

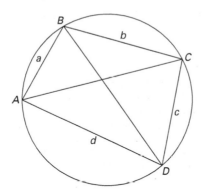

This statement translates into formulas for the lengths of the diagonals AC and BD of the quadrilateral. Since the "sum of the products of the arms adjacent to" diagonal AC, or $ad + bc$, is "mutually divided," that is, divided by the corresponding sum with respect to the second diagonal, or $ab + cd$, and then multiplied by the "sum of the two products of the arms and counter-arms," or $ac + bd$, the result is that

$$AC = \sqrt{\frac{(ac + bd)(ad + bc)}{ab + cd}} \quad \text{and similarly} \quad BD = \sqrt{\frac{(ac + bd)(ab + cd)}{ad + bc}}.$$

A proof of this result too was in the *Yuktibhāṣā*.

8.4   EQUATION SOLVING

The rule for solving quadratic equations seems to have been known in India from at least the end of the fifth century. For Āryabhaṭa, in dealing with arithmetic progressions in two stanzas of his *Āryabhaṭī ya*, provided what amounts to the quadratic formula in a special case:

STANZA II, 19 *The desired number of terms minus one, halved, . . . multiplied by the common difference between the terms, plus the first term, is the middle term. This multiplied by the number of terms desired is the sum of the desired number of terms. Or the sum of the first and last terms is multiplied by half the number of terms.*[15]

This verse presents a formula for the sum S_n of an arithmetic progression with initial term a and common difference d. The formula translates to

$$S_n = n \left[\left(\frac{n-1}{2} \right) d + a \right] = \frac{n}{2}[a + (a + (n-1)d)]. \tag{8.1}$$

STANZA II, 20 *Multiply the sum of the progression by eight times the common difference, add the square of the difference between twice the first term and the common difference, take the square root of this, subtract twice the first term, divide by the common difference, add one, divide by two. The result will be the number of terms.*

In the same circumstances as above, S_n is given and n is to be found. The formula given is

$$n = \frac{1}{2} \left[\frac{\sqrt{8 S_n d + (2a - d)^2} - 2a}{d} + 1 \right]. \tag{8.2}$$

If Equation 8.1 for S_n is rewritten as a quadratic equation in n, namely, $dn^2 + (2a - d)n - 2S_n = 0$, then the value for n in Equation 8.2 follows from the quadratic formula. Although Āryabhaṭa did not explicitly present here a general procedure for solving quadratic equations, Brahmagupta, a century and a quarter later, did so for the equation we would write as $ax^2 + bx = c$. Here the "middle number" is the coefficient b (and also the unknown itself); the *rūpas* is the constant term c and the "square" is the coefficient a.

> Diminish by the middle number the square root of the *rūpas* multiplied by four times the square and increased by the square of the middle number; divide the remainder by twice the square. The result is the middle number.[16]

Brahmagupta's words can easily be translated into the formula

$$x = \frac{\sqrt{4ac + b^2} - b}{2a}.$$

As an example, Brahmagupta presented the solution to the equation $x^2 - 10x = -9$:

> Now to the *rūpas* [−9] multiplied by four times the square [−36], and added to the square [100] of the middle number (making 64), the square root being extracted [8], and lessened by the middle number [−10], the remainder 18 divided by twice the square [2] yields the value of the middle number, 9.[17]

Note here that although the given equation actually has a second positive solution, corresponding to the negative of the square root, Brahmagupta did not mention it. Several hundred years later, Bhāskara II did deal with multiple roots, at least when both are positive. His basic technique for solving quadratic equations was that of completing the square. Namely, he added an appropriate number to both sides of $ax^2 + bx = c$ so that the left side becomes a perfect square: $(rx - s)^2 = d$. He then solved the equation $rx - s = \sqrt{d}$ for x. But, he noted, "if the root of the absolute side of the equation is less than the number, having the negative sign, comprised in the root of the side involving the unknown, then putting it negative or positive, a twofold value is to be found of the unknown quantity."[18] In other words, if $\sqrt{d} < s$, then there are two values for x, namely, $(s + \sqrt{d})/r$ and $(s - \sqrt{d})/r$. Bhāskara did, however, hedge his bets. As he says, "this [holds] in some cases." We consider two examples to see what he meant.

> The eighth part of a troop of monkeys, squared, was skipping in a grove and delighted with their sport. Twelve remaining monkeys were seen on the hill, amused with chattering to each other. How many were they in all?

Bhāskara wrote the equation as $(\frac{1}{8}x)^2 + 12 = x$, then multiplied by 64, and subtracted to get $x^2 - 64x = -768$. Adding 32^2 to each side produced $x^2 - 64x + 1024 = 256$. Taking square roots: $x - 32 = 16$. He then noted that "the number of the root on the absolute side is here less than the known number, with the negative sign, in the root on the side of the unknown." Therefore, 16 can be made positive or negative. So, he concludes, "a two-fold value of the unknown is thence obtained, 48 and 16."

> The fifth part of the troop less three, squared, had gone to a cave; and one monkey was in sight having climbed on a branch. Say how many they were?

The equation becomes $x^2 - 55x = -250$, and Bhāskara found the two roots 50 and 5. "But the second [root] is in this case not to be taken; for it is incongruous. People do not approve a negative absolute number."[19] Here, the negative number is not from the equation itself but from the problem. One cannot subtract three monkeys from one-fifth of five. In the case of quadratic equations, which for us have a positive and a negative root, Bhāskara simply found the positive root. He never gave examples of quadratic equations having two negative roots or no real roots at all, nor did he give examples of quadratic equations having irrational roots. In every example, the square root in the formula is a rational number.

The Indian mathematicians also handled equations in several variables. Thus, Mahāvīra presented a version of the hundred fowls problem in his major treatise, the *Gaṇitasārasaṅgraha*: "Doves are sold at the rate of 5 for 3 coins, cranes at the rate of 7 for 5, swans at the rate of 9 for 7, and peacocks at the rate of 3 for 9. A certain man was told to bring at these rates 100 birds for 100 coins for the amusement of the king's son and was sent to do so. What amount does he give for each?"[20]

Mahāvīra gave a rather complex rule for the solution. Bhāskara, on the other hand, presented the same problem with a procedure showing explicitly why the problem has multiple solutions. He put his unknowns, which we label d, c, s, and p, equal to the number of "sets" of doves, cranes, swans, and peacocks, respectively. From the prices and the numbers of birds he derived the two equations

$$3d + 5c + 7s + 9p = 100$$
$$5d + 7c + 9s + 3p = 100$$

and proceeded to solve them. He solved each equation for d, then equated the two expressions and found the equation $c = 50 - 2s - 9p$. Taking an arbitrary value 4 for p, he reduced the equation to the standard indeterminate form $c + 2s = 14$, for which the solution is $s = t$, $c = 14 - 2t$, with t arbitrary. It follows that $d = t - 2$. Then setting $t = 3$, he calculated that $d = 1$, $c = 8$, $s = 3$, and $p = 4$, hence that the number of doves is 5, of cranes 56, of swans 27, and of peacocks 12, their prices being respectively 3, 40, 21, and 36. He noted further that other choices of t gave different values for the solution. Thus, "by means of suppositions, a multitude of answers may be obtained."[21]

 8.5

INDETERMINATE ANALYSIS

Like the Chinese, Indian mathematicians spent much effort on the solution of congruences, originally probably for much the same reasons.

8.5.1 Linear Congruences

Although we do not know whether the Indians learned the quadratic formula somehow from the Babylonians or from Diophantus, we are fairly certain that Indian mathematicians originated a method for solving linear congruences, because there is no comparable method described anywhere else. In modern notation, the problem was to find N satisfying $N \equiv a \pmod{r}$ and $N \equiv b \pmod{s}$, or to find x and y such that $N = a + rx = b + sy$, or so that $a + rx = b + sy$, or finally, setting $c = a - b$, so that $rx + c = sy$. We first find mention of a method for solving this problem in Āryabhaṭa's work, but Brahmagupta gave somewhat clearer explanations. However, either because of faulty copying over the years or beause the oral tradition never required that every step be written down, there are many places where Brahmagupta's description of his method does not match the steps of his examples. The modern explanations to be presented do, however, convey the main ideas. Note, of course, that Brahmagupta had nothing a modern reader would consider a proof. He just presented an algorithm.

We accompany Brahmagupta's description of his method of *kuṭṭaka* or "pulverizer," taken from chapter 18 of his text, with an example he used: $N \equiv 10 \pmod{137}$ and $N \equiv 0 \pmod{60}$. This problem can be rewritten as the single equation $137x + 10 = 60y$.

> Divide the divisor having the greatest remainder (*agra*) by the divisor having the least remainder; whatever is the remainder is mutually divided; the quotients are to be placed separately one below the other.[22]

Therefore, divide 137 by 60 and continue by dividing the residues. In other words, apply the Euclidean algorithm until the final nonzero remainder is reached:

$$137 = 2 \cdot 60 + 17$$
$$60 = 3 \cdot 17 + 9$$
$$17 = 1 \cdot 9 + 8$$
$$9 = 1 \cdot 8 + 1$$

Then list the quotients one under the other:

$$2$$
$$3$$
$$1$$
$$1$$

Brahmagupta lists 0 for the first quotient, evidently taking the first division as $60 = 0 \cdot 137 + 60$, despite his statement of which divisor is divided into which.

> Multiply the remainder by an arbitrary number such that, when increased by the difference between the two remainders (*agras*), it is eliminated. The multiplier is to be set down as is also the quotient.

The final remainder is 1. Multiply that by some number v so that $1 \cdot v \pm 10$ is exactly divisible by the last divisor, in this case 8. Brahmagupta explained that one uses the $+$ when there are an even number of quotients and the $-$ when there are an odd number. Here, because 0 is one of the quotients, the last equation becomes $1v - 10 = 8w$. Choose $v = 18$ and $w = 1$. The new column of numbers is then

$$0$$
$$2$$
$$3$$
$$1$$
$$1$$
$$18$$
$$1$$

> Beginning from the last, multiply the next to last by the one above it; the product, increased by the last, is the end of the remainders (*agrānta*). [Continue to the top of the column.]

Multiply 18 by 1 and add 1 to get 19. Then replace the term "above," namely, 1, by 19, and remove the last term. Continue in this way (as in the table below) until there are only two terms.

0	0	0	0	0	130
2	2	2	2	297	297
3	3	3	130	130	
1	1	37	37		
1	19	19			
18	18				
1					

The top term, the *agrānta*, is 130. So $x = 130$, $y = 297$, is a solution to the original equation. Brahmagupta, however, wanted a smaller solution, so he first determined N:

> Divide it (the *agrānta*) by the divisor having the least remainder; multiply the remainder by the divisor having the greatest remainder. Increase the product by the greatest remainder; the result is the remainder of the product of the divisors.

Therefore, we divide 130 by 60 and obtain a remainder of 10. Multiplying 10 by 137 and adding the product to 10 then gives 1380 as the value for N modulo the product of 137 and 60, or $N \equiv 1380 \pmod{8220}$. Brahmagupta then solved for y by dividing 1380 by 60 (since $N = 60y$) and calculated a new value for x. Hence, $y = 23$, $x = 10$, is a solution to the equation $137x + 10 = 60y$.

Although we do not know how Brahmagupta justified his procedure to his own students, we will present a modern explanation. Begin with the equation $60y = 137x + 10$, and make step-by-step subsitutions in accordance with the successive quotients appearing in the Euclidean algorithm:

$$60y = 137x + 10 \qquad y = \tfrac{137x+10}{60} = 2x + z \qquad 137x + 10 = 60(2x + z)$$

$$17x = 60z - 10 \qquad x = \tfrac{60z-10}{17} = 3z + u \qquad 17(3z + u) = 60z - 10$$

$$9z = 17u + 10 \qquad z = \tfrac{17u+10}{9} = 1u + v \qquad 9(1u + v) = 17u + 10$$

$$8u = 9v - 10 \qquad u = \tfrac{9v-10}{8} = 1v + w \qquad 8(1v + w) = 9v - 10$$

$$v = 8w + 10$$

Brahmagupta then solved this last equation by inspection: $w = 1$, $v = 18$. The remaining variables are then found by substitution, working up the column of variables.

$$u = 1v + w = 1 \cdot 18 + 1 = 19 \qquad z = 1u + v = 1 \cdot 19 + 18 = 37$$
$$x = 3z + u = 3 \cdot 37 + 19 = 130 \qquad y = 2x + z = 2 \cdot 130 + 37 = 297$$

Although both Brahmagupta in the seventh century and various Chinese authors beginning in the third century were interested in solving systems of linear congruences, a close inspection shows that the two methods were quite different, especially since the Indian author usually dealt with a system of two congruences, while the Chinese authors dealt with a larger system. Even when Brahmagupta did deal with a problem similar to a "Chinese remainder problem," such as, "What number, divided by 6, has a remainder of 5; and divided by 5, a remainder of 4, and by 4, a remainder of 3; and by 3, a remainder of 2?," he solved these congruences two at a time. Namely, he first solved $N \equiv 5 \pmod 6$ and $N \equiv 4 \pmod 5$ to get $N \equiv 29 \pmod{30}$, then solved $N \equiv 29 \pmod{30}$ and $N \equiv 3 \pmod 4$ and so on. It appears, then, that the only similarity between the Indian and Chinese methods is that both made use of the Euclidean algorithm. A more interesting question, then, unanswerable with current evidence, is whether either culture learned the algorithm from the Greeks, whether all three learned it from an earlier culture, or whether the two Asian cultures simply discovered the algorithm independently.

There is good evidence, however, that Brahmagupta and Āryabhaṭa were interested in congruence problems for the same basic reason as the Chinese, namely, for use in astronomy. The Indian astronomical system of the fifth and sixth century had been heavily influenced by Greek astronomy, especially in the notion that the various planets traveled on epicycles that in turn circled the earth. Therefore, Indian astronomers, like their Greek counterparts, needed trigonometry to be able to calculate positions. But a significant idea of Hindu astronomy, similar to one from ancient China but not particularly important in Greece, was that of a large astronomical period at the beginning and end of which all the planets (including the sun and moon) had longitude zero. It was thought that all worldly events would recur with this same period. For Āryabhaṭa, the fundamental period was the *Mahayuga* of 4,320,000 years, the last quarter of which, the *Kaliyuga*, began in 3102 BCE. For Brahmagupta, the fundamental period was the *Kalpa* of 1000 *Mahayugas*.

In any case, to do calculations with heavenly bodies, one had to know their average motion. Since it was difficult to determine these motions empirically, it became necessary to calculate them from current observations and the fact that all the planets were at approximately the same place at the beginning of the period. These calculations were made by solving linear congruences.

8.5.2 The Pell Equation

The ability to solve systems of pairs of linear congruences turned out to be important in the solution of another type of indeterminate equation, the quadratic equation of the form $Dx^2 \pm b = y^2$. Today, the special case where $b = 1$ is usually referred to as Pell's equation (mistakenly named after the seventeenth-century Englishman John Pell). But although there are indications that the Greeks could solve a few of these equations, the general case, first developed in India, was undoubtedly the high point of medieval Indian algebra.

Brahmagupta gave the first explanation of the method of solving these problems. And, as in the case of the *kuttaka*, he introduced rules for dealing with equations of this type, in conjunction with examples. Consider the following:

> He who computes within a year the square of [a number] . . . multiplied by ninety two . . . and increased by one that is a square, he is a calculator.[23]

This equation, $92x^2 + 1 = y^2$, will be solved here in considerably less than a year. Brahmagupta's solution rule began as follows:

> Put down twice the square root of a given square multiplied by a multiplier and increased or diminished by an arbitrary number.

So set down any value, say, 1, and note that if 92 is multiplied by 1^2 and the product added to 8 (the arbitrary number), then the sum is a square, namely, 100. Thus, three numbers x_0, b_0, y_0 have been found satisfying the equation $Dx_0^2 + b_0 = y_0^2$. For convenience, we will write that (x_0, y_0) is a solution for additive b_0. In this case, $(1, 10)$ is a solution for additive 8. Brahmagupta next wrote this solution in two rows as

$$
\begin{array}{ccc}
x_0 & y_0 & b_0 \\
x_0 & y_0 & b_0
\end{array}
$$

or

$$
\begin{array}{ccc}
1 & 10 & 8 \\
1 & 10 & 8.
\end{array}
$$

> The product of the first pair, multiplied by the multiplier, with the product of the last pair, is the [new] last root.

Namely, a new value for the "last root" y is found by setting $y_1 = Dx_0^2 + y_0^2$. In this example, $y_1 = 92(1)^2 + 10^2 = 192$.

> The sum of the thunderbolt products [cross multiplication] is the [new] first root. The additive is equal to the product of the additives.

A new value for the "first root" x is determined as $x_1 = x_0 y_0 + x_0 y_0$ or $x_1 = 2x_0 y_0$, while a new additive is $b_1 = b_0^2$. In other words, $(x_1, y_1) = (20, 192)$ is a solution for additive $b_1 = 64$, or $92(20)^2 + 64 = 192^2$. This result is straightforward to verify, but Brahmagupta in fact considered the more general result, that if (u_0, v_0) is a solution for additive c_0 and (u_1, v_1) is a solution for additive c_1, then $(u_0 v_1 + u_1 v_0, Du_0 u_1 + v_0 v_1)$ is a solution for additive $c_0 c_1$. To check this result, consider the identity

$$D(u_0 v_1 + u_1 v_0)^2 + c_0 c_1 = (Du_0 u_1 + v_0 v_1)^2,$$

given that $Du_0^2 + c_0 = v_0^2$ and $Du_1^2 + c_1 = v_1^2$. We will call this new solution the **composition** of the solutions (u_0, v_0) and (u_1, v_1). Brahmagupta concluded his basic rule: "The two square roots, divided by the [original] additive or subtractive, are the [roots for] additive unity." In the present example, divide 20 and 192 by 8 to get $(\frac{5}{2}, 24)$ as a solution for additive 1. Since, however, one of these roots is not an integer, this was not a satisfactory answer. So Brahmagupta composed this solution with itself to get the integral solution for additive 1, $(120, 1151)$. In other words, $92 \cdot 120^2 + 1 = 1151^2$.

This example, as well as illustrating Brahmagupta's method, shows its limitations. The solution for additive 1 in the general case is the pair $(\frac{x_1}{b_0}, \frac{y_1}{b_0})$. There is no guarantee that these will be integers or even that one can generate integers by combining this solution with itself. Brahmagupta simply gave several more rules and examples, without noting the conditions under which integral solutions exist. First, he noted that composition allows him to get other solutions for any additive, provided he knows one solution for this additive as well as a solution for additive 1. In general, the given equation will have infinitely many solutions.

Second, if he had found a solution (u, v) for additive 4, he showed how to find a solution for additive 1. Namely, if v is odd or u is even, then

$$(u_1, v_1) = \left(u \left(\frac{v^2 - 1}{2} \right), v \left(\frac{v^2 - 3}{2} \right) \right)$$

is the desired solution. In the case where v is even and u is odd,

$$(u_1, v_1) = \left(\frac{2uv}{4}, \frac{Du^2 + v^2}{4} = \frac{2v^2 - 4}{4} \right)$$

is an integral solution. As an example of the first case, Brahmagupta solved $3x^2 + 1 = y^2$ by beginning with the solution $u = 2$, $v = 4$, for $3u^2 + 4 = v^2$.

Brahmagupta gave a similar rule for subtractive 4, as well as rules for solving the Pell equation in other special circumstances. Although his methods were always correct, the text contains no proofs, nor do we learn how Brahmagupta discovered the method. Why the Indian mathematicians were interested in this problem is also a mystery. Some of Brahmagupta's examples use astronomical variables for x and y, but there is no indication that the problems actually came from real-life situations.

In any case, the Pell equation became a tradition in Indian mathematics. It was studied through the next several centuries and was solved completely by the otherwise unknown Acarya Jayadeva (c. 1000). The solution given by Bhāskara II is more easily followed, however.

Bhāskara's goal in his *Līlāvatī* was to show how any equation of the form $Dx^2 + 1 = y^2$ can be solved in integers. He began by recapitulating Brahmagupta's procedure. In particular, he emphasized that once one had found one solution pair, indefinitely many others could be found by composition. More importantly, however, he discussed the so-called cyclic method (*chakravāla*). The basic idea is that by continued appropriate choices of solution pairs for various additives by use of the *kuṭṭaka* method, one eventually reaches one that has the desired additive 1. We present Bhāskara's rule for the general case $Dx^2 + 1 = y^2$ and follow its use in one of his examples, $67x^2 + 1 = y^2$.

> Making the smaller and larger roots and the additive into the dividend, the additive, and the divisor, the multiplier is to be imagined.[24]

Begin as before by choosing a solution pair (u, v) for any additive b. In this example, take $(1, 8)$ as a solution for additive -3. Next, solve the indeterminate equation $um + v = bn$ for m, here $1m + 8 = -3n$. The result is $m = 1 + 3t, n = -3 - t$, for any integer t.

> When the square of the multiplier is subtracted from the "nature" or is diminished by the "nature" so that the remainder is small, that divided by the additive is the new additive. It is reversed if the square of the multiplier is subtracted from the "nature." The quotient of the multiplier is the smaller square root; from that is found the greatest root.

In other words, choose t so that the square of m is as close to D (the "nature") as possible, and take $b_1 = +\frac{D - m^2}{b}$ (which may be negative) for the new additive. The new first root is $u_1 = \frac{um + v}{b}$ while the new last root is $v_1 = \sqrt{Du_1^2 + b_1}$. In the given example, Bhāskara wants m^2 close to 67, so he chooses $t = 2$ and $m = 7$. Then $(D - m^2)/b = (67 - 49)/(-3) = -6$. But, because the subtraction is of the square from the coefficient, the new additive is 6. The new first root is $u_1 = \frac{1 \cdot 7 + 8}{-3} = -5$, but since these roots are always squared, u_1 can be taken as positive. Then $v_1 = \sqrt{67 \cdot 25 + 6} = \sqrt{1681} = 41$, and $(5, 41)$ is a solution for additive 6.

> Then it is done repeatedly, leaving aside the previous square roots and additives. They call this the *chakravāla* (circle). Thus there are two integer square roots increased by four, two or one. The supposition for the sake of an additive one is from the roots with four and two as additives.

Bhāskara here noted that if the above operation is repeated, eventually a solution for additive or subtractive four, two, or one will be reached. As already noted, from a solution with additive or subtractive 4, a solution for additive 1 can be found. This is also easy to do with additive or subtractive 2 and with subtractive 1. Before continuing with the example, however, we need to discuss two questions, neither of which are addressed by Bhāskara. First, why does the method always give integral values at each stage? Second, why does the repetition of the method eventually give a solution pair for additives $\pm 4, \pm 2,$ or ± 1?

To answer the first question, note that Bhāskara's method can be derived by composing the first solution (u, v) for additive b with the obvious solution $(1, m)$ for additive $m^2 - D$. It follows that $(u', v') = (mu + v, Du + mv)$ is a solution for additive $b(m^2 - D)$. Dividing the resulting equation by b^2 gives the solution $(u_1, v_1) = (\frac{mu + v}{b}, \frac{Du + mv}{b})$ for additive $\frac{m^2 - D}{b}$. It is then clear why m must be found so that $mu + v$ is a multiple of b. It is not difficult to prove, although as usual the text does not have a proof, that if $\frac{mu + v}{b}$ is integral, so are $\frac{m^2 - D}{b}$ and $\frac{Du + mv}{b} = \pm\sqrt{Du_1^2 + b_1}$.[25]

The reason that $m^2 - D$ is chosen "small" is so that the second question can be answered. Unfortunately, the proof that the process eventually reaches additive 1 is quite difficult; the first published version only dates to 1929.[26] It may well be that neither Bhāskara nor Jayadeva proved the result. They may simply have done enough examples to convince themselves of its truth. In fact, one can show that the *chakravāla* method leads to the smallest possible solution of the equation and therefore to every solution.

In any case, we continue with Bhāskara's example. Beginning with $67 \cdot 1^2 - 3 = 8^2$, we have derived $67 \cdot 5^2 + 6 = 41^2$. The next step is to solve $5m + 41 = 6n$, with $|m^2 - 67|$ small. The appropriate choice is $m = 5$. Then $(u_2, v_2) = (11, 90)$ is a solution for additive -7, or $67 \cdot 11^2 - 7 = 90^2$. Again, solve $11m + 90 = -7n$. The value $m = 9$ works and $(u_3, v_3) = (27,221)$ is a solution for additive -2, or $67 \cdot 27^2 - 2 = 221^2$. At this point, since additive -2 has been reached, it is only necessary to compose $(27, 221)$ with itself. This gives $(u_4, v_4) = (11934, 97684)$ as a solution for additive 4. Dividing by 2, Bhāskara finally found the desired solution $x = 5967$, $y = 48,842$, to the original equation $67x^2 + 1 = y^2$.

 8.6   COMBINATORICS

The earliest recorded statements of combinatorical rules appear in India, although again without any proofs or justifications. For example, the medical treatise of Susruta, perhaps written in the sixth century BCE, states that 63 combinations can be made out of six different tastes—bitter, sour, salty, astringent, sweet, hot—by taking them one at a time, two at a time, three at a time, and so on.[27] In other words, there are 6 single tastes, 15 combinations of two, 20 combinations of three, and so forth. Other works from the same general time period include similar calculations dealing with such topics as philosophical categories and senses. In all these examples, however, the numbers are small enough that simple enumeration is sufficient to produce the answers. We do not know whether relevant formulas had been developed.

On the other hand, a sixth-century work by Varāhamihira deals with a larger value. It plainly states that "if a quantity of 16 substances is varied in four different ways, the result will be 1820."[28] In other words, since Varāhamihira was trying to create perfumes using 4 ingredients out of a total of 16, he had calculated that there were precisely 1820 ($= C_4^{16}$) different ways of choosing the ingredients. It is unlikely that the author actually enumerated these 1820 combinations, and so we assume that he knew a method to calculate that number.

In the ninth century, Mahāvīra gave an explicit algorithm for calculating the number of combinations:

> The rule regarding the possible varieties of combinations among given things: Beginning with one and increasing by one, let the numbers going up to the given number of things be written down in regular order and in the inverse order (respectively) in an upper and a lower horizontal row. If the product of one, two, three, or more of the numbers in the upper row taken from right to left be divided by the corresponding product of one, two, three, or more of the numbers in the lower row, also taken from right to left, the quantity required in each such case of combination is obtained as the result.[29]

Mahāvīra did not, however, give any proof of this algorithm, which can be translated into the modern formula

$$C_r^n = \frac{n(n-1)(n-2)\ldots(n-r+1)}{r!}.$$

He simply applied the rule to two problems, one about combinations of the tastes—as his predecessor did—and another about combinations of jewels on a necklace, where these may be diamonds, sapphires, emeralds, corals, and pearls.

Bhāskara gave many other calculations using this basic formula and also calculated that the number of permutations of a set of order n was $n!$. He was therefore able to ask and answer the question

> How many are the variations of form of the god Sambhu by the exchange of his ten attributes held reciprocally in his several hands: namely, the rope, the elephant's hook, the serpent, the tabor, the skull, the trident, the bedstead, the dagger, the arrow, and the bow?[30]

Other types of discrete problems also appeared in Indian mathematics. For example, Āryabhaṭa presented the following:

STANZA II, 22 *The sixth part of the triple product of the term count plus one, that sum plus the term count, and the term count, in order, is the total of the series of squares. And the square of the total of the series of natural numbers is the total of the series of cubes.*[31]

These two statements give us formulas for the sums S_n^2, S_n^3, of the first n integral squares and cubes, namely, $S_n^2 = \frac{1}{6}n(n+1)(2n+1)$ and $S_n^3 = (1+2+\cdots+n)^2$. The first of these formulas was in essence known to Archimedes. The second formula is almost obvious, at least as a hypothesis, if one tries a few numerical examples.

As usual, Āryabhaṭa gave no indication of how he discovered or proved these results. But Nīlakaṇṭha (c. 1445–1545), a member of Mādhava's school in Kerala, gave an interesting proof of the first result in his commentary on the *Āryabhaṭīya*:

> Being that this [result on the sum of the squares] is demonstrated if there is equality of the total of the series of squares multiplied by six and the product of the three quantities, their equality is to be shown. A figure with height equal to the term-count, width equal to the term-count plus one, [and] length equal to the term-count plus one plus the term-count is [equal to] the product of the three quantities. But that figure can be made to construct the total of the series of squares multiplied by six.

Nīlakaṇṭha then described the construction of this figure. At each stage k, he used three "dominoes" of thickness 1, width k, and length $2k$ (Fig. 8.9). Thus, the total volume of the dominoes is $6k^2$. From the largest set, he constructed four walls of the desired figure. One of the dominoes forms one wall, a second forms the floor. The third is broken into two pieces,

FIGURE 8.9

Finding the sum of squares using dominoes

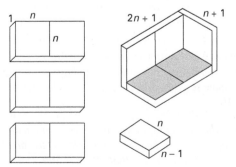

one of length $n + 1$ and one of length $n - 1$, and these form the two ends. The total length of the box is therefore $2n + 1$; its width is $n + 1$; and its height is n. The inside space of the box has length $2n - 1 = 2(n - 1) + 1$, width $n = (n - 1) + 1$, and height $n - 1$. Thus, we can create the walls of this new space with the three dominoes of thickness 1, width $n - 1$, and length $2(n - 1)$. We then continue until the entire box is filled. The result follows.

 8.7   TRIGONOMETRY

During the first centuries of the common era, in the period of the Kushan empire and that of the Guptas, there is strong evidence of the transmission of Greek astronomical knowledge to India, probably along the Roman trade routes. Curiously, Ptolemy's astronomy and mathematics were not transmitted but the work of some of his predecessors instead, in particular, the work of Hipparchus. Just as the needs of Greek astronomy led to the development of trigonometry, the needs of Indian astronomy led to Indian improvements in this field.

8.7.1 Construction of Sine Tables

The earliest known Indian work containing trigonometry is the *Paitāmahasiddhānta*, written in the early fifth century. This is the first of several similar works dealing with astronomy and its associated mathematics written over the next several centuries. To provide a basis for the spherical trigonometrical calculations necessary to solve astronomical problems, the *Paitāmahasiddhānta* contains a table of "half-chords," the literal translation of the Sanskrit term *jyā-ardha* (Sidebar 8.1). Recall that Ptolemy, in order to solve triangles using a table of chords, often had to deal with half the chord of double the angle. It was probably an unknown Indian mathematician who decided that it would be much simpler to tabulate the half-chords of double the angle rather than the chords themselves. Thus, in this work, as in all later Indian astronomical works, it is this half-chord "function" that is used. Now Ptolemy tabulated his chords in a circle of radius 60, while Hipparchus, several centuries earlier, had used a radius of 3438. Because this latter radius was used as the basis of the table in the *Paitāmahasiddhānta*, we surmise that it was Hipparchus's trigonometry rather than Ptolemy's that first reached India. In what follows, we generally use the word "Sine" (with a capital S) to represent the length of the Indian half-chord, given that the half-chord is a line in a circle of radius R, where R will always be stated. We reserve the word "sine" (with a small s) for the modern function (or, equivalently, when the radius of the circle is 1). Thus, $\text{Sin } \theta = R \sin \theta$. (A similar convention will be used for other trigonometric functions, here and in subsequent chapters.)

We consider an early description of the construction of a Sine table, not in the imperfectly preserved *Paitāmahasiddhānta*, but in the *Āryabhaṭīya*. The description of the construction method for a Sine table is given in stanza II, 12, while a table of differences for the sines is given in stanza I, 10.

STANZA II, 12   *By what number the second Sine is less than the first Sine, and by the quotient obtained by dividing the sum of the preceding Sines by the first Sine, by the sum of these two quantities the following Sines are less than the first Sine.*[32]

SIDEBAR 8.1   *The Etymology of "Sine"*

The English word "sine" comes from a series of mistranslations of the Sanskrit *jyā-ardha* (chord-half). Āryabhaṭa frequently abbreviated this term to *jyā* or its synonym *jīva*. When some of the Hindu works were later translated into Arabic, the word was simply transcribed phonetically into an otherwise meaningless Arabic word *jiba*. But since Arabic is written without vowels, later writers interpreted the consonants *jb* as *jaib*, which means bosom or breast. In the twelfth century, when an Arabic trigonometry work was translated into Latin, the translator used the equivalent Latin word *sinus*, which also meant bosom, and by extension, fold (as in a toga over a breast), or a bay or gulf. This Latin word has now become our English "sine."

The "first Sine" s_1 in Indian trigonometry always means the Sine of an arc of $3\frac{3}{4}^{\circ} = 3°45'$, and this Sine, in a circle of radius 3438, is the same as the arc measure in minutes, namely, $s_1 = 225$. The rule in this stanza then allows us to calculate the Sines of each arc in turn in steps of $3°45'$. Thus, to calculate s_2, the Sine of $7°30'$, we subtract 225 from 225 to get 0 (at this stage, the first and second Sines are the same), then divide 225 by 225 to get 1, then subtract $0 + 1 = 1$ from 224 to get 224. That number is the first Sine difference, so $s_2 = 225 + 224 = 449$. To get s_3, subtract 224 from 225 to get 1, then divide 449 by 225, giving 2, then subtract $1 + 2 = 3$ from 225 to get 222 as the next Sine difference. Thus, s_3, the Sine of $11°15'$, is given by $s_3 = 449 + 222 = 671$. In general, then, the nth Sine s_n (the Sine of $n \times 3°45'$) is calculated as

$$s_n = s_{n-1} + \left(s_1 - \frac{s_1 + s_2 + \cdots s_{n-1}}{s_1} \right).$$

All of the Sine differences are listed in

STANZA I, 10   *The twenty-four Sine [differences] reckoned in minutes of arc are 225, 224, 222, 219, 215, 210, 205, 199, 191, 183, 174, 164, 154, 143, 131, 119, 106, 93, 79, 65, 51, 37, 22, 7.*[33]

The values here actually show several slight discrepancies from the values calculated according to the method given above. Perhaps the fractional values of the division process were from time to time distributed among the Sines. In any case, it seems unlikely that the Indians actually originally calculated the Sines by this method. More likely, they calculated them as Hipparchus did: The Sine of $90°$ is equal to the radius $3438'$; the Sine of $30°$ is half the radius, $1719'$; the Sine of $45°$ is $\frac{3438}{\sqrt{2}} = 2431'$; and the Sines of the other arcs are calculated by use of the Pythagorean Theorem and the half-angle formula.

Once the table of Sines from $3°45'$ to $90°$ in steps of $3°45'$ had been constructed, a table of differences and second differences could also have been constructed. If the Indians noticed then that the second differences were proportional to the Sines, it would not have been difficult to construct the rule given in stanza II, 12. Similar Sine tables of roughly the same accuracy were produced in India by many authors over the next several hundred years. Varāhamihira (sixth century) tabulated the Cosine as well as the Sine for his radius of 120 and described the standard relationships between these functions. And the *Sūrya-Siddhānta*, probably written in the seventh century, may have been the source of the Chinese calculation of the Tangent function discussed earlier and even hints at the Secant. For although it does not tabulate these functions, verses 21–22 of chapter 3, in discussing the shadow cast by a gnomon, read, "Of [the sun's meridian zenith distance] find the base Sine and the perpendicular Sine [Cosine]. If then the base Sine and radius be multiplied respectively by the measure of the gnomon in digits, and divided by the perpendicular Sine, the results are the shadow and hypotenuse at mid-day."[34]

8.7.2 Approximation Techniques

Interestingly, no Indian astronomical text until the time of Bhāskara II contained a Sine table for arcs closer together than $3\frac{3}{4}°$. Instead, Indian mathematicians developed methods of approximation. The simplest method, of course, was linear interpolation between the tabulated values. But as early as the seventh century, Brahmagupta had developed a somewhat more accurate interpolation scheme using the second-order differences. In modern notation, if Δ_i represents the ith Sine difference (given in Aryabhata's stanza I, 10), α_i the ith arc, and $h = 3\frac{3}{4}°$ the interval between these arcs, then Brahmagupta's result is that

$$\text{Sin}(\alpha_i + \theta) = \text{Sin}(\alpha_i) + \frac{\theta}{2h}(\Delta_i + \Delta_{i+1}) - \frac{\theta^2}{2h^2}(\Delta_i - \Delta_{i+1}).$$

For example, to calculate $\text{Sin}(20°)$, note that $20 = 18\frac{3}{4} + 1\frac{1}{4}$, where $18\frac{3}{4} = x_5$. The formula then gives

$$\text{Sin}(20) = \text{Sin}\left(18\frac{3}{4} + 1\frac{1}{4}\right) = \text{Sin}\left(18\frac{3}{4}\right) + \frac{1\frac{1}{4}}{2(3\frac{3}{4})}(215 + 210) - \frac{(1\frac{1}{4})^2}{2(3\frac{3}{4})^2}(215 - 210)$$

$$= 1105 + \frac{1}{6}(425) - \frac{1}{18}(5) = 1176$$

to the nearest integer, where the Sine is for a circle of radius 3438.

Brahmagupta unfortunately gave no justification for this interpolation formula, but we note that the right side of the formula is the unique quadratic polynomial in θ that agrees with the left side for $\theta = -3\frac{3}{4}°$, $\theta = 0°$, and $\theta = 3\frac{3}{4}°$. Curiously, Brahmagupta himself also used an algebraic formula to approximate Sines, a formula that seems to have been first given by Bhāskara I in Sanskrit verse in the *Mahābhāskariya*:

> I briefly state the rule [for finding the Sine] without making use of the Sine differences 225 and so on. Subtract the degrees of the [arc] from the degrees of half a circle. Then multiply the remainder by the degrees of the [arc] and put down the result in two places. At one place subtract the result from 40,500. By one-fourth of the remainder [thus obtained] divide the result at the other place as multiplied by the radius. . . . Thus is obtained the [Sine to that radius].[35]

In modern notation, Bhāskara's formula is

$$\text{Sin}\,\theta = R\sin\theta = \frac{R\theta(180 - \theta)}{\frac{1}{4}(40{,}500 - \theta(180 - \theta))} = \frac{4R\theta(180 - \theta)}{40{,}500 - \theta(180 - \theta)}.$$

If we use the formula to calculate the Sine of $\theta = 20°$, we get

$$\text{Sin } 20 = 3438 \cdot \frac{4 \cdot 20 \cdot 160}{40{,}500 - 20 \cdot 160} = 1180$$

to the nearest integer, a value in error by approximately 0.3%.

There are two questions to ask here. First, how was this algebraic formula derived? And second, why did the Indians use an algebraic formula for the Sine when they had an accurate table, derived geometrically, as well as standard interpolation methods? Because as usual the ancient sources give us little help with these questions, we will consider the simplest modern suggestion. This idea is that the inventor noted the close resemblance of the Sine function $R \sin \theta$ to the parabolic function $P(\theta) = R\theta(180 - \theta)/8100$ in the sense that both functions are 0 at $\theta = 0$ and $\theta = 180$ and are equal to R at $\theta = 90$. He then noted that the same is true for the function $F(\theta) = \theta(180 - \theta) \text{ Sin } \theta/8100$. Because $P(30) = (5/9)R$ and $F(30) = (5/18)R$, he proceeded to get a formula giving the correct value $R/2$ for $\theta = 30$ by the use of simple proportions:

$$\frac{P(\theta) - \text{Sin } \theta}{F(\theta) - \text{Sin } \theta} = \frac{\frac{5}{9}R - \frac{1}{2}R}{\frac{5}{18}R - \frac{1}{2}R}.$$

This reduces to the equation

$$\frac{R\theta(180 - \theta) - 8100 \text{ Sin } \theta}{\theta(180 - \theta) \text{ Sin } \theta - 8100 \text{ Sin } \theta} = -\frac{1}{4},$$

which in turn gives us Bhāskara's formula.[36]

The apparent method of producing an approximation formula by beginning with a good guess and then tinkering with it to make it agree with the correct result on a few selected values appears in other parts of Indian mathematics. But since no author says that he is just "tinkering," it is difficult to know not only how the results were obtained but also why. It may simply be that, as usual, mathematicians exercised their creative faculties to produce clever and beautiful results. And because the Sine function was necessary in so many calculations for astronomical purposes, it was a benefit to astronomers to have a very accurate rational approximation to the Sine that saved them the labor of constantly doing interpolations in the published Sine tables. In general, Indian mathematicians never restricted themselves to methods based on a particular formal proof structure. Thus, even though it is certain that they often knew how to "prove" mathematical results, the extant texts often demonstrate that once there was sufficient plausibility to a result, it was just passed down through the generations.

8.7.3 Power Series

Now in the time of Bhāskara I and Brahmagupta, algebraic approximations or interpolation schemes using differences were sufficient for the use to which these Sine values were put in astronomy. But over the next several hundred years, the necessity grew to have more accurate Sine tables. This necessity came out of navigation, for the sailors in the Indian Ocean needed to be able to determine precisely their latitude and longitude. Since observation of the pole star was difficult in the tropics, one had to determine latitude by observation of the solar altitude at noon, μ. A standard formula for determining the latitude ϕ, given in an astronomical work of Bhāskara I, was $R \text{ Sin } \delta = \text{Sin } \phi \text{ Sin } \mu$, where δ is the sun's declination (known from tables or calculations). Determination of longitude was somewhat more difficult, but this could also be accomplished using trigonometry if one knew the distance on the earth's surface of one degree along a great circle. In any case, the more accurate the Sine values, the more accurately one could determine one's location. Thus, mathematicians in south India, in what is now the state of Kerala, developed power series for the Sine, Cosine, and Arctangent, beginning late in the fourteenth century. These series appear in written form in the *Tantrasaṃgraha-vyākhyā* of about 1530, a commentary on a work by Nīlakaṇṭha. Derivations appear in the *Yuktibhāsā*, whose author credits these series to Madhava (1359–1425).

The Indian derivations of these results begin with the obvious approximations to the Cosine and Sine for small arcs and then use a "pull yourself up by your own bootstraps" approach to improve the approximation step-by-step. The derivations all make use of the notion of Sine differences, an idea already used much earlier. In our discussion of the Indian method, we will use modern notation.

We first consider the circle of radius R with a small arc $\alpha = \widehat{AC} \approx AC$ (Fig. 8.10). From the similarity of triangles AGC and OEB, we get

$$\frac{x_1 - x_2}{\alpha} = \frac{y}{R} \quad \text{and} \quad \frac{y_2 - y_1}{\alpha} = \frac{x}{R}$$

$$\text{or} \quad \frac{\alpha}{R} = \frac{x_1 - x_2}{y} = \frac{y_2 - y_1}{x}.$$

In modern terms, if $\angle BOF = \theta$ and $\angle BOC = \angle AOB = d\theta$, these equations amount to

$$\sin(\theta + d\theta) - \sin(\theta - d\theta) = \frac{y_2 - y_1}{R} = \frac{\alpha x}{R^2} = \frac{2Rd\theta}{R} \cos\theta = 2\cos\theta \, d\theta$$

and

$$\cos(\theta + d\theta) - \cos(\theta - d\theta) = \frac{x_2 - x_1}{R} = -\frac{\alpha y}{R^2} = -\frac{2Rd\theta}{R} \sin\theta = -2\sin\theta \, d\theta.$$

(These results, of course, almost give the derivative of the sine and cosine.)

FIGURE 8.10

Derivation of power series for sine and cosine

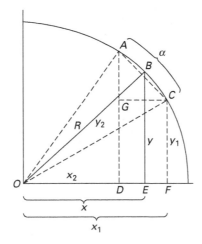

FIGURE 8.11

Differences of y's

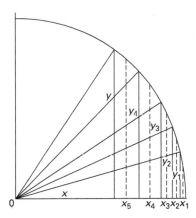

Now, suppose we have a small arc s divided into n equal subarcs, with $\alpha = s/n$. For simplicity, we take $R = 1$, although the Indian mathematicians did not. By applying the previous results repeatedly, we get the following sets of differences for the y's (Fig. 8.11) (where $y_n - y = \sin s$):

$$\Delta_n y = y_n - y_{n-1} = \alpha x_n$$
$$\Delta_{n-1} y = y_{n-1} - y_{n-2} = \alpha x_{n-1}$$
$$\vdots$$
$$\Delta_2 y = y_2 - y_1 = \alpha x_2$$
$$\Delta_1 y = y_1 - y_0 = \alpha x_1.$$

Similarly, the differences for the x's can be written

$$\Delta_{n-1} x = x_n - x_{n-1} = -\alpha y_{n-1}$$
$$\vdots$$
$$\Delta_2 x = x_3 - x_2 = -\alpha y_2$$
$$\Delta_1 x = x_2 - x_1 = -\alpha y_1.$$

We next consider the second differences on the y's:

$$\Delta_2 y - \Delta_1 y = y_2 - y_1 - y_1 + y_0 = \alpha(x_2 - x_1) = -\alpha^2 y_1.$$

In other words, the second difference of the sines is proportional to the negative of the sine. But since $\Delta_1 y = y_1$, we can write this result as

$$\Delta_2 y = y_1 - \alpha^2 y_1.$$

In general, we find that

$$\Delta_k y = y_1 - \alpha^2 y_1 - \alpha^2 y_2 - \cdots - \alpha^2 y_{k-1}.$$

But the sine equals the sum of its differences:

$$y = y_n = \Delta_1 y + \Delta_2 y + \cdots + \Delta_n y$$
$$= ny_1 - [y_1 + (y_1 + y_2) + (y_1 + y_2 + y_3) + \cdots + (y_1 + y_2 + \cdots + y_{n-1})]\alpha^2.$$

Also, $s/n \approx y_1 \approx \alpha$, or $ny_1 \approx s$. Naturally, the larger the value of n, the better each of these approximations is. Therefore,

$$y \approx s - \lim_{n \to \infty} \left(\frac{s}{n}\right)^2 [y_1 + (y_1 + y_2) + \cdots + (y_1 + y_2 + \cdots + y_{n-1})].$$

Next we add the differences of the x's. We get

$$x_n - x_1 = -\alpha(y_1 + y_2 + \cdots + y_{n-1}).$$

But $x_n \approx x = \cos s$ and $x_1 \approx 1$. It then follows that

$$x \approx 1 - \lim_{n \to \infty} \left(\frac{s}{n}\right) (y_1 + y_2 + \cdots + y_{n-1}).$$

To continue the calculation, the Indian mathematicians needed to approximate each y_i and use these approximations to get approximations for $x = \cos s$ and $y = \sin s$. Each new approximation in turn is placed back in the expressions for x and y and leads to a better approximation. Note first that if y is small, y_i can be approximated by is/n. It follows that

$$x \approx 1 - \lim_{n \to \infty} \left(\frac{s}{n}\right) \left[\frac{s}{n} + \frac{2s}{n} + \cdots + \frac{(n-1)s}{n}\right]$$

$$= 1 - \lim_{n \to \infty} \left(\frac{s}{n}\right)^2 [1 + 2 + \cdots + (n-1)]$$

$$= 1 - \lim_{n \to \infty} \frac{s^2}{n^2} \left[\frac{n^2}{2} - \frac{n}{2}\right]$$

$$= 1 - \frac{s^2}{2}.$$

Note that in this calculation, we replaced the sum of the first $n - 1$ integers by a simple expression. To go further, Jyesthadeva needed to know similar formulas for the sums of integral squares, integral cubes, and so on. In particular, he needed to know that

$$\sum_{i=0}^{n-1} i^k = \frac{n^{k+1}}{k+1} \pm \text{lower-order terms.}$$

This result was known in India, as was the result

$$\sum_{p=1}^{n-1} \left(\sum_{i=1}^{p} i^k\right) = n \sum_{i=1}^{n-1} i^k - \sum_{i=1}^{n-1} i^{k+1}$$

from which the earlier result was proved. Since both of these results were discovered several hundred years earlier in the Islamic world, we postpone discussion of them until the next chapter. But we will use these results in what follows. In particular, the former result will be used in the form

$$\lim_{n \to \infty} \frac{\sum_{i=1}^{n-1} i^k}{n^{k+1}} = \frac{1}{k+1}.$$

Thus, to get our new approximation for y, we proceed as follows:

$$y \approx s - \lim_{n \to \infty} \left(\frac{s}{n}\right)^2 \left[\frac{s}{n} + \left(\frac{s}{n} + \frac{2s}{n}\right) + \cdots + \left(\frac{s}{n} + \frac{2s}{n} + \cdots + \frac{(n-1)s}{n}\right)\right]$$

$$= s - \lim_{n \to \infty} \frac{s^3}{n^3} [1 + (1+2) + (1+2+3) + \cdots + (1+2+\cdots+(n-1))]$$

$$= s - \lim_{n \to \infty} \frac{s^3}{n^3} \left[n(1+2+\cdots(n-1)) - (1^2 + 2^2 + \cdots + (n-1)^2)\right]$$

$$= s - s^3 \lim_{n \to \infty} \left[\frac{\sum_{i=1}^{n-1} i}{n^2} - \frac{\sum_{i=1}^{n-1} i^2}{n^3}\right]$$

$$= s - s^3 \left(\frac{1}{2} - \frac{1}{3}\right)$$

$$= s - \frac{s^3}{6}.$$

We thus have a new approximation for y and therefore for each y_i.

To improve the approximation for sine and cosine, we now assume that $y_i \approx (is/n) - (is)^3/(6n^3)$ in the expression for $x = \cos s$ and proceed as before. We use the two sum formulas in the case $k = 3$ to get

$$x \approx 1 - \frac{s^2}{2} + \frac{s^4}{24}.$$

Similarly, we get a new approximation for $y = \sin s$:

$$y \approx s - \frac{s^3}{6} + \frac{s^5}{120}.$$

Because Jyesthadeva considered each new term in these polynomials as a correction to the previous value, he understood that the more terms taken, the more closely the polynomials approach the true values for the sine and cosine. The polynomial approximations can thus be continued as far as necessary to achieve any desired approximation. The Indian authors had therefore discovered the sine and cosine power series.

8.8   TRANSMISSION TO AND FROM INDIA

We are much better informed about Indian mathematics throughout history than we are about the mathematics of China. We know, for example, that India learned trigonometry (and also some astronomy) from Greek sources. We also know that Islamic scholars learned Indian trigonometry when Indian works were brought to Baghdad in the eighth century. And, of course, our decimal place value system traveled from India through Islam to western Europe over a period of several hundreds of years. On the other hand, there is no record of the Indian solution of the Pell equation being known in Europe before European scholars solved it themselves (and in a way different from that of the Indians). Nor do we know how the quadratic formula reached India or whether Islamic scholars learned of it from the Indians. We know that some Indian trigonometric ideas were transmitted to China, but whether the double difference method of finding heights and distances traveled from one of these cultures to the other is not known.

The most interesting question about transmission, however, relates to the power series for the sine and cosine. There is certainly no available documentation showing that any Europeans knew of the Indian developments in this area before the Europeans themselves worked out the power series in the mid-seventeenth century. However, there is some circumstantial evidence. First of all, Europeans, just like the Indians, needed precise trigonometric values for navigation. Secondly, the texts in which these power series were described were easily available in south India. Third, the Jesuits, in their quests to proselytize in Asia, established a center in south India in the late sixteenth century. In general, wherever the Jesuits went, they learned the local languages, collected and translated local texts, and then set up educational institutions to train disciples. But the question remains as to whether, in fact, the Jesuits did find these particular texts and bring them back in some form to Europe. As we will discuss in the chapters on calculus, in the period from 1630 to 1680 some of the basic ideas present in these Indian texts began to appear in European works. In the case of Newton, we can trace his thoughts through his notebooks and therefore have no reason to believe he was aware of Indian material. But for many of the other European mathematicians, we have little documentary evidence of how they discovered and elaborated on their ideas. So at the moment, we can only speculate as to whether Indian trigonometric series were transmitted in some form to Europe by the early seventeenth century.

EXERCISES

1. Use Āryabhaṭa's cube root algorithm to find the cube root of 13,312,053.

2. Show that the construction given in the text for constructing a square equal to the difference of two squares is correct (see Fig. 8.3). Here, $ABCD$ is the larger square with side equal to a, and $PQRS$ the smaller square with side equal to b. We cut off $AK = b$ from AB and draw KL perpendicular to AK intersecting DC in L. With K as center and radius KL, draw an arc meeting AD at M. Thus, show that the square on AM is the required square.

3. Show that the construction given in the text for transforming a rectangle into a square is correct (see Fig. 8.4). The rectangle is $ABCD$. Find L on AB so that $AL = AC$. Then find the midpoint E of LB, and draw EG parallel to LF. Move the rectangle $EBGD$ from where it is to the bottom of the diagram, forming the rectangle $CFHK$. Complete the square by adding the square on FG. Show that using the result of Exercise 2 gives the result.

4. This is the method presented in the text for finding a circle whose area is equal to a given square: In square $ABCD$, let M be the intersection of the diagonals (see Fig. 8.5). Draw the circle with M as center and MA as radius; let ME be the radius of the circle perpendicular to the side AD and cutting AD in G. Let $GN = \frac{1}{3}GE$. Then MN is the radius of the desired circle. Show that if $AB = s$ and $MN = r$, then $\frac{r}{s} = \frac{2+\sqrt{2}}{6}$. Show that this implies a value for π equal to 3.088311755.

5. The *Śulbasūtra* method of "squaring a circle" of diameter d takes the side of the desired square to be $\frac{7}{8} + \frac{1}{8 \times 29} - \frac{1}{8 \times 29 \times 6} + \frac{1}{8 \times 29 \times 6 \times 8}$ times d. Show that this is equivalent to using a value for π equal to 3.088326491.

6. Solve this problem from the *Līlāvatī*: There is a hole at the foot of a pillar nine *hastas* high, and a pet peacock standing on top of it. Seeing a snake returning to the hole at a distance from the pillar equal to three times its height, the peacock descends upon it slantwise. Say quickly, at how many *hastas* from the hole does the meeting of their two paths occur?

7. Brahmagupta asserts that if $ABCD$ is a quadrilateral inscribed in a circle, with side lengths a, b, c, d (in cyclic order) (see Fig. 8.8), then the lengths of the diagonals AC and BD are given by

$$AC = \sqrt{\frac{(ac + bd)(ad + bc)}{ab + cd}}$$

and similarly

$$BD = \sqrt{\frac{(ac + bd)(ab + cd)}{ad + bc}}.$$

Prove this result as follows:

a. Let $\angle ABC = \theta$. Then $\angle ADC = \pi - \theta$. Let $x = AC$. Use the law of cosines on each of triangles ABC and ADC to express x^2 two different ways. Then, since $\cos(\pi - \theta) = -\cos\theta$, use these two formulas for x^2 to determine $\cos\theta$ as a function of a, b, c, and d.

b. Replace $\cos\theta$ in your expression for x^2 in terms of a and b by the value for the cosine determined in part a.

c. Show that
$cd(a^2 + b^2) + ab(c^2 + d^2) = (ac + bd)(ad + bc).$

d. Simplify the expression for x^2 found in part b by using the algebraic identity found in part c. By then taking square roots, you should get the desired expression for $x = AC$. (Of course, a similar argument will then give you the expression for $y = BD$.)

8. Brahmagupta asserts that if $ABCD$ is a quadrilateral inscribed in a circle, as in Exercise 7, then if $s = \frac{1}{2}(a + b + c + d)$, the area of the quadrilateral is given by $S = \sqrt{(s - a)(s - b)(s - c)(s - d)}$ (Fig. 8.12). Prove this result as follows:

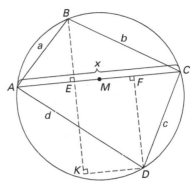

FIGURE 8.12

Area of a quadrilateral inscribed in a circle

a. In triangle ABC, drop a perpendicular from B to point E on AC. Use the law of cosines applied to that triangle to show that $b^2 - a^2 = x(x - 2AE)$.

b. Let M be the midpoint of AC, so $x = 2AM$. Use the result of part a to show that $EM = (b^2 - a^2)/2x$.

c. In triangle ADC, drop a perpendicular from D to point F on AC. Use arguments similar to those in parts a and b to show that $FM = (d^2 - c^2)/2x$.

d. Denote the area of quadrilateral $ABCD$ by P. Show that $P = \frac{1}{2}x(BE + DF)$ and therefore that $P^2 = \frac{1}{4}x^2(BE + DF)^2$.

e. Extend BE to K such that $\angle BKD$ is a right angle, and complete the right triangle BKD. Then $BE + DF = BK$. Substitute this value in your expression from part d; then use the Pythagorean Theorem to conclude that $P^2 = \frac{1}{4}x^2(y^2 - EF^2)$.

f. Since $EF = EM + FM$, conclude that $EF = [(b^2 + d^2) - (a^2 + c^2)]/2x$. Substitute this value into the expression for P^2 found in part e, along with the values for x^2 and y^2 found in Exercise 7. Conclude that

$$P^2 = \frac{1}{4}(ac + bd)^2 - \frac{1}{16}[(b^2 + d^2) - (a^2 + c^2)]^2$$

$$= \frac{1}{16}(4(ac + bd)^2 - [(b^2 + d^2) - (a^2 + c^2)]^2).$$

g. Since $s = \frac{1}{2}(a + b + c + d)$, show that $s - a = \frac{1}{2}(b + c + d - a)$, $s - b = \frac{1}{2}(a + c + d - b)$, $s - c = \frac{1}{2}(a + b + d - c)$, and $s - d = \frac{1}{2}(a + b + c - d)$.

h. To prove the theorem, it is necessary to show that the final expression for P^2 given in part f is equal to the product of the four expressions in part g. It is clear that the denominators are both equal to 16. To prove that the numerators are equal involves a lot of algebraic manipulation. Work carefully and show that the two numerators are in fact equal.

9. Solve the following problem from the Bakhshālī manuscript: One person goes 5 *yojanas* a day. When he has proceeded for seven days, the second person, whose speed is 9 *yojanas* a day, departs. In how many days will the second person overtake the first?

10. Solve the following problem from Mahāvīra: "One night, in a month of the spring season, a certain young lady . . . was lovingly happy along with her husband on . . . the floor of a big mansion, white like the moon, and situated in a pleasure garden with trees bent down with the load of the bunches of flowers and fruits, and resonant with the sweet sounds of parrots, cuckoos and bees which were all intoxicated with the honey obtained from the flowers therein. Then on a love quarrel arising between the husband and the wife, that lady's necklace made up of pearls became sundered and fell on the floor. One-third of that necklace of pearls reached the maid-servant there; one-sixth fell on the bed; then one-half of what remained (and one-half of what remained thereafter and again one-half of what remained thereafter and so on, counting six times in all) fell all of them everywhere; and there were found to remain (unscattered) 1161 pearls. . . . Give out the (numerical) measure of the pearls (in that necklace)."[37]

11. Solve the following problem from Mahāvīra: There are 4 pipes leading into a well. Among these, each fills the well (in order) in 1/2, 1/3, 1/4, and 1/5 of a day. In how much of a day will all of them together fill the well and each of them to what extent?

12. Another problem from Mahāvīra: Of a collection of mango fruits, the king took 1/6; the queen took 1/5 of the remainder, and three chief princes took 1/4, 1/3, 1/2 of what remained at each step; and the youngest child took the remaining three mangoes. O you, who are clever in working miscellaneous problems on fractions, give out the measure of that collection of mangoes.

13. Another problem from Mahāvīra: One-third of a herd of elephants and three times the square root of the remaining part of the herd were seen on a mountain slope; and in a lake was seen a male elephant along with three female elephants constituting the ultimate remainder. How many were the elephants here?

14. Another problem from Mahāvīra: If 3 peacocks cost 2 coins, 4 pigeons cost 3 coins, 5 swans cost 4 coins, and 6 *sārasa* birds cost 5 coins, and if you buy 72 birds for 56 coins, how many of each type of bird do you have?

15. This problem is from Brahmagupta's work on congruences. Given that the sun makes 30 revolutions through the ecliptic in 10,960 days, how many days have elapsed (since the sun was at a given starting point) if the sun has made an integral number of revolutions plus 8080/10,960 of a revolution, that is, "when the remainder of solar revolutions is 8080." If y is the number of days sought and x is the number of revolutions, then, because 30 revolutions take 10,960 days, x revolutions take $(1096/3)x$ days. Therefore, $y = (x + 808/1096)(1096/3)$, or $1096x + 808 = 3y$. Thus, solve $N \equiv 808 \pmod{1096}$ and $N \equiv 0 \pmod 3$.

16. Solve the congruence $N \equiv 23 \pmod{137}$, $N \equiv 0 \pmod{60}$ using Brahmagupta's procedure.

17. Solve $1096x + 1 = 3y$ using Brahmagupta's method. Given a solution to this equation (with "additive" 1), it is easy to find solutions to equations with other additives by simply multiplying. For example, solve $1096x + 10 = 3y$.

18. Prove that Brahmagupta's procedure does give a solution to the simultaneous congruences. Begin by noting that the Euclidean algorithm allows one to express the greatest common divisor of two positive integers as a linear combination of these integers. Note further that a condition for the solution procedure to exist is that this greatest common divisor must divide the "additive." Brahmagupta does not mention this, but Bhāskara and others do.

19. Solve the problem $N \equiv 5 \pmod 6 \equiv 4 \pmod 5 \equiv 3 \pmod 4) \equiv 2 \pmod 3)$ by the Indian procedure and by the Chinese procedure. Compare the methods.

20. Solve the congruence $N \equiv 10 \pmod{137} \equiv 0 \pmod{60}$ by the Chinese procedure and compare your solution step-by-step with the solution by Brahmagupta's method. How do the two methods compare?

21. Solve the indeterminate equation $17n - 1 = 75m$ by both the Indian and Chinese methods explicitly using the Euclidean algorithm. Compare the solutions.

22. Prove that $D(u_0v_1 + u_1v_0)^2 + c_0c_1 = (Du_0u_1 + v_0v_1)^2$ given that $Du_0^2 + c_0 = v_0^2$ and $Du_1^2 + c_1 = v_1^2$.

23. Solve $83x^2 + 1 = y^2$ by Brahmagupta's method. Begin by noting that (1, 9) is a solution for subtractive 2.

24. Show that if (u, v) is a solution to $Dx^2 - 4 = y^2$, then $(u_1, v_1) = (\frac{1}{2}uv(v^2 + 1)(v^2 + 3), (v^2 + 2)[\frac{1}{2}(v^2 + 1)(v^2 + 3) - 1])$ is a solution to $Dx^2 + 1 = y^2$ and that both u_1 and v_1 are integers regardless of the parity of u or v.

25. Solve $13x^2 + 1 = y^2$ by noting that (1, 3) is a solution for subtractive 4 and applying the method of Exercise 24.

26. Show that if (u, v) is a solution to $Dx^2 + 2 = y^2$, then $(u_1, v_1) = (uv, v^2 - 1)$ is a solution to $Dx^2 + 1 = y^2$. Deduce a similar rule if (u, v) is a solution to $Dx^2 - 2 = y^2$.

27. Solve $61x^2 + 1 = y^2$ by Bhāskara's process. The solution is $x = 226,153,980$, $y = 1,766,319,049$.

28. A combinatorics problem from Bhāskara: In a pleasant, spacious, and elegant edifice, with eight doors, constructed by a skillful architect as a palace for the lord of the land, tell me the combinations of apertures taken one, two, three, and so on, at a time.

29. Calculate the fourth, fifth, and sixth Sine differences by using Āryabhaṭa's method. Then determine the fourth, fifth, and sixth Sine values.

30. Use a graphing calculator and/or calculus techniques to show that the algebraic formula of Bhāskara I approximates the Sine between 0 and 180 degrees with an error of no more than 1%. Find the values that are most in error.

31. Show that Bhāskara's algebraic formula for the Sine can be rewritten as an approximation formula for the modern sine in the form

$$\sin x \approx \frac{16x(\pi - x)}{5\pi^2 - 4x(\pi - x)},$$

where x is given in radians. Graph this function on a graphing calculator from 0 to π and compare it with the graph of $\sin x$ on that interval.

32. Use both the interpolation scheme of Brahmagupta and the algebraic formula of Bhāskara I to approximate $\sin(16°)$. Compare the two values to each other and to the exact value. What are the respective errors?

33. Continue the process described for determining the power series for the sine and cosine for two more steps in each case. That is, beginning with $y_i \approx (is/n) - (is)^3/(6n^3)$, show that $x = \cos s \approx 1 - s^2/2 + s^4/24 - s^6/720$ and $y = \sin s \approx s - s^3/6 + s^5/120 - s^7/5040$.

34. Devise a lesson for a number theory course on solving indeterminate equations of the form $rx + c = sy$, using the methods of Brahmagupta.

35. Why would the Indians have thought it better to use an algebraic approximation to the sine function rather than calculate values using geometric methods and methods of interpolation?

REFERENCES AND NOTES

Among the surveys of the general history of Indian mathematics are B. Datta and A. N. Singh, *History of Hindu Mathematics* (Bombay: Asia Publishing House, 1961) (reprint of 1935–1938 original), A. K. Bag, *Mathematics in Ancient and Medieval India* (Varanasi: Chaukhambha Orientalia, 1979), and C. N. Srinivasiengar, *The History of Ancient Indian Mathematics* (Calcutta: World Press Private Ltd., 1967). Other surveys include Volume 2 of D. Chattopadhyaya, ed., *Studies in the History of Science in India* (New Delhi: Editorial Enterprises, 1992); S. N. Sen, "Mathematics," in D. M. Bose, S. N. Sen, and B. V. Subbarayappa, *A Concise History of Science in India* (New Delhi: Indian National Science Academy, 1971), pp. 136–212; chapters 8 and 9 in G. G. Joseph, *The Crest of the Peacock: Non-European Roots of Mathematics* (Princeton: Princeton University Press, 2000); and T. K. Puttaswamy, "The Mathematical Accomplishments of Ancient Indian Mathematicians," in Helaine Selin, ed., *Mathematics across Cultures: The History of Non-Western Mathematics* (Dordrecht: Kluwer Academic Publishers, 2000).

There is a basic study of the mathematical contents of the *Śulbasūtras* by B. Datta: *The Science of the Sulba* (Calcutta: University of Calcutta, 1932), as well as a more general work on Indian geometry: T. A. Sarasvatī, *Geometry in Ancient and Medieval India* (New Delhi: Motilal Banarsidass, 1979). R. H. Gupta has a survey of South Indian mathematics in "South Indian Achievements in Medieval Mathematics," *Gaṇita Bhāratī* 9 (1987), 15–40. The power series methods are discussed in three recent articles: Ranjan Roy, "The Discovery of the Series Formula for π by Leibniz, Gregory, and Nilakantha," *Mathematics Magazine* 63 (1990), 291–306; Victor J. Katz, "Ideas of Calculus in Islam and India," *Mathematics Magazine* 68 (1995), 163–174; and David Bressoud, "Was Calculus Invented in India?" *College Mathematics Journal* 33 (2002), 2–13.

The *Śulbasūtras* are available in English in S. N. Sen and A. K. Bag, eds., *The Śulbasūtras of Baudhāyana, Āpastamba, Kātyāyana and Mānava with Text, English Translation and Commentary* (New Delhi: Indian National Science Academy, 1983). The text of Āryabhaṭa with commentary of Bhāskara is available in an English translation by Agathe Keller: *Expounding the Mathematical Seed: A Translation of Bhāskara I on the Mathematical Chapter of the Āryabhaṭīya* (Basel: Birkhäuser, 2006) as well as in the older translation of Walter E. Clark: *The Āryabhaṭīya of Āryabhaṭa* (Chicago: University of Chicago Press, 1930). The major mathematical texts of Bhāskara II and Brahmagupta were translated by H. T. Colebrooke in *Algebra* with Arithmetic and Mensuration from the Sanskrit of Brah-megupta and Bhāscara (London: John Murray, 1817). More modern translations by Kim Plofker and David Pingree appear in Victor Katz, ed., The Mathematics of Egypt, Mesopotamia, China, India, and Islam: A Sourcebook (Princeton: Princeton University Press, 2007), as do many of the other translations mentioned here. Mahāvīra's Gaṇitāsarasaṅgraha was edited and translated by M. Rangācārya and published in 1912 by Government Press in Madras.

1. Mahāvīra, Gaṇitāsarasaṅgraha, sec. 1.

2. Clark, The Āryabhaṭīya, p. 21.

3. Quoted in D. E. Smith, History of Mathematics (New York: Dover, 1958), vol. 1, p. 166. For more details on Indian mathematical notation, see Saradakanta Ganguli, "The Indian Origin of the Modern Place-Value Arithmetical Notation," American Mathematical Monthly 39 (1932), 251–256, 389–393, and 40 (1933), 25–31, 154–157.

4. Yabuuti Kiyosi, "Researches on the Chiu-chih li," Acta Asiatica 36(1979), 12.

5. For more on this argument, see Lam Lay Yong, "The Conceptual Origins of our Numeral System and the Symbolic Form of Algebra," Archive for History of Exact Sciences 36 (1986), 184–195, and "A Chinese Genesis: Rewriting the History of Our Numeral System," Archive for History of Exact Sciences 38 (1988), 101–108. A more detailed account is in Lam Lay Yong and Ang Tian Se, Fleeting Footsteps: Tracing the Conception of Arithmetic and Algebra in Ancient China (Singapore: World Scientific, 1992).

6. Katz, Sourcebook, p. 403.

7. Ibid., p. 429.

8. Ibid., p. 471.

9. Ibid., p. 453.

10. Ibid., p. 389.

11. Ibid., p. 390. The next four quotations are from p. 391.

12. Ibid., p. 409. Bhhāskara's commentary is on p. 410.

13. Ibid., p. 411. The example problem is also here.

14. Ibid., p. 423. The next quotation is on p. 424.

15. Clark, The Āryabhaṭīya, p. 35. stanza II, 20 is on p. 36.

16. Katz, Sourcebook, p. 431.

17. Colebrooke, Algebra with Arithmetic, pp. 346–347.

18. Ibid., pp. 207–208.

19. Ibid., pp. 215–216. This is the source of the two quadratic problems on monkeys.

20. Mahāvīra, Gaṇitāsarasaṅgraha, p. 134.

21. Colebrooke, Algebra with Arithmetic, p. 235.

22. Katz, Sourcebook, p. 428. The next three quotations are also here.

23. Ibid., p. 433. The next four quotations dealing with the Pell equation are on p. 432.

24. Ibid., p. 473. The next two quotations dealing with the Pell equation are also here.

25. See the proof in C. N. Srinavasiengar, Ancient Indian Mathematics, p. 113.

26. Krishnaswami A. A. Ayyangar, Journal of the Indian Mathematics Society 18 (1929), 232–245. A detailed discussion of the entire process of solving the Pell equation is found in C. O. Selenius, "Rationale of the Chakravāla Process of Jayadeva and Bhāskara II," Historia Mathematica 2 (1975), 167–184.

27. Gurugovinda Chakravarti, "Growth and Development of Permutations and Combinations in India," Bulletin of Calcutta Mathematical Society 24 (1932), 79–88. More information can be found in N. L. Biggs, "The Roots of Combinatorics," Historia Mathematica 6 (1979), 109–136.

28. The quotation is from the Bṛhat Saṁhitā, chapter 77, rule 20, as translated in J. K. H. Kern, "The Brhatsamhita of Varahamihira," Journal of Royal Asiatic Society (1875), 81–134.

29. Mahāvīra, Gaṇitāsarasaṅgraha, p. 150.

30. Colebrooke, Algebra with Arithmetic, p. 124.

31. Katz, Sourcebook, p. 493. The next quotation is on p. 494.

32. Clark, The Āryabhaṭa, p. 29.

33. Ibid., p. 19.

34. E. Burgess, "Translation of the Sūrya-Siddhānta, a Textbook of Hindu Astronomy," Journal of the American Oriental Society 6 (1860), 141–498, p. 252.

35. Quoted in R. C. Gupta, "Bhaskara I's Approximation to Sine," Indian Journal of History of Science 2 (1967), 121–136, p. 122.

36. This suggestion is discussed in R. C. Gupta, "On Derivation of Bhāskara I's Formula for the Sine," Gaṇita Bhāratī 8 (1986), 39–41. A more detailed discussion of approximation methods in general is found in the doctoral dissertation of Kim Plofker: Mathematical Approximation by Transformation of Sine Functions in Medieval Sanskrit Astronomical Texts (Brown University, 1995).

37. Mahāvīra, Gaṇitāsarasaṅgraha, p. 73.

The Mathematics of Islam

You know well . . . for which reason I began searching for a number of demonstrations proving a statement due to the ancient Greeks . . . and which passion I felt for the subject . . . so that you reproached me my preoccupation with these chapters of geometry, not knowing the true essence of these subjects, which consists precisely in going in each matter beyond what is necessary. . . . Whatever way he [the geometer] may go, through exercise will he be lifted from the physical to the divine teachings, which are little accessible because of the difficulty to understand their meaning . . . and because of the circumstance that not everybody is able to have a conception of them, especially not the one who turns away from the art of demonstration.

—Preface to the *Book on Finding the Chords in the Circle* by al-Bīrūnī, c. 1030[1]

It is told that as a student, Omar Khayyam made a compact with two fellow students, Niẓām al Mulk and Ḥassan ibn Sabbah, to the effect that the one who first achieved a high position and great fortune would help the other two. It was Niẓām who in fact became the grand vizier of the Seljuk Sultan Jalāl al-Dīn Malik-shāh and proceeded to fulfill his promise. Ḥassan received the position of court chamberlain, but after he attempted to supplant his friend in the sultan's favor, he was banished from the court. Omar, on the other hand, declined a high position, accepting instead a modest salary that permitted him to have the leisure to study and write.

9.1   INTRODUCTION TO MATHEMATICS IN ISLAM

In the first half of the seventh century, a new civilization came out of Arabia. Under the inspiration of the prophet Muḥammad, the new monotheistic religion of Islam quickly attracted the allegiance of the inhabitants of the Arabian Peninsula. In less than a century after Muḥammad's capture of Mecca in 630, the Islamic armies conquered an immense territory as they propagated the new religion first among the previously polytheistic tribes of the Middle East and then among the adherents of other faiths. Syria and then Egypt were wrested from the Byzantine empire. Persia was conquered by 642, and soon the victorious armies had reached as far as India and parts of central Asia. In the west, North Africa was quickly overrun, and in 711 Islamic forces entered Spain. Their forward progress was eventually halted at Tours by the army of Charles Martel in 732. Already, however, the problems of conquest were being replaced by the new problems of governing the immense new empire. Muḥammad's successors, the caliphs, originally set up their capital in Damascus, but after about a hundred years of wars, including great victories but also some substantial defeats, the caliphate split up into several parts. In the eastern segment, under the Abbasid caliphs, the growth of luxury and the cessation of wars of conquest created favorable conditions for the development of a new culture.

In 766 the caliph al-Manṣūr founded his new capital of Baghdad, a city that soon became a flourishing commercial and intellectual center. The initial impulses of Islamic orthodoxy were soon replaced by a more tolerant atmosphere, and the intellectual accomplishments of all residents of the caliphate were welcomed. The caliph Hārūn al-Rashīd, who ruled from 786 to 809, established a library in Baghdad. Manuscripts were collected from various academies in the Near East that had been established by scholars fleeing from the persecutions of the ancient academies in Athens and Alexandria. These manuscripts included many of the classic Greek mathematical and scientific texts. A program of translation into Arabic was soon begun. Hārūn's successor, the caliph al-Ma'mūn (813–833), established a research institute, the Bayt al-Ḥikma (House of Wisdom), which was to last over 200 years. To this institute were invited scholars from all parts of the caliphate to translate Greek and Indian works as well as to conduct original research. By the end of the ninth century, many of the principal works of Euclid, Archimedes, Apollonius, Diophantus, Ptolemy, and other Greek mathematicians had been translated into Arabic and were available for study to the scholars gathered in Baghdad. Islamic scholars also absorbed the ancient mathematical traditions of the Babylonian scribes, still evidently available in the Tigris-Euphrates Valley, and in addition learned the mathematics of the Hindus.

The Islamic scholars during the first few hundred years of Islamic rule did more than just bring these sources together. They amalgamated them into a new whole and, in particular, as the opening quotation indicates, infused their mathematics with what they felt was divine inspiration. Creative mathematicians of the past had always carried investigations well beyond the dictates of immediate necessity, but in Islam many felt that this was a requirement of God. Islamic culture in general regarded "secular knowledge" not as in conflict with "holy knowledge," but as a way to it. Learning was therefore encouraged, and those who had demonstrated sparks of creativity were often supported by the rulers (usually both secular and religious authorities) so that they could pursue their ideas as far as possible. The mathematicians responded by always invoking the name of God at the beginning and end of their works and even occasionally referring to Divine assistance throughout the texts. Furthermore, since the rulers were naturally interested in the needs of daily life, the Islamic mathematicians, unlike their Greek predecessors, nearly all contributed not only to theory but also to practical applications.[2]

By the eleventh century, however, the status of mathematical thought in Islam was beginning to change. It appears that, even when mathematics was being highly developed in Islam, the areas of mathematics more advanced than basic arithmetic were classified as "foreign sciences," in contrast to the "religious sciences," including religious law and speculative theology. To many Islamic religious leaders, the foreign sciences were potentially subversive to the faith and certainly superfluous to the needs of life, either here or hereafter. And although the earliest Islamic leaders encouraged the study of the foreign sciences, over the centuries the support for such study lessened as more orthodox religious leaders came to the fore. More and more, the institutions of higher learning throughout the Islamic world, the *madrasas*, tended to concentrate on the teaching of Islamic law. A scholar in charge of one of these schools could, of course, teach the foreign sciences, but if he did, he could be the subject of a legal ruling from traditionalists, a ruling that would in fact be based on the law establishing the school specifying that nothing inimical to the tenets of Islam could be taught. Thus, although there were significant mathematical achievements in Islam through the fifteenth century, gradually science became less important.

Given the influence of Islam on science in general, and mathematics in particular, the mathematics of this period will be referred to here as "Islamic" rather than "Arabic," even though not all of the mathematicians were themselves Moslems. Nevertheless, it was the Arabic language that was generally in use in the Islamic domains, and hence the works to be discussed were all written in that language. A complete history of mathematics of medieval Islam cannot yet be written, since so many of these Arabic manuscripts lie unstudied and even unread in libraries throughout the world. The situation has been improving recently as more and more texts are being edited and translated, but political difficulties continue to block access to many important collections. Still, the general outline of mathematics in Islam is known. In particular, Islamic mathematicians fully developed the decimal place value number system to include decimal fractions, systematized the study of algebra and began to consider the relationship between algebra and geometry, brought the rules of combinatorics from India and reworked them into an abstract system, studied and made advances on the major Greek geometrical treatises of Euclid, Archimedes, and Apollonius, and made significant improvements in plane and spherical trigonometry (Fig. 9.1).

FIGURE 9.1

The Arab contribution to science on a Tunisian stamp

 9.2

DECIMAL ARITHMETIC

The decimal place value system had spread from India at least as far as Syria by the mid-seventh century. It was certainly available in Islamic lands by the time of the founding of the House of Wisdom. In fact, in 773 an Indian scholar visited the court of al-Manṣūr in Baghdad, bringing with him a copy of an Indian astronomical text, quite possibly Brahmagupta's *Brāhmasphuṭasiddhānta*. The caliph ordered this work translated into Arabic. Besides containing the Indian astronomical system, this work included at least some indication of the Hindu number system. The Moslems, however, already had a number system with which

SIDEBAR 9.1   *Arabic Names*

In our initial reference to a particular Islamic mathematician, we give his complete name, although afterwards we abbreviate it for reasons of space. Note that the Arabic name not only includes the given name of the person, but also may include his lineage to one or more generations ("ibn" means "son of"), the place of his or his ancestors' birth, the name of his son ("abū" means "father of"), and one or more appellations indicating some particular characteristic. For example, al-Uqlīdisī means having to do with Euclid. Namely, the mathematician in question was probably a copyist of Arabic versions of Euclid's works.

those who needed to use mathematics were quite content. In fact, there were two systems in use. The merchants in the marketplace generally used a form of finger reckoning, which had been handed down for generations. In this system, calculations were generally carried out mentally. Numbers were expressed in words, and fractions were expressed in the Babylonian scale of sixty. When numbers had to be written, a ciphered system was used in which the letters of the Arabic alphabet denoted numbers. Many Arabic arithmetic texts in which one or the other of these systems was discussed were written between the eighth and the thirteenth centuries.

Gradually, the knowledge of the Hindu system began to seep into Islamic mathematics. The earliest available arithmetic text that deals with the Hindu numbers is the *Kitāb al-jam'wal tafrīq bi ḥisāb al-Hind* (*Book on Addition and Subtraction after the Method of the Indians*) by Muḥammad ibn Mūsā al-Khwārizmī (c. 780–850), an early member of the House of Wisdom (Sidebar 9.1). Unfortunately, there is no extant Arabic manuscript of this work, only several different Latin versions made in Europe in the twelfth century. In his text, al-Khwārizmī introduced nine characters to designate the first nine numbers and, as the Latin versions tell us, a circle to designate zero. He demonstrated how to write any number using these characters in our familiar place value notation. He then described the algorithms of addition, subtraction, multiplication, division, halving, doubling, and determining square roots, and gave examples of their use. The algorithms, however, were usually set up to be performed on the dust board, a writing surface on which sand was spread. Thus, calculations were generally designed to have figures erased at each step as one proceeded to the final answer. Al-Khwārizmī sometimes expressed fractions in the Egyptian mode as sums of unit fractions and other times used sixtieths. In the latter case, he used the old Babylonian place value system for fractions, noting, for example, that the product of 7 minutes (i.e., sixtieths) by 6 minutes will be 42 seconds (i.e., 3600ths) and the product of 7 seconds by 9 minutes will be 63 thirds (i.e., 216,000ths). It is thus important to note that one of the most important features of our place value system, decimal fractions, was still missing. Nevertheless, al-Khwārizmī's work was important not only in the Islamic world but also because it introduced many Europeans to the basics of the decimal place value system (Sidebar 9.2).

Numerous other arithmetic works were written in Arabic over the next centuries explaining the Indian methods, both on their own and in connection with the older systems already mentioned. The earliest extant Arabic arithmetic, the *Kitāb al-fuṣūl fi-l-ḥisāb al-hindī*, (*The Book of Chapters on Hindu Arithmetic*) of Abu l-Ḥasan al-Uqlīdisī, was written in 952 in Damascus. The author made clear one of the major reasons for what he knew would be the ultimate success of the Indian numbers:

SIDEBAR 9.2   *Mathematical Words from Arabic*

Al-Khwārizmī's arithmetic text was probably the source of three English mathematical words. One of the Latin manuscripts of this work begins with the words "Dixit Algorismi," or, "al-Khwārizmī says." The word "algorismi," through some misunderstanding, soon became a term referring to various arithmetic operations and, ultimately, the English word "algorithm." Our word "zero" probably derives from the Arabic *sifr*, which was Latinized into "zephirum." The word *sifr* itself was an Arabic translation of the Sanskrit word *sūnyā*, meaning "empty." An alternate medieval translation of *sifr* into "cifra" led to our modern English "cipher."

> Most scribes will have to use it [the Indian method] because it is easy, quick and needs little precaution, little time to get the answer, and little keeping of the heart busy with the working that he has to see between his hands, to the extent that if he talks, that will not spoil his work; and if he leaves it and busies himself with something else, when he turns back to it, he will find it the same and thus proceed, saving the trouble of memorizing it and keeping the heart busy with it. This is not the case in the other (arithmetic) which requires finger bending and other necessaries. Most calculators will have to use it [the Indian method] with numbers that cannot be managed by the hand because they are big.[3]

Al-Uqlīdisī's text, like that of al-Khwārizmī, dealt with the various algorithms of arithmetic. But there were two major innovations. First, the author showed how to perform arithmetic calculations on paper. As he noted, some think it "ugly to see the [dust board] in the hands of the scribe . . . sitting in the market places [so] . . . we have substituted for it something that will not require [the dust board]." For example, al-Uqlīdisī gave the following procedure for multiplying 3249 by 2735. He wrote the first number above the second, multiplied each digit of the first by the entire second number, then added the resulting terms together. For example, the first line of the calculation is 6 21 9 15$(= 2 \cdot 3, \ 7 \cdot 3, \ 3 \cdot 3, \ 5 \cdot 3)$.

$$
\begin{array}{rrrrr}
 & & 3249 & & \\
 & & 2735 & & \\
 & & & & \\
6 & 21 & 9 & 15 & \\
4 & 14 & 6 & 10 & \\
 & 8 & 28 & 12 & 20 \\
 & & 18 & 63 & 27 & 45
\end{array}
$$

The result, 8,886,015, is found by careful adding of the columns, keeping track of the various places. Thus, the second digit from the right in the answer comes from adding the 0 and 7 of 20 and 27 to the 4 in 45. The third digit from the right comes from adding the "carry" (1) from the previous addition to the 2 in 20, the 2 in 27, the 0 in 10, the 2 in 12, and the 3 in 63. In any case, all the numbers are written down and preserved so one can check them.

Second, al-Uqlīdisī treated decimal fractions, the earliest recorded instance of these fractions outside of China. This treatment is in al-Uqlīdisī's section on halving: "In what is drawn on the principle of numbers, the half of one in any place is 5 before it. Accordingly, if we halve an odd number we set the half as 5 before it, the units place being marked by a sign ' above it, to denote the place. The units place becomes tens to what is before it. Next, we halve the five as is the custom in halving whole numbers. The units place becomes hundreds in the second time of halving. So it goes always."[4] The central idea of decimal fractions is clear here. In dealing with numbers less than one, one operates on them in exactly the same manner as on whole numbers. It is only after performing the operation that one worries about the decimal place. Al-Uqlīdīsī provided as an example the halving of 19 five times. In order, he gets 9′5, 4′75, 2′375, 1′1875, and 0′59375. He read the latter number as 59,375 of a hundred thousand. Similarly, in a section on increasing numbers, he noted that to find one-tenth of a number, one simply repeats it "one place down." So to increase 135 by one-tenth of itself five times, he wrote

$$1 \quad 3 \quad 5$$
$$1 \quad 3 \quad 5$$

The sum is 148′5. One-tenth of this is 14′85; the new sum is 163′35. Continuing this process another three times gives the final answer of 217′41885.

Although al-Uqlīdīsī used decimal fractions, it is not clear that he completely grasped their meaning. The only divisions he deals with are by two and ten; he did not try to calculate the decimal form of 14/3, for example. By contrast, al-Samaw'al ibn Yaḥyā ibn Yahūda al-Maghribī (c. 1125–1174), in his *Treatise on Arithmetic* of 1172, showed that he fully understood decimal fractions in the context of approximation. He began by describing the basic idea: "Given that proportional places, starting with the place of the units, follow one another indefinitely according to the tenth proportion, we therefore suppose that on the other side [of the units] the place of the parts [of ten follow one another] according to the same proportion, and the place of units lies half-way between the place of the integers whose units are transferred in the same way indefinitely, and the place of indefinitely divisible parts."[5]

As an example, al-Samaw'al divided 210 by 13, and noted that the division did not come out even, but could be carried as far as desired. He wrote the result to five places as 16 plus 1 part of 10 plus 5 parts of 100 plus 3 parts of 1000 plus 8 parts of 10,000 plus 4 parts of 100,000. Similarly, he calculated the square root of 10 verbally to be 3 plus 1 part of 10 plus 6 parts of 100 plus 2 parts of 1000 plus 2 parts of 10,000 plus 7 parts of 100,000 plus 7 parts of 1,000,000 (3.162277). Unlike his predecessor, he still used words to describe the various places. Nevertheless, he understood the value of using decimal fractions for approximating rational numbers or irrational numbers. In fact, when al-Samaw'al calculated higher roots by a method similar to that used in China, he explicitly noted the purpose of the successive steps of the algorithm: "And thus we operate to determine the side of a cube, of a square-square, a square-cube and other [powers]. This method enables us . . . to obtain an infinite number of answers, each one being more precise and closer to the truth than the preceding one."[6] Al-Samaw'al evidently realized that, in theory at least, one can calculate an infinite decimal expansion of a number, and that the finite decimals of this expansion "converge" to the exact value, a value not expressible in any finite form.

But even with this important work, the development of the place value system was not complete. It is in the work of Ghiyāth al-Dīn Jamshīd al-Kāshī (d. 1429) in the early fifteenth century that we first see both a total command of the idea of decimal fractions and a convenient notation for them, namely, a vertical line to separate the integer part of a number from the decimal fraction part (Fig. 9.2). We can then say that the Hindu-Arabic place value system was complete.

FIGURE 9.2

Al-Kāshī on an Iranian stamp

9.3   ALGEBRA

The most important contributions of the Islamic mathematicians lie in the area of algebra. They took the material already developed by the Babylonians, combined it with the classical Greek heritage of geometry, and produced a new algebra, which they proceeded to extend. By the end of the ninth century, the chief Greek mathematical classics were well known in the Islamic world. Islamic scholars studied them and wrote commentaries on them. The most important idea they learned from their study of these Greek works was the notion of proof. They absorbed the idea that one could not consider a mathematical problem solved unless one could demonstrate that the solution was valid. How does one demonstrate this, particularly for an algebra problem? The answer seemed clear. The only real proofs were geometric. After all, it was geometry that was found in Greek texts, not algebra. Hence, Islamic scholars generally set themselves the tasks of justifying algebraic rules, either the ancient Babylonian ones or new ones they themselves discovered, and justifying them through geometry.

9.3.1   The Algebra of al-Khwārizmī and ibn Turk

One of the earliest Islamic algebra texts, written about 825 by al-Khwārizmī, was entitled *Al-kitāb al-muhtaṣar fī hisāb al-jabr wa-l-muqābala* (*The Condensed Book on the Calculation of al-Jabr and al-Muqabala*), a book that ultimately had even more influence than his arithmetical work. The term *al-jabr* can be translated as "restoring" and refers to the operation of "transposing" a subtracted quantity on one side of an equation to the other side where it becomes an added quantity. The word *al-muqābala* can be translated as "comparing" and refers to the reduction of a positive term by subtracting equal amounts from both sides of the equation. Thus, the conversion of $3x + 2 = 4 - 2x$ to $5x + 2 = 4$ is an example of *al-jabr*, while the conversion of the latter to $5x = 2$ is an example of *al-muqābala*. The word "algebra" is a corrupted form of the Arabic *al-jabr*. When al-Khwārizmī's work and other similar treatises were translated into Latin, no translation was made of the word *al-jabr*, which thus came to be taken for the name of this science.

Al-Khwārizmī explained in his introduction why he came to write his text:

> That fondness for science, by which God has distinguished the Imam al-Ma'mūn, the Commander of the Faithful, . . . that affability and condescension which he shows to the learned, that promptitude with which he protects and supports them in the elucidation of obscurities and in the removal of difficulties, has encouraged me to compose a short work on calculating by *al-jabr* and *al-muqābala*, confining it to what is easiest and most useful in arithmetic, such as men constantly require in cases of inheritance, legacies, partition, law-suits, and trade, and in all their dealings with one another, or where the measuring of lands, the digging of canals, geometrical computation, and other objects of various sorts and kinds are concerned.[7]

Al-Khwārizmī was interested in writing a practical manual, not a theoretical one. Nevertheless, he had already been sufficiently influenced by the introduction of Greek mathematics into the House of Wisdom that even in such a manual he felt constrained to give geometric proofs of his algebraic procedures. The geometric proofs, however, are not Greek proofs. They appear to be, in fact, very similar to the Babylonian geometric arguments out of which the algebraic algorithms grew. Again, like his oriental predecessors, al-Khwārizmī gave numerous examples and problems, but the Greek influence showed through in his systematic

BIOGRAPHY

Muḥammad ibn Mūsā al-Khwārizmī (c. 780–850)

Al-Khwārizmī, or perhaps some of his ancestors, came from Khwarizm, the region south of the Aral Sea now part of Uzbekistan and Turkmenistan (Fig. 9.3). Al-Khwārizmī was one of the first scholars in the House of Wisdom founded by the caliph al-Ma'mūn, and also was one of the astronomers called to cast a horoscope for the dying caliph al-Wāthiq in 847. The story is told that although al-Khwārizmī assured the caliph he would live another 50 years, in fact the caliph died 10 days later. Perhaps al-Khwārizmī felt it was not good policy to be the bearer of bad news to one's ruler. Besides the contributions to mathematics detailed in the text, al-Khwārizmī wrote a work on geography in which he developed a map of the Islamic world much superior to that known from the work of Ptolemy.

FIGURE 9.3

Al-Khwārizmī on a stamp from the former Soviet Union

classification of the problems he intended to solve, as well as in the very detailed explanations of his methods.

Al-Khwārizmī began by noting that "what people generally want in calculating . . . is a number,"[8] the solution of an equation. Thus, the text was to be a manual for solving equations. The quantities he dealt with were generally of three kinds, the square (of the unknown), the root of the square (the unknown itself), and the absolute number (the constant in the equation). He then noted that there are six types of equations that can be written using these three kinds of quantities:

1. Squares are equal to roots ($ax^2 = bx$).
2. Squares are equal to numbers ($ax^2 = c$).
3. Roots are equal to numbers ($bx = c$).
4. Squares and roots are equal to numbers ($ax^2 + bx = c$).
5. Squares and numbers are equal to roots ($ax^2 + c = bx$).
6. Roots and numbers are equal to squares ($bx + c = ax^2$).

One reason for this sixfold classification is that Islamic mathematicians, unlike the Hindus, did not deal with negative numbers at all. Coefficients, as well as the roots of the equations, must be positive. The types listed are the only types that have positive solutions. Our standard form $ax^2 + bx + c = 0$ would make no sense for al-Khwārizmī, because if the coefficients are all positive, the roots cannot be.

Al-Khwārizmī's solutions to the first three types of equations were straightforward. We only need note that 0 is not considered as a solution to the first type. His rules for the compound types of equations were more interesting. We present his solution to type 4. Because al-Khwārizmī used no symbols, we will follow him in writing everything out in words, including the numbers of his example: "What must be the square which, when increased by ten of its own roots, amounts to thirty-nine? The solution is this: you halve the number of roots, which in the present instance yields five. This you multiply by itself; the product is twenty-five. Add this to thirty-nine; the sum is sixty-four. Now take the root of this which is eight, and subtract from it half the number of the roots, which is five; the remainder is three. This is the root of the square which you sought for."[9]

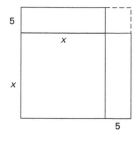

FIGURE 9.4

Al-Khwārizmī's geometric justification for the solution of $x^2 + 10x = 39$

Al-Khwārizmī's verbal description of his procedure was essentially the same as that of the Babylonian scribes. Namely, in modern notation, the solution of $x^2 + bx = c$ is

$$x = \sqrt{\left(\frac{b}{2}\right)^2 + c} - \frac{b}{2}.$$

Al-Khwārizmī's geometric justification of this procedure also demonstrated his Babylonian heritage. Beginning with a square representing x^2, he added two rectangles, each of width five ("half the number of roots") (Fig. 9.4). The sum of the area of the square and the two rectangles is then $x^2 + 10x = 39$. One now completes the square with a single square of area 25 to make the total area 64. The solution $x = 3$ is then easily found. This geometric description corresponds to the Babylonian description of the solution of $x^2 + \frac{2}{3}x = \frac{7}{12}$. (See Section 1.2.4 and Figure 1.21.)

Although al-Khwārizmī's geometric descriptions of his method appear to have been taken over from Babylonian sources, he or his (unknown) predecessors in this field succeeded in changing the focus of quadratic equation solving away from the actual finding of sides of squares into that of finding numbers satisfying certain conditions. For example, he explained the term "root" not as a side of a square but as "anything composed of units which can be multiplied by itself, or any number greater than unity multiplied by itself, or that which is found to be diminished below unity when multiplied by itself."[10] Also, his procedure for solving quadratic equations of type 4, when the coefficient of the square term is other than one, was the arithmetical method of first multiplying or dividing appropriately to make the initial coefficient one, and then proceeding as before. Al-Khwārizmī even admitted somewhat later in his text, when he was discussing the addition of the "polynomials" $100 + x^2 - 20x$ and $50 + 10x - 2x^2$, that "this does not admit of any figure, because there are three different species, i.e., squares and roots and numbers, and nothing corresponding to them by which they might be represented. . . . [Nevertheless], the elucidation by words is easy."[11]

Finally, al-Khwārizmī's presentation of the method and geometric description for type 5, squares and numbers equal to roots, shows that, unlike the Babylonians, he could deal with an equation with two positive roots, at least numerically. In this case, $x^2 + c = bx$, his verbal description of the solution procedure easily translates into our formula

$$x - \frac{b}{2} \pm \sqrt{\left(\frac{b}{2}\right)^2 - c}. \tag{9.1}$$

In fact, he stated that one could employ either addition or subtraction to get a root and also noted the condition on the solution: "If the product [of half the number of roots with itself] is less than the number connected with the square, then the instance is impossible; but if the product is equal to the number itself, then the root of the square is equal to half of the number of roots alone, without either addition or subtraction."[12] The geometric demonstration in this case, which reminds us of the Babylonian description for the system $x + y = b$, $xy = c$ (see Section 1.2.4 and Figure 1.20), only dealt with the subtraction in Equation 9.1. In Figure 9.5, square $ABCD$ represents x^2, whereas rectangle $ABNH$ represents c. Therefore, HC represents b. Bisect HC at G, extend TG to K so that $GK = GA$, and complete the rectangle $GKMH$. Finally, choose L on KM so that $KL = GK$ and complete the square $KLRG$. It is then clear that rectangle $MLRH$ equals rectangle $GATB$. Since the area of

FIGURE 9.5

Al-Khwārizmī's geometric justification for the solution of $x^2 + c = bx$

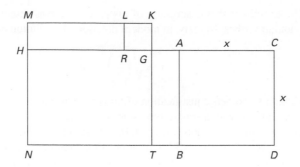

square $KMNT$ is $(\frac{b}{2})^2$, while that square less square $KLRG$ equals rectangle $ABNH$ or c, it follows that square $KLRG$ equals $(\frac{b}{2})^2 - c$. Since the side of that square is equal to AG, it follows that $x = AC = CG - AG$ is given by Equation 9.1 using the minus sign. Although al-Khwārizmī briefly noted that CR could also represent a solution, he did not demonstrate this by a diagram, nor did he deal in his diagram with the special conditions mentioned in his verbal description.

Al-Khwārizmī's text contains the word "condensed" in the title, thus leading one to believe that there were other books at the time discussing algebraic procedures and their attendant geometric justifications in more detail. There is, however, only a fragment of such a work now extant, the section "Logical Necessities in Mixed Equations" from a longer work *Kitāb al-jabr wa'l muqābala* by 'Abd al-Ḥamīd ibn Wāsi ibn Turk al-Jīlī, a contemporary of al-Khwārizmī about whom very little is known. The sources even differ as to whether ibn Turk was from Iran, Afghanistan, or Syria.

In any case, the extant chapter of ibn Turk's book deals with quadratic equations of al-Khwārizmī's types 1, 4, 5, and 6 and includes a much more detailed geometric description of the method of solution than is found in al-Khwārizmī's work. In particular, in the case of type 5, ibn Turk gave geometric versions for all possible cases. His first example is the same as al-Khwārizmī's, namely, $x^2 + 21 = 10x$, but he began the geometrical demonstration by noting that G, the midpoint of CH, may be either on the line segment AH, as in al-Khwārizmī's diagram, or on the line segment CA of Figure 9.6. In this case, squares and rectangles are completed, similar in form to those in Figure 9.5, but the solution $x = AC$ is now given as $CG + GA$, thus using the plus sign in Equation 9.1. In addition, ibn Turk discussed what he called the "intermediate case," where the root of the square is exactly equal to half the number of roots. His example for this situation is $x^2 + 25 = 10x$; the geometric diagram then simply consists of a rectangle divided into two equal squares.

Ibn Turk further noted that "there is the logical necessity of impossibility in this type of equation when the numerical quantity . . . is greater than [the square of] half the number of roots,"[13] as, for example, in the case $x^2 + 30 = 10x$. Again, he resorted to a geometric argument. Assuming that G is located on the segment AH, we know as before that the square $KMNT$ is greater than the rectangle $HABN$ (Fig. 9.7). But the conditions of the problem show that the latter rectangle equals 30 while the former only equals 25. A similar argument works in the case where G is located on CA.

FIGURE 9.6

Ibn Turk's geometric jus-
tification for one case of
$x^2 + c = bx$

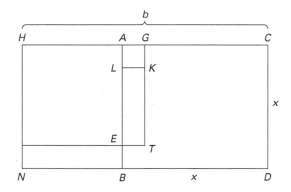

FIGURE 9.7

Ibn Turk's geometric justification of the impossibility of
solving $x^2 + 30 = 10x$

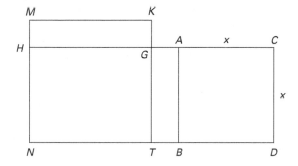

Although the section on quadratic equations of ibn Turk's algebra is the only part still extant, al-Khwārizmī's text contains much else of interest, including an introduction to manipulation with algebraic expressions, explained by reference to similar manipulations with numbers. For example, he noted that if $a \pm b$ is multiplied by $c \pm d$, then four multiplications are necessary. Although none of his numbers are negative, he certainly knew the rules for dealing with multiplication and signs. As he stated, "If the units [b and d in our notation] . . . are positive, then the last multiplication is positive; if they are both negative, then the fourth multiplication is likewise positive. But if one of them is positive and one negative, then the fourth multiplication is negative."[14]

Al-Khwārizmī's text continued with a large collection of problems, many of which involve these manipulations, and most of which result in a quadratic equation. For example, one problem states, "I have divided ten into two parts, and having multiplied each part by itself, I have put them together, and have added to them the difference of the two parts previously to their multiplication, and the amount of all this is fifty-four."[15] It is not difficult to translate this problem into the equation $(10 - x)^2 + x^2 + (10 - x) - x = 54$. The author reduced this to the equation $x^2 + 28 = 11x$ and then used his rule for this equation of type 5 to get $x = 4$. He ignored here the second root, $x = 7$, for then the sum of the two squares would be 58 and the conditions of the problem could not be met. In another example, al-Khwārizmī dealt with a nonrational root: "I have divided ten into two parts; I have multiplied the one by ten and the other by itself, and the products were the same."[16] The equation here is $10x = (10 - x)^2$ and the solution is $x = 15 - \sqrt{125}$. Here he again ignored the root with the positive sign, because $15 + \sqrt{125}$ could not be a "part" of 10.

Despite al-Khwārizmī promising in his preface that he would write about what is "useful," very few of his problems leading to quadratic equations deal with any "practical" ideas. Many of them are similar to the previous examples and begin with "I have divided ten into two parts." Among the few problems written in "real-world" terms is the following: "You divide one dirhem among a certain number of men. Now you add one man more to them, and divide again one dirhem among them. The quota of each is then one-sixth of a dirhem less than at the first time."[17] If x represents the number of men, the equation becomes $\frac{1}{x} - \frac{1}{x+1} = \frac{1}{6}$, which reduces to $x^2 + x = 6$, for which the solution is $x = 2$. An entire section of the text is devoted to elementary problems of mensuration, which will be discussed later, and a brief section is devoted to the "rule of three," but neither of these provides any practical uses of quadratic equations either. Finally, the second half of the text is entirely devoted to problems of inheritance. Dozens of complicated situations are presented, for the solution of which one needs to be familiar with Islamic legacy laws. The actual mathematics needed, however, is never more complicated than the solution of linear equations. One can only conclude that although al-Khwārizmī was interested in teaching his readers how to solve mathematical problems, and especially how to deal with quadratic equations, he could not think of any real-life situations that required these equations. Things apparently had not changed in this regard since the time of the Babylonians.

9.3.2   The Algebra of Thābit ibn Qurra and Abū Kāmil

Within 50 years of the works by al-Khwārizmī and ibn Turk, the Islamic mathematicians had decided that the necessary geometric foundations to the algebraic solution of quadratic equations should be based on the work of Euclid rather than on the ancient traditions. Perhaps the earliest of these justifications was given by Thābit ibn Qurra (836–901). Thābit was born in Ḥarrān (now in southern Turkey), was discovered there by one of the scholars from the House of Wisdom, and was brought to Baghdad in about 870, where he himself became a great scholar. Among his many writings on mathematical topics is a short work entitled *Qawl fī taṣḥīḥ masā'il al-jabr bi l-barāhīn al-handasīya* (*On the Verification of Problems of Algebra by Geometrical Proofs*). To solve the equation $x^2 + bx = c$, for example, Thābit used Figure 9.8, where AB represents x, square $ABCD$ represents x^2, and BE represents b. It follows that the rectangle $DE = AB \times EA$ represents c. If W is the midpoint of BE, Euclid's *Elements* II–6 implies that $EA \times AB + BW^2 = AW^2$. But since $EA \times AB$ and BW^2 are known (equaling, respectively, c and $(b/2)^2$), it follows that AW^2 and therefore AW are known. Then $x = AB = AW - BW$ is determined. Thābit noted explicitly that the geometric procedure of *Elements* II–6 is completely analogous to the procedure of "the algebraists," that

FIGURE 9.8

Thābit ibn Qurra's geometric justification for the solution of $x^2 + bx = c$

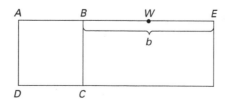

is, the algorithm stated by al-Khwārizmī, and therefore provides the necessary justification. Thābit also showed how to use this same proposition to solve $x^2 = bx + c$ and how to use *Elements* II–5 to solve $x^2 + c = bx$.

Similar justifications of these solutions using *Elements* II were given by the Egyptian mathematician Abū Kāmil ibn Aslam (c. 850–930) in his own algebra text, *Kitāb fī al-jabr wa'l-muqābala*: "I shall explain their rule using geometric figures clarified by wise men of geometry and which are explained in the Book of Euclid."[18] Abū Kāmil, however, unlike Thābit, proved Euclid's results anew in the course of his discussion and also presented numerical examples, in fact, the same initial numerical examples as al-Khwārizmī. Like his predecessor, Abū Kāmil followed his discussion of the various forms of quadratic equations by a treatment of various algebraic rules and then a large selection of problems. But he made some advances over the earlier mathematician by considering many more complicated identities and more complex problems, including in particular manipulations with surds.

Abū Kāmil was not at all worried about dealing with "irrationals." He used them freely in his problems, many of which, like those of al-Khwārizmī, start with "divide 10 into 2 parts." For example, consider problem 37: "If one says that 10 is divided into two parts, and one part is multiplied by itself and the other by the root of 8, and subtract the quantity of the product of one part times the root of 8 from . . . the product of the other part multiplied by itself, it gives 40."[19] The equation in this case is $(10 - x)(10 - x) - x\sqrt{8} = 40$. After rewriting it in the form $x^2 + 60 = 20x + \sqrt{8x^2}(= (20 + \sqrt{8})x)$, Abū Kāmil carried out the algorithm for the case squares and numbers equal roots to conclude that

$$x = 10 + \sqrt{2} - \sqrt{42 + \sqrt{800}}$$

and that $10 - x$, the "other part," is equal to

$$\sqrt{42 + \sqrt{800}} - \sqrt{2}.$$

Abū Kāmil even applied substitutions to simplify problems and could deal with equations of degree higher than 2 as long as they were quadratic in form. Problem 45 illustrates both ideas: "One says that 10 is divided into two parts, each of which is divided by the other, and when each of the quotients is multiplied by itself and the smaller is subtracted from the larger, then there remains 2."[20] The equation is

$$\left(\frac{x}{10 - x}\right)^2 - \left(\frac{10 - x}{x}\right)^2 = 2.$$

Abū Kāmil made a new "thing" y equal to $\frac{10-x}{x}$ and derived the new equation $\frac{1}{y^2} = y^2 + 2$. Multiplying both sides by y^2 gave him the quadratic equation in y^2: $(y^2)^2 + 2y^2 = 1$ for which the solution is $y^2 = \sqrt{2} - 1$. Hence,

$$y = \sqrt{\sqrt{2} - 1}.$$

Then

$$\frac{10 - x}{x} = \sqrt{\sqrt{2} - 1},$$

and Abū Kāmil proceeded to solve for x by first squaring both sides of this equation. The final result is

$$x = 10 + \sqrt{50} - \sqrt{50 + \sqrt{20,000} - \sqrt{5,000}}.$$

Abū Kāmil could even solve systems of equations. Consider problem 61: "One says that 10 is divided into three parts, and if the smallest is multiplied by itself and added to the middle one multiplied by itself, it equals the largest multiplied by itself, and when the smallest is multiplied by the largest, it equals the middle multiplied by itself."[21] In modern symbols, we are asked to find $x < y < z$, where

$$x + y + z = 10, \qquad x^2 + y^2 = z^2, \quad \text{and} \quad xz = y^2.$$

Presumably noticing that the three equations are all homogeneous, Abū Kāmil used the ancient method of false position. Namely, he initially ignored the first equation and set $x = 1$ in the second and third equations to get $1 + y^2 = y^4$. Since this is an equation in quadratic form, he could solve it:

$$z = y^2 = \frac{1}{2} + \sqrt{\frac{5}{4}} \quad \text{and} \quad y = \sqrt{\frac{1}{2} + \sqrt{\frac{5}{4}}}.$$

Now, returning to the first equation, he noted that the sum of his three "false" values was

$$1\frac{1}{2} + \sqrt{\frac{5}{4}} + \sqrt{\frac{1}{2} + \sqrt{\frac{5}{4}}},$$

instead of 10. To find the correct values, he needed to divide 10 by this value and multiply the quotient by the "false" values. Since the false value of x was 1, this just meant that the correct value for x was

$$x = \frac{10}{1\frac{1}{2} + \sqrt{\frac{5}{4}} + \sqrt{\frac{1}{2} + \sqrt{\frac{5}{4}}}}.$$

To simplify this was not a trivial procedure, but Abū Kāmil began by multiplying the denominator by x and setting the product equal to 10. He ultimately turned this equation into a quadratic equation and succeeded in determining that

$$x = 5 - \sqrt{\sqrt{3,125} - 50}.$$

To find y and z by multiplying the false values by this quotient would have been even more difficult, so he chose to find z by beginning the problem anew with the false value $z = 1$. Of course, once he found z, he could determine y by subtraction.

When considering Abū Kāmil's algebra, remember that, like all Islamic algebra texts of his era, it was written without symbols. Thus, the algebraic manipulation that modern symbolism makes almost obvious is carried out completely verbally. (Of course, in our final example, the procedure is by no means "obvious," even with symbolism.) More importantly, however, Abū Kāmil was willing to use the algebraic algorithms that had been systematized by the time of al-Khwārizmī with any type of positive "number." He made no distinction between operating with 2 or with $\sqrt{8}$ or even with $\sqrt{\sqrt{2}-1}$. Since these algorithms came from geometry, on one level that is not surprising. After all, it was the Greek failure to find a "numerical" representation of the diagonal of a square that was one of the reasons for their use of the geometric algebra of line segments and areas. But in dealing with these quantities, Abū Kāmil interpreted all of them in the same way. It did not matter whether a magnitude was technically a square or a fourth power or a root or a root of a root. For Abū Kāmil, the solution of a quadratic equation was not a line segment, as it would be in the interpretation of the appropriate propositions of the *Elements*. It was a "number," even though Abū Kāmil could not perhaps give a proper definition of that term. He therefore had no compunction about combining the various quantities that appeared in the solutions, using general rules. Abū Kāmil's willingness to handle all of these quantities by the same techniques helped pave the way toward a new understanding of the concept of number that was just as important as al-Samaw'al's use of decimal approximations.

9.3.3   Al-Karajī, al-Samaw'al, and the Algebra of Polynomials

The process of relating arithmetic to algebra, begun by al-Khwārizmī and Abū Kāmil, continued in the Islamic world with the work of Abū Bakr al-Karajī (d. 1019) and al-Samaw'al over the next two centuries. These latter mathematicians were instrumental in showing that the techniques of arithmetic could be fruitfully applied in algebra and, reciprocally, that ideas originally developed in algebra could also be important in dealing with numbers.

Little is known of the life of al-Karajī other than that he worked in Baghdad around the year 1000 and wrote many mathematics works as well as works on engineering topics. In the first decade of the eleventh century, he composed a major work on algebra entitled *al-Fakhrī* (*The Marvelous*). The aim of *al-Fakhrī*, and of algebra in general according to al-Karajī, was "the determination of unknowns starting from knowns."[22] In pursuit of this aim, he made use of all the techniques of arithmetic, converted into techniques of dealing with unknowns. He began by making a systematic study of the algebra of exponents. Although earlier writers, including Diophantus, had considered powers of the unknown greater than the third, al-Karajī was the first to fully understand that these powers can be extended indefinitely. In fact, he developed a method of naming the various powers x^n and their reciprocals $\frac{1}{x^n}$. Each power was defined recursively as x times the previous power. It followed that there was an infinite sequence of proportions,

$$1 : x = x : x^2 = x^2 : x^3 = \ldots,$$

and a similar one for reciprocals,

$$\frac{1}{x} : \frac{1}{x^2} = \frac{1}{x^2} : \frac{1}{x^3} = \frac{1}{x^3} : \frac{1}{x^4} = \ldots.$$

Once the powers were understood, al-Karajī could establish general procedures for adding, subtracting, and multiplying monomials and polynomials. In division, however, he only used monomials as divisors, partly because he was unable to incorporate rules for negative numbers into his theory and partly because of his verbal means of expression. Similarly, although he developed an algorithm for calculating square roots of polynomials, it was only applicable in limited circumstances.

BIOGRAPHY

Al-Samaw'al (c. 1125–1174)

Al-Samaw'al was born in Baghdad to well-educated Jewish parents. His father was in fact a Hebrew poet. Besides giving him a religious education, they encouraged him to study medicine and mathematics. Because the House of Wisdom no longer existed in Baghdad, he had to study mathematics independently and therefore traveled to various other parts of the Middle East. He wrote his major mathematical work, *Al-Bāhir*, when he was only nineteen. His interests later turned to medicine, and he became a successful physician and author of medical texts. The only extant one is entitled *The Companion's Promenade in the Garden of Love*, a treatise on sexology and a collection of erotic stories. When he was about forty, he decided to convert to Islam. To justify his conversion to the world, he wrote an autobiography in 1167 stating his arguments against Judaism, a work that became famous as a source of Islamic polemics against the Jews.

Al-Karajī was more successful in continuing the work of Abū Kāmil in applying arithmetic operations to irrational quantities. In particular, he explicitly interpreted the various classes of incommensurables in *Elements* X as classes of "numbers" on which the various operations of arithmetic were defined, but then noted that there were indefinitely many other classes composed of three or more surds. Like Abū Kāmil, he gave no definition of "number," but just dealt with the various surd quantities using numerical rather than geometrical techniques. As part of this process, he developed various formulas involving surds, such as

$$\sqrt{A+B} = \sqrt{\frac{A + \sqrt{A^2 - B^2}}{2}} + \sqrt{\frac{A - \sqrt{A^2 - B^2}}{2}}$$

and

$$\sqrt[3]{A} + \sqrt[3]{B} = \sqrt[3]{3\sqrt[3]{A^2 B} + 3\sqrt[3]{AB^2} + A + B}.$$

Further work in dealing with algebraic manipulation was accomplished by al-Samaw'al, who, in particular, introduced negative coefficients. He expressed his rules for dealing with these coefficients quite clearly in his algebra text *Al-Bāhir fi'l-ḥisāb* (*The Shining Book of Calculation*):

> If we subtract an additive number from an empty power $[0x^n - ax^n]$, the same subtractive number remains; if we subtract the subtractive number from an empty power $[0x^n - (-ax^n)]$, the same additive number remains. If we subtract an additive number from a subtractive number, the remainder is their subtractive sum; if we subtract a subtractive number from a greater subtractive number, the result is their subtractive difference; if the number from which one subtracts is smaller than the number subtracted, the result is their additive difference.[23]

Given these rules, al-Samaw'al could easily add and subtract polynomials by combining like terms. To multiply, of course, he needed the law of exponents. Al-Karajī had in essence used this law, as had Abū Kāmil and others. However, since the product of, for example, a square and a cube was expressed in words as a square-cube, the numerical property of adding exponents could not be seen. Al-Samaw'al decided that this law could best be expressed by using a table consisting of columns, each column representing a different power of either a number or an unknown. In fact, he also saw that he could deal with powers of $\frac{1}{x}$ as easily as with powers of x. In his work, the columns are headed by the Arabic letters standing for the numerals, reading both ways from the central column labeled 0. We will simply use the Arabic numerals themselves. Each column then has the name of the particular power or reciprocal power. For example, the column headed by a 2 on the left is named "square," that headed by a 5 on the left is named "square-cube," that headed by a 3 on the right is named "part of cube," and so on. To simplify matters we will just use powers of x. In his initial explanation of the rules, al-Samaw'al also put a particular number under the 1 on the left, such as 2, and then the various powers of 2 in the corresponding columns:

7	6	5	4	3	2	1	0	1	2	3	4	5	6	7
x^7	x^6	x^5	x^4	x^3	x^2	x	1	x^{-1}	x^{-2}	x^{-3}	x^{-4}	x^{-5}	x^{-6}	x^{-7}
128	64	32	16	8	4	2	1	$\frac{1}{2}$	$\frac{1}{4}$	$\frac{1}{8}$	$\frac{1}{16}$	$\frac{1}{32}$	$\frac{1}{64}$	$\frac{1}{128}$

Al-Samaw'al now used the chart to explain what we call the law of exponents, $x^n x^m = x^{m+n}$: "The distance of the order of the product of the two factors from the order of one of the two factors is equal to the distance of the order of the other factor from the unit. If the factors are in different directions then we count (the distance) from the order of the first factor towards the unit; but, if they are in the same direction, we count away from the unit."[24] So, for example, to multiply x^3 by x^4, count four orders to the left of column 3 and get the result as x^7. To multiply x^3 by x^{-2}, count two orders to the right from column 3 and get the answer x^1. Using these rules, al-Samaw'al could easily multiply polynomials in x and $\frac{1}{x}$ as well as divide such polynomials by monomials.

Al-Samaw'al was also able to divide polynomials by polynomials using a similar chart. In this chart, which reminds us of the Chinese counting board as used in solving polynomial equations, each column again stands for a given power of x or of $\frac{1}{x}$. But now the numbers in each column represent the coefficients of the various polynomials involved in the division process. For example, to divide $20x^2 + 30x$ by $6x^2 + 12$, he first set the 20 and the 30 in the columns headed by x^2 and x, respectively, and the 6 and 12 below these in the columns headed respectively by x^2 and 1. Since there is an "empty order" for the divisor in the x column, he placed a 0 there. He next divided $20x^2$ by $6x^2$, getting 3 1/3, putting that number in the units column on the answer line. The product of 3 1/3 by $6x^2 + 12$ is $20x^2 + 40$. The next step is subtraction. The remainder in the x^2 column is naturally 0. In the x column the remainder is 30, while in the units column the remainder is -40. Al-Samaw'al now presented a new chart in which the 6, 0, 12, are shifted one place to the right, and the directions are given to divide that into $30x - 40$. The initial quotient of $30x$ by $6x^2$ is $5 \cdot 1/x$, so a 5 is placed in the answer line in the column headed by $\frac{1}{x}$, and the process is continued. We display here al-Samaw'al's first two charts for this division problem.

x^2	x	1	$\frac{1}{x}$	$\frac{1}{x^2}$	$\frac{1}{x^3}$
		$3\frac{1}{3}$			
20	30				
6	0	12			

x^2	x	1	$\frac{1}{x}$	$\frac{1}{x^2}$	$\frac{1}{x^3}$
		$3\frac{1}{3}$	5		
30	−40				
6	0	12			

In this particular example, the division was not exact. Al-Samaw'al continued the process through eight steps to get

$$3\frac{1}{3} + 5\left(\frac{1}{x}\right) - 6\frac{2}{3}\left(\frac{1}{x^2}\right) - 10\left(\frac{1}{x^3}\right) + 13\frac{1}{3}\left(\frac{1}{x^4}\right) + 20\left(\frac{1}{x^5}\right) - 26\frac{2}{3}\left(\frac{1}{x^6}\right) - 40\left(\frac{1}{x^7}\right).$$

To show his fluency with the multiplication procedure, he then checked the answer by multiplying it by the divisor. Because the product differed from the dividend by terms only in $\frac{1}{x^6}$ and $\frac{1}{x^7}$, he called the result given "the answer approximately." Nevertheless, he also noted that there is a pattern to the coefficients of the quotient. In fact, if a_n represents the coefficient of $\frac{1}{x^n}$, the pattern is given by $a_{n+2} = -2a_n$. He then proudly wrote out the next 21 terms of the quotient, ending with $54{,}613\frac{1}{3}(\frac{1}{x^{28}})$.

Given that al-Samaw'al thought of extending division of polynomials into polynomials in $\frac{1}{x}$, and thought of partial results as approximations, it is not surprising that he would divide whole numbers by simply replacing x by 10. As already noted, al-Samaw'al was the first to explicitly recognize that one could approximate fractions more and more closely by calculating more and more decimal places. The work of al-Karajī and al-Samaw'al was thus extremely important in developing the idea that algebraic manipulations and manipulations with numbers are parallel. Virtually any technique that applies to one can be adapted to apply to the other.

9.3.4 Induction, Sums of Powers, and the Pascal Triangle

Another important idea introduced by al-Karajī and continued by al-Samaw'al and others was that of an inductive argument for dealing with certain arithmetic sequences. Thus, al-Karajī used such an argument to prove the result on the sums of integral cubes already known to Āryabhaṭa (and even, perhaps, to the Greeks). Al-Karajī did not, however, state a general result for arbitrary n. He stated his theorem for the particular integer 10:

$$1^3 + 2^3 + 3^3 + \cdots + 10^3 = (1 + 2 + 3 + \cdots + 10)^2.$$

His proof, nevertheless, was clearly designed to be extendable to any other integer.

Consider the square $ABCD$ with side $1 + 2 + 3 + \cdots + 10$ (Fig. 9.9). Setting $BB' = DD' = 10$, and completing the gnomon $BCDD'C'B'$, al-Karajī calculated the area of the

FIGURE 9.9

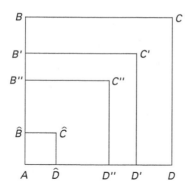

Al-Karajī's proof of the formula for the sum of the integral cubes

gnomon to be

$$2 \cdot 10(1 + 2 + \cdots + 9) + 10^2 = 2 \cdot 10 \cdot \frac{9 \cdot 10}{2} + 10^2 = 9 \cdot 10^2 + 10^2 = 10^3.$$

Since the area of square $ABCD$ is the sum of the areas of square $AB'C'D'$ and the gnomon, it follows that $(1 + 2 + \cdots + 10)^2 = (1 + 2 + \cdots + 9)^2 + 10^3$. A similar argument then shows that $(1 + 2 + \cdots + 9)^2 = (1 + 2 + \cdots + 8)^2 + 9^3$. Continuing in this way to the final square $A\hat{B}\hat{C}\hat{D}$ of area $1 = 1^3$, al-Karajī proved his theorem from the equality of square $ABCD$ to square $A\hat{B}\hat{C}\hat{D}$ plus the sum of the gnomons of areas 2^3, 3^3, ..., 10^3.

Al-Karajī's argument included the two basic components of a modern argument by induction, namely, the truth of the statement for $n = 1$ $(1 = 1^3)$ and the deriving of the truth for $n = k$ from that for $n = k - 1$. Of course, this second component is not explicit since, in some sense, al-Karajī's argument is in reverse. That is, he starts from $n = 10$ and goes down to 1 rather than proceeding upward. Nevertheless, his argument in *al-Fakhrī* is the earliest extant proof of the sum formula for integral cubes.

The formulas for the sums of the integers and their squares had long been known, while the formula for the sum of cubes is easy to discover if one considers a few examples. To give an argument for their validity that generalizes to enable one to find a formula for the sum of fourth powers, however, is more difficult. Nonetheless, this was accomplished early in the eleventh century in a work by the Egyptian mathematician Abū 'Alī al-Ḥasan ibn al-Ḥasan ibn al-Haytham (965–1039). That he did not generalize his result to find the sums of higher powers is probably due to his needing only the formulas for the second and fourth powers in his computation of the volume of a paraboloid, to be discussed in Section 9.5.5.[25]

The central idea in ibn al-Haytham's proof of the sum formulas was the derivation of the equation

$$(n + 1) \sum_{i=1}^{n} i^k = \sum_{i=1}^{n} i^{k+1} + \sum_{p=1}^{n} \left(\sum_{i=1}^{p} i^k \right). \tag{9.2}$$

Ibn al-Haytham did not state this result in general form but only for particular integers, namely, $n = 4$ and $k = 1,\ 2,\ 3$. His proof, however, which, like al-Karajī's, used inductive

BIOGRAPHY

Ibn al-Haytham (965–1039)

Ibn al-Haytham, known in Europe as Alhazen and one of the most influential of Islamic scientists, was born in Basra, now in Iraq, but spent most of his life in Egypt after he was invited by the caliph al-Ḥakim to work on a Nile control project (Fig. 9.10). Although the project never came to fruition, ibn al-Haytham did produce in Egypt his most important scientific work, the *Optics*, in seven books. The *Optics* was translated into Latin in the early thirteenth century and was studied and commented on in Europe for several centuries thereafter. Ibn al-Haytham's fame as a mathematician chiefly rests on his treatment of "Alhazen's problem"—to find the point or points

on some reflecting surface at which the light from one of two points outside that surface is reflected to the other. In the fifth book of the *Optics*, he attempted to solve the problem for a variety of surfaces—spherical, cylindrical, and conical, concave and convex. Although he was not completely successful, his accomplishments showed him to be in full command of both the elementary and advanced geometry of the Greeks. In the final years of his life, ibn al-Haytham earned his living by copying annually, among others, Euclid's *Elements*, Apollonius's *Conics*, and Ptolemy's *Almagest*.

FIGURE 9.10

Ibn al-Haytham's work on optics is honored on a stamp from Pakistan

reasoning, is immediately generalizable to any values of n and k. We consider his proof for $k = 3$ and $n = 4$:

$$(4 + 1)(1^3 + 2^3 + 3^3 + 4^3) = 4(1^3 + 2^3 + 3^3 + 4^3) + 1^3 + 2^3 + 3^3 + 4^3$$
$$= 4 \cdot 4^3 + 4(1^3 + 2^3 + 3^3) + 1^3 + 2^3 + 3^3 + 4^3$$
$$= 4^4 + (3 + 1)(1^3 + 2^3 + 3^3) + 1^3 + 2^3 + 3^3 + 4^3.$$

But, because Equation 9.2 is assumed true for $n = 3$, we have

$$(3 + 1)(1^3 + 2^3 + 3^3) = 1^4 + 2^4 + 3^4 + (1^3 + 2^3 + 3^3) + (1^3 + 2^3) + 1^3.$$

Thus, Equation 9.2 is proved for $n = 4$. It is straightforward to rewrite this argument into a modern proof by induction on n.

Ibn al-Haytham used Equation 9.2 to derive formulas for the sums of integral powers, formulas that are stated in all generality. Thus, for $k = 2$ and $k = 3$, we have

$$\sum_{i=1}^{n} i^2 = \left(\frac{n}{3} + \frac{1}{3}\right) n \left(n + \frac{1}{2}\right) = \frac{n^3}{3} + \frac{n^2}{2} + \frac{n}{6}$$

$$\sum_{i=1}^{n} i^3 = \left(\frac{n}{4} + \frac{1}{4}\right) n(n + 1)n = \frac{n^4}{4} + \frac{n^3}{2} + \frac{n^2}{4}.$$

We will not consider the proofs of these results here, but only the proof of the analogous result for fourth powers. This result, although stated (at the end) in all generality, is only proved for the case $n = 4$, $k = 3$. But we can consider this to represent the method of "generalizable example," a method we have seen Euclid use earlier. In any case, ibn al-Haytham proved the formula for fourth powers by substituting the formulas for cubes and squares into Equation 9.2:

$$(1^3 + 2^3 + 3^3 + 4^3)5 = 1^4 + 2^4 + 3^4 + 4^4 + (1^3 + 2^3 + 3^3 + 4^3)$$
$$+ (1^3 + 2^3 + 3^3) + (1^3 + 2^3) + 1^3$$
$$= 1^4 + 2^4 + 3^4 + 4^4 + \left(\frac{4^4}{4} + \frac{4^3}{2} + \frac{4^2}{4}\right) + \left(\frac{3^4}{4} + \frac{3^3}{2} + \frac{3^2}{4}\right)$$
$$+ \left(\frac{2^4}{4} + \frac{2^3}{2} + \frac{2^2}{4}\right) + \left(\frac{1^4}{4} + \frac{1^3}{2} + \frac{1^2}{4}\right)$$
$$= 1^4 + 2^4 + 3^4 + 4^4 + \frac{1}{4}(1^4 + 2^4 + 3^4 + 4^4)$$
$$+ \frac{1}{2}(1^3 + 2^3 + 3^3 + 4^3) + \frac{1}{4}(1^2 + 2^2 + 3^2 + 4^2)$$
$$= \frac{5}{4}(1^4 + 2^4 + 3^4 + 4^4) + \frac{1}{2}(1^3 + 2^3 + 3^3 + 4^3)$$
$$+ \frac{1}{4}(1^2 + 2^2 + 3^2 + 4^2)$$

$$1^4 + 2^4 + 3^4 + 4^4 = \frac{4}{5}(1^3 + 2^3 + 3^3 + 4^3)(4 + \frac{1}{2}) - \frac{1}{5}(1^2 + 2^2 + 3^2 + 4^2)$$
$$= \frac{4}{5}\left(4 + \frac{1}{2}\right)\left(\frac{4}{4} + \frac{1}{4}\right)4(4 + 1)4 - \frac{1}{5}\left(\frac{4}{3} + \frac{1}{3}\right)4\left(4 + \frac{1}{2}\right)$$

and finally

$$1^4 + 2^4 + 3^4 + 4^4 = \left(\frac{4}{5} + \frac{1}{5}\right)4\left(4 + \frac{1}{2}\right)\left[(4 + 1)4 - \frac{1}{3}\right].$$

From this result for the case $n = 4$, ibn al-Haytham simply stated his general result in words we can translate into the modern formula:

$$\sum_{i=1}^{n} i^4 = \left(\frac{n}{5} + \frac{1}{5}\right)n\left(n + \frac{1}{2}\right)\left[(n + 1)n - \frac{1}{3}\right].$$

Another inductive argument, this time in relation to the binomial theorem and the Pascal triangle, is found in al-Samaw'al's *Al-Bāhir*, where he refers to al-Karajī's treatment of these subjects. Because the particular work of al-Karajī's in which this discussion occurs is no longer extant, we consider al-Samaw'al's version. The binomial theorem is the result

$$(a + b)^n = \sum_{k=0}^{n} C_k^n a^{n-k} b^k,$$

where n is a positive integer and the values C_k^n are the binomial coefficients, the entries in the Pascal triangle. Naturally, al-Samaw'al, having no symbolism, wrote this formula in words in each individual instance. For example, in the case $n = 4$ he wrote, "For a number divided into two parts, its square-square [fourth power] is equal to the square-square of each part, four times the product of each by the cube of the other, and six times the product of the squares of each part."[26] Al-Samaw'al then provided a table of binomial coefficients to show how to generalize this rule for greater values of n:

x	x^2	x^3	x^4	x^5	x^6	x^7	x^8	x^9	x^{10}	x^{11}	x^{12}
1	1	1	1	1	1	1	1	1	1	1	1
1	2	3	4	5	6	7	8	9	10	11	12
	1	3	6	10	15	21	28	36	45	55	66
		1	4	10	20	35	56	84	120	165	220
			1	5	15	35	70	126	210	330	495
				1	6	21	56	126	252	462	792
					1	7	28	84	210	462	924
						1	8	36	120	330	792
							1	9	45	165	495
								1	10	55	220
									1	11	66
										1	12
											1

His procedure for constructing this table is the familiar one, that any entry comes from adding the entry to the left of it to the entry just above that one. He then noted that one can use the table to read off the expansion of any power up to the twelfth of "a number divided into two parts."

With this table in mind, let us see how al-Samaw'al demonstrated the quoted result for $n = 4$. Assume the number c is equal to $a + b$. Since $c^4 = cc^3$ and c^3 is already known to be given by $c^3 = (a + b)^3 = a^3 + b^3 + 3ab^2 + 3a^2b$, it follows that $(a + b)^4 = (a + b)(a + b)^3 = (a + b)(a^3 + b^3 + 3ab^2 + 3a^2b)$. By using repeatedly the result $(r + s)t = rs + rt$, which al-Samaw'al quoted from Euclid's *Elements* II, he found that this latter quantity equals $(a + b)a^3 + (a + b)b^3 + (a + b)3ab^2 + (a + b)3a^2b = a^4 + a^3b + ab^3 + b^4 + 3a^2b^2 + 3ab^3 + 3a^3b + 3a^2b^2 = a^4 + b^4 + 4ab^3 + 4a^3b + 6a^2b^2$. The coefficients here are the appropriate ones from the table, and the expansion shows that the new coefficients are formed from the old ones exactly as stated in the table construction. Al-Samaw'al next quoted the result for $n = 5$ and asserted his general result: "He who has understood what we have just said, can prove that for any number divided into two parts, its quadrato-cube [fifth power] is equal to the sum of the quadrato-cubes of each of its parts, five times the product of each of its parts by the square-square of the other, and ten times the product of the square of each of them by the cube of the other. And so on in ascending order."[27] Like the proofs of al-Karajī and ibn al-Haytham, al-Samaw'al's argument contained the two basic elements of an inductive proof. He began with a value for which the result is known, here $n = 2$, and then used the result for a given integer to derive the result for the next. Although al-Samaw'al did not have any way of stating, and therefore proving, the general binomial theorem, to modern readers there is only a short step from al-Samaw'al's argument to a full inductive proof of the binomial theorem, provided that in the statement of that theorem the coefficients themselves are defined inductively, essentially as al-Samaw'al did define them, as $C_m^n = C_{m-1}^{n-1} + C_m^{n-1}$. In any case, the Pascal triangle, both in Islam and, as we have noted, in China, was used

BIOGRAPHY

Al-Khayyāmī (1048–1131)

Al-Khayyāmī was born in Nishapur, Iran, in 1048 shortly after the area was conquered by the Seljuk Turks. He was able during most of his life to enjoy the support of the Seljuk rulers. In fact, he spent many years at the observatory in Isfahan at the head of a group working to reform the calendar. At various times, as ruler replaced ruler, he fell into disfavor, but he was able ultimately to garner enough support to write many mathematical and astronomical works, as well as poetry and philosophical works. In fact, he is best known in the West for the collection of poems known as the *Rubaiyat*. In the preface of his great algebra work, he complained how difficult it had been for him to work, but then thanks the ruler who provided him with the necessary support:

> I had not been able to find time to complete this work, or to concentrate my thoughts on it, hindered as I had been by troublesome obstacles. . . . Most of our contemporaries

are pseudo-scientists who mingle truth with falsehood, who are not above deceit and pedantry, and who use the little that they know of the sciences for base material purposes only. When they see a distinguished man intent on seeking the truth, one who prefers honesty and does his best to reject falsehood and lies, avoiding hypocrisy and treachery, they despise him and make fun of him. When God favored me with the intimate friendship of His Excellency, our glorious and unique Lord, the supreme judge, the Imām, Sayid Abū-Ṭāhir . . . after I had despaired of meeting such a man . . . who combined in himself profound power in science with firmness of action . . . my heart was greatly rejoiced to see him. . . . My power was strengthened by his liberality and his favors. In order that I might come nearer to his sublime position I found myself obliged to take up again the work which the vicissitudes of time had caused me to abandon in summarizing what I had verified of the essence of philosophical theories.[28]

to develop an algorithm to calculate roots of numbers. In the Islamic case, this algorithm is documented from the time of al-Samaw'al, while there are strong indications that it was known at least a century earlier.

9.3.5 Omar Khayyam and the Solution of Cubic Equations

There was another strand of development in algebra in the Islamic world alongside its arithmetization and the development of inductive ideas, namely, the application of geometry. By the end of the ninth century, Islamic mathematicians, having read the major Greek texts, had noticed that certain geometric problems led to cubic equations, equations that could be solved through finding the intersection of two conic sections. Such problems included the doubling of the cube and Archimedes' splitting of a sphere into two parts whose volumes are in a given ratio. Several Islamic mathematicians during the tenth and eleventh centuries also solved certain cubic equations by taking over this Greek idea of intersecting conics. But it was the mathematician and poet 'Umar ibn Ibrāhīm al-Khayyāmī (1048–1131) (usually known in the West as Omar Khayyam), who first systematically classified and then proceeded to solve all types of cubic equations by this general method.

Al-Khayyāmī announced his project in a brief treatise entitled *On the Division of a Quadrant of a Circle*, in which he proposed to divide a quadrant $ABCD$ at a point G such that, with perpendiculars drawn to two diameters as in Figure 9.11, we have $AE : GH = EH : HB$. Using the method of analysis, he assumed that the problem was solved and then constructed

FIGURE 9.11

Omar Khayyam's quadrant
problem

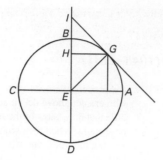

the tangent GI to the circle at G. After a few steps, he found that the right triangle EGI
had the property that its hypotenuse EI was equal to the sum of one of the sides EG and the
perpendicular GH from the right angle to the hypotenuse. He concluded that if he could find
such a right triangle, he could complete the synthesis of the problem.

In order to find the right triangle, he needed algebra. So he tried a particular case, with
$EH = 10$ and $GH = x$. Therefore, $GE^2 = x^2 + 100$. But $GE^2 = EI \cdot EH$, so

$$\frac{GE^2}{10} = \frac{x^2}{10} + 10 = \frac{EI \cdot EH}{10} = EI.$$

But since $EI = EG + GH$, we have $\frac{x^2}{10} + 10 = EG + GH = EG + x$. Therefore,

$$x^2 + 100 = EG^2 = \left(\frac{x^2}{10} + 10 - x\right)^2.$$

Simplifying this equation gave him a cubic equation in x: $x^3 + 200x = 20x^2 + 2000$. Noting
that this equation could not be solved by "plane geometry," al-Khayyāmī proceeded to
solve it by the intersection of a hyperbola and a semicircle whose modern equations are
$xy = \sqrt{20{,}000}$ and $x^2 - 30x + y^2 - \sqrt{800}y + 400 = 0$, respectively. Given the solution, he
finally could construct the right triangle that enabled him to solve the original problem.

With this example in mind, al-Khayyāmī then analyzed all possible cubic equations in
his algebra text, the *Risāla fi-l-barāhīn 'ala masā'il al-jabr wa'l-muqābala* (*Treatise on
Demonstrations of Problems of al-jabr and al-muqabala*). Although the author suggested
that the reader be familiar with Euclid's *Elements* and *Data* and the first two books of
Apollonius's *Conics*, nevertheless, the text addressed algebraic, not geometric, problems.
In fact, al-Khayyāmī would have liked to provide algebraic algorithms for solving cubic
equations, analogous to al-Khwārizmī's three algorithms for solving quadratic equations. As
he wrote, "When, however, the object of the problem is an absolute number, neither we, nor
any of those who are concerned with algebra, have been able to solve this equation—perhaps
others who follow us will be able to fill the gap."[29] It was not until the sixteenth century in
Italy that al-Khayyāmī's hope was realized.

Al-Khayyāmī began his work, in the style of al-Khwārizmī, by giving a complete classification of equations of degree up to three. Since for al-Khayyāmī, as for his predecessors, all

numbers were positive, he had to list separately the various forms that might possess positive roots. Among these were fourteen not reducible to quadratic or linear equations, including, of course, the form analyzed by Archimedes much earlier and the form needed in the quadrant problem. These types of equations were in three groups: one binomial equation, $x^3 = d$; six trinomial equations, $x^3 + cx = d$, $x^3 + d = cx$, $x^3 = cx + d$, $x^3 + bx^2 = d$, $x^3 + d = bx^2$, and $x^3 = bx^2 + d$; and seven tetranomial equations, $x^3 + bx^2 + cx = d$, $x^3 + bx^2 + d = cx$, $x^3 + cx + d = bx^2$, $x^3 = bx^2 + cx + d$, $x^3 + bx^2 = cx + d$, $x^3 + cx = bx^2 + d$, and $x^3 + d = bx^2 + cx$. For each of these forms, the author described the conic sections necessary for its solution, proved that his solution was correct, and finally discussed the conditions under which there may be no solutions or more than one solution.

That al-Khayyāmī gave this classification is strong evidence of the major change in mathematical thinking that had happened in the over 1300 years since Archimedes. Unlike the case with the Greek genius, al-Khayyāmī was no longer interested in solving a specific geometric problem, even though his interest in the subject was sparked by such a problem. He was interested in finding general methods for solving all sorts of problems that could be expressed in the form of equations. Though he did not use our symbolic notation, but just used words, there is no question that al-Khayyāmī was doing algebra, not geometry. And this is true even though every one of his equations was conceived as an equation between solids. For example, in his solution of $x^3 + cx = d$ or, as he puts it, the case where "a cube and sides are equal to a number," since x represents a side of a cube, c must represent an area (expressible as a square), so that cx is a solid, while d itself represents a solid.

To construct the solution to this equation, al-Khayyāmī set AB equal in length to a side of the square c, or $AB = \sqrt{c}$ (Fig. 9.12). He then constructed BC perpendicular to AB so that $BC \cdot AB^2 = d$, or $BC = d/c$. Next, he extended AB in the direction of Z and constructed a parabola with vertex B, axis BZ, and parameter AB. In modern notation, this parabola has the equation $x^2 = \sqrt{c}\,y$. Similarly, he constructed a semicircle on the line BC. Its equation is

$$\left(x - \frac{d}{2c}\right)^2 + y^2 = \left(\frac{d}{2c}\right)^2 \quad \text{or} \quad x\left(\frac{d}{c} - x\right) = y^2.$$

The circle and the parabola intersect at a point D. It is the x coordinate of this point, here represented by the line segment BE, which provides the solution to the equation.

FIGURE 9.12

Al-Khayyāmī's construction for the solution of $x^3 + cx = d$

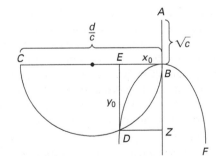

Al-Khayyāmī proved that his solution is correct by using the basic properties of the parabola and the circle. If $BE = DZ = x_0$ and $BZ = ED = y_0$, then first,

$$x_0^2 = \sqrt{c}\, y_0 \qquad \text{or} \qquad \frac{\sqrt{c}}{x_0} = \frac{x_0}{y_0},$$

because D is on the parabola, and second,

$$x_0\left(\frac{d}{c} - x_0\right) = y_0^2 \qquad \text{or} \qquad \frac{x_0}{y_0} = \frac{y_0}{\frac{d}{c} - x_0},$$

because D is on the semicircle. It follows that

$$\frac{c}{x_0^2} = \frac{x_0^2}{y_0^2} = \frac{y_0^2}{\left(\frac{d}{c} - x_0\right)^2} = \frac{y_0}{\frac{d}{c} - x_0}\frac{x_0}{y_0} = \frac{x_0}{\frac{d}{c} - x_0}$$

and then that $x_0^3 = d - cx_0$, so x_0 is the desired solution. Al-Khayyāmī noted here, without any indication of a proof, that this class of equations always has a single solution. In other words, the parabola and circle always intersect in one point other than the origin. The origin, though, does not provide a solution to the problem. Al-Khayyāmī's remark reflects the modern statement that the equation $x^3 + cx = d$ always has exactly one positive solution.

Al-Khayyāmī treated each of his fourteen cases in the same manner. In those cases in which a positive solution did not always exist, he noted that there were zero, one, or two solutions, depending on whether the conic sections involved do not intersect or intersect at one or two points. His one failure in this analysis is in the case of the equation $x^3 + cx = bx^2 + d$, where he did not discover the possibility of three (positive) solutions. In general, however, he did not relate the existence of one or two solutions to conditions on the coefficients. Even when he did, in the case $x^3 + d = bx^2$, it was only in a limited way. In that equation, he noted that if $\sqrt[3]{d} = b$, there was no solution. For if x were a solution, then $x^3 + b^3 = bx^2$, so $bx^2 > b^3$ and $x > b$. Since $x^3 < bx^2$, it is also true that $x < b$, a contradiction. Similarly, there was no solution if $\sqrt[3]{d} > b$. The condition $\sqrt[3]{d} < b$, however, does not guarantee a solution. Al-Khayyāmī noted again that there may be zero, one, or two (positive) solutions, depending on how many times the conics for this problem (a parabola and a hyperbola) intersect.

9.3.6   Sharaf al-Dīn al-Ṭūsī and Cubic Equations

Al-Khayyāmī's methods were improved on by Sharaf al-Dīn al-Ṭūsī (d. 1213), a mathematician born in Tus, Persia. Like his predecessor, he began by classifying the cubic equations into several groups. His groups differed from those of al-Khayyāmī, because he was interested in determining conditions on the coefficients that determine the number of solutions. Therefore, his first group consisted of those equations that could be reduced to quadratic ones, plus the equation $x^3 = d$. The second group consisted of the eight cubic equations that always have at least one (positive) solution. The third group consisted of those types that may or may not have (positive) solutions, depending on the particular values of the coefficients. These include $x^3 + d = bx^2$, $x^3 + d = cx$, $x^3 + bx^2 + d = cx$, $x^3 + cx + d = bx^2$, and $x^3 + d = bx^2 + cx$.

For the second group of equations, his method of solution was the same as al-Khayyāmī's, the determination of the intersection point of two appropriately chosen conic sections. Yet he went beyond al-Khayyāmī by always giving a careful discussion as to why the two conics in fact intersected. It is in the third group, however, that he made his most original contribution.

Consider Sharaf al-Dīn's analysis of $x^3 + d = bx^2$, typical of his analysis of the five equations in this group. He began by putting the equation in the form $x^2(b - x) = d$. He then noted that the question of whether the equation has a solution depends on whether the "function" $f(x) = x^2(b - x)$ reaches the value d or not (Fig. 9.13). He therefore carefully proved that the value $x_0 = \frac{2b}{3}$ provides the maximum value for $f(x)$, that is, for any x between 0 and b, $x^2(b - x) \le (\frac{2b}{3})^2(\frac{b}{3}) = \frac{4b^3}{27}$. It follows that if $\frac{4b^3}{27}$ is less than the given d, there can be no solutions to the equation. If $\frac{4b^3}{27}$ equals d, there is only one solution, $x = \frac{2b}{3}$. Finally, if $\frac{4b^3}{27}$ is greater than d, there are two solutions, x_1 and x_2, where $0 < x_1 < \frac{2b}{3}$ and $\frac{2b}{3} < x_2 < b$.

FIGURE 9.13

Modern graphic interpretation of Sharaf al-Dīn al-Ṭūsī's analysis of the cubic equation $x^3 + d = bx^2$

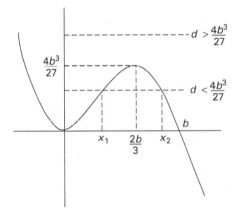

It is curious that Sharaf al-Dīn did not say how he found this particular value for x_0. Perhaps he guessed it by analogy to the fact already known to the Greeks (*Elements* VI–28) that $x = \frac{b}{2}$ provides the maximum value for the expression $x(b - x)$, or by a close study of problem 4 of Archimedes' *On the Sphere and Cylinder II*, in which Archimedes also found that $x_0 = \frac{2b}{3}$ provided the maximum value for the function $x^2(b - x)$. One historian has even suggested that Sharaf al-Dīn found this maximum by considering the conditions on x under which $f(x) - f(y) > 0$ for both $y < x$ and $y > x$, that is, in essence, by calculating a zero of the "derivative" of $f(x)$.[30]

Now, knowing the condition under which solutions exist, Sharaf al-Dīn proceeded to solve the equation by reducing it to a form already known, in this case the equation $x^3 + bx^2 = k$, where $k = \frac{4b^3}{27} - d$. He demonstrated that if a solution X to that equation is found geometrically by the intersection of two conic sections, then the larger solution x_2 to the given equation is $x_2 = X + \frac{2b}{3}$. To find the remaining root x_1, the author provided a new method. He found the positive solution Y to the quadratic equation $x^2 + (b - x_2)x = x_2(b - x_2)$ and then demonstrated, again geometrically, that $x_1 = Y + b - x_2$ is the other positive root of the original equation. Hence, the root of the new polynomial is related to that of the old by this change of variable formula. It is clear, therefore, that Sharaf al-Dīn had a solid understanding of the nature of cubic equations and the relationship of their roots and coefficients. Unlike his predecessors, he was able to see that the various types of cubic equations were related. Solutions of one type could be conveniently used in solving a second type. Note also that although he in effect used the discriminant of a cubic equation, here $\frac{4b^3}{27} - d$, to determine whether positive solutions existed, he was not able to use it algebraically to determine the numerical solutions.

On the other hand, Sharaf al-Dīn was interested in finding numerical solutions to these cubic equations. The example he gave in the case discussed was $x^3 + 14,837,904 = 465x^2$. By the method above, he first calculated that $\frac{4b^3}{27} = 14,895,500$ and $k = \frac{4b^3}{27} - d = 57,596$. It followed that there were two solutions x_1, x_2, with $0 < x_1 < 310$ and $310 < x_2 < 465$. To find x_2, he needed to solve $x^3 + 465x^2 = 57,596$. He found that 11 is a solution and therefore that $x_2 = \frac{2b}{3} + 11 = 310 + 11 = 321$. To find x_1, he needed to solve the quadratic equation $x^2 + 144x = 46,224$. The (positive) solution is an irrational number approximately equal to 154.73, a solution he found by a numerical method related to the Chinese method discussed in Chapter 7. The solution x_1 to the original equation is then 298.73.

9.4   COMBINATORICS

As we have seen, the basic formulas for combinations and permutations were known in India by the ninth century and probably even earlier. Islamic mathematicians too were interested in such questions. For example, al-Khalīl ibn Aḥmad (717–791), a lexicographer interested in classifying the words in the Arabic language, calculated the number of words one could get by taking 2, 3, 4, or 5 letters out of the Arabic alphabet of 28 letters. And al-Samaw'al, in discussing methods for solving large systems of equations, actually wrote down in a systematic fashion all 210 combinations of 10 unknowns taken 6 at a time in his *Al-Bahir*. He did not, however, indicate how to calculate the number for other cases. It is only in the thirteenth century that we see evidence of the derivation of the basic combinatorial formulas. We will consider the contributions of several Islamic mathematicians to this work.

9.4.1   Counting Combinations

Early in the thirteenth century, Aḥmad al-Ab'dari ibn Mun'im discussed the calculation of the number of combinations of r things from a set of n by looking at this number in terms of combinations of $r - 1$ things. Little is known about ibn Mun'im, but he probably lived at the Almohade court in Marrakech (now in Morocco) during the reign of Mohammed ibn Ya'kub al-Nasir (1199–1213). Although the Almohade dynasty originally ruled over a large empire including much of North Africa and Spain, al-Nasir was defeated by a coalition of Christian kings at the battle of Las Navas de Tolosa in Spain in 1212 and lost many of his Spanish domains.

Ibn Mun'im was basically examining the old question of the number of possible words that could be formed out of the letters of the Arabic alphabet. But before dealing with that question, he considered a different problem: how many different pom-poms of one, two, three, and so on, colors can one make out of ten different colors of silk. He calculated these carefully. First of all, he noted that with only one color, there are ten possibilities, that is,

$C_1^{10} = 10$. To calculate the possibilities for two colors, ibn Mun'im listed the pairs in order (where c_i represents the ith color):

$$(c_2, c_1); \ (c_3, c_1), \ (c_3, c_2); \ldots (c_{10}, c_1), \ (c_{10}, c_2), \ldots, (c_{10}, c_9).$$

Then he noted that

$$C_2^{10} = C_1^1 + C_1^2 + \cdots + C_1^9 = 1 + 2 + \cdots + 9 = 45$$

and proceeded to generalize this result for any number of colors of silk: "The number of pom-poms of two colors is then equal to the sum of the successive whole numbers from one to the number that is one less than the number of colors."[31]

To calculate C_3^{10}, ibn Mun'im proceeded analogously.

> As for determining the number of pom-poms of three colors, it is obtained by combining the third color with the first and the second, then by combining the fourth color with each pair of colors among the three colors preceding which are the first, the second, and the third, then by the combination of the fifth color with each pair of colors among the four colors preceding, then by the combination of the sixth color with each pair among the five colors preceding, and so on, until [the combination of] the tenth color, with each pair of colors among the nine colors preceding.

In other words, for each c_k with $k = 3, 4, \ldots, 10$, ibn Mun'im considered the pairs from the previous calculation, which have all indices less than k; for example,

$$(c_3, (c_2, c_1)); \ (c_4, (c_2, c_1)), \ (c_4, (c_3, c_1)), \ (c_4, (c_3, c_2)); \ (c_5, (c_2, c_1)), \ldots.$$

> But each pair of colors is a pom-pom of the second line. For this reason, we write: one, in the first case of the third line opposite the third color, and this will be the pom-pom composed of the first, second and third color; then, we write, in the next case, which is opposite the fourth color, the number of pom-poms obtained by the combination of the fourth color with each pair among the colors preceding, and it is equal to the number of pom-poms of two colors composed of colors preceding the fourth color, and it is also equal to the sum of the content of the two first cases of the second line, and it is three. We then write three in the second case of the third line. And we write in the third case of the third line—this case being that opposite the fifth color—[the number] of pom-poms [obtained] by the combination of the fifth color with the pairs of colors preceding the fifth color. And it is also the sum of the content of the three first cases of the second line. And it is six. We [therefore] write six in the third case of the third line. . . . The sum of the cases of the third line is then equal to the set of pom-poms of three colors each, [obtained] beginning with the [given] colors.

The word "line" here refers to the table in which ibn Mun'im presented these results (Fig. 9.14). The first line of the table is a row of "1s" (which we can think of as representing C_0^k), while the second line lists the numbers $1, 2, \ldots, 10 (= C_1^1, C_1^2, \ldots, C_1^{10})$. Ibn Mun'im's argument thus showed that a given number in the third line (representing the number of pairs with index less than a given number, that is, the numbers $C_2^k = 1, 3, 6, \ldots, 36$) is calculated by summing the numbers in the previous line up to one less than the given number. And then, in the last sentence, he asserted that C_3^{10} is the sum $1 + 3 + 6 + \cdots + 36 = C_2^2 + C_2^3 + C_2^4 + \cdots + C_2^9$. Ibn Mun'im thus developed this table, the Pascal triangle, line by line, in the process showing that

$$C_k^n = C_{k-1}^{k-1} + C_{k-1}^k + C_{k-1}^{k+1} + \cdots + C_{k-1}^{n-1},$$

FIGURE 9.14

Table for numbers of possible pom-poms

Sum											
1	1	Line of pom-poms of ten colors									
10	9	1	Line of pom-poms of nine colors								
45	36	8	1	Line of pom-poms of eight colors							
120	84	28	7	1	Line of pom-poms of seven colors						
210	126	56	21	6	1	Line of pom-poms of six colors					
252	126	70	35	15	5	1	Line of pom-poms of five colors				
210	84	56	35	20	10	4	1	Four colors			
120	36	28	21	15	10	6	3	1	Three colors		
45	9	8	7	6	5	4	3	2	1	Two colors	
10	1	1	1	1	1	1	1	1	1	One color	
all	10th	9th	8th	7th	6th	5th	4th	3rd	2nd	1st	

for $n \leq 10$ and $k \leq n$. In fact, he even noted that "if the number of colors you have is larger than ten, you add columns to the table until the number of its colors is equal to that of your colors."[32]

Returning now to the question of words, ibn Mun'im first dealt with the question of permutations without repetition:

> The problem is: We want to determine a canonical procedure to determine the number of permutations of the letters of a word of which the number of letters is known and which does not repeat any letter. If the word has two letters, it is clear that there will be two permutations, since the first letter may be made the second and the second the first. If we augment this by one letter and consider a three letter word, it is clear that, in each of the permutations of two letters of a two letter word, the third letter may be before the two letters, between the two letters, or in the final position. The letters of a three letter word therefore have six permutations. If the word is now augmented by another letter to make a four letter word, the fourth letter will be in each of the six permutations [in one of four positions]. The four letter word will thus have twenty four permutations.[33]

Thus, ibn Mun'im noted, the number of permutations of the letters of a word of any length is found by multiplying one by two, by three, by four, by five, and so on, up to the number of letters of the word.

After next considering how to calculate permutations with repetitions, ibn Mun'im dealt with the technical details of how Arabic words are created, including the use of vowel signs. Although it is certainly not feasible to determine the total number of possible Arabic words, given the ways such words can be constructed, he concluded his treatise with various examples. For instance, he calculated explicitly the number of words of nine letters, each word having two nonrepeated letters, two letters repeated twice, and one letter repeated three times. The number turns out to have 16 decimal digits.

9.4.2 Combinatorics and Number Theory

In the late thirteenth century, the question of the combinatorial formulas was taken up by Kamāl al-Dīn al-Fārisī (d. 1320), who lived in Persia, this time in connection with factorization of integers and the concept of amicable numbers. Recall that in Book IX of the *Elements*, Euclid had shown how to find perfect numbers, numbers that equaled the sum of their proper divisors. Later Greek mathematicians had generalized this idea and defined the notion of **amicable numbers**, a pair of numbers each of which equaled the sum of the proper divisors of the other. Unfortunately, the Greeks had only discovered one such pair, 220 and 284, and had not been able to find a general theorem that produced such pairs. It was Thābit ibn Qurra who first discovered and proved such a theorem, here stated in modern notation:

IBN QURRA'S THEOREM *For $n > 1$, let $p_n = 3 \cdot 2^n - 1$, $q_n = 9 \cdot 2^{2n-1} - 1$. If p_{n-1}, p_n, and q_n are prime, then $a = 2^n p_{n-1} p_n$ and $b = 2^n q_n$ are amicable.*

As the simplest example, we can take $n = 2$; then $p_1 = 5$, $p_2 = 11$, and $q_2 = 71$ are all prime, and the resultant pair of numbers is 220 and 284. Although other Islamic mathematicians studied ibn Qurra's result, it was not until the late thirteenth century that a second pair of amicable numbers, 17,296 and 18,416, was found by al-Fārisī, in connection with his own study of the theorem. (This pair is commonly attributed to Pierre Fermat in the seventeenth century; another pair, attributed to René Descartes, was also discovered earlier by an Islamic mathematician. See Exercise 24.)

Al-Fārisī's work on ibn Qurra's theorem was through combinatorial analysis, in this case the combinations of the prime divisors of a number. It is these combinations that determine all of the proper divisors of a number. For example, if $n = p_1 p_2 p_3$, where each p_i is prime, then the divisors of n are 1, p_1, p_2, p_3, $p_1 p_2$, $p_1 p_3$, $p_2 p_3$, and $p_1 p_2 p_3$. Thus, there are $C_0^3 + C_1^3 + C_2^3 + C_3^3$ divisors in all. Therefore, a knowledge of the relationships among the combinatorial numbers was necessary for a complete study of the divisors of integers.

Al-Fārisī was able to work out these relationships in some detail, using an argument similar to that of ibn Mun'im. In fact, he also developed the "Pascal" triangle and was able to relate columns not only to numbers of combinations but also to figurate numbers—triangular, pyramidal, and higher-order solids—while at the same time giving an algebraic proof of ibn Qurra's theorem.

9.4.3 Ibn al-Bannā and the Combinatorial Formulas

Al-Fārisī, like his predecessors, developed the results for combinations by taking sums. It was a direct successor of ibn Mun'im in Morocco, Abu-l-'Abbas Ahmad al-Marrakushi ibn al-Bannā (1256–1321), also of Marrakech, who was able to derive the standard multiplicative formula for finding combinations, the formula that was stated much earlier in India. In addition, he dealt with combinatorics in the abstract, not being concerned with what kinds of objects were being combined.

Ibn al-Bannā began by using a counting argument to show that $C_2^n = n(n-1)/2$: An element a_1 is associated with each of $n-1$ elements, a_2 is associated with each of $n-2$ elements, and so on, so C_2^n is the sum of $n-1, n-2, n-3, \ldots, 2, 1$. He then showed that to find the value C_k^n, "we always multiply the combination that precedes the combination sought by the number that precedes the given number, and whose distance to it is equal to the number of combinations sought. From the product, we take the part that names the number of combinations."[34] We can translate ibn al-Bannā's words into the modern formula

$$C_k^n = \frac{n - (k - 1)}{k} C_{k-1}^n.$$

To prove this result, ibn al-Bannā began with C_3^n. To each set of two elements from the n elements, one associates one of the $n - 2$ remaining elements. One obtains then $(n - 2)C_2^n$ different sets. But because $C_2^3 = 3$, each of these sets is repeated three times. For example, $\{a, b, c\}$ occurs as $\{\{a, b\}, c\}$, $\{\{a, c\}, b\}$, and as $\{\{b, c\}, a\}$. Therefore, $C_3^n = \frac{n-2}{3} C_2^n$ as claimed. For the next step, we know that $C_3^4 = 4$. It follows that if we associate to each set of three elements one of the $n - 3$ remaining elements, the total $(n - 3)C_3^n$ is four times larger than C_4^n, or $C_4^n = \frac{n-3}{4} C_3^n$. A similar argument holds for other values of k. Putting these results together, it follows that

$$C_k^n = \frac{n(n - 1)(n - 2) \cdots (n - (k - 1))}{1 \cdot 2 \cdot 3 \cdots k},$$

the standard formula for the number of ways to pick k elements out of a set of n. Using this result and the result of ibn Mun'im that the number of permutations of a set of n objects was $n!$, ibn al-Bannā showed by multiplication that the number P_k^n of permutations of k objects from a set of n is

$$P_k^n = n(n - 1)(n - 2) \ldots (n - (k - 1)).$$

Ibn al-Bannā's proof of the formula for C_k^n as well as ibn Mun'im's proof of the permutation rule are, like earlier proofs of al-Karajī and al-Samaw'al, in inductive style. That is, the author began with a known result for a small value and used it to build up step-by-step to higher values. But neither ibn al-Bannā nor any of his predecessors explicitly stated an induction principle to be used as a basis for proofs. Such a statement was first made by Levi ben Gerson, a younger contemporary of ibn al-Bannā, and will be considered in Chapter 10.

9.5   GEOMETRY

Islamic mathematicians dealt at an early stage with practical geometry, but later worked on various theoretical aspects of the subject, including the parallel postulate of Euclid, the concept of an irrational magnitude, and the exhaustion principle for determining volumes of solids.

9.5.1   Practical Geometry

The earliest extant Arabic geometry, like the earliest algebra, is due to al-Khwārizmī, and occurred as a separate section of his algebra text. A brief reading makes it clear that in his geometry, even more so than in his geometric demonstrations in algebra, al-Khwārizmī was not at all influenced by theoretical Greek mathematics. His text is an elementary compilation of rules for mensuration such as might be needed by surveyors, containing no axioms or proofs.

We begin with al-Khwārizmī's rules for the circle:

> In any circle, the product of its diameter, multiplied by three and one-seventh, will be equal to the circumference. This is the rule generally followed in practical life, though it is not quite exact. The geometricians have two other methods. One of them is, that you multiply the diameter by itself, then by ten, and hereafter take the root of the product; the root will be the circumference. The other method is used by the astronomers among them. It is this, that you multiply the diameter by sixty-two thousand eight hundred thirty-two and then divide the product by twenty thousand. The quotient is the circumference. Both methods come very nearly to the same effect. . . . The area of any circle will be found by multiplying half of the circumference by half of the diameter, since, in every polygon of equal sides and angles, . . . the area is found by multiplying half of the perimeter by half of the diameter of the middle circle that may be drawn through it. If you multiply the diameter of any circle by itself, and subtract from the product one-seventh and half of one-seventh of the same, then the remainder is equal to the area of the circle.[35]

The first of the approximations for π given here is the Archimedean one, $3\frac{1}{7}$, familiar to Heron. The approximation of π by $\sqrt{10}$, attributed to "geometricians," was used in India. Interestingly, however, it is less exact than the "not quite exact" value of 3 1/7. The earliest known occurrence of the third approximation, 3.1416, was also in India, in the work of Āryabhaṭa. The attribution of this value to astronomers is probably connected with its use in the Indian astronomical works that were translated into Arabic.

Al-Khwārizmī gave several other procedures for calculating areas and volumes. For example, to calculate the area of a rhombus, the reader is instructed to multiply the length of one diagonal by half the length of the other. To determine the volume of the frustum of a pyramid, we are not given a formula, as in the *Moscow Papyrus*, but are told to calculate the height to the top of the completed pyramid by using similar triangles, then to subtract the volume of the upper pyramid from that of the lower. And, interestingly, rather than presenting Heron's formula for calculating the area of a triangle with three sides known, al-Khwārizmī dropped a perpendicular from one vertex to the opposite side, then used the Pythagorean Theorem twice to calculate the height of the triangle, and finally multiplied this height by half the base.

As we noted earlier, during the generations following al-Khwārizmī, Islamic mathematicians began to absorb the basic principles of Greek mathematics, including creation of correct geometrical constructions. But since they were always interested in practical applications, in particular in how artisans could create interesting geometrical patterns, they became proficient in doing theoretical constructions that could easily be translated into real-life constructions. We consider here an example taken from the *Book on the Geometrical Constructions Necessary to the Artisan* by Muḥammad Abū al-Wafā' al-Būzjānī (940–997), in which the author was looking at the problem of constructing a large square out of three identical squares. As he wrote, "A number of geometers and artisans have erred in the matter of these squares and their assembling. The geometers [have erred] because they have little practice in constructing, and the artisans [have erred] because they lack knowledge of proofs."[36] Further, he noted, "I was present at some meetings in which a group of geometers and artisans participated. They were asked about the construction of a square from three squares. A geometer easily constructed a line such that the square of it is equal to the three squares, but none of the artisans was satisfied with what he had done. The artisan wants to divide those squares into

BIOGRAPHY

Abū al-Wafā' (940–998)

Born in Buzjan, in the Khorasan region of what is now Iran, Muḥammad Abū al-Wafā al-Būzjānī lived during the time of the Buyid Islamic dynasty in western Iran and Iraq. The high point of this dynasty was during the reign of 'Adud ad-Dawlah, who supported a number of mathematicians at his court in Baghdad. His son Sharaf ad-Dawlah continued his father's policy, and Abū al-Wafā was employed in designing and building an observatory. His *Book on What Is Necessary from the Science of Arithmetic for Scribes and Businessmen* provided an introduction to various practical mathematical ideas, including mensuration, taxes, units of money, and payments to soldiers.

Interestingly, it is virtually the only book in medieval Islam in which negative numbers appear, in the context of debts. But Abū al-Wafā's main contribution was in the simplification and extension of the spherical trigonometry that Islamic scientists had learned from Greek sources. Among his other accomplishments, he was responsible for the earliest proof of the rule of four quantities, which then served as a basis for developing the basic ideas of spherical trigonometry. He was also the first to discover and prove the spherical law of sines as well as the law of tangents.

pieces from which one square can be assembled." (The geometer's construction simply used the Pythagorean Theorem twice to construct the square root of 3. But this construction is not a "physical" construction.)

Abū al-Wafā' then presented one of the incorrect methods of the artisans, in order that "the correct ones may be distinguished from the false ones and someone who looks into this subject will not make a mistake by accepting a false method, God willing [Fig. 9.15]. . . . But this figure which he constructed is fanciful, and someone who has no experience in the art or in geometry may consider it correct, but if he is informed about it he knows that it is false." He went on to note that the angles are correct, and that it looks like a good construction. But, in fact, the side of the proposed large square is equal to the side of the smaller square plus

FIGURE 9.15

Incorrect construction by artisans of one square from three

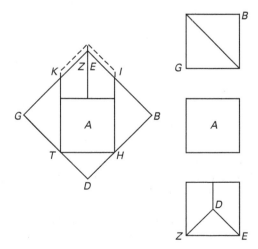

half the diagonal. A quick calculation showed that, in effect, the square of $1 + \frac{\sqrt{2}}{2}$ is not equal to 3.

Abū al-Wafā' finally presented a geometrically correct construction, with proof (Fig. 9.16). He bisected two of the squares along their diagonals. Each of those was applied to one side of the third square; one of the angles of the triangle, which is half a right angle, is placed at one angle of the square and the hypotenuse of the triangle at the side of the square. Then the right angles of the triangles are connected by straight lines. These become the side of the desired square. From each of the original triangles, a small triangle is cut off by the straight line; these are transferred to the "empty" triangles within the square. To prove that this construction is correct, Abū al-Wafā' needed to prove that the triangles extending past the square were congruent to the "vacant" triangles inside the square. But this followed by the angle-angle-side triangle congruence theorem.

FIGURE 9.16

Correct construction of one square from three

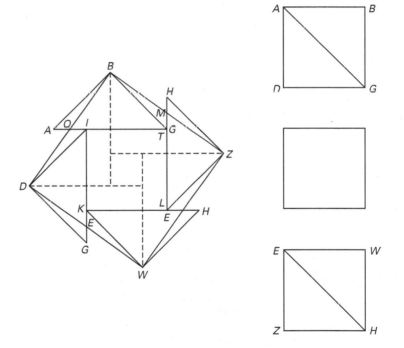

9.5.2 Geometrical Constructions

The preceding construction was designed for the use of artisans. But Islamic geometers were also interested in constructions that simply extended what the Greeks had done. As we saw in Chapter 3, Euclid presented a construction of a regular pentagon in Book IV of the *Elements*. Abū Kāmil showed, using algebra, how to construct an equilateral pentagon in a given square, each of whose sides is equal to 10. To construct the pentagon, he assumed that it was accomplished and used analysis to determine what the length of a side must be (Fig. 9.17). But rather than using the author's words, we will use modern symbols to help

FIGURE 9.17

Abū Kāmil's construction of
a pentagon in a square

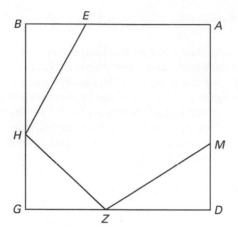

the reader understand the construction. Thus, if the pentagon is $AEHZM$, we set $x = AE$. Then $BE = 10 - x$. Also, since triangle GHZ is isosceles with $HZ = x$, sides GH and GZ will both equal $\sqrt{x^2/2}$. Thus, $HB = 10 - \sqrt{x^2/2}$. It follows that

$$x^2 = HE^2 = HB^2 + BE^2 = \left(10 - \sqrt{x^2/2}\right)^2 + (10 - x)^2$$

$$= \frac{3}{2}x^2 - \left(20 + \sqrt{200}\right)x + 200.$$

Abū Kāmil solved this equation to get $x = 20 + \sqrt{200} - \sqrt{200 + \sqrt{320,000}}$ as the length of a side of the pentagon. Presumably, he could now construct this length.

Abū al-Wafā' also gave a construction of a pentagon with a special condition, the condition being that the compass was a "rusty compass," one with a fixed opening. Thus, to construct a regular pentagon on a line segment AB, he erected a perpendicular BG at B of length equal to AB, found the midpoint D of AB, connected DG and found point S on DG such that $DS = AB$ (Fig. 9.18). He next constructed a perpendicular to DG at the midpoint K of DS, which met AB extended at E. Then, using both A and E as centers, he constructed circles of radius AB, which cut each other at M. Next, he connected BM and extended it to Z so that $MZ = AB$. Finally, he joined AZ, drew circles centered on A and Z with length AB, which met at H, and drew circles centered on B and Z with the same length, which met at T. The vertices A, B, T, Z, H, are now, he claimed, the vertices of an equilateral pentagon.

Islamic mathematicians also worked on constructions requiring the use of conic sections or other devices beyond those of Euclid. For example, Aḥmad ibn Muḥammad ibn 'Abd al-Jalīl al-Sijzī in the late tenth century gave a detailed construction of a heptagon, and an anonymous author a few years later wrote a treatise giving a construction of a regular 9-gon.[37]

FIGURE 9.18

Construction of a pentagon
with a "rusty compass"

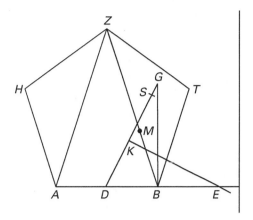

FIGURE 9.18

Construction of a pentagon
with a "rusty compass"

9.5.3 The Parallel Postulate

One of the other pure geometric ideas that recurs in Islamic geometry is that of parallel
lines and the provability of Euclid's fifth postulate. Even in Greek times, mathematicians
were disturbed with this postulate. Many attempts were made to prove it from the others.
So too in the Islamic world. One of the attempts to deal with this question was in the work
of ibn al-Haytham entitled *Maqāla fī sharḥ muṣādarāt kitāb Uqlīdis* (*Commentary on the
Premises of Euclid's Elements*), in which he attempted to reformulate Euclid's theory of
parallels. He began by redefining the concept of parallel lines, deciding that Euclid's own
definition of parallel lines as two lines that never meet was inadequate. His "more evident"
definition included the assumption of the constructibility of such lines. Namely, he wrote that
if a straight line moves so that one end always lies on a second straight line and so that it
always remains perpendicular to that line, then the other end of the moving line will trace
out a straight line parallel to the second line. In effect, this definition characterized parallel
lines as lines always equidistant from one another and also introduced the concept of motion
into geometry. Later commentators, including al-Khayyāmī, were unhappy with this. They
doubted the "self-evidence" of a line moving and always remaining perpendicular to a given
line, and therefore they could not understand how one could base a proof on this idea. As
they knew, Euclid had only used motion in generating new objects from old, as a sphere is
generated by rotating a semicircle. Nevertheless, ibn al-Haytham used this idea in his "proof"
of the fifth postulate.

The crucial step in ibn al-Haytham's proof is the following:

LEMMA *If two straight lines are drawn at right angles to the two endpoints of a fixed
straight line, then every perpendicular line dropped from the one line to the other is equal to
the fixed line.*

In Figure 9.19, GA and DB are drawn at right angles to AB, and a perpendicular is dropped
from G to the line DB. It must be proved that GD is equal to AB. Ibn al-Haytham's proof
was by contradiction. He first assumed that $GD > AB$. He then extended GA past A so that
$AE = AG$ and, similarly, BD past B. From the point E, a perpendicular is dropped to the

FIGURE 9.19

Ibn al-Haytham's proof of
his lemma dealing with the
parallel postulate

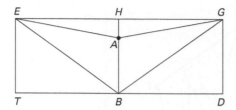

line DB extended, meeting it at T. Then the lines GB and BE are drawn. Triangles EAB and GAB are congruent by side-angle-side. Therefore, $\angle GBA = \angle EBA$, so $\angle GBD = \angle EBT$, and $GB = BE$. It follows that the triangles EBT and GBD are congruent and therefore that $GD = ET$. Now, using his concept of motion, ibn al-Haytham imagined line ET moving along line TD and remaining always perpendicular to it. When T coincides with B, point E will be outside line AB, since $ET > AB$. We call ET at this particular time HB. Of course, when ET reaches GD, the two lines will coincide. It now follows from the definition of parallelism that line GHE is a straight line parallel to DBT. By construction, GAE is also a straight line, so there would be two different straight lines with the same endpoints, and therefore two straight lines would enclose a space. This, of course, is impossible. A similar contradiction resulted from the assumption that $GD < AB$. Hence, the proof is complete.

Because $GD = AB$, it follows easily that $\angle AGD$, like the three other angles of quadrilateral $ABDG$, is a right angle. One can then easily demonstrate Euclid's postulate. Of course, what ibn al-Haytham did not realize was that his original definition of parallel lines already implicitly contained that postulate. In any case, his result made clear the reciprocal relationship between the parallel postulate and the fact that the angle sum of any quadrilateral is four right angles.

Al-Khayyāmī was also interested in this question of parallelism. In his *Sharḥ mā ashkala min musādarāt kitāb Uqlīdis* (*Commentary on the Problematic Postulates of the Book of Euclid*), he began with the principle that two convergent straight lines intersect, and it is impossible for them to diverge in the direction of convergence. By convergent lines, he meant lines that approached one another. Given this postulate, al-Khayyāmī proceeded to prove a series of eight propositions, culminating in Euclid's fifth postulate. He began by constructing a quadrilateral with two perpendiculars of equal length, AC and BD, at the two ends of a given line segment AB and then connecting the points C and D (Fig. 9.20). He proceeded to prove that the two angles at C and D were both right angles by showing that the two other possibilities, that they were both acute or both obtuse, led to contradictions. If they were acute, CD would be longer than AB, whereas if they were obtuse, CD would be shorter than AB. In each case, he showed that the lines AC and BD would diverge or converge on both sides of AB, and this would contradict his original postulate. Al-Khayyāmī was now able to demonstrate Euclid's fifth postulate. In some sense, his treatment was better than ibn al-Haytham's because he explicitly formulated a new postulate to replace Euclid's rather than have the latter hidden in a new definition.

About a century after al-Khayyāmī, another mathematician, Naṣīr al-Dīn al-Ṭūsī (1201–1274) subjected the works of his predecessor to detailed criticism and then attempted his own

FIGURE 9.20

Al-Khayyāmī's quadrilateral:
$AC = BD$, $AC \perp AB$, and
$BD \perp AB$. Are the angles
at C and D acute, obtuse, or
right?

BIOGRAPHY

Naṣīr al-Dīn al-Ṭūsī (1201–1274)

Naṣīr al-Dīn, from Tus in Iran, completed his formal education in Nishapur, Persia, then a major center of learning, and soon gained a great reputation as a scholar (Fig. 9.21). The thirteenth century, however, was a time of great turmoil in Islamic history. The only places of peace in Iran were the forts controlled directly by the Ismāʿili rulers. Fortunately, Naṣīr al-Dīn persuaded one of these rulers to allow him to work at such a fort. After the Mongol leader Hūlāgū defeated the Ismāʿilis in 1256, Naṣīr al-Dīn was able to transfer his allegiance. He served Hūlāgū as a scientific adviser and gained his approval to construct an observatory at Maragha, a town about fifty miles south of Tabriz. It was here that Naṣīr al-Dīn spent the rest of his life as head of a large group of astronomers. During that time, he computed a new set of very accurate astronomical tables and developed an astronomical model that Copernicus may have adapted to design his heliocentric system.

FIGURE 9.21

Naṣīr al-Dīn al-Ṭūsī

FIGURE 9.22

Naṣīr al-Dīn al-Ṭūsī's hypothesis on parallels and perpendiculars

proof of the fifth postulate in his book written in about 1250 entitled *Al-risāla al-shāfiya ʿan al-shakk fi-l-khuṭūṭ al-mutawāziya* (*Discussion Which Removes Doubt about Parallel Lines*). He considered the same quadrilateral as al-Khayyāmī and also tried to derive a contradiction from the hypotheses of the acute and obtuse angles. But in a manuscript probably written by his son Ṣadr al-Dīn in 1298, based on Naṣīr al-Dīn's later thoughts on the subject, there is a new argument based on another hypothesis, also equivalent to Euclid's, that if a line GH is perpendicular to CD at H and oblique to AB at G, then the perpendiculars drawn from AB to CD are greater than GH on the side on which GH makes an obtuse angle with AB and less than GH on the other side (Fig. 9.22).[38]

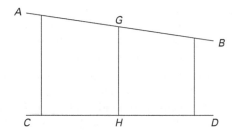

9.5.4 Incommensurables

Another geometric topic of interest to Islamic mathematicians was that of incommensurables. In fact, many Arabic commentaries were written on the topic of Euclid's *Elements*, Book X. Recall that Islamic algebraists early on began to use irrational quantities in their work with equations, ignoring the Euclidean distinction between number and magnitude. There were, however, several commentators who made some attempt to reconcile this use and to put it into a theoretical framework consistent with the Euclidean work.

In the *Risāla fi'l-maqādir al-mushtaraka wa'l-mutabāyana* (*Treatise on Commensurable and Incommensurable Magnitudes*), written sometime around 1000, Abū ʿAbdallāh al-Ḥasan ibn al-Baghdādī attempted to reconcile the operational rules already being used for irrational quantities with the main principles of the *Elements* and thus to prove that the contemporary methods of computation were valid. He was quite aware that these numerical methods of computation were simpler than the geometric modes of Euclid: "It is easier . . . to assume a number and to base oneself on it than to make a similar assumption concerning a magnitude."[39] Because he knew of Aristotle's and Euclid's fundamental distinction between number and magnitude, he began by relating the two concepts by establishing a correspondence between numbers and line segments in what appears to be a modern way. Namely, given a unit magnitude a, each "whole number" n corresponds to an appropriate multiple na of the unit magnitude. Parts of this magnitude, such as $\frac{m}{n}a$, then correspond to parts of a number $\frac{m}{n}$. Ibn al-Baghdādī considered any magnitude expressible this way as a rational magnitude. He showed that these magnitudes relate to one another as numbers to numbers, as in *Elements* X–5. Magnitudes that are not "parts" are considered irrational magnitudes. In effect, ibn al-Baghdādī attempted to imbed the rational numbers into a number line. But he also wanted to connect irrational magnitudes to "numbers."

Ibn al-Baghdādī made the connection through the idea of a root. The root of a number n was the middle term x in the continuous proportion $n : x = x : 1$. Such a root may or may not exist. He then defined the root of a magnitude na similarly as the mean proportional between the unit magnitude a and the magnitude na. This quantity is always constructible by straightedge and compass, so it necessarily exists. It may, of course, be either rational or irrational. Since "rational numbers" correspond to "rational magnitudes," and since the latter always have roots, which may or may not be rational, he could consider roots of the former to continue this correspondence. In particular, he noted that for magnitudes, roots and squares were of the same geometric type. In other words, the root of a magnitude expressed as a line segment was another line segment, just as the square of a line segment could be expressed as a line segment. Ibn al-Baghdādī, like some of his Islamic predecessors, hence moved away from the Greek insistence on homogeneity and toward the notion that all "quantities" can be expressed in the same way, essentially as "numbers."

Ibn al-Baghdādī concluded his book by dealing extensively with the various types of irrational magnitudes treated by Euclid in Book X. As a result of that discussion, he was able to prove a result on the "density" of irrational magnitudes, namely, that between any two rational magnitudes there exist infinitely many irrational magnitudes. For example, he considered the magnitudes represented by the consecutive numbers 2 and 3. The squares of these magnitudes are represented by 4 and 9. Between those magnitudes are magnitudes represented by the numbers 5, 6, 7, and 8. Their roots, $\sqrt{5}$, $\sqrt{6}$, $\sqrt{7}$, and $\sqrt{8}$, which ibn al-Baghdādī called magnitudes of the first order of irrationality, lie between 2 and 3. Similarly, the squares of 4 and 9, namely, 16 and 81, also represent magnitudes, as do the squares 25, 36, 49, and 64. Corresponding to the integers 17, 18, . . . , 24 are magnitudes of the first order of irrationality $\sqrt{17}$, $\sqrt{18}$, . . . , $\sqrt{24}$ as well as magnitudes of the second order of irrationality $\sqrt{\sqrt{17}}$, $\sqrt{\sqrt{18}}$, . . . , $\sqrt{\sqrt{24}}$. The latter magnitudes lie between the original magnitudes 2 and 3. Ibn al-Baghdādī noted that one can continue in this way to find as many magnitudes as one wants, of various higher orders of irrationality, between the two original ones. Ibn al-Baghdādī's work thus demonstrated that Islamic authors understood the arguments of their

Greek predecessors in keeping separate the realms of magnitude and number, but also wanted to break the bonds imposed by this dichotomy so that they could justify their increasing use of "irrationals" in computation.

9.5.5 Volumes and the Method of Exhaustion

One final area of geometry we will discuss also demonstrates that Islamic authors understood the works of the Greeks and wanted to go beyond them, namely, the work in calculating volumes of solids via the method of exhaustion pioneered by Eudoxus and used so extensively by Archimedes. It turns out that although Islamic mathematicians read Archimedes' work *On the Sphere and the Cylinder*, they did not have available his work *On Conoids and Spheroids* in which Archimedes showed how to calculate the volume of the solid formed by revolving a parabola about its axis. Thus, Thābit ibn Qurra found his own proof, which was quite long and complicated, and some 75 years later Abū Sahl al-Kūhī (10th century), from the region south of the Caspian Sea, simplified Thābit's method and solved some similar problems on volumes and analogous problems on the centers of gravity. Al-Kūhī in turn was criticized shortly afterward by ibn al-Haytham for not solving the paraboloid problem in all generality, that is, for not considering the volume of the solid formed by revolving a segment of a parabola about a line perpendicular to its axis. It is this latter problem that ibn al-Haytham proceeded to solve himself.

In modern terminology, ibn al-Haytham proved that the volume of the solid formed by rotating the parabola $x = ky^2$ around the line $x = kb^2$ (which is perpendicular to the axis of the parabola) is 8/15 of the volume of the cylinder of radius kb^2 and height b. His formal argument was a typical exhaustion argument. Namely, he assumed that the desired volume was greater than 8/15 of that of the cylinder and derived a contradiction, then assumed that it was less and derived another contradiction. But the essence of ibn al-Haytham's argument involved "slicing" the cylinder into n disks, each of thickness $h = \frac{b}{n}$, the intersection of each with the paraboloid providing an approximation to the volume of a slice of the paraboloid (Fig. 9.23). The ith disk in the paraboloid has radius $kb^2 - k(ih)^2$ and therefore has volume $\pi h(kh^2n^2 - ki^2h^2)^2 = \pi k^2h^5(n^2 - i^2)^2$. The total volume of the paraboloid is therefore approximated by

$$\pi k^2 h^5 \sum_{i=1}^{n-1} (n^2 - i^2)^2 = \pi k^2 h^5 \sum_{i=1}^{n-1} (n^4 - 2n^2 i^2 + i^4).$$

But ibn al-Haytham already knew formulas for the sums of integral squares and integral fourth powers. Using these, he could calculate that

$$\sum_{i=1}^{n-1} (n^4 - 2n^2 i^2 + i^4) = \frac{8}{15}(n-1)n^4 + \frac{1}{30}n^4 - \frac{1}{30}n = \frac{8}{15}n \cdot n^4 - \frac{1}{2}n^4 - \frac{1}{30}n$$

and therefore that

$$\frac{8}{15}(n-1)n^4 < \sum_{i=1}^{n-1} (n^2 - i^2)^2 < \frac{8}{15}n \cdot n^4.$$

FIGURE 9.23

Revolving a segment of
a parabola around a line
perpendicular to its axis

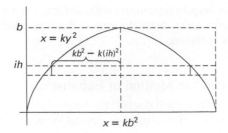

But the volume of a typical slice of the circumscribing cylinder is $\pi h (kb^2)^2 = \pi k^2 h^5 n^4$, and therefore the total volume of the cylinder is $\pi k^2 h^5 n \cdot n^4$, while the volume of the cylinder less its "top slice" is $\pi k^2 h^5 (n-1) n^4$. Therefore, the inequality shows that the volume of the paraboloid is bounded between 8/15 of the cylinder less its top slice and 8/15 of the entire cylinder. Since the top slice can be made as small as desired by taking n sufficiently large, it follows that the paraboloid is exactly 8/15 of the cylinder as asserted.

 9.6

TRIGONOMETRY

An Indian *Siddhanta* was brought to Baghdad late in the eighth century and translated into Arabic. Thus, Islamic scholars were made aware of the trigonometric knowledge of the Hindus, which had earlier been adapted from the Greek version of Hipparchus. They were also soon aware of Ptolemy's trigonometry as detailed in his *Almagest* when that work was translated into Arabic as well. As in other areas of mathematics, the Islamic mathematicians absorbed what they found from other cultures and gradually infused the subject with new ideas.

Like the situation in both Greece and India, trigonometry in Islam was intimately tied to astronomy, so in general mathematical texts on trigonometry were written as chapters of more extensive astronomical works. The mathematicians were particularly interested in using trigonometry to solve spherical triangles because Islamic law required that Moslems face the direction of Mecca when they prayed. To determine the appropriate direction at one's own location required an extensive knowledge of the solution of such triangles on the sphere of the earth. The solution of both plane and spherical triangles was also important in the determination of the correct time for prayers. These times were generally defined in relation to the onset of dawn and the end of twilight as well as the length of daylight and the altitude of the sun on a given day, notions that again required spherical trigonometry to determine accurately.

9.6.1 The Trigonometric Functions

Recall that Ptolemy used only one trigonometric "function," the chord, in his trigonometric work, while the Hindus modified that into the more convenient sine. Early in Islamic trigonometry, both the chord and the sine were used concurrently, but eventually the sine won out. (The Islamic sine of an arc, like that of the Hindus, was the length of a particular line in a circle of given radius R. We will keep to our notation of "Sine" to designate the

Islamic sine function, with a similar convention for the other functions.) It is not entirely clear who introduced the other functions, but we do know that Abū 'Abdallāh Muḥammad ibn Jābir al-Battānī (c. 855–929) used the "sine of the complement to 90°" (our cosine) in his astronomical work designed to be an improvement on the *Almagest*. Because he did not use negative numbers, he defined the cosine only for arcs up to 90°. For arcs between 90° and 180°, he used the Versine, defined as Versin $\alpha = R + \text{Sin}(\alpha - 90°)$. But because al-Battānī did not make use of the tangent, his formulas were no less clumsy than those of Ptolemy.

The tangent, cotangent, secant, and cosecant functions made their appearance in Islamic works in the ninth century, perhaps earliest in the work of Aḥmad ibn 'Abdallāh al-Marwazī Habas al-Ḥāsib (c. 770–870), although the tangent function had already been used in China in the eighth century. We consider here, however, the discussion of these functions by Abu l-Rāyḥān Muḥammad ibn Aḥmad al-Bīrūnī (973–1055) in his *Exhaustive Treatise on Shadows*. "An example of the direct shadow [cotangent] is: Let A be the body of the sun and BG the gnomon perpendicular to EG, which is parallel to the horizon plane, and ABE the sun's ray passing through the head of the gnomon BG [Fig. 9.24a]. . . . EG is that which is called the direct shadow such that its base is G and its end E. And EB, the line joining the two ends of the shadow and the gnomon, is the hypotenuse of the shadow [cosecant]."[40] The tangent and secant are defined similarly by using a gnomon parallel to the horizon plane. In Figure 9.24b, GE is called the "reversed shadow" (tangent) while BE is called the "hypotenuse of the reversed shadow" (secant).

FIGURE 9.24

Al-Bīrūnī's definition of tangent, cotangent, secant, and cosecant. (a) GE is the cotangent of angle E and EB is the cosecant. (b) GE is the tangent of angle B and BE is the secant.

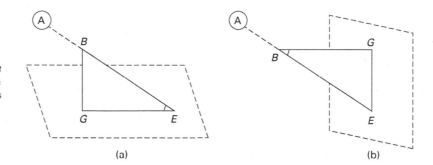

(a)   (b)

Al-Bīrūnī demonstrated the various relationships among the trigonometric functions. For example, he showed that the "ratio of the gnomon to the hypotenuse of the shadow is as the ratio of the sine of the altitude to the total sine."[41] By the "total sine," al-Bīrūnī meant the Sine of a 90° arc, namely, the radius R of the circle on which the arcs are being taken. The formula can then be translated as

$$\frac{g}{g \csc \alpha} = \frac{\text{Sin } \alpha}{R}$$

(where g is the length of the gnomon) or as

$$\csc \alpha = \frac{1}{\sin \alpha}.$$

BIOGRAPHY

Al-Bīrūnī (973–1055)

Al-Bīrūnī was born in Khwarizm, near a town now named Biruni in Uzbekistan, and began scientific studies early in life under the guidance of Abū Naṣr Manṣūr ibn ʿIrāq, a prominent astronomer from the region (Fig. 9.25). Political strife in his homeland compelled him to flee in 995, but two years later he was back in Kāth, the principal city of Khwarizm, to observe a lunar eclipse. He had previously arranged that Abu'l-Wafā' would observe the same eclipse in Baghdad, so that the time difference of the two occurrences would enable him to calculate the difference in longitude of the two sites. In 1017,

Khwarizm was conquered by Sultan Maḥmūd of Ghazna, in Afghanistan, who soon ruled an extensive empire that included parts of northern India. Al-Bīrūnī was taken to the sultan's court, from where he traveled to India and where he wrote a major work on all aspects of Indian culture, including such varied topics as the caste system, Hindu religious philosophy, the rules of chess, notions of time, and calendric procedures. Al-Bīrūnī wrote over 140 works in all, the majority of which were in mathematics, astronomy, and geography.

FIGURE 9.25

Al-Bīrūnī on a Syrian stamp

Al-Bīrūnī noted further that "if we are given the shadow at a certain time, and we want to find the altitude of the sun for that time, we multiply the shadow by its equal and the gnomon by its equal and we take [the square root] of the sum, and it will be the cosecant. Then we divide by it the product of the gnomon by the total sine, and there comes out the Sine of the altitude. We find its corresponding arc in the Sine table and there comes out the altitude of the sun at the time of that shadow."[42] In modern notation, al-Bīrūnī used the relationship

$$\sqrt{g^2 \cot^2 \alpha + g^2} = g \csc \alpha \quad (\text{or } \cot^2 \alpha + 1 = \csc^2 \alpha)$$

and then the previous formula in the form $g R / g \csc \alpha = \text{Sin}\, \alpha$ to determine the Sine function based on the particular value of the radius R used. He then consulted his Sine table in reverse to determine α. Al-Bīrūnī similarly gave rules equivalent to $\tan^2 \alpha + 1 = \sec^2 \alpha$ and $\tan \alpha = \frac{\sin \alpha}{\cos \alpha}$ and presented a table for the tangent and cotangent in which he used the relationship $\cot \alpha = \tan(90° - \alpha)$.

It is perhaps surprising that with the wealth of trigonometric knowledge collected in his text, al-Bīrūnī only used it for dealing with astronomical problems. For determining terrestrial heights and distances, he described nontrigonometric methods. For example, to determine the height of a minaret where the base is accessible, he suggested that, "if surveyed at a time when the altitude of the sun equals an eighth of a revolution [45°], there will be between the end of the shadow and the foot of the vertical a distance equal to [its height]."[43] If the base is not accessible, however, al-Bīrūnī described a procedure similar to the Chinese and Indian procedures discussed in Chapters 7 and 8. Unlike his Indian and Chinese predecessors, however, he gave a description in his text of his reasoning, using the idea of similar triangles.

9.6.2 Trigonometric Tables

With the trigonometric functions defined, it was, of course, necessary to calculate tables of these functions so they could be used to solve problems of astronomy and geography. The key to getting accurate tables was, as we have seen earlier, a method for determining the chord (or the Sine) of $1°$ or $\frac{1}{2}°$. Recall that Ptolemy was able to get accuracy to three sexagesimal places by an approximation procedure. Islamic mathematicians over the centuries worked out ways to improve this accuracy substantially. One early contributor to this effort was Abū al-Wafā'. He first calculated the Sines of $\frac{12}{15}°$, $\frac{15}{32}°$, and $\frac{18}{32}°$ by the application of the half-angle formula and the sum formula (as in Ptolemy's *Almagest*). He then used the formula $\text{Sin}(\alpha + \beta) - \text{Sin}\,\alpha < \text{Sin}\,\alpha - \text{Sin}(\alpha - \beta)$ (Fig. 9.26), which is essentially the result that successive differences of sines decrease as the arcs increase. Thus, to determine $\text{Sin}\,\frac{1}{2}°$, he used his formula twice to get

$$\text{Sin}\,\frac{18°}{32} - \text{Sin}\,\frac{17°}{32} < \text{Sin}\,\frac{17°}{32} - \text{Sin}\,\frac{16°}{32} < \text{Sin}\,\frac{16°}{32} - \text{Sin}\,\frac{15°}{32}.$$

Then $\text{Sin}\,\frac{16°}{32} - \text{Sin}\,\frac{15°}{32}$, the largest of these three differences, is greater than $1/3$ of their sum, the known value $\text{Sin}\,\frac{18°}{32} - \text{Sin}\,\frac{15°}{32}$. Therefore,

$$\text{Sin}\,\frac{1°}{2} > \text{Sin}\,\frac{15°}{32} + \frac{1}{3}\left(\text{Sin}\,\frac{18°}{32} - \text{Sin}\,\frac{15°}{32}\right).$$

FIGURE 9.26

Successive differences of sines decrease as sines increase

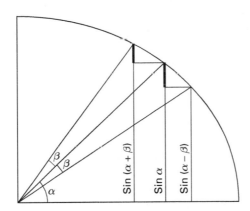

The formula also implies that

$$\text{Sin}\,\frac{16°}{32} - \text{Sin}\,\frac{15°}{32} < \text{Sin}\,\frac{15°}{32} - \text{Sin}\,\frac{14°}{32} < \text{Sin}\,\frac{14°}{32} - \text{Sin}\,\frac{13°}{32} < \text{Sin}\,\frac{13°}{32} - \text{Sin}\,\frac{12°}{32},$$

and by an argument similar to that above, Abū al-Wafā' found that

$$\text{Sin}\,\frac{1°}{2} < \text{Sin}\,\frac{15°}{32} + \frac{1}{3}\left(\text{Sin}\,\frac{15°}{32} - \text{Sin}\,\frac{12°}{32}\right).$$

He therefore had bounded $\text{Sin}\,\frac{1}{2}°$ by two values differing by only 5 units in the fourth sexagesimal place. His final value for the Sine was

$$\text{Sin}\,\frac{1°}{2} = 0;31,24,55,54,55,$$

in error only in the fifth sexagesimal place.

Abū al-Wafā's method was used by various mathematicians for over four centuries. But it was eventually superseded by a completely different method, due to al-Kāshī in the early fifteenth century. The latter started with a version of the triple-angle formula Sin $3\theta = 3$ Sin $\theta - 0;0,4(\text{Sin}^3 \theta)$, valid for his Sine table based on a circle of radius 60. Putting $\theta = 1°$ gave a cubic equation for $x = $ Sin $1°$, namely, $3x - 0;0,4x^3 = $ Sin $3°$. Al-Kāshī rewrote this in the form

$$x = \frac{900 \text{ Sin } 3° + x^3}{2700}.$$

Recall that Sin $3°$ can be calculated to whatever accuracy needed by use of the difference and half-angle formulas. Al-Kāshī in fact used as his value for Sin $3°$ the sexagesimal 3:8,24,33,59,34,28,15. His equation, written in sexagesimal notation, was therefore

$$x = \frac{47,6;8,29,53,37,3,45 + x^3}{45,0}.$$

He proceeded to solve this equation by an iterative procedure, given that he knew that the solution was a value close to 1. Writing the equation symbolically as $x = \frac{q+x^3}{p}$, and assuming that the solution is given as $x = a + b + c + \cdots$, where the various letters represent successive sexagesimal places, one begins with the first approximation $x_1 = \frac{q}{p} \approx a \, (= 1)$. To find the next approximation $x_2 = a + b$, solve for b by setting

$$x_2 = \frac{q + x_1^3}{p} \quad \text{or} \quad a + b = \frac{q + a^3}{p}.$$

Then

$$b \approx \frac{q - ap + a^3}{p} \, (= 2).$$

Similarly, if $x_3 = a + b + c$, set

$$x_3 = \frac{q + x_2^3}{p} \quad \text{or} \quad a + b + c = \frac{q + (a+b)^3}{p}$$

and find

$$c \approx \frac{q - (a+b)p + (a+b)^3}{p} \, (= 49).$$

Al-Kāshī did not justify this iterative approximation procedure, but he evidently knew it converged more rapidly than the solution procedures for cubic equations used by his predecessors. In this case, he calculated $x = 1;2,49,43,11,14,44,16,26,17$, a result equivalent to a decimal value for sin $1°$ of 0.017452406437283571, quite a feat for the days before electronic calculators and well beyond any practical need. Al-Kāshī's patron, Ulūgh Beg (Fig. 9.27), himself an astronomer who ruled a domain in central Asia from his capital of Samarkand, used this work to calculate Sine and Tangent tables for every minute of arc to five sexagesimal places, a total of 5400 entries in each table!

FIGURE 9.27

Ulūgh Beg on a stamp from Turkey

9.6.3 Spherical Trigonometry

The major goal of trigonometry in Islam, as it had been in Greece and India, was to solve astronomical problems, and these mostly required the solution of spherical triangles. Recall that the basic result used by Ptolemy to derive results in spherical trigonometry was Menelaus's theorem. Islamic mathematicians were certainly familiar with this result, but they felt that it was frequently too complicated to use, because it required finding an appropriate Menelaus configuration. Thus, they sought, and found, simpler results, which would serve as the basis for spherical trigonometry. The most important of these results, the rule of four quantities and the spherical law of sines, were discovered independently by two contemporaries of al-Bīrūnī: Abū al-Wafā' and Abū Naṣr Manṣūr ibn 'Iraq (d. 1030), one of al-Bīrūnī's teachers. We will follow the work of the former as presented in his astronomical handbook called, like Ptolemy's work, the *Almagest*.

The rule of four quantities, often called the "theorem that dispenses" (i.e., with the Menelaus configuration), is the following:

THEOREM *If ABC and ADE are two spherical triangles with right angles at B, D, respectively, and a common acute angle at A, then* Sin BC : Sin CA = Sin DE : Sin EA *[Fig. 9.28].*

FIGURE 9.28

The rule of four quantities

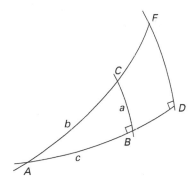

Abū al-Wafā' proved this by drawing various lines from the points in the triangles to the center Z of the sphere and determining the necessary sines in terms of the resulting plane triangles in space. He then proved, as an immediate corollary of this theorem, one of the special cases of Menelaus's theorem discussed in Chapter 5:

THEOREM *If ABC is a right spherical triangle with right angle at B, then* $\frac{\text{Sin } A}{R} = \frac{\text{Sin } a}{\text{Sin } b}$.

To prove this, extend the hypotenuse AC and the base AB to points E and D, respectively, such that both AD and AE are quadrants of a great circle. Then the great circle arc from E to D is perpendicular to both AD and AE, and we can apply the theorem. Because sin DE = sin A, our result is proved. This corollary was in essence used by Ptolemy in many of his calculations. Abu' al-Wafā' also gave proofs of other special cases of the Menelaus theorem, including the results $\frac{\cos a}{\cos b} = \frac{1}{\cos c}$ and $\frac{\sin c}{\tan a} = \frac{1}{\tan A}$. In addition, he gave a proof of the spherical law of sines:

THEOREM *In any spherical triangle ABC,* $\frac{\text{Sin } a}{\text{Sin } A} = \frac{\text{Sin } b}{\text{Sin } B} = \frac{\text{Sin } c}{\text{Sin } C}$. *[Note that since ratios are involved, we could use sines instead of Sines.]*

FIGURE 9.29

Abū al-Wafā's proof of the
sine theorem

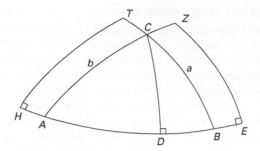

Given the spherical triangle ABC, let CD be an arc of a great circle perpendicular to AB (Fig. 9.29). Extend AB and AC to AE and AZ, both quadrants, and extend BA and BC to BH and BT, also both quadrants. Then A is a pole for the great circle EZ and B a pole for the great circle TH. Because the angles at E and H are right angles, it follows that the triangles ADC and AEZ are spherical right triangles with a common angle at A, while the triangles BDC and BHT are spherical right triangles with a common angle at B. By the rule of four quantities, we have

$$\frac{\text{Sin } DC}{\text{Sin } b} = \frac{\text{Sin } ZE}{\text{Sin } ZA} \quad \text{and} \quad \frac{\text{Sin } DC}{\text{Sin } a} = \frac{\text{Sin } TH}{\text{Sin } TB}.$$

But because A and B are poles of ZE and TH, respectively, arc ZE equals $\angle A$ and arc TH equals $\angle B$. Thus, the equations can be rewritten in the form

$$\frac{\text{Sin } DC}{\text{Sin } b} = \frac{\text{Sin } A}{R} \quad \text{and} \quad \frac{\text{Sin } DC}{\text{Sin } a} = \frac{\text{Sin } B}{R}.$$

Thus, Sin A Sin $b = $ Sin B Sin a, and the sine theorem is proved.

Using the sine theorem, al-Bīrūnī was able to show how to determine the *qibla*, the direction of Mecca relative to one's own location in which a Moslem must face during prayer. One of al-Bīrūnī's solutions to this problem, taken from his *Book of the Determination of Coordinates of Localities*, is outlined here.[44] Assume that M is the position of Mecca and that P is one's current location (Fig. 9.30). Let arc AB represent the equator and T the North Pole, and draw meridians from T through P and M, respectively. The *qibla* is then $\angle TPM$

FIGURE 9.30

The problem of the *qibla*

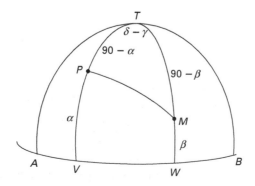

on the earth's sphere. Assuming the latitudes α, β, and the longitudes γ, δ, of P and M, respectively, are known, then arcs TP and TM are known ($90° - \alpha$, $90° - \beta$, respectively), and also $\angle PTM$ ($= \delta - \gamma$) is known. Unfortunately, the sine theorem by itself is not sufficient to solve the triangle PTM because no single angle and side opposite are known. Al-Bīrūnī, however, used the theorem repeatedly on a series of triangles.

We follow al-Bīrūnī's method, taking as an example point P to be Jerusalem (latitude 31°47′ N, longitude 35°13′ E). Mecca itself has latitude 21°45′ N, longitude 39°49′ E. Let the circle $KSQN$ represent the horizon circle of the point P (or its local zenith) as viewed from above and M the zenith of Mecca (Fig. 9.31). If S is the south point of the horizon (P is northwest of M), N the north point, and the arcs PMK and NPS are drawn, the arc NK represents the *qibla*. Let circle CFD represent the horizon circle of Mecca and circle MHJ the horizon circle of F and draw circle MTL through the north celestial pole T. The data of this problem give $TN = \alpha = 31°47′$, $TL = \beta = 21°25′$, $MT = 90° - \beta = 68°35′$, and $\angle MTH = \delta - \gamma = 4°36′$. Since MT, $\angle MTH$, and $\angle THM - 90°$ are known, the sine theorem for triangle MTH (where for simplicity, we use modern sines) shows that

$$\sin MH = \frac{\sin MT \sin \angle MTH}{\sin \angle THM} = .07466.$$

FIGURE 9.31

Al-Bīrūnī's solution of the *qibla*

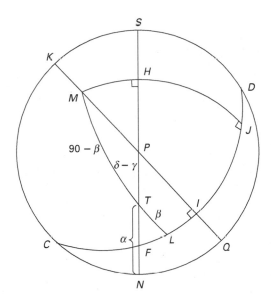

Therefore, $MH = 4°17′$ and $HJ = 90° - MH = 85°43′$. Because $\angle TFL = HJ$, and TL and $\angle TLF = 90°$ are known, the sine theorem applied to triangle TFL determines

$$\sin TF = \frac{\sin TL \sin \angle TLF}{\sin \angle TFL} = .36617,$$

so $TF = 21°29′$, and therefore $FN = \alpha - TF = 10°18′$ and $PF = 90° - FN = 79°42′$. Next, apply the rule of four quantities to triangles FPI and FHJ. Again, because PF,

$FH = 90°$, and HJ are known, $\sin PI$ is determined as

$$\sin PI = \frac{\sin PF \sin HJ}{\sin FH} = .98114,$$

so $PI = 78°51'$ and $IQ = 90° - PI = 11°9'$. But C is the pole of circle $KMPIQ$. Therefore, $\angle FCN \ (= IQ)$ is known. Finally, apply the sine theorem to triangle CFN. Again, three quantities are known, namely, $\angle FCN$, $\angle CFN \ (= \angle TFL)$, and FN, so the fourth quantity, NC, is determined. Thus,

$$\sin NC = \frac{\sin \angle CFN \sin FN}{\sin \angle FCN} = .92204,$$

$NC = 67°14'$ and the *qibla* $NK = NC + CK = 67°14' + 90° = 157°14'$.

Interestingly, as we noted earlier, al-Bīrūnī could solve problems that we would solve with trigonometry by using other methods. So, in this same text, he showed how to determine the distance between two given points on the earth's surface without trigonometry, even though this problem, like the *qibla* problem, amounts to solving a spherical triangle, given two sides and the angle between them. If the two locations are A and B, then let $NCAH$ be the meridian through A, beginning at the North Pole N and ending at H on the equator (Fig. 9.32). Similarly, let $NBDK$ be the meridian through B. Since the latitudes of A and B are assumed known (i.e., arcs HA and KB, respectively), the arcs NA and NB are also known, as is the angle at N representing the difference in the longitudes of A and B. Since C is on the meridian through A at the same latitude as B, and D is on the meridian through B at the same latitude as A, we also know that arc $NC =$ arc NB and arc $ND =$ arc NA. The arcs BC and AD, parts of latitude circles and parallel to the equator, are not, however, arcs of great circles. But we know that the length of arc BC is equal to that of arc HK multiplied by the cosine of the latitude (equal to $\sin NB$) and therefore crd $BC =$ crd $HK \cdot \sin NB$. Similarly, crd $AD =$ crd $HK \cdot \sin NA$.

Al-Bīrūnī next noted that the plane isosceles trapezoid $ACBD$ can be inscribed in a circle, and thus Ptolemy's theorem applies (see Chapter 5). Therefore, crd $AD \cdot$ crd $BC +$ crd $AC \cdot$ crd $BD =$ crd $AB \cdot$ crd DC. But the first two chords have just been calculated. Also, crd $AC =$ crd $BD =$ the chord of the difference in latitudes of the two points, and crd $AB =$ crd DC. It follows that

$$\text{crd } AB = \sqrt{\text{crd } AD \cdot \text{crd } BC + \text{crd}^2 AC}.$$

Of course, once the chord of the arc AB is known, the arc itself can be determined. And, since the radius of the earth is assumed known, the distance AB is then known as well.[45]

9.6.4   Al-Tūsī and the Systematization of Trigonometry

With all the work on spherical trigonometry, it is not surprising that eventually someone would write a treatise just dealing with trigonometry, rather than as an adjunct to astronomy. What is surprising is that the first such treatise appeared in Spain, the *Determination of the Magnitudes of Arcs on the Surface of a Sphere*, written by Abū abd Allāh Muhammad ibn Mu'ādh al-Jayyānī (989–1080), who spent most of his life in Cordoba. Al-Jayyānī began with Menelaus's theorem, although without proof, and then went on to prove the sine theorem, the

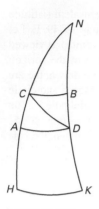

FIGURE 9.32

Distances on the earth

theorem of the four quantities, and several other results. He then used these to systematically provide methods for solving spherical triangles.

But it was in the thirteenth century, in Persia, that there appeared the most comprehensive treatise on both spherical and plane trigonometry written in the Islamic world, the *Treatise on the Complete Quadrilateral*, by Naṣir al-Dīn al-Ṭūsī. This work was unabashedly mathematical, beginning in Book I with a detailed treatment of the composition of ratios, followed in Book II by a long discussion of the various possible cases of Menelaus's theorem on what al-Ṭūsī called the plane sector figure (what we called the Menelaus configuration in Chapter 5). Book III treats plane trigonometry. Al-Ṭūsī began by solving right triangles, then solved arbitrary triangles in a way reminiscent of Ptolemy's method of drawing appropriate circles and using the chords. However, he noted that there was a simpler way to solve triangles, through the use of sines. Thus, he presented, for the first time, a statement and proof of the law of sines for plane triangles:

THEOREM   *In any plane triangle, the ratio of the sides is equal to the ratio of the sines of the angles opposite to those sides. That is, in triangle ABC, we have $AB : AC = \mathrm{Sin}(\angle ACB) : \mathrm{Sin}(\angle ABC)$. [Note that since we are considering a ratio, it is irrelevant whether we use Sines or sines.]*

We consider the first of al-Ṭūsī's two proofs, in the case where all angles are acute (Fig. 9.33). Extend CB to E and BC to H so that $CE = BH = 60$, the radius of the circle in which the Sines are calculated. Then describe circular arcs with radius CE centered on C and radius BC centered on B. Extend CA to its intersection D with the arc centered on C and drop a perpendicular DF to CE. Similarly, drop the perpendicular TK from the intersection T of the extension of BA with the arc centered on B. Then $DF = \mathrm{Sin}(\angle ACB)$ and $TK = \mathrm{Sin}(\angle ABC)$. If we now draw AL perpendicular to BC, we have, by similarity, $AB : AL = TB : TK$ and $AC : AL = DC : DF$. Since $DC = TB$, we conclude that $AB : AC = DF : TK$, the desired result.

FIGURE 9.33

The plane law of sines

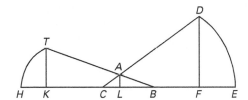

With the law of sines, al-Ṭūsī could now solve easily the triangles where two angles and a side are given or where two sides and an angle opposite one of them are given. For the cases where two sides and the included angle are given, he divided the triangle into two right triangles that he could then solve. Where three sides are given, he also dropped a perpendicular and then applied Proposition II–13 of Euclid's *Elements*, the equivalent of using the modern law of cosines. Al-Ṭūsī concluded Book III by demonstrating the important result that if the sum or difference of two arcs is given along with the ratio of their sines, then the arcs are determined.

In Book IV, al-Ṭūsī studied the properties of the spherical sector figure in great detail, applying this in the final Book V to the detailed study of spherical triangles. He presented several proofs of the spherical law of sines, giving attributions to earlier Islamic mathematicians, and, again with attributions, several proofs of the law of tangents:

THEOREM   *If ABC is a spherical triangle with angle C a right angle, then the ratio of the Sine of arc AC to the radius R equals the ratio of the Tangent of arc BC to the Tangent of angle A.*

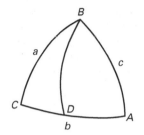

FIGURE 9.34

Solving a spherical triangle with two sides and the included angle known

In modern notation, assuming $R = 1$, we can write this as $\sin b = \tan a / \tan A$, equivalent to Equation 5.3 (but recall that Ptolemy himself did not use the tangent function, or even the sine function). Al-Ṭūsī went on to prove numerous other relationships among the sides and angles of a spherical right triangle, such as $\cos c = \cos a \cos b$ and $\cos A = \cos a \sin B$. He then systematically used his results to solve all possible cases of spherical triangles, frequently giving more than one possible approach. For example, if the right angle C as well as angle A and side c in the spherical right triangle ABC are known, then the law of sines shows that $\text{Sin } a = \text{Sin } A \text{ Sin } c/R$, so a is known. Then side b and angle B are easily found. Therefore, to solve a general spherical triangle ABC with sides b and c and angle A known, al-Ṭūsī used a procedure similar to the method in the analogous plane case (Fig. 9.34): Drop a perpendicular BD from B to side b and apply the right triangle result just quoted to triangle ABD to find BD and AD. Since we now know DC and DB in right triangle BDC, we can use the cosine relationship above to find side $BC = a$. We find the remaining two angles by using either the law of tangents or the law of sines.

As a final example, we consider al-Ṭūsī's procedure for solving a spherical triangle when the three sides are known. So let ABC be the triangle with given arcs AB, AC, and BC (Fig. 9.35). Extend AB and AC to quadrants AD and AE, respectively. Then draw the great circle through DE and extend it to meet BC extended at F. Because the angles at D and E are right, the rule of four quantities implies that $\text{Sin } CF : \text{Sin } BF = \text{Sin } CE : \text{Sin } BD$. Because $CE = 90 - CA$ and $BD = 90 - BA$ are known, the ratio of the Sine of CF to the Sine of BF is known. In addition, $BF - CF = BC$ is known. Therefore, both arcs BF and CF can be found by a result from Book III. Then arcs DF and EF can be found by using results

FIGURE 9.35

Solving a spherical triangle with all sides known

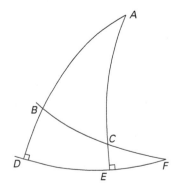

on right spherical triangles. It follows that $DE = DF - EF$ can be found and therefore that angle A (= arc DE) can be found. The remaining angles can then be found by the law of sines.[46]

9.7 TRANSMISSION OF ISLAMIC MATHEMATICS

By the fifteenth century, Islamic scientific civilization was in a state of decline. There were few other scientists of consequence in the years following. But even before the fifteenth century, mathematical activity had resumed in Europe. A central factor of this revival was the work of the translators of the twelfth century, who made available to Europeans a portion of the Islamic mathematical corpus, most importantly both the arithmetic and the algebra works of al-Khwārizmī. The work of Abū Kāmil also became available in Europe, chiefly through the inclusion of numerous problems from his work in Leonardo of Pisa's *Liber Abbaci* (1202) and the fifteenth-century translation of this work into Hebrew in Italy. As far as is known, the more advanced algebraic materials of al-Samaw'al and others did not reach Europe before or during the Renaissance.

We have already noted that both the idea and the notation for decimal fractions were present in the work of al-Kāshī. The system, including decimal fractions, also appeared around this time in a Byzantine textbook, with the method described as "Turkish," that is, Islamic. This textbook was brought to Venice in 1562, but even before that the same notation appeared occasionally in European works. So, although traditionally the complete decimal system in Europe is ascribed to Simon Stevin in the late sixteenth century, it does appear that at least some aspects of it traveled to Europe from the Islamic world before that time.

In combinatorics, there is no known Renaissance translation into a European language of the work of ibn Mun'im or ibn al-Bannā. On the other hand, as we will see in the next chapter, ideas very closely related to theirs were developed in southern France in the fourteenth century by Levi ben Gerson, who in all probability was aware of Islamic advances in this area. However, we do not know whether the work of Levi had any influence on combinatorics in Europe later on. The Pascal triangle itself first appeared in Europe in the thirteenth century. Unfortunately, we can only speculate as to whether this idea traveled to Europe from the Islamic world (or from China) or was discovered there independently.

The only manuscript that we know of containing ibn al-Haytham's work on the volume of a paraboloid of revolution was acquired by the library of the India Office in England in the nineteenth century. Thus, although results similar to ibn al-Haytham's on the sum of integral powers began to appear in Europe in the seventeenth century, we have no way of knowing whether anyone in Europe was aware, either directly or indirectly, of that particular treatise of the Egyptian mathematician. Certainly, however, Europeans were aware of ibn al-Haytham's major work on optics, a work that was translated into Latin early and had major influence on European work on that subject.

But in the study of the parallel postulate, we do know that an important Islamic work on this subject, the 1298 work of Ṣadr al-Dīn al-Ṭūsī, was in fact published in Rome in 1594 in Arabic, but with a Latin title page. As far as we know, it was never formally translated into Latin. However, John Wallis in England was certainly aware of its contents and wrote about the ideas there in developing his own ideas on the parallel postulate. Gerolamo Saccheri also knew of this work and used some of its ideas in his own work, which ultimately led to the development of non-Euclidean geometry in the nineteenth century.

Similarly, although Naṣir al-Dīn al-Ṭūsī's trigonometry work did not reach Europe during the Renaissance, we know that some of the earlier Islamic work on spherical trigonometry did. In particular, the sine theorem and the theorem of the four quantities, along with some of their corollaries, appeared in Spain, in the work of Abū Muḥammad Jābir ibn Aflaḥ al-Ishbīlī (early twelfth century). This work was translated into Latin late in the twelfth century and provided Europeans with one of the earliest versions of the Islamic advances on the trigonometry of Ptolemy. In fact, Regiomontanus, the first European author to write a work on pure trigonometry, clearly took much of his material on spherical trigonometry directly from the book of Jābir.

It will take considerably more research to determine, then, how much of medieval and Renaissance European mathematics was influenced, either directly or indirectly, by the mathematics of Islam.

EXERCISES

1. Multiply 8023 by 4638 using the method of al-Uqlīdisī.

2. Al-Khwārizmī gives the following rule for his sixth case, $bx + c = x^2$: Halve the number of roots. Multiply this by itself. Add this square to the number. Extract the square root. Add this to the half of the number of roots. That is the solution. Translate this rule into a formula. Give a geometric argument for its validity using Figure 9.36, where $x = AB$, $b = HC$, c is represented by rectangle $ABRH$, and G is the midpoint of HC.

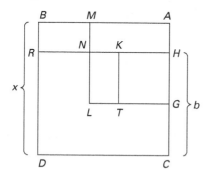

FIGURE 9.36

Al=Khwārizmī's justification for the solution rule for $bx + c = x^2$

3. Solve the following problems due to al-Khwārizmī:

   a. $x^2 + (10 - x)^2 = 58$

   b. I have divided 10 into two parts, and have divided the first by the second, and the second by the first and the sum of the quotients is 2 1/6. Find the parts.

4. Solve $\frac{1}{2}x^2 + 5x = 28$ by multiplying first by 2 and then using al-Khwārizmī's procedure. Similarly, solve $2x^2 + 10x = 48$ by first dividing by 2.

5. Prove that al-Khwārizmī's procedure for solving equations of the form $x^2 + c = bx$ is correct using Euclid's *Elements*, II–5.

6. Solve the following problems of Abū Kāmil:

   a. Suppose 10 is divided into two parts and the product of one part by itself equals the product of the other part by the square root of 10. Find the parts.

   b. Suppose 10 is divided into two parts, each one of which is divided by the other, and the sum of the quotients equals the square root of 5. Find the parts. (Abū Kāmil solves this in two ways, once directly for x, and a second time by first setting $y = \frac{10-x}{x}$.)

7. Solve the following problems of Abū Kāmil:

   a. $[x - (2\sqrt{x} + 10)]^2 = 8x$ (First substitute $x = y^2$.)

   b. $(x + \sqrt{\frac{1}{2}x})^2 = 4x$ (Abū Kāmil does this three different ways; he first solves directly for x, next substitutes $x = y^2$, and finally substitutes $x = 2y^2$.)

8. Complete the solution of Abū Kāmil's problem in three variables given in the text by now beginning with the assumption that $z = 1$.

9. Solve the following problem in three variables due to Abū Kāmil: $x < y < z$, $x^2 + y^2 = z^2$, $xz = y^2$, $xy = 10$. (Begin by setting $y = \frac{10}{x}$, $z = \frac{100}{x^3}$, and substituting in the first equation.)

10. Complete al-Samaw'al's procedure of dividing $20x^2 + 30x$ by $6x^2 + 12$ to get the result stated in the text. Prove that the coefficients of the quotient satisfy the rule $a_{n+2} = -2a_n$ where a_n is the coefficient of $\frac{1}{x^n}$.

11. Give a complete inductive proof of the result

$$\sum_{i=1}^{n} i^3 - \left(\sum_{i=1}^{n} i\right)^2$$

and compare with al-Karaji's proof.

12. Use ibn al-Haytham's procedure to derive the formula for the sum of the fifth powers of the integers:

$$1^5 + 2^5 + \cdots + n^5 = \frac{1}{6}n^6 + \frac{1}{2}n^5 + \frac{5}{12}n^4 - \frac{1}{12}n^2.$$

13. Give a formal proof of Equation 9.2 by induction on n.

14. Show, using the formulas for sums of fourth powers and squares, that

$$\sum_{i-1}^{n-1}(n^4 - 2n^2i^2 + i^4) = \frac{8}{15}(n-1)n^4 + \frac{1}{30}n^4 - \frac{1}{30}n$$

$$= \frac{8}{15} \cdot n \cdot n^4 - \frac{1}{2}n^4 - \frac{1}{30}n.$$

15. Using Figure 9.11, show that if $AE : GH = EH : HB$ and if IG is tangent to the circle at G, then $EG + GH = EI$.

16. Show that one can solve $x^3 + d = cx$ by intersecting the hyperbola $y^2 - x^2 + \frac{d}{c}x = 0$ with the parabola $x^2 = \sqrt{c}y$. Sketch the two conics. Find sets of values for c and d for which these conics do not intersect, intersect once, and intersect twice.

17. Show that $x^3 + cx = bx^2 + d$ is the only one of al-Khayyāmī's cubics that could have three positive solutions. Under what conditions do these three positive solutions exist? How many positive solutions does the equation $x^3 + 200x = 20x^2 + 2000$ have? (The solution of this equation enabled al-Khayyāmī to solve his quadrant problem.)

18. Show that one can solve $x^3 + d = bx^2$ by intersecting the hyperbola $xy = d$ and the parabola $y^2 + dx - db = 0$. Assuming that $\sqrt[3]{d} < b$, determine the conditions on b and d that give zero, one, or two intersections of these two conics.

Compare your answer with Sharaf al-Dīn al-Tūsī's analysis of the same problem.

19. Show using calculus that $x_0 = \frac{2b}{3}$ does maximize the function $x^2(b - x)$. Then use calculus to analyze the graph of $y = x^3 - bx^2 + d$ and confirm Sharaf al-Dīn's conclusion on the number of positive solutions to $x^3 + d = bx^2$.

20. Show, as did Sharaf al-Dīn al-Tūsī, that if x_2 is the larger positive root to the cubic equation $x^3 + d = bx^2$, and if Y is the positive solution to the equation $x^2 + (b - x_2)x = x_2(b - x_2)$, then $x_1 = Y + b - x_2$ is the smaller positive root of the original cubic.

21. Analyze the possibilities of positive solutions to $x^3 + d = cx$ by first showing that the maximum of the function $x(c - x^2)$ occurs at $x_0 - \sqrt{\frac{c}{3}}$. Use calculus to consider the graph of $y = x^3 - cx + d$ and determine the conditions on the coefficients giving it zero, one, or two positive solutions.

22. Show that 17,296 and 18,416 are amicable by using ibn Qurra's theorem.

23. Show that 1184 and 1210 are amicable numbers that are not a consequence of the theorem of Thābit ibn Qurra.

24. Find a pair of amicable numbers different from those in the text. (*Hint*: Try the case $n = 7$ of ibn Qurra's theorem.)

25. Demonstrate the following equalities, typical examples of material on irrationals occurring in works of Islamic commentators on *Elements* X:

   a. $\sqrt{\sqrt{8} \pm \sqrt{6}} = \sqrt[4]{4\frac{1}{2}} \pm \sqrt[4]{\frac{1}{2}}$.

   b. $\sqrt[4]{12} \pm \sqrt[4]{3} = \sqrt{\sqrt{27} \pm (\sqrt{24})} = \sqrt[4]{51 \pm \sqrt{2592}}$.

26. Abū Sahl al-Kūhī knew from his own work on centers of gravity and the work of his predecessors that the center of gravity divides the axis of certain plane and solid figures in the following ratios:

Triangle: $\frac{1}{3}$	Tetrahedron: $\frac{1}{4}$
Segment of a parabola: $\frac{2}{5}$	Paraboloid of revolution: $\frac{2}{6}$
	Hemisphere: $\frac{3}{8}$

   Noting the pattern, he guessed that the corresponding value for a semicircle was 3/7. Show that al-Kūhī's first five results are correct, but that his guess for the semicircle implies that $\pi = 3$ 1/9. (Al-Kūhī realized that this value contradicted Archimedes' bounds of 3 10/71 and 3 1/7, but concluded that there was an error in the transmission of Archimedes' work.)

27. Calculate the first four sexagesimal places of the approximation to $x = \text{Sin } 1°$ following the method indicated in the

text. Your calculation should show why the iteration method works.

28. In the tenth century, the mathematician 'Abd al-'Aziz al-Qabīsī described a trigonometric method, using only the sine, for determining the height and distance of an inaccessible object. One sights the summit A from two locations C, D, and determines, using an astrolabe (an angle-measuring instrument usually used for astronomical purposes), the angles $\alpha_1 = \angle ACB$ and $\alpha_2 = \angle ADB$ (Fig. 9.37). If $CD = d$, then the height $y = AB$ and the distance $x = BC$ are given by

$$y = \frac{d \sin \alpha_2}{\sin(90 - \alpha_2) - \frac{\sin(90 - \alpha_1) \sin \alpha_2}{\sin \alpha_1}},$$

$$x = \frac{y \sin(90 - \alpha_1)}{\sin \alpha_1}.$$

Prove that al-Qabīsī's formula is correct.

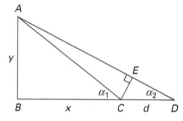

FIGURE 9.37

Al-Qabī's method for determining height and distance by way of two angle determinations

29. Use al-Bīrūnī's procedure to determine the *qibla* for Rome (latitude 41°53′ N, longitude 12°30′ E).

30. Show that the radius r_α of a latitude circle on the earth at $\alpha°$ is given by $r_\alpha = R \cos \alpha$, where R is the radius of the earth.

31. The latitudes of Philadelphia and Ankara, Turkey, are the same (40°), with the first at longitude 75° W and the second at longitude 33° E. Calculate the distance between Philadelphia and Ankara along the latitude circle, by first calculating the radius of that circle, using 25,000 miles for the circumference of the earth. Then calculate the distance along a great circle, by noting that the chord connecting the two cities can be thought of as a chord of that circle as well as a chord of the latitude circle. (*Hint:* You will have to convert the chords to the appropriate sines to make this calculation.)

32. Show directly, without the use of Ptolemy's theorem, that in an isosceles trapezoid, the square on a diagonal is equal

to the sum of the product of the two parallel sides plus the square on one of the other sides.

33. Use al-Bīrūnī's nontrigonometric procedure for calculating distances on the earth to find the great circle distance between New York (latitude 41° N, longitude 74° W) and London (latitude 52° N, longitude 0°). Assume that the circumference of the earth is 25,000 miles.

34. Al-Battānī developed a formula equivalent to what is today called the spherical law of cosines:

$$\cos a = \cos b \cos c + \sin b \sin c \cos A.$$

Use this formula to determine the *qibla* for Rome. (Al-Battānī did not himself do this.)

35. Use the spherical law of cosines (previous exercise) to determine the great circle distance between New York and London (whose coordinates are given in Exercise 33). Again, assume that the circumference of the earth is 25,000 miles.

36. Al-Bīrūnī devised a method for determining the radius r of the earth by sighting the horizon from the top of a mountain of known height h. That is, al-Bīrūnī assumed that one could measure α, the angle of depression from the horizontal at which one sights the apparent horizon (Fig. 9.38). Show that r is determined by the formula

$$r = \frac{h \cos \alpha}{1 - \cos \alpha}.$$

Al-Bīrūnī performed this measurement in a particular case, determining that $\alpha = 0°34′$ as measured from the summit of a mountain of height 652;3,18 cubits. Calculate the radius of the earth in cubits. Assuming that a cubit equals 18″, convert your answer to miles and compare to a modern value. Comment on the efficacy of al-Bīrūnī's procedure.

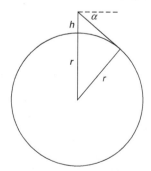

FIGURE 9.38

Al-Bīrūnī's method for calculating the earth's radius

37. Show how to determine arcs α and β if $\alpha + \beta = \gamma$ is given as well as $\sin\alpha / \sin\beta = r$.

38. Use al-Ṭūsī's method to solve the spherical triangle with known sides of 40° and 50° and with the angle between those sides equal to 25°.

39. Use al-Ṭūsī's method to solve a spherical triangle with sides 60°, 75°, and 31°.

40. Al-Ṭūsī demonstrated a method to solve a spherical triangle if all three angles are known. Suppose the three angles of triangle ABC are given (Fig. 9.39), where we assume that all three sides of the triangle are less than a quadrant. We extend each side of the triangle two different ways to form a quadrant. That is, we extend AB to AD and BH; AC to AE and CG; and BC to BK and FC, where all of the six new arcs are quadrants. We then draw great circle arcs through D and E, F and G, and H and K to form the new spherical triangle LMN. Now the vertices of the original triangle are the poles of the three sides of the new triangle. Then, for example, $MD = EN = 90° - DE = 90° - A$, or $MN = 180° - A$. Thus, the three sides of triangle LMN are known, and therefore the triangle can be solved by the procedure sketched in the text. But we also know that the vertices of triangle LMN are the poles of the original triangle. So, for example, $BF = CK = 90° - BC$, and $L = FK = 180° - BC$. We therefore can determine the sides of the original triangle. Use this procedure to solve the triangle ABC, where $A = 75°$, $B = 80°$, and $C = 85°$.

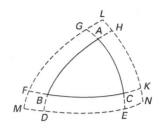

FIGURE 9.39

Solving a spherical triangle when the three angles are given

41. Why did it take many centuries after its introduction for the decimal place value system to become the system of numeration universally used in the Islamic world?

42. Outline a lesson teaching the quadratic formula using geometric arguments in the style of al-Khwārizmī.

43. Compare and contrast the geometric proofs of the quadratic formulas of al-Khwārizmī and Thābit ibn Qurra. Which method would be easier to explain?

44. Design a lesson deriving the multiplicative formula for C_k^n based on the work of ibn al-Bannā.

45. Design a lesson for a trigonometry class showing the application of the rules for solving spherical triangles to various interesting problems.

46. Given ibn al-Haytham's "integration" to determine the volume of a paraboloid of revolution and his general rule for determining the sums of kth powers of integers, why did Islamic mathematicians not discover that the area under the curve $y = x^n$ was $\frac{x^n}{n+1}$ for an arbitrary positive integer n? What needed to happen in Islamic civilization for Islamic mathematicians to discover calculus?

REFERENCES AND NOTES

The best general work on Islamic mathematics is still Adolf P. Youschkevitch, *Les Mathématiques Arabes (VIIIe–XVe Siècles)* (Paris: J. Vrin, 1976). This is one part of a more comprehensive work on medieval mathematics (in Russian) and was translated into French by M. Cazenave and K. Jaouiche. Another excellent work in English that treats various important ideas of Islamic mathematics is J. Lennart Berggren, *Episodes in the Mathematics of Medieval Islam* (New York: Springer Verlag, 1986). This work, although not a history of Islamic mathematics as such, treats certain important mathematical ideas considered by Islamic mathematicians at a level accessible to uni-

versity mathematics students. A collection of papers by Roshdi Rashed has been recently edited and translated into English as *The Development of Arabic Mathematics: Between Arithmetic and Algebra* (Dordrecht: Kluwer, 1994). This collection presents much of Rashed's recent research and provides a new look at the arithmetical and algebraic aspects of Islamic mathematics. Rashed has also written three major articles for the *Encyclopedia of the History of Arabic Science* (London: Routledge, 1996), which together provide an excellent summary of the history of much of Islamic mathematics. These are "Algebra" (Vol. 2, pp. 349–375), "Combinatorial Analysis, Numerical Analysis, Diophantine Analysis and Number Theory" (Vol. 2, pp. 376–417), and "Infinitesimal Determinations, Quadrature of Lunules and Isoperimetric Problems" (Vol. 2, pp. 418–446). These articles are complemented by Marie-Thérèse Debarnot's article in the same *Encyclopedia*, "Trigonometry" (Vol. 2, pp. 495–538), and additional articles on other topics. A general survey of Islamic work in the sciences, which includes material on mathematics, is Edward S. Kennedy, "The Arabic Heritage in the Exact Sciences," *Al-Abhath* 23 (1970), 327–344. This article, and many other useful articles on Islamic science, can also be found in D. A. King and M. H. Kennedy, eds., *Studies in the Islamic Exact Sciences* (Beirut: American University of Beirut Press, 1983).

Many of the major works of Islamic mathematics have been translated into English or French. Al-Uqlīdisī's arithmetic has been translated as A. S. Saidan, *The Arithmetic of Al-Uqlīdisī* (Boston: Reidel, 1978). Al-Khwārizmī's *Algebra* was edited and translated by Frederic Rosen as *The Algebra of Muhammed ben Musa* (London: Oriental Translation Fund, 1831) (reprinted Hildesheim: Olms, 1986). There have also been translations of this work into English from Latin: Louis Karpinski, *Robert of Chester's Latin Translation of the Algebra of al-Khowarizmi* (Ann Arbor: University of Michigan Press, 1930) and Barnabas Hughes, "Gerard of Cremona's Translation of al-Khwarizmi's *Al-Jabr*: A Critical Edition," *Mediaeval Studies* 48 (1986), 211–263. Abū Kāmil's *Algebra* is available in Martin Levey, *The Algebra of Abū Kāmil, Kitāb fī'l-muqābala, in a Commentary by Mordecai Finzi* (Madison: University of Wisconsin Press, 1966). This is an edition and English translation of the fifteenth-century Hebrew translation of Abū Kāmil's work, and provides a detailed discussion of the relation of his algebra to previous Greek and Islamic work. For al-Khayyāmī, see Daoud S. Kasir, *The Algebra of Omar Khayyam* (New York: Columbia Teachers College, 1931). This English translation of al-Khayyāmī's work on algebra includes a detailed discussion of his contributions to the subject. Al-Khayyāmī's algebra is also available in R. Rashed and B. Vahabzadeh, *Omar Khayyam, The Mathematician* (New York: Bibliotheca Persica Press, 2000). This edition also contains translations of his *Treatise on the Division of a Quadrant of a Circle* and his *Commentary on the Difficulties of Certain Postulates of Euclid's Work*. Sharaf al-Dīn's work containing his contribution to the solution of cubic equations has been translated into French by Roshdi Rashed as *Sharaf al-Dīn al-Ṭūsī, oeuvres mathématiques: Algèbre et Géométrie au XIIᵉ Siècle* (Paris: Société d'Édition Les Belles Lettres, 1986). Rashed gives not only an edited Arabic text and a translation, but also an extensive commentary on the mathematics involved, relating it in particular to various modern concepts. Ibn Mun'im's work is available in Ahmed Djebbar, *L'Analyse Combinatoire au Maghreb: L'Exemple d'Ibn Mun'im*

(XIIᵉ–XIIIᵉ s., (Orsay: Université de Paris Sud, Publications Mathématiques D'Orsay, 1985). Al-Bīrūnī's trigonometry is in E. S. Kennedy, *The Exhaustive Treatise on Shadows by Abū al-Rayhān al-Bīrūnī* (Aleppo: University of Aleppo, 1976). The French translation of the trigonometry text of al-Ṭūsī has recently been reprinted: *Traité du Quadrilatère Attribute à Nassiruddin-el-Toussy*, Alexandre Pacha Caratheodory, trans. (Constantinople, 1894), now available as Volume 47 of the series *Islamic Mathematics and Astronomy* (Frankfurt am Main, Institute for the History of Arabic-Islamic Science, 1998). Excerpts from some of these translations and numerous other translations of sections of Islamic works are available in Victor J. Katz, ed., *The Mathematics of Egypt, Mesopotamia, China, India, and Islam: A Sourcebook* (Princeton: Princeton University Press, 2007).

Finally, a summary and bibliography of recent work on Islamic mathematics is J. Lennart Berggren, "Mathematics and Her Sisters in Medieval Islam: A Selective Review of Work Done from 1985 to 1995," *Historia Mathematica* 24 (1997), 407–440.

1. Jens Høyrup, "The Formation of 'Islamic Mathematics': Sources and Conditions," *Science in Context* 1 (1987), 281–329, pp. 306–307.

2. Ibid. This article contains a detailed development of these ideas. The author contends that there was an Islamic "miracle," comparable to the Greek one, involving the integration of mathematical theory and practice, which was also crucial for the creation of modern science.

3. Saidan, *Al-Uqlīdisī*, p. 35.

4. Ibid., p. 110. Katz, *Sourcebook*, p. 532.

5. Al-Samaw'al is quoted in Roshdi Rashed, *Arabic Mathematics*, p. 116.

6. Ibid., p. 123.

7. Rosen, *Algebra of Muhammed ben Musa*, p. 3. Katz, *Sourcebook*, p. 542.

8. Rosen, p. 5; Katz, p. 543.

9. Rosen, p. 8; Katz, p. 543.

10. Karpinski, *Robert of Chester's Latin Translation*, p. 69.

11. Rosen, pp. 33–34.

12. Ibid., p. 12. Katz, *Sourcebook*, p. 544.

13. A. Sayili, *Logical Necessities in Mixed Equations by 'Abd al-Hamīd ibn Turk and the Algebra of his Time* (Ankara: Türk Tarih Kurumu Basimevi, 1962), p. 166.

14. Rosen, p. 22. Katz, *Sourcebook*, p. 546.

15. Rosen, p. 43.

16. Ibid., p. 51.

17. Ibid., p. 63.

18. Levey, *Abū Kāmil*, p. 32.

19. Ibid., p. 144.

20. Ibid., p. 156.

21. Ibid., p. 186.

22. Franz Woepcke, *Extrait du Fakhrī, traité d'algèbre par Aboù Bekr Mohammed ben Alhaçan al-Karkhī* (Paris: L'imprimerie Impériale, 1853), p. 45. (This translation is available in a 1982 reprint by Georg Olms Verlag, Hildesheim.) Note that al-Karaji's name is here given as al-Karkhī. It is not known which transliteration is correct, since available Arabic manuscripts give both versions.

23. Adel Anbouba, "Al Samaw'al," in the *Dictionary of Scientific Biography* (New York: Scribners, 1970–1980), vol. XII, p. 92.

24. Berggren, *Mathematics of Medieval Islam*, p. 114.

25. See Katz, *Sourcebook*, pp. 587–595 for a translation of parts of ibn al-Haytham's work on the volume of a paraboloid.

26. Rashed, *Arabic Mathematics*, p. 65. Also see Katz, *Sourcebook*, pp. 552–554.

27. Ibid. See also J. Lennart Berggren, "Proof, Pedagogy, and the Practice of Mathematics in Medieval Islam," *Interchange* 21 (1990), 36–48, and M. Yadegari, "The Binomial Theorem: A Widespread Concept in Medieval Islamic Mathematics," *Historia Mathematica* 7 (1980), 401–406.

28. Kasir, *Omar Khayyam*, p. 44. For more on Omar Khayyam, see D. J. Struik, "Omar Khayyam, Mathematician," *Mathematics Teacher* 51 (1958), 280–285, and B. Lumpkin, "A Mathematics Club Project from Omar Khayyam," *Mathematics Teacher* 71 (1978), 740-744.

29. Ibid., p. 49.

30. For more details on Sharaf al-Dīn al-Ṭūsī's work, see J. Lennart Berggren, "Innovation and Tradition in Sharaf al-Dīn al-Ṭūsī's *al Mu'ādalāt*," *Journal of the American Oriental Society* 110 (1990), 304–309; Jan P. Hogendijk, "Sharaf al-Dīn al-Ṭūsī on the Number of Positive Roots of Cubic Equations," *Historia Mathematica* 16 (1989), 69–85; and Roshdi Rashed, *Arabic Mathematics*, chapter 3.

31. Katz, *Sourcebook*, p. 662. The next two quotations from ibn Mun'im are also here.

32. Ibid., p. 664.

33. Djebbar, *Ibn Mun'im*, pp. 55–56.

34. Rashed, *Arabic Mathematics*, p. 300.

35. Rosen, pp. 71–72.

36. Katz, *Sourcebook*, p. 612. The next two quotations from Abū al-Wafā' are also here.

37. See Katz, *Sourcebook*, pp. 580–586 for details of these constructions.

38. For more information on non-Euclidean geometry in the Islamic world, see chapter 2 of B. A. Rosenfeld, *A History of Non-Euclidean Geometry: Evolution of the Concept of a Geometric Space*, translated by Abe Shenitzer (New York: Springer Verlag, 1988), and chapter 3 of Jeremy Gray, *Ideas of Space: Euclidean, Non-Euclidean, and Relativistic*, second edition (Oxford: Clarendon Press, 1989). See also D. E. Smith, "Euclid, Omar Khayyam, and Saccheri," *Scripta Mathematica* 3 (1935), 5–10.

39. Ibn al-Baghdādī is quoted in Galina Matvievskaya, "The Theory of Quadratic Irrationals in Medieval Oriental Mathematics," in D. A. King and G. Saliba, eds., *From Deferent to Equant: A Volume of Studies in the History of Science in the Ancient and Medieval Near East in Honor of E. S. Kennedy* (New York: New York Academy of Sciences, 1987), 253–277, p. 267.

40. Kennedy, *Treatise on Shadows*, p. 64. See also E. S. Kennedy, "An Overview of the History of Trigonometry," in *Historical Topics for the Mathematics Classroom* (Reston, VA: National Council of Teachers of Mathematics, 1989), 333 359.

41. Ibid., p. 89.

42. Ibid., p. 90.

43. Ibid., p. 255.

44. See Berggren, *Mathematics of Medieval Islam*, section 6.8, or Katz, *Sourcebook*, pp. 634–637, for more details.

45. See Katz, *Sourcebook*, pp. 632–634 for more details.

46. See Katz, *Sourcebook*, pp. 637–657 for a translation of large sections of the *Treatise on the Complete Quadrilateral*.

Mathematics in Medieval Europe

Who wishes correctly to learn the ways to measure surfaces and to divide them, must necessarily thoroughly understand the general theorems of geometry and arithmetic, on which the teaching of measurement . . . rests. If he has completely mastered these ideas, he . . . can never deviate from the truth.

—Introduction to the *Liber embadorum*, Plato of Tivoli's Latin translation of the Hebrew *Treatise on Mensuration and Calculation* by Abraham bar Ḥiyya, 1116[1]

Coming to Pisa in 1225 on orders of the Holy Roman Emperor Frederick II (1194–1250), Leonardo found that his king was interested in mathematics: "After being brought to Pisa by Master Dominick to the feet of your celestial majesty, most glorious prince, I met Master John of Palermo; he proposed to me a question that had occurred to him, pertaining not less to geometry than to arithmetic . . . When I heard recently from a report from Pisa and another from the Imperial Court that your sublime majesty deigned to read the book I composed on numbers [the *Liber Abbaci*] and that it pleased you to listen to several subtleties touching on geometry and numbers, I recalled the question proposed to me at your court by your philosopher. I took upon myself the subject matter and began to compose in your honor this work, which I wish to call *The Book of Squares*. I have come to request indulgence if in any place it contains something more or less than right or necessary, for to remember everything and be mistaken in nothing is divine rather than human; and no one is exempt from fault nor is everywhere circumspect."[2]

10.1

INTRODUCTION TO THE MATHEMATICS
OF MEDIEVAL EUROPE

The Roman Empire in the West collapsed in 476 under the onslaught of various "barbarian" tribes. Feudal societies were soon organized in parts of the old empire, and the long process of the development of the European nation-states began. For the next five centuries, however, the general level of culture in Europe was very low. Serfs worked the land and few of the barons could read or write, let alone understand mathematics. In fact, there was little practical need for the subject, because the feudal estates were relatively self-sufficient and trade was almost nonexistent, especially after the Moslem conquest of the Mediterranean sea routes.

Despite the lack of mathematical activity, the early Middle Ages had inherited from antiquity the notion that the **quadrivium**—arithmetic, geometry, music, and astronomy—was required study for an educated man, even in the evolving Roman Catholic culture. Thus, St. Augustine (354–430) had written in his *City of God* that "we must not despise the science of numbers, which, in many passages of Holy Scripture, is found to be of eminent service to the careful interpreter. Neither has it been without reason numbered among God's praises: 'Thou hast ordered all things in number, and measure, and weight.'"[3] Yet the only texts available for the study of these subjects were brief introductions, especially those by the Roman scholar Boethius (480–524) and the seventh-century bishop, Isidore of Seville (560–636). Thus, the outline of the mathematical quadrivium was in place, but it was only a shell, nearly devoid of substance.

Virtually the only schools in existence were connected with the monasteries, many of which were founded by monks from Ireland, the first country not originally part of the Roman Empire to adopt Christianity. While much of continental Europe was in turbulence, these monks copied Greek and Latin manuscripts and thus preserved much ancient learning. Students from all over Europe came to study there. Then, from the sixth to the eighth centuries, missionaries went out from Ireland to the continent to establish new centers of learning from which, several centuries later, new intellectual developments eventually sprung forth.

Even in the earliest part of the Middle Ages, however, there was a significant mathematical problem to be considered: the determination of the calendar. In particular, the Church debated whether Easter should be determined using the Roman solar calendar or the Jewish lunar calendar. The two reckonings could be reconciled, but only by those with some mathematical knowledge. Thus, Charlemagne, even before his coronation in 800 as Holy Roman Emperor, formally recommended that the mathematics necessary for Easter computations be part of the curriculum in Church schools.

To help him in establishing more schools, Charlemagne brought in Alcuin of York (735–804) as his educational adviser. Alcuin, who had studied with an Irish teacher and was assisted in Charlemagne's court by several Irish clerics, generally sent to England and Ireland when he needed books. We do not have much direct information about Alcuin's knowledge of mathematics, but a collection of fifty-three arithmetical problems from his time, entitled *Propositiones ad acuendos juvenes* (*Propositions for Sharpening Youths*), is generally attributed to him. The problems of the collection often require some ingenuity for solving, but do not depend on any particular mathematical theory or rules of procedure.

FIGURE 10.1

Gerbert d'Aurillac, Pope
Sylvester II

In the tenth century, a revival of interest in mathematics began with the work of Gerbert d'Aurillac (945–1003), who became Pope Sylvester II in 999 (Fig. 10.1). In his youth, Gerbert studied in Spain, where he probably learned some of the mathematics of the Moslems. Later, under the patronage of Otto II, the Holy Roman Emperor, Gerbert reorganized the cathedral school at Rheims and successfully reintroduced the study of mathematics. Besides teaching basic arithmetic and geometry, Gerbert dealt with the mensuration rules of the Roman surveyors and the basics of astronomy. He also taught the use of a counting board, divided into columns representing the (positive) powers of 10, in each of which he would place a single counter marked with the western Arabic form of one of the numbers 1, 2, 3, . . . , 9. Zero was represented by an empty column. Gerbert's work represents the first appearance in the Christian West of the Hindu-Arabic numerals, although the absence of the zero and the lack of suitable algorithms for calculating with these counters showed that Gerbert did not understand the full significance of the Hindu-Arabic system.

Despite the limited mathematical sources available to Europeans at the turn of the millennium, scholars did know that there was an ancient tradition in mathematics due to the Greeks, but it was virtually inaccessible to them at the time. This heritage, as well as a portion of the mathematics developed in the Islamic world, was only brought into western Europe through the work of translators. European scholars discovered the major Greek scientific works (primarily in Arabic translation) beginning in the twelfth century and started the process of translating these into Latin. Much of this work was accomplished at Toledo in Spain, which at the time had only recently been retaken by the Christians from the former Moslem rulers. Here could be found repositories of Islamic scientific manuscripts as well as people straddling the two cultures. In particular, there was a flourishing Jewish community, many of whose members were fluent in Arabic. The translations then were often made in two stages, first by a Spanish Jew from Arabic into Spanish, and then by a Christian scholar from Spanish into Latin. The list of the translations of major mathematical works (with their dates) is fairly extensive (Sidebar 10.1).

Among the earliest of the translating teams were John of Seville and Domingo Gundisalvo, who were active in the first half of the twelfth century. John was born a Jew, his original name probably being Solomon ben David, but converted to Christianity, while Gundisalvo was a philosopher and Christian theologian. The most important of their mathematical translations was of an elaboration of al-Khwārizmī's work on arithmetic. They also translated a large number of astronomical works, including commentaries on the work of Ptolemy, and numerous medical and philosophical works.

A contemporary of John of Seville was Adelard of Bath (1075–1164), who was born in Bath and spent much of his early years traveling in France, southern Italy, Sicily, and the Near East, the latter two places in particular having many Arabic treatises available. Adelard was responsible for the first translation from the Arabic of Euclid's *Elements*. He also translated the astronomical tables of al-Khwārizmī in 1126. This translation contains the first sine tables available in Latin as well as the first tangent tables, the latter having been added to al-Khwārizmī's work by an eleventh-century editor. Another Englishman, Robert of Chester, who lived in Spain for several years, translated the *Algebra* of al-Khwārizmī in 1145, thus introducing to Europe the algebraic algorithms for solving quadratic equations. Interestingly enough, in the same year, Plato of Tivoli translated from the Hebrew the *Liber*

SIDEBAR 10.1 *Translators and Their Translations*

James of Venice (fl. 1128–1136)

> *Topics, Prior Analytics, Posterior Analytics* of Aristotle

Adelard of Bath (fl. 1116–1142)

> *Astronomical Tables* of al-Khwārizmī
> *Elements* of Euclid
> *Liber ysagogarum Alchorismi*, the arithmetical work of al-Khwārizmī

John of Seville and Domingo Gundisalvo (fl. 1135–1153)

> *Liber alghoarismi de practica arismetrice*, an elaboration of al-Khwārizmī's *Arithmetic*

Plato of Tivoli (fl. 1134–1145)

> *Spherica* of Theodosius (c. 100 BCE)
> *De Motu Stellarum* of al-Battānī, which contains important material on trigonometry
> *Measurement of a Circle* of Archimedes
> *Liber embadorum* of Abraham bar Ḥiyya

Robert of Chester (fl. 1141–1150)

> *Algebra* of al-Khwārizmī
> Revision of al-Khwārizmī's astronomical tables for the meridian of London

Gerard of Cremona (fl. 1150–1185)

> *Posterior Analytics* of Aristotle
> *De Sphaera Mota* of Autolycus
> *Elements* of Euclid
> *Data* of Euclid
> *Measurement of a Circle* of Archimedes
> *Spherica* of Theodosius
> *Almagest* of Ptolemy
> *De Figuris Sphaericis* of Menelaus
> *Algebra* of al-Khwārizmī
> *Elementa Astronomica* by Jābir ibn Aflaḥ

Wilhelm of Moerbeke (fl. 1260–1280)

> *On Spirals* of Archimedes
> *On the Equilibrium of Planes* of Archimedes
> *Quadrature of the Parabola* of Archimedes
> *Measurement of a Circle* of Archimedes
> *On the Sphere and Cylinder* of Archimedes
> *On Conoids and Spheroids* of Archimedes
> *On Floating Bodies* of Archimedes

Note: This listing contains works whose translation can definitely be attributed to a given translator. There are Latin translations of other works known to have been made in the twelfth and thirteenth centuries, including parts of Apollonius's *Conics* and the *Algebra* of Abū Kāmil, whose translators are currently unknown.

embadorum (*Book of Areas*) by the Spanish-Jewish scholar Abraham bar Ḥiyya, a work that also contained the Islamic rules for solving quadratic equations.

The most prolific of all the translators was Gerard of Cremona (1114–1187), an Italian who worked primarily in Toledo and is credited with the translation of more than 80 works. Undoubtedly, not all of these are due to him alone. It is known that one of his assistants was Galippus, a Spanish Christian who had been allowed to practice Christianity under Moslem rule, but the names of his other assistants have been lost to history. Among Gerard's works was a new translation of Euclid's *Elements* from the Arabic of Thābit ibn Qurra and the first translation of Ptolemy's *Almagest* from the Arabic in 1175.

By the end of the twelfth century, then, many of the major works of Greek mathematics and a few Islamic works were available to Latin-reading scholars in Europe. During the next centuries, these works were assimilated and new mathematics began to be created by the Europeans themselves. It is well to note, however, that some Spanish-Jewish scholars had earlier read the Arabic works in the original and had produced works on their own, in Hebrew.

During the twelfth century, in fact, the cultural exchange among the three major civilizations of Europe and the Mediterranean basin, the Jewish, Christian, and Islamic, was very intense. The Islamic supremacy of previous centuries was on the wane, and the other two were gaining strength. By the end of the next century, the genius of western Christendom had manifested itself, while various physical limitations on the lives of the Jews began to lessen the Jewish contribution.

This chapter will discuss both Jewish and Christian contributions of the twelfth through the fourteenth centuries. We will first consider geometry and trigonometry, next developments in combinatorics, next the algebra that grew out of the introduction of Islamic algebra into Europe, and finally, some of the mathematics of kinematics that stemmed from the study of Aristotle's works in the medieval universities.

 10.2   ## GEOMETRY AND TRIGONOMETRY

Euclid's *Elements* was translated into Latin early in the twelfth century. Before then, of course, Arabic versions were available in Spain. And so, when Abraham bar Ḥiyya (d. 1136) of Barcelona wrote his *Hibbur ha-Meshihah ve-ha-Tishboret* (*Treatise on Mensuration and Calculation*) in 1116 to help French and Spanish Jews with the measurement of their fields, he began the work with a summary of some important definitions, axioms, and theorems from Euclid. Not much is known of the life of Abraham bar Ḥiyya, but from his Latin title of *savasorda*, a corruption of the Arabic words meaning "captain of the bodyguard," it is likely that he had a court position, probably one in which he gave mathematical and astronomical advice to the Christian monarch.

10.2.1 Abraham bar Ḥiyya's *Treatise on Mensuration*

Like most of those who dealt with geometry over the next few centuries, Abraham was not so much interested in the theoretical aspects of Euclid's *Elements* as in the practical application of geometric methods to measurement. Nevertheless, he took over the Islamic tradition of proof, absorbed from the Greeks, and gave geometric justifications of methods for solving the algebraic problems he included as part of his geometrical discussions. In particular, Abraham included in his work the major results of *Elements* II on geometric algebra and used them to demonstrate methods of solving quadratic equations. In fact, Abraham's work was the first in Europe to give the Islamic procedures for solving such equations.

For example, Abraham posed the question, "If from the area of a square one subtracts the sum of the (four) sides and there remains 21, what is the area of the square and what is the length of each of the equal sides?"[4] We can translate Abraham's question into the quadratic equation $x^2 - 4x = 21$, an equation he solves in the familiar way by halving 4 to get 2, squaring this result to get 4, adding this square to 21 to get 25, taking the square root to get 5, and then adding that to the half of 4 to get the answer 7 for the side and the answer 49 for the area. Abraham's statement of the problem was not geometrical, in that he wrote of subtracting a length (the sum of the sides) from an area. But in his geometric justification, he restated the problem to mean the cutting off of a rectangle of sides 4 and x from the original square of unknown side x to leave a rectangle of area 21. He then bisected the side of length 4 and applied *Elements* II–6 to justify the algebraic procedure. Thus, Abraham evidently had learned his algebra not from al-Khwārizmī (whose *Algebra* was translated into Latin in the same year as Abraham's work), but from an author such as Abū Kāmil, who used Euclidean justifications. Abraham similarly presented the method and Euclidean proof for examples of the two other Islamic classes of mixed quadratic equations, $x^2 + 4x = 77$ and $4x - x^2 = 3$. In the latter case, he gave both positive solutions. Abraham also solved such quadratic problems as the systems $x^2 + y^2 = 100$, $x - y = 2$, and $xy = 48$, $x + y = 14$.

Abraham's most original contribution, however, is found in his section on measurements in circles. He began by giving the standard rules for finding the circumference and area of a circle, first using 3 1/7 for π but then noting that if one wants a more exact value, as in dealing with the stars, one should use $3\frac{8\frac{1}{2}}{60}(= 3\frac{17}{120})$. Curiously, in the Hebrew version of the text, but not in the Latin, there is a justification of the area formula $A = \frac{C}{2}\frac{d}{2}$ by use of indivisibles. Namely, one thinks of the circle as made up of concentric circles of indivisible threads (Fig. 10.2). If one then slices this circle from the center to the circumference and unfolds it into a triangle, the base of the triangle is the original circumference and the height is the radius. The area formula follows immediately.

FIGURE 10.2

Circle unfolded into a triangle

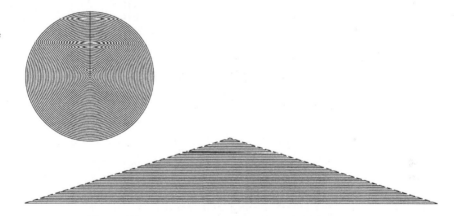

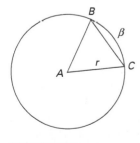

FIGURE 10.3

Area of segment $B\beta C =$ Area of sector $AB\beta C -$ Area of triangle ABC; Area of sector $= r\frac{\beta}{2}$

To measure areas of segments of circles, Abraham noted that one must first find the area of the corresponding sector by multiplying the radius by half the length of the arc (Fig. 10.3). One then subtracts the area of the triangle formed by the chord of the segment and the two radii at its ends. But how does one calculate the length of the arc, assuming one knows the length of the chord? Abraham's answer is, by the use of a table relating chords and arcs. And so for the first time in Europe there appeared what one can call a trigonometric table (Fig. 10.4). Unlike the table of sines of al-Khwārizmī, which appeared in Latin translation shortly after Abraham's book and which used degrees to measure arcs and a circle radius of 60, Abraham's table was a table of arcs to given chords using what seemed to Abraham more convenient measures. Namely, he used a radius of 14 parts, so the semicircumference would be integral (44), and then gave the arc (in parts, minutes, and seconds) corresponding to each integral value of the chord from 1 to 28. So to determine the length of the arc of a segment of a circle, given the chord s and the distance h from the center of the chord to the circumference, Abraham first determined the diameter d of the circle by the formula

FIGURE 10.4

Arc-chord table of Abraham
bar Ḥiyya

Partes Cordatum	Arcus		
	Partes	Min.	Sec.
1	1	0	2
2	2	0	8
3	3	0	26
4	4	0	55
5	5	1	44
6	6	2	54
7	7	4	42
8	8	7	11
9	9	9	56
10	10	13	42
11	11	18	54
12	12	24	38
13	13	31	9
14	14	40	0
15	15	50	10
16	17	2	16
17	18	16	36
18	19	33	27
19	20	53	26
20	22	17	10
21	23	45	6
22	25	19	24
23	27	0	0
24	28	49	56
25	31	26	37
26	33	20	52
27	36	27	32
28	44	0	0

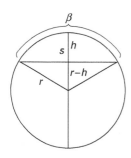

FIGURE 10.5

Length of arc
$\beta = \frac{d}{28}$ arc-chord $\left(\frac{28}{d}s\right)$,
where $d = 2r = \frac{s^2}{4h} + h$

$d = s^2/4h + h$ (Fig. 10.5). Then he multiplied the given chord by $\frac{28}{d}$ (to convert to a circle of diameter 28), consulted his table to determine the corresponding arc α, and multiplied α by $\frac{d}{28}$ to find the actual arc length.

10.2.2   Practical Geometries

Abraham's Hebrew text was one of the earliest of many practical geometrical works to appear in medieval Europe. An early Latin one appeared in the 1120s, probably written by Hugh of St. Victor (1096–1141), a theologian and master of the abbey of St. Victor in Paris. This text, designed for surveyors, is on a much simpler level than Abraham's. Apparently, knowledge of trigonometry had not yet reached Paris nor was there any mention of Euclid in Hugh's work. But Hugh did make use of the astrolabe, the sighting device developed by Islamic astronomers from earlier Greek models and brought through Spain into western Europe. Thus, Hugh's methods of measurement involved the use of the alidade, an altitude-sighting device attached to the astrolabe, which enabled one to measure the ratio of height to distance of an object sighted (Fig. 10.6). If this ratio r is known, and the distance d of the object is also known,

FIGURE 10.6

Astrolabe with alidade OA. One holds the line OB horizontal and sights the distant object along OA. Then r gives the ratio of the height to the distance of that object.

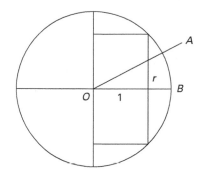

FIGURE 10.7

Measuring height of a distant object using two sightings according to Hugh of St. Victor

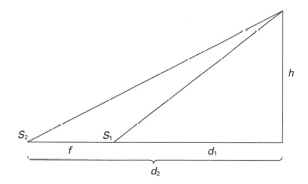

then the height h is given by $h = rd$. Like his predecessors in India, China, and the Islamic world, Hugh also knew that it is not always possible to measure the distance d of a distant object. In that case, two measurements were needed (Fig. 10.7). At point S_1, one finds the ratio r_1 of height h to distance d_1, while at point S_2, one finds the ratio r_2 of h to d_2. It then follows that $d_2 = (r_1/r_2)d_1$. But since $d_2 - d_1 = f$ can be measured, Hugh could calculate d_1 as

$$d_1 = \frac{f}{\frac{r_1}{r_2} - 1}$$

and then evaluate h by $h = r_1 d_1$.[5]

By late in the twelfth century, however, trigonometry and knowledge of Euclid had reached Paris, as exemplified in the anonymous practical geometry generally known by the first three words of the manuscript, *Artis cuiuslibet consummatio* (*The Perfection of Any Art*). This work, originally written in Latin but translated into French in the thirteenth century, opens with a rather poetic introduction:

> The perfection of any art, seen as a whole, depends on two aspects: theory and practice. Anyone deprived of either of these aspects is labeled semiskilled. Truly the modern Latins . . . [by] neglecting the practice fail to reap where they sowed the richest fruits as if picking a spring flower without waiting for its fruit. What is sweeter when once the qualities of numbers have been known through arithmetic than to recognize their infinite dispositions by subtle calculation,

the root, origin, and source necessarily available for every science? What is more pleasant when once the proportion of sounds has been known through music than to discern their harmonies by hearing? What is more magnificent when once the sides and angles of surfaces and solids have been proved through geometry than to know and investigate exactly their quantities? What is more glorious or excellent when once the motion of the stars has been known through astronomy than to discover the eclipses and secrets of the art? We prepare for you therefore a pleasant treatise and delightful memoir on the practice of geometry so that we may offer to those who are thirsty what we have drunk from the most sweet source of our master.[6]

To be truly educated, the author seems to be saying, one not only must study the theoretical aspects of the quadrivium, but also must understand how these subjects are used in the real world. *Artis cuiuslibet consummatio* intends to show, then, the practical aspects of one of the quadrivial subjects, namely, geometry.

The book is divided into four parts: area measurement, height measurement, volume measurement, and calculation with fractions. The last section is designed to help the reader with the computations necessary in the earlier parts. The first part, on areas, begins with the basic procedures for finding the areas of triangles, rectangles, and parallelograms, most of which are justified by an appeal to Euclidean propositions. The author followed this with a section on the areas of various equilateral polygons, all of the formulas for which are incorrect. Instead of being formulas for areas of pentagons, hexagons, heptagons, and so on, of side n, the formulas are always those for the nth pentagonal, hexagonal, heptagonal number. For example, the procedure given for finding the area of a pentagon of side n amounts to using the formula

$$A = \frac{3n^2 - n}{2}.$$

The author may well have been influenced by the material on figurate numbers derived from the work of Nicomachus.

The section of the book on heights showed the author's knowledge of trigonometry. For example, the procedure for measuring the altitude of the sun using the shadow of a vertical gnomon of length 12 is given: "Let the shadow be multiplied by itself. Let 144 be added to the product. Let the root of the whole sum be taken. And then let the shadow be multiplied by 60. Let the product be divided by the root found. The result will be the sine; let its arc be found. Let the arc be subtracted from 90; the remainder will be . . . the altitude of the sun."[7] Namely, if the shadow is designated by s, the altitude α is given by

$$\alpha = 90 - \text{arcSin}\left(\frac{60s}{\sqrt{s^2 + 144}}\right),$$

where, as in most of the Islamic trigonometric works, the Sines were computed using a radius of 60 (Fig. 10.8). Similarly, the author calculated the shadow from the altitude by using

$$s = \frac{12\,\text{Sin}(90 - \alpha)}{\text{Sin}\,\alpha}.$$

These two problems demonstrate that the author knew the use of a table of sines but probably did not know of the tables of cosines, tangents, or cotangents, even though these had already been developed in the Islamic world. It was only the earliest of the Hindu and Islamic improvements on Greek work that were available.

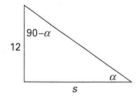

FIGURE 10.8

The calculation of the altitude of the sun given the shadow, and conversely, from the *Artis cuiuslibet consummatio*: $\alpha = 90 - \arcsin\left(\frac{60s}{\sqrt{s^2+144}}\right)$; $s = \frac{12\sin(90-\alpha)}{\sin\alpha}$

For surveying, the author returned to the ancient methods. To measure the height of a tower, not only did he not use trigonometric methods, he reverted to probably the oldest (and simplest) method available: "Wait until the altitude of the sun is 45 degrees . . . ; then the shadow lying in the plane of any body will be equal to its body."[8] If the tower is inaccessible, the author used the ancient methods requiring two sightings similar to those in the Chinese and Indian sources. As in almost all of the Indian, Islamic, and medieval European sources, even when trigonometric methods were known, they were applied solely to heavenly triangles, not to earthly ones.

These two twelfth-century Latin geometries give us an idea of the state of geometrical knowledge in northern Europe of the time. Greek geometrical traditions were just beginning to be reestablished, but practical geometrical methods, also dating to ancient times and not all strictly correct, were used for actually computing geometrical quantities of use in daily life. In southern Europe, however, the Islamic influence was stronger and Euclidean traditions of proof are more in evidence, as in the work of Abraham bar Ḥiyya. Another example is provided by the geometrical work of one of the first Italian mathematicians, Leonardo of Pisa (c. 1170–1240).

10.2.3 Leonardo of Pisa's *Practica Geometriae*

Leonardo's *Practica geometriae* (1220) is more closely related to the work of Abraham bar Ḥiyya than to the *Artis cuiuslibet consummatio* or the work of Hugh of St. Victor. In fact, some of the sections appear to be taken almost directly from the *Liber embadorum*. Leonardo's work is, however, somewhat more extensive. As in the earlier book, Leonardo began with a listing of various definitions, axioms, and theorems of Euclid, including especially the propositions of Book II. So in his section on measuring rectangles, in which he includes the standard methods for solving quadratic equations, he was able to quote Euclid in justification of his procedures. He provided more examples than Abraham, including equations in which the coefficient of the square term is greater than 1. For example, to solve the equation three squares and four roots equal 279 ($3x^2 + 4x = 279$), he divided by 3 and reduced the equation to $x^2 + 1\frac{1}{3}x = 93$ before applying the standard method. Also, many of his problems involve the diagonal of a rectangle and thus deal with the sums of the squares of the sides.

Leonardo, again like Abraham, wrote a section on circles in which he quoted the standard 22/7 for π. But Leonardo, in addition, showed how to calculate this value by the procedure of Archimedes. He found that the ratio of the perimeter of a 96-sided polygon circumscribed about a circle to the diameter of the circle is 1440 to 458 1/5, and the ratio of the perimeter of an inscribed 96-sided polygon to the diameter is 1440 to 458 4/9. Noting that 458 1/3 is approximately halfway between 458 1/5 and 458 4/9, he asserted that the ratio of circumference to diameter is close to 1440 : 458 1/3 = 864 : 275. Because 864 : 274 10/11 = 3 1/7 : 1, Leonardo had rederived the Archimedean value.

Leonardo also calculated areas of segments and sectors of circles. To do this, he too needed a table of arcs and chords. Strangely enough, although he defined the Sine of an arc in the standard way, he did not give a table of Sines, but one of chords, and in fact reproduced the Ptolemaic procedure for determining the chord of half an arc from that of the whole arc. His chord table, though, was not Ptolemaic. In fact, it may well be original to Leonardo because it is based on a radius of 21. Like the value 14 of Abraham, this was chosen so the semicircumference of the circle is integral, but unlike Abraham's table, this table is a direct chord table (Fig. 10.9). For each integral arc from 1 to 66 rods (and also from 67 to 131), the table gives the corresponding chord, in the same measure, with fractions of the rods not in sixtieths, but in the Pisan measures of feet (6 to the rod), unciae (18 to the foot), and points (20 to the uncia). Leonardo then demonstrated how to use the chord table to calculate arcs to chords in circles of radius other than 21.

Like Abraham bar Ḥiyya, Leonardo used the table of chords only to calculate areas of circular sectors and segments. When, later in the same chapter, he calculated the lengths of the sides and diagonals of a regular pentagon inscribed in a circle, he did not use what seems to us the obvious method of consulting his table of chords. He returned to Euclid and quoted appropriate theorems from Book XIII relating the sides of a hexagon, pentagon, and decagon to enable him actually to perform the calculations. And toward the end of the book, when he wanted to calculate heights, again he did not use trigonometry. He used the old methods of similar triangles, starting with a pole of known height to help sight the top of the unknown object, then measuring the appropriate distances along the ground.

10.2.4   Trigonometry

That trigonometry in the medieval period was not used to measure earthly triangles is further demonstrated by two fourteenth-century trigonometry works, one by the Englishman Richard of Wallingford (1291–1336) and the other by the French Jew Levi ben Gerson (1288–1344). Yet both of these texts had something new, especially in the methods of calculating accurate tables.

Richard of Wallingford was a monk who spent the final nine years of his life as the abbot at St. Albans monastery. The *Quadripartitum*, a four-part work on the fundamentals of trigonometry, was written while he was still a student at Oxford, probably around 1320. Perhaps 10 years later, Richard revised and shortened this work in another treatise entitled *De Sectore*. The goal of both works, like that of most texts on trigonometry, was to teach the methods required for the solution of problems in spherical trigonometry, which in turn was required for astronomy. It appears that the chief source of the *Quadripartitum* was the *Almagest* of Ptolemy, modified to incorporate the Hindu Sines in addition to the more ancient chords. But by the time Richard revised the work, he had become familiar with Jābir's trigonometry. In fact, in his section on spherical trigonometry, he presented virtually the whole of Jābir's treatment right after Ptolemy's version based on the theorem of Menelaus.

Richard's treatment of the theorem of Menelaus, both in its plane and spherical versions, was extremely detailed. Because this theorem is concerned with ratios among the various sides in the Menelaus configuration, Richard needed first to consider the basics of the theory of proportions. His study of proportions is closely related to the work of several contemporaries in the universities and will be considered in Section 10.5.1. Here we only note that in his treatment of Menelaus's theorem, Richard considered all the possible cases of the Menelaus configuration and proved the result anew each time. While modern readers might consider his work tedious, he evidently felt that such detail was necessary for the less mathematically experienced readers for whom he was writing. One also sees here, as well as in the beginning sections of the book on the basic results of plane trigonometry, Richard's commitment to strictly Euclidean rigor of argument as he exhausts all the cases. Recall that even though mathematical knowledge was at a low ebb during the early Middle Ages, the basic notion of

FIGURE 10.9

Chord table of Leonardo of Pisa

Arcus pertice	Arcus pertice	Corde pertice	Ar pedes	Cuu vncie	M puncta	Arcus pertice	Arcus pertice	Corde pertice	Ar pedes	CV vncie	VM puncta
1	131	0	5	17	17	34	98	30	2	6	17
2	130	1	5	17	13	35	97	31	0	8	5
3	129	2	5	17	4+	36	96	31	4	8	7
4	128	3	5	17	2	37	95	32	2	5	15
5	127	4	4	12	10	38	94	33	0	1	9
6	126	5	5	16	7+	39	93	34	3	13	0
7	125	6	5	14	5	40	92	35	1	4	15
8	124	7	5	12	9	41	91	35	4	12	10
9	123	8	5	8	16	42	90	36	2	0	0
10	122	9	5	7	8	43	89	36	5	3	5
11	121	10	5	4	2	44	88	37	2	4	6
12	120	11	4	17	18	45	87	37	5	3	2
13	119	12	4	13	6	46	86	38	1	17	15
14	118	13	4	7	16	47	85	38	4	12	13
15	117	14	4	1	0	48	84	38	1	4	0
16	116	15	3	11	18	49	83	39	3	11	15
17	115	16	3	3	12	50	82	39	5	17	2
18	114	17	2	12	8	51	81	40	2	2	1
19	113	18	2	0	15	52	80	40	4	2	10
20	112	19	1	8	12	53	79	40	0	0	11
21	111	20	0	13	18	54	78	40	1	14	5
22	110	21	0	0	0	55	77	41	3	7	8
23	109	21	5	2	16	56	76	41	4	16	2
24	108	22	4	4	5	57	75	41	0	4	12
25	107	23	3	4	8	58	74	41	1	8	1
26	106	24	2	3	2	59	73	41	2	9	0
27	105	25	1	0	6	60	72	41	3	7	14
28	104	25	5	16	2	61	71	41	4	9+	2
29	103	26	4	8	0	62	70	41	4	15	10
30	102	27	3	0	3	63	69	41	5	6	9
31	101	28	1	9	7	64	68	41	5	12	17
32	100	28	5	16	4	65	67	41	5	6	14
33	99	29	4	3	9	66	66	42	0	0	0

BIOGRAPHY

Leonardo of Pisa (c. 1170–1240)

Leonardo, often known today by the name Fibonacci (son of Bonaccio) given to him by Baldassarre Boncompagni, the nineteenth-century editor of his works, was born around 1170. His father was a Pisan merchant who had extensive commercial dealings in Bugia on the North African coast (now Bejaia, Algeria). Leonardo spent much of his early life there learning Arabic and studying mathematics under Moslem teachers. Later he traveled throughout the Mediterranean, probably on business for his father. At each location, he met with Islamic scholars and absorbed the mathematical knowledge of the Islamic world. After his return to Pisa in about 1200, he spent the next 25 years writing works in which he incorporated what he had learned. The ones that have been preserved include the *Liber abbaci* (1202, 1228), the *Practica geometriae* (1220), and the *Liber quadratorum* (1225). Leonardo's importance was recognized both at the court of Frederick II, as noted in the opening story, and also in the city of Pisa, which in 1240 granted him a yearly stipend in thanks for his teaching and other services to the community.

a mathematical proof survived and was reinvigorated, as, for example, by Richard, once the need for more mathematics had established itself.

Although much of Richard's work was derived from earlier trigonometries, he did present a new method of calculating Sin 1°, the value that determined the accuracy of the Sine tables. Thus, after considering both Ptolemy's method from the *Almagest* and the method of Abū al-Wafā', he extended the latter to smaller and smaller arcs. Namely, beginning with the Sine of $\frac{3}{16}^\circ$, calculated by the half-angle formula, he suggested continuing to use that formula to find the Sines of $\frac{3}{32}^\circ$ and $\frac{3}{64}^\circ$. The latter enables one to determine, by the sum formula, $\text{Sin}\,\frac{63}{64}^\circ = \text{Sin}(\frac{3}{64} + \frac{15}{16})^\circ$. Similarly, one can find $\text{Sin}(1 - \frac{1}{256})^\circ$ and $\text{Sin}(1 - \frac{1}{1024})^\circ$ and "proceed in this way even to the 9000th part of a degree, or even to the infinitely small, if by working minutely you wish to do so."[9]

The trigonometrical work of Levi ben Gerson was roughly contemporaneous with the *Quadripartitum*. It formed part of an astronomical treatise that in turn formed part of a major philosophical work, *Sefer Milḥamot Adonai* (*Wars of the Lord*). Levi's trigonometry was based chiefly on Ptolemy, though again, like Richard, Levi generally used Sines rather than chords. Also, like Richard, Levi spent some time dealing with accuracy of his tables. In particular, he noted that tables with intervals of 1° have errors of about 15 minutes of arc when one uses linear interpolation to find arcs corresponding to given Sines, if the arcs are close to 90°. And this large an error was unacceptable. Hence, Levi determined his own tables in steps of $\frac{1}{4}^\circ$.

Levi's main departure from Ptolemy, and also from Richard, is that he gave detailed procedures for solving plane triangles. He first presented the standard methods for solving right triangles and then proceeded to general triangles. In the case where three sides are known, Levi solved the triangle by dropping a perpendicular from one vertex to the opposite side (or opposite side extended), and then applying the version of the law of cosines of *Elements* II–12 and II–13. The same method works also where two sides and the included angle are known. For the case where two sides and the angle opposite one of them are known,

Levi used (with proof) the law of sines. He did not, however, mention the possible ambiguity of this case. Of course in any particular problem, one of the unknown angles is required to be acute or obtuse, so a single solution of the triangle can be found. Finally, Levi noted that the case where two angles and a side are known can also be solved using the law of sines.

Certainly, Levi's methods were not new. Although his procedures were somewhat different from those of Jābir, the methods were available in other Islamic trigonometries. Nevertheless, Levi's brief treatise provided one of the earliest treatments of the basic methods for solving plane triangles available in Europe. But as in the Islamic works and the practical geometry texts, the methods Levi presented were used only for solving astronomical triangles, never for solving earthly ones.

 10.3

COMBINATORICS

We have already discussed the interest in combinatorics in Indian and Islamic sources. In medieval Europe, there was also interest in such questions, primarily in the Jewish community. The earliest Jewish source on this topic seems to be the mystical work *Sefer Yetsirah* (*Book of Creation*), written sometime before the eighth century and perhaps as early as the second century. In it the unknown author calculated the various ways in which the 22 letters of the Hebrew alphabet can be arranged. He was interested in this calculation because the Jewish mystics believed that God had created the world and everything in it by naming these things (in Hebrew, of course): "God drew them, combined them, weighed them, interchanged them, and through them produced the whole creation and everything that is destined to be created. . . . Two stones [letters] build two houses [words], three build six houses, four build twenty-four houses, five build one hundred and twenty houses, six build seven hundred and twenty houses, seven build five thousand and forty houses."[10] Evidently, the author understood that the number of possible arrangements of n letters was $n!$. An Italian rabbi, Shabbetai Donnolo (913–970), derived this factorial rule very explicitly in a commentary on the *Sefer Yetsirah*:

> The first letter of a two-letter word can be interchanged twice, and for each initial letter of a three-letter word the other letters can be interchanged to form two two-letter words—for each of three times. And all the arrangements there are of three-letter words correspond to each one of the four letters that can be placed first in a four-letter word: a three-letter word can be formed in six ways, and so for every initial letter of a four-letter word there are six ways—altogether making twenty-four words, and so on.[11]

10.3.1 The Work of Abraham ibn Ezra

Although the author of the *Sefer Yetsirah* briefly mentioned how to calculate the number of combinations of letters taken two at a time, a more detailed study of combinations was carried out by Rabbi Abraham ben Meir ibn Ezra (1090–1167), a Spanish-Jewish philosopher, astrologer, and biblical commentator. It was in an astrological text that ibn Ezra discussed the number of possible conjunctions of the seven "planets" (including the sun and the moon). It was believed that these conjunctions would have a powerful influence on human life. Ibn Ezra thus calculated C_k^7 for each integer k from 2 to 7 and noted that the total was 120. He began with the simplest case, that the number of binary conjunctions was 21. This number was equal to the sum of the integers from one to six and could be calculated by ibn Ezra's rule for the sum of the integers from one up to a particular number: multiply that number by its half and by half of unity. In modern terms, we can write this as

$$C_2^n = \sum_{i=1}^{n-1} i = (n-1)\left(\frac{n-1}{2}\right) + (n-1)\left(\frac{1}{2}\right) = \frac{n(n-1)}{2}.$$

To calculate ternary combinations, ibn Ezra explained, "We begin by putting Saturn with Jupiter and with them one of the others. The number of the others is five; multiply 5 by its half and by half of unity. The result is 15. And these are the conjunctions of Jupiter."[12] Namely, there are five ternary combinations involving Jupiter and Saturn, four involving Jupiter and Mars, but not Saturn, and so on. Hence, there are $C_2^6 = 15 (= 5 \cdot \frac{5}{2} + 5 \cdot \frac{1}{2})$ ternary conjunctions involving Jupiter. Similarly, to find the ternary conjunctions involving Saturn but not Jupiter, ibn Ezra needed to calculate the number of choices of two planets from the remaining five: $C_2^5 = 10$. He then found the ternary conjunctions involving Mars, but neither Jupiter nor Saturn, and finally concluded with the result

$$C_3^7 = C_2^6 + C_2^5 + C_2^4 + C_2^3 + C_2^2 = 15 + 10 + 6 + 3 + 1 = 35.$$

Ibn Ezra next calculated the quaternary conjunctions by analogous methods. The conjunctions involving Jupiter require choosing three planets from the remaining six. Those with Saturn but not Jupiter require choosing three from five. So finally, $C_4^7 = C_3^6 + C_3^5 + C_3^4 + C_3^3 = 20 + 10 + 4 + 1 = 35$. Ibn Ezra then just stated the results for the conjunctions involving five, six, and seven planets. Essentially, he had given an argument for the case $n = 7$, easily generalizable to the general combinatorial rule:

$$C_k^n = \sum_{i=k-1}^{n-1} C_{k-1}^i.$$

In a later work, ibn Ezra essentially repeated the calculations above for C_2^7 and C_3^7 and then noted that by symmetry $C_4^7 = C_3^7$ and $C_5^7 = C_2^7$, something not explicitly mentioned by either ibn Mun'im or ibn al-Bannā in their own similar derivations somewhat later. Also, in a work on arithmetic in 1146, ibn Ezra introduced the Hebrew-speaking community to the decimal place value system. He used the first nine letters of the Hebrew alphabet to represent the first nine numbers and then instructed his readers on the meaning of place value, the use of the zero (which he wrote as a circle), and the various algorithms for calculation in the Hindu-Arabic system.

10.3.2   Levi ben Gerson and Induction

Early in the fourteenth century, Levi ben Gerson gave careful, rigorous proofs of various combinatorial formulas in a major work, the *Maasei Hoshev* (*The Art of the Calculator*) (1321, with a second edition in 1322). Levi's text is divided into two parts, a first theoretical part in which every theorem receives a detailed proof, and a second "applied" part in which explicit instructions are given for performing various types of calculation. (Levi used ibn Ezra's "Hebrew" place value system in this part.) Levi's theoretical first section began with a quite modern justification for considering theory at all:

Because the true perfection of a practical occupation consists not only in knowing the actual performance of the occupation but also in its explanation, why the work is done in a particular way, and because the art of calculating is a practical occupation, it is clear that it is pertinent to concern oneself with its theory. There is also a second reason to inquire about the theory in this field. Namely, it is clear that this field contains many types of operations, and each type itself concerns so many different types of material that one could believe that they cannot all belong to the same subject. Therefore, it is only with the greatest difficulty that one can achieve understanding of the art of calculating, if one does not know the theory. With the knowledge of the theory, however, complete mastery is easy. One who knows it will understand how to apply it in the various cases which depend on the same foundation. If one is ignorant of the theory, one must learn each kind of calculation separately, even if two are really one and the same.[13]

Of course, as in any mathematical work, the reader must know the prerequisites, in this case Books VII, VIII, and IX of Euclid's *Elements*, "since it is not our intention in this book to repeat [Euclid's] words." But Levi did insist on giving careful, Euclidean-style proofs of all his results. The most important aspects of Levi's work are the combinatorial theorems. It is here that he used, somewhat more explicitly than his Islamic predecessors, the essentials of the method of mathematical induction, what he calls the process of "rising step by step without end." In general, when Levi used such a proof, he first proved the **inductive step**, the step that allows one to move from k to $k + 1$, next noted that the process begins at some small value of k, and then finally gave the complete result. Nowhere did he state the modern principle of induction, but it does appear that he knew how to use it. In fact, he used it initially in connection with two of the earliest theorems in the book, theorems that deal with associativity and commutativity of multiplication.

PROPOSITION 9 *If one multiplies a number which is the product of two numbers by a third number, the result is the same as when one multiplies the product of any two of these three numbers by the third.*

PROPOSITION 10 *If one multiplies a number which is the product of three numbers by a fourth number, the result is the same as when one multiplies the product of any three of these four numbers by the fourth.*

In modern notation, the first result states that $a(bc) = b(ac) = c(ab)$, while the second extends that result to four factors. The proof of Proposition 9 simply involves counting the number of times the various factors of the product appear in that product. In the proof of Proposition 10, Levi noted that $a(bcd)$ contains bcd a times. Since by Proposition 9, bcd can be thought of as $b(cd)$, it follows that the product $a(bcd)$ contains acd b times, or, $a(bcd) = b(acd)$, as desired. Levi then generalized these two results to any number of factors: "By the process of rising step by step without end, this is proved; that is, if one multiplies a number which is the product of four numbers by a fifth number, the result is the same as when one multiplies the product of any four of these by the other number. Therefore, the result of multiplying any product of numbers by another number contains any of these numbers as many times as the product of the others."[14] We see here the essence of the principle of mathematical induction. Levi used the principle again in proving that $(abc)d = (ab)(cd)$ and concluded that one can use the same proof to demonstrate the result without end: Any number contains the product of two of its factors as many times as the product of the remaining factors.

BIOGRAPHY

Levi ben Gerson (1288–1344)

Levi was born probably in the village of Bagnols-sur-Cèze in the south of France and spent most of his life in the nearby town of Orange. He was not only a mathematician but also an astronomer, philosopher, and biblical commentator. Not much is known of his life, except that he maintained contact with many important Christians, at the request of some of whom he composed a set of astronomical tables. His various works show that he was acquainted with the major Greek philosophical, astronomical, and mathematical writings, as well as with significant parts of the Islamic mathematical tradition. His best-known contribution to astronomy is his invention of the Jacob Staff (Fig. 10.10), which was used for centuries to measure the angular separation between heavenly bodies. In particular, it was popular with sixteenth-century European sailors who used it for navigation purposes.

FIGURE 10.10

Jacob Staff, invented by Levi ben Gerson

Levi was certainly not consistent about applying his induction principle. The middle of the text contains many theorems dealing with sums of various sequences of integers, theorems that could be proved by induction. But for many of these, Levi used other methods. For example, in proving that the sum of the first n integers equals $\frac{1}{2}n(n+1)$ (where n is even), he used the idea that the sums of the first and last integers, the second and next to last, and so on, are each equal to $n+1$. The same result when n is odd is proved by noting that those same sums are equal to twice the middle integer. In his proof of the formula for the sum of the first n integral cubes, however, he did use induction, in a way reminiscent of al-Karajī's proof of the same result. The basic inductive step is

PROPOSITION 41   *The square of the sum of the natural numbers from 1 up to a given number is equal to the cube of the given number added to the square of the sum of the natural numbers from 1 up to one less than the given number. [In modern notation, the theorem says that $(1 + 2 + \cdots + n)^2 = n^3 + (1 + 2 + \cdots + (n - 1))^2$.]*

We present Levi's proof in modern notation. First, $n^3 = n \cdot n^2$. Also, $n^2 = (1 + 2 + \cdots + n) + (1 + 2 + \cdots + (n - 1))$. (This result is Levi's Proposition 30.) Then

$$n^3 = n[(1 + 2 + \cdots + n) + (1 + 2 + \cdots + (n - 1))]$$
$$= n^2 + n[2(1 + 2 + \cdots + (n - 1))].$$

But

$$(1 + 2 + \cdots + n)^2 = n^2 + 2n(1 + 2 + \cdots + (n - 1)) + (1 + 2 + \cdots (n - 1))^2.$$

It follows that $n^3 + (1 + 2 + \cdots + (n - 1))^2 = (1 + 2 + \cdots + n)^2$.

Levi next noted that although 1 has no number preceding it, "its third power is the square of the sum of the natural numbers up to it." In other words, he gave the first step of a proof by induction for the result stated as

PROPOSITION 42   *The square of the sum of the natural numbers from 1 up to a given number is equal to the sum of the cubes of the numbers from 1 up to the given number.*

Levi's proof is not quite what we would expect of a proof by induction. Instead of arguing from n to $n + 1$, he argued, as did al-Karajī, from n to $n - 1$. He noted that, first of all, $(1 + 2 + \cdots + n)^2 = n^3 + (1 + 2 + \cdots + (n - 1))^2$. The final summand is, also by the previous proposition, equal to $(n - 1)^3 + (1 + 2 + \cdots + (n - 2))^2$. Continuing in this way, Levi eventually reached $1^2 = 1^3$, and the result is proved. We note further that, although the proposition is stated in terms of an arbitrary natural number, in his proof Levi wrote only a sum of five numbers in his first step rather than the n used in our adaptation. The five are represented by the five initial letters of the Hebrew alphabet. Like many of his predecessors, Levi had no way of writing the sum of arbitrarily many integers and so used the method of generalizable example. Nevertheless, the idea of a proof by induction is evident in Levi's demonstration.

Inductive proofs are also evident in the final section of the theoretical part of the *Maasei Hoshev*, that on permutations and combinations. Levi's first result in this section showed that the number of permutations of a given number n of elements is what we call $n!$:

PROPOSITION 63 *If the number of permutations of a given number of different elements is equal to a given number, then the number of permutations of a set of different elements containing one more number equals the product of the former number of permutations and the given next number.*

Symbolically, the proposition states that $P_{n+1} = (n + 1)P_n$ (where P_k stands for the number of permutations of a set of k elements). This result provides the inductive step in the proof of the proposition $P_n = n!$, although Levi did not mention that result until the end. His proof of proposition 63 was very detailed. Given a permutation, say, *abcde*, of the original n elements and a new element f, he noted that *fabcde* is a permutation of the new set. Because there are P_n such permutations of the original set, there are also P_n permutations of the new set beginning with f. Also, if one of the original elements, for example, e, is replaced by the new element f, there are P_n permutations of the set a, b, c, d, f and therefore also P_n permutations of the new set with e in the first place. Because any of the n elements of the original set, as well as the new element, can be put in the first place, it follows that the number of permutations of the new set is $(n + 1)P_n$. Levi finished the proof of Proposition 63 by showing that all of these $(n + 1)P_n$ permutations are different. He then concluded, "Thus it is proved that the number of permutations of a given set of elements is equal to that number formed by multiplying together the natural numbers from 1 up to the number of given elements. For the number of permutations of 2 elements is 2, and that is equal to $1 \cdot 2$, the number of permutations of 3 elements is equal to the product $3 \cdot 2$, which is equal to $1 \cdot 2 \cdot 3$, and so one shows this result further without end."[15] Namely, Levi mentioned the beginning step and then noted that with the inductive step already proved, the complete result is also proved.

After proving, using a counting argument, that $P_2^n = n(n - 1)$ (where P_k^n represents the number of permutations of k elements in a set of n), Levi proved that $P_k^n = n(n - 1)(n - 2) \cdots (n - k + 1)$ by induction on k. As before, he stated the inductive step as a theorem:

PROPOSITION 65 *If a certain number of elements is given and the number of permutations of order a number different from and less than the given number of elements is a third number, then the number of permutations of order one more in this given set of elements is equal to* *the number which is the product of the third number and the difference between the first and the second numbers.*

Modern symbolism replaces Levi's convoluted wording with a brief phrase: $P^n_{j+1} = (n - j)P^n_j$. Levi's proof is quite similar to that of Proposition 63. At the end, he stated the complete result: "It has thus been proved that the permutations of a given order in a given number of elements are equal to that number formed by multiplying together the number of integers in their natural sequence equal to the given order and ending with the number of elements in the set."[16] To clarify this statement, Levi first gave the initial step of the induction by quoting his previous result in the case $n = 7$, that is, the number of permutations of order 2 in a set of 7 is equal to $6 \cdot 7$. Then the number of permutations of order 3 is equal to $5 \cdot 6 \cdot 7$ (since $5 = 7 - 2$). Similarly, the number of permutations of order 4 is equal to $4 \cdot 5 \cdot 6 \cdot 7$, "and so one proves this for any number."

In the final three propositions of the theoretical part of *Maasei Ḥoshev*, Levi completed his development of formulas for permutations and combinations. Proposition 66 showed that $P^n_k = C^n_k P^k_k$, while Proposition 67 simply rewrote this as $C^n_k = P^n_k / P^k_k$. Since he had already given formulas for both the numerator and denominator of this quotient, Levi thus had demonstrated the standard formula for C^n_k:

$$C^n_k = \frac{n(n - 1) \cdots (n - k + 1)}{1 \cdot 2 \cdots k}.$$

And finally, Proposition 68 demonstrated that $C^n_k = C^n_{n-k}$.

Levi gave examples of many of these results in the second section of his book. For example, he noted that to determine the sum of the cubes of the numbers from 1 to 6, one first calculates that the sum of the numbers themselves is 21 and therefore the sum of the cubes is the square of 21, namely, 441. Or to find the number of permutations of five elements out of a set of eight, P^8_5, one multiplies $4 \cdot 5 \cdot 6 \cdot 7 \cdot 8$ to get 6720. Then the number of combinations of five elements out of eight, namely, C^8_5, is that number divided by $1 \cdot 2 \cdot 3 \cdot 4 \cdot 5$, or 120. The result is 56.

Finally, at the end of the second section, Levi presented a number of interesting problems, most seemingly "practical," and most of which could be solved through a knowledge of ratio and proportion. These problems, including some familiar ones, become progressively harder, but Levi gave a detailed explanation of the solution to each (Sidebar 10.2). Two of these problems are included in the exercises to this chapter.

10.4   MEDIEVAL ALGEBRA

Although the theory of combinatorics appears to have developed in Europe through the Jewish tradition, the writers on algebra in medieval Europe were direct heirs to Islamic work.

10.4.1   Leonardo of Pisa's *Liber Abbaci*

One of the earliest European writers on algebra was Leonardo of Pisa, most famous for his masterpiece, the *Liber abbaci*, or *Book of Calculation*. (The word *abbaci*, from *abacus*, does not refer to a computing device but simply to calculation in general.) The first edition

---

SIDEBAR 10.2   *Did Anyone Read the Works of Levi ben Gerson?*

Although the *Maasei Hoshev* was the earliest work in Europe to consider the combinatorial formulas in detail as well as to provide examples of proof by mathematical induction, it does not seem to have had any influence in the subsequent centuries. As far as can be determined, there are no references to this work in any later European mathematical work, and in fact the combinatorial formulas themselves do not appear anywhere in Europe for the next 200 years, nor is proof by induction used before the work of Pascal in the mid-seventeenth century. So what happened to Levi's book? Did anyone read it?

The simple answer to the last question is, yes. There are today about a dozen manuscript copies of the work extant, in libraries throughout Europe as well as one copy in New York, most written in the fifteenth and sixteenth centuries. For a medieval manuscript, that is not a trivial number of copies. And for some copies, we know the name of the copyist or the original owner. In fact, the copy in London was at one time owned by Mordecai Finzi of Mantua, the fifteenth-century Jewish scientist who translated the work of Abu Kāmil into Hebrew. The more important question then is, Did anyone read the *Maasei Hoshev* who used it to continue work in the field of combinatorics?

As far as Finzi is concerned, there is no record of his ever having written about combinatorics. On the other hand, we do know that Marin Mersenne wrote about combinatorics in his works on music theory in the mid-1630s. And his methodology bears some resemblance to the work of Levi. Could he have read Levi's work or learned of it through one of his many correspondents? For that to have happened, there would have had to be a copy of the manuscript in Paris available for someone who both read Hebrew and understood mathematics. In fact, all of these conditions are satisfied. A copy of the *Maasei Hoshev* was brought to Paris around 1620 by Achille Harlay de Sancy, the French ambassador to Constantinople.

De Sancy donated the manuscript—as part of his large collection of Hebrew manuscripts—to the library of the Oratorian priests, whose Paris house he joined as well. There the manuscript remained until the Oratorian houses were closed in the 1790s during the French Revolution. Now Mersenne was certainly in contact with many priests at the Oratory, and we also know that some of them read Hebrew and were trained in mathematics. But there the trail ends for now. The manuscript itself has no notes on it, nor are there library records from the Oratory that tell us who may have looked at the manuscript. So we may never know the answer to the question posed above.

---

of this work appeared in 1202, while a slightly revised one was published in 1228. The many surviving manuscripts testify to the wide readership the book enjoyed. The sources for the *Liber abbaci* were largely in the Islamic world, which Leonardo visited during many journeys, but he enlarged and arranged the material he collected through his own genius. The book contained not only the rules for computing with the new Hindu-Arabic numerals, but also numerous problems of various sorts in such practical topics as calculation of profits, currency conversions, and measurement, supplemented by the now standard topics of current algebra texts such as mixture problems, motion problems, container problems, the Chinese remainder problem, and, at the end, various forms of problems solvable by use of quadratic equations. Interspersed among the problems is a limited amount of theory, such as methods for summing series, geometric justifications of the quadratic formulas, and even a brief discussion of negative numbers.

Leonardo used a great variety of methods in his solution of problems. Often, in fact, he used special procedures designed to fit a particular problem rather than more general methods. One basic method used often is the old Egyptian method of "false position" in which a convenient, but wrong, answer is given first and then adjusted appropriately to get the correct result. Similarly, he used the method of "double false position," a method that has its origins in China but was also used in medieval Islam. Leonardo also used the methods of al-Khwārizmī for solving quadratic equations. For many of the problems, it is possible to cite Leonardo's sources. He often took problems verbatim from such Islamic mathematicians as al-Khwārizmī, Abū Kāmil, and al-Karajī, many of which he found in Arabic manuscripts discovered in his travels. Some of the problems seem ultimately to have come from China or India, but Leonardo probably learned these in Arabic translations. The majority of the problems, however, are of his own devising and show his creative abilities. A few of Leonardo's problems and solutions should give the flavor of this most influential mathematical work.

Leonardo began his text by introducing the Hindu-Arabic numerals: "The nine Indian figures are 9, 8, 7, 6, 5, 4, 3, 2, 1. With these nine figures, and with the sign 0, which the Arabs call *zephir* (cipher), any number whatsoever is written, as is demonstrated below."[17] He then showed precisely that, giving the names to the various places in the place value system (for integers only). Leonardo next dealt with various algorithms for adding, subtracting, multiplying, and dividing whole numbers and common fractions. His notation for mixed numbers differed from ours in that he wrote the fractional part first, but his algorithms are generally close to the ones we use today. For example, to divide 83 by $5\frac{2}{3}$ (or, as he writes, 2/3 5), Leonardo multiplied 5 by 3 and added 2, giving 17. He then multiplied 83 by 3, giving 249, and finally divided 249 by 17, giving $14\frac{11}{17}$. To add $1/5 + 3/4$ to $1/10 + 2/9$, Leonardo multiplied the first two denominators, 4 and 5, to get 20, then multiplied this by the denominator 9 to get 180. A multiplication by 10 was unnecessary since 10 is already a factor of 180. Then $1/5 + 3/4$ times 180 is 171, while $1/10 + 2/9$ times 180 is 58. The sum of these two, 229, is then divided by 180 to get the final result, $1\frac{49}{180}$. Leonardo wrote the answer as $\frac{1\ 6\ 2}{2\ 9\ 10}1$, by which he meant $1 + \frac{1}{2\cdot9\cdot10} + \frac{6}{9\cdot10} + \frac{2}{10}$. This latter notation perhaps derives from the Pisan monetary system. Because 1 pound is divided into 20 *solidi* and each *solidus* is divided into 12 *denarii*, it was convenient for him, for example, to write 17 pounds, 11 *solidi*, 5 *denarii* as $\frac{5\ 11}{12\ 20}17$. Notations aside, Leonardo was able to use his procedures effectively to show his readers how to perform the intricate calculations needed to convert among the many currencies in use in the Mediterranean basin during his time.

Leonardo presented several versions of the classic problem of buying birds. In the first, he asked how to buy 30 birds for 30 coins, if partridges cost 3 coins each, pigeons 2 coins each, and sparrows 2 for 1 coin. He began by noting that he could buy 5 birds for 5 coins by taking 4 sparrows and 1 partridge. Similarly, 2 sparrows and 1 pigeon give him 3 birds for 3 coins. By multiplying the first transaction by 3 and the second by 5, he procured 12 sparrows and 3 partridges for 15 coins and 10 sparrows and 5 pigeons also for 15 coins. Adding these two transactions gave the desired answer: 22 sparrows, 5 pigeons, 3 partridges.

Another classic problem is that of the lion in the pit: The pit is 50 feet deep. The lion climbs up 1/7 of a foot each day and then falls back 1/9 of a foot each night. How long will it take him to climb out of the pit? Leonardo here used a version of "false position." He assumed the answer to be 63 days, since 63 is divisible by both 7 and 9. Thus, in 63 days the lion will climb up 9 feet and fall down 7, for a net gain of 2 feet. By proportionality, then, to climb 50 feet, the lion will take 1575 days. (By the way, Leonardo's answer is incorrect. At the end of 1571 days, the lion will be 8/63 of a foot from the top. On the next day, he will reach the top.)

Leonardo's example of the Chinese remainder problem asked to find a number that when divided by 2 had remainder 1, by 3 had remainder 2, by 4 had remainder 3, by 5 had remainder 4, by 6 had remainder 5, and by 7 had remainder 0. To solve this, he noted that 60 was evenly divisible by 2, 3, 4, 5, and 6. Therefore, $60 - 1 = 59$ satisfied the first five conditions as did any multiple of 60, less 1. Thus, he had to find a multiple of 60 that had remainder 1 on division by 7. The smallest such number is 120, and therefore 119 is the number sought. (Interestingly, this problem was also posed by ibn al-Haytham two centuries earlier.)

Negative numbers appear in one of Leonardo's many problems dealing with a purse found by a number of men. In this particular problem, there are 5 men. The amount the first has together with the amount in the purse is $2\frac{1}{2}$ times the total of the amounts held by the other four. Similarly, the second man's amount together with the amount in the purse is $3\frac{1}{3}$ times the total held by the others. Analogously, the fraction is $4\frac{1}{4}$ for the third man, $5\frac{1}{5}$ for the fourth man, and $6\frac{1}{6}$ for the fifth man. Leonardo worked out the problem and discovered that the only way this can be solved is for the first man to begin with a debt of 49,154. In a few other problems, he also gave negative answers, and even demonstrated an understanding of the basic rules for adding and subtracting with these numbers.

Leonardo used many methods to solve his problems, but in later chapters of the book he tended toward methods that are explicitly algebraic. In fact, Leonardo credited the Arabs with what he called the "direct" method of solution, a method that involves setting up an equation and then simplifying it according to standard rules. For example, suppose two men have some money, and one said to the other: If you will give me 7 of your *denarii*, then I will have five times as much as you. The other said, if you give me 5 *denarii*, then I will seven times as much as you. Leonardo started by assuming that the second man has "thing" plus 7 *denarii*. Then the first man has five things minus 7. If the first then gives 5 to the second, he will have five things minus 12, while the second man will have thing plus 12. The equation is then "one thing and 12 *denarii* are seven times five things minus 12 *denarii*." Leonardo then solved the equation to find that "thing" is $2\frac{14}{17}$ *denarii*, and therefore that the second man began with $9\frac{14}{17}$ *denarii*, while the first began with $7\frac{2}{17}$ *denarii*.

Leonardo also dealt comfortably with determinate and indeterminate problems in more than two unknowns. For example, suppose there are four men such that the first, second, and third together have 27 *denarii*, the second, third, and fourth together have 31, the third, fourth, and first have 34, while the fourth, first, and second have 37. To determine how much each man has requires solving a system of four equations in four unknowns. Leonardo accomplished this expeditiously by adding the four equations together to determine that four times the total sum of money equals 129 *denarii*. The individual amounts are then easily calculated. On the other hand, in a similar question reducible to the four equations $x + y = 27$, $y + z = 31$, $z + w = 34$, $x + w = 37$, Leonardo first noted that this system is impossible since the two different ways of calculating the total sum of money give two different answers, 61 and 68. However, if one changes the fourth equation to $x + w = 30$, one can simply choose x arbitrarily ($x \leq 27$) and calculate y, z, and w by using the first, second, and third equations, respectively.

The most famous problem of the *Liber abbaci*, the rabbit problem, is tucked inconspicuously between a problem on perfect numbers and the problem just discussed: "How many pairs of rabbits are created by one pair in one year? A certain man had one pair of rabbits together in a certain enclosed place, and one wishes to know how many are created from the pair in one year when it is the nature of them in a single month to bear another pair, and in the second month those born to bear also."[18] Leonardo proceeded to calculate: After the first month there will be two pairs, after the second, three. In the third month, two pairs will produce, so at the end of that month there will be five pairs. In the fourth month, three pairs will produce, so there will be eight. Continuing in this fashion, he showed that there will be 377 pairs by the end of the twelfth month. Listing the sequence 1, 2, 3, 5, 8, 13, 21, 34, 55, 89, 144, 233, 377 in the margin, he noted that each number is found by adding the two previous numbers, and "thus you can do it in order for an infinite number of months." This sequence, calculated recursively, is known today as a **Fibonacci sequence**. It turns out that it has many interesting properties unsuspected by Leonardo, not the least of which is its connection with the Greek problem of dividing a line in extreme and mean ratio.

In his final chapter, Leonardo demonstrated his complete command of the algebra of his Islamic predecessors as he showed how to solve equations that reduce ultimately to quadratic equations. He discussed in turn each of the six basic types of quadratic equations, as given by al-Khwārizmī, and then gave geometric proofs of the solution procedures for each of the three mixed cases. He followed the proofs with some 50 pages of examples, most taken from the works of al-Khwārizmī and Abū Kāmil, including the familiar ones beginning with "divide 10 into two parts." In particular, he included the latter's problem of three equations in three unknowns, discussed in Chapter 9.

The content of the *Liber abbaci* contained no particular advance over mathematical works then current in the Islamic world. In fact, as far as the algebra was concerned, Leonardo was only presenting tenth-century Islamic mathematics and ignoring the advances of the eleventh and twelfth centuries. The chief value of the work, nevertheless, was that it did provide Europe's first comprehensive introduction to Islamic mathematics. Those reading it were afforded a wide variety of methods to solve mathematical problems, methods that provided the starting point from which further progress could ultimately be made.

10.4.2   The *Liber Quadratorum*

Another briefer work by Leonardo, the *Liber quadratorum* (*Book of Squares*) of 1225, is much more theoretical than the *Liber abbaci*. This is a book on number theory, in which Leonardo discussed the solving in rational numbers of various equations involving squares. The book originated in a question posed to Leonardo by a Master John of Palermo, a member of the entourage of the Holy Roman Emperor Frederick II, whom Leonardo met as described in the opening of this chapter. According to Leonardo, Master John proposed the question, to "find a square number from which, when five is added or subtracted, always arises a square number. Beyond this question, the solution of which I have already found, I saw, upon reflection, that this solution itself and many others have origin in the squares and the numbers which fall between the squares."[19] The initial problem, to find x, y, z, so that $x^2 + 5 = y^2$ and $x^2 - 5 = z^2$, is solved as the seventeenth of the 24 propositions of the book, but Leonardo first developed various properties of square numbers and sums of square numbers. Interestingly, John of Palermo was not only a mathematician but also an Arabic scholar, who may well have been familiar with this problem from the work of al-Karajī.

To solve Master John's problem, Leonardo introduced what he called **congruous** numbers, numbers n of the form $ab(a + b)(a - b)$ when $a + b$ is even and $4ab(a + b)(a - b)$ when

NONBIOGRAPHY

Jordanus de Nemore

Although Jordanus has been recognized as one of the best mathematicians of the Middle Ages, there is virtually no available evidence about his life, other than that he appears to have been connected with the University of Paris in the early decades of the thirteenth century. Some years ago, he was identified with Jordanus de Saxonia, the second Master General of the Dominican order, but recent scholarly work has shown that this identification is impossible. The translator of

De numeris datis, Barnabas Hughes, concludes that Jordanus is *sine patre, sine matre, sine genealogia*. He also notes in a letter, however, that "the only explanation that appealed to me [as to why no biographical information is extant] was that the name is a pseudonym. But why a *nom de plume*? Could it be that Jordanus was really a woman? Shades of Hypatia! Thirteenth century women were good for writing poems, songs and prayers; but science?"[21]

$a + b$ is odd. He showed that congruous numbers are always divisible by 24 and that integral solutions of $x^2 + n = y^2$ and $x^2 - n = z^2$ can be found only if n is congruous. The original problem is therefore not solvable in integers. Nevertheless, since $720 = 12^2 \cdot 5$ is a congruous number (with $a = 5$ and $b = 4$) and since $41^2 + 720 = 49^2$ and $41^2 - 720 = 31^2$, it follows by dividing both equations by 12^2 that $x = 41/12$, $y = 49/12$, $z = 31/12$, provides a solution in rational numbers to $x^2 + 5 = y^2$, $x^2 - 5 = z^2$. Leonardo's methodology is different from that of al-Karajī, although he does get the same answer, but it is similar to that in other Islamic treatises on number theory of the same time period, including the work of Abū Ja-far al-Khāzin (tenth century).[20]

It is clear that Leonardo mastered the Islamic mathematics he had learned in his travels and passed what he knew on to his European successors. With respect to the number theory of the *Liber Quadratorum*, Leonardo had no successor until several centuries later when Diophantus's *Arithmetica* was again available in Europe. On the other hand, the practical material in the *Liber abbaci* and the *Practica geometriae* was picked up by Italian surveyors and masters of computation (*maestri d'abbaco*), who were influential in the next several centuries in bringing a renewed sense of mathematics into Italy. It took a full 300 years, however, for this renewed mathematical knowledge to increase to the point where conditions in Italy were advanced enough for new mathematics to be created.

10.4.3 Jordanus de Nemore

One of the first mathematicians to make some advance over the work of Leonardo was a contemporary, Jordanus de Nemore. Although we know virtually nothing about the author himself, it is believed that he taught in Paris around 1220. His writings include several works on arithmetic, geometry, astronomy, mechanics, and algebra, and it appears that he worked to create a Latin version of the quadrivium, based upon a theoretical work on arithmetic. Jordanus's *Arithmetica* was far different from the demonstrationless arithmetic of Boethius, which was then circulating widely in Europe. Jordanus's work was firmly based on a Euclidean model, with definitions, axioms, postulates, propositions, and careful proofs. Also like Euclid, Jordanus did not give any numerical examples.

The *Arithmetica*, a work in 10 books, dealt with such topics found in Euclid as ratio and proportion, prime and composite numbers, the Euclidean algorithm, and the geometrical algebra propositions of *Elements*, Book II. It also considered much material not found in Euclid, including figurate numbers and a detailed study of named ratios, due to Nicomachus. Most interesting, however, are a few items not found in the Greek sources. For example, in Book VI Jordanus solved a problem virtually identical to the central problem of Leonardo's *Book of Squares*.

PROPOSITION VI–12   *To find three square numbers whose continued differences are equal.*

In modern symbols, Jordanus wanted to determine y^2, x^2, and z^2 so that $y^2 - x^2 = x^2 - z^2$. His solution amounted to setting

$$y = \frac{a^2}{2} + ab - \frac{b^2}{2}, \qquad x = \frac{a^2 + b^2}{2}, \qquad z = \frac{b^2}{2} + ab - \frac{a^2}{2},$$

where a, b, have the same parity. In contrast to Leonardo, Jordanus was just interested in integral solutions, but did not give any example. Nevertheless, it is straightforward to see that the difference of the squares in Jordanus's theorem is a congruous number according to Leonardo's definition.

And in Book IX, Jordanus displayed the "Pascal" triangle, for the first time in a European work. The construction of the triangle is the standard one:

> Put 1 at the top and below two 1's. Then the row of two 1's is doubled so that the first 1 is in the first place and another 1 in the last place as in the second row; and 1,2,1 will be in the third row. The numbers are added two at a time, the first 1 to the 2 in the second place, and so on through the row until a final 1 is put at the end. Thus the fourth row has 1,3,3,1. In this way subsequent numbers are made from pairs of preceding numbers.[22]

Most of the medieval manuscripts of the *Arithmetica* display a version of the triangle at this point, some even up to the tenth line. Jordanus then used the triangle explicitly in Proposition IX–70 to construct series of terms in given ratios. For example, if $a = b = c = d = 1$, then the numbers $e = 1a = 1$, $f = 1a + 1b = 2$, $g = 1a + 2b + 1c = 4$, and $h = 1a + 3b + 3c + 1d = 8$ form a continued proportion with constant ratio 2. Similarly, $k = 1e = 1$, $\ell = 1e + 1f = 3$, $m = 1e + 2f + 1g = 9$, and $n = 1e + 3f + 3g + 1h = 27$ form a continued proportion with ratio 3.

Jordanus's *Arithmetica* was widely read, at least judging from the number of extant manuscripts. Similarly, his major work on algebra, *De numeris datis (On Given Numbers)*, also had a large circulation in medieval Europe. *De numeris datis* is an analytic work on algebra, based on but differing in spirit from the Islamic algebras that had made their way into Europe by the early thirteenth century. It appears to be modeled on Euclid's *Data*, available to Jordanus in a Latin translation of Gerard of Cremona, in that it presents problems in which certain quantities are given and then shows that other quantities are therefore also determined. The problems in *De numeris datis*, however, are algebraic rather than geometric. Jordanus's proofs are also algebraic, or, perhaps, arithmetical. In fact, one of his aims is apparently to base the new algebra on arithmetic, the most fundamental of the subjects of the quadrivium, rather than on geometry, and especially on his own work on the subject. He also organized his book in a logical fashion and, in a major departure from his Euclidean model, and even from his own *Arithmetica*, provided numerical examples for most of his theoretical results.

Although many of the actual problems and the numerical examples were available in the Islamic algebras, Jordanus adapted them to his own purposes. In particular, he made the major change of using letters to stand for arbitrary numbers. Jordanus's algebra was no longer entirely rhetorical. That is not to say that his symbolism was modern-looking. He picked his letters in alphabetical order with no distinction between letters representing known quantities and those representing unknowns and used no symbols for operations. Sometimes a single number was represented by two letters. At other times the pair of letters ab represented the sum of the two numbers a and b. The basic arithmetic operations were always written in words. And Jordanus did not use the new Hindu-Arabic numerals. All of his numbers were written as Roman numerals. Nevertheless, the idea of symbolism, so crucial to any major advance in algebraic technique, was found, at least in embryonic form, in Jordanus's work.

To understand Jordanus's contribution, we consider a few of the text's more than 100 propositions, organized into four books. Like Euclid, Jordanus wrote each proposition in a standard form. The general enunciation was followed by a restatement in terms of letters. By use of general rules, the letters representing numbers were manipulated into a canonical form from which the general solution could easily be found. Finally, a numerical example was calculated following the general outlines of the abstract solution. The canonical forms themselves are among the earliest of the propositions.

PROPOSITION I–1 *If a given number is divided into two parts whose difference is given, then each of the parts is determined.*

Jordanus's proof was straightforward: "Namely, the lesser part and the difference make the greater. Thus the lesser part with itself and the difference make the whole. Subtract therefore the difference from the whole and there will remain double the lesser given number. When divided [by two], the lesser part will be determined; and therefore also the greater part. For example, let 10 be divided in two parts of which the difference is 2. When this is subtracted from 10 there remains 8, whose half is 4, which is thus the lesser part. The other is 6."[23]

In modern symbolism, Jordanus's problem amounted to the solution of the two equations $x + y = a$, $x - y = b$. Jordanus noted first that $y + b = x$, so that $2y + b = a$ and therefore $2y = a - b$. Thus, $y = \frac{1}{2}(a - b)$ and $x = a - y$.

Jordanus used this initial proposition in many of the remaining problems of Book I. For example, consider

PROPOSITION I–3 *If a given number is divided into two parts, and the product of one by the other is given, then of necessity each of the two parts is determined.*

This proposition presented one of the standard Babylonian problems: $x + y = m$, $xy = n$. Jordanus's method of solution, however, is different from the classic Babylonian solution, and, in addition, he used symbolism as indicated: Suppose the given number abc is divided into the parts ab and c. Suppose ab multiplied by c is d and abc multiplied by itself is e. Let f be the quadruple of d, and g be the difference of e and f. Then g is the square of the difference between ab and c. Its square root b is then the difference between ab and c. Since now both the sum and difference of ab and c are given, both ab and c are determined according to the first proposition. Jordanus's numerical example used 10 as the sum of the two parts and 21 as the product. He noted that 84 is quadruple 21, that 100 is 10 squared, and that 16 is their difference. Then the square root of 16, namely, 4, is the difference of the two parts of 10. By the proof of the first proposition, 4 is subtracted from 10 to get 6. Then 3 is the desired smaller part while 7 is the larger.

Jordanus's solution, translated into modern symbolism, used the identity $(x - y)^2 = (x + y)^2 - 4xy = m^2 - 4n$ to determine $x - y$ and reduce the problem to Proposition I–1. The solution is then $y = \frac{1}{2}(m - \sqrt{m^2 - 4n})$, $x = m - y$. Jordanus's method appears to be new with him, and he continued to use his own methods throughout the work. Even his solution of problems in Book I equivalent to pure quadratic equations used methods different from the standard ones used by the Islamic algebraists. Nevertheless, the numerical examples themselves have a familiar look. In fact, every proposition in Book I deals with a number divided into two parts, and in every example but one the number to be divided is 10. The solution methods may differ somewhat from those in the Islamic texts, but it is clear that al-Khwārizmī's problems live on!

Many of the propositions in the remaining three books of Jordanus's treatise dealt with numbers in given proportion. They demonstrated his fluency in dealing with the rules of proportion found in Books V and VII of Euclid's *Elements*, most of which are also found in his own *Arithmetica*. Consider

PROPOSITION II–18   *If a given number is divided into however many parts, whose continued proportions are given, then each of the parts is determined.*

Because Jordanus, like his contemporaries, had no way to express arbitrarily many "parts," he dealt in his proof with a number divided into three parts: $a = x + y + z$. Then $x : y = b$ and $y : z = c$ are both known ratios. Jordanus noted that the ratio $x : z$ is also known. It follows that the ratio of x to $y + z$ is known and therefore also that of a to x. Since a is known, x and then y and z can be determined. His example enables us to follow his verbal description. The number 60 is divided into three parts, of which the first is double the second and the second is triple the third. That is, $x + y + z = 60$, $x = 2y$, $y = 3z$. Then $x = 6z$ and therefore $y + z = \frac{2}{3}x$. So $60 = \left(1\frac{2}{3}\right)x$, and $x = 36$, $y = 18$, $z = 6$. One notes that Jordanus easily inverted ratios if necessary and also knew how to combine them.

Among the propositions in Book IV are three giving the three standard forms of the quadratic equation, all presented with algebraic rather than geometric justifications. For these problems, however, Jordanus did use the standard Islamic algorithms, but with his own symbolism. Consider

PROPOSITION IV–9   *If the square of a number added to a given number is equal to the number produced by multiplying the root and another given number, then two values are possible.*

Thus, Jordanus asserted that there are two solutions to the equation $x^2 + c = bx$. He then gave the procedure for solving the equation: Take half of b, square it to get f, and let g be the difference of x and $\frac{1}{2}b$, that is, $g = \pm(x - \frac{1}{2}b)$. Then $x^2 + f = x^2 + c + g^2$ and $f = c + g^2$. Jordanus concluded by noting that x may be obtained by either subtracting g from $b/2$ or by adding g to $b/2$, that is,

$$x = \frac{b}{2} \pm \sqrt{\left(\frac{b}{2}\right)^2 - c}.$$

His example made his symbolic procedure clearer. To solve $x^2 + 8 = 6x$, he squared half of 6, giving 9, and then subtracted 8 from it, leaving 1. The square root of 1 is 1, and this is the difference between x and 3. Hence, x can be either 2 or 4.

Among the other quadratic problems Jordanus solved in Book IV are the systems $xy = a$, $x^2 + y^2 = b$, and $xy = a$, $x^2 - y^2 = b$. In each case, as in all the previous cases, the given example resulted in a positive integral answer. While Jordanus often used fractions as part of his solution, he carefully arranged matters so that final answers were always whole numbers. If, in fact, he had read Abū Kāmil's *Algebra*, which was available in Latin, Jordanus would have seen nonintegral, and even nonrational, solutions to this type of problem. He nevertheless rejected such solutions when he made up his examples. Given his very formal style, however, Jordanus may still have been under the influence of Euclid and may have felt that irrational "numbers" simply did not belong in a work based on arithmetic. Hence, although *De numeris datis* represented an advance from the Islamic works in the use of analysis, in the constant striving toward generality, and in some symbolization, it returned to the strict Greek separation of number from magnitude, an idea from which Jordanus's Islamic predecessors had already departed. Thus, it appears that although Jordanus certainly made use of the new Islamic material available in Europe, his goal was to provide his readers with a mathematics based as much as possible on Greek principles.

10.5 THE MATHEMATICS OF KINEMATICS

The algebraic work of Jordanus de Nemore was not developed further in the thirteenth century, even though a group of followers had appeared in Paris by the middle of that century. Perhaps Europe was not then ready to resume the study of pure mathematics. By early in the fourteenth century, however, certain other aspects of mathematics began to develop in the universities of Oxford and Paris out of attempts to clarify certain remarks in Aristotle's physical treatises (Sidebar 10.3).

10.5.1 The Study of Ratios

One of the new mathematical ideas came from the effort to derive a relationship among the force F applied to an object, its resistance R, and its velocity V. A basic postulate of medieval physics was that F must be greater than R for motion to be produced. (The medieval philosophers did not attempt to measure these quantities in any particular units.) The simplest relationship among these quantities implied by Aristotle's own words may be expressed by the statement that F/R is proportional to V. This mathematical relationship, however, quickly leads to a contradiction of the postulate. For if F is left fixed, the continual doubling of R is equivalent to the continual halving of V. Halving a positive velocity keeps it positive, but the doubling of R eventually makes R greater than F, thus contradicting the notion that F must be greater than R for motion to take place.

Thomas Bradwardine (1295–1349) of Merton College, Oxford, in his 1328 *Tractatus de proportionibus velocitatum in motibus* (*Treatise on the Proportions of Velocities in Movements*), proposed a solution to this dilemma, that is, a "correct" interpretation of Aristotle's

SIDEBAR 10.3 *The Medieval Universities*

It was during the late twelfth century that Europe saw the beginnings of the institutions that were to have immense influence in the development of science in general and mathematics in particular: the universities. We cannot assign any specific date for the origins of the earliest universities. They were formed as societies, or guilds, of masters and pupils and appeared on the scene when there was enough learning available in western Europe to justify their existence. The earliest of these institutions were in Paris, Oxford, and Bologna. In Paris, the university grew out of the cathedral school of Notre Dame. The masters and students gradually grouped themselves into the four faculties of arts, theology, law, and medicine. Although there is evidence of the existence of the university in the late twelfth century, the first official charter dates from 1200. The University at Oxford emerged not out of a church school but from a group of English students who had returned from Paris. Again, although the university certainly existed in the late twelfth century, the first official document dates from 1214. At Bologna, the university began as a law school, perhaps as early as the eleventh century. The Italian university differed from its northern counterparts, however, since it was a guild of students, rather than one of professors, that initially constituted the organization. The students elected the professors and other officials. Student control was somewhat weakened, however, because salaries were paid by the Bolognese municipality and the faculty conducted the examinations.

The curriculum in arts at all of the universities was based on the ancient trivium of logic, grammar, and rhetoric and the quadrivium of arithmetic, geometry, music, and astronomy (Fig. 10.11). This study in the faculty of arts provided the student with preparation for the higher faculties of law, medicine, or theology. The centerpiece of the arts curriculum was the study of logic, and the primary texts for this were the logical works of Aristotle, all of which had by this period been translated into Latin. The masters felt that logic was the appropriate first area of study since it taught the methods for all philosophic and scientific inquiry. Gradually, other works of Aristotle were also added to the curriculum. For several centuries, the great philosopher's works were the prime focus of the entire arts curriculum. Other authors were studied insofar as they allowed one to better understand this most prolific of the Greek philosophers. In particular, mathematics was studied only as it related to the work of Aristotle in logic or the physical sciences. The mathematical curriculum itself—the quadrivium—usually consisted of arithmetic, taken from such works as Boethius's adaptation of Nicomachus or a medieval text on rules for calculation; geometry, taken from Euclid and one of the practical geometries; music, taken also from a work of Boethius; and astronomy, taken from Ptolemy's *Almagest* and some more recent Latin translations of Islamic astronomical works.

FIGURE 10.11

The quadrivium—arithmetic, geometry, music, and astronomy—on two stamps from the Netherlands Antilles

remarks. The rule noted above implies that for two forces F_1, F_2, two resistances R_1, R_2, and two velocities V_1, V_2, the equation

$$\frac{F_2}{R_2} = \frac{V_2}{V_1}\frac{F_1}{R_1}$$

is satisfied. Bradwardine suggested that this should be replaced by the relationship expressed in modern notation as

$$\frac{F_2}{R_2} = \left(\frac{F_1}{R_1}\right)^{\frac{V_2}{V_1}}.$$

In other words, the multiplicative relation should be replaced by an exponential one. This solution indeed removed the absurdity noted above. Given initially that $F > R$ (or $F/R > 1$), halving the velocity in this situation is equivalent to taking roots of the ratio $\frac{F}{R}$. Consequently, F/R will remain greater than 1, and R will never be greater than F. Neither Bradwardine nor anyone else in this period, however, attempted to give any experimental justification for this relationship. The scholars at Merton wanted a mathematical explanation of the world, not a physical one. As it turned out, Bradwardine's idea was discarded as a physical principle by the middle of the next century, but the mathematics behind it led to important new ideas. To deal with these required a systematic study of ratios, in particular, of the idea of compounding (or multiplying) ratios.

Until the fourteenth century, compounding was performed in the classical Greek style. Thus, to deal with the ratio compounded of $a : b$ and $c : d$, one needed to find a magnitude e such that $c : d = b : e$. Then the desired compound ratio would be $a : e$. Gradually, however, the more explicit notion of multiplication of ratios was introduced. For example, Bradwardine's contemporary at Oxford, Richard of Wallingford, defined ratios as well as their compounding and dividing in part II of his *Quadripartitum*:

1. A **ratio** is a mutual relation between two quantities of the same kind.
2. When one of two quantities of the same kind divides the other, what results from the division is called the **denomination** of the ratio of the dividend to the divisor.
3. A ratio [is said to be] **compounded** of ratios when the product of the denominations gives rise to some denomination.
4. A ratio [is said to be] **divided** by a ratio when the quotient of the denominations gives rise to some denomination.[24]

There are several important notions here. First, Richard emphasized that ratios can be taken only between quantities of the same kind. This Euclidean idea meant that velocity could not be treated as a ratio of distance to time. Second, the word *denomination* in these definitions referred to the "name" of the ratio in "lowest terms," as given in the terminology due to Nicomachus now standard in Europe. For example, the ratio $3 : 1$ was called a triple ratio, while that of $3 : 2$ was called the sesqualter. Finally, definitions 3 and 4 showed that for Richard, unlike for Euclid, multiplication (of numbers) was involved in compounding, and the inverse notion of division could also be applied. Thus, although he compounded the ratios $4 : 16$, $16 : 2$, and $2 : 12$ to get $4 : 12$, he noted that since the first ratio is a subquadruple $(1 : 4)$, the second an octuple $(8 : 1)$, and the third a subsextuple $(1 : 6)$, they can be compounded by first dividing 8 by 4 to get 2, and then dividing 2 by 6 to get $1 : 3$ (a subtriple) as the final result. Thus, one can actually use the standard algorithm for multiplying fractions to "compound" ratios.

Nicole Oresme (1320–1382), a French cleric and mathematician associated with the University of Paris, undertook a very detailed study of ratios in his *Algorismus proportionum (Algorithm of Ratios)* and his *De proportionibus proportionum (On the Ratios of Ratios)*. In addition to performing compounding in the traditional manner, Oresme noted explicitly that one can also compound ratios by multiplying the antecedents and then multiplying the consequents. Thus, $4 : 3$ compounded with $5 : 1$ is $20 : 3$. The connecting link between the two methods is presumably that $a : b$ can be expressed as $ac : bc$, $c : d$ as $bc : bd$, and hence the compound of $a : b$ with $c : d$ as the compound of $ac : bc$ with $bc : bd$ or as $ac : bd$. In any case, given a way of multiplying two ratios, Oresme also noted that one could reverse the procedure and divide two ratios. Thus, the quotient of $a : b$ by $c : d$ was the ratio $ad : bc$.

Now that the product of any two ratios had been defined, Oresme discussed the product of a given ratio with itself. Thus, $a : b$ compounded with itself n times gives what would now be written as $(a:b)^n$. More importantly, given any ratio, Oresme devised a language for discussing what are now called "roots" of that ratio. Thus, since $2:1$ is a double ratio, Oresme called that ratio, which when compounded twice with itself equals $2:1$, half of a double ratio. (In modern terminology, this is the ratio $(2:1)^{\frac{1}{2}}$). Similarly, he called $(3:1)^{\frac{3}{4}}$ three fourth parts of a triple ratio. Oresme next developed an arithmetic for these ratios. For example, to multiply $(2:1)^{\frac{1}{3}}$ by $3:2$, Oresme cubed the second ratio to get $27:8$, multiplied this by $2:1$ to get $27:4$, and then took the cube root of the ratio considered as a fraction to get $(6\frac{3}{4})^{\frac{1}{3}}$. Similarly, to divide $(2:1)^{\frac{1}{2}}$ by $4:3$, he divided $2:1$ by the square of $4:3$, namely, $16:9$, to get $9:8$ and then took the square root of that, $(9:8)^{\frac{1}{2}}$. In some sense, then, Oresme's works show for the first time operational rules for dealing with exponential expressions with fractional exponents.

Oresme even attempted to deal with what we would call irrational exponents. He felt intuitively that "every ratio is just like a continuous quantity with respect to division," that is, that one could take any possible "part" of such a ratio. So, "there will be some ratio which will be part of a double ratio and yet will not be half of a double nor a third part or fourth part or two-thirds part, etc., but it will be incommensurable to a double and, consequently [incommensurable] to any [ratio] commensurable to this double ratio."[25] Because Oresme had no notation for irrational exponents, he could only convey his sense of them negatively. That is, he felt that ratios of the form $(2:1)^r$ should exist even when r was not a rational number. "And further, by the same reasoning there could be some ratio incommensurable to a double and also to a triple ratio and [consequently incommensurable] to any ratios commensurable to these. . . . And there might be some irrational ratio which is incommensurable to any rational ratio. Now the reason for this seems to be that if some ratio is incommensurable to two [rational ratios] and some ratio is incommensurable to three rational ratios and so on, then there might be some ratio incommensurable to any rational ratio whatever. . . . However, I do not know how to demonstrate this."[26] What Oresme was apparently expressing, in terms of modern ideas, was that since the number line is continuous and since, for example, the fractional powers of 2 do not exhaust all (real) numbers, there must be (nonfractional) powers of 2 equal to the real numbers not already included. In fact, somewhat later in the text he states a theorem to the effect that irrational ratios are much more prevalent than rational ones:

PROPOSITION III–10   *It is probable that two proposed unknown ratios are incommensurable because if many unknown ratios are proposed it is most probable that any [one] would be incommensurable to any [other].*

Although Oresme had no formal way of proving this result, he noted that if one considers all the integral ratios from $2:1$ up to $101:1$, there are 4950 ways of comparing these two by two in terms of exponents (always comparing a greater ratio to a smaller), but only 25 ways with rational exponents. For example, $4:1 = (2:1)^2$ and $8:1 = (4:1)^{\frac{3}{2}}$. On the other hand, there is no rational exponent r such that $3:1 = (2:1)^r$. Oresme then used a probability argument to conclude that astrology must be fallacious. His argument is that with great probability the ratio of any two unknown ratios, for example, those that represent various celestial motions, will be irrational. Since, therefore, there can be no exact repetitions of planetary conjunctions or oppositions, and since astrology rests on such endless repetitions, the whole basis of that "science" is false.

10.5.2 Velocity

The efforts to turn Aristotle's ideas on motion into quantitative results also resulted in new mathematics. In particular, these ideas were developed by Bradwardine and another scholar, William Heytesbury, at Merton College in the early fourteenth century. Recall that Greek mathematicians, including Autolycus and Strato, had dealt with the notion of uniform velocity and, to some extent, accelerated motion, but never considered velocity or acceleration as independent quantities that could be measured. Velocities were only dealt with by comparing distances and times, and therefore, in essence, only average velocities (over certain time periods) could be compared.

The fourteenth century, however, saw the beginning of the notion of velocity, and in particular instantaneous velocity, as measurable entities. Thus, Bradwardine in his *Tractatus de continuo* (*Treatise on the Continuum*) (c. 1330) defined the "grade" of motion as "that part of the matter of motion susceptible to 'more' and 'less.'"[27] Bradwardine then showed how to compare velocities: "In the case of two local motions which are continued in the same or equal times, the velocities and distances traversed by these [movements] are proportional, i.e., as one velocity is to the other, so the space traversed by the one is to the space traversed by the other. . . . In the case of two local motions traversing the same or equal spaces, the velocities are inversely proportional to the time, that is, as the first velocity is to the second, so the time of the second velocity is to the time of the first."[28] In other words, if two objects travel at (uniform) velocities v_1, v_2, respectively in times t_1, t_2, and cover distances s_1, s_2, then (1) if $t_1 = t_2$, then $v_1 : v_2 = s_1 : s_2$, and (2) if $s_1 = s_2$, then $v_1 : v_2 = t_2 : t_1$. Bradwardine thus considered uniform velocity itself as a type of magnitude, capable of being compared with other velocities.

Heytesbury, only a few years later in his *Regule solvendi sophismata* (*Rules for Solving Sophisms*, 1335), gave a careful definition of instantaneous velocity for a body whose motion is not uniform: "In nonuniform motion . . . the velocity at any given instant will be measured by the path which would be described by the . . . point if, in a period of time, it were moved uniformly at the same degree of velocity with which it is moved in that given instant, whatever [instant] be assigned."[29] Having given this explicit definition, Heytesbury noted by example that if two points have the same instantaneous velocity at a particular instant, they do not necessarily travel equal distances in equal times, because their velocities may well differ at other instants.

Heytesbury also dealt with acceleration in this same section: "Any motion whatever is uniformly accelerated if, in each of any equal parts of the time whatsoever, it acquires an equal increment of velocity. . . . But a motion is nonuniformly accelerated . . . when it acquires . . . a greater increment of velocity in one part of the time than in another equal part. . . . And since any degree of velocity whatsoever differs by a finite amount from zero velocity . . . , therefore any mobile body may be uniformly accelerated from rest to any assigned degree of velocity."[30] This statement provides not only a very clear definition of uniform acceleration but also, in nascent form at least, the notion of velocity changing with time. In other words, velocity is being described by Heytesbury as a "function" of time.

How does one determine the distance traveled by a body being uniformly accelerated? The answer, generally known today as the **mean speed rule**, was first stated by Heytesbury in this same work: "When any mobile body is uniformly accelerated from rest to some given degree [of velocity], it will in that time traverse one-half the distance that it would traverse if, in that same time it were moved uniformly at the [final] degree [of velocity]. . . . For that motion, as a whole, will correspond to . . . precisely one-half that degree which is its terminal velocity."[31] In modern notation, if a body is accelerated from rest in a time t with a uniform acceleration a, then its final velocity is $v_f = at$. What Heytesbury is saying is that the distance traveled by this body is $s = \frac{1}{2}v_f t$. Substituting the first formula in the second gives the standard modern formulation $s = \frac{1}{2}at^2$.

Heytesbury gave a proof of the mean speed theorem by an argument from symmetry, taking as his model a body d accelerating uniformly from rest to a velocity of 8 in one hour. (The number 8 does not represent any particular speed, but is just used as the basis for his example.) He then considered three other bodies, a moving uniformly at a speed of 4 throughout the hour, b accelerating uniformly from 4 to 8 in the first half hour, and c decelerating uniformly from 4 to 0 in that same half hour. First, he noted that body d goes as far in the first half hour as does c and as far in the second half hour as does b. Therefore, d travels as far in the whole hour as the total of b and c in the half hour. Second, he argued that since b increases precisely as much as c decreases, together they will traverse as much distance in the half hour as if they were both held at the speed of 4. This latter distance is the same that a travels in the whole hour. It follows that d goes exactly as far as does a in the hour, and the mean speed theorem is demonstrated, at least to Heytesbury's satisfaction. He then proved the easy corollary, that the body d traverses in the second half hour exactly three times the distance it covered in the first half hour.

Other scholars at Merton College in the same time period began to explore the idea of representing velocity, as well as other varying quantities, by line segments. The basic idea seems to come, in effect, from Aristotle, because such notions as time, distance, and length (of line segments) were conceived of as magnitudes in the Greek philosopher's distinction between the two types of quantities. All were infinitely divisible, and hence it was not unreasonable to attempt to represent the somewhat abstract idea of velocity, now itself being quantified, by the concrete geometric idea of a line segment. Velocities of different "degrees" would thus be represented by line segments of different lengths. Oresme carried this idea to its logical conclusion by introducing a two-dimensional representation of velocity changing with respect to time. In fact, in his *Tractatus de configurationibus qualitatum et motuum* (*Treatise on the Configuration of Qualities and Motions*) of about 1350, Oresme even generalized this idea to other cases where a given quantity varied in intensity over either distance or time. Oresme began by explaining why one can use lines to represent such quantities as velocity:

> Every measurable thing except numbers is imagined in the manner of continuous quantity. There-fore, for the mensuration of such a thing, it is necessary that points, lines, and surfaces, or their properties, be imagined. For in them, as [Aristotle] has it, measure or ratio is initially found, while in other things it is recognized by similarity as they are being referred by the intellect to them [the geometrical entities]. Although indivisible points, or lines, are nonexistent, still it is necessary to feign them mathematically for the measures of things and for the understanding of their ratios. Therefore, every intensity which can be acquired successively ought to be imagined by a straight line perpendicularly erected on some point.[32]

From these straight lines Oresme constructed what he called a **configuration**, a geomet-rical figure consisting of all the perpendicular lines drawn over the base line. In the case of velocities, the base line represented time, while the perpendiculars represented the veloci-ties at each instant. The entire figure represented the whole distribution of velocities, which

FIGURE 10.12

Uniform velocity

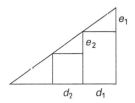

FIGURE 10.13

Uniformly difform velocity, where $d_1 : d_2 = e_1 : e_2$

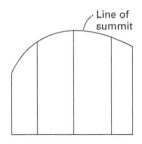

FIGURE 10.14

Difformly difform velocity, or nonuniform acceleration

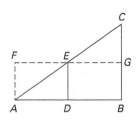

FIGURE 10.15

Proof of mean speed theorem due to Oresme

Oresme interpreted as representing the total distance traveled by the moving object. Oresme did not use what we call coordinates. There was no particular fixed length by which a given degree of velocity was represented. The important idea was only that "equal intensities are designated by equal lines, a double intensity by a double line, and always in the same way if one proceeds proportionally."[33]

For Oresme, then, a uniform quality, for example, a body moving with uniform velocity, is represented by a rectangle, for at each point the velocity is the same (Fig. 10.12). The area of the rectangle represents the total distance traveled. The distance traveled by a body beginning at rest and then moving with constant acceleration, representing what Oresme calls a "uniformly difform" quality, one whose intensity changes uniformly, is the area of a right triangle (Fig. 10.13). As Oresme noted, "A quality uniformly difform is one in which if any three points [of the subject line] are taken, the ratio of the distance between the first and the second to the distance between the second and the third is as the ratio of the excess in intensity of the first point over that of the second point to the excess of that of the second point over that of the third point, calling the first of those three points the one of greatest intensity."[34] This equality of ratios naturally defines a straight line, the hypotenuse of the right triangle. Finally, a "difformly difform" quality, such as nonuniform acceleration, is represented by a figure whose "line of summit" is a curve that is not a straight line (Fig. 10.14). In other words, Oresme in essence developed the idea of representing the functional relationship between velocity and time by a curve. In fact, he noted, "the aforesaid differences of intensities cannot be known any better, more clearly, or more easily than by such mental images and relations to figures."[35] In other words, this geometrical representation of varying quantities provided the best way to study them.

Given this representation of the motion of bodies, it was easy for Oresme to give a geometrical proof of the mean speed theorem. For if triangle ABC represents the configuration of a body moving with a uniformly accelerated motion from rest, and if D is the midpoint of the base AB, then the perpendicular DE represents the velocity at the midpoint of the journey and is half the final velocity (Fig. 10.15). The total distance traveled, represented by triangle ABC, is then equal to the area of the rectangle $ABGF$, precisely as stated by the Mertonians.

Oresme's geometric technique reappeared some 250 years later in the work of Galileo. The difference between the two lay mainly in that Galileo assumed that uniform acceleration from rest was the physical rule obeyed by bodies in free fall, while Oresme was studying the subject only abstractly. This abstraction is evident in Oresme's consideration of cases involving velocities increasing without bound. For example, he considered the case where the velocity of an object during the first half of the time interval AB, taken equal to 1 unit, is equal to 1, that in the next quarter is equal to 2, that in the next eighth is 3, in the next sixteenth 4, and so on, and proceeded to calculate the total distance traveled. In effect, he was summing the infinite series

$$\frac{1}{2} \cdot 1 + \frac{1}{4} \cdot 2 + \frac{1}{8} \cdot 3 + \cdots + \frac{1}{2^n} \cdot n + \cdots.$$

His result was that the sum, representing the total distance, is 2, or, as he put it, "precisely four times what is traversed in the first half of the [time]."[36] His proof, given geometrically, is very elegant. He drew a square of base CD equal to AB (=1) and divided it "to infinity

FIGURE 10.16

Oresme's summation of
$\frac{1}{2} \cdot 1 + \frac{1}{4} \cdot 2 + \frac{1}{8} \cdot 1 + \frac{1}{4} \cdot 3 + \cdots + \frac{1}{2^n} \cdot n + \cdots$

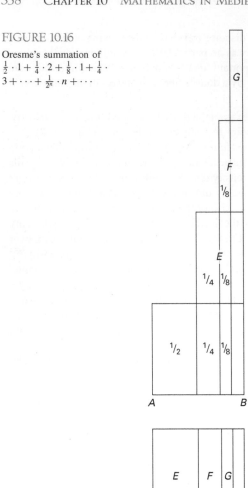

into parts continually proportional according to the ratio 2 to 1" (Fig. 10.16). Namely, E represents half of the square, F one-quarter, G one-eighth, and so on. The rectangle E was placed over the right half of the square on AB, F atop the new configuration over its right quarter, G atop the right eighth, and so on. It is then evident that the total area of the new configuration, which represents the total distance traveled, is not only equal to the sum of the infinite series but also equal to the sum of the areas of the two original squares.

Oresme's idea of representing velocities, as well as other qualities, geometrically, was continued in various works by others over the next century. However, no one was able to extend the representation of distances to situations more complex than Oresme's uniformly difform qualities. Eventually, even this idea was lost. Much the same fate befell the ideas of the other major European mathematicians of the medieval period. Their works were not studied and their new ideas had to be rediscovered centuries later. This lack of "progress" is evident in the stagnant mathematical curricula at the first universities as well as at the many new ones founded in succeeding centuries. With the works of Aristotle continuing to be the basis of the curriculum, the only mathematics studied was that which found its use in helping the student to understand the works of the great philosopher. Although an Oresme might carry these ideas further, such men were rare. In addition, the ravages of the Black Death and the Hundred Years War caused a marked decline in learning in France and England. It was therefore in Italy and Germany that a few of the ideas of the medieval French and English mathematicians would generate new ideas in the Renaissance.

EXERCISES

1. This problem and the next two are from Alcuin's *Propositions for Sharpening Youths*.[37] A cask is filled to 100-*metreta* capacity through three pipes. One-third of its capacity plus 6 *modii* flows in through one pipe; one-third of its capacity flows in through another pipe; but only one-sixth of its capacity flows in through the third pipe. How many *sextarii* flow in through each pipe? (Here a *metreta* is 72 *sextarii* and a *modius* is 200 *sextarii*.)

2. A man must ferry a wolf, a goat, and a head of cabbage across a river. The available boat, however, can carry only the man and one other thing. The goat cannot be left alone with the cabbage, nor the wolf with the goat. How should the man ferry his three items across the river?

3. A hare is 150 paces ahead of a hound that is pursuing him. If the hound covers 10 paces each time the hare cover 6, in how many paces will the hound overtake the hare?

4. Use Abraham bar Ḥiyya's table to find the length of the arc cut off by a chord of length 6 in a circle of diameter $10\frac{1}{2}$.

5. Find the area of the circle segment determined by the chord in Exercise 4.

6. Find the length of the chord that cuts off an arc of length $5\frac{1}{2}$ in a circle of diameter 33.

7. If a chord of length 8 has distance 2 from the circumference, find the diameter of the circle.

8. The *Artis cuiuslibet consummatio* claimed that the formula $A = \frac{3n^2 - n}{2}$ gave the area of a pentagon of side n. Show, instead, that it provides a formula for the nth pentagonal number. Calculate the area of regular pentagons with sides of length $n = 1, 2, 3$, and compare your answer to the value of the $(n + 1)$st pentagonal number. How close an approximation does the given formula provide?

9. Suppose you sight a tower from two stations as in Figure 10.7. At station S_1, you calculate that the ratio of height to distance d_1 is $2 : 5$. At station S_2, you calculate the ratio of height to distance d_2 to be $2 : 7$. If the distance between the two stations is 50 feet, how high is the tower?

10. Use Leonardo's table of chords to solve the following: Suppose a given chord in a circle of diameter 10 is 8 rods, 3 feet, $16\frac{2}{7}$ *unciae*. Find the length of the arc cut off by the chord.

11. From Leonardo's *Practica geometriae:* Given the quadrilateral inscribed in a circle with $ab = ag = 10$ and $bg = 12$, find the diameter ad of the circle (Fig. 10.17).

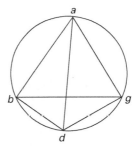

FIGURE 10.17

Determining the diameter of a circle in the work of Leonardo of Pisa

12. Develop a formula, as did Richard of Wallingford, to calculate the chord of the sum of three arcs. Translate this into a formula for the sine of the sum of three arcs.

13. Prove this theorem from Levi ben Gerson's *Trigonometry:* If all sides of any triangle whatever are known, its angles are also known. Start by dropping a perpendicular from one

vertex to the opposite side (or opposite side extended), and show how one can calculate the angles.

14. Prove the general combinatorial rule by induction:

$$C_k^n = \sum_{i=k-1}^{n-1} C_{k-1}^i.$$

15. Prove Proposition 30 from the *Maasei Ḥoshev:* $(1 + 2 + \cdots + n) + (1 + 2 + \cdots + (n-1)) = n^2.$

16. Prove Proposition 32 of the *Maasei Ḥoshev:*

$$1 + (1+2) + (1+2+3) + \cdots + (1+2+\cdots+n)$$
$$= \begin{cases} 1^2 + 3^2 + \cdots + n^2 & n \text{ odd}; \\ 2^2 + 4^2 + \cdots + n^2 & n \text{ even}. \end{cases}$$

17. Prove Proposition 33 of the *Maasei Ḥoshev:*

$$(1 + 2 + 3 + \cdots + n) + (2 + 3 + \cdots + n)$$
$$+ (3 + \cdots + n) + \cdots + n$$
$$= 1^2 + 2^2 + \cdots + n^2.$$

18. Prove Proposition 34 of the *Maasei Ḥoshev:*

$$[(1 + 2 + \cdots + n) + (2 + 3 + \cdots + n) + \cdots + n]$$
$$+ [1 + (1+2) + \cdots + (1+2+\cdots+(n-1))]$$
$$= n(1 + 2 + \cdots + n).$$

19. Use the three previous results to prove the following:

$$1^2 + 2^2 + \cdots + n^2 = [n - \frac{1}{3}(n-1)][1 + 2 + \cdots + n].$$

20. One of the problems from the *Maasei Ḥoshev:* A barrel has various holes: The first hole empties the full barrel in 3 days; the second hole empties the full barrel in 5 days; another hole empties the full barrel in 20 hours; and another hole empties the full barrel in 12 hours. All the holes are opened together. How much time will it take to empty the barrel?

21. Another problem from the *Maasei Ḥoshev:* A merchant sells four drugs. The cost of the first drug is 2 *dinars* per *litra*; the cost of the second is 3 *dinars* per *litra*; the cost of the third is 12 *dinars* per *litra*; the cost of the fourth is 20 *dinars* per *litra*. How many *litras* should one buy of each of the drugs so that the cost for each is the same?

22. Prove that the difference of the squares of two consecutive triangular numbers is a cube. (Factor the difference of squares as a sum times a difference and use the result of Exercise 15.) (This result is from Jordanus's *Arithmetica*. Jordanus uses this result, as did al-Karajī and Levi ben Gerson, to prove that the sum of the integral cubes from 1 to *n* is the square of the *n*th triangular number.)

23. Recall that Jordanus used the Pascal triangle in Proposition IX–70 of the *Arithmetica* to determine series of numbers in continued proportion. Namely, beginning with the series 1, 1, 1, 1, . . . , he derived first the series 1, 2, 4, 8, . . . , and by using those terms, he derived the series 1, 3, 9, 27, ... Now use this latter series in the same way to derive the series 1, 4, 16, 64, ... Formulate and prove by induction a generalization of this result.

24. This problem and the next six are taken from the *Liber abbaci*. One roll of saffron is sold for 3 bezants and $7\frac{1}{4}$ mils (where there are 10 mils in a bezant). How much are 17 rolls and $5\frac{1}{2}$ ounces worth (where there are 12 ounces to the roll)?

25. A Genoese *solidus* is sold for $21\frac{1}{2}$ Pisan *denarii*. How much are 7 Genoese *solidi* and 5 *denarii* worth in Pisan money? (Recall that 1 *solidus* equals 12 *denarii*.)

26. If an Imperial *solidus* is sold for $31\frac{1}{2}$ Pisan *denarii*, and a Genoese *solidus* is worth $19\frac{3}{4}$ Pisan *denarii*, then how many Genoese pounds will one have for 17 Imperial pounds, 11 *solidi*, and 5 *denarii*? (One pound equals 20 *solidi*. Note that the exchange rate between Pisan and Genoese money is different in this exercise from that stated in the previous exercise.)

27. If 7 rolls of pepper are worth 4 bezants, and 9 pounds of saffron are worth 11 bezants, how much saffron will be had for 23 rolls of pepper?

28. If a lion eats one sheep in 4 hours, a leopard eats one sheep in 5 hours, and a bear eats one sheep in 6 hours, how long would it take the three animals together to devour one sheep? (Begin by supposing that the answer is 60, the least common multiple of 4, 5, 6.)

29. Two men have some *denarii*. The first said to the second, if you will give me one of your *denarii*, then mine will equal yours. The other responded, and if you will give me one of your *denarii*, then I will have ten times as many as you. How many does each man have?

30. Solve this problem discussed in the text: There are five men with money who have found a purse with additional money. The amount the first has together with the amount in the purse is $2\frac{1}{2}$ times the total of the amounts held by the other four. Similarly, the second man's amount together with the amount in the purse is $3\frac{1}{3}$ times the total held by the others. Analogously, the fraction is $4\frac{1}{4}$ for the third man, $5\frac{1}{5}$ for the fourth man, and $6\frac{1}{6}$ for the fifth man. Find the amounts of money that each man had originally as well as the amount in the purse. (Note that Leonardo found that the first man actually had a debt of 49,154.)

31. The Fibonacci sequence (the sequence of rabbit pairs) is determined by the recursive rule $F_0 = F_1 = 1$ and $F_n = F_{n-1} + F_{n-2}$. Show that

$$F_{n+1} \cdot F_{n-1} = F_n^2 - (-1)^n$$

and that

$$\lim_{n \to \infty} \frac{F_n}{F_{n-1}} = \frac{1 + \sqrt{5}}{2}.$$

32. Prove that Leonardo's "congruous" numbers are always divisible by 24.

33. From the *Book of Squares:* Find a square number for which the sum of it and its root is a square number and for which the difference of it and its root is similarly a square number. (In modern notation, find x, y, z, such that $x^2 + x = z^2$ and $x^2 - x = y^2$. Leonardo began his solution by using the congruous number 24 in solving $a^2 + 24 = b^2$, $a^2 - 24 = c^2$; he then divided everything by 24.)

34. This problem and the next two are from Jordanus's *De numeris datis*. If the sum of the product of the two parts of a given number and of their difference is known, then each of them is determined. Namely, solve the system $x + y = a$, $xy + x - y = b$. Use Jordanus's example where $a = 9$ and $b = 21$.

35. If the sum of the two quotients formed by dividing the two parts of a given number by two different known numbers is given, then each of the parts is determined. Namely, solve the system $x + y = a$, $x/b + y/c = d$. Jordanus sets $a = 10$, $b = 3$, $c = 2$, and $d = 4$.

36. If the sum of two numbers is given together with the product of their squares, then each of them is determined. Jordanus's example is $x + y = 9$, $x^2 y^2 = 324$.

37. Use Oresme's technique to divide a sesquialterate ratio $(3 : 2)$ by a third part of a double ratio $(2 : 1)^{\frac{1}{3}}$.

38. Show that there are in fact 4950 ways of comparing (by ratio) the 100 integral ratios from $2 : 1$ up to $101 : 1$ and that precisely 25 will have rational exponents.

39. Show that under the assumptions of the mean speed theorem, if one divides the time interval into four equal subintervals, the distances covered in each interval will be in the ratio $1 : 3 : 5 : 7$. Generalize this statement to a division of the time interval into n equal subintervals and prove your result.

40. From Oresme's *Tractatus de configurationibus qualitatum et motuum*: Show geometrically that the sum of the series

$$48 \cdot 1 + 48 \cdot \frac{1}{4} \cdot 2 + 48 \cdot \left(\frac{1}{4}\right)^2 \cdot 4 + \cdots$$
$$+ 48 \left(\frac{1}{4}\right)^n \cdot 2^n + \cdots$$

is equal to 96.

41. Solve the following problem of Oresme: Divide the line AB of length 1 (representing time) proportionally to infinity in a ratio of $2 : 1$; that is, divide it so the first part is one-half, the second one-quarter, the third one-eighth, and so on. Let there be a given finite velocity (say, 1) in the first interval, a uniformly accelerated velocity (from 1 to 2) in the second, a constant velocity (2) in the third, a uniformly accelerated velocity (from 2 to 4) in the fourth, and so on (Fig. 10.18). Show that the total distance traveled is 7/4.

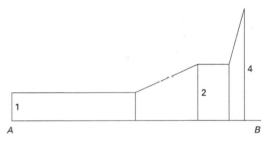

FIGURE 10.18

A problem of Oresme

42. Prove the result of Oresme: $1 + \frac{1}{2} + \frac{1}{3} + \frac{1}{4} + \cdots$ becomes infinite. (This series is usually called the harmonic series.)

43. Determine what mathematics was necessary to solve the Easter problem. What was the result of the debate in the Church? How is the date of Easter determined today? (Note that the procedure in the Roman Catholic Church is different from that in the Eastern Orthodox Church.)

44. Compare Levi ben Gerson's use of "induction" to that of al-Karajī. Should the methods of either be considered "proof by induction"? Discuss.

45. Write a lesson demonstrating proof by induction using some of Levi ben Gerson's examples.

46. Write a lesson developing some of the basic combinatorial rules using the methods of Abraham ibn Ezra and Levi ben Gerson.

47. Explain in detail why the area of one of Oresme's configurations should represent the total distance traveled by a moving object.

REFERENCES AND NOTES

Among the best general sources on the mathematics of medieval Europe are Marshall Clagett, *Mathematics and Its Applications to Science and Natural Philosophy in the Middle Ages* (Cambridge: Cambridge University Press, 1987), and David C. Lindberg, ed., *Science in the Middle Ages* (Chicago: University of Chicago Press, 1978). In particular, chapter 5 ("Mathematics") by Michael S. Mahoney and chapter 7 ("The Science of Motion") by John E. Murdoch and Edith D. Sylla in the latter work provide good surveys. An excellent collection of original source materials translated into English is Edward Grant, ed., *A Source Book in Medieval Science* (Cambridge: Harvard University Press, 1974). Another source book dealing mainly with aspects of mechanics is Marshall Clagett, *The Science of Mechanics in the Middle Ages* (Madison: University of Wisconsin Press, 1961). Basic biographical information on mathematicians of the Middle Ages can be found in George Sarton, *Introduction to the History of Science* (Huntington, NY: Robert E. Krieger, 1975), especially volumes II and III. Finally, a systematic survey of mathematics in the Middle Ages not only in Europe but also in Islam, China, and India is Adolf P. Juschkewitsch, *Geschichte der Mathematik im Mittelalter* (Leipzig: Teubner, 1964), a German translation of the Russian original of 1961.

The Latin version of Abraham bar Ḥiyya's *Treatise on Mensuration* is available in a German translation in Maximilian Curtze, *Der Liber Embadorum des Abraham bar Chijja Savasorda in der Übersetzung des Plato von Tivoli, Abhandlungen zur Geschichte der Mathematischen Wissenschaften* 12 (1902), 1–183. There is no complete modern translation directly from the Hebrew. There is also a German translation of Levi ben Gerson's *Maasei Ḥoshev*: Gerson Lange, *Sefer Maasei Choscheb. Die Praxis der Rechners. Ein Hebräisch-Arithmetisches Werk des Levi ben Gerschom aus dem Jahre 1321* (Frankfurt: Louis Golde, 1909). And although there is no English translation of the entire work, there is a new English translation of the problems at the end: Shai Simonson, "The Missing Problems of Gersonides: A Critical Edition," *Historia Mathematica* 27 (2000), 243–302, 384–431. Richard of Wallingford's trigonometry is available in John North, *Richard of Wallingford: An Edition of His Writings with Introductions, English Translations and Commentary* (Oxford: Clarendon Press, 1976). Levi ben Gerson's trigonometry is included in Bernard Goldstein, *The Astronomy of Levi ben Gerson (1288–1344)* (New York: Springer-Verlag, 1985). The *Sefer Yetzirah* is available in English: Isidor Kalisch, ed. and trans., *The Sepher Yezirah (The Book of Formation)* (Gillette, NJ: Heptangle Books, 1987). Leonardo of Pisa's *Liber abbaci* has recently been translated into English by L. E. Sigler, *Fibonacci's Liber Abaci: A Translation into Modern English of*

Leonardo Pisano's Book of Calculation (New York: Springer, 2002). There is also an English translation of the *Liber quadratorum: Leonardo Pisano Fibonacci, The Book of Squares*, edited and translated by L. E. Sigler, (Boston: Academic Press, 1987). Both the *Arithmetica* and the *De numeris datis* of Jordanus are available in English. The first is in H. L. L. Busard, *Jordanus de Nemore, De elementis arithmetice artis* (Stuttgart: Franz Steiner Verlag, 1991), while the second is in Barnabas Hughes, *Jordanus de Nemore: De numeris datis* (Berkeley: University of California Press, 1981). Hughes also discusses the sources of the work and gives a modern symbolic translation of each proposition. Oresme's work is available in Marshall Clagett, *Nicole Oresme and the Medieval Geometry of Qualities and Motions. A Treatise on the Uniformity and Difformity of Intensities Known as Tractatus de configurationibus qualitatum* (Madison: University of Wisconsin Press, 1968). Besides the Latin text and a complete English translation, this edition contains a detailed commentary as well as translations of some related works. Other works of Oresme are available in English in Edward Grant, *Nicole Oresme: De proportionibus proportionum & Ad pauca respicientes* (Madison: University of Wisconsin Press, 1966).

1. Curtze, *Liber embadorum*, p. 11.

2. Sigler, *The Book of Squares*, p. 3.

3. St. Augustine, *City of God*, XI, 30, quoting Wisdom of Solomon, 11:20.

4. Curtze, *Liber embadorum*, p. 35.

5. For a translation and discussion of Hugh of St. Victor's work, see Frederick A. Homann, *Practical Geometry, Attributed to Hugh of St. Victor* (Milwaukee: Marquette University Press, 1991).

6. Stephen Victor, *Practical Geometry in the High Middle Ages. Artis cuiuslibet consummatio and the Pratike de Geometrie* (Philadelphia: American Philosophical Society, 1979), pp. 109–111. This work contains much general information on the tradition of practical geometry in the Middle Ages as well as the translation of the two texts of the title. Another practical geometry is discussed in H. L. Busard, "The *Practica geometriae* of Dominicus de Calvasio," *Archive for History of Exact Sciences* 2 (1965), 520–575. Other similar works are discussed in Gillian Evans, "The 'Sub-Euclidean' Geometry of the Earlier Middle Ages, Up to the Mid-Twelfth Century," *Archive for History of Exact Sciences* 16 (1976), 105–118.

7. Victor, *Practical Geometry*, p. 221.

8. Ibid., p. 295.

9. North, *Wallingford*, p. 53.

10. Kalisch, *The Sepher Yezirah*, p. 23.

11. Quoted in Nachum L. Rabinovitch, *Probability and Statistical Inference in Ancient and Medieval Jewish Literature* (Toronto: University of Toronto Press, 1973), p. 144. This book provides an interesting look at the beginnings of various central ideas in probability and statistics in the discussions of the rabbis of ancient and medieval times on various points of Jewish law. An older work on Jewish mathematics is M. Steinschneider, *Mathematik bei den Juden* (Hildesheim: Georg Olms, 1964), a reprint of original articles of 1893–1901.

12. J. Ginsburg, "Rabbi ben Ezra on Permutations and Combinations," *The Mathematics Teacher* 15 (1922), 347–356, p. 351.

13. Lange, *Sefer Maasei Choscheb*, p. 1. The quotation in the next paragraph is also here. For more details on Levi's ideas on induction, see Nachum L. Rabinovitch, "Rabbi Levi ben Gershon and the Origins of Mathematical Induction," *Archive for History of Exact Sciences* 6 (1970), 237–248.

14. Ibid., p. 8.

15. Ibid., p. 49.

16. Ibid., p. 51.

17. Sigler, *Fibonacci's Liber Abaci*, p. 17.

18. Ibid., p. 404.

19. Sigler, *The Book of Squares*, p. 3.

20. For details on al-Khāzin, see Roshdi Rashed, *The Development of Arabic Mathematics*, chapter IV. For more information on the relationship of Leonardo's work to that of his Islamic predecessors, see Roshdi Rashed, "Fibonacci et les Mathématiques Arabes," in *Le Scienze alla Corte di Federico II* (Paris: Brepols, 1994), pp. 145–160.

21. The letter is quoted in Jens Høyrup, "Jordanus de Nemore, 13th Century Mathematical Innovator: An Essay on Intellectual Context, Achievement, and Failure," *Archive for History of Exact Sciences* 38 (1988), 307–363. This article provides a broad overview of the work of Jordanus. The article by Wilbur Knorr, "On a Medieval Circle Quadrature: De Circulo Quadrando," *Historia Mathematica* 18 (1991), 107–128, makes a new attempt to place Jordanus in the context of early thirteenth-century Paris.

22. Barnabas Hughes, "The Arithmetical Triangle of Jordanus de Nemore," *Historia Mathematica* 16 (1989), 213–223.

23. Hughes, *Jordanus de Nemore*, p. 57. Many of the following propositions have been translated directly from the Latin to better give the flavor of Jordanus's words than Hughes's more modern translation.

24. These definitions are taken from North, *Richard of Wallingford*, p. 59.

25. Grant, *Nicole Oresme*, p. 161.

26. Ibid., pp. 161–163.

27. Clagett, *The Science of Mechanics*, p. 230. A detailed introduction to the work of Bradwardine and his contemporaries is found in A. G. Molland, "The Geometrical Background to the 'Merton School': An Exploration into the Application of Mathematics to Natural Philosophy in the Fourteenth Century," *British Journal for the History of Science* 4 (1968), 108–125.

28. Ibid., pp. 230–231.

29. Grant, *A Source Book in Medieval Science*, p. 238. This quotation from William Heytesbury is also found in Clagett, *The Science of Mechanics*, p. 236. A detailed study of the work of Heytesbury is Curtis Wilson, *William Heytesbury: Medieval Logic and the Rise of Mathematical Physics* (Madison: University of Wisconsin Press, 1956).

30. Grant, *Source Book*, p. 238, and Clagett, *The Science of Mechanics*, p. 237.

31. Grant, *Source Book*, p. 239, and Clagett, *The Science of Mechanics*, p. 271.

32. Clagett, *Nicole Oresme*, pp. 165–167.

33. Ibid., p. 167. A study of the use of lines to express functions is in Edith Sylla, "Medieval Concepts of the Latitude of Forms: The Oxford Calculators," *Archives d'Histoire Doctrinal et Littérarie du Moyen Âge* 40 (1973), 223–283.

34. Ibid., p. 193.

35. Ibid.

36. Ibid., p. 415.

37. David Singmaster, "Some Early Sources in Recreational Mathematics," in Cynthia Hay, ed., *Mathematics from Manuscript to Print: 1300–1600* (Oxford: Clarendon Press, 1988), p. 199.

Mathematics around the World

The number of sciences is great, and it may be still greater if the public mind is directed towards them at such times as they are in the ascendancy and in general favor with all, when people not only honor science itself, but also its representatives. To do this is, in the first instance, the duty of those who rule over them, of kings and princes. For they alone could free the minds of scholars from the daily anxieties for the necessities of life, and stimulate their energies to earn more fame and favor, the yearning for which is the pith and marrow of human nature. The present times, however, are not of this kind. They are the very opposite, and therefore it is quite impossible that a new science or any new kind of research should arise in our days. What we have of sciences is nothing but the scanty remains of bygone better times.

—Al-Birūnī, from his survey of Hindu astronomy, c. 1030[1]

Spanish conquistadors arrived in the New World beginning at the end of the fifteenth century. Among the peoples they encountered were the Mayans, who centuries earlier had developed a base-20 place value system with which they could do rather sophisticated astronomical calculations. Many of their discoveries were written down in codices stored in libraries in the Mayan heartland. However, the Spaniards, in their zeal to convert the "heathens" to Christianity, decided that these writings were heretical and must be burned. Thus, most of the codices were destroyed so that only in recent times have scholars been able to reconstruct some of the mathematical and astronomical knowledge of the Mayans.

Having studied in some detail the mathematics of China, India, the Islamic world, and Europe up to about the year 1300, it is useful to compare the mathematics known in these places at that time. We then consider the issue of why it was that modern mathematics developed in Europe rather than elsewhere. We can also ask the question as to what mathematical ideas were known in other parts of the world at this time. This question is difficult to answer in much detail, given the current state of our knowledge. Nevertheless, we will present in the second half of this chapter a selection of mathematical ideas that are known to have been developed in the Americas, in sub-Saharan Africa, and in the Pacific.

MATHEMATICS AT THE TURN OF THE FOURTEENTH CENTURY

As the reader has certainly noted, there are many commonalities in the mathematics of the civilizations we have studied. We look at these as well as the possibilities of transmission from one culture to another.

11.1.1 Common Ideas of Mathematics

We begin with geometry. Practical geometry, that is, the measure of fields, the determination of unknown distances and heights, the calculation of volumes, and so on, was performed by much the same techniques in the four societies studied. All of them knew how to calculate areas and volumes, at least approximately, and all knew and used the Pythagorean Theorem when dealing with right triangles. Even the techniques of determining the height of a distant tower were nearly the same.

As far as theoretical geometry was concerned, it was in the world of Islam that the heritage of classical Greek geometry was preserved and studied and in which further advances were made. It was there that questions were raised and answered about the exact volumes of certain solids and about the locations of centers of gravity, using both heuristic methods for arriving at answers and the technique of exhaustion for giving proofs. It was there that questions were raised and answers attempted about Euclid's parallel postulate. It was there that questions were raised and new ideas developed about the classical Greek separation of number and magnitude. And it was there that the Greek idea of proof from stated axioms was most fully understood and developed.

Although Europe had always had at least a version of Euclid's *Elements* available, the beginning of the fourteenth century saw only the bare beginnings of a renewed interest in Euclid and other Greek geometers, stimulated by the appearance of a mass of translations of this material in the twelfth and thirteenth centuries. But although the idea of proof survived, there was still no new work in theoretical geometry. Neither India nor China had been exposed to classical Greek geometry, as far as is known, but that is not to say that they had no notion of proof. In the works of the Chinese mathematicians and their numerous commentators, there are always derivations of results. These derivations are not, however, based on explicitly stated axioms. They are, on the other hand, examples of logical arguments. In India, written derivations from early times have not survived; but beginning in the fourteenth century, we see many attempts to write out explicit derivations of mathematical results.

Related to geometry is the subject of trigonometry, developed in the Hellenic world as a part of the study of astronomy. By the year 1300, trigonometry was in active use in India, Islam, and Europe, generally for the same purpose of studying the heavens. The subject was modified and extended as it traveled from one country to another, but those interested in the heavens in those three civilizations were all fluent in at least some version of the subject. It appears that only China was lacking trigonometry, even though Indian scholars had introduced the elements of the subject in their visits in the eighth century. It is probable that trigonometry was simply not useful to the Chinese in their own astronomical and calendrical calculations.

In certain aspects of algebra, on the other hand, the Chinese were the first to develop techniques that were later used elsewhere. For example, they had from early times constructed efficient methods of solving systems of linear equations. By the fourteenth century, they had developed their early root finding techniques, which involved the use of the Pascal triangle, into a detailed procedure for solving polynomial equations of any degree. They also worked out the basis of what is today called the Chinese remainder theorem, a procedure for solving simultaneous linear congruences.

Linear congruences were also solved in India, by a method different from that of the Chinese but still using the Euclidean algorithm. Indian scholars were even prouder, however, of the techniques they developed for solving the quadratic indeterminate equations known today as the Pell equations. Although there is fragmentary evidence that certain simple cases of these equations were studied in Greece, the general case that was developed in India was studied nowhere else until eighteenth-century Europe. The Indian mathematicians were also familiar with the standard techniques of solving quadratic equations, but since there is no documentation of how they thought about the method, we do not know whether they developed the technique independently or absorbed it from the ancient Babylonians. A third possibility is that they learned it somehow from the work of Diophantus, who in turn was probably aware of the methods at least indirectly from the Babylonians.

For Islam, of course, there is copious documentation of an interest in algebra. Not only did Islamic mathematicians study the quadratic equation in great detail, giving geometric justifications for the various algebraic procedures involved in the solution, they studied cubic equations as well. For these equations, Islamic mathematicians developed a solution method involving conic sections and gained some understanding of the relationship of the roots to the coefficients of these equations. In addition, they knew a method of solving polynomial equations numerically, similar to the Chinese method, ultimately based on the Pascal triangle.

The Pascal triangle also appeared in Islamic mathematics in connection both with the binomial theorem and with the study of combinatorics. The latter field first appeared in India, but the proofs were apparently first worked out in detail in North Africa. Islamic mathematicians who dealt with these two aspects of the Pascal triangle also developed proof techniques closely resembling our modern proof by induction. Such techniques were further worked out in Europe by Levi ben Gerson. Furthermore, Islamic algebraists developed in great detail the methods for manipulating algebraic expressions, especially those involving surds, and thereby began the process of negating the classical Greek separation of number and magnitude.

By the turn of the fourteenth century, algebraic techniques were only beginning their appearance in Europe. Those techniques that were available were clearly based on the Islamic work, although Jordanus de Nemore considered the material from a somewhat different point of view. He also introduced a form of symbolism in his algebraic work, something missing entirely in Islamic algebra but also present, in different form, in India and China. On the other hand, European algebra of this time period, like its Islamic counterpart, did not consider negative numbers at all. India and China, however, were very fluent in the use of negative quantities in calculations, even if they were still hesitant about using them as answers to mathematical problems.

The one mathematical subject present in Europe in this time period that was apparently not considered in the other areas was the complex of ideas surrounding motion. It was seemingly only in Europe that mathematicians studied the mathematical question of the meaning of instantaneous velocity and therefore were able to develop the mean speed rule. Thus, the seed was planted that ultimately grew into one branch of the subject of calculus nearly three centuries later. On the other hand, the Indians beginning in the fourteenth century did begin to consider infinitesimally small quantities as they worked out their own ideas related to calculus.

11.1.2   Possible Transmission of Ideas

It appears that the level of mathematics in China, India, the Islamic world, and Europe was comparable at the turn of the fourteenth century. Although there were specific techniques available in each culture that were not available in others, there were many mathematical ideas and methods common to two or more. The question then arises as to whether the ideas developed independently in the four areas or whether there was transmission among them.

For certain ideas, the lines of transmission are clear. We have already seen how trigonometry moved from Greece to India to Islam and back to Europe, with each culture modifying the material to meet its own requirements. Also, the decimal place value system, with its beginnings in China or India (or perhaps on the border between them) moved to Baghdad in the eighth century and then to Europe (via both Italy and Spain) in the eleventh and twelfth centuries.

But for other common ideas, the situation is less clear. For example, in trigonometry the tangent function is useful in relating the length of a shadow to the altitude of the sun. The first tabulation of the tangent function was in China in the early eighth century, where Yi Xing developed this idea probably with the aid of Indian computations of the sine. The next appearance of a tabulated tangent function was in Islam. Was this notion carried there by Chinese technicians captured in the Battle of the Talas River in 751, which established Islamic hegemony in western Central Asia? In any case, although tangent tables were brought to Europe early in the twelfth century with the translation of an edited version of al-Khwārizmī's astronomical tables, the tangent function is not found in the early European trigonometry works.

What about the Pascal triangle? It appears in Islam in the early eleventh century and then in China perhaps in the middle of that century. Was there transmission? There was certainly contact in this period between Islam and China along the famous silk route. Recall also that al-Bīrūnī in this time period was at the court of Sultan Maḥmūd of Ghazna, where he studied the culture of India, and there was always some contact, particularly via Buddhism, between India and China. So how did the Pascal triangle arrive in Europe by the early thirteenth century? Did

European mathematicians learn of this arrangement of numbers from Arabic manuscripts of which we are today unaware? Or was the knowledge of this material transmitted by scholars traveling between the Islamic world and Europe?

Consider as well the methods of determining heights and distances in all four of these cultures. The earliest documented appearance of the basic method of two sightings is in China in the third century. By the thirteenth century, the method was being used in Europe. Similarly, recreational problems such as that of the hundred fowls or that of the faucets emptying (or filling) a tank appear in Chinese, Islamic, and early medieval European work. Did these problems travel, and, if so, how? Again, the silk route comes to mind as a method of transmission. Or, on a more specific level, there was a group of Jewish merchants known as the Radhanites who regularly traveled from southern France to China, via Damascus and India, in the ninth century, carrying eunuchs, fur, and swords to the East and returning with musk, spices, and medicinal plants. Did they learn of Chinese mathematics and bring it back to Europe or to any of their way stations, or, conversely, did they take Islamic or Indian mathematics to China? This question applies with perhaps more relevance to the question of the Jacob's Staff, the surveying device first described in Europe by Levi ben Gerson early in the fourteenth century but available in China by the eleventh century. Was this carried from China by Jewish merchants?

The answers to many of the questions of transmission can at this point only be speculation. Any documentation of such transmission remains to be discovered. But transmission or not, it does appear that the common mathematical ideas were adapted to meet the mathematical needs of each civilization.

11.1.3 Why Did Modern Mathematics Develop in Europe?

A question about which there has been wide discussion is why modern mathematics (and modern science in general) developed in western Europe rather than in the Islamic world or India or China. Because the technical achievements of these four civilizations around 1300 were comparable, many scholars have sought for the answer in the religious and cultural backgrounds against which the achievements took place. For example, in China there was essentially only one "university," and this was a bureaucratic subdivision of the imperial administration. Thus, unless the imperial administration encouraged mathematical developments—and this was a rare occurrence—there were few places where someone mathematically trained could develop his ideas. As we have noted, the Chinese mathematicians that we know about were isolated, both temporally and geographically, and some of them probably did not know of the existence of others. Thus, although there was some development of new ideas, there was never enough interest emanating from the government to encourage new thinking. In fact, in general, Chinese education was devoted to memorization and commentary on the ancient classics.

In India, there was no central government over the entire subcontinent, and thus no central system of education. On the other hand, there was always opportunity for advanced study as scholars established schools in various sites. It was at one of these "schools," in Kerala, that, as we have seen, there were developments that nearly led to what we call the calculus. It is clear that the mathematicians in the Kerala school understood some basic ideas of the differential calculus as well as the notion of power series and the critical result on sums of integral powers necessary for beginning the development of integral calculus. But although the mathematicians worked on these ideas for a period of several centuries, the ideas evidently never spread out of southwestern India nor did they develop into any general theories. And after the sixteenth century, it appears that the Kerala school itself disappeared.

Islam, like India, had numerous institutions of higher learning and numerous scholars interested in mathematics. Thus, mathematical traditions developed, and, even more so than in India, one could point to the establishment of mathematical "schools," groups of scholars who worked on similar problems using related techniques. So the great mystery is why there was never a breakthrough to modern mathematics under Islam, or why, for example, it was not Islamic mathematicians who developed the calculus or developed heliocentric astronomy. But, in fact, Islamic mathematics suffered a period of decline after the thirteenth century, and significant ideas of earlier time periods were lost.

Although it is difficult to solve this mystery definitively, it would appear that one of the central factors is that, even when mathematics was being highly developed in Islam, the areas of mathematics more advanced than basic arithmetic were classified as "foreign sciences," in contrast to the "religious sciences," including religious law and speculative theology. To many Islamic religious leaders, the foreign sciences were potentially subversive to the faith and certainly superfluous to the needs of life, either here or hereafter. And although the earliest Islamic leaders encouraged the study of the foreign sciences, over the centuries the support for such study lessened as more orthodox religious leaders came to the fore. Even al-Bīrūnī noted this, as is clear from the chapter's opening quotation.

Although there were certainly mathematical achievements in Islam after the time of al-Bīrūnī, the rate of such achievements began to drop. And although there continued to be institutions of higher learning throughout the Islamic world, the *madrasas*, these tended to concentrate on the teaching of Islamic law. A scholar in charge of one of these schools could, of course, teach the foreign sciences, but if he did, he could be the subject of a legal ruling from traditionalists, a ruling that would in fact be based on the law establishing the school, specifying that nothing inimical to the tenets of Islam could be taught.

Interestingly, Catholic leaders in Europe also issued decrees forbidding the teaching of certain subjects. In fact, at several times, it was officially forbidden to teach aspects of Aristotle's works that were apparently in conflict with church doctrine. However, it appears that scholars in the universities at Paris and elsewhere to a large extent ignored the church decrees. The universities in Europe, unlike the *madrasas* of Islam, were corporate bodies, having legally defined autonomy. And if the faculty decided to discuss scientific topics and develop new mathematical ideas surrounding them, it was in general not easy for church leaders to ban such work. Thus, the path was opened in Europe for the development of modern mathematics and, of course, modern science.

That is not to say that the Islamic achievements had no effect. On the contrary, we have already noted various aspects of Islamic mathematical work that were transmitted to Europe during the twelfth and thirteenth centuries. In certain areas of mathematics, Europe's first ideas were direct consequences of original Islamic work or of Islamic modifications of Greek or Hindu work. In subsequent chapters, we will detail some of the influences of Islam on European mathematics. Nevertheless, the locus of the history of mathematics after the fourteenth century was primarily in Europe, so, for better or worse, the remaining chapters of this book concentrate on the mathematical achievements that took place there.

11.2   MATHEMATICS IN AMERICA, AFRICA, AND THE PACIFIC

There were mathematical ideas in the world in civilizations different from the four major medieval societies already considered. Unfortunately, most of the other civilizations were nonliterate, so written documentation is not available. Thus, any description of the mathematics of these societies necessarily comes from artifacts or from the descriptions of ethnologists. Much research on mathematics in various societies has been carried out in recent years, but there are still many unanswered questions. We can only present here a brief sketch of what is known about the mathematical ideas of various societies. References to the current literature are provided, however, so that the interested reader can pursue these matters further.

11.2.1   The Mayans

We will begin in the Americas with the Mayans, the society in the New World about whose mathematics the most is known, primarily because the Mayans did have a written language. Mayan civilization flourished in southern Mexico, Guatemala, Belize, and Honduras and reached its high point between the third and ninth centuries. Thereafter, the Mayans came under the influence of other peoples of Mexico, and many of their cultural centers fell into ruin. Nevertheless, a strong Mayan culture still existed when the Spaniards arrived in the early sixteenth century. Although the Spaniards conquered the Mayans rather quickly in a physical sense, they never succeeded in completely destroying the Mayan culture. Today there still remain approximately two and a half million speakers of Mayan languages who have managed to preserve some aspects of their ancient way of life.

Like many ancient civilizations, the Mayans had a priestly class who studied mathematics and astronomy and kept the calendar (Fig. 11.1). The records of the priests were written down and preserved on a bark paper or carved into stone monuments. Unfortunately, as we noted

FIGURE 11.1

Detail from a Mayan ceramic vessel (c. 750 CE) depicting two mathematicians. The one on the left is a man, while the one in the upper right corner is a woman. The mathematicians are identified by scrolls with number symbols emerging from their armpits.[2]

FIGURE 11.2

Dresden codex on a stamp of the German Democratic Republic

in the chapter opening, the Spanish conquerors destroyed most of the documents they found. And because modern-day Mayans cannot read the ancient hieroglyphics, it has been a long and tedious process to decipher the few documents that remain, in particular, the Dresden codex (named for the library that owns it), which dates from the twelfth century and deals with aspects of the Mayan calendar (Fig. 11.2). Nevertheless, scholars today understand the basics of the classic Mayan calendrical and numeration systems. The documents, however, provide only the results of calculations. There is no record of the methods by which the calculations were made. Some of what follows in the description of Mayan mathematics is therefore speculative.

The Mayan numeration system was a mixed system, like the Babylonian. It was a place value system with base 20 on one level, but for the representation of numbers less than twenty, it was a grouping system with base 5. The Mayans used only two symbols to represent numbers, a dot (·) to represent 1 and a line (—) to represent 5. These were grouped in the appropriate way to represent numbers up to 19. Thus, ____ represented 8, and ▦ represented 17. For numbers larger than 19, a place value system was used. The first place represented the units, the second place the 20s, the third place the 400s, and so on. Unlike the Babylonians, however, the Mayans did have a symbol for 0, namely, ⊖, used to designate an "empty" place. Mayan numbers were generally written vertically, with the highest place value at the top, but for convenience we write them horizontally, using the same conventions we used to represent the Babylonian numbers. Thus, 3,5 represents $3 \times 20 + 5$, or 65.

For calendrical purposes, the Mayans modified their numeration system slightly, using the third place from the bottom to represent 360s, rather than 400s, with every other place still representing 20 times the place before. It is this system we use in what follows, because it was in calendrical calculations that the Mayans used numbers most extensively. In this calendrical numeration system, then, 2,3,5 represents $2 \times 360 + 3 \times 20 + 5$, or 785, while 2,0,12,15 represents $2 \times 7200 + 12 \times 20 + 15$ or 14,655.

For the Babylonians, there is quite extensive evidence of the methods of calculation used. The natural question to ask with regard to the Mayans, then, is how they made calculations using their place value system. Unfortunately, all that exists in the Mayan documents are the results of various computations, nearly all of them correct, without any record of the methods themselves. It is surmised that for addition and subtraction the Mayans used some sort of counting board device to collect the dots and lines in each place and move any excess over 20 into the next place. To do multiplication, presumably all one needs to know are three basic facts: $1 \times 1 = 1$, $1 \times 5 = 5$, and $5 \times 5 = 1,5$. To perform any other multiplication, one needs only the distributive law and a way to keep track of the places, but, naturally, multiplication tables up to 19×19 would make computation easier. There is no record, however, of the use of such tables.

The most important use of computation for the Mayan priests appears to have been for calendrical computations. To understand these, one must first understand the basic Mayan calendar system. The Mayans used two different calendars at the same time. First, there was the 260-day almanac, which was the product of two cycles—one of length 13 and the other of length 20. Namely, any day in the almanac is specified by a pair (t, v), where t is a day number from 1 to 13 and v is one of twenty day names. For example, because the list of twenty day names begins with *Imix* and *Ik*, the day 1 *Imix* is written as $(1, 1)$, while the day 5 *Ik* is written as $(5, 2)$. The second calendar was the 365-day year. This calendar year was divided into 18 months of 20 days and an extra period of five days. For our purposes, however, it is sufficient to designate a day in the 365-day year by its number y. Thus, because *Muan* is the fifteenth month of the calendar, the second day of Muan is designated by $y = 282$. The three cycles of 13 day numbers, 20 day names, and 365 days of the year were traversed independently, and thus the complete cycle of triples (t, v, y) was repeated after $13 \cdot 20 \cdot 73 = 18,980$ days, or 52 calendar years, or 73 almanacs (18,980 is the least common multiple of 260 and 365). This entire cycle is generally called the "calendar round."

The two basic calendrical problems the Mayans needed to solve were, first, given a date (as a triple) and a specified number of days later, to determine the new date, and, second, given two Mayan dates (as triples), to determine the least number of days between them. If the specified number of days is denoted, in base-20 calendrical notation, by m, n, p, q, r, where $0 \leq m, n, p, r \leq 19$, $0 \leq q \leq 17$, the first question can be written in modern notation as given an initial date (t_0, v_0, y_0), determine the date (t, v, y), which is m, n, p, q, r days later. To begin this process, we first add r days, then $20q \equiv 7q$ days, then $360p \equiv -4p$ days, then $7200n \equiv -2n$ days, and finally $144,000m \equiv -m$ days, where all equivalences are modulo 13. Therefore, $t \equiv t_0 - m - 2n - 4p + 7q + r \pmod{13}$. We can show further that

$$v = v_0 + r \pmod{20}$$

and

$$y = y_0 + 190m - 100n - 5p + 20q + 4 \pmod{365}.$$

For example, if the given date is $(4, 15, 120)$, the new date $0, 2, 5, 11, 18$ days later is $(10, 13, 133)$. The second calendrical problem requires one first to determine the smallest intervals between the dates in each of the three component cycles, second to combine the first two to determine the smallest interval in the almanac, and third to combine this value with the third to determine the number of days in the calendar round.

There is no evidence of the exact method used by the Mayan priests to solve either of these problems. But it seems virtually certain that they must have gone through a computation similar to that indicated by the algebraic formulas above, without, of course, the algebraic notation itself. In any case, the priests were able to use their base-20 notation to solve the problems they needed to keep track of their calendar and thus to provide the Mayan government with the correct days on which to celebrate festivals, make appropriate sacrifices, plant the maize, or accomplish whatever other tasks were necessary to run the Mayan kingdom (Fig. 11.3).[3]

FIGURE 11.3

Mayan observatory, El Caracol, at Chichen Itza in the Yucatán Peninsula of Mexico

11.2.2   The Incas

About 2000 miles south of the Mayan heartland was another major civilization of about four million people, the Inca, which flourished in what is now Peru and surrounding areas from about 1400 to 1560. The Incas did not have a written language but did possess a logical numbering system of recording in the knots and cords of what are called *quipus*. The *quipus* were the means by which the Inca leadership monitored its domains. They necessarily received and sent many messages daily, including details of items that were needed in storehouses, taxes that were owed, numbers of workers needed for certain public works projects, and so on. The messages were encoded on the *quipus* and sent to their destination

FIGURE 11.4

Inca runner carrying *quipu*

by a series of runners (Fig. 11.4). Of necessity, the messages had to be concise and compact, so *quipu* makers were trained in Cuzco, the capital, to design and create the *quipus* on which the messages were carried.

A *quipu* is a collection of colored knotted cords, in which the colors, the placement of the cords, the knots on the individual cords, the placement of the knots, and the spaces between the knots all contribute to the meaning of the recorded data. Every *quipu* has a main cord, thicker than the others, to which are attached other cords, called pendant cords, to which may be attached further cords, called subsidiary cords. Sometimes there is a top cord, a cord placed near the center of several pendant cords and tied so that when the *quipu* lies flat it falls in a direction opposite the pendant cords. Data is recorded on the cords (other than the main cord) by a system of knots. The knots are clustered together in groups separated by spaces and represent numbers using a base-10 place value system with the highest value place closest to the main cord. Thus, the cord with three knots near the top and nine knots near the bottom represents the number 39. As additional help for reading the numbers, the knots representing units are generally larger knots than those representing higher powers of ten. The largest number so far discovered on a *quipu* is 97,357. Zeros are generally represented by a particularly wide space (Fig. 11.5).

FIGURE 11.5

Quipu from the Museo Nácional de Antropologia y Arqueologia in Lima, Peru (Smithsonian Institution, catalog no. 289613, department of anthropology)

The pendant cords on *quipus* are themselves generally clustered in groups, sometimes with each group consisting of the same set of distinct colors. It is assumed that each color refers to a particular class of data that is being recorded on the *quipu*. In addition, one often has a top cord associated with a group of cords on which is recorded the sum of the numbers on the individual cords of the group. Sometimes certain pendant cords record sums of the numbers on other such cords. Sometimes it appears that the knots on a particular cord do not represent data at all but are simply labels. In any case, the *quipus* are not calculating tools but records. The calculations on which these records are based must have been done elsewhere, probably with some sort of counting board.

What is not generally known is exactly what kinds of data a given *quipu* records. One particular *quipu* is known to be a record of census data for a region of seven provinces, the people in each of which were classified in one of two groups, each of which was further divided into two subgroups. Thus, the individual pieces of data recorded on certain of the cords were the number of households in each province belonging to each of the subgroups. Other cords then represented the sums of various pieces of this data, with one cord finally giving the grand total of the number of households in the entire region.

In modern terminology, a *quipu* can be thought of as a particular type of graph known as a tree (see Chapter 24). Certainly, the *quipu* makers had to ask themselves the types of questions often associated with the study of trees, including how many different trees can be constructed with a given number of edges. And since the Inca official associated numbers with each edge of the tree, as well as colors, the questions that had to be answered in designing these objects so that they could be useful were not trivial ones.[4]

11.2.3   The North American Indians

In the Inca civilization, as in the Mayan, there was a professional class of "mathematicians," people who had to deal with the mathematics of the culture on a regular basis to help the civilization maintain itself. But in the other cultures to be discussed next, such a class did not exist. In fact, these peoples had no category in their lives called "mathematics." Nevertheless, there are certain aspects of their culture that today we recognize as being mathematical. The people involved did not distinguish and classify these aspects as we do. The mathematical ideas were simply part of what they needed to conduct their lives, to farm, to build, to worship. This mathematics of a group of people, used on a regular basis, is what today is often called "ethnomathematics," the study of which allows us to see the importance of mathematical ideas to various such groups.[5]

Perhaps the most sophisticated civilization in what is now the United States in pre-Columbian times was that of the Anasazi, who lived in the Four Corners area of the Southwest from about 600 BCE to around 1300 CE. The high point of their civilization was reached in the years after 1000 in which they constructed elaborate pueblos and ceremonial structures at various sites, the most prominent being at Mesa Verde (in southwestern Colorado) and at Chaco Canyon (in northwestern New Mexico). It appears that viewing areas were set up in many of their structures for the same purposes as sight lines at the temple at Stonehenge, namely, to determine the occurrences of certain important astronomical events including the summer and winter solstice and even the 18.6-year cycle of moonrise positions.

Although the Anasazi have left us no documented records of their mathematics, we can speculate about what mathematics was necessary for their lives by considering the archaeological remnants of their civilization. For example, one important notion in the Anasazi religion, coming from the myths of their origins, is that of the four cardinal directions. It was evidently important to the Anasazi to align their major buildings in these directions and even to build their roads that way. One of the major roads out of Chaco Canyon was aligned due north and built that way for many miles, irrespective of topographical obstacles. And the great ceremonial structure in Chaco Canyon, Casa Rinconada, is a 63-foot diameter circle, the roof of which was originally supported by four pillars forming the corners of a square aligned exactly along the cardinal directions. The question, then, is how the Anasazi determined the direction of true north (see Fig. 5.28). One possibility, because they were certainly aware of the daily and yearly motions of the sun, is that they used the same techniques as the Roman surveyors of a millennium earlier, to draw a circle centered on a pole, then record the curve of the endpoints of the pole's shadow throughout the day and determine the two points where the curve intersects the circle. The line connecting those two points is an east-west line to which a perpendicular bisector can be drawn to determine a north-south line.[6]

Other North American Indians built carefully aligned structures and even entire urban areas, thus displaying a knowledge of astronomy and geometry. For example, the Cahokian mounds in East St. Louis, Illinois, built by the civilization known as the Mississippian during the period from 900 to 1200, display not only alignments to important celestial events but also evidence of detailed city planning. Similarly, the Bighorn Medicine Wheel, near the summit of Medicine Mountain in the Bighorn Mountains of northern Wyoming, and the Moose Mountain Medicine Wheel, in southeastern Saskatchewan, were probably constructed by Plains Indians to determine the summer solstice. The Moose Mountain Medicine Wheel probably dates from over 2000 years ago, while the Bighorn Medicine Wheel is of much more recent origin.

11.2.4 Sub-Saharan Africa

Like the North American Indians, most African cultures of the past did not leave written records, so it is not possible to say with any degree of certainty when and how mathematical ideas were created in these cultures. What makes matters even worse for the historian is that few artifacts are even available from which mathematical ideas can be inferred, partly because comparatively little archaeology has been done in Africa south of the Sahara. One major ancient structure that is only now being studied in detail is Great Zimbabwe, a massive stone complex 17 miles south of Nyanda, Zimbabwe, which was probably built in the twelfth century. It is evident that the empire that built this complex required mathematics to deal with the administrative and engineering requirements of the construction as well as with the trade, taxes, and calendars required to keep the empire functioning. Similarly, the bureaucracies of the West African states of the medieval period, including Ghana, Mali, and Songhai, also required mathematics like their colleagues elsewhere in the world. Because the influence of Islam penetrated to much of west Africa, and because an Islamic university was in existence in Timbuktu from the fourteenth century at least until 1600, scholars of that region were probably exposed to some of the mathematics of Islam. However, we have no direct information on mathematics or mathematicians of this time and place.

Until more archaeological finds have been discovered, we can find out about the mathematics of the peoples of Africa by considering the reports of the ethnographers who studied these civilizations in the nineteenth and twentieth centuries and pull out of their studies what we consider mathematical ideas. In addition, in certain African cultures we can study indigenous practices that have been relatively unaffected by colonialism to also discover mathematical ideas. Such studies have been made in recent years in Mozambique and surrounding areas of southern Africa.[7] Unfortunately, neither of these methods permits us to determine the sources and the dates of the ideas.

One mathematical idea that appears in the Bushoong culture in Zaire and in the Tshokwe culture of northeastern Angola is the graph theoretical idea of tracing out certain figures in a continuous curve without lifting one's finger from the sand. In Western mathematics, this idea was first dealt with by Leonhard Euler in 1736 (see Chapter 20). The Bushoong children, who first showed their diagrams to a European ethnologist in 1905, not only were evidently aware of the conditions that ensured that the graph could be drawn continuously but also knew the procedure that permitted its drawing most expeditiously. For the Tshokwe, figure drawing is not a children's game, but part of a storytelling tradition among the elders. As part of their storytelling, dots are used to represent humans, and the rather complex curves are drawn including certain dots within the figure and excluding certain others. In fact, the procedure for drawing is to set out a rectangular grid of dots on which the curve is superimposed (Fig. 11.6). Without a special study of the diagrams, it is not easy to determine which dots are inside and which are outside, but the detailed drawing rules that the Tshokwe follow enable them to construct the curves quickly in one continuous motion.[8]

FIGURE 11.6

Examples of Tshokwe graphs

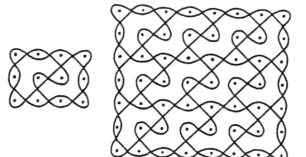

Another mathematical idea that occurs in many African cultures is that of a geometric pattern, as used in cloth weaving or decorative metal work. There are numerous examples from all over Africa of patterned strips, using the seven possible strip patterns, as well as most of the seventeen other plane patterns.[9] In fact, the Bakuba people from Zaire use all seven strip patterns in their cloth, as well as at least twelve of the plane patterns. The artists of Benin (Nigeria) decorate their bronze castings with all of the strip patterns and some of the other plane patterns as well (Fig. 11.7). And the Tellem weavers, from what is now Mali, beginning in the eleventh century explored numerous ways of combining strips of two different colors as they wove cotton cloth to make tunics and other items of clothing.[10]

FIGURE 11.7

Examples of strip patterns from Benin

Mathematical games and puzzles occur in Africa, too. For example, the board game known variously as *wari*, *omweso*, and *mankala* is played throughout Africa and is quite useful in teaching children counting and strategy. Similarly, the familiar puzzle story of a person attempting to transport three objects, *A*, *B*, *C*, across a river, but only being able to take one at a time and not being able to leave either *A* or *C* alone with *B*, occurs in several African cultures. Among the Bamileke (Cameroon), the objects are a tiger, a sheep, and a big spray of reeds. A different problem, where the person can take two objects at a time, is found, among other places, in Algeria (where the objects are a jackal, a goat, and a bundle of hay), in Liberia (a cheetah, a fowl, and some rice), and in Zanzibar (a leopard, a goat, and some tree leaves). Recall that this puzzle also occurred in eighth-century Europe in Alcuin's *Propositiones*.

In Madagascar, off the southeastern coast of Africa, the diviners used manipulations of the seeds of a fano tree to give advice about planting, traveling, adopting a child, and so on. The manipulations of the seeds use an algorithm that we can interpret as a particular operation in Boolean algebra. The process begins with the diviner taking a fistful of seeds, dividing them arbitrarily into four piles, and then removing as many multiples of two as possible from each pile so that the remainders are either one or two. These four remainders become the first column of an array. Note that there are sixteen possibilities for this column, since each entry may be either a one or a two. The process of taking seeds, making piles, and reducing the piles is repeated three more times to get three more columns so that the diviner ends up with a four-by-four array. We label the four columns as C_1, C_2, C_3, and C_4, where these are ordered from right to left, while the rows are labeled C_5, C_6, C_7, and C_8, from top to bottom. The diviner then uses the XOR (exclusive or) operation from Boolean algebra ($\oplus$) componentwise to create eight new columns. (For the purpose of this algebra, we can think of one seed as representing "odd" or "1" and two seeds as representing "even" or "0." The XOR operation then gives "even" when combining two elements of the same parity and "odd" when combining two elements of opposite parity.) For example, column 11 (C_{11}) is created as $C_4 \oplus C_3$. Thus, if $C_4 = (2, 1, 1, 2)$ and $C_3 = (1, 1, 2, 1)$, then $C_{11} = (2 \oplus 1, 1 \oplus 1, 1 \oplus 2, 2 \oplus 1) = (1, 2, 1, 1)$. Similarly, if $C_2 = (2, 2, 1, 1)$ and $C_1 = (1, 2, 1, 1)$, then C_{12}, defined as $C_2 \oplus C_1$, is $(1, 2, 2, 2)$. The final arrangement of the sixteen columns, together with additional manipulations of them, are the basis for the diviner's predictions or answers to his client's questions.

Of course, in performing his algorithm the diviner does not think he is doing Boolean algebra. But when he is finished with creating his columns, he does several checks to see that he has not made any errors. In other words, he knows that certain relationships must be present in his final arrangement, assuming he carried out the algorithm correctly. For example, the final arrangement must always produce at least two identical columns. How did the originators of this process discover this result, and did they develop some sort of proof of it? We do not know the answer to these questions, but we do know that some diviners did interest themselves in certain special final arrangements and therefore in determining what original layouts lead to these arrangements. That is, the diviners acted as "mathematicians."[11]

11.2.5   The South Pacific

Moving to the South Pacific, we find the idea of tracing figures continuously in the sand also in Malekula in the Republic of Vanuatu, an island chain some 1200 miles northeast of Australia. The drawing of figures here is imbedded in Malekula religious life. In fact, passage to the Land of the Dead requires being able to draw these figures accurately. The Malekulans devised standard algorithms for tracing their quite complicated figures using symmetry operations on a few basic drawings. Thus, one can analyze the Malekulan figures using some of the language of modern-day group theory.

Group theory is also convenient in analyzing the kin relationships in Malekula. In fact, the elders explained these relationships to an anthropologist using diagrams that can easily be transformed into a group table. The basic idea is that the society is divided into six sections, and men of one section can only marry women of a different section, while their children belong to still another section. If a given male belongs to the section we label as e (identity), his mother belongs to section m and his father to section f. Then the mother of his father is in section mf and the father of his mother is in section fm. It turns out that the kin rules are such that all the possible "products" of m and f form the dihedral group of order 6, that is, the group of six elements generated by the elements m, f, with the relations $m^3 = e$, $f^2 = e$, and $(mf)(mf) = e$. Marriage can only take place between A and B, if B belongs to the section of the mother of the father of A, or, equivalently, if A belongs to the section of the mother of the father of B. A similar kin relationship group structure of order 8 occurs among the Warlpiri of northern Australia.[12]

In another part of the South Pacific, the Marshall Islands, stick charts are an element of the navigation tradition. Some of these stick constructions contain idealized shapes, which were used to train navigators in the principles of wave motion and especially in the interaction of the waves with the land masses. Other models are essentially maps of the entire Marshall Island archipelago or some subset of it. In both cases, it is clear that the navigators have constructed mathematical models, that is, representations of the most important elements in the complex interaction of wind and water necessary to sail from one island to another. These models have been passed down through the generations, so that new generations of navigators always have the necessary knowledge (Fig. 11.8).[13]

On the island of Bali, in Indonesia, there is a fascinating calendar that raises questions similar to those of the Mayan calendar. This calendar is based on 10 arbitrary cycles, of 10 days, 9 days, 8 days, and so on, down to 1 day. A year in this calendar is 210 days, a number evenly divisible by all the cycle lengths except 4, 8, and 9. Special adjustments are

FIGURE 11.8

A stamp portraying stick charts from the Marshall Islands

made for those cycles so they fit reasonably into this calendar. As in the Mayan calendar, to do calculations in this calendar requires the solution of simultaneous congruences. In this case, we know that the Balinese use a wooden board known as the *tika*, on which an array of seven rows and thirty columns is carved or painted. Various symbols are placed in many of the boxes of the array, representing important days in the calendar. Manipulations on the *tika* then enable the answers to typical calendrical questions to be found. So an unanswered question about the Balinese calendar is how the *tika* was originally constructed.[14]

This brief trip through the world of ethnomathematics shows us that two of the central ideas of mathematics, logical thought and pattern analysis, occur in societies around the world. And although most societies did not have the formal "mathematics" of the literate civilizations of China, India, Islam, or Europe, mathematics was, and is, a force in the lives of people in all parts of the globe.

EXERCISES

1. Complete the reasoning to show why the formulas for determining the Mayan date a specified number of days later than a given date are valid.

2. Given the Mayan date (8, 10, 193), determine the Mayan date that is 0,2,3,5,10 days later.

3. Find an algorithm for deciding the minimum number of days between two Mayan dates (t_0, v_0, y_0) and (t_1, v_1, y_1). It might be easier to first ignore the 365-day cycle altogether and simply determine the minimum number of days between the two almanac dates of (t_0, v_0) and (t_1, v_1). For help with this problem, consult the works of Closs, Lounsbury, or Ascher mentioned in note 3.

4. Show that the minimum number of days between the two Mayan dates of (8, 20, 13) and (6, 18, 191) is 1,8,15,18 (= 10,398). Because these two dates are the birth and death dates of Pacal, a Mayan king, and because it is known from other sources that Pacal's age at death was more than sixty and less than one hundred years, determine the number of days of Pacal's life and his age at death. (Recall that the number of days calculated by the algorithm in Exercise 3 is determined only up to a multiple of one calendar round, 18,980 days or 52 calendar years.)

5. Work out the group table of the kin structure in Malekulan society. For a woman in each of the six sections, determine the section of her husband, her mother, her father, and her children.

6. The full set of algorithms for finding columns 9 through 16 in the diviner's array from Madagascar are as follows: $C_9 = C_8 \oplus C_7$, $C_{10} = C_6 \oplus C_5$, $C_{11} = C_4 \oplus C_3$, $C_{12} = C_2 \oplus C_1$, $C_{13} = C_9 \oplus C_{10}$, $C_{14} = C_{11} \oplus C_{12}$, $C_{15} = C_{13} \oplus$ C_{14}, and $C_{16} = C_{15} \oplus C_1$. Assume that $C_1 = (1, 1, 2, 2)$, $C_2 = (2, 2, 1, 2)$, $C_3 = (1, 1, 1, 2)$, and $C_4 = (1, 2, 2, 2)$. Recall that C_5, C_6, C_7, and C_8 are the four rows of the array formed by placing the first four columns in the order C_4, C_3, C_2, C_1. Calculate all of the columns $C_i, 5 \le i \le 16$.

7. Using the rules in Exercise 6 and the columns calculated there, show that $C_{13} \oplus C_{16} = C_{14} \oplus C_1 = C_{11} \oplus C_2$. Then use the properties of the XOR operation to show that these equalities are always true. The Madagascar diviners knew this result and used it to check their use of their algorithm.

8. It turns out that in the Balinese calendar, to specify a day, it is sufficient to specify the position of the day in just the five, six, and seven day weeks. Thus, we specify a day by the notation (a_5, b_6, c_7), where $1 \le a \le 5$, $1 \le b \le 6$, and $1 \le c \le 7$. Find the minimum number of days between the day $(2_5, 3_6, 5_7)$ and $(5_5, 2_6, 4_7)$.

9. Write a report on the seven possible strip patterns and seventeen possible plane patterns of symmetry. Find examples of each in wallpaper patterns or in fabric patterns. Consult D. K. Washburn and D. W. Crowe, *Symmetries of Culture* (Seattle: University of Washington Press, 1988).

10. Learn the game *mankala* and design a lesson for young children using the game to teach various arithmetic concepts. See Laurence Russ, *Mancala Games* (Algonac, Mich.: Reference Publications, 1984); H. J. R. Murray, *A History of Board Games Other Than Chess* (Oxford: Clarendon Press, 1952); or M. B. Nsimbi, *Omweso, a Game People Play in Uganda* (Los Angeles: African Studies Center, UCLA, 1968) for details.

11. Read the paper by Anna Sofaer, Rolf M. Sinclair, and Joey B. Donahue, "Solar and Lunar Orientations of the Major Architecture of the Chaco Culture of New Mexico," in *Proceedings of the Colloquio Internazionale Archeologia e Astronomia* (Venice, 1990) and some of the references cited in the paper. Do these articles convince you that the Anasazi used mathematical tools in the orientation and basic design of their buildings? What other types of evidence would be worth considering?

12. Chapter 6 of Marcia Ascher and Robert Ascher, *Code of the Quipu*, cited in note 4, deals with *quipus* in terms of the mathematical structure known as a tree. Read the chapter and do some of the exercises. What kinds of analyses of trees did Inca *quipu* makers have to perform in order to create *quipus* appropriate for various purposes?

REFERENCES AND NOTES

Books that discuss the question of why modern science arose in the West rather than in Islam or China include Toby E. Huff, *The Rise of Early Modern Science: Islam, China, and the West* (Cambridge: Cambridge University Press, 1993), and H. Floris Cohen, *The Scientific Revolution: A Historiographical Inquiry* (Chicago: University of Chicago Press, 1994). More information on the study of science in Islam can be found in J. L. Berggren, "Islamic Acquisition of the Foreign Sciences: A Cultural Perspective," in F. Jamil Ragep and Sally P. Ragep, eds., *Tradition, Transmission, Transformation* (Leiden: E. J. Brill, 1996), pp. 263–284.

Two of the best books on ethnomathematics are by Marcia Ascher: *Ethnomathematics: A Multicultural View of Mathematical Ideas* (Pacific Grove, Cal.: Brooks/Cole, 1991) and *Mathematics Elsewhere: An Exploration of Ideas across Cultures* (Princeton: Princeton University Press, 2002). The classic work in the field, which gave a great impetus to the study of mathematics in Africa, is Claudia Zaslavsky, *Africa Counts: Number and Pattern in African Culture* (Boston: Prindle, Weber and Schmidt, 1973). Another important work that contains several chapters dealing with the mathematics of various cultures around the world is Helaine Selin, ed., *Mathematics across Cultures: The History of Non-Western Mathematics* (Dordrecht: Kluwer Academic Publishers, 2000).

1. Quoted in Cohen, *The Scientific Revolution*, p. 367.

2. This picture is taken from Persis B. Clarkson, "Classic Maya Pictorial Ceramics: A Survey of Content and Theme," in Raymond Sidrys, ed., *Papers on the Economy and Architecture of the Ancient Maya* (Los Angeles: Institute of Archaeology, UCLA, 1978), 86–141. Clarkson identified this person as a female scribe. Michael Closs identified her as a mathematician because of the number scroll coming from her armpit.

3. See Michael Closs, "The Mathematical Notation of the Ancient Maya," in *Native American Mathematics*, Michael P. Closs, ed. (Austin: University of Texas Press, 1986), 291–369; Floyd G. Lounsbury, "Maya Numeration, Computation, and Calendrical Astronomy," in *Dictionary of Scientific Biography* (New York: Scribners, 1978), Vol. 15, 759–818; and chapter 3 of Ascher, *Mathematics Elsewhere*, for more details on Mayan mathematical techniques.

4. See Marcia Ascher and Robert Ascher, *Code of the Quipu: A Study in Media, Mathematics, and Culture* (Ann Arbor: University of Michigan Press, 1980) for more details on *quipus*. This work was reprinted under the title *Mathematics of the Incas: Code of the Quipu* (New York: Dover Publications, 1997). It provides a mathematical analysis of various techniques of *quipu* making and also provides exercises to help students learn the relevant mathematical ideas.

5. A discussion of the general idea of ethnomathematics is found in Marcia Ascher and Robert Ascher, "Ethnomathematics," *History of Science* 24 (1986), 125–144, and Bill Barton, "Making Sense of Ethnomathematics: Ethnomathematics Is Making Sense," *Educational Studies in Mathematics* 31 (1996), 201–233. See also Ubiratan D'Ambrosio, *Etnomatemática: Arte ou Técnica de Explicar e Conhecer* (Sao Paulo: Editora Ática S.A., 1990) for a more philosophical study of the notion of ethnomathematics. *For the Learning of Mathematics* 14(2) (1994) was a special issue devoted to ethnomathematics in mathematics education, edited by Marcia Ascher and Ubiratan D'Ambrosio. The *Newsletter* of the International Study Group on Ethnomathematics provides current information in this field.

6. For more details on the Anasazi, consult William Ferguson and Arthur Rohn, *Anasazi Ruins of the Southwest in Color* (Albuquerque: University of New Mexico Press, 1986). For details on the astronomy of the Anasazi and other North American Indians, see Ray A. Williamson, *Living the Sky: The Cosmos of the American Indian* (Norman: University

of Oklahoma Press, 1987), and E. C. Krupp, ed., *In Search of Ancient Astronomies* (New York: McGraw-Hill, 1978).

7. See Paulus Gerdes, *Geometry from Africa: Mathematical and Educational Explorations* (Washington: Mathematical Association of America, 1999) for details on many geometrical ideas present in Mozambican culture and other cultures of southern Africa.

8. More details on the graphing procedures of the Tshokwe are found in the book of note 7 as well as in Paulus Gerdes, "On Mathematical Elements in the Tchokwe 'Sona' Tradition," *For the Learning of Mathematics* 10 (1990), 31–34, from which Figure 11.6 is taken. Also, consult Ascher, *Ethnomathematics*, chapter 2.

9. See Zaslavsky, *Africa Counts*, chapter 14. This chapter, entitled "Geometric Symmetries in African Art," was written by D. W. Crowe. The patterns in Figure 11.7 are taken from that book and used by permission.

10. See Gerdes, *Geometry from Africa*, chapter 1, for more details.

11. See Ascher, *Mathematics Elsewhere*, chapter 1, for more details on the logic of divination.

12. See Ascher, *Ethnomathematics*, chapter 3, for more details.

13. For a detailed survey of the stick charts from the Marshall Islands, see Marcia Ascher, "Models and Maps from the Marshall Islands: A Case in Ethnomathematics," *Historia Mathematica* 22 (1995), 347–370, as well as Ascher, *Mathematics Elsewhere*, chapter 4.

14. See Ascher, *Mathematics Elsewhere*, chapter 3, for more information on the calendar in Bali.

CHAPTER 12

Algebra in the Renaissance

But of number, cosa [unknown], and cubo [cube of the unknown], however they are compounded . . . , nobody until now has formed general rules, because they are not proportional among them. . . . And therefore, until now, for their equations, one cannot give general rules except that, sometimes, by trial, . . . in some particular cases. And therefore when in your equations you find terms with different intervals without proportion, you shall say that the art, until now, has not given the solution to this case, . . . even if the case may be possible.

—From the *Summa de arithmetica, geometrica, proportioni et proportionalita* of Luca Pacioli, 1494[1]

His account of the discovery of the rule for the algebraic solution of a cubic equation is given in chapter 11 of Girolamo Cardano's *Ars Magna*: "Scipio Ferro of Bologna well-nigh thirty years ago [c. 1515] discovered this rule and handed it on to Antonio Maria Fior of Venice, whose contest with Niccolò Tartaglia of Brescia gave Niccolò occasion to discover it. He [Tartaglia] gave it to me in response to my entreaties, though withholding the demonstration. Armed with this assistance, I sought out its demonstration in [various] forms. This was very difficult."[2]

Many changes began to take place in the European economy in the fourteenth century that eventually had an effect on mathematics. The general cultural movement of the next two centuries, known as the Renaissance, also had its impact, particularly in Italy, so it is in that country that we begin our discussion of Renaissance mathematics.

The Italian merchants of the Middle Ages generally were what today we might call venture capitalists. They traveled themselves to distant places in the East, bought goods that were wanted back home, and returned to Italy to sell them in the hope of making a profit. These traveling merchants needed very little mathematics other than the ability to determine their costs and revenues for each voyage. By the early fourteenth century, a commercial revolution spurred originally by the demands of the Crusades had begun to change this system greatly. New technologies in shipbuilding and greater safety on the shipping lanes helped to replace the traveling merchants of the Middle Ages with the sedentary merchants of the Renaissance. These "new men" were able to remain at home in Italy and hire others to travel to the various ports, make the deals, act as agents, and arrange for shipping. Thus, international trading companies began to develop in the major Italian cities, companies that had a need for more sophisticated mathematics than did their predecessors. These new companies had to deal with letters of credit, bills of exchange, promissory notes, and interest calculations. Double-entry bookkeeping began as a way of keeping track of the various transactions. Business was no longer composed of single ventures but of a continuous flow of goods consisting of many shipments from many different ports en route simultaneously. The medieval economy, based in large part on barter, was gradually being replaced by a money economy.

The Italian merchants needed a new facility in mathematics to be able to deal with the new economic circumstances, but the mathematics they needed was not the mathematics of the quadrivium, the mathematics studied in the universities. They needed new tools for calculating and problem solving. To meet this need, a new class of "professional" mathematicians, the *maestri d'abbaco,* or abacists, appeared in early fourteenth-century Italy. These professionals wrote the texts from which they taught the necessary mathematics to the sons of the merchants in new schools created for this purpose.

The first section of this chapter therefore discusses the mathematics of the abacists in Italy and, in particular, their algebra. Because the commercial revolution soon spread to other parts of Europe as well, the next section deals with late fifteenth- and early sixteenth-century algebra in France, England, Germany, and Portugal. But because the major new discoveries in algebra in this time period took place in Italy, partly in response to Luca Pacioli's statement in the chapter's opening quotation that, as of 1494, cubic equations were in general unsolvable algebraically, we go back to Italy to tell the marvelous story of the ultimate discovery of such a solution in the work of Scipione del Ferro, Niccolò Tartaglia, Girolamo Cardano, and Rafael Bombelli.

All of these algebraists based their work on the Islamic algebras first translated into Latin in the twelfth century. But by the middle of the sixteenth century, virtually all of the surviving works of Greek mathematics, newly translated into Latin from the Greek manuscripts that had been stored in Constantinople, were available to European mathematicians. The last sections of this chapter are thus devoted to the works of François Viète, who used his understanding of Greek mathematics to entirely revamp the study of algebra, and Simon Stevin, who once and for all eliminated the Aristotelian distinction between number and magnitude, in effect giving us our current concept of "number."

12.1

THE ITALIAN ABACISTS

The Italian abacists of the fourteenth century were instrumental in teaching the merchants the "new" Hindu-Arabic decimal place value system and the algorithms for using it. As is usual when a new system replaces an old traditional one, there was great resistance to the change. For many years, account books were still kept in Roman numerals. It was believed that the Hindu-Arabic numerals could be altered too easily, and thus it was risky to depend on them alone in recording large commercial transactions. (The current system of writing out the amounts on checks in words dates from this time.) The advantages of the new system, however, eventually overcame the merchants' initial hesitation. The old counting board system required accountants to carry around not only a board but also a bag of counters, while the new system required only pen and paper and could be used anywhere. In addition, using a counting board required that preliminary steps in the calculation be eliminated as one worked toward the final answer. With the new system, all the steps were available for checking when the calculation was finished. (Of course, these advantages would have meant nothing had not a steady supply of cheap paper been recently introduced.) The abacists instructed entire generations of middle-class Italian children in the new methods of calculation, and these methods soon spread throughout the continent.

In addition to the algorithms of the Hindu-Arabic number system, the abacists taught their students methods of problem solving using the tools of both arithmetic and Islamic algebra. The texts written by the abacists, of which several hundred different ones still exist, are generally large compilations of problems along with their solutions.[3] These include not only genuine business problems of the type the students would have to solve when they joined their fathers' companies but also plenty of recreational problems typical of the kind found in modern elementary algebra texts. There were also sometimes geometrical problems as well as problems dealing with elementary number theory, the calendar, and astrology. The solutions in the texts were written in great detail with every step fully described, but, in general, no reasons were given for the various steps, nor any indication of the limitations of a particular method. Perhaps the teachers did not want to disclose their methods in written form, fearing that then there would no longer be any reason to hire them. In any case, it seems clear that these abacus texts were designed not only for classroom use but also to serve as reference manuals for the merchants themselves. A merchant could easily find and readily follow the solution of a particular type of problem without the necessity of understanding the theory behind the solution.

The following are examples of the types of problems found in these texts, most of which can be solved by using the ancient methods of the rule of three or false position:

> The gold florin is worth 5 *lire*, 12 *soldi*, 6 *denarii* in Lucca. How much (in terms of gold florins) are 13 *soldi*, 9 *denarii* worth? (One needs to know that 20 *soldi* make up 1 *lira* and 12 *denarii* make 1 *soldo*.)

> The *lira* earns 3 *denarii* a month in interest. How much will 60 *lire* earn in 8 months? (This is a problem in simple interest. Problems in compound interest also appeared where the period of compounding was generally one year.)

> A field is 150 feet long. A dog stands at one corner and a hare at the other. The dog leaps 9 feet in each leap while the hare leaps 7. In how many feet and leaps will the dog catch the hare?

Although these texts were strictly practical, they did have significant influence on the development of mathematics, because they instilled in the Italian merchant class a facility with numbers without which future advances could not be made. Furthermore, some of these texts also brought to this middle class the study of Islamic algebra as a basic part of the curriculum. During the fourteenth and fifteenth centuries, the abacists extended the Islamic methods in several directions. In particular, they introduced abbreviations and symbolism, developed new methods for dealing with complex algebraic problems, and expanded the rules of algebra into the domain of equations of degree higher than the second. More important than the introduction of a few new techniques, however, was the general teaching of how algebra could be used to solve practical problems. With a growing competence in algebra brought about by the study of these abacus texts, it was only natural that European scholars would attempt to apply these techniques to solve more theoretical problems arising from the rediscovery of many of the classic Greek mathematical texts. This combination of algebra and Greek geometry was to lead in the seventeenth century to the new analytic techniques that serve as the basis of modern mathematics.

12.1.1 Algebraic Symbolism and Techniques

Recall that Islamic algebra was entirely rhetorical. There were no symbols for the unknown or its powers nor for the operations performed on these quantities. Everything was written out in words. The same was generally true in the works of the early abacists and in the earlier Italian work of Leonardo of Pisa. Early in the fifteenth century, however, some of the abacists began to substitute abbreviations for unknowns. For example, in place of the standard words *cosa* (thing), *censo* (square), *cubo* (cube), and *radice* (root), some authors used the abbreviations *c*, *ce*, *cu*, and *R*. Combinations of these abbreviations were used for higher powers. Thus, *ce di ce* or *ce ce* stood for *censo di censo* or fourth power (x^2x^2); *ce cu* or *cu ce*, designating *censo di cubo* and *cubo di censo*, respectively, stood for fifth power (x^2x^3); and *cu cu*, designating *cubo di cubo*, stood for sixth power (x^3x^3). By the end of the fifteenth century, however, the naming scheme for higher powers had changed, and authors used *ce cu* or *censo di cubo* to designate the sixth power ($(x^3)^2$) and *cu cu* or *cubo di cubo* to represent the ninth power ($(x^3)^3$). The fifth power was then designated as *p.r.* or *primo relato* and the seventh power as *s.r.* or *secondo relato*.

Near the end of the fifteenth century, Luca Pacioli introduced the abbreviations $\overline{p}$ and $\overline{m}$ to represent plus and minus (*più* and *meno*). (These particular abbreviations probably came from a more general practice of using the bar over a letter to indicate that some letters were missing.) As with other innovations, however, there was no great movement on the part of all the writers to use the same names or the same abbreviations. This change was a slow one. New symbols gradually came into use in the fifteenth and sixteenth centuries, but modern algebraic symbolism was not fully formed until the mid-seventeenth century.

Even without much symbolism, the Italian abacists, like their Islamic predecessors, were competent in handling operations on algebraic expressions. For example, Paolo Gerardi, in his *Libro di ragioni* of 1328, gave the rule for adding the fractions $100/x$ and $100/(x + 5)$:

> You place 100 opposite one *cosa* [x], and then you place 100 opposite one *cosa* and 5. Multiply crosswise as you see indicated, and you say . . . 100 times the one *cosa* that is across from it makes 100 *cose*. And then you say 100 times one *cosa* and 5 makes 100 *cose* and 500 in number. Now you must add one with the other which makes 200 *cose* and 500 in number. Then multiply one

cosa times 1 *cosa* and 5 in number, making 1 *censo* [x^2] and 5 *cose*. Now you must divide 200 *cose* and 500 in number by one *censo* and 5 *cose* [$(200x + 500/(x^2 + 5x)]$.[4]

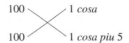

Similarly, the rules of signs were also written out in words and even justified, here in a late fourteenth-century manuscript by an unknown author:

> Multiplying minus times minus makes plus. If you would prove it, do it thus: You must know that multiplying 3 and 3/4 by itself will be the same as multiplying 4 minus 1/4 [by itself]. Multiplying 3 and 3/4 by 3 and 3/4 gives 14 and 1/16. To multiply 4 minus 1/4 by 4 minus 1/4 . . . , say 4 by 4 is 16; now multiply across and say 4 times minus one quarter makes minus 4 quarters, which is minus 1, and 4 times minus 1/4 makes minus 1, so you have minus 2. Subtract this from 16 and it leaves 14. Now take minus 1/4 times minus 1/4. That gives 1/16, so one has the same as the other [multiplication].[5]

In general, the abacus manuscripts have lists of products and quotients of monomials written out, using the abbreviations for the powers of the unknown given above. But one fifteenth-century manuscript makes explicit the rules of exponents after having named the first nine powers of the unknown:

> If you wish to multiply these names [of the powers], . . . multiply the quantities [the coefficients of the powers] one into the other; then add together the degrees of the names and see the degree which is named. . . . If you wish to divide one of those names by another, it is necessary that what you wish to divide has a degree greater than that by which you wish to divide it. Make so: divide the quantities one by another. Afterwards, subtract the quantity of the degrees of those names one from another, . . . and [if] so many degrees remain, that quantity will be of so many degrees.[6]

Antonio de' Mazzinghi (1353–1383), one of the few abacists about whom any biographical details are known, taught in the *Bottega d'abbaco* at the monastery of S. Trinita in Florence. His algebraic problems survive in several fifteenth-century manuscripts. Antonio was expert in devising clever algebraic techniques for solving complex problems. In particular, he explicitly used two different names for the two unknown quantities in many of these problems. For example, consider the following: "Find two numbers such that multiplying one by the other makes 8 and the sum of their squares is 27."[7] The abacist began the solution by supposing that the first number is *un cosa meno la radice d'alchuna quantità* (a thing minus the root of some quantity), while the second number equals *una cosa più la radice d'alchuna quantità* (a thing plus the root of some quantity). The two words *cosa* and *quantità* then serve in his rhetorical explication of the problem as the equivalent of our symbols x and y, that is, the first number is equal to $x - \sqrt{y}$, the second to $x + \sqrt{y}$.

12.1.2 Higher-Degree Equations

The third major innovation of the Italian abacists was the extension of Islamic quadratic equation–solving techniques to higher-degree equations. In general, all of the abacists began their treatments of algebra by presenting al-Khwārizmī's six types of linear and quadratic equations and showing how each can be solved. But Maestro Dardi of Pisa in a work of 1344 extended this list to 198 types of equations of degree up to four, some of which involved radicals.[8] Most of the equations can be solved by a simple reduction to one of the standard forms, although in each case Dardi gave the solution anew, presenting both a numerical example and a recipe for solving the particular type of equation. For example, he noted that the equation $ax^4 = bx^3 + cx^2$ has the solution given by

$$x = \sqrt{\left(\frac{b}{2a}\right)^2 + \frac{c}{a}} + \frac{b}{2a},$$

that is, it has the same solution as the standard equation $ax^2 = bx + c$. (Note that 0 is never considered as a solution.) Similarly, the equation $n = ax^3 + \sqrt{bx^3}$ can be solved for x^3 by reducing it to a quadratic equation in $\sqrt{x^3}$.

More interesting than these quadratic equations are four examples of irreducible cubic and quartic equations. Dardi's cubic equation was $x^3 + 60x^2 + 1200x = 4000$. His rule tells us to divide 1200 by 60 (giving 20), cube the result (which gives 8000), add 4000 (giving 12,000), take the cube root ($\sqrt[3]{12,000}$), and finally subtract the quotient of 1200 by 60. Dardi's answer, which is correct, was that $x = \sqrt[3]{12,000} - 20$. If we write this equation using modern notation and then give Dardi's solution rule, we obtain the solution to the equation $x^3 + bx^2 + cx = d$ in the form

$$x = \sqrt[3]{\left(\frac{c}{b}\right)^3 + d} - \frac{c}{b}.$$

It is easy enough to see that this solution is wrong in general, and Dardi even admitted as much. How then did Dardi figure out the correct solution to his particular case? We can answer this question by considering the problem that illustrates the rule, a problem in compound interest: A man lent 100 *lire* to another and after 3 years received back a total of 150 *lire* in principal and interest, where the interest was compounded annually. What was the interest rate? Dardi set the rate for 1 *lira* for 1 month at x *denarii*. Then the annual interest on 1 *lira* is $12x$ *denarii* or $(1/20)x$ *lire*. So the amount owed after 1 year is $100(1 + x/20)$ and after 3 years is $100(1 + x/20)^3$. Dardi's equation therefore is

$$100\left(1 + \frac{x}{20}\right)^3 = 150 \quad \text{or} \quad 100 + 15x + \frac{3}{4}x^2 + \frac{1}{80}x^3 = 150$$

or, finally,

$$x^3 + 60x^2 + 1200x = 4000.$$

Because the left side of this equation comes from a cube, it can be completed to a cube once again by adding an appropriate constant. In general, because $(x + r)^3 = x^3 + 3rx^2 + 3r^2x + r^3$, to complete $x^3 + bx^2 + cx$ to a cube, we must find r satisfying two separate conditions, $3r = b$ and $3r^2 = c$, conditions that can only be satisfied when $b^2 = 3c$. In Dardi's example, with $b = 60$ and $c = 1200$, the condition is satisfied and $r = c/b = 20$.

Dardi gave a similar rule for solving special quartic equations, while Piero della Francesca (c. 1420–1492), more famous as a painter than as an abacist, extended these rules to fifth- and sixth-degree equations in his own *Trattato d'abaco*. Neither man stated explicitly that the rules apply only to the cases reducible to the form $h(1 + x)^n = k$, where $n = 4$, 5, 6. There is another (anonymous) manuscript of this period, which suggests that the equation $x^3 + px^2 = q$ can be solved by setting $x = y - \frac{p}{3}$ where y is a solution of $y^3 = 3(\frac{p}{3})^2 + [q - 2(\frac{p}{3})^3]$. This is correct as far as it goes, but the author has only managed to replace one cubic equation by another. In the numerical example presented, he solved the new equation by trial, but this could also have been done with the original. Nevertheless, although the abacists did not manage to give a complete general solution to the cubic equation, they, like their Islamic predecessors, wrestled with the problem and arrived at partial results, as noted in the opening quotation from the work of Luca Pacioli (1445–1517).

FIGURE 12.1

Pacioli on an Italian stamp

Pacioli, one of the last of the abacists, was ordained as a Franciscan friar in the 1470s and taught mathematics at various places in Italy during the remainder of his career. He became so famous as a teacher that there is a painting of him by Jacopo di Barbari now hanging in the Naples Museum, which shows him teaching geometry to a young man tentatively identified as Guidobaldo, the son of his patron, the Duke of Urbino (Fig. 12.1). As part of his teaching, Pacioli composed three different abacus texts for his students. He regretted what he believed to be the low ebb to which teaching had fallen. Because he felt that one of the problems was the scarcity of available subject material, he gathered mathematical materials for some twenty years and in 1494 completed the most comprehensive mathematics text of the time, and one of the earliest mathematics texts to be printed. This was the *Summa de arithmetica, geometrica, proportioni et proportionalita*, a 600-page work written in the Tuscan dialect rather than in Latin. It contained not only practical arithmetic but also much of the algebra already discussed, the first published treatment of double-entry bookkeeping, and a section on practical geometry. There was little that was original in this work. In fact, a large number of the algebra problems are taken directly from della Francesco's treatise, while the practical geometry is very similar to that of Leonardo of Pisa. Nevertheless, its comprehensiveness and the fact that it was the first such work to be printed made it into a widely circulated and influential text, extensively studied by sixteenth-century Italian mathematicians. It became the common base from which these men were able to extend the range of algebra. Before considering these advances, however, we first turn to contemporaneous developments elsewhere in Europe. It is not only from Italy that our algebra comes.

12.2 ALGEBRA IN FRANCE, GERMANY, ENGLAND, AND PORTUGAL

The medieval economy was also changing in northern Europe during the fourteenth and fifteenth centuries, although developments were generally a bit behind those in Italy. And so mathematics texts began to appear there to meet the new needs of the society. We will consider here the work of Nicolas Chuquet in France, Christoff Rudolff, Michael Stifel, and Johannes Scheubel in Germany, Robert Recorde in England, and Pedro Nunes in Portugal. There is much similarity among their works in algebra and also similarities between these works and the Italian algebra of the fifteenth century, so it is clear that these mathematicians all had some knowledge of the contemporaneous work elsewhere in Europe, even though explicit reference to the work of others is generally limited or lacking entirely. But each of them also seems to have some original material. It appears that the knowledge of Islamic algebra had spread widely in Europe by the fifteenth century. Each person attempting to write new works used this material and works in algebra from elsewhere in Europe, adapted them to fit the circumstances of his own country, and introduced some of his own new ideas. By the late sixteenth century, with the spread of printing, new ideas could circulate more rapidly throughout the continent, and those generally felt to be most important were absorbed into a new European algebra.

12.2.1   France: Nicolas Chuquet

Nicolas Chuquet (d. 1487) was a French physician who wrote his mathematical treatise in Lyon near the end of his life. Lyon in the late fifteenth century was a thriving commercial community with a growing need, as in the Italian cities, for practical mathematics. It was probably to meet this need that Chuquet composed his *Triparty* in 1484, a work on arithmetic and algebra in three parts, followed by three related works containing problems in various fields in which the rules established in the *Triparty* are used. These supplementary problems show many similarities to the problems in Italian abacus works, but the *Triparty* itself is on a somewhat different level in that it is a text in mathematics itself. Most of the mathematics in it was certainly known to the Islamic algebraists and also to Leonardo of Pisa. Nevertheless, since it is the first detailed algebra in fifteenth-century France, we will consider some of its important ideas.

The first part of the *Triparty* is concerned with arithmetic. Like the Italian works, it began with a treatment of the Hindu-Arabic place value system and detailed the various algorithms for the basic operations of arithmetic, both with whole numbers and with fractions. One of Chuquet's procedures with fractions was a rule "to find as many numbers intermediate between two neighboring numbers as one desires."[9] His idea was that to find a fraction between two fractions, one simply adds the numerators and adds the denominators. Thus, between 1/2 and 1/3 is 2/5, and between 1/2 and 2/5 is 3/7. Chuquet gave no proof that the rule is correct, but he did apply it to deal with finding roots of polynomials. For example, to find the root of $x^2 + x = 39\frac{13}{81}$, Chuquet began by noting that 5 is too small to be a root, while 6 is too large. He then proceeded to find the correct intermediate value by checking, in turn, $5\frac{1}{2}$, $5\frac{2}{3}$, $5\frac{3}{4}$, and $5\frac{4}{5}$ and determining that the root must be between the two last values. Applying his rule to the fractional parts, he next checked $5\frac{7}{9}$, which turns out to be the correct answer.

In part two of the *Triparty*, Chuquet applied the rule to the calculation of square roots of numbers that are not perfect squares. Noting that 2 is too small and 3 too large to be the square root of 6, he began the next stage of his approximation procedure by determining that $2\frac{1}{3}$ is too small and $2\frac{1}{2}$ too large. His next several approximations were, in turn, $2\frac{2}{5}$, $2\frac{3}{7}$, $2\frac{4}{9}$, $2\frac{5}{11}$, and $2\frac{9}{20}$. At each stage he calculated the square of the number chosen and, depending on whether it is larger or smaller than 6, determined between which two values to use his rule of intermediates. He noted that "by this manner one may proceed, . . . until one approaches very close to 6, a little more or a little less, and until it is sufficient. And one should know that the more one should continue in this way, the nearer to 6 one would approach. But one would never attain it precisely. And from all this follows the practice, in which the good and sufficient root of 6 is found to be $2\frac{89}{198}$, which root multiplied by itself produces 6 plus 1/39,204."[10] Chuquet evidently was aware of the irrationality of $\sqrt{6}$ and had developed a new recursive algorithm to calculate it to whatever accuracy may be desired. He had therefore taken another step on the road to denying the usefulness of the Greek dichotomy between the discrete and the continuous, the final elimination of which was to occur about a century later.

Chuquet also displayed in the second part of his work the standard methods for calculating the square and cube roots of larger integers, one integral place at a time, but as is usual in the discussion of these methods, he did not take the method below the unit. He showed no knowledge of the idea of a decimal fraction. If the standard method did not give an exact root, one could choose between calculating using common fractions by his method of intermediates or (and this is the method he preferred) simply not bothering to calculate at all and leaving the answer in the form $R^2 6$ or $R^3 12$, his notation for our $\sqrt{6}$ and $\sqrt[3]{12}$. Chuquet also used the Italian $\overline{p}$ and $\overline{m}$ for plus and minus, but introduced an underline to indicate grouping. Thus, what we would write as $\sqrt{14 + \sqrt{180}}$, Chuquet wrote as $R^2\underline{14\overline{p}R^2 180}$. He proceeded to use this notation with complete understanding through the rest of this second part as he displayed a solid knowledge of computations with radical expressions, both simple and compound, including the necessary rules for dealing with positives and negatives in addition, subtraction, multiplication, and division.

The third part of the *Triparty* was more strictly algebraic, as Chuquet showed how to manipulate with polynomials and how to solve various types of equations. As part of his discussion of polynomials, he introduced an exponential notation for the powers of the unknown, which made calculation somewhat easier than the Italian abbreviations. For example, he wrote 12^2 for what we write as $12x^2$ and, introducing actual negative numbers for the first time in a European work, wrote $\overline{m}12^{\overline{2m}}$ for $-12x^{-2}$. He even noted that the exponent 0 is to be used when one is dealing with numbers themselves. He then showed how to add, subtract, multiply, and divide these expressions (*diversities*) involving exponents (*denominations*) using the standard modern rules, even when one of the exponents is negative. Thus, "whoever would multiply 8^3 by $7^{\overline{1m}}$ it is first necessary to multiply 8 by 7 coming to 56, then he must add the denominations, that is to say $3\overline{p}$ with $1\overline{m}$ coming to 2. Thus, this multiplication comes to 56^2, and so should others be understood."[11] Not only did he give this rule, similar to that of one of his Italian contemporaries, but he also justified it. He wrote down in two parallel columns the powers of 2 (beginning with $1 = 2^0$ and ending with $1,048,576 = 2^{20}$) and the corresponding denomination and then noted that multiplication in the first column corresponded to addition in the second. For example, 128 (which corresponds to 7) multiplied by 512 (which corresponds to 9) gives 65,536 (which corresponds to 16). Because the addition rule of exponents works for numbers, he simply extended it to his diversities. But although he showed that he understood the meaning of negative exponents, his table for numbers did not include them, and, in fact, unlike al-Samaw'al, he made little use of them in what follows.

Chuquet also had a few innovations in his equation-solving techniques. First, he generalized al-Khwārizmī's rules to equations of any degree that are of quadratic type, thus going somewhat further than the Italian abacists. For example, he gave the solution of the equation $cx^m = bx^{m+n} + x^{m+2n}$ as

$$x = \sqrt[n]{\sqrt{(b/2)^2 + c} - (b/2)}.$$

Second, he noted that a particular system of two equations in three unknowns has multiple solutions. To solve the system $x + y = 3z$, $x + z = 5y$, he first picked 12 for x and then found $y = 3\frac{3}{7}$ and $z = 5\frac{1}{7}$. Then he picked 8 for y and calculated $x = 28$ and $z = 12$. "Thus," he concluded, "it appears that the number proposed alone determines the varying answer."[12] Finally, although he was not consistent about this, Chuquet was willing under some circumstances to consider negative solutions to equations, again for the first time in Europe. For example, he solved the problem $\frac{5}{12}(20 - \frac{11}{20}x) = 10$ to get $x = -7\frac{3}{11}$. He then checked the result carefully and concluded that the answer is correct. In other problems, however, he rejected negative solutions as "impossible," and he never considered 0 to be a solution.

The three supplements to the *Triparty* contained hundreds of problems in which the techniques of that work were applied. Many of the problems were commercial, of the same type found in the Italian abacus works, while others were geometrical, both practical and theoretical. This work may have been intended as a text, although probably not in a university, but, unfortunately, the *Triparty* was never printed and exists today only in manuscript form. Some parts of it were incorporated into a work of Estienne de la Roche (probably one of Chuquet's students) in 1520, but neither this work nor Chuquet's itself had much influence.

12.2.2 Germany: Christoff Rudolff, Michael Stifel, and Johannes Scheubel

Algebra in Germany first appeared late in the fifteenth century, probably due to the same reasons that led to its development in Italy somewhat earlier. It is likely, in fact, that many of the actual techniques were also imported from Italy. The very name given to algebra in Germany, the Art of the *Coss*, reveals its Italian origin. *Coss* was simply the German form of the Italian *cosa*, or thing, the name usually given to the unknown in an algebraic equation. Two of the most important Cossists in the first half of the sixteenth century were Christoff Rudolff (sixteenth century) and Michael Stifel (1487–1567).

Christoff Rudolff wrote his *Coss*, the first comprehensive German algebra, in Vienna in the early 1520s. It was published in Strasbourg in 1525. As usual, the book began with the basics of the place value system for integers, giving the algorithms for calculation as well as a short multiplication table. In a section dealing with progressions, Rudolff included a list of nonnegative powers of 2 alongside their respective exponents, just as Chuquet had done. He also noted that multiplication in the powers corresponded to addition in the exponents. He then extended this idea to powers of the unknown, again as Chuquet had done. Although Rudolff did not have the exponential notation of his French predecessor, he did have a system of abbreviations of the names of these powers, where his naming scheme was similar to the Italian multiplicative one (Sidebar 12.1).

To help the reader understand these terms, Rudolff gave as examples the powers of various numbers. He then showed how to add, subtract, multiply, and divide expressions formed from these symbols. Because it is not obvious how to multiply these symbols, unlike the situation in Chuquet's system, Rudolff presented a multiplication table for use with them, which showed, for instance, that ℞ times ʒ was ♅. To simplify matters, he then included numerical values for his symbols. Thus, *radix* was labeled as 1, *zensus* as 2, *cubus* as 3, and so on, and he noted that in multiplying expressions one could simply add the corresponding numbers to find the correct symbol. In this section Rudolff also dealt with binomials, terms connected

SIDEBAR 12.1 *Rudolff's System for Powers of the Unknown*

dragma	φ	radix	$x \leftrightarrow x$
zensus	$\mathcal{Z} \leftrightarrow x^2$	cubus	$\mathcal{C} \leftrightarrow x^3$
zens de zens	$\mathcal{ZZ} \leftrightarrow x^4$	sursolidum	$ß \leftrightarrow x^5$
zensicubus	$\mathcal{ZC} \leftrightarrow x^6$	bissursolidum	$bß \leftrightarrow x^7$
zenszensdezens	$\mathcal{ZZZ} \leftrightarrow x^8$	cubus de cubo	$c\,\mathcal{C} \leftrightarrow x^9$

by an operation sign, and included, for the first time in an algebra text, the current symbols of $+$ and $-$ to represent addition and subtraction. These signs had been used earlier in an arithmetic work of 1518 of Heinrich Schreiber (Henricus Grammateus), Rudolff's teacher at the University of Vienna. Even earlier they had appeared in a work of Johann Widman of 1489. There, however, they represented excess and deficiency rather than operations.

Rudolff also introduced in his *Coss* the modern symbol $\sqrt{}$ for square root. He modified this symbol somewhat to indicate cube roots and fourth roots but did not use modern indices. He did, however, give a detailed treatment of operations on surds, showing how to use conjugates in division as well as how to find the square roots of surd expressions such as $\sqrt{27 + \sqrt{200}}$. He also introduced a symbol for "equals," namely, a period, as in $12x.2\ (x = 2)$. Often, however, he relied on the German *gleich*.

The second half of Rudolff's *Coss* was devoted to the solving of algebraic equations, but Rudolff used his own eight-fold classification rather than the standard six-fold one. The rule for the solution of each type of equation was given in words and then illustrated with examples. Although Rudolff dealt with equations of higher degree than two in his classes, like Chuquet he included only those that could be solved by reduction to a quadratic equation or by simple roots. Thus, for example, one of his classes was that now written as $ax^n + bx^{n-1} = cx^{n-2}$. The solution given was the standard

$$x = \sqrt{\left(\frac{b}{2a}\right)^2 + \frac{c}{a}} - \frac{b}{2a}.$$

His sample equations illustrating this class included $3x^2 + 4x = 20$ and $4x^7 + 8x^6 = 32x^5$, both of which have the solution $x = 2$. Like the other authors, however, Rudolff did not deal with either negative roots or zero as a root.

After presenting the rules, Rudolff, as is typical, gave several hundred examples of problems that could be solved using the rules. Many are commercial problems dealing with buying and selling, exchange, wills, and money, or recreational problems, including a version of the old 100-birds-for-100-coins problem. Most of the problems, especially the more practical ones, were given as examples of Rudolff's first class of equations, $ax^n = bx^{n-1}$, for which the solution is $x = \frac{b}{a}$. The problems needing a version of the quadratic formula are generally artificial ones, including the ubiquitous "divide 10 into two parts such that . . . ." At the end of the text, Rudolff presented three irreducible cubic equations with their answers

BIOGRAPHY

Michael Stifel (1487–1567)

Michael Stifel was ordained as a priest in 1511. Reacting to various clerical abuses, he became an early follower of Martin Luther. In the 1520s he became interested in what he called *wortrechnung* (word calculus), the interpretation of words through the numerical values of the letters involved. Through interpreting certain Biblical passages using his numerical methods, he finally came to the belief that the world would end on October 18, 1533. He assembled his congregation in the church on that morning, but to his great dismay, nothing happened. He was subsequently discharged from his parish and for a time placed under house arrest. Because he had now been cured of prophesying, however, he was given another parish in 1535 through the intervention of Luther. Subsequently, he devoted himself to the study of mathematics at the University of Wittenberg and soon became an expert in algebraic methods, publishing his *Deutsche Arithmetica* in 1545, one year after the *Arithmetica integra*. Later in life, however, he resumed his *wortrechnung* and wrote two books on the subject.

but without giving a method of solution. He simply noted that others who come later will continue the algebraic art and teach how to deal with these. Curiously, on the final page there is a drawing of a cube of side $3 + \sqrt{2}$ divided into eight rectangular prisms. Whether Rudolff intended this diagram to be a hint for the solution of the cubic equation is not known.

Michael Stifel brought out a new edition of Rudolff's text in 1553, nine years after he had published his own, the *Arithmetica integra*.[13] In this latter work, Stifel used the same symbols as Rudolff for the powers of the unknowns, but he was more consistent in using the correspondence between these letters and the integral "exponents." He went further than Rudolff in writing out a table of powers of 2 along with their exponents, which included the negative values -1, -2, and -3 as corresponding to $\frac{1}{2}$, $\frac{1}{4}$, and $\frac{1}{8}$, respectively, but he was probably not aware of Chuquet's similar work with negative exponents.

Although Stifel, like most of his contemporaries, did not accept negative roots to equations, he was the first to compress the three standard forms of the quadratic equation into the single form $x^2 = bx + c$, where b and c were either both positive or of opposite parity. The solution, expressed in words, was then equivalent to

$$x = \frac{b}{2} \pm \sqrt{\left(\frac{b}{2}\right)^2 + c},$$

where the negative sign was only possible in the case where b was positive and c negative. In that case, as long as $(\frac{b}{2})^2 + c > 0$, there were two positive solutions. Combining the three cases of the quadratic into one does not seem a major advance, but in the context of the sixteenth century it was significant. It was another step toward the extension of the number concept, although two centuries were to pass before all algebra texts adopted his procedure.

Stifel's work was also the first European work both to present the Pascal triangle of binomial coefficients and to make use of the table for finding roots (Table 12.1). (The triangle itself had been published earlier on the title page of Peter Apianus's *Arithmetic* of 1527, but Apianus made no use of the triangle in his book.) Stifel noted that he had discovered these

TABLE 12.1   *The Pascal Triangle.*

Stifel's Version of the Pascal Triangle

1						
2						
3	3					
4	6					
5	10	10				
6	15	20				
7	21	35	35			
8	28	56	70			
9	36	84	126	126		
10	45	120	210	252		
11	55	165	330	462	462	
12	66	220	495	792	924	
13	78	286	715	1,287	1,716	1,716
14	91	364	1,001	2,002	3,003	3,432
15	105	455	1,365	3,003	5,005	6,435	6,435
16	120	560	1,820	4,368	8,008	11,440	12,870
17	136	680	2,380	6,188	12,376	19,448	24,310

Scheubel's Version of the Pascal Triangle

2													
3	3												
4	6	4											
5	10	10	5										
6	15	20	15	6									
7	21	35	35	21	7								
8	28	56	70	56	28	8							
9	36	84	126	126	84	36	9						
10	45	120	210	252	210	120	45	10					
11	55	165	330	462	462	330	165	55	11				
12	66	220	495	792	924	792	495	220	66	12			
13	78	286	715	1,287	1,716	1,716	1,287	715	286	78	13		
14	91	364	1,001	2,002	3,003	3,432	3,003	2,002	1,001	364	91	14	
15	105	455	1,365	3,003	5,005	6,435	6,435	5,005	3,003	1,365	455	105	15
16	120	560	1,820	4,368	8,008	11,440	12,870	11,440	8,008	4,368	1,820	560	120	16

coefficients and the root finding procedure only with great difficulty, as he had been unable to find any written accounts of them. Thus, although these coefficients had been used for that purpose in China and in Islamic countries several centuries earlier, the knowledge of this procedure evidently only reached Stifel indirectly.

Other texts by German authors over the next several decades also made use of the Pascal triangle to find roots. For example, Johannes Scheubel (1494–1570) displayed the triangle in his *De numeris et diversis rationibus* of 1545 with the standard instructions for calculating its entries. Scheubel's book, written in Latin, was evidently aimed at a different audience than the books of Rudolff and Stifel. In particular, he made little effort to include "practical" applications of the material. But he did spend many pages working through the method of extracting higher roots using the entries in the Pascal triangle. Although Scheubel's *De numeris* was not an algebra text, in 1552 Scheubel published such a text, again in Latin. This work, *Algebrae compendiosa facilisque descriptio* was printed, however, in France and was the first algebra work printed there, with the exception of de la Roche's version of Chuquet's *Triparty*.

12.2.3 England: Robert Recorde

The *Arithmetica integra* and Stifel's 1553 revision of Rudolff's *Coss* were very important in Germany, influencing textbook writers well into the next century and helping to develop in Germany, as had already been done in Italy, mathematical awareness in the middle classes. They also had influence in England, where they were the major source of the first English algebra, *The Whetstone of Witte*, published in 1557 by the first English author of mathematical works in the Renaissance, Robert Recorde (1510–1558) (Fig. 12.2).

The Whetstone of Witte had little that was original in technique, because it was based on the German sources and even used the German symbols for powers of the unknown, but there are a few points of interest in the text, which taught algebra to an entire generation of English scientists. First, Recorde created the modern symbol for equality: "To avoid the tedious repetition of these words—is equal to—I will set as I do often in work use, a pair of parallels, or gemow [twin] lines of one length, thus ====, because no 2 things can be more equal."[14] Second, he modified and extended the German symbolization of powers of

---

BIOGRAPHY

Robert Recorde (1510–1558)

Robert Recorde graduated from Oxford in 1531 and was licensed in medicine soon thereafter. Although he probably practiced medicine in London in the late 1540s, his only known positions were in the civil service, positions in which he was not notably successful. On the other hand, he did write several successful mathematics textbooks besides *The Whetstone of Witte*, including *The Ground of Arts* (1543) on arithmetic,

The Pathway to Knowledge (1551) on geometry, and *The Castle of Knowledge* (1556) on astronomy. His works show that he was especially interested in pedagogy. In particular, his books were set in the form of a dialogue between master and pupil, in which each step in a particular technique was carefully explained.

FIGURE 12.2

Title page of Robert Recorde's *The Whetstone of Witte* (1557). (*Source*: Smithsonian Institution Libraries, Photo No. 92-338)

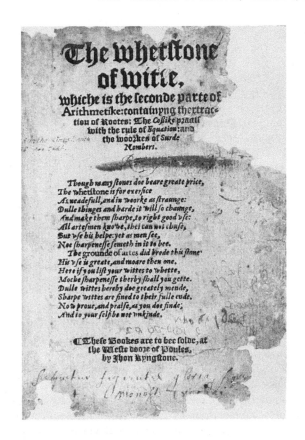

the unknown to powers as high as the 80th, setting the integer of the power next to each symbol and noting that multiplication of these symbols corresponded to addition of the corresponding integers. In fact, he showed how to build the symbol for any power out of the square $\mathfrak{z}$, the cube $\mathscr{ce}$, and various sursolids (prime powers higher than the third) *ß (where * stands for a letter designating the order of the prime). The fifth power is written ß, the seventh power as ^{b}ß (second sursolid), and the eleventh power as ^{c}ß (third sursolid). Then, for instance, the 9th power is written $\mathscr{ce}\mathscr{ce}$ (cube of the cube), the 20th power as $\mathfrak{z}\mathfrak{z}$ß (square of the square of the fifth power), and the 21st power as $\mathscr{ce}^b$ß (cube of the seventh power). Finally, to help students remember the various rules of operation, he gave them in poetic form. His verse giving the procedure for multiplying and dividing expressions of the form ax^n, where the power n is called the "quantity" of the expression, included the standard rule of signs for those operations as well as the rule of exponents:

> Who that will multiplie,
> Or yet divide trulie:
> Shall like still to have more,
> And mislike lesse in store.
> Their quantities doe kepe soche rate,
> That .M. doeth adde; and .D. abate.

BIOGRAPHY

Pedro Nunes (1502–1578)

P edro Nunes studied at the University of Salamanca but received a degree in medicine in Lisbon in 1525 (Fig. 12.3). He made several contributions to the science of navigation and became chief cosmographer to the king of Portugal and professor of mathematics at the University of Coimbra, writing his *Libro de Algebra* in 1532. Many of his students later held high positions at court. Although Nunes was of Jewish origin, he was never persecuted by the Inquisition, probably because one

of his students became the Inquisitor General, Cardinal Don Henrique. Nunes's algebra text was originally written in Portuguese, but because he felt that it would have more influence if it were available in Spanish, he translated it into Spanish some thirty years later and had it printed in the Netherlands in 1567. He wrote his astronomical works chiefly in Latin, however. In addition to his scientific work, Nunes was also a poet of some note.

FIGURE 12.3

Pedor Nunes on a Portugese stamp. Notice the algebra problem in the background.

12.2.4 Portugal: Pedro Nunes

Mathematics was also necessary in Portugal, where already in the fifteenth century navigators were extending the European knowledge of the rest of the world. It was here that Pedro Nunes (1502–1578) wrote his *Libro de Algebra* in 1532.[15] Nunes was influenced by his reading of the work of Pacioli. His notation clearly came from the Italian writers, but it appears that he had no knowledge of his German contemporaries. Thus, he used Italian abbreviations for the various powers of the unknown—*co* for *cosa*, *ce* for *censo*, *cu* for *cubo*—as well as $\overline{p}$ and $\overline{m}$ for plus and minus. In his text, he dealt with the procedures for combining algebraic expressions, for solving equations, and for dealing with radicals and proportions. He included dozens of problems, but unlike those in most of the other algebra texts mentioned, his were all abstract. Neither commercial nor recreational problems were included. He did, however, include a section on applications of algebraic techniques to geometry.

To give the flavor of Nuncs's text, we consider how he solved one of the standard problems, to find two numbers whose product and the sum of whose squares is known. The product is given as 10 and the sum of the squares as 30. Nunes solved the problem three different ways to demonstrate various algebraic techniques. We will use modern notation, with, for example, x rather than Nunes's *co*. First, letting x be the smaller of the two numbers, he took $\frac{10}{x}$ as the larger, squared each, and found the equation $x^2 + \frac{100}{x^2} = 30$. Multiplying by x^2, he reduced this to a quadratic equation in x^2 whose solution by one of al-Khwārizmī's formulas is $x^2 = 15 \pm \sqrt{125}$. Thus, the two desired numbers are $\sqrt{15 - \sqrt{125}}$ and $\sqrt{15 + \sqrt{125}}$. For his second method, he noted that the two numbers cannot be equal, then represented the squares of the two numbers respectively as $15 - x$ and $15 + x$. The numbers themselves being the square roots of these quantities, he derived the equation $\sqrt{15 - x}\sqrt{15 + x} = 10$, which easily reduces to $x = \sqrt{125}$. Therefore, the solution is the same as before. Nunes's third solution made use of the identity $(a + b)^2 = a^2 + 2ab + b^2$. The square of the sum of the two numbers is 50, so their sum is $\sqrt{50}$. The two numbers, therefore, are $\frac{1}{2}\sqrt{50} - x$ and

$\frac{1}{2}\sqrt{50} + x$. Multiplying these together gives the equation

$$12\frac{1}{2} - x^2 = 10,$$

whence $x = \sqrt{2\frac{1}{2}}$. The two numbers in this case are then $\sqrt{12\frac{1}{2}} - \sqrt{2\frac{1}{2}}$ and $\sqrt{12\frac{1}{2}}$ $+ \sqrt{2\frac{1}{2}}$. Nunes now needed to show that this pair of numbers is the same as the pair found earlier. He accomplished this by comparing their respective squares, even though he realized that the equality of squares does not necessarily imply the equality of the roots. Although he was not sure how to avoid this difficulty, he was convinced that the solutions are in fact the same, because both pairs do satisfy the original equation.

 12.3

THE SOLUTION OF THE CUBIC EQUATION

Fra Luca Pacioli noted in 1494 that there was not yet an algebraic solution to the general cubic equation, but throughout the fifteenth and early sixteenth centuries many mathematicians were working on this problem. Finally, sometime between 1500 and 1515, Scipione del Ferro (1465–1526), a professor at the University of Bologna, discovered an algebraic method of solving the cubic equation $x^3 + cx = d$. Recall that, since most mathematicians still did not deal with negative numbers even as coefficients of equations, there were 13 different types of mixed irreducible cubic equations depending on the relative positions of the (positive) quadratic, linear, and constant terms. So del Ferro had only begun the process of "solving" the cubic equation with his solution of one of these cases.

In modern academia, professors announce and publish new results as quickly as possible to ensure priority, so it may be surprising to learn that del Ferro did not publish, nor even publicly announce, his major breakthrough. But academic life in sixteenth-century Italy was far different from that of today. There was no tenure. University appointments were mostly temporary and subject to periodic renewal by the university senate. One of the ways a professor convinced the senate that he was worthy of continuing in his position was by winning public challenges. Two contenders for a given position would present each other with a list of problems, and, in a public forum some time later, each would present his solutions to the other's problems. Often, considerable amounts of money, aside from the university positions themselves, were dependent on the outcome of such a challenge. As a result, if a professor discovered a new method for solving certain problems, it was to his advantage to keep it secret. He could then pose these problems to his opponents, secure in the knowledge that he would prevail.

Before he died, del Ferro disclosed his solution to his pupil, Antonio Maria Fiore (first half of the sixteenth century), and to his successor at Bologna, Annibale della Nave (1500–1558). Although neither of these men publicized the solution, word began to circulate among Italian mathematicians that this old problem had been, or soon would be, solved. Another mathematician, Niccolò Tartaglia of Brescia (1499–1557), in fact, boasted that he too had discovered the solution to a form of the cubic, $x^3 + bx^2 = d$. In 1535, Fiore challenged Tartaglia to a public contest, hoping to win on the strength of his knowledge of the earlier case. Each of his thirty submitted problems dealt with that class of cubic equations. For example, one of the problems read, "A man sells a sapphire for 500 ducats, making a profit of the cube root of his capital. How much is this profit?" $[x^3 + x = 500]$.[16] But Tartaglia, the better mathematician, worked long and hard on that case, and, as he later wrote, on the night of February 12, 1535 he discovered the solution. Since Fiore was unable to solve many of Tartaglia's questions covering other areas of mathematics besides the cubic, Tartaglia was declared the winner, in this case of 30 banquets prepared by the loser for the winner and his friends. (Tartaglia, probably wisely, declined the prize, accepting just the honor of the victory.)

Word of the contest and the new solutions of the cubic soon reached Milan, where Gerolamo Cardano (1501–1576) was giving public lectures in mathematics, supported by a grant from the will of the scholar Tommasso Piatti for the instruction of poor youths. Cardano wrote to Tartaglia, asking that Tartaglia show him the solution so it could be included, with full credit, in the arithmetic text Cardano was then writing. Tartaglia initially refused, but after many entreaties and a promise from Cardano to introduce him and his new inventions in artillery to the Milanese court, he finally came to Milan in early 1539. Tartaglia, after extracting an oath from Cardano that he would never publish Tartaglia's discoveries— Tartaglia planned to publish them himself at some later date—divulged the secrets of three different forms of the cubic equation to Cardano in the form of a poem. Here is the verse explaining $x^3 + cx = d$:

> When the cube and its things near
> Add to a new number, discrete,
> Determine two new numbers different
> By that one; this feat
> Will be kept as a rule
> Their product always equal, the same,
> To the cube of a third
> Of the number of things named.
> Then, generally speaking,
> The remaining amount
> Of the cube roots subtracted
> Will be your desired count.[17]

12.3.1   Gerolamo Cardano and the *Ars Magna*

Cardano kept his promise not to publish Tartaglia's result in his arithmetic book, which soon appeared. In fact, he sent Tartaglia a copy off the press to show his good faith. Cardano then began to work on the problem himself, probably assisted by his servant and student, Lodovico Ferrari (1522–1565). Over the next several years, he worked out the solutions and their justifications to all of the various cases of the cubic. Ferrari managed to solve the fourth-degree equation as well. Meanwhile, Tartaglia still had not published anything on the cubic. Cardano did not want to break his solemn oath, but he was eager that the solutions should be made available. Acting on rumors of the original discovery by del Ferro, he and Ferrari journeyed to Bologna and called on della Nave. The latter graciously gave the two permission to inspect del Ferro's papers, and they were able to verify that del Ferro had discovered the solution first. Cardano no longer felt an obligation to Tartaglia. After all, he would not be publishing Tartaglia's solution, but one discovered some 20 years earlier by a

---

BIOGRAPHY

Gerolamo Cardano (1501–1576)

Cardano was trained as a physician, but was denied admission to the Milan College of Physicians because of his illegitimate birth. For several years, therefore, he practiced medicine in a small town near Padua, returning to Milan in 1533, where he treated occasional private patients as well as lectured in mathematics and wrote a textbook on arithmetic. Finally, he convinced the College of Physicians to change its mind. Cardano soon became the most prominent physician in Milan and in great demand throughout Europe. His most important patient was probably John Hamilton, the Archbishop of Scotland, who in 1551 requested Cardano's services to help him overcome steadily worsening attacks of asthma. Cardano, after spending a month observing the archbishop's symptoms and habits, decided he had a severe allergy to the feathers in the bed he slept in. Thus, Cardano recommended

that the bedding be changed to silk and the pillow to linen. The archbishop's health improved immediately, and he remained extremely grateful to Cardano for the rest of his life, offering money and other assistance whenever Cardano might need it. Cardano was not so successful in his attempt to cast a horoscope for the young king Edward VI on his return from Scotland through England. He predicted a long life, but unfortunately, the sixteen-year-old king died shortly thereafter. Cardano's own life was filled with many tragedies, including the death of his wife in 1546 and the execution of his son for the murder of his own wife in 1560. The final blow came in 1570 when he was brought before the Inquisition on a charge of heresy. Fortunately, the sentence was relatively lenient. Cardano spent his last few years in Rome, where he wrote his autobiography *De Propria Vita*.

---

man now deceased. So in 1545 Cardano published his most important mathematical work, the *Ars magna, sive de regulis algebraicis* (*The Great Art, or On the Rules of Algebra*), chiefly devoted to the solution of cubic and quartic equations (Fig. 12.4). Tartaglia, of course, was furious when Cardano's work appeared. He felt he had been cheated of the rewards of his labor, even though Cardano did mention that Tartaglia was one of the original discoverers of the method. Tartaglia's protests availed him nothing. In an attempt to recoup his prestige, he had another public contest, this time with Ferrari, but was defeated. To this day, the formula providing the solution to the cubic equation is known as **Cardano's formula**.

We now consider the details of the cubic formula, as presented in poetry by Tartaglia (above) and in prose by Cardano in the *Ars magna*, for the equation $x^3 + cx = d$, or, as Cardano put it, "cube and first power equal to number": "Cube one-third the coefficient of the thing; add to it the square of one-half the constant of the equation; and take the square root of the whole. You will duplicate this and to one of the two you add one-half the number you have already squared and from the other you subtract one-half the same. . . . Then subtracting the cube root of the [second] from the cube root of the [first], the remainder [or] that which is left is the value of the thing."[18] Tartaglia's poem tells us (lines 3–4) to find two numbers u^3, v^3, such that $u^3 - v^3 = d$ and (lines 6–8) $u^3 v^3 = (\frac{c}{3})^3$. Then (lines 10–12), $x = u - v$. Cardano himself then simply explained how to solve the standard problem of finding two quantities, u^3 and v^3, whose difference and product are known:

$$u^3 = \sqrt{\left(\frac{d}{2}\right)^2 + \left(\frac{c}{3}\right)^3} + \frac{d}{2},$$

FIGURE 12.4

Title page of Gerolamo Cardano's *Ars magna*. (*Source*: Smithsonian Institution Libraries, Photo No. 76-15322)

while the solution for v^3 differs only by a sign. Then, since x is the difference of u and v, we get Cardano's formula for this case:

$$x = \sqrt[3]{\sqrt{\left(\frac{d}{2}\right)^2 + \left(\frac{c}{3}\right)^3} + \frac{d}{2}} - \sqrt[3]{\sqrt{\left(\frac{d}{2}\right)^2 + \left(\frac{c}{3}\right)^3} - \frac{d}{2}}.$$

Cardano proved his result by a geometric argument involving cubes applied to a specific example. The essence of his proof can be seen more easily through the following algebraic argument: If $x = u - v$, where $u^3 - v^3 = d$ and $u^3 v^3 = (\frac{c}{3})^3$, then

$$\begin{aligned} x^3 + cx &= (u - v)^3 + c(u - v) \\ &= u^3 - 3u^2 v + 3uv^2 - v^3 + 3uv(u - v) \\ &= u^3 - v^3 - 3uv(u - v) + 3uv(u - v) \\ &= d. \end{aligned}$$

To clarify his rule, Cardano presented the example $x^3 + 6x = 20$. The formula tells us, since $\frac{c}{3} = 2$ and $\frac{d}{2} = 10$, that

$$x = \sqrt[3]{\sqrt{108} + 10} - \sqrt[3]{\sqrt{108} - 10}.$$

(In this formula, the square and cube root symbols as well as the operation symbols are modern. Cardano himself wrote this as $\Re$ v : cub $\Re$ 108 p : 10 m : $\Re$ v : cub $\Re$ 108 m : 10.) There is an obvious question about the answer given here. It is clear to us that the solution to the equation $x^3 + 6x = 20$ is $x = 2$. The answer given by the formula is in fact equal to 2, but that is certainly not evident. Cardano himself noted this a few pages later, but did not show how to transform the answer given by the formula into the value 2.

Similarly, in the chapter "On the Cube Equal to the Thing and Number," that is, $x^3 = cx + d$, Cardano presented and proved a rule differing little from his rule for cube and thing equal to number:

$$x = \sqrt[3]{\frac{d}{2} + \sqrt{\left(\frac{d}{2}\right)^2 - \left(\frac{c}{3}\right)^3}} + \sqrt[3]{\frac{d}{2} - \sqrt{\left(\frac{d}{2}\right)^2 - \left(\frac{c}{3}\right)^3}}.$$

After presenting as examples the equations $x^3 = 6x + 40$ and $x^3 = 6x + 6$, he noted the difficulty here if $(\frac{c}{3})^3 > (\frac{d}{2})^2$. In that case, one could not take the square root. To circumvent the difficulty, Cardano described other methods for special cases. As we will see later, it was Rafael Bombelli who showed how to deal with the square roots of negative numbers in the Cardano formula.

Cardano's discussion of the solutions of the various cases of the cubic are in Chapters 11–23 of the *Ars magna*. But the text opens with some general results, including a discussion of the number of roots a given equation could have, whether the roots are positive (true) or negative (fictitious), and how the roots of one equation determine the roots of a related equation. For example, Cardano noted that equations of the form $x^3 + cx = d$ always have one positive solution and no negative ones. Conversely, the number and sign of the roots of the equation $x^3 + d = cx$ depend on the coefficients. If $\frac{2c}{3}\sqrt{\frac{c}{3}} = d$, then this equation has one positive root $r = \sqrt{\frac{c}{3}}$ and one negative one $-s = -2\sqrt{\frac{c}{3}}$. If $\frac{2c}{3}\sqrt{\frac{c}{3}} > d$, then there are two positive roots r and s and one negative root $-t$, where $t = r + s$. In addition, t is the positive root of the equation $x^3 = cx + d$. Parenthetically, Cardano noted that in the first case, one could consider the positive root r as two separate roots, for the negative root equals $-2r$. Finally, if $\frac{2c}{3}\sqrt{\frac{c}{3}} < d$, then there are no positive roots. There is one negative root $-s$, where s is the positive root of $x^3 = cx + d$. Sharaf al-Dīn al-Ṭūsī had given a similar discussion of the roots of this equation (and certain others) some 300 years earlier and had arrived at the same criteria for the existence of positive roots, but whether Cardano used the same method of considering maximums is unknown. Cardano did provide more information than his Islamic predecessor, however, since he considered negative roots. Thus, he was also able to understand, if not prove, that when there are three real roots to a cubic equation, their sum is equal to the coefficient of the x^2 term.

Cardano's pupil Lodovico Ferrari succeeded in finding the solution to the fourth-degree equation. Cardano presented this solution briefly near the end of the *Ars magna*, where he listed the 20 different types of quartic equations, outlined a basic procedure, and calculated a few examples. This basic procedure begins with a linear substitution that eliminates the term in x^3, leaving an equation of the form $x^4 + cx^2 + e = dx$, for instance. To solve this equation, second-degree and constant terms are added to both sides to turn each side into a perfect square. One then takes square roots and calculates the answer. We illustrate the procedure with one of Cardano's examples: $x^4 + 3 = 12x$. If we add $2bx^2 + b^2 - 3$ to both sides (where b is to be determined), the left side becomes $x^4 + 2bx^2 + b^2$, a perfect square, while the right side becomes $2bx^2 + 12x + b^2 - 3$. For this latter expression to be a perfect square, we must have $2b(b^2 - 3) = (12/2)^2$ or $2b^3 = 6b + 36$. Therefore, we need to solve a cubic equation in b. (This equation is now called the resolvent cubic of the given quartic.) Cardano, of course, had a rule for solving this equation, but in this case it is clear that $b = 3$ is a solution. Thus, the added polynomial is $6x^2 + 6$, and the original equation is transformed into $x^4 + 6x^2 + 9 = 6x^2 + 12x + 6$. Taking square roots gives $x^2 + 3 = \sqrt{6}(x + 1)$, the solutions to which are easily found to be

$$x = \sqrt{1\frac{1}{2}} \pm \sqrt{\sqrt{6} - 1\frac{1}{2}}.$$

Are these the only roots of the quartic? One can attempt to find others by taking negative square roots in the equation $x^4 + 6x^2 + 9 = 6x^2 + 12x + 6$, but that leads to complex values for x that Cardano ignored. In other examples, he did use both sets of roots. One could also look for other roots by using a second solution of the resolvent cubic. Cardano evidently considered this possibility but only teased the reader about what happens: "I need not say whether having found another value for b . . . we would come to two other solutions [for x]. If this operation delights you, you may go ahead and inquire into this for yourself."[19]

Much else of interest is found in Cardano's masterpiece, including a solid understanding of the use of negative numbers as solutions to problems and the first appearance of complex numbers, not in connection with cubic equations but in connection with a quadratic problem. This problem is simply to divide 10 into two parts (here we go again!) such that the product is 40. By standard techniques of solving quadratic equations, Cardano showed that the two parts must be $5 + \sqrt{-15}$ and $5 - \sqrt{-15}$. Although he checked that these answers in fact satisfy the conditions of the problem, he was not entirely happy with the solution, for, as he wrote, "So progresses arithmetic subtlety the end of which, as is said, is as refined as it is useless."[20] Cardano thus left off the discussion and wrote no more about complex numbers.

12.3.2   Rafael Bombelli and Complex Numbers

Cardano's *Ars magna* was extremely influential, marking the first substantive advance over the Islamic algebra so long studied in Europe. The author himself was quite proud of his work. At the end of the text there appears in large type, "WRITTEN IN FIVE YEARS, MAY IT LAST AS MANY THOUSANDS." Nevertheless, the book itself was difficult to read. Its arguments were often prolix and not easily followed, and its organization left much to be desired. To improve the teaching of the subject, and to clear up some of the difficulties still remaining, Rafael Bombelli (1526–1572) some 15 years later decided to write a more systematic text in Italian to enable students to master the material on their own. Although only the first three of the five parts were published in Bombelli's lifetime, and although in the questions concerning multiple roots of cubic equations he did not achieve as much as Cardano, his *Algebra* nevertheless marks the high point of the Italian algebra of the Renaissance.

BIOGRAPHY

Rafael Bombelli (1526–1572)

Bombelli was educated as an engineer and spent much of his adult life working on engineering projects in the service of his patron, a Roman nobleman who was a favorite of Pope Paul III. The largest project in which he was involved was the reclamation of the marshes in the Val di Chiana into arable land. Today that valley, extending southeast for about sixty miles between the Arno and the Tiber, is still one of the most fertile in central Italy. Bombelli later served as a consultant on a proposed project for the draining of the Pontine Marshes near Rome. During a lull in the reclamation work caused by a war in the area, he was able to work on his algebra treatise at his patron's villa in Rome sometime between 1557 and 1560. Other professional engagements delayed the printing of it, however, and it did not appear until shortly before his death in 1572.

Bombelli's *Algebra* was more in the tradition of Luca Pacioli's *Summa* and the German *Coss* works than was Cardano's book. Bombelli began the book with elementary material and gradually worked up to the solving of cubic and quartic equations. Like Cardano, he gave a separate treatment to each class of cubics, but he expanded on Cardano's brief treatment of quartics by giving a separate section to each of those classes as well. After dealing with the theoretical material, he presented the student with a multitude of problems using the techniques developed in the earlier chapters. He had originally intended to include practical problems similar to those of the earlier abacus works, but after studying a copy of Diophantus's *Arithmetica* at the Vatican Library, he decided to replace these with abstract numerical problems taken from Diophantus and other sources.

Recall that algebraic symbolism was gradually replacing the strictly verbal accounts of the Moslems and of the earliest Italian algebraists. Cardano had used some symbolism, but Bombelli's was a bit different. For example, he used $R.q.$ to denote the square root, $R.c.$ to denote the cube root, and similar expressions to denote higher roots. He used $\lfloor\ \rfloor$ as parentheses to enclose long expressions, as in $R.c.\lfloor 2\ p\ R.q.21\rfloor$, but kept the standard Italian abbreviations of p for plus and m for minus. His major notational innovation was the use of a semicircle around a number n to denote the nth power of the unknown. Thus, $x^3 + 6x^2 - 3x$ would be written as

$$1\overset{3}{\smile}p\ \ 6\overset{2}{\smile}m\ \ 3\overset{1}{\smile}.$$

Writing powers numerically rather than in the German form of symbols allowed him easily to express the exponential laws for multiplying and dividing monomials.

Late in the first part of the *Algebra*, Bombelli introduced "another sort of cube root much different from the former, which comes from the chapter on the cube equal to the thing and number; . . . this sort of root has it own algorithms for various operations and a new name."[21] This root is the one that occurs in the cubic equations of the form $x^3 = cx + d$ when $(\frac{d}{2})^2 - (\frac{c}{3})^3$ is negative. Bombelli proposed a new name for these numbers, which are neither positive (*più*) nor negative (*meno*), that is, the modern imaginary numbers. The numbers written today as bi, $-bi$, respectively, Bombelli called *più di meno* (plus of minus) and *meno di meno* (minus of minus). For example, he wrote $2 + 3i$ as $2\ p\ di\ m\ 3$ and $2 - 3i$

as 2 *m di m* 3. Bombelli presented the various laws of multiplication for these new (complex) numbers, such as *più di meno* times *più di meno* gives *meno* and *più di meno* times *meno di meno* gives *più* $((bi)(ci) = -bc, bi(-ci) = bc)$.

To illustrate his rules, Bombelli gave numerous examples of the four arithmetic operations on these new numbers. Thus, to find the product of $\sqrt[3]{2 + \sqrt{-3}}$ and $\sqrt[3]{2 + \sqrt{-3}}$ one first multiplies $\sqrt{-3}$ by itself to get -3, then 2 by itself to get 4, then adds these two to get 1 for the "real" part. Next, one multiplies 2 by $\sqrt{-3}$ and doubles the result to get $\sqrt{-48}$. The answer is $\sqrt[3]{1 + \sqrt{-48}}$. To divide 1000 by $2 + 11i$, Bombelli multiplied both numbers by $2 - 11i$. He then divided the new denominator, 125, into 1000, giving 8, which in turn he multiplied by $2 - 11i$ to get $16 - 88i$ as the result. Bombelli, although he noted that "the whole matter seems to rest on sophistry rather than on truth,"[22] nevertheless presented here for the first time the rules of operation for complex numbers. It seems clear from his discussion that he developed these rules strictly by analogy to the known rules for dealing with real numbers. Arguing by analogy is a common method of making mathematical progress, even if one is not able to give rigorous proofs. Of course, because Bombelli did not know what these numbers "really" were, he could give no such proofs.

Proofs notwithstanding, with the rules for dealing with complex numbers now available, Bombelli could discuss how to use Cardano's formula for the case $x^3 = cx + d$ whether $(\frac{d}{2})^2 - (\frac{c}{3})^3$ is positive or negative. He first considered the example $x^3 = 6x + 40$. Cardano's procedure gives $x = \sqrt[3]{20 + \sqrt{392}} + \sqrt[3]{20 - \sqrt{392}}$, even though it is obvious that the answer is $x = 4$. Bombelli showed how one can see that the sum of the two cube roots is in fact 4. He assumed that $20 + \sqrt{392}$ equals the cube of a quantity of the form $a + \sqrt{b}$ for some numbers a and b, or $\sqrt[3]{20 + \sqrt{392}} = a + \sqrt{b}$. This implies that $\sqrt[3]{20 - \sqrt{392}} = a - \sqrt{b}$. Multiplying these two equations together gives $\sqrt[3]{8} = a^2 - b$ or $a^2 - b = 2$. Furthermore, cubing the first equation and equating the parts without square roots gives $a^3 + 3ab = 20$. Bombelli did not attempt to solve this system of two equations in two unknowns by a general argument. Rather, he noted that the only possible integral value for a is $a = 2$. Fortunately, $b = 2$ then provides the other value in each equation, so Bombelli had shown that $\sqrt[3]{20 + \sqrt{392}} = 2 + \sqrt{2}$ and $\sqrt[3]{20 - \sqrt{392}} = 2 - \sqrt{2}$. It follows that the solution to the cubic equation is $x = (2 + \sqrt{2}) + (2 - \sqrt{2}) = 4$ as desired.

For the equation $x^3 = 15x + 4$, the Cardano formula gives

$$x = \sqrt[3]{2 + \sqrt{-121}} + \sqrt[3]{2 - \sqrt{-121}},$$

although again it is clear that the answer is $x = 4$. Bombelli used his newfound knowledge of complex numbers to apply the same method as above. He first assumed that $\sqrt[3]{2 + \sqrt{-121}} = a + \sqrt{-b}$. Then $\sqrt[3]{2 - \sqrt{-121}} = a - \sqrt{-b}$, and a short calculation leads to the two equations $a^2 + b = 5$ and $a^3 - 3ab = 2$. Again, Bombelli carefully showed that $a = 2$ was the only possibility. Then $b = 1$ provides the other solution and the desired cube root is $2 + \sqrt{-1}$. It follows that the solution to the cubic equation is $x = (2 + \sqrt{-1}) + (2 - \sqrt{-1})$ or $x = 4$.

Bombelli presented several more examples of the same type, where in each case he was able somehow to calculate the appropriate values of a and b. He did note, however, that this was not possible in general. If one attempts to solve the system in a and b by a general method, such as substitution, one is quickly led back to another cubic equation. Bombelli also showed that complex numbers could be used to solve quadratic equations that previously had been thought to have no solution. For example, he used the standard quadratic formula to show that $x^2 + 20 = 8x$ has the solutions $x = 4 + 2i$ and $x = 4 - 2i$. Although he could not answer all questions about the use of complex numbers, his ability to use them to solve certain problems provided mathematicians with the first hint that there was some sense to dealing with them. Since mathematicians were still not entirely happy with using negative numbers—Cardano called them fictitious and Bombelli did not consider them as roots at all—it is not surprising that it took many years before they were entirely comfortable with using complex numbers.

Bombelli was the last of the Italian algebraists of the Renaissance. His *Algebra*, however, was widely read in other parts of Europe. Two men, one in France and one in the Netherlands, just before the turn of the seventeenth century used both Bombelli's work and some newly rediscovered Greek mathematical works to take algebra into new directions.

12.4 VIÈTE, ALGEBRAIC SYMBOLISM, AND ANALYSIS

The European algebraists of the sixteenth century had achieved about as much as possible in their continuation of the Islamic algebra of the Middle Ages. They were now expert in algebraic manipulations, even though their symbolism still left something to be desired, and they knew how to solve any polynomial equation of degree up to four. The solutions, however, were given in the form of rules of procedure. Most of these authors used some symbolization for the unknown and its powers, but there were no symbols for the coefficients. Thus, the best that could be done to illustrate a procedure was to use numerical examples. None of these algebra texts contain a written formula, like the quadratic formula found in every current elementary algebra textbook. To be able to write down such formulas required a new approach to symbols.

There was another major trend in mathematics during the sixteenth century in Italy besides this continuation of Islamic algebra. As part of the general revival of knowledge of classical antiquity, there developed a great interest in retrieving all of the Greek mathematical works to be found. The basic works of Euclid, Archimedes, and Ptolemy had been translated several centuries earlier from the Arabic. Since the translators were not expert mathematicians, however, their versions were not always completely understandable. In the sixteenth century, however, a concerted effort was made to retranslate these works as well as to translate other Greek mathematical works from the original Greek, these new translations to be prepared by mathematicians. The most important figure in this mathematical renaissance was the Italian geometer Federigo Commandino (1509–1575), who singlehandedly prepared Latin translations of virtually all of the known works of Archimedes, Apollonius, Pappus, Aristarchus, Autolycus, Heron, and others. Commandino's mathematical talents allowed him to conquer many of the obscurities that centuries of copyists had introduced into the Greek manuscripts. Thus, he included with each translation extensive mathematical commentaries, clarifying difficulties and providing references from one treatise to other related ones.

With the entire corpus of Greek mathematical works that had survived the destruction of the major libraries in late antiquity now available to Europeans, the question of how the Greeks discovered their theorems began to be addressed in earnest. In particular, with the availability of Pappus's *Mathematical Collection* and especially Book VII, *On the Domain of Analysis*,

European mathematicians began to search for the "methods of analysis" used by the ancient Greeks. Most Greek mathematics texts were models of synthetic reasoning, beginning with axioms and proceeding step by step to increasingly complex results. The texts generally gave little clue as to how the results were found or how one might find similar results. Pappus's work was the only one available to provide some hints of the Greek method of geometrical analysis. Not only did Pappus discuss the basic procedure of analysis, but he also gave in Book VII a guide to the major Greek texts, which provided the tools enabling one to use analysis in the discovery of new results or in the solving of new problems. Unfortunately, besides Euclid's *Data* and Apollonius's *Conics*, the treatises referred to by Pappus were no longer available. With a mixture of curiosity and frustration, the Europeans then studied the available Greek texts to try to ferret out the Greek methods and to reconstruct the lost texts of the *Domain of Analysis* using the hints and descriptions of Pappus.

René Descartes in 1629 best expressed these feelings in Rule IV of his *Rules for the Direction of the Mind*:

> But when I afterwards bethought myself how it could be that the earliest pioneers of Philosophy in bygone ages refused to admit to the study of wisdom any one who was not versed in Mathematics, evidently believing that this was the easiest and most indispensable mental exercise and preparation for laying hold of other more important sciences, I was confirmed in my suspicion that they had knowledge of a species of Mathematics very different from that which passes current in our time. I do not indeed imagine that they had a perfect knowledge of it . . . but I am convinced that certain primary germs of truth implanted by nature in human minds . . . had a very great vitality in that rude and unsophisticated age of the ancient world. . . . Indeed I seem to recognize certain traces of this true Mathematics in Pappus and Diophantus. . . . But my opinion is that these writers then with a sort of low cunning, deplorable indeed, suppressed this knowledge. Possibly they acted just as many inventors are known to have done in the case of their discoveries; i.e., they feared that their method being so easy and simple would become cheapened on being divulged, and they preferred to exhibit in its place certain barren truths, deductively demonstrated with show enough of ingenuity, as the results of their art, in order to win from us our admiration for these achievements, rather than to disclose to us that method itself which would have wholly annulled the admiration accorded. Finally, there have been certain men of talent who in the present age have tried to revive this same art. For it seems to be precisely that science known by the barbarous name of Algebra, if only we could extricate it from that vast array of numbers and inexplicable figures by which it is overwhelmed, so that it might display the clearness and simplicity which we imagine ought to exist in a genuine Mathematics.[23]

12.4.1 François Viète and *The Analytic Art*

One of the first "men of talent" who attempted to identify the Greek analysis with the new algebra, and who tried to display this new algebra with "clearness and simplicity," was François Viète (1540–1603). In the closing years of the sixteenth century, he composed the several treatises collectively known as *The Analytic Art*, in which he effectively reformulated the study of algebra by replacing the search for solutions to equations by the detailed study of the structure of these equations, thus developing the earliest consciously articulated theory of equations.

Viète began his program in his *In artem analyticem isagoge* (*Introduction to the Analytic Art*) of 1591 with an announcement of what he wanted to accomplish:

BIOGRAPHY

François Viète (1540–1603)

Viète was born in Fontenay-le-Comte, a village in western France near the Bay of Biscay. After receiving a law degree from the University of Poitiers, he returned to his native village to begin the practice of that profession. His legal reputation grew through his association with a prominent local family, and he was called to Paris by King Henri III for private advice, confidential negotiations, and finally in 1580 a seat on the privy council. One of his tasks for Henri, after the latter moved the court to Tours in 1589, was to act as a cryptanalyst of intercepted messages between the king's enemies. He was so successful at this that he was denounced by some who thought that the decipherment could only have been made by sorcery. Because he continued to work for Henri III and his successor Henri IV, Viète's mathematical work could only be an avocation.

There is a certain way of searching for the truth in mathematics that Plato is said first to have discovered. Theon called it analysis, which he defined as assuming that which is sought as if it were admitted and working through the consequences of that assumption to what is admittedly true. . . . Although the ancients propounded only two kinds of analysis, zetetics and poristics, to which the definition of Theon best applies, I have added a third, which may be called rhetics or exegetics. It is properly zetetics by which one sets up an equation or proportion between a term that is to be found and the given terms, poristics by which the truth of a stated theorem is tested by means of an equation or proportion, and exegetics by which the value of the unknown term in a given equation or proportion is determined. Therefore the whole analytic art, assuming this three-fold function for itself, may be called the science of correct discovery in mathematics.[24]

Recall from the discussion in Chapter 6 that Pappus called the two kinds of analysis of the Greeks problematic analysis and theorematic analysis. Viète renamed these methods, but entirely altered their meaning, meanwhile adding a new type. For Viète, problematic analysis became **zetetic analysis**, the procedure by which one transforms a problem into an equation linking the unknown and the various knowns; theorematic analysis became **poristic analysis**, the procedure exploring the truth of a theorem by appropriate symbolic manipulation; and, finally, **exegetics** is the art of transforming the equation found by zetetics to find a value for the unknown. It is not entirely surprising that Viète tried to identify Greek analysis with algebra. The procedure for solving an equation assumes that one can treat the unknown x as if it were known by using the basic rules of operation. At the end of the series of operations, one then has the unknown expressed ($x =$) in terms of the knowns. This is in some sense the same procedure that Pappus described of assuming what is sought as known and then proceeding through its consequences to something already given. But Viète's use of the terms is not the same as that of Pappus. The actual burden of "finding" the unknown is borne by the new third kind of analysis, exegetics, rather than by the two earlier kinds. In any case, we can understand clearly Viète's goal, as stated in the last sentence of the quotation of the previous paragraph as well as in the final paragraph of the *Introduction*: "Finally, the analytic art endowed with its three forms of zetetics, poristics, and exegetics, claims for itself the greatest problem of all, which is *to leave no problem unsolved*."[25]

To solve problems effectively, better symbolism was necessary, so in the *Introduction*, Viète began as follows: "Numerical logistic is that which employs numbers; symbolic logistic that which uses symbols, as, say, the letters of the alphabet."[26] Viète thus manipulated letters as well as numbers. "Given terms are distinguished from unknown by constant, general and easily recognized symbols, as (say) by designating unknown magnitudes by the letter *A* and the other vowels *E*, *I*, *O*, *U*, and *Y* and given terms by the letters *B*, *G*, *D* and the other consonants."[27] While Viète's convention differed from the modern one in distinguishing knowns from unknowns, he was now able to manipulate completely with these letters. Furthermore, these letters did not need to stand for numbers only. They could stand for any quantity to which one could apply the basic operations of arithmetic. Viète had not entirely broken away from his predecessors, however. He continued to use words or abbreviations for powers rather than exponents, as suggested by Bombelli and Chuquet. Rather than using A^2, B^3, or C^4, Viète wrote *A quadratum*, *B cubus*, or *C quadrato–quadratum*, the first and third of which he sometimes abbreviated to *A quad* or *C quad–quad*. He therefore had to give verbal rules for multiplying and dividing powers—for example, *latus* (side) times *quadratum* equals *cubus*, and *quadratum* times itself equals *quadrato-quadratum*. In numerical examples, Viète often used a different scheme of symbolization, namely, *N* stood for *numerus* (number), *Q* for *quadratus* (square), and *C* for *cubus* (cube).

For operations, Viète adopted the German forms $+$ and $-$ for addition and subtraction, although sometimes he still used words. For multiplication, he generally used the word *in*, while for division, he used the fraction bar. Hence,

$$\frac{A \; in \; B}{C \; quadratum}$$

means, in modern notation, AB/C^2. Viète wrote square roots using the symbol ℓ. for *latus*: ℓ.64 meant the square root of 64 and ℓ.c.64 stood for the cube root of 64. Sometimes, however, he used *R* for *radix* to symbolize square root. Like most of his predecessors, Viète insisted on the law of homogeneity, that all terms in a given equation must be of the same degree. So to make sense of the equation we would write as $x^3 + cx = d$, Viète insisted that *c* be a plane (so that *cx* is a solid) and that *d* be a solid. He wrote this equation as *A cubus + C plano in A aequetus D solido*. Note that he did not use a symbol for "equals" but a word (*aequetus*).

While Viète had come only part way toward modern symbolism, the crucial step of allowing letters to stand for numerical constants enabled him to break away from the style of examples and verbal algorithms of his predecessors. He could now treat general examples, rather than specific ones, and give formulas rather than rules. In addition, eliminating the possibility of actually carrying out numerical computations involving the symbolic constants made it possible to focus on the procedures of the solution rather than the solution itself. One could see that the solution procedures could be applied to quantities other than numbers, such as, for example, line segments or angles. Further, solving equations symbolically made the structure of the solution more evident. Instead of replacing $5 + 3$ by 8, for example, one kept the expression $B + D$ in the displayed formula so that at the end of the argument one could consider its relationship to the original constants. Viète was thus able in some circumstances to discover how the roots of an equation were related to the expressions from which the equation was constructed.

We will consider a few of Viète's problems and methods of solutions in the various treatises that make up *The Analytic Art*. We begin with his rule, in the *Introduction*, that "an equation is not changed by *antithesis*:"

> Let it be given that A square minus D plane is equal to G square minus B in A. I say that A square plus B in A is equal to G square plus D plane and that by this transposition under opposite signs of conjunction the equation is not changed. For since A square minus D plane is equal to G square minus B in A, add D plane plus B in A to both sides. Then by common agreement A square minus D plane plus D plane plus B in A is equal to G square minus B in A plus D plane + B in A. The negative affection on each side of this equation cancels the positive: on one side the affection D plane vanishes; on the other the affection B in A. This leaves A square plus B in A equal to G square plus D plane.[28]

In modern notation, what Viète wrote is simply that if one adds $D + BA$ to each side of the equation $A^2 - D = G^2 - BA$, one gets the transposed version $A^2 + BA = G^2 + D$, and this new equation has the same meaning as the original one. Viète's expressions are still very wordy, but the basic new symbolism is present. We will see later how Thomas Harriot rewrote Viète's rule.

In the *Prior Notes on Symbolic Logistic*, probably written at the same time as the *Introduction* but not published until 1631, Viète showed how to operate on symbolic quantities. He derived many of the standard algebraic identities, most of which were previously known in verbal form at least but were here written for the first time using symbols. For example, Viète noted that $A - B$ times $A + B$ equals $A^2 - B^2$ and also wrote out the expansion of $(A + B)^n$ for each integer n from 2 to 6, as well as the products of $A - B$ with $A^2 + AB + B^2$, $A^3 + A^2B + AB^2 + B^2$, ... to get $A^3 - B^3$, $A^4 - B^4$, ...

In another section of the *Prior Notes*, Viète applied his algebra to trigonometry. Using the identity $(BG + DF)^2 + (DG - BF)^2 = (B^2 + D^2)(F^2 + G^2)$, he showed that given two right triangles, one with base D, perpendicular B, and hypotenuse Z, the other with base G, perpendicular F, and hypotenuse X, a new right triangle can be constructed with base $DG - BF$, perpendicular $BG + DF$, and hypotenuse ZX. The angle at the base of this new triangle is then the sum of the angles at the base of the two original triangles (assuming that sum is less than 90°). It follows that if one starts with two identical triangles with base D, perpendicular B, and hypotenuse Z, the new triangle, with base $D^2 - B^2$, perpendicular $2BD$, and hypotenuse Z^2, has base angle double that of the original. The results here are equivalent to the familiar double-angle formulas of trigonometry. Viète then performed the same construction using the "double-angle" triangle and the original one to get the "triangle of the triple angle." This triangle has base $D^3 - 3B^2D$, perpendicular $3BD^2 - B^3$, and hypotenuse Z^3. These formulas for the base, perpendicular, and hypotenuse are equivalent to the modern triple-angle formulas for cosine and sine. Viète continued his construction to generate formulas for the quadruple and quintuple angles as well.

In the *Five Books of Zetetics* (1591), Viète used his symbolic methods of calculation to deal with a large number of algebraic problems drawn from a variety of sources, both ancient and contemporary. In each problem, he used "analysis," representing the unknowns by letters, then operating on the unknowns and the knowns until it was clear how to express the former in terms of the latter. He began with the same problem with which both Diophantus and Jordanus de Nemore began their texts: Given the difference between two numbers and their sum, to find the numbers. Viète's procedure was straightforward: Letting B be the difference, D the sum, and A the smaller of the two numbers, he noted that $A + B$ is the greater. Then the sum of the numbers is $2A + B$, which equals D. Hence, $2A = D - B$ and $A = (1/2)D - (1/2)B$. The other number is then $E = (1/2)D + (1/2)B$. Having written down the solution in symbols, Viète then restated it in words: "Half the sum of the numbers minus half the difference equals the lesser number, plus that difference, the greater."[29] He concluded with an example: If B is 40 and D is 100, then A is 30 and E is 70. This format is typical of Viète's work. Although he had introduced symbolic methods, he often restated his answers in words as if to convince skeptical readers that the new symbolic method can always be translated back into the more familiar verbal mode of expression.

It is enlightening to compare the same problem in Diophantus, Jordanus, and Viète to see the differences. Diophantus, although stating the problem generally, in fact solved it only for a particular numerical example, the same one that Viète used. Jordanus solved it generally but in words: "Subtract the difference from the whole and there will remain double the lesser given number." Viète solved it totally symbolically. This problem exemplifies the change in algebra over 1350 years.

In the second book of zetetics, Viète dealt with products of unknowns as well as various powers. He showed that if one knows the product of two values and the sum of their squares, or if one knows the product of two values and their sum, or if one knows the sum of two values and the difference between their squares, one can find the unknown values. Several of the results of this book are important in Viète's later treatment of cubic equations. For example, problem 20 asks to find two values given their sum and the sum of their cubes. In this case, Viète set G equal to the sum of the unknown values, D equal to the sum of the cubes, and A equal to the product of the unknowns. From the formula for the expansion of the cube of a binomial, Viète then derived the result $G^3 - D = 3GA$, or, in modern notation,

$$(r + s)^3 - (r^3 + s^3) = 3(r + s)rs. \tag{12.1}$$

Thus, the product $A = rs$ is now known, along with the sum, and one can find the two unknown values.

12.4.2 Viète's Theory of Equations

The central work in Viète's theory of equations is found in the *Two Treatises on the Recognition and Emendation of Equations*. For example, we see how Viète solves the quadratic equation, which he writes as "A quad $+ B2$ in A equals Z plane" (or, in more modern notation, $A^2 + 2BA = Z$. Viète set $A + B$ to be E, or $E - B$ to be A. Then $(E - B)^2 + 2B(E - B) = Z$, which reduces to $E^2 = Z + B^2$. Therefore, A is equal to $\sqrt{Z + B^2} - B$. In Viète's notation, this is

$$A \text{ is } \ell.\overline{Z \text{ plane} + B \text{ quad}} - B,$$

the first occurrence of what we can really call the "quadratic formula." However, we should also note that Viète gave two other versions of the formula, one for the case $A^2 - 2BA = Z$ and one for the case $2DA - A^2 = Z$, and, of course, only in the latter case does one have the possibility of two (positive) solutions.

The treatment of cubics in *Two Treatises* is far more extensive, because there are many more types. We will just consider the equation $x^3 = cx + d$, where, in what follows, we will generally use modern notation. Recall that Cardano's formula gives complex numbers in this case when $(d/2)^2 < (c/3)^3$. Under those conditions, Viète decided to apply trigonometric reasoning. He began by rewriting the equation as $x^3 - 3b^2x = b^2d$, in keeping with homogeneity. The inequality then becomes $(b^2d/2)^2 < (b^2)^3$, which reduces to $2b > d$. Recall that Viète had earlier developed the triple-angle formulas, which reduce to $\cos 3\alpha = 4\cos^3 \alpha - 3\cos^2 \alpha$ or

$$\cos^3 \alpha - \frac{3}{4}\cos \alpha = \frac{1}{4}\cos 3\alpha.$$

By setting $x = r\cos \alpha$ and substituting into his version of the cubic, Viète converted that equation to $r^3 \cos^3 \alpha - 3b^2 r \cos \alpha = b^2d$ or

$$\cos^3 \alpha - \frac{3b^2}{r^2}\cos \alpha = \frac{b^2d}{r^3}.$$

Comparing the two equations involving $\cos^3 \alpha$ shows first that

$$\frac{3b^2}{r^2} = \frac{3}{4} \quad \text{or} \quad r = 2b$$

and second that

$$\frac{1}{4}\cos 3\alpha = \frac{b^2d}{r^3} = \frac{b^2d}{8b^3} \quad \text{or} \quad \cos 3\alpha = \frac{d}{2b}.$$

The inequality $2b > d$ for the coefficients ensures that this final equation makes sense. Thus, if α satisfies $\cos 3\alpha = d/2b$ and $r = 2b$, then $x = r\cos \alpha$ is a solution to our original cubic equation. For example, if $x^3 - 300x = 432$, then $b = 10$ and $d = 432/100$. It follows that $\cos 3\alpha = 432/2000$. By consulting tables, one determines that $\cos \alpha = 0.9$ and thus $x = 2b\cos \alpha = 18$.

For the same equation $x^3 = cx + d$ when $(d/2) < (c/3)^3$, Viète presented an algebraic solution. Rewriting the equation in the form $x^3 - 3bx = 2d$ (where b is plane and d is solid), Viète put the inequality in the form $b^3 < d^2$. Then, referring to Equation 12.1, he noted that x must be the sum of two numbers whose product is b. Therefore, $y(x - y) = b$, or $xy - y^2 = b$, or finally $x = \frac{b+y^2}{y}$. Substituting this expression into $x^3 - 3bx = 2d$ and multiplying all terms by y^3 produces a quadratic equation in y^3: $2dy^3 - (y^3)^2 = b^3$. The solutions to this are $y^3 = d \pm \sqrt{d^2 - b^3}$, so the reason for the inequality condition is evident. Because the desired root x is the sum of the two values for y, the final result is the formula slightly modified from the description of Cardano,

$$x = \sqrt[3]{d + \sqrt{d^2 - b^3}} + \sqrt[3]{d - \sqrt{d^2 - b^3}}.$$

Of course, this formula is in modern notation. Viète's actual version of Cardano's formula stated that the solution to the equation

A cube $-$ B plane 3 in A equals Z solid 2

is given by

$$A \text{ is } \ell.c.\overline{Z \text{ solid}} + \ell.\overline{Z \text{ solidsolid}} - B \text{ planeplaneplane}$$

$$+ \ell.c.\overline{Z \text{ solid}} - \ell.\overline{Z \text{ solidsolid}} - B \text{ planeplaneplane}.$$

Although Viète did not consider negative or complex roots to equations, he did deal to some extent with the relationship of the roots to the coefficients. For example, Viète was aware that the quadratic equation $bx - x^2 = c$ could have two positive roots. To discover the relationship between these two roots, x_1 and x_2, he equated the two expressions $bx_1 - x_1^2$ and $bx_2 - x_2^2$. Then $x_1^2 - x_2^2 = bx_1 - bx_2$, and, dividing through by $x_1 - x_2$, he found that $x_1 + x_2 = b$, that is, "b is the sum of the two roots being sought." Substituting $x_1 + x_2$ for b in the equation $bx_1 - x_1^2 = c$, he found the other relationship $x_1 x_2 = c$, or "c is the product of the two roots being sought."[30]

Viète tried the same device for the cubic equation $bx - x^3 = d$, which he also knew could have two positive roots, x_1, x_2. In this case the results were not so simple. Viète found that the coefficient b is equal to $x_1^2 + x_2^2 + x_1 x_2$, while the constant d is $x_1^2 x_2 + x_2^2 x_1$. What about cubic equations having three positive roots? Such an equation, of course, must have a term of the second degree. Viète's normal method of solving it would replace the equation by an equation without a second-degree term through the use of a linear substitution, and the new equation would have at most two positive roots. So Viète would not normally be able to find a third root. In fact, in an example he gave of such a reduction, using the equation $x^3 - 18x^2 + 88x = 80$, he only calculated the integer root of the reduced equation $20y - y^3 = 16$ and thus gave only one root for the original equation. Nevertheless, at the very end of the second treatise on equations, Viète stated four propositions without proof, one for each degree of equation from two through five, expressing the coefficients of the equation as elementary symmetric functions of the roots. Thus, for the third-degree equation, "If $x^3 - x^2(b + d + g) + x(bd + bg + dg) = bdg$, x is explicable by any of the three b, d or g," and for the fourth, "If $x(bdg + bdh + bgh + dgh) - x^2(bd + bg + bh + dg + dh + gh) + x^3(b + d + g + h) - x^4 = bdgh$, x is explicable by any of the four b, d, g, or h."[31] Viète considered these theorems "elegant and beautiful" and a "crown" of his work.

12.5　SIMON STEVIN AND DECIMAL FRACTIONS

Contemporary with Viète, but living most of his life in the Netherlands, was another mathematician who made a substantial contribution to a major change in mathematical thinking around the turn of the seventeenth century, Simon Stevin (1548–1620). Stevin's major mathematical contribution was the creation of a well-thought-out notation for decimal fractions, for the use of which he proved himself a strong advocate. He also played a fundamental role in changing the basic concepts of "number" and in erasing the Aristotelian distinction between number and magnitude. These contributions are set forth in his most important mathematical works, *De Thiende* (*The Art of Tenths*), known in its French version as *La Disme*, and *l'Arithmétique*, a work containing both arithmetic and algebra, both published in 1585.[32]

BIOGRAPHY

Simon Stevin (1548–1620)

S tevin was born into a wealthy family in Bruges, in what is now Belgium, but eventually left the area, then under Spanish rule, for the new Republic of Holland (Fig. 12.5). Much of his adult life was spent in the service of Maurice of Nassau, the Stadhouder of Holland. Stevin served Maurice as engineer, tutor in mathematics and ballistics, and adviser in various other mathematically dependent fields such as finance and navigation. From 1593 until his death, Stevin served the Dutch government as quartermaster general of the army, responsible for organizing military camps. At the request of Maurice, he organized a school of engineering at the University of Leiden, where Dutch, rather than the traditional Latin, was the language of instruction. He was responsible for meeting the growing need of the Dutch nation for technically trained engineers, merchants, surveyors, and navigators. To aid in this endeavor, Stevin wrote textbooks in Dutch for several of the subjects taught at Leiden.

FIGURE 12.5

Simon Stevin on a Belgian stamp

Decimal fractions were not used in Europe in the late Middle Ages or in the Renaissance. The various arithmetic texts written throughout Europe from the thirteenth through the sixteenth centuries, although they invariably discussed the Hindu-Arabic place value system, used it only for integers. If fractions were needed, they were written as common fractions or, in many trigonometric works, as sexagesimal fractions. Both Rudolff and Viète hinted at decimal fractions in works written during the sixteenth century, but without making a major impact. For example, Viète noted that if one needs to calculate the square root of 2 to a high degree of accuracy, one should add as many zeros as necessary, and calculate the square root of, for example, 20,000,000,000,000,000,000,000,000,000,000,000,000. He showed that root to be 141,421,356,237,309,505, and thus the square root of 2 was approximately $1\frac{41,421,356,237,309,505}{100,000,000,000,000,000}$. And Rudolff did write decimal fractions by using a vertical line to separate the whole number part from the fraction part. Thus, he used 413|4375 to represent 413.4375. But neither Rudolff nor Viète nor any of the others who had used some notation for fractions with denominator a power of 10 demonstrated a clear understanding of the concept of such a fraction. In Islam, however, recall that al-Samaw'al had understood the concept, while al-Kashi had also developed a convenient notation for decimal fractions.

Stevin was probably not influenced by the Islamic development, when he too put idea and notation together in his *De Thiende*. In the preface Stevin made clear its purpose: "It teaches (to speak in a word) the easy performance of all reckonings, computations, and accounts, without broken numbers [common fractions], which can happen in man's business, in such sort as that the four principles of arithmetic, namely addition, subtraction, multiplication, and division, by whole numbers may satisfy these effects."[33] Thus, Stevin promised to show that all operations using his new system could be performed exactly as if one were using whole numbers. That, of course, is the basic advantage of decimal fractions. Stevin began *De Thiende* by defining *thiende* as arithmetic based on geometric progression by tens, using the Hindu-Arabic numerals, and by calling a whole number a *commencement* with the notation ⓪. Thus, 364 is to be thought of as 364 commencements and is written as 364⓪. His major definition, in which he describes his terminology and notation for decimal fractions, is the third: "And each tenth part of the unity of the commencement we call the *prime*, whose sign is ①, and each tenth part of the unity of the prime we call the *second*, whose sign is ② and so of the other; each tenth part of the unity of the precedent sign, always in order one further."[34] To explain this, he gave examples: 3① 7② 5③ 9④ means 3 primes, 7 seconds, 5 thirds, and 9 fourths, or 3/10, 7/100, 5/1000, 9/10,000, or, altogether 3759/10,000. Similarly, $8\frac{937}{1000}$ is written as 8⓪ 9① 3② 7③. Stevin made the point that no fractions were used in his notation and that except in the case of ⓪, there were only single digits to the left of the signs (circled digits). The numbers written according to these rules Stevin named **decimal numbers**.

Stevin proceeded in the second part of this brief pamphlet to show how the basic operations are performed on decimal numbers. The important idea, naturally, is that operations are performed exactly as on whole numbers, with the proviso that one must take into account the appropriate signs. Thus, in addition and subtraction the numbers must be lined up with all ①s, for example, under one another. For multiplication, Stevin noted that once the multiplication in integers is performed, the sign of the rightmost digit is determined by adding the signs of the rightmost digits of the multiplicands. For division, one similarly subtracts the rightmost sign of the divisor from the rightmost sign of the dividend. He also gave rules for determining the sign when finding square and cube roots. Thus, although Stevin's notation differs somewhat from what we use today, he clearly set out the basic rules and rationale for using decimal fractions in calculation. The concluding section of *De Thiende* consisted of pleas to use his new decimal system for calculations in various trades. He suggested using a known basic unit in each case as the commencement and then applying his system for fractions of that unit. His suggestion, however, was not generally carried out until 200 years later, when the French revolutionary government introduced the metric system.

How the idea of decimal fractions in *De Thiende* is connected with a change in the basic concept of number is demonstrated in Stevin's other mathematical work of 1585, *l'Arithmétique*. Certainly, many authors over the centuries had been treating irrational quantities as "numbers," that is, had been dealing with them using the same rules and concepts as with whole numbers. Gradually, the Euclidean distinction between number and magnitude, between discrete quantity and continuous quantity, had broken down. It was Stevin who first stated this breakdown explicitly. Thus, he began *'l'Arithmétique* with two definitions:

1. Arithmetic is the science of numbers.
2. Number is that which explains the quantity of each thing.

Thus, at the very beginning of the work, Stevin made the point that number represents quantity, any type of quantity at all. Number is no longer to be only a collection of units, as defined by Euclid. Stevin even wrote in capitals at the top of the page, THAT UNITY IS A NUMBER. The Greeks had rejected this notion. To them, unity was not a number, but only the generator of number, as the point was the generator of a line. Through the centuries, this idea had been argued. As late as 1547, one of the questions that Ferrari sent to Tartaglia as part of the challenge competition mentioned earlier was whether unity was a number. Tartaglia complained that the question did not have to do with mathematics but with metaphysics. He then hedged his answer by asserting that unity was a number "in potential" but not one "in actuality." Stevin, by contrast, was very sure of himself. His basic philosophical argument was that since the part is of the same matter as the whole, and since unity is a part of a multitude of units (that is, a "number"), then unity must itself be a number. The mathematical argument is simply that one can operate on unity just as on other "numbers." In particular, one can divide unity into as many parts as desired. The Euclidean special role of unity as the basis of "collections of units" and therefore as the basis of the distinction between the discrete and the continuous no longer made sense to Stevin. He boldly asserted this particular idea as well, that "number is *not* discontinuous quantity."[35] Any quantity, including the unit, can be divided "continuously." In some sense, this is the basis of the idea of a decimal fraction. One can continue the signs as far as one likes to determine any division of unity, however fine.

Stevin further explained what number should mean by giving several special definitions. For example, "number explaining the value of the geometric quantity is called a geometric number and receives the name conforming to the species of the quantity that it explains." A "square number" represents a square, and a "cube number" represents a cube. Stevin pointed out, however, that any (positive) number is a square number, and thus the root of any square number is also a number: "The part is of the same matter that is the whole. The root of 8 is part of its square 8. Therefore $\sqrt{8}$ is of the same matter that is 8. But the matter of 8 is number. Therefore the matter of $\sqrt{8}$ is number. And, by consequence, $\sqrt{8}$ is a number."[36] The decimal number system of *De Thiende* then enabled Stevin to represent $\sqrt{8}$ to any accuracy desired, just as it enabled one to represent 8 itself.

Stevin did distinguish between pairs of numbers that are commensurable (have a common measure) and incommensurable (do not have a common measure). But all of these quantities are numbers in his sense. Thus, there is no real point in Euclid's multitude of distinctions of classes of irrational lines in Book X. All of these lines are represented as numbers and can be dealt with by the standard arithmetical operations. Stevin noted that we can even consider more kinds of lines than Euclid did by simply taking more roots and combinations of roots. And all of these lines (or numbers) can be calculated using his decimal arithmetic.

From the current vantage point, where the discrete Euclidean "numbers" have long been incorporated into the continuous number line, it is somewhat difficult to understand the fundamental contribution of Stevin. But Euclid had always been the center of the study of mathematics. His ideas always had to be confronted. If one wanted to change his notions, one needed to make strong and continued arguments. Many mathematicians who read Euclid did ignore his distinctions, both in the Islamic world and in Europe. In particular, the algebraists studied in this chapter tended to manipulate with all quantities in the same way. Others more philosophically inclined, however, were somewhat bothered by this generally cavalier attitude toward Euclid's work. These mathematicians needed to be convinced that there was no longer any mathematical necessity for the distinction. Naturally, Stevin alone did not do this. Not until the nineteenth century was the work of embedding "discrete arithmetic" into "continuous magnitude" completed. Nevertheless, Stevin stood at a watershed of mathematical thinking. Ultimately, he was so successful that it is difficult to understand how things were done before him.

EXERCISES

1. The gold florin is worth 5 *lire* 12 *soldi*, 6 *denarii* in Lucca. How much (in florins) are 13 *soldi*, 9 *denarii* worth? (Note that 20 *soldi* make 1 *lira* and 12 *denarii* make 1 *soldo*.)

2. If 8 *braccia* of cloth are worth 11 florins, what are 97 *braccia* worth?

3. I have 25 pounds of silver alloy that contain 8 ounces of pure silver per pound and 16 pounds that have $9\frac{1}{2}$ ounces of silver per pound. How much copper must be added to the total so that I can make coins containing $7\frac{1}{2}$ ounces of silver per pound?

4. This problem is from the *Treviso Arithmetic*, the first printed arithmetic text, dated 1478: The Holy Father sent a courier from Rome to Venice, commanding him that he should reach Venice in 7 days. And the most illustrious Signoria of Venice also sent another courier to Rome, who should reach Rome in 9 days. And from Rome to Venice is 250 miles. It happened that by order of these lords the couriers started their journeys at the same time. It is required to find in how many days they will meet, and how many miles each will have traveled.[37]

5. This problem and the next two are from the work of Piero della Francesca. Three men enter into a partnership. The first puts in 58 ducats, the second 87; we do not know how much the third puts in. Their profit is 368, of which the first gets 86. What shares of profit do the second and third receive and how much did the third invest?

6. Of three workmen, the second and third can complete a job in 10 days. The first and third can do it in 12 days, while the first and second can do it in 15 days. In how many days can each of them do the job alone?

7. A fountain has two basins, one above and one below, each of which has three outlets. The first outlet of the top basin fills the lower basin in two hours, the second in three hours, and the third in four hours. When all three upper outlets are shut, the first outlet of the lower basin empties it in three hours, the second in four hours, and the third in five hours. If all the outlets are opened, how long will it take for the lower basin to fill?

8. Solve this problem from the work of Antonio de' Mazzinghi. Find two numbers such that multiplying one by the other makes 8 and the sum of their squares is 27. (Put the first number equal to $x + \sqrt{y}$ and the second equal to $x - \sqrt{y}$; then the two equations are $x^2 - y = 8$ and $2x^2 + 2y = 27$.)

9. Divide 10 into two parts such that if one squares the first, subtracts it from 97, and takes its square root, then squares the second, subtracts it from 100, and takes its square root, the sum of the two roots is 17. (This problem is also from the work of Antonio de' Mazzinghi. Mazzinghi set the parts u, v equal to $5 + x$ and $5 - x$, respectively, and derived an equation in x.)

10. Macstro Dardi gave a rule to solve the fourth-degree equation $x^4 + bx^3 + cx^2 + dx = e$ as $x = \sqrt[4]{(d/b)^2 + e} - \sqrt{d/b}$. His problem illustrating the rule is the following: A man lent 100 *lire* to another and after 4 years received back 160 *lire* for principal and (annually compounded) interest. What is the interest rate? As in the text's example, set x as the monthly interest rate in *denarii* per *lira*. Show that this problem leads to the equation $x^4 + 80x^3 + 2400x^2 + 32,000x = 96,000$ and that the solution found by "completing the fourth power" is given by the stated formula.

11. Piero della Francesa presented the problem to divide 10 into two parts such that if their product is divided by their difference, the result is $\sqrt{18}$. To solve this, he used a rule for solving the fourth-degree equation $ax + bx^2 + cx^4 = d + ex^3$, namely, $x = \sqrt[4]{(b/4c)^2 + (d/c)} + (e/4c) - \sqrt{a/2e}$. Show that this formula works in this case, but not in general. How did Piero derive the formula?

12. The equation $6x^3 = 43x^2 + 79x + 30$ is solved in the *Summa* of Luca Pacioli as follows: "Add the number to the *cose* to form a number, and then you get one *cubo* equal to $7\frac{1}{6}$ *censi* plus $18\frac{1}{6}$, after you have reduced to one *cubo* [divided all the terms by 6]. Then divide the *censi* in half and multiply this half by itself, and add it onto the number. It will be $31\frac{1}{144}$ and the *cosa* is equal to the root of this plus $3\frac{7}{12}$, which is half of *censi*."[38] Show that Pacioli's answer is incorrect. What was he thinking of in presenting his rule?

13. Carry Chuquet's approximation procedure for $\sqrt{6}$ further. That is, since $2\frac{4}{9} < \sqrt{6}$, $2\frac{5}{11} > \sqrt{6}$, and $2\frac{9}{20} > \sqrt{6}$, the next approximation is $2\frac{13}{29}$. Continue the procedure until you reach Chuquet's final value of $2\frac{89}{198}$.

14. Use Chuquet's approximation procedure to calculate his values for $\sqrt{5}$, namely, $2\frac{161}{682}$.

15. Find two numbers in the proportion 5 : 7 such that the square of the smaller multiplied by the larger gives 40.

16. Find a number that, when multiplied by 20 and then having 7 added to the product, has the sum in the proportion 3 : 10 with the number formed by multiplying the original number

by 30 and subtracting 9. (Chuquet notes that the problem is impossible. Why?)

17. In a vessel full of wine there are three taps such that if one opens the largest it will empty the vessel in 3 hours, if one opens the middle tap it will empty it in 4 hours, and if one uses the smallest tap it will empty it in 6 hours. How long would it take to empty the vessel if all three taps were open? (This problem and the next are also from Chuquet's work.)

18. A man makes a will and dies leaving his wife pregnant. His will disposes of 100 *écus* such that if his wife has a daughter, the mother should take twice as much as the daughter, but if she has a son, he should have twice as much as the mother. [Sexist problem!] The mother gives birth to twins, a son and a daughter. How should the estate be split, respecting the father's intentions?

19. Express $\sqrt{27 + \sqrt{200}}$ as $a + \sqrt{b}$. (This problem and the next two are from Rudolf's *Coss*.)

20. I am owed 3240 *florins*. The debtor pays me 1 *florin* the first day, 2 the second day, 3 the third day, and so on. How many days does it take to pay off the debt?

21. Divide 10 into two parts such that their product is $13 + \sqrt{128}$.

22. This problem is from Stifel's *Arithmetica integra*. In the sequence of odd numbers, the first odd number equals 1^5. After skipping one number, the sum of the next four numbers $(5 + 7 + 9 + 11)$ equals 2^5. After skipping the next three numbers, the sum of the following nine numbers $(19 + 21 + 23 + 25 + 27 + 29 + 31 + 33 + 35)$ equals 3^5. At each successive stage, one skips the next triangular number of odd integers. Formulate this power rule of fifth powers in modern notation and prove it by induction.

23. The basis of Stifel's procedure for finding higher-order roots (as well as that of Scheubel and others of his time) was the appropriate binomial expansion, or, more specifically, the entries in the appropriate row of the "Pascal" triangle. For example, to find the fourth root of 1,336,336, one first notes that the answer must be a two-digit number beginning with 3. One then subtracts $30^4 = 810,000$ from the original number to get remainder 526,336. Recalling that the entries in the fourth row of the triangle are 1, 4, 6, 4, 1, and guessing that the next digit is 4, one checks this by successively subtracting from that remainder $4 \times 30^3 \times 4 = 432,000$, $6 \times 30^2 \times 4^2 = 86,400$, $4 \times 30 \times 4^3 = 7680$, and $4^4 = 256$. In this case, the result is 0, so the desired root is 34. Write a brief report explaining this procedure in detail and use it to calculate the fourth root of 10,556,001.

24. There is a certain army composed of dukes, earls, and soldiers. Each duke has under him twice as many earls as there are dukes. Each earl has under him four times as many soldiers as there are dukes. The 200th part of the number of soldiers is 9 times as many as the number of dukes. How many of each are there? (This problem and the next two are from Recorde's *The Whetstone of Witte*.)

25. A gentleman, willing to prove the cunning of a bragging arithmetician, said thus: I have in both hands 8 crowns. But if I count the sum of each hand by itself severally and add to it the squares and the cubes of the both, it will make in number 194. Now tell me, what is in each hand?

26. There is a strange journey appointed to a man. The first day he must go $1\frac{1}{2}$ miles, and every day after the first he must increase his journey by $\frac{1}{6}$ of a mile, so that his journey shall proceed by an arithmetical progression. And he has to travel for his whole journey 2955 miles. In what number of days will he end his journey?

27. Show that if r, s, are two positive roots of $x^3 + d = cx$, then $t = r + s$ is a root of $x^3 = cx + d$.

28. Show that if t is a root of $x^3 = cx + d$, then $r = t/2 + \sqrt{c - 3(t/2)^2}$ and $s = t/2 - \sqrt{c - 3(t/2)^2}$ are both roots of $x^3 + d = cx$. Apply this rule to solve $x^3 + 3 = 8x$.

29. Prove that the equation $x^3 + cx = d$ always has one positive solution and no negative ones.

30. Use Cardano's formula to solve $x^3 + 3x = 10$.

31. Use Cardano's formula to solve $x^3 = 6x + 6$.

32. Consider the equation $x^3 = cx + d$. Show that if $(c/3)^3 > (d/2)^2$ (and thus that Cardano's formula involves imaginary quantities), then there are three real solutions.

33. Solve $x^3 + 21x = 9x^2 + 5$ completely by first using the substitution $x = y + 3$ to eliminate the term in x^2 and then solving the resulting equation in y.

34. Use Ferrari's method to solve the quartic equation $x^4 + 4x + 8 = 10x^2$. Begin by rewriting this as $x^4 = 10x^2 - 4x - 8$ and adding $-2bx^2 + b^2$ to both sides. Determine the cubic equation that b must satisfy so that each side of the resulting equation is a perfect square. For each solution of that cubic, find all solutions for x. How many different solutions to the original equation are there?

35. The dowry of Francis's wife is 100 *aurei* more than Francis's own property, and the square of the dowry is 400 more than the square of his property. Find the dowry and the property. (Note the negative answer for Francis's property; Cardano interpreted this as a debt.)

36. Find two numbers x, y, with $x > y$ such that $x + y = y^3 + 3yx^2$ and $x^3 + 3xy^2 = x + y + 64$. (This problem and the next are from Ferrari's contest with Tartaglia. Tartaglia's solution is

$$x = \sqrt[3]{4 + \sqrt{15\frac{215}{216}}} + \sqrt[3]{4 - \sqrt{15\frac{215}{216}}} + 2$$

while $y = x - 4$.)

37. Divide 8 into two parts x, y, such that $xy(x - y)$ is a maximum. (Note that this was posed in the days before calculus.)

38. It is obvious that 3 is a root of $x^3 + 3x = 36$. Show that the Cardano formula gives $x = \sqrt[3]{\sqrt{325} + 18} - \sqrt[3]{\sqrt{325} - 18}$. Using Bombelli's methods, show that this number is in fact equal to 3.

39. Express $\sqrt[3]{52 + \sqrt{-2209}}$ in the form $a + b\sqrt{-1}$.

40. Given a right triangle with base D, perpendicular B, and hypotenuse Z, and a second right triangle with base G, perpendicular F, and hypotenuse X, show that the right triangle constructed in the text in Viète's work with base $DG - BF$, perpendicular $BG + DF$, and hypotenuse ZX has its base angle equal to the sum of the base angles of the original triangles.

41. Given the product of two numbers and their ratio, to find the roots: Let A, E, be the two roots, $AE = B$, $A : E = S : R$. Show that $R : S = B : A^2$ and $S : R = B : E^2$. Viète's example has $B = 20$, $R = 1$, $S = 5$. Show in this case that $A = 10$ and $E = 2$. (Jordanus has the same problem but with different numbers.)

42. Given the difference between two numbers and the difference between their cubes, find the numbers. Let E be the sum of the numbers, B the difference between them, and D the difference between the cubes. Show that $E^2 = \frac{4D - B^3}{3B}$.

Once E^2 is known, so is E and then the numbers themselves. Find the solution when $B = 6$ and $D = 504$. (Diophantus has the same problem twice, once in Book IV with these numerical values and once in Book B.)

43. Write 13.395 and 22.8642 in Stevin's notation. Use his rules to multiply the two numbers together and to divide the second by the first.

44. Given the two numbers 237 ⓪ 5① 7② 8③ and 59 ⓪ 7① 3② 9③, subtract the second from the first.

45. Why is Cardano's formula no longer generally taught in a college algebra course? Should it be? What insights can it bring to the study of the theory of equations?

46. Outline a lesson introducing the study of complex numbers via the problems with Cardano's formula giving a real root as the sum of two complex values. Discuss the merits of such an approach.

47. Compare the various notations for unknowns used by the mathematicians discussed in the text. Write a brief essay on the importance of a good notation for increasing a student's understanding in algebra.

48. The first printed mathematics book is the so-called *Treviso Arithmetic* of 1478, by an unknown author. Write a brief essay on its contents and its importance. Consult Frank J. Swetz, *Capitalism and Arithmetic*, from note 37.

49. Why was the knowledge of mathematics necessary for the merchants of the Renaissance? Did they really need to know the solutions of cubic equations? What, then, was the purpose of the detailed study of these equations in the works of the late sixteenth century?

50. Compare the symbolism of Jordanus and Viète. In what way is Viète's work an advance on that of Jordanus?

51. Explain why mathematicians of the sixteenth century equated the new algebra with the Greek analysis as described by Pappus.

REFERENCES

General works on the history of algebra during the Renaissance include Paul Lawrence Rose, *The Italian Renaissance of Mathematics: Studies on Humanists and Mathematicians from Petrarch to Galileo* (Geneva: Droz, 1975); Warren van Egmond, "The Commercial Revolution and the Beginnings of Western Mathematics in Renaissance Florence, 1300–1500," Dissertation (University of Indiana, 1976); and R. Franci and L. Toti Rigatelli, "Towards a History of Algebra from Leonardo of Pisa to Luca Pacioli," *Janus* 72 (1985), 17–82. Chapter 2 of B. L. Van der Waerden, *A History of Algebra from al-Khwarizmi to Emmy Noether* (New York: Springer, 1985) also provides a good introduction to the material. There is an extensive study of all the German algebra works of the Renaissance by P. Treutlein, "Die Deutsche Coss," *Abhandlungen zur Geschichte der mathematischen Wissenschaften* 2 (1879), 1–124. A more recent work, a collection of articles by experts on individual German algebraists and *Rechenmeisters* is Rainer Gebhardt and Helmuth Albrecht, eds., *Rechenmeister und Cossisten der frühen Neuzeit* (Annaberg-Buchholz: Schriften des Adam-Ries-Bundes, 1996).

Maestro Dardi's algebra work has been translated as Warren van Egmond, "The Algebra of Master Dardi of Pisa," *Historia Mathematica* 10 (1983), 399–421. Many other important fourteenth-century abacus manuscripts are summarized in R. Franci and L. Toti Rigatelli, "Fourteenth-Century Italian Algebra," in Cynthia Hay, ed., *Mathematics from Manuscript to Print: 1300–1600* (Oxford: Clarendon Press, 1988), pp. 11–29. There are no English versions available of Pacioli's *Summa* or the works of Rudolff or Stifel, but there is a photographic reprint of Robert Recorde, *The Whetstone of Witte*, published by Da Capo Press, New York, in 1969. Chuquet's manuscript has been translated and edited in Graham Flegg, Cynthia Hay, and Barbara Moss, *Nicolas Chuquet, Renaissance Mathematician* (Boston: Reidel, 1985). In addition to Chuquet's text, this volume contains a discussion of the author's life and his place in the history of mathematics. Cardano's *Ars magna* is available as Girolamo Cardano, *The Great Art, or The Rules of Algebra*, translated and edited by T. Richard Witmer (Cambridge: MIT Press, 1968). Many aspects of the controversy surrounding the discovery of the solution to the cubic, together with many of the original documents, are in Martin A. Nordgaard, "Sidelights on the Cardan-Tartaglia Controversy," *Mathematics Magazine* 13 (1938), 327–346. Bombelli's algebra text is only available in an Italian reprint: Rafael Bombelli, *Algebra* (Milan: Feltrinelli, 1966). On the other hand, virtually all of Viète's works have been translated into English in François Viète, *The Analytic Art*, translated and edited by T. Richard Witmer (Kent, Ohio: Kent State University Press, 1983). An English translation of Stevin's *De Thiende* is available in Henrietta O. Midonick, ed., *The Treasury of Mathematics* (New York: Philosophical Library, 1965), pp. 735–750. The translation was made by Robert Norton in 1608 and is also reprinted in *The Principal Works of Simon Stevin*, volume II, edited by Dirk J. Struik (Amsterdam: Swets and Zeitlinger, 1958).

1. Quoted in R. Franci and L. Toti Rigatelli, "Towards a History of Algebra," pp. 64–65.

2. Girolamo Cardano, *The Great Art*, p. 96.

3. Warren van Egmond, "Commercial Revolution." This dissertation examines the works of the *maestri d'abbaco*, including both the arithmetic and the algebra contained in them. It also discusses the great importance of these works for reintroducing the basic ideas of mathematics into the general culture. The three quoted problems are taken from this work.

4. R. Franci and L. Toti Rigatelli, "Towards a History of Algebra," p. 31. This article provides a detailed look at the works of various *maestri d'abbaco* and analyzes their relationship to the works of Leonardo of Pisa and Luca Pacioli.

5. R. Franci and L. Toti Rigatelli, "Fourteenth-Century Italian Algebra," p. 16.

6. R. Franci and L. Toti Rigatelli, "Towards a History of Algebra," p. 49.

7. Warren van Egmond, "Commercial Revolution," p. 266. For more details on Mazzinghi, see R. Franci and L. Toti Rigatelli, "Towards a History of Algebra," and B. L. van der Waerden, *History of Algebra*.

8. See Warren van Egmond, "The Algebra of Master Dardi of Pisa," for more details.

9. Graham Flegg, *Nicolas Chuquet*, p. 90.

10. Ibid., p. 105.

11. Ibid., p. 151.

12. Ibid., p. 177.

13. The work of Michael Stifel is discussed in Joseph Hofmann, *Michael Stifel 1487?–1567: Leben, Wirken und Bedeutung für die Mathematik seiner Zeit* (Wiesbaden: Franz Steiner Verlag, 1968).

14. Recorde, *The Whetstone of Witte*. The pages are unnumbered.

15. Information on Nunes's *Algebra* can be found in H. Bosmans, "Sur le 'Libro de Algebra' de Pedro Nuñez," *Bibliotheca Mathematica* (3) 8 (1907), 154–169, and H. Bosmans, "L'Algebre de Pedro Nuñez," *Annaes Scientificos da Academia Politechnica do Porto* 3 (1908), 222–271. A more general treatment of his life and work is by Rodolpho Guimaräes, *Sur la vie et l'oeuvre de Pedro Nuñes* (Coimbra: Imprimerie de l'Université, 1915). The text of the *Libro de Algebra* was photographically reprinted by the Academia das Ciências de Lisboa in 1946.

16. John Fauvel and Jeremy Gray, ed., *The History of Mathematics: A Reader* (London: Macmillan, 1987), p. 254. Many of the problems involved in the challenges between Fiore and Tartaglia and between Ferrari and Tartaglia, as well as Tartaglia's description of his discussion with Cardano, are translated in chapter 8A of this book.

17. Translated from the Italian by my daughter, Sharon Katz Cooper, with the assistance of other students at Princeton University.

18. Cardano, *The Great Art*, pp. 98–99. This work will repay a careful reading as there are many gems included in it not discussed in this text. An English language biography of Cardano is Oystein Ore, *Cardano: The Gambling Scholar* (Princeton: Princeton University Press, 1953). Cardano's autobiography is available as *Cardano, The Book of My*

Life, translated by J. Stoner (New York: Dover, 1962). See also Richard Feldman, "The Cardano-Tartaglia Dispute," *Mathematics Teacher* 54 (1961), 160–163, and James Bidwell and Bernard Lange, "Girolamo Cardano: A Defense of His Character," *Mathematics Teacher* 64 (1971), 25–31.

19. Ibid., p. 250.

20. Ibid., p. 220.

21. Rafael Bombelli, *Algebra*, p. 133. For more on Bombelli, see three articles of S. A. Jayawardene: "The Influence of Practical Arithmetics on the Algebra of Rafael Bombelli," *Isis* 64 (1973), 510–523; "Unpublished Documents Relating to Rafael Bombelli in the Archives of Bologna," *Isis* 54 (1963), 391–395; and "Rafael Bombelli, Engineer-Architect: Some Unpublished Documents of the Apostolic Camera," *Isis* 56 (1965), 298–306.

22. Ibid.

23. René Descartes, *Rules for the Direction of the Mind*, translated by Elizabeth S. Haldane and G. R. T. Ross, *Great Books* edition, (Chicago: Encyclopedia Britannica, 1952), pp. 6–7.

24. François Viète, *The Analytic Art*, pp. 11–12. All further quotations are taken from this edition, but sometimes have been amended better to give the sense of the original Latin. There is an alternative translation of the *Two Treatises on the Recognition and Emendation of Equations*, somewhat more faithful to the original—and therefore somewhat harder for a modern reader to understand—by Robert Schmidt, published in *The Early Theory of Equations: On Their Nature and Constitution* (Annapolis: Golden Hind Press, 1986). There is much in Viète's work that we have not considered in the text. Many of his methods could well be adapted to modern use. Unfortunately, there is no English study of these methods.

25. Ibid., p. 32.

26. Ibid., p. 17.

27. Ibid., p. 24.

28. Ibid., pp. 25–26.

29. Ibid., p. 84.

30. Ibid., p. 210.

31. Ibid., p. 310.

32. The second half of Charles Jones's dissertation "On the Concept of One as a Number" (University of Toronto, 1978) is primarily devoted to a study of Stevin's ideas on decimals and what Jones calls "the breakdown of the Greek number concept." I have summarized some of his arguments in the text.

33. Henrietta Midonick, ed., *The Treasury of Mathematics*, p. 737. More on Stevin can be found in Dirk J. Struik, *The Land of Stevin and Huygens: A Sketch of Science and Technology in the Dutch Republic during the Golden Century* (Dordrecht: Reidel, 1981).

34. Ibid., p. 740.

35. Charles Jones, "Concept of One," p. 239.

36. Ibid., p. 248.

37. Frank J. Swetz, *Capitalism and Arithmetic: The New Math of the 15th Century* (La Salle, Il.: Open Court, 1987), p. 158. This work provides a complete translation of the *Treviso Arithmetic* as well as a detailed commentary on the mathematics and the social setting in which the book was produced.

38. R. Franci and L. Toti Rigatelli, "Towards a History of Algebra," p. 65.

13

Mathematical Methods in the Renaissance

There is (gentle reader) nothing (the work of God only set apart) which so much beautifies and adorns the soul and mind of man as does the knowledge of good arts and sciences. . . . Many . . . arts there are which beautify the mind of man; but of all other none do more garnish and beautify it, than those arts which are called mathematical, unto the knowledge of which no man can attain, without the perfect knowledge and instruction of the principles, grounds, and Elements of Geometry.

—John Dee, *The Mathematical Preface*[1]

In late February, 1616, Cardinal Bellarmine, an advisor to the Pope and a cardinal of the Inquisition, sent two officers to bring Galileo to his home for a discussion. Three months later the cardinal gave Galileo an affidavit certifying what happened at that time: "Sig. Galileo . . . was . . . told of the declaration made by his holiness [Pope Paul V] and published by the Congregation of the Index, that [to say] the earth moves around the sun and that the sun stands still in the center of the universe without motion from east to west is contrary to Sacred Scripture and therefore may not be defended or held."[2]

Algebra was not the only mathematical concern of the Renaissance. In fact, geometry, as John Dee wrote in the opening quotation, was still the central aspect of mathematics. As part of the general revival of interest in classical learning, Renaissance scholars studied the Greek geometry texts. First, naturally, they studied Euclid, whose *Elements*, in various Latin versions, was a major part of the mathematics curriculum at the European universities of the time. It was expected that anyone having any pretense of learning would be familiar with Euclid's work.

Because there were many who did not know Latin and did not attend the universities, vernacular versions of the *Elements* began to appear in the sixteenth century. Tartaglia prepared an Italian version in 1543; Johannes Scheubel and Wilhelm Holzmann (Xylander) translated major portions into German in 1558 and 1562; Pierre Forcadel did the same in French in 1564–1566; and Rodrigo Camorano made a Spanish translation of the first six books in 1576. The most impressive of the vernacular versions, however, was the English translation of Henry Billingsley in 1570. Its nearly 1000 pages included all thirteen original books of the *Elements* as well as three additional books traditionally ascribed to Euclid. It also contained numerous additions and notes from various ancient and modern authors. The printer evidently spared no pains in the production of this work. For example, the material on solid geometry from Book XI contains "pop-up" diagrams, pasted onto the relevant pages, which enabled the reader actually to construct the three-dimensional figures (Fig. 13.1).

FIGURE 13.1

Page from Billingsley's translation of Euclid's *Elements*, containing a pop-up diagram. (*Source*: Library of Congress)

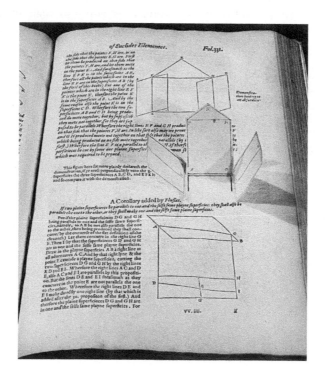

The most noteworthy part of the Billingsley *Elements*, however, is the *Mathematical Preface* by the sixteenth-century English scientist and mystic John Dee (1527–1608). Dee was well qualified to write a preface to the translation of Euclid. He had acquired a wide knowledge of various fields in which mathematics was employed and wanted to convince those about to work their way through this great geometrical work of its value. Thus, he gave detailed descriptions of some 30 different fields that need mathematics and the relationships among them, organized into what he called a "groundplat," or chart. Dee's framework offered an overview of "applied mathematics" in the Renaissance, the subject of this chapter (Fig. 13.2).

A careful student of Greek mathematics, Dee began his preface by noting that "of Mathematical things are two principal kinds; namely Number, and Magnitude."[3] The science of number is called arithmetic, that of magnitude, geometry. These are the two principal divisions of the mathematical arts. Dee noted that arithmetic originally meant the study of whole numbers, but arithmeticians have "extended their name farther than to numbers whose least part is a unit."[4] Various other kinds of numbers have been introduced, including common fractions, sexagesimal (or astronomical) fractions, and radical numbers (roots). Arithmetic has also been extended into the "Arithmetical Art of Equation," that is, algebra. It is, however, the application of geometry, the "science of magnitude," to which most of his preface is devoted.

Dee gave a brief history of geometry to justify its name (meaning "land measuring") which, he wrote, is too "base and scant for a science of such dignity and ampleness." The name "has been suffered to remain, that it might carry with it a perpetual memory of the first and notablest benefit, by that science, to common people showed, which was, when bounds . . . of land and ground were lost and confounded or, that ground bequeathed, were to be assigned . . . or . . . that commons were distributed into severalties. For, where, upon these and such like occasions, some by ignorance, some by negligence, some by fraud, and some by violence, did wrongfully limit, measure, encroach, or challenge . . . those lands and grounds, great loss, disquietness, murder and war did (full oft) ensue. Till, by God's mercy and man's industry, the perfect science of lines, planes, and solids . . . gave unto every man, his own."[5] It is good to know that this science at its origins prevented war and helped to dispense justice.

Dee divided the applications of geometry into two classes: "vulgar" geometry, which includes the various sciences of measurement such as stereometry, the measure of solids, and geography, the study of the methods for creating maps; and the "methodical arts," "which, declining from the purity, simplicity, and immateriality of our principal science of magnitudes, do yet nevertheless use the great aid, direction, and method of the said principal science."[6] Among these methodical arts are perspective, astronomy, music, astrology, statics, architecture, navigation, anthropography (the study of the geometry of the human body), trochilike (the study of circular motions), menadry (the study of simple machines), and Zography (the study of painting). This chapter surveys some of these fields, discussing both Dee's analysis and the actual work of some of the practitioners in the sixteenth and early seventeenth centuries.

FIGURE 13.2

Dee's "groundplat" from the
preface to the Billingsley
translation of Euclid's *Elements*. (*Source*: Smithsonian
Institution Libraries, Photo
No. 93-345)

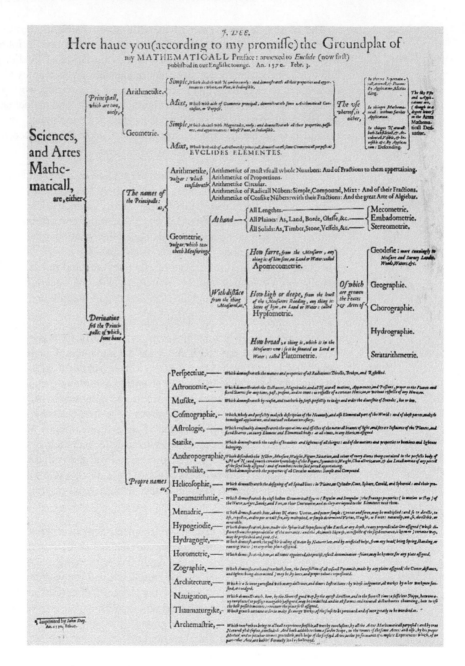

BIOGRAPHY

John Dee (1527–1608)

Dee took his BA from Cambridge University in 1545 and shortly thereafter journeyed to the continent, where he studied with various mathematicians, learned much about such fields as geography, astronomy, and astrology, and lectured on Euclid in Paris. Returning to England, he served as a court astrologer to Queen Elizabeth. His own writings encompassed such varied topics as logic, astronomy, perspective, burning mirrors, and astrology, but later in life he became enamored of the mystical elements in mathematics. Thus, he studied and wrote about how various symbols could be combined in certain figures, the proper understanding of which would enable the reader to understand the hidden secrets of the physical world. Like some of his contemporaries, he also involved himself with gematria, the study of the numerical values of words, and alchemy. Ultimately, his mysticism and accusations that he was involved with the practice of "black magic" caused him to lose his royal patronage. He died in poverty.

 13.1

PERSPECTIVE

According to Dee, "Perspective is an Art Mathematical which demonstrates the manner and properties of all radiations direct, broken and reflected." This art explains why "walls parallel . . . approach afar off" and why "roof and floor parallels, the one to bend downward, the other to rise upward, [approach each other] at a little distance from you."[7] Closely related to perspective is the art of Zography, "which teaches and demonstrates how the intersection of all visual pyramids, made by any plane assigned . . . may be by lines . . . represented."[8] It is these two arts with which a painter must be well acquainted in order that "in winter he can show you the lively view of summer's joy and riches and in summer exhibit the countenance of winter's doleful state and nakedness. Cities, towns, forts, woods, armies, yea whole kingdoms . . . can he, with ease bring with him, home (to any man's judgment) as patterns lively of the things rehearsed."[9]

13.1.1   The Creation of a Mathematical Theory of Perspective

Although there was some use of perspective in ancient times, it was only in the Renaissance that painters began in earnest to attempt to give visual depth to their works. The earliest painters accomplished this through trial and error, but by the fifteenth century, artists were attempting to derive a mathematical basis for displaying three-dimensional objects on a two-dimensional surface. Clearly, objects that are farther away from the observer must be made smaller to give the picture realism. The question then becomes how small a given object should be. The answer to this question, painters ultimately realized, had to come from geometry. Filippo Brunelleschi (1377–1446) was the first Italian artist to make a serious study of the geometry of perspective, but Leon Battista Alberti (1404–1472) wrote the first text on the subject, the *Della Pittura* of 1435 (Fig. 13.3). Alberti noted in this treatise that the first requirement of a painter is to know geometry. Thus, he presented a geometrical result showing how to represent a set of squares in the *ground plane* on the plane of the canvas, the *picture plane*.

FIGURE 13.3

Leon Battista Alberti on an Italian stamp

The picture plane may be thought of as pierced by rays of light from the various objects in the picture to the artist's eye, whose position is called the *station point*. Hence, the picture plane is a section of the projection from the eye (point A') to the scene to be pictured (Fig. 13.4). The perpendicular from the station point to the picture plane intersects the latter in a point V called the *center of vision* or the *central vanishing point*. The horizontal line AV through the central vanishing point is called the *vanishing line* or *horizon line*. All horizontal lines in the picture perpendicular to the picture plane must be drawn to intersect at the vanishing point. All other sets of parallel horizontal lines will intersect at some point on the vanishing line.

FIGURE 13.4

Alberti's rule for perspective drawing of a tiled floor

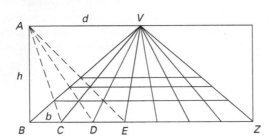 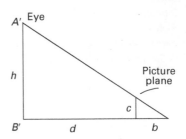

To represent a square tiled pavement (a checkerboard) in the ground plane with sides parallel to and perpendicular to the picture plane, Alberti began by marking off a set of equally spaced points $B, C, D, E, \ldots$ on the line of intersection BZ of the picture plane and the ground plane, the *ground line*. He connected these to the central vanishing point, thus giving one set of sides. To determine the set of sides parallel to the ground line, Alberti invented the following method. Mark the point A on the vanishing line at distance d from the vanishing point V, where d is the distance of the eye from the picture plane. Then draw lines connecting A to each of the points $B, C, D, E, \ldots$. The lines parallel to BZ through the intersections of BV with $AC, AD, AE, \ldots$ will then represent the lines of pavement parallel to the ground line.

To demonstrate that this construction is correct, we will use algebra, although Alberti himself did not give any demonstration. Suppose that the eye is situated at a height h above the ground line. If a line in the ground plane is parallel to the ground line and at a distance b behind it, then its position in the picture plane should be at a distance c above the ground line, where c is determined by the proportion $c : b = h : d + b$ derived from the similar triangles in Figure 13.4. Thus, $c = hb/(d + b)$. Now, if AB and the ground line BZ are taken as coordinate axes, the equation of the line connecting B and V is

$$y = \frac{h}{d}x,$$

while the equation of the line connecting $A = (0, h)$ and $C = (b, 0)$ is

$$y = -\frac{h}{b}x + h.$$

The y coordinate of the intersection of the two lines is then $hb/(d + b)$ as desired. One can easily demonstrate the correctness for as many parallel lines as desired.[10]

This checkerboard construction is at the heart of the system of focused perspective used by artists from the fifteenth century to the present day. Alberti himself did not discuss any more advanced perspective constructions, but Piero della Francesca (1420–1492), in his work *De prospectiva pingendi* (*On Perspective for Painting*), written sometime between 1470 and 1490, gave a detailed discussion of how to draw various two- and three-dimensional geometrical objects in focused perspective. Della Francesca, besides being an artist, was a competent mathematician, problems from whose abacus tract appeared in Chapter 12. His text on perspective included a drawing showing the calculations the artist made in preparing a painting in focused perspective.[11]

13.1.2 Dürer and the Teaching of Perspective

FIGURE 13.5

Self-portrait of Albrecht Dürer

Another artist-mathematician of this same period was the German Albrecht Dürer (1471–1528), who spent several years in Italy studying works on perspective before finally writing his own major treatise, the *Underweysung der Messung mit Zirckel und Richtscheyt in Linien, Ebnen, und gantzen Corporen* (*Treatise on Mensuration with the Compass and Ruler in Lines, Planes, and Whole Bodies*) (Fig. 13.5). Published in 1525, this work was the earliest geometric text written in German.[12] Thus, Dürer had to create a new German vocabulary for scientific terms, including abstract mathematical concepts. Whenever possible, he used the expressions handed down from generation to generation by artisans. For example, *der neue Mondschein* (crescent) denoted the intersection of two circles; *Gabellinie* (fork line) meant hyperbola; and *Eierlinie* (egg line) meant ellipse.

Dürer believed that he needed to instruct the German artists in many of the preliminary geometrical ideas involved in drawing before they could approach perspective, dealt with at the end of the *Underweysung der Messung*. In his opinion, German painters were equal to any in practical skill and imagination, but they were well behind the Italians in rational knowledge. "And since geometry is the right foundation of all painting, I have decided to teach its rudiments and principles to all youngsters eager for art."[13] Therefore, the work was eminently practical. Dürer showed how to apply geometric principles to the representation of objects on canvas (Fig. 13.6).

The first of the four books of the *Underweysung der Messung* dealt with the representation of space curves. Dürer's idea was to project the curve onto both the yz plane and the xy plane in order to determine its nature. Unfortunately, this is not always a straightforward task. Consider his construction of an ellipse from its definition as a section of a right circular cone (Fig. 13.7). Dürer first projected the cone with its cutting plane onto the yz plane. He divided the line segment fg representing the diameter of the ellipse into 12 equal parts, and drew both vertical and horizontal lines through the division points. At each of the eleven points i, the horizontal line represented part of the diameter of the circular section C_i made by a horizontal cutting plane. The two points of intersection of this circle with the ellipse are symmetrically located on the ellipse with respect to its diameter and therefore determine the width w_i of the ellipse there. The projection of the cone onto the xy plane then consisted of this series of concentric circles C_i. The continuation of each vertical line became a chord in the corresponding circle whose length is w_i. Dürer thus had a rough projection of the ellipse.

The outline of the ellipse is, however, not symmetric about its minor axis, since this projection is not taken from a direction perpendicular to the plane of the ellipse itself. But when Dürer attempted to draw the ellipse from its projection, he simply transferred the line

FIGURE 13.6

St. Jerome in His Study.
Here, Dürer illustrates his
application of the theory of
perspective. (*Source*: The
Nelson-Atkins Museum of
Art, Kansas City, Missouri,
58-70/21, gift of Robert B.
Fizzell)

segment representing the axis of the ellipse to a new vertical line fg, divided it at the same points i, drew horizontal line segments through each of width w_i, and then sketched the curve through the ends of these line segments. Dürer's drawing was therefore in error, because the curve is wider at the bottom than at the top. A possible reason Dürer did not realize that the ellipse should be symmetric about its minor axis is that the centerline of the cone, around the projection of which all the circles were drawn, does not pass through the center of the ellipse. Although one can prove that $w_i = w_{12-i}$ ($i = 1, 2, 3, 4, 5$) by an analytic argument, Dürer probably believed that the ellipse was in fact egg-shaped—he did call it an *Eierlinie*—because the cone itself widens toward the bottom.[14]

After describing the construction and representation by projections of other space curves, Dürer continued in the second book of the *Underweysung* to describe methods for constructing various regular polygons, both exact ones using the classical tools of straightedge and compass and approximate ones taken from the tradition of artisans. Thus, the work, which was published in Latin several years after its German edition, served both to introduce the artisans to the Greek classics and to familiarize professional mathematicians with the practical geometry of the workshop. The third book of the text was purely practical, showing how geometry could be applied in such varied fields as architecture and typography. Here Dürer suggested new types of columns and roofs as well as the methods of accurately constructing

FIGURE 13.7

Dürer's construction of an ellipse by projection

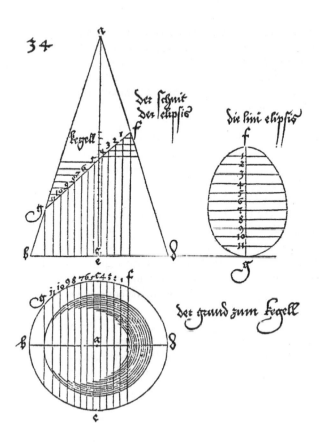

both Roman and Gothic letters. In the final book, Dürer returned to more classical problems and dealt with the geometry of three-dimensional bodies. In particular, he presented a construction of the five regular solids by paper folding, a method still found in texts today, as well as similar procedures for certain semiregular solids. He also presented other problems of construction, including that of doubling the cube, before concluding the work with the basic rules for the perspective drawing of these solid figures.

13.1.3 The Conic Sections

One of the reasons that Dürer had such trouble drawing the ellipse was that he had probably never seen Apollonius's text on the conic sections. He only knew that they were curves formed by cutting a cone with a plane. But exactly what they looked like was evidently not clear to him. Johannes Kepler (1571–1630), on the other hand, was very familiar with Apollonius's work. He realized that he needed a detailed understanding of the subject as he was working on his treatise on optics as well as on his astronomy (see below). In thinking about the subject, he was able to take more out of Apollonius's works than the author had put in. He described his thoughts on the subject in his *Optics* of 1604.

For example, he realized that all the conic sections were really part of the same family of curves. As he wrote, "there exists among these curves the following order, by reason of their properties: it passes from the straight line through an infinity of hyperbolas to the parabola, and thence through an infinity of ellipses to the circle. For the most obtuse of all hyperbolas is a straight line; the most acute, a parabola. Likewise, the most acute of all ellipses is a parabola; the most obtuse, a circle."[15] In other words, the parabola is the boundary curve between the family of hyperbolas and the family of ellipses. Even though it is infinite, like the hyperbola, "the more it is extended, the more it becomes parallel to itself, and does not expand the arms like a hyperbola." The straight line is one of the outer limits of the family of curves. Therefore, the further the hyperbola is extended, the more it becomes like a straight line. Similarly, since the circle is the other outer limit, the further one goes from the center of an ellipse, the more it looks like a circle.

Kepler's use of analogy extended to his discussion of the foci of the conic sections, points which Apollonius had referred to but which Kepler named. The circle has one focus, at the center, thus, as far from the curve as possible. In an ellipse there are two foci, and as the ellipses get more acute, the foci get further apart. Since the "most acute" ellipse is a parabola, the second focus of that curve is at an infinite distance from the first. The more obtuse the hyperbola, the closer the foci are to the curve, until in the case of the straight line, the foci merge again to lie on the line itself.

Finally, Kepler described constructions of the hyperbola and the ellipse via threads tied around pins at the foci. He then noted that since the parabola is in the middle between the hyperbola and ellipse, there should be an analogous construction of that curve. After searching for one, he finally found it and described it too in his text.

13.2   NAVIGATION AND GEOGRAPHY

Two related aspects of mathematics discussed by Dee and extremely important to the world of the sixteenth century were geography and navigation.

13.2.1   Problems of Navigation

As Dee wrote, "The art of Navigation demonstrates how, by the shortest good way, by the aptest direction, and in the shortest time, a sufficient ship, between any two places (in passage navigable) assigned, may be conducted; and in all storms and natural disturbances chancing, how to use the best possible means, whereby to recover the place first assigned."[16] In the fifteenth and sixteenth centuries, Europeans were exploring the rest of the world, and methods of navigation were of central importance. The country that could employ new techniques well had great advantages in the quest for new colonies and their attendant natural resources.

The major question of navigation on the seas was the determination of the ship's latitude and longitude at any given time. The first of these was not too difficult. One's latitude, in the northern hemisphere, was equal to the altitude of the north celestial pole, and this was marked, approximately, by Polaris, the pole star. A good approximation of the latitude was found simply by taking the altitude of that star, although because in the fifteenth century Polaris was about $3\frac{1}{2}^{\circ}$ from the pole, appropriate adjustments needed to be made. An alternate method of finding latitude, especially when sailing close to or south of the equator, was by observation of the sun. As noted in Chapter 5, the zenith distance of the sun at local noon is equal to the latitude minus the sun's declination. Navigators of the fifteenth century had accurate tables of the declination for any day of the year, so they needed only to take a reading of the sun's altitude at noon. This altitude was, of course, the highest altitude of the day and could be determined by finding the shortest shadow of a standard pole.

The determination of longitude was much more difficult. Knowing the difference between the longitudes of two places is equivalent to knowing the difference between their local times, because 15° of longitude is equivalent to one hour. Theoretically, if one had a clock set to the time at a place of known longitude and could determine when, on that clock, local noon occurred at one's current location, the difference in time would enable one to make a determination of longitude. Alternatively, one could compare the known time of an astronomical event, such as an eclipse of the moon, at the place of known longitude with its local time at one's current location. Unfortunately, these methods could not work given the current state of knowledge of the moon's motion or of the accuracy of timekeeping devices. The clocks in use were simply not precise enough, especially if operated on the moving decks and in the changing temperatures of a ship at sea. And when Columbus attempted to determine longitude on his second voyage to America in 1494 using an eclipse of the moon, his error was about 18°. As late as 1707, four British warships ran aground at the Scilly Isles near the southwestern tip of England with the loss of 2000 men because the admiral and his navigators had misjudged the ships' longitude. The British government then offered a reward of £20,000 (approximately $12,000,000 in today's currency) for a method of accurately determining longitude at sea. The money (at least, most of it) was ultimately paid to the English watchmaker John Harrison (1693–1776) after his series of increasingly accurate timepieces survived numerous trials both on land and at sea and won praise from Captain James Cook on his voyages to the South Pacific (Fig. 13.8). Around the same time, lunar tables accurate enough for use in determining longitude began to be produced following a century of detailed observations at the Royal Greenwich Observatory.[17]

Given the difficulties of finding one's location at sea, it is not surprising that seamen often used methods of "guesstimation" rather than mathematical astronomy. While scholars were aware that a great circle route was the shortest distance between two points, sailors generally preferred to sail to the latitude of their destination as quickly as possible and then head due east or west until they reached land.

FIGURE 13.8

Harrison's final timepiece on a British stamp

13.2.2 Mapmaking in the Renaissance

Whatever the method of navigating, however, the seamen needed accurate maps. Dee called the making of these maps Geography: "Geography teaches ways by which in sundry forms (as spherical, plane, or other) the situation of cities, towns, villages, forts, castles, mountains, woods, havens, rivers, creeks, and such other things, upon the outface of the earthly globe . . . may be described and designed in commensurations analogical to nature and verity, and most aptly to our view, may be represented."[18]

Maps had been drawn since antiquity. Because it is impossible to make an absolutely correct map on a flat piece of paper, the mapmaker always had to make some choice of the particular qualities of the projection desired. The mapmaker could choose to preserve areas or shapes or directions or distances. The larger the portion of the earth's surface to be represented, the more difficult it is for the map to have several of these qualities, even approximately. As we have seen in Chapter 5, Ptolemy made two different choices of projection in the world maps he described in his *Geography*. And in his regional maps, Ptolemy simply used a rectangular grid for the meridians and parallels. Because the spacing of the meridians depends on the latitude, he chose a scale in the two directions so that it corresponded approximately to the ratio of the length of one degree of longitude on the middle parallel of the map to one degree of latitude. As we have noted in Chapter 5 and the Exercises there, this ratio is equal to $\cos \phi$, where ϕ is the middle latitude. For example, because Ptolemy's map of Europe reaches from latitude 42° to latitude 54°, the given ratio should be approximately $\cos 48° = .6691$, or $2 : 3$.

During the early Renaissance, before Ptolemy's *Geography* achieved wide circulation, the maps used by seamen were generally constructed in the simplest possible way, by using a rectangular grid for parallels and meridians, with the same scale on each. Because the distances between the meridians were the same at all latitudes, and because the true distance depends on the cosine of the latitude, shapes on these maps had the appearance of being elongated in the horizontal direction. Thus, shape was not preserved, and more important for the sailor, lines of constant compass bearing, called **rhumbs**, were not represented by straight lines. When such maps were of relatively small areas, the rhumb lines were straight enough and were often drawn in for each of eight or sixteen compass directions. But as long sea voyages became increasingly common, improvements were required.

One of the first Renaissance mathematicians to attempt to apply mathematics to the improvement of mapmaking methods was Pedro Nunes, in his *Tratado da sphera* of 1537. He discovered that on a sphere a rhumb line or **loxodrome**, as it is now called, becomes a spiral terminating at the pole. Using globes for navigation, however, was inconvenient because they could not be made large enough. Nunes therefore attempted to develop a map in which loxodromes were straight lines. For accuracy, however, it was necessary that the meridians converge near the poles. Although Nunes was able to design a device that enabled sailors to measure the number of miles in a degree along each parallel, he was not able to solve the problem he had set.

FIGURE 13.9

Mercator on a Belgian stamp

By 1569, Nunes's problem was solved from a slightly different point of view by Gerard Mercator (1512–1594) (Fig. 13.9), with a new projection known ever since as Mercator's projection. Both parallels and meridians were represented by straight lines on this map. To compensate for the "incorrect" spacing of the meridians, therefore, Mercator increased the spacing of the parallels toward the poles. He claimed that on his new map rhumb lines were now straight and a navigator could simply lay a straightedge on his map between his origin and his destination to determine the constant compass bearing to follow. Mercator did not explain the mathematical principle he followed for increasing the distance between the parallels, and some believe that he did it by guesswork alone. Not until the work of Edward Wright (1561–1615), *On Certain Errors in Navigation* (1599), did an explanation of Mercator's methods appear in print.

Because the ratio of the length of a degree of longitude at latitude ϕ to one at the equator is equal to $\cos \phi$, if meridians are straight lines, the distances between them at latitude ϕ are stretched by a factor of $\sec \phi$. For loxodromes to be straight on such a map, the vertical distances must also be stretched by the same factor. Because $\sec \phi$ varies at each point along a meridian, the stretching factor needs to be considered for each small change of latitude. If we denote by $D(\phi)$ the distance on the map between the equator and the parallel of latitude ϕ, the change dD in $D(\phi)$ caused by a small change $d\phi$ in ϕ is determined by $dD = \sec\phi d\phi$. Because the same factor applies horizontally as well, any "small" region on the globe will be represented on the map by a "small" region of the same shape. The angle at which a line crosses a meridian on the globe will be transformed into that same angle on the map and loxodromes will be straight. It follows from this argument that, in modern terminology, the map distance between the equator and the parallel at latitude ϕ is given by

$$D(\phi) = \int_0^\phi \sec\phi\,d\phi, \tag{13.1}$$

where the radius of the globe is taken as 1. Wright, of course, did not use integrals. He took for his $d\phi$ an angle of $1'$ and computed a table of what he called "meridional parts" by adding the products $\sec\phi\,d\phi$ for latitudes up to $75°$. $D(\phi)$ can be calculated by calculus techniques as

$$\ln(\sec\phi + \tan\phi) \quad \text{or} \quad \ln\left(\tan\left(\frac{\phi}{2} + \frac{\pi}{4}\right)\right).\text{[19]}$$

John Dee met Mercator on one of his trips to the continent. He returned with several of Mercator's globes and probably conferred with Wright concerning the mathematical details of Mercator's projection. Thus, he was involved in the process of making maps "analogical to nature." Mercator's map, although well suited for navigation, was unfortunately not "analogical to nature" for regions far from the equator. The spacing out of the parallels greatly increased the relative size of such regions. The popularity of the map led generations of students to believe, for example, that Greenland is larger than South America. Nevertheless, its simplicity of use made it the prime sea chart during the age of European exploration.

13.3   ASTRONOMY AND TRIGONOMETRY

According to Dee, "Astronomy is an art mathematical which demonstrates the distance, magnitudes, and all natural motions, appearances, and passions proper to the planets and fixed stars, for any time past, present and to come, in respect of a certain horizon, or without respect of any horizon. By this art we are certified of the distance of the starry sky and of each planet from the center of the earth, and of the greatness of any fixed star seen, or planet, in respect of the earth's greatness."[20] Thus, the purpose of astronomy is to predict the motions of the heavenly bodies as well as to determine their sizes and distances. A related art is Cosmography, "the whole and perfect description of the heavenly, and also elemental part of the world, . . . and mutual collation necessary."[21] As Dee noted further, cosmography explains the relationship of heavenly to earthly events, allowing us to determine "the rising and setting of the sun, the lengths of days and nights . . . with very many other pleasant and necessary uses."

13.3.1 **Regiomontanus**

Since astronomy and cosmography in the Renaissance, like astronomy in earlier periods, were heavily dependent on trigonometry, we begin with a discussion of the first "pure" trigonometry text written in Europe, the *De Triangulis Omnimodis* (*On Triangles of Every*

Kind) of Johannes Müller (1436–1476), generally known as Regiomontanus because he was born near Königsberg in Lower Franconia. (*De Triangulis* was written about 1463 but not published until 70 years later.)

Regiomontanus had made a new translation of Ptolemy's *Almagest* directly from the Greek and, after completing it, realized that there was a need for a compact systematic treatment of the rules governing the relationships of the sides and angles in both plane and spherical triangles that would improve on Ptolemy's seemingly ad hoc approach. He considered such a treatment a necessary prerequisite to the study of the *Almagest*: "You, who wish to study great and wonderful things, who wonder about the movement of the stars, must read these theorems about triangles. Knowing these ideas will open the door to all of astronomy and to certain geometric problems."[22]

Regiomontanus presented his material in *On Triangles* in careful geometric fashion, beginning with definitions and axioms. He proved each theorem by using the axioms, results from Euclid's *Elements*, or earlier results in the text. Most theorems are accompanied by diagrams and many are followed by examples illustrating the material. Regiomontanus based his trigonometry on the sine of an arc, defined as the half-chord of double the arc, but he did note that one can also consider the sine as depending on the corresponding central angle. Like his European predecessors, he made no use of the tangent function, even though he must have been aware of tables of tangents that had appeared in Europe, mostly taken from Islamic astronomical works. He did, however, use the cosine (written as sine of the complement) and the versine (radius minus the cosine). In any case, Regiomontanus was able to solve all of the standard problems of trigonometry using just the sine, an extensive table for which, based on a radius (or total sine) of 60,000, he appended to the text. (As in Chapter 9, we will use the notation "Sin" to denote Regiomontanus's sine function.)

The first half of Regiomontanus's text deals with plane triangles, the second half with spherical ones. Among his results are various methods for solving triangles. Conceptually, there is nothing particularly new in his methods, but unlike earlier European authors on trigonometry, Regiomontanus often provided clear and explicit examples of his procedures. For example, Theorem I–27 shows how to determine the angles of a right triangle if two sides are known, while Theorem I–29 shows how to determine the unknown sides of a right triangle, if one of the two acute angles and one side is given. In both cases, Regiomontanus used his sine table. His example for the second of these theorems assumes that one acute angle is $36°$ and that the hypotenuse equals 20. Thus, the other angle is $54°$, and the two sides would be 35,267 and 48,541, respectively, if the hypotenuse were 60,000. Using proportions, Regiomontanus calculated that because the hypotenuse is 20, these sides are equal to $11\frac{3}{4}$ and $16\frac{11}{60}$, respectively.

In Theorem I–49, Regiomontanus solved an arbitrary triangle when two sides and the included angle are known. Supposing AB and BC are known together with the included angle ABC, Regiomontanus used the same procedure as Levi ben Gerson (Fig. 13.10). He dropped a perpendicular AD to BC or BC extended. In the right triangle ABD, one of the acute angles and a side is known. By Theorem I–29, the remaining sides and angle can be calculated. Two sides of the right triangle ADC are now known and Theorem I–27 and the Pythagorean Theorem provide the missing side and angles.

For the so-called ambiguous case, where two sides and an angle opposite one of them are known, Regiomontanus improved on Levi ben Gerson by providing several possibilities. He

FIGURE 13.10

De Triangulis: Theorem I–49

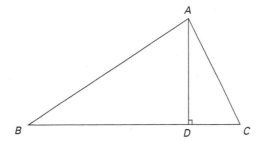

first dealt with the case where the given angle ACB opposite the known side AB is obtuse, AC also being known, by dropping a perpendicular AD to AC extended (Fig. 13.11). The triangle is then solved as in I–49. In his treatment of the case where the given angle is acute, however, he noted that "there is not enough [information given] to find the [other] side and the remaining angles."[23] For with an acute angle ABC given opposite side AC, there are two possible triangles that can be constructed, one of which has an acute angle opposite AB, the other an obtuse angle. He showed how to find the unknown side and angles in each case, but failed to note the possibility that there might not exist any solution, probably because he always considered that the particular triangle, the unknown parts of which were sought, did exist.

FIGURE 13.11

De Triangulis: The ambiguous case

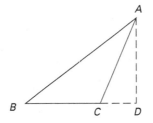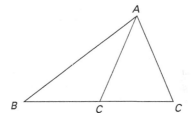

In Theorem II–1 Regiomontanus proved the law of sines: "In every rectilinear triangle the ratio of [one] side to [another] side is as that of the right sine of the angle opposite one of [the sides] to the right sine of the angle opposite the other side."[24] Because Regiomontanus's sines are lines in a circle of a given radius, his proof of the theorem for the triangle ABG requires circles drawn with centers B and G having equal radii BD and GA, respectively (Fig. 13.12). Drawing perpendiculars to BG from A and D, intersecting that line at K and H, respectively, Regiomontanus then noted that DH is the Sine of $\angle ABG$ while AK is the Sine of $\angle AGB$, using circles of the same radius. Since $BD = GA$, $\angle ABG$ is opposite side GA, and $\angle AGB$ is opposite side AB, the similarity of triangles ABK and DBH provides the desired result. Regiomontanus now used this result to solve anew the case of a triangle with two sides and the angle opposite one of them known.

In the remainder of Book II of *On Triangles*, Regiomontanus showed how to determine various parts of triangles if certain information is given, such as the ratio of the sides or the length of the perpendicular from a vertex to the opposite side. In two of these theorems, rather

FIGURE 13.12

Proving the law of sines

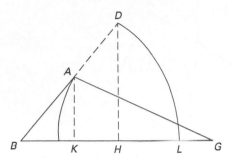

than using geometric arguments, he uses arguments from algebra, what he called "the art of thing and square," because he claimed that no "geometric" proof of his result was available. Thus, to find the sides AB, AG, of a triangle given that the base $BG = 20$, the perpendicular $AD = 5$, and the ratio $AB : AG = 3 : 5$, Regiomontanus set segment DE equal to BD and, for algebraic simplicity, used $2x$ to represent the unknown segment EG (Fig. 13.13). Then $BE = 20 - 2x$, $BD = 10 - x$, and $DG = 10 + x$. Since $AB^2 = BD^2 + AD^2$ and $AG^2 = DG^2 + AD^2$, and since the ratio $AD^2 : AG^2 = 9 : 25$, Regiomontanus concluded that

$$\frac{(10 - x)^2 + 25}{(10 + x)^2 + 25} = \frac{9}{25}.$$

This equation reduces easily to $16x^2 + 2000 = 680x$. Regiomontanus stopped his solution here, noting only that "what remains [to be done], the rules of the art [of algebra] show."[25]

FIGURE 13.13

Regiomontanus's use of algebra

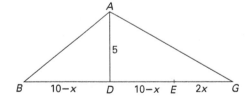

Book III of *On Triangles* provides a basic introduction to spherical geometry, including especially many results on great circles. This discussion is preliminary to the standard material on spherical trigonometry contained in the final two books of the text. Regiomontanus included here the rule of four quantities, and then derived from it the law of sines for both right and arbitrary spherical triangles. He followed these with three other important results, two involving right triangles ABC, with the right angle at C, and one about arbitrary spherical triangles. Theorem IV–18 is the result that $\text{Cos } B = \text{Sin } A \text{ Cos } b$, evidently first proved by Jābir ibn Aflah;[26] Theorem IV–19 is $\text{Cos } c = \text{Cos } a \text{ Cos } b$, the spherical equivalent of the Pythagorean Theorem, essentially known to Ptolemy; and Theorem IV–20 shows that $\frac{\text{Sin } B_2}{\text{Sin } B_1} = \frac{\text{Cos } C}{\text{Cos } A}$, where the perpendicular BD from B to AC divides angle B into two angles B_1 and B_2 (Fig. 13.14).

FIGURE 13.14

On Triangles: Theorem IV–20

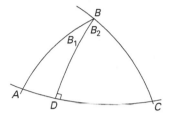

Regiomontanus then gave a detailed discussion of how to solve spherical triangles, given various pieces of information, being very careful to distinguish cases where sides were less than or greater than quadrants and where certain angles were acute or obtuse. After showing how to use the law of sines as well as Theorems IV–18 and IV–19 to solve right triangles, he then proceeded through the various cases of arbitrary triangles. His usual technique is to drop a perpendicular from one vertex to the opposite side (or the opposite side extended) and then use results on right triangles. For example, Theorem IV–20 gave him a simpler procedure than that of Naṣir al-Dīn for solving the spherical triangle, all of whose angles are given. For then the ratio of Sin B_1 to Sin B_2 was known as well as the sum of B_1 and B_2. It followed that both B_1 and B_2 could be found. Therefore, all angles in the right triangles ABD and BDC were known, so by Theorem IV–18 and the law of sines, the sides of both of these triangles could be found.

In Book V, Regiomontanus proved the result known to al-Battānī that gave him an alternate method for solving certain triangles:

$$\frac{\text{Versin } A}{\text{Versin } a - \text{Versin}(b - c)} = \frac{R^2}{\text{Sin } b \, \text{Sin } c},$$

a result equivalent to the spherical law of cosines: $\cos a = \cos b \cos c + \sin b \sin c \cos A$. In Book IV, he had provided two rather complicated methods for solving a spherical triangle with three sides given, one of them being the same as that of Naṣir al-Dīn, but the law of cosines then gave him a much simpler method.

Even though Regiomontanus's book was not published until 1533, it nevertheless was extremely influential in the development of European trigonometry and astronomy. In general, the other authors who wrote trigonometry texts in the last two-thirds of the sixteenth century modeled themselves on Regiomontanus's work, although they did improve his tables and introduce the other trigonometric functions, all, like the sine, defined as lengths of certain lines depending on a given arc in a circle of a fixed radius.[27] The radius was generally of size 10^n or 6×10^n, with n tending to be larger toward the end of the sixteenth century. The large radius enabled all values to be given in integers, since decimal fractions were still not in use.

George Joachim Rheticus (1514–1574) was the first to define the trigonometric functions directly in terms of angles of a right triangle, holding one of the sides fixed at a large numerical value. Rheticus thus called the sine the "perpendiculum" and the cosine the "basis" of the triangle with fixed hypotenuse. Other authors gave other names to the trigonometric functions. The first author to use the modern terms "tangent" and "secant" was Thomas Finck (1561–1656) in his *Geometria rotundi libra XIV* of 1583. He called the three cofunctions "sine complement," "tangent complement," and "secant complement." Many of these trigonometry texts gave various numerical examples to illustrate methods of solving plane and spherical triangles, but not until the work of Bartholomew Pitiscus (1561–1613) in 1595 did there appear any problem in such a text explicitly involving the solving of a real plane triangle on earth. Pitiscus, in fact, invented the term "trigonometry." He titled his book *Trigonometriae sive, de dimensione triangulis, Liber* (*Book of Trigonometry, or the Measurement of Triangles*).

Pitiscus intended in the text to show how to measure triangles and, in appendix 2 on Altimetry, he gave trigonometric methods for determining the height BC of a distant tower. In Figure 13.15, a quadrant is used to measure $\angle AKM = \angle ABC = 60°20'$. The distance AC from the observer to the tower is measured as 200 feet. Pitiscus set up the proportion Sin $60°20' : AC = $ Sin $29°40' : BC$ and then calculated that $BC = 113\frac{80,204}{86,892}$, or, approximately, 114 feet. This calculation used Pitiscus's sine table, calculated to a radius of 100,000. He gave a second procedure using his tangent table, in which the required proportion is $AC : 100,000 = BC : \tan 29°40'$. The major difference between Pitiscus's methods and current ones is that he always adjusted for the fact that his trigonometric values are lengths of certain lines in a particular circle. The trigonometric ratios in use today had yet to arrive.

FIGURE 13.15

Measuring the height of a tower

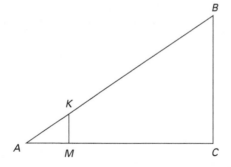

13.3.2   Nicolaus Copernicus and the Heliocentric System

The trigonometry of the fifteenth century exemplified by Regiomontanus's work, even without trigonometric ratios, provided the mathematics necessary to attack the astronomical problems of the day. Some of these problems, discussed in Regiomontanus's edition of the *Almagest*, were ones involving fundamental questions about Ptolemy's system, still the accepted view of the nature of the universe. Islamic and Jewish astronomers through the centuries had noted certain discrepancies between Ptolemy's predictions and their own observations and had made various adjustments to some of Ptolemy's details. But the Christian view of the universe at the beginning of the Renaissance was still based on the views of Aristotle and Ptolemy to the effect that the universe was composed of a system of nested spheres centered on the earth and that it was the rotation of these spheres, to which were attached the planets, that caused the appearances in the heavens. The various additional parts of the model, such as epicycles and eccenters, were all somehow embedded in the various spheres. This basic view of the universe can perhaps most easily be seen in Dante's *Divine Comedy* (1328), which describes the poet's journey through each of the celestial spheres holding the planets and stars up to the final immovable sphere containing the throne of God.

By the fifteenth century, however, astronomers were having very serious difficulties accepting Ptolemy's system in detail. One type of error was pointed out by Regiomontanus, who noted that Ptolemy's theory of the moon required the observed size to vary considerably more than it really does. More importantly, because even small errors tended to accumulate over the centuries, astronomers found many occasions when Ptolemy's predictions of planetary positions or lunar eclipses were greatly in error. And as European explorers set out on voyages around the globe, they needed improved navigational techniques that could come only through correct astronomical tables. In addition, through these explorations, Europeans found so many parts of the world previously unknown to them that they realized that Ptolemy's *Geography* was also in error. The way was prepared for believing that the fundamentals of his astronomy could be wrong.

The Catholic Church was also aware by the early Renaissance that the Julian calendar, used since the time of the Roman Empire, had serious inadequacies. In particular, since the true solar year was $11\frac{1}{4}$ minutes less than the $365\frac{1}{4}$ days on which that calendar was based, the cumulative errors threatened to change the relationship of the calendar months to the seasons. For instance, according to Church law, Easter was to be celebrated on the first Sunday after the first full moon following the vernal equinox. The equinox was always reckoned as March 21, but by the sixteenth century it actually took place about March 11. Without correction, Easter would eventually arrive in the summer rather than in the spring. When calendar reform became an official Church project, however, the astronomers advised that existing astronomical observations were inadequate and did not yet permit an accurate, mathematically based calendar change.

Among the astronomers who refused an invitation to participate in the reform of the calendar was Nicolaus Copernicus (1473–1543), who, having studied Ptolemy's system in great detail and having become aware of all its inaccuracies, came to the conclusion that it was impossible to patch up the earth-centered approach any longer. "[The astronomers] have not been able to discover or deduce from [their hypotheses] the chief thing, that is the form of the universe, and the clear symmetry of its parts. They are just like someone including in a picture hands, feet, head, and other limbs from different places, well painted indeed, but not modeled from the same body, and not in the least matching each other, so that a monster would be produced from them rather than a man."[28] To redo the "painting" and eliminate the monster, Copernicus decided to read all the opinions of the ancients to determine whether anyone had proposed a system of the universe different from the earth-centered one. Having discovered that some Greek philosophers had proposed a sun-centered (**heliocentric**) system in which the earth moves, Copernicus explored the consequences of reforming the system under that assumption: "Thus assuming the motions which I attribute to the Earth, . . . I eventually found by long and intensive study that if the motions of the wandering stars are referred to the circular motion of the Earth and calculated according to the revolution of each star, not only do the phenomena agree with the result, but also it links together the arrangement of all the stars and spheres, and their sizes, and the very heaven, so that nothing can be moved in any part of it without upsetting the other parts and the whole universe."[29]

Copernicus's fundamental treatise in which he expounded his system of the universe was *De revolutionibus orbium celestium* (*On the Revolutions of the Heavenly Spheres*), a book that represented the work of a lifetime but that was only published in 1543, the year of his death. This book sets forth the first mathematical description of the motions of the heavens based on

BIOGRAPHY

Nicolaus Copernicus (1473–1543)

Copernicus was born in Torun in East Prussia into the family of a wealthy merchant and was sent at the age of 18 to study at the University of Cracow (Fig. 13.16). Upon leaving Cracow, he was appointed to a clerical post through the influence of his uncle, the Bishop of Ermland. He therefore not only received a salary but also was permitted to travel to Italy to study at Bologna and Padua over the next ten years. It is assumed that in Italy he learned of some of the work of the Islamic astronomers from Maragha, including al-Tūsī, although there is no definite evidence of this. Finally returning home, he spent the remainder of his life in Ermland, serving as a Canon of Frauenburg Cathedral. The job not being a particularly demanding one, he was generally free to concentrate on his study of astronomy and was able to complete his manuscript of *De revolutionibus* by about 1530. He was unwilling, however, to publish the work. In about 1514, he had already written a brief outline of his system, *The Commentariolus*, which was circulated to various scholars. But it was not until George Rheticus, professor of mathematics at the University of Wittenberg, arrived in Frauenburg in 1539 to learn firsthand about Copernicus's system, that Copernicus was finally persuaded to allow his masterwork to be published.

FIGURE 13.16

Copernicus and his system on a Hungarian stamp

the assumption that the earth moves, for, as Copernicus noted in his preface, "Mathematics is written for mathematicians."[30] *De revolutionibus*, following very closely the model of Ptolemy's *Almagest*, is a very technical work in which the author uses detailed mathematical calculation, based on the assumption that the sun is at the center of the universe and buttressed by the results of observations taken by Copernicus and his predecessors, to describe the orbits of the moon and the planets and to show how these orbits are reflected in the positions observed in the skies. Copernicus sketched his theory very briefly in the first book of *De revolutionibus* and presented the simplified diagram of the sun in the center of seven concentric spheres, one each for the six planets, including the earth, and one for the fixed sphere of the stars (Fig. 13.17).

Copernicus, like his predecessors, conceived of the system of the universe as a series of nested spheres containing the planets, rather than as empty space through which the planets

FIGURE 13.17

Copernicus's system of the universe

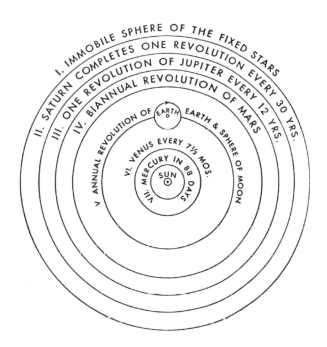

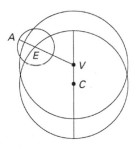

FIGURE 13.18

Ptolemy's equant: The planet *A* moves on the epicycle centered at *E*; *E* travels around the circle centered at *C* in such a way that the radius vector *VE* moves uniformly around *V*, the equant point.

travel in circles. Of course, Copernicus had no physics to keep the planets in their orbits. Spheres for Copernicus, however, as for Aristotle, had a natural motion that needed no other physical basis: "The movement of a sphere is a revolution in a circle, expressing its shape by the very action, in the simplest of figures, where neither beginning nor end is to be found, nor can the one be distinguished from the other, as it moves always in the same place."[31] In fact, one of Copernicus's aims in his reform of Ptolemy's work was to return to one of the classic principles of astronomy, that all heavenly motion must be composed of uniform motion of circles about their centers. Copernicus believed Ptolemy had violated that principle by accounting for certain aspects of a planet's motion through the use of the equant, a point within the planet's orbit around which the radius vector to the center of the planet's epicycle revolved uniformly (Fig. 13.18). The uniform motion in that case was not about the center of the circle on which the epicycle traveled. Islamic astronomers at Maragha, led by Nasir al-Dīn al-Ṭūsī, had also been bothered by this problem. Copernicus adapted their solution in his own work, although it is not known how Copernicus learned of the Islamic work. The Islamic astronomers, of course, had not taken Copernicus's major step of challenging the centrality and immovability of the earth.

Copernicus himself did not—and could not—present any real evidence for either the earth's daily rotation on its axis or its yearly revolution about the sun. For the first motion, he simply argued that it is more reasonable to assume that the relatively small earth rotates rather than the immense sphere of the stars. For the second motion, his argument was in essence that the qualitative behavior of the planets can more easily be understood by attributing part of their motion to the earth's own yearly revolution. Thus, retrogression can be explained in

FIGURE 13.19

Retrograde motion for a planet
outside the orbit of the earth.
The observed positions of the
planet against the sphere of the
stars are marked, in order, 1,
2, 3, 4, 5, 6, 7. Retrogression
takes place between 3 and 5.

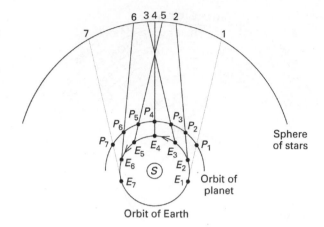

terms of the combined orbital motions of the earth and the planet rather than by an epicycle
(Fig. 13.19). The observed variation in the planets' distances from the earth also is more
easily understood in terms of the two orbits. Copernicus answered the objection to the earth's
motion around the sun, that it would cause the fixed stars to appear different at different times
of the year (the so-called annual parallax), by assuming that the radius of the earth's orbit
is so much smaller than the radius of the sphere of the stars that no such parallax could be
observed. Thus, one of the effects of Copernicus's theory was to vastly increase the size of
the universe.

After his basic introduction to the new system, Copernicus followed his mentor Ptolemy
by presenting an outline of the plane and spherical trigonometry necessary to solve the mathematical problems presented by the movements of the celestial bodies. Despite the advances
in trigonometric technique now available in Europe following the work of Regiomontanus,
Copernicus's own treatment stays very close to that of the second-century astronomer, even
to the use of chords. He did, however, make some concessions to the 1400 years of work
since the time of Ptolemy. First, he used 100,000 for his circle radius (now that Arabic numerals were in general use) rather than the 60 used by the ancients. Second, his table did
not give the chords of the various arcs but instead half the chords of twice the arcs "because
the halves come more frequently into use in demonstration and calculation than the whole
chords do,"[32] but he did not use the now common term "sine." Third, unlike Ptolemy, Copernicus did present specific methods for solving the various cases of both plane and spherical
triangles, rather than developing them ad hoc. However, his methods did not involve either
the plane or spherical law of sines. In general, he drew appropriate perpendiculars and then
dealt with solving right triangles. Still, his procedures were sufficient for the astronomical
work to follow.

In the remaining books of his treatise, Copernicus used his new sun-centered model, along
with both ancient and modern observations, to calculate the basic parameters of the orbits
of the moon and the planets. Reading these later books reveals that moving the center of the
universe away from the earth did not simplify Ptolemy's picture very much. Copernicus found
that simply placing the planets on sun-centered spheres did not satisfy the requirements of

FIGURE 13.20

Earth revolves around C_E, which rotates on an epicycle centered at O, where O revolves around the sun

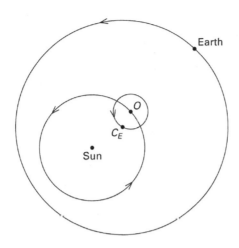

FIGURE 13.21

The 400th anniversary of the Gregorian calendar on a Vatican stamp

observation. Thus, he, like Ptolemy, introduced various mathematical devices. For example, Copernicus's calculations showed that the center of the earth's (circular) orbit was not the sun, but a point C_E in space that revolved on an epicycle whose center O revolved about the sun (Fig. 13.20). Similarly, the centers of the various planetary orbits were located neither at the sun nor even at the center of the earth's orbit. In the end, the full system as described in *De revolutionibus* was of the same order of complexity as that of Ptolemy.

The mathematical details of Copernicus's work made it unreadable to all but the best astronomers of his day, its primary audience. Over the next several decades, these mathematicians found that calculations of astronomical phenomena were simplified by applying Copernicus's theory and techniques. It was unnecessary to believe in the movement of the earth to use these techniques. Therefore, many people, both astronomers and educated laymen, took Copernicus's work merely as a mathematical hypothesis and not as a physical theory. In fact, the foreword to the printed text of *De revolutionibus*, written by Andreas Osiander, the Lutheran theologian who saw the book through the press, claimed that Copernicus's views on the earth's motion should not be taken as true but only as a hypothesis for calculation, "since the true laws cannot be reached by the use of reason."[33]

During the latter half of the sixteenth century, however, various churchmen, particularly Protestant clerics deeply involved in the fierce conflict with the Roman Catholic Church, began to express severe opposition to Copernicus's ideas because they explicitly contradicted various Biblical passages asserting the earth's stability. These Protestant leaders believed that the Roman Church had departed greatly from the views expressed in the Bible. They vehemently rejected any doctrines seen as deviating from the literal words of Scripture. During this same period, the Catholic Church itself had little to say about Copernicus's work. *De revolutionibus*, in fact, was taught at various Catholic universities, and the astronomical tables derived from it provided the basis for the reform of the calendar promulgated for the Catholic world by Pope Gregory XIII in 1582 (Fig. 13.21). Ironically, it was not until the seventeenth century, after most astronomers were convinced of the earth's movement by new evidence and a better heliocentric theory than that of Copernicus, that the Catholic Church brought its full power to bear against the heresy represented by the moving earth.

13.3.3   Tycho Brahe

One astronomer who used Copernicus's work as the basis for astronomical calculations was Erasmus Reinhold (1511–1553). In 1551 he issued the first complete set of astronomical tables prepared in Europe for over three centuries, generally called the Prutenic tables after his patron, the Duke of Prussia. These tables were markedly superior to the older ones, partly because they were based on more and better data. Nevertheless, they were not intrinsically more accurate than tables based on Ptolemy's work. There were still errors of a day or more in the prediction of lunar eclipses.

One way to improve the results no matter how one calculated the tables, however, was to have better observations. Tycho Brahe (1546–1601) was one astronomer who devoted much of his life to making these observations. To do so obviously required excellent instruments, which in return required funds. It was fortunate that in 1576 he was able to convince King Frederick II of Denmark to allow him the use of the island of Hveen near Copenhagen and to provide him with funds for building a magnificent observatory and also for hiring the assistants necessary to provide year-round observations using the newly constructed instruments (Fig. 13.22). Brahe was the first astronomer to realize the necessity for making continuous observations of the various planets. Although he eventually left Denmark and moved to Prague to work for the Austrian emperor Rudolph II, he was able to accumulate enormous amounts of data over a 25-year period, generally accurate to within a couple of minutes of arc—an accuracy far in excess of the best work of any of the ancients.

Two of Brahe's most important series of observations convinced him that the Ptolemaic system with its Aristotelian philosophy could not be correct. First, beginning in late 1572 he tracked for 16 months a new object that had appeared in the heavens, a nova. Because this object did not move with respect to the sphere of the stars—Brahe demonstrated this by very precise observation—he concluded that it belonged to the region of the fixed stars. Hence, despite Aristotle, change was possible in the heavens, and therefore one distinction between the earth and the heavens was removed. The possibility of change in the heavens was further confirmed by his observation of a comet in 1577. Again, by a comparison of the parallax of the comet with that of the moon and the planets, he concluded that the comet lay beyond the moon and that it revolved around the sun at a distance greater than that of Venus. Since its distance from the sun apparently varied greatly during the course of his observations, Brahe further concluded that the heavens could not be filled by solid spheres carrying the planets. There must in fact be space between the planets in which another heavenly object could travel.

FIGURE 13.22

Tycho Brahe's observatory, with one of his quadrants and the nova of 1572, on a stamp from Ascension Island

13.3.4   Johannes Kepler and Elliptical Orbits

Brahe was primarily an excellent observer rather than a theoretician. He did devise a model of the universe "intermediate" between that of Ptolemy and Copernicus, in which all of the planets except the earth traveled around the sun while the whole system revolved around the central immovable earth, but he was not able to elaborate it mathematically. Johannes Kepler (1571–1630), who worked with Brahe for the final two years of his life in Prague, was the astronomer able to use the mass of Brahe's observations to construct a new heliocentric theory that could accurately predict heavenly events without the elaborate machinery of epicycles.

It was perhaps his theological training, combined with a philosophical bent, that provided Kepler with the goal from which he never wavered, of discovering the mathematical rules God

BIOGRAPHY

Johannes Kepler (1571–1630)

Kepler was born in Weil-der-Stadt in southwest Germany and studied in the University of Tübingen where he became acquainted with Copernicus's theory and convinced himself that in essence it represented the correct system of the world. Although he had originally planned to become a Protestant minister, fate intervened, and he was recommended by the University to fill a job as mathematics professor at the Protestant school in the Austrian town of Graz. When the school was closed several years later and all Protestant officials were exiled because of the Counter Reformation, an exception was made in Kepler's case. He was allowed to return and to spend time thinking about mathematics and astronomy. Kepler knew that to work out in complete detail a correct version of Copernicus's theory, he had to have access to the observations of Tycho Brahe. He therefore began a correspondence with the Dane, which finally resulted in his being appointed his assistant in Prague by Emperor Rudolph II. Although Brahe died about 18 months after Kepler's arrival, Kepler had by this time learned enough about Brahe's work to be able to use the material in working on his own major project. He was himself appointed as Imperial Mathematician to succeed Tycho Brahe and spent the next 11 years in Prague (Fig. 13.23).

FIGURE 13.23

Kepler and his system of the universe on a Hungarian stamp

used for creating the universe. As he stated in his earliest work, the *Mysterium cosmographicum* (*The Secret of the Universe*) of 1596, "Quantity was created in the beginning along with matter."[34] In a note to the second edition of 1621, Kepler clarified what he meant: "Rather the ideas of quantities are and were coeternal with God, and God himself. . . . On this matter the pagan philosophers and the Doctors of the Church agree."[35] Throughout his life, Kepler attempted through both philosophical analysis and prodigious calculation to demonstrate the numerical relationships with which God had created the universe. His goal appeared to be nothing less than to reconfirm on a higher level the Pythagorean doctrine that the universe is made up of number. Taking as his starting point Copernicus's placing of the sun at the center of the universe, he was able to discover the three laws of planetary motion, today known as **Kepler's laws**, and many other relationships that we tend to dismiss as mystical.

Kepler discussed one of these relationships in great detail in the *Mysterium cosmographicum*: Why are there precisely six planets? Because "God is always a geometer," the Supreme Mathematician wanted to separate the planets with the regular solids. Euclid had proved that there could be only five such solids, so Kepler took this as the reason God chose to provide just six planets. He then worked out the idea that between each pair of spheres containing the orbits of adjacent planets there was inscribed one of the regular solids (Fig. 13.24). Thus, inside the sphere of Saturn was to be inscribed a cube, which in turn circumscribed the orbit of Jupiter. Similarly, between the orbits of Jupiter and Mars was a tetrahedron, between Mars and Earth a dodecahedron, between Earth and Venus an icosahedron, and between Venus and Mercury an octahedron. These solids lay in the interspherical spaces, and their sizes provided a measure of the relationship between the sizes of the various planetary orbits. For example, Kepler noted that the diameter of Jupiter's orbit is triple that of Mars, while the ratio of the diameter of the sphere circumscribed about the tetrahedron is triple that of the sphere inscribed in the tetrahedron. Not all of the values came out exactly correct. There was still some discrepancy. But even this fact did not bother Kepler too much. He gave various reasons why

FIGURE 13.24

Kepler's regular solids representing the orbits of the planets, on a plate from the *Mysterium cosmographicum* of 1621. (*Source*: Courtesy of the Department of Special Collections, Stanford University Libraries)

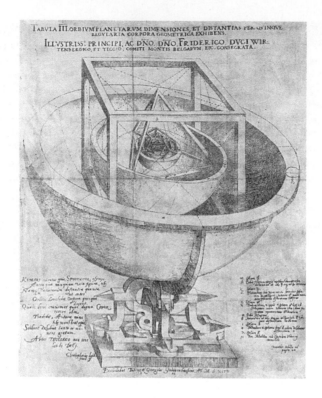

the values could not be expected to be exact, including the fact that even the data from the Prutenic tables was not entirely accurate. He was so convinced of the correctness of his basic idea that such discrepancies were of little moment. Kepler's views on this matter were not a mere function of his youth. In fact, he returned time and again to this basic proposition, each time attempting to adduce new reasons for its correctness.

Kepler also interested himself in the ratios of the sizes of the planets themselves. In his *Epitome astronomiae Copernicanae* (*Epitome of Copernican Astronomy*) (1618), he noted, "Nothing is more in concord with nature than that the order of magnitudes [of the sizes] should be the same as the order of the spheres."[36] In other words, Mercury should be the smallest planet and Saturn the largest. But what ratios should exist among these sizes? He presented several possibilities. Since Saturn is approximately 10 times as far from the sun as the earth, he claimed that either Saturn's diameter is ten times that of the earth, or its surface area is ten times that of the earth, or its mass is ten times that of the earth. To choose among these possibilities, he referred to certain new telescopic observations and picked the third option. It followed that the ratios of the diameters are as the cube roots of the distance, while the ratios of the surface areas are as the square of the cube roots, provided that the density of Saturn was the same as that of the earth. Kepler had little evidence to check whether his theoretical statements were true—there was no way he could measure the mass of a planet— so the assertion remained simply a theory. But as in the case of the sizes of the orbits, Kepler was convinced that a simple numerical relationship had to hold.

Kepler was well schooled in music. Thus, he would have been very familiar with the Pythagorean ratios of string lengths, which give consonant harmonies: a ratio of $1:2$ is that of an octave, $2:3$ a fifth, $3:4$ a fourth, and so on. In his *Harmonices mundi* (*Harmonies of the World*) (1619), Kepler attempted to assign these harmonic ratios to various numbers connected with the different planets. First, he tried the periods of revolution, but these did not give any harmonic ratios. Next, he tried the volumes of the planets, the greatest and smallest solar distances, the extreme velocities, and the variations in time needed by a planet to cover a unit length of its orbit. Nothing appeared to work. Finally, after a lengthy argument, Kepler hit on the "right" numbers, the apparent daily angular movements of the planets as seen from the sun. Thus, the daily movement of Saturn at aphelion (the point on its orbit farthest from the sun) is $1'46''$, while its daily movement at perihelion (the point closest to the sun) is $2'15''$. The ratio between these two values is approximately $4:5$, a major third. The corresponding ratio for Mars is $26'14'' : 38'1''$ or approximately $2:3$, a fifth. Not only did Kepler find consonances between the extreme movements of the individual planets, but he also found them between movements of different planets. The ratio of Saturn at perihelion to Jupiter at aphelion turned out to be $1:2$, an octave. Further, when he transposed a particular set of these relations into a common key, Kepler found a major scale beginning with the aphelion of Saturn and a minor one beginning with Saturn's perihelion. Kepler included in his book the various notes "played" by the planets, both singly and together, concluding with several multipart harmonies: "Accordingly the movements of the heavens are nothing except a certain everlasting polyphony. . . . Hence it is no longer a surprise that man, the ape of his Creator, should finally have discovered the art of singing polyphonically . . . in order that he might play the everlastingness of all created time in some short part of an hour by means of an artistic concord of many voices and that he might to some extent taste the satisfaction of God the Workman with His own works, in that very sweet sense of delight elicited from this music which imitates God."[37]

The reader may well wonder if Kepler could indeed be thought of as a scientist, given his propensity toward what we might today call mysticism. The answer, however, is a resounding yes. Kepler was responsible for some of the most important astronomical discoveries of his time. There is a direct line from his three laws of planetary motion to the fundamental work of Newton on the laws of motion.

Kepler announced in his *Mysterium cosmographicum* that among his goals was to discover the "motion of the circles" of the planets, that is, to determine their orbits. He gave numerous arguments in that work for the basic correctness of the Copernican system, but by the end of the century he realized that Copernicus's mathematical details did not give the complete solution to the problem. For example, Copernicus still treated the earth as special rather than as just another planet. To correct Copernicus's work, Kepler knew that he needed better observational data—data that could come only from Tycho Brahe. With these finally in hand by 1601, Kepler could proceed to determine the exact details of the planetary orbits. He began with the case of Mars, because that planet's orbit had always been the most difficult to comprehend. If he could understand Mars's orbit, Kepler believed, he could understand them all.

In his *Astronomia nova* (*New Astronomy*) of 1609, Kepler described his eight years of detailed calculations, false starts, stupid mistakes, and continued perseverance to calculate the orbit of Mars. He first decided that he needed accurate parameters for the earth's own

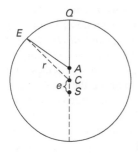

FIGURE 13.25

Kepler's assumption for the earth's orbit

orbit, because his overriding Copernican theory was that the motion of Mars was viewed from a moving earth. Kepler took the earth's orbit as a circle with radius r centered on a point C, with the sun at a point S making $CS = e = 0.018r$ (Fig. 13.25). (The orbit of the earth is very close to circular, so the assumption that it was a circle did not lead Kepler astray.) Furthermore, there was another point A on the diameter CS, with $AC = CS$, such that $\angle EAQ$ varied uniformly with time, where Q is the earth's aphelion. In other words, he reintroduced the equant that Copernicus had rejected. Because the earth moved with a constant angular velocity on its orbit with respect to A, its linear velocity necessarily changed as its distance from the sun changed. Kepler showed that the earth's velocity near aphelion and perihelion varied inversely with its distance from the sun, and then generalized this result to the rest of the orbit. (Unfortunately, that rule turned out to be incorrect. As Kepler understood later, it is the component of the planet's velocity perpendicular to the radius vector that varies inversely with the distance from the sun.)

Unlike Ptolemy or Copernicus, however, Kepler was interested not only in the pure mathematics of the celestial motions, that is, in "saving the appearances," but in the physics as well. He was trying to describe the actual orbit of the earth through space and so wanted to know what caused the earth to move, what kept it in its orbit, and why the velocity changed with the distance to the sun. Having read the work of William Gilbert, *On Magnets* (1601), Kepler settled on the fact that some force emanating from the sun acts on the planet and sweeps it around in its orbit. He could understand this force acting on the earth as it moved around in a circle much better than he could see it acting on a planet moving on an epicycle. It also made sense that, like magnetic force, the sun's force weakened with distance, so that the planet's velocity was smaller at a greater distance. This change from a mathematical to a physical point of view was one of the reasons Kepler felt comfortable reintroducing the equant as well as rejecting epicycles.

Returning to the motion of Mars, Kepler began by using his earlier assumption of the circularity of its orbit, because it at least provided approximately correct results. His aim was to calculate the relation between the length of arc QP traveled by the planet after aphelion Q and the time it takes to traverse that arc. He knew that the planet moved more slowly the farther it was from the sun. Since the calculation of the exact relationship between velocity and arc, however, was beyond his capabilities, Kepler resorted to approximation. His assumption, now taken for Mars as well as for the earth, that the planet's velocity varied inversely with the length of the radius vector, implied that the time required to pass over an (infinitesimal) arc was proportional to that vector. The time could therefore be represented by the radius vector, with appropriate choice of units. Kepler then argued that the total time required to pass over a finite arc QP could be thought of as the sum of the radius vectors making up that part of the circle, or as the area swept out by the radius vector (Fig. 13.26). Kepler realized that such an infinitesimal argument was not rigorous, but he stated it anyway as a law based on the incorrect circular orbit and the incorrect velocity law: The radius vector sweeps out equal areas in equal times. This law is generally referred to as **Kepler's second law**, because it is today regarded as a supplement to the first law. Interestingly enough, Kepler made no attempt to prove it differently even when he discovered that the correct planetary orbit was an ellipse.

That the shape of the orbit is an ellipse is the content of Kepler's first law. Kepler informed us fully how he discovered this law as well. Having worked out the orbit of the earth, he made

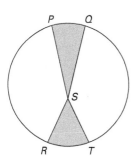

FIGURE 13.26

Kepler's second law: A planet sweeps out equal areas in equal times. The time of the planet's motion from Q to P is equal to that from R to T when area SPQ is equal to area SRT.

various calculations of the distances of Mars to the supposed center of its orbit and found that they were larger near aphelion and perihelion and smaller during the remainder of the orbit. Thus, the circularity of the orbit was impossible. Kepler concluded that the orbit had to be an oval of some sort. It was somewhat strange to reject the comforting circularity of the Greeks and replace it with a rather vaguely shaped oval, because such a curve would seemingly destroy all possibility of the "harmony of the spheres" for which Kepler had been scarching. Nevertheless, Kepler began the long process of calculating the exact shape of the oval.

After two years of calculation, the result appeared to Kepler virtually by accident. To aid in certain computations, he had been approximating the oval by an ellipse. He noted that the distance AR between the circumference of the circle and the end of the minor axis of the ellipse was equal to 0.00429 (the radius of the circle being set at 1), which turned out to be $(1/2)e^2$ where $e = CS$ was the distance between the center of the circle and the sun (Fig. 13.27). It followed that the ratio

$$CA : CR = 1 : \left(1 - \frac{e^2}{2}\right) \approx 1 + \frac{e^2}{2} = 1.00429.$$

What struck Kepler about this number was that he had seen it before. It was equal to the secant of $5°18'$, the value of the angle ϕ between the directions AC and AS, where A is the point on the circle 90° away from the aphelion point Q. The secant in this case is the ratio of the length of the radius vector to its projection onto a diameter. Realizing that $CA : CR \approx SA : SB \approx SA : CA$, Kepler then had the brilliant inspiration that when the angle between CQ and a direction CP had any value β (and not just 90°), the ratio of the distance SP to that actual sun-Mars distance was also the ratio of SP to its perpendicular projection PT on a diameter. In other words, he realized that the actual sun-Mars distance was PT, where $PT = PC + CT = 1 + e \cos \beta$. The remaining question for Kepler was how to lay off this distance. Kepler first decided to lay it off with one end at the sun and the other on the radius vector PC, that is, to make $SV = PT$. Unfortunately, the curve so traced turned out to be not quite in accord with observation.

FIGURE 13.27

Kepler's derivation of the elliptical orbit

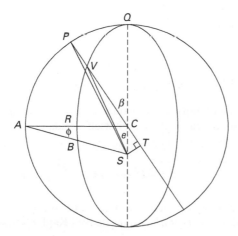

Kepler finally discovered the correct result, that the distance given by $\rho = 1 + e \cos \beta$ should be laid off from the sun so that the endpoint is on a line perpendicular to the line CQ, where β is the angle that CQ makes with the line from C to the intersection W of that perpendicular with the auxiliary circle (Fig. 13.28). (One should note that the difference between Kepler's first idea and this one is extremely small, producing discrepancies of at most about 5′ of arc.) Kepler was able to demonstrate that the curve he now produced was an ellipse, using an argument summarized here in modern notation. Assume that an ellipse is centered at C with $a = 1$ and $b = 1 - \frac{e^2}{2}$, where $e = CS$. This ellipse can be thought of as being formed from the circle of radius 1 by reducing all the ordinates perpendicular to QC in the proportion b. If v represents the angle at S subtended by the arc RQ, then $\rho \cos v = e + \cos \beta$ and $\rho \sin v = b \sin \beta$. Squaring the two equations and adding gives

$$\rho^2 = e^2 + 2e \cos \beta + \cos^2 \beta + \left(1 - \frac{e^2}{2}\right)^2 \sin^2 \beta$$

$$= e^2 + 2e \cos \beta + 1 - e^2 \sin^2 \beta + \frac{e^4}{4} \sin^2 \beta.$$

FIGURE 13.28

Kepler's proof that the curve of the orbit is an ellipse

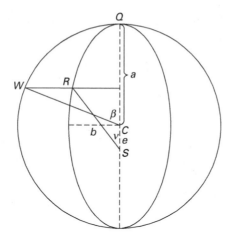

Neglecting the term in e^4 then produces the result

$$\rho^2 = 1 + 2e \cos \beta + e^2 \cos^2 \beta = (1 + e \cos \beta)^2.$$

Thus, the equation of the ellipse can be written as $\rho = 1 + e \cos \beta$, exactly the same equation as already derived for the curve of the orbit itself. In addition, the distance c of the center of the ellipse from the focus is given by

$$c^2 = 1 - b^2 = 1 - \left(1 - \frac{e^2}{2}\right)^2 = e^2,$$

if again the term in e^4 is ignored. It follows that the sun is at one focus of the ellipse and that e is the eccentricity (Fig 13.29). We have now derived **Kepler's first law** of planetary motion:

A planet travels in an ellipse around the sun with the sun at one focus. Kepler himself, having derived the law for the case of Mars, merely checked it briefly for the other planets before asserting its general validity.[38]

Kepler's third law appeared for the first time in the *Harmonice mundi*, stated as an empirical fact. In some sense it was a culmination of the work begun earlier in the *Mysterium cosmographicum* because it provided another answer to the general questions Kepler had asked regarding the size and motions of the orbits: "It is absolutely certain and exact that the ratio which exists between the periodic times of any two planets is precisely the ratio of the $\frac{3}{2}$th power of the mean distances [of the planet to the sun]."[39] Kepler discovered the law by studying more of Tycho Brahe's measurements, but never gave any derivation of it from other principles.

Kepler's three laws of planetary motion had great consequences in the development of astronomical as well as physical theory. Their discovery provides an excellent example of the procedures used by scientists. They need some theory to begin with, but then must always compare the results of the theory with the results of observation. If they have confidence in their observations, and these do not agree with the predictions of their theory, they must modify the theory. Kepler did this often until he finally reached theoretical results agreeing with his observations. He spent years performing the necessary calculations. Toward the end of his life, however, the invention of logarithms greatly simplified Kepler's calculations as well as those of other astronomers.

FIGURE 13.29

Kepler's working out of elliptical orbits on a German stamp

 13.4   LOGARITHMS

The idea of the logarithm probably had its source in the use of certain trigonometric formulas, which transformed multiplication into addition or subtraction. Recall that if one needed to solve a triangle using the law of sines, a multiplication and a division were required. Because sines were generally calculated to seven or eight digits (using a circle of radius 10,000,000 or 100,000,000), these calculations were long and errors were often made. Astronomers realized that it would be simpler and reduce the number of errors if one could replace the multiplications and divisions by additions and subtractions. To accomplish this task, sixteenth-century astronomers often used formulas such as $2 \, \text{Sin} \, \alpha \, \text{Sin} \, \beta = \text{Cos}(\alpha - \beta) - \text{Cos}(\alpha + \beta)$. Thus, if one wanted to multiply 4,378,218 by the sine of $27°15'22''$, one determined α such that $\text{Sin} \, \alpha = 2,189,109$, set $\beta = 27°15'22''$ and used a table to determine $\text{Cos}(\alpha - \beta)$ and $\text{Cos}(\alpha + \beta)$. The difference of these two latter values was then the desired product, found without any actual multiplying.

A second, more obvious, source of the idea of a logarithm was probably found in the work of such algebraists as Stifel and Chuquet, who both displayed tables relating the powers of 2 to the exponents and showed that multiplication in one table corresponded to addition in the other. But because these tables had increasingly large gaps, they could not be used for the necessary calculations. Around the turn of the seventeenth century, however, two men working independently, the Scot John Napier (1550–1617) and the Swiss Jobst Bürgi (1552–1632) came up with the idea of producing an extensive table that would allow one to multiply any desired numbers together (not just powers of 2) by performing additions. Napier published his work first.

SIDEBAR 13.1   *The Modern Notation for Decimal Fractions*

Napier is primarily responsible for the introduction of our modern notation for decimal fractions. Stevin had detailed the idea, along with a suggestion for a notation. But Napier, near the beginning of the *Constructio*, after noting that accuracy of computation requires the use of large numbers like 10,000,000 as the base for a table of sines, wrote: "In computing tables, these large numbers may again be made still larger by placing a period after the number and adding ciphers. . . . In numbers distinguished thus by a period in their midst, whatever is written after the period is a fraction, the denominator of which is unity with as many ciphers after it as there are figures after the period."[40] For example, he wrote, 25.803 is the same as $25\frac{803}{1000}$ and 9999998.0005021 means $9999998\frac{5021}{10,000,000}$. The publication of Napier's tables, in which these decimal fractions appeared, soon resulted in their general use throughout Europe. It had taken about 400 years since the introduction into Europe of the Hindu-Arabic numbers for the complete decimal place value system to be generally accepted.

13.4.1   The Idea of the Logarithm

Napier's logarithmic tables first appeared in 1614 in a book entitled *Mirifici logarithmorum canonis descriptio* (*Description of the Wonderful Canon of Logarithms*). This work contained only a brief introduction, showing how the tables were to be used. His second work on logarithms, describing the theory behind the construction of the tables, *Mirifici logarithmorum canonis constructio* (*Construction of the Wonderful Canon of Logarithms*) appeared in 1619, two years after his death. In this latter work appears his imaginative idea of using geometry to construct a table for the improvement of arithmetic.

Realizing that astronomers' calculations involved primarily trigonometric functions, especially sines, Napier aimed to construct a table by which multiplications of these sines could be replaced by addition. For the definition of logarithms Napier conceived of two number lines, on one of which an increasing arithmetic sequence, 0, b, $2b$, $3b$, . . . is represented, and on the other a sequence whose distances from the right endpoint form a decreasing geometric sequence, ar, a^2r, a^3r, . . . , where r is the length of the second line (Fig. 13.30). (Napier chose r to be 10,000,000, because that was the radius for his table of sines, and a to be a number smaller than but very close to 1 (Sidebar 13.1).) The points on this second line can be marked 0, $r - ar$, $r - a^2r$, $r - a^3r$, . . . , with these values representing sines of certain angles.

FIGURE 13.30

Napier's moving points

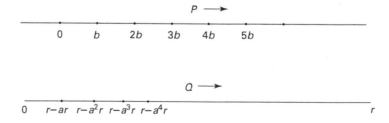

Napier now considered points P and Q moving to the right on each line as follows: P moves on the upper line "arithmetically" (that is, with constant velocity). Thus, P covers each equal interval $[0, b]$, $[b, 2b]$, $[2b, 3b]$, ... in the same time. Q moves on the lower line "geometrically." Its velocity changes so that it too covers each (decreasing) interval $[0, r - ar]$, $[r - ar, r - a^2r]$, $[r - a^2r, r - a^3r]$, ... in the same time. The distances traveled in each interval form a decreasing geometric sequence $r(1 - a)$, $ar(1 - a)$, $a^2r(1 - a)$, ..., each member of which is the same multiple of the distance of the left endpoint of the interval to the right end of the line. Because distances covered in equal times have the same ratios as the velocities, it follows that the point's velocity over each interval is proportional to the distance of the beginning of that interval from the right end of the line. It appears that Napier initially thought of the velocity of the lower point as changing abruptly when it passed each marked point, remaining constant in each of the given intervals. In his definition of logarithm, however, Napier smoothed out these changes by considering the second point's velocity as changing continuously (without, naturally, using that terminology). Thus, a point moves geometrically if its velocity is always proportional to its distance from the right end of the line. For Napier, "the logarithm of a given sine is that number which has increased arithmetically with the same velocity throughout as that with which radius began to decrease geometrically, and in the same time as radius has decreased to the given [number]."[41] In other words, if the upper point P began to move from 0 with constant velocity equal to that with which the lower point Q also began to move (geometrically) from 0, and if P has reached y when Q has reached a point whose distance from the right endpoint (radius) is x, then y is said to be the **logarithm** of x.

In modern calculus notation, Napier's idea is reflected in the differential equations

$$\frac{dx}{dt} = -x, \qquad x(0) = r; \qquad \frac{dy}{dt} = r, \qquad y(0) = 0.$$

The solution to the first equation is $\ln x = -t + \ln r$, or $t = \ln \frac{r}{x}$. Combining this with the solution $y = rt$ of the second equation shows that Napier's logarithm y (here written as $y = \text{Nlog } x$) may be expressed in terms of the modern natural logarithm as $y = \text{Nlog } x = r \ln \frac{r}{x}$. Napier's logarithm is thus closely related to the natural logarithm. It does not, however, share the common properties of the natural logarithm since, for example, its value decreases when the value of x increases.

13.4.2   The Use of Logarithms

Although Napier's definition is somewhat different from the modern one, he nevertheless was able to derive important properties of logarithms analogous to those of our modern logarithm as well as to show how to construct a table of logarithms of sines. He began by noting that the definition implied immediately that $\text{Nlog } r = 0$, for the upper point will not have moved at all. Napier in fact realized that he could have assigned 0 to be the logarithm of any fixed number, but, he wrote, "it was best to fit it to the whole sine, that the addition or subtraction of the logarithm which is most frequent in all calculations, might never after be any trouble to us."[42] Similarly, if $\frac{\alpha}{\beta} = \frac{\gamma}{\delta}$, then $\text{Nlog } \alpha - \text{Nlog } \beta = \text{Nlog } \gamma - \text{Nlog } \delta$. This result also follows from the definition because the geometrical motion of the lower point implies that its time to travel from α to β equals its time to travel from γ to δ. From this result follow rules enabling one to use logarithms in calculation. For example, if $x : y = y : z$, then $2 \text{ Nlog } y = \text{Nlog } x + \text{Nlog } z$, and if $x : y = z : w$, then $\text{Nlog } x + \text{Nlog } w = \text{Nlog } y + \text{Nlog } z$. On the other hand, Napier did not show us how to calculate the logarithm of a product, probably because he was not interested in pure multiplications as such. He constructed his logarithms with trigonometry in mind, and many calculations involved in the solving of triangles require the finding of a fourth proportional, for which his rules indeed apply.

As an example of this type of calculation, consider the right triangle whose hypotenuse c and leg a are known. The problem is to find the angle α opposite the given leg. Napier made use of the basic trigonometric relation

$$\frac{\text{Sin } \alpha}{r} = \frac{a}{c},$$

where $r = 10^7$ is the radius of the circle in which the sines are defined. Napier used his table and the rule for proportions given above to calculate Nlog Sin α = Nlog a − Nlog c + Nlog r. Because Nlog $r = 0$, he had found the logarithm of the sine of α in terms of the logarithms of the sides. Reading his table in reverse gives the desired angle. Note that although Napier's table is a table of logarithms of sines, he used it to calculate the logarithms of the numerical lengths needed in this problem by looking in the table for a sine that was close enough to the desired number, making appropriate adjustments for the number of digits in one or the other, and then taking the logarithm of that sine value.

As another example, consider Napier's use of the law of sines

$$\frac{\text{Sin } \alpha}{a} = \frac{\text{Sin } \beta}{b},$$

to solve a plane triangle given two sides a, b and the angle α opposite side a. The logarithm property gave Nlog Sin β = Nlog Sin α + Nlog a − Nlog b. Thus, Napier could read his table in reverse to find the two possible values for β, one less than a right angle and one greater. Finally, to solve a triangle given two sides a, b and the included angle γ, Napier did not use the standard method of dropping a perpendicular, because that method is not suited to logarithmic calculation. Instead, he made use of the **law of tangents**:

$$\frac{a + b}{a - b} = \frac{\tan \frac{1}{2}(\alpha + \beta)}{\tan \frac{1}{2}(\alpha - \beta)}.$$

If γ is given, then $\alpha + \beta$ is known. Applying logarithms to this proportion allowed him to find $\tan \frac{1}{2}(\alpha - \beta)$, therefore $\frac{1}{2}(\alpha - \beta)$, and therefore both α and β.

How did Napier calculate the logarithm of a tangent from his table of logarithms of sines? To answer this question, consider the following line from Napier's actual table, which included seven columns for each minute of arc from 0° to 45°:

34°40′   5688011   5642242   3687872   1954370   8224751   55°20′

The first column gives the value of an arc (or angle), while the second gives the sine of that arc. The final column gives the arc that is complementary to that in the first column, while the sixth column gives its sine. It follows that the sixth column gives the sine of the complement of the arc of the first column, that is, the cosine of that arc. The third and fifth columns give Napier's logarithms of the sines in the second and sixth columns, respectively, or, as Napier also notes, the logarithms of the sine complements of the sixth and second columns, respectively. Finally, the middle column represents the difference of the entries in the third and fifth columns, or Napier's logarithm of the tangent of the arc of the first column.

Because the logarithm of 10,000,000 is 0, logarithms of numbers greater than 10,000,000 must be negative and are defined by simply reversing the directions of the moving points in the original definition. These numbers, of course, cannot represent sines but can represent tangents or secants. In this case, the negative of the logarithm in the middle column is the logarithm of the tangent of $55°20'$, while the negative of the logarithm in the third column is the logarithm of the secant of that same angle.

Napier's actual construction of his table of logarithms took him twenty years.[43] And even though this work was done in the era of hand calculations, there were remarkably few errors. Late in his life, however, Napier decided that it would be more convenient to have logarithms whose value was 0 at 1 rather than at 10,000,000. In that case the familiar properties of logarithms, $\log xy = \log x + \log y$ and $\log \frac{x}{y} = \log x - \log y$, would hold. Furthermore, if the logarithm of 10 were set at 1, the logarithm of $a \times 10^n$, where $1 \le a < 10$, would simply be n added to the logarithm of a. Napier died before he could construct a new table based on these principles, but Henry Briggs (1561–1631), who discussed this matter thoroughly with Napier in 1615, began the calculation of such a table. Rather than simply convert Napier's logarithms to these new "common" logarithms by simple arithmetic procedures, however, Briggs worked out the table from scratch. Starting with $\log 10 = 1$, he calculated successively $\sqrt{10}, \sqrt{\sqrt{10}}, \sqrt{\sqrt{\sqrt{10}}}, \ldots$, until after 54 such root extractions he reached a number very close to 1. All of these calculations were carried out to 30 decimal places. Since $\log \sqrt{10} = 0.5000$, $\log \sqrt{\sqrt{10}} = 0.2500, \ldots$, $\log(10^{\frac{1}{2^{54}}}) = \frac{1}{2^{54}}$, he was able to build up a table of logarithms of closely spaced numbers using the laws of logarithms. Briggs's table, completed by Adrian Vlacq in 1628, became the basis for nearly all logarithm tables into the twentieth century. Astronomers very quickly discovered the great advantages of using logarithms for calculations. Logarithms became so important that the eighteenth-century French mathematician Pierre-Simon de Laplace was able to assert that the invention of logarithms, "by shortening the labors, doubled the life of the astronomer."

13.5   KINEMATICS

The final mathematical arts of John Dee that we will consider are those that deal with motion. Thus, "Statics is an art mathematical which demonstrates the causes of heaviness and lightness of all things and of motions and properties to heaviness and lightness belonging,"[44] while "Trochilike . . . demonstrates the properties of all circular motions, simple and compound."[45] The man generally considered to be the founder of modern physics, Galileo Galilei (1564–1642), was responsible in large measure for reformulating the laws of motion considered first by the Greeks and later by certain medieval scholars. But like his predecessors, he proposed to use geometry, not algebra, to explicate his ideas. Although he did work in what today is generally called statics, his most important new ideas, dealing with the "natural" accelerated motion of free fall and the "violent" motion of a projectile, were published in 1638 in his *Discourses and Mathematical Demonstrations Concerning Two New Sciences*. Galileo thus applied mathematics to the study of motion on earth, much as Copernicus and Kepler had applied it to the study of motion in the heavens.

BIOGRAPHY

Galileo Galilei (1564–1642)

Galileo studied at the University of Pisa from 1581 to 1585 ostensibly for a medical degree at the request of his father (Figure 13.31). But he was more interested in mathematics and ultimately left the university without any degree. His training in mathematics was, however, the classical one. He mastered Aristotle and Euclid and read some of the works of Archimedes. Thus, he was well versed in Eudoxian proportion theory, but evidently had little knowledge of the algebra of Cardano or Bombelli or the more recent work of Viète. Nevertheless, he was convinced of the importance of mathematics, particularly geometry, in the study of natural phenomena.

Galileo is today probably most famous for his clash with the Catholic Church over his publication of the *Dialogue Concerning the Two Chief World Systems* (1632), in which he presented the arguments for and against both the Ptolemaic and the Copernican theories of the universe. As noted in the opening of the chapter, Galileo had been warned by church authorities in 1616 that the Church's official position was that the earth did not move and that Galileo must not hold or defend such views. Galileo in his book therefore took some pains to present the Copernican position as a hypothetical one and simply to consider its consequences as well as the failings of the traditional Ptolemaic position. Nevertheless, a careful reading of the text shows that in fact Galileo was convinced of the truth of the earth's motion around the sun—not surprising at this date—and made the defenders of the older position in his *Dialogue* appear foolish.

Still, Galileo believed that scientific and religious truths were compatible. As he had written in 1615, when "we have arrived at any certainties in physics, we ought to utilize these as the most appropriate aids in the true exposition of the Bible and in the investigation of those meanings which are necessarily contained therein, for these must be concordant with demonstrated truths."[46] It is thus unfortunate that the church leaders in the 1630s were equally as stubborn as Galileo and were convinced that any challenge to the current interpretation of the Bible must be confronted directly. Thus, in 1633 Galileo was brought before the Inquisition in Rome and forced to confess his error. He was then sentenced to house imprisonment for the remainder of his life and forbidden to publish any more books. He did manage, however, to publish his most important work, the *Discourses and Mathematical Demonstrations Concerning Two New Sciences* in 1638 by sending the manuscript beyond the reach of the Inquisition to Leiden in the Netherlands, where it was printed by the publishing house of the Elseviers.

Even though the Church banned the *Dialogue*, so many copies were already in circulation that it was impossible for its effect to be negated. Thus, the Italian public, as well as readers elsewhere, soon were convinced that the Copernican system was in fact true and that eventually even the Church would have to acknowledge that its interpretation of certain statements in the Bible must be changed.

FIGURE 13.31

Galileo on an Italian stamp

13.5.1   Accelerated Motion

Galileo wrote the *Two New Sciences* in the form of a dialogue among three people, whose discussion of motion is carried out around the framework of a formal treatise on motion written in the Euclidean format, including a definition, a postulate, and many theorems and proofs. The definition reads, "Motion is equably or uniformly accelerated which, abandoning rest, adds on to itself equal momenta of swiftness in equal times."[47] The definition is essentially the same as that put forth by Heytesbury 300 years earlier. Galileo, however, made two major advances. First, he discovered by 1604 that uniformly accelerated motion is precisely that of a freely falling body, and second, he worked out numerous mathematical consequences of this fact, some of which he could confirm by experiment.[48]

At one time Galileo believed that the velocity of a falling body increased in the ratio of the distances fallen rather than in the ratio of the times elapsed. In the *Two New Sciences*, he gave an argument showing that this first possibility is erroneous. First, he noted that if two different velocities of a given body are proportional to the distances covered while the body has each velocity in turn, then the times for the body to cover those distances are equal. This statement is virtually obvious for velocities constant over the given period of time. Galileo then assumed it to be true also for continuously changing velocities. Thus, "if the speeds with which the falling body passed the space of four *braccia* were the doubles of the speeds with which it passed the first two *braccia*, as one space is double the other space, then the times of those passages are equal."[49] Galileo was here comparing two (infinite) sets of velocities, those at each instant in which the falling body passed a point in the first two *braccia* (an Italian measure of distance) with those at each instant it passed a point in the first four *braccia* twice as far from the point of origin. His statement that the total times are equal is the result of applying the argument for finite times to infinitesimal times and adding up the entire set of these infinitesimal times. Galileo concluded that it is ridiculous that a given fallen body starting from rest could cover both two *braccia* and four *braccia* in the same time, and thus that it is false that speed increases as the distance traveled.

Galileo's argument by comparing two infinite sets provides one of the first such arguments in mathematical history—Archimedes had used a similar argument in his *Method*—but one that he used in other contexts as well. In particular, he used it in his proof of the mean speed rule.

THEOREM *The time in which a certain space is traversed by a moveable in uniformly accelerated movement from rest is equal to the time in which the same space would be traversed by the same moveable carried in uniform motion whose degree of speed is one-half the maximum and final degree of speed of the previous, uniformly accelerated, motion.*[50]

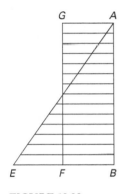

FIGURE 13.32

Galileo's proof of the mean speed theorem

With AB representing the time of travel, EB representing the maximum speed attained by the moveable, and F the midpoint of BE, Galileo constructed right triangle ABE and rectangle $ABFG$, whose areas are equal (Fig. 13.32). There are then one-to-one correspondences between the instants of time represented by points of the line AB and the parallels in the triangle representing the increasing degrees of speed on the one hand, and those instants and the parallels in the rectangle representing the equal speeds at half the final speed on the other. Galileo concluded that "there are just as many momenta of speed consumed in the accelerated motion according to the increasing parallels of triangle AEB as in the equable motion according to the parallels of the [rectangle] GB,"[51] because the deficit above the halfway point is made up by the surplus below it. Since these "momenta" of speed for each instant of time are proportional to the distances traveled in those instants, it follows that the total distance in each case is the same. As before, Galileo used an argument with infinitesimals. One may wonder why he believed in such arguments, given that they violated classic geometric concepts. But use them he did.

As a corollary to the theorem, Galileo proved that in the case of a moveable falling from rest, the distances traveled in any times are as the squares of those times. That is, he showed that if the body falls a distance d_1 in time t_1, and d_2 in time t_2, then $d_1 : d_2 = t_1^2 : t_2^2$. Galileo's result is written in modern notation as $d = kt^2$, but Galileo always used Euclidean proportionality concepts instead of modern "function" concepts. To prove this corollary,

Galileo first noted that for two bodies both traveling at constant, but unequal, velocities, the distances traveled are in the ratio compounded of the ratios of the speeds and the times, or $d_1 : d_2 = (v_1 : v_2)(t_1 : t_2)$. This result is derived from the facts that for equal times distances are proportional to velocities and for equal velocities distances are proportional to times. By the theorem, the distances traveled by the falling body in the two times are the same as if in each case the body had a constant velocity equal to half its final velocity. These halves of the final velocities are also proportional to the times. It follows that in the compound ratio, the ratio of the velocities can be replaced by the ratio of the times, and the corollary is proved.

Galileo stated and proved some 38 propositions on naturally accelerated motion. He was interested in comparing velocities, times, and distances for motion along inclined planes as well as for free fall. Thus, he presented a postulate to the effect that the velocity acquired by an object sliding down an inclined plane (without friction) depends only on the height of the plane and not on the angle of inclination. Using this postulate, he deduced results such as that the times of descent for a given object along two different inclined planes of the same height are to one another as the lengths of the planes, and, conversely, the times of descent over planes of equal lengths are to one another inversely as the square roots of their heights. Galileo also made progress toward solving the **brachistochrone problem**, that is, discovering the path by which an object moves in shortest time from one point to another point at a lower level. He showed that in a given vertical circle, the time taken for a body to descend along a chord from any point to the bottom of the circle, say, DC, is greater than its time to descend along the two chords DB, BC, the first beginning at the same point as the original chord, the second ending at the same bottom point (Fig. 13.33). (Here DC must subtend an arc no greater than 90°.) By extending this result to more and more chords, he concluded, erroneously, that the path of swiftest descent is a circular arc. It was not until the end of the century that several mathematicians deduced that this curve was in fact a cycloid.

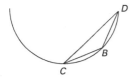

FIGURE 13.33

Galileo and the brachis-tochrone problem

13.5.2 The Motion of Projectiles

In the final part of the *Two New Sciences*, Galileo discussed the motion of projectiles. These motions are compounded from two movements, the horizontal one being of constant velocity and the vertical one being naturally accelerated. As he wrote,

> I mentally conceive of some moveable projected on a horizontal plane all impediments being put aside. Now it is evident . . . that equable motion on this plane would be perpetual if the plane were of infinite extent; but if we assume it to be ended, and [situated] on high, the moveable . . . , driven to the end of this plane and going on further, adds on to its previous equable and indelible motion that downward tendency which it has from its own heaviness. Thus, there emerges a certain motion, compounded from equable horizontal and from naturally accelerated downward [motion], which I call projection.[52]

In other words, Galileo stated part of the fundamental law of inertia, that a body moving on a frictionless horizontal plane at constant velocity will not change its motion, because, as he had noted earlier, "there is no cause of acceleration or retardation."[53] Isaac Newton extended this principle into one of his laws of motion by replacing Galileo's "causes" by his own notion of force. Galileo, however, was not so interested in the law itself as in the path of the projectile. Thus, he proved the following

THEOREM *When a projectile is carried in motion compounded from equable horizontal and from naturally accelerated downward [motions], it describes a semiparabolic line in its movement.*

Galileo discovered this theorem in 1608 in connection with an experiment rolling balls off tables, an experiment that convinced him that the horizontal motion was unaffected by the downward motion due to gravity.[54] His proof in the *Two New Sciences* used this assumption. Galileo drew a careful graph of the path of the object, noting that in equal times the horizontal distances traveled are equal, while in those same times the vertical distances increase in proportion to the squares of the times. Therefore, the curve has the property that for any two points on it, say, F, H, the ratio of the squares of the horizontal distances, $FG^2 : HL^2$, is the same as that of the vertical distances (to the plane), $BG : BL$ (Fig. 13.34). Galileo then concluded from his familiarity with the work of Apollonius that the curve was a parabola as claimed.

FIGURE 13.34

Galileo and the parabolic motion of a projectile

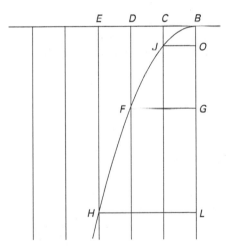

Galileo continued this discussion of projectile motion by proving that for objects fired at an angle to the horizontal, as from a cannon, the path would also be parabolic. In fact, he calculated several tables giving the height and distance traveled by such a projectile as functions of the initial angle of elevation, showing, for example, that the maximum range is achieved by an initial angle of 45°. Tartaglia had already established this latter rule in his *Nova scientia* of 1537, without, however, knowing that the path was a parabola. He even anticipated Galileo in determining that projectiles shot at complementary angles would have equal ranges.

But Tartaglia, unlike Galileo, had not started with any basic principles of physics. Nor did he have the solid understanding of the idea of a mathematical model that Galileo had. As Galileo wrote, "No firm science can be given of such events of heaviness, speed, and shape which are variable in infinitely many ways. Hence, to deal with such matters scientifically, it is necessary to abstract from them. We must find and demonstrate conclusions abstracted from the impediments, in order to make use of them in practice under those limitations that experience will teach us. . . . Indeed, in projectiles that we find practicable, . . . the deviations from exact parabolic paths will be quite insensible."[55] Galileo thus stated his firm belief in the application of mathematics to physics. One must always form a mathematical model by considering only the most important ideas in a given situation. Only after deriving mathematically the consequences of one's model and comparing these to experiment can one decide whether adjustments to the model are necessary. Galileo, like Kepler, followed these basic precepts of mathematical modeling of physical phenomena. Kepler, because he was dealing with astronomical phenomena, could only compare his theoretical results to observation. Galileo, on the other hand, conducted experiments to verify (or refute) the results of his reasoning. The detailed explication of this process of mathematical modeling, even more than his actual physical theorems, formed Galileo's most fundamental contribution to the mutual development of mathematics and physics. His ideas came into full flower as the scientific revolution of the seventeenth century reached its climax in the work of Isaac Newton.

EXERCISES

1. Make a perspective drawing of a checkerboard. First, establish a reasonable distance for the vanishing line and vanishing point and then construct the horizontal lines using the rules given in the text.

2. Suppose you are adding a row of telephone poles, all of the same actual height, to the picture of the tiled floor in Figure 13.4. The poles are equally distant from each other (each a distance of one square from the previous one) and are going off into the distance along the line from E to V in the figure. If the height on your canvas of the pole right at the picture plane is p, what should be the heights on your canvas of the remaining poles?

3. Prepare a small collection of Renaissance paintings, including some from Piero, Alberti, and Dürer. Mark the vanishing point and vanishing line and indicate some lines in the painting that meet at the vanishing point.

4. Kepler gave the following construction for a hyperbola with foci at A and B and with one vertex at C: Let pins be placed at A and B. To A let a thread with length AC be tied and to B a thread with length BC. Let each thread be lengthened by an amount equal to itself. Then grasp the two threads together with one hand (starting at C) and little by little move away from C, paying out the two threads. With the other hand, draw the path of the join of the two threads at the fingers. Show that the path is a hyperbola.

5. This problem provides details on constructing a Mercator chart to represent the region between the equator and 30° N

latitude and between 75° and 85° W longitude. Draw a line 10 cm long to represent the equator between those meridians. Divide it into intervals of 1 cm and draw the meridians perpendicular to the line of the equator. Then 1° of longitude is taken as 1 cm. To find the distance on the map to the 10° parallel, note first that since 1 cm corresponds to 1° on a great circle, the radius of the corresponding sphere must be $\frac{180}{\pi}$. One must therefore multiply this value by $D(10°)$, computed by Equation 13.1. Similarly, to calculate the distance to the parallel at 20° from that at 10°, find $D(20°) - D(10°)$ and multiply by the radius. Also, calculate the distance of the 30° parallel from the equator. To make the chart somewhat more precise, determine the distances of the parallels at 5°, 15°, and 25°.

6. Modify the calculation of Exercise 5 for placing parallels in order to map a region between 80° and 100° W longitude and between 40° and 60° N latitude, assuming that 1 cm on the parallel of 40° N corresponds to 1° of longitude.

7. Complete the solution of the problem from *On Triangles* discussed in the text of finding two sides AB, AG, of triangle ABG given that $BG = 20$, the perpendicular $AD = 5$, and the ratio $AB : AG = 3 : 5$ (see Fig. 13.13).

8. In triangle ABC, suppose the ratio $\angle A : \angle B = 10 : 7$ and the ratio $\angle B : \angle C = 7 : 3$. Find the three angles and the ratio of the sides. (This problem and the next two are also from *On Triangles*.)

9. In triangle ABC with AD perpendicular to BC, suppose $AB - AC = 3$, $BD - DC = 12$, and $AD = 30$. Find the three sides.

10. Show that if the sum of two arcs is known and the ratio of their sines is known, then each arc may be found. In particular, suppose the sum of the two arcs is 40° and ratio of the sine of the larger part to that of the smaller is 7 : 4. Determine the two arcs. (Although Regiomontanus only used sines, it is probably easier to do this using cosines and tangents as well.)

11. Suppose the three angles of a spherical triangle are 90°, 70°, and 50°. Find the lengths of the sides.

12. Show that Regiomontanus's versine formula is equivalent to the spherical law of cosines: $\cos a = \cos b \cos c + \sin b \sin c \cos A$.

13. The following problem is from Pitiscus's *Trigonometry*: Find the area of the field $ABCDE$ given the following measurements: $AB = 7$, $BC = 9$, $AC = 13$, $CD = 10$, $CE = 11$, $DE = 4$, and $AE = 17$. Begin by drawing $BF \perp AC$, $CG \perp AE$, and $DH \perp CE$ (Fig. 13.35).

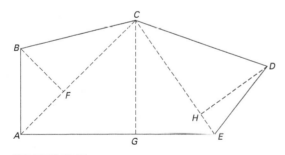

FIGURE 13.35

A problem in area from Pitiscus's *Trigonometry*

14. This problem is from Copernicus's *De revolutionibus*. Given the three sides of an isosceles triangle, to find the angles. Circumscribe a circle around the triangle and draw another circle with center A and radius $AD = \frac{1}{2}AB$ (Fig. 13.36). Then show that each of the equal sides is to the base as the radius is to the chord subtending the vertex angle. All three angles are then determined. Perform the calculations with $AB = AC = 10$ and $BC = 6$.

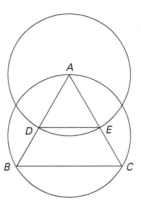

FIGURE 13.36

To find the angles of an isosceles triangle

15. Prove that
$$\sin \alpha \sin \beta = \frac{1}{2}[\cos(\alpha - \beta) - \cos(\alpha + \beta)].$$

16. Use the formula of Exercise 15 to multiply 4,378,218 by the sine of 27°15′22″. Check your answer by using your calculator in the standard way.

17. Given that the period of the earth is 1 year, and given that Mars's mean distance from the sun is 1.524 times that of the earth's mean distance, use Kepler's third law to determine the period of Mars.

18. According to Kepler's second law, at what point in the planet's orbit will the planet be moving the fastest?

19. Using the definition of the function Nlog presented in the text, determine $\mathrm{Nlog}(xy)$ and $\mathrm{Nlog}(x/y)$ in terms of $\mathrm{Nlog}\, x$ and $\mathrm{Nlog}\, y$.

20. Use the law of tangents to solve a triangle with sides 10 and 13 and included angle 35°.

21. Prove: If the same moveable is carried from rest on an inclined plane and also along a vertical of the same height, the times of movements will be to one another as the lengths of the plane and of the vertical. A corollary is that the times of descent along differently inclined planes of the same height are to one another as the lengths.

22. Prove: The times of motion of a moveable starting from rest over equal planes unequally inclined are to each other inversely as the square root of the ratio of the heights of the planes.

23. Show that a projectile fired at an angle α from the horizontal follows a parabolic path.

24. Galileo states that if a projectile fired at an angle α from the horizontal at a given initial speed reaches a distance of 20,000 if $\alpha = 45°$, then with the same initial speed it will reach a distance of 17,318 if $\alpha = 60°$ or $\alpha = 30°$. Check this statement.

25. Galileo states that if a projectile fired at a given initial speed at an angle α to the horizontal reaches a maximum height of 5000 if $\alpha = 45°$, then with the same initial speed it will reach a height of 2499 when $\alpha = 30°$ and a height of 7502 when $\alpha = 60°$. Check this statement.

26. Given that the distances traveled in any times by a body falling from rest are as the squares of the times, show that the distances traveled in successive equal intervals are as the consecutive odd numbers 1, 3, 5, ...

27. Find out how the use of logarithms was mechanized in the seventeenth century by the invention of the slide rule. Give examples of the various types of slide rules used. When did the slide rule itself become obsolete and why?

28. Compare Galileo's and Kepler's attitudes toward the interaction of experiment (or observation) and theory in developing a new body of knowledge.

29. Read Kepler's biography in Arthur Koestler's *The Sleepwalkers*, referred to in References and Notes for this chapter. Koestler also discusses the lives of Copernicus and Galileo. How believable is Koestler's "sleepwalking" hypothesis to explain the discovery of the new ideas in astronomy? Comment.

30. Look up the eccentricities of the orbits of Mars and the other visible planets. Compare these with the eccentricity of the earth to see why Kepler was able to assume the circularity of the earth's orbit. Considering these eccentricities, why did Kepler study Mars in detail rather than Mercury?

31. Look up a treatment of geometrical perspective in a modern text on techniques of painting. How does it compare to the discussion of Alberti?

32. Look up the recent review of Galileo's case by the Roman Catholic Church. Has the Church revised its opinion that Galileo disobeyed orders?

✦ REFERENCES AND NOTES

There are several books that discuss aspects of the applied mathematics of the Renaissance in greater detail. Julian Lowell Coolidge, *The Mathematics of Great Amateurs*, second edition (Oxford: Clarendon Press, 1990), contains three chapters dealing with the mathematics of the artists of the Renaissance as well as a chapter on John Napier. J. V. Field, *The Invention of Infinity: Mathematics and Art in the Renaissance* (Oxford: Oxford University Press, 1997), describes in detail the relationship between art and mathematics during this time period. Thomas S. Kuhn, *The Copernican Revolution* (Cambridge: Harvard University Press, 1957), and E. J. Dijksterhuis, *The Mechanization of the World Picture* (Princeton: Princeton University Press, 1986), each contain sections on the developments in astronomy. The latter work, in particular, has mathematical details. Arthur Koestler, *The Sleepwalkers* (New York: Penguin, 1959), provides a lively treatment of the history of astronomy from Greek times to the time of Galileo, including biographies of Copernicus, Brahe, Kepler, and Galileo. Many of his interpretations are controversial. Finally, Ernan McMullin, ed., *Galileo, Man of Science* (Princeton Junction: Scholar's Bookshelf, 1988), is a collection of essays on various aspects of Galileo's scientific achievements.

Barnabas Hughes has prepared a modern edition of Regiomontanus's trigonometry: *Regiomontanus on Triangles* (Madison: University of Wisconsin Press, 1967). This work has the original Latin of *De triangulis omnimodis* as well as an English translation, an introduction, and extensive notes. A recent translation of Copernicus's *De revolutionibus* is A. M. Duncan, trans., *Copernicus: On the Revolutions of the Heavenly Spheres* (New York: Barnes and Noble, 1976). Kepler's *Optics* is available in the translation of William H. Donahue (Santa Fe: Green Lion Press, 2000). The *Mysterium cosmographicum* is available as A. M. Duncan, trans., *The Secret of the Universe* (New York: Abaris, 1981). This is a translation of the second edition, so it includes not only the original text of 1596 but also Kepler's notes added in 1621. It also contains copies of the original diagrams. The *New Astronomy* is also now available in a translation by William H. Donahue (Cambridge: Cambridge University Press, 1992), while sections of the *Epitome of Copernican Astronomy* and the *Harmonice mundi*, both translated by Charles Glenn Wallis, appear in volume 16 of the *Great Books* (Chicago: Encyclopedia Britannica, 1952). Napier's two volumes on logarithms, the *Descriptio* and the *Constructio*, were translated in the seventeenth century. Photographic reprints of both are available as John Napier, *Descriptio*, translated by Edward Wright (New York: Da Capo Press, 1969), and John Napier, *Constructio*, translated by William R. MacDonald (London: Dawsons, 1966). The former includes Napier's original table, while the latter shows in detail the ingenious interpolation schemes Napier used to guarantee eight-place accuracy. Finally, Galileo's *Two New Sciences* was translated by Stillman Drake (Madison: University of Wisconsin Press, 1974). Drake also wrote an introduction, extensive notes, and a glossary of Galileo's technical terms.

1. John Dee, *Mathematical Preface* (New York: Science History Publications, 1975), p. i, "The Translator to the Reader." This work is a photographic reproduction of the original with an introduction by Allen G. Debus, who discusses Dee's life and influence. The pages are unnumbered just as in the original. A discussion of Dee's philosophy and of the mathematical preface is also found in F. A. Yates, *Theatre of the World* (Chicago: Chicago University Press, 1969).

2. Stillman Drake, *Galileo at Work* (Chicago: University of Chicago Press, 1978), pp. 347–348. There is an enormous literature on Galileo and his conflict with the Church. One of the more interesting volumes on this subject is Pietro Redondi, *Galileo Heretic*, Raymond Rosenthal, trans. (Princeton: Princeton University Press, 1987), but a glance at the appropriate shelf in any university library will provide many more.

3. John Dee, *Mathematical Preface*, p. 3.

4. Ibid., p. 5.

5. Ibid., p. 13.

6. Ibid., p. 19.

7. Ibid.

8. Ibid., p. 38.

9. Ibid.

10. For more on Alberti's work on perspective, see J. and P. Green, "Alberti's Perspective: A Mathematical Comment," *Art Bulletin* 64 (1987), 641–645.

11. For more information on Leon Battista Alberti and Piero della Francesca, see Julian Lowell Coolidge, *Mathematics of Great Amateurs*, chapter 3.

12. An English version of this work is available as W. Strauss, trans., *The Painter's Manual* (New York: Abaris, 1977).

13. Erwin Panofsky, "Dürer as a Mathematician," in James R. Newman, ed., *The World of Mathematics* (New York: Simon and Schuster, 1956), vol. 1, 603–621, pp. 611–612. This section is taken from Erwin Panofsky, *Albrecht Dürer* (Princeton: Princeton University Press, 1945), a work that discusses Dürer's life and art in great detail.

14. For more information, see Roger Herz-Fischler, "Dürer's Paradox or Why an Ellipse Is Not Egg-Shaped," *Mathematics Magazine* 63 (1990), 75–85.

15. Kepler, *Optics*, p. 107.

16. Dee, *Mathematical Preface*, p. 42.

17. For more on John Harrison and the problem of longitude, see Dava Sobel, *Longitude: The True Story of a Lone Genius Who Solved the Greatest Scientific Problem of His Time* (New York: Walker and Company, 1995).

18. Dee, *Mathematical Preface*, p. 15.

19. More details can be found in V. Frederick Rickey and Philip M. Tuchinsky, "An Application of Geography to Mathematics: History of the Integral of the Secant," *Mathematics Magazine* 53 (1980), 162–166. See also Florian Cajori, "On an Integration Ante-dating the Integral Calculus," *Bibliotheca Mathematica* (3) 14 (1914), 312–319.

20. Dee, *Mathematical Preface*, p. 20.

21. Ibid., p. 23.

22. Hughes, *Regiomontanus on Triangles*, p. 27.

23. Ibid., p. 101.

24. Ibid., p. 109.

25. Ibid., p. 119.

26. The relationship between the work of Jābir and Regiomontanus is discussed in detail by Richard Lorch in "Jābir ibn Aflaḥ and the Establishment of Trigonometry in the West," in Richard Lorch, *Arabic Mathematical Sciences: Instruments, Texts, Transmission* (Aldershot, UK: Ashgate Publishing Ltd., 1995).

27. For more details on the history of trigonometry see M. C. Zeller, "The Development of Trigonometry from Regiomontanus to Pitiscus," (Dissertation, University of Michigan, 1944, and Ann Arbor: Edwards Bros., 1946). This work has an extensive table comparing various aspects of trigonometrical notation and terminology in European texts to the beginning of the seventeenth century. See also J. D. Bond, "The Development of Trigonometric Methods Down to the Close of the XVth Century," *Isis* 4 (1921), 295–323.

28. Copernicus, *On the Revolutions*, p. 25. An older translation of *De revolutionibus*, by Charles Glenn Wallis, appears in volume 16 of the *Great Books* (Chicago: Encyclopedia Britannica, 1952).

29. Ibid., p. 26.

30. Ibid., p. 27.

31. Ibid., p. 38.

32. Ibid., p. 60.

33. Ibid., p. 22.

34. Kepler, *The Secret of the Universe*, p. 67.

35. Ibid., p. 73.

36. Kepler, *Epitome of Copernican Astronomy*, p. 878.

37. Kepler, *The Harmonies of the World*, p. 1048. It is an interesting exercise to play on a piano the notes Kepler assigns to the various planets and so to understand his version of the "Music of the Spheres."

38. For more details on Kepler's discovery of the elliptical path, see Curtis Wilson, "How Did Kepler Discover His First Two Laws," *Scientific American* 226 (March, 1972), 92–106; Eric Aiton, "How Kepler Discovered the Elliptical Orbit," *Mathematical Gazette* 59 (1975), 250–260; and Dijksterhuis, *Mechanization of the World*, pp. 303–323.

39. Kepler, *Harmonies*, p. 1020.

40. Napier, *Constructio*, p. 8. Many essays on Napier's work are found in C. G. Knott, ed., *Napier Tercentenary Memorial Volume* (London: Longmans, Green and Co., 1915). A general discussion of logarithms is found in E. M. Bruins, "On the History of Logarithms: Bürgi, Napier, Briggs, de Decker, Vlacq, Huygens," *Janus* 67 (1980), 241–260, and in F. Cajori, "History of the Exponential and Logarithmic Concepts," *American Mathematical Monthly* 20 (1913), 5–14, 35–47, 75–84, 107–117.

41. Napier, *Constructio*, p. 19.

42. Napier, *Descriptio*, p. 6. This work includes many examples of the use of logarithms in solving both plane and spherical triangles.

43. See C. H. Edwards, *The Historical Development of the Calculus* (New York: Springer-Verlag, 1979), chapter 6, for details of Napier's table construction methods.

44. Dee, *Mathematical Preface*, p. 25.

45. Ibid., p. 34.

46. Stillman Drake, ed., *Discoveries and Opinions of Galileo* (New York: Doubleday, 1957), p. 183.

47. Galileo, *Two New Sciences*, p. 154. For other essays on various aspects of Galileo's scientific life, see Stillman Drake, *Galileo Studies: Personality, Tradition and Revolution* (Ann Arbor: University of Michigan Press, 1970) and the classic work by Alexandre Koyré, *Galileo Studies*, translated by John Mepham (Atlantic Highlands, NJ: Humanities Press, 1978).

48. See Stillman Drake, "Galileo's Discovery of the Law of Free Fall," *Scientific American* 228 (May, 1973), 84–92, for more information. Drake has analyzed certain of Galileo's manuscript notes that provide details on the discovery.

49. Galileo, *Two New Sciences*, p. 160.

50. Ibid., p. 165.

51. Ibid.

52. Ibid., p. 217.

53. Ibid., p. 196.

54. Stillman Drake, "Galileo's Discovery of the Parabolic Trajectory," *Scientific American* 232 (March, 1975), 102–110, has many details on Galileo's manuscript notes detailing his experiments.

55. Galileo, *Two New Sciences*, p. 225.

Algebra, Geometry, and Probability in the Seventeenth Century

Whenever two unknown magnitudes appear in a final equation, we have a locus, the extremity of one of the unknown magnitudes describing a straight line or a curve.

—Pierre de Fermat's *Introduction to Plane and Solid Loci,* 1637[1]

So that he could improve his chances at gambling, Antoine Gombaud, the chevalier de Méré, asked Blaise Pascal two questions on betting around 1652. The first was on the number of tosses of two dice necessary to have at least an even chance of getting a double six and the second on the equitable division of stakes in a game interrupted before its conclusion. It was out of Pascal's answers that the theory of probability grew. The two men had been introduced by the Duke of Roannez, who had a salon in Paris early in the 1650s that provided a meeting place for mathematicians among others.

In the early seventeenth century, the pace of mathematical development began to accelerate. By now printing was well established, and communication, both through letter and through the printed word, was becoming much more rapid. The ideas of one mathematician were passed on to others, to be criticized, commented upon, and finally extended. In this chapter, we will survey some of the newly developing areas of mathematics.

Viète's ideas on the use of algebra in analysis were critical in the new developments. Starting in around 1610, William Oughtred, Thomas Harriot, Albert Girard, and others began to turn Viète's notation into recognizably modern notation, while at the same time further developing his theory of equations. Then in the 1630s Viète's analysis was applied to geometry and reformulated into the new subject of **analytic geometry**. The two central figures in the development of this field, which was to prove vital in the subsequent invention of the calculus, are Pierre de Fermat and René Descartes. Both of these men played central roles in other areas of mathematics as well. Descartes continued the work of developing a theory of equations. Fermat was involved, in his correspondence with Blaise Pascal, in the early development of probability theory, the first textbook on which was written by Christian Huygens in 1656. Fermat also was responsible for the first new work in number theory since Leonardo of Pisa, while Pascal, along with Girard Desargues, made some of the earliest contributions to the subject of projective geometry.

 14.1

THE THEORY OF EQUATIONS

Algebraic methods for solving cubic and quartic equations were discovered in Italy in the sixteenth century and improved on somewhat by Viète near the turn of the seventeenth century. But Cardano was hampered by a lack of a convenient notation and Viète always restricted himself to positive solutions. Thus, even though the former gave various examples of relationships among the roots of a single cubic equation and between roots of related equations and the latter was able to express algebraically the relationship between the coefficients and the solutions of equations of degree up to five, provided all values were positive, the general theory was still incomplete.

14.1.1   William Oughtred and Thomas Harriot

By the early seventeenth century, two English mathematicians, William Oughtred (1575–1660) and Thomas Harriot (1560–1621), had made careful studies of Viète's work and were converted to the method of symbolical reasoning he had introduced. Both of them attempted in their own work to go beyond Viète and to make algebraic arguments even more symbolic. Oughtred was a cleric, who evidently spent most of his time on mathematics and mathematics teaching. His major work, the *Clavis mathematicae* (*Key of Mathematics*), first appeared in 1631 and had several subsequent editions, both in Latin and in English. The *Clavis* introduced English readers to Viète's symbolic algebra, and, in particular, attempted to show, as did his French predecessor, that algebra could really be considered as the "analytical art, . . . in which by taking the thing sought as known, we find out that we seek."[2] In other words, Oughtred felt that mathematical problems, including geometric ones, should be translated into symbolic equations and then solved by the methods of algebra.

BIOGRAPHY

Thomas Harriot (1560–1621)

Harriot entered the service of Sir Walter Raleigh after finishing his undergraduate studies at Oxford and went on an expedition to Virginia in 1585 as an expert on cartography. Besides learning how to smoke there, a habit that ultimately led to his death from cancer, he also wrote a report on the colony and its native inhabitants: *A Briefe and True Report of the New Found Land of Virginia*. After he returned to England, he found a patron and benefactor, Sir Henry Percy, even though Percy was imprisoned from 1605 to 1621 for his supposed involvement in the Gunpowder Plot.

Among Harriot's friends was Nathaniel Torporley (1564–1632), who in the 1590s became Viète's amanuensis. It was through Torporley, then, that Harriot became acquainted with Viète's work, often before the works were actually published. And thus, beginning in about 1600, Harriot made this work his own, writing hundreds of manuscript pages with his transformation of Viète's algebra into a recognizably modern form. By 1610, he had organized some of these pages into a *Treatise on Equations*. Unfortunately, despite the entreaties of many of his friends, Harriot never published this or any other of his mathematical discoveries. His executors did publish some parts of the *Treatise on Equations* in the *Artis analyticae praxis* (*Practice of the Analytic Art*), which appeared in 1631. But because some of the manuscript pages had been scattered in various collections and because the executors did not fully understand Harriot's ideas, this book does not fairly represent Harriot's discoveries. Thus, even today, some of his most important work remains only in manuscript, although it appears that many of these manuscripts circulated in England during and after his lifetime. In fact, it is quite probable that Charles Cavendish (1591–1645), another member of the aristocracy, even carried many of Harriot's ideas to his mathematical acquaintances on the Continent. It was only in recent years that Jacqueline Stedall recovered the manuscript of the *Treatise on Equations* and published it. It is clear that this work would have had enormous influence had Harriot published it himself.[3]

Oughtred introduced many symbols, including the $\times$ to represent multiplication, but in part because there were so many and they often confused his students and his typesetters, few of his symbols have lasted. For variables, constants, and their powers, however, he basically kept to Viète's plan, only making the notation a bit shorter by using abbreviations. For example, he used A_q for the square of A and A_c for its cube. But he did show how to use algebra to solve problems, as when he rewrote Euclid's Proposition II-11 as an equation. He let A stand for the greater of the two segments in which the line of length B was to be cut. Then $B - A$ was the lesser segment, so the rectangle contained by the whole and the lesser segment was $B_q - BA$. Since this was required to be equal to the square on A, he had $A_q = B_q - BA$, or, $A_q + BA = B_q$. He then used his version of the quadratic formula to solve the equation: $A = \sqrt{} : \frac{1}{4}B_q + B_q : -\frac{1}{2}B$. Note that Oughtred here used juxtaposition to indicate multiplication, although he did not do this consistently, sometimes reverting to Viète's *in*. He also used a colon where we use parentheses.

Harriot's *Treatise on Equations* was written much earlier than Oughtred's work and evidently circulated in manuscript in England (and perhaps on the Continent) for several decades. The contributions of Harriot that we discuss come from this treatise, even though some were not in the *Praxis*. Harriot took over from Viète the idea of using vowels for unknowns and consonants for knowns, although he used small letters instead of Viète's capital letters. He also consistently used juxtaposition for multiplication, an idea that had shown up occasionally earlier in the work of both Oughtred and Michael Stifel. Thus, Harriot wrote *ba* where Viète wrote "B in A," and *aaaa* in place of Viète's "A square-square." He also used Recorde's equal sign, the now standard signs < and > for "less than" and "greater than," and our usual signs for square root and cube root. Harriot was thus able to simplify considerably Viète's rules. For example, he replaced Viète's justification of transposition (p. 411) by the following:

Let $aa - dc = gg - ba$. To be added to each $+ba + dc$. Whence $aa + ba = gg + dc$.

This looks quite modern, but note that where Viète used the expression "D plane," Harriot, although he used symbols, still felt constrained by the notion of homogeneity and thus replaced this by *dc* rather than by a single letter.

Harriot also realized that equations could be generated from their roots $b, c, d, \ldots$ by multiplying together expressions of the form $a - b, a - c, a - d. \ldots$ Thus, he was led to the basic relationship between the roots and the coefficients of the equation, even sometimes in the case of negative and imaginary roots, although he never seems to have stated this explicitly as a theorem and was not consistent in dealing with negatives and imaginaries. For example, Harriot multiplied together $a - b$, $a - c$, and $a - d$ to get the equation $aaa - baa - caa + bca - daa + bda + cda - bcd = 0$, from which he noticed that the only roots are b, c, and d. And the sum of these is the negative of the coefficient of the square term. But when he multiplied together $a - b$, $a - c$, and $a + d$, he only found that the roots were b and c. However, in a later numerical example, considering the equation $12 = 8a - 13aa + 8aaa - aaaa$, he saw first that two of the roots are 2 and 6. Noting that the sum of these real roots is already equal to 8, the coefficient of a^3, he stated that there could not be any further real roots, because they would make the sum greater than 8, the coefficient of the cubic term. However, since he realized that there ought to be four solutions, he used a substitution to remove the cubic term: $a = 2 - e$. The new equation in e was $-20e + 11ee - eeee = 0$, whose real roots are 0 and -4. After reducing this to a cubic, $eee - 11e = -20$, with one root equal to -4, he concluded that the sum of the remaining roots must be 4 and their product 5. The two values satisfying this are $e = 2 + \sqrt{-1}$ and $e = 2 - \sqrt{-1}$ and therefore the complex roots of the original equation are $a = 2 - (2 + \sqrt{-1}) = -\sqrt{-1}$ and $a = 2 - (2 - \sqrt{-1}) = +\sqrt{-1}$.

With his symbolic notation, Harriot could, like Viète, write "formulas" for the solution of quadratic and cubic equations, formulas that are much closer than Viète's to our modern formulas. Thus, he solved the quadratic equation $aa + bb = 2ca$ by transposing the bb and $2ca$ terms and then completing the square to get $aa - 2ca + cc = cc - bb$. There are then two roots of the left side, $a - c$ and $c - a$. If $a - c = \sqrt{cc - bb}$, then $a = c + \sqrt{cc - bb}$, while if $c - a = \sqrt{cc - bb}$, then $a = c - \sqrt{cc - aa}$. For cubic equations, he was able to write down a formula representing Cardano's method (Sidebar 14.1). Cardano, recall, had numerous procedures, one for each type of cubic equation. But Harriot always used substitution to reduce cubics with square terms to those without one and therefore only needed three different formulas. Similarly, he solved quartic equations using the method of Ferrari, but also showed in this case how to eliminate the cubic term from any quartic by a substitution.

SIDEBAR 14.1 *Changes in Notation*

The best way to see the rapid changes in notation over about a century is to look at how Cardano, Viète, Harriot, and Descartes wrote solutions to cubic equations. First, remember that Cardano could only give a solution to a particular cubic, because he had no way of writing a general cubic. Thus, he wrote the solution to "cube equals six things and 40," or $x^3 = 6x + 40$, as

$$\mathfrak{R} \, \text{v} : \text{cu.20 p} : \mathfrak{R} \, 392 \, \text{p} : \mathfrak{R} \, \text{v} : \text{cu.20 m} : \mathfrak{R} \, 392.$$

Viète wrote that the solution to the equation

$$A \text{ cube} - B \text{ plane 3 in } A \text{ equals } Z \text{ solid 2}$$

is given by

$$A \text{ is } l.c. \; \overline{Z \text{ solid} + l.Z \text{ solidsolid } - B \text{ planeplaneplane}}$$

$$+ l.c. \; \overline{Z \text{ solid} - l.Z \text{ solidsolid } - B \text{ planeplaneplane}}.$$

Then, we have Harriot's solution to $2ccc = 3bba + aaa$:

$$a = \sqrt[3]{\sqrt{bbbbbb + cccccc} + ccc}$$

$$- \sqrt[3]{\sqrt{bbbbbb + cccccc} - ccc}.$$

It is only a short step from Harriot's notation to that of Descartes. Here is Descartes' solution to the equation $z^3 = -pz + q$:

$$z = \sqrt{\text{C.} + \frac{1}{2}q + \sqrt{\frac{1}{4}qq + \frac{1}{27}p^3}}$$

$$- \sqrt{\text{C.} - \frac{1}{2}q + \sqrt{\frac{1}{4}qq + \frac{1}{27}p^3}}.$$

14.1.2 Albert Girard and the Fundamental Theorem of Algebra

Albert Girard (1595–1632) was much clearer than Harriot on the relationship between the roots and coefficients of a polynomial in his 1629 work *Invention nouvelle en l'algèbre* (*A New Discovery in Algebra*) and also gave the first explicit statement of the fundamental theorem of algebra. Girard was probably born in St. Mihiel in the French province of Lorraine but spent much of his life in the Netherlands, where he studied at Leiden and served as a military engineer in the army of Frederick Henry of Nassau. Although he wrote a work on trigonometry and edited the works of Stevin, his most important contributions are to algebra. In *A New Discovery in Algebra*, Girard clearly introduced the notion of a fractional exponent ("the numerator is the power and the denominator the root"[4]) but also used the current notation for higher roots (e.g., $\sqrt[3]{}$ for cube root) as an alternative to the exponent 1/3. However, the fractional exponent was not attached directly to an unknown. For example, Girard wrote $(\frac{3}{2})49$ to denote the cube of the square root of 49, or 343, and, taking his cue from Bombelli, $49(\frac{3}{2})$ to mean what is today written as $49x^{3/2}$. But since he did not use letters to represent coefficients, he could only write out particular cubic equations, for example, such as: let 1(3) be equal to 6(1) + 40.

Furthermore, Girard was among the first to note the geometric meaning of a negative solution to an equation: "The minus solution is explicated in geometry by retrograding; the minus goes backward where the plus advances."[5] He then gave an example of a geometric problem whose algebraic translation has two positive and two negative solutions and noted on the relevant diagram that the negative solutions were to be interpreted as being laid off in the direction opposite that of the positive ones.

Not only did Girard understand the meaning of negative solutions to equations, he also systematized the work of Viète and Harriot and explicitly considered **factions**, today called the elementary symmetric functions of n variables: "When several numbers are proposed, the entire sum may be called the first *faction*; the sum of all the products taken two by two may be called the second *faction*; the sum of all the products taken three by three may be called the third *faction*; and always thus to the end, but the product of all the numbers is the last *faction*. Now, there are as many *factions* as proposed numbers."[6] He pointed out that for the numbers 2, 4, 5, the first faction is 11, their sum; the second is 38, the sum of all products of pairs; while the third is 40, the product of all three numbers. He also noted that the Pascal triangle of binomial coefficients, which Girard called the "triangle of extraction," tells how many terms each of the factions contains. In the case of four numbers the first faction contains four terms, the second, six, the third, four, and the fourth and last, one.

Girard's basic result in the theory of equations was the following theorem, to which he gave no proof:

THEOREM   *Every algebraic equation . . . admits of as many solutions as the denomination of the highest quantity indicates. And the first faction of the solutions is equal to the [coefficient of the second highest] quantity, the second faction of them is equal to the [coefficient of the third highest] quantity, the third to the [fourth], and so on, so that the last faction is equal to the [constant term]—all this according to the signs that can be noted in the alternating order.*[7]

What Girard meant by the last statement about signs is that one first needs to arrange the equation so that the degrees alternate on each side of the equation. Thus, $x^4 = 4x^3 + 7x^2 - 34x - 24$ should be rewritten as $x^4 - 7x^2 - 24 = 4x^3 - 34x$. The roots of this equation being 1, 2, -3, and 4, the first faction is equal to 4, the coefficient of x^3; the second to -7, the coefficient of x^2; the third to -34, the coefficient of x; and the fourth to -24, the constant term. Similarly, the equation $x^3 = 167x - 26$ can be rewritten as $x^3 - 167x = 0x^2 - 26$. Because -13 is one solution, his result implies that the product of the two remaining roots is 2, while their sum is 13. To find these simply requires solving a quadratic equation. The answers are $6\frac{1}{2} + \sqrt{40\frac{1}{4}}$ and $6\frac{1}{2} - \sqrt{40\frac{1}{4}}$.

In the first part of the theorem, Girard was asserting the truth of the fundamental theorem of algebra, that every polynomial equation has a number of solutions equal to its degree (denomination of the highest quantity). As his examples show, he acknowledged that a given solution could occur with multiplicity greater than one. He also fully realized that in his count of solutions he would have to include imaginary ones (which he called impossible). So in his example $x^4 + 3 = 4x$, he noted that the four factions are 0, 0, 4, 3. Because 1 is a solution of multiplicity 2, the two remaining solutions have the property that their product is 3 and their sum is -2. It follows that these are $-1 \pm \sqrt{-2}$. In answer to the anticipated question of the value of these impossible solutions, Girard answered that "they are good for three things: for the certainty of the general rule, for being sure that there are no other solutions, and for its utility."[8]

What the "utility" of impossible solutions is, Girard did not explain. Nor did he show how he derived the theorem. Given that he considered solutions with multiplicity greater than one, however, it appears he, like Harriot, must have understood that equations of degree n come from multiplying together n expressions of the form $x - r_i$, where some of the r_i may be identical. It was Descartes, however, who made this procedure precise, as we will see below.

 14.2   # ANALYTIC GEOMETRY

Analytic geometry was born in 1637 of two fathers, René Descartes (1596–1650) and Pierre de Fermat (1601–1665). Naturally, there had been a gestation period, but early in that year Fermat sent to his correspondents in Paris a manuscript entitled *Ad locos planos et solidos isagoge* (*Introduction to Plane and Solid Loci*) while at about the same time Descartes was readying for the printer the galley proofs of his *Discours de la méthode pour bien conduire sa raison et chercher la vérité dans les sciences* (*Discourse on the Method for Rightly Directing One's Reason and Searching for Truth in the Sciences*) with its three accompanying essays, among which was *La Géométrie* (*The Geometry*). Both Fermat's *Introduction* and Descartes' *Geometry* present the same basic techniques of relating algebra and geometry, the techniques whose further development culminated in the modern subject of analytic geometry. Both men came to the development of these techniques as part of the effort of rediscovering the "lost" Greek techniques of analysis. Both were intimately familiar with the Greek classics, in particular with the *Domain of Analysis* of Pappus, and both tested their new ideas against the four-line-locus problem of Apollonius and its generalizations. But Fermat and Descartes developed distinctly different approaches to their common subject, differences rooted in their differing points of view toward mathematics.

14.2.1 Fermat and the *Introduction to Plane and Solid Loci*

Fermat began his study of mathematics with the normal university curriculum at Toulouse, which probably covered little more than an introduction to Euclid's *Elements*. But after completing his baccalaureate degree and before beginning his legal education, he spent several years in Bordeaux studying mathematics with former students of Viète, who during the late 1620s were engaged in the editing and publishing of their teacher's work. Fermat became familiar both with Viète's new ideas for symbolization in algebra and with his program of discovering and elucidating the secret analysis of the Greek mathematicians. In Bordeaux, Fermat began his own project of using Pappus's annotations and lemmas in the *Domain of Analysis* to restore the *Plane Loci* of Apollonius. Fermat tried to reconstruct the original work along with Apollonius's reasoning in the discovery of the various theorems. His study of Viète naturally led Fermat to attempt to replace Apollonius's geometric analysis with an algebraic version. It is this algebraic version of Apollonius's locus theorems that provided the beginnings of Fermat's analytic geometry.

For example, Fermat considered the following result: If, from any number of given points, straight lines are drawn to a point, and if the sum of the squares of the lines is equal to a given area, the point lies on a circumference [circle] given in position. The theorem deals with an indeterminate number of points, but Fermat's treatment of the simplest case, that of two points, contained the germs of the two major ideas of analytic geometry, the correspondence between geometric loci and indeterminate algebraic equations in two or more variables, and

BIOGRAPHY

Pierre de Fermat (1601–1665)

Fermat was born into a moderately wealthy family in Beaumont-de-Lomagne in the south of France, where his father was a leather merchant and minor local official (Fig. 14.1). He received his undergraduate education at the University of Toulouse and took a bachelor of civil law degree in 1631 at Orléans. He then returned to Toulouse, in the vicinity of which he spent the remainder of his life practicing law. He was a member of various official bodies in Toulouse, including the chambers of the Parlement, a body charged with both administrative and legal functions. Although Fermat served as a jurist for many years, he was evidently never a brilliant lawyer, probably because he spent much of his time on his first love, mathematics. Due to the state of his health and the press of his legal work, however, he never traveled far from his home. Thus, all of his mathematical work was communicated to others via his extensive correspondence.

Fermat always considered mathematics a hobby, a refuge from the continual disputes with which he had to deal as a jurist. He therefore refused to publish any of his discoveries, because to do so would have forced him to complete every detail and to subject himself to possible controversies in another arena. In many cases it is not known what, if any, proofs Fermat constructed nor is there always a systematic account of certain parts of his work. Fermat often tantalized his correspondents with hints of his new methods for solving certain problems. He would sometimes provide outlines of these methods, but his promises to fill in gaps "when leisure permits" frequently remained unfulfilled. Nevertheless, a study of his manuscripts, published by his son 14 years after his death, as well as his many letters, enables scholars today to have a reasonably complete picture of Fermat's methods.[9]

FIGURE 14.1

Fermat on a French stamp

the geometric framework for this correspondence, a system of axes along which lengths are measured.

Fermat took the two given points A, B and bisected the line AB at E. With IE as radius (I yet to be determined) and E as center, he described a circle (Fig. 14.2).[10]. He then showed that any point P on this circle satisfied the conditions of the theorem, namely, that $AP^2 + BP^2$ equals the given area M, provided that I is chosen so that $2(AE^2 + IE^2) = M$. The important ideas in this proof were that the locus, the circle, was determined by the sum of the squares of two variable quantities, AP and BP, and that the point I was determined in terms of its "coordinate" measured from the "origin" E.

FIGURE 14.2

Fermat's analysis of a special case of a theorem of Apollonius

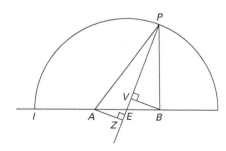

The idea of using an origin from which to determine the (horizontal) coordinate of a point was also apparent in Fermat's treatment of the case of Apollonius's theorem with several noncollinear points. In that situation, he used a base line such that all the points lie on one side and dropped perpendiculars from the given points to the line. Not only did he use the horizontal coordinates GH, GL, GK in his analysis of the problem, but also the vertical coordinates GA, HB, LD, KC, measured along perpendiculars to the base line (Fig. 14.3). In fact, he showed that the horizontal coordinate GM of the center O of the desired circle is given by $GM = \frac{1}{4}(GH + GL + GK)$, while the vertical coordinate $MO = \frac{1}{4}(GA + HB + LD + KC)$. The radius OP is determined by the equation

$$M = AO^2 + BO^2 + CO^2 + DO^2 + 4OP^2,$$

where M is the given area.

FIGURE 14.3

A second special case of
Apollonius's theorem

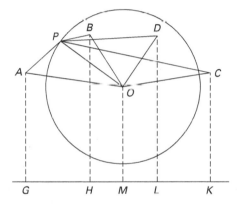

In his treatment of the general case of Apollonius's theorem, however, Fermat did not express the circle by means of an equation, probably because he was trying to write the text as Apollonius would have. But two years after he finished his reconstruction, he set down his new ideas on analytic geometry in his *Introduction to Plane and Solid Loci,* with this chapter's opening quotation sounding the central theme. Fermat asserted that if in solving algebraically a geometric problem one ends up with an equation in two unknowns, the resulting solution is a locus, either a straight line or a curve, the points of which are determined by the motion of one endpoint of a variable line segment, the other endpoint of which moves along a fixed straight line.

Fermat's chief assertion in this brief introduction was that if the moving line segment makes a fixed angle with the fixed line, and if neither of the unknown quantities occurs to a power greater than the square, then the resulting locus is a straight line, a circle, or one of the other conic sections. He proceeded to prove his result by a treatment of each of the various possible cases. Let us consider first the case of the straight line: "Let NZM be a straight line given in position, with point N fixed. Let NZ be equal to the unknown quantity A, and ZI, the line drawn to form the angle NZI, the other unknown quantity E. If D times A equals B times E, the point I will describe a straight line given in position."[11] Fermat thus began with a

FIGURE 14.4

Fermat's analysis of the
equation D times A equals
B times E

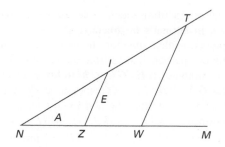

single axis NZM and a linear equation (Fig. 14.4). (Fermat used Viète's convention of vowels for unknowns and consonants for knowns.) He wanted to show that this equation, which is written in modern notation as $dx = by$, represents a straight line. Because $D \cdot A = B \cdot E$, also $B : D = A : E$. Because $B : D$ is a known ratio, the ratio $A : E$ is determined, and so also the triangle NZI. Thus, the line NI is, as Fermat wrote, given in position. Fermat dismissed as "easy" the necessary completion of the argument, to show that any point T on NI determines a triangle TWN with $NW : TW = B : D$.

Although the basic notions of modern analytic geometry are apparent in Fermat's description, Fermat's ideas differed somewhat from current ones. First, Fermat used only one axis. The curve is thought of, not as made up of points plotted with respect to two axes, but as generated by the motion of the endpoint I of the variable line segment ZI as Z moves along the given axis. Fermat often took the angle between ZI and ZN as a right angle, although there was no particular necessity for so doing. Second, to Fermat, as to Viète and most others of the time, the only proper solutions to algebraic equations were positive. Thus, Fermat's "coordinates" ZN and ZI, solutions to his equation $D \cdot A = B \cdot E$, represented positive numbers. Hence, Fermat drew only the ray emanating from the origin into the first quadrant.

Fermat's restriction to the first quadrant is quite apparent also in his treatment of the parabola: "If Aq equals D times E, point I lies on a parabola."[12] Fermat intended to show that the equation $x^2 = dy$ (in modern notation) determines a parabola. He began with the basic two line segments NZ and ZI, in this case at right angles. Drawing NP parallel to ZI, he then asserted that the parabola with vertex N, axis NP, and latus rectum D is the parabola determined by the given equation (Fig. 14.5). Fermat was, of course, assuming that his readers were very familiar with Apollonius's *Conics*. For the parabola, Apollonius's construction showed that the rectangle contained by D and NP was equal to the square on PI (or NZ), a statement translated into algebra by the equation $dy = x^2$. Although Fermat knew what a parabola looked like, his diagram only included part of half of it. He did not deal with negative lengths along the axis.

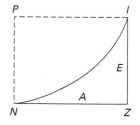

FIGURE 14.5

Fermat's analysis of the
equation Aq equals D times
E

Fermat proceeded to determine the curves represented by five other quadratic equations in two variables. In modern notation, $xy = b$ and $b^2 + x^2 = ay^2$ represent hyperbolas, $b^2 - x^2 = y^2$ represents a circle, $b^2 - x^2 = ay^2$ represents an ellipse, and $x^2 \pm xy = ay^2$ represents a straight line. In each case, his argument assumed the construction of a particular conic section according to Apollonius's procedures and showed that this conic had the desired equation. Finally, Fermat sketched a method of reducing any quadratic equation to one of his seven canonical forms, by showing how to change variables. For example, he asserted that any equation containing ax^2 and ay^2 along with bx and/or cy can be reduced to the canonical equation of a circle, provided that the angle between the axis and the line tracing out the curve is a right angle. Thus, the equation

$$p^2 - 2hx - x^2 = y^2 + 2ky$$

can be transformed by first adding k^2 to both sides so the right side becomes a square. Setting

$$r^2 = h^2 + k^2 + p^2 \quad \text{or} \quad r^2 - h^2 = k^2 + p^2,$$

Fermat could rewrite the original equation as

$$r^2 - (x + h)^2 = (y + k)^2,$$

the canonical equation of a circle if $x + h$ is replaced by x' and $y + k$ by y'. Fermat also dealt with equations containing an xy term by an appropriate change of variable.

Fermat was able to determine the locus corresponding to any quadratic equation in two variables and show that it had to be a straight line, a circle, or a conic section. To conclude his *Introduction*, Fermat noted that one could apply his methods to the following generalization of the four-line-locus problem: "If, given any number of lines in position, lines be drawn from one and the same point to each of them at given angles, and if [the sum of] the squares of all of the drawn lines is equal to a given area, the point will lie on a [conic section] given in position."[13] Fermat, however, left the actual solution to the reader.

In an appendix to the *Introduction*, Fermat began to explore the application of his ideas to the solutions of cubic and quartic equations. Thus, like Archimedes and Omar Khayyam, he showed how to find the solutions by intersecting two conic sections. But since, unlike his predecessors, he could write out the conics using algebra, he was able to simplify the process considerably. For example, to solve the equation $x^3 + bx^2 = bc$, he set each side equal to bxy, "in order that by division of this solid, on the one hand by x and on the other by b, the matter is reduced to quadratic loci."[14] In other words, the two new equations were $x^2 + bx = by$ and $c = xy$, the first a parabola and the second a hyperbola. Then the x coordinate of the intersection of the two curves (and Fermat was interested only in a single intersection) determined a solution to the original equation.

14.2.2 Descartes and the *Geometry*

Fermat's brief treatise created a stir when it reached Paris. A circle of mathematicians, centered on Marin Mersenne (1588–1648), had been gathering regularly to discuss new ideas in mathematics and physics. Mersenne acted as the recording and corresponding secretary of the group, and as such, received material from various sources, copied it, and distributed it widely. Mersenne thus served as France's "walking scientific journal." Fermat had begun a regular correspondence with Mersenne in 1636, but because many of his manuscripts were brief and lacking in detail, Mersenne often forwarded to Fermat requests to amplify his work. Nevertheless, the *Introduction* was received positively and established Fermat's name as a first-class mathematician. The manuscript had, however, reached Paris—and then Descartes—just prior to the publication of Descartes' own version of analytic geometry. One can only imagine Descartes' chagrin at seeing material similar to his own appearing before his own work reached its intended audience.

BIOGRAPHY

René Descartes (1596–1650)

Descartes was born at La Haye (now La Haye-Descartes) near Tours into a family of the old French nobility. Because he was sickly throughout his youth, he was permitted during his school years to rise late. He thus developed the habit of spending his mornings in meditation. His thoughts led him to the conclusion that little he had learned in school was certain. In fact, he became so full of doubts that he decided to abandon his studies. As he reported in his *Discourse on Method*, "I used the rest of my youth to travel, to see courts and armies, to frequent people of differing dispositions and conditions, to store up various experiences, to prove myself in the encounters with which fortune confronted me, and everywhere to reflect upon the things that occurred, so that I could derive some profit from them."[15] Thus, he participated in several campaigns during the Thirty Years War before settling in Holland in 1628 to begin his lifelong goal of creating a new philosophy suited to discovering truth about the world. He resolved to accept as true only ideas so clear and distinct that they would cause no

doubt and then to follow the model of mathematical reasoning through simple, logical steps to discern new truths. He soon wrote a major treatise on physics, but at the last minute, having heard of Galileo's condemnation by the Church, decided not to publish it for fear that a small doctrinal error might lead to the banning of his entire philosophy. He was soon persuaded, however, that he should share his new ideas with the world. In 1637 he published his *Discourse on Method*, along with three essays on optics, meteorology, and geometry designed to show the efficacy of the "method" (Fig. 14.6).

Descartes' international reputation was enhanced with the publication of several other philosophical works, and in 1649 he was invited by Queen Christina of Sweden to come to Stockholm to tutor her. He reluctantly accepted. Unfortunately, his health could not withstand the severity of the northern climate, especially since Christina required him, contrary to his long-established habits, to rise at an early hour. Descartes soon contracted a lung disease, which led to his death in 1650.

FIGURE 14.6

Descartes and his *Discours de la Méthode* on a French stamp

Descartes' analytic geometry was, nevertheless, somewhat different from that of Fermat. To understand it, one must realize that the *Geometry* was written to demonstrate the application to geometry of Descartes' methods of correct reasoning discussed in the *Discourse*, reasoning based on self-evident principles. Like Fermat, Descartes had studied the works of Viète and saw in them the key to the understanding of the analysis of the Greeks. But rather than dealing with the relationship of algebra to geometry through the study of loci, Descartes was more concerned with demonstrating this relationship through the geometric construction of solutions to algebraic equations. In some sense, then, he was merely following in the ancient tradition, a tradition that had been continued by such Islamic mathematicians as al-Khayyāmī and Sharaf al-Dīn al-Ṭūsī. But Descartes did take the same crucial step as Fermat, a step his Islamic predecessors failed to take, of using coordinates to study this relationship between geometry and algebra.

The *Geometry* begins, "Any problem in geometry can easily be reduced to such terms that a knowledge of the lengths of certain straight lines is sufficient for its construction."[16] In the first of the three books of this work, Descartes found these lengths by the use of lines and circles, the standard Euclidean curves. But Descartes made these Euclidean techniques appear modern in his clear use of algebraic techniques. For example, to find the solution of the quadratic equation $z^2 = az + b^2$, he constructed a right triangle NLM with $LM = b$ and

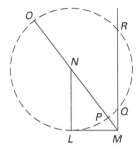

FIGURE 14.7

Descartes' construction of the solution to a quadratic equation

$LN = \frac{1}{2}a$ (Fig. 14.7). Prolonging the hypotenuse to O, where $NO = NL$, and constructing the circle centered on N with radius NO, he concluded that OM is the required value z because the value of z is given by the standard formula

$$z = \frac{1}{2}a + \sqrt{\frac{1}{4}a^2 + b^2}.$$

Under the same conditions, MP is the solution to $z^2 = -az + b^2$, while if MQR is drawn parallel to LM, then MQ and MR are the two solutions to $z^2 = az - b^2$.

Descartes noted, however, that "often it is not necessary thus to draw the lines on paper, but it is sufficient to designate each by a single letter."[17] As long as it is known what operations are possible geometrically, it is feasible just to perform the algebraic operations, and state the result as a formula. In these algebraic operations Descartes took another major step. He represented the terms aa (or a^2) and a^3 as line segments, rather than as geometric squares and cubes. Thus, he could also consider higher powers without worrying about their lack of geometric meaning. Descartes made only a brief bow to the homogeneity requirements carefully kept by Viète by noting that any algebraic expression could be considered to include as many powers of unity as necessary for this purpose, but in fact he freely added algebraic expressions whatever the power of the terms. Furthermore, Descartes replaced Viète's (and Harriot's) vowel-consonant distinction of unknowns and knowns with the current usage of letters near the end of the alphabet for unknowns and those near the beginning for knowns.

Descartes concluded his first book with a detailed discussion of Apollonius's problem of the four lines. It is here that he introduced a coordinate axis to which all the lines as well as the locus of the solution is referred. The problem requires the finding of points from which lines drawn to four given lines at given angles satisfy the condition that the product of two of the line lengths bears a given ratio to the product of the other two. Descartes noted that "since there are always an infinite number of different points satisfying these requirements, it is also required to discover and trace the curve containing all such points."[18]

Using Figure 14.8, Descartes noted that matters are simplified if all lines are referred to two principal ones. Thus, he set x as the length of segment AB along the given line EG and y as the length of segment BC along the line BC to be drawn, where C is one of the points satisfying the requirements of the problem. The lengths of the required line segments, CB, CH, CF, and CD (drawn to the given lines EG, TH, FS, and DR, respectively), can each be expressed as a linear function of x and y. For example, because all angles of the triangle ARB are known, the ratio $BR : AB = b$ is also known. It follows that $BR = bx$ and $CR = y + bx$. Because the three angles of triangle DRC are also known, so is the ratio $CD : CR = c$, and therefore $CD = cy + bcx$. Similarly, setting the fixed distances $AE = k$ and $AG = \ell$ and the known ratios $BS : BE = d$, $CF : CS = e$, $BT : BG = f$, and $CH : TC = g$, one shows in turn that $BE = k + x$, $BS = dk + dx$, $CS = y + dk + dx$, $CF = ey + dek + dex$, $BG = \ell - x$, $BT = f\ell - fx$, $CT = y + f\ell - fx$, and finally $CH = gy + fg\ell - fgx$. Because the problem involves comparing the products of certain pairs of the line lengths, it follows that the equation expressing the desired locus is a quadratic equation in x and y. Furthermore, as many points of the locus as desired can be constructed, because if any value of y is given, the value of x is expressed in the form of a determinate quadratic equation whose solution has already been provided. The required curve can then be drawn. In book two of the *Geometry*, Descartes returned to this problem and showed that the curve given by the

FIGURE 14.8

The problem of the four-line locus

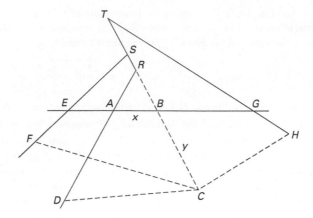

quadratic equation in two variables is either a circle or one of the conic sections, depending on the values of the various constants involved.

Now Pappus (and Apollonius) already knew this result. What Descartes wanted to do was show that his new methods enabled him to go further and solve the problem for arbitrarily many lines. Thus, he presented a five-line problem: Suppose four of the lines are equally spaced parallel lines and the fifth perpendicular to each of the others. Suppose that the lines drawn from the required point meet the given lines at right angles. Then it is required to find the locus of all points such that the product of the lengths of the lines drawn to three of the parallel lines is equal to the constant spacing times the product of the lengths of the lines drawn to the remaining two lines. To solve this problem, let the four parallel lines be L_1, L_2, L_3 and L_4, the perpendicular line be L_5, and the constant spacing be a. Let P be a point satisfying the problem, and let d_1, d_2, d_3, d_4, and d_5 be the perpendicular distances from P to the five line segments (Fig. 14.9). Letting $d_5 = x$ and $d_3 = y$, we calculate that $d_1 = 2a - y$, $d_2 = a - y$, and $d_4 = y + a$. The conditions of the problem that $d_1d_2d_4 = ad_3d_5$ then give $(2a - y)(a - y)(y + a) = ayx$, or $y^3 - 2ay^2 - a^2y + 2a^3 = axy$.

With the equation known, Descartes now intended to construct the desired curve, both point by point and as a single unit. In fact, in order to establish his main thesis about constructibility of solutions to geometric problems, he had to decide what curves were acceptable in geometric constructions. He based his definition of such curves on Euclid's postulates 1 and 3 on drawing straight lines and circles and another axiom (see below) that probably came out of his study of Pappus's problem, beginning around 1632.[19] So let us consider a four-line problem where three of the lines L_1, L_2, and L_3 are parallel and equidistant (with distance a), while the fourth line L_0 is perpendicular to the other three. The problem is to determine the curve whose points P satisfy $d_0d_1 = \alpha d_2d_3$, where d_i is the perpendicular distance of P to L_i (Fig. 14.10). It is straightforward to determine the equation of the curve. Namely, let $P = (x, y)$, where $x = d_3$ is the distance of P from L_3 and $y = d_0$ is the distance of P from L_0. Then $d_1 = a - x$ and $d_2 = a + x$. Thus, the equation of the curve is

$$y(a - x) = \alpha(a + x)x.$$

FIGURE 14.9

The five-line problem, with
ruler and parabola

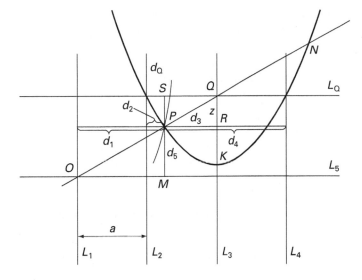

FIGURE 14.10

The four-line problem, with
ruler and straight line

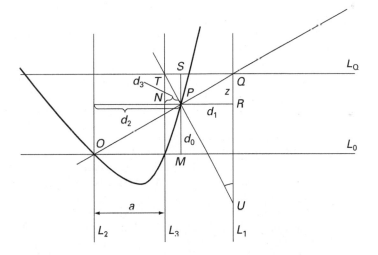

But Descartes may well have looked at this problem in a slightly different way, beginning by taking the intersection point O of L_2 and L_0 and drawing the straight line OP, whose extension intersects L_1 at Q. Now, draw line L_Q through Q parallel to L_0, and set $z = QR = PS$. By similarity, $z : d_1 = d_0 : d_2$, or $zd_2 = d_0d_1$. It follows, by comparing this with the defining property of the locus, that $z = \alpha d_3$. Because z is the distance of P from L_Q, we know that P is on the locus of points whose distances from L_Q and L_3 have a constant ratio. Such a locus is a straight line that passes through the intersection T of L_Q and L_3. Now as P varies, Q moves up and down along L_1. But since L_1 is a fixed distance from L_3, the distance QU and the angle at U must also remain constant, where U is the intersection of the straight line through T with L_1. It follows that this entire straight line moves up and down with Q.

Therefore, as the "ruler" OPQ pivots around O, the desired four-line locus is traced by the intersection of this ruler with the moving straight line UPT.

We can now use the definition of the four-line locus as an intersection to give a new derivation of the equation of the desired curve. Thus, set $QU = c$. Because $QT = a$, the similarity of triangles TQU and TNP implies that $QR = TN = cx/a$. The similarity of triangles QRP and PMO then implies that $QR : (a - x) = y : (a + x)$, or

$$\frac{cx}{a}(a + x) = (a - x)y.$$

Because $\alpha = z/d_3 = TN/PN = c/a$, this equation is the same as the one we found by the original Pappus condition. The curve defined by this equation is in fact a hyperbola.

It was probably from considerations of this type of problem that Descartes stated his third construction axiom: "Two or more lines can be moved, one upon the other, determining by their intersection other curves."[20] That is, he accepted in geometry precisely those curves traced by some continuous motion generated by certain machines. In the *Geometry*, he gave several examples of instruments designed to trace such curves. In particular, because such a machine could generate a parabola, he created a new machine to solve the original five-line problem by replacing the straight line TPU by a new "line," namely, a parabola (see Fig. 14.9). That is, the point P in this case lies on the ruler OPQ pivoting at O (forcing Q to move along L_3) but also on a parabola moving so that its axis remains along L_3. With the parameter of the parabola chosen appropriately, Descartes showed that this new machine generated the curve $y^3 - 2ay^2 - a^2y + 2a^3 = axy$, the solution to the five-line problem.

14.2.3   Descartes and Equation Solving

Besides constructing (or "tracing") the solution curve as a whole, Descartes also wanted to be able to construct points on it. To do so required a geometric solution of a cubic equation. Thus, in his third book, Descartes discussed the solution of algebraic equations. He began by quoting—almost—Girard's result that "every equation can have as many distinct roots as the number of dimensions of the unknown quantity in the equation."[21] Descartes used "can have" rather than Girard's "admits of" because he only considered distinct roots and because, at least initially, he did not want to consider imaginary roots. Later on, however, he noted that roots are sometimes imaginary and that "while we can always conceive of as many roots for each equation as I have already assigned [that is, as many as the dimension], yet there is not always a definite quantity corresponding to each root so conceived of."[22]

Descartes showed explicitly how equations are built up from their solutions. Thus, if $x = 2$ or $x - 2 = 0$ and if also $x = 3$ or $x - 3 = 0$, Descartes noted that the product of the two equations is $x^2 - 5x + 6 = 0$, an equation of dimension 2 with the two roots 2 and 3. Again, if this latter equation is multiplied by $x - 4 = 0$, there results an equation of dimension 3, $x^3 - 9x^2 + 26x - 24 = 0$, with the three roots 2, 3, and 4. Multiplying further by $x + 5 = 0$, an equation with a "false" root 5, produces a fourth-degree equation with four roots, three "true" and one "false." Descartes concluded that "it is evident from the above that the sum of an equation having several roots [that is, the polynomial itself] is always divisible by a binomial consisting of the unknown quantity diminished by the value of one of the true roots, or plus the value of one of the false roots. In this way, the degree of an equation can be lowered. Conversely, if the sum of the terms of an equation is not divisible by a binomial consisting of the unknown quantity plus or minus some other quantity, then this latter quantity is not a root of the equation."[23] This is the earliest statement of the modern **factor theorem**. In his usual fashion, Descartes did not give a complete proof, just writing that the result is "evident."

Similarly, Descartes also stated without proof the result today known as **Descartes' Rule of Signs**: "An equation can have as many true [positive] roots as it contains changes of sign, from + to − or from − to +; and as many false [negative] roots as the number of times two + signs or two − signs are found in succession."[24] As illustration, the equation $x^4 - 4x^3 - 19x^2 + 106x - 120 = 0$ has three changes of sign and one pair of consecutive minus signs. Thus, it can have up to three positive roots and one negative one. In fact, the roots are 2, 3, 4, and −5.

Descartes was, however, primarily interested in the construction of solutions to equations, so toward the end of the third book he demonstrated explicitly some construction methods for equations of higher degree. In particular, for equations of degree 3 or 4, he used the intersection of a parabola and a circle, both of which meet his criteria for constructible curves. Descartes' methods are similar to those of al-Khayyāmī, but, unlike his Islamic predecessor, Descartes realized that certain intersection points represented negative (false) roots of the equation and also that "if the circle neither cuts nor touches the parabola at any point, it is an indication that the equation has neither a true nor a false root, but that all the roots are imaginary."[25]

Descartes showed further how to solve equations of degree higher than the fourth by intersecting a circle with a curve constructed by one of his machines. Although he only briefly sketched his methods and applied them to a few examples, Descartes believed that it was "only necessary to follow the same general method to construct all problems, more and more complex, ad infinitum; for in the case of a mathematical progression, whenever the first two or three terms are given, it is easy to find the rest."[26] Over the remainder of the century, various mathematicians attempted to generalize Descartes' methods to find other geometrical means for constructing the solutions to various types of equations. Geometric methods, however, proved inadequate to gaining a complete understanding of the nature of such solutions. It turned out that algebraic methods as well as the new ideas of calculus were better suited for solving even the kinds of geometrical problems to which Descartes applied his construction techniques.

14.2.4 Descartes and Geometric Curves

How did Descartes decide on his defining characteristic of "geometric" curves as ones traced by continuous motions? It appears that his basic reason for defining such curves was that "all points of those curves . . . must bear a definite relation to all points of a straight line, and that this relation must be expressed by means of a single equation."[27] In other words, any such curve is expressible as an algebraic equation. But then, because Descartes could also construct solutions to polynomial equations, he saw that such a curve could be constructed pointwise. Finally, he was convinced that a curve for which any point could be constructed could also be generated by continuous motion. In other words, Descartes was convinced, although he never wrote down an actual proof, that curves defined by algebraic equations in two variables were precisely those whose construction could be realized by an appropriate machine.

There are probably several reasons why Descartes defined "geometric" curves by continuous motion rather than directly as curves having an algebraic equation. First, he was interested in reforming the study of geometry. Defining acceptable curves by a completely algebraic criterion would have reduced his work to algebra. Second, because he wanted to be able to construct the solution points of geometric problems, he needed to be able to determine intersections of algebraic curves. It was evident to him that defining curves by continuous motion would explicitly determine intersection points. It was not at all evident that curves defined by algebraic equations had intersection points. Because he was studying geometry, Descartes could not adopt the algebraic definition as an axiom. Finally, Descartes evidently was not convinced that an algebraic equation was the best way to define a curve. Nowhere in the *Geometry* did he begin with an equation. Unlike Fermat, Descartes always described a curve geometrically and then, if appropriate, derived its equation. An equation for Descartes was thus only a tool in the study of curves and not the defining criterion.

It must also be asked, on the other hand, why Descartes rejected curves not definable geometrically. He was certainly aware of curves without algebraic equations. An ancient example was the quadratrix, defined by a combination of a rotary and a linear motion (see Fig. 4.6). What bothered Descartes about such a curve, as it had also bothered the ancients, was that the two motions had no exact, measurable relation, because one could not precisely determine the ratio of the circumference of the circle to its radius. As Descartes wrote, "the ratios between straight and curved lines are not known, and I believe cannot be discovered by human minds, and therefore no conclusion based upon such ratios can be accepted as rigorous and exact."[28] Unfortunately for Descartes, the first determination of the exact lengths of various curves in the 1650s as well as the study of areas under his nongeometric (or transcendental) curves soon undermined Descartes' basic distinction between acceptable and nonacceptable curves in geometry.

14.2.5   Descartes versus Fermat

It is clear that both Fermat and Descartes understood the basic connection between a geometric curve and an algebraic equation in two unknowns. Both used as their basic tool a single axis along which one of the unknowns was measured rather than the two axes used today, and neither insisted that the lines measuring the second unknown intersect the single axis at right angles. Both used as their chief examples the familiar conic sections, although both were also able to construct curves whose equations were of degree higher than two. And both recognized a new relationship of algebra to geometry. Recall that algebra grew out of some simple manipulation of geometric shapes. Then, during the medieval period and the Renaissance, algebra gradually freed itself from geometry. But now, algebra returned to the service of geometry. It became a much more flexible tool that could be used not only to determine solutions of equations but also to find entire curves. It therefore was available for use in the study of motion, a study central in the development of calculus.

Descartes and Fermat came at the subject of analytic geometry, however, from different viewpoints. Fermat gave a very clear statement that an equation in two variables determines a curve. He always started with the equation and then described the curve. Descartes, on the other hand, was more interested in geometry. For him, the curves were primary. Given a geometric description of a curve, he was able to come up with the equation. Thus, Descartes was forced to deal with algebraic equations considerably more complex than those of Fermat. It was this very complication of Descartes' equations that led him to discover methods of dealing with polynomial equations of high degree.

Descartes and Fermat emphasized the two different aspects of the relationship between equations and curves. Unfortunately, Fermat never published his work. Although it was presented clearly and circulated through Europe in manuscript, it never had the influence of a published work. Descartes' work, conversely, proved very difficult to read. It was published in French, rather than the customary Latin, and had so many gaps in arguments and complicated equations that few mathematicians could fully understand it. Descartes was actually proud of the gaps. He wrote at the end of the work, "I hope that posterity will judge me kindly, not only as to the things which I have explained, but also as to those which I have intentionally omitted so as to leave to others the pleasure of discovery."[29] But a few years after the publication of the *Geometry*, Descartes changed his mind somewhat. He encouraged other mathematicians to translate the work into Latin and to publish commentaries to explain what he had intended. It was only after the publication of the Latin version by Frans van Schooten (1615–1660), a professor at the engineering school in Leiden, first in 1649 with commentary by van Schooten himself and by Florimond Debeaune (1601–1652) and then with even more extensive commentaries and additions in 1659–1661, that Descartes' work achieved the recognition he desired.

14.2.6 The Work of Jan de Witt

One of the additions to van Schooten's 1659–1661 edition of Descartes' *Geometry* was a treatise on conic sections by Jan de Witt (1623–1672). In his student days, de Witt had studied with van Schooten, who had known Descartes and had studied Fermat's works during a sojourn in Paris. Through van Schooten, de Witt became acquainted with the works of both of the inventors of analytic geometry. In 1646, at the age of 23, he composed the *Elementa curvarum linearum* (*Elements of Curves*) in which he treated the subject of conic sections from both a synthetic and an analytic point of view. The first of the two books of the *Elements* was devoted to developing the properties of the various conic sections using the traditional methods of synthetic geometry. In the second book, the first systematic treatise on conic sections using the new method, de Witt extended Fermat's ideas into a complete algebraic treatment of the conics beginning with equations in two variables. Although the methodology was similar to that of Fermat, de Witt's notation was the modern one of Descartes.

For example, in theorem I, de Witt proved that the equation $y = \frac{bx}{a}$ has for its locus a straight line. De Witt's proof is similar to that of Fermat, except that he explicitly showed by use of similarity that any point with coordinates x, y on his constructed line satisfies the relationship $a : b = x : y$. Like Fermat, de Witt only dealt with positive values in both his constants and his coordinates, so the desired line only appears as a ray emanating from the origin A. But he went further by next showing that several other equations also determine straight lines: $y = \frac{bx}{a} + c$, $y = \frac{bx}{a} - c$, $y = c - \frac{bx}{a}$, $y = c$, and $x = c$. Only that part of the line lying in the first quadrant was drawn, however, in each case.

Again, de Witt proceeded like Fermat to show that $y^2 = ax$ represents a parabola. He also showed the graphs of parabolas determined by such equations as $y^2 = ax + b^2$, $y^2 = ax - b^2$, $y^2 = b^2 - ax$ and the equations formed from these by interchanging x and y. As before, de

BIOGRAPHY

Jan de Witt (1623–1672)

Jan de Witt was a talented mathematician who, because of his family background, could devote little time to mathematics (Fig. 14.11). Born into a politically active Dutch family, he became a leader in his hometown of Dort and, after the death of Prince William II of Orange, was appointed in 1653 to the position of grand pensionary of Holland, in effect the prime

minister. He guided Holland through difficult times over the next 19 years, successfully balancing the conflicting demands of England and France. When France invaded Holland in 1672, however, the people called William III to return to power. Violent demonstrations ensued against de Witt, and he was murdered by an infuriated mob.

FIGURE 14.11

Jan de Witt on a Dutch stamp

Witt showed only that part of the graph for which both x and y are positive. But he also considered in detail the more complicated equation

$$y^2 + \frac{2bxy}{a} + 2cy = bx - \frac{b^2x^2}{a^2} - c^2.$$

Setting $z = y + bx/a + c$ reduced this equation to

$$z^2 = \frac{2bc}{a}x + bx,$$

or, with $d = \frac{2bc}{a} + b$, to $z^2 = dx$, an equation that de Witt knew represents a parabola. He then showed how to use this transformation to draw the locus. If the coordinates of D are (x, y) using AE as the x axis and AF as the y axis, set $BE = AG = c$ and extend DB to C such that $GB : BC = a : b$ or $BC = bx/a$ (Fig. 14.12). It follows that $DC = y + c + bx/a = z$. Also, setting $GB : GC = a : e$ gives $GC = ex/a$. In modern terminology, de Witt had used the transformation $x = (a/e)x'$; $y = z - (b/e)x' - c$ to convert from the oblique axes AE and AF to the perpendicular axes GC and GF; thus, the point D with coordinates (x, y) has new coordinates (x', z) related to the original ones via this transformation. In the new coordinates, the equation of the curve is $z^2 = (da/e)x'$, a parabola with vertex G, axis GC, and latus rectum of length da/e. De Witt could thus draw this parabola, or, more particularly,

FIGURE 14.12

De Witt's construction of the parabola $y^2 + \frac{2bxy}{a} + 2cy = bx - \frac{b^2x^2}{a^2} - c^2$

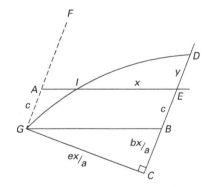

the part ID above the original axis AE, because only that part can serve as the desired locus. De Witt completed the proof by noting that given an arbitrary point D on this locus, the basic property of a parabola implies that the square on DC equals the rectangle on $GC(=ex/a)$ and the latus rectum da/e. Thus, $z^2 = dx$, and substituting $z = y + c + bx/a$ gives back the original equation.

De Witt similarly gave detailed treatments of both the ellipse and the hyperbola, presenting standard forms such as

$$\frac{ey^2}{g} = f^2 - x^2 \quad \text{(ellipse)}$$

and

$$\frac{ey^2}{g} = x^2 - f^2 \quad \text{and} \quad xy = f^2 \quad \text{(hyperbolas)}$$

first and then showing how other equations can be reduced to one of these by appropriate substitutions. Although de Witt did not state the conditions on the original equation that determine whether the locus is a parabola, ellipse, or hyperbola, it is easy enough to discover these by analyzing his examples. He concluded his work by noting that any quadratic equation in two variables can be transformed into one of the standard forms and therefore represents a straight line, circle, or conic section. Although both Fermat and Descartes had sketched this same result, it was de Witt who provided all the details to solve the locus problem for quadratic equations.

 14.3 ELEMENTARY PROBABILITY

The modern theory of probability is usually considered to begin with the correspondence of Pascal and Fermat in 1654, partially in response to the gambling questions de Méré raised to Pascal (noted in the opening of the chapter). But because gambling is one of the oldest leisure activities, it would seem that from earliest times people had considered the basic ideas of probability, at least on an empirical basis, and, in particular, had some vague conception of how to calculate the odds of the occurrence of any given event in a gambling game. Dice from several ancient cultures have been found. Although it is not always known what the purpose of these objects was, there are strong indications that they were used for predicting the future and for gaming. Unfortunately, no written evidence survives from any of these civilizations about how the various games were played and whether any calculations of odds were made.

A little more is known about such calculations in Jewish sources dating back to the early years of the common era, although these were concerned with the application of various Jewish laws rather than with games. The Talmud, the Jewish work recording the discussions of the rabbis in their interpretations of Jewish law, contains applications of, for example, the laws of addition and multiplication for determining the probability of events compounded of events of known probability. Such probabilities are then used in determining, at least approximately, what we consider the "expectation" of an event in order to justify various decisions. For example, in a marriage contract the *ketubah* is the amount that a husband must pay the wife if he divorces her or dies before her. If she dies first, on the other hand, he inherits all her possessions. But the wife's rights to her *ketubah* are a saleable commodity, so the question becomes how much one should pay the wife for these rights. The rabbis of talmudic times (third–sixth centuries) did not actually try to estimate this value, but Maimonides, writing in the twelfth century, noted that this value depended on such conditions as the health of the wife and whether there was peace between husband and wife. Similar types of calculations occur in Roman law, where there are discussions of the appropriate pricing of a life annuity, the rights to an inheritance, or a maritime loan. However, in none of these cases are there records of detailed numerical calculations.[30]

14.3.1   The Earliest Beginnings of Probability Theory

In Europe in the late Middle Ages, some elementary probabilistic ideas connected with dice playing were spelled out. For example, there are several documents that calculate the number of different ways two or three dice can fall, 21 ways in the case of two dice and 56 in the case of three. These numbers are correct, assuming one only counts the different sets of dots that can occur, without examining the order in which they happen. Thus, in the case of two dice, there is one way to roll a 2, one way to roll a 3, two ways to roll a 4 (2, 2 and 1, 3), two ways to roll a 5 (1, 4 and 2, 3), and so on. In modern terms, these ways are not "equiprobable" (equally likely) and could not serve as the basis for calculating odds in play. But counting the ways the dice could fall most likely came from the earlier use of dice in divination, where it was the actual dice faces showing that determined the future and where odds were not involved. The earliest known comment that the 56 ways three dice fall are not equiprobable occurs in an anonymous Latin poem *De vetula* written sometime between 1200 and 1400: "If all three [dice] are alike there is only one way for each number; if two are alike and one different there are three ways; and if all are different there are six ways."[31] An analysis of the situation according to the stated rule then shows that the total number of ways for three dice to fall is 216 (Fig. 14.13).

By the sixteenth century, the idea of equiprobable events was beginning to be understood, and thus it became possible for actual probability calculations to be made. The earliest systematic attempt to make these calculations is in the *Liber de ludo aleae* (*Book on Games of Chance*) written about 1526 by Cardano, although not published during his lifetime. Besides counting accurately the number of ways two or three dice can fall, Cardano demonstrated an understanding of the basic notions of probability. Thus, having counted that there are 11 different throws of two dice in which a 1 occurs, 9 additional ones in which a 2 occurs, and 7 more in which a 3 occurs, he calculated that for the problem of throwing a 1, 2, or 3 there are 27 favorable occurrences and 9 unfavorable ones, and therefore the odds are 3 : 1. It follows that a fair wager would be 3 coins for the one betting on getting a 1, 2, or 3 versus 1 coin for the player betting against, because in four throws they would expect to come out even.

Cardano also was aware of the multiplication rule of probabilities for independent events, but in his book he recorded his initial confusion as to what exactly should be multiplied. Thus, he calculated that the chances of at least one 1 appearing in a toss of three dice is 91 out of 216, so the odds against are 125 to 91. To determine the odds against throwing at least one 1 in two successive rolls, he squared the odds and calculated the result as 15,625 to 8281, or approximately 2 to 1. After consideration of the matter, however, he noted that this reasoning must be false, because if the chances of a given event are even (odds of 1 : 1), the reasoning would imply that the chances would still be even of the given event occurring twice or three times in succession. This, he noted, is "most absurd. For if a player with two

FIGURE 14.13

Page from *De vetula* showing all 56 ways three dice can fall. (*Source*: The Houghton Library, Harvard University)

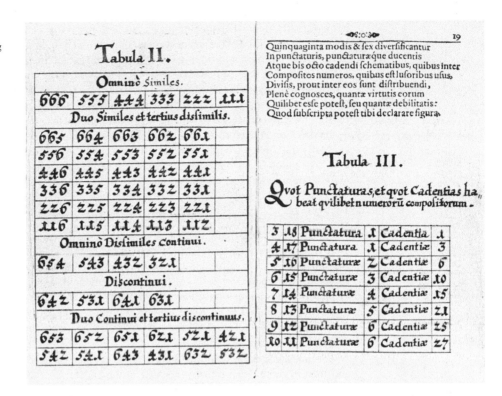

dice can with equal chances throw an even and an odd number, it does not follow that he can with equal fortune throw an even number in each of three successive casts."[32] Cardano then proceeded to correct his error. After careful calculation in some easy cases, he realized that it is the probabilities that must be multiplied and not the odds. Thus, by counting in a case where the odds for success are 3 to 1, or the probability of success is $\frac{3}{4}$, he showed that for two successive plays, there are 9 chances of repeated success and 7 otherwise. Therefore, the probability of succeeding twice is $\frac{9}{16}$, while the odds are 9 to 7 in favor. He then generalized and noted that for n repeated trials in a situation with f possible outcomes and s successes, the correct odds in favor are s^n to $f^n - s^n$.

Cardano also discussed the problem de Méré would pose to Pascal of determining how many throws must be allowed to provide even odds for attaining two sixes on a pair of dice, a problem that evidently was popular for years. Cardano argued that since there is 1 chance in 36 of throwing two sixes, on average such a result will occur once every 36 rolls. Therefore, the odds are even that one will occur in half that number of rolls, or 18. He similarly argued that in dealing with one die, there are even odds that a 2 would appear in 3 rolls. Cardano's reasoning implied that in 6 rolls of one die a 2 is certain or in 36 rolls of two dice a double 6 is certain, but he did not realize his error.

The problem of de Méré on the division of the stakes had also been considered earlier in Italy, in particular, in the *Summa* of Luca Pacioli. Pacioli's version of the problem has two players playing a fair game that was to continue until one player had won six rounds. The game actually stops when the first player has won five rounds and the second three. Pacioli's answer to the division of the stakes was that they should be split in the ratio of 5 to 3. Tartaglia, in his *Generale trattato* written some 60 years later, noted that this answer must be wrong, for the reasoning implied that if the first player had won one round and the second none when the game was suspended, the first player would collect all of the stakes, an obviously unjust result. Tartaglia argued that since the difference between the two scores was two games, one-third of the number needed to win, the first player should take one-third of the second's share of the stake, and therefore the total stake should be divided in the ratio of 2 to 1. Tartaglia was evidently not entirely confident of his answer either, for he concluded that "the resolution of such a question is judicial rather than mathematical, so that in whatever way the division is made there will be cause for litigation."[33]

14.3.2   Blaise Pascal, Probability, and the Pascal Triangle

The ideas of Cardano and Tartaglia on probability, however, were not taken up by others of their time and were forgotten. It was only in the decade surrounding 1660 that probability entered European thought, and then in two senses: first as a way of understanding stable frequencies in chance processes and second as a method of determining reasonable degrees of belief. The work of Blaise Pascal (1623–1662) exemplified both of these senses. In his mathematical answer to de Méré's division problem, Pascal dealt with a game of chance, while in his decision-theoretic argument for belief in God there is no concept whatever of chance.

Pascal described his solution to the division problem in several letters to Fermat in 1654 and then in more detail a few years later at the end of his *Traité du triangle arithmétique* (*Treatise on the Arithmetical Triangle*). He began with two basic principles to apply to the division. First, if the position of a given player is such that a certain sum belongs to him whether he wins or loses, he should receive that sum even if the game is halted. Second, if the position of the two players is such that if one wins, a certain sum belongs to him and if he loses, it belongs to the other, and if both players have equally good chances of winning, then they should divide the sum equally if they are unable to play.

Pascal next noted that what determines the split of the stakes is the number of games remaining and the total number that the rules say either player must win to obtain the entire stake. Therefore, if they are playing for a set of two games with a score of 1 to 0, or for a set of three games with the score 2 to 1, or for a set of eleven games with the score 10 to 9, the results of the division of the stakes at the time of interruption should all be the same. In all these cases, the first player needs to win one more game, while the second player needs two.

As an example of Pascal's principles, suppose that the total stake in the contest is 80 dollars. First, if each player needs one game to win and the contest is stopped, simply divide the 80 dollars in half, so each gets 40. Second, suppose that the first player needs one game to win and the second player two. If the first player wins the next game, he will win the 80 dollars. If he loses, then both players will need one game, so by the first case, the first player will win 40 dollars. If they stop the contest now, the first player is therefore entitled to the 40 dollars he would win in any case plus half of the remaining 40, that is, to 60 dollars, the mean of the two possible amounts he could win. Similarly, if the first player needs one game to win while the second player needs three games, there are two possibilities for the next game. If the first player triumphs, he wins the 80 dollars, while if he loses, the situation is the same

BIOGRAPHY

Blaise Pascal (1623–1662)

Pascal was born in Clermont Ferrand, France, and showed his mathematical precocity very early. His father, Étienne, introduced him as a young man to the circle around Mersenne. Thus, the young Pascal was soon acquainted with the major mathematical developments in France, including the work of Fermat. He began his own mathematical and scientific researches before he was 20. Among his accomplishments were the invention of a calculating machine and the investigation of the action of fluids under the pressure of air. After 1654, however, his scientific interests were overshadowed by an increasing interest in religious matters. Never in good health, he died at the age of 39 after a violent illness (Fig. 14.14).

FIGURE 14.14

Pascal on a French stamp

as in the second case, in which he is entitled to 60 dollars. It follows that if the next game is not played, the first player should receive 60 dollars plus half of the remaining 20, that is, 70 dollars, the mean of his two possible winnings.

The general solution to the division problem, it turns out, requires some of the properties of Pascal's triangle. Before considering Pascal's solution, therefore, we must first look at his construction and use of what he called the **arithmetical triangle**, the triangle of numbers that had been used in various parts of the world already for more than 500 years. Pascal's *Treatise on the Arithmetical Triangle*, famous also for its explicit statement of the principle of mathematical induction, began with his construction of the triangle starting with a 1 in the upper left-hand corner and then using the rule that each number is found by adding together the number above it and the number to its left (Fig. 14.15). In the discussion of Pascal's results, however, it is clearer to use the modern table and modern notation to identify the various entries in the triangle (Table 14.1). The standard binomial symbol $\binom{n}{k}$ is used to name the kth entry in the nth row (where the initial column and initial row are each numbered 0). The basic construction principle is then that

$$\binom{n}{k} = \binom{n-1}{k} + \binom{n-1}{k-1}.$$

Pascal began his study by considering how various entries are related to sums of others. His proofs were usually by the method of "generalizable example," because, like his forebears, he had no good way of symbolizing general terms. For example, Pascal's "third consequence" (of the definition of the triangle) states that any entry is the sum of all the elements in the preceding column up to the preceding row:

$$\binom{n}{k} = \sum_{j=k-1}^{n-1} \binom{j}{k-1}.$$

In this case, Pascal took as his example the particular entry $\binom{4}{2}$, which, by the method of construction, is equal to $\binom{3}{1} + \binom{3}{2}$. Because $\binom{3}{2} = \binom{2}{1} + \binom{2}{2}$ and $\binom{2}{2} = \binom{1}{1}$, the result followed.

Pascal's proof of the eighth consequence, that the sum of the elements in the nth row is equal to 2^n, is by mathematical induction, where the inductive step, going from k to $k+1$, is

FIGURE 14.15

Pascal's version of the
arithmetical triangle

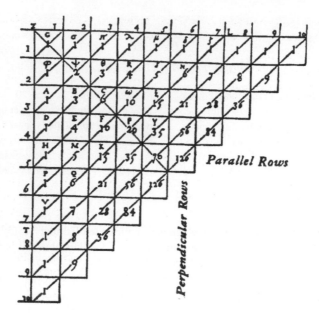

TABLE 14.1   *The Modern Arithmetical Triangle.*

Row	Column								
	0	1	2	3	4	5	6	7	8
0	1								
1	1	1							
2	1	2	1						
3	1	3	3	1					
4	1	4	6	4	1				
5	1	5	10	10	5	1			
6	1	6	15	20	15	6	1		
7	1	7	21	35	35	21	7	1	
8	1	8	28	56	70	56	28	8	1

accomplished in the seventh consequence: the sum of the elements of any row is double that of the preceding row. The proof of that proposition is again by the method of generalizable example. Pascal considered a particular row, the third, noted that the first and last entries are equal to the first and last entries in the second row and that every other entry in the third row is equal to the sum of two entries in the second row. Thus, the sum of the entries in the third row included each element of the second row twice. Pascal completed the demonstration of the eighth consequence by simply noting that the 0th row has a single 1 in it, thus its sum equals 2^0, while each succeeding row is double the previous one.

Curiously, it is only in the proof of the twelfth consequence that Pascal stated the principle of mathematical induction explicitly, not in all generality but just in the context of the specific result to be proved:

$$\binom{n}{k} : \binom{n}{k+1} = (k+1) : (n-k).$$

Pascal noted that "although this proposition has an infinity of cases, I shall demonstrate it very briefly by supposing two lemmas," namely, the two basic parts of an induction argument. "The first, which is self evident, [is] that this proportion is found in the [first row], for it is perfectly obvious that $[\binom{1}{0} : \binom{1}{1} =]1 : 1$. The second [is] that if this proportion is found in any [row], it will necessarily be found in the following [row]. Whence it is apparent that it is necessarily in all the [rows]. For it is in the second [row] by the first lemma; therefore by the second lemma it is in the third [row], therefore in the fourth, and to infinity."[34] Although this is a clear statement of the induction principle for the specific case at hand, and of the reason for its use in demonstrating a general result, Pascal again did not prove the second lemma generally but only showed that the truth of the lemma in the third row implied its truth in the fourth. Thus, to demonstrate that $\binom{4}{1} : \binom{4}{2} = 2 : 3$, he first noted that $\binom{3}{0} : \binom{3}{1} = 1 : 3$ and therefore that $\binom{4}{1} : \binom{3}{1} = \left(\binom{3}{1} + \binom{3}{0}\right) : \binom{3}{1} = 4 : 3$. Next, since $\binom{3}{1} : \binom{3}{2} = 2 : 2$, it follows that $\binom{4}{2} : \binom{3}{1} = \left(\binom{3}{2} + \binom{3}{1}\right) : \binom{3}{1} = 4 : 2$. The desired result comes from dividing the first of these two proportions by the second. Pascal was aware that this proof is not general, for he completed it by noting that "the proof is the same for all other [rows], since it requires only that the proportion be found in the preceding [row], and that each [entry] be equal to the [entry above it and the entry to the left of that one], which is everywhere the case."[35] In any case, this twelfth consequence enabled Pascal to demonstrate easily, by compounding ratios, that

$$\binom{n}{k} : \binom{n}{0} = (n-k+1)(n-k+2)\cdots n : k(k-1)\cdots 1$$

or, since $\binom{n}{0} = 1$, that

$$\binom{n}{k} = \frac{n(n-1)\cdots(n-k+1)}{k!}.$$

Having set out the basic properties of the arithmetical triangle, Pascal showed how to apply it in several areas. He demonstrated, using an argument by induction, that $\binom{n}{k}$ equals the number of combinations of k elements in a set of n elements. He showed that the row entries in the triangle are the binomial coefficients, that is, that the numbers in row n are the coefficients of the powers of a in the expansion $(a+1)^n$. But one of the more important applications of the triangle, Pascal believed, was to the problem of the division of stakes, which he solved via the following

THEOREM *Suppose that the first player lacks r games of winning the set while the second player lacks s games, where both r and s are at least 1. If the set of games is interrupted at*

this point, the stakes should be divided so that the first player gets that proportion of the total as $\sum_{k=0}^{s-1} \binom{n}{k}$ is to 2^n, where $n = r + s - 1$ (the maximum number of games left).

The theorem asserts that the probability of the first player winning is the ratio of the sum of the first s terms of the binomial expansion of $(1 + 1)^n$ to the total 2^n. One can consider the first term of the expansion as giving the number of chances for the first player to win n points, the second the number of chances to win $n - 1$, and so on, while the sth term gives the number of chances to win $n - (s - 1) = r$ points. Since one may as well assume that in fact exactly n more games must be played, these coefficients give all of the ways the first player can win.

Pascal proved the theorem by induction, beginning with the case where $n = 1$, or $r = s = 1$, the case where the stakes should be evenly split. The assertion of the theorem is that they should be divided so that the first player gets the proportion $\binom{1}{0}$ to 2, or 1/2, and therefore the result is true for $n = 1$. The next step is to assume that the result is true when the maximum number of games left is m and prove it for the case where the maximum number of games left is $m + 1$, where the first player lacks r games and the second s games. As before, Pascal's proof of this inductive step is by a generalizable example, taking $m = 3$. But we will provide the complete proof, using modern notation. Consider the two possibilities if the players were to play one more game. If the first player wins, he would then lack $r - 1$ games while the second player would still lack s games. Since $r - 1 + s - 1 = m$, the induction hypothesis shows that the first player should get that proportion of the stakes that $\sum_{k=0}^{s-1} \binom{m}{k}$ is to 2^m. On the other hand, if the first player loses the next game, the induction hypothesis shows that he should be awarded that proportion of the stakes that $\sum_{k=0}^{s-2} \binom{m}{k}$ is to 2^m. Thus, by Pascal's basic principles, the award to the first player if that next game is not played should be the mean of those two values, namely, the proportion of the stakes that

$$\sum_{k=0}^{s-1} \binom{m}{k} + \sum_{k=0}^{s-2} \binom{m}{k}$$

is to $2 \cdot 2^m$. The sum of binomial coefficients can be rewritten as

$$\binom{m}{0} + \sum_{k=1}^{s-1} \binom{m}{k} + \sum_{k=1}^{s-1} \binom{m}{k-1}.$$

By the rule for construction of the arithmetic triangle, and because $\binom{m}{0} = \binom{m+1}{0}$, this sum is in turn equal to

$$\sum_{k=0}^{s-1} \binom{m+1}{k}.$$

Because $2 \cdot 2^m = 2^{m+1}$, the award to the first player is precisely as asserted by the theorem for the case $n = m + 1$, and the proof is complete.

Pascal had thus answered completely de Méré's problem of division. In his correspondence with Fermat, the two men discussed the same problem when there were more than two players and found themselves in agreement on the solution. Pascal also mentioned briefly the other problem, of determining the number of throws of two dice for which there are even odds that a pair of sixes will occur. He noted that in the analogous problem for one die, the odds for throwing a six in four throws are 671 to 625, but did not show his method of calculating the result. De Méré evidently believed that since four throws were sufficient to guarantee at least even odds in the case of one die (where there are six possible outcomes), the same ratio 4 : 6 would hold no matter how many dice were thrown. Because there were 36 possibilities in tossing two dice, he thought that the correct value should be 24. He probably posed the question to Pascal because this value did not seem to be empirically correct. Pascal noted that the odds are against success in 24 throws, but did not detail in his letters or in any other work the theory behind this statement.

Pascal's decision-theoretic argument in favor of belief in God demonstrates the second side of probabilistic reasoning, a method of coming to a "reasonable" decision. Either God is or God is not, according to Pascal. One has no choice but to "wager" on which of these statements is true, where the wager is in terms of one's actions. In other words, a person may act either with complete indifference to God or in a way compatible with the (Christian) notion of God. Which way should one act? If God is not, it does not matter much. If God is, however, wagering that there is no God will bring damnation while wagering that God exists will bring salvation. Because the latter outcome is infinitely more desirable than the former, the outcome of the decision problem is clear, even if one believes that the probability of God's existence is small: the "reasonable" person will act as if God exists.

FIGURE 14.16

Christian Huygens on a Dutch stamp

14.3.3 Christian Huygens and the Earliest Probability Text

Pascal's argument in favor of belief in God is certainly valid, given his premises. (Whether one accepts the premises is a different matter.) In fact, his notion of somehow calculating the "value" of a particular action became the basis for the first systematic treatise on probability, written in 1656 by Christian Huygens (1629–1695), a student of van Schooten (Fig. 14.16). Huygens became interested in the question of probability during a visit to Paris in 1655 and wrote a brief book on the subject, the *De ratiociniis in aleae ludo* (*On the Calculations in Games of Chance*), which appeared in print in 1657.

Huygens's work contained only 14 propositions and concluded with five exercises for the reader. The propositions included ones dealing with both of de Méré's problems, but Huygens also gave detailed discussions of the reasoning behind the solutions, in particular how to calculate in a game of chance: "Although in a pure game of chance the results are uncertain, the chance that one player has to win or to lose depends on a determined value."[36] Huygens's "value" is similar to Pascal's notion in his wager, but in the case of games of chance, Huygens could calculate it explicitly. In modern terms, the "value" of a chance is the **expectation**, the average amount that one would win if one played the game many times. It is this amount that a player would presumably pay to have the privilege of playing an equitable game. For example, Huygens's first proposition is: "To have equal chances of winning [amounts] a or b is worth $(a + b)/2$ to me."[37] This proposition is the same as one of the principles Pascal stated in solving the division problem. Huygens, however, gave a proof. He postulated two players each putting in a stake of $(a + b)/2$ with each player having the same chance of winning. If the first wins, he receives a and his opponent b. If the second wins, the payoffs are reversed. Huygens considered this an equitable game. In modern terminology, since the probability of winning each of a or b is 1/2, the expectation for each player is $(1/2)a + (1/2)b$, Huygens's "value" of the chance.

Huygens generalized this result in his third proposition: "To have p chances to win a and q chances to win b, the chances being equivalent, is worth $(pa + qb)/(p + q)$ to me."[38] In other words, if $p + q = r$, if the probability of winning a is p/r, and if the probability of winning b is q/r, then the expectation is given by $(p/r)a + (q/r)b$. Huygens proved this result by embedding the problem in a symmetric game played by $p + q$ players arranged in a circle, each of whom puts in the same stake x and each of whom has the same chance of winning.[39] If a given player wins, he takes the entire stake and pays b to each of the $q - 1$ players to his left and a to each of the p players to his right, retaining the remainder. To make this remainder equal to b, it must be true that

$$(p + q)x - (q - 1)b - pa = b \quad \text{or} \quad x = \frac{pa + qb}{p + q}.$$

But now it is clear that each player has q chances of winning b and p chances of winning a, so the game is equitable, and each player should be willing to risk the stated stake.

Huygens took as an axiom that each player in an equitable game would be willing to risk the calculated fair stake and would not be willing to risk more. In fact, however, as the history of gambling shows, that assumption is, at the very least, debatable. It is not at all clear that the fair stake defined by Huygens is the most a given person is willing to pay for the chance to participate in a game. The success of state-run lotteries, not to mention the gambling palaces in Las Vegas and Atlantic City, testifies to precisely the opposite. Nevertheless, Huygens based the remainder of his treatise on the results of his third proposition, and even today the concept of expectation is considered a useful one.

Huygens's discussion of de Méré's problem of division was similar to Pascal's, but he gave a more extensive analysis of the problem of the dice in proposition 11. He showed how to determine the number of times two dice should be thrown, so that one would be willing to wager $(1/2)a$ in order to win a if two sixes appear in that many plays. Huygens proceeded in stages. Supposing that one wins a when two sixes turn up, he argued that on the first throw one has 1 chance of winning a and 35 chances of winning 0, so the value of a chance on one throw is $(1/36)a$. If the player fails on the first throw, he takes a second, whose value is naturally the same $(1/36)a$. Hence, for the first throw the player has 1 chance of winning a and 35 chances of taking the second throw, which is worth $(1/36)a$. The value of his chance of throwing a double six on the two throws is, by the third proposition,

$$\frac{1a + 35(1/36)a}{1 + 35} \quad \text{or} \quad \frac{71}{1296}a.$$

Huygens next moved to the case of four throws. If the player gets a double six on one of the first two plays, he wins a; if not, he has a second pair of chances, the value of which is $(71/1296)a$. Since there are 71 chances of winning a on the first pair of plays and therefore 1225 chances of not winning (out of 1296), there are 1225 chances of reaching the second pair of plays, whose value is also $(71/1296)a$. Again, by the third proposition, the value of the player's chance on a double six in four throws is

$$\frac{71a + 1225(71/1296)a}{1296} \quad \text{or} \quad \frac{178,991}{1,679,616}a.$$

Because this value is still considerably less than the desired $(1/2)a$, Huygens had to continue the process. Although he did not present any further calculations, he noted that one next considers 8 throws, then 16, and then 24 and 25. The results show that in 24 throws the player is at a very slight disadvantage on the bet of $(1/2)a$ while for 25 he has a very slight advantage.[40]

At the conclusion of his little treatise, Huygens presented as exercises some problems of drawing different-colored balls from urns, problems of the type that today appear in every elementary probability text. These problems were discussed by many mathematicians over the next decades, especially since Huygens's text was the only introduction available to the theory of probability until the early eighteenth century. Even then, its influence continued because James Bernoulli incorporated it into his own more extensive work on probability, the *Ars conjectandi* of 1713.

14.4   NUMBER THEORY

Fermat, involved in the beginnings of analytic geometry and probability, also made contributions to number theory, contributions that were virtually ignored during his lifetime and indeed until the middle of the next century. One of the reasons for this was probably his deep secrecy about his methods. Thus, although many of his results are known, because he announced them proudly in letters to his various correspondents and presented them with challenges to solve similar problems, there is virtually no record of any of his proofs and only vague sketches of some of his methods.

Fermat's earliest interest in number theory grew out of the classical concept of a perfect number, one equal to the sum of all of its proper divisors. Book IX of Euclid's *Elements* contains a proof that if $2^n - 1$ is prime, then $2^{n-1}(2^n - 1)$ is perfect. The Greeks had, however, only been able to discover four perfect numbers—6, 28, 496, and 8128—because it was difficult to determine the values of n for which $2^n - 1$ is prime. Fermat discovered three propositions that could help in this regard, propositions he communicated to Mersenne in a letter in June of 1640. The first of these results was that if n is not itself prime, then $2^n - 1$ cannot be prime. The proof of this result just exhibited the factors: If $n = rs$, then

$$2^n - 1 = 2^{rs} - 1 = (2^r - 1)(2^{r(s-1)} + 2^{r(s-2)} + \cdots + 2^r + 1).$$

The basic question therefore reduced to asking for which primes p is $2^p - 1$ prime. Such primes are today called **Mersenne primes** in honor of Fermat's favorite correspondent.

Fermat's second proposition was that if p is an odd prime, then $2p$ divides $2^p - 2$, or p divides $2^{p-1} - 1$. His third was that, with the same hypothesis, the only possible divisors of $2^p - 1$ are of the form $2pk + 1$. Fermat indicated no proofs of these results in his letter, but only gave a few numerical examples. He confirmed that $2^{37} - 1$ was composite by testing its divisibility by numbers of the form $74k + 1$ until he found the factor $223 = 74 \cdot 3 + 1$. But in a letter written a few months later to Bernard Frenicle de Bessy (1612–1675), he stated a more general theorem of which these two propositions are easy corollaries. This theorem, today known as **Fermat's Little Theorem**, is, in modern terminology, that if p is any prime and a any positive integer, then p divides $a^p - a$. It is often written in the form $a^p \equiv a \pmod{p}$, or, adding the condition that a and p are relatively prime, in the form $a^{p-1} \equiv 1 \pmod{p}$. It follows that if n is the smallest positive integer such that p divides $a^n - 1$, then n divides $p - 1$ and, in addition, that all powers k such that p divides $a^k - 1$ are multiples of n. Fermat gave no indication in any of his writings how he discovered or proved this result. In any case, the second of the propositions in the letter to Mersenne is simply the case $a = 2$ of the theorem (where $p > 2$). The third proposition requires only a bit more work. Suppose q is a prime divisor of $2^p - 1$. The theorem then implies that p divides $q - 1$ or that $q - 1 = hp$ for some integer h. Since $q - 1$ is even, 2 must divide hp and therefore must divide h. It follows that $h = 2k$ and $q = 2kp + 1$ as asserted.

Fermat's Little Theorem turned out to be an extremely important result in number theory with many applications. But his work on another aspect of primality showed that even Fermat could be mistaken. In his correspondence, he repeatedly asserted that the so-called Fermat numbers, those of the form $2^{2^n} + 1$, were all prime. As late as 1659, he wrote that he had found a proof. It is not difficult to show that the numbers of this form for $n = 0, 1, 2, 3$, and 4 are prime. But Leonhard Euler discovered in 1732 that 641 was a factor of $2^{2^5} + 1$, and, in fact, no prime Fermat numbers have been found beyond $2^{2^4} + 1$. How did Fermat make such an error? It is quite likely that his attempted proof was of the type he outlined in another area of his number-theoretic work, the **method of infinite descent**, and that he simply believed that the methods used for integers up to 4 would work for larger ones.

The method of infinite descent demonstrates the nonexistence of positive integers having certain properties by showing that the assumption that one integer has such a property implies that a smaller one has the same property. By continuing the argument, one gets an infinite decreasing sequence of positive integers, an impossibility. Fermat used this method in the only number-theoretic proof that he actually wrote out in detail, the proof that it is impossible to find an integral right triangle whose area is a square. In other words, it is impossible to find integers x, y, z, w such that $x^2 + y^2 = z^2$ and $(1/2)xy = w^2$.

Fermat knew that any Pythagorean triple (x, y, z) with the numbers relatively prime could be generated by a pair of relatively prime numbers p, q of opposite parity, with $p > q$, by setting $(x, y, z) = (2pq, p^2 - q^2, p^2 + q^2)$. Now suppose that there existed an integral right triangle whose area was a square. Then $(1/2)xy = pq(p^2 - q^2)$ would be a square. Because the factors of this product are relatively prime, each of them must also be a square. So $p = d^2$, $q = f^2$, and $p^2 - q^2 = d^4 - f^4 = c^2$. Next, Fermat noted that since $c^2 = (d^2 + f^2)(d^2 - f^2)$ and since d and f are relatively prime, $d^2 + f^2$ and $d^2 - f^2$ must both also be squares, say, $d^2 + f^2 = g^2$ and $d^2 - f^2 = h^2$. Subtracting the second of the two equations from the first gives $2f^2 = g^2 - h^2 = (g + h)(g - h)$. Because g^2 and h^2 are both odd and relatively prime, $g + h$ and $g - h$ are both even and can have no common factor other than 2. It follows that $g + h$ can be written as $2m^2$ and $g - h$ as n^2 (or vice versa), where n is even and m odd. So $g = m^2 + n^2/2$, $h = m^2 - n^2/2$, and $d^2 = (1/2)(g^2 + h^2) = (m^2)^2 + (n^2/2)^2$. But then m^2 and $\frac{n^2}{2}$ are sides of a new right triangle whose area $\frac{m^2 n^2}{4}$ is also a square. Since the hypotenuse d of this new triangle is smaller than the hypotenuse of the original triangle, the method of infinite descent implies that the original assumption must be false.

One can pull out of this argument an argument by infinite descent showing that one cannot find three positive integers a, b, c such that $a^4 - b^4 = c^2$. It follows that one also cannot express a fourth power as a sum of two other fourth powers. Fermat wrote a generalization of this result—that "one cannot split a cube into two cubes, nor a fourth power into two fourth powers, nor in general any power beyond the square *in infinitum* into two powers of the same name,"[41]—as a marginal note to Diophantus's Problem II–8 in his copy of the 1621 Latin edition of the *Arithmetica*. This generalization is the content of what has become known as **Fermat's Last Theorem** (see Fig. 6.2). In modern terms, this conjecture asserts that there do not exist nonzero integers a, b, c, and $n > 2$, such that $a^n + b^n = c^n$. This result, of which Fermat claimed he had "a truly marvelous demonstration . . . which this margin is too narrow to contain," provided mathematicians since the seventeenth century with a major challenge. In 1995, Andrew Wiles (1953–) of Princeton University gave the first proof of Fermat's Last Theorem, a proof based on the work of many other mathematicians in the late twentieth century and using techniques from algebraic geometry unavailable to Fermat. Thus, most historians believe that Fermat erred in his own claim of a proof, probably because he erroneously assumed that the method of infinite descent, which works in the case $n = 4$, would generalize to larger values of n.

Although Fermat's claim in the case of the Fermat numbers was wrong and his assertion of the truth of Fermat's Last Theorem was premature, most of his claims of results in number theory announced in his correspondence or scribbled in the margins of his copy of Diophantus have proved true. But although he tried on many occasions to stimulate other European mathematicians to work on his various number-theoretic problems, his pleas fell on deaf ears. It took until the next century before a successor could be found to continue the work in number theory begun by the French lawyer.

 14.5

PROJECTIVE GEOMETRY

The fate of being ignored also befell Girard Desargues (1591–1661), a French engineer and architect whose most original contributions to mathematics were in the field of projective geometry. As part of his professional interests, he wanted to continue the study of perspective begun by the Renaissance artists. Having mastered the geometrical work of the Greeks, especially that of Apollonius, he proposed to unify the various methods, not by algebraicizing them as did Fermat, but by subsuming them under new synthetic techniques of projection. In particular, he attempted in his *Brouillon projet d'une atteinte aux événemens des rencontres d'un cone avec un plan* (*Rough Draft of an Attempt to Deal with the Outcomes of the Meetings of a Cone with a Plane*) of 1639 to unify the study of conics by use of projective techniques. It was well known, for example, that a circle viewed obliquely appears as an ellipse. Because viewing obliquely is equivalent to projecting the circle from a certain point not in its plane onto another plane, Desargues wanted to study those properties of conics that are invariant under projections.

As part of his study, Desargues had to consider points at infinity, the points, like the vanishing point in a drawing in perspective, where parallel lines meet. "Every straight line is, if necessary, taken to be produced to infinity in both directions." When several straight lines are either parallel or intersect at the same point, Desargues wrote that they belong to the same **ordinance**. "Thus any two lines in the same plane belong to the same ordinance, whose *butt* [intersection point] is at a finite or infinite distance."[42] The collection of all the points at infinity makes the line at infinity. It follows that every plane must be considered to extend to infinity in all directions. In addition, because the cylinder could be considered a cone with vertex at infinity, Desargues treated cones and cylinders simultaneously. Thus, two plane sections of a cone are related by a projective transformation, projection from the vertex, and two plane sections of a cylinder are related by projection from the point at infinity. Since the circle is a plane section of a cone (or cylinder), Desargues was able to regard all conics as projectively equivalent to a circle. The ellipses are those projections of circles that do not meet the line of infinity in their plane; the parabolas are those that just touch it; and the hyperbolas are those that cut the line at infinity. Thus, any property of the circle invariant under projections could easily be proved to be a property of all conics.

Desargues' most famous result, however, occurs not in the *Rough Draft* but in the appendix to a practical work by a friend, Abraham Bosse (1602–1676), *Maniére universelle de M. Desargues pour practiquer la perspective* (*Universal Method of M. Desargues for Using Perspective*): "When the lines HDa, HEb, CED, ℓga, ℓfb, $H\ell K$, DgK, EfK, in different planes or in the same plane, having any order or direction whatsoever, meet in like points, the points c, f, g lie in one line cfg."[43] In modern terminology, Desargues was considering two triangles, KED and $ab\ell$, which are related by a projection from the "like" point H (Fig. 14.17). In other words, the lines joining pairs of corresponding vertices meet at H. The conclusion is then that the intersection points g, f, c of pairs of corresponding sides, here DK, $a\ell$; EK, $b\ell$; and DE, ab, all lie on the same line. Desargues proved the result by applying Menelaus's theorem.

FIGURE 14.17

Desargues' theorem

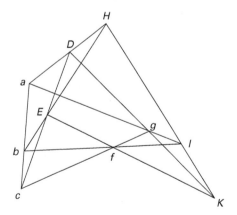

Desargues' work was not well received, partly because he invented and used so many new technical terms that few could follow it and partly because mathematicians were just beginning to appreciate Descartes' analytic unification of geometry and were not ready to consider a new synthetic version. Apparently, the only contemporary mathematician to appreciate his work was Pascal, who published in 1640 a brief *Essay on Conics* in which he credited Desargues with introducing him to projective methods. This work contains a version of the theorem ever since known by Pascal's name:

FIGURE 14.18

Pascal's hexagon theorem

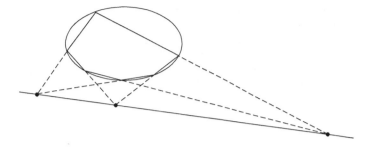

THEOREM   *If a hexagon is inscribed in a conic, then the opposite sides intersect in three collinear points (Fig. 14.18).*

Because this theorem is meant to be a statement in projective geometry, among the possible cases are those where some of the pairs of opposite sides are parallel and have their intersection point at infinity. This is naturally the case when the hexagon is a regular one inscribed in a circle. Pascal gave no proof of his theorem in his brief essay. He merely claimed its truth first for circles and then for arbitrary conics. Presumably, he meant to prove the general result by following Desargues' outline. Pascal promised to reveal more of his results along with his methods in a more complete work on conics, a work he wrote in the mid-1650s. Unfortunately, this larger work was never published, and all manuscript copies have subsequently disappeared. In fact, projective methods in geometry were effectively ignored until early in the nineteenth century.

EXERCISES

1. Find the fourth-degree polynomial in a generated by multiplying $b + a$, $c - a$, and $df = aa$. What are the roots of this polynomial? (This problem and the next three are from Harriot's *Treatise on Equations*.)

2. Viète has the following rule: If to A plane$/B$, there should be added Z square$/G$, the sum will be $(G$ in A plane $+ B$ in Z square$)/G$. Rewrite this in Harriot's notation.

3. Consider the cubic equation $aaa - 3raa = 2xxx$. Show that if one sets $a = e + r$, the resulting cubic equation in e does not have a square term. For example, show that the equation $aaa - 6aa = 400$ can be reduced to the equation $eee - 12e = 416$. Find a solution of the last equation for e and therefore find a solution for the equation in a.

4. Consider the cubic equation $aaa - 3raa + ppa = 2xxx$. Show that the substitution $a = e + r$ reduces this to an equation without a square term. As an example, reduce the equation $aaa - 18aa + 87a = 110$ to a cubic equation in e without a square term. Find all three solutions to the

equation in e and therefore find the solutions to the original equation in a.

5. Solve $x^3 = 300x + 432$ using Girard's technique, given that $x = 18$ is one solution.

6. Solve $x^3 = 6x^2 - 9x + 4$ using Girard's technique. First, determine one solution by inspection.

7. Show that in the equation $x^4 + Bx^2 + D = Ax^3 + Cx$, A is the sum of the roots, $A^2 - 2B$ is the sum of the squares of the roots, $A^3 - 3AB + 3C$ is the sum of the cubes of the roots, and $A^4 - 4A^2B + 4AC + 2B^2 - 4D$ is the sum of the fourth powers of the roots.

8. This problem illustrates Girard's geometric interpretation of negative solutions to polynomial equations. Let two straight lines DG, BC intersect at right angles at O (Fig. 14.19). Determine A on the line bisecting the right angle at O so that $ABOF$ is a square of side 4. Draw ANC as in the diagram so that $NC = \sqrt{153}$. Find the length FN. (Girard notes that if $x = FN$, then $x^4 =$

$8x^3 + 121x^2 + 128x - 256$ and so there are four possible solutions, each of which can be calculated. The two positive solutions are represented by FN and FD, while the two negative ones are represented by FG and FH, the latter two taken in the opposite direction from the former two.)

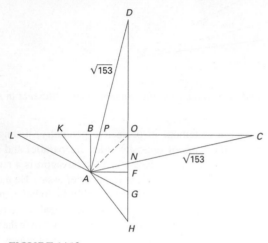

FIGURE 14.19

A problem from Girard

9. Assuming that $xy = c$ represents a hyperbola with the x and y axes as asymptotes, show that $xy + c = rx + xy$ also represents a hyperbola. Find its asymptotes.

10. Determine the locus of the equation $b^2 - 2x^2 = 2xy + y^2$. (*Hint*: Add x^2 to both sides.)

11. Show that $b^2 + x^2 = ay$ represents a parabola. Draw that portion lying in the first quadrant.

12. Determine the equation of the circle that solves the problem from Apollonius's *Plane Loci* for the case of two points. (Set the coordinates of A and B in Figure 14.2 to be $(-a, 0)$ and $(a, 0)$, respectively.)

13. Determine the equation of the circle that solves the problem from Apollonius's *Plane Loci* for the case of four non-collinear points (x_i, y_i) $(i = 1, 2, 3, 4)$.

14. Descartes was able to construct the product and quotient of two quantities via the use of similar triangles. Suppose AB is taken equal to 1, and we want to multiply BD by BC (Fig. 14.20). Join AC and draw DE parallel to CA. Show that BE is the product of BD and BC. Similarly, given two lengths BE and BD, construct the quotient length.

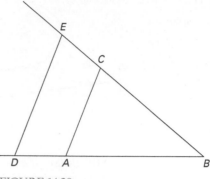

FIGURE 14.20

The product and quotient of two quantities

15. Show that MQ and MR in Figure 14.20 represent the two solutions to the equation $z^2 = ax - b^2$.

16. How can one use Figure 14.20 to represent the negative solution to $z^2 = az + b$?

17. Using the various constants mentioned in the text in the discussion of the general four-line problem, determine the equation of the locus that solves the problem in the special case where the product of the first two lines equals the product of the second two. What type of curve is this?

18. Show that the equation that solves Descartes' special four-line problem, $y(a - x) = \alpha(a + x)x$, is the question of a hyperbola. In terms of Figure 14.10, determine the asymptotes of the hyperbola.

19. Using Figure 14.9, show that the curve that solves the five-line problem of the text is generated by the intersection of the ruler OPQ with the parabola PKN, with parameter a, the distance between the parallel vertical lines. Point Q on the axis of the parabola L_3 is chosen so that $KQ = a$. As in the text, set $d_3 = y$ and $d_5 = x$. Use the similarity of triangles OMP and PRQ to show that $OM : MP = PR : RQ$. Translate this proportion into algebra to find RQ and then RK. Then use the fact that RK is on the axis of a parabola with parameter a to show that the equation of the desired curve is $y^3 - 2ay^2 - a^2y + 2a^3 = axy$.

20. This problem illustrates one of Descartes' machines (Fig. 14.21). Here GL is a ruler pivoting at G. It is linked at L with a device $CNKL$ that allows L to be moved along AB, always keeping the line KN parallel to itself. The intersection C of the two moving lines GL and KN determines a curve. To find the equation of the curve, begin by setting $CB = y$, $BA = x$, and the constants $GA = a$, $KL = b$, and $NL = c$. Then find BK, BL, and AL in terms of x, y, a, b, and c. Finally, use the similarity relation

$CB : BL = GA : AL$ to show that the equation is

$$y^2 = cy - \frac{c}{b}xy + ay - ac.$$

Descartes stated, without proof, that this curve is a hyperbola. Show that he was correct.

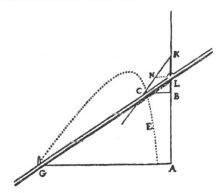

FIGURE 14.21

Descartes' curve-drawing instrument

21. Although Descartes claimed that "any problem" in geometry could be solved by applying his methods, he got into some trouble when he suggested to Princess Elizabeth of Bohemia that she apply his method to solve the "Apollonian problem" of finding a circle tangent to each of three given circles in the plane. Assume the construction completed (Fig. 14.22). Let $A = (0, 0)$, $B = (a, 0)$, and $C = (c, d)$ be the centers of the given circles with radii r, s, t, respectively. Let $D = (x, y)$ be the center of the constructed circle with unknown radius z. Find equations relating the three unknowns to the known quantities and show that solutions can be constructed according to Descartes' principles. Could Descartes have actually completed this construction?

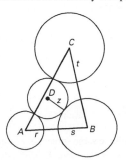

FIGURE 14.22

Apollonius's problem of the three circles

22. To solve the fourth-degree equation $x^4 - px^2 - qx - r = 0$, Descartes considered the cubic equation in y^2: $y^6 - 2py^4 + (p^2 + 4r)y^2 - q^2 = 0$. If y is a solution, show that the original polynomial factors into two quadratics: $r_1(x) = x^2 - yx + \frac{1}{2}y^2 - \frac{1}{2}p - \frac{q}{2y}$, $r_2(x) = x^2 + yx + \frac{1}{2}y^2 - \frac{1}{2}p + \frac{q}{2y}$, each of which can be solved. Apply this method to solve the equation $x^4 - 17x^2 - 20x - 6 = 0$. Note that the corresponding equation in y, $y^6 - 34y^2 + 313y^2 - 400 = 0$, has the solution $y^2 = 16$.

23. Solve the equation $x^3 - \sqrt{3}x^2 + \frac{26}{27}x - \frac{8}{27\sqrt{3}} = 0$ by first substituting $y = \sqrt{3}x$ and then $z = 3y$ to get an equation in z with integral coefficients.

24. In de Witt's substitution $z = y + \frac{b}{a}x + c$, which simplifies the equation

$$y^2 + \frac{2bxy}{a} + 2cy = bx - \frac{b^2x^2}{a^2} - c^2,$$

he has rotated one of the axes through an angle α. Find the sine and cosine of α.

25. Show that de Witt's equation

$$y^2 + \frac{2bxy}{a} + 2cy = \frac{fx^2}{a} + ex + d$$

represents a hyperbola. (Use the substitution $z = y + \frac{b}{a}x + c$ and show that this substitution, when combined with a substitution of the form $x' = \beta x$, converts the original oblique x-y coordinate system into a new x'-z coordinate system based on perpendicular axes.) Sketch the curve.

26. Prove by induction on n that

$$\binom{n}{k} = \sum_{j=k-1}^{n-1} \binom{j}{k-1}$$

for all k less than n.

27. Prove that

$$\binom{n}{k} : \binom{n}{k+1} = (k+1) : (n-k).$$

28. Prove that

$$\binom{n}{k} : \binom{n-1}{k} = n : (n-k).$$

29. Pascal stated that the odds in favor of throwing a six in four throws of a single die are 671 to 625. Show why this is true.

30. Show that the odds against at least one 1 appearing in a throw of three dice is $125 : 91$. (This answer was stated by Cardano.)

31. Determine the appropriate division of the stakes in a game between two players if the first player is lacking three games to win and the second four.

32. Suppose three players play a fair series of games under the condition that the first player to win three games wins the stakes. If they stop play when the first player needs one game while the second and third players each need two games, find the fair division of the stakes. (This problem was discussed in the correspondence of Pascal with Fermat.)

33. For a roll of three dice, show that both a 9 and a 10 can be achieved in six different ways. Nevertheless, show that the probability of rolling a 10 is higher than that of rolling a 9. (A discussion of this idea is found in a fragment of a work of Galileo.)

34. If two players play a game with two dice with the condition that the first player wins if the sum thrown is 7, the second wins if the sum is 6, and the stakes are split if there is any other sum, find the expectation (value of the chance) of each player.

35. If I play with another player throwing two dice alternately under the condition that I win when I have thrown a 7 and he wins when he throws a 6, and if he throws first, what is the ratio of my chance to his?

36. There are 12 balls in an urn, 4 of which are white and 8 black. Three blindfolded players, A, B, C draw a ball in turn, first A, then B, then C. The winner is the one who first draws a white ball. Assuming that each (black) ball is replaced after being drawn, find the ratio of the chances of the three players.

37. There are 40 cards, 10 from each suit. A wagers B that he will draw four cards and get one of each suit. What are the fair amounts of the wagers of each?

38. Prove that if p is prime, then $2^p \equiv 2 \pmod{p}$ by writing $2^p = (1 + 1)^p$, expanding by the binomial theorem, and noting that all of the binomial coefficients $\binom{p}{k}$ for $1 \leq k \leq p - 1$ are divisible by p. Prove $a^p \equiv a \pmod{p}$ by induction on a, using this result and the fact that $(a + 1)^p \equiv a^p + 1 \pmod{p}$.

39. For a proof of the Fermat Little Theorem in the case where a and p are relatively prime, consider the remainders of the numbers 1, a, a^2, . . . on division by p. These remainders must ultimately repeat (why?), and so $a^{n+r} \equiv a^r \pmod{p}$ or $a^r(a^n - 1) \equiv 0 \pmod{p}$ or $a^n \equiv 1 \pmod{p}$. (Justify each of these alternatives.) Take n as the smallest positive integer satisfying the last congruence. By applying the division algorithm, show that n divides $p - 1$.

40. Construct a tangent to a point P on a conic section by using Pascal's hexagon theorem. Consider the tangent line as passing through two neighboring points at P. Then pick four other points on the conic and apply the theorem.

41. The best-known quotation from Descartes is "I think, therefore I am," from the *Discourse on Method*. The context is Descartes' resolve only to accept those ideas that are self-evidently true. There is a well-known joke based on this quote: Descartes goes into a restaurant. The waiter asks him, "Would you like tonight's special?" He replies, "I think not," and disappears. Comment on the logical validity of this joke.[44]

42. Compare the analytic geometries of Descartes, Fermat, and de Witt. Adapt the formulation of one of these authors to give a presentation of the subject to a precalculus class.

43. Outline a lesson in the theory of equations using Descartes' algebraic techniques to teach such results as the factor theorem and the methods of solving polynomial equations of degree higher than two.

44. Outline a lesson in elementary probability theory using the ideas of Cardano. Include material on justification of the various rules involved and on the possible mistakes one can make.

45. Outline a lesson on the principle of mathematical induction using material from Pascal's *Treatise on the Arithmetical Triangle*.

46. Compare Pascal's use of mathematical induction to the use of it by ibn al-Haytham, al-Samaw'al, and Levi ben Gerson.

REFERENCES AND NOTES

The first two chapters of Helena Pycior, *Symbols, Impossible Numbers, and Geometric Entanglements* (Cambridge: Cambridge University Press, 1997) discuss the contributions of Oughtred and Harriot to the development of symbolic algebra. A more recent book dealing with the same subject is Jacqueline A. Stedall, *A Discourse Concerning Algebra: English Algebra to 1685* (Oxford: Oxford University Press, 2002). Carl Boyer has written the only general history of analytic geometry: *History of Analytic Geometry* (New York: Scripta Mathematica, 1956), but recently reprinted. This work covers the subject from its beginnings in ancient times up through the nineteenth century. The best general histories of probability, which deal to some extent with the seventeenth century and earlier, are F. N. David, *Games, Gods, and Gambling* (New York: Hafner, 1962), and Ian Hacking, *The Emergence of Probability* (Cambridge: Cambridge University Press, 1975). The latter book is more philosophical, while the former discusses the relevant texts in more detail. A general history of number theory from a historical point of view is André Weil, *Number Theory: An Approach through History from Hammurapi to Legendre* (Boston: Birkhäuser, 1983).

Oughtred's *Clavis* was translated into English in 1647 by Robert Wood, under the title of *The Key of the Mathematics, New Forged and Filed*. Although there is no recent reprint, there is a study of it in Florian Cajori, *William Oughtred: A Great Seventeenth Century Teacher of Mathematics* (Chicago: Open Court, 1916) and in Jacqueline Stedall, *Discourse*, chapter 3. Harriot's *Praxis* was translated into English by Muriel Seltman as part of her MSc unpublished dissertation at the University of London. The *Treatise on Equations* is now available as Jacqueline Stedall, *The Greate Invention of Algebra: Thomas Harriot's Treatise on Equations* (Oxford: Oxford University Press, 2003). Other works discussing Harriot's manuscripts include Muriel Seltman, "Harriot's Algebra: Reputation and Reality," in Robert Fox, ed., *Thomas Harriot: An Elizabethan Man of Science* (Aldershot: Ashgate, 2000), and J. A. Lohne, "Essays on Thomas Harriot," *Archive for History of Exact Sciences* 20 (1979), 189–312. Girard's *A New Discovery in Algebra* is available in a translation by Ellen Black in *The Early Theory of Equations: On Their Nature and Constitution* (Annapolis: Golden Hind Press, 1986). Fermat's *Introduction to Plane and Solid Loci* has been translated by Joseph Seidlin and is in David Eugene Smith, *A Source Book in Mathematics* (New York: Dover, 1959), vol. 2, pp. 389–396. The standard modern version of Descartes' *Geometry* is David Eugene Smith and Marcia L. Latham, trans., *The Geometry of René Descartes* (New York: Dover, 1954). This book contains the original French and the English translation on facing pages as well as Descartes' original notation and diagrams. The first book of Jan de Witt's *Elements of Curves* was recently translated by A. W. Grootendorst and Miente Bakker as *Jan de Witt's Elementa curvarum linearum, Liber Primus* (New York: Springer, 2000). Pascal's *Treatise on the Arithmetical Triangle*, in a translation by Richard Scofield, is available in *Great Books of the Western World* (Chicago: Encyclopedia Britannica, 1952), vol. 33. This edition also contains the letters between Pascal and Fermat on probability. Huygens's *On the Calculations in Games of Chance* is in his *Oeuvres Completes* (The Hague: 1888–1950), vol. 14. The work of Desargues and Pascal on projective geometry is

discussed in J. V. Field and J. J. Gray, *The Geometrical Work of Girard Desargues* (New York: Springer, 1987).

1. Smith, *Source Book*, vol. 2, p. 389.

2. Quoted in Cajori, *William Oughtred*, p. 20.

3. See Stedall, *Greate Invention*, for more details on Harriot's life and work.

4. Black, *The Early Theory of Equations*, p. 107. This book also contains treatises of Viète and Debeaune.

5. Ibid., p. 145.

6. Ibid., p. 138.

7. Ibid., p. 139.

8. Ibid., p. 141.

9. The best study of the life and mathematical work of Fermat is Michael S. Mahoney, *The Mathematical Career of Pierre de Fermat 1601–1665* (Princeton: Princeton University Press, 1973). This book contains a detailed analysis of Fermat's work, not only in analytic geometry but also in the various aspects of the calculus to be treated in Chapter 15.

10. Ibid., pp. 102–103.

11. Smith, *Source Book*, p. 390.

12. Ibid., p. 392.

13. Mahoney, *Mathematical Career*, p. 91.

14. Ibid., p. 126.

15. Smith and Latham, trans., *Geometry*, p. 2.

16. René Descartes, *Discourse on Method, Optics, Geometry, and Meteorology* translated by Paul J. Olscamp (Indianapolis: Bobbs-Merrill Co., 1965), p. 9. This version contains not only the *Geometry* but also the other two essays. The *Optics* in particular contains more of Descartes' mathematics as he demonstrates how to construct certain curves, lenses in the shape of which will have prescribed optical properties. For a general study of Descartes' work in mathematics and physics, see J. F. Scott, *The Scientific Work of René Descartes* (London: Taylor and Francis, 1952).

17. Smith and Latham, trans., *Geometry*, p. 5.

18. Ibid., p. 22.

19. See the very interesting paper of H. J. M. Bos, "On the Representation of Curves in Descartes' Géométrie," *Archive for History of Exact Sciences* 24 (1981), 295–338. Bos discusses Descartes' general program for geometry, which is outlined in the *Geometry*. Also see H. J. M. Bos, *Redefining Geometrical Exactness: Descartes' Transformation of the Early Modern Concept of Construction* (New York: Springer, 2001). This book discusses Descartes' construction procedures in great detail and puts them into the

context of the seventeenth century. See also A. Molland, "Shifting the Foundations: Descartes' Transformation of Ancient Geometry," *Historia Mathematica* 3 (1976), 21–49. See E. G. Forbes, "Descartes and the Birth of Analytic Geometry," *Historia Mathematica* 4 (1977), 141–161, for an argument that analytic geometry should be credited to Marino Ghetaldi (1566–1627), a pupil of Viète.

20. Smith and Latham, trans., *Geometry*, p. 43. For more information on Descartes' curve-drawing devices, see David Dennis, "René Descartes' Curve-Drawing Devices: Experiments in the Relations between Mechanical Motion and Symbolic Language," *Mathematics Magazine* 70 (1997), 163–175.

21. Ibid., p. 159.

22. Ibid., p. 175.

23. Ibid., p. 159.

24. Ibid., p. 160.

25. Ibid., p. 200.

26. Ibid., p. 240.

27. Ibid., p. 48.

28. Ibid., p. 91.

29. Ibid., p. 240.

30. The early contributions of the Jews to probability are discussed in Nachum L. Rabinovitch, *Probability and Statistical Inference in Ancient and Medieval Jewish Literature* (Toronto: University of Toronto Press, 1973).

31. Quoted in David, *Games, Gods*, p. 33.

32. Oystein Ore, *Cardano, the Gambling Scholar* (Princeton: Princeton University Press, 1953), p. 203. Gerolamo Cardano's *The Book on Games of Chance* appears in translation by Sydney Henry Gould at the end of this biographical work. Ore discusses Cardano's work in detail and argues more strongly than either David or Hacking for crediting Cardano with originating many of the central ideas of probability.

33. Quoted in Oystein Ore, "Pascal and the Invention of Probability Theory," *American Mathematical Monthly* 67 (1960), 409–419, p. 414.

34. Pascal, *Treatise on the Arithmetical Triangle*, p. 452. For more on Pascal, see Harold Bacon, "The Young Pascal," *Mathematics Teacher* 30 (1937), 180–185, Morris Bishop, *Pascal, The Life of Genius* (New York: Reynal and Hitchcock, 1936), and Jean Mesnard, *Pascal, His Life and Works* (New York: Philosophical Library, 1952).

35. Pascal, *Treatise on the Arithmetical Triangle*, p. 452.

36. Huygens, *On the Calculations in Games of Chance*, p. 61.

37. Ibid., p. 62.

38. Ibid., p. 64.

39. Huygens's version of his game is incomplete. This improvement is found in Olav Reiersol, "Notes on Some Propositions of Huygens in the Calculus of Probability," *Nordisk Matematisk Tidskrift* 16 (1968), 88–91.

40. For a modern discussion of de Méré's dice problem, see Jane B. Pomeranz, "The Dice Problem—Then and Now," *College Mathematics Journal* 15 (1984), 229–237.

41. Quoted in Mahoney, *Mathematical Career*, p. 344. For more on Fermat's method of infinite descent, see Howard Eves, "Fermat's Method of Infinite Descent," *Mathematics Teacher* 53 (1960), 195–196. See also Michael Mahoney, "Fermat's Mathematics: Proofs and Conjectures," *Science* 178 (1972), 30–36.

42. Field and Gray, *Geometrical Work*, pp. 69–70. For more on Desargues, see N. A. Court, "Desargues and his Strange Theorem," *Scripta Mathematica* 20 (1954), 5–13, 155–164.

43. Smith, *Source Book*, pp. 307–308.

44. Told to me by my son Ari.

15
CHAPTER

The Beginnings of Calculus

All other properties of curves [besides those concerning quadrature] depend only on the angles that these curves make with other lines. But the angle formed by two intersecting curves can be as easily measured as the angle between two straight lines, provided that a straight line can be drawn making right angles with one of these curves at its point of intersection with the other. This is my reason for believing that I shall have given here a sufficient introduction to the study of curves when I have given a general method of drawing a straight line making right angles with a curve at an arbitrarily chosen point upon it. And I dare say that this is not only the most useful and most general problem in geometry that I know, but even that I have ever desired to know.

—From Descartes' *Geometry*[1]

To indicate the extent of his research in finding areas, Fermat wrote to Roberval on September 22, 1636: "I have squared infinitely many figures composed of curved lines; as, for example, if you would imagine a figure like the parabola but such that the cubes of the ordinates are proportional to the abscissas. This figure approaches the parabola and differs only in that, whereas in the parabola one takes the ratios of the squares, I take in this figure that of the cubes; it is for that reason that M. de Beaugrand, to whom I showed the proposition, calls it a 'solid parabola.' . . . I have had to follow a path other than that of Archimedes in the quadrature of the parabola and that I would never have solved it by the latter means."[2]

Building on the work of many mathematicians over the centuries who considered the problems of determining the areas of regions bounded by curves and of finding the maximum or minimum values of certain functions, two geniuses of the last half of the seventeenth century, Isaac Newton and Gottfried Leibniz, created the machinery of the calculus, the foundation of modern mathematical analysis and the source of application to an increasing number of other disciplines. The maximum-minimum problem and the area problem, along with the related problems of finding tangents and determining volumes, had been attacked and solved for various special cases over the years. But virtually every solution had required an ingenious construction. No one had developed an algorithm that would enable these problems easily to be solved in new situations.

New situations did not often occur in either the Greek or Islamic setting, since those mathematicians had few ways of describing new curves or solids for which to calculate tangents, areas, or volumes. But with the advent of analytic geometry in the first half of the seventeenth century, the possibility suddenly opened up of constructing all sorts of new curves and solids. After all, any algebraic equation determined a curve, and a new solid could be formed, for example, by rotating a curve around any line in its plane. With an infinity of new examples to deal with, mathematicians of the seventeenth century sought for and discovered new ways of finding maximums, constructing tangents, and calculating areas and volumes. These mathematicians were not, however, concerned with functions. They were concerned with curves, defined by some relation between two variables. And in the process of finding tangents, they often considered other geometric aspects of the curves. Figure 15.1 illustrates some of the quantities connected with a point on a given curve: the abscissa x, the ordinate y, the arclength s, the subtangent t, the tangent τ, the normal n, and the subnormal v.

FIGURE 15.1

Quantities connected with a curve: x is the abscissa, y the ordinate, s the arclength, t the subtangent, τ the tangent, n the normal, and v the subnormal

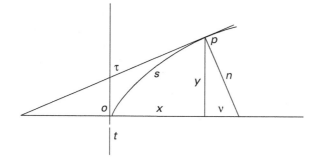

In this chapter we will explore first the various methods used to construct tangents and find extrema and next the methods developed to determine areas and volumes. Finally, we will discuss the ways of accomplishing what Descartes said could not be done, the determination of lengths of curves, and see how these methods led to the first inkling of the inverse relationship between areas and tangents.

15.1 TANGENTS AND EXTREMA

In 1615, Kepler wrote his *Nova stereometria doliorum vinariorum* (*New Solid Geometry of Wine Bottles*), in which he showed that Austrian wine merchants had a rather accurate way of determining how much wine remained in a given barrel. As part of this study of various solid shapes, he proved that the largest parallelepiped that can be inscribed in a given sphere is a cube. In fact, he actually tabulated the volumes of parallelepipeds inscribed in a sphere of radius 10 for all integral altitudes from 1 to 20. It was therefore clear to him that near the maximum value of approximately 1540, the volume changed little with small changes in the altitude: "Near the maximum, the decrements on both sides are initially imperceptible."[3]

15.1.1 Fermat's Method of *Adequality*

Fermat, in the late 1620s, was able to turn Kepler's idea into an algorithm, but he was stimulated to consider the question by a study of Viète's work relating the coefficients to the roots of a polynomial: "While I was pondering Viète's method . . . and was exploring more accurately its use in discovering the structure of . . . equations, there came to mind a new method to be derived from it for finding maxima and minima, by means of which some doubts pertaining to *diorismos* [conditions], which have caused trouble to ancient and modern geometry, are most easily dispatched."[4]

Recall that Viète had shown that the sum of the two roots x_1, x_2 of $bx - x^2 = c$ was b by equating $bx_1 - x_1^2$ and $bx_2 - x_2^2$ and dividing through by $x_1 - x_2$. The equation $bx - x^2 = c$ comes from the geometric problem of dividing a line of length b into two parts whose product is c. Fermat knew from Euclid that the maximum possible value of c was $\frac{b^2}{4}$ and also that for any number less than the maximum, there were two possible values for x whose sum was b. But what happened as c approached its maximum value? The geometrical situation made it clear to Fermat that even for this maximum value, the equation had two solutions, each of the same value: $x_1 = \frac{b}{2}$ and $x_2 = b - x_1 = b - \frac{b}{2} = \frac{b}{2}$. This insight gave Fermat his method for maximizing a polynomial $p(x)$: Set $p(x_1) = p(x_2)$. Then divide through by $x_1 - x_2$ to find the relationship between the coefficients and any two roots of the polynomial. Finally, set the two roots equal to one another and solve.

From $bx_1 - x_1^2 = bx_2 - x_2^2$, Fermat derived the fact that $b = x_1 + x_2$, an equation holding for any two roots. Setting $x_1 = x_2 (= x)$ gives $b = 2x$. Thus, the maximum occurs when $x = \frac{b}{2}$. Similarly, to maximize $bx^2 - x^3$, Fermat set $bx_1^2 - x_1^3 = bx_2^2 - x_2^3$ and derived $b(x_1^2 - x_2^2) = x_1^3 - x_2^3$ and $bx_1 + bx_2 = x_1^2 + x_1 x_2 + x_2^2$. He then set $x_1 = x_2 (= x)$ and determined that $2bx = 3x^2$, from which he concluded that $x = \frac{2b}{3}$ provides the maximum value. He knew that this value was a maximum from the geometry of the situation. More generally, in other situations he used geometry to determine which answers gave maximums and which minimums when there were two or more solutions to his final equation.

But Fermat's method raised a significant methodological question. How can one divide through by $x_1 - x_2$ and then set that value equal to 0? For Fermat, the geometric situation showed that the roots were distinguishable even when their difference was 0. Thus, he never felt he was dividing by 0. He simply assumed that the relationships worked out using Viète's methods were perfectly general (for example, $x_1 + x_2 = b$) and thus held for any particular values of the variables, even those at the maximum.

Fermat did realize, however, that if the polynomial $p(x)$ were somewhat complicated, the division by $x_1 - x_2$ might be rather difficult. Thus, he modified his method to avoid this. Instead of considering the two roots as x_1 and x_2, he wrote them as x and $x + e$. Then, after equating $p(x)$ with $p(x + e)$—Fermat actually used the term *adequate*, which he had read in Diophantus—he had only to divide by e or one of its powers. In the resulting expression, he then removed any term that contained e to get an equation enabling the maximum to be found. Thus, using his original example of $p(x) = bx - x^2$, Fermat *adequated* $bx - x^2$ with $b(x + e) - (x + e)^2 = bx - x^2 + be - 2ex - e^2$. (We will write this as $bx - x^2 \approx bx - x^2 + be - 2ex - e^2$.) Canceling common terms gave him $be \approx 2ex + e^2$ and, on dividing by e, he found $b \approx 2x + e$. Removing the term that contains e gave Fermat his known result: $x = \frac{b}{2}$. In his description of this procedure, which was probably written before 1630 but only reached Paris in late 1636, Fermat wrote that "we can hardly expect a more general method."[5]

In this same document, Fermat showed how the method of *adequality* can be adapted to determine a tangent to a curve, in particular to a parabola. Because Fermat discovered this method before he invented analytic geometry, he used a geometric description of the parabola. In 1638, however, once the possibility opened up of defining curves by algebraic equations rather than through geometric properties, Fermat could explain his method more easily. (Descartes, in fact, had strongly criticized his geometric explanation.) To draw a tangent line at B to a curve represented in modern notation by $y = f(x)$, pick an arbitrary point A on the tangent line and drop perpendiculars AI and BC to the axis (Fig. 15.2). Fermat's idea was then to *adequate* FI/BC with EI/CE, where F is the intersection of AI with the curve. If $CI = e$, $CD = x$, and $CE =$ the subtangent t, this *adequality* can be written as

$$\frac{f(x + e)}{f(x)} \approx \frac{t + e}{t}$$

or $tf(x + e) \approx (t + e)f(x)$. By applying his rules of canceling common terms, dividing through by e, and then removing any remaining terms containing e, Fermat could calculate the relation between t and x that determined the tangent line. For example, if the curve is the parabola $f(x) = \sqrt{x}$, then Fermat's method gives the *adequality* $t\sqrt{x + e} \approx (t + e)\sqrt{x}$. Squaring both sides and simplifying gives $t^2 e \approx 2etx + e^2 x$. If we divide through by e and then remove the term still containing e, we get the result $t = 2x$. This is, of course,

FIGURE 15.2

Fermat's method for determining subtangents

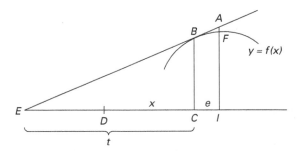

Apollonius's result (Proposition I–33) that the subtangent to the parabola at a point is double the abscissa.

In response to a challenge from Descartes, Fermat modified his method to deal with curves expressed in the form $f(x, y) = 0$. In fact, he justified his method to Descartes by showing him how the method could be used to find the tangent to the curve Descartes proposed to him, $x^3 + y^3 = pxy$.[6]

15.1.2 Descartes and the Method of Normals

One of the reasons Descartes was critical of Fermat was that Fermat had discovered the same new mathematics as Descartes, independently of the great philosopher. As indicated in the opening quotation, Descartes was immensely proud of his own discovery of a method of drawing a normal to a curve at any point, from which, naturally, one could easily determine the tangent as well.

Descartes derived his idea for drawing a normal from the realization that a radius of a circle is always normal to the circumference. Thus, the radius of a circle tangent to a given curve at the given point will be normal to that curve as well. To construct a circle tangent to a curve required an idea similar to that of Fermat, namely, that the two intersection points of a circle with the curve near the given point will become one if the circle is in fact tangent. To carry out this procedure at a point C of a curve given by $y = f(x)$, assume that P is the center of the required circle, take an arbitrary point A on the axis through P, and set $CP = n$ and $PA = v$ (Fig. 15.3). If $C = (x, y)$, then $PM = v - x$ and the equation of the circle is $n^2 = y^2 + (v - x)^2$ or $n^2 = [f(x)]^2 + v^2 - 2vx + x^2$. Descartes then used this equation to determine v, which in turn determined the point P. As he noted, "if the point P fulfills the required conditions, the circle about P as center and passing through the point C will touch but not cut the curve CE; but if this point P be ever so little nearer to or farther from A than it should be, this circle must cut the curve not only at C but also in another point. Now if this circle cuts [the curve also at E], the equation . . . must have two unequal roots. . . . The nearer together the point C and E are taken, however, the less difference there is between the roots; and when the points coincide, the roots are exactly equal."[7] In other words, for P to be the center of a tangent circle, the equation $[f(x)]^2 + v^2 - 2vx + x^2 - n^2 = 0$ must have a double root. As Descartes knew from his study of roots of equations, this meant that the polynomial had a factor of $(x - x_0)^2$ where x_0 is the double root. Setting then $[f(x)]^2 + v^2 - 2vx + x^2 - n^2 = (x - x_0)^2 q(x)$ and equating the coefficients of like powers of x, Descartes could solve for v in terms of x_0.

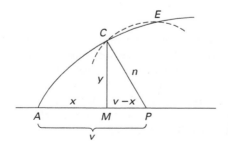

FIGURE 15.3

Descartes' method for finding normals

As usual in the *Geometry*, Descartes provided quite difficult examples of his procedure. We therefore present a simple example to clarify his method, namely, that of determining a normal to the parabola $y = x^2$ at the point (x_0, x_0^2). In this case, the polynomial with the double root is $(x^2)^2 + v^2 - 2vx + x^2 - n^2$. Because this is a fourth-degree polynomial, it must be equated to $(x - x_0)^2 q(x)$, where $q(x)$ has degree two. Thus,

$$x^4 + x^2 - 2vx + v^2 - n^2 = (x - x_0)^2(x^2 + ax + b)$$

or

$$x^4 + x^2 - 2vx + v^2 - n^2 = x^4 + (a - 2x_0)x^3 + (b - 2x_0a + x_0^2)x^2 + (ax_0^2 - 2bx_0)x + bx_0^2.$$

Equating coefficients gives

$$a - 2x_0 = 0$$
$$b - 2x_0a + x_0^2 = 1$$
$$ax_0^2 - 2bx_0 = -2v$$
$$bx_0^2 = v^2 - n^2.$$

Solving the first three equations for v by setting $a = 2x_0$ and $b = 2ax_0 - x_0^2 + 1$ gives $v = 2x_0^3 + x_0$ as the (horizontal) coordinate of the desired point P. (Since v determines n, the fourth equation is not necessary.) Because Descartes was interested just in constructing the normal, he stopped the procedure here with the point P determined. But we note further that the slope of the normal line is

$$\frac{-y_0}{v - x_0} = \frac{-x_0^2}{2x_0^3} = \frac{-1}{2x_0},$$

and therefore the slope of the tangent line is $2x_0$, a familiar result.

15.1.3   The Algorithms of Hudde and Sluse

By the late 1630s, Gilles Persone de Roberval (1602–1675) had discovered a kinematic method of determining tangents by considering a curve to be generated by a moving point. But his method depended on the geometric description of the curve and thus could not meet the need for a simple algebraic algorithm to determine tangents. The procedures of Fermat, and especially of Descartes, often led to such complicated algebra that these methods too could not provide the desired ease of calculation. But a study of these methods led two other mathematicians, Johann Hudde (1628–1704) and René François de Sluse (1622–1685), to discover simpler algorithms in the 1650s.

Hudde was one of van Schooten's students, who, like de Witt, became active in political life in the Netherlands. His contributions to mathematics were made in the late 1650s, when two of his papers appeared in van Schooten's 1659 edition of Descartes' *Geometry*. In *De maximis et minimis* (*On Maximums and Minimums*), Hudde described his algorithm for simplifying the calculations necessary to determine a double root to a polynomial equation, necessary for Descartes' method of finding normals. **Hudde's rule**, for which he only sketched a proof, states that if a polynomial $f(x) = a_0 + a_1x + a_2x^2 + \cdots + a_nx^n$ has a double root $x = \alpha$, and if $p, p + b, p + 2b, \ldots, p + nb$ is an arithmetic progression, then the polynomial $pa_0 + (p + b)a_1 x + (p + 2b)a_2 x^2 + \cdots + (p + nb)a_n x^n$ also has the root $x = \alpha$. In modern terminology, the new polynomial can be expressed as $pf(x) + bxf'(x)$. Hudde's result follows immediately, because if $f(x)$ has a double root, then $f'(x)$ has the same root. Although his rule permitted the arbitrary choice of an arithmetic progression, Hudde most often used the progression with $p = 0$, $b = 1$. In this case the new polynomial is $xf'(x)$, a result that helped to bring out the computational importance of what we now call the derivative.

As a first example of the rule, consider the problem of determining the normal to the parabola $y = x^2$, where it is necessary to find the relationship between the coefficient v and the double root x_0 of the polynomial $x^4 + x^2 - 2vx + v^2 - n^2$. Using Hudde's rule with $p = 0$ and $b = 1$ gives the new polynomial equation $4x^4 + 2x^2 - 2vx = 0$ or $4x^3 + 2x^2 - 2v = 0$. Because x_0 is a solution of this equation, it follows as before that $v = 2x_0^3 + x_0$ and therefore that the slope of the tangent line is $(v - x_0)/x_0^2 = 2x_0$. An easy generalization of this example makes it possible to show that the slope of the tangent line to $y = x^n$ at (x_0, x_0^n) is nx_0^{n-1}, a result extremely difficult to find using Descartes' procedure.

Hudde also applied his rule to the determination of extreme values, using Fermat's idea that if a polynomial $f(x)$ has an extreme value M, then the polynomial $g(x) = f(x) - M$ has a double root. Thus, to maximize $x^2(b - x)$, use the rule with $p = 0$, $b = 1$ on the polynomial $-x^3 + bx^2 - M$. The new polynomial equation is $-3x^3 + 2bx^2 = 0$, the nonzero root of which, $x = \frac{2b}{3}$, gives the desired maximum. Hudde in addition used his rule to find the tangents to curves determined by equations of the form $f(x, y) = 0$, but Sluse gave an even simpler algorithm for this case.

Sluse was born and spent most of his life in Liège in what is now Belgium and, like Hudde, had little time for mathematics. He nevertheless carried on an extensive correspondence with mathematicians all over Europe. His algorithm for determining the subtangent t (and of course the tangent) to a curve given by a polynomial equation $f(x, y) = 0$ was probably discovered in the 1650s but only appeared in print in a letter to Henry Oldenburg (1615–1677) in England in 1673. The algorithm begins with the elimination of constant terms. One then leaves all terms with x on the left and transfers all terms with y to the right with appropriate change of sign. Thus, any term containing both x and y will now appear on each side of the equation. Next, one multiplies each term on the right by its exponent of y and each term on the left by its exponent of x. Finally, one replaces one x in each term on the left by t and solves the resultant equation for t. For example, given the equation $x^5 + bx^4 - 2q^2 y^3 + x^2 y^3 - b^2 = 0$, one eliminates the constant term and transfers all the terms in y to get $x^5 + bx^4 + x^2 y^3 = 2q^2 y^3 - x^2 y^3$. Multiplying by the appropriate exponents and replacing one x in each term on the left by t gives $5x^4 t + 4bx^3 t + 2txy^3 = 6q^2 y^3 - 3x^2 y^3$. The subtangent t is therefore given by

$$t = \frac{6q^2 y^3 - 3x^2 y^3}{5x^4 + 4bx^3 + 2xy^3}$$

and the slope of the tangent by

$$\frac{y}{t} = \frac{5x^4 + 4bx^3 + 2xy^3}{6q^2 y^2 - 3x^2 y^2}.$$

In modern terms, it is easy enough to see that Sluse has calculated

$$t = -\frac{yf_y(x, y)}{f_x(x, y)} \quad \text{or} \quad \frac{dy}{dx} = -\frac{f_x(x, y)}{f_y(x, y)}.$$

Sluse, however, gave no written justification of his method or hinted how he discovered it. The best guess is that he generalized it from a study of many examples. In any case, the importance of the rules of Hudde and Sluse is that they provided general algorithms by which one could routinely construct tangents to curves given by polynomial equations. It was no longer necessary to develop a special technique for each particular curve. Anyone could now determine the tangent.

AREAS AND VOLUMES

Both Greek and Islamic mathematicians had been able to determine areas and volumes of certain regions bounded by curved lines or surfaces. The texts available, however, generally gave only the result with a proof based on the method of exhaustion. (Archimedes's *Method* was not available in the seventeenth century.) Such results gave seventeenth-century mathematicians few clues as to how to determine the areas bounded by the many new curves now available for study or the volumes of solid regions generated by revolving these curves around lines in the plane. The only clear idea passed down from Greek times was that somehow the given region needed to be broken up into very small regions, whose individual areas or volumes were known.

15.2.1 Infinitesimals and Indivisibles

Recall that Kepler used the procedure of adding up small regions in his discovery of the laws of planetary motion. And in his *Nova stereometria*, he calculated the area of a circle of radius AB by first noting that "the circumference . . . has as many parts as points, namely, an infinite number; each of these can be regarded as the base of an isosceles triangle with equal sides AB so that there are an infinite number of triangles in the area of the circle, all having their vertices at the center A."[8] Kepler then stretched the circumference of the circle out into a straight line, upon each point of which, "arranged one next to the other," he placed triangles equal to the ones in the circle, all having the altitude AB (Fig. 15.4). It follows that the area of the triangle ABC "consisting of all those triangles, will be equal to all the sectors of the circle and therefore equal to the area of the circle which consists of all of them." Therefore, the area

FIGURE 15.4

Kepler's method of determining the area of a circle

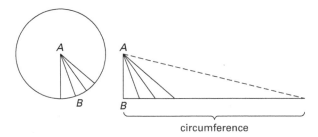

circumference

of the circle is one-half of the radius multiplied by the circumference, or, as Kepler put it, the area of the circle is to the square on the diameter as 11 to 14. Similarly, Kepler calculated the volume of a ring (torus) by slicing it into "an infinite number of very thin disks,"[9] each of which is thinner toward the center and thicker toward the outside. Kepler never claimed, however, that his method was rigorous, noting only that "we could obtain absolute and in all respects perfect demonstrations from these books of Archimedes themselves, were we not repelled by the thorny reading thereof."[10]

Kepler's use of "very thin" disks or very small triangles illustrate what came to be called the **method of infinitesimals**. Galileo, in contrast, used the **method of indivisibles**, in which a given geometric object is considered to be made up of objects of dimension one less. Thus, like Archimedes in *The Method*, he considered plane figures as made up of lines and solid figures as made up of surfaces. Nor did he believe he needed an argument by reductio ad absurdum to justify their use. As he wrote:

> I say it is most true and necessary that the line be composed of points, and the continuum of indivisibles. . . . Recognize clearly that the continuum is divisible into parts always divisible only because it is constituted of indivisibles. For if the division and subdivision must be able to go on forever, it must necessarily be that the multitude of the parts is such that one can never go beyond it, and therefore the parts are infinite [in number], otherwise the subdivision would come to an end, and if they are infinite, they must be without magnitude, because an infinity of parts endowed with magnitude compose an infinite magnitude."[11]

Infinity, however, has strange characteristics, one of which, a possible equality between a point and a line, Galileo illustrated in his calculation of the volume of a "soup bowl." We begin with a half-sphere AFB resting in a cylinder and also consider a cone whose vertex is at the point C on the diameter AB with base equal to AB (Fig. 15.5). If we now remove the sphere from the solid cylinder, what remains is the solid region called the soup bowl. To calculate the volume of the soup bowl, Galileo considered a horizontal slice along GK (of the vertical slice of all the solids in the figure). We have $IC^2 = IP^2 + PC^2$. But $IC = AC = GP$; therefore, $GP^2 = IP^2 + PC^2$. But $PC = PH$ and hence $GP^2 = IP^2 + PH^2$, or, $GP^2 - IP^2 = PH^2$. Because the cone and the soup bowl are generated by revolving the various segments around the central axis CF and because circles are as the squares on their diameters, it follows that the slice of the soup bowl is equal in area to the slice of the cone. By the principle used earlier by Heron and Zu Geng, that two solids with equal cross sections at corresponding heights have the same volume, Galileo concluded that the volume of the

FIGURE 15.5

Galileo's soup bowl cut out of a cylinder

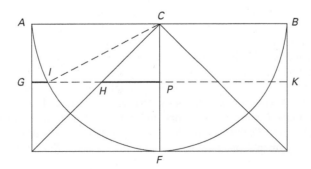

BIOGRAPHY

Bonaventura Cavalieri (1598–1647)

Cavalieri began his study of mathematics in Pisa while a member of a small religious order and there began a correspondence with Galileo that lasted nearly until Galileo's death. Probably through the latter's influence, he obtained a professorship at Bologna in 1629 and succeeded in having the appointment renewed every three years until his own death.

Besides the works mentioned in the text, Cavalieri published many other books on mathematics, including a work on astrology, and also investigated lenses and mirrors. His fame rests, however, on the method of indivisibles discussed in the *Geometria*, a work that was widely known, although, due to its difficulty, probably little studied.

soup bowl equals the volume of the cone. But what concerned Galileo was not so much the volumes, but the fact that the equality at each level must also be true at the top of the figure, in which case the cone is equal to a point and the soup bowl to a complete circle. Thus, a point is equal to a line. As Galileo wrote, "Now why should these not be called equal, if they are the last remnants and vestiges left by equal magnitudes?"[12]

It was Bonaventura Cavalieri (1598–1647), a disciple of Galileo, who first developed a complete theory of indivisibles, elaborated in his *Geometria indivisibilibus continuorum nova quadam ratione promota* (*Geometry, Advanced in a New Way by the Indivisibles of the Continua*) of 1635 and his *Exercitationes geometricae sex* (*Six Geometrical Exercises*) of 1647. The central concept of Cavalieri's work was that of *omnes lineae*, or "all the lines" of a plane figure F, to be written as $\mathcal{O}_F(\ell)$. By this, Cavalieri meant the collection of intersections of the plane figure with a perpendicular plane moving parallel to itself from one side of the given figure to the other. These intersections are lines, and it is the collection of such lines, thought of as a single magnitude, that Cavalieri dealt with throughout his work. Cavalieri's lines in some sense made up the given figure, but he was careful to distinguish $\mathcal{O}_F(\ell)$ from F itself. He was also able to generalize the idea by considering higher-dimensional objects such as "all the squares" or "all the cubes" of a given figure. One can think of "all the squares" of a triangle, for example, as representing a pyramid, each of whose cross sections is a square of side the length of a particular line in the triangle.

The basis for Cavalieri's computations was a result to this day known as **Cavalieri's principle**, a two-dimensional version of the principle Galileo used in the soup bowl calculation: "If two plane figures have equal altitudes and if sections made by lines parallel to the bases and at equal distances from them are always in the same ratio, then the plane figures are also in this ratio."[13] Cavalieri proved this result by an argument using superposition. It followed that if there were a fixed ratio between corresponding lines of the two figures F and G, then $\mathcal{O}_F(\ell) : \mathcal{O}_G(\ell) = F : G$. For example, suppose the rectangle F of length a and width b is divided by its diagonal into two triangles T, S (Fig. 15.6). Since each line segment BM in triangle T corresponds to one and only one equal line segment HE in triangle S, then $\mathcal{O}_T(\ell) = \mathcal{O}_S(\ell)$. On the other hand, since every line segment BA of the rectangle is made up of one segment from triangle S and one from triangle T, $\mathcal{O}_F(\ell) = \mathcal{O}_T(\ell) + \mathcal{O}_S(\ell)$. It follows

FIGURE 15.6

Cavalieri's method of "all the lines" in a triangle and rectangle

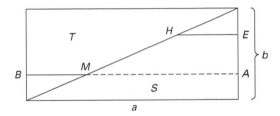

that $\mathcal{O}_F(\ell) = 2\mathcal{O}_T(\ell)$, or, all the lines of the square are double all the lines of the triangle. In modern notation, this result is equivalent to

$$ab = 2 \int_0^b \frac{a}{b} t \, dt$$

or, more simply, to

$$b^2 = 2 \int_0^b t \, dt.$$

Cavalieri was similarly able to demonstrate that "all the squares" of the rectangle F are triple "all the squares" of each triangle, or, in modern notation, that

$$a^2 b = 3 \int_0^b \frac{a^2}{b^2} t^2 \, dt \quad \text{or} \quad b^3 = 3 \int_0^b t^2 \, dt.$$

By 1647, he had demonstrated analogous results for certain higher powers and was able to infer that the area under the "higher parabola" $y = x^k$ inscribed in a rectangle is $\frac{1}{k+1}$ times the area of the rectangle or that

$$\int_0^b x^k \, dx = \frac{1}{k+1} b^{k+1}.$$

This result was also discovered by Fermat, Pascal, Roberval, and Torricelli in the same time period.

15.2.2 Torricelli and the Infinitely Long Solid

Evangelista Torricelli (1608–1647), another disciple of Galileo, also worked with indivisibles, but he cautioned that their uncritical use could lead to paradoxes. For example, suppose one uses mutually perpendicular indivisibles to calculate the areas of the two triangles in the rectangle $ABCD$ (Fig. 15.7). In this case, since the lines FE are always to the lines EG as AB is to BC, it would seem to follow that the triangle ABD is to the triangle DBC in that same ratio, an absurd result. Torricelli's solution to this paradox was essentially to revert to infinitesimals, namely, to consider that the "indivisible" line segments in fact had a thickness. In this particular case, the vertical line segments were thicker than the horizontal ones in the ratio AB to BC, so that when one took all of them together, the triangles ABD and DBC did in fact have the same area. Although much of Torricelli's work was not published in his lifetime, it did circulate in Italy in the work of his own students. Thus, it was known that

FIGURE 15.7

Torricelli's paradox using
indivisibles

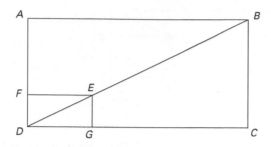

he had solved the problem of determining areas under and tangents to the curves $y = x^{\frac{m}{n}}$.
Interestingly, unlike many of his contemporaries, he generally gave complete classical proofs
of his results by reductio ad absurdum arguments.

Torricelli's most surprising discovery, however, was announced in 1643. He showed that
the volume of the infinitely long solid formed by rotating the hyperbola $xy = k^2$ around
the y axis from $y = a$ to $y = \infty$ was finite and in fact that the sum of its volume and that
of the cylinder of radius k^2/a and altitude a was equal to the volume of the cylinder of
altitude $\frac{k^2}{a}$ and radius equal to the semidiameter $AS = \sqrt{2}k$ of the hyperbola (Fig. 15.9).
Torricelli used a method similar to the cylindrical shell method taught today, but expressed
in terms of indivisibles, analogous to the lines of his friend Cavalieri. First, he showed that
the lateral surface area of any cylinder inscribed in his infinite hyperbolic solid, such as
$POMN$, was equal to the area of the circle of radius AS. (In modern terms, this is simply
that $2\pi x(k^2/x) = \pi(\sqrt{2}k)^2$.) Next, he noted that the infinite solid (including its base cylinder)
can be considered to be composed of all these cylindrical surfaces, to each of which there
corresponds one of the circles making up the cylinder $ACHI$. It follows that the infinite solid
is equal to the cylinder $ACHI$.

Torricelli wrote that "it may seem incredible that although this solid has an infinite length,
nevertheless none of the cylindrical surfaces we considered has an infinite length, but all of
them are finite."[14] Because he believed that this result was "incredible," however, he decided
to present a second proof, this one by exhaustion, to lend more strength to this result.

FIGURE 15.8

Torricelli on an Italian stamp

BIOGRAPHY

Evangelista Torricelli (1608–1647)

Torricelli studied mathematics in Rome with Benedetto
Castelli (1578–1643), a pupil of Galileo, and in 1641 was
able to study with Galileo himself at his house in Arcetri.
He stayed there until Galileo's death and was soon thereafter appointed to Galileo's old position of mathematician and
philosopher to the Grand Duke of Tuscany. Torricelli remained

in Florence for the rest of his life, continuing Galileo's work
on motion and grinding lenses for more powerful telescopes.
He is probably most famous for his discovery of the principle
of the barometer in 1643. He died of typhoid fever in 1647
(Fig. 15.8).

FIGURE 15.9

Torricelli's infinite hyperbolic
solid

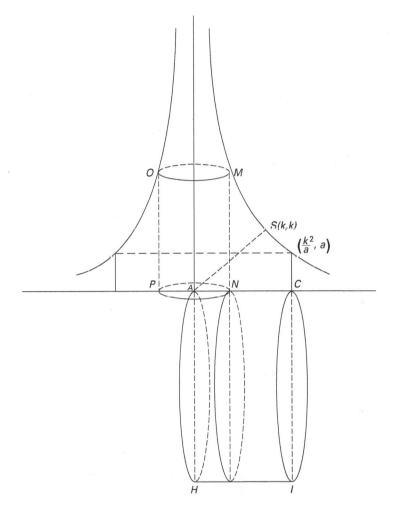

15.2.3 Fermat and the Area under Parabolas and Hyperbolas

As noted in the chapter opening, Fermat claimed on September 22, 1636, that he had been able to square "infinitely many figures composed of curved lines," in particular, that he could calculate the area of a region under any higher parabola $y = px^k$, but that he used methods different from those of Archimedes. In particular, rather than use Archimedes's triangles, Fermat would use simpler figures. Roberval, writing back in October, claimed that he too had found the same result, using a formula for the sums of powers of the natural numbers: "The sum of the square numbers is always greater than the third part of the cube which has for its root the root of the greatest square, and the same sum of the squares with the greatest square removed is less than the third part of the same cube; the sum of the cubes is greater than the fourth part of the [fourth power] and with the greatest cube removed, less than the fourth part, etc."[15] In other words, finding the area of the region bounded by the parabola

$y = px^k$, the x axis, and a given vertical line depends on the formula

$$\sum_{i=1}^{N-1} i^k < \frac{N^{k+1}}{k+1} < \sum_{i=1}^{N} i^k.$$

It is easy enough to see why this formula is fundamental, by considering the graph of $y = px^k$ over the interval $[0, x_0]$. Divide the base interval into N equal subintervals, each of length x_0/N, and erect over each subinterval a rectangle whose height is the y coordinate of the right endpoint (Fig. 15.10). The sum of the areas of these N circumscribed rectangles is then

$$p\frac{x_0^k}{N^k}\frac{x_0}{N} + p\frac{(2x_0)^k}{N^k}\frac{x_0}{N} + \cdots + p\frac{(Nx_0)^k}{N^k}\frac{x_0}{N} = \frac{px_0^{k+1}}{N^{k+1}}\left(1^k + 2^k + \cdots + N^k\right).$$

Similarly, one can calculate the sum of the areas of the inscribed rectangles, those whose height is the y coordinate of the left endpoint of the corresponding subinterval. If A is the area under the curve between 0 and x_0, then

$$\frac{px_0^{k+1}}{N^{k+1}}(1^k + 2^k + \cdots + (N-1)^k) < A < \frac{px_0^{k+1}}{N^{k+1}}(1^k + 2^k + \cdots + N^k).$$

FIGURE 15.10

The area under $y = px^k$ according to Fermat and Roberval

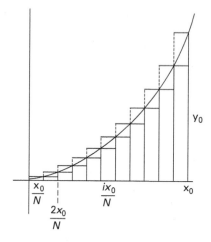

The difference between the outer expressions of this inequality is simply the area of the rightmost circumscribed rectangle. Because x_0 and $y_0 = px_0^k$ are fixed, this difference may be made less than any assigned value simply by taking N sufficiently large. It follows from the inequality cited by Roberval that both the area A and the value

$$\frac{px_0^{k+1}}{k+1} = \frac{x_0 y_0}{k+1}$$

are "squeezed" between two values whose difference approaches 0. Thus, Fermat (and Roberval) found that

$$A = \frac{x_0 y_0}{k + 1}.$$

The obvious question then is how either of these two men discovered the formula for the sums of powers, a formula that was in essence known to ibn al-Haytham 600 years earlier. Fermat claimed that he had a "precise demonstration" and doubted that Roberval had one. However, as is typical in Fermat's work, all we have is his own general statement in terms of numbers, pyramidal numbers, and the other numbers that occur as columns in Pascal's triangle: "The last side multiplied by the next greater makes twice the triangle. The last side multiplied by the triangle of the next greater side makes three times the pyramid. The last side multiplied by the pyramid of the next greater side makes four times the triangulotriangle. And so on by the same progression in infinitum."[16] Fermat's statement, which we write as

$$N\binom{N+k}{k} = (k+1)\binom{N+k}{k+1},$$

is equivalent to Pascal's twelfth consequence. Using the properties of Pascal's triangle, it is then not difficult to derive for each k in turn (beginning with $k = 1$) an explicit formula for the sum of the kth powers. This formula will be of the form

$$\sum_{i=1}^{N} i^k = \frac{N^{k+1}}{k+1} + \frac{N^k}{2} + p(N)$$

where $p(N)$ is a polynomial in N of degree less than k. A careful study of the form of $p(N)$ enables one then to derive Roberval's inequality.

It is not known whether Fermat actually proved the general result quoted or merely tried a few values of k and assumed it would be true for any value. And it is also not known how Fermat derived formulas for the sums of powers of integers. Probably, Fermat was not aware of the work of Johann Faulhaber (1580–1635), a *Rechenmeister* from Ulm, who by 1631 had developed explicit formulas for the sums of kth powers of integers through $k = 17$.[17] And Pascal himself, writing in 1654, may not have been aware of Fermat's results either, when he gave an explicit derivation for sums of powers from properties of his triangle and noted that "those who are even a little familiar with the doctrine of indivisibles will not fail to see that one may use this result for the determination of curvilinear areas. This result permits one immediately to square all types of parabolas and an infinity of other curves."[18]

In any case, Fermat was not completely satisfied with his method of finding areas because it only worked for higher parabolas. He could not see how to adapt it for curves of the form $y^m = px^k$ or for "higher hyperbolas" of the form $y^m x^k = p$. In modern terms, this method for finding areas under $y = px^k$ only worked if k was a positive integer. Fermat wanted a method that would work if k were any rational number, positive or negative. Although he only announced such a method in his *Treatise on Quadrature* of about 1658, it seems clear that he discovered this new procedure in the 1640s.

To apply his earlier method to the question of determining the area under $y = px^{-k}$ to the right of $x = x_0$ required dividing either the x axis or the line segment $x = x_0$ from

0 to $y_0 = px_0^{-k}$ into finitely many intervals and summing the areas of the inscribed and circumscribed rectangles. Using the latter procedure, however, would give Fermat an infinite rectangle as the difference between his circumscribed and inscribed rectangles, one for which it was not at all clear that the area could be made as small as desired. On the other hand, there was no way of dividing the (infinite) x axis into finitely many intervals ultimately to be made as small as one wishes. Fermat's solution to his dilemma was to divide the x axis into infinitely many intervals, whose lengths were not equal but formed a geometric progression, and then to use the known formula for summing such a progression to add up the areas of the infinitely many rectangles.

Fermat began by partitioning the infinite interval to the right of x_0 at the points $a_0 = x_0$, $a_1 = \frac{m}{n} x_0$, $a_2 = (\frac{m}{n})^2 x_0$, . . . , $a_i = (\frac{m}{n})^i x_0$, . . . where m and n $(m > n)$ are positive integers (Fig. 15.11). The intervals $[a_{i-1}, a_i]$ will ultimately be made as small as desired by taking $\frac{m}{n}$ sufficiently close to 1. Fermat next circumscribed rectangles above the curve over each small interval. The first circumscribed rectangle has area

$$R_1 = \left(\frac{m}{n} x_0 - x_0 \right) y_0 = \left(\frac{m}{n} - 1 \right) x_0 \frac{p}{x_0^k} = \left(\frac{m}{n} - 1 \right) \frac{p}{x_0^{k-1}}.$$

The second rectangle has area

$$R_2 = \left[\left(\frac{m}{n} \right)^2 x_0 - \left(\frac{m}{n} \right) x_0 \right] \frac{p}{(\frac{m}{n} x_0)^k} = \left(\frac{m}{n} \right) \left(\frac{m}{n} - 1 \right) x_0 \left(\frac{n}{m} \right)^k \frac{p}{x_0^k} = \left(\frac{n}{m} \right)^{k-1} R_1.$$

Similarly, the third rectangle has area

$$R_3 = \left(\frac{n}{m} \right)^{2(k-1)} R_1.$$

It follows that the sum of all the circumscribed rectangles is

$$R = R_1 + \left(\frac{n}{m} \right)^{k-1} R_1 + \left(\frac{n}{m} \right)^{2(k-1)} R_1 + \cdots$$

$$= R_1 \left[1 + \left(\frac{n}{m} \right)^{k-1} + \left(\frac{n}{m} \right)^{2(k-1)} + \cdots \right]$$

or, using the formula for the sum of a geometric series,

$$R = \frac{1}{1 - (\frac{n}{m})^{k-1}} R_1 = \frac{1}{1 - (\frac{n}{m})^{k-1}} \left(\frac{m}{n} - 1 \right) \frac{p}{x_0^{k-1}} = \frac{1}{\frac{n}{m} + (\frac{n}{m})^2 + \cdots + (\frac{n}{m})^{k-1}} \frac{p}{x_0^{k-1}}.$$

Fermat could have made a similar calculation for the inscribed rectangles, but decided it wasn't necessary. He let the area of the first rectangle "go to nothing," or, in modern terminology, found the limiting value of his sum, by letting $\frac{n}{m}$ approach 1. The value of R then approaches $\frac{1}{k-1} \frac{p}{x_0^{k-1}}$, and therefore, the desired area A is given by

$$A = \frac{1}{k-1} x_0 y_0.$$

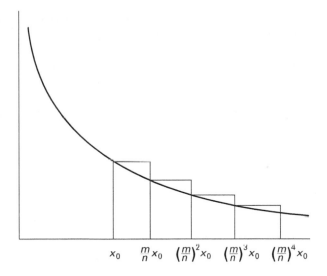

$$x_0 \quad \frac{m}{n}x_0 \quad \left(\frac{m}{n}\right)^2 x_0 \quad \left(\frac{m}{n}\right)^3 x_0 \quad \left(\frac{m}{n}\right)^4 x_0$$

Fermat quickly noticed that this division of the axis into infinite intervals could also be applied to find the known area under the parabolas $y = px^k$ from $x = 0$ to $x = x_0$. He simply divided this finite interval $[0, x_0]$ into an infinite set of subintervals by beginning from the right: $a_0 = x_0$, $a_1 = \frac{n}{m}x_0$, $a_2 = (\frac{n}{m})^2 x_0, \ldots, a_i = (\frac{n}{m})^i x_0, \ldots$ where here $n < m$, and proceeded as above to show that this area is equal to $\frac{1}{k+1}x_0 y_0$. In the other cases Fermat wanted to solve, namely, the areas under the curves $x^k y^m = p$ and $y^m = px^k$, the method had to be modified slightly to avoid having the geometric series involve fractional powers (Sidebar 15.1). But Fermat did succeed in showing that the area under the "hyperbola" $x^k y^m = p$ to the right of $x = x_0$ is $\frac{m}{k-m}x_0 y_0$, while that under the "parabola" $y^m = px^k$ from 0 to x_0 is $\frac{m}{k+m}x_0 y_0$.

15.2.4 Wallis and Fractional Exponents

Another mathematician who derived the same "integration" formulas as Fermat was John Wallis (1616–1703). Wallis, the first mathematician actually to explain fractional exponents and use them consistently, had read about Cavalieri's work but was never able to find a copy of his books. Thus, although he used indivisibles, he took an approach somewhat different from Cavalieri's in his *Arithmetica infinitorum* (*Arithmetic of Infinites*) of 1655. To determine the ratio of the area under $y = x^2$ between $x = 0$ and $x = x_0$ to the circumscribed rectangle whose area was $x_0 y_0$, he noted that the ratio of the corresponding line segments over a given abscissa x was $x^2 : x_0^2$. But since there were infinitely many such abscissas, Wallis needed to calculate the ratio of the sum of the infinitely many antecedents to the sum of the infinitely many consequents. Taking his abscissas in arithmetic progression $0, 1, 2, \ldots$, Wallis wanted to determine what in modern terminology would be

$$\lim_{n \to \infty} \frac{0^2 + 1^2 + 2^2 + \cdots + n^2}{n^2 + n^2 + n^2 + \cdots + n^2}.$$

SIDEBAR 15.1   *Did Fermat Invent the Calculus?*

By the mid-1640s, Fermat had determined the area under any curve of the form $y = x^k$ (except, of course, $y = x^{-1}$, a curve for which Fermat realized his method did not apply) and also had been able to construct the tangent to such a curve. Since he had solved the two major problems of the calculus, at least in these significant special cases, why should he not be considered as the inventor of the calculus? The answer must be that Fermat did not realize the inverse relationship between the two problems, partly because he did not understand that the two basic operations of the calculus, what we call the derivative and the integral, each determine new functions to which one can again apply these operations. A student today, seeing that the derivative of $y = x^k$ was the function $y' = kx^{k-1}$ and also that the area under $y = x^k$ from 0 to x was the function $\frac{x^{k+1}}{k+1}$ would probably immediately recognize the inverse property. Fermat did not, because he was not asking the questions that would lead him to it. For Fermat, construction of a tangent meant exactly that: find the length of the subtangent and then draw the line from the point on the curve to the appropriate point on the axis. Thus, he did not generally consider the slope of the tangent line, our derivative. In dealing with $y = x^k$, he would find that the subtangent t equaled $\frac{x}{k}$ rather than that the slope of the tangent equaled kx^{k-1}. Similarly, to find an area under a curve meant for Fermat to find a suitable rectangle equal in area to the given curvilinear region. In other words, the area under $y = x^k$ from 0 to x_0 equaled the area of the rectangle whose width was x_0 and whose height was $\frac{1}{k+1} y_0$. He never considered the area from a fixed coordinate to a variable one as determining a function, expressible as a new curve. Thus, although Fermat was able to solve the two basic problems of the calculus in many instances, he did not ask the "right" questions. It was others who were able to see what Fermat missed.

To calculate this ratio, he tried various cases:

$$\frac{0+1}{1+1} = \frac{1}{2} = \frac{1}{3} + \frac{1}{6}$$

$$\frac{0+1+4}{4+4+4} = \frac{5}{12} = \frac{1}{3} + \frac{1}{12}$$

$$\frac{0+1+4+9}{9+9+9+9} = \frac{14}{36} = \frac{1}{3} + \frac{1}{18}$$

and, in general,

$$\frac{0^2 + 1^2 + 2^2 + \cdots + n^2}{n^2 + n^2 + n^2 + \cdots + n^2} = \frac{1}{3} + \frac{1}{6n}.$$

Wallis concluded that if the number of terms was infinite, that is, if the lines "filled up" the desired areas, the ratio would be exactly 1/3. After calculating the analogous ratio for cubes to be 1/4, Wallis took the leap by what he called "induction" to the conclusion that for any positive integer k,

$$\frac{0^k + 1^k + 2^k + \cdots + n^k}{n^k + n^k + n^k + \cdots + n^k} = \frac{1}{k+1}$$

if there were an infinite number of terms.

Wallis's next step was to generalize this result to other powers by using analogy. Thus, he noted that given any arithmetic sequence of powers, say, 2, 4, 6, . . . , the consequents of the

BIOGRAPHY

John Wallis (1616–1703)

Although Wallis studied mathematics in his university days in Cambridge, much of his early life was spent in preparing for an ecclesiastical career. Nevertheless, his interest in various scientific questions led him to be involved in the first informal meetings in the 1640s in London of that group of men who formed the Royal Society in 1662. These weekly meetings were devoted to the discussion of "Philosophical Inquiries," including matters of anatomy, geometry, astronomy, and mechanics that were currently undergoing detailed investigations in England as well as on the continent. Wallis's early interest in mathematics being revived about 1647, he was appointed two years later to the vacancy in the Savilian chair of mathematics at Oxford caused by the incumbent finding himself on the wrong side in the English civil war. It was at Oxford that Wallis wrote his mathematical works, which included, besides the *Arithmetica infinitorum*, tracts on algebra, conic sections, and mechanics.

corresponding ratio of areas was also an arithmetic sequence, namely, 3, 5, 7, ... It followed that if the consequent of the ratio was 1, the sequence of powers must have index 0, that is, that m^0 must be 1 for every m. Furthermore, he noted that the sequence of second powers, with consequent 3, was composed of the square roots of the sequence of fourth powers, with consequent 5, and that 3 is the arithmetic mean between 1 and 5, the consequents of the series of powers 0 and 4. Wallis then made another bold generalization. Taking the series of terms $\sqrt{0}$, $\sqrt{1}$, $\sqrt{2}$, . . . , whose terms are the square roots of the series 0, 1, 2, . . . , he decided that the consequent of the corresponding ratio should be the arithmetic mean between 1 and 2, the consequents of the series of powers 0 and 1. In other words, the ratio

$$\frac{\sqrt{0} + \sqrt{1} + \sqrt{2} + \cdots \sqrt{n}}{\sqrt{n} + \sqrt{n} + \sqrt{n} + \cdots + \sqrt{n}}$$

must ultimately be equal to $\frac{1}{1\,1/2} = \frac{2}{3}$. In addition, the power of this series should be the arithmetic mean between 0 and 1, namely, 1/2, or, as Wallis put it, the index of $\sqrt{x}$ is 1/2. Wallis similarly concluded that the index of $\sqrt[3]{x}$ must be 1/3 and that of $\sqrt[3]{x^2}$ must be 2/3, while the consequents of their corresponding ratios must be the two arithmetic means between 1 and 2, namely, $1\frac{1}{3}$ and $1\frac{2}{3}$. Then, defining a fractional power for an arbitrary positive fraction p/q as the index of the qth root of the pth power, Wallis brought all these generalizations together into a theorem: "Proposition 64: If there is considered an infinite series of quantities, beginning from a point or 0, continually increasing according to any power either simple or composite, then the ratio of all of them to a series of the same number of terms equal to the greatest, is that of unity to the index of that power increased by one."[19]

Although Wallis applied this result to solve the area problem for a curve of the form $y = x^{p/q}$, that is, he found

$$\int_0^1 x^{p/q}\, dx = \frac{1}{p/q + 1},$$

he did not prove that his answer was correct except in the case where the index was $1/q$. He was, however, a firm believer in the power of analogy. Thus, he generalized his ideas of indexes to both negative and irrational numbers and showed that these indexes obeyed our familiar laws of exponents. His methods fell short when he attempted to generalize his theorem and the solution of the area problem to curves of the form $y = x^{-k}$. His basic rule told him that the corresponding ratio in the case of exponent -1 should be $1/(-1+1) = 1/0$, while in the case of exponent -2 the ratio should be $1/(-1)$. It was reasonable to assume that the area under the hyperbola $y = 1/x$ was in some sense $1/0$ or infinity, but what did it mean that the area under the curve $y = 1/x^2$ was $1/(-1)$? Since for indices 3, 2, 1, 0, the corresponding ratios were 1/4, 1/3, 1/2, and 1/1, and these values formed an increasing sequence, he assumed that the ratio $1/(-1)$ for index -2 should be greater than the ratio $1/0$ for index -1. But what it meant for $1/(-1)$ to be greater than infinity, Wallis could never quite figure out.

Passing over this problem, but also realizing that his method could be applied to finding areas under curves given by sums of terms of the form $ax^{p/q}$, Wallis next attempted to generalize his methods to the more complicated problem of determining arithmetically the area of a circle of radius 1, namely, of finding the area under the curve $y = \sqrt{1 - x^2} = (1 - x^2)^{1/2}$. To use his technique of arguing by analogy, he actually attacked a more general problem, to find the ratio of the area of the unit square to the area enclosed in the first quadrant by the curve $y = (1 - x^{1/p})^n$. The case $p = 1/2$, $n = 1/2$, is the case of the circle, where the ratio is $4/\pi$. It was easy enough for Wallis to calculate by his known methods the ratios in the cases where p and n were integral. For example, if $p = 2$ and $n = 3$, the area under $y = (1 - x^{1/2})^3$ from 0 to 1 is that under $y = 1 - 3x^{1/2} + 3x - x^{3/2}$, that is, $1 - 2 + 3/2 - 2/5 = 1/10$. Since the area of the unit square is 1, the ratio here is $1 : 1/10 = 10$. Wallis thus constructed the following table of these ratios, where for $p = 0$ he simply used the area under $y = 1^n$:

$p \backslash n$	0	1	2	3	4	5	6	7	...
0	1	1	1	1	1	1	1	1	...
1	1	2	3	4	5	6	7	8	...
2	1	3	6	10	15	21	28	36	...
3	1	4	10	20	35	56	84	120	...
⋮	⋮	⋮	⋮	⋮	⋮	⋮	⋮	⋮	⋱

Wallis clearly recognized Pascal's arithmetical triangle in his table. What he wanted was to be able to interpolate rows corresponding to $p = 1/2$, $p = 3/2$, ... and columns corresponding to $n = 1/2$, $n = 3/2$, ... from which he could find the desired value, which he wrote as □ when both parameters equaled 1/2. The basic formulas for rows in Pascal's triangle enabled him to interpolate the values $a_{p,n}$ when p was integral, where $a_{p,n}$ designates the entry in row p, column n. Thus, because $a_{2,n} = \frac{(n+1)(n+2)}{2}$ for integral n, Wallis could determine $a_{2,n}$ for fractional n. For example, $a_{2,\frac{1}{2}} = \frac{15}{8}$ and $a_{2,\frac{3}{2}} = \frac{35}{8}$. In general, Wallis found that $a_{p,n} = \frac{p+n}{n} a_{p,n-1}$, for p integral, and thus decided to use this same rule for the row $p = 1/2$. First, he noted that $a_{1/2,0} = 1$, because all other entries in column 0 were equal to 1. It followed that

$$a_{1/2,1} = \left(\frac{1/2+1}{1}\right) \cdot 1 = 3/2, \qquad a_{1/2,2} = \left(\frac{1/2+2}{2}\right) \cdot \frac{3}{2} = \frac{5}{4} \cdot \frac{3}{2} = \frac{15}{8},$$

$$a_{1/2,3} = \frac{7}{6} \cdot \frac{5}{4} \cdot \frac{3}{2} = \frac{105}{48}, \qquad \dots .$$

Similarly, since $a_{1/2,1/2} = \square$, he had

$$a_{1/2,3/2} = \left(\frac{1/2+3/2}{3/2}\right)\square = \frac{4}{3}\square, \qquad a_{1/2,5/2} = \frac{6}{5} \cdot \frac{4}{3}\square, \dots ,$$

and the row $p = 1/2$ was

$$1 \quad \square \quad \tfrac{3}{2} \quad \tfrac{4}{3}\square \quad \tfrac{15}{8} \quad \tfrac{8}{5}\square \quad \dots .$$

By this point in his book, Wallis had already come to believe that the value of $\square$ "cannot be forced out in numbers according to any method of notation so far accepted."[20] So Wallis attempted to bound the value by using ratios in row $p = 1/2$. Because it was evident that the ratios of alternate terms continually decreased, that is, $a_{1/2,k+2} : a_{1/2,k} > a_{1/2,k+4} : a_{1/2,k+2}$ for all k, he made the assumption that this was true for ratios of adjoining terms as well. It followed that

$$\square : 1 > 3/2 : \square, \quad \text{so } \square > \sqrt{3/2};$$

that

$$3/2 : \square > 4/3 \sqcap : 3/2, \quad \text{so } \square < (3/2)\sqrt{3/4} = \lfloor (3 \times 3)/(2 \times 4)\rfloor\sqrt{4/3};$$

and, similarly, that

$$\square > [(3 \times 3)/(2 \times 4)\sqrt{5/4}, \quad \square < (3 \times 3 \times 5 \times 5)/(2 \times 4 \times 4 \times 6)\sqrt{6/5}, \dots .$$

Wallis thus finally asserted that the fraction

$$\frac{3 \times 3 \times 5 \times 5 \times 7 \times 7 \times \cdots}{2 \times 4 \times 4 \times 6 \times 6 \times 8 \times \cdots}$$

"continued indefinitely is itself precisely the required number $\square = 4/\pi$."[21]

15.2.5 Roberval and the Cycloid

Although an infinite product was not perhaps the kind of area result Wallis had hoped for, other mathematicians of the period also had to be satisfied with answers not strictly arithmetical in their consideration of different curves. Roberval, for example, around 1637 determined the area under a **cycloid**, the curve traced by a point attached to the rim of a wheel rolling along a line. Roberval defined this curve as follows: "Let the diameter AB of the circle AGB move along the tangent AC, always remaining parallel to its original position, until it takes the position CD, and let AC be equal to the semicircle AGB [Fig. 15.12]. At the same time, let the point A move on the semicircle AGB in such a way that the speed of AB along AC may be equal to the speed of A along the semicircle AGB. Then, when AB has reached the position CD, the point A will have reached the position D. The point A is carried along by two motions—its own on the semicircle AGB and that of the diameter along AC."[22]

FIGURE 15.12

Roberval's determination of
the area bounded by a cycloid

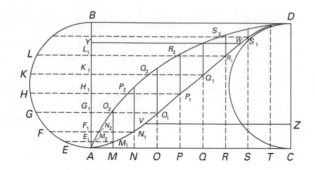

Roberval began his calculation by dividing the axis AC and the semicircle AGB into infinitely many equal parts. Along the semicircle these parts are $AE = EF = FG = \cdots$ while along the axis they are $AM = MN = NO = \cdots$. Furthermore, since the motion that generates the cycloid is composed of equal motions along the semicircle and the axis, Roberval set $AE = AM, EF = MN, \ldots$. Because the point A will be at E when the base of the diameter is at M, the point M_2, whose horizontal distance from M is the same as that of E from the point E_1 on the axis, is a point on the cycloid. Similarly, the point N_2 whose horizontal distance from N is the same as the distance of F from the point F_1 on the axis, is also a point on the curve, as are the points O_2, P_2, $\ldots$ indicated in Figure 15.12. Roberval then constructed a new curve, the companion of the cycloid, through the points M_1, N_1, O_1, $\ldots$, where M_1 has the same x coordinate as M and the same y coordinate as E, and so on. In modern notation, this curve is given by $x(t) = at$, $y(t) = a(1 - \cos t)$, or, in nonparametric form, as $y = a(1 - \cos \frac{x}{a})$, where a is the radius of the circle. The cycloid itself is given by $x(t) = a(t - \sin t)$, $y(t) = a(1 - \cos t)$.

To determine the area under half of one arch of the cycloid, Roberval first demonstrated that the area between the cycloid and its companion is equal to that of half the generating circle. This follows from Cavalieri's principle, because $M_1M_2 = EE_1$, $N_1N_2 = FF_1$, $\ldots$, and the corresponding pairs of lines are each at the same altitude. To finish his calculation, Roberval noted that to each line VZ in region $ACDM_1$, there corresponds an equal line WY in AM_1DB. Therefore, again by Cavalieri's principle, the companion curve to the cycloid bisects the rectangle $ABCD$. Because the area of the rectangle is equal to the product of half the circumference of the circle with the diameter (or $2\pi a^2$), the area under the companion curve is equal to that of the generating circle (πa^2). It follows that the area under half of one arch of the cycloid is equal to 3/2 that of the generating circle or that the area under an entire arch is three times that of the circle.

15.2.6 Pascal and the Sine Curve

Roberval's companion to the cycloid was in effect a Cosine curve, although Roberval did not identify it as such. But in the same work he did draw, probably for the first time, a curve identified as a curve of Sines, although it only consisted of the sines of one quadrant of the circle. In addition, Roberval was able to determine that the area under this curve was equal to the square of the radius defining the particular Sines. Pascal, some twenty years later, in a small treatise entitled *Traité des sinus du quart de cercle* (*Treatise on the Sines of a Quadrant of a Circle*) was able to find the area under any portion of that curve. Consider the quadrant ABC of the circle, and let D be any point from which the Sine DI is drawn to the radius AC (Fig. 15.13). Pascal then drew a "small" tangent EDE' and perpendiculars ER, $E'R$ to the radius. His claim was that "the sum of the sines of any arc of a quadrant is equal to the portion of the base between the extreme sines, multiplied by the radius."[23] By the "sum of the sines," Pascal meant the sum of the infinitesimal rectangles formed by multiplying each Sine by the infinitesimal arc represented by the tangent EE'. Recall that Pascal's Sine is our sine multiplied by the radius. Therefore, his result in modern notation is

$$\int_{\alpha}^{\beta} r \sin \theta \, d(r\theta) = r(r \cos \alpha - r \cos \beta).$$

FIGURE 15.13

Pascal's area under the sine curve

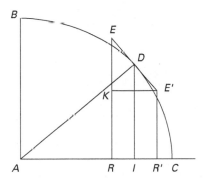

For his proof, Pascal noted that triangles EKE' and DIA are similar; hence, $DI : DA = E'K : EE' = RR' : EE'$ and therefore $DI \cdot EE' = DA \cdot RR'$. In other words, the rectangle formed from the Sine and the infinitesimal arc (or tangent) is equal to the rectangle formed by the radius and the part of the axis between the ends of the arc, or $r \sin \theta d(r\theta) = r(r \cos(\theta + d\theta) - r \cos(\theta)) = r(d(r \cos \theta))$. Adding these rectangles between the two given angles produces the cited result. Although this result proved important, and although Pascal generalized it immediately to give formulas for the integrals of powers of the Sine, the most significant aspect of Pascal's work was the appearance of the "differential triangle" EKE'. Leibniz's study of this particular work of Pascal was instrumental in his own realization of the connection between the area problem and the tangent problem.

15.2.7 The Area under the Rectangular Hyperbola

Our final example of a mid-seventeenth-century solution to the area problem is the work of the Belgian mathematician Gregory of St. Vincent (1584–1667) on the area under the hyperbola $xy = 1$ (Fig. 15.14). In his *Opus geometricum* of 1647, Gregory showed that if (x_i, y_i) for $i = 1, 2, 3, 4$ are four points on this hyperbola such that $x_2 : x_1 = x_4 : x_3$, then the area under the hyperbola over $[x_1, x_2]$ equals that over $[x_3, x_4]$ (Fig. 15.15). To prove this, divide the interval $[x_1, x_2]$ into subintervals at the points a_i, $i = 0, \ldots, n$. Because $x_2 : x_1 = x_4 : x_3$, it follows that $x_3 : x_1 = x_4 : x_2 = v$ or $x_3 = vx_1$, $x_4 = vx_2$. One can therefore conveniently subdivide the interval $[x_3, x_4]$ at the points $b_i = va_i$, $i = 0, \ldots, n$. If rectangles are then inscribed in

FIGURE 15.14

The Frontispiece of Gregory
of St. Vincent's *Opus
geometricum.* Gregory
claimed that he had squared
the circle. (*Source:* Special
Collections Division, USMA
Library, West Point, New
York)

ANTVERPLÆ, APVD IOANNEM ET IACOBVM MEVRSIOS ANNO M.DC.XLVI

and circumscribed about the hyperbolic areas A_j over $[a_j, a_{j+1}]$ and B_j over $[b_j, b_{j+1}]$, it is straightforward to calculate the corresponding inequalities:

$$(a_{j+1} - a_j)\frac{1}{a_{j+1}} < A_j < (a_{j+1} - a_j)\frac{1}{a_j} \quad \text{and} \quad (b_{j+1} - b_j)\frac{1}{b_{j+1}} < B_j < (b_{j+1} - b_j)\frac{1}{b_j}.$$

Substituting the values $b_j = va_j$ into the second set of inequalities gives

$$(a_{j+1} - a_j)\frac{1}{a_{j+1}} < B_j < (a_{j+1} - a_j)\frac{1}{a_j}.$$

Thus, both hyperbolic regions are squeezed between rectangles of the same areas. Because both intervals can be divided into subintervals as small as desired, it follows that the two hyperbolic areas are equal.

FIGURE 15.15

Greory of St. Vincent and
the area under the hyperbola
$xy = 1$

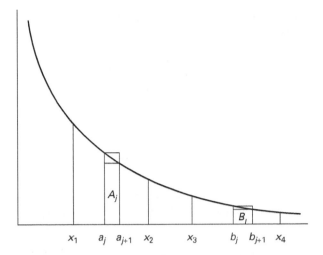

Gregory had not actually calculated the area under the hyperbola. However, when the Belgian Jesuit Alfonso Antonio de Sarasa (1618–1667) read Gregory's work in 1649, he immediately noticed that this calculation implied that the area $A(x)$ under the hyperbola from 1 to x had the logarithmic property $A(\alpha\beta) = A(\alpha) + A(\beta)$. After all, the ratio $\beta : 1$ equals the ratio $\alpha\beta : \alpha$, so the area from 1 to β equals the area from α to $\alpha\beta$. And because the area from 1 to $\alpha\beta$ is the sum of the areas from 1 to α and from α to $\alpha\beta$, the logarithmic property follows immediately. Thus, if one could calculate the areas under portions of the hyperbola $xy = 1$, one could calculate logarithms.

In 1668, Nicolaus Mercator (1620–1687) followed up on de Sarasa's hint in his *Logarithmotechnica* (*Logarithmic Teachings*). Having also learned from earlier mathematicians results on sums of integral powers, he decided to calculate $\log(1 + x)$ (the area A under the hyperbola $y = \frac{1}{1+x}$ from 0 to x) by using such sums. To do this, he divided the interval $[0, x]$ into n subintervals of length x/n and approximated A by the sum

$$\frac{x}{n} + \frac{x}{n}\left(\frac{1}{1 + \frac{x}{n}}\right) + \frac{x}{n}\left(\frac{1}{1 + \frac{2x}{n}}\right) + \cdots + \frac{x}{n}\left(\frac{1}{1 + \frac{(n-1)x}{n}}\right).$$

Since each term $\frac{1}{1 + (kx/n)}$ is the sum of the geometric series $\sum_{j=0}^{\infty}(-1)^j(\frac{kx}{n})^j$, it follows that

$$A \approx \frac{x}{n} + \frac{x}{n}\sum_{j=0}^{\infty}(-1)^j\left(\frac{x}{n}\right)^j + \frac{x}{n}\sum_{j=0}^{\infty}(-1)^j\left(\frac{2x}{n}\right)^j + \cdots + \frac{x}{n}\sum_{j=0}^{\infty}(-1)^j\left(\frac{(n-1)x}{n}\right)^j$$

$$= n\frac{x}{n} - \frac{x^2}{n^2}\sum_{i=1}^{n-1}i + \frac{x^3}{n^3}\sum_{i=1}^{n-1}i^2 + \cdots + (-1)^j\frac{x^{j+1}}{n^{j+1}}\sum_{i=1}^{n-1}i^j + \cdots$$

$$= x - \frac{\sum_{i=1}^{n-1}i}{n \cdot n}x^2 + \frac{\sum_{i=1}^{n-1}i^2}{n \cdot n^2}x^3 + \cdots + (-1)^j\frac{\sum_{i=1}^{n-1}i^j}{n \cdot n^j}x^{j+1} + \cdots$$

Mercator knew that if n is infinite, the coefficient of x^{k+1} in this expression is equal to $\frac{1}{k+1}$. Therefore,

$$\log(1+x) = x - \frac{x^2}{2} + \frac{x^3}{3} - \frac{x^4}{4} + \cdots,$$

a power series in x that enabled actual values of the logarithm to be calculated easily.

RECTIFICATION OF CURVES AND THE FUNDAMENTAL THEOREM

Descartes stated in his *Geometry* that the human mind could discover no rigorous and exact method of determining the ratio between curved and straight lines, that is, of determining exactly the length of a curve. Only two decades after Descartes wrote those words, however, several human minds proved him wrong. Probably the first rectification of a curve was that of the semicubical parabola $y^2 = x^3$ by the Englishman William Neile (1637–1670) in 1657 acting on a suggestion of Wallis. This was followed within the next two years by the rectification of the cycloid by Christopher Wren (1632–1723), the architect of St. Paul's Cathedral and much else in London, and the reduction of the rectification of the parabola to finding the area under a hyperbola by Huygens. The most general procedure, however, was that by Hendrick van Heuraet (1634–1660?), which appeared in van Schooten's 1659 Latin edition of Descartes' *Geometry*.

15.3.1 The Work of van Heuraet

Van Heuraet began his paper *De transmutatione curvarum linearum in rectas* (*On the Transformation of Curves into Straight Lines*) by showing that the problem of constructing a line-segment equal in length to a given arc is equivalent to finding the area under a certain curve. Let P be an arbitrary point on the arc MN of the curve α (Fig. 15.16). The length PS of the normal line from P to the axis can be determined by Descartes' method. Taking an arbitrary line segment σ, van Heuraet defined a new curve α' by the ratio $P'R : \sigma = PS : PR$, where P' is the point on α' associated with P. (The σ is included so that both ratios are ratios

FIGURE 15.16

Van Heuraet's rectification of a curve

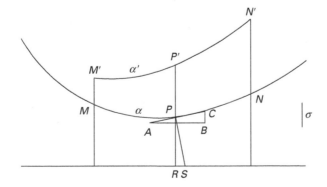

BIOGRAPHY

Hendrick van Heuraet (1634–1660?)

Van Heuraet was born in Haarlem in the Netherlands and went to Leiden to study mathematics under van Schooten in 1653.[24] His inheritance from his cloth-merchant father upon the latter's death the previous year made him rather wealthy, so he could afford to study and travel without worrying about his means of support. His early mathematical work showed such great promise that van Schooten published not only his treatise on rectification but also his work on inflection points. As far as is known, however, van Heuraet died at an early age. There is no extant record of his activities after early 1660.

of lines.) Drawing the differential triangle ACB with AC tangent to α at P, he noted that $PS : PR = AC : AB$. In modern notation, if $AC = ds$ and $AB = dx$, then van Heuraet's ratios yield $P'R : \sigma = ds : dx$ or $\sigma \, ds = P'R \, dx$. Because the sum of the infinitesimal tangents— or equivalently, the infinitesimal pieces of the arc—over the curve MN gives the length of MN, van Heuraet concluded that $\sigma \cdot$(length of MN) = area under the curve α' between M' and N'. Thus, if it is possible to derive the equation of α' from that of α and to calculate the area under it, the length of MN can also be calculated. Using modern notation, with $z = P'R = \sigma \frac{ds}{dx} = \sigma \sqrt{1 + (\frac{dy}{dx})^2}$, van Heuraet's procedure can be written as

$$\sigma \cdot \text{(length of } MN) = \int_a^b z \, dx = \int_a^b \sigma \sqrt{1 + \left(\frac{dy}{dx}\right)^2} \, dx,$$

where a and b represent the x coordinates of M and N, essentially the modern arclength formula.

Van Heuraet illustrated his procedure with one of the few curves for which the area under the associated curve can actually be calculated, the semicubical parabola $y^2 = x^3$ that had been considered by Neile somewhat earlier. Using Descartes' normal method, he calculated that the equation that must have a double root is $x^3 + x^2 - 2vx + v^2 - n^2 = 0$. Using Hudde's rule for finding the double root, he multiplied the terms of this equation by 3, 2, 1, 0 to get $3x^3 + 2x^2 - 2vx = 0$. Therefore, $v - x = \frac{3x^2}{2}$, and $PS = \sqrt{\frac{9}{4}x^4 + x^3}$. Setting $\sigma = 1/3$, van Heuraet defined the new curve α' by taking

$$z = P'R = \sigma \cdot \frac{PS}{PR} = \frac{1}{3} \frac{\sqrt{(9/4)x^4 + x^3}}{\sqrt{x^3}} = \sqrt{\frac{1}{4}x + \frac{1}{9}}$$

or, equivalently, $z^2 = (1/4)x + 1/9$. Van Heuraet easily identified this curve as a parabola, the area under which he knew how to calculate. The length of the semicubical parabola from $x = 0$ to $x = b$ then equals this area divided by σ, that is,

$$\sqrt{\left(b + \frac{4}{9}\right)^3} - \frac{8}{27}.$$

BIOGRAPHY

James Gregory (1638–1675)

After studying at Marischal College in Aberdeen, Gregory left Scotland in 1663 and spent the next five years abroad, studying in Italy under Torricelli's pupil, Stefano degli Angeli, in Padua. It was there that he wrote his first two mathematical works. In 1668, he returned to the chair of mathematics at St. Andrews, where he spent much of his time teaching elementary mathematics. His correspondence with John Collins in London was his sole contact with the rest of the mathematical world. And some of these letters contained surprising discoveries, whose origins are still unknown.

For example, in a letter of December 19, 1670, Gregory wrote that the arc whose Sine is B (where the radius of the circle is R) is expressible as[25]

$$B + \frac{B^3}{6R^2} + \frac{3B^5}{40R^4} + \frac{5B^7}{112R^6} + \frac{35B^9}{1152R^8} + \cdots.$$

In modern terminology, Gregory's series is the series for $\frac{1}{R} \arcsin \frac{B}{R}$, which, if $R = 1$, can be written as

$$\arcsin x = x + \frac{x^3}{6} + \frac{3x^5}{40} + \frac{5x^7}{112} + \frac{35x^9}{1152} + \cdots$$

Similarly, in a letter of February 15, 1671, Gregory included, among others, the series for the arc a given the tangent t and vice versa, written in modern notation as[26]

$$\arctan x = x - \frac{x^3}{3} + \frac{x^5}{5} - \frac{x^7}{7} + \frac{x^9}{9} - \cdots$$

$$\tan y = y + \frac{y^3}{3} + \frac{2y^5}{15} + \frac{17y^7}{315} + \frac{3233y^9}{181,440} + \cdots.$$

In 1673, he was forced to leave St. Andrews because of political problems, but was soon thereafter able to assume a professorship at Edinburgh. Unfortunately, he was blinded by a stroke in October, 1675, and died shortly afterward.

After remarking that one can similarly explicitly determine the lengths of the curves $y^4 = x^5$, $y^6 = x^7$, $y^8 = x^9$, . . . , van Heuraet concluded the paper with the more difficult rectification of an arc of the parabola $y = x^2$, a length that depends on the determination of the area under the hyperbola $z = \sqrt{4x^2 + 1}$. That problem, in 1659, had not yet been satisfactorily solved. Nevertheless, van Heuraet's methods soon became widely known. In particular, the use of the differential triangle and the association of a new curve to the given curve helped to lead others to the ideas relating the tangent problem to the area problem.

15.3.2   Gregory and the Fundamental Theorem

Among the mathematicians who related the tangent problem to the area problem were Isaac Barrow (1630–1677) and James Gregory (1638–1675), both of whom decided to organize the material relating to tangents, areas, and rectification gathered in their travels through France, Italy, and the Netherlands and to present it systematically. Not surprisingly, then, the *Lectiones geometricae* (*Geometrical Lectures*) (1670) of Barrow and the *Geometriae pars universalis* (*Universal Part of Geometry*) (1668) of Gregory contained much of the same material presented in similar ways. In effect, both of these works were treatises on material today identified as calculus, but with presentations in the geometrical style each author had learned in his university study. Neither was able to translate the material into a method of computation useful for solving problems.

As an example, consider how Gregory presented the fundamental theorem of calculus, the result linking the ideas of area and tangent. This result was the natural outcome of Gregory's study of the general problem of arclength as he found it in the work of van Heuraet. Consider a monotonically increasing curve $y = y(x)$ with two other curves associated with it, the normal curve $n(x) = y\sqrt{1 + (dy/dx)^2}$ and $u(x) = cn/y = c\sqrt{1 + (dy/dx)^2}$, where c is a given constant. Now constructing the differential triangle dx, dy, ds at a given point, Gregory argued from its similarity with the triangle formed by the ordinate y, the subnormal v, and the normal n that $y : n = dx : ds = c : u$ and thus that both $u\,dx = c\,ds$ and $n\,dx = y\,ds$ (Fig. 15.17). Summing the first equation over the curve showed Gregory, as it had van Heuraet, that the arclength $\int ds$ can be expressed in terms of the area under the curve $\frac{1}{c}(u(x))$. The sum of the second equation enabled Gregory to show that the area under $n = n(x)$ is equal, up to a constant multiple, to the area of the surface formed by rotating the original curve around the x axis. Gregory proved both of these results by a careful Archimedean argument using inscribed and circumscribed rectangles and a double reductio ad absurdum.

FIGURE 15.17

Gregory's differential triangle

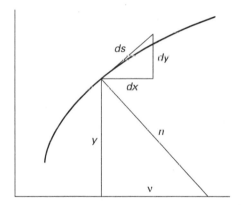

Having now shown that arclength can be found by an area, Gregory made a fundamental advance by asking the converse question. Can one find a curve $u(x)$ whose arclength s has a constant ratio to the area under a given curve $y(x)$? In modern notation, Gregory was asking whether it is possible to determine u such that

$$c \int_0^x \sqrt{1 + \left(\frac{du}{dx}\right)^2}\, dx = \int_0^x y\, dx.$$

But this means that $c^2(1 + (du/dx)^2) = y^2$ or that $du/dx = (1/c)\sqrt{y^2 - c^2}$. In other words, Gregory had to determine a curve u, the slope of whose tangent is equal to a given function. Letting $z = \sqrt{y^2 - c^2}$, Gregory simply defined $u(x)$ to be the area under the curve z/c from the origin to x. He then had to show that the slope of the tangent to this curve is given by z/c. What he in fact demonstrated, again by a reductio argument, is that the line connecting a point K on the u curve to the point on the axis at a distance cu/z from the x coordinate of K is tangent to the curve at K.

Gregory's crucial advance, then, was the abstraction of the idea of area under a specific curve between two given x values into the idea of area as a function of a variable. In other words, he constructed a new curve whose ordinate at any value x was equal to the area under the original curve from a fixed point up to x. Once this idea was conceived, it turned out that it was not difficult to construct the tangent to this new curve and show that its slope at x was always equal to the original ordinate there.

15.3.3   Barrow and the Fundamental Theorem

Gregory had the idea of constructing a new curve for the particular purpose of finding arclength. Isaac Barrow, on the other hand, stated a more general version of part of the fundamental theorem as proposition 11 in lecture X of his *Geometrical Lectures*:

THEOREM   *Let ZGE be any curve of which the axis is AD and let ordinates applied to this axis, AZ, PG, DE, continually increase from the initial ordinate AZ. Also let AIF be a curve such that if any straight line EDF is drawn perpendicular to AD, cutting the curves in the points E, F, and AD in D, the rectangle contained by DF and a given length R is equal to the intercepted space $ADEZ$. Also let $DE : DF = R : DT$ and join FT. Then TF will be tangent to AIF (Fig. 15.18).*[27]

FIGURE 15.18

Barrow's version of the fundamental theorem

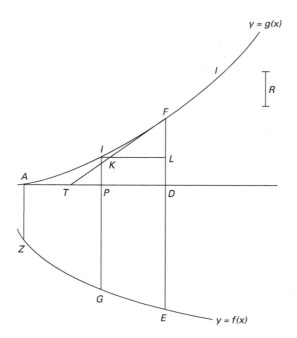

Like Gregory, Barrow began with a curve ZGE, written in modern notation as $y = f(x)$, and constructed a new curve $AIF = (g(x))$ such that $Rg(x)$ is always equal to the area bounded by $f(x)$ between a fixed point and the variable point x. In modern notation, $Rg(x) = \int_a^x f(x)\,dx$. Barrow then proved that the length $t(x)$ of the subtangent to $g(x)$

is given by $Rg(x)/f(x)$, or that

$$g'(x) = \frac{g(x)}{t(x)} = \frac{f(x)}{R} \quad \text{or} \quad \frac{d}{dx} \int_a^x f(x)\, dx = f(x).$$

Barrow proved this result by showing that the line TF always lies outside the curve AIF. If I is any point on the curve $g(x)$ on the side of F toward A, and if IG is drawn parallel to AZ and KL parallel to AD, the nature of the curve shows that $LF : LK = DF : DT = DE : R$ or $R \cdot LF = LK \cdot DE$. Because $R \cdot IP$ equals the area of $APZG$, it follows that $R \cdot LF$ equals the area of $PDEG$. Therefore, $LK \cdot DE = $ area $PDEG < PD \cdot DE$. Hence, $LK < PD$ or $LK < LI$ and the tangent line is below the curve at I. A similar argument applies for a point I on the side of F away from A.

In theorem 19 of lecture XI, Barrow proved the second part of the fundamental theorem, namely, that

$$\int_a^b Rf'(x)\, dx = R(f(b) - f(a)),$$

by showing a correspondence between infinitesimal rectangles in the region under the curve $Rf'(x)$ and those in the (large) rectangle $R(f(b) - f(a))$.

How did Barrow discover the inverse relationship of the tangent and the area problems? Barrow did not tell us explicitly, but a careful reading of the early parts of the *Geometrical Lectures* shows that he often thought of curves as being generated by the motion of a moving point. Thus, he demonstrated that the slope of the tangent line at a point P to a curve so generated is equal to the velocity of the moving point at P. Furthermore, he also represented the varying velocities of a point by the varying ordinates of a curve whose axis represents time. "Hence, if through all points of a line representing time are drawn straight lines so disposed that no one coincides with another (i.e., parallel lines), the plane surface that results as the aggregate of the parallel straight lines, when each represents the degree of velocity corresponding to the point through which it is drawn, exactly corresponds to the aggregate of the degrees of velocity, and thus most conveniently can be adapted to represent the space

---

BIOGRAPHY

Isaac Barrow (1630–1677)

Barrow entered Trinity College, Cambridge, in 1643, receiving his BA in 1648 and his MA in 1652. Because he had royalist sympathies, he was ousted from the university in 1655 and prevented from assuming a professorship. He took the opportunity to tour the continent for four years and learn mathematics in France, Italy, and the Netherlands. He returned to Cambridge at the time of the Restoration, took holy orders, and became the Regius professor of Greek. In 1662, he accepted concurrently the Gresham Professorship of Geometry in London and the following year became the first Lucasian Professor of Mathematics at Cambridge. After presenting several courses of lectures over the next few years, in elementary mathematics, geometry, and optics, he resigned his position in 1669 to become the royal chaplain in London. In 1673, he returned to Trinity College as master and two years later was appointed vice chancellor of the University, but he died in 1677, probably due to an overdose of drugs.

traversed also."[28] This idea of representing distance as the area under the velocity curve goes back to Galileo and Oresme, but, combined with the notion of velocity as the slope of a tangent, it could have easily led to Barrow's understanding of the inverse relationship of the differential and integral processes.

During the years just preceding the publication of his *Geometrical Lectures*, Barrow was the Lucasian Professor of Mathematics at Cambridge University. It is not known whether Isaac Newton ever attended any of Barrow's lectures, but he may well have been influenced by Barrow's idea of curves being generated by motion. Newton in fact suggested a few improvements to Barrow's book, in particular, that Barrow include an algebraic method of calculating tangents based on the differential triangle. This method consists of drawing the differential triangle NMR at a point M on a given curve and calculating the ratio of $MP = y$ to $PT = t$ by using the corresponding ratio of $MR = a$ to $NR = e$ in the infinitesimal triangle (Fig. 15.19). Thus, if the curve is $y^2 = x^3$, Barrow replaced y by $y + a$, x by $x + e$, and found that $(y + a)^2 = (x + e)^3$ or $y^2 + 2ay + a^2 = x^3 + 3x^2e + 3xe^2 + e^3$. He then removed all terms containing a power of a or e or a product of the two, "for these terms have no value," and determined that $y^2 + 2ay = x^3 + 3x^2e$. Next, "rejecting all terms consisting of letters denoting known or determined quantities . . . for these terms, brought over to one side of the equation, will always be equal to zero," he was left with $2ay = 3x^2e$. In the final step, he substituted y for a and t for e to get the ratio $y : t$. In this case, the result is $y : t = 3x^2 : 2y$. Barrow further noted that "if any indefinitely small arc of the curve enters the calculation, an indefinitely small part of the tangent or of any straight line equivalent to it is substituted for the arc."[29] Barrow made no attempt to justify this method, a modification of Fermat's method of *adequality*, but only noted that he frequently used it in his own calculations (Sidebar 15.2).

FIGURE 15.19

Barrow's differential triangle

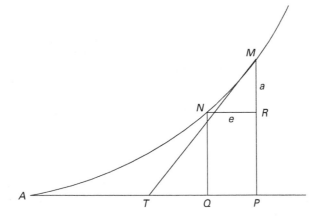

The work of Barrow and Gregory can be thought of as a culmination of all of the seventeenth-century methods of area and tangent calculations. But neither of these men in 1670 could mold these methods into a true computational and problem solving tool. In the five years before that date, however, Isaac Newton, communicating with practically no one from his rooms at Cambridge, was already using his intense powers of concentration to consolidate and extend the work of all his predecessors into the subject we today call the calculus.

SIDEBAR 15.2 *Did Barrow Invent the Calculus?*

Given that Barrow knew the algebraic procedures for calculating tangents and areas and was also aware of the fundamental theorem, should he be considered one of the inventors of the calculus? The answer must be no. Barrow presented all of his work in a classic geometric form. It does not appear that he was aware of the fundamental nature of the two theorems presented in the text. Barrow did not mention that they are particularly important; he just presented them as two among many geometrical results dealing with tangents and areas. And Barrow never used them to calculate areas. Perhaps if Newton had not come along, Barrow would have seen the uses to which these theorems could be put. But because he realized that Newton's abilities outshone his own, and because he was more concerned with pursuing theological interests, Barrow abandoned the study of mathematics to his younger colleague and left to him the invention of the calculus.

EXERCISES

1. Show that the largest parallelepiped that can be inscribed in a sphere is a cube. Determine the dimensions of the cube and its volume if the sphere has radius 10.

2. Show that the largest circular cylinder that can be inscribed in a sphere is one in which the ratio of diameter to altitude is $\sqrt{2} : 1$ (Kepler).

3. Show that Fermat's two methods of determining a maximum or minimum of a polynomial $p(x)$ are both equivalent to solving $p'(x) = 0$.

4. Use one of Fermat's methods to find the maximum of $bx - x^3$. How would Fermat decide which of the two solutions to choose as his maximum?

5. Justify Fermat's first method of determining maxima and minima by showing that if M is a maximum of $p(x)$, then the polynomial $p(x) - M$ always has a factor $(x - a)^2$, where a is the value of x giving the maximum.

6. Use Fermat's tangent method to determine the relation between the abscissa x of a point B and the subtangent t that gives the tangent line to $y = x^3$.

7. Modify Fermat's tangent method to be able to apply it to curves given by equations of the form $f(x, y) = c$. Begin by noting that if $(x + e, \bar{y})$ is a point on the tangent line near to (x, y), then $\bar{y} = \frac{t+e}{t} y$. Then *adequate* $f(x, y)$ to $f(x + e, \frac{t+e}{t} y)$. Apply this method to determine the subtangent to the curve $x^3 + y^3 = pxy$.

8. Show that in modern notation, Fermat's method of finding the subtangent t to $y = f(x)$ determines t as $t = f(x)/f'(x)$. Show similarly that the modified method of Exercise 7 is equivalent in modern terms to determining t as $t = -y(\partial f/\partial y)/(\partial f/\partial x)$.

9. Use Fermat's method to determine the subtangent to the ellipse $x^2/a^2 + y^2/b^2 = 1$. Compare your answer with that of Apollonius in Chapter 4.

10. Use Descartes' circle method to determine the subnormal to $y = x^{3/2}$.

11. Use Descartes' circle method to determine the slope of the tangent line to $y^2 = x$.

12. Use Hudde's rule applied to Descartes' method to show that the slope of the tangent line to $y = x^n$ at (x_0, x_0^n) is nx_0^{n-1}.

13. Maximize $3ax^3 - bx^3 - \frac{2b^2a}{3c}x + a^2b$ using Hudde's rule. (This example is taken from Hudde's *De maximis et minimis*.)

14. Apply Sluse's rule to find the subtangent to Fermat's equation $x^3 + y^3 - pxy = 0$.

15. Apply Sluse's rule to find the tangent to the circle $x^2 + y^2 = bx$.

16. Derive Sluse's rule for the special case $f(x, y) = g(x) - y$ from Fermat's rule for determining the subtangent to $y = g(x)$. Derive Sluse's general rule from the modification of Fermat's rule discussed in Exercise 7.

17. Given that the volume of a cone is $(1/3)hA$, where h is the height and A the area of the base, use Kepler's method to divide a sphere of radius r into infinitely many infinitesimal cones of height r, and then add up their volumes to get a formula for the volume of the sphere.

18. Show that Fermat's rule,

$$N\binom{N+k}{k} = (k+1)\binom{N+k}{k+1},$$

is equivalent to

$$N \sum_{j=k-1}^{N+k-1} \binom{j}{k-1} = (k+1)\frac{N(N+1)\cdots(N+k)}{(k+1)!}$$

and also to

$$\sum_{j=1}^{N} \frac{j(j+1)\cdots(j+k-1)}{k!} = \frac{N(N+1)\cdots(N+k)}{(k+1)!}.$$

19. Set $k = 3$ in the last formula of Exercise 18 and derive the formula for the sums of cubes from this result and the known formulas for the sum of the integers and the sums of the squares.

20. Discover the fifth-degree polynomial formula for $\sum_{j=1}^{N} j^4$ by using the formulas of Exercise 18.

21. Fermat included the following result in a letter to Roberval dated August 23, 1636: If the parabola with vertex A and axis AD is rotated around the line BD perpendicular to its axis, the volume of this solid has the ratio 8 : 5 to the volume of the cone of the same base and vertex (Fig. 15.20). Prove that Fermat is correct and show that this result is equivalent to the result on the volume of this same solid discovered by ibn al-Haytham, discussed in Chapter 9.

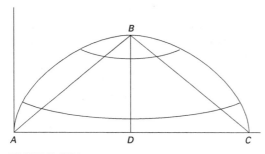

FIGURE 15.20

Fermat's problem of revolving a parabola around a line perpendicular to its axis

22. Determine the area under the curve $y = px^k$ from $x = 0$ to $x = x_0$ by dividing the interval $[0, x_0]$ into an infinite set of subintervals, beginning from the right with the points $a_0 = x_0$, $a_1 = \frac{n}{m}x_0$, $a_2 = (\frac{n}{m})^2 x_0$, ..., where $n < m$, and proceeding as in Fermat's derivation of the area under the hyperbola.

23. Using Wallis's method, interpolate the row $p = 3$ in his ratio table for $n = 1/2$, $n = 3/2$, and $n = 5/2$.

24. Using Wallis's method, calculate the values $n = 0, 1/2, 1, 3/2, 2, 5/2$ in row $p = \frac{3}{2}$ of his ratio table.

25. Find the length of arc of the curve $y^4 = x^5$ from $x = 0$ to $x = b$.

26. Show that to find the length of an arc of the parabola $y = x^2$ one needs to determine the area under the hyperbola $y^2 - 4x^2 = 1$.

27. Gregory derived various formulas for calculating the subtangents of curves composed of other curves by addition, subtraction, and the use of proportionals. In particular, suppose that four functions are related by the proportion $u : v = w : z$. Show that the subtangent t_z is given by the formula

$$t_z = \frac{t_u t_v t_w}{t_u t_v + t_u t_w - t_v t_w}.$$

Derive the product and quotient rules for derivatives from this formula, given that if a function u is a constant, then its subtangent t_u is infinite.

28. Use Barrow's a, e method to determine the slope of the tangent line to the curve $x^3 + y^3 = c^3$.

29. Barrow was perhaps the first to calculate the slope of the tangent to the curve $y = \tan x$ using his a, e method. Suppose DEB is a quadrant of a circle of radius 1 and BX the tangent line at B (Fig. 15.21a). The tangent curve AMO (Fig. 15.21b) is defined to be the curve such that if AP is equal to arc BE, then PM is equal to BG, the tangent of arc BE. Use the differential triangle to calculate the slope of the tangent to curve AMO as follows: Let $CK = f$ and $KE = g$. Since $CE : EK = $ arc $EF : LK = PQ : LK$,

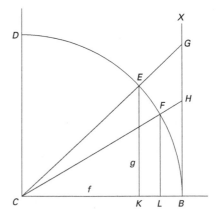

FIGURE 15.21A

Barrow's calculation of the tangent line to the tangent function

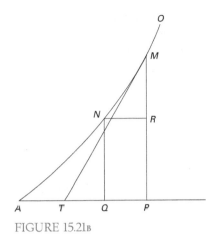

FIGURE 15.21B

it follows that $1 : g = e : LK$ or $LK = ge$ and $CL = f + ge$. Then $LF = \sqrt{1 - f^2 - 2fge} = \sqrt{g^2 - 2fge}$. Because

$CL : LF = CB : BH$, one can transfer the ratio in the circle to that on the tangent curve. Demonstrate finally that $PT = t = \frac{BG \cdot CB^2}{CG^2} = \frac{BG \cdot CK^2}{CE^2}$, and show that this result can be translated into the familiar formula $d(\tan x)/dx = \sec^2 x$. Given this result, can one say that Barrow has differentiated a trigonometric function? Why or why not?

30. Compare the efficacy of the tangent method of Fermat and the circle method of Descartes to determine the slope of the tangent line to the curve $y = x^n$. Note the kinds of calculations needed in each instance.

31. Outline a lesson introducing the concept of integration via the method of Fermat applied to curves whose equations are of the form $y = x^n$ for n a positive integer.

32. Outline a lesson introducing the determination of arclength using the method of van Heuraet. How does this differ from the method normally presented in calculus texts?

REFERENCES AND NOTES

There are three current works on the history of calculus, which together provide a fairly complete treatment of the material. The earliest is Carl Boyer, *The History of the Calculus and Its Conceptual Development* (New York: Dover, 1959). This work deals primarily with the central concepts underlying the calculus. Although it generally considers ideas only as they prefigure modern ones rather than considering them as part of their own time, the book is still an excellent treatment that covers most of the ideas studied not only in this chapter but also, insofar as they are relevant to calculus, in several earlier and later chapters. Margaret E. Baron, *The Origins of the Infinitesimal Calculus* (Oxford: Pergamon Press, 1969), traces more of the methods actually used in the calculus up to the time of Newton and Leibniz. It is very strong on the first part of the seventeenth century in particular and provides many examples that are useful in understanding how various mathematicians actually solved the problems they encountered. The third book, C. H. Edwards, *The Historical Development of the Calculus* (New York: Springer, 1979), also is devoted to showing exactly how mathematicians calculated, but unlike the previous work it covers in detail the contributions of Newton and Leibniz as well the work of their eighteenth- and nineteenth-century successors. A more general discussion of seventeenth-century mathematics is found in D. T. Whiteside, "Patterns of Mathematical Thought in the Later Seventeenth Century," *Archive for History of Exact Sciences* 1 (1960–62), 179–388.

Although few of the works cited in this chapter are available in their entirety in English, excerpts from many, including works of Kepler, Cavalieri, Fermat, Torricelli, Roberval, and Pascal, are available in D. J. Struik, ed., *A Source Book in Mathematics, 1200–1800* (Cambridge: Harvard University Press, 1969). Mahoney's biography of Fermat, cited in Chapter 14, contains many details of Fermat's work on calculus. Descartes' work on normals is available in the English edition of *The Geometry*, cited in Chapter 14. John Wallis's *Arithmetica infinitorum* has been translated into English by Jacqueline Stedall: *The Arithmetic of Infinitesimals: John Wallis 1656* (New York: Springer, 2004). James Gregory's *Geometriae pars universalis* is available in English online, in a translation by Andrew Leahy: http://math.knox.edu/aleahy/gregory/WORKING/gpu.html. Isaac Barrow's *Geometrical Lectures* are available in English in J. M. Child, *The Geometrical Lectures of Isaac Barrow* (Chicago: Open Court, 1916). Child, in his commentary, seems to credit Barrow with most of the invention of the calculus by translating his geometrical work into modern analytical terms, a step Barrow himself never took. Nevertheless, there is much of interest in Barrow's lectures.

1. David Eugene Smith and Marcia L. Latham, trans., *The Geometry of René Descartes* (New York: Dover, 1954), p. 95.

2. Michael S. Mahoney, *The Mathematical Career of Pierre de Fermat 1601–1665* (Princeton University Press, 1973), p. 220.

3. Johann Kepler, *Gesammelte Werke*, ed. by M. Caspar (Munich: Beck, 1960), vol. IX, p. 85.

4. Quoted in Mahoney, *Mathematical Career*, p. 148.

5. Struik, ed., *Source Book*, p. 223.

6. For a discussion of the dispute between Fermat and Descartes regarding Fermat's method, and the role Mersenne played in exacerbating it, see Mahoney, *Mathematical Career*, chapter IV.

7. Smith and Latham, trans., *The Geometry*, pp. 100–104.

8. Struik, *Source Book*, p. 194.

9. Ibid., p. 196.

10. Quoted in Edwards, *Historical Development*, p. 103.

11. Quoted in François De Gandt, *Force and Geometry in Newton's* Principia, (Princeton: Princeton University Press, 1995), pp. 172–173. This book traces back some of the basic physical and mathematical ideas that served as a basis for Newton's masterwork of 1687.

12. Ibid., p. 176.

13. Kirsti Andersen, "Cavalieri's Method of Indivisibles," *Archive for History of Exact Sciences* 31 (1985), 291–367, p. 316. The discussion of Cavalieri is largely based on this paper.

14. Paolo Mancosu and Ezio Vailati, "Torricelli's Infinitely Long Solid and Its Philosophical Reception in the Seventeenth Century," *Isis* 82 (1991), 50–70, p. 54.

15. Mahoney, *Mathematical Career*, p. 221.

16. Ibid., p. 230.

17. For a discussion of attempts to develop formulas for the sums of integral powers, see Ivo Schneider, "Potenzsummenformeln im 17. Janrhundert," *Historia Mathematica* 10 (1983), 286–296.

18. Blaise Pascal, *Oeuvres*, ed. by L. Brunschvicg and P. Boutroux (Paris: Hachette, 1909), vol. III, p. 365.

19. Stedall, *Arithmetic of Infinitesimals*, p. 56. For more on the work of John Wallis, see J. F. Scott, *The Mathematical Work of John Wallis* (New York: Chelsea, 1981).

20. Stedall, *Arithmetic of Infinitesimals*, p. 162.

21. Ibid., p. 166

22. Evelyn Walker, *A Study of the Traité des Indivisibles of Gilles Persone de Roberval* (New York: Columbia University, 1932), p. 174.

23. Struik, *Source Book*, p. 239.

24. The best treatment of van Heuraet is in J. A. van Maanen, "Hendrick van Heuraet (1634–1660?): His Life and Mathematical Work," *Centaurus*, 27 (1984), 218–279.

25. H. W. Turnbull, ed., *James Gregory Tercentenary Memorial Volume* (London: G. Bell and Sons, 1939), p. 148.

26. Ibid., p. 170.

27. J. M. Child, *Geometrical Lectures*, p. 117. Also see Florian Cajori, "Who Was the First Inventor of the Calculus?" *American Mathematical Monthly* 26 (1919), 15–20, for a review of this book.

28. Ibid., p. 39.

29. Ibid., p. 120.

Newton and Leibniz

*The prime occasion from which arose my
discovery of the method of the Characteristic
Triangle, and other things of the same sort,
happened at a time when I had studied
geometry for not more than six months. . . .
At that time I was quite ignorant of
Cartesian algebra and also of the method
of indivisibles; indeed I did not know the
correct definition of the center of gravity. For,
when by chance I spoke of it to Huygens, I
let him know that I thought that a straight
line drawn through the center of gravity
always cut a figure into two equal parts. . . .
Huygens laughed when he heard this, and
told me that nothing was farther from the
truth. So I, excited by this stimulus, began
to apply myself to the study of the more
intricate geometry.*

—From a 1680 letter from Gottfried
Leibniz to Ehrenfried Walter
von Tschirnhaus (1651–1708)[1]

On October 24, 1676, Newton sent his second (and last)
letter to Leibniz (the *Epistola posterior*), via Henry Oldenburg.
Because he was fearful of giving away too many secrets, he
concealed the basic goal of his version of the calculus by an anagram:
6accdæ13eff7i3l9n4o4qrr4s8t12ux. It is doubtful that Leibniz was able to
read it as "Data æquatione quotcunque fluentes quantitates involvente,
fluxiones invenire; et vice versa" (given an equation involving any
number of fluent quantities, to find the fluxions, and vice versa).[2]
Nevertheless, Leibniz responded enthusiastically to Newton's letter, as
he had also responded to the *Epistola prior*, giving details of his own
work on the calculus and encouraging further dialogue. Newton never
replied again.

Two of the greatest mathematical geniuses of all time, Isaac Newton and Gottfried Leibniz, were contemporaries in the last half of the seventeenth century. Independently of each other, they developed general concepts—for Newton the fluxion and fluent, for Leibniz the differential and integral—that were related to the two basic problems of calculus, extrema and area. They developed notations and algorithms, which allowed the easy use of these concepts, and they understood and applied the inverse relationship of their two concepts. Finally, they used these two concepts in the solution of many difficult and previously unsolvable problems. In this chapter, we first discuss the work of Newton, then the work of Leibniz, and then conclude with a study of the contents of the earliest textbooks in calculus, both those based on Leibniz's work and those based on Newton's.

 16.1

ISAAC NEWTON

Isaac Newton, according to biographer Richard Westfall, was "one of the tiny handful of supreme geniuses who have shaped the categories of the human intellect, a man not finally reducible to the criteria by which we comprehend our fellow beings."[3] Because calculus was only one of the many areas in which he made major contributions to our understanding of the world around us, and because his collected mathematical papers, newly edited by Derek Whiteside, fill up eight thick volumes, we can only present here a brief glimpse of the reasons why he is considered such a "supreme genius." But what will be apparent in the next several pages is that over the course of a brief few years in the 1660s, Newton succeeded in consolidating and generalizing all the material on tangents and areas developed by his seventeenth-century predecessors into the magnificent problem solving tool exhibited in the thousand-page calculus textbooks of our own day.

Having mastered through self-study the entire achievement of seventeenth-century mathematics, Newton spent the two years from late 1664 to late 1666 working out his basic ideas on calculus, partly in his room at Cambridge and partly back at his home in Woolsthorpe. More work followed in the next several years, although there were breaks in his mathematical study during that time that he devoted to other topics, including optics, mechanics, and alchemy. On at least three occasions, Newton wrote up his research into a form suitable for publication. Unfortunately, for various reasons, Newton never published any of these three papers on calculus. Nevertheless, the so-called October 1666 tract on fluxions, the *De analysi per aequationes numero terminorum infinitas* (*On Analysis by Equations with Infinitely Many Terms*) of 1669, and the *Tractatus de methodis serierum et fluxionum* (*A Treatise on the Methods of Series and Fluxions*) of 1671 all circulated to some extent in manuscript in the English mathematical community and demonstrated the great power of Newton's new methods. Because the latter treatise summarizes and deepens the results of the two earlier ones, we will use it as the framework for the study of Newton's calculus, referring to the others as necessary.

16.1.1 Power Series

Newton clearly believed that his methods had greatly expanded the power of the "new analysis" that he had found in his readings. In particular, he believed that the idea of power series, the topic with which the 1671 treatise began, was central to expanding the field of

BIOGRAPHY

Isaac Newton (1642–1727)

Newton was born on December 25, 1642, at Woolsthorpe, near Grantham, some 100 miles north of London, to a mother already widowed in October. When he was three years old, his mother remarried and left young Isaac in the care of his grandmother until she returned to Woolsthorpe in 1653 upon the death of her second husband. In 1655, Newton was sent to Grantham to attend the local grammar school. It was here that he mastered Latin, the mainstay of the classical school curriculum, and also was introduced to the study of mathematics by the somewhat unusual schoolmaster Henry Stokes. Not only did Newton learn basic arithmetic, he also studied such advanced topics as plane trigonometry and geometric constructions, thus putting him far ahead of his fellow students on his matriculation at Trinity College, Cambridge, in 1661.

Mathematics, however, was not generally part of the course of study at Cambridge, even after the appointment of Barrow as Lucasian Professor of Mathematics in 1663. In fact, the university had few requirements at all. If one stayed in residence for four years and paid one's fees, one received a bachelor's degree. On the other hand, because in 1663 Newton started to explore on his own the mathematics he had been introduced to at school, it was to his advantage that the university did not particularly care what he studied. He mastered Euclid so that he could understand trigonometry, then the *Clavis mathematicae* (*Key to Mathematics*) of William Oughtred (1574–1660), then Descartes' *Geometry* in van Schooten's Latin edition along with the hundreds of pages of commentary, Viète's collected works, and finally Wallis's *Arithmetica infinitorum*. Because Isaac Barrow was giving his first series of Lucasian lectures on the foundations of mathematics in 1664, in all probability the older mathematician encouraged the younger, perhaps even lending him books from his own mathematics collection. To devote himself fully to research, however, Newton needed the security of university financial support. This was assured through a scholarship in 1664, a fellowship in 1667, and the appointment as Lucasian professor in 1669, all probably through the influence of Barrow.

Apparently, one of the central reasons for Newton's success in his development not only of the calculus but also of the basic principles of optics and mechanics was his intense facility for concentration. As John Maynard Keynes wrote, "I believe that the clue to his mind is to be found in his unusual powers of continuous concentrated introspection. . . . His peculiar gift was the power of holding in his mind a purely mental problem until he had seen straight through it. . . . I believe that Newton could hold a problem in his mind for hours and days and weeks until it surrendered to him its secret."[4] Newton's powers of concentration are exemplified by many stories told of him, similar to stories about Archimedes. For example, "when he had friends to entertain at his chamber, if he stept into his study for a bottle of wine, and a thought came into his head, he would sit down to paper and forget his friends."[5] In fact, "thinking all hours lost, that were not spent in his studies, . . . he seldom left his chamber, unless at Term Time, when he read in the schools, as being Lucasian professor." But when he lectured, "so few went to hear him, and fewer that understood him, that ofttimes he did in a manner, for want of hearers, read to the walls."[6] Perhaps Newton was not a success as a professor, but as the central figure in the Scientific Revolution, his works continue to exert their influence on our lives (Fig. 16.1).

FIGURE 16.1

Newton on a stamp from the Soviet Union

analysis. And Newton was especially struck by the analogy between the infinite decimals of arithmetic and the infinite degree "polynomials" that we call power series:

> Since the operations of computing in numbers and with variables are closely similar . . . I am amazed that it has occurred to no one (if you except N. Mercator with his quadrature of the hyperbola) to fit the doctrine recently established for decimal numbers in similar fashion to variables, especially since the way is then open to more striking consequences. For since this doctrine in species has the same relationship to Algebra that the doctrine in decimal numbers

has to common Arithmetic, its operations of Addition, Subtraction, Multiplication, Division and Root-extraction may easily be learnt from the latter's provided the reader be skilled in each, both Arithmetic and Algebra, and appreciate the correspondence between decimal numbers and algebraic terms continued to infinity. . . . And just as the advantage of decimals consists in this, that when all fractions and roots have been reduced to them they take on in a certain measure the nature of integers, so it is the advantage of infinite variable-sequences that classes of more complicated terms (such as fractions whose denominators are complex quantities, the roots of complex quantities and the roots of affected equations) may be reduced to the class of simple ones: that is, to infinite series of fractions having simple numerators and denominators and without the all but insuperable encumbrances which beset the others.[7]

Newton proceeded, then, at the beginning of the *Treatise*, to show by example the advantage of infinite variable sequences, or power series, which he considered simply as generalized polynomials with which he could operate just as with ordinary polynomials. Thus, for example, the fraction $1/(1 + x)$ can be written as the series

$$1 - x + x^2 - x^3 + x^4 - x^5 + \cdots$$

by simply using long division to divide $1 + x$ into 1. Similarly, one can use the standard arithmetic algorithm for determining square roots to calculate the roots of polynomials as power series. Applying this method to $\sqrt{1 + x^2}$, Newton easily calculated the result as

$$1 + \frac{x^2}{2} - \frac{x^4}{8} + \frac{x^6}{16} - \frac{5x^8}{128} + \frac{7x^{10}}{256} - \cdots$$

The reduction of "affected equations," that is, the solving of an equation $f(x, y) = 0$ for y in terms of a power series in x, is somewhat more difficult, Newton believed, because the method of solving equations $f(y) = 0$ numerically was not completely familiar. Thus, Newton explained his method of solving such equations in terms of the example $y^3 - 2y - 5 = 0$. He noted first that the integer 2 can be taken as an initial approximation to a root. He then set $y = 2 + p$ and substituted in the original equation to get the new equation $p^3 + 6p^2 + 10p - 1 = 0$. Because p is small, Newton could neglect p^3 and $6p^2$ and solve $10p - 1 = 0$ to get $p = 0.1$. It follows that $y = 2.1$ is the second approximation to the root. The next step is to set $p = 0.1 + q$ and substitute that into the equation for p. In the resulting equation, $q^3 + 6.3q^2 + 11.23q + 0.061 = 0$, the two highest-degree terms are again neglected. The linear equation is then solved to get $q = -0.0054$, yielding a new approximation for y as 2.0946. One can continue this method as far as desired. Newton himself stopped after one more step with the value $y = 2.09455148$. He then adapted the numerical equation solving method to algebra and calculated several examples. Thus, the solution of $y^3 + a^2y + axy - 2a^3 - x^3 = 0$ is given as

$$y = a - \frac{x}{4} + \frac{x^2}{64a} + \frac{131x^3}{512a^2} + \frac{509x^4}{16,384a^3} + \cdots$$

while that of $y^5/5 - y^4/4 + y^3/3 - y^2/2 + y - z = 0$ is

$$y = z + \frac{1}{2}z^2 + \frac{1}{6}z^3 + \frac{1}{24}z^4 + \frac{1}{120}z^5 + \cdots$$

16.1.2 The Binomial Theorem

Newton's discovery of power series came out of his reading of Wallis's *Arithmetica infinitorum*, especially the section on determining the area of a circle. In fact, he got out of Wallis's work more than Wallis had put in. In considering areas, Wallis had always looked for a specific numerical value, or the ratio of two such values, because he wanted to determine the area under a curve between two fixed values, say, 0 and 1. Newton realized that one could see further patterns if one calculated areas from 0 to an arbitrary value x, namely, if one considered area under a curve as a function of the varying endpoint of the interval. Thus, in looking at the same problem as Wallis of calculating the area of a circle, he considered a sequence of curves similar to those of Wallis, that is, the curves $y = (1 - x^2)^n$. But Newton then tabulated the values under these curves as functions of the variable x. For example, using modern notation,

$$\int_0^x (1 - x^2)^0 \, dx = x$$

$$\int_0^x (1 - x^2)^1 \, dx = x - \frac{1}{3}x^3$$

$$\int_0^x (1 - x^2)^2 \, dx = x - \frac{2}{3}x^3 + \frac{1}{5}x^5$$

$$\int_0^x (1 - x^2)^3 \, dx = x - \frac{3}{3}x^3 + \frac{3}{5}x^5 - \frac{1}{7}x^7$$

$$\int_0^x (1 - x^2)^4 \, dx = x - \frac{4}{3}x^3 + \frac{6}{5}x^5 - \frac{4}{7}x^7 + \frac{1}{9}x^9.$$

Newton then tabulated not numerical areas, but the coefficients of the various powers of x. This tabulation is illustrated in Table 16.1.

TABLE 16.1   *Newton's original table of coefficients from calculating areas.*

$n = 0$	$n = 1$	$n = 2$	$n = 3$	$n = 4$	$\ldots$	times
1	1	1	1	1	$\ldots$	x
0	1	2	3	4	$\ldots$	$-\frac{x^3}{3}$
0	0	1	3	6	$\ldots$	$\frac{x^5}{5}$
0	0	0	1	4	$\ldots$	$-\frac{x^7}{7}$
0	0	0	0	1	$\ldots$	$\frac{x^9}{9}$

Like Wallis, Newton realized that Pascal's triangle was here and so he attempted to interpolate. In fact, to solve the problem of the area of the circle, he needed the values in the column corresponding to $n = \frac{1}{2}$. To find these values, he rediscovered Pascal's formula $\binom{n}{k} = \frac{n(n-1)(n-2)\cdots(n-k+1)}{k!}$ for positive integer values of n and decided to use the same formula even when n was not a positive integer. Thus, the entries in the column $n = 1/2$ would be

$$\binom{\frac{1}{2}}{0} = 1, \quad \binom{\frac{1}{2}}{1} = \frac{1}{2}, \quad \binom{\frac{1}{2}}{2} = \frac{\frac{1}{2}(\frac{1}{2}-1)}{2} = -\frac{1}{8}, \quad \binom{\frac{1}{2}}{3} = \frac{\frac{1}{2}(\frac{1}{2}-1)(\frac{1}{2}-2)}{6} = \frac{1}{16}, \ldots$$

Newton could now fill in the table for columns corresponding to $n = k/2$ for any positive integral k. He realized further that in the original table each entry was the sum of the number to its left and the one above that. If, in his table with extra columns interpolated, he revised that rule slightly to read that each entry should be the sum of the number two columns to its left and the one above that, the new entries found by the binomial coefficient formula satisfied that rule as well. Not only did this give Newton confidence that his interpolation was correct, but it also convinced him to add columns to the left corresponding to negative values of n. The sum rule made it clear to him that in the column $n = -1$ the first number had to be 1, while the next number had to be -1, since $1 + (-1) = 0$, and 0 was the second entry in the column $n = 0$. Similarly, the third number in the $n = -1$ column was 1, the fourth -1, and so on. Of course, the binomial coefficient formula gave these same alternating values of 1 and -1. Newton's interpolation for calculating the area under $y = (1 - x^2)^n$ from 0 to x is shown in Table 16.2.

TABLE 16.2   *Newton's expanded table of coefficients from calculating areas.*

$n=-1$	$n=-\frac{1}{2}$	$n=0$	$n=\frac{1}{2}$	$n=1$	$n=\frac{3}{2}$	$n=2$	$n=\frac{5}{2}$	$\ldots$	times
1	1	1	1	1	1	1	1	$\ldots$	x
-1	$-\frac{1}{2}$	0	$\frac{1}{2}$	1	$\frac{3}{2}$	2	$\frac{5}{2}$	$\ldots$	$-\frac{x^3}{3}$
1	$\frac{3}{8}$	0	$-\frac{1}{8}$	0	$\frac{3}{8}$	1	$\frac{15}{8}$	$\ldots$	$\frac{x^5}{5}$
-1	$-\frac{5}{16}$	0	$\frac{3}{48}$	0	$-\frac{1}{16}$	0	$\frac{5}{16}$	$\ldots$	$-\frac{x^7}{7}$
1	$\frac{35}{128}$	0	$-\frac{15}{384}$	0	$\frac{3}{128}$	0	$-\frac{5}{128}$	$\ldots$	$\frac{x^9}{9}$
$\vdots$	$\vdots$	$\vdots$	$\vdots$	$\vdots$	$\vdots$	$\vdots$	$\vdots$	$\ddots$	$\vdots$

Newton soon realized that, first, there was no necessity of dealing only with fractions with denominator 2. The multiplicative rule for $\binom{n}{k}$ would apply for any fractional value of n, positive or negative. Second, as he noted in his letter to Leibniz of October 24, 1676, the terms $(1 - x^2)^n$ for n integral "could be interpolated in the same way as the areas generated by them; and that nothing else was required for this purpose but to omit the denominators 1, 3, 5, 7, etc., which are in the terms expressing the areas"[8] (and, of course, reduce the corresponding powers by 1). Finally, there was no reason to limit himself to binomials of the form $1 - x^2$. With appropriate modification, the coefficients of the power series for $(a + bx)^n$ for any value of n could be calculated using the formula for the binomial coefficients. Thus, Newton had discovered, although hardly proved, the general binomial theorem:

$$(a + bx)^n = \binom{n}{0}a^n + \binom{n}{1}a^{n-1}bx + \binom{n}{2}a^{n-2}b^2x^2 + \binom{n}{3}a^{n-3}b^3x^3 + \cdots$$

Newton was, however, completely convinced of its correctness because it provided him in several cases with the same answer that he had derived in other ways. For example, Newton noted that the series derived from $1/(1 + x)$ by division was the same as that derived from the binomial theorem using exponent -1:

$$(1 + x)^{-1} = 1 + (-1)x + \frac{(-1)(-2)}{2!}x^2 + \frac{(-1)(-2)(-3)}{3!}x^3 + \cdots$$

$$= 1 - x + x^2 - x^3 + \cdots$$

Using his knowledge that the area under $y = 1/(1 + x)$ was the logarithm of $1 + x$ and the area under $y = x^n$ was $x^{n+1}/(n + 1)$, Newton found the power series for $\log(1 + x)$ by "integrating" the above series term by term:

$$\log(1 + x) = x - \frac{x^2}{2} + \frac{x^3}{3} - \frac{x^4}{4} + \cdots$$

This was, of course, the same series that Mercator had found. Newton then proceeded to use the series to calculate the logarithms of 1 ± 0.1, 1 ± 0.2, 1 ± 0.01, and 1 ± 0.02 to over fifty decimal places. Using appropriate identities, such as $2 = (1.2 \times 1.2)/(0.8 \times 0.9)$ and $3 = (1.2 \times 2)/0.8$, as well as the basic properties of logarithms, Newton was able to calculate the logarithms of many small positive integers.

Knowledge of the binomial theorem let Newton deal with many other interesting series. For example, he worked out the series for $y = \arcsin x$ using a geometrical argument: Suppose the circle AEC has radius 1 and $BE = x$ is the sine of the arc $y = AE$, or $y = \arcsin x$ (Fig. 16.2). The area of the circular sector APE is known to be $\frac{1}{2}y = \frac{1}{2}\arcsin x$. On the other hand, it is also equal to the area under $y = \sqrt{1 - x^2}$ from 0 to x less $\frac{1}{2}x\sqrt{1 - x^2}$. By his earlier calculation, Newton knew that

$$\sqrt{1 - x^2} = (1 - x^2)^{1/2} = 1 - \frac{1}{2}x^2 - \frac{1}{8}x^4 - \frac{1}{16}x^6 - \cdots$$

It follows by integrating term by term and multiplying the above series by x that

$$y = \arcsin x = 2\int_0^x \sqrt{1 - x^2}\, dx - x\sqrt{1 - x^2} = x + \frac{1}{6}x^3 + \frac{3}{40}x^5 + \frac{5}{112}x^7 + \cdots$$

FIGURE 16.2

Newton's power series for $y = \arcsin x$

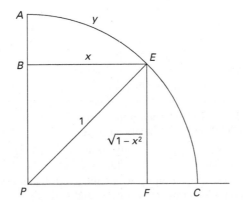

Newton could then solve this "equation" for $x = \sin y$ by his method for affected equations. Thus, in *De analysi* there occurs for the first time in European mathematics the series

$$x = \sin y = y - \frac{1}{6}y^3 + \frac{1}{120}y^5 - \frac{1}{5040}y^7 + \cdots$$

as well as the series for $x = \cos y$ that Newton derived by calculating $\sqrt{1 - (\sin y)^2}$.

In dealing with power series today, one always considers the question of convergence. It would seem that Newton did not worry much about this problem. Near the end of *De analysi* he wrote, "Whatever common analysis performs by equations made up of a finite number of terms, . . . , this method [of series] may always perform by infinite equations. To be sure, deductions in the latter are no less certain than in the other, nor its equations less exact, even though we, mere men possessed only of finite intelligence, can neither designate all their terms nor so grasp them as to ascertain exactly the quantities we desire from them."[9] Nevertheless, Newton clearly realized the limitations of his methods, at least intuitively. Although he never gave a formal treatment of the question of convergence, he did note, for example, in the course of the calculations that gave him the area under the hyperbola $y = 1/(1 + x)$, that the first few terms of this logarithm series "will be of some use and sufficiently exact provided x be considerably less than [1]."[10]

16.1.3   Algorithms for Calculating Fluxions

Series were of fundamental importance to Newton's calculus. He used them in dealing with every algebraic or transcendental relation not expressible as a finite polynomial in one variable. But there was naturally much more in his *Treatise on Methods*, beginning with the problems that he only indicated in his second letter to Leibniz via the anagram, the two central problems, the solutions to which would resolve all the difficulties about curves faced by his predecessors:

1. Given the length of the space continuously [that is, at every time], to find the speed of motion at any time proposed.
2. Given the speed of motion continuously, to find the length of the space described at any time proposed.[11]

For Newton, the basic ideas of calculus had to do with motion. Every variable in an equation was to be considered, at least implicitly, as a distance dependent on time. Of course, this idea was not new with Newton, but he did make the idea of motion fundamental: "I consider quantities as though they were generated by continuous increase in the manner of a space over which a moving object describes its course."[12] The constant increase of time itself Newton considered virtually an axiom, for he gave no definition of time. What he did define was the concept of fluxion: The **fluxion** $\dot{x}$ of a quantity x dependent on time (called the **fluent**) was the speed with which x increased via its generating motion. In his early works, Newton did not attempt any further definition of speed. The concept of continuously varying motion was, Newton believed, completely intuitive.

Newton solved problem 1 by a perfectly straightforward algorithm that determined the relationship of the fluxions $\dot{x}$ and $\dot{y}$ of two fluents x and y related by an equation of the form $f(x, y) = 0$: "Arrange the equation by which the given relation is expressed according to the dimensions of some fluent quantity, say, x, and multiply its terms by any arithmetical progression and then by $\frac{\dot{x}}{x}$. Carry out this operation separately for each one of the fluent quantities and then put the sum of all the products equal to nothing, and you have the desired equation."[13] As an example, Newton presented the equation $x^3 - ax^2 + axy - y^3 = 0$. First considering this as a polynomial of degree 3 in x, Newton multiplied using the progression 3, 2, 1, 0 to get $3x^2\dot{x} - 2ax\dot{x} + ay\dot{x}$. Next, considering the equation as a polynomial of degree 3 in y and using the same progression, he calculated $ax\dot{y} - 3y^2\dot{y}$. Putting the sum equal to nothing gave the desired relationship $3x^2\dot{x} - 2ax\dot{x} + ay\dot{x} + ax\dot{y} - 3y^2\dot{y} = 0$. In terms of a ratio, this result is $\dot{x} : \dot{y} = (3y^2 - ax) : (3x^2 - 2ax + ay)$.

There are several important ideas to note in Newton's rule for calculating fluxions. First, Newton was not calculating derivatives, for he did not in general start with a function. What he did calculate is the differential equation satisfied by the curve determined by the given equation. In other words, given $f(x, y) = 0$ with x and y both functions of t, Newton's procedure produced what is today written as

$$\frac{\partial f}{\partial x}\frac{dx}{dt} + \frac{\partial f}{\partial y}\frac{dy}{dt} = 0.$$

Second, Newton used Hudde's notion of multiplying by an arbitrary arithmetic progression. In practice, however, Newton generally used the progression starting with the highest power of the fluent. Third, if x and y are considered as functions of t, the modern product rule for derivatives is built into Newton's algorithm. Any term containing both x and y is multiplied twice and the two terms added.

Newton justified his rule, in effect, via infinitesimals. He first defined the **moment** of a fluent quantity to be the amount by which it increases in an "infinitely small" period of time. Thus, the increase of x in an infinitesimal time o is the product of the speed of x by o, or $\dot{x}o$. It follows that after this time interval, x will become $x + \dot{x}o$ and similarly y will become $y + \dot{y}o$. "Consequently, an equation which expresses a relationship of fluent quantities without variance at all times will express that relationship equally between $x + \dot{x}o$ and $y + \dot{y}o$ as between x and y; and so $x + \dot{x}o$ and $y + \dot{y}o$ may be substituted in place of the latter quantities, x and y, in the said equation."[14]

Newton explained further through the example $x^3 - ax^2 + axy - y^3 = 0$ given earlier. Substituting $x + \dot{x}o$ for x and $y + \dot{y}o$ for y, the new equation becomes

$$(x^3 + 3x^2\dot{x}o + 3x\dot{x}^2o^2 + \dot{x}^3o^3) - (ax^2 + 2ax\dot{x}o + a\dot{x}^2o^2)$$
$$+ (axy + ay\dot{x}o + ax\dot{y}o + a\dot{x}\dot{y}o^2) - (y^3 + 3y^2\dot{y}o + 3y\dot{y}^2o^2 + \dot{y}^3o^3) = 0.$$

"Now by hypothesis $x^3 - ax^2 + axy - y^3 = 0$, and when these terms are erased and the rest divided by o there will remain

$$3x^2\dot{x} + 3x\dot{x}^2o + \dot{x}^3o^2 - 2ax\dot{x} - a\dot{x}^2o + ay\dot{x} + ax\dot{y} + a\dot{x}\dot{y}o - 3y^2\dot{y} - 3y\dot{y}^2o - \dot{y}^3o^2 = 0.$$

But further, since o is supposed to be infinitely small so that it be able to express the moments of quantities, terms which have it as a factor will be equivalent to nothing in respect of the others. I therefore cast them out and there remains $3x^2\dot{x} - 2ax\dot{x} + ay\dot{x} + ax\dot{y} - 3y^2\dot{y} = 0$, as . . . above."[15]

Although this calculation is only an example and not a proof, Newton noted that it is immediately generalizable: "It is accordingly to be observed that terms not multiplied by o

will always vanish, as also those multiplied by o of more than one dimension; and that the remaining terms after division by o will always take on the form they should have according to the rule. This is what I wanted to show."[16] In other words, Newton assumed that the reader understood that the coefficient of $x^{n-1}\dot{x}o$ in the expansion of $(x + \dot{x}o)^n$ is n itself. But note also that Newton's only justification of his step of "casting out" any terms in which o appears was that they are "equivalent to nothing in respect of the others." There is no limit argument here. There is only the intuitive notion of the properties of these infinitesimal increments of time.

As already noted, the product rule for derivatives is essentially built into Newton's algorithm. On the other hand, Newton's approach to the modern chain rule was via substitution. For example, to determine the relationship of the fluxions in the equation $y - \sqrt{a^2 - x^2}$, he put z for the square root and dealt with the two equations $y - z = 0$ and $z^2 - a^2 + x^2 = 0$. The first gave $\dot{y} - \dot{z} = 0$ while the second gave $2z\dot{z} + 2x\dot{x} = 0$, or $\dot{z} = -x\dot{x}/z$. Thus, the relationship between the fluxions of x and y is

$$\dot{y} + \frac{x\dot{x}}{\sqrt{a^2 - x^2}} = 0.$$

A similar approach works in dealing with quotients.

16.1.4 Applications of Fluxions

With the calculation of fluxions accomplished, Newton used them to solve various problems. For example, Newton found maxima and minima by setting the relevant fluxion equal to zero, because "when a quantity is greatest or least, at that moment its flow neither increases nor decreases; for if it increases, that proves that it was less and will at once be greater than it now is, and conversely so if it decreases."[17] Again he used the equation $x^3 - ax^2 + axy - y^3 = 0$ as his example for determining the greatest value of x. Setting $\dot{x} = 0$ in the equation involving the fluxions, he found that $-3y^2\dot{y} + ax\dot{y} = 0$ or $3y^2 = ax$. This equation must then be solved simultaneously with the original one to find the desired value for x. Similarly, to find the maximum value of y, one sets $\dot{y} = 0$ and uses the resulting equation $3x^2 - 2ax + ay = 0$. Newton's discussion of this method was brief, however, and he gave no criteria for determining whether the values found are maxima or minima. Presumably, that determination can be made from the context in any given problem.

To draw tangents, Newton used Barrow's differential triangle. Thus, if x changes to $x + \dot{x}o$ while y changes to $y + \dot{y}o$, then the ratio $\dot{y}o : \dot{x}o = \dot{y} : \dot{x}$ of the sides of this triangle is the slope of the tangent line, thought of as the direction of instantaneous motion of the particle describing the curve. This ratio is in turn equal to that of the ordinate y to the subtangent t. Since to draw the tangent means to find the subtangent, Newton simply noted that $t = y(\dot{x}/\dot{y})$. As a slight simplification in this calculation and others, Newton sometimes set $\dot{x} = 1$. This is equivalent to considering x as flowing uniformly, or as itself representing time.

A final example of Newton's use of fluxions is his calculation of the curvature of a curve, a problem that "has the mark of exceptional elegance and of being pre-eminently useful in the science of curves."[18] Newton defined curvature in terms of a circle, that is, he noted that a circle has the same curvature everywhere and that the curvatures of two circles are inversely proportional to their radii. In modern terminology, the curvature of a circle of radius r is defined to be $\kappa = 1/r$. For an arbitrary curve, Newton defined the curvature at a point to be

FIGURE 16.3

Newton's method for finding curvature

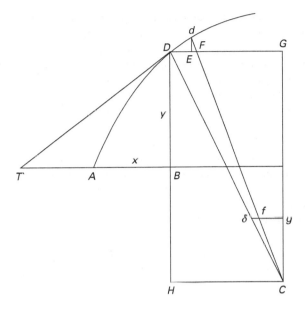

the curvature of the circle tangent to the curve at the point that has the further property that no other tangent circle can be drawn between the curve and the circle. The definition means that this **osculating circle** at a point D also passes through any point d infinitely close to D. To find the curvature at D, one simply needs to find the radius of this circle, that is, the distance DC to the intersection point of the normals to the curve at D and d, respectively (Fig. 16.3). Drawing the tangent line dDT to the curve through D and d, completing the rectangle $DGCH$, taking g on GC so that $Cg = 1$, and setting $AB = x$, $BD = y$, and $g\delta = z$, Newton concluded from the similarity of triangle DBT to triangle $Cg\delta$ that $Cg : g\delta = TB : BD$ or that $1 : z = \dot{x} : \dot{y}$. Since d is infinitely close to D, it follows that $\delta f = \dot{z}o$, $DE = \dot{x}o$, and $dE = \dot{y}o$. Furthermore, since one can consider DdF a right triangle, $DE : dE = dE : EF$ and therefore $DF = DE + EF = DE + (dE^2/DE) = \dot{x}o + (\dot{y}^2o/\dot{x})$. Thus, $Cg : CG = \delta f :$ $DF = \dot{z}o : (\dot{x}o + \dot{y}^2o/\dot{x})$ and $CG = (\dot{x}^2 + \dot{y}^2)/\dot{x}\dot{z}$. Assuming that $\dot{x} = 1$, Newton concluded that $\dot{y} = z$, $CG = (1 + z^2)/\dot{z}$, $DG = (CG \cdot BD)/BT = z \cdot CG = (z + z^3)/\dot{z}$, and finally that

$$DC = \sqrt{CG^2 + DG^2} = \frac{(1 + z^2)^{\frac{3}{2}}}{\dot{z}}.$$

Newton's result is naturally equivalent to the modern version, that the curvature of $y = f(x)$ equals

$$\frac{y''}{(1 + y'^2)^{\frac{3}{2}}},$$

since his z is our y' and his $\dot{z}$ our y''.

16.1.5   Procedures for Finding Fluents and Areas

Problem 2 of the *Treatise on Methods* asks to find the distance, given the velocity, that is, the fluent given the fluxion. Newton's first method of attack was to reverse the procedure for finding the fluxion. "Since this problem is the converse of the preceding [discussed in 16.1.3], it ought to be resolved the contrary way: namely by arranging the terms multiplied by $\dot{x}$ according to the dimensions of x and dividing by $\frac{\dot{x}}{x}$ and then by the number of dimensions, . . . by carrying out the same operation in the terms multiplied by . . . $\dot{y}$, and, with redundant terms rejected, setting the total of the resulting terms equal to nothing."[19] As an example, he used his earlier problem. Starting with $3x^2\dot{x} - 2ax\dot{x} + ay\dot{x} - 3y^2\dot{y} + ax\dot{y} = 0$, he divided the terms having $\dot{x}$ by $\frac{\dot{x}}{x}$ (or, what amounts to the same thing, removed the $\dot{x}$ and raised the power of x by 1), then divided each term again by the new power of x to get $x^3 - ax^2 + axy$. Doing the analogous operation on the terms containing $\dot{y}$, he found $-y^3 + axy$. Noting that axy occurs twice, he removed one of those terms to produce the final equation $x^3 - ax^2 + axy - y^3 = 0$.

Newton naturally realized that this procedure does not always work. He suggested, in fact, that one always check the result. But if the problem cannot be solved by this simple "antiderivative" approach, Newton generally used the method of power series. Since the fluent equation determined by the fluxional equation $\dot{y} = x^n\dot{x}$ or $\dot{y}/\dot{x} = x^n$ is $y = x^{n+1}/n + 1$, he suggested that when $\dot{y}/\dot{x}$ depends only on x, one should express the ratio by a power series and apply that rule to each term. For example, the equation $\dot{y}^2 = \dot{x}\dot{y} + x^2\dot{x}^2$ can be rewritten as $\dot{y}^2/\dot{x}^2 = \dot{y}/\dot{x} + x^2$. This quadratic equation in $\dot{y}/\dot{x}$ can be solved to give $\dot{y}/\dot{x} = \frac{1}{2} \pm \sqrt{1/4 + x^2}$. By applying the binomial theorem, one gets the two series

$$\frac{\dot{y}}{\dot{x}} = 1 + x^2 - x^4 + 2x^6 - 5x^8 + \cdots \quad \text{and} \quad \frac{\dot{y}}{\dot{x}} = -x^2 + x^4 - 2x^6 + 5x^8 + \cdots$$

The solutions to the original problem are then easily found to be

$$y = x + \frac{1}{3}x^3 - \frac{1}{5}x^5 + \frac{2}{7}x^7 + \cdots \quad \text{and} \quad y = -\frac{1}{3}x^3 + \frac{1}{5}x^5 - \frac{2}{7}x^7 + \cdots$$

The solution method is more complicated if $\dot{y}/\dot{x}$ is given by an equation in both x and y, but even then Newton's basic idea is to express the given equation in terms of a power series.

Of course, Newton's procedure of finding the fluent, once he had a power series, amounted to using the now standard procedure of raising the exponent by 1 and dividing by that new exponent. But Newton also realized very early in his research that this problem of finding the fluent is equivalent to finding the area under a curve from its equation. In this context, he discovered and used what we call the fundamental theorem of calculus. For Newton, this theorem was virtually self-evident. Because he thought of the curve AFD as being generated by the motions of x and y, it followed that the area $AFDB$ was generated by the motion of the moving ordinate BD (Fig. 16.4). It was therefore obvious that the fluxion of the area was in fact the ordinate multiplied by the fluxion of BD. That is, if z represents the area under the curve, then $\dot{z} = y\dot{x}$, or $\dot{z}/\dot{x} = y$. This equation translates immediately into part of the modern fundamental theorem, that if $A(x)$ represents the area under $y = f(x)$ from 0 to x, then $dA/dx = f(x)$. Newton noted that the area z can be found explicitly from the equation $\dot{z}/\dot{x} = y$ by using the techniques already discussed of finding fluents. But as he wrote a few pages later, "hitherto we have exposed the quadrature of curves defined by less simple equations by the technique of reducing them to equations consisting of infinitely many simple terms. However, curves of this kind may sometimes be squared by means of finite equations also."[20]

To square curves by finite equations (that is, to find the area under them), one needs what is today referred to as a table of integrals. So Newton provided one. The first entry in the table he produced for the *Treatise on Methods* is the simple one—the area under $y = ax^{n-1}$ is $\frac{a}{n}x^n$— but the other entries are considerably more complex. In this brief excerpt from Newton's table, the function z on the right represents the area under the function y on the left:

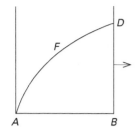

A

FIGURE 16.4

Newton and the fundamental theorem

$$y = \frac{ax^{n-1}}{(b+cx^n)^2} \qquad z = \frac{(a/nb)x^n}{b+cx^n} \tag{16.1}$$

$$y = ax^{n-1}\sqrt{b+cx^n} \qquad z = \frac{2a}{3nc}(b+cx^n)^{3/2} \tag{16.2}$$

$$y = ax^{2n-1}\sqrt{b+cx^n} \qquad z = \frac{2a}{nc}\left(-\frac{2}{15}\frac{b}{c} + \frac{1}{5}x^n\right)(b+cx^n)^{3/2} \tag{16.3}$$

$$y = \frac{ax^{2n-1}}{\sqrt{b+cx^n}} \qquad z = \frac{2a}{nc}\left(-\frac{2}{3}\frac{b}{c} + \frac{1}{3}x^n\right)\sqrt{b+cx^n} \tag{16.4}$$

In contrast to a modern table of integrals, Newton's table listed no transcendental functions, no sines or cosines or even logarithms. Although Newton knew the power series of these functions, he never treated them on an equal basis with algebraic functions. He did not operate with the sine, cosine, or logarithm algebraically, by combining them with polynomials and other algebraic expressions. Newton did, however, extend his table to functions whose integrals today would be expressed in terms of transcendental functions by using an appropriate substitution and then expressing the integral in terms of areas bounded by certain conic sections, areas that could be calculated by the use of power series techniques. Three examples are

$$y = \frac{ax^{n-1}}{e+fx^n} \qquad z = \frac{1}{n}s \qquad u = x^n \quad v = \frac{a}{e+fu} \tag{16.5}$$

$$y = \frac{ax^{n-1}}{\sqrt{e+fx^n+gx^{2n}}} \qquad z = \frac{8ags - 4agxv - 2afx}{4neg - nf^2} \qquad u = x^n \quad v = \sqrt{e+fu+gu^2} \tag{16.6}$$

$$y = \frac{ax^{2n-1}}{\sqrt{e+fx^n+gx^{2n}}} \qquad z = \frac{-4afs + 2afuv + 4aev}{4neg - nf^2} \qquad u = x^n \quad v = \sqrt{e+fu+gu^2} \tag{16.7}$$

where, as before, z is the area under the function y, but where s is the area under the conic section with coordinates u, v, or, in modern terms, $s = \int v\,du$.

There were some curves, however, for which Newton's table could not give an answer, namely, curves defined geometrically, as, for example, the cycloid. In such a case, Newton proceeded geometrically as well. Given the cycloid ADF traced by the point A as the circle ALE rolls along EF (Fig. 16.5), he noted that the tangent DT to the cycloid at an arbitrary point D is always parallel to AL, where L is the intersection with the circle of the line through D parallel to EF. This is because the motion generating the cycloid is composed of the equal motions of the diameter AE moving along EF and the motion of A around the circle. Because

FIGURE 16.5

Newton's calculation of the area bounded by an arch of a cycloid

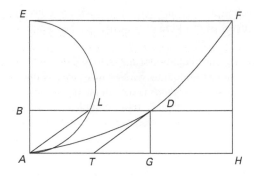

$\dot{y} : \dot{x}$ is the slope of the tangent line, and this is equal to $DG : GT$ or to $DG : BL$, it follows that $(DG)\dot{x} = (BL)\dot{y}$. But $(DG)\dot{x}$ is the fluxion of the area ADG and $(BL)\dot{y}$ is the fluxion of the area ALB. It follows that the area AHF below the half-arch of the cycloid equals the area of the semicircle $ALEB$. Because the area of rectangle $AEFH$ is twice that of the generating circle, it follows that the area above one complete arch of the cycloid is equal to three times that of the circle, the same result discovered by Roberval over three decades earlier.

There is much else in Newton's *Treatise on Methods*, including techniques equivalent to the modern rules of substitution (as in the previous examples of integrals) and integration by parts as well as the method of determining arclength. This never-published text thus included virtually all of the important ideas found in the first several chapters of any modern calculus text, as well as some that are thought too advanced for such. One missing idea, however, is that of a limit. That is not to say that Newton never considered that idea. He did, but only published it in connection with his masterpiece, the *Philosophiae naturalis principia mathematica* (*Mathematical Principles of Natural Philosophy*) of 1687, the work in which Newton formulated his laws of motion and used them along with his theory of gravity and the mathematics of the heavens to derive the "system of the world" (Fig. 16.6).

FIGURE 16.6

The 300th anniversary of Newton's *Principia* honored by a British stamp

16.1.6   The Synthetic Method of Fluxions

Perhaps one of the reasons Newton did not publish his *Treatise on Methods* was that by the mid-1670s, he was somewhat unhappy with his use of "analysis" in developing the ideas of calculus. He had been studying the ancient Greek texts and believed that mathematical "truth" must be based on the tenets of proof that had been developed in Greece. Thus, although he had confidence in the efficacy of the algebraic methods of the "moderns," he began to reformulate his ideas more in keeping with the "geometry of the ancients." And therefore, when he began to compose the *Principia*, he decided to write his text in the form of a synthetic geometric treatise.

That he wrote the book at all is perhaps surprising, given Newton's aversion to publishing. But in the summer of 1684, the young English astronomer Edmond Halley (1656–1741) traveled to Cambridge to pose a critical question to Newton: "What would be the curve that would be described by the planets supposing the force of attraction towards the sun to be reciprocal to the square of their distance from it?"[21] Newton's immediate answer was that he had already "calculated" the answer to this problem and that the curve would be an ellipse. When Halley pressed Newton for details, the Cambridge professor promised to send them along shortly. Several months went by, but in November of 1684 Halley received a 10-page treatise from Newton, not only purporting to answer the original question but also sketching a reformulation of astronomy in terms of forces. Halley was so impressed with this work, *De motu corporum in gyram* (*On the Motion of Bodies in an Orbit*), that he hurried back to Cambridge to attempt to persuade Newton to publish. Evidently, Halley did not have to work very hard on Newton; Newton was already on the way to revising and expanding this short treatise into his magnum opus on physics.

To develop his physics, however, Newton needed to establish a mathematical framework. So he reformulated his ideas on fluxions from the analytic methods he used earlier into a more synthetic method that he called "the method of first and ultimate ratios" and then used this method in proving 11 important lemmas in section 1 of the *Principia*. For example, we have:

LEMMA 1   *Quantities, and also ratios of quantities, which in any finite time constantly tend to equality, and which before the end of that time approach so close to one another that their difference is less than any given quantity, become ultimately equal.*[22]

The proof was obvious. If the quantities are ultimately unequal, they differ by a positive value D and therefore do not approach nearer to equality than D, a contradiction.

In Lemma 2, Newton presented the situation where there is a set of inscribed rectangles in a curvilinear area and a corresponding set of circumscribed rectangles around that area. Newton claimed that the ultimate ratios that the inscribed figure, the circumscribed figure, and the curvilinear figure itself bear to one another as the width of the rectangles is diminished and their number increased indefinitely is the ratio of equality. Newton's proof is similar to one we would give today. He showed that the difference between the areas of the circumscribed and inscribed figures is the area of a single rectangle, which, because its width "is diminished indefinitely, becomes less than any rectangle." As another example, consider

LEMMA 9   *If the straight line AE and the curve ABC, both given in position, intersect each other at a given angle A, and if BD and CE are drawn as ordinates to the straight line AE at another given angle and meet the curve in B and C, and if then points B and C simultaneously approach point A, I say that the areas of the triangles ABD and ACE will ultimately be to each other as the squares of the sides [Fig. 16.7].*

For the proof, Newton assumed that Ad and Ae are always proportional to AD and AE, the curve Abc is similar to ABC, and the line Ag is tangent to both curves at A. Then, with the length Ae remaining fixed, as B and C come together at A and the angle cAg vanishes, the curvilinear areas Abd and Ace will coincide with the similar rectilinear areas Afd and Age and therefore will be in the squared ratio of the sides Ad and Ae. But since the areas ABD and ACE are always proportional to Afd and Age (because the curves Abc and ABC are similar) and because sides AD and AE are proportional to Ad and Ae, it follows that the areas ABD and ACE also are ultimately as the squares of the sides AD and AE.

FIGURE 16.7

Principia, Lemmas 9 and 10

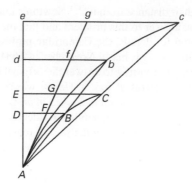

Newton then used Lemma 9 to demonstrate one of Galileo's important results in

LEMMA 10   *The spaces which a body describes when urged by any finite force, whether that force is determinate and immutable or is continually increased or continually decreased, are at the very beginning of the motion in the square ratio of the times.*

Newton's proof used Figure 16.7, in which he represented the times by AD and AE and the velocities by DB and DC. Areas ABD and ACE then represent the distances (spaces), so Lemma 9 produces the conclusion.

Newton believed that his method of first and ultimate ratios replaced the lengthy ancient proofs by reductio ad absurdum, as well as proofs by "indivisibles," but he realized that he had to convince his readers. So in the scholium to section 1, he wrote:

> Since the hypothesis of indivisibles is problematical and this method is therefore accounted less geometrical, I have preferred to make the proofs of what follows depend on the ultimate sums and ratios of vanishing quantities and the first sums and ratios of nascent quantities, that is, on the limits of such sums and ratios. . . . For the same result is obtained by these as by the method of indivisibles, and we shall be on safer ground using principles that have been proved. Accordingly, whenever in what follows I consider quantities as consisting of particles or whenever I use curved line-elements in place of straight lines, I wish it always to be understood that I have in mind not indivisibles but evanescent divisibles, and not sums and ratios of definite parts but the limits of such sums and ratios, and that the force of such proofs always rests on the method of the preceding lemmas.
>
> It may be objected that there is no such thing as an ultimate proportion of vanishing quantities, inasmuch as before vanishing the proportion is not ultimate, and after vanishing it does not exist at all. But by the same argument it could equally be contended that there is no ultimate velocity of a body reaching a certain place at which the motion ceases; for before the body arrives at this place, the velocity is not the ultimate velocity, and when it arrives there, there is no velocity at all. But the answer is easy: to understand the ultimate velocity as that with which a body is moving, neither before it arrives at its ultimate place and the motion ceases, nor after it has arrived there, but at the very instant when it arrives, that is, the very velocity with which the body arrives at its ultimate place and with which the motion ceases. And similarly the ultimate ratio of vanishing quantities is to be understood not as the ratio of quantities before they vanish or after they have vanished, but the ratio with which they vanish. . . . There exists a limit which their velocity can attain at the end of the motion, but cannot exceed. This is the ultimate velocity. . . . And since this limit is certain and definite, the determining of it is properly a geometrical problem. . . .

It can also be contended that if the ultimate ratios of vanishing quantities are given, their ultimate magnitudes will also be given; and thus every quantity will consist of indivisibles, contrary to what Euclid had proved. . . . But this objection is based on a false hypothesis. Those ultimate ratios with which quantities vanish are not actually ratios of ultimate quantities, but limits which the ratios of quantities decreasing without limit are continually approaching, and which they can approach so closely that their difference is less than any given quantity, but which they can never exceed and can never reach before the quantities are decreased indefinitely.[23]

A translation of Newton's words into an algebraic statement would give a definition of limit close to, but not identical with, the modern one. Newton never made such a translation. Nevertheless, it seems clear that Newton intuitively knew what he was doing in using "limits" to calculate fluxions. To see this, we consider his final tract on fluxions, the *De quadratura curvarum* (*On the Quadrature of Curves*) of 1691 (published in 1704), where we read: "Fluxions are in the first ratio of the nascent augments or in the ultimate ratio of the evanescent part, but they may be expounded by any lines that are proportional to them." Newton then showed how to calculate the fluxion of x^n, where x flows uniformly:

In the time that the quantity x comes in its flux to be $x + o$ [here o can be thought of as the "nascent augment"], the quantity x^n will come to be $(x + o)^n$, that is, by the method of infinite series,

$$x^n + nox^{n-1} + \frac{n^2 - n}{2}o^2x^{n-2} + \cdots;$$

and so the augments o and $nox^{n-1} + \frac{n^2-n}{2}o^2x^{n-2} + \cdots$ are one to the other as 1 and $nx^{n-1} + \frac{n^2-n}{2}ox^{n-2} + \cdots$. Now let those augments come to vanish [so now o is the "evanescent part"] and their ultimate ratio will be 1 to nx^{n-1}; consequently the fluxion of the quantity x is to the fluxion of the quantity x^n as 1 to nx^{n-1}.[24]

This demonstration is not very different from the earlier calculation of fluxions, except that Newton now did not write of just casting out terms which are "equivalent to nothing in respect of the others." But in another manuscript, *Geometria curvilinea*, probably written a decade earlier but never published, Newton calculated the fluxions of the sine, tangent, and secant by the same method, a calculation he had never made by his original "analytic" method:

THEOREM   *In a given circle the fluxion of an arc is to the fluxion of its sine as the radius to its cosine; to the fluxion of its tangent as its cosine is to its secant; and to the fluxion of its secant as its cosine to its tangent.*

To demonstrate the result about the tangent, Newton considered a circle with center C and radius AC, where AB is the given arc, and AT a straight line tangent to it at A (Fig. 16.8). He then drew the secant CT, meeting the arc at B, and drew AS perpendicular to CT. Note that AS is the sine of AB, while CS is the cosine. Now let the arc and tangent flow until they become Ab and At, respectively. Since the area of sector CBb is $\frac{1}{2}CA \times Bb$ and that of triangle CTt is $\frac{1}{2}CA \times Tt$, the ratio of arc Bb to segment Tt equals that of sector CBb to triangle CTt. To measure this ratio, Newton drew a new triangle Cpq between the lines CT and Ct similar to triangle CTt, but equal in area to sector CBb. The ratio of the two similar triangles is as the squares of their sides; thus, $Bb : Tt = Cp^2 : CT^2$. Now as the "augments" vanish, t and T will come together, as will p and q, and so Cp will become equal to CB. Thus, the ultimate ratio of Bb to Tt is CB^2 to CT^2. But $CT : CA = CA : CS$, and $CB = CA$,

FIGURE 16.8

Newton's derivation of the
fluxion of the tangent

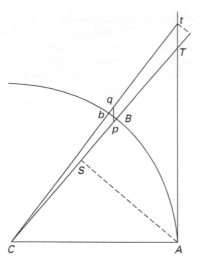

so $CB^2 = CT \cdot CS$. Therefore, $CB^2 : CT^2 = (CT \cdot CS) : CT^2 = CS : CT$. In other words, the fluxion of the arc to the fluxion of the tangent, that is, the ultimate ratio of Bb to Tt, is equal to the cosine to the secant, as claimed. Newton proved the other two statements of the theorem by using this result and the basic trigonometric relationships.

16.1.7   Newton and Celestial Physics

To see how Newton used his ideas of limits in the *Principia*, we will study his derivation of versions of Kepler's laws of planetary motion. But given that the *Principia* is written as a synthetic geometric treatise, we begin by giving Newton's axioms, the three laws of motion:

1. Every body perseveres in its state of being at rest or of moving uniformly straight forward, except insofar as it is compelled to change its state by forces impressed.
2. A change in motion is proportional to the motive force impressed and takes place along the straight line in which that force is impressed.
3. To any action there is always an opposite and equal reaction.

Newton used these laws immediately, beginning with the following proposition (from section 2 of Book I).

PROPOSITION 1   *The areas which bodies made to move in orbits described by radii drawn to an unmoving center of forces . . . are proportional to the times.*

This result is, of course, Kepler's second law. Newton began his proof by dividing the time into equal finite parts. Suppose that in the first part, the body by its "inherent force" moves along the straight line segment AB (Fig. 16.9). If nothing were to impede it, it would move in the next time interval along an equal segment Bc in the same direction, according to Newton's first law. Thus, if lines are drawn from A, B, and c to the center S, the triangles ASB and BSc will have equal areas. But since the body is being drawn to the center, Newton assumed that when it reaches B the centripetal force acts and causes the body to change its path so it moves in the direction BH. Now draw a line from c parallel to BS, meeting BH at C. By the parallelogram law of combining forces, which Newton had worked out earlier, the body will be found at C at the end of the second time interval. If we now connect the center S to C and c, then triangle BSC is equal in area to triangle BSc and therefore to triangle ASB. Because this argument can be repeated for other equal time intervals, and because one can combine the equal triangles into larger regions, it follows that in this situation of force acting discretely, "any sums $SADS$ and $SAFS$ of the areas are to each other as the times of description." But Newton knew, of course, that the force acts continuously. Thus, he concluded his proof with the following: "Let the number of triangles be increased and their width decreased indefinitely, and their ultimate perimeter will be a curved line; and thus the centripetal force by which the body is continually drawn back from the tangent of this curve will act uninterruptedly, while any areas described, $SADS$ and $SAFS$, which are always proportional to the times of description, will be proportional to those times in this case."[25]

FIGURE 16.9

Newton's determination of the area law

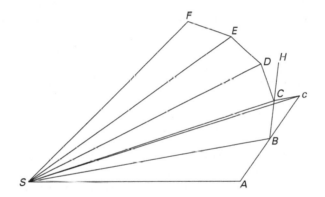

To deal with the central forces by geometrical methods, Newton needed a geometrical representation of such a force, even when the force changes its magnitude and direction continuously. This he accomplished in Proposition 6 and its corollaries, in which a body is orbiting about a center S in any curve (Fig. 16.10): If PX is tangent to the curve at P, if QT is perpendicular to PS at any other point Q on the orbit, and if QR is drawn to PX parallel to PS, then the centripetal force will be inversely as the solid $(SP^2 \times QT^2)/QR$, "provided that the magnitude of that solid is always taken as that which it has ultimately when the points P and Q come together."[26] Recall that, by Lemma 10, the distance that a body travels under even a variable force is, at the very beginning of motion, as the square of the time. In this case, that distance is QR. But since QR also represents the change in motion of the body, it is proportional to the force (by Newton's second law). Therefore, QR is proportional to the force and the square of the time. But by Proposition 1, the time is proportional to the area swept out, namely, the triangle SPQ, and the area of that triangle is $\frac{1}{2}SP \cdot QT$. Thus, QR is proportional both to the centripetal force and to $(SP \cdot QT)^2$. Therefore, the force F is proportional to QR and inversely proportional to $(SP \cdot QT)^2$, as claimed. In modern terms, we can think of QR as a vector representing the acceleration caused by the initial force applied. The length of this vector is the magnitude of the acceleration, which, since

FIGURE 16.10

Determination of a geometrical method to measure the force

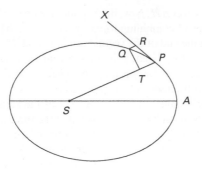

$d = \frac{1}{2}at^2$, is itself proportional to the distance and inversely to the square of the time. Since force is proportional to acceleration and time to the area swept out, Newton's result follows.

With a geometrical representation of the force at hand, Newton could now calculate the force for various specific orbits. We consider Proposition 11, the case of most interest, where the orbit is elliptical and the force is directed toward a focus S (Fig. 16.11). In this case, Newton demonstrated that the force is inversely proportional to the square of the distance of the body from the focus. Using the same notation as in Proposition 6, let DK, PG be conjugate diameters of the ellipse, where DK is parallel to PR, and let QV be drawn parallel to RP meeting PG at V. Furthermore, let SP cut DK at E and QV at Y, thus completing the parallelogram $QYPR$. As usual, let a and b represent the lengths of the semimajor and semiminor axes of the ellipse, respectively, and p represent the parameter. Newton first showed that $EP = a$. For if H is the second focus of the ellipse, and if HI is drawn parallel to EC, then $ES = EI$ and $\angle PIH = \angle YPR = \angle ZPH = \angle PHI$. Thus, $EP = (PS + PI)/2 = (PS + PH)/2 = 2a/2 = a$. He then set out five proportions:

$$p \times QR : p \times PV = QR : PV = PY : PV = PE : PC = a : PC \tag{16.8}$$

$$p \times PV : GV \times PV = p : GV \tag{16.9}$$

$$GV \times PV : QV^2 = PC^2 : CD^2 \tag{16.10}$$

$$QV^2 : QY^2 = M : N \tag{16.11}$$

$$QY^2 : QT^2 = EP^2 : PF^2 = a^2 : PF^2 = CD^2 : b^2 \tag{16.12}$$

Proportion 16.8 follows from the similarity of triangles PVY and PCE and from the fact that $EP = a$, while Proportion 16.9 is simply the cancellation law. Newton knew Proportion 16.10 from his study of conic sections; this is Apollonius's Proposition I–21 referred to a pair of conjugate axes. (See Chapter 4, Exercise 23d.) Proportion 16.11 is simply a definition of M and N, while the final proportion (16.12) depends on the similarity of triangles QTY and PFE and Apollonius's Proposition VII–31, that the rectangles constructed on any pair of conjugate diameters are equal. (See Chapter 4, Exercise 23e.) If one multiplies these proportions together, recalling that $b^2 = pa/2$, the result is that $p \times QR : QT^2 = (2PC : GV) \times (M : N)$. But as the points P and Q "come together," the two ratios on the right be-

FIGURE 16.11

An elliptical orbit entails an inverse square force law

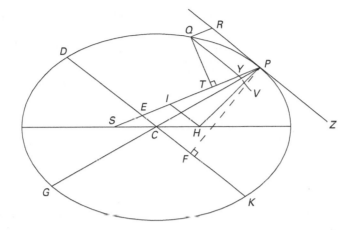

come ratios of equality. It follows that $p \times QR = QT^2$ and, on multiplying both sides by SP^2/QR, that

$$p \times SP^2 = \frac{SP^2 \times QT^2}{QR}.$$

Since by Proposition 6 the centripetal force is inversely proportional to the expression on the right, it is also inversely proportional to $p \times SP^2$. Because p is a constant, Newton had demonstrated his result that the force is inversely proportional to the square of the distance SP of a point from the focus.

After proving this result, Newton proved analogous results for hyperbolas (Proposition 12) and parabolas (Proposition 13) and then concluded with the following:

COROLLARY   *From the last three propositions, it follows that if any body P departs from the place P along any straight line PR with any velocity whatever and is at the same time acted upon by a centripetal force that is inversely proportional to the square of the distance of places from the center, this body will move in some one of the conics having a focus in the center of forces; and conversely.*

This statement was criticized by many early readers of the *Principia*, since they wondered how Newton could conclude that an inverse square force law implied a conic section orbit from the converse of that result proved in Propositions 11–13. So Newton added a brief argument to the corollary in the second and third editions (1713 and 1726) to the effect that a conic section can be constructed through a given point with a given focus, given tangent, and given curvature, that motion along this conic satisfies an inverse square force law, and that because the force and velocity together determine the curvature, this conic is the unique solution to the initial value problem implied by the inverse square force law. Although sketchy, this argument is quite correct and can be expanded into a full formal proof of the corollary.[27] In fact, Newton provided other results in the *Principia* showing how to complete this proof, especially in Proposition 41 and its corollaries. The proofs here assumed "the quadratures of curvilinear figures," such as those provided in his integral table in the *Treatise on Methods*. In fact, integral 16.6 from Newton's table (with $n = -1$) is the basic tool necessary in showing that an inverse square force does imply a conic section orbit.

Proposition 11 and the corollary are closely related to Kepler's second law. Newton also proved a result closely related to Kepler's third law:

PROPOSITION 15   *The squares of the periodic times in ellipses are as the cubes of the major axes.*

Using the same notation and diagram as in Proposition 11, we recall that by Proposition 1, areas swept out are proportional to the time elapsed. Therefore, if Δt is the time taken in each ellipse to sweep out the infinitesimal area PSQ, the entire area of the ellipse is to the periodic time T ultimately as the area of triangle $PSQ (= \frac{1}{2} QT \cdot PS)$ is to Δt. Because the area of the ellipse is proportional to ab, we know that ab is ultimately proportional to the product of T and $QT \cdot PS$. Also, for each of the elliptical orbits, the parameter p equals QT^2/QR. But QR, representing the force, is inversely proportional to SP^2. So p is ultimately proportional to $(QT \cdot PS)^2$, or $\sqrt{p}$ is ultimately proportional to $QT \cdot PS$. It follows that ab is proportional to $p^{1/2}T$. But since $b^2 = pa/2$, we have that ab is also proportional to $a^{3/2} p^{1/2}$. Therefore, T is proportional to $a^{3/2}$, or T^2 is proportional to a^3, as claimed.

It has frequently been asserted that Newton developed the calculus in order to work out his system of the world in the *Principia*, including not only Kepler's laws but also the law of universal gravitation, discussed in Book III. The evidence from his mass of manuscripts shows that this is not the case, that in fact the calculus was developed well before the physics. Nevertheless, as should be clear from the discussion above, Newton did use the ideas and the methodology of the calculus, although not always the analytic machinery, to derive many physical results. In his physical arguments based on geometry, he usually proceeded in three steps. First, he established a result for finite regions. Then, he assumed that the result will remain true in infinitesimal regions of the same type. And finally, he used the infinitesimal result to conclude something about the original figure. In Proposition 1 of the *Principia*, for example, having proved the area rule for finite triangles, he simply assumed that the result will also hold for infinitesimal triangles. He then seems to assert that because the area law is true infinitesimally and because the infinitely many, infinitely small triangles make up the region encompassed by the orbit, the area law must hold true for the entire region. Similarly, in Proposition 6, he showed that a force could be measured by a particular solid, that this result held infinitesimally, and finally, in the following applications, that the infinitesimal result could be converted into a result dealing with finite distances. And in his fluxion arguments, whether algebraic or geometric, he used the same three steps. He first found a result using finite quantities, then asserted that it is true even if certain of the quantities are infinitesimally small, and finally applied the new result to a finite situation.

But occasionally in the *Principia* and elsewhere, Newton showed that he could, if necessary, translate his geometry into analysis. As noted earlier, Proposition 41 of Book I can be explicated analytically by referring to an integral, and the same is true, for example, of Proposition 91. There Newton was calculating the total attraction on a point outside of an ellipsoid of revolution, assuming that the mass of the ellipsoid is distributed homogeneously and that the attracting force is inverse square. Although the text itself only contains geometry, Newton discussed this matter with Roger Cotes (1682–1716), the editor of the second edition of the *Principia*, and showed that the geometric method followed from integral 16.7 in our short version of Newton's table of integrals. And in a draft of *De quadratura* around 1690,

Newton even wrote, "let y be the height or distance of the body from the center [of force toward which it gravitates]. Then if the body ascends or descends straight up or down, its speed will be $\dot{y}$ and gravity $\ddot{y}$. For the fluxion of the height is as the body's speed and the fluxion of the speed is as the body's gravity."[28] This statement is certainly very close to our usual formulation $F = ma$ of Newton's second law. In other manuscripts of the same time period, Newton used such a fluxional approach to solve several problems.

Thus, although Newton did not invent the calculus to do celestial mechanics, he did use the ideas and results of his theory of fluxions as the mathematical underpinning of his most important physical work. And even though he did not publish any of his treatises on fluxions until late in his life, the ideas he generated in his rooms at Cambridge and at home in Woolsthorpe in the mid-1660s proved critical when he began to work out his system of the world in the mid-1680s. It is also important to note that, contrary to what Newton himself tried to have us believe in his reminiscences toward the end of his life, the basic ideas of the *Principia* were not developed in the 1660s. He had certainly begun to think about the problem of gravity then, but it was only in the 1680s that he was able to tie together the mathematical and physical ideas into his great work of 1687.

The *Principia*, arguably the most important text of the Scientific Revolution, was the work that defined the study of physics for the next 200 years. It secured Newton's reputation and ultimately led to his becoming Master of the Mint in 1696 and president of the Royal Society in 1703. On the other hand, Newton's calculus had relatively little influence because only parts appeared in print and, even those, many years after they were written. In fact, it was work accomplished some eight to ten years after Newton's own discoveries that constituted the basis of the first publication of the ideas of the calculus, work done by the co-inventor of the calculus, Gottfried Wilhelm Leibniz (1646–1716).

 16.2

GOTTFRIED WILHELM LEIBNIZ

As indicated in the chapter opening, Christian Huygens brought Leibniz to the frontiers of mathematical research during his stay in Paris from 1672 to 1676 by encouraging him to read such material as van Schooten's edition of Descartes' *Geometry* and the works of Pascal that included the differential triangle. Leibniz was then able to begin the investigations that led to his own invention of the differential and integral calculus toward the end of that period. It was only 10 years later, however, that he began to publish his results in short notes in the *Acta eruditorum*, the German scientific journal that he helped to found. The presentation here of Leibniz's calculus is taken from those notes, a work entitled *Historia et origo calculi differentialis* (*History and Origin of the Differential Calculus*), which Leibniz wrote in 1714 as a response to the assertion by English mathematicians that he had stolen his methods from Newton, and from the manuscripts of his early work in Paris in which he kept virtually a running log of his thoughts on the new calculus.

16.2.1 Sums and Differences

Leibniz's idea, out of which his calculus grew, was the inverse relationship of sums and differences in the case of sequences of numbers. He noted that if A, B, C, D, E was an increasing sequence of numbers, and L, M, N, P was the sequence of differences, then

---

BIOGRAPHY

Gottfried Wilhelm Leibniz (1646–1716)

The second inventor of the calculus, Gottfried Wilhelm Leibniz, was born in Leipzig to the third wife of the vice chairman of the faculty of philosophy at the University of Leipzig. Although his father died when he was only six, the young Leibniz had already been inculcated with a desire to read and study. During his youth, he taught himself Latin and plowed through the Latin classics as well as the philosophical and theological works in his father's extensive library. In 1661, he entered the University of Leipzig where he spent most of his time studying philosophy. He did attend introductory lectures on Euclid, but commented in later life about the low level of mathematics teaching at Leipzig. Leibniz received his bachelor's degree in 1663 and his master's degree in 1664, but although he prepared a dissertation for the degree of Doctor of Law, the university refused to award it to him, probably because of some political problems in the faculty. Leibniz thus left Leipzig and received his degree in 1667 from the University of Altdorf in Nuremberg.

Meanwhile, Leibniz had been introduced to advanced mathematics during a brief stay at the University of Jena in 1663 and began to work out the details of what he hoped would be his most original contribution to philosophy, the development of an alphabet of human thought, a way of representing all fundamental concepts symbolically and a method of combining these symbols to represent more complex thoughts. Although Leibniz never completed this project, his initial ideas are contained in his *Dissertatio de arte combinatoria* (*Dissertation on the Combinatorial Art*) of 1666, in which he worked out for himself Pascal's arithmetic triangle as well as the various relations among the quantities included. This interest in finding appropriate symbols to represent thoughts and ways of combining these, however, ultimately led him to the invention of the symbols for calculus we use today.

Soon after Leibniz finished his university studies, he entered upon a career first in diplomacy for the Elector of Mainz and during much of his later life as a counselor to the Duke of Hanover. Although there were various periods of his life when his job kept him extremely busy, he was nevertheless able to find time to pursue his ideas on mathematics and to carry on a lively correspondence on the subject with colleagues all over Europe (Fig. 16.12).

FIGURE 16.12

Leibniz on a German stamp

$E - A = L + M + N + P$, that is, "the sums of the differences between successive terms, no matter how great their number, will be equal to the difference between the terms at the beginning and the end of the series."[29] It followed that difference sequences were easily summed. Thus, Leibniz considered not only the arithmetical triangle of Pascal, in which each column consists of the sums of elements of the preceding column, or conversely, each column consists of the differences of the succeeding column, but also a new triangle of fractions with similar properties, which he called his "harmonic triangle":

$$\frac{1}{1}$$
$$\frac{1}{2} \quad \frac{1}{2}$$
$$\frac{1}{3} \quad \frac{1}{6} \quad \frac{1}{3}$$
$$\frac{1}{4} \quad \frac{1}{12} \quad \frac{1}{12} \quad \frac{1}{4}$$
$$\frac{1}{5} \quad \frac{1}{20} \quad \frac{1}{30} \quad \frac{1}{20} \quad \frac{1}{5}$$
$$\frac{1}{6} \quad \frac{1}{30} \quad \frac{1}{60} \quad \frac{1}{60} \quad \frac{1}{30} \quad \frac{1}{6}$$
$$\frac{1}{7} \quad \frac{1}{42} \quad \frac{1}{105} \quad \frac{1}{140} \quad \frac{1}{105} \quad \frac{1}{42} \quad \frac{1}{7}$$

Each column in this harmonic triangle is formed by taking quotients of the first column with the corresponding columns of the arithmetical triangle. For example, the elements 1/3, 1/12, 1/30, . . . of the third column arise from dividing 1/1, 1/2, 1/3, . . . by 3, 6, 10, . . . , the elements in the third column of Pascal's triangle. Because each column consists of differences of the elements in the column to its left, it follows that the sum of the elements in each column up to a certain value can be found by Leibniz's principle as the difference of the first and last values in the column immediately preceding. For example, $1/2 + 1/6 + 1/12 = 1/1 - 1/4$. Leibniz noted in addition that this rule could be extended to infinite sums because the more terms taken, the smaller the last value of the preceding sequence became. He was therefore able to derive such results as

$$\frac{1}{3} + \frac{1}{12} + \frac{1}{30} + \cdots + \frac{1}{n(n+1)(n+2)/2} + \cdots = \frac{1}{2}.$$

By multiplying this sequence by 3, Leibniz was able to rewrite it as the sum of reciprocals of the pyramidal numbers:

$$\frac{1}{1} + \frac{1}{4} + \frac{1}{10} + \cdots = \frac{3}{2}.$$

Leibniz's actual results here were not new. Their importance lay in what the possibility of summing difference sequences implied when the idea was transferred to geometry. Thus, Leibniz considered a curve defined over an interval divided into subintervals and erected ordinates y_i over each point x_i in the division. If one forms the sequence $\{\delta y_i\}$ of differences of these ordinates, its sum, $\sum_i \delta y_i$, is equal to the difference $y_n - y_0$ of the final and initial ordinates. Similarly, if one forms the sequence $\{\sum y_i\}$, where $\sum y_i = y_0 + y_1 + \cdots + y_i$, the difference sequence $\{\delta \sum y_i\}$ is equal to the original sequence of the ordinates. Leibniz extrapolated these two rules to handle the situation where there were infinitely many ordinates. He considered the curve as a polygon with infinitely many sides, at each intersection point of which an ordinate y is drawn to the axis. If the infinitesimal difference in ordinates is designated by dy, and if the sum of infinitely many ordinates is designated by $\int y$, the first rule translates into $\int dy = y$ while the second gives $d \int y = y$. Geometrically, the first means simply that the sum of the **differentials** (infinitesimal differences) in a segment equals the segment. (Leibniz assumed here that the initial ordinate equals 0.) The second rule does not have an obvious geometric interpretation, because the sum of infinitely many finite terms may well be infinite. So Leibniz replaced the finite ordinate y with an infinitesimal area $y \, dx$, where dx was the infinitesimal part of the x axis determined by the intersection points of the sides of the infinite-sided polygon. Thus, $\int y \, dx$ could be interpreted as the area under the curve, and the rule $d \int y \, dx = y \, dx$ simply meant that the differences between the terms of the sequence of areas $\int y \, dx$ are the terms $y \, dx$ themselves.

As part of his quest for the appropriate notation to represent ideas, Leibniz introduced the two notations d and $\int$ to represent his generalization of the idea of difference and sum. The latter is simply an elongated form of the letter S, the first letter of the Latin *summa*, while the former is the first letter of the Latin *differentia*. For Leibniz, both dy and $\int y$ were variables. In other words, d and $\int$ were operators that assigned an infinitely small variable and an infinitely large variable, respectively, to the finite variable y. But dy is always thought of as an actual difference, that between two neighboring values of the variable y, while $\int y$ is conceived of as an actual sum of all values of the variable y from a certain fixed value to the given one. Since dy is a variable, it too can be operated on by d to give a second-order differential, written as $d\,dy$, or even one of higher order. It is perhaps difficult for a modern reader to conceive of these infinitesimal differences and infinite sums, but Leibniz and his followers became extremely adept at using these concepts in developing methods for solving many types of problems.

16.2.2   The Differential Triangle and the Transmutation Theorem

One of the earliest applications Leibniz made of the concept of a differential was to the idea of the differential triangle, a version of which he had seen in his reading of Pascal and, perhaps, of Barrow. The differential triangle, the infinitesimal right triangle whose hypotenuse ds connects two neighboring vertices of the infinite-sided polygon representing a given curve, is similar to the triangle composed of the ordinate y, the tangent τ, and the subtangent t, so $ds : dy : dx = \tau : y : t$ (Fig. 16.13). Because ratios are involved in the idea of a tangent, Leibniz generally made one of these three differentials a constant. In other words, in choosing how to represent a curve as a polynomial with infinitely many sides, he could make the polygon have equal sides (ds constant or $d\,ds = 0$), the projection of the sides on the x axis equal (dx constant or $d\,dx = 0$), or the projection of the sides on the y axis equal (dy constant or $d\,dy = 0$). In some sense, the variable chosen to have a constant differential can be thought of as the independent variable. In any case, it was through manipulations of the differentials in the differential triangle, using his basic rules for manipulating with differentials, that Leibniz found the central techniques for his version of the calculus.

FIGURE 16.13

Leibniz's differential triangle

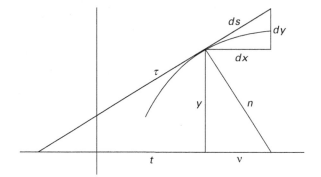

Pascal had used the differential triangle in a circle of radius r to show that, in Leibniz's language, $y\,ds = r\,dx$. Leibniz realized that this rule could be generalized to any curve if one replaced the radius by the normal line n, because the triangle made up of the ordinate, normal, and subnormal v was similar to the differential triangle. Therefore, $y : dx = n : ds$ or $y\,ds = n\,dx$. Because $2\pi y\,ds$ can be interpreted as the surface area of the surface formed by rotating ds around the x axis, this formula replaced a surface area calculation with an area calculation. Similarly, Leibniz noted that $dx : dy = y : v$ or $y\,dy = v\,dx$. Because he realized that $\int y\,dy$ represented a triangle whose area was $(1/2)b^2$, where b was the final value of the ordinate y, he had the result that $\int v\,dx = (1/2)b^2$. Therefore, to find the area under a curve with ordinate z, it was sufficient to find a curve y whose subnormal v was equal to z. But since $v = y\frac{dy}{dx}$, this was equivalent to solving the equation $y(dy/dx) = z$. In other words, an area problem was reduced to what Leibniz called an inverse problem of tangents.

Although these particular rules did not lead Leibniz to any previously unknown result, a generalization of this method gave him his **transmutation theorem** and led him to his arithmetical quadrature of the circle, a series expression for $\pi/4$. In the curve $OPQD$, where P and Q are infinitesimally close, he constructed the triangle OPQ (Fig. 16.14). Extending $PQ = ds$ into the tangent to the curve, drawing OW perpendicular to the tangent, and setting h and z as in the figure, he showed, using the similarity of triangle TWO to the differential triangle, that $dx : h = ds : z$ or that $z\,dx = h\,ds$. The left side of the second equation is the area under the rectangle $UVRS$ while the right side is twice the area of the triangle OPQ. It follows that the sum of all the triangles, namely, the area bounded by the curve $OPQD$ and the line OD, equals half the area under the curve whose ordinate is z, or $\int y\,dx - (1/2)OG \cdot GD = \frac{1}{2}\int z\,dx$. Denoting OG by x_0 and GD by y_0, Leibniz's transmutation theorem can now be stated as

$$\int_0^{x_0} y\,dx = \frac{1}{2}\left(x_0 y_0 + \int_0^{x_0} z\,dx\right).$$

Because $z = y - PU = y - x(dy/dx)$, and because Leibniz could calculate tangents by using the rules of Hudde or Sluse, the transmutation theorem enabled him to find the area under the original curve, provided that $\int z\,dx$ was simpler to compute than $\int y\,dx$.

FIGURE 16.14

Leibniz's transmutation theorem

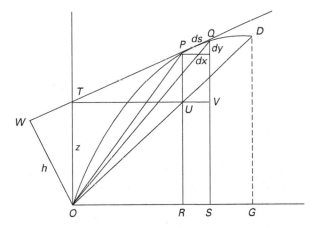

For example, Leibniz applied this result to calculate the area of a quarter of the circle of radius 1 given by $y^2 = 2x - x^2$. In this case

$$z = y - x\left(\frac{1-x}{y}\right) = \frac{x}{y} = \sqrt{\frac{x}{2-x}}$$

or

$$z^2 = \frac{x}{2-x} \quad \text{or, finally,} \quad x = \frac{2z^2}{1+z^2}.$$

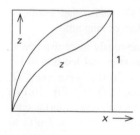

FIGURE 16.15

The transmutation function
for the circle: $z^2 = \frac{x}{2-x}$
or $x = \frac{2z^2}{1+z^2}$

By the transmutation theorem, $\int y\, dx$ (or $\pi/4$) is equal to $(1/2)(1 + \int z\, dx)$. Since it is clear from Figure 16.15 that $\int z\, dx = 1 - \int x\, dz$, Leibniz concluded that

$$\int y\, dx = 1 - \int \frac{z^2}{1+z^2}\, dz.$$

By an argument analogous to Mercator's, he then showed that

$$\frac{z^2}{1+z^2} = z^2(1 - z^2 + z^4 - z^6 + \cdots)$$

and hence that

$$\int y\, dx = 1 - \frac{1}{3}z^3 + \frac{1}{5}z^5 - \frac{1}{7}z^7 + \cdots.$$

Leibniz's formula for arithmetical quadrature, $\pi/4 = 1 - 1/3 + 1/5 - 1/7 + \cdots$, followed immediately.

16.2.3   The Calculus of Differentials

Leibniz discovered his transmutation theorem and the arithmetical quadrature of the circle in 1674. During the next two years, he discovered all of the basic ideas of his calculus of differentials. He only first published some of these results in "A New Method for Maxima and Minima as well as Tangents, which is neither impeded by fractional nor irrational Quantities, and a remarkable Type of Calculus for them," a brief article appearing in 1684 in the *Acta eruditorum*. In this paper, Leibniz was reluctant to define his differentials dx as infinitesimals because he believed there would be great criticism of these quantities, which had not been rigorously defined. Thus, he introduced dx as an arbitrary finite line segment. If y was the ordinate of a curve for which x was the abscissa, and if τ was the tangent to the curve at a point with t the subtangent, then dy was defined to be that line such that $dy : dx = y : t$. He then stated some basic rules of operation. If a is a constant, then $da = 0$; $d(v \pm y) = dv \pm dy$; $d(vw) = v\, dw + w\, dv$; and $d(v/y) = (\pm v\, dy \mp y\, dv)/y^2$. (The signs in the quotient rule depend, according to Leibniz, on whether the slope of the tangent line is positive or negative.)

Leibniz had discovered the product and quotient rules in 1675. In fact, in a manuscript of November 11 of that year he wrote, "Let us now examine whether $dx\, dy$ is the same thing as $d(xy)$, and whether dx/dy is the same thing as $d(x/y)$."[30] To check his conjecture for the product rule, he did an example where $y = z^2 + bz$ and $x = cz + d$. First, he calculated dy as the difference of the y values at $z + dz$ and at z. So $dy = (z + dz)^2 + b(z + dz) - z^2 - bz = (2z + b)\, dz + (dz)^2$. Since $(dz)^2$ is infinitely less than dz, he discarded that term and concluded that $dy = (2z + b)\, dz$. Similarly, $dx = c\, dz$ and $dx\, dy = (2z + b)c(dz)^2$. He then wrote, "but you get the same thing if you work out $d(xy)$ in a straightforward manner." Unfortunately, Leibniz did not here "work it out." Later in the manuscript, however, he realized his error in another example by showing that $d(x^2)$ is not the same as $(dx)^2$. Ten days later, he wrote the correct version of the product rule, later giving a simple proof by a difference argument: "$d(xy)$ is the same thing as the difference between two successive xy's; let one of these be xy, and the other $(x + dx)(y + dy)$; then we have $d(xy) = (x + dx)(y + dy) - xy = x\, dy + y\, dx + dx\, dy$. The omission of the quantity $dx\, dy$, which is infinitely small in comparison with the rest, . . . will leave $x\,dy + y\,dx$."[31] He proved the quotient rule similarly.

In the 1684 paper, Leibniz continued by giving, also without proof, the power rule $d(x^n) = nx^{n-1}dx$ and the rule for roots $d\sqrt[b]{x^a} = (a/b)\sqrt[b]{x^{a-b}}dx$, noting that the first law includes the second if a root is written as a fractional power. The chain rule is almost obvious using Leibniz's notation. For example, to calculate the differential of $z = \sqrt{g^2 + y^2}$, where g is a constant, Leibniz set $r = g^2 + y^2$ and noted that $dr = 2y\,dy$ and $dz = d\sqrt{r} = dr/2\sqrt{r}$. Substituting the first equation into the second, he concluded that

$$dz = \frac{2y\,dy}{2z} = \frac{y\,dy}{z}.$$

To demonstrate the usefulness of his new calculus, Leibniz discussed how to determine maxima and minima. Thus, he noted that dv will be positive when v is increasing and negative when v is decreasing, since the ratio of dv to the always positive dx gives the slope of the tangent line. It follows that $dv = 0$ when v is neither increasing nor decreasing. At that place, the ordinate will be a maximum (if the curve is concave down) or a minimum (if it is concave up). The tangent there will be horizontal. The question of concavity, Leibniz noted further, depends on the second differentials $d\,dv$: "When with increasing ordinates v its increments or differences dv also increase (that is, when dv is positive, $d\,dv$, the difference of the differences, is also positive, and when dv is negative, $d\,dv$ is also negative), then the curve is [concave up], in the other case [concave down]. Where the increment is maximum or minimum, or where the increments from decreasing turn into increasing, or the opposite, there is a *point of inflection*,"[32] that is, when $d\,dv = 0$.

As the final problem of his 1684 paper, Leibniz presented an example of one of "the most difficult and most beautiful problems of applied mathematics, which without our differential calculus or something similar no one could attack with any such ease."[33] This is the problem that had been posed by Debeaune to Descartes in 1639 to find a curve whose subtangent is a given constant a. If y is the ordinate of the proposed curve, the differential equation of the curve is $y(dx/dy) = a$ or $a\,dy = y\,dx$. Leibniz set dx as constant, equivalent to having the abscissas form an arithmetical progression. The equation then can be written as $y = k\,dy$, where k is constant. It follows that the ordinates y are proportional to their increments dy, or that the y's form a geometric progression. Since the relationship of a geometric progression in y to an arithmetic progression in x is as numbers are to their logarithms, Leibniz concluded that the desired curve will be a "logarithmic" curve. (It is now called an "exponential" curve—but, after all, today's exponential and logarithmic curves are the same curves referred to different axes.) It follows from Leibniz's discussion that, since $x = \log y$, $d(\log y) = a(dy/y)$, where the constant a depends on the particular logarithm used.

Leibniz did not consider the logarithm further in 1684, but after discussion with Johann Bernoulli (1667–1748) some years later, he returned in 1695 to the question of the differential not only of the logarithm but also of the exponential function. In a paper of that year, he responded to criticism by Bernard Nieuwentijdt (1654–1718) that his methods would not suffice to calculate the differential of the exponential expression $z = y^x$ (where x and y are both variables).[34] A direct calculation of the differential gives $dz = (y + dy)^{x+dx} - y^x$. Applying the binomial theorem and discarding powers of dy higher than the first as well as multiples $dx\,dy$ produces the equation $dz = y^{x+dx} + xy^{x+dx-1}\,dy - y^x$, a differential equation that is not homogeneous and cannot apparently be simplified further, even in the special case where $y = b$ is constant and therefore $dy = 0$. To circumvent this difficulty, Leibniz, following a suggestion of Bernoulli in 1694, attacked the problem differently by taking logarithms of both sides of the equation $z = y^x$ to get $\log z = x \log y$. The differential of this equation is then

$$a\,\frac{dz}{z} = xa\,\frac{dy}{y} + \log y\,dx.$$

It follows that

$$dz = \frac{xz}{y}\,dy + \frac{z \log y}{a}\,dx \quad \text{or} \quad d(y^x) = xy^{x-1}\,dy + \frac{y^x \log y}{a}\,dx.$$

If $x = r$ is constant, Leibniz noted, this rule reduces to the power rule $d(y^r) = ry^{r-1}\,dy$.

Two years later, Johann Bernoulli published a paper entitled "Principles of the Exponential Calculus," in which he generalized Leibniz's results to find relationships of the differentials in such equations as $y = x^x$, $x^x + x^c = x^y + y$, and $z = x^{y^v}$. He also stated explicitly the standard result on differentials of logarithms in the case where $a = 1$ that "the differential of the logarithm, however composed, is equal to the differential of the [function] divided by the [function]."[35] For example, he wrote,

$$d\left(\log \sqrt{x^2 + y^2}\right) = \frac{x\,dx + y\,dy}{x^2 + y^2}.$$

16.2.4   The Fundamental Theorem and Differential Equations

Recall that Leibniz began his research into what became his calculus with the idea that sums and differences are inverse operations. It followed that the fundamental theorem of calculus was completely obvious. He amplified this idea, however, in a manuscript of about 1680 in which he noted, first, that "I represent the area of a figure by the sum of all the rectangles contained by the ordinates and the differences of the abscissae," or as $\int y\,dx$, and, second, that "I obtain the area of a figure by finding the figure of its summatrix or quadratrix; and of this indeed the ordinates are to the ordinates of the given figure in the ratio of sums to differences."[36] That is, to find the area under a curve with ordinates y, one needs to find a curve with ordinates z such that $y = dz$. Leibniz made this idea more explicit in a 1693 paper in the *Acta eruditorum* where he showed that "the general problem of quadratures can be reduced to the finding of a curve that has a given law of tangency."[37] As he demonstrated, if, given the curve with ordinates y, one can find a curve z such that $dz/dx = y$ (a curve with a given law of tangency), then $\int y\,dx = z$, or, in modern notation, assuming that $z(0) = 0$,

$$\int_0^b y\,dx = z(b).$$

But Leibniz, like Newton, was not so much interested in finding areas as in solving differential equations, especially since it turned out that important physical problems could be expressed in terms of such equations. And Leibniz, also like Newton, used power series methods to solve such equations. His technique, however, was different. For example, consider the

FIGURE 16.16

Leibniz's derivation of the
differential equation for the
sine

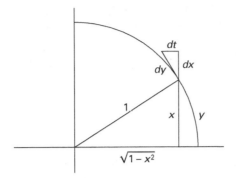

equation expressing the relationship between the arc y and its sine x in a circle of radius 1 as discussed by Leibniz in 1693.[38] The differential triangle with sides dy, dt, and dx is similar to the large triangle with corresponding sides 1, x, $\sqrt{1-x^2}$ (Fig. 16.16), so

$$dt = \frac{x\,dx}{\sqrt{1-x^2}}.$$

By the Pythagorean Theorem, $dx^2 + dt^2 = dy^2$. Substituting into this the value of dt and simplifying gave Leibniz the differential equation relating the arc and the sine: $dx^2 + x^2dy^2 = dy^2$. Considering dy as constant, he applied his operator d to this equation and concluded that $d(dx^2 + x^2dy^2) = 0$ or, using the product rule, that $2dx(d\,dx) + 2x\,dx\,dy^2 = 0$. Leibniz simplified this into the second-order differential equation

$$d^2x + xdy^2 = 0 \quad \text{or} \quad \frac{d^2x}{dy^2} = -x,$$

the familiar differential equation of the sine. (Note that Leibniz's method of manipulating with second-order differentials explains our seemingly strange placement of the 2s in the modern notation for second derivative.)

Given the differential equation, Leibniz next assumed that x could be written as a power series in y: $x = by + cy^3 + ey^5 + fy^7 + gy^9 + \cdots$, with the coefficients to be determined. It was obvious to him that there could be no even-degree terms and that, since $\sin 0 = 0$, the constant term was also 0. Differentiating this series twice gives $\frac{d^2x}{dy^2} = 2 \cdot 3cy + 4 \cdot 5ey^3 + 6 \cdot 7fy^5 + 8 \cdot 9gy^7 + \cdots$, a power series to be equated to the power series expressing $-x$. The identity of the coefficients then gives a series of simple equations:

$$2 \cdot 3c = -b$$
$$4 \cdot 5e = -c$$
$$6 \cdot 7f = -e$$
$$8 \cdot 9g = -f$$
$$\vdots$$

Setting $b = 1$ as the second initial condition, Leibniz solved these easily to get $c = -1/3!$, $e = 1/5!$, $f = -1/7!$, $g = 1/9!$, . . . , and thus derived the sine series, a series he had discovered by 1676:

$$x = \sin y = y - \frac{1}{3!}y^3 + \frac{1}{5!}y^5 - \frac{1}{7!}y^7 + \frac{1}{9!}y^9 + \cdots .$$

Leibniz had by the early 1690s discovered most of the ideas present in current calculus texts, but had never written out a complete, coherent treatment of the material. Nevertheless, like Newton after the 1670s, he wanted to justify his work by appealing to Greek standards. He gave two separate justifications. First, he attempted to relate infinitesimals to Archimedean exhaustion: "For instead of the infinite or the infinitely small, one takes quantities as large, or as small, as necessary in order that the error be smaller than the given error, so that one differs from Archimedes' style only in the expression, which are more direct in our method and conform more to the art of invention."[39] Thus, he seemed to think, like Kepler, that any argument using infinitesimals can be replaced by a perfectly rigorous argument in the style of the Greeks. But if one always had to give those arguments, one would never be able to gain new insights. Leibniz's second approach made use of a law of continuity: "If any continuous transition is proposed terminating in a certain limit, then it is possible to form a general reasoning, which covers also the final limit."[40] In other words, if one determined that a particular ratio is true in general, when, for example, the quantities dx, dy are finite, the same ratio will be true in the limiting case, when these quantities are themselves equal to 0. This justification is, in fact, very similar to Newton's own notion of a limit. But justified or not, the technique of manipulating with these infinitesimal differentials became a very useful one, particularly for Leibniz's immediate followers, Johann Bernoulli and Jakob Bernoulli (1655–1705). They seemed to accept infinitesimals as actual mathematical entities and through their use achieved many important results both in calculus itself and in its applications to physical problems.

A few words about the priority controversy between Leibniz and Newton are in order here.[41] It should be clear that although the two men discovered essentially the same rules and procedures that today are collectively called the calculus, their approaches to the subject were entirely different. Newton's approach was through the ideas of velocity and distance while Leibniz's was through those of differences and sums. Nevertheless, since Newton's work was not published until the early eighteenth century, although it was well known in England much earlier, the successes of Leibniz and the Bernoulli brothers in applying their version caused certain English mathematicians to accuse Leibniz of plagiarism, particularly because he had read some of Newton's material during his brief visits to London in the 1670s and had received two letters from Newton through Henry Oldenburg, the secretary of the Royal Society, in which Newton himself discussed some of his results. Conversely, precisely because Newton had not published, the Bernoullis accused Newton of plagiarism from Leibniz. In 1711, the Royal Society, of which Newton was then the president, appointed a commission to look into the charges. Naturally, the commission found Leibniz guilty as charged. The unfortunate result of the controversy was that the interchange of ideas between English and Continental mathematicians virtually ceased. As far as the calculus was concerned, the English all adopted Newton's methods and notation, while on the Continent, mathematicians used those of Leibniz. It turned out that Leibniz's notation and his calculus of differentials proved easier to work with. Thus, progress in analysis was faster on the Continent. To its ultimate detriment, the English mathematical community deprived itself for nearly the entire eighteenth century of the great progress.

 ## 16.3 FIRST CALCULUS TEXTS

The differences between the English and Continental approaches appear vividly in the first calculus texts to appear, those of the Marquis de l'Hospital (1661–1704) in France in 1696 and those of Charles Hayes (1678–1760) and Humphry Ditton (1675–1715) in England in 1704 and 1706, respectively. We conclude this chapter with a brief study of certain aspects of these initial texts to give the reader an idea of what an early student of the calculus would need to master.

16.3.1 L'Hospital's *Analyse des Infiniment Petits*

Guillaume François l'Hospital was born into a family of the nobility and served in his youth as an army officer. In about 1690, he became interested in the new analysis that was just then beginning to appear in journal articles by Leibniz as well as the Bernoulli brothers. Unfortunately, these articles were often brief to the point of obscurity, at least where the methods were concerned. Because Johann Bernoulli was spending time in Paris in 1691, l'Hospital asked him to provide, for a good fee, lectures on the new subject. Bernoulli agreed and some of the lectures were given. After about a year, Bernoulli left Paris to become a professor at the University of Groningen in the Netherlands. Because l'Hospital wanted the instruction to continue, they came to an agreement that for a large monthly salary, Bernoulli would not only continue sending l'Hospital material on the calculus, including any new discoveries he might make, but also give no one else access to them. In effect, Bernoulli was working for l'Hospital. By 1696, l'Hospital decided that he understood differential calculus well enough to publish a text on it, and since he had paid well for Bernoulli's work, he felt no compunction about using much of the latter's organization and discoveries in the new mathematics. Although Bernoulli was somewhat unhappy that his work was being published by another with only a bare acknowledgment, he kept silent on the matter. Since l'Hospital died before he could publish a work on integral calculus, Bernoulli eventually published his own lectures on that material.

L'Hospital began in this first extremely successful calculus text, entitled *Analyse des infiniment petits pour l'intelligence des lignes courbes* (*Analysis of Infinitely Small Quantities for the Understanding of Curves*), by defining variable quantities as those that continually increase or decrease and then giving his fundamental definition of a differential: "The infinitely small part by which a variable quantity increases or decreases continually is called the differential of that quantity." He then presented two postulates to govern his use of these differentials:

1. Grant that two quantities, whose difference is an infinitely small quantity, may be taken (or used) indifferently for each other; or (which is the same thing) that a quantity which is increased or decreased only by an infinitely small quantity may be considered as remaining the same.

2. Grant that a curve may be considered as the assemblage of an infinite number of infinitely small straight lines; or (which is the same thing) as a polygon of an infinite number of sides, each infinitely small, which determine the curvature of the curve by the angles they make with each other.[42]

For l'Hospital, then, there was no question about the existence of infinitesimals. They exist; they can be represented by elements of the differential triangle; and calculations can be made using the various rules that he presented. The rules were generally stated and proved the same way that Leibniz did originally, but although Leibniz and the Bernoullis had begun to consider transcendental curves in their own work, l'Hospital dealt virtually exclusively with algebraic curves. He only mentioned briefly the logarithmic curve, defined as one whose subtangent $y\frac{dx}{dy}$ is constant, and did not consider anything resembling a trigonometric curve.

L'Hospital's treatment of maxima and minima was slightly more general than that of Leibniz. He noted that the differential dy will be positive if the ordinates are increasing and negative if they are decreasing, but showed further that dy can change from positive to negative, and the ordinates from increasing to decreasing, in two possible ways, if dy passes through 0 or through infinity. As part of this discussion, he presented diagrams illustrating four possibilities, two where the tangent line is horizontal and two where there are cusps and the tangent line is vertical, as well as examples illustrating these possibilities. Thus, to find the maximum of $y - a = a^{\frac{1}{3}}(a - x)^{\frac{2}{3}}$, he calculated

$$dy = -\frac{2\sqrt[3]{a}\,dx}{3\sqrt[3]{a-x}}.$$

Since $dy = 0$ is impossible, he set dy equal to infinity. This implied that $3\sqrt[3]{a-x} = 0$ or that $x = a$. L'Hospital gave no particular method for distinguishing between maxima and minima, but the nature of the extremum is generally clear from the conditions of the problem. For example, consider the now standard problem of finding among all rectangular parallelepipeds with a given volume a^3 and with one side equal to a given line b the one with the least surface area. Since the sides of the parallelepiped are b, x, and $\frac{a^3}{bx}$, the problem reduces to finding the minimum of $y = bx + a^3/x + a^3/b$. L'Hospital concluded that this minimum occurs when $x = \sqrt{a^3/b}$.

L'Hospital naturally discussed second-order differences and concluded like Leibniz that points of inflection occur when $d^2y = 0$, assuming dx is taken as constant. He also developed the formula

$$r = \frac{(dx^2 + dy^2)^{\frac{3}{2}}}{dx\,d^2y}$$

for determining the radius of curvature of a given curve by a method similar to Newton's. But L'Hospital's *Analyse* is probably most famous as the source of l'Hospital's rule—which should probably be renamed Bernoulli's rule—for calculating limits of quotients in the case where the limits of both numerator and denominator are zero:

PROPOSITION   *Let AMD be a curve ($AP = x$, $PM = y$, $AB = a$) such that the value of the ordinate y is expressed by a fraction, of which the numerator and denominator each become 0 when $x = a$, that is to say, when the point P corresponds to the given point B. It is required to find what will then be the value of the ordinate BD [Fig. 16.17].*[43]

FIGURE 16.17

L'Hospital's diagram illustrating l'Hospital's rule. Notice that the function g is drawn below the x axis, but the quotient function, represented by curve AMD, is above the x axis. Think of all values of the functions involved as representing positive quantities.

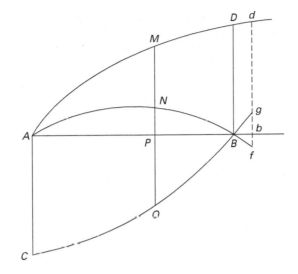

Supposing that $y = p/q$, l'Hospital simply noted that for an abscissa b infinitely close to B, the value of the ordinate y will be given by

$$y + dy = \frac{p + dp}{q + dq}.$$

But since this ordinate is infinitely close to y, and since at B both p and q are 0, l'Hospital simply noted that $y = dp/dq$. In other words, "if the differential of the numerator be found, and that be divided by the differential of the denominator, after having made $x = a \ldots$, we shall have the value of the ordinate $\ldots$ sought." (Note that no limits are involved in the statement or proof of this theorem.) L'Hospital did not believe here in trivial examples. His first, communicated to him some years earlier by Bernoulli, is the function

$$y = \frac{\sqrt{2a^3x - x^4} - a\sqrt[3]{a^2x}}{a - \sqrt[4]{ax^3}},$$

where the value is to be found when $x = a$. A straightforward calculation of differentials gave him the answer $y = (16/9)a$.

16.3.2 The Works of Ditton and Hayes

Turning to the English writers, we find in Ditton's *An Institution of Fluxions* and Hayes's *A Treatise of Fluxions* a somewhat different type of text. The two authors were not well known. They had both studied the new calculus, however, and were teaching it to their own students, for whom they felt that an English language text would be useful. Although they had both read the continental authors, they naturally preferred the fluxional approach of Newton to the differential approach of Leibniz. Thus, Ditton wrote that quantities are not to be imagined as "the aggregates or sums total of an infinite number of little constituent elements but as the result of a regular flux, proceeding incessantly, from the first moment of its beginning to that of perfect rest. A line is described not by the apposition of little lines or parts, but by the continual motion of a point. . . . The fundamental principle upon which the method of fluxions is built is more accurate, clear, and convincing than those of differential calculus."[44] Ditton then attempted to convince the reader that the discarding of certain terms because they are "nothing" in the differential calculus was not as valid as the removal of terms in the fluxional calculus because they were multiplied by a quantity that "does at last really vanish." Whether these philosophical arguments convinced the students or not, both Hayes and Ditton gave a clear treatment, similar to that of Newton, of the basics of both branches of the calculus.

It is in the detailed calculus of the logarithm and exponential functions, taken presumably from Bernoulli's paper, as well as in the treatment of the integral calculus, that these books differ in content from l'Hospital's work. Thus, both proved Bernoulli's theorem that the fluxion of the logarithm of any quantity is equal to the fluxion of the quantity divided by the quantity. (In fluxional notation, this result is $\dot{\ell}(x) = \frac{\dot{x}}{x}$, where $\ell(x)$ represents the logarithm.) Ditton gave a proof using power series: Since $\ell(1 + x) = x - \frac{1}{2}x^2 + \frac{1}{3}x^3 - \frac{1}{4}x^4 + \cdots$, it follows that $\dot{\ell}(1 + x) = \dot{x} - x\dot{x} + x^2\dot{x} - x^3\dot{x} + \cdots = \dot{x}(1 - x + x^2 - x^3 + \cdots) = \dot{x}/(1 + x)$, a result equivalent to the desired theorem.

With the logarithm taken care of, both authors turned to the exponential function $y = a^x$, treating it in a way directly translated from Leibniz's procedure. To determine the fluxion of y, the authors noted that $\ell(y) = x\ell(a)$ and, taking fluxions of both sides, calculated that $\dot{y}/y = \dot{x}\ell(a) + x\dot{\ell}(a)$. Because a is constant, the fluxion of its logarithm is 0. It follows that $\dot{y} = y\dot{x}\ell(a) = a^x\ell(a)\dot{x}$.

Hayes also considered the curve determined by the exponential function. Recall that the logarithmic curve is the curve whose ordinates are in geometric progression when its abscissas are in arithmetic proportion or as the curve whose subtangent is constant. Because the subtangent of any curve y is given by $y(\dot{x}/\dot{y})$, and because the subtangent of the curve $y = a^x$ is given by

$$y\frac{\dot{x}}{y\dot{x}\ell(a)} = \frac{1}{\ell(a)},$$

the curve defined by $y = a^x$ must be the logarithmic curve. Furthermore, Hayes calculated the area under it by first noting that, in general, the fluxion of area is $y\dot{x}$. Because in this case the subtangent $y(\dot{x}/\dot{y})$ to the curve is a constant c, it follows that $y\dot{x} = c\dot{y}$ and therefore, the fluxion of the area is $c\dot{y}$ and the area itself must be cy. Hayes's conclusion was that the area under the logarithmic curve between any two abscissas is proportional to the difference between the corresponding ordinates, a result not explicit in the work of either Leibniz or Bernoulli.

Ditton treated other aspects of the integral calculus in detail, including rectification of curves, areas of curved surfaces, volumes of solids, and centers of gravity. But his text, like those of Hayes and l'Hospital, had no treatment of the calculus of the sine or cosine. There was an occasional mention of these trigonometric relations as part of certain problems, but there is nowhere at the turn of the eighteenth century any treatment of the calculus of these functions. This was not to come until the work of Leonhard Euler in the 1730s.

EXERCISES

1. Calculate a power series for $\sqrt{1+x}$ by applying the square root algorithm to $1+x$.

2. Calculate a power series for $1/(1-x^2)$ by using long division.

3. Square Newton's power series for $(1-x^2)^{1/2}$ and show that the resultant power series is equal to $1-x^2$. (You need to convince yourself that every coefficient beyond that for x^2 is equal to 0.)

4. Use Newton's method to solve the equation $x^2 - 2 = 0$ to a result accurate to eight decimal places. How many steps does this take? Compare the efficacy of this method with that of the Chinese square root algorithm.

5. Solve $y^3 + y - 2 + xy - x^3 = 0$ for y as a power series in x. Begin by finding the value of y when $x = 0$, that is, by solving $y^3 + y - 2 = 0$. Since $y = 1$ is a solution, assume that $y = 1 + p$ is a solution to the original equation. Substitute this value for y and get $1 + 3p + 3p^2 + p^3 + 1 + p - 2 + x + px - x^3 = 0$. Removing all terms of degree higher than 1 in x and p, solve $4p + x = 0$ to get $p = -\frac{1}{4}x$. Thus, $1 - (1/4)x$ are the first two terms of the desired power series for y. To go further, substitute $p = -(1/4)x + q$ in the equation for p and continue as before. Show that the next term in the series is $(1/64)x^2$.

6. Use Newton's method of Exercise 5 to solve the equation $(1/5)y^5 - (1/4)y^4 + (1/3)y^3 - (1/2)y^2 + y - z = 0$ for y. Begin with the first approximation $y = z$. Next, substitute $y = z + p$ into the series, delete nonlinear terms in p, and solve to get $y = z + (1/2)z^2$ as the second approximation. Continue in this way to get two more terms, $(1/6)z^3$ and $(1/24)z^4$, of this series.

7. Calculate, using the power series for $\log(1 + x)$, the values of the logarithm of 1 ± 0.1, 1 ± 0.2, 1 ± 0.01, 1 ± 0.02 to eight decimal places. Using the identities presented in the text and others of your own devising, calculate a logarithm table of the integers from 1 to 10 accurate to eight decimal places.

8. Calculate the relationship of the fluxions in the equation $x^3 - ax^2 + axy - y^3 = 0$ using multiplication by the progression 4, 3, 2, 1. What do you notice? What would happen if you used a different progression?

9. Find the relationship of the fluxions using Newton's rules for the equation $y^2 - a^2 - x\sqrt{a^2 - x^2} = 0$. Put $z = x\sqrt{a^2 - x^2}$.

10. Solve the fluxional equation $\dot{y}/\dot{x} = 2/x + 3 - x^2$ by first replacing x by $x + 1$ and then using power series techniques.

11. Find the curvature of the ellipse $x^2 + 4y^2 = 1$ by using Newton's procedure.

12. Check the third value in Newton's integral table (integral 16.3) by showing that the derivative of

$$z = \frac{2a}{nc}\left(-\frac{2}{15}\frac{b}{c} + \frac{1}{5}x^n\right)(b + cx^n)^{3/2}$$

is $y = ax^{2n-1}\sqrt{b + cx^n}$.

13. Use modern techniques to integrate $y = \frac{ax^{n-1}}{e + fx^n}$ and compare your answer with Newton's answer in integral 16.5: $z = \frac{1}{n}s$, where $u = x^n$ and s is the area under the hyperbola $v = \frac{a}{e + fu}$.

14. Find the derivative of $z = \frac{8ags - 4agxv - 2afx}{4neg - nf^2}$, where $u = x^n$ and s is the area under the curve $v = \sqrt{e + fu + gu^2}$. This should equal Newton's value of $y = \frac{ax^{n-1}}{\sqrt{e + fx^n + gx^{2n}}}$, from integral 16.6.

15. Use a modern table of integrals to find the antiderivative of

$$y = \frac{ax^{2n-1}}{\sqrt{e + fx^n + gx^{2n}}}.$$

Show that your answer is equivalent to Newton's answer in integral 16.7: $z = \frac{-4afs + 2afuv + 4aev}{4neg - nf^2}$, where $u = x^n$ and s is the area under the curve $v = \sqrt{e + fu + gu^2}$.

16. Find the ratio of the fluxion of x to the fluxion of $1/x$ using Newton's "synthetic" method of fluxions.

17. Find the ratio of the fluxion of x to the fluxion of $1/x^n$ by using Newton's "synthetic" method of fluxions.

18. Derive Newton's result that the fluxion of an arc to the fluxion of its secant is as its cosine to its secant. Use the result on the tangent already demonstrated and the fluxional relationship derived from $CT^2 + AT^2 + AC^2$ (see Fig. 16.8), noting that AC, the radius of the circle, is fixed. Translate this result into a standard modern result on the derivative of the secant.

19. Derive Newton's result that the fluxion of an arc is to the fluxion of its sine as the radius to its cosine. Use the result on the tangent proved in the text, the result of Exercise 18, and fluxional relationships coming from the geometry of the situation.

20. Suppose in a simplified solar system that all planets revolved uniformly in circles with the sun at the center. If the centripetal force is inversely as the square of the radius,

show that the squares of the periodic times of the planets are as the cubes of the radii. (This is a special case of Kepler's third law.)

21. Construct Leibniz's harmonic triangle by beginning with the harmonic series 1/1, 1/2, 1/3, 1/4, . . . and taking differences. Develop a formula for the elements in this triangle.

22. Show that the sum of the denominators in row n of the harmonic triangle is given by $n2^{n-1}$.

23. Use the harmonic triangle to derive the result
$$\frac{1}{4} + \frac{1}{20} + \frac{1}{60} + \cdots + \frac{1}{n(n-1)(n-2)(n-3)/6} = \frac{1}{3}.$$

24. Given the curve $y^q = x^p$ ($q > p > 0$), show using the transmutation theorem that
$$\int_0^{x_0} y\, dx = \frac{qx_0y_0}{p+q}.$$

Note that from $y^q = x^p$, it follows that $q\, dy/y = p\, dx/x$, and therefore that $z = y - x\, dy/dx = [(q-p)/q]y$.

25. Prove the quotient rule $d(\frac{x}{y}) = \frac{y\, dx - x\, dy}{y^2}$ by an argument using differentials.

26. Derive the rule $d(x^3) = 3x^2\, dx$ using differentials.

27. Derive the general power rule $d(x^n) = nx^{n-1}$ for n a positive integer using differentials.

28. Find the relationship between the differentials in the equation $y = x^x$.

29. Find the relationship between the differentials in the equation $x^x + x^c = x^y + y$.

30. Derive the power series for the logarithm by beginning with the differential equation $dy = \frac{1}{x+1}\, dx$, assuming that y is a power series in x with undetermined coefficients, and solving simple equations to determine each coefficient in turn.

31. Derive the power series that determines the number $x + 1$ given its logarithm y, as Leibniz puts it, that is, the power series for the exponential function, by the method of undetermined coefficients. Begin with the differential equation $x + 1 = dx/dy$.

32. Apply l'Hospital's rule to his example
$$y = \frac{\sqrt{2a^3x - x^4} - a\sqrt[3]{a^2x}}{a - \sqrt[4]{ax^3}}$$
to find the value when $x = a$.

33. Use the method of Hayes and Ditton to calculate the fluxion of $y = x^x$. Compare with Exercise 28.

34. Compare and contrast the "calculuses" of Newton and Leibniz in terms of their notation, their ease of use, and their foundations.

35. Outline a series of lessons on power series using the ideas of Newton. Is it useful to introduce such series early in a calculus course? Why or why not?

36. Could one structure a calculus course along the lines of Newton's *Treatise on Methods*? How would this differ from the normal organization of a calculus course? How would this organization compare to that of one of the new reform calculus courses?

37. Is the notion of a differential as an infinitesimal a useful idea to present in a modern calculus class, either a standard one or a reform one? Would it make the derivation of the basic rules of calculus easier? Why or why not?

38. Outline a lesson on the general binomial theorem following the argument from analogy of Newton.

39. Why are Newton and Leibniz considered the inventors of the calculus rather than some of the mathematicians considered in Chapter 15?

✿ REFERENCES AND NOTES

The three works on the history of calculus mentioned in the notes to Chapter 15 also deal with the work of Newton and Leibniz. Isaac Newton's published treatises on fluxions, all published many years after they were written, are available in facsimile edition in Derek T. Whiteside, ed., *The Mathematical Works of Isaac Newton*, vol. 1 (New York: Johnson Reprint Corporation, 1964). Whiteside is also the editor of the eight-volume set of all of Newton's surviving mathematical manuscripts, including the material published in the previous reference: *The Mathemat-* *ical Papers of Isaac Newton* (Cambridge: Cambridge University Press, 1967–1981). Although the originals of many of the papers are in Latin, Whiteside has translated most of them into English. These volumes repay careful browsing. Newton's letters are in H. W. Turnbull, ed., *The Correspondence of Isaac Newton* (Cambridge: Cambridge University Press, 1960). This seven-volume set contains virtually all the extant letters to and from Newton, as well as other related material. A new translation of *The Principia* has recently been made, complete with a guide to its study:

Isaac Newton, *The Principia: Mathematical Principles of Natural Philosophy*, trans. by I. Bernard Cohen and Anne Whitman (Berkeley: University of California Press, 1999). Excerpts from Leibniz's first published calculus treatises are in D. J. Struik, ed., *A Source Book in Mathematics, 1200–1800* (Cambridge: Harvard University Press, 1969). J. M. Child, *The Early Mathematical Manuscripts of Leibniz* (Chicago: Open Court, 1920) contains the edited translations of many of Leibniz's mathematical manuscripts up to about 1680. The commentaries must be read with care, because Child seems to be most interested in showing that much of Leibniz's work is derived from that of Isaac Barrow.

There are numerous recent works dealing with Newton. The best biography is Richard Westfall, *Never at Rest* (Cambridge: Cambridge University Press, 1980), which covers in stimulating detail not only Newton's mathematical achievements but also his work in various other areas of science. An excellent summary of Newton's mathematical achievements is V. Frederick Rickey, "Isaac Newton: Man, Myth, and Mathematics," *College Mathematics Journal* 18 (1987), 362–389. A summary of Newton's early work is found in two articles of Derek T. Whiteside: "Isaac Newton: Birth of a Mathematician," *Notes and Records of the Royal Society* 19 (1964), 53–62, and "Newton's Marvelous Years: 1666 and All That," *Notes and Records of the Royal Society* 21 (1966), 32–41. For studies of the *Principia* and related works, see François De Gandt, *Force and Geometry in Newton's Principia*, (Princeton: Princeton University Press, 1995); Dana Densmore, *Newton's Principia: The Central Argument: Translation, Notes, and Expanded Proofs* (Santa Fe: Green Lion Press, 1995); and Niccolò Guicciardini, *Reading the Principia: The Debate on Newton's Mathematical Methods for Natural Philosophy from 1687 to 1736* (Cambridge: Cambridge University Press, 1999). The best recent works on Leibniz include Eric Aiton, *Leibniz, A Biography* (Bristol: Adam Hilger Ltd, 1985), a general work covering his entire scientific career, and Joseph E. Hofmann, *Leibniz in Paris, 1672–1676* (Cambridge: Cambridge University Press, 1974), covering in great detail the years in which Leibniz invented his version of the calculus.

1. Child, *Early Mathematical Manuscripts*, p. 215.

2. Turnbull, ed., *Correspondence*, vol. II, p. 115 and p. 153. In interpreting the anagram, one must keep in mind that in Latin, the letters "u" and "v" are interchangeable.

3. Westfall, *Never at Rest*, p. x.

4. John Fauvel, Raymond Flood, Michael Shortland, and Robin Wilson, eds., *Let Newton Be! A New Perspective on His Life and Works* (Oxford: Oxford University Press, 1988), p. 15.

5. Westfall, *Never at Rest*, p. 191.

6. Ibid., pp. 192 and 209.

7. Whiteside, *Mathematical Papers*, vol. III, pp. 33–35. A brief presentation of Newton's calculus is in Philip Kitcher, "Fluxions, Limits and Infinite Littleness: A Study of Newton's Presentation of the Calculus," *Isis* 64 (1973), 33–49.

8. Turnbull, *Correspondence*, vol. II, p. 131. The letter in which this discussion of the binomial theorem occurs, the *Epistola posterior*, was sent by Newton to Henry Oldenburg, who in turn forwarded it to Leibniz. It and the *Epistola prior* of June 13, 1676, were later quoted as central pieces of evidence in the Royal Society report convicting Leibniz of plagiarism.

9. Whiteside, *Mathematical Papers*, II, pp. 241–243.

10. Ibid., II, p. 213.

11. Ibid., III, p. 71.

12. Ibid., III, p. 73.

13. Ibid., III, p. 75.

14. Ibid., III, p. 81.

15. Ibid.

16. Ibid.

17. Ibid., III, p. 117.

18. Ibid., III, p. 151. For more on curvature, see Julian L. Coolidge, "The Unsatisfactory Story of Curvature," *American Mathematical Monthly* 59 (1952), 375–379.

19. Whiteside, *Mathematical Papers*, III, p. 83.

20. Ibid., III, p. 237.

21. Westfall, *Never at Rest*, p. 403.

22. Newton, *Principia*, trans. by Cohen and Whitman, p. 433.

23. Ibid., pp. 441–443.

24. Whiteside, *Mathematical Papers*, VIII, pp. 126–129.

25. Newton, *Principia*, trans. by Cohen and Whitman, p. 445.

26. Ibid., p. 454.

27. For a discussion of Newton's proof of the result that an inverse square force implies an elliptical orbit, see Bruce Pourciau, "Reading the Master: Newton and the Birth of Celestial Mechanics," *American Mathematical Monthly* 104 (1997), 1–19.

28. Whiteside, *Mathematical Papers*, VII, pp. 128–129. This is referred to in an as yet unpublished book of Niccolo Guicciardini.

29. Child, *Early Mathematical Manuscripts*, pp. 30–31. This is part of the "Historia et Origo."

30. Ibid., p. 100.

31. Ibid., p. 143.

32. Struik, *Source Book*, p. 275.

33. Ibid., p. 279.

34. G. W. Leibniz, *Mathematische Schriften*, C. I. Gerhardt, ed., (Hildesheim: Georg Olms Verlag, 1971), vol. V, pp. 320–328.

35. Johann Bernoulli, *Opera omnia* (Hildesheim: Georg Olms Verlag, 1968), vol. I, pp. 179–187, p. 183.

36. Child, *Early Mathematical Manuscripts*, p. 138.

37. G. W. Leibniz, *Mathematische Schriften*, vol. V, pp. 294–301. An English version is in Struik, *Source Book*, p. 282.

38. Ibid., vol. V, pp. 285–288.

39. H. J. M. Bos, "Differentials, Higher-Order Differentials and the Derivative in the Leibnizian Calculus," *Archive for History of Exact Sciences* 14 (1974), 1–90, p. 56. The original paper from which this quotation is taken may be consulted in Leibniz, *Mathematische Schriften*, vol. V, p. 350. Bos's paper is an excellent study of the general idea of a differential.

40. Ibid.

41. The controversy is discussed in great detail in A. R. Hall, *Philosophers at War: The Quarrel between Newton and Leibniz* (Cambridge: Cambridge University Press, 1980).

42. Struik, *Source Book*, p. 313. Struik presents a translation of some sections of l'Hospital's *Analyse*. For more details on l'Hospital's work, see Carl Boyer, "The First Calculus Textbook," *Mathematics Teacher* 39 (1946), 159–167.

43. Ibid., pp. 315–316. See also Dirk Struik, "The Origin of l'Hospital's Rule," *Mathematics Teacher* 56 (1963), 257–260.

44. Humphry Ditton, *An Institution of Fluxions* (London: Botham, 1706), p. 1.

Analysis in the Eighteenth Century

CHAPTER

17

Jean Bernoulli, public professor of mathematics, pays his best respects to the most acute mathematicians of the entire world. Since it is known with certainty that there is scarcely anything which more greatly excites noble and ingenious spirits to labors which lead to the increase of knowledge than to propose difficult and at the same time useful problems through the solution of which . . . they may attain to fame and build for themselves eternal monuments among posterity, so I should expect to deserve the thanks of the mathematical world if . . . I should bring before the leading analysts of this age some problem upon which . . . they could test their methods, exert their powers, and, in case they brought anything to light, could communicate with us in order that everyone might publicly receive his deserved praise from us.

—Proclamation made public at
Gröningen, the Netherlands,
January 1697[1]

Realizing only in 1739 that it was necessary to treat the sine and cosine as functions on the same level as the exponential function, Leonhard Euler presented a paper on March 30 of that year to the Academy of Sciences at St. Petersburg in which appears, for the first time, a discussion of the calculus of the trigonometric functions. Until that time there was no sense of the sine and cosine functions being expressed, like the algebraic functions, as formulas involving letters and numbers, whose relationship to other such formulas could be studied using the techniques of the calculus. But in that year he first understood that the sine and cosine were solutions to certain differential equations coming from the theory of vibrations. He made his discovery known through letters to other mathematicians and finally published the material in detail in his *Introductio* in 1748.

The driving force in the continued development of calculus in the eighteenth century was the desire to solve physical problems, the mathematical formulation of which was often in terms of equations among fluxions or among differentials. Although mathematicians in Britain as well as on the Continent participated in this effort, the flexibility of Leibniz's notation seemed to give Continental mathematicians an advantage here, and the method of differential equations soon outstripped methods using fluxions. Thus, Continental mathematicians thought it important to translate Newton's geometrical analysis of the *Principia* into the more algebraic analysis of differentials and thus derive many of Newton's results by their own methods. But mathematicians also posed and solved many new problems arising from applications of Newton's laws of motion. Gradually, the emphasis changed from the study of curves, which was central to both Newton's and Leibniz's mathematics, to the study of analytical expressions involving one or more variable quantities as well as certain constants, that is, functions of one or several variables. The relationship between the differentials of these variables and the variable dependent on them, determined by some physical situation, led to a differential equation whose solution explicitly determined the desired function. In fact, new classes of functions were discovered and analyzed through the differential equations that they satisfied.

The major figure in the development of analysis in the eighteenth century was the most prolific mathematician in history, Leonhard Euler. Much of this chapter is devoted to his work in the theory of differential equations, the calculus of variations, and multivariable calculus as well as to his three influential textbooks in analysis. The chapter begins, however, with some of the challenge problems set by the Bernoullis for the mathematicians of Europe, problems whose solutions helped to establish new ideas in mathematics that were later developed by Euler and others. Because influential ideas and techniques also appeared in the works of Thomas Simpson and Colin Maclaurin in England and that of Maria Gaetana Agnesi in Italy, these texts are considered as well. In particular, because Maclaurin wrote his text partly to answer the criticisms of George Berkeley regarding the foundations of calculus, his response to this criticism is discussed. The chapter concludes with the attempt by Joseph Louis Lagrange to eliminate all reference to infinitesimal quantities or even to limits and to base the calculus on the notion of a power series.

17.1   DIFFERENTIAL EQUATIONS

It was the brothers Jakob and Johann Bernoulli (often known as Jacques and Jean, or James and John) who were among the first in Europe to understand the new techniques of Leibniz and to apply them to solve new problems. For example, in 1659 Huygens had discovered using infinitesimals, and then proved geometrically, that the curve along which an object descending under the influence of gravity takes the same amount of time to reach the bottom, from whichever point on the curve its descent begins, was a cycloid. Huygens then used this idea in his invention of a pendulum clock. He realized that if the pendulum were constrained to move in a cycloidal arc, it would keep time perfectly, whatever the size of the oscillation. Jakob Bernoulli in 1690 was able to prove Huygens's result analytically by setting up the differential equation for this curve of equal time, the **isochrone**.

Having succeeded in the isochrone problem, Jakob then proposed a new one, to determine the shape of the catenary, the curve assumed by a flexible but inelastic cord hung freely

BIOGRAPHY

Jakob (1654–1705) and Johann Bernoulli (1667–1748)

Jakob Bernoulli (Fig. 17.1) taught himself mathematics, spending time in France, the Netherlands, and England where he acquainted himself with the works of various scientists and mathematicians. On his return to his native Basel in 1683, he also studied van Schooten's edition of Descartes' *Geometry* and was ultimately appointed to the University of Basel, first as a lecturer in experimental physics and in 1687 as a professor of mathematics, a position he held to the end of his life. Meanwhile, his younger brother Johann, having tried unsuccessfully to please his father by becoming a businessman, enrolled at the university to study medicine, again to meet his father's wishes. Nevertheless, he spent much time studying mathematics with Jakob. Together, they mastered the early works of Leibniz and were soon able to make contributions on their own.

Johann Bernoulli developed Leibniz's techniques in more detail in various articles in the early 1690s, thus enabling him, through the help of Huygens, to be offered the chair of mathematics at the University of Gröningen in the Netherlands, a position he held until his brother's death allowed him to succeed to the mathematics chair in Basel in 1705.

FIGURE 17.1

Jakob Bernoulli on a Swiss stamp

between two fixed points. Galileo had thought that this curve was a parabola. Jakob himself was unable to solve the problem, but in the *Acta eruditorum* for June 1691 there appeared solutions by Leibniz, Huygens, and Johann Bernoulli. Johann was immensely proud that he had surpassed his older brother, reporting that the solution had cost him a night of sleep. His solution, which later appeared with more details in the lectures he gave to l'Hospital, began with the differential equation $dy/dx = s/a$ derived from an analysis of the forces acting to keep the cord in position, where s represents arclength. Because $ds^2 = dx^2 + dy^2$, squaring the original equation gives

$$ds^2 = \frac{s^2 dy^2 + a^2 dy^2}{s^2} \quad \text{or} \quad ds = \frac{\sqrt{s^2 + a^2}\, dy}{s}$$

or, finally,

$$dy = \frac{s\, ds}{\sqrt{s^2 + a^2}}.$$

An integration then shows that $y = \sqrt{s^2 + a^2}$ or that $s = \sqrt{y^2 - a^2}$. Bernoulli concluded that

$$dx = \frac{a\, dy}{s} = \frac{a\, dy}{\sqrt{y^2 - a^2}}.$$

He was not able to express this integral in closed form, but was able to construct the desired curve by making use of certain conic sections. In modern terminology, this equation can be solved in the form $x = a \ln(y + \sqrt{y^2 - a^2})$ or in the form $y = a \cosh \frac{x}{a}$. For Bernoulli in 1691, however, as for his contemporaries, an answer in terms of areas under, or lengths of, known curves was sufficient.

Over the next several years both brothers posed other problems involving differential equations and, along with Leibniz, made much progress in developing methods of solution. In particular, in 1691 Leibniz found the technique of separating variables, that is, of rewriting a differential equation in the form $f(x)dx = g(y)dy$ and then integrating both sides to give the solution. He also developed the technique for solving the homogeneous equation $dy = f(y/x)dx$ by substituting $y = vx$ and then separating variables. By 1694, Leibniz had in addition solved the general first-order linear differential equation $m\,dx + ny\,dx + dy = 0$, where m and n are both functions of x. (In modern notation, this is the equation $dy/dx + ny = -m$.) He defined p by the equation $dp/p = n\,dx$ and substituted to get $pm\,dx + y\,dp + p\,dy = 0$. Because the last two terms on the left side are equal to $d(py)$, an integration gives $\int pm\,dx + py = 0$. This equation, giving the answer in terms of an area, provided Leibniz with the desired solution.

17.1.1   The Brachistochrone Problem

Probably the most significant of the problems proposed by Johann Bernoulli, in terms of its ultimate consequences for mathematics, was that of the **brachistochrone**, the curve of quickest descent. He first proposed it in the June 1696 issue of the *Acta eruditorum* as a "New Problem Which Mathematicians Are Invited to Solve: If two points A and B are given in a vertical plane, to assign to a mobile particle M the path AMB along which, descending under its own weight, it passes from the point A to the point B in the briefest time."[2] Bernoulli noted that the required curve was not a straight line, but a curve "well known to geometers." He had requested the solutions by the end of 1696, but in early January of 1697, acting on a suggestion of Leibniz, he extended the deadline to Easter and sent the problem, as mentioned in the opening quotation, to those who had not seen the note in the *Acta*. He offered a prize "neither of gold nor silver. . . . Rather, since virtue itself is its own most desirable reward and fame is a powerful incentive, we offer the prize . . . compounded of honor, praise, and approbation; thus we shall crown, honor, and extol, publicly and privately, in letter and by word of mouth, the perspicacity of our great Apollo."[3] Among those to whom the challenge was sent was Newton, who, Bernoulli believed, had stolen Leibniz's methods and would not be able himself to solve this problem.

When Newton received the letter from Bernoulli at about 4:00 p.m. on January 29, he was very tired after a difficult day at the mint. Nevertheless, he stayed up until he had solved the problem by 4:00 the next morning. Bernoulli was forced to acknowledge Newton's talents. Leibniz was sufficiently embarrassed by the incident that he wrote to the Royal Society denying that he had been involved in it. In any case, Newton's solution was published in the May 1697 issue of the *Acta* along with the solutions of Leibniz, Jakob Bernoulli, and Johann himself.

We will consider Johann's solution here. He began by noting that, according to Galileo, the velocity acquired by a falling body is proportional to the square root of the distance fallen. Second, he recalled Snell's law, that when a light ray passes from a thinner to a denser medium, the ray is bent so that the sine of the angle of incidence is to the sine of the angle of refraction inversely as the densities of the media and therefore directly as the velocities in those media. This law had been derived by Fermat as an application of the principle that the path traversed by the light ray must take the least time. Bernoulli then assumed that the vertical plane of the problem was composed of infinitesimally thick layers whose densities varied. The brachistochrone, the path of least time, was thus the curved path of a light ray whose direction changed continually as it passed from one layer to the next. At every point the sine of the angle between the tangent to the curve and the vertical axis was proportional to the velocity, and the velocity was in turn proportional to the square root of the distance fallen.

Now, denoting the desired brachistochrone curve by AMB and the curve representing the velocity at each point by AHE, Bernoulli set x and y to be the vertical and horizontal coordinates, respectively, of the point M measured from the origin A and t to be the horizontal coordinate of the corresponding velocity point H (Fig. 17.2). With m a point infinitesimally close to M, he represented the infinitesimals Cc, Mm, and nm by dx, ds, and dy, respectively. From the fact that the sine of the angle of refraction nMm is $dy : ds$, which is in turn proportional to the velocity t, Bernoulli derived the equation $dy : t = ds : a$, or $a\,dy = t\,ds$, or $a^2\,dy^2 = t^2\,ds^2 = t^2\,dx^2 + t^2\,dy^2$, or finally,

$$dy = \frac{t\,dx}{\sqrt{a^2 - t^2}}.$$

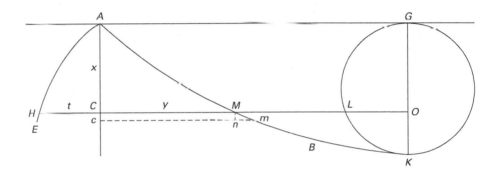

FIGURE 17.2

Johann Bernoulli's brachistochrone problem

Because the curve AHE is a parabola with equation $t^2 = ax$ or $t = \sqrt{ax}$, substitution of this value for t produces the differential equation of curve AMB:

$$dy = dx\sqrt{\frac{x}{a-x}}.$$

Bernoulli immediately recognized this equation as defining a cycloid. To prove this analytically, he noted that

$$dx\sqrt{\frac{x}{a-x}} = \frac{a\,dx}{2\sqrt{ax-x^2}} - \frac{(a-2x)\,dx}{2\sqrt{ax-x^2}}.$$

Given that $y^2 = ax - x^2$ is the equation of a circle GLK, and that the first term on the right is the differential of arclength along this circle, an integration of the equation gives $CM = $ arc $GL - LO$. Because $MO = CO - CM = CO - $ arc $GL + LO$, and with the assumption that CO is equal to half the circumference of the circle, it follows that $MO = $ arc $LK + LO$ or that $ML = $ arc LK. It is then immediate that the curve AMK is a cycloid as asserted. Bernoulli expressed a pleased amazement that this curve was the same as the isochrone. He noted that this was true only because velocity is proportional to the square root of the distance and not to any other power. "We may conjecture," he continued, "that nature wanted it to be thus. For, as nature is accustomed to proceed always in the simplest fashion, so here she accomplished two different services through one and the same curve."[4]

Although Johann Bernoulli's solution of the brachistochrone problem was very ingenious, his brother's solution was more easily generalizable. Jakob reasoned that if the entire curve is that along which a point moves in shortest time, then any infinitesimal segment of the curve will have the same property. Using a geometrical argument, he was able to derive the differential equation $ds = \sqrt{a}\,dy/\sqrt{x}$ for the curve, an equation easily converted to Johann's equation.[5] Jakob's method provided the beginning of a new field of mathematics, the calculus of variations, in which a curve is sought that satisfies some maximum or minimum property (see Section 17.1.5).

On the other hand, Johann's solution to the problem led to an investigation of the properties of families of curves, which in turn led to some fundamental new concepts in the theory of functions of several variables. Given that one could construct a family $\{C_\alpha\}$ of cycloids, each being the curve of fastest descent from a given point A to a point B_α, Johann Bernoulli posed and solved the problem of finding a new family $\{D_\beta\}$ of curves, called **synchrones**, each point of which was the place reached by particles in a given time t_β descending from A along the various cycloids C_α (Fig. 17.3). In physical terms, if the cycloids represent light rays, then the synchrones represent the wave fronts, the simultaneous positions of the various light pulses emitted from A at the same instant. Bernoulli realized that optical theory (developed by Huygens) predicted that the synchrones would intersect the brachistochrone cycloids in right angles. Thus, his geometric problem was to find a family $\{D_\beta\}$ of curves orthogonal to a family $\{C_\alpha\}$ of cycloids with a given vertex. Johann was able to construct the synchrones with little difficulty, but he challenged others to solve the general problem of finding orthogonal trajectories to a given family of transcendental curves. This type of problem soon led to the development of ideas about partial differentiation, which is discussed in Section 17.2.1.

FIGURE 17.3

Two orthogonal families of curves

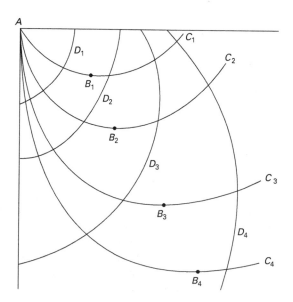

17.1.2 Translating Newton's Synthetic Method of Fluxions into the Method of Differentials

Under an appropriate translation, Newton's *Principia* was full of differential equations. And so Leibniz, the Bernoullis, and other Continental mathematicians involved in the development of Leibniz's calculus of differentials, in order to show that their methods were as good as, or better than, Newton's methods, made a major effort to translate Newton's synthetic method of fluxions into the methods of the differential and integral calculus. (As we noted in Chapter 16, Newton himself had translated some of the results of the *Principia* into the analytical method of fluxions.)

For example, we consider the derivation of Kepler's area law by Jacob Hermann (1678–1733), a student of Jacob Bernoulli in Basel. Recall that this law was derived by Newton as proposition 1 of the *Principia*. Hermann proved it anew in his *Phoronomia* of 1716. He assumed that the trajectory of the body was a plane curve ANB, where D is the center of force, ds is the infinitesimal element of arc, and NC and nc are tangents to the curve at N and n, respectively, (Fig. 17.4). The line DC ($= p$) is perpendicular to the tangent NC ($= q$). Lines ON and On are perpendicular to the tangents at the "neighboring" points N and n; thus, they meet at the center O of the **osculating circle**, the circle that best approximates the curve ANB near n. The radius $ON = \rho$ of that circle is what is known as the **radius of curvature** of curve ANB at N.

FIGURE 17.4

Hermann's proof of Newton's proposition 1

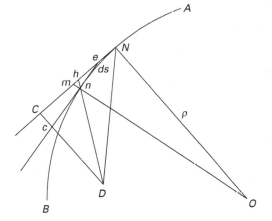

Hermann began with two basic principles for an arbitrary force G acting on a unit mass. First, since under those circumstances force and acceleration are equal, he knew that (1) $Gt = v$, where v is the velocity and t is the time. Second, he took Galileo's law in the form (2) $2\ell/G = t^2$, where ℓ is the distance fallen from rest in time t. Now, let the central force F of the problem be split into two components F_T, the force acting along the tangent, and F_N, the force acting along the normal. (Note that all forces may be assumed constant in an infinitesimal time interval.) By (1), $F_T\, dt = -dv$, where dt is the infinitesimal time it takes the body to go from N to n and $-dv$ is the corresponding change in the velocity. Therefore, $F_T v\, dt = -v\, dv$, and, since $v\, dt = ds$, we get (3) $F_T\, ds = -v\, dv$.

Next, by (2) we have $2\,d\alpha/F = dt^2 = ds^2/v^2$, where $d\alpha = \ell n$ is an infinitesimal Galilean fall from the tangent to the curve. Thus, $d\alpha = ds^2 F/2v^2$. Since triangles Nmn and QnN are similar (Fig. 17.5), we get $nm : Nn = Nn : NQ = Nn : 2ON$, or $d\beta^2 = ds^2/2\rho$, where $d\beta = nm$ is the infinitesimal change in position along the radius. (Note that here mn is perpendicular to NC rather than being the prolongation of On, while ln is parallel to DN. This ambiguity, common in the work of Hermann and others calculating with infinitesimals, does not affect the final result because the difference in the lengths of the infinitesimal line segments in question is a higher-order infinitesimal, which is neglected.) We now have

$$\frac{F}{F_N} = \frac{\ell n}{nm} = \frac{d\alpha}{d\beta} = \frac{ds^2 F}{2v^2} \cdot \frac{2\rho}{ds^2} = \frac{F\rho}{v^2},$$

so it follows that (4) $F_N \rho = v^2$, a standard result relating acceleration and velocity around a circle.

FIGURE 17.5

Hermann's proof of Newton's proposition 1, continued

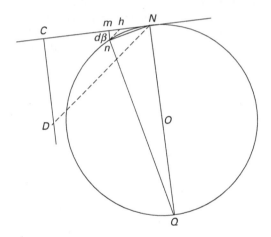

We know that $F_T : F_N = NC : DC = q : p$. Because triangles Cec and NOn are similar, again neglecting higher-order infinitesimals, we also have $Cc : Ce = Nn : NO$. But since Ce and CN differ by a higher-order infinitesimal, we can replace Ce by CN in this proportion. Since $Cc = dp$ and $Nn = ds$, we then have $dp/q = ds/\rho$. Dividing equation 3 by equation 4 and simplifying gives

$$\frac{F_T\,ds}{F_N \rho} = \frac{-v\,dv}{v^2} \quad \text{or} \quad \frac{dv}{v} = -\frac{F_T\,ds}{F_N \rho} = -\frac{q}{p}\frac{dp}{q} = -\frac{dp}{p}.$$

It follows that $p\,dv + v\,dp = 0$ or $d(pv) = 0$ or, finally, $pv = 2k$, where k is a constant. But $p\,ds$ is twice the area of triangle DNn, or twice the infinitesimal area dA swept out by the line from the central force to the moving body. Therefore,

$$2k = pv = p\frac{ds}{dt} = \frac{p\,ds}{dt} = \frac{2dA}{dt}$$

and

$$\frac{dA}{dt} = k,$$

that is, the rate of change of area is constant. This is the content of Kepler's law of areas, so Hermann had now proved Newton's proposition 1.

Hermann also used differentials to prove the result that Newton stated in his corollary to propositions 11–13 of Book I of the *Principia*, namely, that an inverse-square force law implies a conic section orbit. In Figure 17.6 (slightly modified from Fig. 16.10 by completing right triangles SPI and QRB and then drawing lines QH, QK, RG, and KP as shown), a body is moving along curve $APQL$ under the attraction of a central force F at S, inversely proportional to the square of the distance SP. We set S to be the origin of the coordinate system. As in Figure 16.10, PQ is infinitesimal. We therefore set $PQ = ds$, $SI = x$, and $PI = y$. Then $SP = \sqrt{x^2 + y^2}$, $QH = PK = dx$, $PH = GB = dy$, and $KG - QR = -ddx = -d^2x$.

FIGURE 17.6

Hermann's proof of Kepler's first law

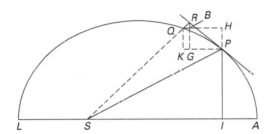

Because triangles QRB and SPI are similar, it follows that $QB : QR = SI : SP$, or $-d^2x : QR = x : \sqrt{x^2 + y^2}$. Therefore,

$$QR = \frac{-d^2x\sqrt{x^2 + y^2}}{x}.$$

Now Newton had shown, and Hermann used this result, that the force F is proportional to QR and inversely proportional to the square of the area of the infinitesimal triangle SQP. By standard techniques, that area is given by $\frac{1}{2}(y\,dx - x\,dy)$. Because F is also inversely proportional to $SP^2 = x^2 + y^2$, we get

$$\frac{-ad^2x\sqrt{x^2 + y^2}}{x(y\,dx - x\,dy)^2} = \frac{1}{x^2 + y^2} \quad \text{or} \quad -ad^2x = \frac{x(y\,dx - x\,dy)^2}{(x^2 + y^2)^{3/2}}$$

as the second-order differential equation that Hermann needed to solve.

Now Hermann used the earlier area result, which implied that $y\,dx - x\,dy$ is constant. He therefore could perform two integrations. First, rewriting the differential equation in the form

$$-ad^2x = (y\,dx - x\,dy)\frac{xy\,dx - x^2\,dy}{(x^2 + y^2)^{3/2}},$$

he claimed that the integral was

$$-a\,dx = \frac{-y}{\sqrt{x^2+y^2}}(y\,dx - x\,dy) = \frac{xy\,dy - y^2\,dx}{\sqrt{x^2+y^2}}.$$

Although Hermann did not show how he found this integral, it is straightforward to show that the differential of this latter equation gives the original equation. Then, rewriting this result in the form

$$-\frac{ab\,dx}{x^2} = \frac{bxy\,dy - by^2\,dx}{x^2\sqrt{x^2+y^2}},$$

where b is a constant, he integrated again to get

$$\frac{ab}{x} \pm c = \frac{b\sqrt{x^2+y^2}}{x}, \quad \text{or} \quad a \pm \frac{cx}{b} = \sqrt{x^2+y^2},$$

with c the arbitrary constant of integration. Hermann knew that this last equation was the equation of a conic section. In fact, it is a parabola if $b = c$, an ellipse if $b > c$, and a hyperbola if $b < c$. Hermann had therefore proved the result that Newton only sketched in his corollary to propositions 11–13 of the *Principia*.

Interestingly enough, Johann Bernoulli criticized Hermann's result, because he felt it was not general enough, namely, that Hermann did not introduce an arbitrary constant of integration in his first integral. But as others pointed out, such a constant would have just changed the axis of the curve, and Hermann simply made the assumption that that axis would be the x axis. Bernoulli himself eventually gave another proof of the same result using differential calculus, but from a slightly different point of view.

17.1.3 Differential Equations and the Trigonometric Functions

Ideas involving the solution of differential equations were what led Leonhard Euler (1707–1783) to his invention of the modern notion of the sine and cosine functions in the 1730s. Recall that Newton had been able to calculate the fluxion of the sine and that Leibniz had derived the differential equation for the sine by a geometrical argument and then had solved it to get its power series representation. But in the early years of the eighteenth century, physical problems that led to such equations were typically solved geometrically. For example, the equation

$$dt = \frac{c\,ds}{\sqrt{c^2 - s^2}}$$

would be solved for t as $t = c \arcsin \frac{s}{c}$ by consideration of the geometrical situation, rather than as $s = c \sin \frac{t}{c}$. One of the problems that led to an equation of this type was that of determining the motion of an object subject to a force proportional to its distance from a given point, considered by Johann Bernoulli in 1728. The sine, however, was still considered as a line in a circle of a given radius rather than as a function in modern terms and was not a part of these solutions.

In contrast to the "missing" sine function, the exponential function was well known to Euler by 1730. In fact, in a manuscript on differential calculus that he wrote for use as a text shortly after arriving in St. Petersburg, Euler noted that there were two classes of functions, algebraic and transcendental. The latter class consisted solely of the exponential and logarithmic functions, the properties of which he proceeded to discuss. Thus, he knew that the differential equation satisfied by $y = e^{ax}$ was $dy = ae^{ax}dx$, or, to put this the other way around, that the solution to $dy = ay\,dx$ was $y = e^{ax}$. In various papers in the early 1730s, Euler made use of this property of the exponential to solve other differential equations. For example, he noted that the equation

$$dz - 2zdv + \frac{z\,dv}{v} = \frac{dv}{v}$$

could be solved by multiplying through by the "integrating factor" $e^{-2v}v$ to give $e^{-2v}v\,dz - 2e^{-2v}zv\,dv + e^{-2v}z\,dv = e^{-2v}\,dv$. Because the left side is the differential of $e^{-2v}vz$, the solution of the equation is

$$e^{-2v}vz = C - \frac{1}{2}e^{-2v} \quad \text{or} \quad 2vz + 1 = Ce^{2v}.$$

Higher-order equations could also be solved by exponential functions, but by the mid-1730s Euler realized that these functions were not sufficient. For example, in 1735 Daniel Bernoulli wrote to Euler to discuss a problem on the vibrations of an elastic band. The problem led to the fourth-order equation $k^4(d^4y/dx^4) = y$. Both Bernoulli and Euler realized that $e^{\frac{x}{k}}$ was a solution, but Bernoulli wrote that this solution is "not general enough for the present business."[6] Euler was able to solve the equation using power series methods, but did not recognize a sine or cosine in the resulting answer.

It was not until 1739 that Euler realized that the sine function would enable closed-form solutions of such higher order equations to be given. On March 30 of that year, Euler presented a paper to the St. Petersburg Academy of Sciences in which he solved the differential equation of motion of a sinusoidally driven harmonic oscillator, that is, of an object acted on by two forces, one proportional to the distance and one varying sinusoidally with the time. The very statement of the problem is perhaps the earliest use of the sine as a function of time, and the resulting differential equation,

$$2a\,d^2s + \frac{s\,dt^2}{b} + \frac{a\,dt^2}{g}\sin\frac{t}{a} = 0$$

(where s represents position and t time), is the earliest use of that function in such an equation. There are two aspects of Euler's solution that are of interest. First, as a special case, he deleted the sine term and solved the equation $2a\,d^2s + \frac{s\,dt^2}{b} = 0$, or, after multiplying through by $b\,ds$, the equation $2ab\,ds\,d^2s = -s\,ds\,dt^2$. An integration with respect to s then gave $2ab\,ds^2 = (C^2 - s^2)\,dt^2$, or

$$dt = \frac{\pm\sqrt{2ab}\,ds}{\sqrt{C^2 - s^2}},$$

the differential equation for the arcsine (with the positive sign) or the arccosine (with the negative sign). Euler, since he was interested in the motion rather than the time, solved the arccosine equation for s instead of t: $s = C\cos(t/\sqrt{2ab})$, the first such explicit analytic solution on record. Second, to solve the general case, Euler postulated a solution of the form

---

BIOGRAPHY

Leonhard Euler (1707–1783)

Born in Basel, Switzerland, Euler showed his brilliance early, graduating with honors from the University of Basel when he was fifteen. Although his father preferred that he prepare for the ministry, Euler managed to convince Johann Bernoulli to tutor him privately in mathematics. The latter soon recognized his student's genius and persuaded Euler's father to allow him to concentrate on mathematics. In 1726 Euler was turned down for a position at the University, partly because of his youth. A few years earlier, however, Peter the Great of Russia, on the urging of Leibniz, had decided to create the St. Petersburg Academy of Sciences as part of his efforts to modernize the Russian state. Among the earliest members of the Academy, appointed in 1725, were Nicolaus II (1695–1726) and Daniel Bernoulli (1700–1782), two of Johann's sons with whom Euler had developed a friendship. Although there was no position in mathematics available in St. Petersburg in 1726, they nevertheless recommended him for the vacancy in medicine and physiology, a position Euler immediately accepted. (He had studied these fields during his time at Basel.)

In 1733, due to Nicolaus's death and Daniel's return to Switzerland, Euler was appointed the Academy's chief mathematician. Late in the same year he married Catherine Gsell with whom he subsequently had 13 children. The life of a foreign scientist was not always carefree in Russia at the time.

Nevertheless, Euler was able generally to steer clear of controversies, until the problems surrounding the succession to the Russian throne in 1741 convinced him to accept the invitation of Frederick II of Prussia to join the Berlin Academy of Sciences, founded by Frederick I also on the advice of Leibniz. He soon became director of the Academy's mathematics section and, with the publication of his texts in analysis as well as numerous mathematical articles, became recognized as the premier mathematician of Europe. In 1755 the Paris Academy of Sciences named him a foreign member, partly in recognition of his winning their biennial prize competition 12 times.

Ultimately, however, Frederick tired of Euler's lack of philosophical sophistication. When the two could not agree on financial arrangements or on academic freedom, Euler returned to Russia in 1766 at the invitation of Empress Catherine the Great, whose succession to the throne marked Russia's return to the westernizing policies of Peter the Great. With the financial security of his family now assured, Euler continued his mathematical activities even though he became almost totally blind in 1771. His prodigious memory enabled him to perform detailed calculations in his head. Thus, he was able to dictate his articles and letters to his sons and others virtually until the day of his sudden death in 1783 while playing with one of his grandchildren (Fig. 17.7).

FIGURE 17.7

Euler on a Swiss stamp

$s = u \cos(t/\sqrt{2ab})$, where u is a new variable. He then substituted that solution into the equation and solved for u. This manipulation showed that Euler was already familiar with the basic differentiation rules for the sine and cosine. These had, in effect, been known to Newton and Leibniz and, as will be noted below, had already appeared in several printed sources.

There is more to the story of the sine and cosine. On May 5, 1739, Euler wrote to Johann Bernoulli, noting that he had solved in finite terms the third-order equation $a^3 d^3 y = y dx^3$:

> Although it appears difficult to integrate, needing a triple integration and requiring the quadrature of the circle and hyperbola, it may be reduced to a finite equation; the equation of the integral is
>
> $$y = be^{x/a} + ce^{-(x/2a)} \sin \frac{(f+x)\sqrt{3}}{2a}$$
>
> . . . where b, c, f are arbitrary constants arising from the triple integration.[7]

Euler did not reveal how he discovered this solution, but a good guess would be that he used the known exponential solution $y = e^{x/a}$ to reduce the order of the equation. In this technique, which Euler had used earlier, one multiplies the original equation $a^3 d^3 y - y dx^3 = 0$ by $e^{-(x/a)}$ and assumes that this is the differential of $e^{-(x/a)}(A \, d^2 y + B \, dy \, dx + C y \, dx^2)$. It is then straightforward to show that a new solution of the original equation also satisfies the second-order equation $a^2 \, d^2 y + a \, dy \, dx + y \, dx^2 = 0$. To solve this latter equation requires a different Eulerian technique. Namely, one guesses a solution to be of the form $y = u e^{\alpha x}$ and substitutes this for y in the equation. Again, a bit of manipulation shows that the term $du \, dx$ can be eliminated by setting $\alpha = -1/2a$. The equation then reduces to $a^2 \, d^2 u + \frac{3}{4} u \, dx^2 = 0$, an equation of the same form as the one Euler solved in March. In this case, the solution is $u = C \sin((x + f)\sqrt{3}/2a)$, from which the general solution to the original third-order solution follows. Note that this reconstruction uses both the quadrature of the hyperbola (the exponential function, related naturally to the logarithm function defined as the area under a hyperbola) and the quadrature of the circle (the sine function, related to the arcsine function whose definition involves the area of the circle), as well as three integrations.

Since the sine and exponential functions had been used in the solution of the same differential equation, it was clear that Euler now considered the sine, and by extension, the other trigonometric functions, as functions in the same sense as the exponential function. But even more interesting, it was the very introduction of these functions into calculus that led Euler to the solution method for the class of linear differential equations with constant coefficients, that is, equations of the form

$$y + a_1 \frac{dy}{dx} + a_2 \frac{d^2 y}{dx^2} + a_3 \frac{d^3 y}{dx^3} + \cdots + a_n \frac{d^n y}{dx^n} = 0.$$

In a letter to Johann Bernoulli on September 15, 1739, Euler noted that "after treating this problem in many ways, I happened on my solution entirely unexpectedly; before that I had no suspicion that the solution of algebraic equations had so much importance in this matter."[8] Euler's "unexpected" solution was to replace the given differential equation by the algebraic equation

$$1 + a_1 p + a_2 p^2 + a_3 p^3 + \cdots + a_n p^n = 0$$

and factor this "characteristic polynomial" into its irreducible real linear and quadratic factors. For each linear factor $1 - \alpha p$, one takes as solution $y = A e^{x/\alpha}$, while for each irreducible quadratic factor $1 + \alpha p + \beta p^2$ one takes as solution

$$e^{-\frac{\alpha x}{2\beta}} \left(C \sin \frac{x\sqrt{4\beta - \alpha^2}}{2\beta} + D \cos \frac{x\sqrt{4\beta - \alpha^2}}{2\beta} \right).$$

The general solution is then a sum of the solutions corresponding to each factor. As an example, Euler solved the equation proposed by Daniel Bernoulli some four years earlier:

$$y - k^4 \frac{d^4 y}{dx^4} = 0.$$

The corresponding algebraic equation $1 - k^4 p^4$ factors as $(1 - kp)(1 + kp)(1 + k^2 p^2)$. Thus, the solution is

$$y = Ae^{-\frac{x}{k}} + Be^{\frac{x}{k}} + C \sin \frac{x}{k} + D \cos \frac{x}{k}.$$

Euler did not say how he arrived at his algebraic solution method. But since several months earlier he had discovered that trigonometric functions were involved in the solution to the equation $y - a^3 \frac{d^3 y}{dx^3} = 0$, one can surmise that he merely generalized that method. For given one solution of the equation, of the form $y = e^{x/\alpha}$, the reduction procedure indicated there provides in essence a factorization of the characteristic polynomial as $1 - a^3 p^3 = (1 - ap)(1 + ap + a^2 p^2)$. The general factorization method indicated in Euler's September letter would then have followed easily. In particular, it would have been clear that the sine and cosine terms come from the irreducible quadratic factors.

Johann Bernoulli was somewhat bothered by Euler's solution. He noted that the irreducible quadratic factors of the characteristic polynomial could be factored over the complex numbers, and thus that Euler's method led to complex roots of this polynomial being related to real solutions involving sines and cosines. Euler finally convinced Bernoulli late in 1740 that $2 \cos x$ and $e^{ix} + e^{-ix}$ were identical because they satisfied the same differential equation, and therefore that using imaginary exponentials amounted to the same thing as using sines and cosines.

17.1.4   Logarithms of Negative and Complex Numbers

The relationship between complex exponential functions on the one hand and sines and cosines on the other, expressed by the formulas

$$e^{ix} = \cos x + i \sin x \quad \text{and} \quad e^{-ix} = \cos x - i \sin x$$

coming from the equality of $2 \cos x$ and $e^{ix} + e^{-ix}$ soon led to the resolution of a controversy over the status of logarithms of negative numbers, a controversy that began with a series of letters between Johann Bernoulli and Leibniz in 1712–1713. Bernoulli had insisted that $\ln(-x) = \ln x$, because the rectangular hyperbolas used to define the logarithm had two branches. Leibniz, however, thought that the logarithm of a negative number was "impossible." Later, in letters to Euler in 1727, Bernoulli used arguments using differentials to bolster his case. Namely, because $d \ln(-x) = \frac{-1}{-x} \, dx = \frac{1}{x} \, dx = d \ln x$, then $\ln(-x) = \ln x$. Of course, because equality of differentials only implies equality of the quantities up to a constant, this argument also required that $\ln(-1) = 0$. But, after all, $0 = \ln 1 = \ln(-1)^2 = 2 \ln(-1)$, so that also seemed to be true.

Euler countered Bernoulli with various arguments, but in particular referred him to a paper that Bernoulli himself had written in 1702 in which he had noted that the differential $a \, dz/(b^2 + z^2)$ was not only the differential of a sector of a circle but also could be expressed as the sum of the two complex differentials $\frac{1}{2} a \, dz/(b^2 + bzi)$ and $\frac{1}{2} a \, dz/(b^2 - bzi)$. Although this result seems surprising, Bernoulli noted that "one sees that imaginary logarithms can be taken for real circular sectors because of the compensation which imaginary quantities make on being added together of destroying themselves in such a way that their sum is always real."[9] Given that the area of a sector of central angle θ in a circle with radius a is given by

$(a^2/2)\theta$, Euler was able to use Bernoulli's result, with $b = a$, to derive the formula that the area of the sector is

$$\frac{a^2}{4i} \ln \frac{x + yi}{x - yi},$$

where $x = \cos\theta$ and $y = \sin\theta$. To understand this derivation, note that if t is the tangent of θ in the circle of radius a (Fig. 17.8), then $t : a = x : y$ and $\theta = \arctan(t/a) = \arctan(y/x)$. Then

$$\frac{a^2}{2}\theta = \frac{a^2}{2} \arctan \frac{t}{a}$$

$$= \frac{a^2}{2} \int \frac{a\, dt}{a^2 + t^2} = \frac{a^2}{2} \int \left[\frac{1/2}{a + ti} + \frac{1/2}{a - ti} \right] dt$$

$$= \frac{a^2}{4} \left[\frac{1}{i} \ln(a + ti) - \frac{1}{i} \ln(a - ti) \right] = \frac{a^2}{4i} \ln \left[\frac{a + ti}{a - ti} \right]$$

$$= \frac{a^2}{4i} \ln \left[\frac{a + (ay/x)i}{a - (ay/x)i} \right] = \frac{a^2}{4i} \ln \left[\frac{x + yi}{x - yi} \right]$$

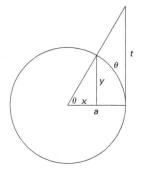

FIGURE 17.8

Sine, cosine, and tangent of an arc (or angle) θ in a circle of radius a

Using this result, Euler noted that the formula gave the area of the first quadrant of the circle when $x = 0$. Therefore,

$$\frac{\pi a^2}{4} = \frac{a^2}{4i} \ln(-1).$$

Thus, not only could $\ln(-1)$ not equal 0, but in fact $\ln(-1) = \pi i$. This result is equivalent to $e^{\pi i} + 1 = 0$, usually called Euler's Identity, although evidently Euler never wrote this down explicitly. But this was not the end of the story. In fact, Euler soon realized that there were infinitely many logarithms to any number, whether real or complex. And in 1747, he read a paper to the Berlin Academy explaining his ideas in full detail, using ideas similar to the derivation above.

Euler started by looking at an arbitrary arc ϕ in a circle of radius 1, with cosine x and sine y, so that $x = \sqrt{1 - y^2}$. Since $\phi = \arcsin y$, we have

$$d\phi = \frac{dy}{\sqrt{1 - y^2}}.$$

Now Euler again went to the complex number domain by letting $y = iz$, so that

$$d\phi = \frac{i\, dz}{\sqrt{1 + z^2}}.$$

Euler could easily integrate this differential:

$$\phi = \int \frac{i\, dz}{\sqrt{1 + z^2}} = i \ln(\sqrt{1 + z^2} + z) + C = i \ln \left(\sqrt{1 - y^2} + \frac{y}{i} \right) + C.$$

Because $\phi = 0$ when $y = 0$, it was clear that $C = 0$. Therefore,

$$\phi = i \ln(x - iy) = \frac{1}{i} \ln[(x - iy)^{-1}] = \frac{1}{i} \ln(x + iy).$$

But because the formula must hold for any arc ϕ with the same sine and cosine, he found that

$$\phi \pm 2n\pi = \frac{1}{i} \ln(x + iy) \quad \text{or} \quad \ln(x + iy) = i(\phi \pm 2n\pi).$$

As he then wrote, "from this it is clear that the same number $x + iy$ corresponds to infinitely many logarithms, which are all included in this general formula $i(\phi \pm 2n\pi)$, where in place of n we can put any whole number we wish."[10] He then rewrote the formula as

$$\ln(\cos \phi + i \sin \phi) = i(\phi \pm 2n\pi)$$

from which the exponential formulas follow immediately. Euler proceeded to use his formulas to find the infinitely many logarithms of many particular real and imaginary quantities. For example, $\ln(1) = \pm 2n\pi i$, $\ln(-1) = (1 \pm 2n)\pi i$, and $\ln(i) = (\frac{1}{2} \pm 2n)\pi i$. As he noted further, positive numbers have exactly one real logarithm, while all the logarithms of negative numbers are imaginary. He could then conclude that now "all the difficulties which might be encountered in this matter will disappear entirely, and the doctrine of the logarithms will be safeguarded from all attacks."[11]

17.1.5 The Calculus of Variations

Euler contributed heavily to another area of analysis related to differential equations, the calculus of variations. This subject grew out of consideration of problems in which the goal was to find a curve that maximizes or minimizes a particular integral. For example, the problem of the brachistochrone was to find a curve that minimized the time of descent under gravity, that is, a curve that minimized

$$I = \int dt = \int \frac{ds}{v} = \int \frac{\sqrt{1 + y'^2}\, dx}{\sqrt{2gy}},$$

where dt is the element of time, ds is the element of arclength, g is the acceleration due to gravity, $y = y(x)$ represents the curve, and $v = \sqrt{2gy}$ is the velocity of the body falling under the influence of gravity.

After studying this problem and many similar ones, Euler was able to combine them all into a general theory, that of determining a curve y that minimizes or maximizes an integral of the form $I(y) = \int_a^b F(x, y, y')\, dx$. This theory appeared in a few early papers and then in his classic work of 1744, *Methodus inveniendi lineas curvas maximi minimive proprietate gaudentes* (*Method of Finding Curved Lines That Show Some Property of Maximum or Minimum*). Euler's central idea was to use a polygonal approximation to the integral to develop a necessary condition for an extremal value. We give a brief outline of his method in modern notation.

Partition $[a, b]$ into n equal subintervals $[x_{i-1}, x_i]$, where $x_i = x_{i-1} + \Delta x$, $i = 1, \ldots, n$, and consider the polygonal curve connecting the points (x_i, y_i), where $y_i = y(x_i)$ (Fig. 17.9). Then $I(y)$ can be approximated by

$$I(y) \approx I(y_1, y_2, \ldots, y_{n-1}) = \sum_{i=0}^{n-1} F\left(x_i, y_i, \frac{y_{i+1} - y_i}{\Delta x}\right) \Delta x.$$

FIGURE 17.9

Euler's polygonal approximation to $y = y(x)$

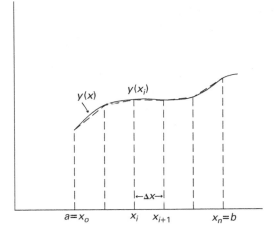

For the curve y to be an extremum, the derivative of I with respect to each y_i must be 0; that is, any change in the "corners" of the curve must result in the integral losing its extremal property. Thus, for each i from 1 to $n - 1$, we have the equation

$$
\begin{aligned}
0 &= \frac{1}{\Delta x} \frac{\partial I}{\partial y_i} \\
&= \frac{\partial F}{\partial y}\left(x_i, y_i, \frac{y_{i+1} - y_i}{\Delta x}\right) + \frac{\partial F}{\partial y'}\left(x_i, y_i, \frac{y_{i+1} - y_i}{\Delta x}\right)\left(\frac{-1}{\Delta x}\right) \\
&\quad + \frac{\partial F}{\partial y'}\left(x_{i-1}, y_{i-1}, \frac{y_i - y_{i-1}}{\Delta x}\right)\left(\frac{1}{\Delta x}\right) \\
&= \frac{\partial F}{\partial y}\left(x_i, y_i, \frac{y_{i+1} - y_i}{\Delta x}\right) - \left(\frac{1}{\Delta x}\right)\left[\frac{\partial F}{\partial y'}\left(x_i, y_i, \frac{\Delta y_i}{\Delta x}\right) - \frac{\partial F}{\partial y'}\left(x_{i-1}, y_{i-1}, \frac{\Delta y_{i-1}}{\Delta x}\right)\right] \\
&= \frac{\partial F}{\partial y}\left(x_i, y_i, \frac{\Delta y_i}{\Delta x}\right) - \frac{\Delta\left(\frac{\partial F}{\partial y'}\right)}{\Delta x}.
\end{aligned}
$$

When there are infinitely many subintervals and corners, these equations can be replaced by the single differential equation

$$
\frac{\partial F}{\partial y} - \frac{d}{dx}\left(\frac{\partial F}{\partial y'}\right) = 0,
$$

the fundamental necessary condition of the variational problem today known as the Euler equation. Euler himself used differential notation to write his equation: If F is a function of x, y, and $p = dy/dx$, so that $dF = M\,dx + N\,dy + P\,dp$, then the integral will have an extreme value when $N\,dy - p\,dP = 0$. But because $dy = p\,dx$, this can be rewritten as $N\,dx - dP = 0$ or as $N - dP/dx = 0$.

To illustrate the use of Euler's equation, we give two examples. First, it is well known that the shortest distance between two points is a straight line. To find this via the calculus of variations, we must minimize ds for any curve between the two points; that is, we must find y, which minimizes $I = \int \sqrt{1 + y'^2} \, dx$. Since $F(x, y, y') = \sqrt{1 + y'^2}$, and therefore $\partial F / \partial y = 0$, it follows that the Euler equation reduces to

$$\frac{d}{dx}\left(\frac{\partial F}{\partial y'}\right) = 0 \quad \text{or} \quad \frac{\partial F}{\partial y'} = c \quad \text{or} \quad \frac{y'}{\sqrt{1 + y'^2}} = c.$$

But this last equation reduces to $y' = a$, for some constant a, and therefore the desired curve is of the form $y = ax + b$, that is, a straight line.

Our second example, again one given by Euler, is the brachistochrone problem. In this case, since the function $F = \sqrt{1 + y'^2}/\sqrt{2gy}$ is only a function of y and y', it is easiest to modify Euler's equation to handle that situation. Note first that in that case

$$\frac{dF}{dx} = y'\frac{\partial F}{\partial y} + y''\frac{\partial F}{\partial y'}.$$

By Euler's equation,

$$\frac{\partial F}{\partial y} = \frac{d}{dx}\left(\frac{\partial F}{\partial y'}\right), \quad \text{so} \quad \frac{dF}{dx} = y'\frac{d}{dx}\left(\frac{\partial F}{\partial y'}\right) + y''\frac{\partial F}{\partial y'} = \frac{d}{dx}\left(y'\frac{\partial F}{\partial y'}\right).$$

It follows that

$$\frac{d}{dx}\left(F - y'\frac{\partial F}{\partial y'}\right) = 0 \quad \text{or} \quad F - y'\frac{\partial F}{\partial y'} = c.$$

In the case of interest, the last equation becomes

$$\frac{\sqrt{1 + y'^2}}{\sqrt{2gy}} - \frac{y'^2}{\sqrt{2gy}\sqrt{1 + y'^2}} = c \quad \text{or} \quad \frac{1}{\sqrt{2gy}\sqrt{1 + y'^2}} = c \quad \text{or} \quad \frac{1}{2gy(1 + y'^2)} = c^2.$$

If we set $a = 1/2gc^2$, this last equation can be rewritten as $1 + y'^2 = a/y$ or as $y'^2 = (a - y)/y$. Thus,

$$dy = \sqrt{\frac{a - y}{y}}\,dx \quad \text{or} \quad dx = dy\sqrt{\frac{y}{a - y}},$$

the same equation, with x and y interchanged, which Johann Bernoulli found for the cycloid.[12]

Euler eventually realized that his derivation of the Euler equation was not completely adequate nor did it give a sufficient condition for an extremum. But after he received a letter in 1755 from Joseph-Louis Lagrange (1736–1813) giving a better method for deriving his equation, he praised Lagrange, presented Lagrange's method to the Berlin Academy, and left it to the younger man to develop the field further.

17.2 THE CALCULUS OF SEVERAL VARIABLES

Functions of several variables had appeared in problems beginning in the late seventeenth century. In this section we consider both the differential and integral calculus of such functions.

17.2.1 The Differential Calculus of Functions of Two Variables

Although we have already seen partial derivatives appearing in several contexts, the earliest appearance of this notion was not, as one might expect from a modern perspective, in terms of surfaces defined by functions of two variables, but in terms of families $\{C_\alpha\}$ of curves. Such families had been initially considered in the early 1690s by Leibniz and then later by Johann Bernoulli. In the basic situation there are two infinitesimally close curves from a given family, C_α and $C_{\alpha+d\alpha}$, intersected by a third curve D defined geometrically in terms of that family. For example, D may be orthogonal to all members of the family. In such a situation, to find the differential equation of D or construct its tangent, it was necessary to consider three different differentials of the ordinate y. Let P, P' be points on C_α and Q, Q' points on $C_{\alpha+d\alpha}$ (Fig. 17.10). One differential of y is that between two points on a single curve of the family, say, $y(P') - y(P)$. This is the differential of y with x variable and α constant, designated by $d_x y$. A second differential is that between the y values of two corresponding points on the neighboring curves, say $y(Q) - y(P)$. This is the differential of y with α variable and x constant, designated by $d_\alpha y$. Differentiation using this differential was referred to as differentiation from curve to curve. Finally, there is a third differential, $y(Q') - y(P)$, denoted dy, which is the differential along the curve D.

FIGURE 17.10

Partial differentiation in a
family of curves

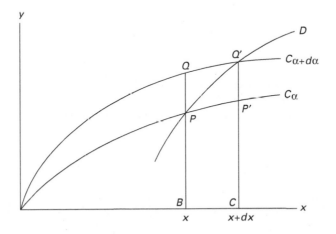

For Leibniz there was no difficulty in calculating either of the first two differentials, provided the curves were given algebraically. He could use his standard rules with either α or x being treated as a constant, in effect taking the partial derivatives with respect to x or α, respectively. Unfortunately, many interesting curves were not given algebraically but in terms of an integral. For example, the family of brachistochrone cycloids above was given by

$$C_\alpha = y(x, \alpha) = \int_0^x \sqrt{\frac{x}{\alpha - x}} \, dx.$$

(The modern notation of limits on the integral sign is being used here for clarity. Neither Leibniz nor the Bernoullis used such symbolism.) The question for Leibniz, brought home to him in a letter from Johann Bernoulli in July 1697, was how to perform the differentiation, with α variable, in a case like this.

After thinking about the problem for a few days while at the same time trying to meet with Czar Peter the Great of Russia, on his way from Berlin to the Netherlands, Leibniz succeeded in solving it. His solution went back to his own basic ideas of the calculus, namely, that the differential is the extrapolation of a finite difference and the integral that of a finite sum. Since for two sets of finite quantities, the sum of the differences of the parts is equal to the difference of the sums of the parts, Leibniz discovered what is called the **interchangeability theorem** for differentiation and integration, namely,

$$d_\alpha \int_b^x f(x, \alpha) \, dx = \int_b^x d_\alpha f(x, \alpha) \, dx.$$

(The modern formulation of this theorem is in terms of derivatives with respect to α rather than differentials.) Leibniz was then able to apply this result to determining the tangent to the curve D that cuts off equal arcs on the family of logarithmic curves $y(x, \alpha) = \alpha \log x$.

Because the arclength differential ds for the logarithmic curves is given by

$$ds = \frac{\sqrt{x^2 + \alpha^2} \, dx}{x},$$

the desired curve D is determined by the condition

$$s(x, \alpha) = \int_1^x \frac{\sqrt{x^2 + \alpha^2}}{x} \, dx = K,$$

where K is a constant. To determine the tangent, Leibniz needed to calculate the ratio of QB to QB' in the differential triangle of Figure 17.11. The first value is straightforward: $QB = d_\alpha \alpha \log x = \log x_0 \, d\alpha$, where x_0 is the x coordinate of B. The arc QB' equals the difference $AB - AQ$ since $AB = AB'$. Therefore,

$$QB' = d_\alpha s = d_\alpha \int_1^{x_0} \frac{\sqrt{x^2 + \alpha^2}}{x} \, dx.$$

By the interchangeability theorem, this integral is equal to

$$\int_1^{x_0} d_\alpha \frac{\sqrt{x^2 + \alpha^2}}{x} \, dx = \int_1^{x_0} \frac{\alpha}{x\sqrt{x^2 + \alpha^2}} \, d\alpha \, dx.$$

Because the limits of integration involve only x, however,

$$QB' = \alpha \, d\alpha \int_1^{x_0} \frac{dx}{x\sqrt{x^2 + \alpha^2}}.$$

FIGURE 17.11

Finding the tangent to the curve D, which cuts off equal arcs on the logarithmic curves $y = \alpha \log x$

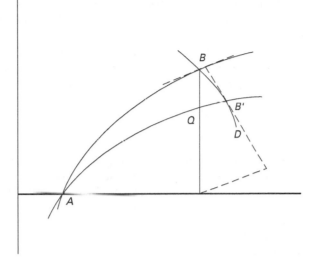

QB', now expressed by the area under a particular curve, could be considered as known; hence, so was the desired ratio.

Leibniz sent his solution to Bernoulli, who was certain that Leibniz's result would enable him to make significant progress in solving the problem of orthogonal trajectories for curves defined by integrals involving the parameter. Unfortunately, although he found the differential equation of the desired family of curves, the technical difficulties in integrating it were overwhelming, and Bernoulli was never able to develop a truly general method using Leibniz's result. Nevertheless, the interchangeability theorem became one of the foundations of the differential calculus of functions of two variables.

17.2.2 Total Differentials

Two other important aspects of the calculus of functions of two variables, the notion of a total differential and the equality of the mixed second-order differentials, were discovered by Nicolaus Bernoulli (1687–1759), a nephew of both Johann and Jakob, who lived in the shadow of his more famous uncles and chose to publish very little in mathematics. In an article in the *Acta eruditorum* of June 1719, Nicolaus discussed the orthogonal trajectory problem, but he did not provide the demonstrations until years later and then only in manuscript form. In that manuscript, Nicolaus asserted that the differential dy associated with a family of curves with parameter α is given by $dy = p\, dx + q\, d\alpha$, an expression he called the **complete differential equation** of the family, today known as the **total differential**. In this equation, p and q are functions of x, y, and α. Although Bernoulli did not say how he derived this result, one possible method was via a geometric argument. The differential dy along the curve D crossing the given family can be expressed by

$$dy = y(Q') - y(P) = [y(Q') - y(Q)] + [y(Q) - y(P)]$$
$$= d_x(y + d_\alpha y) + d_\alpha y = d_x y + d_x d_\alpha y + d_\alpha y.$$

Since the middle term, a second-order differential, is infinitely small compared to the other terms, the result is that $dy = d_x y + d_\alpha y$, or $dy = p\,dx + q\,d\alpha$ (see Fig. 17.10). As we will see later, it was Euler who eventually made the differential coefficients p and q themselves the fundamental concept of his calculus, replacing the differentials of Leibniz and the Bernoullis.

The equality of mixed second-order differentials followed from another geometrical argument based on Figure 17.10, one in which the line segment $P'Q'$ is found in two different ways. On the one hand, $P'Q' = PQ + d_x PQ = d_\alpha y + d_x d_\alpha y$. On the other, $P'Q' = d_\alpha C P'$, which is in turn equal to $d_\alpha (BP + d_x BP) = d_\alpha BP + d_\alpha d_x BP = d_\alpha y + d_\alpha d_x y$. Comparing the two expressions yields the result $d_x d_\alpha y = d_\alpha d_x y$. Interestingly enough, Nicolaus Bernoulli did not consider this argument as a proof but simply as an illustration of a result "which I thought to be obvious to anybody from the mere notion of differentials."[13] He used this result in his own work on orthogonal trajectories, but because his main arguments did not appear in his published work, they had little influence. The theorem on equality of mixed partial differentials was thus proved anew about 10 years later by Euler, who noted that this result and the interchangeability theorem, which both he and Nicolaus Bernoulli derived from the equality theorem, provided the basis for a theory of partial derivatives. Euler used it to give his own solutions to the same problems discussed by his predecessors of finding new curves defined geometrically in terms of given families of curves.

Euler developed another important idea in the solution of differential equations around 1740, one that had been found independently by Alexis-Claude Clairaut (1713–1765) in 1739—the condition for solvability of the homogeneous linear differential equation $P\,dx + Q\,dy = 0$, where P and Q are functions of x and y. If $P\,dx + Q\,dy$ is the total differential of a function $f(x, y)$, then the equality of the mixed second-order differentials shows that $\frac{\partial P}{\partial y} = \frac{\partial Q}{\partial x}$. More importantly, however, Clairaut demonstrated that this condition was sufficient for $P\,dx + Q\,dy$ to be a total differential. He asserted, in fact, that under that condition, the function $f(x, y)$ was given by $\int P\,dx + r(y)$, where r was a function of y to be determined. The differential of $\int P\,dx + r(y)$ was, by Leibniz's result, equal to $P\,dx + dy \int \frac{\partial P}{\partial y}\,dx + dr$. But since

$$\frac{\partial P}{\partial y} = \frac{\partial Q}{\partial x} \quad \text{and} \quad \int \frac{\partial Q}{\partial x}\,dx = Q + s(y),$$

the differential can be rewritten as $P\,dx + Q\,dy + dr + s\,dy$. Therefore, if r is chosen so that $dr = -s\,dy$, the differential becomes $P\,dx + Q\,dy$ as desired. Clairaut had thus reduced the original two-variable problem to an ordinary differential equation in one variable, an equation he assumed to be solvable. He also easily extended this "exactness" result to homogeneous linear equations in more than two variables.

17.2.3 Double Integrals

The subject of double integrals had its beginnings in Leibniz's solution to a challenge problem of Vincenzo Viviani (1622–1703) of April 4, 1692. Viviani, hiding behind the anagram D. Pio Lisci Pusillo Geometra (which stood for *Postremo Galileo Discipulo* (the last student of Galileo)), proposed the problem of determining four equal "windows" on a hemispherical surface such that the remainder of the surface was equal in area to a region constructible

---

BIOGRAPHY

Alexis Clairaut (1713–1765)

Clairaut, born in Paris, was a child genius who mastered l'Hospital's *Analyse des infiniment petits* by the age of 10 and read a paper at the Paris Academy of Sciences two years later. The research on curves that led to his book of 1731 was begun when he was 13 and ultimately led to his election to the Academy at the age of 18. Clairaut soon turned to celestial mechanics and later to pedagogy. His five major works in the former field proved quite influential, and his two texts on geometry and algebra (the first of which is discussed in Chapter 20) were attempts to introduce a historical, or "natural," approach to the teaching of these subjects.

---

by straightedge and compass. Leibniz solved the problem on May 27, 1692, the same day he received it. In doing so, he had to calculate areas of various regions on a hemisphere, for which he integrated expressions involving products of two differentials by integrating first with respect to one variable, with the second held constant, and then with respect to the second.

This problem and similar ones were solved somewhat later by the Bernoullis and l'Hospital, among others, but it was not until 1731 that a systematic attempt to calculate volumes of certain regions as well as the areas of their bounding surfaces was published by Clairaut in his *Recherches sur les courbes a double courbure* (*Research on Curves of Double Curvature*). Although Clairaut demonstrated that surfaces could in general be represented by a single equation in three variables, he most often considered cylindrical surfaces generated by a curve in one of the coordinate planes. Thus, to calculate the volume of a region between two cylinders given by $y = f(x)$, $z = g(y)$, he showed that the element of volume was given by $dx \int z \, dy$, then used his equations to rewrite z and dy in terms of x so that he could integrate $z \, dy$. With the volume element now given entirely in terms of x, he was able to integrate again to calculate the desired volume. Similarly, he expressed an element of surface area by $dx \int \sqrt{dx^2 + dy^2}$ and performed analogous calculations.

In a work on the calculus of variations in 1760, Joseph-Louis Lagrange (1736–1813) also had to deal with volumes and surface areas. Lagrange simply wrote $\iint z \, dx \, dy$ for the volume and $\iint dx \, dy \sqrt{1 + P^2 + Q^2}$ for the surface area, where the equation of the surface is given by $z = f(x, y)$ and where $dz = P \, dx + Q \, dy$. These notations occur without much discussion in both his letters of the 1750s to Euler and in his early papers, although he did note that the double integral signs indicated that two integrations must be performed successively.

It was only in a paper of 1769 that Euler gave the first detailed explanation of the concept of a **double integral**. For Euler, as for Leibniz and the Bernoullis, an "integral" was what we call an antiderivative. The use of these for finding areas was just an application. So Euler began by generalizing this notion of an integral to two variables. Thus, $\iint Z \, dx \, dy$ was to mean a function of two variables that when twice differentiated, first with respect to x alone, second with respect to y alone, gave $Z \, dx \, dy$ as the differential. For example, Euler noted that $\iint a \, dx \, dy = axy + X + Y$, where X is a function of x and Y a function of y. A more complicated example is provided by

$$\iint \frac{dx\,dy}{x^2 + y^2}.$$

The first integration can be performed with respect to either variable. Integrating each way, Euler found the values

$$\int \frac{dx}{x} \arctan \frac{y}{x} + X \quad \text{and} \quad \int \frac{dy}{y} \arctan \frac{x}{y} + Y.$$

Because the only way to perform the second integration, in either case, is by writing the integrand as a power series, Euler did so and showed that both integrations led to the same final result:

$$\iint \frac{dx\,dy}{x^2 + y^2} = X + Y - \frac{y}{x} + \frac{y^3}{9x^3} - \frac{y^5}{25x^5} + \cdots.$$

Given then the idea of a double integral as a double antiderivative, Euler generalized the concept of finding the area via a single integration to that of finding volumes using this double integration. His basic idea, like that of Leibniz, was to integrate with respect to one variable first, keeping the other constant, and then deal with the second one. His first example is to find the volume under one octant of the sphere of radius a whose equation is $z = \sqrt{a^2 - x^2 - y^2}$. Taking an element of area $dx\,dy$ in the first quadrant of the circle in the xy plane, Euler noted that the volume of the solid column above that infinitesimal rectangle is $dx\,dy\sqrt{a^2 - x^2 - y^2}$ (Fig. 17.12). To determine this volume, Euler first integrated with respect to y, holding x constant, to get

$$\left[\frac{1}{2}y\sqrt{a^2 - x^2 - y^2} + \frac{1}{2}(a^2 - x^2) \arcsin \frac{y}{\sqrt{a^2 - x^2}} \right] dx$$

as the volume under that piece of the sphere over the rectangle whose width is dx and whose length is y. Replacing y by $\sqrt{a^2 - x^2}$, Euler calculated the volume of the same piece up to that value of y to be $\frac{\pi}{4}(a^2 - x^2)\,dx$. Integrating with respect to x then gave $\frac{\pi}{4}(a^2 x - \frac{1}{3}x^3)$

FIGURE 17.12

The volume of a sphere

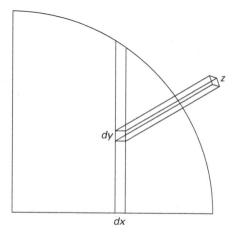

as the volume from the y axis to x, and replacing x by a gave the total volume of the octant as $\frac{\pi}{6}a^3$ and that of the whole sphere as $\frac{4\pi}{3}a^3$.

After showing further how to calculate volumes of solid regions bounded above by the sphere and below by various areas of the plane, Euler noted that a double integral may also be used to calculate surface area. He gave, without much discussion, the element of surface area of the sphere as

$$\frac{a\,dx\,dy}{\sqrt{a^2 - x^2 - y^2}},$$

presumably knowing the general formula given earlier by Lagrange. Furthermore, he also noted that $\iint dx\,dy$ over a region A was precisely the area of A.

17.2.4 Change of Variables in Multiple Integration

The most interesting part of Euler's paper on double integrals was his discussion of what happens to a double integral if the variables are changed. In other words, if x and y are given as functions of two new variables t and v, Euler wanted to determine how to transform the integral of $Z\,dx\,dy$ into a new integral with area element $dt\,dv$.

Euler realized that if the given transformation of variables is a translation followed by a rotation, that is, if

$$x = a + mt + v\sqrt{1 - m^2}$$
$$y = b + t\sqrt{1 - m^2} - mv,$$

where m is the cosine of the angle θ of the rotation, then the area elements $dx\,dy$ and $dt\,dv$ should be equal. But when he performed the obvious formal calculation

$$dx = m\,dt + dv\sqrt{1 - m^2}$$
$$dy = dt\sqrt{1 - m^2} - m\,dv$$

and multiplied the two equations together, he arrived at

$$dx\,dy = m\sqrt{1 - m^2}dt^2 + (1 - 2m^2)dt\,dv - m\sqrt{1 - m^2}dv^2,$$

a result that he noted was obviously wrong. Clearly, a similar calculation would also be wrong if t, v were related to x, y by a more complicated transformation. What Euler wanted to do, then, was to develop a method that in the above situation gave $dx\,dy = dt\,dv$ and in the more general case $dx\,dy = W\,dt\,dv$, where W is some function of t and v.

Euler's idea was to deal with one variable at a time, just as was done in double integration. Thus, he first introduced the new variable v such that y was a function of x and v. Then $dy = P\,dx + Q\,dv$. Assuming x fixed, Euler obtained $dy = Q\,dv$ and $\iint dx\,dy = \iint Q\,dx\,dv = \int dv \int Q\,dx$. Similarly, now letting x be a function of t and v so that $dx = R\,dt + S\,dv$ and holding v constant, he calculated that $\int dv \int Q\,dx = \int dv \int QR\,dt = \iint QR\,dt\,dv$. Thus, his initial solution to his problem was that $dx\,dy = QR\,dt\,dv$. This answer was not completely satisfactory, however, because Q may depend on x and because the method is not symmetric. Euler therefore continued by considering y as a function of t and v and calculating dy anew:

$$dy = P\, dx + Q\, dv = P(R\, dt + S\, dv) + Q\, dv = PR\, dt + (PS + Q)\, dv.$$

Because it was also true that $dy = T\, dt + V\, dv$, it followed that $PR = T$ and $PS + Q = V$ or that $QR = VR - ST$. Euler's final answer was that $dx\, dy = (VR - ST)dt\, dv$. He noted further that one must, in fact, use the absolute value of $VR - ST$ because area is a positive quantity. In modern notation Euler's result, expressed for double integrals of functions, is that

$$\iint f(x, y)dx\, dy = \iint f(x(t, v), y(t, v)) \left| \frac{\partial x}{\partial t} \frac{\partial y}{\partial v} - \frac{\partial x}{\partial v} \frac{\partial y}{\partial t} \right| dt\, dv,$$

where the domains over which the integrals are taken are related by the given functional relationship between (x, y) and (t, v).

As is typical of eighteenth-century mathematical proofs, Euler's argument was formal. And Lagrange used a similar formal argument to derive a change of variable formula in an integration in three dimensions. Neither man used any notion of a limit or an infinitesimal approximation or even worried whether there were points at which the relevant derivatives might not exist. But although Euler in particular was immensely successful in developing new mathematics through the use of such arguments, the lack of an axiomatic foundation on the model of Greek geometry bothered some of his contemporaries. Thus, a debate developed as to the proper basis of the chief concepts of the calculus, a debate we consider in Section 17.4.

17.2.5 Partial Differential Equations: The Wave Equation

Because many physical situations involve functions of two or more variables, it became necessary in the eighteenth century to deal with partial differential equations. Among the originators of this theory were Euler, Jean Le Rond d'Alembert (1717–1783), and Daniel Bernoulli. Here we discuss only one particular type of partial differential equation—the wave equation—because the debate over the subject of vibrating strings, from which the equation was derived, led not only to certain methods of solution but also to a new understanding of the notion of a function.

The discussion on the subject of vibrating strings began with a paper of d'Alembert written in 1747, in which he proposed a solution to the problem of the shape of a taut string placed in vibration. Because the position of a point on the string varies both with its abscissa and with time, this shape is determined by a function $y = y(t, x)$ of two variables. D'Alembert considered the string to be composed of an infinite number of infinitesimal masses and then used Newton's laws to derive the partial differential equation for y now called the **wave equation** and given in modern notation as

$$\frac{\partial^2 y}{\partial t^2} = c^2 \frac{\partial^2 y}{\partial x^2}.$$

D'Alembert then solved the equation for the special case $c^2 = 1$ in the form $y = \Psi(t + x) + \Gamma(t - x)$, where Ψ and Γ are arbitrary twice-differentiable functions. As d'Alembert pointed out, "this equation includes an infinity of curves."[14] The elaboration of that statement led to much controversy.

BIOGRAPHY

Jean d'Alembert (1717–1783)

D'Alembert was abandoned as an infant on the steps of a Parisian church by his mother, who had just renounced her nun's vows and feared retribution. He was soon adopted into a poor family, but his wealthy father provided him with a substantial annuity and helped him gain admission to the Collège Mazarin, where he received a classical education. Although he became a lawyer in 1738, his true interest was in mathematics, a subject he learned on his own. After publishing several papers, particularly in the area of differential equations, he was admitted to the Paris Academy in 1741 and soon became one of the leading mathematicians of Europe. His works include not only major treatises on dynamics and fluid mechanics but also, after 1750, many sections of the French *Encyclopédie*, the 28-volume work that aimed to set forth the basic principles of all the arts and sciences (Fig. 17.13).

FIGURE 17.13

D'Alembert on a French stamp

D'Alembert himself first discussed the case where $y = 0$ when $t = 0$ for every x, that is, where the string is in the equilibrium position for $t = 0$. He further required that $y = 0$ for $x = 0$ and $x = l$ for all t, that is, that the string be held fixed at the two ends of the interval $[0, l]$. The first requirement shows that $\Psi(x) + \Gamma(-x) = 0$, while the second gives the results $\Psi(t) + \Gamma(t) = 0$ and $\Psi(t + l) + \Gamma(t - l) = 0$. It follows that $\Gamma(t - x) = -\Psi(t - x)$ or that $y = \Psi(t + x) - \Psi(t - x)$; that $\Psi(-x) = -\Gamma(x) = \Psi(x)$ or that Ψ is an even function; and that $\Psi(t + l) = \Psi(t - l)$ or that Ψ is periodic of period $2l$. Furthermore, because the initial velocity is given by $\partial y / \partial t$ at $t = 0$, that is, by $v = \Psi'(x) - \Psi'(-x)$, and because the derivative of an even function is odd, d'Alembert concluded that "the expression for the initial velocity . . . must be such that when reduced to a series it includes only odd powers of x. Otherwise . . . the problem would be impossible."[15] In a paper written shortly afterward, d'Alembert generalized the solution to the case where the initial position of the string was given by $y(0, x) = f(x)$ and the initial velocity by $v(0, x) = g(x)$. In this case, he obtained the result that the solution is only possible if $f(x)$ and $g(x)$ are odd functions of period $2l$ and, in order that one could operate with these, that each function is given by a single analytic expression that is twice differentiable. For d'Alembert, then, a function was exactly that—an analytic expression, or what today is called a *formula*. Thus, even though $f(x)$ and $g(x)$ are given just on $[0, l]$, the function $y = \Psi(u)$ determining them must itself be given as a formula defined for all values of u. No other type of function could occur as a solution to a physical problem. As d'Alembert concluded in another paper three years later, "in any other case, the problem cannot be solved, at least by my method, and I am not certain whether it will not surpass the power of known analysis. . . . Under this supposition we find the solution of the problem only for the cases in which all the different figures of the vibrating string can be comprehended by one and the same equation [formula]. It seems to me that in all other cases it will be impossible to give y a general form."[16]

Two years after d'Alembert's initial paper, Euler published his own solution to the same problem, getting the same formal result as d'Alembert although with a somewhat different derivation. But Euler differed from d'Alembert in what kinds of initial position functions f could be permitted. First, he announced that f could be any curve defined on the interval $[0, l]$, even one that was not determined by an analytic expression. It could be a curve drawn by hand.

Thus, it was not necessary for the function to be differentiable at every point. Second, it was only the definition on the initial interval that was important. One could make the curve odd and periodic by simply defining it on $[-l, 0]$ by $f(-x) = -f(x)$ and then extending it to the entire real line by using $f(x \pm 2l) = f(x)$. After all, Euler reasoned contrary to d'Alembert, as far as the physical situation was concerned, the initial shape of the string could be arbitrary. Even if there were isolated points where the function was not differentiable, one could still consider the curve a solution to the differential equation, because behavior at isolated points was not, Euler believed, relevant to the general behavior of a function over an interval. Euler also felt that it was not essential to have a single formula defining the function everywhere.

Both d'Alembert and Euler, although they carried on their analyses in terms of general "functions," always had in mind the examples of the sine and cosine. The former was odd and periodic while the latter was even and periodic. In fact, in 1750 d'Alembert derived the solution $y = (A \cos Nt)(B \sin Nx)$ to the wave equation by the technique of separation of variables, that is, by assuming $y = f(t)g(x)$ and then differentiating. Nevertheless, it was the third participant in the debate, Daniel Bernoulli, who explicitly referred to combinations of sines and cosines in his attempt to bring the debate back to the reality of physical strings.

Bernoulli, whose positions at the University of Basel encompassed medicine, metaphysics, and natural philosophy, and whose chief work was in hydrodynamics and elasticity, wrote in a paper of 1753, "The calculations of Messrs. d'Alembert and Euler, which certainly contain all that analysis can have at its deepest and most sublime, . . . show at the same time that an abstract analysis which is accepted without any synthetic examination of the question under discussion is liable to surprise rather than enlighten us. It seems to me that we have only to pay attention to the nature of the simple vibrations of the strings to foresee without any calculation all that these two great geometers have found by the most thorny and abstract calculations that the analytical mind can perform."[17]

Bernoulli's more physical solution of the problem was to explore the idea that a vibrating string potentially represents an infinity of tones, each superimposed upon the others, and each being separately represented as a sine curve. It followed, although Bernoulli did not write out the result in this generality, that the movement of a vibrating string can be represented by the function

$$y = \alpha \sin \frac{\pi x}{l} \cos \frac{\pi t}{l} + \beta \sin \frac{2\pi x}{l} \cos \frac{2\pi t}{l} + \gamma \sin \frac{3\pi x}{l} \cos \frac{3\pi t}{l} + \cdots,$$

where the sum is infinite. The initial position function, over which Euler and d'Alembert quarreled, is then represented by the infinite sum

$$y(0, x) = \alpha \sin \frac{\pi x}{l} + \beta \sin \frac{2\pi x}{l} + \gamma \sin \frac{3\pi x}{l} + \cdots.$$

Interestingly enough, Euler had written these series in 1750, probably intending only a finite sum, as an example of a possible solution to the equation. Bernoulli believed that the latter series could represent any arbitrary initial position function $f(x)$ with the appropriate choice of constants $\alpha, \beta, \gamma, \ldots$ but could give no mathematical argument for the correctness of his view, only writing somewhat later that his representation provides an infinite number of constants that can be used for adjusting the curve to pass through an infinite number of specified points. His view was challenged by Euler, who not only could not see any way of determining these coefficients but also realized that for a function to be represented by such a trigonometric series it had to be periodic. By this argument, though, Euler showed himself to be caught between the older notion of function as formula and the newer view, which he was instrumental in helping evolve. Euler, after all, was willing to allow the arbitrary curve $f(x)$ defined over the interval $[0, l]$ to be extended by periodicity to the entire real line. But this was an example of what may be called geometric periodicity. It took no account of the algebraic expression by which f may be expressed. On the other hand, Euler's argument against Bernoulli was based on the algebraic periodicity of the trigonometric functions themselves on the whole real line. Euler had only an inkling of the modern notion of the domain of a function with the concomitant possibility that functions can be represented by different expressions on various parts of their domain.

The debate over the kinds of functions acceptable as solutions to the wave equation was continued through the next decades by these three mathematicians without any of them being convinced by any of the others. Although other mathematicians also entered the debate, it was not until the early years of the nineteenth century that a resolution of the problem was worked out through a complete analysis of the nature of trigonometric series.

 17.3

CALCULUS TEXTS

Although there were a few calculus texts written around the turn of the eighteenth century, including those discussed in Chapter 16, the middle third of the century saw many more, including both texts in the vernacular, for the education of laymen, and an important series of texts in Latin, designed for those with a university education. These books attempted to organize the topics just discussed, with the Continental texts dealing with the calculus of differentials and the British texts with the calculus of fluxions. To some extent, the ideas were mutually translatable, but it gradually became clear that it was much easier both to calculate and to discover new ideas using the Leibnizian approach. We now look at the highlights of a few of these texts, concentrating on the ideas that were special in each one.

17.3.1 Thomas Simpson's *Treatise of Fluxions*

In England a growing demand for mathematical knowledge by the middle class was met in part by a group of private teachers who wrote texts to supplement their instruction. A typical example was Thomas Simpson (1710–1761), whose earliest text, *A New Treatise of Fluxions*, was published in 1737 by subscription of his private students.

Simpson's *Treatise* is basically Newtonian in approach, making much use of infinite series to solve, in particular, problems in integration. It is replete with problems, many of which have become familiar to today's students. Thus, in an early section on maxima and minima, Simpson showed how to find the greatest parallelogram inscribed in a triangle, the smallest isosceles triangle that circumscribes a given circle, and the cone of least surface area with a given volume. Also included in this section is perhaps the earliest solution to the problem of determining the maximum of a function of several variables, namely, $w = (b^3 - x^3)(x^2z - z^3)(xy - y^2)$. Although Simpson did not use the language of partial derivatives, he did calculate the relationships of the fluxions of w to that of each of x, y, and z separately, holding the other two variables constant, before setting $\dot{w}$ equal to 0 in each case and solving the resulting equations simultaneously. In a later section of the book, buried in a

---

BIOGRAPHY

Thomas Simpson (1710–1761)

Born in the village of Market Bosworth, not far from Birmingham, Simpson was raised by his father to become a weaver. Thomas's thirst for a better education, however, led to a rift with his father, and he was forced to leave home. By the age of 25 he had learned mathematics on his own, including material on the calculus from an English translation of l'Hospital's work. In 1735 he moved to London, where he joined the Mathematical Society at Spitalfields, a weaving community in what is now a residential suburb. The rules of this society made it the duty of every member "if he be asked any mathematical or philosophical question by another member, to instruct him in the plainest and easiest manner he is able."[18] Through his activity in the society, Simpson became a teacher of mathematics and was soon able to give up weaving and bring his family to London. His reputation in mathematics, enhanced by the publication of several textbooks, finally enabled him in 1743 to secure a position as professor of mathematics at the Royal Military Academy at Woolwich, a school founded to provide military cadets with sufficient mathematical education to succeed as engineers. Shortly thereafter, Simpson was elected to the Royal Society.

---

problem on navigation, is probably the first publication in a text of the rule for differentiating the sine. In Simpson's words, "the fluxion of any circular arch is to the fluxion of its sine, as radius to the cosine."[19] The proof, given over 20 years earlier in a paper by Roger Cotes (1682–1716), the editor of the second edition of Newton's *Principia*, uses similar triangles but is different from Newton's own proof from the 1680s. If z denotes the arc of a circle of radius An centered on A, $x = Ab$ the sine of z, and bn its cosine, then the differential triangle nrm, whose hypotenuse nr represents $\dot{z}$ (the fluxion of the arc) and whose side mr represents $\dot{x}$ (the fluxion of the sine), is similar to triangle Anb (Fig. 17.14). It follows that $\dot{z} : \dot{x} = An : bn$, the result quoted. In modern notation, taking $An = 1$, this can be written as $d(\sin z)/dz = \cos z$.

FIGURE 17.14

The fluxion of the sine

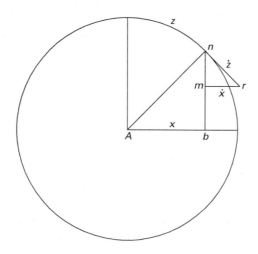

BIOGRAPHY

Colin Maclaurin (1698–1746)

Maclaurin, unlike Thomas Simpson, had both a university education and a university career. Born in Kilmodan, a village in western Scotland, he entered the University of Glasgow at the age of 11 and soon mastered the university mathematics curriculum. At the age of 19, Maclaurin was appointed to a chair of mathematics at the University of Aberdeen, but shortly thereafter left for a three-year tour of Europe as tutor to the son of a wealthy lord. The authorities at Aberdeen were not particularly happy with his absence and soon after his return forced him to resign. Meanwhile, Newton had recommended him for a position at Edinburgh, where he remained, teaching subjects ranging from Euclid and elementary algebra to fluxions and Newton's *Principia*, for the rest of his life. In 1745, Maclaurin helped to fortify Edinburgh against the forces of Bonnie Prince Charlie, but when the city fell, he left for York. He fell ill there and never recovered, dying at the age of 48.

Simpson is most famous today for the rule for numerical integration by parabolic approximation that bears his name. This rule appears not in his calculus text but in his *Mathematical Dissertations on a Variety of Physical and Analytical Subjects* of 1743. It is, however, not original to Simpson, having appeared in the works of other authors even in the seventeenth century.

17.3.2 Colin Maclaurin's *Treatise of Fluxions*

The name of the Scottish mathematician, Colin Maclaurin (1698–1746), is also known to today's students from a concept in the calculus text not original to him, the Maclaurin series. The series is found in Maclaurin's *A Treatise of Fluxions*, which appeared in 1742 partly in response to the criticism of the foundations of the theory of fluxions voiced by George Berkeley (1685–1753) eight years earlier. (A discussion of this criticism is found in Section 17.4.1.) Book I of this work treated the foundations of the Newtonian calculus from a geometric point of view. But in Book II, Maclaurin had a different agenda, to demonstrate the rules of fluxions and their applications in an algebraic and algorithmic manner. Maclaurin thus provided details of the entire range of problems to which the calculus was being applied. He discussed maxima and minima and points of inflection; he found tangent lines and asymptotes; he determined curvature; and he gave a complete account of the brachistochrone problem. Maclaurin calculated areas under curves given by y in terms of x by showing that the fluxion of this area was $y\dot{x}$ and then using one of several methods to determine the fluent of this expression. Similarly, he calculated volumes and surface areas of solids of revolution by first determining their fluxions. He used an elementary form of multiple integration to study the gravitational attraction of ellipsoids. Finally, in dealing with logarithms, Maclaurin began with Napier's original definition in terms of motion. It was then easy for him to determine the standard fluxional properties of the logarithm and to use these properties to calculate the fluxions of exponential functions.

Maclaurin's work contained somewhat more about trigonometric functions than Simpson's or any earlier work. For example, in addition to Simpson's theorem on sines, he showed geometrically that the fluxion of an arc is to the fluxion of its tangent in the duplicate ratio of the radius to the cosine and that the fluxion of an arc is to the fluxion of its secant as the square on the radius is to the rectangle determined by the secant and the tangent. Although these results can be translated into modern calculus theorems ($\frac{d}{dx} \tan x = \frac{1}{\cos^2 x}$; $\frac{d}{dx} \sec x = \sec x \tan x$), for Maclaurin they only gave ratios of fluxions in relation to those of the line segments representing the trigonometric functions. These results were not applied analytically as were those on the calculus of logarithmic and exponential functions. Maclaurin's results involving inverse trigonometric functions, however, were so applied. They appeared in the context of finding fluents (or integrals) of given fluxions. Thus, Maclaurin noted that the fluent of $\dot{y}/(a^2 + y^2)$ was the arc whose tangent was y in a circle of radius a, while the fluent of $a\dot{y}/\sqrt{a^2 - y^2}$ was the arc whose sine was y. More interestingly, however, Maclaurin, like Johann Bernoulli earlier, also realized that a minor change in the function changed the fluent from a circular arc to a logarithm. For example, a change in the sign of y^2 in the first problem changes the fluent to $(1/2a) \log[(a + y)/(a - y)]$. Although this result showed Maclaurin that circular arcs could be represented by imaginary logarithms, he was unable to derive the consequences that Euler did from the same idea.

The series named for Maclaurin also occurs in Book II: Suppose that y is expressible as a series in z, say, $y = A + Bz + Cz^2 + Dz^3 + \cdots$. If $E, \dot{E}, \ddot{E}, \ldots$ are the values of y and its fluxions of various orders when z vanishes, then the series can be expressed in the form

$$ y = E + \dot{E}z + \frac{\ddot{E}z^2}{1 \times 2} + \frac{\dddot{E}z^3}{1 \times 2 \times 3} + \cdots $$

(with the assumption that $\dot{z} = 1$). Maclaurin's proof was easy, given his assumption that y can be written in a power series. Namely, he first set $z = 0$ to get $A = E$. Next, he took the fluxion of the series and again set $z = 0$. It follows that $B = \dot{E}/\dot{z} = \dot{E}$. He continued to take fluxions and set $z = 0$ to complete the result. Maclaurin noted that this theorem had already been discovered by Brook Taylor and published in his *Methodus incrementorum* (*Method of Increments*) in 1715.

Maclaurin worked out many examples of these series, including the series for the sine and cosine in a circle of radius a. For example, if $y = \cos z$ (in a circle of radius a), then $\frac{\dot{y}}{\dot{z}} = \sqrt{a^2 - y^2}/a$. It follows that

$$ \frac{\dot{y}^2}{\dot{z}^2} = \frac{a^2 - y^2}{a^2} $$

and that

$$ \frac{2\dot{y}\ddot{y}}{\dot{z}^2} = -\frac{2y\dot{y}}{a^2} \quad \text{or} \quad \frac{\ddot{y}}{\dot{z}^2} = -\frac{y}{a^2}. $$

Therefore, since $y = a$ when $z = 0$, we get $E = a$, $\dot{E} = 0$ and $\ddot{E} = -\frac{1}{a}$. The first three terms of the series for $y = \cos z$ are then $y = a + 0z - \frac{1}{2a}z^2$. More terms are easily found without any necessity for calculating derivatives of sines and cosines.

Maclaurin also used his series for developing the standard derivative tests for determining maxima and minima: "When the first fluxion of the ordinate vanishes, if at the same time its second fluxion is positive, the ordinate is then a minimum, but is a maximum if its second fluxion is then negative."[20] If the ordinate $AF = E$ and two values of the abscissa, one to the

FIGURE 17.15

Maclaurin and the second
derivative test

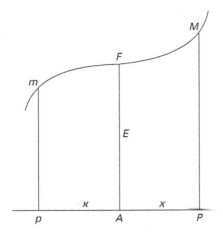

right of A (designated x) and one the same distance to the left (designated $-x$), are given
(Fig. 17.15), the Maclaurin series shows that the corresponding ordinates are

$$PM = E + \dot{E}x + \frac{\ddot{E}x^2}{2} + \frac{\dddot{E}x^3}{6} + \cdots$$

and

$$pm = E - \dot{E}x + \frac{\ddot{E}x^2}{2} - \frac{\dddot{E}x^3}{6} + \cdots.$$

Assuming that $\dot{E} = 0$ and that x is small enough, Maclaurin concluded that both of these
ordinates will exceed the ordinate $AF = E$ when $\ddot{E}$ is positive (so that AF is a minimum)
and both will be less than AF when $\ddot{E}$ is negative (so that AF is a maximum). Furthermore,
Maclaurin concluded that if $\ddot{E}$ also vanishes, and if $\dddot{E}$ does not, then either $PM > AF$ and
$pm < AF$ or vice versa so that AF is neither a maximum nor a minimum.

Maclaurin ended his text with probably the earliest analytic proof of part of the fundamental theorem of calculus, at least for the special case of power functions. (He had given a more
general geometric proof earlier in the text.) "Supposing n to be any [positive] integer, . . .
if the area upon the base AP or x is always equal to x^n, then the ordinate PM or y shall
be always equal to nx^{n-1}."[21] Maclaurin began by taking an increment $o = Pp$ of the base
x and noting that $PM \times Pp = yo < $ area $PMmp = (x + o)^n - x^n$ (Fig. 17.16). Because
$(x + o)^n - x^n < n(x + o)^{n-1}o$ by an algebraic result he had proved earlier, it followed that
$yo < n(x + o)^{n-1}o$. Similarly, using a value of the abscissa to the left of P, Maclaurin found
that $yo > n(x - o)^{n-1}o$ and therefore that $n(x - o)^{n-1} < y < n(x + o)^{n-1}$. Rather than use
a modern limit argument to show that $y = nx^{n-1}$, Maclaurin used a *reductio* argument. If
$y > nx^{n-1}$, then $y = nx^{n-1} + r < n(x + o)^{n-1}$ for any increment o. But if o is chosen to
be $(x^{n-1} + \frac{r}{n})^{1/(n-1)} - x$, a brief calculation shows that $y = n(x + o)^{n-1}$, a contradiction.
A similar contradiction occurs if y is assumed less than nx^{n-1}, and Maclaurin's proof was
complete.

FIGURE 17.16

Maclaurin and the fundamental theorem of calculus

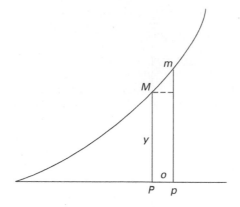

17.3.3   Maria Agnesi's *Instituzioni Analitiche*

Maclaurin's *Treatise of Fluxions* was read on the European continent, especially after it was translated into French in 1749. The year before, however, the first important successor to l'Hospital's text appeared in Europe, the *Instituzioni analitiche ad uso della gioventu italiana* (*Foundations of Analysis for the Use of Italian Youth*), by Maria Gaetana Agnesi (1718–1799). Agnesi's text, not surprisingly, showed the influence of Leibniz and his followers rather than Newton. Thus, it was written in the language of differentials and infinitesimals rather than that of fluxions. (Interestingly enough, the English translator (in 1801) replaced all of the dx's by $\dot{x}$'s, although he often kept the word "differential" rather than replace it by "fluxion.") Because Agnesi wrote it originally to instruct her younger brothers in the subject, the text explained concepts clearly and provided numerous examples. Thus, in her section on maxima and minima, Agnesi presented such problems as that of cutting a line at a point so that the product of the length of one segment and the square of the other is maximal and that of finding the line segment of minimum length that passes through one vertex of a rectangle and intersects the extensions of both of the opposite sides. She even showed how to find the point of maximal curvature on the logarithmic curve, that defined by $a\frac{dy}{y} = dx$.

For Agnesi, as also for Johann Bernoulli earlier and Leonhard Euler later, the integral calculus is the inverse of the differential calculus, that is, the method of determining, from a given differential expression, the quantity of which that expression is the differential. Thus, the symbol $\int y\,dx$ means an antiderivative. But the symbolism of the integrand, that $y\,dx$ represents the area of an infinitesimal rectangle, led Agnesi, virtually as an afterthought, to note that areas under curves can be calculated by this same inverse process.

Agnesi was especially thorough in her treatment of the logarithmic (exponential) curve. She noted that the ordinary rule for integration leads from $dx = ay^{-1}dy$ to $x = ay^{-1+1}/(-1+1)$ or $ay^0/0$ and that this "teaches us nothing." Thus, she dealt with this curve in other ways. She showed first that the curve whose ordinates increase geometrically while the abscissas increase arithmetically has the differential equation $dx = ay^{-1}dy$ and then that one can make computations by using appropriate infinite series. She also showed how to find the area under this curve, both over a finite interval and over an infinite one stretching to the left from a fixed abscissa x. She calculated this "improper integral" (today written as $\int_{-\infty}^{x} e^{t/a}\,dt$) to be

BIOGRAPHY

Maria Agnesi (1718–1799)

Agnesi was the eldest child of a wealthy Milanese merchant, who encouraged his daughter to pursue scientific interests by hiring various distinguished professors to tutor her. By the age of 11, she was fluent in seven languages and in her teens was able to dispute important matters in such fields as mechanics, logic, zoology, and mineralogy with the best scholars of the day. Having studied the major mathematical works of the time, she began to instruct her younger brothers in the subject. She soon decided that her work, including material on algebra as well a complete treatise on the differential and integral calculus, should be published to benefit all Italian youth. This text was so clearly written that a committee of the French Academy, in authorizing its translation into French in 1749, noted that "there is no other book, in any language, which would enable a reader to penetrate as deeply, or as rapidly, into the fundamental concepts of analysis."[22] And John Colson, the Lucasian Professor of Mathematics at Cambridge in the middle of the eighteenth century, was so impressed with the book that he learned Italian for the sole purpose of translating the work into English so that British youth would have the same benefits as those of Italy.

The Pope, too, recognized Agnesi's talents and appointed her to the chair of mathematics at the University of Bologna, but Agnesi never assumed the position. Soon after her father's death in 1752, she withdrew from all scientific pursuits and spent the rest of her life in religious studies and social work among the poor.

ay, where y is the ordinate corresponding to x, and also calculated the volume of the solid generated by revolving the curve around the x axis. But as in most of the texts of the first half of the eighteenth century, there was little concerning trigonometric functions.

Curiously, Agnesi's name, like those of the two other textbook authors discussed above, is attached to a small item in her book not even original to her. As an example in analytic geometry, she described geometrically a curve whose equation she determined to be

$$y = \frac{a\sqrt{a - x}}{\sqrt{x}},$$

a curve that had earlier been named *la versiera*, derived from the Latin meaning "to turn." Unfortunately, the word *versiera* was also the abbreviation for the Italian word *avversiera*, meaning "wife of the devil." Because the English translator rendered this word as "witch," the curve has ever since been referred to as the "witch of Agnesi."

17.3.4   Euler's *Introductio*

The three texts discussed so far were all written in the vernacular. It was a series of Latin texts, however, that proved more important for the future. These were the works of Euler, the two volumes of the *Introductio in analysin infinitorum* (*Introduction to Analysis of the Infinite*) (published in 1748 but written several years earlier), the *Institutiones calculi differentialis* (*Methods of the Differential Calculus*) (1755), and the three volumes of the *Institutiones calculi integralis* (*Methods of the Integral Calculus*) (1768–1770).

The *Introductio*, Euler's "precalculus" text, was an attempt to develop those topics "which are absolutely required for analysis" so that the reader "almost imperceptibly becomes acquainted with the idea of the infinite."[23] (Here we will just deal with ideas from this text important for analysis, leaving the discussions of purely algebraic material and the analytic geometry for later chapters.) The "idea of the infinite" is certainly critical to the study of the calculus itself. But, since for us and for Euler, analysis is concerned with functions, Euler began his work with that topic. In fact, making functions the central topic of the book represented a change in viewpoint in the history of analysis. Newton and Leibniz, in their development of calculus, dealt with "curves." And recall that the title of l'Hospital's calculus text was, after all, *Analysis of Infinitely Small Quantities for the Understanding of Curves*. Euler, in his studies of differential equations in the 1730s, had gradually come to the conclusion that the notion of "function" should be the basis of analysis. And so chapter 1 of the *Introductio* opens with a definition of the term: "A **function** of a variable quantity is an analytic expression composed in any way whatsoever of the variable quantity and numbers or constant quantities."[24] The first point to note about Euler's definition is that the word "function" means an "analytic expression," that is, a formula. Thus, Euler's definition is very similar to that of d'Alembert. Next, Euler was dealing with a "variable quantity," that is, "one which can take on any value." In fact, Euler frequently included complex numbers as well as real numbers under the term "any value." Finally, Euler's statement as to how his formulas are to be formed, "in any way whatsoever," can only be understood by considering his further discussion.

Although Euler had thus changed the central idea of analysis from curves to functions, he of course understood the relationship between the two concepts. Namely, any function can be translated into geometry to determine a curve in the plane. But curves and functions were not identical, partly because curves were not always represented by a single function. In fact, Euler divided curves into two classes, continuous and discontinuous: "A **continuous** curve is one such that its nature can be expressed by a single function of x. If the curve is of such a nature that for its various parts . . . different functions of x are required for its expression, that is, after one part . . . is defined by one function of x, then another function is required to express the [next] part . . . , then we call such a curve **discontinuous**. . . . This is because such a curve cannot be expressed by one constant law, but is formed from several continuous parts."[25] Thus, in the debate between d'Alembert and Euler over the initial conditions of the wave equation, it turned out that the former would only allow "continuous" curves while the latter was willing to admit discontinuous ones.

Euler also divided the set of functions into two basic classes, algebraic and transcendental. The former are formed from the variables and constants by addition, subtraction, multiplication, division, raising to a power, extraction of roots, and the solution of an equation. The latter are those defined by exponentials, logarithms, and, more generally, by integrals. Because integrals could not be discussed in a precalculus work, the transcendental functions discussed in the *Introductio* are limited to the special cases of trigonometric, exponential, and logarithmic functions. Furthermore, as the remainder of the *Introductio* showed, "in any way whatsoever" includes the notions of infinite series, infinite products, and infinite continued fractions. Thus, an important tool in Euler's discussion of functions is that of a power series.

Euler was convinced that any function could be expressed, except perhaps at isolated points, by a power series, but gave no proof. Rather, he attempted to convince the reader of this truth by showing how to expand any algebraic function as well as various transcendental functions into such a series. His methods for algebraic functions were not new, being a combination of Newton's methods using division (in the case of rational functions) and the binomial theorem (for functions expressible in terms of any powers). And there is no discussion of convergence.

The central chapters of Book I of the *Introductio*, and those that were to prove most influential, are the chapters dealing with the exponential, logarithmic, and trigonometric functions, for it is there that Euler introduced the notations and concepts that were to make obsolete all the discussions of such functions in earlier texts. All modern treatments of these functions are in some sense derived from those of Euler. Thus, Euler defined exponential functions as powers in which exponents are variable and then—and this is a first—defined logarithms in terms of these. Namely, if $a^z = y$, Euler defined z to be the logarithm of y with base a. He then derived the basic properties of the logarithm function from those of the exponential. Using these properties, Euler showed how to compute a logarithm, $\log_{10} 5$. He began with $A = 1$, $B = 10$, calculated $C = \sqrt{AB} = \sqrt{10}$, and then noted that $\log_{10} C = \frac{1}{2}(\log_{10} A + \log_{10} B) = 0.5$. Similarly, he calculated $D = \sqrt{BC}$ so that $\log_{10} D = 0.75$. He continued in this manner for 26 steps (when the letters ran out) until he finally reached $Z = 5.000000$ and $\log_{10} Z = 0.6989700$. Of course, as Euler was to show shortly, there are more efficient ways of calculating logarithms.

After dealing with various functional aspects of the logarithm and exponential, Euler next developed their power series for an arbitrary base a by use of the binomial theorem. His technique made important use of both "infinitely small" and "infinitely large" numbers, concepts whose use today is frowned upon. Nevertheless, Euler rarely erred. For example, he noted that since $a^0 = 1$, it follows that $a^\omega = 1 + \psi$, where both ω and ψ are infinitely small. Therefore, ψ must be some multiple of ω, depending on a, and

$$a^\omega = 1 + k\omega \quad \text{or} \quad \omega = \log_a(1 + k\omega).$$

Euler noted next that for any j, $a^{j\omega} = (1 + k\omega)^j$, and, expanding the right side by the binomial theorem, that

$$a^{j\omega} = 1 + \frac{j}{1}k\omega + \frac{j(j-1)}{1 \cdot 2}k^2\omega^2 + \frac{j(j-1)(j-2)}{1 \cdot 2 \cdot 3}k^3\omega^3 + \cdots.$$

If j is taken to equal z/ω, where z is finite, then j is infinitely large and $\omega = z/j$. The series now becomes

$$a^z = 1 + \frac{1}{1}kz + \frac{1(j-1)}{1 \cdot 2j}k^2z^2 + \frac{1(j-1)(j-2)}{1 \cdot 2j \cdot 3j}k^3z^3 + \cdots.$$

Because j is infinitely large, $(j-n)/j = 1$ for any positive integer n. The expansion then reduces to the series

$$a^z = 1 + \frac{kz}{1} + \frac{k^2z^2}{1 \cdot 2} + \frac{k^3z^3}{1 \cdot 2 \cdot 3} + \cdots,$$

where k depends on the base a. Euler also noted that the equation $\omega = \log_a(1 + kw)$ implies that if $(1 + kw)^j = 1 + x$, then $\log_a(1 + x) = j\omega$. Since $k\omega = (1 + x)^{1/j} - 1$, it follows that

$$\log_a(1 + x) = \frac{j}{k}(1 + x)^{\frac{1}{j}} - \frac{j}{k}.$$

Another clever use of the binomial theorem finally allowed Euler to derive the series

$$\log_a(1+x) = \frac{1}{k}\left(\frac{x}{1} - \frac{x^2}{2} + \frac{x^3}{3} + \cdots\right).$$

The choice of $k = 1$ or, equivalently, $a = e$, gives the standard power series for e^z and $\ln(1 + x)$. The latter series, and one that is easily derived from it, namely,

$$\ln\left(\frac{1+x}{1-x}\right) = \frac{2x}{1} + \frac{2x^3}{3} + \frac{2x^5}{5} + \cdots,$$

can then be used, as promised earlier, to calculate logarithms efficiently to the base e—and Euler proudly displayed values for the logarithms of the first 10 positive integers to 25 decimal places.

Euler's treatment of "transcendental quantities which arise from the circle" is the first textbook discussion of the trigonometric functions that deals with these quantities as functions having numerical values, rather than as lines in a circle of a certain radius. Euler did not, in fact, give any new definition of the sine and cosine. He merely noted that he would always consider the sine and cosine of an arc z to be defined in terms of a circle of radius 1. All basic properties of the sine and cosine, including the addition and periodicity properties, are assumed known, although Euler did derive some relatively complicated identities. More importantly, he derived the power series for the sine and cosine through use of the binomial theorem and complex numbers.

From the easily derived identity $(\cos z \pm i \sin z)^n = \cos nz \pm i \sin nz$, Euler concluded that

$$\cos nz = \frac{(\cos z + i \sin z)^n + (\cos z - i \sin z)^n}{2}$$

and, by expanding the right side, that

$$\cos nz = (\cos z)^n - \frac{n(n-1)}{1 \cdot 2}(\cos z)^{n-2}(\sin z)^2$$
$$+ \frac{n(n-1)(n-2)(n-3)}{1 \cdot 2 \cdot 3 \cdot 4}(\cos z)^{n-4}(\sin z)^4 + \cdots.$$

Again letting z be infinitely small, n infinitely large, and $nz = v$ finite, it follows from $\sin z = z$ and $\cos z = 1$ that

$$\cos v = 1 - \frac{v^2}{1 \cdot 2} + \frac{v^4}{1 \cdot 2 \cdot .3 \cdot 4} - \cdots.$$

Similarly, Euler derived the power series for the sine.

Virtually as an aside, Euler then derived the formulas relating complex exponentials to sines and cosines: $e^{\pm iv} = \cos v \pm i \sin v$, and used these to develop the classic power series for the arctangent:

$$\arctan t = \frac{t}{1} - \frac{t^3}{3} + \frac{t^5}{5} - \frac{t^7}{7} + \cdots.$$

Noting that this series implies that $\pi/4 = 1 - 1/3 + 1/5 - 1/7 + \cdots$ but that this series "hardly converges," he then manipulated with the tangent function to give a much more rapidly converging series for $\pi/4$, namely, $\pi/4 = \arctan(1/2) + \arctan(1/3)$. Curiously, however, he did not use the formulas to show how to calculate logarithms of negative or imaginary quantities. Evidently, his calculation of these quantities, discussed in Section 17.1.4, was completed after the writing of the *Introductio*.

The remainder of Volume 1 of the *Introductio* includes much else about infinite processes, including infinite products as well as infinite series. For example, Euler considered the equality

$$\prod_p \frac{1}{1 - \frac{1}{p^n}} = \sum_m \frac{1}{m^n},$$

where the product is taken over all primes and the sum over all positive integers. This product and sum, both generalized to the case where n is any complex number s, are today called the **Riemann zeta function** of the variable s, the study of which has led to much new mathematics. It is also in terms of products and factors that the hyperbolic functions appear, although they are not named. Thus, Euler showed that

$$\frac{e^x - e^{-x}}{2} = \frac{x}{1} + \frac{x^3}{1 \cdot 2 \cdot 3} + \frac{x^5}{1 \cdot 2 \cdot 3 \cdot 4 \cdot 5} + \cdots$$

may be factored as

$$\frac{e^x - e^{-x}}{2} = x \left(1 + \frac{x^2}{\pi^2}\right) \left(1 + \frac{x^2}{4\pi^2}\right) \left(1 + \frac{x^2}{9\pi^2}\right) \cdots$$

and similarly

$$\frac{e^x + e^{-x}}{2} = 1 + \frac{x^2}{1 \cdot 2} + \frac{x^4}{1 \cdot 2 \cdot 3 \cdot 4} + \cdots = \left(1 + \frac{4x^2}{\pi^2}\right) \left(1 + \frac{4x^2}{9\pi^2}\right) \left(1 + \frac{4x^2}{25\pi^2}\right) \cdots.$$

(It was Johann Heinrich Lambert (1728–1777) who introduced the names hyperbolic sine ($\sinh x$) and hyperbolic cosine ($\cosh x$) for these functions in 1768.)

Euler further showed that replacing the x in the above series by an imaginary number zi gives the ordinary sine and cosine series for z as well as their infinite product representation. He then used the relationship between roots and coefficients of a polynomial equation (extended to power series) to calculate the infinite sums

$$\sum_{n=1}^{\infty} \frac{1}{n^{2k}}$$

and

$$\sum_{n=1}^{\infty} \frac{1}{(2n-1)^{2k}}.$$

The simplest of these formulas gave a solution to a long-standing question of Johann Bernoulli as to the sum of the reciprocal squares of the integers:

$$\sum_{n=1}^{\infty} \frac{1}{n^2} = \frac{\pi^2}{6}.$$

But Euler also showed, for example, that the sum of the reciprocal fourth powers equals $\frac{\pi^4}{90}$ and that the sum of the reciprocal squares of the odd integers is $\frac{\pi^2}{8}$. Similarly, Euler derived Wallis's infinite product formula

$$\frac{\pi}{2} = \frac{2 \cdot 2 \cdot 4 \cdot 4 \cdot 6 \cdot 6 \cdots}{1 \cdot 3 \cdot 3 \cdot 5 \cdot 5 \cdot 7 \cdots}$$

by using infinite product representations of the sine and the cosine functions.

17.3.5   Euler's *Differential Calculus*

Although Volume 1 of the *Introductio* was largely concerned with series, Euler considered this material as a prerequisite for the calculus. He discussed the calculus itself in his *Institutiones calculi differentialis* of 1755. That work began with his definition of the differential calculus: "It is a method for determining the ratios of the vanishing increments that any functions take on when the variable, of which they are functions, is given a vanishing increment."[26] Euler had already given a definition of "function" in the *Introductio*, but here he generalized it somewhat: "Those quantities that . . . undergo a change when others change, are called functions of these quantities. This definition applies rather widely and includes all ways in which one quantity can be determined by others."[27] Thus, Euler no longer required a function to be an "analytic expression." The reason for this change is perhaps connected to the controversy over the vibrating string problem discussed in Section 17.2.5. Euler was, of course, well aware of the many applications of the differential calculus to geometry. He wrote, however, that in this regard "I have nothing new to offer, and this is all the less to be required, since in other works I have treated this subject so fully."[28] Thus, he decided to keep the *Differential Calculus* as a work of pure analysis, so that there was no need for any diagrams. Similarly, Euler did not deal with the relationship of the subject to physics.

Because calculus has to do with ratios of "evanescent increments," Euler started with a discussion of increments in general, that is, with finite differences. Given a sequence of values of the variable, say, x, $x + \omega$, $x + 2\omega$, . . . and the corresponding values of the function y, y', y'', . . . , Euler considered various sequences of finite differences. The first differences are $\Delta y = y' - y$, $\Delta y' = y'' - y'$, $\Delta y'' = y''' - y''$, . . . ; the second differences are $\Delta\Delta y = \Delta y' - \Delta y$, $\Delta\Delta y' = \Delta y'' - \Delta y'$, . . . ; third and higher differences are defined analogously. For example, if $y = x^2$, then $y' = (x + \omega)^2$ and $\Delta y = 2\omega x + \omega^2$, $\Delta\Delta y = 2\omega^2$, while the third and higher differences are all 0. Using various techniques, including expansion in series, Euler calculated the differences for all of the standard elementary functions. Furthermore, using the sum Σ to denote the inverse of the Δ operation, he derived various formulas for that operation as well. Thus, because $\Delta x = \omega$, it followed that $\Sigma\omega = x$ and that $\Sigma 1 = x/\omega$. Similarly, because $\Delta x^2 = 2\omega x + \omega^2$, it followed that $\Sigma(2\omega x + \omega^2) = x^2$ and that

$$\Sigma x = \frac{x^2}{2\omega} - \Sigma\frac{\omega}{2} = \frac{x^2}{2\omega} - \frac{x}{2}.$$

Euler then easily developed rules for Σ from the corresponding rules for Δ. Rather than discuss the rules for finite differences, however, it is more useful to discuss Euler's rules for differentials.

"The analysis of the infinite . . . is nothing but a special case of the method of differences . . . wherein the differences are infinitely small, while previously the differences were assumed to be finite."[29] Euler's rules for calculating with these infinitely small quantities, the differentials, produce the standard formulas of the differential calculus. For example, if $y = x^n$, then $y' = (x + dx)^n = x^n + nx^{n-1}dx + \frac{n(n-1)}{1 \cdot 2}x^{n-2}dx^2 + \cdots$. Thus, $dy = y' - y = nx^{n-1}dx + \frac{n(n-1)}{1 \cdot 2}n^{n-2}dx^2 + \cdots$. "In this expression the second term and all succeeding terms vanish in the presence of the first term."[30] Thus, $d(x^n) = nx^{n-1}dx$. It should be noted here that Euler intended his argument to apply not just to positive integral powers of x but to arbitrary powers. The binomial theorem, after all, applies to all powers. Thus, the expansion of $(x + dx)^n$ does not necessarily represent a finite sum; it may well represent an infinite series. Euler therefore noted immediately that $d(\frac{1}{x^m}) = -\frac{m\,dx}{x^{m+1}}$ and, more generally, that $d(x^{\mu/\nu}) = (\mu/\nu)x^{(\mu-\nu)/\nu}dx$.

Euler did not give an explicit statement of the modern chain rule but dealt with special cases as the need arose. Thus, if p is a function of x whose differential is dp, then $d(p^n) = np^{n-1}dp$. Euler's derivation of the product rule was virtually identical to that of Leibniz, but his derivation of the quotient rule was more original. He expanded $1/(q + dq)$ into the power series

$$\frac{1}{q + dq} = \frac{1}{q}\left(1 - \frac{dq}{q} + \frac{dq^2}{q^2} - \cdots\right),$$

neglected the higher-order terms, and then wrote

$$\frac{p + dp}{q + dq} = (p + dp)\left(\frac{1}{q} - \frac{dq}{q^2}\right) = \frac{p}{q} - \frac{p\,dq}{q^2} + \frac{dp}{q} - \frac{dp\,dq}{q^2}.$$

It follows, since the second-order differential $dp\,dq$ vanishes with respect to the first-order ones, that

$$d\left(\frac{p}{q}\right) = \frac{p + dp}{q + dq} - \frac{p}{q} = \frac{dp}{q} - \frac{p\,dq}{q^2} = \frac{q\,dp - p\,dq}{q^2}.$$

The differential of the logarithm requires the power series derived in the *Introductio*. If $y = \ln x$, then

$$dy = \ln(x + dx) - \ln(x) = \ln\left(1 + \frac{dx}{x}\right) = \frac{dx}{x} - \frac{dx^2}{2x^2} + \frac{dx^3}{3x^3} - \cdots.$$

Dispensing with the higher-order differentials immediately gave Euler the formula $d(\ln x) = \frac{dx}{x}$. The approach to the arcsine function is through complex numbers. Substituting $y = \arcsin x$ into the formula $e^{iy} = \cos y + i \sin y$ gives $e^{iy} = \sqrt{1 - x^2} + ix$. It follows that $y = \frac{1}{i}\ln(\sqrt{1 - x^2} + ix)$ and therefore that

$$dy = d(\arcsin x) = \frac{1}{i}\frac{1}{\sqrt{1 - x^2} + ix}\left(\frac{-x}{\sqrt{1 - x^2}} + i\right)dx = \frac{dx}{\sqrt{1 - x^2}}.$$

Rather than use this result to calculate the differential of the sine, Euler began anew by calculating $d(\sin x) = \sin(x + dx) - \sin x = \sin x \cos dx + \cos x \sin dx - \sin x$. Euler then recalled his series expansions of the sine and cosine and, again rejecting higher-order terms, noted that $\cos dx = 1$ and $\sin dx = dx$. It follows that $d(\sin x) = \cos x\,dx$ as desired. (To be fair to Euler, at this point he also noted that this result could be easily derived—without using power series—from the previous calculation of the differential of the arcsine.)

Euler's chapter on the differentiation of functions of two or more variables did not record any of the struggles discussed earlier in the development of this idea. He merely noted that in dealing with such a function, the variables can change independently. The central concepts in this chapter, as in the case of functions of one variable, are that of the differential and the differential coefficient. Euler showed, chiefly through the use of examples, that if V is a function of the two variables x and y, then dV, the change in V resulting from the changes x to $x + dx$ and y to $y + dy$, is given by $dV = p\,dx + q\,dy$, where p, q are the differential coefficients resulting from leaving y and x constant, respectively. There is naturally no difficulty in calculating p or q since, in modern terms, $p = \partial V/\partial x$ and $q = \partial V/\partial y$. One merely applies the rules already derived, treating one or the other variable as a constant. Euler showed further, by an algebraic argument involving differentials, that the "mixed partial derivatives" are equal.

Euler's text had many other features, including an introduction to differential equations, in which he showed how to generate these from a given equation in two variables, a discussion of the Taylor series, a chapter on various methods of converting functions to power series, an extensive discussion on finding the sums of various series, including those for the sums of the various powers of the integers, and a variety of ways of finding the roots of equations numerically. The remainder of the discussion here, however, will center on Euler's two chapters on finding maxima and minima. Recall that there are no diagrams in the text and therefore no pictures of curves possessing maxima or minima. Everything is done analytically. But Euler began the discussion by distinguishing between an absolute maximum, a value greater than any other of the function, and a local maximum, a value of y taken at $x = f$ that is greater than any other value of y for x "near" f on either side.

Euler derived the basic criteria for a function to have a maximum or minimum value at $x = \alpha$, in terms of the first and second derivatives, by use of the Maclaurin series, in a way virtually identical to that of Maclaurin himself. But Euler provided many more examples than his Scottish predecessor and often sought to generalize. Thus, after considering maxima and minima for several specific polynomials, he discussed in some detail the case of an arbitrary polynomial $y = x^n + Ax^{n-1} + Bx^{n-2} + \cdots + D$. After dealing with several cases of rational functions, he considered the more general rational function

$$\frac{(\alpha + \beta x)^m}{(\gamma + \delta x)^n}.$$

Following his discussion of the lack of a power series for $x^{2/3}$ around 0, and therefore, the necessity of formulating some different criteria for a maximum or minimum, he dealt with the more general case $x^{2pz/(2q-1)}$. Most of Euler's examples are of algebraic functions, but he concluded with a few examples using transcendental functions, including the functions $x^{1/x}$ and $x \sin x$, both of which required detailed numerical work to arrive at an exact solution for an extreme value.

For functions V of two variables, Euler began by considering the special case of functions of the form $X + Y$, where X is a function solely of x and Y of y. In that case, a pair of values (x_0, y_0) such that x_0 is a maximum for X and y_0 a maximum for Y clearly gives a maximum for $X + Y$. For the more general case, Euler realized, by holding each variable constant in turn, that an extreme value of V can only occur when the differential $dV = P\,dx + Q\,dy = 0$, therefore only when both $P = \partial V/\partial x = 0$ and $Q = \partial V/\partial y = 0$. The question of determining whether a point (x_0, y_0) where both first partial derivatives vanish produces a maximum, a minimum, or neither is more difficult, and Euler failed to give complete results. He claimed, in fact, that if $\frac{\partial^2 V}{\partial x^2}$ and $\frac{\partial^2 V}{\partial y^2}$ are both positive at (x_0, y_0), then the function V has a minimum there, and if they are both negative, there is a maximum. Euler gave several examples illustrating the method, including $V = x^3 + ay^2 - bxy + cx$. He noted that an extreme value would occur when

$$x = \frac{b^2 \pm \sqrt{b^4 - 48a^2c}}{12a},$$

as long as $b^4 - 48a^2c > 0$. Furthermore, since $\frac{\partial^2 V}{\partial x^2} = 6x$ and $\frac{\partial^2 V}{\partial y^2} = 2a$, he claimed that when $a > 0$ and both possible values of x are positive, then the two extreme values are both minima. In particular, in Euler's special case where $a = 1$, $b = 3$, and $c = 3/2$, his criteria imply that $V = x^3 + y^2 - 3xy + (3/2)x$ has a minimum both when $x = 1$, $y = 3/2$ and when $x = 1/2$, $y = 3/4$. Unfortunately, Euler was wrong; the latter point is not a minimum but a saddle point.

17.3.6   Euler's *Integral Calculus*

The final part of Euler's trilogy in analysis, the *Institutiones calculi integralis*, began with a definition of integral calculus. It is the method of finding, from a given relation of differentials of certain quantities, the quantities themselves. Namely, for Euler as it was for Agnesi and Johann Bernoulli, integration is the inverse of differentiation rather than the determination of an area. Thus, the first part of the work dealt with techniques for integrating (finding antiderivatives of) functions of various types, while the remainder of the text dealt with the solutions of differential equations. Although Euler began his section on techniques with such standard results as

$$\int ax^n \, dx = \frac{a}{n+1}x^{n+1} + C$$

for $n \neq -1$ and

$$\int \frac{a\,dx}{x} = a \ln x + C = \ln cx^a,$$

he quickly moved on to many types of integrals, some being familiar and others of types not usually covered in today's texts. Thus, he noted that to integrate any rational function, it sufficed to integrate functions of the form

$$\frac{A}{(a+bx)^n} \quad \text{and} \quad \frac{A+Bx}{(a^2 - 2abx\cos\zeta + b^2x^2)^n},$$

having found it convenient to express an irreducible quadratic in a trigonometric form. The first type of integral is straightforward. For the second, he began with the special case $n = 1$:

$$\int \frac{(A + Bx)\, dx}{a^2 - 2abx \cos \zeta + b^2 x^2}$$

$$= \frac{B}{2b^2} \ln(a^2 - 2abx \cos \zeta + b^2 x^2) + \frac{Ab + Ba \cos \zeta}{ab^2 \sin \zeta} \arctan \frac{bx - a \cos \zeta}{a \sin \zeta}.$$

In this example, as in the others discussed below, Euler considered various special cases before generalizing. And then, once he had his general integral results, he often specialized again, frequently calculating the same integral in more than one way.

To integrate functions involving square roots, Euler used substitution, although not our modern trigonometric substitutions. For example, to integrate

$$\frac{dx}{\sqrt{\alpha + \beta x + \gamma x^2}},$$

he considered two cases, depending on whether the quadratic polynomial factored into two real factors or not. In the first case, he assumed the factorization was $(a + bx)(f + gx)$. Then, "to remove the irrationality," he set $(a + bx)(f + gx) = (a + bx)^2 z^2$, or $(f + gx) = (a + bx)z^2$. Solving for x gives $x = (az^2 - f)/(g - bz^2)$, and therefore

$$dx = \frac{2(ag - bf)z\, dz}{(g - bz^2)^2} \quad \text{and} \quad \frac{dx}{\sqrt{(a + bx)(f + gx)}} = \frac{2\, dz}{g - bz^2}.$$

Assuming $g > 0$, we then have that if $b > 0$, the integral is

$$\frac{1}{\sqrt{bg}} \ln \left(\frac{\sqrt{g} + z\sqrt{b}}{\sqrt{g} - z\sqrt{b}} \right),$$

while if $b < 0$, the integral is

$$\frac{2}{\sqrt{bg}} \arctan \left(\frac{z\sqrt{b}}{\sqrt{g}} \right).$$

An analogous substitution works in the case where the original quadratic polynomial is irreducible over the real numbers.

Euler next considered integration by the use of infinite series, Newton's favorite technique. To integrate functions involving logarithms, he invoked the technique of what we call integration by parts. As he described this technique, if the function V can be factored as $V = PQ$, and if the integral $P\, dx = S$ is known, then from $P\, dx = dS$, we get $V\, dx = PQ\, dx = Q\, dS$. Thus, since $d(QS) = Q\, dS + S\, dQ$, we have $\int V\, dx = QS - \int S\, dQ$. He immediately applied this rule to integrating functions of the form $x^n \ln x$:

$$\int x^n \ln x\, dx = \frac{1}{n+1} x^{n+1} \ln x - \int \frac{1}{n+1} x^{n+1}\, d(\ln x) = \frac{1}{n+1} x^{n+1} \left(\ln x - \frac{1}{n+1} \right).$$

In another chapter, Euler dealt with numerous procedures for integration of powers of trigonometric functions. For example, he used the substitution $\cos \phi = \frac{1-x^2}{1+x^2}$, $\sin \phi = \frac{2x}{1+x^2}$ to convert rational functions involving sines and cosines to ordinary rational functions. Thus, he showed that, in the case where $a > b$,

$$\int \frac{d\phi}{a + b \cos \phi} = \int \frac{2\,dx}{a + b + (a - b)x^2}$$

$$= \frac{2}{\sqrt{a^2 - b^2}} \arctan \frac{(a - b)x}{\sqrt{a^2 - b^2}} = \frac{1}{\sqrt{a^2 - b^2}} \arctan \frac{\sin \phi \sqrt{a^2 - b^2}}{a \cos \phi + b}.$$

Although normally Euler just calculated antiderivatives, occasionally he calculated what we would call a "definite integral." Thus, he first demonstrated the reduction formula:

$$\int \frac{x^{m+1}\,dx}{\sqrt{1 - x^2}} = \frac{m}{m + 1} \int \frac{x^{m-1}\,dx}{\sqrt{1 - x^2}} - \frac{1}{m + 1} x^m \sqrt{1 - x^2}.$$

Then, noting that the second term on the right vanished at both $x = 0$ and $x = 1$ and that

$$\int_0^1 \frac{dx}{\sqrt{1 - x^2}} = \frac{\pi}{2} \quad \text{and} \quad \int_0^1 \frac{x\,dx}{\sqrt{1 - x^2}} = 1,$$

he concluded that

$$\int_0^1 \frac{x^{2n}\,dx}{\sqrt{1 - x^2}} = \frac{1 \cdot 3 \cdot 5 \cdots (2n - 1) \cdot \pi}{2 \cdot 4 \cdot 6 \cdots 2n \cdot 2} \quad \text{and} \quad \int_0^1 \frac{x^{2n+1}\,dx}{\sqrt{1 - x^2}} = \frac{2 \cdot 4 \cdot 6 \cdots 2n}{3 \cdot 5 \cdot 7 \cdots (2n + 1)}.$$

(Note that Euler himself did not write limits of integration; they are put there for clarity.)

After considering these various techniques of integration, Euler moved on to deal with methods of solving differential equations. Euler solved the general first-order linear equation $dy + Py\,dx = Q\,dx$ (or, in modern terms, $y' + Py = Q$) by separation of variables to get

$$y = e^{-\int P\,dx} \int e^{\int P\,dx} Q\,dx.$$

As examples of this, he solved $dy + y\,dx = ax^n\,dx$ for various values of n. For $n = 3$, he found that $y = Ce^{-x} + x^3 - 3x^2 + 6x - 6$. He showed how to integrate $P\,dx + Q\,dy$ in the "exact" case where $\partial P/\partial y = \partial Q/\partial x$, using Clairaut's idea of 1739, again following the general discussion with numerous examples. He demonstrated how to find integrating factors in the case where $P\,dx + Q\,dy$ is not exact, detailing the method through various examples. He considered many cases of second and higher-order differential equations, including the linear case with constant coefficients, which required the solving of a polynomial equation. Finally, Euler concluded the book with a long discussion of partial differential equations.

The *Integral Calculus*, like the *Differential Calculus*, is a text in pure analysis, so much so that Euler did not even deal with applications to geometry, let alone physics. This is perhaps especially surprising in the *Integral Calculus* since the original motivation for the solution of differential equations came from physical questions, questions that in fact led Euler to some of these methods of solution in the 1730s and 1740s. So the modern reader will also be surprised that in the *Differential Calculus* there are no tangent lines or normal lines, no tangent planes, no study of curvature—all topics with which Euler was fully conversant in 1740 but that only appear in some of his geometrical works. And in the *Integral Calculus* there is no mention of the vibrating string problem or various other vibration problems that had led Euler to "invent" the trigonometric functions in the 1730s, nor is there any calculation of areas, nor any material on lengths of curves, or volumes, or surface areas of solids. It follows that the fundamental theorem of calculus, central to modern works, does not appear. Euler was certainly familiar with using antiderivatives to calculate area and in fact used such ideas in various papers. On the other hand, since there does not appear in his work any clear notion of the area under a curve as a function, he did not consider the derivative of such an area function.

Most historians consider Euler's analysis texts as the most influential texts of the eighteenth century. But it is difficult to quantify this influence. Certainly, they were among the works to which Laplace referred when he wrote, "Read Euler; read Euler. He is the master of us all." But to figure out who actually did read them is difficult. Certainly, the *Introductio* was read frequently. For not only were Euler's notation and methods taken up in numerous analysis texts that followed, but also the book itself saw several reprints even during the eighteenth century, as well as translations toward the end of that century into both French (twice) and German. On the other hand, the *Differential Calculus* only had a single German translation— in 1790—while there were no eighteenth-century translations of the *Integral Calculus* at all; a German one finally appeared in 1828–1830.

Certainly, the techniques that Euler developed to determine derivatives and integrals continued to appear in other texts, but his use of infinitesimals as a basis for the calculus was gradually replaced by the idea of a limit, as we discuss later in this chapter. However, it was the French Revolution and the influence of Napoleon that really changed everything. Suddenly, with the aristocracy removed in France, and greatly weakened elsewhere, there was a great need for educating a new class of students who were entering the sciences. And it was this need that inspired the writing of many new texts in the vernacular, texts that replaced those of Euler and were the direct ancestors of the texts of today.

 17.4

THE FOUNDATIONS OF CALCULUS

The eighteenth century saw extensive development of the techniques of the calculus. Many texts were written explaining the calculus to those who wanted to learn, while multitudes of papers appeared in which new procedures and methods of solving various kinds of physical problems were demonstrated. But in the minds of some, there was a nagging doubt as to the foundations of the subject. Most mathematicians had read Euclid's *Elements* and regarded it as a model of how mathematics should be done. Yet there was no logical basis for the central procedures of calculus. In general, the practitioners themselves did not worry much about this. Newton, Leibniz, Euler, and the others had a strong intuitive feel for the subject and knew when what they were doing was correct. Even in the light of modern standards, these great mathematicians rarely made errors. Nevertheless, the explanations they themselves gave of the foundations for their procedures left something to be desired.

17.4.1 George Berkeley's *The Analyst*

The most important criticism of both infinitesimals and fluxions was made by the Irish philosopher Bishop George Berkeley (1685–1753) in a 1734 tract entitled *The Analyst*

FIGURE 17.17

Bishop Berkeley on an Irish stamp

(Fig. 17.17). The work was addressed "to an infidel mathematician," generally supposed to be the astronomer Edmond Halley, who financed the publication of Newton's *Principia* and helped see it through the press. Berkeley presumably considered him an infidel because he had persuaded a mutual friend that the doctrines of Christianity were inconceivable. Berkeley's aim in *The Analyst* was not to deny the utility of the calculus or the validity of its many new results, but to show that mathematicians had no valid arguments for the procedures they invoked.

Thus, "the Method of Fluxions is the general key by help whereof the modern mathematicians unlock the secrets of Geometry, and consequently of Nature." The fluxions themselves, however, "are said to be nearly as the increments of the flowing quantities [moments], generated in the least equal particles of time; and to be accurately in the first proportion of the nascent, or in the last of the evanescent increments. . . . By moments we are not to understand finite particles . . . [but] only the nascent principles of finite quantities."[31] What are these "nascent principles"? Berkeley noted that even though "the minutest errors are not to be neglected in mathematics"—a quotation from Newton himself—the actual finding of these fluxions determined by the nascent principles involved precisely that kind of neglect.

Berkeley demonstrated his point by analyzing the calculation of the fluxion of x^n:

In the same time that x by flowing becomes $x + o$, the power x^n becomes $(x + o)^n$, i.e., by the method of infinite series

$$x^n + nox^{n-1} + \frac{n^2 - n}{2}o^2x^{n-2} + \cdots,$$

and the increments o and $nox^{n-1} + \frac{n^2-n}{2}o^2x^{n-2} + \cdots$ are one to another as 1 to $nx^{n-1} + \frac{n^2-n}{2}ox^{n-2} + \cdots$. Let now the increments vanish, and their last proportion will be 1 to nx^{n-1}. But it should seem that this reasoning is not fair or conclusive. For when it is said, let the increments vanish, i.e., let the increments be nothing, or let there be no increments, the former supposition that the increments were something, or that there were increments, is destroyed, and yet a consequence of that supposition, i.e., an expression got by virtue thereof, is retained.[32]

Berkeley thus questioned how one can take a nonzero increment, do calculations with it, and then in the end set it equal to zero. He noted further that the methods of the Continental mathematicians were no better. Rather than considering "flowing quantities and their fluxions, they consider the variable finite quantities as increasing or diminishing by the continual addition or subtraction of infinitely small quantities." And these lead to exactly the same kinds of problems. In particular, Berkeley claimed he cannot conceive of infinitely small quantities. "But to conceive a part of such infinitely small quantity that shall be still infinitely less than it, and consequently though multiplied infinitely shall never equal the minutest finite quantity, is, I suspect, an infinite difficulty to any man whatsoever."[33] Thus, second-order differentials, and similarly, fluxions of fluxions form an "obscure mystery. The incipient celerity of an incipient celerity, the nascent augment of a nascent augment, i.e., of a thing which hath no magnitude—take it in what light you please, the clear conception of it will, if I mistake not, be found impossible."[34] Since Halley could not comprehend the arguments of theology, Berkeley counterattacked by noting that "he who can digest a second or third fluxion, a second or third difference, need not, methinks, be squeamish about any point in divinity."[35]

17.4.2 Maclaurin's Response to Berkeley

Berkeley's criticisms of the foundations of calculus were valid. The question of when a value was zero and when it was not zero extended back even to the work of Fermat, and neither Newton nor Leibniz were ever quite able to resolve it. Nevertheless, several English mathematicians sprang to Newton's defense under Berkeley's attack. The most important response was that of Maclaurin in his *Treatise of Fluxions*. As he noted in his preface, Maclaurin wanted to "deduce those Elements [of the theory of Fluxions] after the Manner of the Ancients from a few unexceptionable Principles by Demonstrations of the strictest Form."[36] He noted that he would not use any indivisible or infinitely small part of time or space as part of the demonstration, "the supposition of an infinitely little magnitude being too bold a Postulate for such a Science as Geometry."[37] Maclaurin therefore had to consider finite lengths and times as his basic elements, for "no quantities are more clearly conceived than the limited parts of space and time."[38] These spaces and times then determine (average) velocity. But because it is instantaneous velocity that is the basic concept necessary to the theory of fluxions, Maclaurin attempted a definition of this as well: "The velocity of a variable motion at any given term of time is not to be measured by the space that is actually described after that term in a given time, but by the space that would have been described if the motion had continued uniformly from that term."[39] This definition is reminiscent of that given by Heytesbury in the fourteenth century. From a modern point of view, however, Maclaurin, by giving such a definition, missed the fundamental idea of instantaneous velocity as a limit of average velocities as the time interval approaches zero. In any case, given the definition of variable velocity, Maclaurin presented axioms for the use of this definition and then proceeded to prove numerous theorems in the "manner of the ancients," using each time a double reductio ad absurdum.

In particular, because one of Maclaurin's aims was to show that "infinitesimals" in the arguments of Newton can always be replaced by finite quantities, he demonstrated that even the differential triangle can be derived rigorously:

PROPOSITION   *Let ET be the tangent of the curve FE at E and, EI being parallel to the base AD, let IT be parallel to the ordinate DE. Then the fluxions of the base, ordinate, and curve shall be measured by the lines EI, IT, and ET, respectively [Fig. 17.18].*[40]

Maclaurin proved this result by first noting that if the curve is concave up, then the increase of the ordinate in the time of a given increase of the base is greater than that which would have been generated had the motion of the ordinate been uniform. To show that this latter increase, proportional to the fluxion of the ordinate DE, is precisely equal to IT, he made the initial assumption that it was greater than IT and showed by use of his axioms on velocity that a contradiction results. A similar contradiction followed from the contrary assumption. The entire proof was then repeated in the case where the curve was concave down. From a modern point of view, the problem with Maclaurin's proof lay in his definition of a tangent, a concept generally accepted by his contemporaries as "self-evident." True to his belief in the methods of the ancients, however, Maclaurin presented the ancient definition, that a tangent to a curve is a straight line that touches the curve in such a way that no other straight line can be inserted between the curve and the line.

But despite his definition taking instantaneous velocity as the basic concept and his geometrical definition of a tangent, Maclaurin was well aware of Newton's use of the notion

FIGURE 17.18

Maclaurin's differential
triangle

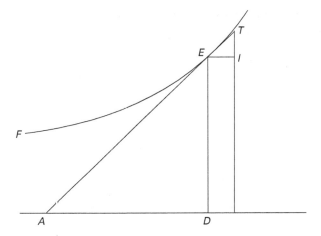

of "ultimate proportion of evanescent quantities" or "limits." Thus, he wrote about the ratio
that is the limit of the various proportions that finite simultaneous increments of two variable
quantities bear to one another as the two increments decrease until they vanish. He noted
that to discover this limit one must first determine the ratio of increments in general and
then reduce to the simplest terms so that a part of the result might be independent of the
value of the increments themselves. The desired limit then readily appears if one supposes
the increments to "decrease until they vanish."[41] For example, to find the ratio of the fluxion
of x^2 to the fluxion of ax, Maclaurin calculated the ratio of the increments (as x increases to
$x + o$) to be $(2xo + o^2) : ao$ or $(2x + o) : a$. "This ratio of $2x + o$ to a continually decreases
while o decreases and is always greater than the ratio of $2x$ to a while o is any real increment,
but it is manifest that it continually approaches to the ratio of $2x$ to a as its limit."[42]

Maclaurin vehemently denied Berkeley's contention that the method of first supposing a
finite increment and then letting that increment vanish is contradictory. In fact, he noted, this
method allows one to determine the ratio of the increments when the increments are finite and
to determine how the ratio varies with the increment. One can then easily determine what limit
the ratio approaches as the increments are diminished. As a final response to Berkeley, he
even redefined the tangent as a limit: "The tangent . . . is the . . . line that limits the position
of all the secants that can pass through the point of contact, though strictly speaking it be no
secant, [just as] a ratio may limit the variable ratios of the increments, though it cannot be
said to be the ratio of any real increments."[43]

The problem with Maclaurin's treatment of the calculus, as noted by many of his contemporaries, was not that he failed rigorously to derive the rules but that he really did this in the
"manner of the ancients." In particular, he used the method of exhaustion and its accompanying reductio ad absurdum argument. The use of such a method imposed a heavy toll on the
reader. For example, the first 590 pages of this 754-page work do not contain any notation
of fluxions. Every new idea was derived geometrically with great verbosity. And in the eighteenth century, few were willing to read through these detailed arguments. Maclaurin himself
realized that the advantages of the new calculus were that it enabled old problems to be solved
in expeditious fashion and new discoveries to be made with ease. "But when the principles

and strict method of the ancients, which had hitherto preserved the evidence of this science entire, were so far abandoned, it was difficult for the Geometricians to determine where they should stop."[44] Nevertheless, although Maclaurin's great efforts answered Berkeley's objections, they were not appreciated by most eighteenth-century mathematicians, people who saw themselves as breaking new ground rather than extending the methods of the ancients.

17.4.3 Euler and d'Alembert

On the European continent, too, some justification of the procedures of the calculus was necessary. In his *Differential Calculus*, Euler developed the idea that the ratios involved in the calculation of derivatives were in fact simply versions of the ratio $0 : 0$. For Euler, infinitely small quantities were quantities actually equal to 0, because the latter is the only quantity smaller than any given quantity. "There is really not such a great mystery lurking in this idea as some commonly think and thus have rendered the calculus of the infinitely small suspect to so many."[45] But although two zeroes are equal in such a way that their difference is always zero, Euler insisted that the ratio of two zeroes, which depends on the origin of the quantities that are becoming zero, must be calculated in each specific case. As an example, he noted that $0 : 0 = 2 : 1$ is a correct statement because the first quantity on each side of the equal sign is double the second quantity. In fact, then, the ratio $0 : 0$ may be equal to any finite ratio at all. Therefore, "in the calculus of the infinitely small, we deal precisely with geometric ratios of infinitely small quantities."[46]

Euler continued his discussion in this vein, noting further, for example, that "the infinitely small vanishes in comparison with the finite and hence can be neglected," hence "the objection brought up against the analysis of the infinite, that it lacks geometric rigor, falls to the ground under its own weight, since nothing is neglected except that which is actually nothing."[47] Similarly, the infinitely small quantity dx^2 will vanish with respect to dx and can thus be neglected because $(dx \pm dx^2)/dx = 1 \pm dx = 1$.

Interestingly enough, d'Alembert, in the article "Différentiel," which he wrote for the *Encyclopédie* in 1754, combined the ideas of both Euler and Maclaurin. He agreed with Euler that there was no absurdity in considering the ratio $0 : 0$ because it may in fact be equal to any quantity at all. But the central idea of the differential calculus is that dy/dx is the limit of a certain ratio as the quantities involved approach 0. The "most precise and neatest possible definition of the differential calculus" is that it "consists in algebraically determining the limit of a ratio, for which we already have the expression in terms of lines, and in equating those two expressions."[48] As an example of what he meant, d'Alembert calculated the slope of the tangent line to the parabola $y^2 = ax$ by first determining the slope of a secant through the two points (x, y) and $(x + u, y + z)$. This slope, the ratio $z : u$, is easily seen to be equal to $a : (2y + z)$. "This ratio is always smaller than $a : 2y$; but the smaller z is, the greater the ratio will be and, since one may choose z as small as one pleases, the ratio $a : (2y + z)$ can be brought as close to the ratio $a : 2y$ as we like. Consequently $a : 2y$ is the limit of the ratio $a : (2y + z)$."[49] It follows that $dy/dx = a/2y$. D'Alembert's wording was virtually identical to that of Maclaurin. He went somewhat further, however, by giving an explicit definition of the term "limit" in his *Encyclopédie* article on that notion: "One magnitude is said to be the **limit** of another magnitude when the second may approach the first within any given magnitude, however small, though the second magnitude may never exceed the magnitude it approaches."[50] His idea, although apparently geometric rather than arithmetic, was not followed up by his eighteenth-century successors. Through the remainder of the century, most of the works on calculus attempted to explain the basis of the subject in terms of infinitesimals, fluxions, or the ratios of zeroes.

17.4.4 Lagrange and Power Series

It was Lagrange near the end of the eighteenth century who attempted to give a precise definition of the derivative by eliminating all reference to infinitesimals, fluxions, zeroes, and even limits, all of which he believed lacked proper definitions. He sketched his new ideas about derivatives in a paper of 1772 and then developed them in full in his text of 1797, the full title of which expressed what he intended to do: *The Theory of Analytic Functions, containing the principles of the differential calculus, released from every consideration of the infinitely small or the evanescent, of limits or of fluxions, and reduced to the algebraic analysis of finite quantities.* How could Lagrange accomplish the reduction of calculus purely to algebraic analysis? He did so by formalizing the idea used by most of his predecessors without question that any function can be represented as a power series. For Lagrange, if $y = f(x)$ is any function, then $f(x + i)$, where i is an indeterminate, can "by the theory of series" be expanded into a series in i:

$$f(x + i) = f(x) + pi + qi^2 + ri^3 + \cdots,$$

where $p, q, r \ldots$ are new functions of x independent of i. Lagrange then showed that the ratio dy/dx can be identified with the coefficient $p(x)$ of the first power of i in this expansion. He therefore had a new definition of this basic concept of the calculus. Since the function p is "derived" from the original function f, Lagrange named it a *fonction dérivée* (from which comes the English word *derivative*) and used the notation $f'(x)$. Similarly, the derivative of f' is written f'', that of f'' is written f''', and so on. Lagrange easily showed that $q = (1/2) f''$, $r = (1/6) f'''$, ...

Lagrange's argument for the expansion of a function f—generally thought of as a formula—into a power series began with the assertion that $f(x + i) = f(x) + iP$, where $P(x, i)$ is defined by

$$P(x, i) = \frac{f(x + i) - f(x)}{i}.$$

Lagrange assumed further that one can separate out from P that part p that does not vanish at $i = 0$. Thus, $p(x)$ is defined as $P(x, 0)$ and then

$$Q(x, i) = \frac{P(x, i) - p(x)}{i}$$

or $P = p + iQ$. It follows that $f(x + i) = f(x) + ip + i^2 Q$. Repeating the argument for Q, he wrote $Q = q + iR$ and substituted again. As an example of the procedure, Lagrange used

BIOGRAPHY

Joseph-Louis Lagrange (1736–1813)

Lagrange was born in Turin into a family of French descent. His father wanted him to study law, but he was attracted to mathematics in school and at the age of 19 became a professor of mathematics at the Royal Artillery School in Turin. At about the same time, having read Euler's book on the calculus of variations, he wrote to the latter explaining a better method he had recently discovered for deriving the central equation of the subject. Euler praised Lagrange greatly and arranged to present his paper to the Berlin Academy. Frederick II was also impressed with Lagrange's work, and, when Euler left Berlin to return to St. Petersburg, Frederick appointed Lagrange to fill Euler's post at the Academy of Sciences. After Frederick's death, Lagrange accepted the invitation of Louis XVI to come

to Paris, where he spent the rest of his life, there publishing in 1788 his most important work, *Analytical Mechanics*, a work that extended the mechanics of Newton, the Bernoullis, and Euler and emphasized the fact that problems in mechanics can generally be solved by reducing them to the theory of ordinary and partial differential equations. In 1792 he married 17-year-old Renée Le Monnier, who brought renewed joy to his life. Because of his generally introverted personality, he was able to survive the excesses of the French Revolution, in fact being treated with honor, but the death of several of his colleagues disturbed him greatly. After the Terror, he took an active role in improving university education in France and was ultimately honored by Napoleon for his life's work (Fig. 17.19).

FIGURE 17.19

Lagrange on a French stamp

$f(x) = 1/x$. Since $f(x + i) = 1/(x + i)$, he calculated

$$P = \frac{1}{i}\left(\frac{1}{x+i} - \frac{1}{x}\right) = -\frac{1}{x(x+i)} \qquad p = -\frac{1}{x^2}$$

$$Q = \frac{1}{i}\left(-\frac{1}{x(x+i)} + \frac{1}{x^2}\right) = \frac{1}{x^2(x+i)} \qquad q = \frac{1}{x^3}$$

$$\vdots$$

Thus, the series becomes

$$\frac{1}{x+i} = \frac{1}{x} - \frac{i}{x^2} + \frac{i^2}{x^3} - \frac{i^3}{x^4} + \cdots.$$

At each stage of the expansion, the terms iP, i^2Q, . . . can be considered as the error terms resulting from representing $f(x + i)$ by terms up to that point. Furthermore, Lagrange claimed, the value of i can always be taken so small that any given term of this series is greater than the sum of the remaining terms, that is, that the remainders are always sufficiently small so that in fact the function is represented by the series. This result is what Lagrange used often later on. He also used a somewhat different form of his expansion result containing what is now called the **Lagrange form of the remainder** in the Taylor series. Namely, he showed that for any given positive integer n, one can write

$$f(x+i) = f(x) + if'(x) + \frac{i^2 f''(x)}{2} + \cdots + \frac{i^n f^{(n)}(x)}{n!} + \frac{i^{n+1} f^{(n+1)}(x+j)}{(n+1)!}$$

for some value j between 0 and i. Although this new form is, perhaps, no more convincing to the modern reader than his earlier one, Lagrange himself was satisfied that his principle of a power series representation for every function was correct. After all, he claimed, it enabled him to derive anew all of the basic results of the calculus without any consideration of infinitesimals, fluxions, or limits.

One of these basic results is part of what is known today as the **fundamental theorem of calculus**, that if $F(x)$ represents the area under the curve $y = f(x)$ from a fixed ordinate, then $F'(x) = f(x)$. (It should be noted that Lagrange had no definition of area. He simply assumed that the area under a curve $y = f(x)$ is a well-determined quantity.) Lagrange began his proof, reminiscent of Maclaurin's proof of the same result for power functions, by noting that $F(x + i) - F(x)$ represents that portion of the area between the abscissas x and $x + i$. Keeping to Euler's dictum that in a text on analysis one should not include diagrams, Lagrange nevertheless wrote that even without a figure one can easily convince oneself that if $f(x)$ is monotonically increasing, then $if(x) < F(x + i) - F(x) < if(x + i)$, with the inequalities reversed if $f(x)$ is monotonically decreasing (Fig. 17.20).

FIGURE 17.20

Lagrange and the fundamental theorem of calculus: $if(x) < F(x + i) - F(x) < if(x + i)$

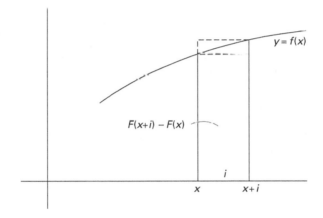

Now expanding both $f(x + i)$ and $F(x + i)$, Lagrange determined that

$$f(x + i) = f(x) + if'(x + j)$$

and

$$F(x + i) = F(x) + iF'(x) + \frac{i^2}{2}F''(x + j),$$

where $0 < j < i$ (although the value of j may not be the same in both expansions). It follows that $if(x) < iF'(x) + \frac{i^2}{2}F''(x + j) < if(x) + i^2 f(x + j)$ and therefore that

$$\left| i[F'(x) - f(x)] + \frac{i^2}{2}F''(x + j) \right| < i^2 f'(x + j),$$

where the absolute value sign is necessary to take care of both the increasing and decreasing cases. Lagrange concluded that because the inequality held no matter how small i was taken to be, it must be true that $F'(x) = f(x)$. He even calculated that if the conclusion were not true, the inequality would fail for

$$i < \frac{F'(x) - f(x)}{f'(x+j) - \frac{1}{2}F''(x+j)}.$$

To finish his proof Lagrange removed the condition that $f(x)$ be monotonic on the original interval $[x, x+i]$. For if it is not, there is a maximum or minimum of f on that interval and i can be chosen small enough so that the extreme value falls outside of the new interval $[x, x+i]$.

It is curious, of course, that despite Lagrange's claim that this work will only use "algebraic analysis of finite quantities," in this very important proof, as well as in his remainder arguments, he used the notion of a limit. In other sections of the work, where he found tangent lines, curvature, and maxima and minima, among other geometric quantities, Lagrange used limits in the same way, along with his central concept of the expansion of the functions involved in a power series. And, in fact, these very arguments were used in the nineteenth-century treatments of calculus that used limits explicitly.

Most of the early objections to Lagrange's new foundation for the calculus were aimed at his new notations and the length of some of his calculations rather than at his assertion that any function can be expanded in a power series. Mathematicians in general continued to use the earlier differential methods, especially since Lagrange's book assured them that because there was a correct basis to the entire subject, any method that worked would be legitimate. Even Lagrange in some of his other work continued to employ the notation of differentials rather than that of derivatives. It was not until the second decade of the nineteenth century that various mathematicians pointed out that there existed differentiable functions that do not have a power series representation and thus that Lagrange's basic concept was not tenable. The story of the new attempts to supply a foundation to the ideas of calculus is therefore continued in Chapter 22.

EXERCISES

1. Derive Johann Bernoulli's differential equation for the catenary, $dy/dx = s/a$, as follows: Let the lowest point of the hanging cord be the origin of the coordinate system, and consider a piece of the chord of length s over the interval $[0, x]$. Let $T(x)$ be the (vector) tension of the cord at the point $P = (x, y)$. Let α be the angle that $T(x)$ makes with the horizontal and let ρ be the density of the cord. Show that the equilibrium of horizontal forces gives the equation $|T(0)| = |T(x)| \cos \alpha$, while that of the vertical forces gives $\rho s = |T(x)| \sin \alpha$. Since $dy/dx = \tan \alpha$, Bernoulli's equation can be derived by dividing the second equation by the first.

2. Using modern techniques, put Johann Bernoulli's solution to the catenary problem,

$$dy = \frac{a\,dx}{\sqrt{x^2 - a^2}},$$

into a closed-form expression.

3. Derive Johann Bernoulli's differential equation for the brachistochrone $dy = dx\sqrt{\frac{x}{a-x}}$ from Jakob Bernoulli's equation $ds = \frac{\sqrt{a}\,dy}{\sqrt{x}}$.

4. Suppose $A = a_1 + a_2 + \cdots + a_n$ and $B = b_1 + b_2 + \cdots + b_n$. Show that $\sum(b_i - a_i) = B - A$, or, that the sum of the differences of the parts is equal to the difference of the sums of the parts.

5. Translate Leibniz's solution of $m\,dx + ny\,dx + dy = 0$ into modern terms by noting that $dp/p = n\,dx$ is equiv-

alent to $\ln p = \int n \, dx$ or to $p = e^{\int n \, dx}$. Solve $-3x \, dx + (1/x)y \, dx + dy = 0$ by using Leibniz's procedure.

6. The modern way to derive Kepler's area law is to break the force into its radial and transverse components, rather than the tangential and normal components used by Hermann, and use polar coordinates whose origin is the center of force. Assume then that the center of force is at the origin of a polar coordinate system. Using vector notation, set $\mathbf{u}_r = \mathbf{i} \cos \theta + \mathbf{j} \sin \theta$ and $\mathbf{u}_\theta = -\mathbf{i} \sin \theta + \mathbf{j} \cos \theta$. Show that $d\mathbf{u}_r/d\theta = \mathbf{u}_\theta$ and $d\mathbf{u}_\theta/d\theta = -\mathbf{u}_r$. Then show that if $\mathbf{r} = r\mathbf{u}_r$, then the velocity $\mathbf{v}$ is given by $r(d\theta/dt)\mathbf{u}_\theta + (dr/dt)\mathbf{u}_r$. Show next that the radial component a_r and the transverse component a_θ of the acceleration are given by

$$a_r = \frac{d^2 r}{dt^2} - r\left(\frac{d\theta}{dt}\right)^2 \quad \text{and} \quad a_\theta = r\frac{d^2\theta}{dt^2} + 2\frac{dr}{dt}\frac{d\theta}{dt}.$$

Since the force is central, $a_\theta = 0$. Multiply the differential equation expressing that fact by r and integrate to get $r^2 \frac{d\theta}{dt} = k$, where k is a constant. Show finally that $r^2 \frac{d\theta}{dt} = \frac{dA}{dt}$, where A is the area swept out by the radius vector. This proves Kepler's law of areas.

7. Show that the area of the infinitesimal triangle SQP of Figure 17.6 can be written as $\frac{1}{2}(y \, dx - x \, dy)$.

8. Convert Newton's geometrical description of the central force as proportional to QR and inversely proportional to the square of the area of triangle SQP (Fig. 17.6) into differentials and compare this with the component a_r of the force derived in Exercise 6.

9. Show that the differential equation

$$\frac{d^2 r}{dt^2} - r\left(\frac{d\theta}{dt}\right)^2 = -\frac{k}{r^2}$$

derived by assuming that the component a_r of the force from Exercise 6 is inversely proportional to r^2 is equivalent to the differential equation Hermann derived using the inverse square property of the central force.

10. Show that the equation $a \pm cx/b = \sqrt{x^2 + y^2}$ is a parabola if $b = c$, is an ellipse if $b > c$, and is a hyperbola if $b < c$.

11. Show that the equation of any conic section may be written in the form $\sqrt{x^2 + y^2} = \alpha x + \beta$, where the origin is a focus of the conic. How do α and β determine the nature of the conic section?

12. Show that $y = e^{x/a}$ is a solution to the differential equation $a^3 d^3 y - y \, dx^3 = 0$. Next, assume that the product $e^{-(x/a)}(a^3 d^3 y - y \, dx^3)$ is the differential of

$$e^{-(x/a)}(A \, d^2 y + B \, dy \, dx + Cy \, dx^2)$$

and show that a new solution of the original equation must also satisfy $a^2 d^2 y + a \, dy \, dx + y \, dx^2 = 0$. (*Hint:* Calculate the differential and equate the two expressions. It may be easier if you rewrite the equations in modern notation using derivatives.)

13. Given that $y = e^x$ is a solution of $y''' - 6y'' + 11y' - 6y = 0$, show by a method analogous to that of Exercise 12 that any other solution must satisfy $y'' - 5y' + 6y = 0$.

14. Solve the differential equation of Exercise 13 by using Euler's procedure of factoring the characteristic polynomial.

15. Show that if $y = ue^{\alpha x}$ is assumed to be a solution of $a^2 d^2 y + a \, dy \, dx + y \, dx^2 = 0$, then if $\alpha = -1/2a$, conclude that u is a solution to $a^2 d^2 u + (3/4)u \, dx^2 = 0$.

16. Solve $a^2 d^2 u + (3/4)u \, dx^2 = 0$. First multiply by du and integrate once to get $4a^2 \, du^2 = (K^2 - 3u^2)dx^2$ or

$$dx = \frac{2a}{\sqrt{K^2 - 3u^2}}du.$$

Integrate a second time to get

$$x = \frac{2a}{\sqrt{3}} \arcsin \frac{\sqrt{3}u}{K} - f.$$

Rewrite this equation for u in terms of x as

$$u = C \sin\left(\frac{(x+f)\sqrt{3}}{2a}\right).$$

17. Find the natural logarithms of the three cube roots of 1 and of the five fifth roots of 1.

18. Find the curve joining two points in the upper half-plane, which, when revolved around the x axis, generates a surface of minimal surface area. If $y = f(x)$ is the equation of the curve, then the desired surface area is $I = 2\pi \int y \, ds = 2\pi \int y\sqrt{1 + y'^2} \, dx$. So use the Euler equation in the modified form $F - y'(\partial F/\partial y') = c$, where $F = y\sqrt{1 + y'^2}$. (*Hint:* Begin by multiplying the equation through by $\sqrt{1 + y'^2}$.)

19. Determine a procedure for finding the differential equation of a family of orthogonal trajectories to a given family $f(x, y, \alpha) = 0$. (Use the fact that orthogonal lines have negative reciprocal slopes.) Use your procedure to find the family orthogonal to the family of hyperbolas $x^2 - y^2 = a^2$.

20. Determine and solve the differential equation for the family of synchrones, the family orthogonal to the family of brachistochrones.

21. Solve the differential equation $(2xy^3 + 6x^2 y^2 + 8x) \, dx + (3x^2 y^2 + 4x^3 y + 3) \, dy = 0$ using Clairaut's method.

22. Use Clairaut's technique of multiple integration to calculate the volume of the solid bounded by the cylinders $ax = y^2$, $by = z^2$ and the coordinate planes. First determine the volume element $dx \int z \, dy$ by converting the integrand to a function of x and integrating. Then integrate the volume element with appropriate limits. Compare this method to the standard modern method.

23. Suppose that x and y are given in terms of t and u by the functions

$$x = \frac{t}{\sqrt{1+u^2}}, \qquad y = \frac{tu}{\sqrt{1+u^2}}.$$

Show that the change-of-variable formula is given by

$$dx \, dy = \frac{t \, dt \, du}{1+u^2}.$$

24. Suppose that the solution to the wave equation $\frac{\partial^2 y}{\partial t^2} = \frac{\partial^2 y}{\partial x^2}$ is given by $y = \Psi(t+x) - \Psi(t-x)$. Show that the initial conditions $y(0, x) = f(x)$, $y'(0, x) = g(x)$ and the condition $y(t, 0) = y(t, l) = 0$ for all t lead to the requirements that $f(x)$ and $g(x)$ are odd functions of period $2l$ (d'Alembert).

25. Suppose that $y = F(t)G(x) = \Psi(t+x) - \Psi(t-x)$ is a solution to the wave equation $\frac{\partial^2 y}{\partial t^2} = \frac{\partial^2 y}{\partial x^2}$. Show by differentiating twice that $\frac{F''}{F} = \frac{G''}{G} = C$, where C is some constant, and therefore that $F = ce^{t\sqrt{C}} + de^{-t\sqrt{C}}$ and $G = c'e^{x\sqrt{C}} + d'e^{-x\sqrt{C}}$. Apply the condition $y(t, 0) = y(t, l) = 0$ to show that C must be negative, and hence derive the solution $F(t) = A \cos Nt$, $G(x) = B \sin Nx$ for the appropriate choice of A, B, and N (d'Alembert).

26. Find the isosceles triangle of smallest area that circumscribes a circle of radius 1 (Simpson).

27. Find the cone of least surface area with given volume V (Simpson).

28. Show that $w = (b^3 - x^3)(x^2 z - z^3)(xy - y^2)$ has a maximum when $x = \frac{1}{2}b\sqrt[3]{5}$, $y = \frac{1}{4}b\sqrt[3]{5}$, and $z = \frac{b\sqrt[3]{5}}{2\sqrt{3}}$ (Simpson).

29. Calculate the first four nonzero terms of the power series for $y = \cos z$ using Maclaurin's technique without explicitly using the derivatives of the cosine or sine. Assume that the radius of the circle is 1.

30. Find the point of maximal curvature on the curve defined by the differential equation $a(dy/y) = dx$ (Agnesi).

31. Given a rectangle, find the line of minimum length that passes through one vertex and through the extensions of the two opposite sides (Agnesi).

32. Sketch a particular example of the "witch of Agnesi," the curve given by $y^2 = \frac{4(2-x)}{x}$. Show that it is symmetric about the x axis and asymptotic to the y axis. Show that the area between the curve and its asymptote is 4π.

33. Assume that after the flood the human population was 6 and that 200 years later the population was 1,000,000. Find the annual rate of growth of the population (Euler). (*Hint:* If the annual rate of growth is $1/x$, then the equation for the problem is

$$6\left(\frac{1+x}{x}\right)^{200} = 1{,}000{,}000.)$$

34. Show that k in the series given in the text for a^z and $\log_a(1+x)$ is given by $k = \ln a$ (Euler).

35. Derive the power series for $\ln(1+x)$ from the equation $\ln(1+x) = j(1+x)^{1/j} - j$ by using the binomial theorem and assuming that j is infinitely large (Euler).

36. Replace x by ix in the expansion in the text for $(e^x - e^{-x})/2$ to get both the power series for the sine and a representation of the sine as an infinite product. By using the relationship between the roots and coefficients of a polynomial (extended to power series), show that

$$\sum_{k=1}^{\infty} \frac{1}{k^2} = \frac{\pi^2}{6} \quad \text{and} \quad \sum_{k=1}^{\infty} \frac{1}{k^4} = \frac{\pi^4}{90}.$$

37. Determine all the relative extrema for $V = x^3 + y^2 - 3xy + (3/2)x$, and for each one determine whether it is a maximum or minimum. Compare your answer with that of Euler.

38. If $y = \arctan x$, show that $\sin y = x/\sqrt{1+x^2}$ and $\cos y = 1/\sqrt{1+x^2}$. Then, if $p = x/\sqrt{1+x^2}$, show that $\sqrt{1-p^2} = 1/\sqrt{1+x^2}$. Since $y = \arcsin p$, it follows that $dy = dp/\sqrt{1-p^2}$ and $dp = dx/(1+x^2)^{3/2}$. Conclude that

$$dy = \frac{dx}{1+x^2} \quad \text{(Euler)}.$$

39. Calculate dy for $y = a^x$ by noting that $dy = a^{x+dx} - a^x = a^x(a^{dx} - 1)$ and then expanding $a^{dx} - 1$ into the power series $\ln a \, dx + \frac{(\ln a)^2 dx^2}{2} + \cdots$ (Euler).

40. Calculate dy for $y = \tan x$ by using the addition formula

$$\tan(x + dx) = \frac{\tan x + \tan dx}{1 - \tan x \tan dx} \quad \text{(Euler)}.$$

41. Use Euler's procedure for integrating rational functions to find $\int \frac{(1+x)dx}{1-x+x^2}$. First show that $\cos \zeta = \frac{1}{2}$.

42. Use Euler's procedure for integrating functions involving square roots to determine $\int \frac{dx}{\sqrt{2+5x+3x^2}}$. First, factor $2 + 5x + 3x^2$.

43. Translate d'Alembert's definition of a limit into algebraic language and compare with the modern definition of a limit.

44. Use Lagrange's technique to calculate the quantities p, q, r for the function $f(x) = \sqrt{x}$ and thus determine the first three terms of its power series representation.

45. Show why Lagrange's power series representation fails for the case $f(x) = e^{-1/x^2}$.

46. Given that $f(x + i) = f(x) + pi + qi^2 + ri^3 + \cdots$, show that $p = f'(x)$, $q = f''(x)/2!$, $r = f'''(x)/3!$, ...

47. Did eighteenth-century mathematicians prove or use the fundamental theorem of calculus in the sense it is used today? What concepts must be defined before one can even consider this theorem? How are these concepts dealt with by eighteenth-century mathematicians? Did these mathematicians consider the fundamental theorem as "fundamental"?

48. Develop a lesson to enhance students' understanding of the fundamental theorem of calculus by using the work of Maclaurin and Lagrange.

49. Compare Euler's trilogy of precalculus and calculus texts to a modern series. What items are common? What does Euler have that is missing in today's texts, and conversely? Could one use Euler's texts today?

50. Trace the development of the concept of the limit from Newton through Maclaurin to d'Alembert. How do their formulations agree? How do they compare with the modern formulation of this concept?

51. Develop several lessons teaching the basic methods for solving various classes of differential equations using the formulations of Leibniz, Clairaut, and Euler.

REFERENCES AND NOTES

Three books that cover aspects of the history of analysis in the eighteenth century are S. B. Engelsman, *Families of Curves and the Origins of Partial Differentiation* (Amsterdam: Elsevier, 1984); Umberto Bottazzini, *The Higher Calculus: A History of Real and Complex Analysis from Euler to Weierstrass* (New York: Springer-Verlag, 1986); and Ivor Grattan-Guinness, *The Development of the Foundations of Mathematical Analysis from Euler to Riemann* (Cambridge: MIT Press, 1970). Two good survey articles dealing with this material are H. J. M Bos, "Calculus in the Eighteenth Century—the Role of Applications," *Bulletin of the Institute of Mathematics and Its Applications* 13 (1977), 221–227, and Craig Fraser, "The Calculus As Algebraic Analysis: Some Observations on Mathematical Analysis in the 18th Century," *Archive for History of Exact Sciences* 39 (1989), 317–336.

Johann Bernoulli's solution to the brachistochrone problem is available in David Eugene Smith, *A Source Book in Mathematics* (New York: Dover, 1959), vol. 2, pp. 644–655. D. J. Struik, *A Source Book in Mathematics, 1200–1800* (Cambridge: Harvard University Press, 1969), provides translations of the important parts of the papers involved in the controversy over vibrating strings. The book also includes selections from Berkeley's *Analyst*, from Euler's own discussion of the metaphysics of the calculus, and from Lagrange's *Theory of Analytic Functions*. Maria Agnesi's *Foundations of Analysis* was translated into English as *Analytical Institutions*, John Colson, trans. (London: Taylor and Wilks, 1801). Euler's *Introductio* has been translated in two volumes as *Introduction to Analysis of the Infinite*, John D.

Blanton, trans. (New York: Springer-Verlag, 1988, 1990). The first volume of Euler's *Differential Calculus* is also available: *Foundations of Differential Calculus*, John D. Blanton, trans. (New York: Springer-Verlag, 2000).

1. David Eugene Smith, *A Source Book in Mathematics*, vol. 2, p. 646.

2. Ibid., p. 645. (The original problem as well as Johann Bernoulli's solution are presented here.)

3. Ibid., p. 647.

4. Ibid., p. 654.

5. Jakob Bernoulli's solution to the brachistochrone problem appears in Struik, *A Source Book*, pp. 396–399.

6. C. Truesdell, "The Rational Mechanics of Flexible or Elastic Bodies: 1638–1788," in Leonhard Euler, *Opera omnia* (Leipzig, Berlin, and Zurich: Societas Scientarum Naturalium Helveticae, 1911–), (2) 11, part 2, p. 166. This article contains many details on aspects of Euler's work. It is an introduction to a section of Euler's collected works, now totaling over 70 volumes, which has been appearing in four series since 1911 and is still not complete. In general, references to Euler's papers here and in the next few chapters will include a reference to their location in this set as well as their Eneström numbers. A more detailed treatment of the invention of the calculus of the sine and cosine functions is found in Victor J. Katz, "The Calculus of the Trigonometric Functions," *Historia Mathematica* 14 (1987), 311–324.

More information on the work of Euler is found in several new books published in honor of his 300th birthday. These include Robert E. Bradley, Lawrence A. D'Antonio, C. Edward Sandifer, eds., *Euler at 300: An Appreciation* (Washington: MAA, 2007), C. Edward Sandifer, *How Euler Did It* (Washington: MAA, 2007), C. Edward Sandifer, *The Early Math of Leonhard Euler* (Washington: MAA, 2007), William Dunham, ed., *The Genius of Euler: Reflections* (Washington: MAA, 2007), as well as the book mentioned in note 10.

7. G. Eneström, "Der Briefwechsel zwischen Leonhard Euler und Johann I Bernoulli," *Bibliotheca Mathematica* (3) 6 (1905), 16–87, p. 31.

8. Ibid., p. 46.

9. John Fauvel and Jeremy Gray, *The History of Mathematics: A Reader* (London: Macmillan, 1987), p. 439.

10. This paper (E807) may be found in its entirety, translated by Todd Doucet, in *Convergence*, at http://convergence.mathdl.org. It is also in *Opera omnia* (2) 19, pp. 417–438. This quotation is from §26, p. 14. An analysis of this paper and other details of Euler's work on logarithms is found in Robert Bradley, "Euler, D'Alembert and the Logarithm Function," in Robert Bradley and C. Edward Sandifer, eds., *Leonhard Euler: Life, Work, and Legacy* (Amsterdam: Elsevier, 2007), pp. 255–278.

11. Ibid., §34, p. 19.

12. This treatment of Euler's version of the calculus of variations is due to Steven Schot, of American University, in some unpublished lectures. A detailed history of the calculus of variations can be found in Herman H. Goldstine, *A History of the Calculus of Variations from the 17th through the 19th Century* (New York: Springer Verlag, 1980).

13. Engelsman, *Families of Curves*, p. 106. This work is the best presentation of the history of partial differentiation and the source of much of the material for this section.

14. D. J. Struik, *Source Book*, p. 355. A discussion of the entire history of the vibrating string problem can be found in Jerome R. Ravetz, "Vibrating Strings and Arbitrary Functions," a chapter in *The Logic of Personal Knowledge: Essays Presented to Michael Polanyi on his Seventieth Birthday, 11 March 1961* (London: Routledge and Paul, 1961), 71–88, as well as in chapter 1 of Bottazzini, *Higher Calculus*, and in C. Truesdell, "Rational Mechanics." A comprehensive study of the work of d'Alembert is Thomas Hankins, *Jean d'Alembert: Science and the Enlightenment* (Oxford: Clarendon Press, 1970).

15. Quoted in Truesdell, "Rational Mechanics," p. 239.

16. Struik, *Source Book*, p. 361.

17. Ibid.

18. Frances Marguerite Clarke, *Thomas Simpson and His Times* (New York: Columbia University, 1929), p. 16. This work is the only biography of Thomas Simpson and provides many details of his life and work.

19. Thomas Simpson, *A New Treatise of Fluxions* (London: Gardner, 1737), p. 179.

20. Colin Maclaurin, *A Treatise of Fluxions* (Edinburgh: Ruddimans, 1742), p. 694.

21. Ibid., p. 753. For more details on Maclaurin's calculus and its influence, see Judith Grabiner, "Was Newton's Calculus a Dead End? The Continental Influence of Maclaurin's *Treatise of Fluxions*," *American Mathematical Monthly* 104 (1997), 393–410.

22. Edna Kramer, "Maria Agnesi," *Dictionary of Scientific Biography*, vol. 1, p. 76. A more recent work on Agnesi is C. Truesdell, "Maria Gaetana Agnesi," *Archive for History of Exact Sciences* 40 (1989), 113–147.

23. Euler, *Introduction to Analysis of the Infinite, Book I* (E101), p. v. This work is the first volume of a recent English translation of Euler's *Introductio*. It repays a careful reading as it has much information that could not be summarized in the text. See also Carl Boyer, "The Foremost Textbook of Modern Times (Euler's *Introductio in analysin infinitorum*)," *American Mathematical Monthly* 58 (1951), 223–226.

24. Euler, *Introduction, Book I*, p. 3. For more on the notion of functions, see A. P. Youshkevitch, "The Concept of Function Up to the Middle of the Nineteenth Century," *Archive for History of Exact Sciences* 16 (1982), 37–85.

25. Euler, *Introduction, Book II* (E102), p. 6.

26. Euler, *Foundations of Differential Calculus* (E212), p. vii.

27. Ibid., p. vi.

28. Ibid., p. xi.

29. Ibid., p. 64.

30. Ibid., p. 77.

31. George Berkeley, *The Analyst*, in James Newman, *The World of Mathematics* (New York: Simon and Schuster, 1956), vol. 1, pp. 288–289. Selections of Berkeley's work are reprinted here and also in Struik, *Source Book*, pp. 333–338.

32. Ibid., pp. 291–292.

33. Struik, *Source Book*, p. 335.

34. Newman, *World of Mathematics*, p. 289.

35. Ibid., p. 290.

36. Maclaurin, *Treatise of Fluxions*, p. 1. A brief biography of Colin Maclaurin appears in H. W. Turnbull, *Bicentenary of the Death of Colin Maclaurin* (Aberdeen, University Press, 1951). See also C. Tweedie, "A Study of the Life and Writings of Colin Maclaurin," *Mathematical Gazette* 8 (1915), 132–151, and 9 (1916), 303–305, and H. W. Turnbull, "Colin Maclaurin," *American Mathematical Monthly* 54 (1947), 318–322.

37. Maclaurin, *Treatise of Fluxions*, p. 4.

38. Ibid., p. 53.

39. Ibid., p. 55.

40. Ibid., p. 181.

41. Ibid., p. 420.

42. Ibid., p. 421.

43. Ibid., p. 423.

44. Ibid., p. 38.

45. Euler, *Foundations of Differential Calculus*, p. 51.

46. Ibid.

47. Ibid., p. 52.

48. Struik, *Source Book*, p. 345. This selection is from the article "Différentiel" by d'Alembert in the *Encyclopédie*.

49. Ibid., pp. 343–344.

50. From the article "Limite," in the *Encyclopédie*, quoted in Bottazzini, *The Higher Calculus*, p. 37.

Probability and Statistics in the Eighteenth Century

The only thing needed for correctly forming conjectures on any matter is to determine the numbers of these cases accurately and then to determine how much more easily some can happen than others. But here we come to a halt, for this can hardly ever be done. Indeed, it can hardly be done anywhere except in games of chance. . . . But what mortal, I ask, may determine, for example, the number of diseases, as if they were just as many cases, which may invade at any age the innumerable parts of the human body and which imply our death? And who can determine how much more easily one disease may kill than another? . . . Likewise who will count the innumerable cases of the changes to which the air is subject every day and on this basis conjecture its future constitution after a month, not to say after a year?

—Jakob Bernoulli's *Ars conjectandi*, 1713[1]

Years after Euler was asked by Frederick the Great on September 15, 1749, to investigate the possible profits that running a lottery could provide as well as the possible hazards that it could engender, Euler made a detailed study of lotteries and presented it to the Berlin Academy of Sciences. He had already answered the king's questions about a specific lottery, but then generalized this to other lotteries. He not only developed the mathematics of such lotteries but also made recommendations as to the pricing of tickets and the amount of profit to be realized.

In this chapter, we consider eighteenth-century developments in probability and statistics. Jakob Bernoulli took up Huygens's work in probability and extended it, ultimately proving what is today known as the Law of Large Numbers. Abraham de Moivre carried this work even further by applying his knowledge of series to develop the curve of normal distribution and some of its properties. Thomas Bayes and Pierre-Simon de Laplace then attempted to answer questions of inverse probability, that is, how to determine probability from a consideration of certain empirical data. Their answers have proved somewhat controversial ever since. Meanwhile, various mathematicians applied some of the basic ideas of probability to such questions as observational errors, the pricing of annuities, and the mathematics of lotteries.

 18.1   # THEORETICAL PROBABILITY

The early work on probability discussed in Chapter 14 was chiefly concerned with the question of determining expectations and the associated probabilities in cases arising from various types of games or other gambling questions. But Pascal's idea of a "fair" distribution of the stakes in an interrupted game and Huygens's interest in "equitable" games show that probability in its beginnings was closely related to the notion of an **aleatory** contract, a contract providing for the exchange of a present certain value for a future uncertain one. Such contracts included annuities and maritime insurance policies in which a certain sum of money was paid now in exchange for an unknown sum to be returned at a later date under certain conditions. For the contract to be "fair," the mathematicians argued, one needed to be able somehow to quantify the risk involved.

In the case of certain types of games, the early practitioners were able to work out efficient ways of counting successes and failures and thus to determine the expectation or probability a priori. In most realistic situations, however, it was much more difficult to quantify risk, that is, to determine the degree of belief that a "reasonable man" would have. How could one determine a "reasonable" price to pay for insurance? As indicated in the opening quotation, Jakob Bernoulli, in his study of the subject over some 20 years, wanted to be able to quantify risk in situations where it was impossible to enumerate all possibilities. To do this, he proposed to ascertain probabilities a posteriori by looking at the results observed in many similar instances, that is, by considering some statistics. "If, for example, there once existed three hundred people of the same age and body type as Titius now has, and you observed that two hundred of them died before the end of a decade, while the rest lived longer, you could safely enough conclude that there are twice as many cases in which Titius also may die within a decade as there are cases in which he may live beyond a decade."[2]

18.1.1 Jakob Bernoulli and the *Ars Conjectandi*

It seemed reasonably obvious to Bernoulli that the more observations one made of a given situation, the better one would be able to predict future occurrences. But he wanted to give a "scientific proof" of this principle, a proof he finally found before his death in 1705. Bernoulli presented this scientific proof in his Law of Large Numbers and placed it in the fourth and final part of his important text on probability, the *Ars conjectandi* (*Art of Conjecturing*), a work not published until 1713.

The first three parts of the *Ars conjectandi* were more in the spirit of earlier work on probability. Part One contained a reprint of Huygens's 1657 *De Ratiociniis in aleae ludo*, with substantial added commentary, often aimed at giving more general results or better ways of solving the problems. For example, consider how Bernoulli solved Huygens's proposition 11 (Section 14.3.3), which asked to determine the number of throws of two dice necessary to give an even chance of throwing a double six. Bernoulli first generalized the problem by considering the case where there are c possible outcomes of a throw, of which b do not give the desired outcome. Thus, in n throws, the total number of cases will be c^n, while the number of cases in which the desired outcome does not appear will be b^n. It follows that the desired outcome will occur in each of the remaining $c^n - b^n$ cases. Thus, for an even chance of this outcome, we must have $b^n = c^n - b^n$, or $c^n = 2b^n$. Using logarithms, we get

$$n \log c = \log 2 + n \log b \quad \text{or} \quad n = \frac{\log 2}{\log c - \log b}.$$

In Huygens's particular case, $c = 36$ and $b = 35$, thus $n = \log 2/(\log 36 - \log 35)$, a quotient more than 24 and less than 25, as Huygens had himself concluded.

Part Two of the *Ars conjectandi* developed anew various laws of permutations and combinations because, Bernoulli claimed, the most frequent error in answering probability questions is the "insufficient enumeration" of all possibilities. Therefore, "the Art called Combinatorics should be judged . . . most useful, because it remedies this defect of our minds and teaches us how to enumerate all possible ways in which several things can be combined, transposed, or joined with each other, so that we may be sure that we have omitted nothing that can contribute to our purpose."[3] Among the applications of these laws that Bernoulli discussed here was a generalization of Pascal's ideas on the division of stakes in an interrupted game to the case where the chances of the two players winning a game are not equal, or, more generally, to the case of an experiment in which the chances of success or failure are not equal. Bernoulli showed that if the chance of success is a while that of failure is b (out of $a + b$ trials), then the probability of r successes in n trials is the ratio of $\binom{n}{n-r}a^r b^{n-r}$ to $(a + b)^n$. Similarly, the probability of at least r successes in n trials is the ratio of $\sum_{j=0}^{n-r} \binom{n}{j} a^{n-j} b^j$ to $(a + b)^n$.

A second application of Pascal's arithmetic triangle in this part was Bernoulli's calculation of the sums of integral powers. Bernoulli surpassed ibn al-Haytham and Jyesthadeva in this regard, not only by writing out formulas for the sums of the integral powers up to order 10 but also by noting a pattern that gave him a general result for any power c:

$$\sum_{j=1}^{n} j^c = \frac{1}{c+1}n^{c+1} + \frac{1}{2}n^c + \frac{c}{2}B_2 n^{c-1} + \frac{c(c-1)(c-2)}{2 \cdot 3 \cdot 4}B_4 n^{c-3}$$

$$+ \frac{c(c-1)(c-2)(c-3)(c-4)}{2 \cdot 3 \cdot 4 \cdot 5 \cdot 6}B_6 n^{c-5} + \cdots,$$

where the series ends at the last positive power of n and where $B_2 = \frac{1}{6}$, $B_4 = -\frac{1}{30}$, $B_6 = \frac{1}{42}$, ... These latter quantities, today called the **Bernoulli numbers**, may be calculated by noting that on the first sum in which a given one occurs, it is that number which "completes to unity" the sum of the previous coefficients of the powers of n. Thus, because $\sum j^4 = \frac{1}{5}n^5 + \frac{1}{2}n^4 + \frac{1}{3}n^3 + B_4 n$, $B_4 = 1 - \frac{1}{5} - \frac{1}{2} - \frac{1}{3} = -\frac{1}{30}$.[4]

In Part Three of the *Ars conjectandi,* Bernoulli applied his results to numerous games of chance, both ones that he invented as well as ones that were actually played at the time, but it is Part Four for which the work is most famous. This part is entitled *The Use and Application of the Preceding Doctrine in Civil, Moral, and Economic Matters*, but a glance at the published book shows that this part is much shorter than Bernoulli must have intended. Although he began by setting out various principles and giving examples for using probability to evaluate arguments in criminal trials, even developing some algebraic formulas to help, he never actually applied probability to the "civil, moral, and economic matters" that he promised. Still, since in most real-life situations, the probability of a particular event cannot be determined a priori, as in the throws of a die, he was successful in showing how to determine such probabilities a posteriori. He also realized that in most real situations, absolute certainty (or probability equal to 1) is impossible to achieve. Thus, Bernoulli introduced the idea of **moral certainty**. He decided that for an outcome to be morally certain, it should have a probability no less than 0.999. Conversely, an outcome with probability no greater than 0.001 he considered to be morally impossible. The goal of Bernoulli's theorem, then, was to show that "as the number of observations increases, so the probability increases of obtaining the true ratio between the numbers of cases in which some event can happen and not happen, such that this probability may eventually exceed any given degree of certainty."[5] It was to determine from experimental evidence the true probability of an event, with moral certainty, that Bernoulli formulated his theorem, the Law of Large Numbers, which occupies the major portion of Part Four.

To understand the discussion of the theorem, one should keep in mind one of Bernoulli's examples. Suppose there is an urn containing 3000 white and 2000 black pebbles, although that number is unknown to the observer. The observer wants to determine the proportion of white to black by taking out, in turn, a certain number of pebbles and recording the outcome, at each step always replacing each pebble before taking out the next. Thus, in what follows, an observation is the removal of one pebble and a success is that the pebble is white. Assume then that N observations are made, that X of these are successes, and that $p = r/(r + s)$ is the (unknown) probability of a success. (Here r is the total of successful cases and s the total of unsuccessful ones. In the example, $p = 3/5$.)

In modern terminology, the theorem states that given any small fraction $\epsilon = \frac{1}{r+s}$ and any large positive number c, a number $N = N(c)$ may be found so that the probability that X/N differs from p by no more than ϵ is greater than c times the probability that X/N differs from p by more than ϵ. In symbols, this result can be written as

$$P\left(\left|\frac{X}{N} - p\right| \leq \epsilon\right) > cP\left(\left|\frac{X}{N} - p\right| > \epsilon\right).$$

In other words, the probability that X/N is "close" to p is very much greater than the probability that it is not "close." Bernoulli's statement can easily be converted into the standard modern formulation: Given any $\epsilon > 0$ and any positive number c, there exists an N such that

$$P\left(\left|\frac{X}{N} - p\right| > \epsilon\right) < \frac{1}{c+1}.$$

Because Bernoulli considered the basic statement of the theorem virtually intuitive, he felt his main contribution would be to determine the value $N(c)$ from which he could recover the true probability $p = r/(r + s)$ with "moral certainty," that is, with $c = 1000$. In fact, he showed that, if $t = r + s$, then $N(c)$ could be taken to be any integer greater than the larger of

$$mt + \frac{st(m - 1)}{r + 1} \quad \text{and} \quad nt + \frac{rt(n - 1)}{s + 1},$$

where m, n are integers such that

$$m \geq \frac{\log c(s - 1)}{\log(r + 1) - \log r} \quad \text{and} \quad n \geq \frac{\log c(r - 1)}{\log(s + 1) - \log s}.$$

In his example, Bernoulli calculated that for $r = 30$ and $s = 20$, the second expression was larger and therefore $N = 25,550$ for $c = 1000$. In other words, Bernoulli's result enabled him to know that 25,550 observations would be sufficient for moral certainty that the relative frequency found would be within 1/50 of the true proportion 3/5. Bernoulli's text ended with this calculation and similar ones for other values of c, perhaps because he was unhappy with this result. For the early 1700s, 25,550 was an enormous number, larger than the entire population of Basel, for example. What the result seemed to say was that nothing reliable could be learned in a reasonable number of observations. Bernoulli may have felt that he had failed in his quest to quantify the measure of uncertainty, especially since his intuition told him that 25,550 was much larger than necessary.[6] So perhaps this was one of the reasons that he did not include the promised applications of his method. Nevertheless, Bernoulli pointed the way toward a more successful attack on the problem by his slightly younger contemporary, Abraham De Moivre (1667–1754).

18.1.2   De Moivre and *The Doctrine of Chances*

De Moivre's major mathematical work was *The Doctrine of Chances*, first published in 1718, with new editions in 1738 and 1756. This probability text is much more detailed than the work of Huygens, partly because of the general advances in mathematics since 1657. He even began with a precise definition of probability, one that had not been given explicitly before: "The Probability of an Event is greater, or less, according to the number of Chances by which it may happen, compared with the whole number of Chances by which it may either happen or fail." Thus, it is clear that "the Probabilities of happening and failing being added together, their Sum will always be equal to Unity."[7] De Moivre could then use his definition in solving problems, such as the dice problem of de Méré. Like Bernoulli, he solved it as part of a more comprehensive problem:

PROBLEM III   *To find in how many trials an event will probably happen, or how many trials will be necessary to make it indifferent to lay on its happening or failing, supposing that a is the number of chances for its happening in any one trial and b the number of chances for its failing.*[8]

De Moivre began his solution, similar to Bernoulli's, by noting that if there are x trials, then $\frac{b^x}{(a+b)^x}$ is the probability for the event failing x consecutive times. Since there are to be even odds as to whether the event happens at least once in x trials, this probability must equal

BIOGRAPHY

Abraham De Moivre (1667–1754)

De Moivre was born in Vitry, a town in France about a hundred miles east of Paris, into a Protestant family. Between the ages of 11 and 14, he was educated in the classics at the Protestant secondary school in Sedan, but after the school was closed in 1681, he studied first in Saumur and then in Paris. At Saumur, he read Huygens's probability text and in Paris he studied physics as well as the standard mathematics curriculum beginning with Euclid. Soon after the revocation of the edict of Nantes in 1685 made life for Protestants in France very difficult, De Moivre was imprisoned for more than two years. When he was freed in April of 1688, he left France for England, never to return. It was in England that De Moivre mastered Newton's theory of fluxions and began his own original work. Although he was elected to the Royal Society in 1697, he never achieved a university position. He made his living by tutoring and by solving problems arising from games of chance and annuities for gamblers and speculators.

1/2, that is, x must satisfy the equation

$$\frac{b^x}{(a+b)^x} = \frac{1}{2} \quad \text{or} \quad (a+b)^x = 2b^x.$$

De Moivre easily solved this equation by taking logarithms:

$$x = \frac{\log 2}{\log(a+b) - \log b}.$$

Furthermore, he noted that if $a : b = 1 : q$, so that the odds against a success are q to 1, then the original equation can be rewritten in the form

$$\left(1 + \frac{1}{q}\right)^x = 2 \quad \text{or} \quad x \log\left(1 + \frac{1}{q}\right) = \log 2.$$

By expanding $\log(1 + \frac{1}{q})$ in a power series, De Moivre concluded that if q is very large, then the first term $1/q$ of the series is sufficient and the solution can be written as $x = q \log 2$ or $x \approx 0.7q$. Thus, to solve de Méré's specific problem of finding how many throws of two dice are necessary to give even odds of throwing two sixes, De Moivre simply noted that $q = 35$, so $x = 24.5$. The required number of throws is therefore between 24 and 25, the same answer found by both Huygens and Bernoulli.

De Moivre often used infinite series to perform his probability calculations. But more important than these calculations themselves is his detailed discussion of approximating the sum of terms of the binomial $(a + b)^n$, printed as an appendix to *The Doctrine of Chances* in its second and third editions, although first written in 1733, and in which appeared for the first time the so-called normal approximation to the binomial distribution. De Moivre's aim in his discussion, like that of Bernoulli, was to estimate probability by means of experiment: "Supposing for instance that an event might as easily happen as not happen, whether after three thousand experiments it may not be possible it should have happened two thousand times and failed a thousand; and that therefore the odds against so great a variation from equality should be assigned, whereby the mind would be the better disposed in the conclusions derived from the experiments."[9] For De Moivre, as for Bernoulli, the method of calculating the relevant probabilities lay in the calculation of certain binomial coefficients. He initially restricted himself to equally likely occurrences and sought to find the probability of $n/2$ successes in n trials, that is, the ratio that the middle term of $(1 + 1)^n$ has to the sum of all the terms, 2^n, for n large and even. He determined that this ratio $\binom{n}{n/2} : 2^n$ approached $\frac{2T(n-1)^n}{n^n\sqrt{n-1}}$ as n became large, where

$$\log T = \frac{1}{12} - \frac{1}{360} + \frac{1}{1260} - \frac{1}{1680} + \cdots = \frac{B_2}{1 \cdot 2} + \frac{B_4}{3 \cdot 4} + \frac{B_6}{5 \cdot 6} + \frac{B_8}{7 \cdot 8} + \cdots,$$

the B_i being the Bernoulli numbers.

De Moivre's derivation of this result showed his great familiarity with infinite series and with logarithms. He began by noting that the middle term $M = \binom{n}{n/2} = n!/(n/2)!^2$, where $n = 2m$, can be written as

$$M = \frac{(m + 1)(m + 2) \cdots (m + (m - 1))(m + m)}{(m - 1)(m - 2) \cdots (m - (m - 1))m}.$$

It follows that $\log M$ can be written as a sum of logarithms of quotients of the factors. Each of these logarithms can then be expanded in a power series in $1/m$. Thus,

$$\log M = \log \frac{m + 1}{m - 1} + \log \frac{m + 2}{m - 2} + \cdots + \log \frac{m + (m - 1)}{m - (m - 1)} + \log 2,$$

and, for example,

$$\log \frac{m + 1}{m - 1} = \log \frac{1 + \frac{1}{m}}{1 - \frac{1}{m}} = 2\left(\frac{1}{m} + \frac{1}{3m^3} + \frac{1}{5m^5} + \cdots\right)$$

and

$$\log \frac{m + 2}{m - 2} = \log \frac{1 + \frac{2}{m}}{1 - \frac{2}{m}} = 2\left(\frac{2}{m} + \frac{8}{3m^3} + \frac{32}{5m^5} + \cdots\right).$$

De Moivre then cleverly noted that the sum of these power series can be determined by adding vertically instead of horizontally. Thus, except for the term $\log 2$, this sum can be expressed as the sum of the following columns, where $s = m - 1$:

$$\text{col. } 1 = \frac{2}{m}(1 + 2 + \cdots + s)$$

$$\text{col. } 2 = \frac{2}{3m^3}\left(1^3 + 2^3 + \cdots + s^3\right)$$

$$\text{col. } 3 = \frac{2}{5m^5}\left(1^5 + 2^5 + \cdots + s^5\right)$$

$$\vdots$$

The columns, because they involve sums of integral powers, can be calculated using Bernoulli's formulas to write each sum as a polynomial in s. De Moivre added the highest-degree terms of each polynomial together, getting a power series that he could express in finite terms as $(2m - 1)\log(2m - 1) - 2m \log m$. Similarly, the sum of the second-highest-degree terms of each polynomial formed the power series expressing the function

(1/2) log(2m − 1). The sums of the third-, fourth-, . . . , highest-degree terms are more difficult to determine, but De Moivre showed that in the limit as m approaches infinity, these become 1/12, −1/360, ... Remembering the extra term log 2, De Moivre concluded that the logarithm of M is

$$\left(2m - \frac{1}{2}\right) \log(2m - 1) - 2m \log m + \log 2 + \frac{1}{12} - \frac{1}{360} + \cdots$$

and, subtracting off log $2^n = \log 2^{2m} = 2m \log 2$, that the logarithm of the ratio $M : 2^n$ is

$$n \log(n - 1) - \frac{1}{2} \log(n - 1) - n \log n + \log 2 + \frac{1}{12} - \frac{1}{360} + \cdots.$$

Thus, $M : 2^n = \frac{2T(n-1)^n}{n^n \sqrt{n-1}}$ as stated.

Because De Moivre wanted to be able to calculate with this ratio, he showed by use of the series for log T that 2T is approximately 2.168 = 2 21/125. He also determined, by a method similar to that described above, that for m large,

$$\log m! = \sum_{k=1}^{m} \log k \approx \left(m + \frac{1}{2}\right) \log m - m + \log B \quad \text{or} \quad m! \approx Bm^{m+\frac{1}{2}}e^{-m},$$

where log $B − 1 = \log T$, a formula today named for James Stirling (1692–1770). Stirling was responsible for calculating that $B = \sqrt{2\pi}$, probably by an argument similar to De Moivre's but starting from Wallis's product for π. It followed then that log $T = 1 - \frac{1}{2} \log 2\pi$, or that $T = e/\sqrt{2\pi}$. Since De Moivre knew that if n is large, then $\frac{(n-1)^n}{n^n} = (1 - \frac{1}{n})^n$ approximates e^{-1}, he concluded that the ratio of the middle term M of $(1 + 1)^n$ to the sum 2^n is equal to $2/\sqrt{2\pi n}$.

To deal with terms other than the middle terms, De Moivre generalized his method somewhat and concluded that if Q is a term of the binomial expansion $(1 + 1)^n$ at a distance t from the middle term M, then

$$\log \frac{M}{Q} = \left(m + t - \frac{1}{2}\right) \log(m + t - 1) + \left(m - t + \frac{1}{2}\right) \log(m - t + 1)$$

$$- 2m \log m + \log \frac{m + t}{m},$$

where m = n/2. He concluded, again approximating the logarithms by power series, that for n large,

$$\log \left(\frac{Q}{M}\right) \approx -\frac{2t^2}{n} \quad \text{or} \quad Q \approx Me^{-\frac{2t^2}{n}}.$$

In modern notation, this means that

$$P(X = \frac{n}{2} + t) \approx P(X = \frac{n}{2})e^{-(2t^2/n)} = \frac{2}{\sqrt{2\pi n}}e^{-(2t^2/n)}.$$

De Moivre thought of the various values of $Q = P(X = \frac{n}{2} + t)$ as forming a curve: "If the terms of the binomial are thought of as set upright, equally spaced at right angles to and above a straight line, the extremities of the terms follow a curve. The curve so described has two inflection points, one on each side of the maximal term."[10] He calculated that the inflection points of this curve occurred at a distance $\frac{1}{2}\sqrt{n}$ from the maximum term. Thus, De Moivre had found what today we call the **normal curve**, here seen as an approximation to the binomial distribution.

Given his approximation to the individual terms of the binomial expansion and his representation of Q as a curve, De Moivre was able to calculate the sums of large numbers of such terms by integration and thus improve considerably on Bernoulli's quantification of uncertainty. Thus, to find

$$\sum_{t=0}^{k} P\left(X = \frac{n}{2} + t\right),$$

he approximated this sum by

$$\frac{2}{\sqrt{2\pi n}} \int_{0}^{k} e^{-(2t^2/n)} \, dt$$

and evaluated the integral by writing the integrand as a power series and integrating term by term. For $k = \frac{1}{2}\sqrt{n}$, the series converged rapidly enough for him to conclude that the sum was equal to 0.341344 and therefore that "if it was possible to take an infinite number of experiments, the probability that an event which has an equal number of chances to happen or fail shall neither appear more frequently than $\frac{1}{2}n + \frac{1}{2}\sqrt{n}$ nor more rarely than $\frac{1}{2}n - \frac{1}{2}\sqrt{n}$ times will be expressed by the double sum of the number exhibited . . . , that is, by 0.682688."[11] He therefore could conclude that, in modern terminology, for n large, the probability that the number of occurrences of a symmetric binomial experiment would fall within $\frac{1}{2}\sqrt{n}$ of the middle value $\frac{1}{2}n$ was 0.682688. De Moivre then calculated the corresponding values for various other multiples of $\sqrt{n}$. Thus, "to apply this to particular examples, it will be necessary to estimate the frequency of an event's happening or failing by the square root of the number which denotes how many experiments have been, or are designed to be taken; and this square root . . . will be as it were the *modulus* by which we are to regulate our estimation."[12] For De Moivre, $\sqrt{n}$ was the unit by which distances from the center were to be measured. Thus, the accuracy of a probability estimate increased as the square root of the number of experiments.

The discussion above applies only to cases where the chances of an event happening or failing are equal. But De Moivre did sketch a generalization of his method by showing how to approximate terms in $(a + b)^n$, where $a \neq b$. He concluded that if n is large and if M is the greatest term in the binomial expansion, then, first,

$$\frac{M}{(a + b)^n} \approx \frac{a + b}{\sqrt{2\pi abn}},$$

and, second, if Q is a term at distance t from M, then

$$Q \approx M e^{-\frac{(a+b)^2}{2abn}t^2}.$$

In modern notation, the first result means that

$$P(X = np) \approx \frac{1}{\sqrt{2\pi p(1 - p)n}},$$

while the second means that

$$P(X = np + t) \approx P(X = np)e^{-\frac{t^2}{2np(1-p)}},$$

where X is a binomial distribution with n observations and probability of success $p = a/(a + b)$, with np assumed to be an integer.

De Moivre's ideas could be used to show that far fewer experiments are necessary to achieve the accuracy demanded in Bernoulli's example. For instance, it can be shown that in the case where Bernoulli required 25,550 trials, De Moivre's method required but 6498. De Moivre himself, however, only gave examples in the equiprobable case. Thus, he showed, for example, that 3600 experiments will suffice to give the probability 0.682688 that an event will occur at least 1770 times and no more than 1830 times or the probability 0.99874 that an event will occur at least 1710 times and no more than 1890 times. Unfortunately, although De Moivre's results were in fact more precise than those of Bernoulli, he was not able to apply them. Apparently, he did not even recognize the importance of the curve he had developed other than having the $\sqrt{n}$ serve as a measure for estimating the accuracy of an experiment. Nevertheless, his work was to have profound influence on later developments in the century.

18.2   STATISTICAL INFERENCE

One reason that neither De Moivre's nor Bernoulli's work was immediately applied to real situations was that they did not directly answer the question necessary for applications, the question of statistical inference: Given empirical evidence that a particular event happened a certain number of times in a given number of trials, what is the probability of this event happening in general? De Moivre (and Bernoulli) could only tell how likely it was that observed frequencies approximated a given probability. It was Thomas Bayes and Pierre Laplace who first attempted a direct answer to the question of how to determine probability from observed frequencies.

18.2.1 Bayes and Statistical Inference

Thomas Bayes (1702–1761) gave his answer in his *An Essay towards Solving a Problem in the Doctrine of Chances*, written toward the end of his life and not published until three years after his death. Bayes began the essay with a statement of the basic problem: "Given the number of times in which an unknown event [i.e., an event of unknown probability] has happened and failed. Required the chance that the probability of its happening in a single trial lies somewhere between any two degrees of probability that can be named."[13] In modern notation, if X represents the number of times the event has happened in n trials, x the probability of its happening in a single trial, and r and s the two given probabilities, Bayes's aim was to calculate $P(r < x < s | X)$, that is, the probability that x is between r and s, given X. Bayes proceeded to develop axiomatically, from a definition of probability, the two basic results that he would need about the probabilities of two events E and F. Proposition 3 states that "the probability that two subsequent events will both happen is a ratio compounded of the probability of the 1st, and the probability of the 2nd on supposition the 1st happens." Proposition 5, today generally known as **Bayes's theorem**, is: "If there be two subsequent events, the probability of the 2nd b/N and the probability of both together P/N, and it being first discovered that the 2nd event has happened, from hence I guess that the 1st event has also happened, the probability I am in the right is P/b."[14] In modern notation, letting E be the first event and F the second, proposition 3 can be written as $P(E \cap F) = P(E)P(F|E)$, that is, the probability of both happening is the product of the probability of E with the probability of F given E, while Bayes's theorem itself can be written as $P(E|F) = P(E \cap F)/P(F)$, that is, the probability of E given that F has happened is the quotient of the probability of both happening divided by the probability of F alone. Bayes's basic problem, then, was the calculation of $P(E|F)$, where E is the event "$r < x < s$" and F is the event "X successes in n trials." To apply Bayes's theorem to this calculation, he therefore needed a way of determining the two probabilities $P(E \cap F)$ and $P(F)$.

Bayes naturally knew Bernoulli's result that if the probability of a success were a and that of a failure were b, then the probability of p successes and q failures in $n = p + q$ trials was $\binom{n}{q}a^p b^q$. But whereas Bernoulli could give only a rough approximation to the sum of these terms and De Moivre chiefly considered the equiprobable case where $a = b$, Bayes used De Moivre's approach via area to attack the problem directly. Thus, he began by modeling the probabilities by a certain area:

> I suppose the square table . . . $ABCD$ [of side 1] [Fig. 18.1] to be so made and leveled, that if either of the balls O or W be thrown upon it, there shall be the same probability that it rests upon any one equal part of the plane as another. . . . I suppose that the ball W shall be first thrown, and through the point where it rests a line ot shall be drawn parallel to AD, and meeting CD and AB in t and o; and that afterwards the ball O shall be thrown $p + q$ or n times, and that its resting between AD and ot after a single throw be called the happening of the event M in a single trial.[15]

FIGURE 18.1

Bayes's theorem

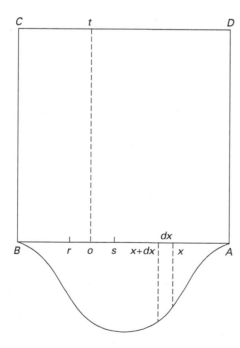

In terms of the basic problem, the position of W determines the probability x. Bayes noted that the probability of the point o falling between any two points r and s is simply the length rs. Similarly, the probability of the event M given that W has been thrown is the length of Ao.

To calculate $P(E \cap F)$, Bayes used proposition 3. Any given probability range for the point o is represented by an interval on the axis AB, say, $[x, x + dx]$, measured from A. Because a particular x represents the probability of the ball landing to the right of ot, $1 - x$ represents the probability of it landing to the left. The probability that the ball will land to the right p times in $p + q = n$ throws is therefore given by $y = \binom{n}{q} x^p (1 - x)^q = \binom{n}{p} x^p (1 - x)^{n-p}$. Bayes drew the curve given by this function below the axis AB and used proposition 3 to conclude that the probability of W lying above the coordinate interval $[x, x + dx]$ and the ball landing p times to the right of W is represented by the area under $[x, x + dx]$ and above the curve. It follows that $P(E \cap F) = P((r < x < s) \cap (X = p))$ is represented by the total area under the interval $[r, s]$ and above the curve, or, in modern notation, by

$$\int_r^s \binom{n}{p} x^p (1 - x)^{n-p} \, dx.$$

Because $P(F) = P(X = p)$ can be thought of as $P((0 < x < 1) \cap (X = p))$, it follows from the above argument that $P(X = p)$ is represented by the entire area under the axis AB and above the curve, or by

$$\int_0^1 \binom{n}{p} x^p (1 - x)^{n-p} \, dx.$$

Bayes's theorem then showed how to calculate $P(E|F)$:

$$P(E|F) = P((r < x < s)|(X = p)) = \frac{\int_r^s \binom{n}{p} x^p (1 - x)^{n-p} \, dx}{\int_0^1 \binom{n}{p} x^p (1 - x)^{n-p} \, dx}.$$

Bayes thus concluded "that in the case of such an event as I there call M, from the number of times it happens and fails in a certain number of trials, without knowing anything more concerning it, one may give a guess whereabouts its probability is, and, by the usual methods computing the magnitudes of the areas there mentioned, see the chance that the guess is right."[16]

Although Bayes's problem was, in fact, formally solved, there were two obstacles to be overcome before one could consider the solution as a practical one. First, does Bayes's physical analogy of rolling balls on a table truly mirror the actual problems to which the theory would be applied? Could nature's choice of an unknown probability x really be the same as the rolling of a ball across a level table? Bayes answered this question by, in effect, restricting the application of the rule to just those circumstances in which for any given number n of trials, all possible outcomes $X = 0$, $X = 1$, $X = 2$, . . . are equally likely, that is, for events concerning which "I have no reason to think that, in a certain number of trials, it should rather happen any one possible number of times than another."[17] But is ignorance about the probabilities in a given situation equivalent to all possible outcomes being equally likely? This question has been debated extensively since Bayes's time.

Second, can one actually calculate the integrals in Bayes's formula? Bayes attempted to do so by expanding the integrands in power series. The integral in the denominator turned out to be $\frac{1}{n+1}$. The integral in the numerator, while not difficult to approximate when either p or $n - p$ is small, turned out to be very difficult otherwise. Richard Price (1723–1791), the friend of Bayes who submittewd his paper to the Royal Society, worked out a few special cases when p is close to n. For example, if $p = n$, then the relevant quotient is

$$\frac{\int_r^s x^n \, dx}{\int_0^1 x^n \, dx} = s^{n+1} - r^{n+1}.$$

So suppose nothing is known about an event M except that it has happened once. The chance that the unknown probability x of M is greater than 1/2, that is, between 1/2 and 1, is then $1^2 - (1/2)^2 = 3/4$. Similarly, if M has happened twice, the probability that x is greater than 1/2 is 7/8; in other words, the odds are 7 to 1 that there is more than an even chance of it happening. In this same situation, the odds are still better than even that the probability of x is greater than 2/3.

FIGURE 18.2

Laplace on a French stamp

18.2.2 The Calculations of Laplace

Bayes's formula did provide a start in answering the basic question of statistical inference. Further progress was made a few years later by Pierre-Simon de Laplace (1749–1827). In 1774, Laplace, using principles similar to those of Bayes, derived essentially the same result involving integrals for determining probability, given empirical evidence. Putting the question in terms of drawing tickets from an urn, he supposed that p white and q black tickets had been

---

BIOGRAPHY

Pierre-Simon de Laplace (1749–1827)

Born in Normandy, Laplace entered the University of Caen in 1766 to begin preparation for a career in the Church. He discovered there his mathematical talents, however, and in 1768 left for Paris to continue his studies. He met with d'Alembert, who was so impressed with him that he secured for Laplace a position in mathematics at the École Militaire, where he taught elementary mathematics to aspiring cadets. Legend has it that he examined, and passed, Napoleon there in 1785. A steady stream of mathematical papers soon began to flow from his pen, winning him election to the Academy of Sciences in 1773. During the Revolution, he served as a member of the Commission on Weights and Measures, but was eventually dismissed for not being a strong republican. He then retired to the country, where he was able to work in relative peace.

Laplace's most important accomplishments were in the field of celestial mechanics. During the period from 1799 to 1825, he produced his five-volume *Traité de mécanique céleste* (*Treatise on Celestial Mechanics*), in which he successfully applied calculus to the motions of the heavenly bodies and showed, among much else, why Newton's law of gravitation implied the long-term stability of the solar system. Laplace also contributed heavily to the field of probability, producing his *Théorie analytique des probabilités* (*Analytic Theory of Probability*) in 1812. Although he was honored by Napoleon, he voted against him in 1814 as a member of the Senate, supporting Louis XVIII instead. Laplace was rewarded with the title of marquis. At his death he was eulogized as "the Newton of France" (Fig. 18.2).

drawn from an urn containing an unknown ratio x of white tickets. Given then any guessed value for x, Laplace showed how to calculate the probability that x differed from $\frac{p}{p+q}$ by as small a value ϵ as one wished. He was in fact able to demonstrate that

$$P\left(\left|x - \frac{p}{p+q}\right| \leq \epsilon | X = p\right) \cong \frac{2(p+q)^{3/2}}{\sqrt{2\pi}\sqrt{pq}} \int_0^\epsilon e^{-[(p+q)^3/2pq]z^2} \, dz$$

$$\cong \frac{2}{\sqrt{2\pi}} \int_0^{\epsilon/\sigma} e^{-(u^2/2)} \, du,$$

where $\sigma^2 = pq/(p+q)^3$. To show that this probability approached 1 as $p+q$ became large, whatever the value of ϵ, Laplace had to integrate $\int_0^\infty e^{-(u^2/2)} \, du$. Using a result of Euler's, he was in fact able to show that this integral equaled $\sqrt{\pi/2}$ and therefore established his result.

To go further in calculating, naturally, Laplace had to evaluate the integral $\int_0^T e^{-(u^2/2)} \, du$ for arbitrary T. This he did in 1785 by deriving two different series for this integral, one that converged rapidly for small T and one for large T. He then applied his results to a genuine problem in statistical inference. During the 26-year period from 1745 to 1770, 251,527 boys and 241,945 girls had been born in Paris. Setting x as the probability of a male birth, he made a straightforward calculation and demonstrated that the probability that $x \leq 1/2$ was 1.15×10^{-42}. He therefore concluded that it was "morally certain" that $x > 1/2$. He then extended his analysis using similar data from London to show that it was also morally certain that the probability of a male birth in London was greater than that in Paris.

18.3 APPLICATIONS OF PROBABILITY

In addition to the standard applications to games and urn problems, probability theory was also applied in the eighteenth century to problems of real concern.

18.3.1 Errors in Observations

One important question considered in the eighteenth century was how to deal with observational errors made in astronomy and other fields. It was certainly known that every observation was subject to error, so that if one wanted to develop a theory, one had to understand the nature of the errors and how to compensate for them. For example, suppose one knows that a particular physical relationship is expressed by a linear function $y = a + bx$. One performs several observations of the phenomenon in question and finds the data points $(x_1, y_1), (x_2, y_2), \ldots, (x_k, y_k)$. Replacing x and y in the equation by these k pairs in turn gives k equations for the two unknown coefficients a, b. The system of k linear equations in two unknowns is thus overdetermined and, in general, has no exact solution. The idea, then, is somehow to determine the "best" approximation to a solution. In geometrical terms, the problem is to find the straight line that is "closest," in some sense, to passing through the k observed points.

This problem of the combination of observations was discussed by various mathematicians in the eighteenth century, primarily in regard to astronomical observations. Among those who attempted solutions to the problem were Leonhard Euler in 1749, Tobias Mayer (1723–1762) in 1750, and Roger Boscovich (1711–1787) in 1760. Euler, in working on a problem involving the mutual gravitational influence of Jupiter and Saturn on each other's orbits, ended up with a system of 75 equations in eight unknowns. He attempted to find the best solution by solving various small sets of his equations and combining the answers. Mayer, on the other hand, in looking at the detailed motion of the moon, had to solve a system of 27 equations in three unknowns. He developed a systematic method of attack by dividing his equations into three groups of nine, adding the equations in each of the groups separately and then solving the resulting system of three equations in three unknowns. What was not entirely clear was exactly what criteria should be used to divide the equations. It was Boscovich, however, who made a significant advance in this problem as he dealt with a question involving the true shape of the earth. He stated the actual criteria that a method of determining the solutions to such systems of equations ought to satisfy, including the important one of minimizing the sum of the absolute values of the errors determined by substituting any particular set of values into the equations. A few years later, Laplace turned Boscovich's method into a detailed algebraic method in his own work on the same problem. Unfortunately, Laplace's method turned out to be difficult to work with and was replaced early in the next century by the method of least squares.

A related question also considered in the eighteenth century was to find a mathematical description of the error function itself. For example, Thomas Simpson in 1755 attempted to show that the error in observations would be diminished by taking the mean of several observations. He did this by assuming, for example, that the probability of errors in seconds of sizes $-5, -4, -3, -2, -1, 0, 1, 2, 3, 4, 5$ in a particular astronomical measurement was respectively proportional to 1, 2, 3, 4, 5, 6, 5, 4, 3, 2, 1. Thus, the probability that a single error does not exceed 1 second is $16/36 = 0.444$ and that it does not exceed 2 seconds is $24/36 = 0.667$. On the other hand, he calculated that the probability that the mean of six errors does not exceed 1 second is 0.725 and that it does not exceed 2 seconds is 0.967, thus showing the advantage of taking means.

Simpson next tried to generalize this result on taking means to more general error functions. But it was Laplace in the 1770s who made a more careful analysis by making explicit assumptions on the conditions an error function $\phi(x)$ should meet. These conditions were, first, that $\phi(x)$ should be symmetric about zero, assuming that it is equally probable that an observation is too big as that it is too small; second, that the curve must be asymptotic to the real axis in both directions, because the probability of an infinite error is zero; and third, that the total area under $\phi(x)$ should be 1, since the area under that curve between any two values represents the probability that the observation has error between those values. Unfortunately, there were many curves that satisfied Laplace's requirements. Through the use of various other arguments, Laplace settled on the curve $y = (m/2)e^{-m|x|}$ for some positive value m. Laplace soon found out, however, that calculations based on this error function led to great difficulties. A better answer was found in the nineteenth century by Gauss.

18.3.2   De Moivre and Annuities

To apply probability to the "real world" required knowledge of outcomes and their events. One particular application, which was of interest to De Moivre, was the application of probability to the pricing of annuities. Annuities had been sold for centuries but were generally considered a "bet" by the annuitant and a loan at interest by the seller. That is, the annuitant, who paid a fixed sum for a guarantee of regular payments until his death, was in effect betting that he would live long enough to collect all of his payment and more. The seller, on the other hand, considered the initial payment as a loan and the payoff to be interest, usually at a rate higher than the legal rate for lending money.

Before mathematicians thought about the subject, the pricing of annuities was set either by the experience of the parties involved or by a need of the seller for cash rather than by a consideration of statistics. For example, an English law of 1540 declared that a government annuity is worth seven years' purchase. That means that the government would sell an annuity for A pounds that would guarantee the annuitant P pounds a year for life, where the P pounds per year was the amount necessary to pay back A pounds, including interest at a fixed rate, in seven years. Apparently, this contract was offered independently of the age or health of the buyer. The relationship between P and A was easy enough to determine, given an interest rate of r. Because $\frac{P}{1+r}$ pounds gives you P pounds in one year, $\frac{P}{(1+r)^2}$ gives you P pounds in two years, and so on, we find that

$$A = \frac{P}{1+r} + \frac{P}{(1+r)^2} + \cdots + \frac{P}{(1+r)^7} = \frac{P}{1+r} \sum_{i=0}^{6} \left(\frac{1}{1+r}\right)^i$$

$$= \frac{P}{1+r}\left(\frac{1-\left(\frac{1}{1+r}\right)^7}{1-\frac{1}{1+r}}\right) = \frac{P[(1+r)^7 - 1]}{r(1+r)^7}.$$

For example, if $P = 1$ and $r = .05$, we calculate that $A = 5.7864$. That is, one could buy a life annuity of one pound per year for 5.7864 pounds, assuming interest at 5%, regardless of one's age.

In modern terms, A is called the **present value of an annuity** of P pounds per year for seven years. Of course, it is easy enough to generalize this idea to an annuity for n years. That is, given an interest rate r, an annuity of P pounds per year for n years has a present value of

$$A = \frac{P[(1+r)^n - 1]}{r(1+r)^n}.$$

Again, if interest is at 5%, the present value of an annuity of one pound per year for 36 years is 16.5468.

What De Moivre wanted to do was to apply probabilities to this calculation to find a way of pricing annuities, fairer to both buyer and seller, given that older annuitants were more likely to die sooner than younger ones. So the question for De Moivre was "how much more likely were older annuitants to die sooner than younger ones?" To answer this question, he needed mortality tables. Information for such tables was already being collected in the seventeenth century. For example, John Graunt (1620–1674) collected and analyzed information from the lists of deaths and their causes that began to be compiled in England in the sixteenth century, originally to keep track of the plague. Graunt published his material in *Natural and Political Observations on the Bills of Mortality* in 1662. Other studies of death rates were done by Jan de Witt in the Netherlands in 1671 and Edmond Halley (using data from Breslau) in 1693. From looking at the data in these studies, De Moivre concluded that, for purposes of pricing annuities, he could assume, first, that the maximum age a person might live was 86 years, and, second, that roughly the same number of people who were k years old now would die in each succeeding year. Putting this in probability terms, if we let $n = 86 - k$, the number of possible years left, the probability of a k-year-old person dying in any given year of the n possible years left would be $1/n$. To put this another way, the probability of a k-year-old person living at least one year is $(n - 1)/n$, at least two years is $(n - 2)/n$, at least three years is $(n - 3)/n$, and so on.

Given his assumptions, De Moivre showed, in problem 1 of his *Treatise of Annuities on Lives* (1724, 1743, 1750), that the "fair" price of a life annuity of one pound per year for a person of age k, given an interest rate r, was

$$Q = \frac{1 - \frac{sA}{n}}{s - 1},$$

where $n = 86 - k$, $s = 1 + r$, and A is the present value of an annuity of one pound for n years. To prove this, De Moivre began by noting that since the probability of having to pay back one pound at the end of the first year is $\frac{n-1}{n}$, we need only invest enough to pay back that amount, namely, $\frac{n-1}{ns}$. Similarly, the probability of having to pay back one pound after two years is $\frac{n-2}{n}$. The amount we need to invest today to do that is $\frac{n-2}{ns^2}$. Therefore, to be sure we have enough to pay back one pound per year for life, we need to invest the sum

$$\frac{n - 1}{ns} + \frac{n - 2}{ns^2} + \frac{n - 3}{ns^3} + \cdots + \frac{n - (n - 1)}{ns^{n-1}}.$$

This sum is then the present value Q of a life annuity of one pound per year for a person of age k years, where $n = 86 - k$.

To transform this formula into the closed form for Q, De Moivre took the present value formula for an annuity of one pound per year for n years:

$$A = \frac{1}{s} + \frac{1}{s^2} + \frac{1}{s^3} + \cdots + \frac{1}{s^n}.$$

Now, we multiply A by $\frac{s}{n}$ and subtract this value from 1:

$$1 - \frac{s}{n}A = \frac{n - 1}{n} - \frac{1}{ns} - \frac{1}{ns^2} - \cdots - \frac{1}{ns^{n-1}}.$$

To get Q, De Moivre needed to multiply this formula by $\frac{1}{s-1}$, which he rewrote as

$$\frac{1}{s} + \frac{1}{s^2} + \frac{1}{s^3} + \cdots.$$

He then multiplied in columns:

$$
\begin{array}{cccccccccc}
\frac{n-1}{ns} & - & \frac{1}{ns^2} & - & \frac{1}{ns^3} & - & \frac{1}{ns^4} & - & \frac{1}{ns^5} & - \cdots \\
& + & \frac{n-1}{ns^2} & - & \frac{1}{ns^3} & - & \frac{1}{ns^4} & - & \frac{1}{ns^5} & - \cdots \\
& & & + & \frac{n-1}{ns^3} & - & \frac{1}{ns^4} & - & \frac{1}{ns^5} & - \cdots \\
& & & & & + & \frac{n-1}{ns^4} & - & \frac{1}{ns^5} & - \cdots \\
& & & & & & & + & \frac{n-1}{ns^5} & - \cdots
\end{array}
$$

If the terms are added vertically, the result is

$$\frac{n-1}{ns} + \frac{n-2}{ns^2} + \frac{n-3}{ns^3} + \frac{n-4}{ns^4} + \frac{n-5}{ns^5} + \cdots,$$

thus proving De Moivre's result.

As an example, because we know that the present value of an annuity of one pound per year for 36 years at 5% interest is 16.5468, we calculate the present value of a life annuity of $1 per year at 5% for a 50-year-old person. We set $A = 16.5468$ and $n = 86 - 50 = 36$. The present value Q is then

$$Q = \frac{1 - \frac{1.05 \cdot 16.5468}{36}}{0.05} = 10.3477.$$

Therefore, to get a life annuity of one pound per year at 5%, a 50-year-old person must pay 10.3477, based on De Moivre's mortality assumptions. De Moivre's treatise contained tables to calculate life annuities at ages from 1 to 86, with interest rates of 3%, $3\frac{1}{2}$%, 4%, 5%, and 6%.

18.3.3 Probability and Lotteries

Just because De Moivre worked out a mathematical way of pricing annuities did not mean that everyone used his methods. For example, a town in the Netherlands raised 100,000 florins by selling 400 annuities at 250 florins each and then dividing its total payment of 4000 florins per year among as many of the annuitants who were still living. A similar result happened with the mathematics of lotteries. Governments had been raising money through lotteries for centuries, although they became particularly popular in the late seventeenth and early eighteenth centuries. Nevertheless, in general, governments made no probabilistic calculations to determine prize money, because there was no good reason to do so. The lotteries were major sources of profit, because the prizes were very low in relation to the odds. For example, the payoff in the French Royal Lottery for someone correctly picking five numbers drawn out of ninety, in the order in which they were drawn, was 1,000,000 times the price paid for the ticket. The odds against winning such a prize, however, are more than five billion to one. Evidently, the French of the eighteenth century, particularly those of the lower classes, like their counterparts today, were willing to gamble on the exceedingly slim possibility of becoming rich. There were occasions where mathematicians wrote learned articles demonstrating why people should not play the lottery at those odds. But the only mathematicians to whom people paid attention were those who sold sure fire methods for picking the winning numbers!

One of the few cases in which a government did consult a mathematician was when Frederick II of Prussia asked Euler to examine a proposal for a lottery that had been submitted to the Prussian monarch. The lottery was a type known as the Genoese lottery, in which five tickets are drawn at random from ninety numbered tickets, and the players may bet on a single number, two numbers, or three numbers appearing among the five drawn. Euler quickly responded to Frederick's request, calculating the probabilities of winning in each case, determining what the price of a ticket should be in each case if the game were equitable, and then showing how much the state would win if the ticket price were set considerably higher. For example, the probability of winning if one bets on a single number is 5/90 = 1/18.

Thus, if the game were equal, a ticket that promised a prize of 100 ecus should sell for an eighteenth of 100, or about 5.56 ecus. In fact, the price of a ticket was set at 8 ecus, so that the state would earn a decent sum. Similarly, the probability of winning if one bets on two numbers is $\frac{5 \cdot 4}{90 \cdot 89} = \frac{2}{801}$. Thus, for the game to be equal, a ticket promising 100 ecus should sell for approximately 0.25 of an ecu. But the proposal set the wager at approximately 0.58 ecus, more than double the "fair" price. Thus, "in order to more encourage the players," the actual prize for this wager was set at 120 ecus rather than 100. In his analysis, however, Euler noted that the results are "only probable" and that it is possible that the state would be forced to disburse more than was paid in. But he further noted that "the more the number of players is great, the more also the real profit will approach the calculated."[18]

For whatever reason, Frederick did not begin this lottery in 1749, but waited until 1763, the same year that Euler wrote a much more detailed analysis of this lottery and calculated probabilities in the general case of drawing t tickets out of a total of n and in which a player can bet on k numbers ($0 < k \leq t$) and win something if he matches i of them for $0 < i \leq k$. We consider the case where $k = 2$ and think of the player as choosing the two numbers one at a time. The probability of the first chosen number being drawn is then $\frac{t}{n}$. The probability of the second chosen number being drawn is $\frac{t-1}{n-1}$, so the probability for both is $\frac{t(t-1)}{n(n-1)}$, in complete analogy with the second example above. On the other hand, to match one number, we note that the probability that the first number drawn is a chosen one and the second is not is $\frac{t}{n} \frac{n-t}{n-1}$, with an identical result for the opposite occurrence. Thus, the probability of winning by matching one number is $\frac{2t(n-t)}{n(n-1)}$. Euler worked out all the details for $k = 1, 2, 3, 4$ and then showed how to determine the general formula for arbitrary k.

Next, Euler detailed several methods of awarding "fair" prizes for each case and then gave some recommendations for actual prizes so that the state would make a reasonable profit on the game. For example, let us suppose in the case $k = 2$ that one awards a prize of a when both numbers picked by the player are drawn and a prize of b when one number is drawn. Then the "advantage" to the player is

$$\frac{t(t-1)}{n(n-1)}a + \frac{2t(n-t)}{n(n-1)}b.$$

If the game were equitable, this value should equal the cost of a ticket, say, 1 ecu. Of course, given that we now have a single equation in two unknowns a and b, Euler realized that there were infinitely many possibilities, even assuming that both summands were positive. That is, if we choose α and β positive numbers such that $\alpha + \beta = 1$, then the two prizes should be

$$a = \frac{\alpha n(n-1)}{t(t-1)} \quad \text{and} \quad b = \frac{\beta n(n-1)}{2t(n-t)}.$$

Euler then gave three possibilities for the choice of α and β and computed the prizes both for general n and t and for the original lottery where $n = 90$ and $t = 5$. First, if $\alpha = \beta = \frac{1}{2}$, then the two specific prizes are $a = 200.25$ and $b = 4.71$. If, however, one wanted to reduce the prize for getting both numbers right, one could choose $\alpha = \frac{1}{3}$ and $\beta = \frac{2}{3}$, or even $\alpha = \frac{1}{5}$ and $\beta = \frac{4}{5}$. In the first case, the prizes are $a = 133.5$ and $b = 6.28$, while in the second, they are $a = 80.1$ and $b = 7.54$. Euler then suggested that, to give the state a profit, the smaller prizes should be reduced by 10% and the larger ones by 20%, to lessen the possibility that the state could lose money. When k is 3, 4, or 5, he suggested cutting the largest prizes by 30%,

40%, and 50%, respectively. As he wrote, "A greater hold back in the small prizes would also be only too apparent and disgust the participants, whereas, in the great prizes, one does not perceive nearly the diminution, seeing that few persons are in a condition to calculate the just value."[19]

EXERCISES

1. Calculate the Bernoulli numbers B_8, B_{10}, and B_{12}. The sequence of Bernoulli numbers is usually completed by setting $B_0 = 1$, $B_1 = -\frac{1}{2}$, and $B_k = 0$ for k odd and greater than 1.

2. Write out explicitly, using Bernoulli's techniques, the formulas for the sums of the first n fourth, fifth, and tenth powers. Then show that the sum of the tenth powers of the first 1000 positive integers is

   91, 409, 924, 241, 424, 243, 424, 241, 924, 242, 500.

   Bernoulli claimed that he calculated this value in "less than half of a quarter of an hour" (without a calculator).

3. Show that if one defines the Bernoulli numbers B_i by setting

   $$\frac{x}{e^x - 1} = \sum_{i=0}^{\infty} \frac{B_i}{i!} x^i,$$

   then the values of B_i for $i = 2, 4, 6, 8, 10, 12$ are the same as those calculated in the text and in Exercise 1.

4. Suppose that a is the probability of success in an experiment and $b = 1 - a$ is the probability of failure. If the experiment is repeated three times, show that the probabilities of the number of successes S being 3, 2, 1, 0, respectively, are given by $P(S = 3) = 1a^3$, $P(S = 2) = 3a^2b$, $P(S = 1) = 3ab^2$, and $P(S = 0) = 1b^3$.

5. Generalize Exercise 4 to the case of n trials. Show that the probability of r successes is $P(S = r) = \binom{n}{n-r} a^r b^{n-r}$.

6. Using the results of Exercise 5, with $a = 1/3$, $b = 2/3$, and $n = 10$, calculate $P(4 \leq S \leq 6)$.

7. Complete Bernoulli's calculation of his example for the Law of Large Numbers by showing that if $r = 30$ and $s = 20$ (so $t = 50$) and if $c = 1000$, then

   $$nt + \frac{rt(n-1)}{s+1} > mt + \frac{st(m-1)}{r+1}$$

   where m, n are integers such that

   $$m \geq \frac{\log c(s-1)}{\log(r+1) - \log r}$$

and

$$n \geq \frac{\log c(r-1)}{\log(s+1) - \log s}.$$

Conclude that in this case the necessary number of trials is $N = 25{,}550$.

8. Use Bernoulli's formula to show that if greater certainty is wanted in the problem of Exercise 7, say, $c = 10{,}000$, then the number of trials necessary is $N = 31{,}258$.

9. In his *Letter to a Friend on Sets in Court Tennis*, written in 1687 but not published until 1713, Jakob Bernoulli analyzed the probabilities at any point in a game or set of court tennis, whose scoring rules are virtually identical with those of tennis today. He determined the odds both when the players were evenly matched and when one player was stronger than the other. If two players A and B are evenly matched in a tennis game with the score $15 : 30$, determine the probability of player A winning. (Remember that one must win by two points.)

10. Continuing from Exercise 9, suppose that player A is twice as strong as player B. Suppose that the score is $30 : 30$. Determine the probability of player A winning. What is the probability of A winning if the score is $15 : 30$?

11. Suppose that the probability of success in an experiment is $1/10$. How many trials of the experiment are necessary to ensure even odds on it happening at least once? Calculate this both by De Moivre's exact method and his approximation.

12. How many throws of three dice are necessary to ensure even odds that three ones will occur at least once?

13. In a lottery in which the ratio of the number of losing tickets to the number of winning tickets is $39 : 1$, how many tickets should one buy to give oneself even odds of winning a prize?

14. Generalize De Moivre's procedure in Problem III (of his text) to solve Problem IV: To find how many trials are necessary to make it equally probable that an event will happen twice, supposing that a is the number of chances for its happening in any one trial and b the number of chances for its failing. (*Hint:* Note that $b^x + xab^{x-1}$ is the

number of chances in which the event may succeed no more than once, while $(a + b)^x$ is the total number of chances.) Approximate the solution for the case where $a : b = 1 : q$, with q large, and show that $x \approx 1.678q$.

15. Show that the sums labeled col. 1, col. 2, and col. 3 in De Moivre's derivation of the ratio $\binom{n}{n/2} : 2^n$ may be written explicitly as

$$\text{col. 1} = \frac{s^2 + s}{m}$$

$$\text{col. 2} = \frac{\frac{1}{2}s^4 + s^3 + \frac{1}{2}s^2}{3m^3}$$

$$\text{col. 3} = \frac{\frac{1}{3}s^6 + s^5 + \frac{5}{6}s^4 - \frac{1}{6}s^2}{5m^5}$$

Determine the corresponding value for col. 4.

16. Add the highest-degree terms of the columns from Exercise 15 to get

$$s\left(\frac{s}{m} + \frac{1}{2 \cdot 3}\frac{s^3}{m^3} + \frac{1}{3 \cdot 5}\frac{s^5}{m^5} + \frac{1}{4 \cdot 7}\frac{s^7}{m^7} + \cdots\right),$$

which, setting $x = s/m$, is equal to

$$s\left(\frac{2x}{1 \cdot 2} + \frac{2x^3}{3 \cdot 4} + \frac{2x^5}{5 \cdot 6} + \frac{2x^7}{7 \cdot 8} + \cdots\right).$$

Show that the series in the parenthesis can be expressed in finite terms as

$$\log\left(\frac{1 + x}{1 - x}\right) + \frac{1}{x}\log(1 - x^2)$$

and therefore that the original series is

$$mx \log\left(\frac{1 + x}{1 - x}\right) + m \log(1 - x^2).$$

Since $s = m - 1$ (or $mx = m - 1$), show therefore that the sum of the highest-degree terms of the columns of Exercise 15 is equal to

$$(m - 1) \log\left(\frac{1 + \frac{m-1}{m}}{1 - \frac{m-1}{m}}\right)$$

$$+ m \log\left[\left(1 + \frac{m - 1}{m}\right)\left(1 - \frac{m - 1}{m}\right)\right],$$

which in turn is equal to $(2m - 1)\log(2m - 1) - 2m \log m$.

17. Show that the sum of the second-highest-degree terms of each column from Exercise 15 is

$$\frac{s}{m} + \frac{s^3}{3m^3} + \frac{s^5}{5m^5} + \frac{s^7}{7m^7} + \cdots,$$

which, since $s = m - 1$, is equal to

$$\frac{1}{2}\log\left(\frac{1 + \frac{s}{m}}{1 - \frac{s}{m}}\right) \quad \text{or} \quad \frac{1}{2}\log(2m - 1).$$

18. Derive De Moivre's result

$$\log\left(\frac{Q}{M}\right) \approx -\frac{2t^2}{n} \quad \text{or equivalently} \quad \log\left(\frac{M}{Q}\right) \approx \frac{2t^2}{n}.$$

(*Hint:* Divide the arguments of the first two logarithm terms in the expression in the text by m. Then simplify and replace the remaining logarithm terms by the first two terms of their respective power series.)

19. De Moivre's result developing the normal curve implies that the probability P_ϵ of an observed result lying between $p - \epsilon$ and $p + \epsilon$ in n trials is given by

$$P_\epsilon = \frac{1}{\sqrt{2\pi np(1 - p)}} \int_{-n\epsilon}^{n\epsilon} e^{-\frac{t^2}{2np(1-p)}} \, dt.$$

Change variables by setting $u = t/\sqrt{np(1 - p)}$ and use symmetry to show that this integral may be rewritten as

$$P_\epsilon = \frac{2}{\sqrt{2\pi}} \int_0^{\frac{\sqrt{n}\epsilon}{\sqrt{p(1-p)}}} e^{-\frac{1}{2}u^2} \, du.$$

Calculate this integral for Bernoulli's example, using $p = .6$, $\epsilon = .02$, and $n = 6498$, and show that in this case $P_\epsilon = 0.999$, a value giving moral certainty. (Use a graphing utility.) Find a value for n that gives $P_\epsilon = 0.99$.

20. Calculate $P(r < x < s | X = n - 1)$ explicitly, using Bayes's theorem. In particular, suppose that you have drawn 10 white and 1 black ball from an urn containing an unknown proportion of white to black balls. If you now guess that this unknown proportion is greater than 7/10, what is the probability that your guess is correct?

21. Show that if an event of unknown probability happens n times in succession, the odds are $2^{n+1} - 1$ to 1 for more than an even chance of its happening again.

22. Imagine an urn with two balls, each of which may be either white or black. One of these balls is drawn and is put back before a new one is drawn. Suppose that in the first two draws white balls have been drawn. What is the probability of drawing a white ball on the third draw?

23. With interest at 4%, what is the present value of an annuity of one pound per year for 50 years?

24. With interest at 4%, what is the present value of a life annuity of one pound per year for someone of age 36?

25. In the French Royal Lottery of the late eighteenth century, five numbered balls were drawn at random from a set of 90 balls. Originally, a player could buy a ticket on any one number or on a pair or on a triple. Later on, one was permitted to bet on a set of four or five as well as on a set given in the order drawn. Show that the odds against winning with a bet on a single number, a pair, and a triple are $17 : 1$, $399.5 : 1$, and $11{,}747 : 1$, respectively. The payoffs on these bets are 15, 270, and 5,500.

26. In Euler's analysis of the lottery for the case $k = 2$, determine the general formulas for the "fair" prizes a and b for matching two numbers and for matching one number, respectively, in terms of n and t, where t tokens are drawn out of a total of n.

27. Work out the probabilities for the case $k = 3$ in the lottery described in the text. That is, assuming that t tokens are drawn out of a total of n, find the probabilities that if a player picks three numbers, he will match all three, match two, or match one of the numbers drawn.

28. For the case of the lottery from Exercise 27, determine the specific probabilities that a player will match three numbers, match two numbers, or match one number in the case $n = 90$ and $t = 5$.

29. In the situation of Exercise 28, find the "advantage" of the player and determine the equation that determines "fair" prizes a, b, and c for each of the three possibilities, assuming a bet of 1 ecu. (Here a is the prize for matching all three numbers, b the prize for matching two numbers, and c the prize for matching one number.)

30. In the situation of Exercise 29, assume that the three summands in the equation are each equal to 1/3. Determine the prizes in that event. Then assume that the first summand (for matching all three numbers) is 1/7, while the other two are each 3/7, and determine the prizes. Finally, assume that the first summand is 1/16, the second (for matching two numbers) is 6/16, and the third (for matching one number) is 9/16 and determine the prizes. (These three cases are all discussed by Euler.)

31. The so-called St. Petersburg Paradox was a topic of debate among those mathematicians involved in probability theory in the eighteenth century. The paradox involves the following game between two players. Player A flips a coin until a tail appears. If it appears on his first flip, player B pays him 1 ruble. If it appears on the second flip, B pays 2 rubles, on the third, 4 rubles, . . . , on the nth flip, 2^{n-1} rubles. What amount should A be willing to pay B for the privilege of playing? Show first that A's expectation, namely, the sum of the probabilities for each possible outcome of the game multiplied by the payoff for each outcome, is

$$\sum_{i=0}^{\infty} \frac{1}{2^i} 2^{i-1}$$

and then that this sum is infinite. Next, play the game 10 times and calculate the average payoff. What would you be willing to pay to play? Why does the concept of expectation seem to break down in this instance?

32. Outline a lesson for a statistics course deriving Bayes's theorem and discussing its usefulness.

REFERENCES AND NOTES

There are several good books on the early history of probability and statistics, including the works of Hacking and David cited in Chapter 14. A newer book that gives great detail on the mathematics of the first studies in probability, although often in modern language, is Anders Hald, *A History of Probability and Statistics and Their Applications before 1750* (New York: John Wiley, 1965). Another good treatment of the early history of statistics, which naturally also includes work on probability, is Stephen M. Stigler, *The History of Statistics* (Cambridge: Harvard University Press, 1986). A more philosophical treatment, concentrating on the ideas behind the notion of probability, is Lorraine Daston, *Classical Probability in the Enlightenment* (Princeton: Princeton University Press, 1988).

The *Ars conjectandi* is available in English as Jacob Bernoulli, *The Art of Conjecturing*, translated by Edith Dudley Sylla (Baltimore: Johns Hopkins University Press, 2006). There is a reprint edition available of Abraham De Moivre, *The Doctrine of Chances*, 3rd ed. (New York: Chelsea, 1967), which is well worth perusing for its many problems and examples. Thomas Bayes, "An Essay towards Solving a Problem in the Doctrine of Chances," has been reprinted with a biographical note by G. A. Barnard in E. S. Pearson and M. G. Kendall, *History of Statistics and Probability*, pp. 131–154. Laplace's earliest memoir on probability is translated in Stephen Stigler, "Laplace's 1774 Memoir on Inverse Probability," *Statistical Science* 1 (1986), 359–378. Euler's article, "Reflections on a Singular Kind of Lottery Named the Genoise Lottery" (E812, written in 1763, but not published until 1862), is available in a translation by Richard Pulskamp at http://cerebro.xu.edu/math/Sources/Euler/.

1. Bernoulli, *The Art of Conjecturing*, p. 327. A brief look at the entire *Ars conjectandi* is in Ian Hacking, "Jacques Bernoulli's Art of Conjecturing," *British Journal for the History of Science* 22 (1971), 209–229.

2. Bernoulli, *The Art of Conjecturing*, p. 327.

3. Ibid., p. 193.

4. Ibid., pp. 214 ff.

5. Ibid., p. 328.

6. See Stigler, *History of Statistics*, pp. 66–70, for more details. See also O. B. Sheynin, "On the Early History of the Law of Large Numbers," *Biometrika* 55 (1968), 459–467, reprinted in E. S. Pearson and M. G. Kendall, *Studies in the History of Statistics and Probability* (London: Griffin, 1970), pp. 231–240, and Karl Pearson, "James Bernoulli's Theorem," *Biometrika* 17 (1925), 201–210.

7. De Moivre, *The Doctrine of Chances*, pp. 1–2. More information on De Moivre can be found in H. M. Walker, "Abraham De Moivre," *Scripta Mathematica* 2 (1934), 316–333, and in Ivo Schneider, "Der Mathematiker Abraham De Moivre (1667–1754)," *Archive for History of Exact Sciences* 5 (1968), 177–317.

8. De Moivre, *Doctrine of Chances*, p. 36.

9. Ibid., p. 242.

10. Quoted in Stigler, *History of Statistics*, p. 76.

11. De Moivre, *Doctrine of Chances*, p. 246.

12. Ibid., p. 248.

13. Thomas Bayes, "An Essay towards Solving a Problem in the Doctrine of Chances," reprinted in Pearson and Kendall, *Studies in the History of Statistics*, p. 136.

14. Ibid., p. 139.

15. Ibid., p. 140.

16. Ibid., p. 143.

17. Ibid. This quotation has led to numerous discussions in the literature as to the types of situations to which Bayes's theorem applies. Two recent analyses of this matter are in Stigler, *History of Statistics*, pp. 122–131, and in Donald A. Gillies, "Was Bayes a Bayesian?," *Historia Mathematica* 14 (1987), 325–346.

18. Letter of Euler to Frederick II, dated 17 September, 1749. Translated by Richard Pulskamp and available at http://cerebro.xu.edu/math/Sources/Euler/. For more on Euler and the lottery, see Robert Bradley, "Euler's Analysis of the Genoese Lottery," *Convergence*, http://convergence.mathdl.org, 2004, and D. R. Bellhouse, "Euler and Lotteries," in Robert Bradley and C. Edward Sandifer, eds., *Leonhard Euler: Life, Work and Legacy* (Amsterdam: Elsevier, 2007), pp. 385–394.

19. Euler, "Reflections on a Singular Kind of Lottery," p. 20. This paper is E812 and is in *Opera omnia* (1) 7, pp. 466–494.

Algebra and Number Theory in the Eighteenth Century

All the pains that have been taken in order to resolve equations of the fifth degree, and those of higher dimensions, . . . or, at least, to reduce them to inferior degrees, have been unsuccessful; so that we cannot give any general rules for finding the roots of equations, which exceed the fourth degree.

—Leonhard Euler's *Algebra*, 1767[1]

Two letters of Euler in 1742 confirmed that he then believed the truth of the fundamental theorem of algebra. On October 1, 1742, he wrote to Nicolaus Bernoulli that every real polynomial could be factored into real linear and quadratic factors, even though Bernoulli claimed to have a counterexample. And then on December 15, in a letter to Christian Goldbach (1690–1764), he noted that complex roots of polynomials always occur in conjugate pairs, while using this result to help in actually factoring Bernoulli's example. Unfortunately, Goldbach was not convinced and gave his own supposed counterexample.

There were few major new developments in algebra in the eighteenth century, in contrast to the work in other fields. The major effort, accomplished by mathematicians whose chief influence was felt elsewhere, was a systematization of earlier material. For example, there were methods to solve systems of linear equations, as well as methods to solve algebraic equations of degree up to four. But these methods were ad hoc, so mathematicians sought more general procedures for which some theoretical analysis was necessary. We first look at these equation-solving ideas in three major algebra texts, one by Newton (compiled from his lectures at Cambridge from 1673 to 1683 but only published in 1707), one by Maclaurin (published in 1748 although probably written in the 1730s), and one by Euler (published in 1767), each of which served to introduce students to the field and set the basis for future work. We then consider some of the extensions of these ideas in the theory of equations later in the century as well as Euler's work in number theory, which was to have important consequences in subsequent years. We conclude with a brief look at mathematics in the Americas in the eighteenth century.

 19.1 ## ALGEBRA TEXTS

In the eighteenth century, algebra meant the solving of equations. Thus, Newton, Maclaurin, and Euler each presented their own ideas on this subject.

19.1.1 Newton's *Arithmetica Universalis*

Newton lectured on algebra for 10 years at Cambridge until finally, in 1683, he decided he should comply with the rules of the Lucasian professorship. Thus, sometime during the winter of 1683–1684, he wrote up the lectures, carefully noting the date that each one was supposedly delivered, and deposited them as required in the university library. Some 20 years later, Newton's successor William Whiston (1667–1752) prepared the lectures for publication, and, although Newton was not entirely happy with the results, they appeared in published form in 1707 as the *Arithmetica universalis* (*Universal Arithmetic*). Despite Newton's misgivings, this book proved very popular, going through numerous editions in Latin, English, and French, into the early nineteenth century.

Newton's text began at a very elementary level, but by the end he had given a rather comprehensive course with many interesting details on the solution of algebraic equations. Consider first Newton's treatment of addition and multiplication:

> Addition: In the case of numbers which are not unduly complicated addition is self-evident. Thus it is clear at first glance that 7 and 9, that is, 7 + 9, make 16 and that 11 + 15 make 26. But in more complicated cases the operation is achieved by writing the numbers in a descending sequence and gathering the sums of the columns separately.[2]

> Multiplication: Simple algebraic terms are multiplied by "drawing" numbers into numbers and variables into variables, and then setting the product positive if both factors be positive or both negative, and negative otherwise.[3]

Newton made no attempt here, or elsewhere, to justify the multiplication rule. He just stated it. Nor did he justify any of his other arithmetic algorithms. Evidently, justification was not necessary for his listeners or, presumably, his readers. All that was necessary were techniques for manipulation. And Newton produced these in abundance, both with numbers and with algebraic expressions. He also covered the basics of solving equations and spent a good bit of time showing how to translate problems into algebra, including much material from geometry. He presented the quadratic formula and also Cardano's cubic formula—although for the latter, he wrote that it is "extremely rare of use."

Many of Newton's "word problems" are very familiar, since versions still appear in algebra texts today:

> If two couriers A and B, 59 miles apart, set out one morning to meet each other, and of these A completes 7 miles in 2 hours and B 8 miles in 3 hours, while B starts his journey 1 hour later than A: how far a distance has A still to travel before he meets B?[4]

> If a scribe can copy out 15 sheets in 8 days, how many scribes of the same output are needed to copy 405 sheets in 9 days?[5]

By the end of his text, however, Newton had also solved much more difficult problems, including problems in physics and astronomy, and had developed Descartes' rule of signs, the relationships between the coefficients of a polynomial and its roots, and formulas to determine the sums of various integral powers of the roots of a polynomial equation. Yet because Newton was no longer heavily involved in mathematics in 1707, he never put the work into a truly polished form. It was his successors, Maclaurin and Euler, who absorbed his insights and reworked his material into texts that were to have even more influence.

19.1.2 Maclaurin's *Treatise of Algebra*

Maclaurin, like Newton, thought of algebra as "a general method of computation by certain signs and symbols which have been contrived for this purpose and found convenient. It is called an Universal Arithmetic and proceeds by operations and rules similar to those in common arithmetic, founded upon the same principles."[6] In other words, for Maclaurin algebra is not "abstract" but simply generalized arithmetic. Thus, because it is necessary to understand arithmetic before one can understand algebra, Maclaurin began his *A Treatise of Algebra in Three Parts* not only with algorithms for calculation but also with attempts to explain the reasoning behind the algorithms. For example, in dealing with negative numbers, he noted that any quantity can enter algebraic computation as either an increment or a decrement. As examples of these two forms, he included such concepts as excess and deficit, value of money due to a man and due by him, a line drawn to the right and one to the left, and elevation above horizon and depression below. He noted that one can subtract a greater quantity from a lesser of the same kind, the remainder in that case always being opposite in kind, but one can only do this if it makes sense. For example, one cannot subtract a greater quantity of matter from a lesser. Nevertheless, Maclaurin always considered a negative quantity to be no less real than a positive one. He thus demonstrated how to calculate with positive and negative quantities. In particular, to show the reason for the rule of signs in multiplying such quantities, he observed that since $+a - a = 0$, also $n(+a - a) = 0$. But the first term of this product, $+na$, is positive. The second term must therefore be negative. Therefore, $-a$ multiplied by $+n$ is negative. Similarly, since $-n(+a - a) = 0$ and the first term of this product is negative, the second term, $(-n)(-a)$, must be positive and equal to $+na$.

Maclaurin continued in the first part of his work to deal with such topics as manipulation with fractions, powers of binomials, roots of polynomials, and sums of progressions. He showed the reader how to calculate terms in the expansion $(a + b)^n$, both for n integral and fractional, the latter calculation naturally resulting in an infinite series. He showed how to solve linear and quadratic equations, including a fair number of "word problems" as examples. In the case of linear equations in more than one unknown, he showed in the cases of two and three equations in the same number of unknowns that the solution can be found by solving for one unknown in terms of the others and substituting. He noted that if there are more unknowns than equations, there may be an infinite number of solutions, while in the opposite case, there may be no solutions at all, but he did not give any examples of either situation.

He did, however, present what he called a "general theorem" for eliminating unknowns in a system of equations, the method known today as **Cramer's rule**, named after the Swiss mathematician Gabriel Cramer (1704–1752) who used it in his 1750 book *Introduction to the Analysis of Algebraic Curves*. If

$$ax + by = c$$
$$dx + ey = f,$$

then solving the first equation for x and substituting gives

$$y = \frac{af - dc}{ae - db}$$

and a similar answer for x. The system of three equations

$$ax + by + cz = m$$
$$dx + ey + fz = n$$
$$gx + hy + kz = p$$

is dealt with by first solving each equation for x, thus reducing the problem to a system in two unknowns, and then using the earlier rule to find

$$z = \frac{aep - ahn + dhm - dbp + gbn - gem}{aek - ahf + dhc - dbk + gbf - gec}.$$

In addition to giving the answer, Maclaurin described the general rule that the numerator consists of the various products of the coefficients of x and y as well as the constant terms, each product consisting of one coefficient from each equation, while the denominator consists of products of the coefficients of all three unknowns. He also explained how to determine the sign of each term. Furthermore, he solved for y and x and showed that the general rule determining each of these values is analogous to the one for z. In particular, each of the three expressions has the same denominator. Maclaurin even extended the rule to systems of four equations in four unknowns, but did not discuss any further generalization.

The numerators and denominators involved in Maclaurin's solutions are, of course, what are known today as **determinants**. But the use of such combinations of coefficients as tools for solving systems of linear equations had appeared somewhat earlier. Leibniz had suggested a similar idea in a letter to l'Hospital in 1693 and had even devised a way of indexing the coefficients of the system by the use of numbers. And halfway around the world, the Japanese mathematician Seki Takakazu (1642–1708) described the use of determinants in a manuscript

BIOGRAPHY

Seki Takakazu (1642–1708)

Seki Takakazu, often called Seki Kōwa because of different ways of reading the Japanese characters making up his name, was born into the family of a samurai retainer of a feudal lord in Fujioka, a town about 50 miles northwest of Tokyo. Seki himself served as an accountant to two feudal lords in Kōfu, now in Yamanashi prefecture, later moving to Tokyo to perform the same service. Although Seki published little, there is evidence in his many manuscripts that he understood much of the basics of the theory of equations, including the notion that

a polynomial equation may have as many roots as the degree. Recall that the Chinese mathematicians, whose material Seki studied, were satisfied with determining only one solution. He introduced determinants in a work of 1683 dealing with setting up and solving equations. The diagrams, one of which is partially visible in the illustration (Fig. 19.1), demonstrate the method of determining the elements to be multiplied together and the signs to be attached to each product.

FIGURE 19.1

Seki Takakazu on a Japanese stamp

10 years earlier, carefully showing by use of diagrams how to decide whether a given term is to be positive or negative.

The second part of Maclaurin's work is a treatise on the solving of polynomial equations, which presents in well-organized form all that had been discovered up to his time. Thus, Maclaurin included not only Cardano's rule for solving cubics and Ferrari's rule for quartics but also Descartes' rule of signs and Newton's methods for approximating numerically the solution to an equation. He noted that the procedure by which equations are generated—multiplying together equations such as $x - a = 0$ or other equations of degree smaller than the given one—shows that no equation can have more roots than the degree of the highest power. Furthermore, "roots become impossible [complex] in pairs" and therefore "an equation of an odd dimension has always one real root."[7] He then discussed the general procedure for finding integral roots of monic polynomials: check all divisors of the constant term as possible roots and, if one such root α is found, divide the polynomial by $x - \alpha$ to reduce the degree.

Maclaurin concluded his text with a discussion of the application of algebraic techniques to geometric problems and, conversely, of the use of geometrical procedures to solve equations. The major difference between the uses of algebra and geometry, he wrote, is that in the former, one can express even impossible roots explicitly, but in the latter such quantities do not appear at all. Maclaurin included in this part detailed rules for constructing solutions to quadratic equations using circles and solutions to cubic and quartic equations using conic sections. Although there was little new mathematically in the text, Maclaurin's work became popular enough with students that it was republished several times during the century.

19.1.3 Euler's *Introduction to Algebra*

An even better introduction to algebra, perhaps, was provided by Euler's *Vollständige Anleitung zur Algebra* (*Complete Introduction to Algebra*). Euler, like Maclaurin, began his text by providing a definition of the subject: "The foundation of all the mathematical sciences must be laid in a complete treatise on the science of numbers, and in an accurate examination of the different possible methods of calculation. This fundamental part of mathematics is called

Analysis, or Algebra. In Algebra, then, we consider only numbers, which represent quantities, without regarding the different kinds of quantity."[8] Later on in the text, he made the definition somewhat more specific: Algebra is "the science which teaches how to determine unknown quantities by means of those that are known."[9] He noted that even ordinary addition of two known quantities can be thought of as fitting under this definition, and so the second, perhaps more common, definition really included the first.

Euler began the text with a discussion of the algebra of positive and negative quantities. His discussion of multiplication was somewhat less formal than that of Maclaurin: "Let us begin by multiplying $-a$ by $+3$. Now, since $-a$ may be considered as a debt, it is evident that if we take that debt three times, it must thus become three times greater, and consequently the required product is $-3a$."[10] Euler then noted the obvious generalization that $-a$ times b will be $-ba$ or $-ab$ and continued to the case of the product of two negatives. Here he simply wrote that $-a$ times $-b$ cannot be the same as $-a$ times b, or $-ab$, and therefore must be equal to $+ab$.

After discussing various other operations, Euler introduced the concept of an imaginary number:

> Since all numbers which it is possible to conceive are either greater or less than 0, or are 0 itself, it is evident that we cannot rank the square root of a negative number amongst possible numbers, and we must therefore say that it is an impossible quantity. In this manner we are led to the idea of numbers, which from their nature are impossible; and therefore they are usually called *imaginary quantities*, because they exist merely in the imagination. All such expressions as $\sqrt{-1}, \sqrt{-2} \ldots$ are consequently impossible, or imaginary numbers, since they represent roots of negative quantities; ... but notwithstanding this these numbers present themselves to the mind; they exist in our imagination, and we still have a sufficient idea of them; since we know that by $\sqrt{-4}$ is meant a number which, multiplied by itself, produces -4; for this reason also, nothing prevents us from making use of these imaginary numbers, and employing them in calculation.[11]

Curiously, Euler did not realize that there may be problems in these calculations. For although he has noted that $\sqrt{-4} \times \sqrt{-4} = -4$, somewhat later he wrote that the general rule for multiplying square roots implies that $\sqrt{-1} \times \sqrt{-4} = \sqrt{(-1)(-4)} = \sqrt{4} = 2$.

Euler continued the text by discussing logarithms, infinite series, and the binomial theorem. He defined logarithms as he did in the *Introductio*: If $a^b = c$, then b is the logarithm of c with base a. Logarithms were then applied in a chapter on calculation of compound interest. Infinite series were introduced in terms of division, with the first example being $\frac{1}{1-a} = 1 + a + a^2 + a^3 + \cdots$. Although Euler did not discuss convergence as such, he asserted that "there are sufficient grounds to maintain that the value of this infinite series is the same as that of the fraction."[12] He then dealt with some examples so that this statement would be "easily understood." Thus, if $a = 1$, the fraction is equal to $1/0$, "a number infinitely great," while the series becomes $1 + 1 + 1 + \cdots$, also infinite, thus confirming the assertion. But, Euler concluded, "the whole becomes more intelligible" if values for a less than 1 are taken. In that case, "the more terms we take, the less the difference [between the fraction and the series] becomes; and consequently, if we continue the series to infinity, there will be no difference at all between its sum and the value of the fraction."[13]

Given that Euler defined algebra in terms of finding unknowns, it is not surprising that a large section of the *Algebra* was devoted to determining such solutions. In fact, he devoted considerable space to explaining how to set up problems as equations. As he wrote, "in algebra, when we have a question to resolve, we represent the number sought by one of the last letters of the alphabet, and then consider in what manner the given conditions can form an equality between two quantities. This equality is represented by a kind of formula, called an *equation*, which enables us finally to determine the value of the number sought, and consequently to resolve the question."[14] He then very systematically took his readers through the algebraic solution of equations of degrees one, two, three, and four, before concluding with the statement given in the opening of this chapter.

The final part of Euler's text was devoted to a subject not found at all in the works of Newton and Maclaurin, the solution of indeterminate equations. Many of the problems solved in this part are, in fact, the problems of Diophantus's *Arithmetica*. But Euler, like Fermat a century earlier, always gave general solutions to the problems rather than the single solution typical of the Greek algebraist. As an example, consider the following problem, virtually the same as Diophantus's Problem II–11:

QUESTION 2   *To find such a number x, that if we add to it any two numbers, for example, 4 and 7, we obtain in both cases a square.*[15]

Diophantus solved this problem by the method of the double equation. Euler used a different technique. Setting $x + 4 = p^2$, he concluded that $x + 7 = p^2 + 3$ is a square whose root is $p + q$. Setting $q = r/s$, it follows that $p^2 + 3 = p^2 + 2pq + q^2$ or that $p = (3 - q^2)/2q$ or finally that

$$x = p^2 - 4 = \frac{9 - 22q^2 + q^4}{4q^2} = \frac{9s^4 - 22r^2s^2 + r^4}{4r^2s^2}.$$

Euler then noted that any choice of integers for r and s gives a solution for x.

Much of this section on indeterminate equations, however, is devoted to some general methods rather than specific problems. Euler dealt especially with techniques for finding solutions, in either rational numbers or integers, to equations of the form $p(x) = y^2$, where $p(x)$ is a polynomial of degree 2, 3, or 4. As a special case, he considered the solution in integers of the equation $Dx^2 + 1 = y^2$ discussed in Chapter 8, the equation whose solution Euler incorrectly attributed to the English mathematician John Pell (1610–1685) and that Fermat claimed to have had. Rather than giving a general method of solution, Euler demonstrated a procedure to be applied in each case separately. He then concluded his discussion by presenting a table in which solutions to the equation are listed for values of D from 2 to 100. Although Euler did not prove that solutions exist for every D, such a proof was given by Lagrange in 1766 and included as an appendix in later editions of the *Algebra*.

   19.2

ADVANCES IN THE THEORY OF EQUATIONS

Eighteenth century mathematicians raised significant questions about various aspects of the theory of equations, questions that were not answered completely until the nineteenth century.

19.2.1   The Fundamental Theorem of Algebra

Recall that both Girard and Descartes had stated versions of what is today called the fundamental theorem of algebra, to the effect that every polynomial equation of degree n with real coefficients has n solutions, either real or complex. But neither of them could actually prove this result. On the other hand, by the beginning of the eighteenth century, doubts began to arise as to the correctness of the theorem. Leibniz, in fact, raised the question as to "whether every algebraic equation . . . can be decomposed into simple or plane real factors,"[16] that is, whether every real polynomial can be factored into real polynomials of degree one or two. (This statement is equivalent to the earlier version of the fundamental theorem, given the result that complex roots of real polynomials always occur in conjugate pairs.) The reason this result was important to Leibniz was to answer the question as to whether every rational function can be integrated via the method of partial fractions. As it turned out, Leibniz believed that he had found a counterexample to the fundamental theorem, namely, the polynomial $x^4 + a^4$, which he factored as

$$x^4 + a^4 = (x^2 + a^2\sqrt{-1})(x^2 - a^2\sqrt{-1})$$
$$= \left(x + a\sqrt{\sqrt{-1}}\right)\left(x - a\sqrt{\sqrt{-1}}\right)\left(x + a\sqrt{-\sqrt{-1}}\right)\left(x - a\sqrt{-\sqrt{-1}}\right).$$

He unfortunately did not know that

$$\sqrt{\sqrt{-1}} = \frac{1 + \sqrt{-1}}{\sqrt{2}}$$

and thus believed that no product of any two of the factors would give a real quadratic divisor of $x^4 + a^4$. In other words, he evidently believed that $\sqrt{\sqrt{-1}}$ was not actually a complex number, but some new kind of number.

Several mathematicians in the first third of the eighteenth century, including Cotes and DeMoivre, showed that Leibniz's example—as well as many others—were in fact factorable. But the first mathematician to publish a purported proof of the fundamental theorem was d'Alembert in 1746. This proof was incomplete, but meanwhile, Euler himself had been thinking about the issue. In chapter 2 of the *Introductio*, he dealt with various kinds of factoring of real polynomials. For example, he demonstrated that complex linear factors of a real polynomial always occur in pairs whose product is real, that if a polynomial is the product of four complex linear factors, then it can also be represented as the product of two real quadratic factors, and, essentially by using the Intermediate Value Theorem, that any polynomial of odd degree has at least one real linear factor.

In chapter 9, Euler analyzed the factoring of certain types of polynomials, including a generalization of Leibniz's example. In particular, he found the irreducible quadratic factors of $a^n \pm z^n$ to be $a^2 - 2az \cos \frac{2k\pi}{n} + z^2$, when the sign is negative, and $a^2 - 2az \cos \frac{(2k+1)\pi}{n} + z^2$, when the sign is positive. In the former case, if one sets $a = 1$, one gets the irreducible factors determining the nth roots of unity. And after finding quadratic factors of certain other classes of polynomials, he wrote that "if there were any doubt that every polynomial can be expressed as a product of real linear and real quadratic factors, then that doubt by this time should be almost completely dissipated."[17] Curiously, however, he did not claim here an actual proof of this version of the fundamental theorem. That he only did in an article in 1749, "Recherches sur les racines imaginaires des équations" ("Investigations on the Imaginary Roots of Equations").

In the article, after discussing a few particular examples, Euler began his general proof of the fundamental theorem by again using the Intermediate Value Theorem to show that any odd-degree polynomial equation has a real root and that any even-degree polynomial with a negative constant term has two real roots. He then proved the result that any fourth-degree polynomial equation can be factored into two real quadratic factors. To do this, he noted that since the cubic term can always be removed by a linear substitution, it sufficed to look at equations of the form $x^4 + Bx^2 + Cx + D = 0$. The two real factors of the polynomial must then be $x^2 + ux + \alpha$ and $x^2 - ux + \beta$, with u, α, and β to be determined. Comparing coefficients shows that the equations for these three unknowns are

$$\alpha + \beta - u^2 = B, \qquad (\beta - \alpha)u = C, \quad \text{and} \quad \alpha\beta = D.$$

It then follows that $\alpha + \beta = B + u^2$ and $\beta - \alpha = C/u$, so that

$$2\beta = u^2 + B + \frac{C}{u} \quad \text{and} \quad 2\alpha = u^2 + B - \frac{C}{u},$$

Because $4\alpha\beta = 4D$, multiplying the last two equations together gives

$$u^4 + 2Bu^2 + B^2 - \frac{C^2}{u^2} = 4D \quad \text{or} \quad u^6 + 2Bu^4 + (B^2 - 4D)u^2 - C^2 = 0.$$

By the result on even-degree polynomials with negative constant terms, Euler knew that the equation for u had two real nonzero roots. Whichever one he chose would then give real values for α and β and thus determine the real factors of the fourth-degree polynomial. (Note that we are assuming here that $C \neq 0$; a much simpler argument will work if $C = 0$.)

In the remainder of the paper, Euler attempted to generalize the proof for fourth-degree polynomials to analogous results for polynomials of degree 2^n, from which he could derive the result for any degree. Unfortunately, although Euler believed that he had succeeded, later mathematicians noted some logical holes in his treatment of these higher-degree cases, and so the fundamental theorem remained unproved for another half century. It was only Carl Gauss who gave proofs of the fundamental theorem that are still considered valid, by making use of the geometrical interpretation of complex numbers that was unknown to Euler.[18]

19.2.2 Euler and Systems of Linear Equations

In his *Algebra*, Euler did not go into great detail on the subject of solving systems of linear equations. He did not present Cramer's rule or any other general procedure. He simply suggested that one solve a system by solving for one unknown in terms of the others, substituting, and thus reducing the system to one with fewer equations and fewer unknowns.

On the other hand, by 1750 Euler was already exploring some more general ideas in the solving of systems of equations. In a paper of that year, his concern was to solve a paradox that Cramer had formulated, based on an earlier suggestion of Maclaurin. This paradox was based on two propositions that everyone believed in the early eighteenth century:

1. An algebraic curve of order n is uniquely determined by $n(n + 3)/2$ of its points.
2. Two algebraic curves of order n and m intersect in nm points.

The first result comes from elementary combinatorics. Basically, a curve of order n, one described by an nth-degree polynomial in two variables, has one coefficient of degree 0, two of degree 1, three of degree 2, four of degree 3, and so on. But since we can divide by any one coefficient, the total number of "independent" coefficients is

$$\sum_{i=1}^{n+1} i - 1 = \frac{(n+1)(n+2)}{2} - 1 = \frac{n(n+3)}{2},$$

and thus that many points are necessary to determine the curve. In fact, it was this problem of determining the coefficients of a curve by a knowledge of certain points on it that had led Cramer to his own discovery of Cramer's rule and also to formulating the paradox. As for the second result, although it was known that the points of intersection of algebraic curves may be multiple or imaginary, examples were known where all mn points were real and distinct. The paradox then came from consideration of the case $n \geq 3$, where it appears from the second proposition that there are n^2 points common to two algebraic curves of order n, while the first proposition implies that $n(n+3)/2$ (which is less than n^2) points should determine a unique curve.

Euler discussed the paradox and concluded that the first result, based on the fact that n linear equations in n unknowns should determine a unique n-fold solution, was not true without restriction. The conviction at the time that n equations determined n unknowns was so strong that no one earlier had really taken the pains to discuss the cases where this did not happen. As noted, Maclaurin had briefly discussed the situation where the number of equations was not equal to the number of unknowns, but had not mentioned the possibilities in the present instance.

In his paper, Euler discussed various examples, without, however, being able to state a definite theorem. For example, he noted that $3x - 2y = 5$, $4y = 6x - 10$ do not determine two unknowns, because if we solve for x and substitute, the equation for y becomes an identity, which does not allow us to determine a value. He also gave a system of four equations in four unknowns in which, after solving for two of the variables in terms of the others and substituting into the remaining two equations, again identities appeared, so that the remaining two values were undetermined. Thus, the four equations do not determine four unknowns. So, he concluded, when it is said that to determine n unknowns it is sufficient to have n equations, it is necessary to add the restriction that these equations are so different that none of them is already "comprised" in the others. Although Euler did not explicitly define "comprise," it seems that, at least intuitively, he understood the concept of the "rank" of a system.

To resolve Cramer's paradox, Euler finally noted that "when two curves of fourth order meet in 16 points, as 14 points, when they lead to different equations, are sufficient to determine one curve of this order, these 16 points will always be such that three or more equations are already comprised in the others. In this way, these 16 points do not determine more than if they were 13 or 12 or even fewer points and in order to determine the curve entirely, one must add to these 16 points one or two others."[19]

Although Euler had solved the immediate problem, it took over a century and a quarter for mathematicians to completely understand the ideas involved in undetermined or inconsistent systems. We thus resume the discussion of these concepts in Chapter 21.

19.2.3 Lagrange and the Solution of Polynomial Equations

As we have noted, Maclaurin and Euler had included methods for solving cubic and quartic polynomial equations in their texts. Other mathematicians of the time period attempted to generalize these methods to solve algebraically polynomial equations of degree five and higher, but without success. But it was Lagrange, in his *Réflexions sur la théorie algébrique des équations* (*Reflections on the Algebraic Theory of Equations*) of 1770, who began a new phase in this work by undertaking a detailed review of these earlier solutions to determine why the methods for cubics and quartics worked. He was not able to find analogous methods for higher-degree equations but was able to sketch a new set of principles for dealing with these equations, which he hoped might ultimately succeed.

Lagrange began with a systematic study of the methods of solution of the cubic equation $x^3 + nx + p = 0$, starting essentially with Cardano's procedure. Setting $x = y - (n/3y)$ transforms this equation into the sixth-degree equation $y^6 + py^3 - (n^3/27) = 0$, which, with $r = y^3$, reduces in turn to the quadratic equation $r^2 + pr - (n^3/27) = 0$. This latter equation has two roots, r_1 and $r_2 = -(\frac{n}{3})^2 \frac{1}{r_1}$. But whereas Cardano took the sum of the real cube roots of r_1 and r_2 as his solution, Lagrange knew that each equation $y^3 = r_1$ and $y^3 = r_2$ had three roots. Thus, there were six possible values for y, namely, $\sqrt[3]{r_1}$, $\omega\sqrt[3]{r_1}$, $\omega^2\sqrt[3]{r_1}$, $\sqrt[3]{r_2}$, $\omega\sqrt[3]{r_2}$, and $\omega^2\sqrt[3]{r_1}$, where $\omega = (-1 + \sqrt{-3})/2$ is a complex root of $x^3 - 1 = 0$, or of $x^2 + x + 1 = 0$. Lagrange could then show that the three distinct roots of the original equation were given by

$$x_1 = \sqrt[3]{r_1} + \sqrt[3]{r_2}$$
$$x_2 = \omega\sqrt[3]{r_1} + \omega^2\sqrt[3]{r_2}$$
$$x_3 = \omega^2\sqrt[3]{r_1} + \omega\sqrt[3]{r_2}.$$

Lagrange next noted that rather than consider x as a function of y, one could reverse the procedure, because the equation for y, which he called the *réduite* or *reduced* equation, was the one whose solutions enabled the original equation to be solved. The idea then was to express those solutions in terms of the original ones. Thus, Lagrange noted that any of the six values for y could be expressed in the form $y = \frac{1}{3}(x' + \omega x'' + \omega^2 x''')$, where (x', x'', x''') was some permutation of (x_1, x_2, x_3). It was this introduction of the permutations of the roots of an equation that provided the cornerstone not only for Lagrange's method but for the methods others were to use in the next century.

In the case of the cubic, there are several important ideas to note. First, the six permutations of the x_i lead to the six possible values for y and thus show that y satisfies an equation of degree six. Second, the permutations of the expression for y can be divided into two sets, one consisting of the identity permutation and the two permutations that interchange all three of the x_i and the second consisting of the three permutations that interchange just two of the x_i. (In modern terminology, the group of permutations of a set of three elements has been divided into two cosets.) For example, if $y_1 = \frac{1}{3}(x_1 + \omega x_2 + \omega^2 x_3)$, then the two nonidentity permutations in the first set change y_1 to $y_2 = \frac{1}{3}(x_2 + \omega x_3 + \omega^2 x_1)$ and $y_3 = \frac{1}{3}(x_3 + \omega x_1 + \omega^2 x_2)$, respectively. But then $\omega y_2 = \omega^2 y_3 = y_1$ and $y_1^3 = y_2^3 = y_3^3$. Similarly, if the results of the permutations of the second set are y_4, y_5, and y_6, it follows that $y_4^3 = y_5^3 = y_6^3$. Thus, because there are only two possible values for $y^3 = \frac{1}{27}(x' + \omega x'' + \omega^2 x''')^3$, the equation for y^3 is of degree 2. Finally, the sixth-degree equation satisfied by y has coefficients that are rational in the coefficients of the original equation. Lagrange considered several other methods of solution of the cubic equation but found in each case the same underlying idea. Each led to a rational expression in the three roots, which took on only two values under the six possible permutations, thus showing that the expression satisfied a quadratic equation.

Lagrange next considered the solutions of the quartic equation. Ferrari's method of solving $x^4 + nx^2 + px + q = 0$ was to add $2yx^2 + y^2$ to each side, rearrange, and then determine a value for y such that the right side of the new equation

$$x^4 + 2yx^2 + y^2 = (2y - n)x^2 - px + y^2 - q$$

was a perfect square. After taking square roots of each side, he could then solve the resulting quadratic equations. The condition that the right side be a perfect square is that

$$(2y - n)(y^2 - q) = \left(\frac{p}{2}\right)^2 \quad \text{or} \quad y^3 - \frac{n}{2}y^2 - qy + \frac{4nq - p^2}{8} = 0.$$

Therefore, the *réduite* is a cubic, which can, of course, be solved. Given the three solutions for y, Lagrange then showed, as in the previous case, that each is a permutation of a rational function of the four roots x_1, x_2, x_3, x_4 of the original equation. In fact, it turned out that $y_1 = \frac{1}{2}(x_1x_2 + x_3x_4)$ and that the 24 possible permutations of the x_i lead to only three different values for that expression, namely, y_1, $y_2 = \frac{1}{2}(x_1x_3 + x_2x_4)$, and $y_3 = \frac{1}{2}(x_1x_4 + x_2x_3)$. The expression must therefore satisfy a third-degree equation, again one with coefficients rational in the coefficients of the original equation.

Having studied the methods for solving cubics and quartics, Lagrange was ready to generalize. First, as was clear from the discussion of cubic equations, the study of the roots of equations of the form $x^n - 1 = 0$ was important. For the case of odd n, Lagrange could show that all the roots could be expressed as powers of one of them. In particular, if n is prime and $\alpha \neq 1$ is one of the roots, then α^m for any $m < n$ can serve as a generator of all of the roots. Second, however, Lagrange realized that to attack the problem of equations of degree n, he needed a way of determining a *réduite* of degree $k < n$. Such an equation must be satisfied by certain functions of the roots of the original equation, functions that take on only k values when the roots are permuted by all $n!$ possible permutations. Because relatively simple functions of the roots did not work, Lagrange attempted to find some general rules for determining such functions and the degree of the equation that they would satisfy.

Lagrange noted that if the values of the roots of the *réduite* are f_1, f_2, . . . , f_k, where each f_i is a function of the n roots of the original equation, then the *réduite* is given by $(t - f_1)(t - f_2) \cdots (t - f_k) = 0$. Although he could not prove that the degree of this equation in general is less than $n!$, he was able to show that its degree k, the number of different values taken by f under the permutations of the variables, always divided $n!$. One can read into this statement Lagrange's theorem to the effect that the order of any subgroup of a group divides the order of the group, but Lagrange never treated permutations as a "group" of operations. He did go on, however, to show how functions of the roots may be related. He proved that if all permutations of the roots that leave one such function u unchanged also leave another such function v unchanged, then v can be expressed as a rational function of u and the coefficients of the original equation. Furthermore, if u is unchanged by permutations that do change v, and if v takes on r different values for each one taken on by u, then v is the root of an equation of degree r whose coefficients are rational in u and the coefficients of the original equation. For example, in the cubic equation $x^3 + nx + p = 0$, the expression $v = \frac{1}{27}(x_1 + \omega x_2 + \omega^2 x_3)^3$ takes on two values under the six permutations of the roots, while $u = x_1 + x_2 + x_3$ is unchanged under those permutations. Then $v^2 + pv - (n^3/27) = 0$ is the equation satisfied by v. (Note here that $u = 0$.)

Lagrange presumably hoped to solve the general polynomial equation of degree n by use of this theorem. Namely, he would start with a symmetric function of the roots, say, $u = x_1 + x_2 + \cdots + x_n$, which was unchanged under all $n!$ permutations, then find a function v, which takes on r different values under these permutations. Thus, v would be a root of an equation of degree r with coefficients rational in the original coefficients (because the given symmetric function u was one of those coefficients). If that equation could be solved, then he could find a new function w, which takes on, say, s values under the permutations that leave v unchanged. Thus, w would satisfy an equation of degree s. He would continue in this way until the function x_1 is reached. Unfortunately, Lagrange was unable to find a general method of determining these intermediate functions such that they were of a form that could be solved by known methods. He was thus forced to abandon his quest. Nevertheless, his work did form the foundation on which all nineteenth-century work on the algebraic solution of equations was based. The story is therefore continued in Chapter 21.

 19.3   ## NUMBER THEORY

Euler worked on and solved many interesting number-theoretical problems during his life, some of which had been suggested by Fermat or grew out of problems Fermat solved. Thus, in 1749 Euler proved Fermat's claim that every prime of the form $4n + 1$ can be written as the sum of two squares and in 1773, after working on the problem for many years, gave a proof that every integer can be expressed as a sum of no more than four squares. (Lagrange had proved this result three years before; Euler's proof was a generalization of his earlier proof for two squares.) We will, however, only discuss Euler's proof of the case $n = 3$ of Fermat's Last Theorem, his study of residues, his generalization of Fermat's Little Theorem, and his discovery of the law of quadratic reciprocity.

19.3.1   Fermat's Last Theorem

Recall that although Fermat had claimed to have a general proof of his "Last Theorem," the only proof that saw the light of day was his proof of the case $n = 4$ by infinite descent. Euler presented a similar proof of this case in his algebra text and then went on to prove the case $n = 3$:

THEOREM   *It is impossible to find any two cubes, whose sum, or difference, is a cube.*

Euler's proof here was also by infinite descent. Thus, he began with the assumption that there were relatively prime integers x, y, z satisfying $x^3 + y^3 = z^3$ and showed that he could find smaller integers that satisfied the same equation, a result leading to an impossibility. There is no loss of generality in assuming that both x and y are odd (and z is even), so Euler set $x + y = 2p$, $x - y = 2q$. Then $x = p + q$, $y = p - q$, and $x^3 + y^3 = 2p(p^2 + 3q^2)$. He then showed that if this latter expression were a cube, he could find a triple f, g, r less than x, y, z satisfying $f^3 + g^3 = r^3$.

If $2p(p^2 + 3q^2)$ were a cube, that cube would be even, so divisible by 8. Therefore, $\frac{1}{4}p(p^2 + 3q^2)$ is also an integral cube. Since the second factor in this expression is odd, so not divisible by 4, it follows that $\frac{1}{4}p$ is an integer. Euler then noted that the two factors must be relatively prime, except in the case where p is divisible by 3. So, assuming this is not the case, it follows that each factor must be a cube. Since $p^2 + 3q^2$ factors as $(p + q\sqrt{-3})(p - q\sqrt{-3})$, Euler asserted, without proof, that each of these factors must itself be a cube. This result is in fact true, but it is certainly not obvious, particularly because the complex numbers of the form $a + b\sqrt{-3}$ do not form a unique factorization domain. In any case, given this result, Euler wrote that $p \pm q\sqrt{-3} = (t \pm u\sqrt{-3})^3$, so $p = t(t^2 - 9u^2)$ and $q = 3u(t^2 - u^2)$, with t odd and u even. In addition, since $\frac{1}{4}p$ is a cube, so also is $2p$. Thus, $2p = 2t(t + 3u)(t - 3u)$ is a cube and, since neither p nor t is divisible by 3, these three factors must be relatively prime. Thus, each of the factors is also a cube, say, $2t = r^3$, $t + 3u = f^3$, $t - 3u = g^3$. But then $f^3 + g^3 = r^3$, with each term less than the corresponding term in our original sum-of-cubes expression.

In the case where $p = 3r$, a similar argument shows that $\frac{2}{3}r = 2u(t + u)(t - u)$ is a cube that is the product of relatively prime factors. Thus, each factor is a cube and we again get a new sum-of-cubes expression with smaller terms than our original one. Therefore, the case $n = 3$ of Fermat's Last Theorem is proved by infinite descent. Euler believed, however, that his proof for $n = 3$ was sufficiently different from his proof for $n = 4$ that there was no hope of generalizing either into a general proof of the theorem.

19.3.2 Residues

Probably around 1750 Euler began to write an elementary treatise on number theory, but after completing 16 chapters he set it aside. The manuscript was discovered after his death and eventually published in 1849 under the title *Tractatus de numerorum doctrina* (*Treatise on the Doctrine of Numbers*). The early chapters contain the calculation of such number-theoretic functions as $\sigma(n)$, the number of divisors of an integer n, and $\phi(n)$, the number of integers prime to n and less than n. The most important part of the treatise, however, beginning in the fifth chapter, was Euler's treatment of the concept of congruence with respect to a given number d, now called the **modulus**. Euler defined the **residue of a with respect to d** as the remainder r on the division of a by d: $a = md + r$. He noted that there are d possible remainders and that therefore all the integers are divided into d classes, each class consisting of those numbers having the given remainder. For example, division by 4 divides the integers into four classes, numbers of the form $4m$, $4m + 1$, $4m + 2$, and $4m + 3$. All numbers in a given class he regarded as "equivalent." Euler further showed that one can define operations on the classes such that if A and B are in the class of residues α and β, respectively, then $A + B$, $A - B$, nA, and AB are in the class of residues $\alpha + \beta$, $\alpha - \beta$, $n\alpha$, and $\alpha\beta$, respectively. Euler thus demonstrated, in modern terminology, that the function assigning an integer to its "residue class" is a ring homomorphism. In fact, it was out of such ideas that the theory of rings eventually developed.

Similarly, basic ideas of group theory are evident in Euler's discussion of residues of a series in arithmetic progression $0, b, 2b, \ldots$. Euler showed that if the modulus d and the number b are relatively prime, then this series contains elements from each of the d different residue classes. Therefore, b has an "inverse" with respect to d, a number p such that the residue of pb equals 1. On the other hand, if the greatest common divisor of d and b is g, then only d/g different residues appear and such an inverse does not exist. For example, the set of multiples of 2 contains elements from nine different residue classes with respect to modulus 9, and 5 is the inverse of 2, while the set of multiples of 3 contains elements from only three distinct residue classes with respect to modulus 9, and no inverse exists for 3.

Euler continued this line of investigation by considering the residues of a geometric series $1, b, b^2, b^3, \ldots$, where b is prime to d. The number n of distinct residues of this series can be no more than $\mu = \phi(d)$. Euler noted that this number n is the smallest number greater than 1 such that b^n has residue 1, because once this power is reached, all subsequent powers simply repeat the same remainders. To show that n is a factor of μ, he used an argument later to be standard in group theory of, in effect, considering the cosets of the subgroup of powers of b in the multiplicative group of residues of d relatively prime to d and showing that the order of the subgroup divides the order of the group. Euler first demonstrated that if r and s are residues, say, of b^ρ and b^σ, then rs is also a residue, of $b^{\rho+\sigma}$. Similarly, r/s is a residue. Thus, if r is a residue and $x < d$ is a nonresidue (a number prime to d not a residue of the series of powers), xr must also be a nonresidue. Therefore, if $1, \alpha, \beta, \ldots$ form the entire set of n residues, then $x, x\alpha, x\beta, \ldots$ form a set of n distinct nonresidues. Because any nonresidue not included in this latter list also leads to a set of n nonresidues, all distinct from the first list, Euler concluded that $\mu = mn$ for some integer m. It follows that $b^\mu = b^{mn}$ has remainder 1 on division by d, or that $b^\mu - 1$ is divisible by d. A special case of this theorem, when d is a prime p, is Fermat's Little Theorem.

19.3.3   Quadratic Reciprocity

The question of residues of powers was of interest to Euler over a long period of his life, and he often did computations using them. Thus, the end of the seventh chapter of the number theory manuscript provided a list of tables of powers of numbers and their residues for all moduli d from 2 to 13. Euler also did extensive computing of prime divisors of expressions of the form $x^2 + ny^2$ and attempted to determine which primes can be written in that form. He published a paper in 1751 that simply contained tables of such results for 16 positive and 18 negative values of n. These calculations led Euler by 1783 to a statement of a theorem equivalent to the **quadratic reciprocity theorem**.

Euler called $p \neq 0$ a **quadratic residue** with respect to a prime q if there exist a and n such that $p = a^2 + nq$, that is, if $x^2 \equiv p \pmod{q}$ has a solution. Note that the condition of being a quadratic residue with respect to q depends only on the residue class of p with respect to q. For example, 1, 4, 9, $5 \equiv 4^2$, and $3 \equiv 5^2$ are quadratic residues with respect to 11, while 2, 6, 7, 8, and 10 are nonresidues. In his paper of 1783, Euler first proved that if $q = 2m + 1$ is an odd prime, then there are exactly m quadratic residues and therefore m nonresidues. Furthermore, he showed that the product and the quotient of two quadratic residues are again quadratic residues. He then determined that -1 is a residue with respect to q if q is of the form

$4n + 1$, while it is a nonresidue if q is of the form $4n + 3$. At the end of the paper, however, after considering more examples, Euler presented four conjectures relating, for two different odd primes q and s, the conditions under which each may or may not be a quadratic residue with respect to the other. These conditions may be written as follows:

1. If $q \equiv 1 \pmod 4$ and q is a quadratic residue with respect to s, then s and $-s$ are both quadratic residues with respect to q.
2. If $q \equiv 3 \pmod 4$ and $-q$ is a quadratic residue with respect to s, then s is a quadratic residue and $-s$ is not with respect to q.
3. If $q \equiv 1 \pmod 4$ and q is not a quadratic residue with respect to s, then s and $-s$ are both nonresidues with respect to q.
4. If $q \equiv 3 \pmod 4$ and $-q$ is not a quadratic residue with respect to s, then $-s$ is a quadratic residue and s is a nonresidue with respect to q.

Euler was not able to prove these results in 1783. They were restated in a somewhat different form by Adrien-Marie Legendre (1752–1833) in a paper of 1785 and in his textbook *Essai sur la théorie des nombres* of 1798, both times, however, with an incomplete proof. The first complete proof was given by Carl Friedrich Gauss in 1801 in his great work *Disquisitiones arithmeticae*, to be considered in Chapter 21.

19.4   MATHEMATICS IN THE AMERICAS

The development of mathematics discussed in this chapter and the several preceding ones all took place in Europe. But the age of discovery in Europe resulted in the colonization of vast areas of the Western Hemisphere, and by the late eighteenth century, the American Revolution resulted in the creation of the United States, with its own new system of colleges. Of course, mathematics in the American colonies and in the new republic was initially nowhere near the level of that in Europe. By the time of the revolution, there were nine colleges in what was to become the United States. Because most were founded with the primary purpose of training clergymen, many of the professors of mathematics were themselves in the clergy and had their chief interest in theology rather than mathematics. Thus, the level of instruction in the subject was even lower than the corresponding instruction in European universities of the time. There were generally courses available in arithmetic, algebra through the solution of simple equations, basic geometry, and trigonometry through the solution of plane triangles by use of logarithms, with application to surveying and astronomy. By the middle of the eighteenth century, courses in fluxions began to be offered, at least at Harvard and Yale. There was also private instruction in mathematics available. Thus, by 1727 Isaac Greenwood (1702–1745), Hollis Professor of Mathematics at Harvard from 1728 to 1738, was available for teaching both the method of fluxions and the Leibnizian differential calculus, as well as navigation, surveying, mechanics, optics, and astronomy. In 1729 he authored the first published arithmetic work written by an American. In general, however, Americans used textbooks imported from England.

Two scientific societies were founded in America during the century, modeled on the similar societies in Europe. These were the American Philosophical Society, founded in 1743, and the American Academy of Arts and Sciences, organized in 1780. But even the journal of

BIOGRAPHY

Benjamin Banneker (1731–1806)

Benjamin Banneker's father was a freed slave, and his mother was the daughter of a former indentured servant from England who had married her own black slave, himself perhaps the son of an African chieftain. His grandmother taught him to read and write and arranged for him to attend a small country school during the winter months. His technical abilities surfaced early in his life, although his circumstances as a free black farmer in Maryland did not allow him to develop his talents. Nevertheless, by the age of 22 he had constructed an accurate clock, mostly out of wood, a clock that continued to operate until it was destroyed in a fire shortly after his death. Banneker was, however, fortunate in that his neighbors included the Ellicott family, a family of businessmen and surveyors from whom he was able to borrow technical books and

a few scientific instruments. Having taught himself the principles of mathematics, surveying, and astronomy, he was invited by Andrew Ellicott to assist him in the survey of the boundaries of the District of Columbia, a task to which Ellicott had been appointed by President Washington in 1791. After his return from this task, Banneker continued his studies to the point that he was able to compile an almanac for 1792, which included daily positions of the sun, moon, and planets, times and descriptions of solar and lunar eclipses, times of rising and setting of the sun, moon, and certain bright stars, and local tide tables. Banneker's almanac proved popular, and he was able to have similar almanacs published yearly through 1797 (Fig. 19.2).

FIGURE 19.2

Benjamin Banneker honored on a U.S. stamp

the latter society was pessimistic about American achievement in mathematics at the time. An editorial in the first volume of its *Memoirs* noted that the mathematical papers appearing would be of interest to very few readers and that, in any case, they contained little in the way of research, being chiefly practical.

Although there were no research mathematicians in America during the eighteenth century, there were astronomers and surveyors of note who at least appreciated the importance of the subject. John Winthrop (1714–1779), who served for more than 40 years as Hollis Professor of Mathematics at Harvard, taught mathematics in the Newtonian tradition, including such topics as "the elements of Geometry, together with the doctrine of Proportion, the principles of Algebra, Conic Sections, Plane and Spherical Trigonometry, with general principles of Mensuration of Planes and Solids, the use of globes, [and] the calculations of the motions and phenomena of the heavenly bodies according to the different hypotheses of Ptolemy, Tycho Brahe, and Copernicus."[20] David Rittenhouse (1732–1796), an astronomer, clockmaker, and surveyor, who performed many detailed astronomical observations and assisted in the establishment of the Mason-Dixon line, published a few mathematical papers, including one dealing with the integration of powers of the sine. And Benjamin Banneker (1731–1806), the first American black to achieve distinction in science, taught himself sufficient mathematics and astronomy to publish a series of almanacs in the 1790s.

There were also two major public figures who provided some encouragement to the study of mathematics in the late eighteenth century. Benjamin Franklin (1706–1790), although not himself a mathematician, did possess a mathematical mind. He was "adept at the systematic and creative ways of thinking about numbers, arrangements, and relationships that characterize mathematical thought."[21] Not only did he create all sorts of magic squares and circles, but also he produced an almanac for 25 years, used coded messages during his service as American commissioner in France, created successful lotteries, and even composed an essay in 1751 in the field of demography. And because he was well known in Europe for some of his scientific experiments and a member of the Royal Society, he was able to provide the beginnings of an interaction between the scientific communities of Europe and the nascent one in the United States.

Finally, Thomas Jefferson (1743–1826), who had studied some mathematics at William and Mary but was largely self-educated in the field, developed during his time as ambassador to France a renewed interest in mathematics and its applications, publishing numerous articles in such fields as surveying, astronomy, spherical trigonometry, and the new metric system. In a letter of 1799 written from his home in Monticello, he expressed his views on the value of mathematics:

> Trigonometry . . . is most valuable to every man. There is scarcely a day in which he will not resort to it for some of the purposes of common life. The science of calculation also is indispensable as far as the extraction of the square and cube roots. Algebra as far as the quadratic equation and the use of logarithms are often of value in ordinary cases. But all beyond these is but a luxury, a delicious luxury indeed, but not to be indulged in by one who is to have a profession to follow for his subsistence. In this light I view the conic sections, curves of the higher orders, perhaps even spherical trigonometry, algebraical operations beyond the second dimension, and fluxions.[22]

At the end of the eighteenth century, and indeed well into the nineteenth, the United States had no need of the "luxuries" of mathematics.

In other parts of the Americas, the major concern for mathematics was also a practical one. For example, Father J. P. DeBonnécamps taught applied mathematics, including hydrography, surveying, and astronomy, at Quebec. And south of the Rio Grande, the decision of the Pope to divide that part of the Americas between Spain and Portugal led to the necessity of competent surveyors in South America. The very first mathematics work in the Americas was written in Mexico in 1560 by Juan Diez Freyle, the *Sumario Compendioso de las Cuentas* (*Brief Summary of Reckoning*), a book that contained extensive tables relating to gold and silver exchange, among other monetary affairs, as well as arithmetic problems relating to the tables and some elementary algebra. Other texts were published in Mexico over the next 200 years, but the most interesting were probably the *Comentarios a las Ordenanzas de Minas*, by Francisco Javier Gamboa, published in 1761, and the *Elementos de Matemáticas*, by Benito Bails, in 1772. The latter book treated analytic geometry and calculus, while the former dealt with the applied mathematics necessary to solve problems related to water and mining, two areas critical to the development of the country.

There were similar books in practical mathematics written in many of the other Latin American colonies, including the two works by the Brazilian José Fernandes Pinto Alpoim (1695–1765) entitled *Exame de Artilheiros* (1744) and *Exame de Bombeiros* (1748). Both of these books dealt with topics in what we would call military mathematics, and both were written in the form of questions and answers. But like in North America, the few universities in Latin America taught little mathematics. It was not until independence in the nineteenth century that there was a chance for mathematics to develop there. Ultimately, both north and south of the Rio Grande, enough students mastered the basics of mathematics that the mathematics of Europe could be further developed in the New World.

EXERCISES

1. Of three workmen, A can finish a given job once in three weeks, B can finish it three times in eight weeks, while C can finish it five times in twelve weeks. How long will it take for the three workmen to complete the job together? (This exercise and the next two are from Newton's *Universal Arithmetic*.)

2. If 12 cattle eat up 3 1/3 acres of meadow in 4 weeks and 21 cattle eat up 10 acres of exactly similar meadow in 9 weeks, how many cattle shall eat up 36 acres in 18 weeks? (*Hint:* The grass continues to grow.)

3. Given the perimeter a and the area b^2 of a right triangle, find its hypotenuse.

4. Suppose that the distance between London and Edinburgh is 360 miles and that a courier for London sets out from the Scottish city running at 10 mph at the same time that one sets out from the English capital for Edinburgh at 8 mph. Where will the couriers meet? (This exercise and the next two are from Maclaurin's *Treatise of Algebra*.)

5. Derive Cramer's rule for three equations in three unknowns from the rule for two equations in two unknowns: Given the system

$$ax + by + cz = m$$
$$dx + ey + fz = n$$
$$gx + hy + kz = p,$$

solve each equation for x in terms of y and z, then form two equations in those variables and solve for z. Finally, determine y and x by substitution.

6. A company dining together find that the bill amounts to $175. Two were not allowed to pay. The rest found that their shares amounted to $10 per person more than if all had paid. How many were in the company?

7. How can one reconcile the algebraic rule $\sqrt{a}\sqrt{b} = \sqrt{ab}$ with the computation $\sqrt{-1}\sqrt{-4} = -2 \neq \sqrt{(-1)(-4)}$? Why do you think Euler erred in his discussion of this matter?

8. Twenty persons, men and women, dine at a tavern. The share of the bill for one man is $8, for one woman $7, and the entire bill amounts to $145. Required, the number of men and women separately. (This exercise and the next two are from Euler's *Introduction to Algebra*.)

9. A horse dealer bought a horse for a certain number of crowns and sold it again for 119 crowns, by which means his profit was as much per cent as the horse cost him. What was the purchase price?

10. Three brothers bought a vineyard for $100. The youngest says that he could pay for it alone if the second gave him half the money which he had; the second says that if the eldest would give him only a third of his money, he could pay for the vineyard singly; lastly, the eldest asks only a fourth part of the money of the youngest, to pay for the vineyard himself. How much money did each have?

11. Factor Leibniz's polynomial $x^4 + a^4$ into two real quadratic polynomials. (*Hint:* Add and subtract $2a^2x^2$.)

12. Factor $x^5 - 1$ into linear and real quadratic factors.

13. Nicolaus Bernoulli claimed that the polynomial $x^4 - 4x^3 + 2x^2 + 4x + 4$ could not be factored into a product of two quadratic polynomials. Show that in fact the factors are

$$\left(x^2 - \left(2 + \sqrt{4 + 2\sqrt{7}}\right)x + \left(1 + \sqrt{4 + 2\sqrt{7}} + \sqrt{7}\right)\right)$$

and

$$\left(x^2 - \left(2 - \sqrt{4 + 2\sqrt{7}}\right)x + \left(1 - \sqrt{4 + 2\sqrt{7}} + \sqrt{7}\right)\right).$$

14. Use Euler's procedure from his proof that all real quartics factor to determine the factorization of $x^4 - 2x^2 + 8x - 3$ as a product of two quadratic polynomials.

15. Find a cubic curve and a quadratic curve that intersect in six real points.

16. Consider the following system of linear equations given by Euler:

$$5x + 7y - 4z + 3v - 24 = 0$$
$$2x - 3y + 5z - 6v - 20 = 0$$
$$x + 13y - 14z + 15v + 16 = 0$$
$$3x + 10y - 9z + 9v - 4 = 0.$$

Show that these four equations are "worth only two," so that they do not determine a unique 4-tuple as a solution.

17. Show that if n is prime, then the roots of $x^n - 1 = 0$ can all be expressed as powers of any such root $\alpha \neq 1$.

18. Let x_1, x_2 be the two roots of the quadratic equation $x^2 + bx + c = 0$. Since $t = x_1 + x_2$ is invariant under the two permutations of the two roots, while $v = x_1 - x_2$ takes on two distinct values, v must satisfy an equation of degree 2 in t. Find the equation. Similarly, x_1 is invariant under the same permutations as $x_1 - x_2$. Thus, x_1 can be expressed rationally in terms of $x_1 - x_2$. Find such a rational expression. Use the rational expression and equation you found to "solve" the original quadratic equation.

19. Determine the three roots x_1, x_2, x_3 of $x^3 - 6x - 9 = 0$. Use Lagrange's procedure to find the sixth-degree equation satisfied by y, where $x = y + 2/y$. Determine all six solutions of this equation and express each explicitly as $\frac{1}{3}(x' + \omega x'' + \omega^2 x''')$, where (x', x'', x''') is a permutation of (x_1, x_2, x_3) and ω is a complex root of $x^3 - 1 = 0$.

20. Show that the expression $x_1 x_2 + x_3 x_4$ takes on only three distinct values under the 24 permutations of four elements.

21. Let x_1, x_2, x_3, x_4 denote the four roots of the quartic equation $x^4 + ax^3 + bx^2 + cx + d = 0$. Set $\alpha = x_1 x_2 + x_3 x_4$, $\beta = x_1 x_3 + x_2 x_4$, $\gamma = x_1 x_4 + x_2 x_3$. Show that $\alpha + \beta + \gamma = b$, that $\alpha\beta + \alpha\gamma + \beta\gamma = ac - 4d$, and that $\alpha\beta\gamma = a^2 d + c^2 - 4bd$. Show that this implies that α, β, and γ are the roots of the cubic equation $y^3 - by^2 + (ac - 4d)y - (a^2 d + c^2 - 4bd) = 0$.

22. Use the results of Exercise 21 to determine the reduced equation for the quartic equation $x^4 - 12x + 3 = 0$. Solve this reduced equation, a cubic. Use the values you obtain for α, β, and γ to solve the original quartic equation.

23. In Euler's proof of the case $n = 3$ of Fermat's Last Theorem, show that $\frac{1}{4}p$ and $p^2 + 3q^2$ are relatively prime if p is not divisible by 3. (Recall that $x = p + q$, $y = p - q$ are odd and relatively prime.)

24. In Euler's proof of the case $n = 3$ of Fermat's Last Theorem, we now consider the situation where $p = 3r$. We then know that $\frac{3}{4}r(9r^2 + 3q^2) = \frac{9}{4}r(3r^2 + q^2)$ must be a cube. Show that the two factors in this expression are relatively prime. It follows that each must be a cube. In particular, $q^2 + 3r^2$ must be a cube. Factor this expression as in the text, using complex numbers of the form $a + b\sqrt{-3}$, and conclude that $q = t(t^2 - 9u^2)$, $r = 3u(t^2 - u^2)$, where t is odd and u is even. Also, since $\frac{9}{4}r$ is a cube, show that $\frac{2}{3}r = 2u(t + u)(t - u)$ is a cube where the factors are relatively prime. Conclude as in the case detailed in the text that we can now

find three integers smaller than the original set for which the sum of their cubes is a cube.

25. Calculate the distinct residues $1, \alpha, \beta, \ldots$ of $1, 5, 5^2, \ldots$ modulo 13. Then pick a nonresidue x of the sequence of powers and determine the coset $x, x\alpha, x\beta, \ldots$. Continue to pick nonresidues and determine the cosets until you have divided the group of all 12 nonzero residues modulo 13 into nonoverlapping subsets, the cosets of the group of powers of 5.

26. Determine the quadratic residues modulo 13.

27. Prove that -1 is a quadratic residue with respect to a prime q if and only if $q \equiv 1 \pmod{4}$.

28. Benjamin Banneker was fond of solving mathematical puzzles and recorded many in his notebook, including his own version of the old hundred fowls problem: A gentleman sent his servant with £100 to buy 100 cattle, with orders to give £5 for each bullock, 20 shillings for each cow, and 1 shilling for each sheep. (Recall that 20 shillings equals £1.) What number of each sort of animal did he bring back to his master?[23]

29. Divide 60 into four parts such that the first increased by 4, the second decreased by 4, the third multiplied by 4, and the fourth divided by 4 shall each equal the same number (Banneker).

30. Suppose a ladder 60 feet long is placed in a street so as to reach a window on one side 37 feet high, and without moving it at the bottom, to reach a window on the other side 23 feet high. How wide is the street? (Banneker)

31. Outline a lesson for an algebra course that uses Maclaurin's technique to teach the principles of Cramer's rule.

32. Prepare a report on the discovery of determinants by Seki Takakazu, the Japanese mathematician.

33. Compare the treatments of multiplication of signed numbers in the texts of Maclaurin and Euler. What laws of arithmetic does each tacitly assume?

REFERENCES AND NOTES

Three recent books on the history of algebra have sections dealing with the eighteenth century. These are Luboš Nový, *Origins of Modern Algebra* (Prague: Academia Publishing House, 1973), Hans Wussing, *The Genesis of the Abstract Group Concept* (Cambridge, MIT Press, 1984), and B. L. van der Waerden, *A History of Algebra* (New York: Springer-Verlag, 1985). A good treatment of the early history of the theory of vector spaces is

Jean-Luc Dorier, "A General Outline of the Genesis of Vector Space Theory," *Historia Mathematica* 22 (1995), 227–261. The early history of number theory is dealt with by André Weil in *Number Theory: An Approach through History from Hammurapi to Legendre* (Boston: Birkhäuser, 1984). A detailed discussion of the history of Fermat's Last Theorem up to the time of Dirichlet, including the work of Euler, is Harold Edwards, *Fermat's*

Last Theorem: A Genetic Introduction to Algebraic Number Theory (New York: Springer, 1977). There is a good treatment of Lagrange's work on the theory of equations, as well as earlier work dealing with the fundamental theorem of algebra, in Jean-Pierre Tignol, *Galois' Theory of Algebraic Equations* (Singapore: World Scientific, 2001).

The algebra treatise of Colin Maclaurin is *A Treatise of Algebra in Three Parts*, 2nd ed. (London: Millar and Nourse, 1756). Euler's original paper in which he discusses Cramer's paradox is E147: "Sur une contradiction apparente dans la doctrine des lignes courbes," *Mémoires de l'Académie des Sciences de Berlin* 4 (1750), 219–223 = *Opera omnia* (3) vol. 26, 33–45. Euler's algebra text, containing his ideas on the solution of polynomial equations and his proofs of the cases $n = 3$ and $n = 4$ of Fermat's Last Theorem is Leonhard Euler, *Elements of Algebra*, translated from the *Vollständige Anleitung zur Algebra* by John Hewlett (New York: Springer-Verlag, 1984). This Springer edition, a reprint of the original 1840 English translation, contains an introduction by C. Truesdell. Its many problems and ingenious methods of solution are worth a careful reading. Euler's article on the fundamental theorem of algebra (E 170) has been translated by Todd Doucet and is available online in *Convergence*: http://convergence.mathdl.org. It is found in *Opera omnia* (1) 6, 78–150. Euler's incomplete manuscript on number theory, E792, the *Tractatus de numerorum doctrina*, is in *Opera omnia* (1) vol. 5, 182–283.

1. Euler, *Elements of Algebra*, p. 286

2. Derek T. Whiteside, ed., *The Mathematical Papers of Isaac Newton* (Cambridge: Cambridge University Press, 1967–1981), vol. V, p. 65.

3. Ibid., p. 73.

4. Ibid., p. 137.

5. Ibid., p. 141.

6. Maclaurin, *A Treatise of Algebra*, p. 1.

7. Ibid., p. 132.

8. Euler, *Elements of Algebra*, p. 2.

9. Ibid., p. 186.

10. Ibid., p. 7.

11. Ibid., p. 43.

12. Ibid., p. 91.

13. Ibid., pp. 91–92.

14. Ibid., p. 187.

15. Ibid., p. 413.

16. Quoted in Tignol, *Galois' Theory*, p. 74.

17. Euler, *Introduction to Analysis of the Infinite*, John D. Blanton, trans. (New York: Springer-Verlag, 1988, 1990) p. 124.

18. For more information on Euler's proof of the fundamental theorem of algebra, consult William Dunham, *Euler: The Master of Us All* (Washington: MAA, 1999), chapter 6.

19. Quoted in Jean-Luc Dorier, "A General Outline of the Genesis of Vector Space Theory," *Historia Mathematica* 22 (1995), 227–261, p. 230.

20. Quoted in Dirk J. Struik, "Mathematics in Colonial America," in Dalton Tarwater, ed., *The Bicentennial Tribute to American Mathematics: 1776–1976* (Washington: Mathematical Association of America, 1976), 1–7, p. 3.

21. Paul Pasles, *Benjamin Franklin's Numbers: An Unsung Mathematical Odyssey* (Princeton: Princeton University Press, 2008), p. 4. This book is the first to deal extensively with Franklin's mathematics and provides wonderful detail on all of his many mathematical ideas, particularly his numerous magic squares.

22. Quoted in David Eugene Smith and Jekuthiel Ginsburg, *A History of Mathematics in America before 1900* (Chicago: Mathematical Association of America, 1934), p. 62.

23. See Silvio A. Bedini, *The Life of Benjamin Banneker* (New York: Scribner's, 1972) for details on Banneker's life and work as well as copies of various relevant documents, including the problems included here.

Geometry in the
Eighteenth Century

I have principally sought such consequences of the [hypothesis of the acute angle] to see if it did not contradict itself. From them all I saw that this hypothesis would not destroy itself at all easily. I will therefore adduce some such consequences. . . . The most remarkable of such conclusions is that if the [hypothesis of the acute angle] holds, we would have an absolute measure of length for each line, for the content of each surface and each bodily space. Now this overturns a theorem that one can unhesitatingly count amongst the fundamentals of Geometry, and which up to now no one has doubted, namely that there is no such absolute measure.

—From J. H. Lambert, *Theory of Parallel Lines*, 1766[1]

Euler wrote to Carl Ehler, the mayor of Danzig, on April 3, 1736, regarding the problem of the bridges of Königsberg, which had evidently been posed to Euler several months earlier: "Thus you see, most noble Sir, how this type of solution bears little relationship to mathematics, and I do not understand why you expect a mathematician to produce it, rather than anyone else, for the solution is based on reason alone, and its discovery does not depend on any mathematical principle."[2] Little did Euler know that his solution was in fact to lead, in the nineteenth and twentieth centuries, to graph theory, a new field of mathematics.

Geometry in the eighteenth century was connected both to algebra, through the relationships codified under the term analytic geometry, and to calculus, through the application of infinitesimal techniques to the study of curves and surfaces. But there was also considerable interest in the continuing problem of Euclid's parallel postulate. Before considering these aspects of geometry, however, we look at one of the textbooks written to introduce students to this field. We conclude with a brief look at the effects of the French Revolution on mathematics and its teaching.

 20.1

CLAIRAUT AND THE *ELEMENTS OF GEOMETRY*

One of the important geometry texts of the eighteenth century, Clairaut's *Éléments de Géométrie (Elements of Geometry)* of 1741, exemplified the author's belief that beginners in the subject should learn the material in what he called a "natural" way. Thus, Clairaut wrote, "I intended to go back to what might have given rise to geometry; and I attempted to develop its principles by a method natural enough so that one might assume it to be the same as that of geometry's first inventors, attempting only to avoid any false steps that they might have had to take."[3] His text had considerable influence on the teaching of geometry well into the nineteenth century, going through 11 editions in France and being translated into Swedish, German, and English.

Clairaut believed that the measurement of fields was the beginning of geometry—after all, the name itself has to do with measurement of the earth—and therefore he would start students off with that basic idea, not with axioms and definitions as in Euclid. He then planned to develop more complex ideas based on analogy to the first principles of measurement, all the time showing how people's natural curiosity enabled them to solve new problems and discover new concepts. He thus hoped to encourage this spirit of discovery in his readers. He realized that he would be criticized for not being "rigorous" in demonstrations, but he felt that it was unnecessary to use abstract reasoning to prove results that anyone of good sense knew to be true.

Clairaut's natural approach began with the concept of the measurement of length by use of a known measure. Because the straight line is the shortest distance from one point to another, the distance between two points is measured by the length of the straight line connecting them. To measure the distance from a point C to a line AB, one simply realizes that the shortest such line "leans" neither toward A nor toward B and is therefore the perpendicular from C to AB. To determine this line, however, one needed a method of constructing a perpendicular. Clairaut thus presented one, using a compass. Given now the notion of a perpendicular, Clairaut could define a rectangle, a four-sided figure with each side perpendicular to the adjoining ones, and a square, a rectangle with four equal sides. Clairaut also gave a "natural" definition of parallel lines as lines such that the distance between them is always the same.

Rectangles are measured by the use of squares of unit side. Thus, Clairaut showed that the area of a rectangle is a product of the measures of the length and width and, since a triangle is always half of a rectangle, that the area of a triangle is half the product of its base and altitude. Because fields do not always have straight sides, however, Clairaut noted that one could measure these by approximating the curved sides by segments of straight lines, then dividing the region into triangles, and measuring each of these. One can always approximate closely enough so that "all sensible error is eliminated."

Clairaut developed in his text most of the important results of Euclid's Books I–IV, VI, XI, and XII but always in what he considered a natural way. For example, given a triangle ABC with known sides, he showed how to construct a new triangle congruent to it. This was necessary, because it might not be possible to measure triangle ABC where it is; the perpendicular to the base might, for example, pass through an obstacle. The construction itself was simple. Clairaut transferred the base AB to a new location DE and then used compasses set at lengths AC and BC, respectively, to determine the point F such that $DF = AC$ and $EF = BC$ (Fig. 20.1). It was then obvious to Clairaut that the constructed triangle is equal in every way to the given triangle. Similarly, to show that a triangle is determined if two sides and the included angle are known, he first showed how to construct an angle equal to a given one by using, in the obvious way, an instrument abc consisting of two rules, ba and bc, which can pivot around b (Fig. 20.2). Then, given a triangle ABC with known angle B and a new line EF equal to BC, he placed the instrument with bc along EF, drew a new line DE at given angle B to line EF, and made DE equal to BA. The triangle DEF, completed by drawing DF, is then congruent to the original one. Clairaut used analogous techniques to construct a triangle similar to a given one and thus "measure" the distance to an inaccessible point.

FIGURE 20.1

Constructing a triangle with sides equal to those in a given triangle

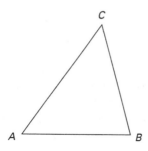
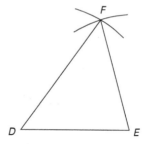

FIGURE 20.2

Clairaut's instrument for constructing angles

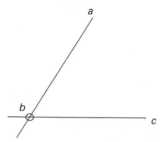

Clairaut noted that geometers did not want to measure by approximation the area of regions bounded by curves as he had suggested. It is more "rigorous" to measure such regions directly if possible. The only such figure he dealt with in this manner, however, was the circle, where he showed that the area is equal to the product of the circumference and half the radius.

Because Clairaut did not want to use the Greek method of exhaustion and the concomitant reductio argument, he decided to use the "fact" that a circle is a polygon with infinitely many sides. Thus, he showed first that the area of any regular polygon inscribed in a circle is equal to the perimeter multiplied by half the apothem and then noted that if there are an infinite number of sides, the area, perimeter, and apothem of the polygon become equal to the area, circumference, and radius, respectively, of the circle. In the section of his book on solid geometry, he similarly considered a square pyramid as being made up of infinitely many slices parallel to the base and argued from this that two pyramids of the same base and the same height have equal volumes. As we have seen, this argument by indivisibles goes back several thousand years. To show, however, that the volumes of two pyramids of the same height are to each other as their bases, Clairaut presented a more rigorous argument, in effect using Archimedean methods. Given this result, he derived the formula $V = (1/3)hB$ for the volume of a pyramid of height h and base B by beginning with the decomposition of a cube into six equal pyramids each with the vertex at the center. He then calculated the volume of a sphere of radius r by noting that the sphere is made up of infinitely many pyramids, each of height r. Because the sum of the area of the bases of these pyramids is the surface area of the sphere, the desired volume is one-third of the radius multiplied by this surface area. To derive the formula for surface area of a sphere, Clairaut used an argument involving infinitesimal cones, areas of which he had already determined through yet another argument using infinitesimals.

 20.2

THE PARALLEL POSTULATE

The eighteenth century saw renewed interest in the attempt to derive "rigorously" Euclid's parallel postulate from the remaining axioms and postulates and thus show that it was unnecessary for Euclid to have assumed his non-self-evident fifth postulate. Among those who wrote on this subject were Girolamo Saccheri (1667–1733) and Johann Lambert.

20.2.1 Saccheri and the Parallel Postulate

Saccheri entered the Jesuit order in 1685 and subsequently taught philosophy in Genoa, Milan, Turin, and then at the University in Pavia, near Milan, where he held the chair of mathematics until his death. In 1697, he published a work in logic containing a study of certain types of false reasoning in which one begins with hypotheses that are incompatible with one another. Ultimately, he was led to the consideration of Euclid's postulates and the study of whether an alternative to Euclid's parallel postulate would be compatible or incompatible with the remaining axioms and postulates. It was this study that Saccheri finally published in 1733 in his *Euclides ab omni naevo vindicatus* (*Euclid Freed of All Blemish*). Saccheri treated the "blemish" of the parallel postulate in the first part of this work. In the second part he considered two other "blemishes," one dealing with the existence of fourth proportionals, the other with the compounding of ratios.

Saccheri's aim in Part One, the only part to be considered here, was "clearly to demonstrate the disputed Euclidean axiom"[4] by assuming that it is false and then deriving it as a logical consequence. Saccheri began with a consideration of the quadrilateral $ABCD$ with two equal sides CA and DB, both perpendicular to the base AD, the same quadrilateral considered some 600 years earlier by al-Khayyāmī (Fig. 20.3). Using only Euclidean propositions not requiring the parallel postulate, Saccheri easily demonstrated that the angles at C and D are equal. There are then three possibilities for these angles: that they are both right, both obtuse, or both acute. Saccheri called these possibilities the *hypothesis of the right angle*, the *hypothesis of the obtuse angle*, and the *hypothesis of the acute angle*, respectively. He then showed that these hypotheses are equivalent, respectively, to the line segment CD being equal to, less than, or greater than the line segment AB. It was "obvious" to Saccheri, as it was to all who considered the question in earlier times, that the only "true" possibility was the hypothesis of the right angle, since it is, in fact, implied by the parallel postulate. The other two hypotheses come from the assumption that the parallel postulate is false. Saccheri intended to derive the parallel postulate from each of these two "false" hypotheses, using only the "self-evident" axioms of Euclid, thus demonstrating that each possibility led to a contradiction.

FIGURE 20.3

Saccheri's quadrilateral

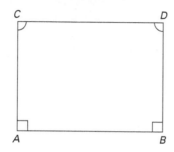

Saccheri began by proving that if either of the hypotheses is true for one quadrilateral, then it is true for all. He then continued with

PROPOSITION VIII   *Given any triangle ABD, right-angled at B; extend DA to any point X, and through A erect HAC perpendicular to AB, the point H being within the angle XAB. I say the external angle XAH will be equal to, or less, or greater than the internal and opposite ADB, according to whether the hypothesis of the right angle, or obtuse angle, or acute angle is true, and conversely.*

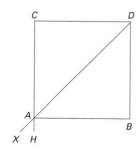

FIGURE 20.4

Saccheri's Propositions VIII and IX

Saccheri's proof made use of various propositions of Euclid's *Elements*, Book I. He began by assuming that AC is equal to BD and connecting CD, thus forming a Saccheri quadrilateral $ABCD$ (Fig. 20.4). By the hypothesis of the right angle, $CD = AB$. It follows that $\angle ADB = \angle DAC = \angle XAH$, and the first case is proved. Under the hypothesis of the obtuse angle, $CD < AB$. Then $\angle XAH = \angle DAC < \angle ADB$ and the second case is proved. Similarly, under the hypothesis of the acute angle, $CD > AB$, and the statement about angles follows. The converse is proved by arguments nearly as brief. This proposition leads to a more important one:

PROPOSITION IX   *In any right triangle, the two acute angles remaining are, taken together, equal to one right angle, in the hypothesis of the right angle; greater than one right angle, in the hypothesis of the obtuse angle; but less in the hypothesis of the acute angle.*

Because under either hypothesis, angles XAH and HAD together equal two right angles, while angle HAB is a right angle, angles XAH and DAB together equal one right angle. The result is then immediate from Proposition VIII. There is, unfortunately, a problem with this theorem that Saccheri evidently did not realize. His statement of the theorem says that both of the nonright angles of the triangle are acute. In fact, this follows from *Elements* I–17 to the effect that any two angles of a triangle are together less than two right angles. As noted in Chapter 3, that theorem depends on an assumption used by Euclid but never explicitly stated—that a straight line can be extended to any given length, an assumption that turns out not to be valid under the hypothesis of the obtuse angle.

Although Saccheri was unaware of the straight line result, he did prove that the hypothesis of the obtuse angle leads to a contradiction of *Elements* I–17 by first demonstrating, in the case of either the hypothesis of the right angle (Proposition XI) or that of the obtuse angle (Proposition XII), that if a line AP intersects PL at right angles and AD at an acute angle, then AD will ultimately intersect PL (Fig. 20.5). He proved this by taking points M_1, M_2, M_3, ... along AD with $AM_1 = M_1M_2 = M_2M_3 = \cdots$ and showing that if N_i is the foot of the perpendicular from M_i to AP for each i, then $AN_1 \leq N_1N_2 \leq N_2N_3 \leq \cdots$. It follows that some N_i will lie beyond the point P and therefore that AD intersects PL at some point between M_{i-1} and M_i.

FIGURE 20.5

Saccheri's Propositions XI and XII that AD and PL will ultimately intersect

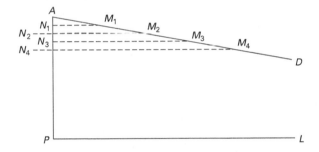

Saccheri could now prove Euclid's parallel postulate under these hypotheses:

PROPOSITION XIII   *If the straight line XA (of given length however great) meets two straight lines AD, XL, making with them on the same side internal angles XAD, AXL less than two right angles [Fig. 20.6], I say that these two, even if neither angle is right, will meet in some point on the side of those angles, and indeed at a finite distance, if either hypothesis holds, of the right angle or of the obtuse angle.*

FIGURE 20.6

Saccheri's Proposition XIII: AD and PL eventually intersect

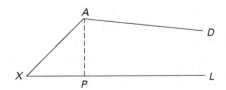

Again, the proof depended on *Elements* I–17. Because one of the angles, say, *AXL*, is acute, one can drop a perpendicular *AP* on *XL*, which, by that proposition, falls on the side of the acute angle *AXL*. Because in either hypothesis the two acute angles *PAX* and *PXA* are not together less than a right angle, if these are subtracted from the sum of the given angles *XAD* and *AXL*, the remaining angle *DAP* will be less than a right angle. Propositions XI and XII then allowed Saccheri to conclude that the two lines will intersect.

But now, because the acute angles of triangle *APX* are, under the hypothesis of the obtuse angle, greater than one right angle, Saccheri could choose an acute angle *PAD*, which, together with those two angles, make up two right angles. By Proposition XII, the line *AD* eventually intersects *XP* extended at, say, *L*. It then follows that two angles of triangle *XAL* themselves sum to two right angles, contradicting I–17. Also, of course, because the parallel postulate had been proved, Saccheri used it to prove, like Euclid in *Elements* Book I, that the three angles of any triangle are together equal to two right angles, contradicting, via Proposition IX, the hypothesis of the obtuse angle itself. As he therefore put it,

PROPOSITION XIV   *The hypothesis of the obtuse angle is absolutely false, because it destroys itself.*

Saccheri next showed that the hypotheses of the right, obtuse, and acute angles are equivalent, respectively, to the results that the sum of the angles of any triangle is equal to, greater than, or less than two right angles and that the sum of the angles of a quadrilateral is equal to, greater than, or less than four right angles. He then proceeded to investigate in more detail the consequences of the hypothesis of the acute angle. Here, however, he was not able to derive the parallel postulate as a consequence. He did, however, derive other intriguing results, for example,

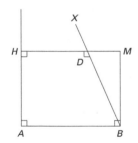

FIGURE 20.7

Saccheri's Proposition XVII: *BD* and *AH* do not ultimately intersect

PROPOSITION XVII   *If the straight line AH is at right angles to any straight line AB, however small, I say that under the hypothesis of the acute angle it cannot be true that every straight line BD intersecting AB in an acute angle will ultimately meet AH produced [Fig. 20.7].*

Suppose *BM* is also perpendicular to *AB*. Drop a perpendicular from *M* to *AH* intersecting that line at *H*. Because the sum of the angles of a quadrilateral is less than four right angles, it follows that angle *BMH* is acute. Similarly, if *BX* is drawn from *B* perpendicular to *HM*, intersecting that line at *D*, then angle *XBA* is also acute. But now *BD* extended cannot intersect *AH* extended because, the angles at *H* and *D* both being right, this would contradict *Elements* I–17.

Because Proposition XVII implies that there are two straight lines in the plane that do not meet, Saccheri could show in Proposition XXIII that for such lines, either they have a common perpendicular or else "they mutually approach ever more toward each other."[5] Furthermore, in the latter case, the distance between the lines becomes smaller than any assigned length; that is, the lines are asymptotic. Saccheri was then able to prove in Proposition XXXII that given a line *BX* perpendicular to a line segment *AB*, there is a certain acute angle *BAX* such that the line *AX* "only at an infinite distance meets *BX*,"[6] while lines making smaller acute angles with *BA* intersect *BX* and those making larger ones all have a common perpendicular with *BX* (Fig. 20.8). Saccheri then concluded with

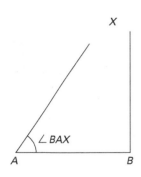

FIGURE 20.8

Saccheri's "angle of parallelism": *AX* and *BX* only meet at an "infinite" point X

BIOGRAPHY

Johann Lambert (1728–1777)

Lambert was a self-educated mathematician and philosopher who had mastered mathematics while assisting his father as a tailor in Alsace. In 1748 he moved to Switzerland as a private tutor for a wealthy family and later toured Europe on an educational journey with his pupils. During this period, Lambert was able to study in the family library and carry out both theoretical and experimental investigations. He never, however, adopted conventional bourgeois attitudes. He was finally proposed for a position at the Prussian Academy of Sciences in Berlin. When he arrived there early in 1764, he was welcomed by Euler, but his strange appearance and behavior delayed his appointment for a year. Eventually overcoming the initial hostility of Frederick II, Lambert produced over 150 works before his untimely death at the age of 49.

PROPOSITION XXXIII *The hypothesis of the acute angle is absolutely false, because it is repugnant to the nature of the straight line.*

Saccheri hardly gave a "proof" of this result. It appears, in fact, that he only ended his quest with this proposition because he had faith that the parallel postulate must be true. He merely wrote that, given the hypothesis of the acute angle, there must exist two straight lines that eventually run together "into one and the same straight line, truly receiving, at one and the same infinitely distant point a common perpendicular in the same plane with them."[7] But then, apparently having second thoughts on the matter, he spent the next 30 pages attempting a further justification of his result, showing along the way that two straight lines cannot enclose a space, that two straight lines cannot have a segment in common, and that there is a unique perpendicular to a given line at a given point—all of which relate to finite straight lines and none of which have much to do with his two straight lines that both join together and have a common perpendicular at infinity. Nevertheless, Saccheri believed that he had accomplished his aim.

20.2.2 Lambert and the Parallel Postulate

Johann Lambert, having studied at least a summary of Saccheri's work, attempted to improve upon it. But his work on the parallel postulate, *Theorie der Parallellinien* (*Theory of Parallel Lines*), finished by 1766, was never published, perhaps because Lambert was not finally happy with the conclusions. In the book, he considered a quadrilateral with three right angles and made three hypotheses as to the nature of the fourth angle, essentially the same three hypotheses as had Saccheri, that it could be right, obtuse, or acute. Again using the principle that a straight line can be of arbitrarily great length, Lambert rejected the second hypothesis. But he had great difficulty in rejecting the third hypothesis. As he noted, in the opening of the chapter, "this hypothesis would not destroy itself at all easily."

Like Saccheri, Lambert began to deduce various consequences from that hypothesis. The most surprising was that in his fundamental quadrilateral the difference between 360° and the sum of the angles was proportional to the area of the quadrilateral, that is, the larger the quadrilateral, the smaller the angle sum. Consider the quadrilateral $ABCD$ with right angles at A, B, and C and an acute angle at D of measure β (Fig. 20.9). At point E between A and B, construct a perpendicular EF to AB. It follows that $\angle CFE$ is also acute. If its measure is α, then the measure of $\angle EFD$ is $180° - \alpha$. But the sum of the angles of quadrilateral $EBFD$ is less than 360°; thus, $90 + 90 + 180 - \alpha + \beta < 360$ or $\beta < \alpha$. Therefore, the angle sum of quadrilateral $ABCD$ is less than that of quadrilateral $AECF$, as stated.

FIGURE 20.9

Lambert's proof that the angle of a quadrilateral decreases with the size of the quadrilateral

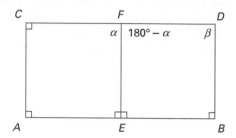

As noted in the opening quotation of this chapter, Lambert concluded from this result that the third hypothesis implies an absolute measure of length, area, and volume. In other words, if we assume that quadrilateral $AEFC$ has $AE = AC$, then $\angle EFC$ is a determined acute angle, an angle that can fit no other such quadrilateral. Thus, the measure α of $\angle EFC$ may be taken as the absolute measure of the quadrilateral. Lambert was not able to deduce this absolute measure, that is, he could not determine what the angle would be if $AE = AC =$ one foot, but he did realize that this hypothesis destroyed entirely the notion of similar figures. It also implied that the difference between 180° and the sum of the angles of a triangle, the **defect** of the triangle, was proportional to the area of the triangle. Lambert realized that a similar result is true under the second hypothesis, with the defect being replaced by the excess of the angle sum over 180°. But he also knew that spherical triangles had this same property that their angle sum was greater than 180° and that the excess was proportional to the area. He then argued by analogy: "I should almost therefore put forward the proposal that the third hypothesis holds on the surface of an imaginary sphere."[8]

Lambert abandoned his study of Euclid's parallel postulate once he felt that he could not successfully refute the hypothesis of the acute angle, even though it seems he was convinced that Euclid's geometry was true of space. Nevertheless, he believed that because the geometry of the obtuse angle hypothesis was reflected in geometry on the sphere, the sphere of imaginary radius would perform the same function for the acute angle hypothesis. Although by 1770 he had introduced the hyperbolic functions as complex analogues of the circular ones, in the sense that $\cosh ix = \cos x$ and $\sinh ix = i \sin x$, he was not able to apply these functions to develop a geometry on the imaginary sphere based on the hypothesis of the acute angle, nor could he give a construction in three-dimensional space of this imaginary sphere. It was only in the early nineteenth century, when analysis of this type could be brought to bear on the alternatives to the parallel postulate, that what is today called non-Euclidean geometry was developed. We tell that story in Chapter 24.

20.3 ANALYTIC AND DIFFERENTIAL GEOMETRY

The central thrust of eighteenth-century geometry was its connection to analysis. We consider this in Euler's work on the analytic geometry of curves in the plane, in the work of Clairaut and Euler on curves in three-space, and finally in Monge's work at the end of the century that systematized the subject texts designed for use in the École Polytechnique.

20.3.1 Euler and Plane Analytic Geometry

Volume 2 of Euler's *Introductio* was devoted to what we now call analytic geometry, beginning with the concept of curves given by functions. As was customary at the time, Euler used only a single axis, not our standard two. The "variable quantity" x (or, the abscissa) is laid out along a horizontal straight line, while the dependent quantity y is simply determined at each point along that horizontal line by erecting a perpendicular (the ordinate) of the appropriate length, above the line if y is positive and below if y is negative. Euler noted that it is also possible to have the ordinate oblique to the axis of abscissas. The curve that represents the function is then constructed by connecting the tips of the perpendicular straight lines y. As Euler wrote, "any function of x is translated into geometry and determines a line, either straight or curved, whose nature is dependent on the nature of the function."[9]

After an initial discussion of curves in general, Euler considered separately curves of first order (i.e., straight lines), curves of second order (i.e., conic sections), curves of third order, and curves of fourth order. Euler gave the general equation of a straight line in the form $\alpha + \beta x + \gamma y = 0$, noting also that the line is actually determined by the two ratios $\beta : \alpha$ and $\gamma : \alpha$. Thus, two points suffice to determine exactly one straight line. Interestingly, Euler gave no geometric interpretations of the coefficients in the equation of a straight line; there is nothing about slope or intercepts. However, he did note that to find where the line intersects the axis, one simply sets $y = 0$ and solves.

A curve of second order is given by the equation $\alpha + \beta x + \gamma y + \delta x^2 + \epsilon xy + \zeta y^2 = 0$ and, for the same reason as before, the curve is really determined by five ratios or, to put it another way, five points completely determine such a curve. With that in mind, Euler discussed various properties of second-order curves in general, including such concepts as conjugate diameters, foci, parameters, vertices, and a method of constructing a tangent, naturally without using calculus. After the generalities, Euler showed how to recognize the three types of conic sections—the ellipse, the parabola, and the hyperbola, noting that the essential difference "lies in the number of branches which go to infinity."[10] The ellipse has no part going to infinity; the parabola has two branches going to infinity, while the hyperbola has four. He then derived the basic properties of these three types, using their equations rather than the sectioning of a cone. Euler also discussed and classified both third-order and fourth-order curves, giving 146 different forms for quartics. He even dealt with exponential (or logarithmic) curves and the trigonometric curves, including examples such as $2y = x^i + x^{-i}$ (or $y = \cos(\ln x)$) and $y = x^x$ and also sketched the earliest graph of $y = \arcsin x$. For certain curves, such as the Archimedean spiral, he made use of polar coordinates, described in a modern fashion. Thus, if s represents the polar angle and z the length of the radius, then the equation of that spiral is $z = as$. Similarly, the equation $z = ae^{s/n}$ represents the logarithmic spiral, whose graph he also displayed.

Euler concluded the *Introductio* with a systematic treatment of the study of quadric surfaces in three-dimensional space. Euler used a single coordinate plane, with only one axis defined on it, and represented the third coordinate by the perpendicular distance from a point to that plane. But he did remark that it was possible to use three coordinate planes and often described a surface by means of its trace in various such planes. He gave the equation for a plane in three-space as $\alpha x + \beta y + \gamma z = a$ but described the meaning of the coefficients only in terms of the cosine of the angle θ between that plane and the xy plane: $\cos \theta = \gamma / \sqrt{\alpha^2 + \beta^2 + \gamma^2}$. In his discussion of the quadric surfaces themselves, Euler began by noting that the general second-degree equation in three variables can be reduced by a change of coordinates to one of the forms $Ax^2 + By^2 + Cz^2 = a^2$, $Ax^2 + By^2 = Cz$, or $Ax^2 = By$. The relationships among the coefficients then determined the type of surface: ellipsoid, elliptic or hyperbolic paraboloid, elliptic or hyperbolic hyperboloid (now called hyperboloids of one and two sheets, respectively), cone, and parabolic cylinder.

20.3.2   Clairaut and Space Curves

The earliest published work on curves in space was the 1731 *Recherches sur les courbes á double courbure* (*Researches on Curves of Double Curvature*) by Clairaut. For Clairaut, a curve in space could only be defined as an intersection of certain surfaces. Thus, he began his study by dealing with various simple cases of surfaces. He showed from the geometric definition that a sphere has equation $x^2 + y^2 + z^2 = a^2$, that a paraboloid is $y^2 + z^2 = ax$, and that, in general, the equation of a surface of revolution formed by revolving a curve $f(x, u) = k$ around the x axis is found by replacing u by $\sqrt{y^2 + z^2}$. He developed the general notion of a cone as a surface formed by connecting an arbitrary curve in the plane to a point outside the plane and showed that every equation in three variables in which each term is of the same degree must be a conic surface. And he proved the general result that an equation in three variables always defines a surface whose properties are determined by the equation.

Clairaut applied the techniques of differential calculus to find tangents and perpendiculars to curves in space and, therefore, considered such curves as being composed of "an infinity of small sides."[11] To determine the tangent line to a curve at a point N is to determine the point t where the extension of the line segment Nn, which connects N to an infinitely close point n on the curve, intersects the xy plane, or, if M is the projection of N onto that plane, to determine the length Mt (Fig. 20.10). Of course, this goal of finding the subtangent Mt is the direct analogue of the standard seventeenth-century method of determining tangents to plane curves. The result is the direct analogue of the result in two dimensions as well, although the procedure in three-space is complicated somewhat by the necessity of keeping all relevant lines in the same plane. Clairaut took Mm as the projection in the xy plane of the infinitesimal side Nn of the curve and extended it to meet the intersection point t. Then the triangle NtM defines the plane in which the tangent line lies. Clairaut too only made use of one axis, the x axis. So if AP is taken to represent the x coordinate of N, the z and y coordinates are represented by the length MN of the perpendicular from N to the xy plane and the length MP of the perpendicular from there to the axis, respectively. If Ap, nm, pm are the corresponding coordinates of n, and, further, if Nh is drawn parallel to Mm and MH is drawn parallel to Ap, then Pp represents dx, nh represents dz, mH represents dy, and

FIGURE 20.10

Clairaut and tangents to space
curves

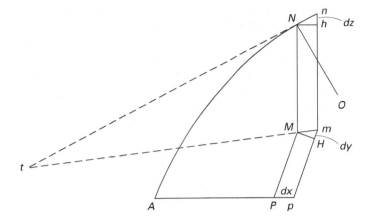

$\sqrt{dx^2 + dy^2}$ represents Mm. Because triangles nNh and NMt are similar, Clairaut derived the proportion $nh : Nh = MN : Mt$. Because $Nh = Mm$, it follows that

$$\frac{dz}{\sqrt{dx^2 + dy^2}} = \frac{z}{Mt} \quad \text{or} \quad Mt = \frac{z\sqrt{dx^2 + dy^2}}{dz}.$$

The tangent Nt itself is then given by

$$Nt = \sqrt{MN^2 + Mt^2} = \frac{z\sqrt{dx^2 + dy^2 + dz^2}}{dz}.$$

Furthermore, the perpendicular NO from the curve to the xz plane, which is also perpendicular to the plane of triangle NtM, is given by

$$NO = \frac{z(dx^2 + dy^2 + dz^2)}{\sqrt{dx^2 + dy^2}}.$$

Clairaut gave several examples of this computation, including the curve determined by the intersection of the two parabolic cylinders $ax = y^2$ and $by = z^2$. In this case $a\,dx = 2y\,dy$ and $b\,dy = 2z\,dz$. It follows that

$$dy = \frac{a\,dx}{2\sqrt{ax}}, \qquad dz = \frac{b\,dy}{2\sqrt{by}} = \frac{ab\,dx}{4\sqrt[4]{b^2 a^3 x^3}}, \quad \text{and} \quad \sqrt{dx^2 + dy^2} = \frac{dx\sqrt{4x + a}}{4x}.$$

Since $z = \sqrt[4]{ab^2 x}$, the subtangent $Mt = \frac{z\sqrt{dx^2 + dy^2}}{dz}$ is given by $Mt = \sqrt{4x + a}$. Similarly, the tangent Nt is found to be $Nt = \sqrt{by + 4x + a}$. An analogous calculation can be made for the perpendicular NO.

20.3.3   Euler and Space Curves and Surfaces

It was not until 1775 that Euler took up the subject of space curves, this time expressing them parametrically via the arclength s.[12] Thus, a curve was given by three equations $x = x(s)$, $y = y(s)$, $z = z(s)$. Taking differentials of each led Euler to the expressions $dx = p\,ds$,

$dy = q\,ds, dz = r\,ds$ from which he derived the result $p^2 + q^2 + r^2 = 1$. The functions $p, q,$ r, the derivatives of the coordinate functions with respect to arclength, are the components of the unit tangent vector to the curve. These components are also called the **direction cosines** of the tangent line (or of the curve itself) at the specified point.

Curvatures of plane curves at a point P had been defined by Newton, among others, as the reciprocal of the radius of the osculating circle meeting the curve at P. To define the curvature of a space curve, Euler used the unit sphere centered at a point $(x(s), y(s), z(s))$. If the "unit vectors" (p, q, r) at the neighboring parameter values s and $s + ds$, both considered to be emanating from that center, differ by an arc on the sphere equal to ds', then the **curvature** κ at that point was defined as $|\frac{ds'}{ds}|$, a value measuring how the curve at any point differs from a great circle on the sphere. Because the vector ds' is given by

$$\left(\frac{dx}{ds}(s+ds) - \frac{dx}{ds}(s), \frac{dy}{ds}(s+ds) - \frac{dy}{ds}(s), \frac{dz}{ds}(s+ds) - \frac{dz}{ds}(s) \right)$$
$$= \left(\frac{d^2x}{ds^2}ds, \frac{d^2y}{ds^2}ds, \frac{d^2z}{ds^2}ds \right),$$

it follows that

$$\kappa = \left| \frac{ds'}{ds} \right| = \sqrt{ \left(\frac{d^2x}{ds^2} \right)^2 + \left(\frac{d^2y}{ds^2} \right)^2 + \left(\frac{d^2z}{ds^2} \right)^2 }.$$

Euler next defined the **radius of curvature** ρ to be the reciprocal of the curvature. Thus,

$$\rho = \frac{ds^2}{\sqrt{(d^2x)^2 + (d^2y)^2 + (d^2z)^2}}.$$

It turned out, although it was not proved until the nineteenth century, that curvature is one of the two essential properties of a space curve. The other quantity is the **torsion**, which measures the rate at which the curve deviates from being a plane curve. If the curvature and torsion are given as functions of arclength along a curve, then the curve is completely determined up to its actual position in space.

Fifteen years earlier, Euler had made a beginning in the differential geometry of surfaces, with a paper entitled *Recherches sur la courbure des surfaces* (*Research on the Curvature of Surfaces*).[13] In that work Euler noted that although the method of finding the curvature of a plane curve at a given point was well known, even to define the curvature of a surface in space at a point was difficult. Each section of a surface by a plane through the given point gives a different curve, and the curvatures of each of these sections may well be different, even if one restricts oneself to plane sections that are perpendicular to the surface. In the paper, Euler calculated these various curvatures and established some relationships among them. First, however, he needed to characterize planes perpendicular to the surface, that is, planes that pass through the normal line to the surface at the given point P. He showed that the plane with equation $z = \alpha y - \beta x + \gamma$ is perpendicular to the surface defined by $z = f(x, y)$ if $\beta \frac{\partial z}{\partial x} - \alpha \frac{\partial z}{\partial y} = 1$. Defining the principal plane to be the plane through P perpendicular both to the surface and to the xy plane, Euler then demonstrated that if a given plane perpendicular to the surface makes an angle ϕ with the principal plane, the curvature of the section formed by that plane is given by $\kappa_\phi = L + M \cos 2\phi + N \sin 2\phi$ where L, M, N depend solely on the partial derivatives of z at P. Taking the derivative of this expression with respect to ϕ, Euler found that the maximum and minimum curvatures occur when $-2M \sin 2\phi + 2N \cos 2\phi = 0$ or when $\tan 2\phi = N/M$. But since $\tan(2\phi + 180°) = \tan 2\phi$, Euler concluded that if a maximum curvature occurs for a given value of ϕ, the minimum occurs at $\phi + 90°$. He was finally able to show that if κ_1 is the maximum curvature and κ_2 the minimum, and if the minimum curvature occurs at the principal plane, then the curvature of any section made by a plane at angle ϕ to the principal plane is given by $\kappa = \frac{1}{2}(\kappa_1 + \kappa_2) - \frac{1}{2}(\kappa_1 - \kappa_2) \cos 2\phi$.

20.3.4 The Work of Monge

Gaspard Monge (1746–1818) systematized the basic results of both analytic and differential geometry and added much new material in several papers beginning in 1771 and finally in two textbooks written for his students at the École Polytechnique at the end of the century. For example, in a paper published in 1784 Monge presented for the first time the point-slope form of the equation of a line: "If one wishes to express the fact that this line [with slope-intercept equation $y = ax + b$] passes through the point M of which the coordinates are x' and y', which determines the quantity b, the equation becomes $y - y' = a(x - x')$, in which a is the tangent of the angle which the straight line makes with the line of x's."[14] Monge's text *Géométrie descriptive* of 1799, on the other hand, did not deal with algebra at all but relied on the basic ideas of pure geometry. Monge outlined many techniques for representing three-dimensional objects in two dimensions. He systematically used projections and other transformations in space to draw in two dimensions various aspects of space figures. He described in detail such concepts as the tangent plane to a surface, the intersection of two surfaces, the notion of a developable surface (a surface that can be flattened out to a plane without distortion), and the curvature of a surface.

In his second text, the *Application de l'analyse á la géométrie* of 1807, which grew out of lecture notes dating from 1795, Monge showed how to apply analysis to geometry. The first part of this work, which used only algebra, contained the earliest detailed presentation of the analytic geometry of lines in two- and three-dimensional space as well as planes in three-dimensional space. Thus, Monge indicated that points in space are to be determined by considering perpendiculars to each of three coordinate planes. A line in space is determined by its projection onto two of these three planes, the equations of the projections onto the xy plane, for example, being given in the slope-intercept form or in the point-slope form. Monge showed how to find the intersection of two lines, as well as how to find a line parallel to a given line through a given point and a line through two given points. He also noted that the lines in the plane with equations $y = ax + \alpha$ and $y = a'x + \alpha'$ are perpendicular if $aa' = -1$.

Monge wrote the equation of a plane both in the form $z = ax + by + c$, where a and b are the slopes of the lines of intersection of this plane with the xz plane and the yz plane, respectively, and in the symmetric form $Ax + By + Cz + D = 0$, where the coefficients A, B, and C determine the direction cosines of the angles between the plane and the coordinate planes. He then proceeded to discuss all of the familiar problems dealing with points, lines, and planes, such as finding the normal line to a plane passing through a given point, finding the shortest distance between two lines, and finding the angle between two lines or between a line and a plane.

The second part of Monge's text was devoted to the study of surfaces. Here he used the entire machinery of calculus to develop analytically all of the topics he had considered in his

BIOGRAPHY

Gaspard Monge (1746–1818)

Monge was born in Beaune, a town about 150 miles south-east of Paris. He was a brilliant student in Lyon and, after preparing a plan of his native town, was invited to the Royal Engineering School at Mézières, where he soon had an opportunity to display his abilities. He was asked to develop a plan for a particular type of fortification. Instead of using the traditional complex method, he employed a new graphical method, a method he subsequently enlarged into the subject of

descriptive geometry. He was therefore promoted to a teaching position from which he was able to influence the scientific training of French military engineers. He was elected to the Academy of Sciences in 1780 and thereafter held many positions of responsibility under the royal, the revolutionary, and finally the imperial governments of France over the next 35 years (Fig. 20.11).

FIGURE 20.11

Monge on a French stamp

Géométrie descriptive. Thus, he considered in detail how to determine from various types of descriptions the partial differential equation that represents a given surface as well as how, in certain cases, to integrate that equation. To develop the equations of the tangent plane and normal line to a surface, Monge began by noting that the differential equation that represents the surface $z = f(x, y)$ near a point (x', y', z') is

$$dz = \frac{\partial z}{\partial x}\, dx + \frac{\partial z}{\partial y}\, dy,$$

where the partial derivatives are evaluated at x' and y'. On the other hand, the equation of any plane through (x', y', z') can be written as $A(x - x') + B(y - y') + C(z - z') = 0$. For this plane to be a tangent plane, any point on it infinitely near the given point must also be on the surface, that is, must satisfy the differential equation of the surface. So, taking $x - x'$ as dx, $y - y'$ as dy, and $z - z'$ as dz, Monge noted that the equation $A\, dx + B\, dy + C\, dz = 0$ must be identical to $dz = \frac{\partial z}{\partial x}\, dx + \frac{\partial z}{\partial y}\, dy$. It follows that $A/C = -\partial z/\partial x$, $B/C = -\partial z/\partial y$, and that the equation of the tangent plane is

$$z - z' = (x - x')\frac{\partial z}{\partial x} + (y - y')\frac{\partial z}{\partial y}.$$

The equations of the normal line to the surface, that is, the normal line to the tangent plane, are then calculated to be

$$x - x' + (z - z')\frac{\partial z}{\partial x} = 0; \qquad y - y' + (z - z')\frac{\partial z}{\partial y} = 0.$$

Monge's general idea of connecting partial differential equations with the geometry of space has had great influence through the years. Perhaps even more important, his teaching at the École Polytechnique influenced an entire generation of French engineers, mathematicians, and scientists, a matter to which we return in Section 20.5.

THE BEGINNINGS OF TOPOLOGY

It was in the mid-1730s that Euler became aware of a little problem coming out of the town of Königsberg, in East Prussia (now in Russia). In the middle of the river Pregel, which ran through the town, there were two islands. The islands and the two banks of the river were connected by seven bridges. The question asked by the townspeople was whether it was possible to plan a stroll that passed over each bridge exactly once. Although, as noted in the chapter opening, Euler did not consider this as a "mathematical" problem, nevertheless, instead of considering this problem in isolation, he attacked and solved the general problem of the existence of such a path, whatever the number of regions and bridges. In a paper published in 1736, he noted first that if one labeled the regions by letters $A, B, C, D, \ldots$, one could then label a path by a series of letters representing the successive regions passed through (Fig. 20.12). Thus, $ABDA$ would represent a path leading from region A to region B and then to D and back to A, regardless of the particular bridges crossed. It followed immediately that a complete path satisfying the desired conditions must contain one more letter than the number of bridges. In the Königsberg case, that number must be eight.

FIGURE 20.12

The seven bridges of Königsberg

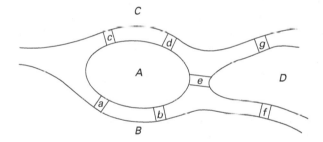

Next, Euler realized that if the number k of bridges leading into a given region is odd, then the letter representing that region must occur $(k + 1)/2$ times. Thus, if there is one bridge leading to region A, then A will occur once; if there are three bridges, A will occur twice; and so on. It does not matter in this case whether the path starts in the region A or in some other region. On the other hand, if k is even, then the letter representing that region will occur $k/2$ times if the path starts outside of the region and $k/2 + 1$ times if the path begins in the region. For example, if there are four bridges leading into region A, then A will occur twice if the route begins outside of A and three times if it begins in A. From the two different determinations of the number of letters that occur in his representation of a particular path, Euler could determine whether a path passing over each bridge exactly once is possible:

> If the total of all the occurrences [calculated above for each region] is equal to the number of bridges plus one, the required journey will be possible, and will have to start from an area with an odd number of bridges leading to it. If, however, the total number of letters is one less than the number of bridges plus one, then the journey is possible starting from an area with an even number of bridges leading to it, since the number of letters will therefore be increased by one.[15]

In the Königsberg case, there is an odd number of bridges leading into each of the regions A, B, C, D, namely, 5, 3, 3, and 3, respectively. The sum of the corresponding numbers, 3, 2,

2, and 2, is 9, which is more than "the number of bridges plus one." It follows that the desired path is impossible. In general, Euler noted that such a path will always be impossible if there are more than two regions approached by an odd number of bridges. If there are exactly two such regions, then the path is possible as long as it starts in one of those regions. Finally, if all the regions are approached by an even number of bridges, then a path crossing each bridge exactly once is always possible. Because Euler believed that once one knows whether a path is possible, the construction of it was straightforward, he never gave a method for actually producing the path.

FIGURE 20.13

Euler and the polyhedron formula on a stamp from Switzerland commemorating his 300th birthday

Euler struggled a bit with another geometrical problem. In 1750, he wrote in a letter to Christian Goldbach (1690–1764), "I cannot yet give an entirely satisfactory proof of the following proposition: In every solid enclosed by plane faces the aggregate of the number of faces and the number of solid angles exceeds by two the number of edges." In other words, given any polyhedron with V vertices (solid angles), E edges, and F faces, then $V + F = E + 2$, or, more familiarly, $V - E + F = 2$ (Fig. 20.13). On the other hand, Euler also wrote in the same letter that "it astonishes me that these general properties of stereometry have not, as far as I know, been noticed by anyone else."[16] In any case, Euler did submit a proof to the St. Petersburg Academy of Sciences in 1751, in which he successively removed tetrahedron-shaped pieces from the given polyhedron in such a way that $V - E + F$ remained unchanged at each stage. Continuing in this process, Euler reached a single tetrahedron, for which $V = F = 4$ and $E = 6$, so that the desired relationship holds. Unfortunately, it was not at all clear that Euler's dissection procedure could be carried out for an arbitrary polyhedron. A completely correct proof was given in 1794 by Adrien-Marie Legendre (1752–1833). Interestingly, Euler noted that the polyhedron problem as well as the Königsberg bridge problem were apparently part of a branch of geometry in which the relations depend on position alone and not at all on magnitudes. It was not until the late nineteenth and early twentieth centuries, however, that these facts and certain others were systematically studied and finally turned into the subject of topology.

 20.5   THE FRENCH REVOLUTION AND MATHEMATICS EDUCATION

The major mathematicians of the eighteenth century were associated, not with universities, but with academies founded under the patronage of various monarchs to gain prestige for their countries as well as to provide a ready source of scientific assistance in both military and civilian projects necessary for the advancement of the nation. The universities in general did not provide an advanced education in mathematics, since even in the eighteenth century they were primarily dominated by philosophers. In France in particular there had been no first-rate mathematicians associated with the University of Paris since the fourteenth century. The only schools that provided a mathematical and scientific education were the military schools, one of whose major functions was to produce military engineers. Thus, early in his career Monge taught at the military school at Mézieres, where he developed his first ideas on descriptive geometry in connection with drafting plans for military fortifications. Similarly, Laplace and Legendre taught for a time at the École Militaire in Paris.

Because the military schools and the universities were centers of Royalist support during the French Revolution, most were closed by 1794 when the revolution reached its most radical phase. Nevertheless, with the attacks on France by the armies of her neighbors as well as the fleeing of some of her best-educated citizens, it was necessary to have schools in which students of non-noble background, who showed a "constant love of liberty and equality and a hatred of tyrants,"[17] could be trained to serve in both military and civilian capacities as engineers and scientists. It was for this purpose that the National Convention on September 28, 1794, founded the École Centrale des Travaux Publiques, soon renamed the École Polytechnique. The school was to be more than just an engineering school, however. It was also to develop well-educated citizens and, in particular, to stimulate the talent necessary to advance science in general.

Monge, who before the revolution had helped to reform the teaching of science in the naval schools, was appointed to the commission responsible for organizing the École Polytechnique. He was therefore instrumental in developing the "revolutionary course," the three-month concentrated survey of the sciences that the first students began in December, which was to serve as a preview of the two- or three-year course of study they would ultimately pursue. The students were to study four basic areas, descriptive geometry, chemistry, analysis and mechanics, and physics, as well as to have a course in engineering drawing. The latter course met for three hours every evening, and the first three each had an hour lecture scheduled every morning followed by an hour of directed study. The physics course, however, only met for four hours in each ten-day *décade*. (The revolutionary calendar divided the month into three ten-day periods, *décades*, rather than into seven-day weeks.)

The first month of the descriptive geometry course taught by Monge himself was to cover essentially the material described in Section 20.3.4. Thus, the students were to study general methods of projection, the determination of tangents and normals to curves and surfaces, the construction of intersections of surfaces, and the notion of a developable surface. They would also consider applications of these ideas to various questions in such fields as building construction and mapmaking. The second month of the course covered architecture and public works, while the third month dealt with fortifications. The course in analysis, also to be taught by Monge, was to start the first month with the solution of polynomial equations up through the fourth degree and then to continue with algebraic and geometric methods for solving systems of equations and a study of the curves and surfaces these equations represent. The second month was to deal with the theory of series, exponential and logarithmic functions, elementary probability, and differential calculus with applications to geometry. The final month was then to deal with the integral calculus, including the finding of lengths, areas, and volumes, and the solution of differential equations.

The syllabus was ambitious. Unfortunately, it could not and did not work. Monge was ill when the school opened for classes on December 21, 1794. His course in descriptive geometry was therefore postponed, while C. J. Ferry taught his analysis course, and C. Griffet-Labaume repeated the course for those students who wanted it during the free hour. Unfortunately, it was clear before the first *décade* was finished that most of the students could not comprehend the course at all. It was soon decided by Lagrange, who was on the governing board of the school, to let Griffet-Labaume teach an elementary course in algebra instead of his repetitions. But even though this new course did not go further in the first month than the representation of plane curves by equations in two variables, fewer than a third of the students remained to the end. Things improved somewhat in the second month, with the continuation of the elementary algebra course, the addition of a course in trigonometry, and Monge's return to teach his course in descriptive geometry, but it was still evident that there was a great gap between the plans and the reality. There were several reasons for the poor beginning of the school, among which was the severe Paris winter, aggravated by a food shortage. But the primary reason was the poor preparation of the students. Students had been examined in their hometowns before being admitted, chiefly for "political correctness," but a political test for entrance could not make up for the lack of any consistent academic standards.

Despite the poor beginning, Monge and others were soon able to make vast improvements at the École Polytechnique, which became the model for colleges of engineering throughout Europe and the United States. National standards for education were established, partly through the creation in 1795 of a new national school for teachers, the École Normale Superieure. But even before these standards were established, the École Polytechnique itself sent examiners to the provinces to ensure that admitted students were well prepared. The course offerings were made somewhat more realistic. The "revolutionary course" was not taught again, only the regular courses, spread over the three-year program. Finally, France's best mathematicians all taught at the school, including Lagrange, Laplace, and Sylvestre Lacroix (1765–1843), and several wrote elementary textbooks for use there. Lacroix, in particular, wrote texts on arithmetic, trigonometry, analytic geometry, synthetic geometry, and differential and integral calculus. Most of these works went through numerous editions and were translated into several languages. In fact, Lacroix's calculus text, translated into English in 1816, was influential in bringing Continental methods to both England and the United States.

FIGURE 20.14

A French stamp commemorating the introduction of the metric system

Besides changing the nature of technical education in France, the revolutionary government was also responsible for the standardization of weights and measures in France and the introduction of the metric system (Fig. 20.14). The Constituent Assembly passed an initial law requiring the standardization in May of 1790, and the Academy of Sciences then formed a committee, including Laplace, Lagrange, and Monge, to consider the subject. The initial recommendation was that the unit of length should be that of a pendulum that beats in seconds; but by March of 1791 the committee decided that the standard length should be the ten millionth part of a quadrant of a great circle on the earth, since that would be more "natural" than using time. The Assembly then enacted a law providing for a new geodesic survey of the meridian of Paris so this length could be accurately determined. With the unit of length defined—it was named the "meter" a year later on the suggestion of Laplace—it was decided that all subdivisions and multiples would be decimal. Furthermore, measures for area and volume were to be defined in terms of the measure for length. Thus, a basic unit for area, the square on a side of 100 meters, was to be called an "are." Similarly, a basic unit of mass, the gram, was defined to be the mass of a cubic centimeter of water at a given temperature (Fig. 20.15).

The members of the committee went even further, devising decimal systems for money, related to weight through the value of gold and silver, and for angles, by dividing the quadrant of a circle into 100 equal parts, now called grads. Finally, they designed the revolutionary calendar to extend the metric system to the realm of history. Laplace went along with this

FIGURE 20.15

Plate from French work of the Commission Temporaire des Poids et Mesures (1793) giving conversions from old standards of weight to the new metric standard. Note that the basic unit of weight was called a *grave* or a *nouveau poid* rather than a kilogram. (*Source*: Smithsonian Institution Libraries, Photo No. 89-8736)

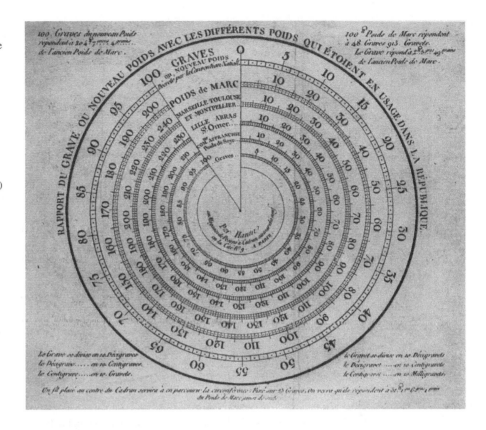

division of the month into three ten-day *décades*, with five extra holidays at the end of the year, even though he realized that this decimalization of the calendar would cause more problems than it would solve, given the incommensurability between the day and the year. Interestingly enough, although the decimalization of weights and measures was accepted over the next century by virtually the entire world, neither the decimalization of angles nor of the calendar lasted more than a dozen years.

Napoleon, having taken control of France in 1799, restored the Gregorian calendar to France in 1806. Meanwhile, he had been able to gain the support of France's important scientists. Monge, in particular, traveled to Egypt with Bonaparte in 1798 and became a strong supporter of the emperor. In return, he was named a senator for life and later a grand officer of the Legion of Honor and count of Péluse. Legendre, Lagrange, and Laplace also received honors, the first being named a chevalier de l'empire, and the others counts. With the fall of Napoleon in 1815, Monge lost all of his positions, spending the remaining three years of his life in intellectual exile. Lagrange had died in 1813, but Laplace and Legendre were able to make peace with the restored monarchy and continue their work unabated.

✦ EXERCISES

1. Let $ABCD$ be a Saccheri quadrilateral as in Figure 20.3 with right angles at A and B. Show, using only Euclidean propositions not requiring the parallel postulate, that $\angle C = \angle D$.

2. Given the hypothesis of the acute angle, both Saccheri and Lambert showed that the sum of the angles of any triangle is less than two right angles. Let the difference between $180°$ and the angle sum of a triangle be the defect of the triangle. Suppose triangle ABC is split into two triangles by line BD (Fig. 20.16). Show that the defect of triangle ABC is equal to the sum of the defects of triangles ABD and BDC.

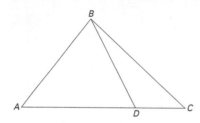

FIGURE 20.16

Calculation of the defect of trianlge ABC

3. Assume that the defect of equilateral triangle DEF is $\beta > 0$. Bisect the sides at points A, B, C and form triangle ABC (Fig. 20.17). Show that triangles AEB, DAC, and CBF are all congruent isosceles triangles and that triangle ABC is equilateral. If the defect of triangle CBF is α and the defect of triangle ABC is γ, show that $\beta = 3\alpha + \gamma$. Also, show that, contrary to the situation in Euclidean geometry, triangle ABC is not congruent to either of the triangles AEB, DAC, or CBF.

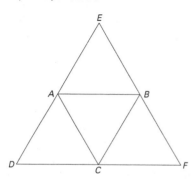

FIGURE 20.17

Calculation of the defect of trianlge DEF

4. Given that the angle sum of a triangle made of great circle arcs on a sphere (a spherical triangle) is greater than two right angles, define the excess of a triangle as the difference between its angle sum and $180°$. Show that if a spherical triangle ABC is split into two triangles by an arc BD from vertex B to the opposite side, then the excess of triangle ABC is equal to the sum of the excesses of triangles ABD and BDC.

5. Given the relationships $\cosh ix = \cos x$ and $\sinh ix = i \sin x$, determine $\cosh x$ and $\sinh x$ in terms of the cosine and sine functions and show that $\cosh^2 x - \sinh^2 x = 1$.

6. Clairaut developed the method of finding the length of a space curve by the use of the integral calculus, namely, by integrating $ds = \sqrt{dx^2 + dy^2 + dz^2}$. Use this result to find the length of the curve given by the intersection of the cylinders $ax = y^2$ and $(9/16)az^2 = y^2$, between the origin and the point (x_0, y_0).

7. Use Clairaut's methods to calculate the subtangent and the tangent to the curve defined as the intersection of the cylinders $x^2 - a^2 = y^2$, $y^2 - a^2 = z^2$.

8. Calculate the length of the perpendicular from a point P on the curve defined by $ax = y^2$, $by = z^2$ to the xz plane, where the perpendicular is also perpendicular to the plane defined by the tangent and subtangent to that curve.

9. Prove that the angle θ between the plane $\alpha x + \beta y + \gamma z = a$ and the xy plane is given by $\cos \theta = \gamma/\sqrt{\alpha^2 + \beta^2 + \gamma^2}$. Determine the cosine of the angles this plane makes with the other two coordinate planes.

10. Show that Euler's result relating the curvature of any section of a surface made by a plane at angle ϕ to the principal plane is equivalent to the modern formulation $\kappa_\phi = \kappa_1 \sin^2 \phi + \kappa_2 \cos^2 \phi$.

11. Show that the plane $z = \alpha y - \beta x + \gamma$ is perpendicular to the surface $z = f(x, y)$ if $\beta \frac{\partial z}{\partial x} - \alpha \frac{\partial z}{\partial y} = 1$. (Show that the plane contains a normal line to the surface.)

12. Find the normal line to the plane $Ax + By + Cz + D = 0$ that passes through the point (x_0, y_0, z_0).

13. Convert Monge's form of the equations of the normal line to the surface $z = f(x, y)$ into the modern vector equation of the line.

14. Show that an "Euler path" over a series of bridges connecting certain regions (a path that crosses each bridge exactly once) is always possible if there are either two or no regions that are approached by an odd number of bridges.

15. Construct Euler paths in the situations of Figure 20.18.

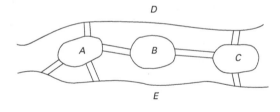

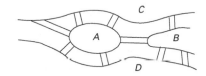

FIGURE 20.18

Bridge path problems

16. Find the numbers of vertices, edges, and faces for each of the five regular polyhedra and confirm that Euler's formula holds in these five cases.

17. Develop a lesson in a course in three-dimensional analytic geometry that uses the work of Monge to derive the equation of the tangent plane to a surface.

REFERENCES AND NOTES

The standard work on the history of non-Euclidean geometry is Roberto Bonola, *Non-Euclidean Geometry, A Critical and Historical Study of Its Development*, translated by H. S. Carslaw (New York: Dover, 1955). More recent works, both of which contain material on Saccheri and Lambert, include Jeremy Gray, *Ideas of Space: Euclidean, Non-Euclidean and Relativistic* (Oxford: Clarendon Press, 1989), and Boris A. Rosenfeld, *A History of Non-Euclidean Geometry: Evolution of the Concept of a Geometric Space*, translated by Abe Shenitzer (New York: Springer-Verlag, 1988). Briefer surveys of the subject include Jeremy Gray, "Non-Euclidean Geometry—A Re-interpretation," *Historia Mathematica* 6 (1979), 236–258, and Torkil Heiede, "The History of Non-Euclidean Geometry," in Victor Katz, ed., *Using History to Teach Mathematics: An International Perspective* (Washington: MAA, 2000), pp. 201–211. A textbook in geometry containing much historical material on non-Euclidean geometry is Marvin Greenberg, *Euclidean and Non-Euclidean Geometries: Development and History*, 3rd ed. (New York: Freeman, 1993). A general survey of the history of differential geometry is found in Dirk J. Struik, "Outline of a History of Differential Geometry," *Isis* 19 (1933), 92–120, while the history of analytic geometry, including material on the eighteenth century, is detailed in Carl Boyer, *History of Analytic Geometry* (New York: Scripta Mathematica, 1956).

The work of Saccheri is available in its entirety in English: Girolamo Saccheri, *Euclides Vindicatus*, translated by G. B. Halstead (New York: Chelsea, 1986). This Chelsea edition is a reprint of the first English translation of 1920 with added notes by Paul Stäckel and Friedrich Engel. Selections from Lambert's work *The Theory of Parallel Lines* are found in John Fauvel and Jeremy Gray, eds., *The History of Mathematics: A Reader* (London: Macmillan, 1987), pp. 517–520. Euler's early work in analytic geometry is in Volume 2 of *Introduction to Analysis of the Infinite*, John D. Blanton, trans., (New York: Springer, 1990). His first work on differential geometry is Leonhard Euler, "Recherches sur la courbure des surfaces," *Mémoires de l'Académie des Sciences de Berlin* 16 (1760), 119–143 = *Opera omnia* (1) 28, 1–22. Monge's work is thoroughly discussed in René Taton, *L'oeuvre scientifique de Monge* (Paris: Presses Universitaires de France, 1951). Norman L. Biggs, E. Keith Lloyd, and Robin J. Wilson, *Graph Theory: 1736–1936* (Oxford: Clarendon Press, 1986) contains a complete translation of Euler's article on the problem of the seven bridges of Königsberg. An English translation by D. Richeson and C. Francese of the paper in which Euler proved the polyhedral formula, "Proof of Some Notable Properties with Which Solids Enclosed by Plane Faces Are Endowed," is online at http://www.eulerarchive.org. This website, the Euler Archive, contains virtually all of Euler's published work, as well as much of his correspondence.

1. Fauvel and Gray, eds., *The History of Mathematics*, p. 517. For more on Lambert, see J. J. Gray and L. Trilling, "Johann Heinrich Lambert, Mathematician and Scientist," *Historia Mathematica* 5 (1978), 13–41.

2. B. Hopkins and R. Wilson, "The Truth about Königsberg," in R. Bradley and C. E. Sandifer, eds., *Leonhard Euler: Life, Work and Legacy* (Amsterdam: Elsevier, 2007), 409–439, p. 413.

3. Alexis-Claude Clairaut, *Eléments de géométrie* (Paris, 1741), preface.

4. Saccheri, *Euclides Vindicatus*, p. 9. For a brief treatment of Saccheri's ideas, see Louis Kattsoff, "The Saccheri Quadrilateral," *Mathematics Teacher* 55 (1962), 630–636.

5. Ibid., p. 117.

6. Ibid., p. 169.

7. Ibid., p. 173.

8. Fauvel and Gray, eds., *History of Mathematics*, p. 520.

9. Euler, *Introductio*, vol. 2, p. 5.

10. Ibid., p. 83.

11. Alexis-Claude Clairaut, *Recherches sur les courbes á double courbure* (Paris: Quillau, 1731), p. 39.

12. Leonhard Euler, "Methodus facilis omnia symptomata linearum curvarum non in eodem plano sitarum investigandi" (E602), *Acta Acad. Sci. Petrop.* 1 (1782), 19–57 = *Opera* (1) 28, 348–381.

13. Leonhard Euler, "Recherches sur la courbure des surfaces," (E333) *Mémoires de l'Académie des Sciences de Berlin* 16 (1760), pp. 119–143 = *Opera* (1) 28, 1–22.

14. Quoted in Boyer, *Analytic Geometry*, pp. 205–206.

15. Biggs, Lloyd, and Wilson, *Graph Theory: 1736–1936*, p. 6.

16. See Euler Archive, http://www.eulerarchive.org, letter from Euler to Goldbach of November 14, 1750.

17. Janis Langins, *La République avait besoin de savants* (Paris: Belin, 1987), p. 123. This book provides a detailed study of the first year of the École Polytechnique and includes copies of many of the relevant documents.

Algebra and Number Theory in the Nineteenth Century

It is greatly to be lamented that this virtue of the real integers that they can be decomposed into prime factors which are always the same for a given integer does not belong to the complex integers [of arbitrary cyclotomic number fields], for were this the case, the entire theory, which is still laboring under many difficulties, could be easily resolved and brought to a conclusion. For this reason, the complex integers we are considering are seen to be imperfect, and there arises the question whether other types of complex numbers can be found . . . which would preserve the analogy with the real integers with respect to this fundamental property.

—Ernst Kummer in "De numeris complexis, qui radicibus unitatis et numeris integris realibus constant," 1847[1]

A s he later reported in a letter to his son, William Rowan Hamilton discovered quaternions late one fall afternoon: "On the 16th day of [October, 1843]—which happened to be a Monday, and a Council day of the Royal Irish Academy—I was walking in to attend and preside, and your mother was walking with me, along the Royal Canal; . . . and although she talked with me now and then, yet an *under-current* of thought was going on in my mind, which gave at last a *result*, whereof it is not too much to say that I felt *at once* the importance. An *electric* circuit seemed to close; and a spark flashed forth . . . Nor could I resist the impulse—unphilosophical as it may have been—to cut with a knife on a stone of Brougham Bridge . . . the fundamental formula with the symbols i, j, k, namely,

$$i^2 = j^2 = k^2 = ijk = -1,$$

which contains the *Solution* of the *Problem*."[2]

Algebra in 1800 meant the solving of equations. By 1900, the term was beginning to encompass the study of various mathematical structures, that is, sets of elements with well-defined operations, satisfying certain specified axioms. It is this change in the notion of algebra that will be explored in this chapter.

The nineteenth century opened with the appearance of the *Disquisitiones Arithmeticae* of Carl Friedrich Gauss, in which the "Prince of Mathematicians" discussed the basics of number theory, not only proving the law of quadratic reciprocity, but also introducing various new concepts that provided early examples of groups and matrices. Gauss's study of higher reciprocity laws soon led to his study of the so-called Gaussian integers, complex numbers of the form $a + bi$ where a and b are ordinary integers. Attempts to generalize the properties of these integers to integers in other number fields led Ernst Kummer to the realization that some of the most important of these properties, such as unique factorization, fail to hold. To recover this property, along with a reasonable new meaning of the term "prime," Kummer created what he called "ideal complex numbers" by 1846, the study of which led Richard Dedekind in the 1870s to define "ideals" in rings of algebraic integers, these ideals having the property of unique factorization into primes.

Gauss's study of the solutions of cyclotomic equations in the *Disquisitiones* as well as the detailed study of permutations by Augustin-Louis Cauchy in 1815 helped with a new attack on the problem of solving algebraic equations of degree higher than 4. It was Niels Henrik Abel who finally proved the impossibility of the solution of a general equation of degree 5 or higher in terms of radicals (in 1827). Soon thereafter, Evariste Galois sketched the relationship between algebraic equations and groups of permutations of the roots, a relationship whose complete development transformed the question of the solvability of equations to one of the study of subgroups and factor groups of the group of the equation. Galois' work was not published until 1846 nor completely understood until somewhat later.

English mathematicians in the first third of the nineteenth century, including George Peacock and Augustus De Morgan, attempted to axiomatize the basic ideas of algebra and to determine exactly how much one can generalize the properties of the integers to other types of quantities. Such study led eventually to the discovery of quaternions by William Rowan Hamilton in 1843, partly in an attempt to determine a physically meaningful algebra in three-dimensional space. Quaternions, however, were four-dimensional objects, and so the physicists could only use the three-dimensional part of the quaternions in any algebraic manipulations. After a debate lasting nearly to the end of the century, the algebra of vectors developed by Oliver Heaviside and Josiah Willard Gibbs won out over the algebra of quaternions to become the language of the physicists. Meanwhile, the algebraic freedom to determine laws of operation, exploited by Hamilton for use in physics, was applied by George Boole to the study of logic. Boole's work turned out to be important in computer design a century later.

Another aspect of today's algebra, the theory of matrices, was also developed in the mid-nineteenth century. Determinants had been used as early as the seventeenth century, but it was only in 1850 that James Joseph Sylvester coined the term **matrix** to refer to a rectangular array of numbers. Soon thereafter, Cayley developed the algebra of matrices. The study of eigenvalues was begun by Cauchy early in the century in his work on quadratic forms and then fully developed by, among others, Karl Weierstrass, Camille Jordan, and Georg Frobenius. The latter in particular organized the theory of matrices into essentially the form it has today.

We conclude the chapter with a survey of how the ideas described earlier began to coalesce into the structural approach to algebra. Thus, in 1854 Arthur Cayley gave the earliest definition of an abstract group, a definition that was not fully exploited until nearly 30 years later in the work of Walther von Dyck and Heinrich Weber. Meanwhile, the study of the "numbers" determined by the solutions of algebraic equations led to the definition of a field of numbers both by Leopold Kronecker and Richard Dedekind and soon after to an abstract definition of a field by Weber. These ideas of structure were, however, not fully developed until the twentieth century.

 21.1

NUMBER THEORY

Legendre had published his work on number theory in 1798, but at the same time a young man in Brunswick, a city in northern Germany, was putting the finishing touches on a number-theoretic work that would ultimately have far more influence than that of Legendre. Carl Friedrich Gauss (1777–1855) noted in the preface to his *Disquisitiones Arithmeticae* (*Investigations in Arithmetic*), published in 1801, that he had written most of it before studying the works of some of his contemporaries. But although some of what he thought he had discovered was already known to Euler or Lagrange or Legendre, Gauss's work contained many new discoveries in the theory of numbers.

21.1.1 Gauss and Congruences

Gauss began in chapter 1 of his book by presenting the modern definition and notation for congruence: The integers b and c are **congruent** relative to the **modulus** a if a divides the difference of b and c. Gauss wrote this as $b \equiv c \pmod{a}$, noting that he adopted the symbol $\equiv$ "because of the analogy between equality and congruence,"[3] and called b and c each a **residue** of the other. He then discussed the elementary properties of congruence. For example, Gauss showed that the linear congruence $ax + b \equiv c \pmod{m}$ is always solvable by use of the Euclidean algorithm, if the greatest common divisor of a and m is 1. He also showed how to solve the Chinese remainder problem and calculate the Euler function $\phi(n)$, which gives the number of integers less than n that are relatively prime to n. In chapter 3, Gauss, like Euler before him, considered residues of powers. Noting that if p is prime and a any number less than p, then the smallest exponent m such that $a^m \equiv 1 \pmod{p}$ is a divisor of $p - 1$, Gauss went on to show that in fact "there always exist numbers with the property that no power less than the $(p - 1)$st is congruent to unity."[4] A number a satisfying this property is called a **primitive root** modulo p. If a is a primitive root, then the powers $a, a^2, a^3, \ldots, a^{p-1}$ all have different residues modulo p and thus exhaust all of the numbers $1, 2, \ldots, p - 1$. In particular, $a^{(p-1)/2} \equiv -1 \pmod{p}$, a property crucial to Gauss's proof of

WILSON'S THEOREM $(p - 1)! \equiv -1 \pmod{p}$.

Let a be a primitive root modulo p. Then

$$(p - 1)! = a^1 a^2 \cdots a^{p-1} = a^{1+2+\cdots+p-1} = a^{p(p-1)/2}.$$

BIOGRAPHY

Carl Friedrich Gauss (1777–1855)

Gauss was born into a family that, like many others of the time, had recently moved into town, hoping to improve its lot from that of impoverished farm workers. One of the benefits of living in Brunswick was that the young Carl could attend school. There are many stories told about Gauss's early-developing genius, among which is one that comes from his mathematics class when he was nine. At the beginning of the year, to keep his 100 pupils occupied, the teacher, J. G. Büttner, assigned them the task of summing the first 100 integers. He had barely finished explaining the assignment when Gauss wrote the single number 5050 on his slate and deposited it on the teacher's desk. Gauss had noticed that the sum in question was simply 50 times the sum 101 of the various pairs 1 and 100, 2 and 99, 3 and 98, . . . and had performed the required multiplication in his head. Impressed by his young student, Büttner arranged for Gauss to have special textbooks, to have tutoring by his assistant Martin Bartels (1769–1836), who himself later became a professor of mathematics in Russia, and to be admitted to a secondary school where he mastered the classical curriculum.

In 1791, the Duke of Brunswick granted Gauss a stipend, which enabled him first to attend the Collegium Carolinium, a new science-oriented academy funded by the Brunswick government to train bureaucrats and military officers, then to matriculate at the University of Göttingen in neighboring Hanover, which already had a reputation in the sciences, and

finally to continue his research back in Brunswick while receiving a PhD from the local University of Helmstedt. Not only did Gauss publish his research in number theory in 1801, with the book being dedicated to his patron the duke, but he also developed at the same time a new method for calculating orbits that enabled several asteroids to be discovered. The patronage of the duke lasted until he was killed in battle against France in 1806 and the duchy was occupied by the French army. Fortunately for science, the French general had been given explicit orders to look out for Gauss's welfare. Thus, Gauss was able to stay in Brunswick until he accepted a position at Göttingen the following year as professor of astronomy and director of the observatory. Gauss remained at Göttingen for the remainder of his life, doing research in pure and applied mathematics as well as astronomy and geodesy.

Gauss was never particularly happy with teaching classes, because most of the students were uninterested in, and not well prepared for, mathematics, but he was willing to work privately with any actively interested student who approached him. Compared to his predecessor Euler and to his French contemporary Cauchy, Gauss ultimately published little, his collected works occupying only (!) 12 volumes. Nevertheless, his mathematical papers in various fields are of such profundity that they have influenced the progress of the subject to the present (Fig. 21.1).[5]

FIGURE 21.1

Gauss on a German stamp

Because $\frac{p(p-1)}{2} \equiv \frac{p-1}{2}$ (mod $p - 1$), it follows that $(p - 1)! \equiv a^{(p-1)/2} \equiv -1$ (mod p). (This theorem had first been stated by John Wilson (1741–1793) and first been proved by Lagrange in 1773.)

The central topic of chapter 4 of the *Disquisitiones* is the law of quadratic reciprocity. Recall that Euler had stated, but not proved, this law, which describes the conditions under which two odd primes are quadratic residues of each other. Gauss considered this theorem so important that he gave six different proofs, the first of which he found in the spring of 1796 after much effort. In the *Disquisitiones*, Gauss gave numerous examples and special cases derived from calculations before presenting his second proof. He showed first, following Euler, that -1 is a quadratic residue of primes of the form $4n + 1$ and a nonresidue of primes of the form $4n + 3$. He next dealt with 2 and -2 and concluded that for primes of the form $8n + 1$, both 2 and -2 are quadratic residues; for primes of the form $8n + 3$, -2 is a quadratic residue and 2 is not; for primes of the form $8n + 5$, both 2 and -2 are nonresidues; and for primes of the form $8n + 7$, 2 is a quadratic residue and -2 is not. The demonstrations of these properties are not difficult. For example, to prove the first result, Gauss chose a primitive root a for the prime $8n + 1$ and noted that $a^{4n} \equiv -1 \pmod{8n + 1}$. This congruence can be rewritten in either of the two forms

$$(a^{2n} + 1)^2 \equiv 2a^{2n} \pmod{8n + 1} \quad \text{or} \quad (a^{2n} - 1)^2 \equiv -2a^{2n} \pmod{8n + 1}.$$

Therefore, both $2a^{2n}$ and $-2a^{2n}$ are squares modulo $8n + 1$. But because a^{2n} is a square, it follows that so are 2 and -2.

Gauss went on to characterize the primes for which 3 and -3 are quadratic residues as well as those for 5 and -5 and 7 and -7. He could then state the general result:

QUADRATIC RECIPROCITY THEOREM   *If p is a prime number of the form $4n + 1$, $+p$ will be a [quadratic] residue or nonresidue of any prime number, which taken positively is a residue or nonresidue of p. If p is of the form $4n + 3$, $-p$ will have the same property.*

The proof is too long to be discussed here, but note that Gauss was unaware of Legendre's suggestive notation for quadratic residues, in which the symbol $(\frac{p}{q})$ is equal to 1 if p is a quadratic residue modulo q and -1 if not, where q is an odd prime. The theorem can then be stated in the following elegant form:

$$\left(\frac{p}{q}\right)\left(\frac{q}{p}\right) = (-1)^{\frac{p-1}{2}\frac{q-1}{2}}.$$

Similar formulas can be written expressing quadratic residue properties of -1 and ± 2 modulo a prime p. Given the product rule for residues, that

$$\left(\frac{a}{p}\right)\left(\frac{b}{p}\right) = \left(\frac{ab}{p}\right),$$

where a, b are prime to p, as well as rules for determining the residue situation when two numbers have common factors, it is possible to determine for any two positive numbers P and Q whether Q is a residue or nonresidue of P. For example, to decide whether 453 is or is not a quadratic residue modulo 1236 ($= 4 \cdot 3 \cdot 103$), Gauss noted first that 453 is a quadratic residue both of 4 and of 3. By the Chinese remainder theorem, it follows that the question is reduced to determining $(\frac{453}{103})$. Using Legendre's notation, we have

$$\left(\frac{453}{103}\right) = \left(\frac{41}{103}\right) = \left(\frac{103}{41}\right)(-1)^{\frac{41-1}{2}\frac{103-1}{2}} = \left(\frac{103}{41}\right) = \left(\frac{-20}{41}\right) = \left(\frac{-1}{41}\right)\left(\frac{2^2}{41}\right)\left(\frac{5}{41}\right).$$

Because each of the three factors on the right is equal to 1, it follows that 453 is a quadratic residue modulo 103 and also modulo 1236. Gauss showed, in fact, that $453 \equiv 297^2 \pmod{1236}$.

Over the next several decades, Gauss attempted to generalize the law of quadratic reciprocity to laws of cubic and quartic reciprocity, that is, to laws determining when numbers are congruent to cubes or fourth powers modulo other numbers. As early as 1805 he realized that "the natural source of a general theory [of quartic residues] is to be found in the expansion of the domain of arithmetic."[6] Thus, he studied the Gaussian integers, the complex numbers $a + bi$, where a and b are ordinary integers, and established certain analogies between the

Gaussian integers and ordinary integers in a paper published in 1832. After noting that there are four units (invertible elements) among the Gaussian integers, 1, -1, i, and $-i$, he defined the **norm** of an integer $a + bi$ to be the product $a^2 + b^2$ of the integer with its complex conjugate $a - bi$. He then called an integer **prime** if it cannot be expressed as the product of two other integers, neither of them units. Because an odd real prime p can be expressed as $p = a^2 + b^2$ if and only if it is of the form $4n + 1$, it follows that such primes, considered as Gaussian integers, are composite, $p = (a + bi)(a - bi)$. Conversely, primes of the form $4n + 3$ are still prime as Gaussian integers. Because $2 = (1 + i)(1 - i)$, 2 is also composite. To determine which other Gaussian integers $a + bi$ are prime, Gauss used the norm and showed that $a + bi$ is prime if and only if its norm is a real prime, which can only be 2 or of the form $4n + 1$. In other words, 2 and primes of the form $4n + 1$ split as the product of two Gaussian primes in the domain of Gaussian integers, while primes of the form $4n + 3$ remain prime there.

With primes defined in the domain of Gaussian integers, Gauss proved the analogue of the fundamental theorem of arithmetic, using theorems similar to those in Euclid's *Elements*, Book VII. First, he easily showed that any integer can be factored as the product of primes. To complete the analogy with ordinary integers, he then proved that this factorization is unique, at least up to unit factors, by first demonstrating, using his description of primes, that if any Gaussian prime p divides the product $qrs \cdots$ of Gaussian primes, then p must itself be equal to one of those primes, or one of them multiplied by a unit. Having established the unique factorization property of the Gaussian integers, Gauss considered congruence modulo Gaussian integers and then quartic reciprocity, stated in terms not of ordinary integers but of Gaussian integers. He was furthermore aware that a law of cubic reciprocity would involve complex numbers of the form $a + b\omega$, where $\omega^3 = 1$ and a, b are ordinary integers. But Gauss never published any investigation of the primes and factorization of numbers of that type.

21.1.2   Fermat's Last Theorem and Unique Factorization

This entire question of factorization in various domains turned out to be related not only to reciprocity but also to the continued attempts to prove Fermat's Last Theorem. The next mathematician after Euler to make any progress on the theorem was Sophie Germain (1776–1831). In fact, in a letter to Gauss written in 1819, she outlined a strategy to accomplish this proof. First she proved the following

LEMMA   *If $x^p + y^p = z^p$ is a solution to the Fermat equation, where p is an odd prime, then every prime number of the form $2Np + 1$ (N being any integer), for which there are no two consecutive pth power residues, necessarily divides one of the numbers x, y, or z.*

She then noted that if for a fixed p one could find infinitely many primes θ satisfying the condition stated, each of these would have to divide one of x, y, or z, and therefore one of these three numbers would be divisible by infinitely many primes, which is absurd. Unfortunately, Germain was not able to find infinitely many primes satisfying this condition, even for a single exponent. But she did, for example, find enough such primes for exponent $p = 5$ to show that any solution would have to have at least 39 decimal digits. In addition, in connection with further work dealing with the size of numbers that could possibly satisfy the Fermat equation, she was able to prove

BIOGRAPHY

Sophie Germain (1776–1831)

Forced to study in private due to the turmoil of the French Revolution and the opposition of her parents, Germain nevertheless mastered mathematics through calculus on her own. She wanted to continue her studies at the École Polytechnique when it opened in 1794, but women were not admitted as students. Nevertheless, she diligently collected and studied the lecture notes from various mathematics classes and even submitted a paper of her own to Lagrange. Mastering Gauss's *Disquisitiones* soon after it was published, she also began a correspondence with him, using the pseudonym of M. Le Blanc. Germain was, in fact, responsible for suggesting to the French general leading the army occupying Brunswick in 1807 that

he insure Gauss's safety. Gauss, naturally, did not then know the name of Sophie Germain, but the misunderstanding was cleared up with an exchange of letters. Perhaps surprisingly for a mathematician brought up in Germany, Gauss was pleased to learn that his correspondent and protector was a woman. As he wrote, "When a person of the sex which, according to our customs and prejudices, must encounter infinitely more difficulties than men to familiarize herself with these thorny researches, succeeds nevertheless in surmounting these obstacles and penetrating the most obscure parts of them, then without doubt she must have the noblest courage, quite extraordinary talents, and a superior genius."[7]

SOPHIE GERMAIN'S THEOREM   *For an odd prime exponent p, if there exists an auxiliary prime θ such that there are no two nonzero consecutive pth powers modulo θ, nor is p itself a pth power modulo θ, then in any solution to the Fermat equation $z^p = x^p + y^p$, one of x, y, or z must be divisible by p^2.*

Unfortunately, although Germain made many valiant efforts to find the infinitely many primes required by her lemma and also believed that she had actually proved Fermat's Last Theorem for even exponents of the form $2p$, where $p = 8n \pm 3$ is prime, none of her work has stood the test of time. Of course, it should be remembered that because she was a woman, she was, in general, isolated from the mathematical community at the time, and so had little opportunity to discuss her work with other mathematicians.

In 1825, Legendre succeeded in proving Fermat's Last Theorem for the case $p = 5$. Seven years later, Peter Lejeune-Dirichlet (1805–1859) proved the theorem for $n = 14$, while it took an additional seven years before the result was proved for $p = 7$ by Gabriel Lamé (1795–1870). These latter proofs were all very long, involved difficult manipulations, and did not appear capable of generalization. If the theorem were to be proved, it seemed that an entirely new approach would be necessary.

Such a new approach to proving Fermat's Last Theorem was announced with great fanfare by Lamé at the Paris Academy meeting of March 1, 1847.[8] Lamé claimed he had solved this long-outstanding problem and gave a brief sketch of the proof. The basic idea began with the factorization of the expression $x^p + y^p$ over the complex numbers as

$$x^p + y^p = (x + y)(x + \alpha y)(x + \alpha^2 y) \cdots (x + \alpha^{p-1} y)$$

where α is a primitive root of $x^p - 1 = 0$. Lamé next planned to show that if x and y are such that the factors in this expression are all relatively prime and if also $x^p + y^p = z^p$, then each of the factors must itself be a pth power. He would then use Fermat's technique of infinite descent to find a solution in smaller numbers. On the other hand, if the factors were not relatively prime, he hoped to show that they had a common factor. Dividing by this factor would then reduce the problem to the first case. When Lamé finished his presentation, Joseph Liouville (1809–1882) took the floor and cast some serious doubt on Lamé's proposal. Basically, he noted that Lamé's idea to conclude that each factor was a pth power because the factors were relatively prime and their product was a pth power depended on the theorem that any integer can be uniquely factored into a product of primes. It was by no means obvious, he concluded, that such a result was true for complex integers of the form $x + \alpha^j y$.

Over the next several weeks, Lamé tried without success to overcome Liouville's objection. But on May 24, Liouville read into the proceedings a letter from Ernst Kummer (1810–1893) that effectively ended the discussion. Namely, Kummer noted not only that unique factorization fails in some of the domains in question, but that three years earlier he had published an article, admittedly in a somewhat obscure publication, in which he had demonstrated this failure in the case $p = 23$. Kummer's article of 1844 was related to the question of higher reciprocity laws, but contained a general study of complex numbers involving roots of unity.

21.1.3   Kummer and Ideal Numbers

The complex numbers studied by Kummer, important in connection with Fermat's Last Theorem and with general reciprocity laws, are called the **cyclotomic integers**. They are the complex numbers of the form

$$f(\alpha) = a_0 + a_1\alpha + a_2\alpha^2 + \cdots + a_{n-1}\alpha^{n-1}$$

where $\alpha \neq 1$ is a root of $x^n - 1 = 0$ and each a_i is an ordinary integer. In particular, we will assume here that n is prime. The numbers formed by replacing α by the other solutions $\alpha^i \neq 1$ of the equation $x^n - 1 = 0$, written $f(\alpha^2)$, $f(\alpha^3)$, . . . , $f(\alpha^{n-1})$, are called the **conjugates** of the given number $f(\alpha)$. The product of all the conjugates is called the **norm** of $f(\alpha)$, written $Nf(\alpha)$, and is an ordinary integer. The norm satisfies the relationship $N[f(\alpha)g(\alpha)] = Nf(\alpha) Ng(\alpha)$. This multiplication property of the norm became one of Kummer's primary tools to deal with factorization of cyclotomic integers, because a factorization of such an integer $f(\alpha)$ implies a factorization of the ordinary integer $Nf(\alpha)$.

To deal with Kummer's argument, we need two definitions. A complex integer in a particular domain is **irreducible** if it cannot be factored into two other integers in that domain, neither of which is a unit. It is **prime** if whenever it divides a product it divides one of the factors. (Note that this definition is different from that of Gauss.) It is not difficult to show that a prime is irreducible. Kummer found, however, that in the domain Γ of cyclotomic integers generated by α, a 23rd root of unity, there were irreducible integers that are not prime. He was then able to demonstrate the failure of unique factorization in Γ.

First, Kummer showed that the norm of any integer in Γ is of the form $(x^2 + 23y^2)/4$. It followed that the primes 47 and 139 are not the norms of any integer because neither $4 \cdot 47 = 188$ nor $4 \cdot 139 = 556$ can be written as a square plus 23 times a square. On the other hand, $47 \cdot 139$ is the norm of $\beta = 1 - \alpha + \alpha^{-2}$. It follows that β divides $47 \cdot 139$. If β were prime, it would have to divide either 47 or 139. But this is impossible because $N(\beta)$ divides neither $N(47) = 47^{22}$ nor $N(139) = 139^{22}$. If β could be factored, the norm of one of these factors would have to be 47, contradicting the result that 47 is not a norm. It follows

BIOGRAPHY

Ernst Kummer (1810–1893)

Kummer, born in Sorau, Germany (now Zary, Poland), a town halfway between Berlin and Wrocław (Breslau), entered the University of Halle in 1828 to study theology. He soon changed his specialty to mathematics and, after receiving his doctorate in 1831, taught mathematics and physics in a gymnasium in Liegnitz (now Legnica) for 10 years before receiving an appointment to the University of Breslau in 1842. It was at Breslau that Kummer devoted himself to research in number theory. In 1855, after Dirichlet left Berlin to succeed Gauss at Göttingen, Kummer was appointed to the vacant position at Berlin, where, along with Karl Weierstrass, he established Germany's first ongoing seminar in pure mathematics. This seminar soon attracted many mathematicians from throughout the world and helped to make Berlin one of the most important world centers of mathematics in the late nineteenth and early twentieth centuries.

that β is irreducible but not prime. An explicit decomposition of $47 \cdot 139$ into two different sets of irreducible factors is then feasible. First, $47 \cdot 139 = N(\beta)$, giving a factorization into 22 irreducible factors each of norm $47 \cdot 139$. Second, if $h(\alpha) = \alpha^{10} + \alpha^{-10} + \alpha^8 + \alpha^{-8} + \alpha^7 + \alpha^{-7}$ and $g(\alpha) - \alpha^{10} + \alpha^{-10} + \alpha^8 + \alpha^{-8} + \alpha^4 + \alpha^{-4}$, then $N\,h(\alpha) = 47^2$ and $N\,g(\alpha) = 139^2$, so $h(\alpha)$, $g(\alpha)$ and all of their conjugates are irreducible. Further, setting $f(\alpha) = h(\alpha)g(\alpha)$, one can show that

$$47 \cdot 139 = f(\alpha)f(\alpha^4)f(\alpha^{-7})f(\alpha^{-5})f(\alpha^3)f(\alpha^{-11})f(\alpha^2)f(\alpha^8)f(\alpha^9)f(\alpha^{-10})f(\alpha^6)$$

where the factors are those conjugates generated by the transformation of α to α^4. This new factorization of $47 \cdot 139$ into 22 irreducibles, half with norm 47^2 and half with norm 139^2, is clearly different from the original one. Interestingly enough, Liouville, before he had heard from Kummer in May of 1847, had written in his notebook a much simpler example of nonunique factorization, in the domain generated by a root of $x^2 = -17$ rather than $x^n = 1$:

$$169 = 13 \cdot 13 = (4 + 3\sqrt{-17})(4 - 3\sqrt{-17}).$$

Having discovered nonunique factorization, Kummer spent the next several years fashioning an answer to the question posed in the opening quotation of this chapter. He devised a new type of complex numbers, "ideal complex numbers," which would uniquely factor into "ideal" prime factors. As an example, consider the domain of the 23rd roots of unity, in which neither 47 nor 139 have prime factors. In terms of "ideal" prime factors, however, because $N(1 - \alpha + \alpha^{-2}) = 47 \cdot 139$, it should be true that each of the 22 irreducible factors on the left would be divisible by two ideal prime factors on the right, one a factor of 47 and the other a factor of 139. To describe such a prime factor, Kummer defined what it meant to be divisible by it. Thus, let P be the prime factor of 47 that divides $\beta = 1 - \alpha + \alpha^{-2}$ and ψ the product of the 21 conjugates of β. Then ψ will be divisible by all of the ideal prime factors of 47 except P, so that $\gamma\psi$ will be divisible by 47 if and only if γ is divisible by P. Hence, an integer γ in the domain of the 23rd roots of unity is **divisible** by the ideal prime factor P if $\gamma\psi$ is divisible by 47. Similarly, γ is divisible m times by P if $\gamma\psi^m$ is divisible by 47^m. Kummer generalized this idea to arbitrary domains of cyclotomic integers and, because an arbitrary ideal number was defined to be the (formal) product of prime ideal numbers, Kummer succeeded, virtually by definition, in restoring unique factorization into primes to the ideal numbers. By further detailed study of these ideal numbers, Kummer was also able to establish certain conditions on a prime integer n under which Fermat's Last Theorem was true for that n and to prove the theorem for all primes less than 100 with the exception of 37, 59, and 67.[9]

In his original paper on ideal complex numbers, Kummer wrote that he intended to generalize his work to domains other than the cyclotomic integers, domains generated by a root of $x^2 - D = 0$, where D is an integer. Kummer never published any such generalization, partly because he could not find the "correct" generalization of the concept of integer to those domains. The obvious generalization appears to be that an integer of that domain would be a complex number of the form $x + y\sqrt{D}$, where x and y are ordinary integers. It turns out, however, that this definition leads to problems. For example, consider the domain of numbers of the form $x + y\sqrt{-3}$, and let $\beta = 1 + \sqrt{-3}$. Then $\beta^3 = -8$. Since 2 does not divide β in this domain, there must be an ideal prime factor P of 2, which divides 2 with a greater multiplicity than it divides β. For simplicity, we write this as $\mu_P(2) > \mu_P(\beta)$. On the other hand, 2^k divides $8\beta^k$ for all k. It follows that $k\mu_P(2) \leq \mu_P(8) + k\mu_P(\beta)$ or $k(\mu_P(2) - \mu_P(\beta)) \leq \mu_P(8)$ for all k. This implies that $\mu_P(2) \leq \mu_P(\beta)$, a contradiction.

21.1.4   Dedekind and Ideals

It was Richard Dedekind (1831–1916) who by 1871 solved this problem of defining the integers. Furthermore, since he was unhappy that "Kummer did not define ideal numbers themselves, but only the divisibility of these numbers,"[10] Dedekind created a new concept to restore unique factorization to his newly defined domains. He worked on this concept over many years, publishing four versions, three as supplements to his edition of Dirichlet's *Lectures on Number Theory* (1871, 1879, 1894), and once, in French, in installments in the *Bulletin des Sciences Mathématiques et Astronomiques* (1876), republished the following year as *Sur la Théorie des Nombres Entiers Algébriques* (*On the Theory of Algebraic Numbers*).

Dedekind began by defining an **algebraic number** as a complex number θ that satisfies any algebraic equation over the rational numbers, that is, an equation of the form

$$\theta^n + a_1\theta^{n-1} + a_2\theta^{n-2} + \cdots a_{n-1}\theta + a_n = 0$$

where the a_i are rational numbers. He then defined an **algebraic integer** to be an algebraic number that satisfied such an equation with all coefficients being ordinary integers. For example, $\theta = \frac{1}{2} + \frac{1}{2}\sqrt{-3}$ is an algebraic integer because it satisfies the equation $\theta^2 - \theta + 1 = 0$, even though it is not of the "obvious" form $x + y\sqrt{-3}$ with x, y integers. Dedekind then showed that the sum, difference, and product of algebraic integers are also algebraic integers. He defined divisibility in the standard way: an algebraic integer α is **divisible** by an algebraic integer β if $\alpha = \beta\gamma$ for some algebraic integer γ. To develop general laws of divisibility, however, Dedekind needed to restrict himself to only a part of the domain of all algebraic integers. Thus, given any algebraic integer θ that satisfies an irreducible equation of degree n, he defined the system of algebraic integers Γ_θ corresponding to θ to be the set of those algebraic integers of the form $x_0 + x_1\theta + x_2\theta^2 + \cdots + x_{n-1}\theta^{n-1}$, where the x_i are rational

BIOGRAPHY

Richard Dedekind (1831–1916)

Dedekind, like Gauss, was born in Brunswick and studied both at the Collegium Carolinium and then at the University of Göttingen. Even after receiving his doctorate under Gauss in 1852, he continued to study with the best of the German mathematicians, both at Göttingen and Berlin. In 1858, he became a professor at the Polytechnikum in Zurich and four years later returned to Brunswick to teach at the Polytechnikum there, the successor to Collegium Carolinium. Although at various times he could have received an appointment to a major German university, Dedekind chose to remain in Brunswick where he felt he had sufficient freedom to pursue his mathematical research. It was his work as editor of Dirichlet's *Vorlesungen über Zahlentheorie* (*Lectures on Number Theory*) (1863 and later editions) that convinced him to publish his own ideas on the subject, ideas he had been developing in his lectures over the years (Fig. 21.2).

FIGURE 21.2

Dedekind and ideal factorization on a stamp from the former DDR

numbers. In any domain Γ_θ, Dedekind could now define, like Kummer, what it meant for an integer to be prime or irreducible.

Dedekind noted that the Gaussian integers Γ_i are a system of algebraic integers and that Gauss had proved that unique factorization into primes is true in this system. But, following the work of Dirichlet, he gave a proof different from that of Gauss. Namely, he first showed that the Euclidean division algorithm is true in this domain in the sense that given any two nonzero Gaussian integers z and m, there always exist two other Gaussian integers q and r such that $z = qm + r$ with $N(r) < N(m)$. (It turns out that such a division algorithm does not generally exist in domains of algebraic integers; a domain in which it is true is called a **Euclidean domain**.) Dedekind then showed, exactly as in the case of ordinary integers, that the repeated use of the division algorithm determines for any two Gaussian integers z and m a greatest common divisor d, which can be written in the form $d = az + bm$. In particular, if an irreducible Gaussian integer p divides a product rs of two Gaussian integers, then it must divide one of the factors. For if p does not divide r, then $1 = ap + br$ or $s = aps + brs$ and p divides s. Unique factorization follows immediately, and Dedekind, like Gauss, could determine all the primes of Γ_i.

Dedekind noted further that if θ is a root of any of the equations $x^2 + x + 1 = 0$, $x^2 + x + 2 = 0$, $x^2 + 2 = 0$, $x^2 - 2 = 0$, or $x^2 - 3 = 0$, then a similar division algorithm is valid in Γ_θ and unique factorization holds. Thus, since

$$\theta = \frac{-1 + \sqrt{-3}}{2}$$

is a root of the first equation, $\Gamma_\theta = \{x + y\theta\}$, with x and y ordinary integers, is a Euclidean domain and has unique factorization, so Kummer's problem with $\{x + y\sqrt{-3}\}$ is resolved. On the other hand, the division algorithm does not apply to the domain determined by a root of $x^2 + 5 = 0$, where the norm of any integer $\omega = x + y\sqrt{-5}$ is $N(\omega) = x^2 + 5y^2$. Dedekind considered the integers $a = 2$, $b = 3$, $b_1 = -2 + \sqrt{-5}$, $b_2 = -2 - \sqrt{-5}$, $d_1 = 1 + \sqrt{-5}$, and $d_2 = 1 - \sqrt{-5}$ and proved, by use of the norm, that each of these integers is irreducible. Furthermore, easy multiplications showed that $ab = d_1 d_2$, $b^2 = b_1 b_2$, and

$ab_1 = d_1^2$. It followed that unique factorization did not hold. Dedekind went further, however. "Imagine for a moment that the . . . preceding numbers are rational integers."[11] Using the general laws of divisibility and assuming that a and b are relatively prime, and also b_1 and b_2, Dedekind deduced that there would exist integers α, γ, and δ such that $a = \alpha^2$, $b = \gamma\delta$, $b_1 = \gamma^2$, $b_2 = \delta^2$, $d_1 = \alpha\gamma$, and $d_2 = \alpha\delta$. Then, for example, $ab = \alpha^2\gamma\delta = d_1 d_2$. These integers, however, do not exist in the given domain. After all, the original integers are all irreducible. To create substitutes for these new integers and thereby restore unique factorization was Dedekind's goal in the creation of his new concept, the ideal, a concept he believed easier to understand than Kummer's ideal numbers.

Dedekind decided that he did not need a new creation like Kummer's ideal numbers, but that "it is sufficient to consider a system of actual numbers."[12] Because Kummer had only defined divisibility by an ideal number, Dedekind took for his "system of actual numbers" the set I of all those integers in Γ_θ that are divisible by the given ideal number, a set he named an **ideal**. Because for any numbers α and β divisible by an ideal number, the sum $\alpha + \beta$ is also divisible, and because $\omega\alpha$ is divisible for any ω in Γ_θ, these conditions are necessary for a set to be an ideal. But because Dedekind could also show that any set satisfying those conditions was the set divisible by some ideal number, he could take those conditions as his definition of an ideal I in a domain Γ_θ of algebraic integers. Furthermore, he defined a **principal ideal** (α) to be the set of all multiples of a given integer α. If α and β are both elements of Γ_θ, then the ideal consisting of all integers of the form $r\alpha + s\beta$, with r, s in Γ_θ, is denoted by (α, β).

Dedekind's next task was to define divisibility of ideals. He noted that if α is divisible by β, or $\alpha = \mu\beta$, the principal ideal generated by α is contained in that generated by β. Conversely, if every multiple of α is also a multiple of β, then α is divisible by β. Dedekind therefore extended this definition to arbitrary ideals: An ideal I is **divisible** by an ideal J if every number in I is contained in J. An ideal P, different from Γ_θ, is said to be **prime**, if it has no divisor other than itself and Γ_θ, that is, if it is contained in no other ideal except Γ_θ itself. For example, the principal ideal (2) in $\Gamma_{\sqrt{-5}}$, although it is generated by an irreducible element, is not prime; it is contained in the prime ideal $(2, 1 + \sqrt{-5})$. (Note that today's definition of prime ideal is somewhat different from Dedekind's.)

Dedekind noted further that there was a natural definition of a **product** of two ideals, namely, that IJ consists of all sums of products of the form $\alpha\beta$ where α is in I and β is in J. It is then obvious that IJ is divisible by both I and J. To complete the relationship between the two notions of product and divisibility, however, he had to prove two further theorems, both of which took him several years of work to achieve:

THEOREM 1   *If the ideal C is divisible by an ideal I, then there exists a unique ideal J such that the product I J is identical with C.*

THEOREM 2   *Every ideal different from Γ_θ is either a prime ideal or may be represented uniquely in the form of a product of prime ideals.*

These two theorems provided Dedekind with his new way of restoring unique factorization to any domain Γ_θ of algebraic integers. Thus, in the domain of algebraic integers of the form $x + y\sqrt{-5}$, the (nonexistent) prime factors α, γ, δ were replaced by prime ideals $A = (2, 1 + \sqrt{-5})$, $G = (3, 1 + \sqrt{-5})$, $D = (3, 1 - \sqrt{-5})$ so that the principal ideals $(a) = (2)$,

$(b) = (3), (d_1) = (1 + \sqrt{-5}), (d_2) = (1 - \sqrt{-5})$ factored as $(a) = A^2, (b) = GD, (d_1) = AG$, and $(d_2) = AD$. The nonunique integer factorization $ab = d_1 d_2$ could then be replaced by the unique factorization into prime ideals: $(a)(b) = A^2 GD = (d_1)(d_2)$.

 21.2   ## SOLVING ALGEBRAIC EQUATIONS

Because the solving of equations was the central concern of algebra before the nineteenth century, it is not surprising that major features of the new forms algebra took in that century grew out of new approaches to this problem. In fact, some of the central ideas of group theory grew out of these new approaches.

21.2.1 Cyclotomic Equations and Constructions

Lagrange had studied in detail the solvability of equations of degree less than 5 and had indicated possible means of attack for equations of higher degree. In the final chapter of his *Disquisitiones Arithmeticae*, Gauss discussed the solution of cyclotomic equations, equations of the form $x^n - 1 = 0$, and the application of these solutions to the construction of regular polygons. Naturally, Gauss knew the solutions of this equation in the form $\cos \frac{2\pi k}{n} + i \sin \frac{2\pi k}{n}$ for $k = 0, 1, 2, \ldots, n - 1$, but his aim in this chapter was to determine these solutions algebraically. Because the solution of the equation for composite integers follows immediately from that for primes, Gauss restricted his attention to the case where n is prime, and because $x^n - 1$ factors as $(x - 1)(x^{n-1} + x^{n-2} + \cdots + x + 1)$, it was the equation

$$x^{n-1} + x^{n-2} + \cdots + x + 1 = 0$$

that provided the focus for his work.

Gauss's plan for the solution of this $(n - 1)$st degree equation was to solve a series of auxiliary equations, each of degree a prime factor of $n - 1$, with the coefficients of each in turn being determined by the roots of the previous equation. Thus, for $n = 17$, where $n - 1 = 2 \cdot 2 \cdot 2 \cdot 2$, he wanted to determine four quadratic equations, while for $n = 73$ he needed three quadratics and two cubics. Gauss knew that the roots of $x^{n-1} + x^{n-2} + \cdots + x + 1$ could be expressed as powers r^i $(i = 1, 2, \ldots, n - 1)$ of any fixed root r. Furthermore, he realized that if g is any primitive root modulo n, the powers $1, g, g^2, \ldots, g^{n-2}$ include all the nonzero residues modulo n. It follows that the $n - 1$ roots of the equation can be expressed as $r, r^g, r^{g^2}, \ldots, r^{g^{n-2}}$ or even as $r^\lambda, r^{\lambda g}, r^{\lambda g^2}, \ldots, r^{\lambda g^{n-2}}$ for any λ less than n. His method of determining the auxiliary equations involved constructing **periods**, certain sums of the roots r^j, which in turn were the roots of the auxiliary equations. An analysis of the particular example $n = 19$ will give the flavor of Gauss's work.

For $n = 19$, the factors of $n - 1$ are 3, 3, and 2. Gauss began by determining three periods of six terms each, each period to be the root of an equation of degree 3. The periods are found by choosing a primitive root modulo 19, here 2, setting $h = 2^3$, and computing

$$\alpha_i = \sum_{k=0}^{5} r^{ih^k} \quad \text{for } i = 1, 2, 4.$$

In modern terminology, the permutations of the 18 roots of $x^{18} + x^{17} + \cdots + x + 1 = 0$ form a cyclic group G determined by the mapping $r \to r^2$, where r is any fixed root. The periods here are the sums that are invariant under the subgroup H of G generated by the mapping $r \to r^h$. These sums contain six elements because $h^6 = 2^{18} \equiv 1 \pmod{19}$, that is, H is a subgroup of G of order 6. Furthermore, for $i = 1, 2, 4$, those mappings of the form $r \to r^{ih^k}$ for $k = 0, 1, \ldots, 5$ are precisely the three cosets of H in the group G. For example, because $H = \{r \to r, r^8, r^{64}, r^{512}, r^{4096}, r^{32768}\}$, it follows that

$$\alpha_1 = r + r^8 + r^7 + r^{18} + r^{11} + r^{12},$$

where the powers are reduced modulo 19. Similarly,

$$\alpha_2 = r^2 + r^{16} + r^{14} + r^{17} + r^3 + r^5 \quad \text{and} \quad \alpha_4 = r^4 + r^{13} + r^9 + r^{15} + r^6 + r^{10}.$$

Gauss then showed that $\alpha_1, \alpha_2, \alpha_4$ are roots of the cubic equation $x^3 + x^2 - 6x - 7 = 0$.

The next step is to divide each of the three periods into three further periods of two terms each, where again the new periods will satisfy an equation of degree 3. These periods,

$$\beta_i = \sum_{k=0}^{1} r^{im^k} \quad \text{for } i = 1, 2, 4, 8, 16, 13, 7, 14, 9,$$

where $m = 2^9$, are invariant under the subgroup M generated by the mapping $r \to r^m$. Because $m^2 \equiv 1 \pmod{19}$, M has order 2 and has nine cosets corresponding to the values of i given. For example,

$$\alpha_1 = \beta_1 + \beta_8 + \beta_7 = (r + r^{18}) + (r^8 + r^{11}) + (r^7 + r^{12}).$$

Given these new periods of length 2, Gauss showed that β_1, β_8, and β_7 are all roots of the cubic equation $x^3 - \alpha_1 x^2 + (\alpha_1 + \alpha_4)x - 2 - \alpha_2 = 0$. It turns out that each of the other β_i can be expressed as a polynomial in β_1. Finally, Gauss broke up each of the periods with two terms into the individual terms of which it is formed and showed that, for example, r and r^{18} are the two roots of $x^2 - \beta_1 x + 1 = 0$. The 16 remaining roots of the original equation are then simply powers of r or can be found by solving eight other similar equations of degree 2.

Because the equations of degree 3 involved in the above example are solvable by the use of radicals, as is the equation of degree 2, Gauss had demonstrated that the roots of $x^{19} - 1 = 0$ are all expressible in terms of radicals. His more general result, applicable to any equation $x^n - 1 = 0$, only showed that a series of equations can be discovered, each of prime degree less than $n - 1$, whose solutions would determine the solution of the original equation. But, Gauss continued:

> Everyone knows that the most eminent geometers have been unsuccessful in the search for a general solution of equations higher than the fourth degree, or (to define the search more accurately) for the reduction of mixed equations to pure equations. [Pure equations are those of the form $x^m - A = 0$, which can be solved by taking an mth root once the solutions of $x^m - 1 = 0$ are known.] And there is little doubt that this problem is not merely beyond the powers of contemporary analysis but proposes the impossible. . . . Nevertheless, it is certain that there are innumerable mixed equations of every degree which admit a reduction to pure equations, and we trust that geometers will find it gratifying if we show that our equations are always of this kind.[13]

Gauss then sketched a proof, although one with a minor gap, that the auxiliary equations involved in his solution of $x^n - 1 = 0$ for n prime can always be reduced to pure equations. He therefore demonstrated, by induction, that these equations are always solvable in radicals. Naturally, if $n - 1$ is a power of 2, all of the auxiliary equations are quadratic and no special proof is necessary. In this case, however, Gauss noted further that the solutions can be constructed geometrically by Euclidean techniques. Because the roots of $x^n - 1 = 0$ can be considered as the vertices of a regular n-gon (in the complex plane), Gauss had proved that such a polygon can be constructed whenever $n - 1$ is a power of 2. The only such primes known to Gauss, and even to us today, are 3, 5, 17, 257, and 65,537. In fact, the story is told that Gauss's discovery of the construction of the regular 17-gon was instrumental in his decision to pursue a career in mathematics. Gauss concluded with a warning: "Whenever $n - 1$ involves prime factors other than 2, we are always led to equations of higher degree. . . . We can show with all rigor that these higher-degree equations cannot be avoided in any way nor can they be reduced to lower-degree equations. . . . We issue this warning lest anyone attempt to achieve geometric constructions for sections [of the circle] other than the ones suggested by our theory (e.g., sections into 7, 11, 13, 19, etc., parts) and so spend his time uselessly."[14]

Interestingly, although Gauss made this assertion, he did not in fact prove that regular n-gons, where $n = 7, 11, 13, 19$, and so on, cannot be constructed. This gap was filled in 1837 by Pierre Wantzel (1814–1848). Wantzel also gave the final resolution of two classical Greek construction problems by showing that any construction problem that does not lead to an irreducible polynomial equation with degree a power of 2 and with constructible coefficients cannot be accomplished using a straightedge and compass. For example, because the problem of doubling a cube of side a requires the solution of the irreducible cubic $x^3 - 2a^3 = 0$, it is impossible to construct this solution with Euclidean tools. Similarly, the problem of trisecting an angle α requires the solution of the irreducible cubic equation expressing $x = \sin(\alpha/3)$ in terms of the known value $a = \sin\alpha$: $4x^3 - 3x + a = 0$. Again, Wantzel's result showed that this construction is impossible with Euclidean tools.

Recall, however, that Greek mathematicians solved both of these problems using conic sections, a result generalized by both Omar Khayyam and Descartes, who gave explicit constructions for solving cubics. Analogously, the American mathematician James Pierpont (1866–1932) proved in 1895 that a regular polygon of n sides (n prime) can be constructed using conic sections if and only if $n - 1$ contains no prime factors other than 2 or 3. For example, a regular 7-gon can be constructed using conics while a regular 11-gon cannot.

The other important construction problem unsolved by the Greeks, that of squaring the circle, was also shown to be impossible. Algebraically, this problem is equivalent to solving $x^2 - \pi = 0$, a quadratic equation. Unfortunately, one coefficient of this quadratic is π, and the Greeks found no way of constructing, using Euclidean tools, a line segment with that length. By the nineteenth century, it had long been suspected that π could not be expressed as the root of any algebraic equation with rational coefficients, that is, that it was a **transcendental** number rather than an algebraic number. It was Liouville who in 1844 was the first actually to display a transcendental number:

$$\frac{1}{10} + \frac{1}{10^{2!}} + \frac{1}{10^{3!}} + \cdots + \frac{1}{10^{n!}} + \cdots = 0.110001000000000000000000100\ldots.$$

He then attempted without success to show that both e and π were also transcendental. A protegé of Liouville, Charles Hermite (1822–1901), finally demonstrated that e was transcendental in 1873, and, on the basis of Hermite's ideas, Ferdinand Lindemann (1852–1939) showed that π was transcendental nine years later. It followed immediately that it was impossible to square the circle with Euclidean tools.

21.2.2 The Theory of Permutations

It is clear that Gauss was convinced that the general equation of degree higher than 4 could not be solved by radicals. Recall that Lagrange had already attempted to find a solution by considering permutations of the roots. To consider the question of higher-degree equations in detail, therefore, it was necessary to understand the theory of permutations. Substantial work on this concept was accomplished early in the nineteenth century by Augustin-Louis Cauchy (1789–1857).

Up to Cauchy's time, the term "permutation" generally referred to an arrangement of a certain number of objects, say, letters. It was Cauchy who first considered the importance of the action of changing from one arrangement to another. He used the word **substitution** to refer to such an action, what one would today call a permutation, that is, a one-to-one function from the given (finite) set of letters to itself. In a series of papers on the subject nearly 30 years after his initial efforts of 1815, Cauchy used the words "substitution" and "permutation" interchangeably to refer to such functions. To avoid confusion, it is the latter word we will generally use, in its modern sense, here.

Besides focusing on the functional aspect of a permutation, Cauchy used a single letter, say, S, to denote a given permutation and defined the product of two such permutations S, T, written ST, to be the permutation determined by first applying S to a given arrangement and then applying T to the resulting one. He named the permutation that leaves a given arrangement fixed the "identity permutation," then noted that the powers S, S^2, S^3, . . . of a given permutation must ultimately result in the identity, and finally defined the **degree** of a permutation S to be the smallest power n such that S^n is equal to the identity. Cauchy even defined what he called a **circular** (cyclic) permutation on the letters $a_1, a_2, \ldots, a_n$ to be one that takes a_1 to a_2, a_2 to a_3, . . . , a_{n-1} to a_n, and a_n to a_1. In 1844, Cauchy introduced the notation $(a_1 a_2 \cdots a_n)$ for such a permutation. At that time he also defined the inverse of a permutation S in the obvious way, using the notation S^{-1}, and introduced the notation 1 for the identity. Further, given any set of substitutions on n letters, he defined what he called the **system of conjugate substitutions** determined by these, today called the **subgroup** generated by the given set, as the collection of all substitutions formed from the original ones by taking all possible products. Finally, he showed that the order of this system (the number of elements in the collection) always divides the order $n!$ of the complete system of substitutions on n letters.

21.2.3 The Unsolvability of the Quintic

The first proposed proof that the general fifth-degree equation could not be solved using radicals appeared in a privately printed treatise by the Italian Paolo Ruffini (1765–1822) in 1798, a treatise whose purported proof no contemporary could understand. In the mid-1820s,

BIOGRAPHY

Niels Henrik Abel (1802–1829)

Abel, born near Stavanger in Norway, unfortunately enjoyed but a brief life. His native abilities in mathematics were discovered by his instructor at the Cathedral School in Oslo, who encouraged Abel to read various advanced mathematics treatises available at the University of Oslo. Becoming interested in the problem of the fifth-degree equation, he believed that he was able, in fact, to solve it using radicals. Because no one in Norway could understand his arguments, he had the paper forwarded to Denmark. Before it could be published, Abel was asked to provide some numerical examples. In searching for these, he realized that his method was incorrect. Although he then proceeded to do research in other areas, in particular the theory of elliptic functions, he continued to work on the solvability question over the next several years, while studying at the University of Oslo, until he managed to prove its impossibility to his own satisfaction. He published the result in a small pamphlet at his own expense in 1824, but the brevity caused by his attempt to save money prevented most

mathematicians from understanding it. Thus, two years later, during his travels through Europe to visit various mathematicians and better prepare himself for a scientific career, he wrote an expanded version that was published in the first volume of the new German mathematics journal, *Journal für die reine und angewandte Mathematik* (*Journal for Pure and Applied Mathematics*), edited by August Crelle, who soon became one of Abel's best friends. When Abel returned to Norway in 1827, he found that there were no positions available to him, the only mathematics professorship at the university having recently been awarded to his former secondary school teacher. Abel struggled to make a living by tutoring and substituting at the university, meanwhile preparing a large number of new mathematical papers. But in January, 1829, he suffered an attack of tuberculosis from which he was not able to recover. He died in April, two days before Crelle wrote to him with the news that an appointment had been secured for him in Berlin (Fig. 21.3).[15]

FIGURE 21.3

Abel honored on a Norwegian stamp

however, Niels Henrik Abel (1802–1829) finally gave a complete proof of the impossibility of such a solution.

Abel's unsolvability proof involved applying results on permutations to the set of the roots of the equation.[16] It is well to note, however, that after proving his unsolvability result, Abel continued his research to attempt to solve the following problems: "1. To find all equations of any given degree that are algebraically solvable. 2. To decide whether a given equation is algebraically solvable or not."[17] Although he was not able in what remained of his life to solve either of these questions in its entirety, he did make progress with a particular type of equation. In a paper published in Crelle's *Journal* in 1829, Abel generalized Gauss's solution method for the equations $x^n - 1 = 0$. For that equation, every root is expressible as a power of one of them. Abel was able to show that "if the roots of an equation of any degree are related so that all of them are rationally expressible in terms of one, which we designate as x, and if, furthermore, for any two of the roots θx and $\theta_1 x$ [where θ and θ_1 are rational functions], we have $\theta\theta_1 x = \theta_1\theta x$, then the equation is algebraically solvable."[18] He demonstrated this result by showing that in this situation, as in the cyclotomic case, one could always reduce the solution to that of auxiliary equations of prime degree. It is because of this result that commutative groups today are often referred to as Abelian.

21.2.4 The Work of Galois

Although Abel could not complete his research program, this was largely accomplished by another genius who died young, Evariste Galois (1811–1832). Galois' thoughts on the subject of solvability of algebraic equations by radicals are outlined in the manuscript he submitted to the French Academy in 1831 and in the letter he wrote to his friend Auguste Chevalier just before the duel that ended his life. In the manuscript, he began by clarifying the idea of rationality. Since an equation has coefficients in a certain domain—for example, the set of ordinary rational numbers—to say that an equation is solvable by radicals means that one can express any root using the four basic arithmetic operations and the operation of root extraction, all applied to elements of this original domain. It turns out, however, that it is usually convenient to solve an equation in steps, as Gauss did in the cyclotomic case. Therefore, once one has solved $x^n = \alpha$, for instance, one has available as coefficients in the next step these solutions, expressible as $\sqrt[n]{\alpha}$, $r\sqrt[n]{\alpha}$, $r^2\sqrt[n]{\alpha}$, ..., where r is an nth root of unity. Galois noted that such quantities are adjoined to the original domain and that any quantity expressible by the four basic operations in terms of these new quantities and the original ones can then also be considered as **rational**. (In modern terminology, one begins with a particular field, then constructs an extension field by adjoining certain quantities not in the original field.) Of course, "the properties and the difficulties of an equation can be altogether different, depending on what quantities are adjoined."[19] Galois also discussed in his introduction the notion of a permutation, using the same somewhat ambiguous language as Cauchy, and used the word "group," although not always in a strictly technical sense, sometimes to refer to a set of permutations that is closed under composition and other times just to refer to a set of arrangements of letters determined by applying certain permutations.

Galois expressed his main result as

PROPOSITION I *Let an equation be given of which a, b, c, ... are the m roots.* [Galois tacitly assumed that this equation is irreducible and that all the roots are distinct.] *There will always be a group of permutations of the letters a, b, c, ... which has the following property: 1. that every function of the roots, invariant under the substitutions of the group, is rationally known; 2. conversely, that every function of the roots which is rationally known is invariant under the substitutions.*[20]

Galois called this group of permutations the **group of the equation**. In modern usage, one normally considers the group of the equation (the **Galois group**) as a group of automorphisms acting on the entire field created by adjoining the roots of the equation to the original field of coefficients. Galois' result is then that the group of the equation is that group of automorphisms of the extension field that leaves invariant precisely the elements of the original field (the elements "rationally known"). Besides giving a brief proof of his result, Galois presented Gauss's example of the cyclotomic polynomial $\frac{x^n - 1}{x - 1}$ for n prime. In that case, supposing that r is one root and g a primitive root modulo n, the roots can be expressed in the form $a = r$, $b = r^g$, $c = r^{g^2}$, ... and the group of the equation is the cyclic group of $n - 1$ permutations generated by the cycle $(abc \ldots k)$. On the other hand, the group of the general equation of degree n, that is, of the equation with literal coefficients, is the group of all $n!$ permutations of n letters.

Having stated the main theorem, Galois explored its application to the solvability question. His second proposition shows what happens when one adjoins to the original field one or all of

BIOGRAPHY

Evariste Galois (1811–1832)

Galois' tragically brief life has been the subject of a fictionalized biography, which included speculation that his death in a duel was engineered by government agents because of his radical political views. The known facts, however, do not support this contention.[21] Galois was born in Bourg-la-Reine, a town not far from Paris in which his father was elected mayor in 1815. He had mixed success in the preparatory school of Louis-le-Grand, especially after discovering his talents in mathematics. Although he published a short paper before he turned 18 and submitted a memoir on the solvability of equations of prime degree to the French Academy at the same time, he nevertheless twice failed the entrance examinations for the École Polytechnique, the first time probably because he had not mastered the basics and the second time perhaps because his father had committed suicide a few days earlier due to a scandal concocted by a reactionary priest. Galois was forced to enroll at the École Normale, whose director locked the students into the building so they could not participate in the political activities leading to the July revolution of 1830. After Galois attacked the director in a December letter for favoring "legitimacy" over "liberty," he was expelled from school and joined a heavily republican division of the National Guard, a division that was soon dissolved because of its perceived threat to the throne occupied by the "bourgeois" King Louis-Phillipe. Now heavily involved in political activity, Galois nevertheless continued his mathematical research, submitting a revised version of his memoir on solvability of equations to the Academy in January 1831. The referee rejected the manuscript some six months later because he could not understand the proofs. He suggested that Galois complete and clarify his theory and resubmit it.

Meanwhile, however, Galois had been arrested twice, the first time for threatening the life of the king and the second time for wearing the uniform of the dissolved National Guard division. For the second offense, he was convicted and sentenced to six months in prison, during which time his hatred of the Academy for their failure to appreciate his work grew to such a degree that he lashed out at France's "official scientists" in a vicious diatribe intended as a preface to the private publication of his work. Before the publication could take place, however, Galois became involved with "an infamous coquette and her two dupes"[22] and, although the exact circumstances have never been clarified, was forced (or chose) to fight a duel in which he was killed, five months before his 21st birthday. On the night before the duel, fearing the worst, he wrote a letter to his friend Auguste Chevalier amplifying and annotating some of his earlier manuscripts. He concluded with the following: "I have often dared in my life to state propositions of which I was not certain. But all that I have written here has been clear in my head for more than a year, and it would not be in my interest . . . to announce theorems of which I do not have complete proof. Publicly beseech Jacobi or Gauss to give their opinion not of the truth, but of the importance of these theorems. After that, I hope, there will be men who will find it profitable to decipher all this mess"[23] (Fig. 21.4).

FIGURE 21.4

Galois on a French stamp

the roots of some auxiliary equation (or of the original equation). Because any automorphism leaving the new field invariant certainly leaves the original field invariant, the group H of the equation over the new field is a subgroup of the group G over the original field. In fact, G can be decomposed either as $G = H + HS + HS' + \cdots$ or as $G = H + TH + T'H + \cdots$ where $S,\ S',\ T,\ T',\ \ldots$ are appropriately chosen permutations. Galois explained this entire procedure in his letter to Chevalier and noted that ordinarily these two decompositions do not coincide. When they do, however, and this always happens when all the roots of an auxiliary equation are adjoined, he called the decomposition **proper**. "If the group of an equation has a proper decomposition so that it is divided into m groups of n permutations, one may solve the given equation by means of two equations, one having a group of m permutations, the other one of n permutations."[24] In modern terminology, a proper decomposition occurs when the subgroup H is **normal**, that is, when the right cosets $\{HS\}$ coincide with the left cosets $\{TH\}$. In these circumstances, the question of solvability reduces to the solvability of two equations each having groups of order less than the original one.

Gauss had already shown that the roots of the polynomial $x^p - 1$ with p prime can be expressed in terms of radicals. It follows that if the pth roots of unity are assumed to be in the original field, then the adjunction of one root of $x^p - \alpha$ amounts to the adjunction of all of the roots. If G is the group of an equation, this adjunction therefore leads to a normal subgroup H of the group G such that the index of H in G (the quotient of the order of G by that of H) is p. Galois also proved the converse, that if the group G of an equation has a normal subgroup of index p, then one can find an element α of the original field (assuming that the pth roots of unity are in that field) such that adjoining $\sqrt[p]{\alpha}$ reduces the group of the equation to H. Galois concluded, both in his manuscript and in his letter, that an equation is solvable by radicals as long as one can continue the process of finding normal subgroups until all of the resulting indices are prime. Galois gave the details of this procedure in the case of the general equation of degree 4, showing that the group of the equation of order 24 has a normal subgroup of order 12, which in turn contains one of order 4, which contains one of order 2, which contains the identity. It follows that the solution can be obtained by first adjoining a square root, then a cube root, and then two more square roots. Galois noted that the standard solution to the quartic equation uses precisely those steps.

Galois then provided two additional results applicable to solving irreducible equations of prime degree. First, he showed that such an equation is solvable by radicals if and only if each of the permutations in the Galois group transforms a root x_k into a root $x_{k'}$, with $k' \equiv ak + b \pmod{p}$ (under a suitable ordering of the roots). For example, the Galois group of an irreducible cubic with one real and two complex roots is the group of all six permutations on three letters. If we use 0, 1, and 2 to identify the three roots, then these six permutations can be expressed as $k \rightarrow ak + b$, with $a = 1, 2$ and $b = 0, 1, 2$. As another example, note that in the case $p = 5$ the group of permutations identified by Galois has 20 elements, while the group of the general fifth-degree polynomial has 120 elements. Therefore, Galois' result shows that the general quintic is not solvable by radicals. Second, in his final theorem, Galois proved that an irreducible equation of prime degree is solvable if and only if all of its roots can be expressed rationally in terms of any two of them. As a consequence, if an irreducible fifth-degree polynomial has three real roots and two complex ones, the condition is not met and the equation is not solvable by radicals.

21.2.5   Jordan and the Theory of Groups of Substitutions

With Galois' death, his manuscripts lay unread until they were finally published in 1846 by Liouville in his *Journal des mathématiques*. Within the next few years, several mathematicians included Galois' material in university lectures or published commentaries on the work. It was not until 1866, however, that Galois theory was included in a text, the third edition of the *Cours d'algebre* of Paul Serret (1827–1898). Four years later, Camille Jordan (1838–1922) published his monumental *Traité des substitutions et des équations algébriques* (*Treatise on Substitutions and Algebraic Equations*), which contained a somewhat revised version of Galois theory, among much else.

It is in Jordan's text, and in some of his papers of the preceding decade that are essentially incorporated in it, that many modern notions of group theory first appear, although always in the context of groups of permutations (substitutions). Thus, Jordan defined a **group** to be a system of permutations of a finite set with the condition that the product (composition) of any two such permutations belongs to the system. He could then show that every group contains a unit element 1 and, for every permutation a another permutation a^{-1} such that $aa^{-1} = 1$. Jordan defined the **transform** of a permutation a by a permutation b to be the permutation $b^{-1}ab$ and the transform of the group $A = \{a_1, a_2, \ldots, a_n\}$ by b to be the group $B = \{b^{-1}a_1b, b^{-1}a_2b, \ldots, b^{-1}a_nb\}$ consisting of all the transforms. If B coincides with A, then A is said to be **permutable** with b. Although Jordan did not explicitly define a normal subgroup of a group, he did define a **simple** group as one that contains no subgroup (other than the identity) permutable with all elements of the group. For a nonsimple group G, there must then exist a **composition series**, a sequence of groups $G = H_0, H_1, H_2, \ldots, \{1\}$ such that each group is contained in the previous one and is permutable with all its elements (that is, is normal) and that no other such group can be interposed in this sequence. Jordan further proved that if the order of G is n and the orders of the subgroups are successively $\frac{n}{\lambda}$, $\frac{n}{\lambda\mu}$, $\frac{n}{\lambda\mu\nu}$, $\ldots$, then the integers $\lambda, \mu, \nu, \ldots$, are unique up to order, that is, that any other such sequence has the same **composition factors**.[25]

Jordan investigated in particular the set of groups that today are referred to as the **classical linear groups**. These are groups of what Jordan called linear substitutions, and what are now written as $n \times n$ matrices operating on vectors in n-space. In general, the field of coefficients for these linear substitutions is the finite field of p elements, although Jordan did not refer to this set as a field. Among the groups he studied are the groups $GL(n, p)$ of all invertible linear transformations on n variables modulo p (the general linear group), $SL(n, p)$, the group of all such transformations with determinant 1 (the special linear group), and $PSL(n, p)$, the quotient group of $SL(n, p)$ by its subgroup of multiples of the identity matrix (the projective special linear group). For example, $SL(2, 5)$ consists of 2×2 matrices with each entry an integer between 0 and 4 and the determinant $ad - bc \equiv 1 \pmod 5$, while $PSL(2, 5)$ consists of equivalence classes of elements from $SL(2, 5)$ modulo the subgroup consisting of the identity matrix I and the matrix $4I$. The order of $SL(2, 5)$ is 120, while that of $PSL(2, 5)$ is 60. Jordan was able to show that $PSL(2, p)$ is a simple group for $p > 3$ and that $PSL(n, p)$ is simple for all $n \geq 3$. It was in fact the existence of these simple groups that led twentieth-century mathematicians to try to find all possible finite simple groups.

Note that if in the two-dimensional space of vectors with coefficients in the field of p elements, we put $(x_1, y_1) \equiv (x_2, y_2)$ if $\frac{x_1}{y_1} = \frac{x_2}{y_2}$ (where $\frac{x_1}{0}$ is defined to be ∞), the set of equivalence classes is called $P_1(p)$, the **one-dimensional projective space** over the field with p elements. In this situation, we can consider the group $PSL(2, p)$ as the group of linear fractional transformations of the form

$$z' = \frac{az + b}{cz + d} \pmod p,$$

with $ad - bc = 1$, where $z = \frac{x}{y}$; that is, $PSL(2, p)$ is a transformation group acting on the projective space $P_1(p)$. This group of linear fractional transformations is often called the **modular group**. Of course, one can generalize this entire construction and consider $PSL(n, p)$ as a group acting on projective space of dimension $n - 1$.

Jordan used some of the group-theoretic concepts he developed to restate some of Galois' results. He defined a **solvable group** to be one that belongs to an equation solvable by radicals. Thus, a solvable group is one that contains a composition series with all composition factors prime. Because a commutative group always has prime composition factors, Jordan could show that an **Abelian equation**, one "of which the group contains only substitutions which are interchangeable among themselves,"[26] is always solvable by radicals. On the other hand, because the alternating group on n letters, which has order $n!/2$, is simple for $n > 4$, it follows immediately that the general equation of degree n is not solvable by radicals. With Jordan's work clarifying that of Galois, it was now evident that the theory of permutation groups was intimately connected with the solvability of equations.

 21.3   SYMBOLIC ALGEBRA

In the nineteenth century, algebra in England was characterized by a new interest in symbolic manipulation and its relation to mathematical truth (see Sidebars 21.1 and 21.2). One of the leaders in this new movement in algebra and, in general, a man interested in the reform of mathematical study in England, was George Peacock (1791–1858). Peacock explained his new symbolical approach to algebra in his *Treatise on Algebra* of 1830, which he extensively enlarged and revised during the period from 1842 to 1845. Peacock's interest in reform can be traced back to questions on the meaning of negative and imaginary numbers, questions that had been raised by several English mathematicians of the late eighteenth century. Negatives and imaginaries were freely used in the eighteenth century (and earlier) and were considered necessary in obtaining all sorts of algebraic results. But mathematicians were unable to explain their meanings in any way other than by various physical analogies. It was this lack of an adequate foundation for these concepts that led Francis Maseres (1731–1824) and William Frend (1757–1841) to write algebra texts specifically renouncing their use. It was clear, however, that this was too radical a step to be generally accepted, given the practical value of negatives and imaginaries in the study of solutions of equations.

21.3.1   Peacock's *Treatise on Algebra*

Peacock took it upon himself to rescue negatives and imaginaries by distinguishing two types of algebra, what he called "arithmetical algebra" and "symbolical algebra." Arithmetical algebra was universal arithmetic, that is, a means of developing the basic principles of the arithmetic of the nonnegative real numbers by use of letters rather than the numbers themselves. Thus, in arithmetical algebra one can write that $a - (b - c) = a + c - b$ but only under the conditions that $c < b$ and $b - c < a$, so that the subtractions can in fact be performed. In symbolical algebra, on the other hand, the symbols (letters) need not have any particular interpretation. Manipulations with the symbols are to be derived from analogous manipulations in arithmetic, but in symbolical algebra it is not necessary to limit their range of applicability. For example, the equation above is universally valid in symbolical algebra.

Peacock's answer to the question, "What *is* a negative number?" is that it is simply a symbol of the form $-a$. One operates with these symbols in the way derived from arithmetic. Since $(a - b)(c - d) = ac - ad - bc + bd$ in arithmetic, provided that $a > b$ and $c > d$, the same rule applies in symbolical algebra without that restriction. It follows then, setting $a = c = 0$,

SIDEBAR 21.1 *Mathematics at Cambridge*

By the mid-eighteenth century, mathematics had become central in the system of Cambridge studies, so much so that the most important examination at Cambridge, the Senate House examination, usually called the *tripos* after the three-legged stool on which candidates originally had sat during the questioning, was primarily devoted to mathematics. After all, the study of mathematics was presumed to develop the mind and thus help to prepare the English gentleman to assume his place in the leadership of the Church or state. The mathematics necessary to pass the tripos exam included the synthetic mathematics of Euclid and Apollonius, together with algebra, trigonometry, fluxions, and the elements of physics as presented in Book I of Newton's *Principia*. The more serious students, however, who hoped to become *wranglers*, that is, to finish at the top of the honors list, studied more advanced mathematics on their own. This material included the remainder of the *Principia* and, increasingly in the early nineteenth century, the work of such French mathematicians as Lagrange, Lacroix, and Laplace. Becoming a wrangler virtually guaranteed one a college fellowship at Cambridge and thus a beginning on a career, especially important to any student not of independent means.

The traditional mode of mathematics instruction throughout the eighteenth century was the synthetic geometric approach, the approach even Newton followed in his *Principia*. It was therefore not easy for Cambridge students to understand the analytical methods being practiced with such success on the Continent. To remedy this situation, several Cambridge undergraduates decided in 1812 to form a new society, the Analytical Society, whose purpose was to advance Continental analytic mathematics in Britain and, in particular, to bring this mathematics into the regular curriculum at Cambridge. Although the Analytical Society only lasted about a year, many of the original members, including George Peacock and Charles Babbage (1792–1871), were influential in the conversion of Cambridge to the new analytic style by the mid-1820s. One of the effects of their work was ultimately to change the role of mathematics at Cambridge from that of the mainstay of a liberal education to that of a profession in its own right, one whose goal was the development of new mathematics as part of the general advancement of knowledge.

that $(-b)(-d) = bd$ and, setting $a = d = 0$, that $(-b)c = -bc$. Similarly, $\sqrt{-1}$ is simply a symbol that obeys the same rules that the square root symbols obey in arithmetic. Therefore, $\sqrt{-1}\sqrt{-1} = -1$. These are examples of what Peacock called the "principle of the permanence of equivalent forms": "Whatever algebraical forms are equivalent, when the symbols are general in form but specific in value, will be equivalent likewise when the symbols are general in value as well as in form."[28] In other words, any law of arithmetic, expressible as an equation, determines a law of symbolical algebra by the removal of any limitations on the interpretation of the symbols involved. As an example not involving negatives or imaginaries, Peacock noted that since

$$(1 + x)^m = 1 + mx + m(m - 1)\frac{x^2}{1 \cdot 2} + \cdots$$

is true in arithmetic if m is rational and $0 < x < 1$, the same equation holds in symbolical algebra, no matter what the values of x and m.

In the first version of his *Algebra* in 1830, Peacock defined symbolical algebra as "the science which treats of the combinations of arbitrary signs and symbols by means of defined though arbitrary laws."[29] Thus, a major focus of the text had to do with the laws of operation of the various symbols of algebra. Still, Peacock did spend some time on the solution of

BIOGRAPHY

George Peacock (1791–1858)

Peacock was born in Denton, a town in Lincolnshire only a few miles from Newton's birthplace. In 1809 he entered Trinity College, Cambridge, and four years later graduated as a second wrangler (see Sidebar 21.1), becoming in turn a fellow at Trinity, a college lecturer, a tutor, and, in 1837, professor of astronomy and geometry. It was his position as moderator of

the tripos exam from 1817 to 1819 that enabled him to introduce Continental mathematics into that exam. Some years later, he participated in the commission that rewrote Cambridge's statutes to remove religious tests as prerequisites for receiving a degree.

equations, even including a treatment of the solution of cyclotomic equations based on the work of Gauss.

Interestingly, Peacock did not avail himself of the "arbitrary laws" of combination he advocated, either in 1830 or in 1845. All the laws in his symbolical algebra were in fact derived by the principle of permanence from the corresponding laws of arithmetic for the same operations. In fact, in 1845 he wrote that "I believe that no views of the nature of Symbolical Algebra can be correct or philosophical which made the selection of its rules of combination arbitrary and independent of arithmetic."[30] But despite his failure to use his asserted freedom to create laws for symbolical algebra, Peacock's statement that the results of this algebra "may be said to exist by convention only"[31] marked the beginning of a new meaning for the entire subject of algebra, a meaning that was soon to be exploited by other English mathematicians.

21.3.2   De Morgan and the Laws of Algebra

Augustus De Morgan (1806–1871) was influenced by his reading of Peacock's treatises, but recognized more clearly than his predecessor that the laws of algebra could be created without using those of arithmetic as a suggestive device. Rather than beginning with the laws of arithmetic, De Morgan believed that one could create an algebraic system by beginning with arbitrary symbols and creating (somehow) a set of laws under which these symbols are operated on. Only afterward would one provide interpretations of these laws. He gave a simple example of such a creation in 1849:

> Given symbols M, N, $+$, and one sole relation of combination, namely, that $M + N$ is the same as $N + M$. Here is a symbolic calculus: how can it be made a significant one? In the following ways, among others. 1. M and N may be magnitudes, and $+$ the sign of addition of the second to the first. 2. M and N may be numbers, and $+$ the sign of multiplying the first by the second. 3. M and N may be lines, and $+$ the direction to make a rectangle with the antecedent for a base, and the consequent for an altitude. 4. M and N may be men and $+$ the assertion that the antecedent is the brother of the consequent. 5. M and N may be nations, and $+$ the sign of the consequent having fought a battle with the antecedent.[32]

Although De Morgan asserted the freedom to create algebraic axioms for his symbols and even realized that the symbols could represent things other than "quantities" or "magnitudes,"

BIOGRAPHY

Augustus De Morgan (1806–1871)

Augustus De Morgan, born in India into the family of an English army officer, studied at Cambridge in the 1820s after some of Peacock's reforms had gone into effect. Thus, he was introduced to the Continental analytic mathematics from the start. Because he graduated in 1827 only as a fourth wrangler (see Sidebar 21.1), partly because other interests interfered with the "cramming" generally necessary to do well in the tripos, he felt that this showing was too poor for him to attempt a career in mathematics and so prepared for the bar. Nevertheless, in 1828 he was selected for the chair in mathematics at the newly created London University, a position he held for most of the rest of his life. De Morgan was a dedicated teacher, regularly giving four courses in each semester, courses ranging from elementary arithmetic to the calculus of variations. He spent much of his creative talents on devising better ways of instruction and wrote not only various mathematics texts but also articles and books on the teaching of mathematics.[33]

he, like Peacock, did not attempt to create any new system obeying laws different from those obeyed by the numbers of arithmetic. In fact, in 1841 he set out what he believed were the rules that were "essential to algebraical process."[34] These rules included the substitution principle (that two expressions connected by an $=$ sign can be substituted for one another), the inverse principle for both addition and multiplication (that $+$ and $-$ as well as $\times$ and $\div$ are "opposite in effect"), the commutative principle for addition and multiplication, the distributive laws (of multiplication over both addition and subtraction), and the exponential laws $a^b a^c = a^{b+c}$ and $(a^b)^c = a^{bc}$. Having presented the laws, which he believed to be "neither insufficient nor redundant," he commented that the "most remarkable point . . . is that the laws of operation prescribe much less of connexion between the successive symbols $a + b$, ab and a^b than a person who has deduced these laws from arithmetical explanation would at first think sufficient."[35] In other words, there is no necessity of deriving the meaning of multiplication from that of addition nor the meaning of exponentiation from that of multiplication. Nevertheless, although one can certainly derive all sorts of algebraic results by using just these principles, such an algebra would have no more meaning than the putting together of a jigsaw puzzle by using the backs of the pieces. True mathematics, De Morgan believed, must have real content. Laying out the axiomatic structure of a system was far less important than the task of interpretation. It was only the interpretation, which De Morgan recognized was outside of the logical framework established by axioms, which gave a mathematical system its meaning and significance.

21.3.3 Hamilton: Complex Numbers and Quaternions

It was the Irish mathematician and physicist William Rowan Hamilton (1805–1865) who was able finally to create a new algebraic system having a genuine interpretation but not conforming to all of the axioms set out by De Morgan. Like Peacock and De Morgan, Hamilton wanted to be able to justify the use of negatives and imaginaries in algebra, concepts he agreed had poor foundations. As he wrote in his fundamental paper of 1837, "Theory of

SIDEBAR 21.2   *The Tripos Examination for 1785*

The question paper was introduced by a memorandum telling the candidates to write distinctly and to observe that "at least as much will depend upon the clearness and precision of the answers as upon the quantity of them."[27]

1. To prove how many regular Solids there are, what are those Solids called, and why there are no more.
2. To prove the Asymptotes of an Hyperbola always external to the Curve.
3. Suppose a body thrown from an Eminence upon the Earth, what must be the Velocity of Projection, to make it become a secondary planet to the Earth?
4. To prove in all the conic sections generally that the force tending to the focus varies inversely as the square of the Distance.

5. Supposing the periodical times in different Ellipses round the same center of force, to vary in the sesquiplicate ratio of the mean distances, to prove the forces in those mean distances to be inversely as the square of the distance.
6. What is the relation between the 3rd and 7th Sections of Newton, and how are the principles of the 3rd applied to the 7th?
7. To reduce the biquadratic equation $x^4 + qx^2 + rx + s = 0$ to a cubic one.
8. To find the fluent of $\dot{x} \times \sqrt{a^2 - x^2}$.
9. To find a number from which if you take its square, there shall remain the greatest difference possible.
10. To rectify [an arbitrary] arc DB of the circle $DBRS$.

Conjugate Functions, or Algebraic Couples; with a Preliminary and Elementary Essay on Algebra As the Science of Pure Time":

> It requires no peculiar scepticism to doubt, or even to disbelieve, the doctrine of Negatives and Imaginaries, when set forth (as it has commonly been) with principles like these: that a *greater magnitude may be subtracted from a less*, and that the remainder is *less than nothing*; that *two negative numbers* or numbers denoting magnitudes each less than nothing, may be *multiplied* . . . and that the product will be a *positive* number . . . and that although the *square* of a number . . . is therefore *always positive*, whether the number be positive or negative, yet that numbers, called *imaginary*, can be found or conceived or determined, and operated on by all the rules of positive and negative numbers, as if they were subject to those rules, *although they have negative squares* and must therefore be supposed to be themselves neither positive nor negative, nor yet null numbers, so that the magnitudes which they are supposed to denote can neither be greater than nothing, nor less than nothing, nor even equal to nothing.[36]

To place algebra, like geometry, on a firm foundation required creating for it certain intuitive principles, and these Hamilton felt could come from the intuition of pure time. What Hamilton meant by "pure time" derives from his reading of Immanuel Kant's *Critique of Pure Reason*; it is "the form of inner sense by which we order all perceptions or sensible intuitions as existing simultaneously or successively."[37] In more modern terminology, Hamilton assumed that there was a set M of "moments," which were ordered by a relation $<$ such that for all A, B in M, either $A = B$ or $A < B$ or $A > B$. Hamilton then defined an equivalence relation on the set of pairs of moments by setting (A, B) equivalent to (C, D) if the following conditions are satisfied: "If the moment B be identical with A, then D must be identical with C; if B be later than A, then D must be later than C, and exactly so much later; and if B be earlier than A, then D must be earlier than C, and exactly so much earlier."[38] To avoid confusion later, it should be noted that Hamilton did not use the modern pair notation in discussing this equivalence. He represented the equivalence class defined by this relation

---

BIOGRAPHY

William Rowan Hamilton (1805–1865)

Hamilton was born in Dublin, but educated in the town of Trim, some 30 miles to the northwest, under the tutelage of his uncle, a scholar of the classics. Because Hamilton showed signs of genius at an early age, his uncle proceeded to turn this precocity to the study of languages. By the time he was 10, William was fluent not only in Latin, Greek, and the modern European languages, but also in Hebrew, Persian, Arabic, and Sanskrit, among others. His uncle would have had him master Chinese as well, except for the difficulty of securing Chinese books in Dublin. From an early age, Hamilton also learned arithmetic, using methods of computation of his own devising, but his mathematical interest was spurred by his contact with an American calculating prodigy who was consistently able to best William in competition. Soon afterward, he discovered Euclidean geometry as well as more modern areas of

mathematics. By the time of his entrance into Trinity College, Dublin, in 1823, he was prepared to deal with the Continental analytic mathematics taught at Trinity in line with the reforms instituted at Cambridge. Hamilton swiftly moved beyond the prescribed curriculum and soon mastered the mathematical texts in use at the École Polytechnique. His first important original work was in optics, rather than in pure mathematics, and, in fact, he is today more famous for his work in dynamics than for his mathematics. It was this work in physics that led to Hamilton's appointment, even before he received his degree, as Astronomer Royal of Ireland in 1827, a position he held for the remainder of his life. His contributions to mathematics and physics led to his being named in 1865 the first foreign associate of the newly created National Academy of Sciences of the United States (Fig. 21.5).

FIGURE 21.5

William Rowan Hamilton and his quaternion formulas on an Irish stamp

first by the suggestive notation $B - A$ and later by a single symbol a, a symbol one can think of as denoting the **time step** from A to B.

It is the time steps that provided the basis of Hamilton's construction of negatives. Namely, if a represents the pair $B - A$, then $\ominus a$ (Hamilton's notation for $-a$) represents the pair $A - B$. (Hamilton created this particular notation from the letter O, the initial letter of the Latin *oppositio* (*opposite*).) Taking a given step a as a unit and using a natural definition of the sum of two steps, Hamilton then proceeded to construct the set of rational numbers. Positive integers are determined by multiples (successive sums) of a with itself while negatives come from multiples of $\ominus a$. Rational numbers are defined through the comparison of two integral multiples of the step a. Hamilton then demonstrated the standard rules for the arithmetic operations on these (positive and negative) multiples. For example, the product of two negative multiples of a must be positive since such a product involves reversing the direction of the step a twice. Having to his own satisfaction answered the objections to negative numbers indicated above, without resorting to quantities "less than nothing," Hamilton attempted next to construct the real numbers from the rationals. Not only was this attempt a failure from a modern perspective, but it also had little effect on the arithmetization of analysis carried out in Germany later in the century. On the other hand, his construction of the complex numbers from the reals in the final part of this same essay is the one often used in textbooks today.

In this final part, Hamilton considered couples, or pairs, of moments, time steps, and numbers. Thus, two pairs (A_1, B_1), (A_2, B_2) of moments determine a pair $(a, b) = (B_1 - A_1, B_2 - A_2)$ of time steps, while the ratio α of the two pairs of steps $(\alpha a, \alpha b)$, (a, b) led Hamilton to conceive that any two pairs of steps would have a ratio expressible as a pair of numbers (α, β). (The original ratio α would then be replaced by the pair $(\alpha, 0)$). It was clear that addition and subtraction of these pairs should be defined by

$$(\alpha, \beta) \pm (\gamma, \delta) = (\alpha \pm \gamma, \beta \pm \delta).$$

Assuming the distributive law for multiplication, Hamilton then argued that a general rule for multiplication can be given by

$$(\alpha, \beta)(\gamma, \delta) = (\alpha\gamma - \beta\delta, \beta\gamma + \alpha\delta).$$

It followed that division should be defined as

$$\frac{(\alpha, \beta)}{(\gamma, \delta)} = \left(\frac{\alpha\gamma + \beta\delta}{\gamma^2 + \delta^2}, \frac{\beta\gamma - \alpha\delta}{\gamma^2 + \delta^2} \right).$$

As Hamilton wrote, "these definitions are really *not arbitrarily chosen, and that though others might have been assumed, no others would be equally proper*,"[39] because from them follow the known laws of operation on complex numbers. For example, $(0, 1)(0, 1) = (-1, 0)$ and therefore, with the identification of $(\alpha, 0)$ with the number α, $\sqrt{-1}$ can be identified with the pair $(0, 1)$. The complex number $\alpha + \beta\sqrt{-1}$ can then simply be *defined* to be the number pair (α, β). The rules above then determine the standard rules of operation of complex numbers. Hamilton thus succeeded in constructing the complex numbers from the reals, bypassing any appeal to "imaginary" numbers, thereby answering the question of what complex numbers *really* are.

Hamilton concluded his essay with the statement that he hoped to develop a "Theory of Triplets and Sets of Moments, Steps and Numbers, which includes this Theory of Couples."[40] He had already learned that the operations on complex numbers have a geometrical interpretation in the two-dimensional plane. But because much of physics took place in three-dimensional space, a system of operations (that is, an algebra) of triplets would prove immensely useful. As he wrote to De Morgan in 1841, "if my view of Algebra be just, it *must* be possible, in *some* way or other, to introduce not only triplets but *polyplets*, so as in some sense to satisfy the symbolical equation $a = (a_1, a_2, \ldots, a_n)$; a being here one symbol, as indicative of one (complex) thought; and $a_1, a_2, \ldots, a_n$ denoting n real numbers positive or negative."[41] The struggle for Hamilton, of course, was not in the addition of his triplets—that was easy—but in the multiplication. Knowing the basic laws for his couples, he wanted his triplets similarly to satisfy the associative and commutative properties of multiplication as well as the distributive law. He wanted division to be always possible (except by 0) and to always have a unique result. He wanted the moduli to multiply, that is, if $(a_1, a_2, a_3)(b_1, b_2, b_3) = (c_1, c_2, c_3)$, then $(a_1^2 + a_2^2 + a_3^2)(b_1^2 + b_2^2 + b_3^2) = c_1^2 + c_2^2 + c_3^2$. Finally, he wanted the various operations to have a reasonable interpretation in three-dimensional space. Hamilton had begun his search for a multiplicative law for triplets as early as 1830. After 13 years of considering the problem, he finally solved it, but not in the way he had hoped, in an experience described in the opening of this chapter.

Hamilton's solution was not to consider triplets at all, but quadruplets, (a, b, c, d), which he wrote, in analogy with the standard notation for complex numbers, as $a + bi + cj + dk$. The basic multiplication laws $i^2 = j^2 = k^2 = ijk = -1$ and the derived rules $ij = k$, $ji = -k$, $jk = i$, $kj = -i$, $ki = j$, and $ik = -j$, when extended by the distributive law to all quadruplets, or **quaternions**, gave this system all of the properties Hamilton sought, with the sole exception of the commutative law of multiplication. In modern terminology, the set of quaternions forms a noncommutative division algebra over the real numbers. Hamilton's system was the first significant system of "quantities" that did not obey all of the standard laws that Peacock and De Morgan had set down. As such, its creation broke a barrier to the consideration of systems violating these laws, and soon the freedom of creation advocated by Peacock became a reality.

Hamilton himself was so taken with his discovery that he spent the remainder of his life writing several tomes on the theory of quaternions. In these works, he justified the necessity for objects with four components in dealing with three-dimensional space by considering "quotients" of vectors. A quotient of a vector **v** by a vector **w** would represent the "quantity" that would turn **w** into **v**. In two dimensions, this quantity was composed of two numerical values, the ratio of the lengths of the two vectors and the angle needed to rotate **w** into **v**. Thus, one could reasonably think of the quotient of two vectors in two dimensions as being again a vector in two dimensions. In the three-dimensional case, however, the rotation itself depends on three numerical values, two to determine the direction of the axis of rotation and a third to give the angle of rotation, and a fourth value is necessary to represent the ratio of the lengths. Therefore, a quotient of two vectors in three dimensions determines a quantity with four components, a quaternion.

Although few physicists used quaternions in their work, Hamilton's idea marked the beginning of today's common use of vector terminology in physical theories. In fact, Hamilton himself noted the convenience of writing a quaternion $Q = a + bi + cj + dk$ in two parts, the real part a and imaginary part $bi + cj + dk$. He named the former the **scalar** part, because all values it can attain are on "one *scale* of progression of number from negative to positive infinity," while he named the latter the **vector** part, because it can be geometrically constructed in three-dimensional space as a "straight line or radius vector."[42] (The word "radius vector" had been part of mathematical vocabulary since the early eighteenth century. It was Hamilton, however, who first used the word "vector" in today's more general sense.) Thus, Hamilton wrote $Q = S.Q + V.Q$, where $S.Q$ is the scalar part and $V.Q$ the vector part. In particular, if we consider the product

$$(ai + bj + ck)(xi + yj + zk)$$
$$= -(ax + by + cz) + (bz - cy)i + (cx - az)j + (ay - bx)k$$

of two quaternions α, β with zero scalar parts, then $S.\alpha\beta$ is the negative of the modern dot product of the vectors α and β, while $V.\alpha\beta$ is the modern cross product.

21.3.4 Quaternions and Vectors

Hamilton's successors in the advocacy of quaternion concepts for use in physics were the Scottish physicists Peter Guthrie Tait (1831–1901) and James Clerk Maxwell (1831–1879) (Fig. 21.6), friends and fellow students at both the University of Edinburgh and Cambridge University. Tait in fact wrote an *Elementary Treatise on Quaternions* in 1867 in which he advocated quaternion methods in physics. Tait's treatise contained equivalents of virtually all the modern laws of operation of the dot and cross product of vectors, although written in quaternion notation. In particular, he showed that $S.\alpha\beta = -T\alpha T\beta \cos\theta$, where $T\alpha$ is the length of α and θ is the angle between α and β, and that $V.\alpha\beta = T\alpha T\beta \sin\theta \cdot \eta$, where η is a unit vector perpendicular to both α and β.

FIGURE 21.6

Maxwell on a stamp from San Marino

Maxwell, in his *Treatise on Electricity and Magnetism*, also advocated Hamilton's ideas. His main purpose, however, as stated in his opening chapter, was "to avoid explicitly introducing the Cartesian coordinates, and to fix the mind at once on a point of space instead of its three coordinates, and on the magnitude and direction of a force instead of its three components."[43] Thus, quaternions and the associated vectors were to be used to represent physical quantities in a more conceptual way than the usual coordinate form. In the treatise itself, Maxwell generally expressed his physical results both ways, in coordinate form and in quaternion form.

It was, however, Josiah Willard Gibbs (1839–1903) (Fig. 21.7) at Yale University and Oliver Heaviside (1850–1925) in England who independently realized, after their reading of Tait and Maxwell, that the full algebra of quaternions was not necessary for discussing physical concepts. It was only the two types of products of vectors, the dot product and the cross product, that were needed. Gibbs published his version of vector analysis privately in 1881 and 1884 and lectured on the subject for many years at Yale, while Heaviside first published his method in papers on electricity in 1882 and 1883. It is to the former, however, that our modern notations of $A \cdot B$ and $A \times B$, for the dot product and the cross product, respectively, are due. With the formal publication of Gibbs's *Vector Analysis* in a 1901 work derived from his lectures, it was clear to the physics community that vectors, rather than quaternions, provided the necessary language for describing physical concepts. Although quaternions were to remain important mathematically, their use in physics soon died a quiet death.

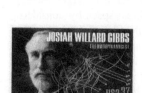

FIGURE 21.7

Josiah Willard Gibbs on a stamp from the United States

21.3.5   Boole and Logic

The algebraic freedom advocated by Peacock and De Morgan was exploited in a different way by the self-taught English logician George Boole (1815–1864). In 1847, Boole published a small book, *The Mathematical Analysis of Logic*, and seven years later expanded it into *An Investigation of the Laws of Thought*, a book that helped bring the study of logic out of metaphysics, where it had remained since the time of Aristotle, and into mathematics. Boole's aim in his *Laws of Thought* was to "investigate the fundamental laws of those operations of the mind by which reasoning is performed; to give expression to them in the symbolical language of a Calculus, and upon this foundation to establish the science of Logic and construct its method."[44]

Because algebra is studied by means of signs, Boole put into his opening proposition the basic signs by which logic would be analyzed:

PROPOSITION 1   *All the operations of Language, as an instrument of reasoning, may be conducted by a system of signs composed of the following elements, viz.:*

   1st. Literal symbols, as x, y, etc., representing things as subjects of our conceptions.

   2nd. Signs of operation, as +, −, ×, standing for those operations of the mind by which the conceptions of things are combined or resolved so as to form new conceptions involving the same elements.

   3rd. The sign of identity, =.

And these symbols of Logic are in their use subject to definite laws, partly agreeing with and partly differing from the laws of the corresponding symbols in the science of Algebra.[45]

Boole next defined the laws of his symbols of language. A letter was to represent a class, or set, of objects. Thus, x could stand for the class of "men" and y for the set of "good things." The combination xy would then stand for the class of things to which both x and y are applicable, that is, the class of "good men." It was obvious to Boole that the commutative law holds for his multiplication: $xy = yx$. Among other laws for multiplication that Boole derived are that $x^2 = x$, since the class to which x and x are applicable is simply that of x, and $xy = x$ in the case where the class represented by x is contained in that represented by y.

Addition for Boole, written $x + y$, represented the conjunction of the two classes represented by x and y, while subtraction, written $x - y$, stood for the class of those things represented by x with the exception of those represented by y. The commutative law of addition then holds as well as the distributive laws $z(x \pm y) = zx \pm zy$. With 0 used to represent the empty class and 1 the universal class, Boole similarly derived the familiar laws $0y = 0$ and $1y = y$ as well as the not-so-familiar one $x(1 - x) = 0$, which "affirms that it is impossible for any being to possess a quality, and at the same time not to possess it."[46]

Because the stated laws agree with the laws of numbers restricted to just the values 0 and 1, Boole decided that his algebra of logic would deal with variables that only take on these values. In particular, he considered functions of one or several logical variables, $f(x)$, $f(x, y), \ldots$, in which the variables can only take on the values 0 and 1. For example, he showed that any such function $f(x)$ can be expanded in the form $f(x) = f(1)x + f(0)(1 - x)$, or, putting $\bar{x} = 1 - x$, in the form $f(x) = f(1)x + f(0)\bar{x}$. Similarly, $f(x, y) = f(1, 1)xy + f(1, 0)x\bar{y} + f(0, 1)\bar{x}y + f(0, 0)\bar{x}\bar{y}$. Thus, the function $1 - x + xy$ can be expanded as $1xy + 0x\bar{y} + 1\bar{x}y + 1\bar{x}\bar{y} = xy + \bar{x}y + \bar{x}\bar{y}$.

Boole then proved that if V is some function, one can interpret the equation $V = 0$ by expanding V according to the above rules and equating to 0 every constituent whose coefficient does not vanish. As an instance of this procedure, Boole considered the definition of "clean beasts" from Jewish law: Clean beasts are those that both divide the hoof and chew the cud. With x, y, z representing clean beasts, beasts that divide the hoof, and beasts that chew the cud, respectively, the definition of clean beasts is given by the equation $x = yz$ or $V = x - yz = 0$. Expanding $x - yz$, Boole found that

$$V = 0xyz + xy\bar{z} + x\bar{y}z + x\bar{y}\bar{z} - \bar{x}yz + 0\bar{x}y\bar{z} + 0\bar{x}\bar{y}z + 0\bar{x}\bar{y}\bar{z}.$$

Equating each nonvanishing term to 0 then gives

$$xy\bar{z} = 0 \qquad x\bar{y}z = 0 \qquad x\bar{y}\bar{z} = 0 \qquad \bar{x}yz = 0.$$

The interpretation of these is the assertion of the nonexistence of certain classes of animals. For example, the first equation asserts that there are no beasts that are clean and divide the hoof, but do not chew the cud.

The algebra of classes developed by Boole, seemingly dormant for many years after Boole published it, is today known as Boolean algebra and has resurfaced as central in the study of the algebra of circuit design, the algebra by which the logic behind modern calculators and computers is developed. Boole would probably be pleased that his calculus of the laws of thought is in fact used in nearly the way he forecast over a century ago.

21.4   MATRICES AND SYSTEMS OF LINEAR EQUATIONS

The idea of a matrix has a long history, dating at least from its use by Chinese scholars of the Han period for solving systems of linear equations. In the eighteenth century, and even somewhat earlier, mathematicians calculated and used determinants of square arrays of numbers, often in the solution of systems of linear equations, even though the square arrays themselves were not singled out for attention. Other work in the nineteenth century led to more formal computations with such arrays and by mid-century to a definition of a matrix and the development of the algebra of matrices. But alongside this formal work, there was a deeper side to the development of the theory of matrices, namely, the work growing out of Gauss's study of quadratic forms, which ultimately led to the concepts of similarity, eigenvalues, diagonalization, and finally the classification of matrices via canonical forms.

21.4.1   Basic Ideas of Matrices

Gauss discussed the theory of quadratic forms, that is, functions of two variables x, y of the form $ax^2 + 2bxy + cy^2$, with a, b, c integers, in chapter 5 of his *Disquisitiones*. In this discussion, he considered the idea of a linear substitution that transforms one form into another. Namely, if $F = ax^2 + 2bxy + cy^2$, then the substitution

$$x = \alpha x' + \beta y'$$
$$y = \gamma x' + \delta y'$$

converts F into a new form F' whose coefficients depend on the coefficients of F and those of the substitution. Gauss noted that if F' is transformed into F'' by a second linear substitution,

$$x' = \epsilon x'' + \zeta y''$$
$$y' = \eta x'' + \theta y'',$$

then the composition of the two substitutions gives a new substitution transforming F into F'':

$$x = (\alpha\epsilon + \beta\eta)x'' + (\alpha\zeta + \beta\theta)y''$$
$$y = (\gamma\epsilon + \delta\eta)x'' + (\gamma\eta + \delta\theta)y''.$$

The coefficient "matrix" of the new substitution is the product of the coefficient matrices of the two original substitutions. Gauss performed an analogous computation in his study of ternary quadratic forms $Ax^2 + 2Bxy + Cy^2 + 2Dxz + 2Eyz + Fz^2$, which in effect gave the rule for multiplying two 3×3 matrices. But although he wrote the coefficients of the substitution in a rectangular array and even used a single letter S to refer to a particular substitution, Gauss did not explicitly refer to this idea of composition as a "multiplication."

In 1815 Cauchy published a fundamental memoir on the theory of determinants, in which he not only introduced the name "determinant" to replace several older terms but also used the abbreviation $(a_{1,n})$ to stand for what he called the "symmetric system,"

$$
\begin{matrix}
a_{1,1} & a_{2,2} & \cdots & a_{1,n} \\
a_{2,1} & a_{2,2} & \cdots & a_{2,n} \\
\vdots & \vdots & \ddots & \vdots \\
a_{n,1} & a_{n,2} & \cdots & a_{n,n}
\end{matrix}
,
$$

to which the determinant is associated. Although many of the basic results on calculating determinants had been known earlier, Cauchy gave the first complete treatment of these in this memoir, including such ideas as the array of minors associated to a given array (the **adjoint**) and the procedure for calculating a determinant by expanding on any row or column. In addition, he followed Gauss in explicitly recognizing the idea of composing two systems $(\alpha_{1,n})$ and $(a_{1,n})$ to get a new system $(m_{1,n})$ defined by the familiar law of multiplication,

$$m_{i,j} = \sum_{k=1}^{n} \alpha_{i,k} a_{k,j}.$$

He then showed that the determinant of the new system was the product of those of the two original ones.

Ferdinand Gotthold Eisenstein (1823–1852), a student of Gauss who visited Hamilton in Ireland in 1843, introduced the explicit notation $S \times T$ to denote the substitution composed of S and T in his discussion of ternary quadratic forms in a paper of 1844, perhaps because of Cauchy's product theorem for determinants. About this notation Eisenstein wrote, "Incidentally, an algorithm for calculation can be based on this; it consists in applying the usual rules for the operations of multiplication, division, and exponentiation to symbolical equations between linear systems; correct symbolical equations are always obtained, the sole consideration being that the order of the factors, i.e., the order of the composing systems, may not be altered."[47] It is interesting, but probably futile, to speculate whether Eisenstein's discussions with Hamilton in 1843 stimulated either to realize the possibility of an algebraic system with a noncommutative multiplication.

21.4.2 Matrix Operations

Eisenstein never developed fully his idea of an algebra of substitutions because of his untimely death at the age of 29. That development was carried out in England by Arthur Cayley (1821–1895) and James Joseph Sylvester (1814–1897) in the 1850s.

In 1850 Sylvester coined the term **matrix** to denote "an oblong arrangement of terms consisting, suppose, of m lines and n columns" because out of that arrangement "we may form various systems of determinants."[48] (The English word *matrix* meant "the place from which something else originates.") Sylvester himself made no use of the term at the time. It was his friend Cayley who put the terminology to use in papers of 1855 and 1858. In the former, Cayley noted that the use of matrices is very convenient for the theory of linear equations. Thus, he wrote

$$(\xi, \eta, \zeta, \ldots) = \begin{pmatrix} \alpha, & \beta, & \gamma, & \cdots \\ \alpha' & \beta', & \gamma', & \cdots \\ \alpha'', & \beta'', & \gamma'', & \cdots \\ \vdots & \vdots & \vdots & \ddots \end{pmatrix} (x, y, z, \ldots)$$

BIOGRAPHY

Arthur Cayley (1821–1895)

Cayley studied mathematics at Trinity College, Cambridge, graduating as senior wrangler (see Sidebar 21.1), but because there was no suitable teaching job available, decided to become a lawyer and was called to the bar in 1849. Although he became skilled in legal work, he regarded the law just as the means of providing him with an income and always reserved a substantial portion of his time for mathematics. In fact, in his 14 years as an attorney he produced close to 300 mathematical papers. In 1863, he was elected to the newly created Sadlerian professorship of mathematics at Cambridge, a position he accepted with eagerness, even though it meant a substantial cut in his earnings.

The duties of the Sadlerian professor were to "teach the principles of pure mathematics" and also "to apply himself to the advancement of that science." In regard to the first duty, Cayley was not very successful. His lectures at the University generally attracted few students, partly because he usually spoke about his latest research. On the other hand, his contributions to mathematics were enormous, comprising nearly 1000 papers in various fields. In addition, he served as a referee for hundreds of papers by others and took great pleasure in encouraging young men just beginning their research.

to represent the square system of equations

$$\xi = \alpha x + \beta y + \gamma z + \cdots$$
$$\eta = \alpha' x + \beta' y + \gamma' z + \cdots$$
$$\zeta = \alpha'' x + \beta'' y + \gamma'' z + \cdots$$
$$\vdots = \cdots + \cdots + \cdots + \cdots.$$

He then determined the solution of this system using what he called the inverse of the matrix:

$$(x, y, z, \ldots) = \begin{pmatrix} \alpha, & \beta, & \gamma, & \cdots \\ \alpha', & \beta', & \gamma', & \cdots \\ \alpha'', & \beta'', & \gamma'', & \cdots \\ \vdots & \vdots & \vdots & \ddots \end{pmatrix}^{-1} (\xi, \eta, \zeta, \ldots).$$

This representation came from the basic analogy of the matrix equation to a simple linear equation in one variable. Cayley, however, knowing Cramer's rule, then described the entries of the inverse matrix in terms of fractions (involving the appropriate determinants).

In 1858, Cayley introduced single-letter notation for matrices and showed not only how to multiply them but also how to add and subtract:

It will be seen that matrices (attending only to those of the same order) comport themselves as single quantities; they may be added, multiplied or compounded together, &c.: the law of the addition of matrices is precisely similar to that for the addition of ordinary algebraical quantities; as regards their multiplication (or composition), there is the peculiarity that matrices are not in general convertible [commutative]; it is nevertheless possible to form the powers (positive or negative, integral or fractional) of a matrix, and thence to arrive at the notion of a rational and integral function, or generally of any algebraical function, of a matrix.[49]

BIOGRAPHY

James Joseph Sylvester (1814–1897)

Sylvester, who was born into a Jewish family in London and studied for several years at Cambridge, was not permitted to take his degree there for religious reasons. Therefore, he received his degree from Trinity College, Dublin. In 1841, he accepted a professorship at the University of Virginia but remained there only a short time. His horror of slavery and an altercation with a student who did not show him the respect he felt he deserved led to his resignation in 1843. After his return to England, he spent 10 years as an attorney and 15 years as professor of mathematics at the Royal Military Academy at Woolwich, before accepting in 1877 the chair of mathematics at the newly opened Johns Hopkins University in Baltimore. While at Hopkins, Sylvester founded the *American Journal of Mathematics* and helped develop a tradition of graduate education in mathematics in the United States. Returning to England in 1884, he finally found a suitable academic position in his native land, the Savilian Chair of Geometry at Oxford.

Cayley then exploited his idea, making constant use of the analogy between ordinary algebraic manipulations and those with matrices, but carefully noting where this analogy fails. Thus, using the formula for the inverse of a 3×3 matrix, he wrote that "the notion of the inverse . . . matrix fails altogether when the determinant vanishes; the matrix is in this case said to be indeterminate. . . . It may be added that the matrix zero is indeterminate; and that the product of two matrices may be zero, without either of the factors being zero, if only the matrices are one or both of them indeterminate."[50]

It was perhaps Cayley's use of the notational convention of single letters for matrices that suggested to him the result known as the **Cayley-Hamilton theorem**. For the case of a 2×2 matrix

$$M = \begin{pmatrix} a & b \\ c & d \end{pmatrix},$$

Cayley stated this result explicitly as

$$\det \begin{pmatrix} a - M & b \\ c & d - M \end{pmatrix} = 0.$$

Cayley first communicated this "very remarkable" theorem in a letter to Sylvester in November of 1857. In 1858, he proved it by simply showing that the determinant $M^2 - (a + d)M^1 + (ad - bc)M^0$ equaled 0 (where M^0 is the identity matrix). Stating the general version in essentially the modern form that M satisfies the equation in λ, $\det(M - \lambda I) = 0$, the **characteristic equation**, Cayley noted that he had "verified" the theorem in the 3×3 case, but wrote further that "I have not thought it necessary to undertake the labour of a formal proof of the theorem in the general case of a matrix of any degree."[51] It was Georg Frobenius (1849–1917) who took advantage of Cayley's notational innovation to give a complete proof some 20 years later.

Cayley's motivation in stating the Cayley-Hamilton theorem was to show that "any matrix whatever satisfies an algebraical equation of its own order" and therefore that "every rational and integral function . . . of a matrix can be expressed as a rational and integral function of an order at most equal to that of the matrix, less unity."[52] Cayley went on to show that one can adapt this result even for irrational functions. In particular, he showed how to calculate $L = \sqrt{M}$, where M is the 2×2 matrix given above. The result is given in the form

$$L = \begin{pmatrix} \frac{a+Y}{X} & \frac{b}{X} \\ \frac{c}{X} & \frac{d+Y}{X} \end{pmatrix}.$$

where $X = \sqrt{a + d + 2\sqrt{ad - bc}}$ and $Y = \sqrt{ad - bc}$. Cayley failed, however, to give conditions under which this result holds. A similar argument, again dependent on manipulation of symbols without any consideration of special cases in which the manipulation fails, enabled Cayley to come to a false characterization of all the matrices L that commute with M. In fact, it was that very question that led Camille Jordan 10 years later to develop a fundamental classification of matrices by means of what today is called the Jordan Canonical Form.

21.4.3   Eigenvalues and Eigenvectors

Jordan's classification depends not on formal manipulation of matrices but on spectral theory, the results surrounding the concept of an eigenvalue. In modern terminology, an **eigenvalue** of a matrix is a solution λ either of the matrix equation $AX = \lambda X$, where A is an $n \times n$ matrix and X is an $n \times 1$ matrix, or of $XA = \lambda X$, where A is $n \times n$ and X is $1 \times n$. An **eigenvector** corresponding to the eigenvalue λ is a vector X, which satisfies the same equation. These concepts, in their origins and later development, were independent of matrix theory per se; they grew out of a study of various ideas that ultimately were included in that theory. Thus, the context within which the earliest eigenvalue problems arose during the eighteenth century was that of the solution of systems of linear differential equations with constant coefficients.

D'Alembert, in works dating from 1743 to 1758, and motivated by the consideration of the motion of a string loaded with a finite number of masses (here restricted for simplicity to three), considered the system

$$\frac{d^2 y_i}{dt^2} + \sum_{k=1}^{3} a_{ik} y_k = 0 \quad i = 1, \ 2, \ 3.$$

To solve this system, he multiplied the ith equation by a constant v_i for each i and added the equations together to obtain

$$\sum_{i=1}^{3} v_i \frac{d^2 y_i}{dt^2} + \sum_{i,k=1}^{3} v_i a_{ik} y_k = 0.$$

If the v_i are then chosen so that $\sum_{i=1}^{3} v_i a_{ik} + \lambda v_k = 0$ for $k = 1, \ 2, \ 3$, that is, if (v_1, v_2, v_3) is an eigenvector corresponding to the eigenvalue $-\lambda$ for the matrix $A = (a_{ik})$, the substitution $u = v_1 y_1 + v_2 y_2 + v_3 y_3$ reduces the original system to the single differential equation

$$\frac{d^2 u}{dt^2} + \lambda u = 0,$$

an equation which, after Euler's work on differential equations, was easily solved and led to solutions for the three y_i. A study of the three equations in which it appears shows that λ is determined by a cubic equation with three roots. D'Alembert realized that for the solutions to make physical sense they had to be bounded as $t \to \infty$. This, in turn, would only be true provided that the three values of λ were distinct, real, and positive.

It was Cauchy who first solved the problem of determining in a special case the nature of the eigenvalues from the nature of the matrix (a_{ik}) itself. In all probability, he was not influenced by d'Alembert's work on differential equations, but by the study of quadric surfaces, a study necessary as part of the analytic geometry that Cauchy was teaching from 1815 at the École Polytechnique. A quadric surface (centered at the origin) is given by an equation $f(x, y, z) = K$, where f is a ternary quadratic form. To classify such surfaces, Cauchy needed to find a transformation of coordinates under which f is converted to a sum or difference of squares. In geometric terms, this problem amounts to finding a new set of orthogonal axes in three-dimensional space by which to express the surface. But Cauchy then generalized the problem to quadratic forms in n variables, the coefficients of which can be written as a symmetric matrix. For example, the binary quadratic form $ax^2 + 2bxy + cy^2$ determines the symmetric 2×2 matrix

$$\begin{pmatrix} a & b \\ b & c \end{pmatrix}.$$

Cauchy's goal was to find a linear substitution on the variables such that the matrix resulting from this substitution was diagonal, a goal he achieved in a paper of 1829. Because the details in the general case are somewhat involved and because the essence of Cauchy's proof is apparent in the two-variable case, it is that case that we will consider here.

To find a linear substitution that converts the binary quadratic form $f(x, y) = ax^2 + 2bxy + cy^2$ into a sum of squares, it is necessary to find the maximum and minimum of $f(x, y)$ subject to the condition that $x^2 + y^2 = 1$. The point at which such an extreme value of f occurs is then a point on the unit circle that also lies on the end of one axis of one member of the family of ellipses (or hyperbolas) described by the equations $f(x, y) = k$. If one takes the line from the origin to that point as one of the axes and the perpendicular to that line as the other, the equation in relation to those axes will only contain the squares of the variables. By the principle of Lagrange multipliers, the extreme value occurs when the ratios $f_x/2x$ and $f_y/2y$ are equal. Setting each of these equal to λ gives the two equations

$$\frac{ax + by}{x} = \lambda \quad \text{and} \quad \frac{bx + cy}{y} = \lambda,$$

which can be rewritten as the system

$$(a - \lambda)x + by = 0$$
$$bx + (c - \lambda)y = 0.$$

Cauchy knew that this system has nontrivial solutions only if its determinant equals 0, that is, if $(a - \lambda)(c - \lambda) - b^2 = 0$. In matrix terminology, this equation is the characteristic equation $\det(A - \lambda I) = 0$, the equation that Cayley dealt with some 30 years later.

To see how the roots of the characteristic equation allow one to diagonalize the matrix, let λ_1 and λ_2 be those roots and (x_1, y_1), (x_2, y_2) be the corresponding solutions for x and y. Thus,

$$(a - \lambda_1)x_1 + by_1 = 0 \quad \text{and} \quad (a - \lambda_2)x_2 + by_2 = 0.$$

If one multiplies the first of these equations by x_2, the second by x_1, and subtracts, the result is the equation

$$(\lambda_2 - \lambda_1)x_1x_2 + b(y_1x_2 - x_1y_2) = 0.$$

Similarly, starting with the two equations involving $c - \lambda_i$, one arrives at the equation

$$b(y_2x_1 - y_1x_2) + (\lambda_2 - \lambda_1)y_1y_2 = 0.$$

Adding these two equations gives $(\lambda_2 - \lambda_1)(x_1x_2 + y_1y_2) = 0$. Therefore, if $\lambda_1 \neq \lambda_2$—and this is surely true in the case being considered, unless the original form is already diagonal—then $x_1x_2 + y_1y_2 = 0$. Because (x_1, y_1), (x_2, y_2) are only determined up to a constant multiple, one can arrange to have $x_1^2 + y_1^2 = 1$ and $x_2^2 + y_2^2 = 1$. In modern terminology, the linear substitution

$$x = x_1u + x_2v$$
$$y = y_1u + y_2v$$

is orthogonal. One easily computes that the new quadratic form arising from this substitution is $\lambda_1u^2 + \lambda_2v^2$ as desired. That λ_1 and λ_2 are real follows from assuming, on the contrary, that they are complex conjugates of one another. In that case, x_1 would be the conjugate of x_2 and y_1 that of y_2, and $x_1x_2 + y_2y_2$ could not be 0. Cauchy had therefore shown that all eigenvalues of a symmetric matrix are real and, at least in the case where they are all distinct, that the matrix can be diagonalized by the use of an orthogonal substitution.

21.4.4   Canonical Forms

The basic arguments of Cauchy's paper provided the beginnings to an extensive theory dealing with the eigenvalues of various types of matrices and with canonical forms. In general, however, throughout the middle of the nineteenth century, these results were all written in terms of forms, not in terms of matrices. Quadratic forms lead to symmetric matrices. The more general case of bilinear forms, functions of $2n$ variables of the form

$$\sum_{i,j=1}^{n} a_{ij}x_iy_j,$$

lead to general square matrices.

The most influential part of the theory of forms was worked out by Camille Jordan in his *Traité des substitutions*. Jordan came to the problem of classification, not through the study of bilinear forms, but through the study of the linear substitutions themselves. He had made a detailed study of Galois' work on solutions of algebraic equations and especially of his work on solving equations of prime power degree. These solutions involved the study of linear substitutions on these roots, substitutions whose coefficients could be considered to be elements of a finite field of order p. Such a substitution on the roots $x_1, x_2, \ldots x_n$ could be expressed in terms of a matrix A. In other words, if X represents the $n \times 1$ matrix of the roots x_i, then the substitution can be written as $X' \equiv AX \pmod{p}$. Jordan's aim was to find what he called a "transformation of indices" so that the substitution could be expressed in terms that were as simple as possible. In matrix notation, that means that he wanted to find an $n \times n$ invertible matrix P so that $PA \equiv DP$, where D is the "simplest possible" matrix. Thus, if $Y \equiv PX$, then $PAP^{-1}Y \equiv PAX \equiv DPX \equiv DY$ and the substitution on Y

is "simple." Using the characteristic polynomial for A, Jordan noted that if all of the roots of $\det(A - \lambda I) \equiv 0$ are distinct, then D could be taken to be diagonal, with the diagonal elements being the eigenvalues. On the other hand, if there are multiple roots, Jordan showed that a substitution can be found so that the resulting D is in block form,

$$
\begin{pmatrix}
D_1 & 0 & 0 & \cdots & 0 \\
0 & D_2 & 0 & \cdots & 0 \\
\vdots & \vdots & \vdots & \ddots & \vdots \\
0 & 0 & 0 & \cdots & D_m
\end{pmatrix},
$$

where each block D_i is a matrix of the form

$$
\begin{pmatrix}
\lambda_i & 0 & 0 & \cdots & 0 & 0 \\
\lambda_i & \lambda_i & 0 & \cdots & 0 & 0 \\
\vdots & \vdots & \vdots & \ddots & \vdots & \vdots \\
0 & 0 & 0 & \cdots & \lambda_i & \lambda_i
\end{pmatrix}
$$

and $\lambda_i \not\equiv 0 \pmod{p}$ is a root of the characteristic polynomial. The canonical form known today as the Jordan canonical form, where the values λ_i off of the main diagonal of the matrix are all replaced by 1s, was introduced by Jordan in 1871, when he realized that his method could be applied to the solution of systems of linear differential equations whose coefficients, instead of being taken from a field of p elements, were either real or complex numbers. Thus, Jordan returned, over a hundred years after the work of d'Alembert, to the origins of the entire complex of ideas associated with the eigenvalues of a matrix.

Jordan, however, did not use Cayley's single-letter notation to represent linear substitutions. It was Frobenius who in 1878 combined the ideas of his various predecessors into the first complete monograph on the theory of matrices. In particular, Frobenius dealt with various types of relations among matrices. For example, he defined two matrices A and B to be **similar** if there were an invertible form P such that $B = P^{-1}AP$ and **congruent** if a P existed with $B = P^t AP$, where P^t is the transpose of P. He showed that when two symmetric matrices were similar, the transforming matrix P could be taken to be **orthogonal**, that is, one whose inverse equaled its transpose. Frobenius then made a detailed study of orthogonal matrices and showed, among other things, that their eigenvalues were complex numbers of absolute value 1. Frobenius concluded his paper by showing the relationship between his symbolical matrix theory and the theory of quaternions. Namely, he determined four 2×2 matrices whose algebra was precisely that of the quantities $1, i, j,$ and k of quaternion algebra.

21.4.5 Solutions of Systems of Equations

Frobenius was also responsible for clarifying the question of the nature of the set of solutions to a system of linear equations, a special case of which Euler had considered many years earlier. Recall that Euler had been bothered because a particular system did not determine a specific value for each of the unknowns. He did realize, of course, that it was the vanishing of the determinant of the system's matrix of coefficients that prevented that system from having a unique solution. By the middle years of the nineteenth century, mathematicians were asking different questions; they wanted to determine not only when a system of m linear equations in n unknowns had solutions but also the size of the set of solutions. From experience with determinants, they learned that if one extracted from such a system a subsystem of k equations with a nonvanishing $k \times k$ determinant in its matrix of coefficients, that subsystem could be solved, although it would not have a unique solution. There were then other conditions on the determinants of the original system that would determine the nature of the solution set to that system, or indeed whether it could be solved at all. In the case where there were more equations than unknowns, it was understood that, in general, there would not be solutions; the system was overdetermined. So the major concern was with systems in which $n \geq m$, a system we will write in Cayley's notation in the form $AX = B$, with A an $m \times n$ matrix.

One of the first to concern himself with the theory behind these notions was Henry J. S. Smith (1826–1883), the Savilian Professor of Geometry at Oxford University. In a paper of 1861 he developed two basic concepts, what he called the index of indeterminates of a homogeneous system of equations $AX = 0$ and the idea of a complete set of independent solutions of such a system, both concepts only discussed in the case where $n > m$ and A had a nonvanishing determinant of order m. The former concept is the excess of the number of unknowns above the number of "really independent equations," that is, $n - m$. Smith then showed that there was a set of $n - m$ solutions $(x_{i1}, x_{i2}, \ldots, x_{in})$, $i = 1, 2, \ldots, n - m$, such that any solution is a linear combination of these and such that the determinants of order m of the matrix (x_{ij}) are not all zero. In fact, he found numerous ways of actually determining this complete set of independent solutions. In addition, Smith noted that to solve a nonhomogeneous system $AX = B$, where $B \neq 0$, one just needed to find one particular solution X^* and then any solution would be expressible in the form $X^* + X$, where X is a solution of the corresponding homogeneous system.

Although Smith used the phrase "really independent equations," he did not consider the case where that number was smaller than the actual number of equations, or, equivalently, the case where the maximal order of a nonzero determinant of the matrix A was less than m. That case was thoroughly covered by Charles L. Dodgson (1832–1898), more commonly known today as Lewis Carroll, in his *An Elementary Treatise on Determinants* of 1867 (Fig. 21.8). There he discussed conditions on both the $m \times n$ matrix A of the linear system $AX = B$ and the $m \times (n + 1)$ **augmented matrix** $(A|B)$ of the system that determined whether the system was consistent or inconsistent. Furthermore, he stated and proved a very general theorem that specified the nature of the set of solutions of an arbitrary system:

FIGURE 21.8

Lewis Carroll (Charles Dodgson) on a stamp from the Republic of Mali

DODGSON'S THEOREM   *If there are m equations, containing n variables ($n \geq m$), and if there are among them r equations which have a nonvanishing order r determinant of their unaugmented matrix; and if when these r equations are taken along with each of the remaining equations successively, each set of r + 1 equations has every order r + 1 determinant of its augmented matrix equal to zero, then the equations are consistent. If any nonvanishing order r determinant of the system of r equations is selected, then the n − r variables whose coefficients are not contained in it may have arbitrary values assigned to them. For each such set of arbitrary values, there is only one set of values for the other variables, and the remaining equations are dependent on these r equations.*[53]

Dodgson's proof of this result was very constructive, and he proceeded to give several examples. Thus, consider the system of four equations in five unknowns:

$$u + v - 2x + y - z = 6$$
$$2u + 2v - 4x - y + z = 9$$
$$u + v - 2x \qquad\qquad = 5$$
$$u - v + x + y - 2z = 0$$

Dodgson noted that there is a nonvanishing order 2 determinant for the first two equations, that there is no nonvanishing order 3 determinant for the first three equations, but that there is a nonvanishing order 3 determinant for the system consisting of equations 1, 2, and 4. Thus, he concludes, those equations are consistent, and, since they contain 5 variables, there are $5 - 3 = 2$ variables to which one can assign arbitrary values. In addition, equation 3 is dependent on equations 1 and 2.

The notion of rank is implicit in Dodgson's theorem, but it was Georg Frobenius (1849–1917) who was able to abstract this concept from the work of his predecessors in 1879: "If in a determinant all minors of order $r + 1$ vanish, but not all of those of order r are zero, then I call r the **rank** of the determinant."[54] A few years earlier, he had also clarified Smith's notion of "really independent equations" and defined the notion of **linear independence**, both for equations and for n-tuples representing solutions to a system: In a homogeneous system, the solutions $(x_{11}, x_{12}, \ldots, x_{1n})$, $(x_{21}, x_{22}, \ldots, x_{2n})$, ..., $(x_{k1}, x_{k2}, \ldots, x_{kn})$ are independent when $c_1 x_{1j} + c_2 x_{2j} + \cdots + c_k x_{kj}$ cannot be zero for $j = 1, 2, \ldots, n$ without all of the c_i being zero. To define independence for equations, Frobenius set up a duality relationship. To a given system of linear homogeneous equations, he associated a new system for which the coefficients of the equations constituted a basis for the solutions of the original system. Thus, n-tuples and equations were similar objects seen from two different points of view. He then demonstrated that if the rank of a system of m equations in n variables was r, one can find a set of $n - r$ independent solutions. Reversing the roles of coefficients and coordinates of the solution set, he then found the associated system, which had rank $n - r$, and showed that this new system itself has an associated system with the same solutions as the original system.

For example, we have already seen that the homogeneous system

$$u + v - 2x + y - z = 0$$
$$2u + 2v - 4x - y + z = 0$$
$$u + v - 2x \qquad\qquad = 0$$
$$u - v + x + y - 2z = 0$$

has rank 3. Thus, it has a set of 2 independent solutions, and a basis for the set of solutions can be taken to be $(1, 3, 2, 0, 0)$ and $(1, -1, 0, 2, 2)$. The associated system is then

$$u + 3v + 2x \qquad\qquad = 0$$
$$u - v \qquad + 2y + 2z = 0.$$

This system has rank 2, with a basis for the set of solutions given by $(1, 1, -2, 0, 0)$, $(3, -1, 0, -2, 0)$, $(3, -1, 0, 0, -2)$. It is then straightforward to show that the system associated with these solutions, namely,

$$u + v - 2x \qquad\qquad = 0$$
$$3u - v \qquad - 2y \qquad = 0,$$
$$3u - v \qquad\qquad - 2z = 0$$

has the same solutions as the original system.

Although Frobenius had completed the study of solutions to systems of equations, as well as the properties of various special classes of matrices, it was not until the beginning of the twentieth century that textbooks appeared in which all of that material was organized in the terminology of matrices. And it was not until the fourth decade of the century that the fundamental relationship of matrices to linear transformations of vector spaces was explicitly recognized. For that to happen, it was necessary for the abstract idea of a vector space to be made explicit. Because that development grew out of certain geometric ideas, its discussion is postponed until Chapter 24.

21.4.6   Systems of Linear Inequalities

Although the theory of linear equations was worked out in detail in the nineteenth century, there was only limited progress in what we would consider a related subject, that of systems of linear inequalities. The first one to deal with these ideas in some detail was Joseph Fourier (1768–1830) in the 1820s. Fourier was interested in various types of problems in which inequalities appeared, including problems in mechanics, probability, elections, and the minimization of errors in a statistical context. In this context, Fourier worked out both algebraic and geometric methods of finding the region of solutions. For example, using a system of six inequalities in two variables, he found the convex polygon of feasible solutions (Fig. 21.9). In addition, Fourier gave indications of an interest in what is today called linear programming, that is, in problems that require the finding of not only the "feasible region" but also an optimal point of some sort. In one example, in the case of three variables, where the linear inequalities defined half-planes in space, he first found the point of intersection furthest from the origin, then descended down from plane to plane, edge by edge, until the desired maximum or minimum value was found. Unfortunately, although Fourier's work was noted by others over the next few years, the technical difficulties involved in actually solving these linear programming problems at the time caused the theory to die out until the twentieth century.

 21.5   GROUPS AND FIELDS—THE BEGINNING OF STRUCTURE

In the first four sections of this chapter, we have seen that the ideas of solving equations and manipulating symbols according to the laws of arithmetic were still central to the notion of algebra in the nineteenth century. Yet at the same time, the notion of abstracting some important ideas out of these various concrete situations was beginning to take hold, and the idea of a mathematical structure started to appear. Thus, Dedekind developed the notion of an ideal. But although this was an abstract concept, it was only defined as a subset of a domain of algebraic integers and not independently as a set in an abstract ring. And Frobenius seemed to understand implicitly the idea of a vector space and a basis but could only deal with this notion in terms of n-tuples. The first mathematical structure to be fully understood was that of a group. In his early work on number theory and the solvability of equations by radicals, Gauss seemed to understand implicitly certain concepts of group theory. As the century wore on, other mathematicians began to bring these structural aspects of the subject into the open.

FIGURE 21.9

Polygon of feasible solutions
to a system of six inequalities
in two variables

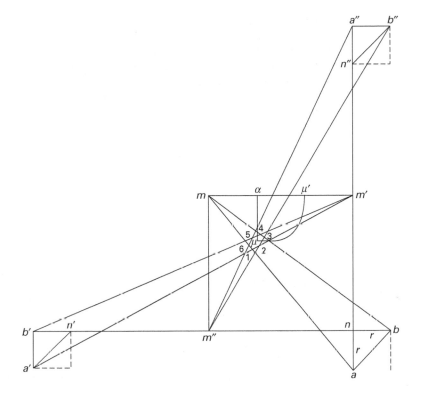

21.5.1 Gauss and Quadratic Forms

As we have already noted, Gauss discussed the theory of quadratic forms in chapter 5 of his *Disquisitiones*. Gauss's primary aim in his discussion of forms was to determine whether a given integer can be represented by a particular form. As a tool in the solution of this problem, he defined equivalence of two forms. A form $f = ax^2 + 2bxy + cy^2$ is **equivalent** to a form $f' = a'x'^2 + 2b'x'y' + cy'^2$ if there exists a linear substitution $x = \alpha x' + \beta y'$, $y = \gamma x' + \delta y'$ with $\alpha\delta - \beta\gamma = 1$ that transforms f into f'. An easy calculation shows that any two equivalent forms have the same discriminant $b^2 - ac$. On the other hand, two forms with the same discriminant are not necessarily equivalent. Gauss was able to show that for any given value D of the discriminant there were finitely many classes of equivalent forms. In particular, there was a distinguished class, the principal class, consisting of those forms equivalent to the form $x^2 - Dy^2$.

To investigate these classes, Gauss presented the rule of composition for forms. That is, given forms f, f', of the same discriminant, Gauss defined, as we have seen in Section 21.4.1, a new form F composed of f, f' (written $F = f + f'$) that had certain desirable properties. First, Gauss showed that if f and g are equivalent and if f' and g' are equivalent, then $f + f'$ is equivalent to $g + g'$. Therefore, the composition operation is an operation on classes. Gauss next showed that the operation of composition is both commutative and associative. Finally,

BIOGRAPHY

Leopold Kronecker (1823–1891)

K ronecker, attempting to get the best mathematics education possible, studied at the Universities of Berlin, Bonn,
and Breslau, before receiving his doctorate in 1845 at Berlin.
For several years thereafter, he managed the family business,
ultimately becoming financially independent. Having carried
on mathematical research as a hobby, he was elected in 1861

to the Berlin Academy and permitted to lecture at the University. In 1880, he took over the editorship of Crelle's *Journal* and
three years later, on Kummer's retirement, became professor of
mathematics in Berlin and, with Karl Weierstrass, codirector
of the influential mathematics seminar there.

Gauss showed that "if any class K is composed with the principal class, the result will be
the class K itself," that for any class K there is a class L (opposite to K) such that the
composite of the two is the principal class, and that "given any two classes K, L of the same
[discriminant], . . . we can always find a class M with the same [discriminant] such that L
is composed of M and K." Given that composition enjoys the basic properties of addition,
Gauss noted that "it is convenient to denote composition of classes by the addition sign, $+$,
and identity of classes by the equality sign."[55]

With the addition sign as the sign of operation, Gauss designated the composite of a class
C with itself by $2C$, the composite of C with $2C$ as $3C$, and so on. Gauss then proved that
for any class C, there is a smallest multiple mC that is equal to the principal class and that, if
the total number of classes is n, then m is a factor of n. Naturally, this result reminded him of
earlier material in the *Disquisitiones*. "The demonstration of the preceding theorem is quite
analogous to the demonstrations [on powers of residue classes] and, in fact, the theory of the
[composition] of classes has a great affinity in every way with the subject treated [earlier]."[56]
He could therefore assert, without proof, various other results coming from this analogy, in
terms of what is now the theory of Abelian groups.

21.5.2   Kronecker and the Structure of Abelian Groups

Gauss, although he recognized the analogy between his two treatments, did not attempt to
develop an abstract theory of groups. This development took many years. In the mid-1840s,
Kummer, in working out his theory of ideal complex numbers, noted that it was analogous in
many respects to Gauss's theory of forms. In particular, Kummer defined an equivalence of
ideal complex numbers that partitioned them into classes whose properties were analogous to
those of Gauss's classes of forms. But it was Kummer's student, Leopold Kronecker (1823–
1891), who finally saw that an abstract theory could be developed out of these analogies.

In a paper of 1870 in which he developed certain properties of the number of classes of
Kummer's ideal complex numbers, Kronecker recalled Gauss's work on quadratic forms:

> The very simple principles on which Gauss's method rests are applied not only in the given
> context but also frequently elsewhere, in particular in the elementary parts of number theory.
> This circumstance shows, and it is easy to convince oneself, that these principles belong to a
> more general, abstract realm of ideas. It is therefore appropriate to free their development from all

unimportant restrictions, so that one can spare oneself from the necessity of repeating the same argument in different cases. This advantage already appears in the development itself, and the presentation gains in simplicity, if it is given in the most general admissible manner, since the most important features stand out with clarity.[57]

Kronecker thus began to develop the simple principles: "Let θ', θ'', θ''', . . . be finitely many elements such that to each pair there is associated a third by means of a definite procedure." Kronecker required that this association, which he first wrote as $f(\theta', \theta'') = \theta'''$ and later as $\theta' \cdot \theta'' = \theta'''$, be commutative and associative and that, if $\theta'' \neq \theta'''$, then $\theta'\theta'' \neq \theta'\theta'''$. From the finiteness assumption, Kronecker then deduced the existence of a unit element 1 and, for any element θ, the existence of a smallest power n_θ such that $\theta^{n_\theta} = 1$.

Finally, Kronecker stated and proved what is now called the fundamental theorem of Abelian groups, that there exists a finite set of elements θ_1, θ_2, . . . , θ_m such that every element θ can be expressed uniquely as a product of the form $\theta_1^{h_1}\theta_2^{h_2} \cdots \theta_m^{h_m}$, where for each i, $0 \leq h_i < n_{\theta_i}$. Furthermore, the θ_i can be arranged so that each n_θ is divisible by its successor and the product of these numbers is precisely the number of elements in the system. Having demonstrated the abstract theorem, Kronecker interpreted the elements in various ways, noting that analogous results in each case had been proved previously by others.

21.5.3 Cayley and the Definition of a Group

Interestingly enough, Kronecker did not give a name to the system he defined, nor did he interpret it in terms of the permutation groups arising from Galois theory, perhaps because he was dealing solely with commutative groups. Kronecker was also probably unaware that, 16 years earlier, Cayley had developed a similar abstract theory based on the notion of groups of substitutions.

Cayley, in his "On the Theory of Groups," noted that the idea of a group of permutations was due to Galois and proceeded to generalize it to any set of operations, or functions, on a set of quantities. He used the symbol 1 to represent the function that leaves all quantities unchanged and noted that for functions, there is a well-defined notion of composition that is associative, although not, in general, commutative. But then Cayley abstracted the basic ideas out of the concrete notion of operations and defined a **group** to be a "set of symbols, 1, α, β, . . . , all of them different, and such that the product of any two of them (no matter in what order), or the product of any one of them into itself, belongs to the set. . . . It follows that if the entire group is multiplied by any one of the symbols, either [on the right or the left], the effect is simply to reproduce the group."[58] From a modern point of view, Cayley has left out a significant portion of the definition, and thus it is not clear why the last statement "follows." But at the same time Cayley wrote this paper, he also wrote another, on the theory of caustics, in which he considered a set of six functions on a particular set that formed, under composition, what he now called a group. Thus, it appears that in writing his initial paper on groups, he was thinking about finite collections of symbols whose composition was always assumed to be associative and, as he even wrote in the introduction to the group theory paper, that " if $\theta = \phi$, then, whatever the symbols α, β may be, $\alpha\theta\beta = \alpha\phi\beta$, and conversely." This "converse" statement is usually referred to as the cancellation law and implies, assuming finiteness, that every symbol has an inverse, another symbol whose product with the first was 1. And then it does "follow" that multiplication by any symbol produces a permutation of the group elements.

To study different abstract groups, Cayley introduced the group table,

	1	α	β	$\cdots$
1	1	α	β	$\cdots$
α	α	α^2	$\beta\alpha$	$\cdots$
β	β	$\alpha\beta$	β^2	$\cdots$
$\vdots$	$\vdots$	$\vdots$	$\vdots$	$\ddots$

in which, as he claimed in his definition, each row and each column contains all the symbols of the group. Furthermore, Cayley noted that each element θ satisfies the symbolic equation $\theta^n = 1$, if n is the number of elements in the group.

Cayley showed by a familiar argument that if n is prime, then the group is necessarily of the form $1, \alpha, \alpha^2, \ldots, \alpha^{n-1}$. If n is not prime, there are other possibilities. In particular, he displayed the group tables of the two possible groups of four elements and the two possible groups of six elements. In another paper of 1859, he described all five groups of order 8 by giving a list of their elements and defining relations as well as the smallest power (index) of each element that equals 1. For example, one of these groups contains the elements $1, \alpha, \beta, \beta\alpha, \gamma, \gamma\alpha, \gamma\beta, \gamma\beta\alpha$ with the relations $\alpha^2 = 1, \beta^2 = 1, \gamma^2 = 1, \alpha\beta = \beta\alpha, \alpha\gamma = \gamma\alpha$, and $\beta\gamma = \gamma\beta$. Each element in this group, except the identity, has index 2.

Although Cayley wrote an article for the *English Cyclopedia* in 1860 in which he explained the term "group," among others, no Continental mathematician over the next few years paid attention to this (nearly) abstract definition. On the other hand, in his lectures on Galois theory in 1856, Dedekind abstracted the identical notion of a finite group out of Galois' own, somewhat imprecise, definition dealing solely with permutations. As Dedekind wrote in his lecture notes, "the results [derived about permutations] are therefore valid for any finite domain of elements, things, concepts $\theta, \theta', \theta'', \ldots$ admitting an arbitrarily defined composition $\theta\theta'$, for any two given elements θ, θ', such that $\theta\theta'$ is itself a member of the domain, and such that this composition satisfies [associativity and right and left cancelability]. In many parts of mathematics, and especially in the theory of numbers and in algebra one often finds examples of this theory."[59]

In 1878, however, Cayley published four new papers on the same subject in which he repeated his definition and results of 1854. In particular, he wrote that "a group is defined by means of the laws of combinations of its symbols."[60] Further, "although the theory [of groups] as above stated is a general one, including as a particular case the theory of substitutions, yet the general problem of finding all the groups of a given order n is really identical with the apparently less general problem of finding all the groups of the same order n which can be formed with the substitutions upon n letters."[61] Cayley took this result, today known as Cayley's theorem, as nearly obvious, noting merely that any element of a group may be thought of as acting upon all the elements of the group by the group operation, such operation inducing a permutation of the group elements. However, Cayley noted, this "does not in any wise show that the best or the easiest mode of treating the general problem is thus to regard it as a problem of substitutions; and it seems clear that the better course is to consider the general problem in itself, and to deduce from it the theory of groups of substitutions."[62] Thus,

Cayley, like Kronecker, realized that problems in group theory could often best be attacked by considering groups in the abstract, rather than in their concrete realizations. In fact, it is often only by dealing with the abstractions that one can make further progress.

21.5.4 The Axiomatization of the Group Concept

Beginning in 1879, many mathematicians began to realize that it was worthwhile to combine Kronecker's and Cayley's definitions into a single abstract group concept. In particular, in 1882 Walther von Dyck (1856–1934) published his "Gruppentheoretische Studien" in which he formulated the basic problem: "To define a group of discrete operations, which are applied to a certain object, while one ignores any special form of representation of the individual operations, regarding these only as given by the properties essential for forming the group."[63] Dyck, although he alluded to the associative and inverse properties, did not give these as defining properties of a group. Instead, he showed how to construct a group by use of generators and relations. Namely, he began with a finite number of operations, $A_1, A_2, \ldots, A_m$, then built the "most general" group G on these elements by considering all possible products of powers of these elements and their inverses. This group, today called the **free group** on $\{A_i\}$, automatically satisfies the modern group axioms. Dyck next specialized to other groups by assuming various relations of the form $F(A_1, A_2, \ldots, A_m) = 1$. In fact, he showed that if the group $\bar{G}$ is formed from operations $\bar{A}_1, \bar{A}_2, \ldots, \bar{A}_m$, which satisfy the given relations, then "all these infinitely many operations of the group G, which are equal to the identity in $\bar{G}$, form a [sub]group H and this . . . commutes with all operations $S, S', \ldots$ of the group G."[64] Dyck then proved that the mapping $A_i \to \bar{A}_i$ defined what he called an **isomorphism** from G onto $\bar{G}$. In modern terminology, Dyck had shown that the subgroup H is normal in G and that $\bar{G}$ is isomorphic to the factor group G/H.

A second paper of the same year, this one by Heinrich Weber (1842–1913) on quadratic forms, was the first to give a complete axiomatic description of a finite group without any reference to the nature of the elements composing it:

A system G of h elements of any sort $\theta_1, \theta_2, \ldots, \theta_h$ is called a group of order h if it satisfies the following conditions:

   I. Through some rule, which is called composition or multiplication, one derives from any two elements of the system a new element of the same system. In signs, $\theta_r \theta_s = \theta_t$.

  II. Always $(\theta_r \theta_s)\theta_t = \theta_r(\theta_s \theta_t) = \theta_r \theta_s \theta_t$.

 III. From $\theta \theta_r = \theta \theta_s$ and from $\theta_r \theta = \theta_s \theta$, there follows $\theta_r = \theta_s$.[65]

From the given axioms and the finiteness of the group, Weber derived the existence of a unique unit element and, for each element, the existence of a unique inverse. He further defined a group to be an **Abelian group** if the multiplication is commutative and then proved the fundamental theorem of Abelian groups by essentially the same method used by Kronecker.

Although the use of the abstract group concept became more common over the next several years, it was not until 1893 that Weber published a definition that included infinite groups. He repeated his three conditions of 1882 and noted that if the group is finite these suffice to ensure that if any two of the three group elements A, B, C are known, there is a unique solution to the equation $AB = C$. On the other hand, this conclusion is no longer valid for infinite groups. In that case, one must assume the existence of unique solutions to $AB = C$

as a fourth axiom. This fourth axiom, even without finiteness, implies a unique identity and unique inverses for every element of the group.

After defining the modern notion of isomorphism of groups, Weber illuminated the basis for his abstract approach: "One can combine all groups isomorphic to one another into a class of groups, which itself is again a group, whose elements are the generic characters which one obtains if one combines the corresponding elements of the individual isomorphic groups into a general concept. The individual isomorphic groups are then to be considered as different representatives of the generic concept, and it is irrelevant which representative one uses to study the properties of the group."[66] Weber produced many examples of groups, including the additive group of vectors in the plane, the group of permutations of a finite set, the additive group of residue classes modulo m, the multiplicative group of residue classes modulo m relatively prime to m, and the group of classes of binary quadratic forms of a given discriminant under Gauss's law of composition.

Interestingly, although Weber incorporated his abstract definition of groups into his 1895 algebra text, *Lehrbuch der Algebra*, he did not present the notion of a group as a central concept of algebra. For Weber, the central concept of the subject was still the solving of polynomial equations, and groups were a tool to help in that process. Thus, he discussed permutation groups in the first volume of the text in connection with his presentation of Galois theory. It was only in the second volume that Weber presented the general definition of a group. And although he was well aware of many instances of groups, he was evidently not entirely sure of the purpose of his general definition but knew that it would stimulate further research:

> The general definition of group leaves much in darkness regarding the nature of the concept. . . . The definition of group contains more than appears at first sight, and the number of possible groups that can be defined given the number of their elements is quite limited. The general laws concerning this question are barely known, and thus every new special group, in particular of a reduced number of elements, offers much interest and invites detailed research.[67]

By the time Weber's text appeared, many mathematicians were already taking up Weber's challenge and attempting to find general theorems about the structure of groups or determine all finite groups of a given order. For example, Frobenius in 1887 reproved abstractly the theorem that if p^n is the largest power of a prime p that divides the order of a finite group, then the group has at least one subgroup of order p^n and, in fact, that the number of such subgroups divides the order of the group and is congruent to 1 modulo p. This theorem had originally been proved by Ludvig Sylow (1832–1918) for permutation groups. Also, Otto Hölder (1859–1937), in his work on Galois theory in 1889, defined the notion of a **factor group** (or **quotient group**) and showed how these are Galois groups of the auxiliary equations that may come up in the process of solving a particular solvable equation. He further showed that in the composition series one gets in accomplishing the solution, the actual factor groups are unique, and not just their order. This result is now called the Jordan-Hölder theorem. Then in several papers in the early 1890s, Hölder studied finite groups of various orders and worked out some structure theorems. In particular, he determined the possible groups of orders p^2, p^3, p^4, pq, pq^2, and pqr, where p, q, r are distinct primes. The cases p^2 and pq had been proved earlier for permutation groups, but Hölder reworked the proofs in a more abstract manner. For example, given a group of order pq with $p > q$, the abstract version of the Sylow theorem shows that there is one subgroup of order p and, if q does not divide

$p - 1$, one subgroup of order q. In that case, the group is cyclic. On the other hand, if q does divide $p - 1$, then there is a second group that is generated by two elements S and T, with $S^q = 1$, $T^p = 1$, and $S^{-1}TS = T^r$, with $r \not\equiv 1 \pmod{p}$ and $r^q \equiv 1 \pmod{p}$.

21.5.5 The Concept of a Field

The story of field theory is much simpler to tell than that of group theory. The notion of a field is certainly implicit in Galois' work around 1830. Recall that Galois discussed what it meant for quantities to be rational and how to adjoin a new element to a given set of rational quantities. For Galois, the notions of the rational number field Q and of an extension field $Q(\alpha)$ generated by either a transcendental quantity or a root of a given equation were intuitively obvious, and there was no need to name this concept. It was Kronecker, beginning in the 1850s, who tried to be more specific in actually constructing these fields. Kronecker believed that algebra and analysis could be put on a more rigorous basis by basing all concepts on constructions beginning with the whole numbers: "God Himself made the whole numbers—everything else is the work of men."[68] Thus, he felt that irrational quantities like $\sqrt{2}$ made no sense unless one could find a definite way of constructing them out of the whole numbers. In terms of fields, then, he wanted to find a method of constructing extension fields of the rational numbers, or indeed of any already determined field, which would not depend on the prior existence of irrational numbers.

Kronecker began with the idea of a domain of rationality determined by certain elements R', R'', ... This domain included all the quantities that were rational functions of R', R'', . . . with integer coefficients. Thus, he assumed the existence of the integers and therefore the rational numbers. He was then able to solve his problem of adjoining $\sqrt{2}$ to a domain of rationality in which $x^2 - 2$ had no root by considering the remainders of polynomials with rational coefficients on division by $x^2 - 2$. Because two polynomials with the same remainder would be considered equal, it was straightforward to define the basic operations on this set of remainders and thereby construct a new domain of rationality. Another way of looking at this construction is simply to consider the new domain of rationality as containing a new element α as well as all rational functions of α, with the condition that α^2 is always replaced by 2.

Dedekind, also beginning in the 1850s, was more concerned with the set of elements itself than with the process of adjunction. Recall that Dedekind was interested in the arithmetic of algebraic integers, complex numbers that could be expressed as roots of algebraic equations. Thus, Dedekind gave the following definition in his supplement to the second edition of Dirichlet's *Vorlesungen* (1871): "A system A of real or complex numbers α is called a *field* if the sum, difference, product and quotient of every pair of these numbers α belongs to the same system A."[69] (He noted that 0 cannot be a denominator in any such quotient and that a field must contain at least one number other than 0.) The smallest such system, of course, is the field of rational numbers, which is contained in every field, while the largest such system is the field of complex numbers, which contains every field. Thus, for Dedekind, unlike Kronecker, the adjunction of an algebraic element to a field always took place in the field of complex numbers. In fact, given any set K of complex numbers, Dedekind defined the field $Q(K)$ to be the smallest field that contains all the elements K.

In a joint paper with Weber of 1882, Dedekind did, however, consider fields that were not subfields of the field of complex numbers, in particular, the field of rational functions over a field of complex numbers. Such a field was defined, as in his earlier definition, to be a system of such functions closed under the four basic operations (except, of course, division by 0). In this work, in fact, the authors exploited the analogy between algebraic integers and the integral polynomials, including the use of the theory of ideals, to formulate results on the factorization of polynomials. Yet even here, as in his earlier work, Dedekind only considered fields containing the field of rational numbers. Neither he nor Kronecker attempted to extend the definition to other types of fields.

Interestingly, as far back as 1830 Galois had published a brief paper that in essence described finite fields. Galois' aim in that paper was to generalize the ideas of Gauss in solving congruences of the form $x^2 \equiv a \pmod{p}$. Galois asked what would happen if, when a solution did not exist, one created a solution, exactly as one created the solution i to $x^2 + 1 = 0$. Thus, designating a solution to an arbitrary congruence $F(x) \equiv 0 \pmod{p}$ by the symbol i (where $F(x)$ is of degree n and no residue modulo p is itself a solution), Galois considered the collection of p^n expressions $a_0 + a_1 i + a_2 i^2 + \cdots + a_{n-1} i^{n-1}$ with $0 \leq a_j < p$ and noted that these expressions can be added, subtracted, multiplied, and divided in the obvious manner. Galois next noted that if α is any of the nonzero elements of his set, some smallest power n of α must be equal to 1 and, by arguments analogous to those of Gauss in the case of residues modulo p, showed that all such elements satisfy $\alpha^{p^n - 1} \equiv 1$ and that there is a primitive root β such that every nonzero element is a power of β. Galois concluded the paper by showing that for every prime power p^n, one can find an irreducible nth-degree congruence modulo p, a root of which generates what is today called the **Galois field of order** p^n. The simplest way to find such a polynomial, Galois remarked, is by trial and error. As an example, he showed that $x^3 - 2$ is irreducible modulo 7 and therefore that the set of elements $\{a_0 + a_1 i + a_2 i^2\}$, with i a zero of that polynomial and $0 \leq a_j < 7$ for $j = 0, 1, 2$, forms the field of order 7^3.

It was Heinrich Weber who finally combined the Dedekind-Kronecker version of a field with the finite systems of Galois into an abstract definition of a field in the same paper of 1893 in which he gave an abstract definition of a group. In fact, he used the notion of group in his definition. Thus, a **field** was a set with two forms of composition, addition and multiplication, under the first of which it was a commutative group and under the second of which the set of nonzero elements formed a commutative group. Furthermore, the two forms of composition are related by the following rules: $a(-b) = -ab$; $a(b + c) = ab + ac$; $(-a)(-b) = ab$; and $a \cdot 0 = 0$. Weber further noted that in a field a product can only be zero when one of the factors is zero. He then gave several examples of fields, including the rational numbers, the finite fields (of which he only cited the residue classes modulo a prime), and the "form fields," the fields of rational functions in one or more variables over a given field F. As in the case of groups, although Weber presented an abstract definition of fields in his textbook of 1895, he essentially used fields there just as a tool in his discussion of Galois theory. Thus, the only fields he considered in the text were subfields of the fields of complex numbers.

We can consider Weber as being on the cusp of what was to become the structural revolution in algebra. He was the first to give fully abstract definitions of groups and fields. Already by his time, group theory was becoming a completely abstract discipline in which one studied a particular type of mathematical structure, a set of elements with a well-defined operation satisfying certain axioms. And Weber certainly understood a central consequence of the structural approach, that two isomorphic groups were in essence the same mathematical object. He also knew that there were many examples of groups and fields and that their unification under a single definition was critical for progress in mathematics. But in putting the ideas of algebra into a form for students, he still reverted to the earlier idea that algebra was about the solving of equations, with groups and fields coming into play as tools in dealing with that issue. Weber's text became the standard text in Germany for the next 30 years. Yet even as it was being used in the universities, research mathematicians were pushing the old definition of algebra into the background and introducing the idea that algebra was the study of mathematical structures. We will thus continue this discussion in Chapter 25.

EXERCISES

1. Prove that if p is prime and $0 < a < p$, then the smallest exponent m such that $a^m \equiv 1 \pmod{p}$ is a divisor of $p - 1$.

2. For the prime $p = 7$, calculate for each integer a with $1 < a < 7$ the smallest exponent m such that $a^m \equiv 1 \pmod{7}$. Show that the theorem in Exercise 1 holds for all a.

3. Determine the primitive roots of $p - 13$, that is, determine numbers a for which $p - 1$ is the smallest exponent such that $a^{p-1} = 1 \pmod{p}$.

4. Complete Gauss's determination that 453 is a quadratic residue modulo 1236 by showing that

   a. If $x^2 \equiv 453 \pmod{4}$, $x^2 \equiv 453 \pmod{3}$, and $x^2 \equiv 453 \pmod{103}$ are all solvable, then so is $x^2 \equiv 453 \pmod{4 \cdot 3 \cdot 103}$.

   b. 453 is a quadratic residue modulo both 4 and 3.

   c. $(\frac{453}{103}) = (\frac{41}{103})$.

   d. $(\frac{5}{41}) = 1$.

5. Show that the Gaussian integer $a + bi$ ($b \neq 0$) is prime if and only if the norm $a^2 + b^2$ is an ordinary prime.

6. Show that if any Gaussian prime p divides the product $abc \cdots$ of Gaussian primes, then p must equal one of those primes, or one of them multiplied by a unit. (*Hint:* Take norms of both sides.)

7. Factor $3 + 5i$ as a product of Gaussian primes.

8. Use Germain's theorem to show that if there is a solution to the Fermat equation for exponent 3, then one of x, y, or z must be divisible by 9. To do this, show, first, that 3 is not a cube modulo 7 and, second, that no two nonzero third-power residues modulo 7 differ by 1.

9. Show that a prime complex integer is irreducible.

10. Show that in the domain of integers of the form $a + b\sqrt{-17}$, Liouville's factorization $169 = 13 \cdot 13 = (4 + 3\sqrt{-17})(4 - 3\sqrt{-17})$ in fact demonstrates that unique factorization into primes fails in that domain. (*Hint:* Use norms to show that each of the four factors is irreducible.)

11. Show that the Gaussian integers form a Euclidean domain. That is, show that, given two Gaussian integers z, m, there exist two others, q, r, such that $z = qm + r$ and $N(r) < N(m)$.

12. Show that the domain of complex integers of the form $a + b\sqrt{-2}$ is Euclidean. First, determine explicitly the analogue of the Euclidean algorithm in this domain.

13. Show that in the domain of complex integers of the form $a + b\sqrt{-5}$, the integers 2, 3, $-2 + \sqrt{-5}$, $-2 - \sqrt{-5}$, $1 + \sqrt{-5}$, $1 - \sqrt{-5}$ are all irreducible.

14. Show that in the domain of complex integers of the form $a + b\sqrt{-5}$, the principal ideal (2) is equal to A^2, where A is the ideal $(2, 1 + \sqrt{-5})$.

15. Determine the cosets of the cyclic subgroup of order 6 of the cyclic group of order 18.

16. Use Gauss's method to solve the cyclotomic equation $x^6 + x^5 + x^4 + x^3 + x^2 + x + 1 = 0$.

17. In the example dealing with Gauss's solution to $x^{19} - 1 = 0$, show that α_1, α_2, and α_4 are roots of the cubic equation $x^3 + x^2 - 6x - 7 = 0$. (*Hint:* Use the relationship between the coefficients of this equation and the symmetric functions of the roots.)

18. In the example dealing with Gauss's solution to $x^{19} - 1 = 0$, show that β_1, β_8, and β_7 are roots of the cubic equation $x^3 - \alpha_1 x^2 + (\alpha_2 + \alpha_4)x - 2 - \alpha_2 = 0$, where the α's and β's are as in the text.

19. In the example dealing with Gauss's solution to $x^{19} - 1 = 0$, show that r and r^{18} are both roots of $x^2 - \beta_1 x + 1 = 0$, where r and β_1 are as in the text.

20. Calculate the Galois group G of the equation $x^3 + 6x = 6$ over the rational numbers. Show that this group has a normal subgroup H such that both H and the index of H in G are primes.

21. Show that the Galois group from Exercise 20 can be expressed in the form $x_k \to x_k'$ with $k' \equiv ak + b \pmod{3}$, where the three roots of the polynomial are x_0, x_1, and x_2 and where $a = 0$, 1 and $b = 0$, 1, 2.

22. Show that the Galois group of the equation $x^5 - 2$ over the rational numbers can be expressed as the group of substitutions of the form $x' \equiv ax + b \pmod{5}$ and therefore has 20 elements.

23. Find a fifth-degree polynomial that is not solvable by radicals.

24. Show that the order of the group $SL(2, p)$ is $p(p^2 - 1)$ and that of $PSL(2, p)$ is $\frac{1}{2}p(p^2 - 1)$.

25. Show that the group $PSL(2, p)$ can be considered as the group of linear fractional transformations $z' = (az + b/cz + d)$ (mod p), with $ad - bc \equiv 1$ (mod p), acting on the one-dimensional projective space $P_1(p)$.

26. Show that Hamilton's laws of operation on number couples (α, β) mirror the analogous laws of operation on complex numbers $\alpha + \beta i$.

27. Let $\alpha = 3 + 4i + 7j + k$ and $\beta = 2 - 3i + j - k$ be quaternions. Calculate $\alpha\beta$ and α/β.

28. Define the modulus $|\alpha|$ of a quaternion $a + bi + cj + dk$ by $|\alpha| = a^2 + b^2 + c^2 + d^2$. Show that $|\alpha\beta| = |\alpha||\beta|$.

29. Determine the general form of the expansion of a function $f(x, y, z)$ of three logical variables into a polynomial in terms of the form $x'y'z'$, where x', for example, represents either x or $\bar{x}$. Use this expansion to expand the function $V = x - yz$.

30. Interpret the remaining three of Boole's equations $x\bar{y}z = 0$, $x\bar{y}\bar{z} = 0$, $\bar{x}yz = 0$ from the case in the text, where x stands for clean beasts, y for beasts that divide the hoof, and z for beasts that chew the cud.

31. Show that if the substitution $x = \alpha x' + \beta y'$, $y = \gamma x' + \delta y'$ with $\alpha\delta - \beta\gamma = 1$ transforms the quadratic form $F = ax^2 + 2bxy + cy^2$ into the form $F' = a'x'^2 + 2b'x'y' + c'y'^2$, then there is an "inverse" substitution of the same form that transforms F' into F.

32. Prove that if the product of two matrices is the zero matrix, then at least one of the factors has determinant 0.

33. Show explicitly the truth of the Cayley-Hamilton theorem that a matrix A satisfies its characteristic equation $\det(A - \lambda I) = 0$ in the case where A is a 2×2 matrix.

34. Show that the matrix

$$L = \begin{pmatrix} \frac{a+Y}{X} & \frac{b}{X} \\ \frac{c}{X} & \frac{d+Y}{X} \end{pmatrix},$$

where $X = \sqrt{a + d + 2\sqrt{ad - bc}}$ and $Y = \sqrt{ad - bc}$, is the square root of the matrix

$$M = \begin{pmatrix} a & b \\ c & d \end{pmatrix}.$$

35. Determine the conditions on the 2×2 matrix M of Exercise 34 so that a square root exists. How many square roots are there?

36. Determine the square root L of an arbitrary 3×3 matrix M. Begin by writing M in its Jordan canonical form. Then use the fact that $L^2 = M$ and that M satisfies its characteristic equation.

37. Use Cauchy's technique to find an orthogonal substitution that converts the quadratic form $2x^2 + 6xy + 5y^2$ into a sum or difference of squares.

38. Solve explicitly the system of linear equations

$$\begin{array}{rcrcrcrcrcl} 2u & + & v & + & 2x & + & y & + & 3z & = & 0 \\ 5u & + & 3v & - & 4x & + & 3y & - & 6z & = & 0 \\ u & + & v & - & 8x & + & y & - & 12z & = & 0 \end{array}.$$

First determine the order of the maximal nonvanishing determinant in the matrix of coefficients.

39. Determine the rank of the matrix of coefficients in the system of equations in Exercise 38. Find a basis for the set of solutions of this system.

40. Using the result of Exercise 39, determine a system of linear equations that is associated with the system of Exercise 38. Find a basis for the set of solutions for this system and then show that the system associated with the new system has the same solution set as the system of Exercise 38.

41. Show that two equivalent quadratic forms have the same discriminant.

42. Describe the five distinct groups of order 8.

43. Show that there are exactly two groups of order p^2, both of which are Abelian.

44. Given a group of order pq, $(p > q)$, with $S^q = 1$ and $T^p = 1$, show that if $S^{-1}TS = T^r$, then $r^q \equiv 1$ (mod p).

45. Create a field of order 5^3 by finding a third-degree irreducible congruence modulo 5.

46. Compare Weber's definition of a group with the standard modern definition. Show that they are equivalent.

47. Compare Weber's definition of a field with the standard modern definition. Can some of Weber's axioms be proved from other ones?

48. Try to create a multiplication for number triples, written, say, in the form $\alpha + \beta i + \gamma j$, which satisfies Hamilton's criteria for a reasonable multiplication. Namely, the multiplication must satisfy the commutative and associative laws, must be distributive over addition, must allow unique division, and must satisfy the modulus multiplication rule. What problems do you run into?

49. Design a lesson introducing the concept of a group through the notion of

    a. a permutation of a finite set;
    b. composition of quadratic forms;
    c. the residue classes modulo a prime p.

50. Compare the advantages and disadvantages of introducing a class to algebraic number fields by Kronecker's method of construction and by Dedekind's method of considering subfields of the complex numbers.

51. Compare De Morgan's version of the laws of algebra with Weber's axioms for a field.

52. Design a lesson explaining negative numbers using either Peacock's principle of permanence of equivalent forms or Hamilton's formulation via pairs of positive numbers.

Which formulation would work better in a classroom? Why?

53. How do students in high school "understand" negative numbers? Do they understand why a negative times a negative is a positive? Is such an understanding necessary?

54. Design a lesson explaining complex numbers using Hamilton's ordered pairs.

55. Outline a lesson on manipulation of matrices following Cayley's 1858 treatment of the subject.

REFERENCES AND NOTES

General references on the history of algebra include the three works mentioned in the references to Chapter 19, the books of Nový, Wussing, and van der Waerden. A work dealing with the history of nineteenth- and twentieth-century algebra specifically is Leo Corry, *Modern Algebra and the Rise of Mathematical Structures* (Basel: Birkhäuser, 1996). Other useful works include two books by Harold Edwards, *Fermat's Last Theorem: A Genetic Introduction to Number Theory* (New York: Springer-Verlag, 1977) and *Galois Theory* (New York: Springer-Verlag, 1984), as well as Michael J. Crowe, *A History of Vector Analysis: The Evolution of the Idea of a Vectorial System* (Notre Dame, IN: University of Notre Dame Press, 1967). An excellent survey of the history of group theory is Israel Kleiner, "The Evolution of Group Theory: A Brief Survey," *Mathematics Magazine* 59 (1986), 198–215. There have been several excellent studies of British algebra, including Joan Richards, "The Art and Science of British Algebra: A Study in the Perception of Mathematical Truth," *Historia Mathematica* 7 (1980), 343–365; Helena Pycior, "George Peacock and the British Origins of Symbolical Algebra," *Historia Mathematica* 8 (1981), 23–45, and Ernest Nagel, " 'Impossible Numbers': A Chapter in the History of Modern Logic," *History of Ideas* 3 (1935), 429–474. These articles contain a much fuller picture of British algebra than could be presented in the text. The history of the theory of matrices is presented in three articles by Thomas Hawkins: "Cauchy and the Spectral Theory of Matrices," *Historia Mathematica* 2 (1975), 1–29; "Another Look at Cayley and the Theory of Matrices," *Archives Internationales d'Histoire des Sciences* 26 (1977), 82–112; and "Weierstrass and the Theory of Matrices," *Archive for History of Exact Sciences* 17 (1977), 119–163. A reading of these articles gives a very detailed picture of the development of matrix theory and related areas of mathematics. A briefer and more elementary survey of the history of matrix theory is R. W. Feldmann, "History of Elementary Matrix Theory," *Mathematics Teacher* 55 (1962), 482–484, 589–590, 657–659, and 56 (1963), 37–38, 101–102, 163–164.

Gauss's *Disquisitiones* is available in English as Carl Friedrich Gauss, *Disquisitiones Arithmeticae*, translated by Arthur A. Clarke (New York: Springer-Verlag, 1986). A perusal of the entire book is well worth the trouble. Some of Germain's work on Fermat's last theorem has been translated into English and is in Reinhard Laubenbacher and David Pengelley, *Mathematical Expeditions: Chronicles by the Explorers* (New York: Springer, 1999), ch. 4. Dedekind's work on ideal theory is in Richard Dedekind, *Theory of Algebraic Integers*, translated by John Stillwell (Cambridge: Cambridge University Press, 1996). Galois' "Memoir on the Conditions for Solvability of Equations by Radicals," translated by Harold Edwards, is found in Edwards, *Galois Theory*.

1. Ernst Kummer, "De numeris complexis, qui radicibus unitatis et numeris integris realibus constant," *Journal de Mathématiques Pures et Appliquées* 12 (1847), 185–212, reprinted in Kummer's *Collected Papers* (New York: Springer-Verlag, 1975), vol. 1, 165–192, p. 182.

2. Michael J. Crowe, *Vector Analysis*, p. 29.

3. Gauss, *Disquisitiones Arithmeticae*, p. 1.

4. Ibid., p. 35.

5. A recent biography of Gauss, which includes a bibliography of his works and a survey of the secondary literature, is W. K. Bühler, *Gauss: A Biographical Study* (New York: Springer-Verlag, 1981).

6. Translated from Gauss, *Untersuchungen über höhere Arithmetik* (New York: Chelsea, 1965), p. 540. This work consists of the German translations by H. Maser of various papers of Gauss on "higher arithmetic" and was originally published in 1889.

7. Quoted in Edwards, *Fermat's Last Theorem*, p. 61. Three recent articles on Sophie Germain are J. H. Sampson, "Sophie Germain and the Theory of Numbers," *Archive for History of Exact Sciences* 41 (1991), 157–161, Amy Dahan

Dalmédico, "Sophie Germain," *Scientific American* (December 1991), 117–122, and Reinhard Laubenbacher and David Pengelley, "'Voici ce que j'ai trouvé:' Sophie Germain's Grand Plan to Prove Fermat's Last Theorem," http://www.math.nmsu.edu/~davidp/germain.pdf (2007).

8. The story of the Paris Academy meeting of March 1, 1847, is told in more detail in Edwards, *Fermat's Last Theorem*, pp. 76–80.

9. For a more detailed discussion of divisors and of Kummer's work on Fermat's Last Theorem, consult chapter 4 of Edwards, *Fermat's Last Theorem*. See also H. M. Edwards, "The Genesis of Ideal Theory," *Archive for History of Exact Sciences* 23 (1980), 321–378.

10. Dedekind, *Theory of Algebraic Numbers*, p. 57.

11. Ibid., p. 88.

12. Ibid., p. 57. For more on the creation of ideals, see H. M. Edwards, "Dedekind's Invention of Ideals," in Esther Phillips, ed., *Studies in the History of Mathematics* (Washington, DC: MAA, 1987), pp. 8–20.

13. Gauss, *Disquisitiones*, p. 445.

14. Ibid., p. 459.

15. Two biographies of Abel are Oystein Ore, *Niels Henrik Abel: Mathematician Extraordinary* (New York: Chelsea, 1974), and Arild Stubhaug, *Niels Henrik Abel and His Times: Called Too Soon by Flames Afar*, Richard Daly, trans. (Berlin: Springer, 2000).

16. See Van der Waerden, *History of Algebra*, pp. 85–88, for details of the proof.

17. Quoted in Wussing, *Abstract Group Concept*, p. 98.

18. Ibid., p. 100.

19. Galois, "Memoir on the Conditions for Solvability of Equations by Radicals," in Edwards, *Galois Theory*, p. 102.

20. A detailed discussion of Galois' propositions and their proofs is in Edwards, *Galois Theory*. A further clarification is in H. M. Edwards, "A Note on Galois Theory," *Archive for History of Exact Sciences* 41 (1991), 163–169.

21. The best recent article on the life of Galois is T. Rothman, "Genius and Biographers: The Fictionalization of Evariste Galois," *American Mathematical Monthly* 89 (1982), 84–106. A recent biography, which includes Galois' mathematical work, is Laura Toti Rigatelli, *Evariste Galois: 1811–1832*, John Denton, trans. (Basel: Birkháuser, 1996). The fictionalized biography is Leopold Infeld, *Whom the Gods Love: The Story of Evariste Galois* (New York: Whittlesey House, 1948). Another biography is found in E. T. Bell, *Men of Mathematics* (New York: Simon and Schuster, 1937).

22. Quoted in Rothman, "Fictionalization of Evariste Galois," p. 97.

23. R. Bourgne and J. P. Azra, eds., *Ecrits et Mémoires Mathématiques d'Evariste Galois* (Paris: Gauthier-Villars, 1962), p. 185.

24. Ibid., p. 175.

25. Camille Jordan, *Traité des substitutions et des équations algébriques* (Paris: Gauthier-Villars, 1870), sec. 54.

26. Ibid., sec. 402. For more information on the history of Galois theory, see B. Melvin Kiernan, "The Development of Galois Theory from Lagrange to Artin," *Archive for History of Exact Sciences* 8 (1971), 40–154.

27. William Rouse Ball, *The Origin and History of the Mathematical Tripos* (Cambridge: Cambridge University Press, 1880), p. 195.

28. George Peacock, *A Treatise on Algebra*, reprint edition (New York: Scripta Mathematica, 1940), vol. 2, p. 59.

29. Quoted in Pycior, "George Peacock," p. 35.

30. Peacock, *Treatise on Algebra*, p. 453.

31. Ibid., p. 449.

32. Augustus De Morgan, "On the Foundation of Algebra, No. II," *Transactions of the Cambridge Philosophical Society* 7 (1839–1842), 287–300, p. 287.

33. Augustus De Morgan, *Trigonometry and Double Algebra* (London: Taylor, 1849), pp. 92–93. This is quoted in Nagel, "Impossible Numbers," pp. 185–186.

34. See Abraham Arcavi and Maxim Bruckheimer, "The Didactical De Morgan: A Selection of Augustus De Morgan's Thoughts on Teaching and Learning Mathematics," *For the Learning of Mathematics* 9 (1989), 34–39. This article is made up of numerous quotations from De Morgan's works illustrating his commitment to better teaching of mathematics.

35. De Morgan, "On the Foundations of Algebra, No. II," pp. 288–289.

36. William Rowan Hamilton, "Theory of Conjugate Functions, or Algebraic Couples; with a Preliminary and Elementary Essay on Algebra as the Science of Pure Time," *Transactions of the Royal Irish Academy* 17 (1837), 293–422, reprinted in Hamilton, *Mathematical Papers* (Cambridge: Cambridge University Press, 1967), vol. 3, 3–96, p. 4. For more information on Hamilton's work, see Jerold Mathews, "William Rowan Hamilton's Paper of 1837 on the Arithmetizatíon of Analysis," *Archive for History of Exact Sciences* 19 (1978), pp. 177–200, and Thomas Hankins, "Algebra of Pure Time: William Rowan Hamilton and the Foundations of Algebra," in P. J. Mackamer and R. G.

Turnbull, eds., *Motion and Time, Space and Matter: Interrelations in the History and Philosophy of Science* (Columbus, OH: Ohio State University Press, 1976), pp. 327–359.

37. Thomas L. Hankins, *Sir William Rowan Hamilton* (Baltimore, MD: Johns Hopkins University Press, 1980), p. 343. This biography is an excellent study of Hamilton's work, not only in mathematics but also in physics.

38. Hamilton, *Mathematical Papers*, vol. 3, p. 10.

39. Ibid., p. 83.

40. Ibid., p. 96.

41. Quoted in Crowe, *Vector Analysis*, p. 27.

42. Ibid., p. 32.

43. Clerk Maxwell, *Treatise on Electricity and Magnetism* (London: Oxford University Press, 1873), pp. 9–10.

44. George Boole, *An Investigation of the Laws of Thought* (New York: Dover, 1958), p. 1. More on Boole is to be found in D. MacHale, *George Boole: His Life and Work* (Dublin: Boole Press, 1985).

45. Ibid., p. 27.

46. Ibid., p. 49.

47. Quoted in Hawkins, "Another Look at Cayley," p. 86. A recent biography of Cayley is Tony Crilly, *Arthur Cayley: Mathematician Laureate of the Victorian Age* (Baltimore, MD: Johns Hopkins University Press, 2005).

48. James Joseph Sylvester, "On a New Class of Theorems," *Philosophical Magazine* (3) 37 (1850), 363–370, reprinted in *Collected Mathematical Papers* (Cambridge: Cambridge University Press, 1904–1912), vol. 1, 145–51, p. 150. A recent biography of Sylvester is Karen Parshall, *James Joseph Sylvester: Jewish Mathematician in a Victorian World* (Baltimore, MD: Johns Hopkins University Press, 2006).

49. Arthur Cayley, "A Memoir on the Theory of Matrices," *Philosophical Transactions of the Royal Society of London* 148 (1858), reprinted in Cayley, *Collected Mathematical Papers*, vol. 2, 475–496, p. 476.

50. Ibid., p. 481.

51. Ibid., p. 483.

52. Ibid.

53. Adapted from Charles L. Dodgson, *An Elementary Treatise on Determinants with Their Application to Simultaneous Linear Equations and Algebraical Geometry* (London: Macmillan, 1867), p. 50.

54. Georg Frobenius, "Über homogene totale Differentialgleichungen," *Journal für die Reine und Angewandte Mathematik* 86 (1879), 1–19, p. 1. This paper is reprinted in Georg Frobenius, *Gesammelte Abhandlungen* (Berlin: Springer-Verlag, 1968), vol. 1, pp. 435–453.

55. Gauss, *Disquisitiones*, pp. 264–265.

56. Ibid., p. 366.

57. Quoted in Wussing, *Abstract Group Concept*, p. 64.

58. Arthur Cayley, "On the Theory of Groups, As Depending on the Symbolic Equation $\theta^n = 1$," *Philosophical Magazine* (4) 7, 40–47, p. 41. This paper is also found in Cayley, *The Collected Mathematical Papers* (Cambridge: Cambridge University Press, 1889–1897), vol. 2, pp. 123–130. For more discussion of Cayley's paper, see Sujoy Chakraburty and Munibur Rahman Choudhury, "Arthur Cayley and the Abstract Group Concept," *Mathematics Magazine* 78 (2005), 269–282.

59. Quoted in Corry, *Modern Algebra*, p. 77.

60. Arthur Cayley, "On the Theory of Groups," *Proceedings of the London Mathematical Society* 9 (1878), 126–133, p. 127. This paper is reprinted in Cayley, *Collected Mathematical Papers*, vol. 10, pp. 324–330.

61. Arthur Cayley, "The Theory of Groups," *American Journal of Mathematics* 1 (1878), 50–52, p. 52. This paper is reprinted in Cayley, *Collected Mathematical Papers*, vol. 10, pp. 401–403.

62. Ibid.

63. Walter Dyck, "Gruppentheoretische Studien," *Mathematische Annalen* 20 (1882), 1–44, p. 1. This paper is discussed further in Wussing, *Abstract Group Concept*, p. 240, and in van der Waerden, *History of Algebra*, p. 152.

64. Ibid., p. 12.

65. Heinrich Weber, "Beweis des Satzes, dass jede eigentlich primitive quadratische Form Unendlich viele Primzahlen fähig ist," *Mathematische Annalen* 20 (1882), 301–329, p. 302. Weber's work is also discussed in the books of Wussing, van der Waerden, and Corry.

66. Heinrich Weber, "Die allgemeinen Grundlagen des Galois'schen Gleichungstheorie," *Mathematische Annalen* 43 (1893), 521–549, p. 524.

67. Quoted in Corry, *Modern Algebra*, p. 42.

68. Quoted in Kurt-R. Biermann, "Kronecker," *Dictionary of Scientific Biography* (New York: Scribners, 1970–1980), vol. 7, pp. 505–509.

69. Richard Dedekind, supplement XI to Dirichlet, *Vorlesungen über Zahlentheorie*, (Braunschweig: Vieweg und Sohn, 1893), p. 452. The quoted edition is the fourth one, but the material also occurs in the second and third editions.

22

Analysis in the Nineteenth Century

In the above mentioned work of M. Cauchy [Cours d'analyse de l'école royale polytechnique] . . . one finds the following theorem: 'If the different terms of the series $u_0 + u_1 + u_2 + u_3 + \cdots$ are functions of one and the same variable quantity x, and indeed are continuous functions with respect to that variable in the neighborhood of a particular value for which the series converges, then the sum s of the series, in the neighborhood of this particular value, is also a continuous function of x.' But it appears to me that this theorem admits exceptions. For example, the series $\sin x - \frac{1}{2}\sin 2x + \frac{1}{3}\sin 3x - \cdots$ is discontinuous for each value $(2m + 1)\pi$ of x, where m is an integer. It is well known that there are many series with similar properties.

—Niels Henrik Abel, in "Untersuchungen über die Reihe $1 + \frac{m}{1}x + \frac{m}{1}\frac{m-1}{2}x^2 + \frac{m}{1}\frac{m-1}{2}\frac{m-2}{3}x^3 + \cdots$," 1826[1]

Cuts were invented by Richard Dedekind on November 24, 1858, as a way of providing an arithmetic definition of the real numbers. During the autumn, Dedekind had to lecture for the first time on the elements of the differential calculus. In preparing for these lectures, he decided that although the traditional geometric approach to some of the basic concepts of the subject had pedagogic value in an introductory course, there was still no truly "scientific" foundation for that part of the calculus having to do with limits of functions. Therefore, he decided to concentrate his energies on providing the basis for an arithmetic definition of the concept of real numbers. Although he soon arrived at his goal, he did not feel completely at ease with his presentation, so did not publish his idea of "Dedekind cuts" until 1872.

Toward the end of the eighteenth century, with the French Revolution's restructuring of mathematics education throughout the European continent and with the increasing necessity for mathematicians to teach rather than just do research, there was an increasing concern with how mathematical ideas should be presented to students and a concomitant increase in concern for "rigor." Recall that Lagrange attempted to base all of calculus on the notion of a power series. And although Lacroix wrote his calculus texts using Lagrange's method, among others, it was soon discovered that not all functions could be expressed by such series.

It was Augustin-Louis Cauchy, the most prolific mathematician of the nineteenth century, who first established the calculus on the basis of the limit concept so familiar today. Although the notion of limits had been discussed much earlier, even by Newton, Cauchy was the first to translate the somewhat vague notion of a function approaching a particular value into arithmetic terms by means of which one could actually prove the existence of limits. Cauchy used his notion of limits in defining continuity (in the modern sense) and convergence of sequences, both of numbers and of functions. Cauchy's notion of convergence, first published in 1821, was also developed in essence by the Czech mathematician Bernhard Bolzano in 1817 and by the Portuguese mathematician José Anastácio da Cunha as early as 1782. Unfortunately, the works of these latter two appeared in the far corners of Europe and were not appreciated, nor even read, in the mathematical centers of France and Germany. Thus, it was out of Cauchy's work that today's notions developed.

One of Cauchy's important results, that the sum of an infinite series of continuous functions is continuous, assuming this sum exists, turned out not to be true. Counterexamples were discovered as early as 1826 in connection with the series of sine and cosine functions now known as Fourier series. These series, although considered briefly by Daniel Bernoulli in the middle of the eighteenth century, were first studied in detail by Joseph Fourier in his work on heat conduction in the early nineteenth century. Fourier's works stimulated Peter Lejeune-Dirichlet to study in more detail the notion of a function and Bernhard Riemann to develop the concept known today as the Riemann integral.

Some unresolved questions in the work of Cauchy and Bolzano as well as the study of points of discontinuity growing out of Cauchy's wrong theorem led several mathematicians in the second half of the century to consider the structure of the real number system. In particular, Richard Dedekind and Georg Cantor each developed methods of constructing the real numbers from the rational numbers and, in this connection, began the detailed study of infinite sets.

In his calculus texts, Cauchy defined the integral as a limit of a sum rather than as an antiderivative, as had been common in the eighteenth century. His extension of this notion of the integral to the domain of complex numbers led him to begin the development of complex analysis by the 1820s. Riemann further developed and extended these ideas in the middle of the century.

Because integration in the complex domain can be thought of as integration over a real two-dimensional plane, Cauchy was also able to state the theorem today known as Green's theorem, relating integration around a closed curve to double integration over the region it bounds. Similar theorems relating integrals over a region to integrals over the boundary of the region were discovered by Mikhail Ostrogradsky and William Thomson. These theorems, today known as the divergence theorem and Stokes's theorem, turned out to have frequent applications in physics.

RIGOR IN ANALYSIS

Silvestre-Francois Lacroix replaced Lagrange at the École Polytechnique in 1799. Two years earlier, the first of the three volumes of his *Traité du calcul différentiel et du calcul intégral* had appeared. Lacroix intended this work to be a survey of the methods of calculus developed since the time of Newton and Leibniz. Thus, Lacroix presented not only Lagrange's view of the derivative of a function $f(x)$ as the coefficient of the first-order term in the Taylor series for f, but also dealt with the definition of dy/dx as a limit in the style of d'Alembert and as a ratio of infinitesimals in the manner of Euler. Lacroix was proud of the book's comprehensiveness and hoped that the true metaphysics of the subject would be found in what the various methods had in common.

For his teaching in Paris, however, Lacroix wrote a shortened one-volume version of the text entitled *Traité élémentaire du calcul différentiel et du calcul intégral*, a text whose continued popularity is attested by its appearance in nine editions between 1802 and 1881. In this work, Lacroix decided to base the differential calculus initially on the notion of a limit, defined in the process of his determination of the limit of a differential quotient. Thus, he showed that if $u = ax^2$ and $u_1 = a(x + h)^2$, then $2ax$ is "the **limit** of the ratio $\frac{u_1 - u}{h}$, or, is the value towards which this ratio tends in proportion as the quantity h diminishes, and to which it may approach as near as we choose to make it."[2] After calculating several other limits of ratios, Lacroix explained that in fact "the differential calculus is the finding of the limit of the ratios of the simultaneous increments of a function and of the variables on which it depends."[3] Thus, Lacroix, following his predecessors Euler and Lagrange, made no attempt at the beginning of the book to motivate the differential calculus in terms of slopes of tangent lines. The differential calculus was part of "analysis" and required no geometrical motivation or diagrams. Tangent lines were simply an application of the calculus and as such were discussed in section 7 of the text, "On the Application of the Differential Calculus to the Theory of Curves."

Despite Lacroix's decision in the *Traité élémentaire* to begin differential calculus with limits, he rapidly moved to establish the Taylor series of a function. Like Lagrange, he believed that all functions could be expressed as series except perhaps at isolated points. He then proceeded to use the Taylor series representation to develop the differentiation formulas for various transcendental functions and the methods for determining maxima and minima, even for functions of several variables. As part of this latter discussion, he corrected Euler's error in giving the conditions for a function of two variables to have an extreme value. In fact, he showed that a sufficient condition for $u(x, y)$ to have an extreme value at a point where both first partial derivatives vanish is that

$$\frac{\partial^2 u}{\partial x^2} \frac{\partial^2 u}{\partial y^2} > \left[\frac{\partial^2 u}{\partial x \partial y}\right]^2$$

at that point.

Lagrange's method of power series also appealed to the reform-minded members of the Cambridge Analytical Society, including George Peacock, Charles Babbage, and John Herschel (1792–1871), who translated Lacroix's *Traité élémentaire* into English in 1816 to provide an analytic text for use at Cambridge. The translators were so disappointed that

BIOGRAPHY

Augustin-Louis Cauchy (1789–1857)

Although Cauchy was the most prolific mathematician of the nineteenth century, he was never easy to deal with. As Abel wrote in a letter to a friend in 1826 during his visit to Paris, "there is no way to get along with him, although he is at present the mathematician who knows best how mathematics ought to be treated. . . . Cauchy is immoderately Catholic and bigoted, a very strange thing for a mathematician. Otherwise he is the only one [in Paris] who at present works in pure mathematics."[5] Born in the capital in the year the French Revolution began, he received an excellent classical education and then studied engineering at the École Polytechnique from 1805 to 1807. While working as an engineer from 1810 to 1813 on various military projects of the Napoleonic government, he showed such a strong interest in pure mathematics that he was encouraged by Laplace and Lagrange to leave engineering. With their help he secured a teaching position at the École Polytechnique and

several years later also at the Collège de France. Upon the appearance of his texts in analysis, he became one of the most respected members of the French mathematical community. He wrote so many mathematical papers that the journal of the Paris Academy was forced to limit the contributions of any one person. Cauchy got around these restrictions by establishing his own journal.

When the July revolution of 1830 led to the overthrow of the last Bourbon king, Cauchy, as an ardent conservative, refused to take the oath of allegiance to the new king and went into a self-imposed exile in Italy and then in Prague. He returned to Paris in 1838 but did not return to his teaching posts until the revolution of 1848 led to the removal of the requirement of an oath of allegiance. It was perhaps because of his political views that the French government did not honor him with a postage stamp until the 200th anniversary of his birth (Fig. 22.1).

FIGURE 22.1

Cauchy honored on a recent French stamp

Lacroix had "substituted the method of limits of d'Alembert in the place of the more correct and natural method of Lagrange"[4] that they provided notes so that the reader could use Lagrange's method rather than that of limits.

In spite of the appeal of Lagrange's method in England, Cauchy, back in France, found that this method was lacking in "rigor." Cauchy in fact was not satisfied with what he believed were unfounded manipulations of algebraic expressions, especially infinitely long ones. Equations involving these expressions were only true for certain values, those values for which the infinite series was convergent. In particular, Cauchy discovered that the Taylor series for the function $f(x) = e^{-x^2} + e^{-(1/x^2)}$ does not converge to the function. Thus, because from 1813 he was teaching at the École Polytechnique, Cauchy began to rethink the basis of the calculus entirely. In 1821, at the urging of several of his colleagues, he published his *Cours d'analyse de l'École Royale Polytechnique* in which he introduced new methods into the foundations of the calculus. We will study Cauchy's ideas on limits, continuity, convergence, derivatives, and integrals in the context of an analysis of this text as well as its sequel of 1823, *Résumé des leçons données á l'École Royale Polytechnique sur le calcul infinitesimal*, for it is these texts, used in Paris, that provided the model for calculus texts for the remainder of the century.

22.1.1 Limits

Cauchy's definition of limit appears near the beginning of his *Cours d'analyse*: "If the successive values attributed to the same variable approach indefinitely a fixed value, such that finally they differ from it by as little as one wishes, this latter is called the **limit** of

SIDEBAR 22.1   *What Is a Limit?*

Leibniz (1684): If any continuous transition is proposed terminating in a certain limit, then it is possible to form a general reasoning, which covers also the final limit.

Newton (1687): The ultimate ratio of evanescent quantities . . . [are] limits towards which the ratios of quantities decreasing without limit do always converge; and to which they approach nearer than by any given difference, but never go beyond, nor in effect attain to, till the quantities are diminished *in infinitum*.

Maclaurin (1742): The ratio of $2x + o$ to a continually decreases while o decreases and is always greater than the ratio of $2x$ to a while o is any real increment, but it is manifest that it continually approaches to the ratio of $2x$ to a as its limit.

D'Alembert (1754): This ratio $[a : 2y + z]$ is always smaller than $a : 2y$, but the smaller z is, the greater the ratio will be and, since one may choose z as small as one pleases, the ratio $a : 2y + z$ can be brought as close to the ratio $a : 2y$ as we like. Consequently, $a : 2y$ is the limit of the ratio $a : 2y + z$.

Lacroix (1806): The limit of the ratio $(u_1 - u)/h$ . . . is the value towards which this ratio tends in proportion as the quantity h diminishes, and to which it may approach as near as we choose to make it.

Cauchy (1821): If the successive values attributed to the same variable approach indefinitely a fixed value, such that they finally differ from it by as little as one wishes, this latter is called the limit of all the others.

all the others."[6] As an example, Cauchy noted that an irrational number is the limit of the various fractions that approach it (see Sidebar 22.1). He also defined an **infinitely small quantity** to be a variable whose limit is zero. Note that Cauchy was not defining the modern concept $\lim_{x \to a} f(x) = b$, for that concept involves two different variables. He seemed to suppress the role of the independent variable entirely. Furthermore, it may appear that Cauchy's definition of a limit is little different from that of d'Alembert. To see, however, what Cauchy meant by his verbal definition and to discover the difference between it and the definitions of his predecessors, it is necessary to consider his use of the definition to prove certain specific results on limits. In fact, Cauchy not only dealt with both the dependent and independent variables, but also translated his statement arithmetically by use of the language of inequalities. As an example, consider Cauchy's proof of the following

THEOREM   *If, for increasing values of x, the difference $f(x + 1) - f(x)$ converges to a certain limit k, the fraction $\frac{f(x)}{x}$ converges at the same time to the same limit.*[7]

Cauchy began by translating the hypothesis of the theorem into an arithmetic statement: Given any value ϵ, as small as one wants, one can find a number h such that if $x \geq h$, then $k - \epsilon < f(x + 1) - f(x) < k + \epsilon$. He then proceeded to use this translation in his proof. Because each of the differences $f(h + i) - f(h + i - 1)$ for $i = 1, 2, \ldots n$ satisfies the inequality, so does their arithmetic mean

$$\frac{f(h + n) - f(h)}{n}.$$

It follows that

$$\frac{f(h + n) - f(h)}{n} = k + \alpha,$$

where $-\epsilon < \alpha < \epsilon$ or, setting $x = h + n$, that

$$\frac{f(x) - f(h)}{x - h} = k + \alpha.$$

But then $f(x) = f(h) + (x - h)(k + \alpha)$ or

$$\frac{f(x)}{x} = \frac{f(h)}{x} + \left(1 - \frac{h}{x}\right)(k + \alpha).$$

Because h is fixed, Cauchy concluded that as x gets large, $f(x)/x$ approaches $k + \alpha$, where $-\epsilon < \alpha < \epsilon$. Because ϵ is arbitrary, the conclusion of the theorem holds. Cauchy also proved the theorem for the cases $k = \pm\infty$ and then used it to conclude, for example, that as x gets large $\log x/x$ converges to 0 and a^x/x $(a > 1)$ has limit ∞.

22.1.2   Continuity

Given the definition of limit, Cauchy could now define the crucial concept of continuity. Recall that for Euler a continuous function was one expressed by a single expression, while a discontinuous one was expressed in different parts of its domain by different expressions. Cauchy realized that such a definition was contradictory. For example, the function $f(x)$, which is equal to x for positive values and $-x$ for negative values, would appear to be discontinuous according to Euler's definition. On the other hand, one can write this same function using the single analytic expression

$$f(x) = \frac{2}{\pi} \int_0^\infty \frac{x^2 \, dt}{t^2 + x^2},$$

so $f(x)$ is continuous. The geometric notion of a continuous curve, one without any breaks, was generally understood, but Cauchy sought to find an analytic definition expressing this idea for functions. Lagrange had attempted such a definition earlier in the specific case of a function "continuous at 0" and having value 0 there: "We can always find an abscissa h corresponding to an ordinate less than any given quantity; and then all smaller values of h correspond also to ordinates less than the given quantity."[8]

Having read Lagrange's work, Cauchy generalized Lagrange's idea and gave his own new definition: "The function $f(x)$ will be, between two assigned values of the variable x, a **continuous function** of this variable if for each value of x between these limits, the numerical [absolute] value of the difference $f(x + \alpha) - f(x)$ decreases indefinitely with α (see Sidebar 22.2). In other words, the function $f(x)$ will remain continuous with respect to x between the given values if, between these values, an infinitely small increment of the variable always produces an infinitely small increment of the function itself."[9] Note that Cauchy presented both an arithmetic definition and one using the more familiar language of infinitely small quantities. But because Cauchy had already defined such quantities in terms of limits, the two definitions meant the same thing. Cauchy demonstrated how to use his definition by showing, for example, that $\sin x$ is continuous (on any interval). For $\sin(x + \alpha) - \sin x = 2 \sin \frac{1}{2}\alpha \cos(x + \frac{1}{2}\alpha)$, and the right side clearly decreases indefinitely with α.

It is interesting that Bernhard Bolzano (1781–1848), a Czech mathematician also familiar with the work of Lagrange, had some four years earlier given a definition of continuity virtually identical to Cauchy's. As part of his plan to prove rigorously the **intermediate**

SIDEBAR 22.2   *Definitions of Continuity*

Euler (1748): A continuous curve is one such that its nature can be expressed by a single function of x. If a curve is of such a nature that for its various parts . . . different functions of x are required for its expression, . . . , then we call such a curve discontinuous.

Bolzano (1817): A function $f(x)$ varies according to the law of continuity for all values of x inside or outside certain limits . . . if [when] x is some such value, the difference $f(x + \omega) - f(x)$ can be made smaller than any given quantity provided ω can be taken as small as we please.

Cauchy (1821): The function $f(x)$ will be, between two assigned values of the variable x, a continuous function of this variable if for each value of x between these limits, the [absolute] value of the difference $f(x + \alpha) - f(x)$ decreases indefinitely with α.

Dirichlet (1837): One thinks of a and b as two fixed values and of x as a variable quantity that can progressively take all values lying between a and b. Now if to every x there corresponds a single, finite y in such a way that, as x continuously passes through the interval from a to b, $y = f(x)$ also gradually changes, then y is called a continuous function of x in this interval.

Heine (1872): A function $f(x)$ is continuous at the particular value $x = X$ if for every given quantity ϵ, however small, there exists a positive number η_0 with the property that for no positive quantity η which is smaller than η_0 does the absolute value of $f(X \pm \eta) - f(X)$ exceed ϵ. A function $f(x)$ is continuous from $x = a$ to $x = b$ if for every single value $x = X$ between $x = a$ and $x = b$, including $x = a$ and $x = b$, it is continuous.

---

value theorem, "that between any two values of the unknown quantity which give results of opposite sign [when substituted in a continuous function $f(x)$] there must always lie at least one real root of the equation $[f(x) = 0]$,"[10] Bolzano needed to give a clear definition of the type of functions for which the theorem would hold. Noting that others had given definitions of continuity in terms of nonmathematical concepts such as time and motion, Bolzano gave what he called a "correct definition": "A function $f(x)$ varies according to the law of continuity for all values of x inside or outside certain limits . . . if [when] x is some such value, the difference $f(x + \omega) - f(x)$ can be made smaller than any given quantity provided ω can be taken as small as we please."[11] Although the two definitions are very similar, there is no convincing evidence that Cauchy had read Bolzano's work when he produced his own definition. Both men, being interested in giving a definition from which certain "obvious" results could be proved, came up with essentially the same idea.

From a modern point of view, of course, neither man defined continuity at a point. They both defined continuity over an interval. It would appear, however, that one can read each definition as defining continuity at each point in the interval. Namely, given a particular value x and an $\epsilon > 0$, one can find a $\delta > 0$ such that $|f(x + \alpha) - f(x)| < \epsilon$ whenever $\alpha < \delta$. As we will see in the next section, though, Cauchy was not entirely clear on what quantities δ depended. This lack of clarity would lead him to an incorrect result.

22.1.3   Convergence

Cauchy's concept of the convergence of series also had predecessors. But it was his definition, accompanied by explicit criteria that could actually be used to test for convergence, that has lasted to the present. Cauchy's definition appears in chapter 6 of his *Cours d'analyse*: "Let

BIOGRAPHY

Bernhard Bolzano (1781–1848)

Bolzano studied mathematics, philosophy, and physics in the university of his hometown of Prague. In 1805 he was appointed there to a chair in the philosophy of religion, a position established by order of Emperor Franz I of Austria to counter the new trends of enlightenment being spread through Europe in the wake of the French Revolution. Bolzano, however, was not a particular sympathizer of the Catholic restoration and expressed his own enlightened views on religion in his lectures. He was finally dismissed from his post in 1819 and put under police supervision on suspicion of heresy. Meanwhile, however, his philosophical training had attracted him to questions about the foundations of analysis, questions he was able to resolve to his satisfaction with new definitions and proofs related to the intuitive ideas of limit and continuity (Fig. 22.2).

FIGURE 22.2

Bolzano on a Czechoslovakian stamp

$s_n = u_0 + u_1 + u_2 + \cdots + u_{n-1}$ be the sum of the first n terms [of a series], n designating an arbitrary integer. If, for increasing values of n, the sum s_n approaches indefinitely a certain limit s, the series will be called **convergent**, and the limit in question will be called the **sum** of the series. On the contrary, if, as n increases indefinitely, the sum s_n does not approach any fixed limit, the series will be **divergent** and will not have a sum."[12] To clarify the definition, Cauchy stated what has become known as the "Cauchy criterion" for convergence. He realized that for a series to be convergent, it was necessary that the absolute values of the individual terms u_n must decrease to zero. This condition, however, was not sufficient. Convergence could only be assured if the various sums $u_n + u_{n+1}, u_n + u_{n+1} + u_{n+2}, u_n + u_{n+1} + u_{n+2} + u_{n+3}, \ldots$ "taken, from the first, in whatever number one wishes, finish by constantly having an absolute value less than any assignable limit."[13] Cauchy offered no proof of this sufficiency condition, because without some arithmetical definition of the real numbers no such proof is possible. He did, however, offer examples. Thus, he showed that the geometric series $1 + x + x^2 + x^3 + \cdots$, with $|x| < 1$, converges, because the sums $x^n + x^{n+1}$, $x^n + x^{n+1} + x^{n+2}, \ldots$, respectively equal to $x^n \frac{1-x^2}{1-x}, x^n \frac{1-x^3}{1-x}, \ldots$, are always between x^n and $x^n/(1-x)$, and the latter values both converge to zero with increasing n.

Interestingly, Cauchy had been preceded in his statement of the Cauchy criterion not only by Bolzano but also by the Portuguese scholar José Anastácio da Cunha (1744–1787), who included this material in his *Principios Mathematicos*, a comprehensive text that stretched from basic arithmetic to the calculus of variations and was published in sections beginning in 1782. Da Cunha's version of convergence and the Cauchy criterion is as follows: "A **convergent** series, so Mathematicians say, is one whose terms are similarly determined, each one by the number of the preceding terms, in such a way that the series can always be continued, and finally it need not matter whether it does or does not because one may neglect without considerable error the sum of any number of terms one might wish to add to those already written or indicated."[14] Da Cunha even used the criterion to demonstrate, like Cauchy, the convergence of the geometric series. Unfortunately, although da Cunha's work was translated into French in 1811, it was apparently little noticed and had little influence.

BIOGRAPHY

José Anastácio da Cunha (1744–1787)

Da Cunha was educated in Lisbon and became a military officer during the French and Spanish invasion of Portugal in 1762. Pursuing the study of ballistics, he wrote a memoir on the subject in 1769, analyzing various manuals used for instruction on the subject. His work brought him to the attention of the Marquis of Pombal, the powerful minister of King José I who had been able to reduce the powers of the Inquisition and remove the Jesuits from their positions of power. Pombal

arranged for da Cunha to be appointed to the chair of geometry at the newly reorganized University of Coimbra in 1773. Upon the death of the king in 1777, however, Pombal lost power, and his protegé da Cunha, having gained a reputation as a free thinker, was arrested by the Inquisition and convicted of heterodox religious opinions. He was pardoned in 1781 and spent the remainder of his life as a professor of mathematics in a Lisbon school organized for the education of poor children.

Bolzano's work, published as well in a distant corner of Europe, also had little contemporary influence. But he did show that the Cauchy criterion implied the **least upper bound principle**, eventually seen to be one of the defining properties of the real number system. Bolzano's convergence definition and his statement of the Cauchy criterion (applied to a series of functions rather than constants) are contained in a

THEOREM   *If a series of quantities $F_1(x)$, $F_2(x)$, $F_3(x)$, . . . , $F_n(x)$, . . . [where each $F_i(x)$ can be thought of as representing the sum of the first i terms of a series] has the property that the difference between its nth term $F_n(x)$ and every later term $F_{n+r}(x)$, however far from the former, remains smaller than any given quantity if n has been taken large enough, then there is always a certain constant quantity, and indeed only one, which the terms of this series approach, and to which they can come as close as desired if the series is continued far enough.*[15]

Bolzano's proof of uniqueness of the limit is straightforward, but his proof of the existence for each x of a number $X(x)$ to which the series converges is faulty because Bolzano, like Cauchy, had no way of defining an arbitrary real number X. Nevertheless, he did show how to determine the X to within any degree of accuracy d. If n is taken sufficiently large so that $F_{n+r}(x)$ differs from $F_n(x)$ by less than d for every r, then $F_n(x)$ is the desired approximation to X.

Bolzano could now also prove the least upper bound property of the real numbers:

THEOREM   *If a property M does not belong to all values of a variable x, but does belong to all values which are less than a certain u, then there is always a quantity U which is the greatest of those of which it can be asserted that all smaller x have property M.*

Bolzano's proof of the existence of this least upper bound U of all numbers having the property M involves the creation of a series to which the convergence criterion can be applied. Because M is not valid for all x, there must exist a quantity $V = u + D$ such that it is false that M is valid for all x smaller than V. Bolzano then considered the quantities $V_m = u + D/2^m$ for each positive integer m. If for all m, it is false that M is valid for all x less than V_m, then u itself must be the desired least upper bound. On the other hand, suppose that M is valid for all x less than $u + D/2^m$ but not for all x less than $u + D/2^{m-1}$. The difference between those two quantities is $D/2^m$, so Bolzano next applied this bisection technique to the interval $[u + D/2^{m-1}, u + D/2^m]$ and determined the smallest integer n so that M is valid for all x less than $u + D/2^m + D/2^{m+n}$ but not for all x less than $u + D/2^m + D/2^{m+n-1}$. Continuing this procedure, Bolzano constructed a sequence u, $u + D/2^m$, $u + D/2^m + D/2^{m+n}$, ... that satisfied his Cauchy criterion and therefore must converge to a value U, which he could easily prove satisfied the conditions of the theorem. (Bolzano's proof was modified slightly by Weierstrass in the 1860s to demonstrate what is now called the Bolzano-Weierstrass theorem, that given any bounded infinite set S of real numbers, there exists a real number r in every neighborhood of which there are other points of S.)

The least upper bound principle easily implies the **intermediate value theorem**, that if $f(x)$ is continuous and if $f(\alpha)$ and $f(\beta)$ have opposite signs, then there is a value of x between α and β for which $f(x) = 0$. For suppose $f(\alpha) < 0$ and $f(\beta) > 0$. Without loss of generality, we can also assume that $f(x) < 0$ for all $x < \alpha$. Then the property M that $f(x) < 0$ is not satisfied by all x but is satisfied by all x smaller than a certain $u = \alpha + \omega$, where $\omega < \beta - \alpha$ (because f is assumed continuous). It follows that there is a value U that is the largest such that $f(x) < 0$ for all $x < U$. Then $f(U)$ cannot be negative, because otherwise, by continuity, there would be a larger value U' with $f(x) < 0$ for all $x < U'$. Similarly, $f(U)$ cannot be positive, because then there would be a smaller value U'' with $f(U'') > 0$ and $f(x)$ would not be negative for all $x < U$. Thus, $f(U) = 0$ and the theorem is proved.

Cauchy's own proof of the same result did not make use of the Cauchy criterion. Instead, it relied implicitly on another axiom of the real number system, that any bounded monotone sequence has a limit, and explicitly on a result he had proved earlier that if $f(x)$ is continuous and the sequence $\{a_i\}$ converges to a, then the sequence $\{f(a_i)\}$ converges to $f(a)$. Cauchy's procedure came from a standard approximation procedure used by Lagrange and earlier mathematicians to approximate the solution to a polynomial equation $f(x) = 0$. Thus, if $f(\alpha) < 0$ and $f(\beta) > 0$, then setting $h = \beta - \alpha$ and m an arbitrary positive integer, Cauchy considered the signs of $f(\alpha + \frac{ih}{m})$ for $i = 1, 2, \ldots, m$. Because there is some pair of consecutive values, say, α_1 and β_1, for which $f(\alpha_1) < 0$ and $f(\beta_1) > 0$, with $\alpha < \alpha_1 < \beta_1 < \beta$, Cauchy next divided the interval $[\alpha_1, \beta_1]$ of length h/m into subintervals of length h/m^2 and repeated the argument. Continuing, he obtained an increasing sequence $\alpha, \alpha_1, \alpha_2, \ldots$ and a decreasing sequence $\beta, \beta_1, \beta_2, \ldots$, each of which must converge to the same limit a. It follows that both sequences $f(\alpha), f(\alpha_1), f(\alpha_2), \ldots$ and $f(\beta), f(\beta_1), f(\beta_2), \ldots$ converge to the common limit $f(a)$. Because the values of the first sequence are all negative and those of the second positive, it must be that $f(a) = 0$ and the theorem is proved. It is noteworthy, however, that this proof was presented in the appendix of Cauchy's text. In the main body of the text, he only gave a geometric argument, noting that the curves $y = f(x)$ and $y = 0$ must cross each other. One can only speculate as to which method Cauchy actually presented in his classes.

Given the Cauchy criterion for the convergence of a series, Cauchy developed in his text various tests by which one could demonstrate convergence in particular cases, beginning with tests for series of positive terms, say, $u_0 + u_1 + u_2 + \cdots$. Cauchy used the **comparison test**, that if a given series is term-by-term bounded by a convergent series, then the given series is itself convergent, without any particular comment. His most common comparison was to a geometric series with ratio less than 1, a series whose convergence Cauchy proved initially by use of the Cauchy criterion. In fact, he used the comparison test to demonstrate the validity of many of his other tests. For example, Cauchy proved the **root test**, that if the limit of the values $\sqrt[n]{u_n}$ is a number k less than 1, then the series converges. Choosing a number U such that $k < U < 1$, Cauchy noted that for n sufficiently large, $\sqrt[n]{u_n} < U$ or $u_n < U^n$. It then follows by comparison with the convergent geometric series $1 + U + U^2 + U^3 + \dots$ that the given series also converges. Similarly, if the limit of the roots is a number greater than 1, then the series diverges.

Cauchy used a similar proof to demonstrate the **ratio test**: If, for increasing values of n, the ratio u_{n+1}/u_n converges to a fixed limit k, the series u_n will be convergent if one has $k < 1$ and divergent if one has $k > 1$.

For series involving positive and negative terms, Cauchy dealt with the idea of absolute convergence (although not with that terminology), adapted the root and ratio test to that case, proved the **alternating series test**, and showed how to calculate the sum and product of two convergent series. He also adapted his various tests to sequences of functions. In particular, he showed how to find the interval of convergence of a power series. Although some of the individual results on series had been known previously, Cauchy was the first to organize them into a coherent theory that allowed him and others to generalize to the case of series of complex numbers and functions.

There is one significant result in the *Cours d'analyse*, however, which, as Abel noted in 1826, is incorrect as stated:

THEOREM 6-1-1   *When the different terms of the series $[\sum_{n=0}^{\infty} u_n]$ are functions of the same variable x, continuous with respect to that variable in the neighborhood of a particular value for which the series is convergent, the sum s of the series is also, in the neighborhood of this particular value, a continuous function of x.*

Cauchy's "proof" of this result is quite simple. We will present his argument in words as Cauchy did and then translate the words into modern symbols. Writing s_n as the sum of the first n terms of the series and s as the sum of the entire series, Cauchy denoted by r_n the remainder $s - s_n$. (Here s, s_n, and r_n are all functions of x.) To prove continuity of s, he needed to show that an infinitely small increment in x leads to an infinitely small increment in $s(x)$, that is, given $\epsilon > 0$,

$$\exists \delta \text{ such that } \forall a, \ |a| < \delta \Rightarrow |s(x+a) - s(x)| < \epsilon. \tag{22.1}$$

Although in certain earlier proofs, Cauchy actually calculated appropriate values for δ, in this case he just attempted an argument using arbitrary infinitely small quantities. Thus, he wrote that an infinitely small increment α of x leads to an infinitely small increment of $s_n(x)$ because the latter is continuous for every n, or

$$\exists \delta \text{ such that } \forall a, \ |a| < \delta \Rightarrow |s_n(x+a) - s_n(x)| < \epsilon. \tag{22.2}$$

Next, because the series converges for any x, r_n will itself be infinitely small for n large enough and will remain so for an infinitely small increment of x, or

$$\exists N \text{ such that } \forall n, \ n > N \Rightarrow |r_n(x)| < \epsilon \text{ and } |r_n(x+a)| < \epsilon. \tag{22.3}$$

Because the increment of s is the sum of those of s_n and r_n, Cauchy concluded that this increment is also infinitely small and therefore that s is itself continuous. We would write

$$|s(x + a) - s(x)| = |s_n(x + a) + r_n(x + a) - s_n(x) - r_n(x)|$$
$$\leq |s_n(x + a) - s_n(x)| + |r_n(x + a)| + |r_n(x)| \leq \epsilon + \epsilon + \epsilon = 3\epsilon. \quad (22.4)$$

What Cauchy failed to notice was that the δ in Equation 22.2 depends on ϵ, x, and n, while the N in Equation 22.3 depends on ϵ, x, and a. Unless we know that there is some value N that will work in Equation 22.3 for all a with $|a| < \delta$, we cannot assert the truth of Equation 22.4 (or Equation 22.1). Cauchy's arguments with infinitely small increments obscured the needed relationships among the various quantities involved. For the proof to be valid, a notion of uniform convergence was necessary, that is, one needed the additional hypothesis that the number N can be chosen independently of x, at least in some fixed interval. We return to this question in Section 22.1.8.[16]

22.1.4 Derivatives

Cauchy's *Cours d'Analyse* provided a treatment of the basic ideas of functions and series. In his 1823 text, *Résumé des leçons données á l'École Royale Polytechnique sur le calcul infinitesimal*, Cauchy applied his new ideas on limits to the study of the derivative and the integral, the two basic concepts of the infinitesimal calculus.

After beginning this text with the same definition of continuity as in the earlier one, Cauchy proceeded in lesson 3 to define the derivative of a function, what he called the *fonction derivée*, as the limit of $[f(x + i) - f(x)]/i$ as i approaches the limit of 0, as long as this limit exists. Just as he did in his definition of continuity, Cauchy defined the concept of a derivative over an interval, in fact an interval in which the function f is continuous. He noted that this limit has a definite value for each value of x, and therefore is a new function of that variable, a function for which he used Lagrange's notation $f'(x)$. The definition itself, although it can be thought of as expressing the quotient of infinitesimal differences as in Euler's work, is more directly taken from that section of Lagrange's *Analytic Functions* in which Lagrange, as part of his power series expansion of f, showed that $f(x + i) = f(x) + if'(x) + iV$, where V is some function that goes to zero with i. Cauchy was able to translate this theorem about derivatives into a definition of the derivative. He then calculated the derivatives of several elementary functions. For example, if $f(x) = \sin x$, then the quotient of the definition reduces to

$$\frac{\sin(1/2)i}{(1/2)i} \cos(x + (1/2)i),$$

whose limit $f'(x)$ is seen to be $\cos x$.

There is, of course, nothing new about Cauchy's calculations of derivatives. Nor is there anything particularly new about the theorems Cauchy was able to prove about derivatives. Lagrange had derived the same results from his own definition of the derivative. But because Lagrange's definition of a derivative rested on the false assumption that any function could be expanded into a power series, the significance of Cauchy's works lies in his explicit use of the modern definition of a derivative, translated into the language of inequalities through his definition of limit, to prove theorems. The most important of these results, in terms of its later use, was in lesson 7:

THEOREM *If the function $f(x)$ is continuous between the values $x = x_0$ and $x = X$, and if we let A be the smallest, B the largest, value of the derivative $f'(x)$ in that interval, then the ratio of the finite differences*

$$\frac{f(X) - f(x_0)}{X - x_0}$$

must be between A and B.[17]

Cauchy's proof of this theorem is the first to use the δ and ϵ so familiar to today's students. Cauchy began by choosing $\epsilon > 0$ and then choosing δ so that for all values of i with $|i| < \delta$ and for any value of x in the interval $[x_0, X]$, the inequality

$$f'(x) - \epsilon < \frac{f(x + i) - f(x)}{i} < f'(x) + \epsilon$$

holds. That such values exist follows from Cauchy's definition of the derivative as a limit. Note, however, that Cauchy used the fact, implicit in his definition of derivative on an interval rather than at a point, that, given ϵ, the same δ works for every x in the interval. In any case, Cauchy next interposed $n - 1$ new values $x_1 < x_2 < \cdots < x_{n-1}$ between x_0 and $x_n = X$ with the property that $x_i - x_{i-1} < \delta$ for each i and applied the above inequality to the subintervals determined by each successive pair of values. It follows that for $i = 1, 2, \ldots, n$,

$$A - \epsilon < \frac{f(x_i) - f(x_{i-1})}{x_i - x_{i-1}} < B + \epsilon.$$

Cauchy then used an algebraic result to conclude that the sum of the numerators divided by the sum of the denominators also must satisfy the same inequality, that is, that

$$A - \epsilon < \frac{f(X) - f(x_0)}{X - x_0} < B + \epsilon.$$

Because this result is true for every ϵ, the conclusion of the theorem follows.

As an immediate consequence of this theorem, Cauchy derived the **mean value theorem for derivatives**. Assuming that $f'(x)$ is continuous in the given interval, an assumption that of course justifies the assumption that it has a smallest value A and a largest value B, Cauchy used the intermediate value theorem to conclude that $f'(x)$ takes on every value between A and B, with x between x_0 and X. In particular, $f'(x)$ takes on the value in this theorem. Thus, the mean value theorem holds, that if $f'(x)$ is continuous on $[x_0, X]$, then there is a value θ between 0 and 1 with

$$\frac{f(X) - f(x_0)}{X - x_0} = f'(x_0 + \theta(X - x_0)).$$

Using the mean value theorem, Cauchy then proved that a function with positive derivative on an interval is increasing, one with negative derivative is decreasing, and one with zero derivative is constant.

22.1.5   Integrals

Cauchy's treatment of the derivative, although using his new definition of limits, was closely related to the treatments in the works of Euler and Lagrange. Cauchy's treatment of the integral, on the other hand, broke entirely new ground. Recall that, in the eighteenth century, integration was defined simply as the inverse of differentiation. Even Lacroix wrote that "the integral calculus is the inverse of the differential calculus, its object being to ascend from the differential coefficients to the function from which they are derived."[18] Although Leibniz had developed his notation to remind one of the integral as an infinite sum of infinitesimal areas, the problems inherent in the use of infinities convinced eighteenth-century mathematicians to take the notion of the indefinite integral, or antiderivative, as their basic notion for the theory of integration. They of course recognized that one could evaluate areas not only by use of antiderivatives but also by various approximation techniques. But it was Cauchy who first took these techniques as fundamental and proceeded to construct a theory of definite integrals upon them.

There are probably several reasons why Cauchy felt compelled to define the integral as the limit of a sum rather than in terms of antiderivatives. First, there were many situations where it was clear that an area under a curve made sense even though it could not be calculated by evaluating an antiderivative at the endpoints of an interval; such was the case in particular for certain piecewise-continuous functions that showed up in Fourier's work on series of trigonometric functions. A second reason may well have developed out of Cauchy's work in developing a theory of integrals of complex functions. Finally, Cauchy may have realized in the course of organizing his material for lectures at the École Polytechnique that there is no guarantee that an antiderivative exists for every function. Cauchy's own explanation of his reason for choosing to define an integral in terms of a sum, however, was that it works. As he wrote in an 1823 article, "it seems to me that this manner of conceiving a definite integral [as the sum of the infinitely small values of the differential expression placed under the integral sign] ought to be adopted in preference . . . because it is equally suitable to all cases, even to those in which we cannot pass generally from the function placed under the $\int$ sign to the primitive function."[19] Furthermore, he noted, "when we adopt this manner of conceiving definite integrals, we easily demonstrate that such an integral has a unique and finite value, whenever, the two limits of the variable being finite, the function under the $\int$ sign itself remains finite and continuous in the entire interval included between these values."[20]

In the second part of his *Resumé*, Cauchy presented the details of a rigorous definition of the integral using sums. Cauchy probably took his definition from the work on approximations of definite integrals by Euler and by Lacroix. But rather than consider this method a way of approximating an area, presumably understood intuitively to exist, Cauchy made the approximation into a definition. Thus, supposing that $f(x)$ is continuous on $[x_0, X]$, he took $n-1$ new intermediate values $x_1 < x_2 < \ldots < x_{n-1}$ between x_0 and $x_n = X$ and formed the sum

$$S = (x_1 - x_0)f(x_0) + (x_2 - x_1)f(x_1) + \cdots + (X - x_{n-1})f(x_{n-1}).$$

Cauchy noted that S depends both on n and on the particular values x_i selected. But, he wrote, "it is important to observe that if the numerical values of the elements $[x_{i+1} - x_i]$ become very small and the number n very large, the method of division will have only an insensible influence on the value of S."[21]

To prove this result, Cauchy noted that if one chose a new subdivision of the interval by subdividing each of the original subintervals, the corresponding sum S' could be rewritten in the form

$$S' = (x_1 - x_0)f(x_0 + \theta_0(x_1 - x_0)) + (x_2 - x_1)f(x_1 + \theta_1(x_2 - x_1)) + \cdots$$
$$+ (X - x_{n-1})f(x_{n-1} + \theta_{n-1}(X - x_{n-1})),$$

where each θ_i is between 0 and 1. By the definition of continuity, this expression can be rewritten as

$$S' = (x_1 - x_0)[f(x_0) + \epsilon_0] + (x_2 - x_1)[f(x_1) + \epsilon_1] + \cdots + (X - x_{n-1})[f(x_{n-1}) + \epsilon_{n-1}]$$
$$= S + (x_1 - x_0)\epsilon_1 + (x_2 - x_1)\epsilon_2 + \cdots + (X - x_{n-1})\epsilon_{n-1}$$
$$= S + (X - x_0)\epsilon'$$

where ϵ' is a value between the smallest and largest of the ϵ_i. Cauchy then argued that if the subintervals are sufficiently small, the ϵ_i and consequently ϵ' will be very close to zero, so that the taking of a subpartition does not change the value of the sum appreciably. Given any two sufficiently small subdivisions, one can take a third that subdivides each. The value of the sum for this third is then arbitrarily close to the values for each of the original two. It follows that "if we let the numerical values of [the lengths of the subdivisions] decrease indefinitely by increasing their number, the value of S ultimately becomes sensibly constant or, in other words, it will end by attaining a certain limit that will depend uniquely on the form of the function $f(x)$ and the extreme values x_0, X attributed to the variable x. This limit is what we call the definite integral [written $\int_{x_0}^{X} f(x)\, dx$]."[22] This definition thus used a generalization of Cauchy's criterion for convergence to sequences not necessarily indexed by the natural numbers.

With the integral now defined in terms of a limit of sums, it was not difficult for Cauchy to prove the **mean value theorem for integrals**, that

$$\int_{x_0}^{X} f(x)\, dx = (X - x_0) f[x_0 + \theta(X - x_0)]$$

for some θ with $0 \leq \theta \leq 1$, and also the additivity theorem for integrals over intervals. He then easily demonstrated the

FUNDAMENTAL THEOREM OF CALCULUS   *If $f(x)$ is continuous in $[x_0, X]$, if $x \in [x_0, X]$, and if*

$$F(x) = \int_{x_0}^{x} f(x)\, dx,$$

then $F'(x) = f(x)$.

To prove the theorem, Cauchy used the mean value theorem and additivity to get

$$F(x + \alpha) - F(x) = \int_{x}^{x+\alpha} f(x)\, dx = \alpha f(x + \theta\alpha).$$

If one divides both sides by α and passes to the limit, the conclusion follows from the continuity of $f(x)$. This version of the fundamental theorem can be considered the first one meeting modern standards of rigor, because it was the first in which $F(x)$ was clearly defined through an existence proof for the definite integral.

Although one of Cauchy's original hypotheses for the existence of a definite integral of $f(x)$ was that $f(x)$ be continuous on the interval of integration, Cauchy also realized that his definition made sense even if that condition were relaxed somewhat. Hence, he showed that if $f(x)$ had finitely many discontinuities in the given interval, the integral could still be defined by breaking the interval into subintervals at the points of discontinuity and defining the integral by a further limit argument. Similarly, if $f(x)$ is continuous on $(a, b]$, Cauchy made the definition

$$\int_a^b f(x)\, dx = \lim_{\epsilon \to 0} \int_{a+\epsilon}^b f(x)\, dx.$$

Cauchy made analogous definitions for integrals of functions over infinite intervals.

Cauchy used a different generalization of his definition of an integral in his presentation of a new method of integrating the differential equation $A\, dx + B\, dy$ in the case where $\partial A/\partial y = \partial B/\partial x$. He showed that the desired integral $f(x, y)$ could be defined by taking definite integrals in the plane from a fixed point (x_0, y_0):

$$f(x, y) = \int_{x_0}^x A(x, y)\, dx + \int_{y_0}^y B(x_0, y)\, dy.$$

Because this definition implies that $\partial f/\partial x = A(x, y)$ and

$$\frac{\partial f}{\partial y} = \int_{x_0}^x \frac{\partial A}{\partial y}(x, y)\, dx + B(x_0, y) = \int_{x_0}^x \frac{\partial B}{\partial x}(x, y)\, dx + B(x_0, y)$$

$$= B(x, y) - B(x_0, y) + B(x_0, y) = B(x, y),$$

this function indeed solves the problem.

The *Cours d'analyse* and the *Resumé* were the bases of Cauchy's first-year course at the École Polytechnique. In the second year, he gave a detailed introduction to the theory of differential equations. Much of that course was concerned with the standard techniques for solving such equations already developed in the eighteenth century, including, of course, the theorem just mentioned. But with his new concern for rigor in analysis, Cauchy wanted to determine the conditions under which one could prove the existence of a solution to $y' = f(x, y)$ satisfying a prescribed initial condition. The approximation technique he used in his proof of beginning with the given initial point (x_0, y_0) and constructing small straight line segments to approximate the desired curve had in essence been used in the eighteenth century. It was Cauchy who was able to demonstrate, however, by use of a version of the Cauchy criterion, that if both $f(x, y)$ and $\partial f(x, y)/\partial y$ are finite, continuous, and bounded in a region of the plane containing (x_0, y_0), then the approximation method produces polygons that converge to a solution curve to the differential equation in some neighborhood of (x_0, y_0) contained in the original region.

There is a curious story connected with Cauchy's treatment of differential equations. Cauchy never published an account of this second-year course, and it is only recently that proof sheets for the first thirteen lectures of the course have come to light. It is not clear why these notes stop at this point, but there is evidence that Cauchy was reproached by the directors of the school. He was told that, because the École Polytechnique was basically an engineering school, he should use class time to teach applications of differential equations rather than to deal with questions of rigor. Cauchy was forced to conform and announced that he would no longer give completely rigorous demonstrations. He evidently then felt that he could not publish his lectures on the material, because they did not reflect his own conception of how the subject should be handled.

22.1.6   Fourier Series and the Notion of a Function

Abel's counterexample to Cauchy's false result on convergence was a **Fourier series**, a series of trigonometric functions of the type Euler and Daniel Bernoulli argued about in the middle of the eighteenth century. It was Joseph Fourier who made a detailed study of these series in connection with his investigation of heat diffusion early in the nineteenth century. He first presented his work to the French Academy in 1807 and later reworked and expanded it into his *Théorie analytique de la chaleur* (*Analytic Theory of Heat*) in 1822. Fourier began by considering the special case of the temperature distribution $v(t, x, y)$ at time t in a rectangular lamina infinite in the positive x direction, of width 2 in the y direction, with the edge $x = 0$ being maintained at a constant temperature 1 and the edges $y = \pm 1$ being kept at temperature 0. By making certain assumptions about the flow of heat, Fourier was able to show that v satisfies the partial differential equation

$$\frac{\partial v}{\partial t} = \frac{\partial^2 v}{\partial x^2} + \frac{\partial^2 v}{\partial y^2}.$$

Fourier then solved this equation under the condition that the temperature of the lamina had reached equilibrium, that is, that $\partial v/\partial t = 0$. Assuming that $v = \phi(x)\psi(y)$ (the method of separation of variables), he differentiated twice with respect to each variable to get $\phi''(x)\psi(y) + \phi(x)\psi''(y) = 0$ or

$$\frac{\phi(x)}{\phi''(x)} = -\frac{\psi(y)}{\psi''(y)} = A$$

for some constant A. The obvious solutions to these equations are $\phi(x) = \alpha e^{mx}$, $\psi(y) = \beta \cos ny$, where $m^2 = n^2 = 1/A$. Physical reasoning dictates that m be negative (for otherwise the temperature would tend toward infinity for x large), so Fourier concluded that the general solution of the original partial differential equation is a sum of functions of the types $v = ae^{-nx} \cos ny$. By using the boundary conditions $v = 0$ when $y = \pm 1$, Fourier showed that n must be an odd multiple of $\frac{\pi}{2}$ and therefore that the general solution is given by the infinite series

$$v = a_1 e^{-(\pi x/2)} \cos\left(\frac{\pi y}{2}\right) + a_2 e^{-(3\pi x/2)} \cos\left(\frac{3\pi y}{2}\right) + a_3 e^{-(5\pi x/2)} \cos\left(\frac{5\pi y}{2}\right) + \cdots.$$

To determine the coefficients a_i, Fourier used the additional boundary condition that $v = 1$ when $x = 0$. Setting $u = \frac{\pi y}{2}$, that implied that the a_i satisfied the equation

$$1 = a_1 \cos u + a_2 \cos 3u + a_3 \cos 5u + \cdots,$$

a single equation for infinitely many unknowns, which Fourier turned into infinitely many equations by differentiating it infinitely often and each time setting $u = 0$. By noting the patterns determined by solving the first several of these equations, Fourier was able to determine that $a_1 = 4/\pi$, $a_2 = -4/3\pi$, $a_3 = 4/5\pi$, ... and therefore solve the partial differential equation. But in the usual spirit of mathematicians, once the original problem was solved he began to consider the mathematical ramifications of his new type of solution. First, he noted that his values for the coefficients implied that

$$\cos u - \frac{1}{3} \cos 3u + \frac{1}{5} \cos 5u - \cdots = \frac{\pi}{4}.$$

BIOGRAPHY

Jean Baptiste Joseph Fourier (1768–1830)

Orphaned at the age of nine, Fourier was placed by the bishop of his hometown of Auxerre, 90 miles southwest of Paris, in the local military school, where he soon displayed a talent for mathematics. Although he hoped to become an army engineer, such a career was not available to him at the time because he was not of noble birth. He therefore took up a teaching position. With the outbreak of the Revolution, Fourier became prominent in local affairs. His defense of victims of the Terror in 1794, however, led to his arrest. Fortunately, after the death of Robespierre, he was released and was appointed in 1795 as an assistant to Lagrange and Monge at the École Polytechnique. Three years later, during Napoleon's Egyptian campaign, he became secretary of the Institut d'Égypte, in which position he was able to conduct extensive research on Egyptian antiquities. On his return to France, he served for 12 years as prefect of the department of Isère in southeastern France and succeeded in accomplishing many public improvements. Fortunately, even with the fall of Napoleon, Fourier was able to be elected to the reconstituted Académie des Sciences and, in 1822, became its perpetual secretary, that is, in modern terms, its executive director, a position he held until his death.

with $u \in (-\frac{\pi}{2}, \frac{\pi}{2})$. But the same series clearly represents 0 for $u = \frac{\pi}{2}$ and represents $-\frac{\pi}{4}$ for $u \in (\frac{\pi}{2}, \frac{3\pi}{2})$. Fourier realized that this result would not be immediately believable to his readers, however: "As these results appear to depart from the ordinary consequences of the calculus, it is necessary to examine them with care and to interpret them in their true sense. One considers the equation $y = \cos u - \frac{1}{3} \cos 3u + \frac{1}{5} \cos 5u - \frac{1}{7} \cos 7u + \cdots$ as that of a line of which u is the abscissa and y the ordinate. One sees . . . that this line must be composed of separate parts $aa, bb, cc, dd, \ldots$ of which each is parallel to the axis and equal to $[\pi]$. These parallels are placed alternatively above and below the axis at a distance of $[\pi/4]$ and are joined by the perpendiculars $ab, cb, cd, ed, \ldots$ which are themselves part of the line."[23] In other words, Fourier considered the infinite cosine series to represent the "square wave" of Figure 22.3. To a modern reader, it is not clear why Fourier drew in the line segments perpendicular to the abscissa, for the value of the series at $u = k\pi/2$ for k odd is always 0. Fourier, however, realizing that the partial sums of this series could always be represented by a curve without breaks, thought that the infinite sum too should be represented by such a curve. The questions of whether this curve represented a "function" in the modern sense or whether the series represented a "continuous" function in Cauchy's sense were not relevant to Fourier's work. He was interested in a physical problem and probably conceived of this

FIGURE 22.3

Fourier's square wave:
$y = \cos u - \frac{1}{3} \cos 3u + \frac{1}{5} \cos 5u - \frac{1}{7} \cos 7u + \cdots$

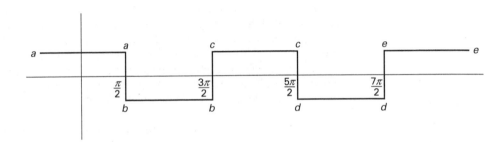

solution in geometrical terms, where he could draw a "continuous" curve without worrying about whether it represented a "function."

Fourier next investigated the representation of various functions by series of trigonometric functions, the most general being the series of the form $c_0 + c_1 \cos x + c_2 \cos 2x + \cdots + d_1 \sin x + d_2 \sin 2x + \cdots$. Usually, however, Fourier limited himself to either a sine series or a cosine series. Furthermore, because he wanted to convince his readers of the validity of his methods, he determined the coefficients of his original series and others by new methods. For example, he showed that if one writes the function $(1/2)\pi f(x)$ as a sine series,

$$\frac{1}{2}\pi f(x) = a_1 \sin x + a_2 \sin 2x + a_3 \sin 3x + \cdots,$$

one can multiply both sides by $\sin nx\, dx$ for each integer n in turn and integrate over the interval $[0, \pi]$. Because

$$\int_0^\pi \sin mx \sin nx \, dx = \begin{cases} 0, & \text{if } m \neq n; \\ \frac{\pi}{2}, & \text{if } m = n, \end{cases}$$

it follows that

$$a_k = \int_0^\pi f(x) \sin kx \, dx,$$

as long as the integrals representing the coefficients make sense, that is, as long as the area under $f(x) \sin kx$ is well defined.

What kinds of functions, then, could be represented by these trigonometric series? To begin to answer this question, Fourier defined what he meant by the term "function": "In general, the function $f(x)$ represents a succession of values or ordinates each of which is arbitrary. An infinity of values being given to the abscissa x, there is an equal number of ordinates $f(x)$. All have actual numerical values, either positive or negative or null. We do not suppose these ordinates to be subject to a common law; they succeed each other in any manner whatever, and each of them is given as if it were a single quantity."[24] Despite this modern-sounding definition, Fourier never considered what would today be called "arbitrary functions." His examples show that he only intended to consider piecewise-continuous functions. And, of course, Fourier only asserted that the series represented the given arbitrary function on the interior of a particular finite interval, such as $[0, \pi]$. The value of the series at the endpoints could easily be calculated separately, while the periodicity properties of the sine function enabled one to extend the original function geometrically to the entire real line (see Sidebar 22.3).

Fourier attempted in certain cases to prove that his expansion actually represented the function by using trigonometric identities to rewrite the partial sum of the first n terms in closed form and then considering the limit as n increased. In general, however, he believed that his explicit calculation of the coefficients in his proposed expansion of an arbitrary function in terms of integrals that represented real areas was a convincing enough argument that the expansion was valid. Thus, for example, he calculated the expansion given later by Abel:

$$\frac{1}{2}x = \sin x - \frac{1}{2}\sin 2x + \frac{1}{3}\sin 3x - \frac{1}{4}\sin 4x + \cdots$$

SIDEBAR 22.3 *What Is a Function?*

Johann Bernoulli (1718): I call a function of a variable magnitude a quantity composed in any manner whatsoever from this variable magnitude and from constants.

Euler (1748): A function of a variable quantity is an analytic expression composed in any way whatsoever of the variable quantity and numbers or constant quantities.

Euler (1755): When quantities depend on others in such a way that [the former] undergo changes themselves when [the latter] change, then [the former] are called functions of [the latter]; this is a very comprehensive idea which includes in itself all the ways in which one quantity can be determined by others.

Lacroix (1810): Every quantity whose value depends on one or more other quantities is called a function of these latter, whether one knows or is ignorant of what operations it is necessary to use to arrive from the latter to the first.

Fourier (1822): In general, the function $f(x)$ represents a succession of values or ordinates each of which is arbitrary. An infinity of values being given to the abscissa x, there is an equal number of ordinates $f(x)$. All have actual numerical values, either positive or negative or null. We do not suppose these ordinates to be subject to a common law; they succeed each other in any manner whatever, and each of them is given as if it were a single quantity.

Heine (1872): A single-valued function of a variable x is an expression which for every single rational or irrational value of x is uniquely defined.

Dedekind (1888): A function ϕ on a set S is a law according to which to every determinate element s of S there belongs a determinate thing which is called the transform of s and denoted by $\phi(s)$.

This series represented $(1/2)x$ on $[0, \pi/2)$ and the function of Figure 22.4 over the entire real line. Abel realized not only that this function violated Cauchy's result on the sum of a series of continuous functions, but also that Fourier's attempts at a proof that the Fourier series converged to the original function were not sufficient.

Cauchy himself attempted a new proof of Fourier's assertion in 1826, but in that proof he assumed that if $(u_k - v_k) \to 0$ and $k \to \infty$, and if $\sum u_k$ is convergent, then the same is true

FIGURE 22.4

Fourier's graph of $y = \sin x - \frac{1}{2}\sin 2x + \frac{1}{3}\sin 3x - \frac{1}{4}\sin 4x + \cdots$

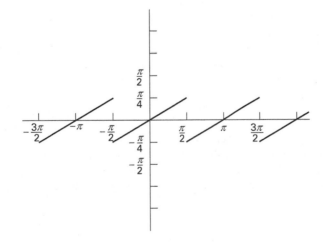

BIOGRAPHY

Gustav Peter Lejeune-Dirichlet (1805–1859)

Because the level of mathematics instruction in his native Germany was generally low during his youth, Dirichlet went to Paris in 1822 to study at the Collége de France. He became a tutor to the children of a famous French general and was thus able to meet many of the most prominent French intellectual figures, including Joseph Fourier, who ultimately proved the strongest influence on his mathematics. Dirichlet returned to Germany in 1825 and three years later was appointed to the faculty of the University of Berlin, a position he held for 27 years. He generally had a very heavy teaching load in Berlin, partly because he taught at the military academy as well as the university. He was therefore willing to accept an invitation to move to Göttingen upon Gauss's death in 1855, where he had more time for research. Unfortunately, his time there lasted only three and a half years until his own death in 1859.

of $\sum v_k$. Dirichlet noted that this assumption was erroneous in a paper of 1829 that contained the counterexample

$$u_k = \frac{(-1)^k}{\sqrt{k}} \qquad v_k = \frac{(-1)^k}{\sqrt{k}} + \frac{1}{k},$$

where it is obvious that the first series converges while the second diverges. Dirichlet then successfully attacked the problem of the convergence of Fourier series with a proof in the style of Cauchy's own analysis.

Rather than try to show that the Fourier series for an "arbitrary" function converged, Dirichlet lowered his sights drastically and found sufficient conditions on the function that would assure this convergence. In particular, he showed that if a function $f(x)$ defined on $[-\pi, \pi]$ was continuous and bounded on that interval, except perhaps for a finite number of finite discontinuities, and in addition had only a finite number of turning values in that interval, then the Fourier series converged at each x in $(-\pi, \pi)$ to the value $\lim_{\epsilon \to 0} \frac{1}{2}[f(x - \epsilon) + f(x + \epsilon)]$. (This value is equal to $f(x)$ if f is continuous at x.) The result held at the endpoints as well if, like Fourier, one interpreted $f(x)$ as being geometrically periodic outside the given interval. Dirichlet chose his conditions in order to be able to integrate products of the given function with trigonometric functions over certain intervals. Recall that Cauchy's new formulation of the definite integral only guaranteed the existence of the integral for functions with finitely many discontinuities. Dirichlet realized that it would be quite difficult to extend this result to functions with infinitely many discontinuities in a given interval. In fact, he provided an example of a function that did not satisfy his original conditions, one that is continuous nowhere: "$f(x)$ equals a determined constant c when the variable x takes a rational value, and equals another constant d, when this variable is irrational. The function thus defined has finite and determined values for every value of x, and meanwhile one cannot substitute it into the series, seeing that the different integrals which enter into that series lose all significance in this case."[25]

22.1.7 The Riemann Integral

In 1853, Georg Bernhard Riemann (1826–1866) attempted to generalize Dirichlet's result by first determining precisely which functions were integrable according to Cauchy's definition of the integral $\int_a^b f(x)\,dx$. He began, in fact, by changing the definition somewhat. Like Cauchy, he divided the interval $[a, b]$ into n subintervals $[x_{i-1}, x_i]$ for $i = 1, 2, \ldots, n$. Setting $\delta_i = x_i - x_{i-1}$, he now considered the sum

$$S = \sum_{i=1}^{n} \delta_i f(x_{i-1} + \epsilon_i \delta_i),$$

where each ϵ_i is between 0 and 1. This sum is more general than Cauchy's because Riemann allowed the argument of the function f to take any value in the relevant subinterval. He then defined the integral to be the limit to which S tends, provided it exists, no matter how the δ_i and ϵ_i are taken. Riemann now asked a question that Cauchy had not: In what cases is a function integrable and in what cases not? Cauchy himself had only shown that a certain class of functions was integrable, but had not tried to find all such functions. Riemann, on the other hand, formulated a necessary and sufficient condition for a finite function $f(x)$ to be integrable: "If, with the infinite decrease of all the quantities δ, the total size s of the intervals in which the variations of the function $f(x)$ are greater than a given quantity σ always becomes infinitely small in the end, then the sum S converges when all the δ become infinitely small"[26] and conversely. (The **variation** of a function in an interval is the difference between the maximum and minimum values of the function on that interval.) As an example of a function defined on [0, 1] that does not satisfy Cauchy's criterion for integrability but is Riemann-integrable, Riemann gave

$$f(x) = \sum_{n=1}^{\infty} \frac{\phi(nx)}{n^2},$$

where $\phi(x)$ is defined to be x minus the nearest integer, or, if there are two equally close integers, is defined to be equal to 0. It turns out that f is continuous everywhere except at the infinitely many points $x = p/2n$ with p and n relatively prime. But because the variation of f near such a point is equal to $\pi^2/8n^2$, the points near which the variation is greater than any $\sigma > 0$ are finite in number and the function satisfies Riemann's integrability criterion.

With a new class of functions now proved to be integrable, Riemann was able to extend Dirichlet's result on convergence of Fourier series. But rather than determine sufficient conditions on a function that ensure the convergence of its series, Riemann attacked the problem in reverse by beginning with a function which is representable by a trigonometric series and attempting to determine what consequences this representation has for the behavior of the function. Riemann was able in this way to find many types of functions that were representable by trigonometric series, but he never answered the entire question to his own satisfaction. Probably this was the reason that Riemann never published the manuscript in which this material appears.

BIOGRAPHY

Karl Weierstrass (1815–1897)

Born in Westphalia, Germany, Weierstrass entered the University of Bonn in 1834, at his father's urging, to study public finance and administration with the aim of a career in the Prussian civil service. His natural bent for mathematics, combined with his interest in the camaraderie of the Bonn taverns, prevented him from doing well in his intended field, so he left Bonn in 1838 with no degree. To earn a living, therefore, he studied for a teaching certificate and, beginning in 1841, taught such subjects as mathematics, physics, German, botany, geography, history, gymnastics, and calligraphy at various gymnasia for the next 14 years. After a series of brilliant mathematics papers in Crelle's *Journal*, he was awarded an honorary doctorate by the University of Königsberg in 1854 and finally received an appointment as an *extraordinarius* (associate professor) at the University of Berlin in 1856, at the same time holding a professorship at the Industry Institute in Berlin. Although health problems caused him to teach while seated, with an advanced student writing on the blackboard, his clear lectures soon won him a European-wide reputation, and his classes were attended each year by hundreds of students. In 1861, he, along with Kummer, introduced the mathematics seminar to the University of Berlin, another factor in making the University the premier university in the world for pure mathematics during the late nineteenth century.

22.1.8 Uniform Convergence

The work of Dirichlet and Riemann made it absolutely clear that a Fourier series can represent a discontinuous function and that therefore Cauchy's theorem on the sum of a series of continuous functions had to be modified. This was accomplished by several mathematicians, but it was Karl Weierstrass (1815–1897) who realized how to ensure that the sum function was continuous over an entire interval as Cauchy had originally concluded. In the course of his lectures at the University of Berlin beginning in the 1850s, Weierstrass made a careful distinction between convergence of a series of numbers and that of a series of functions, a distinction that Cauchy had glossed over. He was then able to identify a crucial property of convergence of functions, the property of uniform convergence over an interval: An infinite series $\sum u_n(x)$ is **uniformly convergent** in $[a, b]$ when, given any $\epsilon > 0$, there exists an N (dependent on ϵ) such that $|r_n(x)| < \epsilon$ for every $n > N$ and for every x in the interval $[a, b]$.

Given Weierstrass's definition, it was simple enough to correct Cauchy's proof for the case where the convergence of the series was uniform. But this definition also had a deeper influence. Not only did Weierstrass make absolutely clear how certain quantities in his definition depended on other quantities, but he also completed the transformation away from the use of terms such as "infinitely small." Henceforth, all definitions involving such ideas were given completely arithmetically. For example, Eduard Heine (1821–1881), a professor at Halle who spent much time in Berlin discussing mathematics with Weierstrass, not only gave a definition of continuity at a point in a paper of 1872 but also reworked Cauchy's definition of continuity over an interval into the following: "A function $f(x)$ is called . . . **uniformly continuous** from $x = a$ to $x = b$ if for any given quantity ϵ, however small, there exists a positive quantity η_0 such that for all positive values η which are less than

BIOGRAPHY

Sofia Kovalevskaya (1850–1891)

While some little girls awoke to delicate flowers on their nursery walls, Sofia Vasilevna Korvin's room was papered with the lecture notes from Mikhail Ostrogradsky's course in calculus. Her father, an army officer, had liked mathematics and allowed Sofia to study with a tutor. She grew to like mathematics too, but could not pursue her studies because she was a woman. Russian universities did not yet permit women to attend officially, and her family was not about to let her go off alone to a European university. Sofia solved this problem by contracting a "marriage of convenience" with Vladimir Kovalevsky, a publisher of scientific and political works and himself an aspiring scientist.

With a husband, Sofia was able to go abroad and study mathematics, first at the University of Heidelberg and then in Berlin with Weierstrass. Even though Weierstrass and others made a strong case for her to the faculty senate, the University of Berlin, unlike that of Heidelberg, refused to admit a woman officially. Sofia studied privately with Weierstrass and, after writing several publishable mathematics papers, the most significant being on the theory of partial differential equations, received her doctorate in 1874 from the University of Göttingen, a university that was willing to grant doctorates *in absentia*.

Returning to Russia, the Kovalevskys became connubial enough to produce a daughter in 1878. A few years later, Sofia resumed the mathematical research she had briefly set aside in favor of domestic and societal concerns. Upon the death of her husband in 1883, Sofia was able to secure a position as a professor at the University of Stockholm, a first for a woman in modern times. She soon became an active member of the European mathematical community, serving as an editor of the Swedish journal *Acta Mathematica* and receiving the Prix Bordin of the French Academy of Sciences in 1888 for her work on the revolution of a solid body about a fixed point.

Life was difficult for Sofia Kovalevskaya as a single mother. As she wrote in a letter to a friend, "All these stupid but unpostponable practical affairs are a serious test of my patience, and I begin to understand why men treasure good, practical housewives so highly. Were I a man, I'd choose myself a beautiful little housewife who would free me from all this."[28] Unfortunately, Kovalevskaya had little time to fulfill her mathematical promise. In early 1891, she contracted pneumonia on a trip to France and Germany and died a few days after her return to Sweden. For years after her untimely death, her grave in Stockholm remained a place of pilgrimage, not only for mathematicians but also for supporters of the rights of women (Fig. 22.5).

FIGURE 22.5

Sofia Kovalevskaya on a Russian stamp

η_0, $f(x \pm \eta) - f(x)$ remains less than ϵ. Whichever value one may give to x, assuming only that x and $x \pm \eta$ belong to the interval from a to b, the same η_0 must effect the required [property]."[27] Heine then went on to prove that a function continuous over a closed interval is uniformly continuous in that interval, making implicit use of what is today called the **Heine-Borel theorem**. He also gave the earliest published proof of the theorem that a continuous function on a closed interval attains a maximum and a minimum.

Because Weierstrass himself did not publish many of his new ideas, it was through the efforts of his followers and his students, including the first woman to earn a doctorate in mathematics, Sofia Kovalevskaya (1850–1891), that Weierstrass's concepts became the standard in mathematical analysis, a standard still in place today.

22.2   ## THE ARITHMETIZATION OF ANALYSIS

Even with the new definitions of continuity and convergence by Bolzano and Cauchy, it was apparent by the middle of the nineteenth century that a crucial step was missing in their proofs of such results as the intermediate value theorem and the existence of a limit of a bounded increasing sequence. Although the new definitions enabled mathematicians to show that a certain sequence of numbers satisfied the Cauchy criterion, there was no way to assert the existence of a limit if one could not specify in advance what type of "number" this limit would be. Cauchy, among others, understood intuitively what the real numbers were. He had even asserted that an irrational number can be considered as a limit of a certain sequence of rational numbers. But he was thereby asserting the a priori existence of such a number, without any argument as to how that assertion could be justified.

22.2.1 Dedekind Cuts

By the middle of the century, several mathematicians were actively considering this matter of exactly what an irrational number is. They were no longer content to assume the existence of such objects as their eighteenth-century predecessors had, particularly because similar such "obvious" assumptions were leading to incorrect conclusions. For example, Weierstrass, in the course of teaching the basics of analysis in Berlin, constructed a function that was everywhere continuous but nowhere differentiable, a function that no one in the previous century would have believed possible. Thus, Weierstrass and Dedekind began the detailed consideration of this question of the meaning of the irrationals. As Dedekind wrote in the introduction to his brief work "Stetigkeit und irrationale Zahlen" ("Continuity and Irrational Numbers"), first worked out in 1858 but only published in 1872,

> As professor in the Polytechnic School in Zürich I found myself for the first time obliged to lecture upon the elements of the differential calculus and felt more keenly than ever before the lack of a really scientific foundation for arithmetic. In discussing the notion of the approach of a variable magnitude to a fixed limiting value . . . I had recourse to geometric evidences. Even now such resort to geometric intuition in a first presentation of the differential calculus, I regard as exceedingly useful, from the didactic standpoint, and indeed indispensable if one does not wish to lose too much time. But that this form of introduction into the differential calculus can make no claim to being scientific, no one will deny. . . . The statement is so frequently made that the differential calculus deals with continuous magnitude, and yet an explanation of this continuity is nowhere given; even the most rigorous expositions of the differential calculus . . . depend upon theorems which are never established in a purely arithmetic manner. . . . It then only remained to discover the true origin [of these theorems] in the elements of arithmetic and thus at the same time to secure a real definition of the essence of continuity.[29]

Dedekind's program of research led to what is now called the "arithmetization of analysis," the deduction of the theorems of analysis first from the basic postulates defining the integers and then from the principles of the theory of sets.

To secure his definition of the "essence of continuity" in a "purely arithmetic manner," Dedekind began by considering the order properties of the set R of rational numbers. The ones he considered most important were, first, that if $a > b$ and $b > c$, then $a > c$; second, that if $a \neq c$, then there are infinitely many rational numbers lying between a and c; and third, that any rational number a divides the entire set R into two classes A_1 and A_2, with A_1 consisting of those numbers less than a and A_2 consisting of those numbers greater than a, the number a itself being assigned to either one of the two classes. These two classes have the obvious property that every number in A_1 is less than every number in A_2. In making the correspondence between rational numbers and points on the line, Dedekind noted that it was clear even to the Greeks that "the straight line L is infinitely richer in point-individuals than the domain R of rational numbers in number-individuals." But one cannot use the geometrical line to define numbers arithmetically. Dedekind's aim, therefore, was "the creation of new numbers such that the domain of numbers shall gain the same completeness, or . . . the same continuity, as the straight line."[30]

To create these new numbers, Dedekind decided to transfer to the domain of numbers the property he considered the essence of continuity of the straight line, namely, that "if all points of the straight line fall into two classes such that every point of the first class lies to the left of every point of the second class, then there exists one and only one point which produces this division."[31] Thus, Dedekind generalized his third property of the rationals into a new definition: "If any separation of the system R into two classes A_1, A_2 is given which possesses only this characteristic property that every number a_1 in A_1 is less than every number a_2 in A_2, then . . . we shall call such a separation a **cut** and designate it by (A_1, A_2)."[32] Every rational number a determines a cut for which a is either the largest of the numbers in A_1 or the smallest of those in A_2, but there are certainly cuts not produced by rational numbers, for example, the cut determined by defining A_2 to be the set of all positive rational numbers whose square is greater than 2 and A_1 to be the set of all other rational numbers. "In this property that not all cuts are produced by rational numbers consists the incompleteness or discontinuity of the domain R of all rational numbers. Whenever, then, we have to do with a cut (A_1, A_2) produced by no rational number, we create a new, an *irrational* number α, which we regard as completely defined by this cut (A_1, A_2); we shall say that the number α corresponds to this cut, or that it produces this cut."[33] Dedekind thus considered α to be a new creation of the mind corresponding to the cut. Other mathematicians, however, felt that it was better to define the real number α to *be* the cut.

In any case, the collection of all such cuts determines the system $\mathcal{R}$ of real numbers. Dedekind was able to show that this system possesses a natural ordering $<$, which satisfies the same properties of order satisfied by the rational numbers. He then proved that the system $\mathcal{R}$ also possesses the attribute of continuity, that if it is broken up into two classes $\mathcal{A}_1$, $\mathcal{A}_2$ such that every number α_1 in $\mathcal{A}_1$ is less than every number α_2 in $\mathcal{A}_2$, then there exists one and only one real number α that is either the greatest number in $\mathcal{A}_1$ or the smallest number in $\mathcal{A}_2$. In fact, α is the real number corresponding to the cut (A_1, A_2), where A_1 consists of all rational numbers in $\mathcal{A}_1$ and A_2 consists of all rational numbers in $\mathcal{A}_2$. It is straightforward to show that α possesses the requisite properties.

Dedekind completed his essay by defining the standard arithmetic operations in his new system $\mathcal{R}$. He was thus able to "arrive at real proofs of theorems (as, e.g., $\sqrt{2}\sqrt{3} = \sqrt{6}$), which to the best of my knowledge have never been established before."[34] One of these theorems was that every bounded increasing sequence $\{\beta_i\}$ of real numbers has a limit. Letting $\mathcal{A}_2$ be the set of all numbers γ such that $\beta_i < \gamma$ for all i, and $\mathcal{A}_1$ all remaining numbers, Dedekind easily showed that the cut $(\mathcal{A}_1, \mathcal{A}_2)$ determines the number α, which is the least number in $\mathcal{A}_2$ as well as the required limit.

BIOGRAPHY

Georg Cantor (1845–1918)

Georg Cantor might have followed in the footsteps of his mother's family of musicians as a violinist and sometimes wondered why he did not. Instead, he became interested in mathematics during his school years in St. Petersburg and studied it at the University of Zürich beginning in 1862 and then in Berlin under Weierstrass a year later. It took him only 10 years to become a full professor at the University of Halle, but he aspired to a better-paying and more prestigious position in Berlin. Kronecker, who opposed Cantor's theory of infinite sets, managed to keep him from the capital. Despite mental illness during his later years, Cantor was able to organize the Association of German Mathematicians in 1890 and was chiefly responsible for setting up the first International Congress of Mathematicians, held in Zürich in 1897.

22.2.2 Cantor and Fundamental Sequences

Dedekind's work on cuts appeared in print in 1872. Interestingly enough, the question of defining the real numbers arithmetically was then so much in the air that at least four other works accomplishing the same goal were published around that same time, all based on the general idea of defining an irrational number in terms of a sequence satisfying Cauchy's convergence criterion. Charles Meray (1835–1911) was the first to publish (in 1869), followed in 1872 by Ernst Kossak (1839–1902), who gave an account of Weierstrass's method, Georg Cantor (1845–1918), and Eduard Heine, who used Cantor's method to derive the basic theorems of the arithmetic of the real numbers. We just discuss Cantor's method here, because it had far-reaching implications for the theory of sets.

Cantor came at the problem of creating the real numbers from a point of view different from that of Dedekind. He was interested in the old problem of the convergence of Fourier series and took up the question of whether a trigonometric series that represents a given function is necessarily unique. In 1870 he managed to prove uniqueness under the assumption that the trigonometric series converged for all values of x. But then he succeeded in weakening the conditions. First, in 1871, he showed that the theorem was still true if either the convergence or the representation failed to hold at a finite number of points in the given interval. Second, in the following year, he was able to prove uniqueness even if the number of these exceptional points was infinite, provided that the points were distributed in a specified way. To describe accurately this distribution of points, Cantor realized that he needed a new way of describing the real numbers.

Beginning like Dedekind with the set of rational numbers, Cantor introduced the notion of a **fundamental sequence**, a sequence $a_1, a_2, \ldots, a_n, \ldots$ with the property that "for any positive rational value ϵ there exists an integer n_1 such that $|a_{m+n} - a_n| < \epsilon$ for $n \geq n_1$, for any positive integer m."[35] Such a sequence, now called a Cauchy sequence, satisfies the criterion Cauchy set out in 1821. For Cauchy, it was obvious that such a sequence converged to a real number b. Cantor, on the other hand, realized that to say this was to commit a logical error, for that statement presupposed the existence of such a real number. Therefore, Cantor used the fundamental sequence to *define* a real number b. In other words, Cantor associated a real number to every fundamental sequence of rational numbers. The rational number r was itself associated to a sequence, the sequence $r, r, \ldots, r, \ldots$, but there were also sequences that were not associated to rationals. For example, the sequence 1, 1.4, 1.41, 1.414, ..., generated by a familiar algorithm for calculating $\sqrt{2}$, was such a fundamental sequence.

Realizing that two fundamental sequences could well converge to the same real number, Cantor went on to define an equivalence relation on the set of such sequences. Thus, the number b associated to the sequence $\{a_i\}$ was said to be equal to the number b' associated to the sequence $\{a_i'\}$ if for any $\epsilon > 0$, there exists an n_1 such that $|a_n - a_n'| < \epsilon$ for $n > n_1$. The set B of real numbers was then the set of equivalence classes of fundamental sequences. It was not difficult to define an order relationship on these sequences as well as to establish the basic arithmetic operations. But Cantor wanted to show that the set he had defined was in some sense the same as the number line. It was clear to Cantor that every point on the line corresponded to a fundamental sequence, but he realized that the converse required an axiom, namely, that to every real number (equivalence class of fundamental sequences) there corresponds a definite point on the line.

Having defined real numbers, Cantor returned to his original question in the theory of trigonometric series. By using his identification of the real numbers with the points on the line, he defined the **limit point** of a point set P to be "a point of the line so placed that in every neighborhood of it we can find infinitely many points of P. . . . By the neighborhood of a point should here be understood every interval which has the point in its interior. Thereafter it is easy to prove that a [bounded] point set consisting of an infinite number of points always has at least one limit point."[36] Cantor denoted the set of these limit points by P', calling this set the first **derived set** of P. Similarly, if P' is infinite, Cantor defined the second derived set P'' to be the set of the limit points of P'. (If P' is finite, the set of its limit points is empty.) Continuing in this way, Cantor defined derived sets of any finite order. He then distinguished two types of bounded point sets. Those of the first species were ones for which the derived set $P^{(n)}$ was empty for some value of n, while those of the second species were those not satisfying this condition. For example, in the interval [0, 1] the point set $\{1, 1/2, 1/3, \ldots\}$ has derived set $\{0\}$ and is therefore of the first species, while the set of rational numbers in that interval has the entire interval as the derived set and is therefore of the second species. Cantor was able to show that point sets of the first species existed for which $P^{(n)}$ was finite for any given n, and was then able to demonstrate that the trigonometric series corresponding to a function was unique provided that either the convergence or the representation failed only at a point set of the first species.

22.2.3 Infinite Sets

The concept of a derived set led Cantor into an entirely new realm. Because he realized that for any set P, it was true that $P' \supseteq P'' \supseteq P''' \supseteq \cdots$, he defined a new set Q to be the intersection of all the $P^{(n)}$. Q is not in general a derived set, in the sense of Cantor's original definition. Nevertheless, because it is "derived" from P, Cantor wrote $Q = P^{(\infty)}$. He then began the process again, defining $P^{(\infty+1)}$ as $(P^{(\infty)})'$, $P^{(\infty+2)} = (P^{(\infty)})''$, $\ldots$, and even

$$P^{(\infty \cdot 2)} = \bigcap_n P^{(\infty+n)}.$$

Thus, Cantor was led to consider what are now called the "transfinite ordinal numbers," those "beyond" the finite ones.

Another question also occurred to Cantor in dealing with these point sets. He knew that the rational numbers were dense but not continuous. It would seem, therefore, that in some sense there should be "more" real numbers than rational numbers. In November 1873, he posed this question in a letter to Dedekind: "Take the collection of all positive whole numbers n and denote it by (n); then think of the collection of all real numbers x and denote it by (x); the question is simply whether (n) and (x) may be corresponded so that each individual of one collection corresponds to one and only one of the other? . . . As much as I am inclined to the opinion that (n) and (x) permit no such unique correspondence, I cannot find the reason."[37]

Dedekind could not answer Cantor's question, but only a month later Cantor was able to show that such a correspondence was impossible. His proof was by contradiction. If the real numbers in the interval (a, b) could be put into one-to-one correspondence with the natural numbers, then one could list these real numbers sequentially: $r_1, r_2, r_3, \ldots, r_n, \ldots$. Cantor then proceeded to find a real number in the interval that was not included in the list. He picked the first two numbers a', b' from the sequence such that $a' < b'$. Similarly, he picked the first two numbers a'', b'' in (a', b') such that $a'' < b''$. Continuing in this way, he determined a nested sequence of intervals (a, b), (a', b'), (a'', b''), $\ldots$. There were then two possibilities. First, the number of such intervals could be finite. In that case, there was certainly a real number in the smallest interval $(a^{(n)}, b^{(n)})$ that was not in the original list. Second, if the number of such intervals were infinite, they determined two bounded monotonic sequences $\{a^{(i)}\}$, $\{b^{(i)}\}$, which had the limits $\bar{a}$, $\bar{b}$, respectively. If $\bar{a} \neq \bar{b}$, then the interval $(\bar{a}, \bar{b})$ surely contained a real number not in the original list. Finally, if $\bar{a} = \bar{b} = \eta$, then η cannot be in the list either. For if it were equal to r_k for some k, it would not be in the intervals past a certain index, while from its definition as a limit it must be in all of the intervals.

In the paper of 1874 that contained the above proof, Cantor also included a proof that the set of all algebraic numbers can be put into one-to-one correspondence with the set of natural numbers. It followed that there were an infinite number of transcendental numbers. More importantly, however, Cantor had established a technique of counting infinite collections and had determined a clear difference in the size (or cardinality) of the continuum of real numbers on the one hand and the set of rational or algebraic numbers on the other. Shortly afterward, in another letter to Dedekind, he asked whether it might be possible to find a one-to-one correspondence between a square and an interval. The obvious answer here was "no," and in fact some of Cantor's colleagues felt that the question was ridiculous. But within three years Cantor discovered that the answer was "yes." He constructed the correspondence by mapping the pair (x, y) represented by the infinite decimal expansions $x = a_1 a_2 a_3 \ldots$, $y = b_1 b_2 b_3 \ldots$ to the point z represented by the expansion $z = a_1 b_1 a_2 b_2 a_3 b_3 \ldots$. There was a slight problem with this correspondence, related to the fact that $0.19999 \ldots$ and $0.20000 \ldots$ represented the same number, but Cantor soon corrected his proof and established that a one-to-one correspondence existed. Although the result surprised the mathematical community, Dedekind pointed out that Cantor's mapping was discontinuous and therefore had little to do with the geometric meaning of dimension. In fact, several mathematicians soon offered proof that no such continuous one-to-one mapping was possible and that, therefore, dimension was invariant under continuous, one-to-one correspondences.

22.2.4   The Theory of Sets

Cantor soon realized that his concept of a one-to-one correspondence could be placed at the foundation of a new theory of sets. In 1879, he used it to begin the study of the cardinality of an infinite set, a concept that he ultimately related to the idea of transfinite ordinals. Two sets A and B were defined to be of the **same power** if there was a one-to-one correspondence between the elements of A and the elements of B. Cantor initially singled out two special cases, those sets of the same power as the set N of natural numbers—these were called denumerable sets—and those sets of the power of the set of real numbers. In his further attempt to understand the properties of the continuum, Cantor was led to establish over the next two decades a detailed theory of infinite sets. Much of this theory was outlined in two papers published in 1895 and 1897 and collectively titled *Beiträge zur Begründung der transfiniten Mengenlehre* (*Contributions to the Founding of Transfinite Set Theory*).

The *Beiträge* began with a definition of a set as "any collection into a whole M of definite and separate objects m of our intuition or our thought."[38] "Every set M has a definite 'power,' which we will also call its 'cardinal number,'" he continued. "We will call by the name 'power' or 'cardinal number' of M the general concept which, by means of our active faculty of thought, arises from the set M when we make abstraction of the nature of its various elements m and of the order in which they are given."[39] By this "abstraction" Cantor meant that the cardinality of an infinite set was the generalization of the concept of "number of elements" for a finite set. Thus, the set of natural numbers and the set of real numbers have different cardinal numbers. That of the set of natural numbers, the set of smallest transfinite cardinality, Cantor called "aleph-zero," written $\aleph_0$, while that of the real numbers is denoted $\mathcal{C}$. Two sets are **equivalent**, or have the same cardinality, if there is a one-to-one correspondence between them. Cantor also defined the notion of $<$ for transfinite cardinals: The cardinality $\overline{\overline{M}}$ of a set M is less than that of a set N if there is no part of M that is equivalent to N while there is a part of N that is equivalent to M. It is then clear that for two sets M and N no more than one of the relations $\overline{\overline{M}} = \overline{\overline{N}}$, $\overline{\overline{M}} < \overline{\overline{N}}$, or $\overline{\overline{N}} < \overline{\overline{M}}$ can occur. But Cantor could not show that at least one of these relations had to occur.

Because $\aleph_0 < \mathcal{C}$, Cantor posed the question whether any other cardinalities were possible for subsets of the real numbers. In 1878, he thought he had answered this question in the negative: "Through a process of induction, which we do not describe further at this point, one is led to the theorem that the number of classes of linear point-sets yielded by this [equivalence] is finite and, indeed, that it is equal to *two*."[40] This conjecture that every subset of the real numbers has cardinality either $\aleph_0$ or $\mathcal{C}$ is called the **continuum hypothesis**. Although Cantor many times believed he had a proof of the result, and at least once believed that he had a proof of its negation, neither he nor anyone else was able to prove or disprove the hypothesis. In fact, the continuum hypothesis was eventually shown to be unprovable by use of any reasonable collection of axioms for the theory of sets.

Although Cantor was unable to answer all of the questions he raised in connection with the theory of infinite sets, his conception of these sets soon achieved both wide acceptance and strong criticism. In particular, Leopold Kronecker believed that any mathematical construct must be capable of being completed in a finite number of operations. Because some of Cantor's constructions did not meet Kronecker's criterion, Kronecker, as an editor of Crelle's

Journal, held up the publication of one of Cantor's articles for so long that Cantor refused to publish again in the *Journal*, even though it was the most influential of the mathematics journals of the time. Nevertheless, although Kronecker and others continued to oppose Cantor's transfinite methods, there were also a growing number of mathematicians who supported his new approach to set theory. The conflict between the two groups, however, continues to this day.

22.2.5   Dedekind and Axioms for the Natural Numbers

Cantor had developed some of the more advanced ideas of set theory and had shown, along with Dedekind, how to construct the real numbers starting from the rational numbers. But it was Dedekind who completed the process of arithmetizing analysis by characterizing the natural numbers, and therefore the rational numbers, in terms of sets. In the work in which he accomplished this task—"Was sind und was sollen die Zahlen?" ("The Nature and Meaning of Numbers")—developed over a 15-year period but only published in 1888, he also provided an introduction to the basic notions of set theory.

To characterize the natural numbers, Dedekind began with the notion that the natural numbers form a set of *things*, or "objects of our thought." Therefore, Dedekind defined the term *systeme*, here translated as *set*: "It very frequently happens that different things, a, b, c, . . . for some reason can be considered from a common point of view, can be associated in the mind, and we say that they form a set S. . . . Such a set S as an object of our thought is likewise a thing; it is completely determined when with respect to every thing it is determined whether it is an element of S or not."[41] Given this necessarily somewhat vague definition, Dedekind proceeded to describe various simple relations involving sets. For example, a set A is a **part** of a set S when every element of A is also an element of S. Also, the set **compounded** out of any sets A, B, C, . . . , denoted $\mathcal{M}(A, B, C, \ldots)$ consists of those elements that are in at least one of the sets A, B, C, . . . , while the set of elements common to A, B, C, . . . , is denoted $\mathcal{G}(A, B, C, \ldots)$. In modern terminology, Dedekind's "part" is our **subset**, $\mathcal{M}(A, B, C, \ldots)$ is the **union** of the sets A, B, C, . . . , and $\mathcal{G}(A, B, C, \ldots)$ is their **intersection**.

A fundamental property of the natural numbers is that each number has a unique successor. In other words, there is a function ψ from the set N of natural numbers to itself given by $\psi(n) = n + 1$. In general, Dedekind defined a function ϕ on S to be a "law according to which to every determinate element s of S there belongs a determinate thing which is called the **transform** of s and denoted by $\phi(s)$."[42] Because different elements of N have different successors, Dedekind was led to the notion of an *ähnlich* (**similar** or **injective**) transformation, one for which "to different elements a, b of the set S there always correspond different transforms $a' = \phi(a)$, $b' = \phi(b)$."[43] In this case there is an inverse transformation $\bar{\phi}$ of the system $S' = \phi(S)$, defined by assigning to every element s' of S' the unique element s, which was transformed into it by ϕ. Two sets R and S are then said to be **similar** to one another if there exists an injective transformation ϕ, defined on R such that $S = \phi(R)$.

The natural numbers also have the property that the image of N under the successor transformation is a proper subset of N itself, with the only element not belonging to that image being the element 1. It is, in fact, in the property that the image is a proper subset that the infinitude of the set N resided: "A set S is said to be infinite when it is similar to a proper part of itself; in the contrary case S is said to be a finite set."[44] But do there exist infinite sets at all? Dedekind was hesitant to prove results about such sets without an argument that they exist, so he gave one: "The totality S of all things which can be objects of my thought is infinite. For if s signifies an element of S, then is the thought s', that s can be object of my thought, itself an element of S."[45] Given the transformation from S to itself defined by $s \to s'$, for which it is clear that the image is not all of S and that the transformation is injective, Dedekind concluded that the set S does satisfy the requirements of his definition.

Dedekind realized that the properties that N was a set possessing an injective successor function whose image was a proper subset of N did not characterize N uniquely. There may well be extraneous elements in any set S that satisfy these properties, elements that are not natural numbers. For example, the set of positive rational numbers satisfies all of the properties. So Dedekind added one more property, that an element belongs to N if and only if it is an element of every subset K of S having the property that 1 belongs to K and that the successor of every element of K is also in K. In other words, N is characterized by being the intersection of all sets satisfying the original properties. N thus contains a base element 1, the successor $\phi(1)$ of 1, the successor $\phi(\phi(1))$ of that element, and so forth, but no other elements.

From his characterization of the natural numbers, Dedekind was able to derive the principle of mathematical induction as well as to give a definition and derive the properties of the order relationship on N and the operations of addition and multiplication. Two other mathematicians, Giuseppe Peano (1858–1932) and Gottlob Frege (1848–1925), also considered the same question of the construction of the natural numbers and the derivation of their important properties in the 1880s. Frege's work appeared in print in 1884 and Peano's in 1889. It was this work, together with the work of Weierstrass and his school, that enabled calculus to be placed on a firm foundation beginning with the basic notions of set theory. Thus, it was shown that calculus has an existence independent of the physical world of motion and curves, the world used by Newton to create the subject in the first place.

 ## 22.3   COMPLEX ANALYSIS

Recall that William Rowan Hamilton had by 1837 developed the theory of complex numbers as ordered pairs of real numbers, thus giving one answer to the question of what this mysterious square root of -1 really was. But mathematicians had been using complex numbers since the sixteenth century and even after Hamilton's work did not generally conceive of them in this abstract form. It was the geometrical representation of these numbers, first published by the Norwegian surveyor Caspar Wessel (1745–1818) in an essay in 1797, that ultimately became the basis for a new way of thinking about complex quantities, a way that soon convinced mathematicians that they could use them without undue worry.

22.3.1 Geometrical Representation of Complex Numbers

Wessel's aim in his "On the Analytical Representation of Direction" was not initially related to complex numbers as such. He felt that certain geometrical concepts could be more clearly understood if there was a way to represent both the length and direction of a line segment in the plane by a single algebraic expression. Wessel made clear that these expressions had to be capable of being manipulated algebraically. In particular, he wanted a way of expressing an arbitrary change of direction algebraically that was more general than the simple use of a negative sign to indicate the opposite direction.

Wessel began by dealing with addition: "Two straight lines are added if we unite them in such a way that the second line begins where the first one ends and then pass a straight line from the first to the last point of the united lines. This line is the sum of the united lines."[46] Thus, whatever the algebraic expression of a line segment was to be, the addition of two had to satisfy this obvious property drawn from Wessel's conception of motion. In other words, he conceived of line segments as representing vectors. It was multiplication, however, that provided Wessel with the basic answer to his question of the representation of direction. To derive this multiplication, he established a number of properties that he felt were essential. First, the product of two lines in the plane had to remain in the plane. Second, the length of the product line had to be the product of the lengths of the two factor lines. Finally, if all directions were measured from the positive unit line, which he called 1, the angle of direction of the product was to be the sum of the angles of direction of the two factors.

Designating by ϵ the line of unit length perpendicular to the line 1, Wessel easily showed that his desired properties implied that $\epsilon^2 = (-\epsilon)^2 = -1$ or that $\epsilon = \sqrt{-1}$. A line of unit length making an angle θ with the positive unit line could now be designated by $\cos\theta + \epsilon\sin\theta$ and, in general, a line of length A and angle θ by $A(\cos\theta + \epsilon\sin\theta) = a + \epsilon b$, where a and b are chosen appropriately (Fig. 22.6). Thus, from Wessel's algebraic interpretation of a geometrical line segment arose the geometrical interpretation of the complex numbers. The obvious algebraic rule for addition satisfied Wessel's requirements for that operation, while the multiplication $(a + \epsilon b)(c + \epsilon d) = ac - bd + \epsilon(ad + bc)$ satisfied his axioms for multiplication. Wessel also derived from his definitions the standard rules for division and root extraction of complex numbers.

FIGURE 22.6

Wessel's geometric interpretation of complex numbers

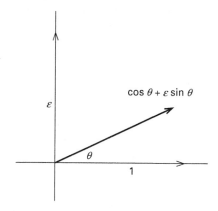

Unfortunately, Wessel's essay remained unread in most of Europe for many years after its publication. The same fate awaited the similar geometric interpretation of the complex numbers put forth by the Swiss bookkeeper Jean-Robert Argand (1768–1822) in a small book published in 1806. It was only because Gauss used the same geometric interpretation of the complex numbers in his proofs of the fundamental theorem of algebra and in his study of quartic residues that this interpretation gained acceptance in the mathematical community

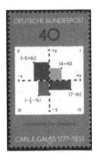

FIGURE 22.7

The Gaussian complex plane on a German stamp

(Fig. 22.7). Gauss was so intrigued with the fundamental theorem—that every polynomial $p(x)$ with real coefficients has a real or complex root—that he published four different proofs of it, in 1799, 1815, 1816, and 1848. Each proof used in some form or other the geometric interpretation of complex numbers, although in the first three Gauss hid this notion by considering the real and imaginary parts of the numbers separately. Thus, in his initial proof, Gauss in essence set $p(x + iy) = u(x, y) + iv(x, y)$ and then noted that a root of p would be an intersection point of the curves $u = 0$ and $v = 0$. He therefore made a detailed study of these curves and, through the use of the intermediate value theorem, showed that the curves must cross. It was only in his final proof in 1848 that Gauss believed mathematicians would be comfortable enough with the geometric interpretation of complex numbers so that he could use it explicitly. In fact, in that proof, similar to his first one, he even permitted the coefficients of the polynomial to be complex.

22.3.2 Complex Integration

By the second decade of the century, Gauss, with his geometric understanding of the meaning of complex numbers, began the development of the theory of complex functions. In a letter of 1811 to his friend Friedrich Wilhelm Bessel (1784–1846), Gauss discussed not only the geometric interpretation of the complex numbers but also the meaning of $\int_\mu^v \phi(x)\, dx$, where the variable x is complex:

> We must assume that x passes through infinitely small increments (each of the form $\alpha + \beta i$) from the value for which the integral is 0 to $x = a + bi$, and then sum all the $\phi(x)\, dx$. In this way the meaning is completely established. But the passage can occur in infinitely many ways; just as one can think of the entire domain of all real magnitudes as an infinite straight line, so one can make the entire domain of all magnitudes, real and imaginary, meaningful as an infinite plane, wherein each point determined by abscissa $= a$ and ordinate $= b$ represents the magnitude $a + bi$ as it were. The continuous passage from one value of x to another $a + bi$ accordingly occurs along a line and is consequently possible in infinitely many ways.[47]

Gauss went on to assert the "very beautiful theorem" that as long as $\phi(x)$ is never infinite within the region enclosed by two different curves connecting the starting and ending points of this integral, then the value of the integral is the same along both curves. Although he did not express himself in those terms, Gauss was considering $\phi(x)$ as an analytic function. In any case, he never published a proof of this result. Such a proof was published in 1825 by Cauchy, however, so the theorem is generally called **Cauchy's integral theorem**.

Cauchy first considered the question of integration in the complex domain in a memoir written in 1814 but not published until 1827. In this work he was mainly interested in the evaluation of definite integrals where one or both of the limits of integration is infinite. To perform such an evaluation, he attempted to make rigorous various procedures developed by Euler and Laplace involving moving the paths of integration into the complex plane. In particular, he used an idea of Euler's to derive the **Cauchy-Riemann equations**. Euler, in a paper written about 1777, asserted that the most important theorem about complex functions was that every function $Z(x + iy)$ that can be written as the sum $M(x, y) + iN(x, y)$ has the property that $Z(x - iy) = M - iN$. In this case it follows that if

$$V = \int Z\, dz = \int (M + iN)(dx + i\, dy) = \int M\, dx - N\, dy + i \int N\, dx + M\, dy = P + iQ,$$

then

$$P - iQ = \int M\,dx - N\,dy - i \int N\,dx + M\,dy.$$

Therefore, $P = \int M\,dx - N\,dy$ and $Q = \int N\,dx + M\,dy$, where, as usual for Euler, the integral signs stand for antidifferentiation. Because P is the integral of the differential $M\,dx - N\,dy$, it follows that

$$\frac{\partial M}{\partial y} = -\frac{\partial N}{\partial x}.$$

Similarly, the expression for Q shows that

$$\frac{\partial M}{\partial x} = \frac{\partial N}{\partial y}.$$

These two equations, the Cauchy-Riemann equations, ultimately became the characteristic property of complex functions.

In his 1821 *Cours d'analyse*, Cauchy dealt with complex quantities, as had Euler, by considering separately the real and imaginary parts. Thus, he considered the "symbolic expressions" $a + ib$ and multiplied them together using normal algebraic rules "as if $\sqrt{-1}$ was a real quantity whose square was equal to -1."[48] He defined a function of a complex variable in terms of two real functions of two real variables and showed what is meant by the various standard transcendental functions in the complex domain. He then generalized most of his results on convergence of series to complex numbers, by using the modulus $\sqrt{a^2 + b^2}$ of the quantity $z = a + ib$ as the analogue of the absolute value of a real number. He also defined continuity for a complex function in terms of the continuity of its two constituent functions.

It was not until 1825, however, having discovered his new definition of a definite integral, that Cauchy was able to deal with complex functions in their own right. In his *Mémoire sur les intégrales définies prises entre des limites imaginaires* (*Memoir on Definite Integrals Taken between Imaginary Limits*), he explicitly defined the definite complex integral

$$\int_{a+ib}^{c+id} f(z)\,dz$$

to be the "limit or one of the limits to which the sum of products of the form $[(x_1 - a) + i(y_1 - b)]f(a + ib)$, $[(x_2 - x_1) + i(y_2 - y_1)]f(x_1 + iy_1), \ldots [(c - x_{n-1}) + i(d - y_{n-1})]f(x_{n-1} + iy_{n-1})$ converge when each of the two sequences $a, x_1, x_2, \ldots, x_{n-1}, c$ and $b, y_1, y_2, \ldots, y_{n-1}, d$ consist of terms that increase or decrease from the first to the last and approach one another indefinitely as their number increases without limit."[49] In other words, Cauchy made Gauss's vague definition explicit by directly generalizing his own definition of a real definite integral and taking partitions of the two intervals $[a, b]$ and $[c, d]$. Cauchy realized, however, as had Gauss, that there were infinitely many different paths of integration beginning at $a + ib$ and ending at $c + id$. It was therefore not clear that this definition made sense.

To demonstrate his integral theorem, which in effect stated that the definition did make sense, Cauchy began by considering a path determined by the parametric equations $x = \phi(t)$, $y = \psi(t)$, where ϕ and ψ are monotonic differentiable functions of t in the interval $[\alpha, \beta]$, with $\phi(\alpha) = a$, $\phi(\beta) = c$, $\psi(\alpha) = b$, and $\psi(\beta) = d$. The two sequences $\{x_j\}$ and $\{y_j\}$ are then determined by taking a single sequence $\alpha, t_1 t_2, \ldots, t_{n-1}, \beta$ and calculating the values of this sequence under ϕ and ψ, respectively. Assuming that the lengths of the various subintervals determined by the t_j are small, Cauchy noted that $x_j - x_{j-1} \approx (t_j - t_{j-1})\phi'(t_j)$ and $y_j - y_{j-1} \approx (t_j - t_{j-1})\psi'(t)$. It follows that the definite integral is the limit of sums of terms of the form $(t_j - t_{j-1})[\phi'(t_j) + i\psi'(t_j)]f[\phi(t_j) + i\psi(t_j)]$ and therefore can be rewritten in the form

$$\int_{a+ib}^{c+id} f(z)\, dz = \int_{\alpha}^{\beta} [\phi'(t) + i\psi'(t)]f[\phi(t) + i\psi(t)]\, dt,$$

or (setting $x' = \phi'(t)$, $y' = \psi'(t)$) as

$$\int_{\alpha}^{\beta} (x' + iy')f(x + iy)\, dt.$$

"Now suppose that the function $f(x + iy)$ remains bounded and continuous as long as x stays between the limits a and c, and y between the limits b and d. In this special case one easily proves that the value of the integral . . . is independent of the nature of the functions $x = \phi(t)$, $y = \psi(t)$."[50] Cauchy's proof of this statement, which requires the existence and continuity of $f'(z)$—and Cauchy had not explicitly defined what was meant by the derivative of a complex function—was based on the calculus of variations. Cauchy varied the curve infinitesimally by replacing the functions ϕ and ψ by $\phi + \epsilon u$, $\psi + \epsilon v$, where ϵ is "an infinitesimal of the first order," and u, v both vanish at $t = \alpha$ and $t = \beta$, and expanded the corresponding change in the integral in a power series in ϵ. Using an integration by parts, Cauchy demonstrated that the coefficient of ϵ in this series is 0 and therefore that an infinitesimal change in the path of integration produces an infinitesimal change in the integral of the order of ϵ^2. Cauchy concluded that a finite change in the path, that is, a change from one path of integration to a second such path, can produce but an infinitesimal change in the integral, that is, no change at all. The integral theorem was therefore proved according to Cauchy's, if not modern, standards.

Cauchy next considered the case where f becomes infinite at some value $z_1 = r + is$ in the rectangle $a \leq x \leq c$, $b \leq y \leq d$. The integrals along two paths that together enclose z_1 are no longer the same. Defining R to be $\lim_{z \to z_1}(z - z_1)f(z)$, Cauchy calculated the difference in the integrals along two paths infinitely close to each other and to the point z_1 to be $2\pi Ri$. To put this another way, the integral of the function around the closed path formed by these two paths is also $2\pi Ri$. For example, if $f(z) = 1/(1 + z^2)$, then f becomes infinite at $z = i$. Because

$$\lim_{z \to i} \frac{z - i}{1 + z^2} = \lim_{z \to i} \frac{z - i}{(z - i)(z + i)} = \frac{1}{2i},$$

it follows that the difference in the values of the integrals of this function over the two paths L_1 and L_2 from -2 to 2 in Figure 22.8 is $2\pi \frac{1}{2i} i = \pi$. In other words, the integral of the function around the closed path consisting of L_2 followed by the negative of L_1 is π.

In a paper written in 1826, Cauchy looked at this question somewhat differently. Given a value z_1 for which $f(z)$ is infinite, Cauchy noted that the expansion of $f(z_1 + \epsilon)$ in powers of ϵ will begin with negative powers. The coefficient of $1/\epsilon$ in this expansion is what Cauchy

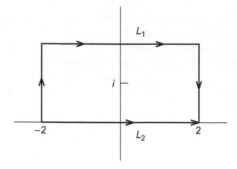

terms the **residue** of $f(z)$ at z_1, denoted by $R(f, z_1)$. Thus, if $(z - z_1)f(z) = g(z)$ is bounded near z_1, then

$$f(z_1 + \epsilon) = \frac{g(z_1 + \epsilon)}{\epsilon} = \frac{1}{\epsilon}g(z_1) + g'(z_1 + \theta\epsilon)$$

for θ a number between 0 and 1. It follows that the residue of $f(z)$ at z_1 is $g(z_1) = \lim_{z \to z_1}(z - z_1)f(z)$, the same value denoted earlier by R. With the residue calculated, Cauchy could then show that the integral of a function around a closed path containing z_1 was $2\pi i g(z_1)$.

Cauchy noted that his theory of residues had applications to such problems as the splitting of rational fractions, the determination of the values of certain definite integrals, and the solution of certain types of equations. For example, he demonstrated that

$$\int_{-\infty}^{\infty} \frac{\cos x}{1 + x^2}\, dx = \pi e^{-1}$$

by extending the interval of integration to a closed path in the complex plane containing the value i for which the integrand becomes infinite. To accomplish this integration, Cauchy first noted that the integral of any odd function over the interval $[-\infty, \infty]$ is zero. Therefore, because $\sin x/(1 + x^2)$ is odd and $e^{ix} = \cos x + i \sin x$ the integrand in this problem can be replaced by $e^{iz}/(1 + z^2)$. Next, he saw that one can use the theory of residues to integrate this function over the path consisting of the line interval from $-R$ to R followed by the half-circle of radius R extending back from R to $-R$, with $R > 1$. Since this path contains the value i, the integral is equal to $2\pi i$ multiplied by the residue of the function at i, and this residue he calculated to be $1/2ie$. He then noted that as the radius of the circle and the corresponding length of the interval grew larger, the part of the integral taken over the half-circle approached 0. The desired result followed.

In 1831, after he had exiled himself from Paris to Turin, Cauchy wrote a series of memoirs expanding on his ideas about complex analysis. In particular, he derived what is now called **Cauchy's integral formula** for circles, a result showing that the value of a complex function inside a closed curve is determined by its values on the curve itself. Cauchy began with the remark that for any positive integer n,

$$\int_{-\pi}^{\pi} e^{ni\theta}\, d\theta = \int_{-\pi}^{\pi} e^{-ni\theta}\, d\theta = 0,$$

while if $n = 0$, then

$$\int_{-\pi}^{\pi} d\theta = 2\pi.$$

It follows immediately that if $f(x) = a_0 + a_1 x + a_2 x^2 + \cdots + a_n x^n$ is any polynomial and $z = Re^{i\theta}$ is a complex number of modulus R, then

$$\int_{-\pi}^{\pi} f(z)\, d\theta = \int_{-\pi}^{\pi} f(1/z)\, d\theta = 2\pi a_0 = 2\pi f(0).$$

To generalize this result to arbitrary functions, assumed finite and continuous inside a circle of radius R around 0, Cauchy first noted that if $z = re^{i\theta}$, then $\frac{d}{dr} f(z) = \frac{1}{ir} \frac{d}{d\theta} f(z)$. He then integrated both sides of this equation over the disk given by $0 \le r \le R$ and $-\pi \le \theta \le \pi$. The left side gives

$$\int_{-\pi}^{\pi} d\theta \int_{0}^{R} \frac{d}{dr} f(z)\, dr = \int_{-\pi}^{\pi} \left[f(Re^{i\theta}) - f(0) \right] d\theta = \int_{-\pi}^{\pi} f(Re^{i\theta})\, d\theta - 2\pi f(0).$$

Integrating the right side in the opposite order gives

$$\int_{0}^{R} dr \int_{-\pi}^{\pi} \frac{1}{ir} \frac{d}{d\theta} f(z)\, d\theta = \int_{0}^{R} \left[\frac{1}{ir} \left(f(re^{i\pi}) - f(re^{-i\pi}) \right) \right] dr = 0.$$

Cauchy's generalization follows:

$$\int_{-\pi}^{\pi} f(Re^{i\theta})\, d\theta = 2\pi f(0).$$

Cauchy next noted that if $f(0) = 0$, then $\int_{-\pi}^{\pi} f(Re^{i\theta})\, d\theta = 0$. So, if one now takes a value a inside the circle of radius R and sets $F(z) = z[f(z) - f(a)]/(z - a)$, then $F(0) = 0$, so

$$\int_{-\pi}^{\pi} F(z)\, d\theta = \int_{-\pi}^{\pi} \frac{z[f(z) - f(a)]}{z - a}\, d\theta = 0.$$

Therefore, by expanding $z/(z - a)$ as a power series, Cauchy found his result:

$$\int_{-\pi}^{\pi} \frac{zf(z)}{z - a}\, d\theta = f(a) \int_{-\pi}^{\pi} \frac{z}{z - a}\, d\theta = f(a) \int_{-\pi}^{\pi} \left(1 + \frac{a}{z} + \frac{a^2}{z^2} + \cdots \right) d\theta = 2\pi f(a).$$

This result is equivalent to the modern version of Cauchy's integral formula. To see that, note that if $z = Re^{i\theta}$, then $dz = iRe^{i\theta}\, d\theta - iz\, d\theta$. Therefore, if C is the circle of radius R centered on the origin,

$$\int_{C} \frac{f(z)}{z - a}\, dz = \int_{-\pi}^{\pi} \frac{f(z)}{z - a} iz\, d\theta = i \int_{-\pi}^{\pi} \frac{zf(z)}{z - a}\, d\theta = 2\pi i f(a).$$

This latter result is the usual statement of Cauchy's formula.

22.3.3 Complex Functions and Line Integrals

There are many other standard results in complex function theory for which Cauchy was at least partially responsible, most being applications of his integral theorem or his calculus of residues. But the discussion of his work will be concluded with a brief analysis of a paper of 1846 which, although it did not mention complex functions at all, led to a new way of proving the integral theorem and also provided the beginning of some fundamental ideas in both vector analysis and topology. This short paper, "Sur les intégrales qui s'étendent a tous les points d'une courbe fermée" ("On the Integrals that Extend to All the Points of a Closed Curve"), contained the bare statement of several theorems, without proofs. Cauchy promised to provide the proofs later, but apparently did not do so. The theorems deal with a function k of several variables x, y, z, ... which is to be integrated along the boundary curve Γ of a surface S lying in a space of an unspecified number of dimensions. The most important results are collected in the following

THEOREM   *Suppose*

$$k = X\frac{dx}{ds} + Y\frac{dy}{ds} + Z\frac{dz}{ds} + \cdots$$

where $X\,dx + Y\,dy + Z\,dz + \cdots$ is an exact differential. [To say that this differential is exact is to say that $\partial X/\partial y = \partial Y/\partial x$, $\partial X/\partial z = \partial Z/\partial x$, $\partial Y/\partial z = \partial Z/\partial y$, ...] Suppose that the function k is finite and continuous everywhere on S except at finitely many points P, P', P'', ... in its interior. If α, β, γ, ... are closed curves in S surrounding these points, respectively, then

$$\int_\Gamma k\,ds = \int_\alpha k\,ds + \int_\beta k\,ds + \int_\gamma k\,ds + \ldots$$

In particular, if there are no such singular points, then

$$\int_\Gamma k\,ds = 0.$$

In the two-dimensional case, where S is a region of the plane and k is an arbitrary differential, then

$$\int_\Gamma k\,ds = \pm \int\int_S \left(\frac{\partial X}{\partial y} - \frac{\partial Y}{\partial x} \right)\,dx\,dy.$$

If k is an exact differential, then $\partial X/\partial y = \partial Y/\partial x$, so the right side, and therefore the left, vanish.

The Cauchy integral theorem follows from the last statement. A complex function $f(z) = f(x + iy)$ can be expressed as $f(x,\,y) = u(x,\,y) + iv(x,\,y)$ and, therefore, since $dz = dx + i\,dy$,

$$f(z)\,dz = \int (u\,dx - v\,dy) + i\int (v\,dx + u\,dy).$$

The Cauchy-Riemann equations then imply that both integrands are exact differentials and therefore that the integral theorem holds.

More interesting than the integral theorem, however, is the appearance in Cauchy's paper both of the concept of a line integral in n-dimensional space (and of the matter-of-fact occurrence of a space of dimension higher than three) and of the statement (in the next-to-the-last sentence) of the theorem today generally known as **Green's theorem**. In fact, results somewhat akin to that theorem appear in an 1828 paper of George Green (1793–1841) dealing with electricity and magnetism, but Cauchy's version is the first printed statement of

BIOGRAPHY

Georg Bernhard Riemann (1826–1866)

Riemann needed his father's permission to switch from the study of theology and philology to the study of mathematics in 1846 when he enrolled at the University of Göttingen. He had started life in the village of Breselenz, about 60 miles southeast of Hamburg, and now he would journey to Berlin because mathematics education was not particularly strong at Göttingen. In Berlin he met Dirichlet, who became his mentor. He returned to Göttingen a few years later to study with Gauss

and received his PhD in 1851. For two years he researched and prepared his lectures qualifying him to teach at Göttingen. In 1857 he was appointed as an associate professor and two years later, on the death of Dirichlet, who had in the meantime come to Göttingen, as full professor. His mathematical work was brilliant, but tuberculosis cut his work short when it claimed his life in the summer of 1866 during one of his several trips to Italy to find a cure.

the result so named in today's textbooks. Finally, the expression of the line integral around the boundary of the surface as a sum of line integrals around isolated singular points, whose values are called **periods**, marked the beginning of the study of the relationships of integrals to surfaces over which they are not everywhere defined. Since Cauchy never published the proof of his 1846 theorem, one can only speculate as to how far he carried all of these new concepts. It was Riemann, however, who restated Cauchy's results a few years later, with full proofs, and extended the result on periods far beyond Cauchy's conception.

22.3.4 Riemann and Complex Functions

Riemann's dissertation, "Grundlagen für eine allgemeine Theorie der Functionen einer veränderlichen complexen Grösse" ("Foundations for a General Theory of Functions of One Complex Variable"), began with a discussion of an important distinction between real and complex functions. Although the definition of function, "to every one of [the] values [of a variable quantity z] there corresponds a single value of the indeterminate quantity w,"[51] can be applied both to the real and complex case, Riemann realized that in the latter case, where $z = x + iy$ and $w = u + iv$, the limit of the ratio dw/dz defining the derivative could well depend on how dz approaches 0. Because for functions defined algebraically one could calculate the derivative formally and not have this problem, Riemann decided to make this existence of the derivative the basis for the concept of a complex function: "The complex variable w is called a function of another complex variable z when its variation is such that the value of the derivative dw/dz is independent of the value of dz."[52] Cauchy, of course, had essentially used this notion in his entire discussion of complex functions but had only made it explicit toward the end of his career.

As a first application of this definition, Riemann showed that such a complex function considered as a mapping from the z plane to the w plane preserves angles. For suppose p' and p'' are infinitely close to the origin P in the z plane, with their images q', q'' infinitely close to the image Q of P. Writing the infinitesimal distance from p' to P both as $dx' + i\,dy'$ and as $\epsilon' e^{i\phi'}$, and that from q' to Q as both $du' + i\,dv'$ and $\eta' e^{i\psi'}$, with similar notations for the other infinitesimal distances, Riemann noted that his condition on the function implies that

$$\frac{du' + i\,dv'}{dx' + i\,dy'} = \frac{du'' + i\,dv''}{dx'' + i\,dy''}$$

or that

$$\frac{du' + i\,dv'}{du'' + i\,dv''} = \frac{\eta'}{\eta''}e^{i(\psi'-\psi'')} = \frac{dx' + i\,dy'}{dx'' + i\,dy''} = \frac{\epsilon'}{\epsilon''}e^{i(\phi'-\phi'')}.$$

It follows that $\eta'/\eta'' = \epsilon'/\epsilon''$ and that $\psi' - \psi'' = \phi' - \phi''$, or, in other words, that the infinitesimal triangles $p'Pp''$ and $q'Qq''$ are similar. Such an angle-preserving mapping is called a **conformal mapping**. In some sense, both Euler and Gauss knew that analytic complex functions had this property, but it was Riemann who gave this argument and who, in addition, was able to demonstrate the **Riemann mapping theorem**, that any two simply connected regions in the complex plane can be mapped conformally on each other by means of a suitably chosen complex function.

Riemann next derived the Cauchy-Riemann equations by determining what the existence of the derivative means in terms of the two functions u and v:

$$\frac{dw}{dz} = \frac{du + i\,dv}{dx + i\,dy} = \frac{\frac{\partial u}{\partial x}\,dx + \frac{\partial u}{\partial y}\,dy + i\left(\frac{\partial v}{\partial x}\,dx + \frac{\partial v}{\partial y}\,dy\right)}{dx + i\,dy}$$

$$= \frac{\left(\frac{\partial u}{\partial x} + i\frac{\partial v}{\partial x}\right)dx + \left(\frac{\partial v}{\partial y} - i\frac{\partial u}{\partial y}\right)i\,dy}{dx + i\,dy}.$$

If this value is independent of how dz approaches 0, then setting dx and dy in turn equal to zero and equating the real and imaginary parts of the two resulting expressions shows that

$$\frac{\partial u}{\partial x} = \frac{\partial v}{\partial y} \quad \text{and} \quad \frac{\partial v}{\partial x} = -\frac{\partial u}{\partial y}.$$

Conversely, if those Cauchy-Riemann equations are satisfied, then the desired derivative is easily calculated to be $\partial u/\partial x + i\,\partial v/\partial x$, a value independent of dz. Riemann made these equations the center of his theory of complex functions, along with the second set of partial differential equations easily derived from them:

$$\frac{\partial^2 u}{\partial x^2} + \frac{\partial^2 u}{\partial y^2} = 0 \quad \text{and} \quad \frac{\partial^2 v}{\partial x^2} + \frac{\partial^2 v}{\partial y^2} = 0.$$

As an example, Riemann gave a detailed proof of the Cauchy integral theorem following the outline provided by Cauchy in 1846. The important idea was Green's theorem, which Riemann stated in the following form:

THEOREM   *Let X and Y be two functions of x and y continuous in a finite region T with infinitesimal area element designated by dT. Then*

$$\int_T \left(\frac{\partial X}{\partial x} + \frac{\partial Y}{\partial y}\right) dT = -\int_S (X\cos\xi + Y\cos\eta)\,ds,$$

where the latter integral is taken over the boundary curve S of T, ξ, η designating the angles the inward-pointing normal line to the curve makes with the x and y axes, respectively.

Riemann proved this by using the fundamental theorem of calculus to integrate $\partial X/\partial x$ along lines parallel to the x axis, getting values of X where the lines cross the boundary of the region. Because $dy = \cos \xi \, ds$ at each of those points, he could integrate with respect to y to get

$$\int \left[\int \frac{\partial X}{\partial x} \, dx \right] dy = - \int X \, dy = - \int X \cos \xi \, ds.$$

The other half of the theorem is proved similarly. Riemann then noted that

$$\frac{dx}{ds} = \pm \cos \eta \quad \text{and} \quad \frac{dy}{ds} = \mp \cos \xi,$$

where the sign depends on whether one gets from the tangent line to the inward normal line by traveling counterclockwise or clockwise. It follows that Green's theorem can be rewritten as

$$\int_T \left(\frac{\partial X}{\partial x} + \frac{\partial Y}{\partial y} \right) dT = \int_S \left(X \frac{dy}{ds} - Y \frac{dx}{ds} \right) ds,$$

from which the Cauchy integral theorem follows easily.

Much of Riemann's dissertation involved the introduction of an entirely new concept in the study of complex functions, the idea of a Riemann surface. In the case of functions of a real variable, it is possible to picture the function by a curve in two-dimensional space. Such a representation is no longer possible for complex functions, because the graph would need to be in a space of four real dimensions. An alternative way of picturing complex functions, then, is to trace the independent variable z along a curve in one plane and consider the curve generated by the dependent variable w in another plane. Riemann realized from the fact that a complex function always had a power series representation that "a function of $x + iy$ defined in a region of the (x, y) plane can be continued analytically in only one way." It follows that once one knows the values in a certain region, one can continue the function and even return to the same z value by, say, a continuous curve. There are then two possibilities. "Depending on the nature of the function to be continued, either this function will always assume the same value for the same value of [z], no matter how it is continued, or it will not."[53] In the first case, Riemann called the function single-valued, while in the second it is multiple-valued. As a simple example of the latter, one can take $w = z^{\frac{1}{2}}$. To study such functions effectively, it was not possible simply to use two planes as indicated above, for one would not know which value the function had for a given point on the first plane. Thus, Riemann came up with a new idea, to use a multiple plane, a covering of the z plane by as many sheets as the function has values. These sheets are attached along a line, say, the negative real axis, in such a way that whenever one moves in a curve across that line one changes from one sheet to another. In this way the multiple-valued function has only one value defined at each point of this Riemann surface. Since it may happen that after several circuits (two in the example above) one returns to a former value, the top sheet of this covering must be attached to the bottom one. It follows that it is not in general possible to construct a physical model of a Riemann surface in three-dimensional space. Nevertheless, the study of Riemann surfaces, initiated by Riemann to deal with multiple-valued complex functions, soon led Riemann and others into the realm of what is today called **topology**. The connection of topology with integration along curves and surfaces, barely touched by Cauchy in 1846, was explored in great detail in the second half of the nineteenth century and the early years of the twentieth.

22.3.5 The Riemann Zeta Function

One complex function that Riemann studied extensively, now called the Riemann zeta function, has had major importance since his time. As we saw in Chapter 17, this function had its start in the formula of Euler,

$$\sum_{n=1}^{\infty} \frac{1}{n^s} = \prod_{p} \frac{1}{\left(1 - \frac{1}{p^s}\right)},$$

where the product on the right ranges over all prime numbers p. The formula results from expanding each factor on the right as

$$\frac{1}{\left(1 - \frac{1}{p^s}\right)} = 1 + \frac{1}{p^s} + \frac{1}{(p^2)^s} + \frac{1}{(p^3)^s} + \cdots$$

and then noting that their product is a sum of terms of the form

$$\frac{1}{(p_1^{n_1} p_2^{n_2} \cdots p_k^{n_k})^s},$$

where the p_i are distinct primes and the n_i are positive integers. Because every positive integer can be expressed uniquely as a product of primes, the sum of all such expressions is exactly the left side of Euler's formula. Euler used this formula for integer values of s. In particular, letting $\zeta(s) = \sum_{n=1}^{\infty} 1/n^s$, he was able to show that $\zeta(2) = \pi^2/6$, $\zeta(4) = \pi^4/90$, and to provide a general method for calculating $\zeta(2n)$. (See Exercise 40 of Chapter 17.) Dirichlet, somewhat later, extended $\zeta(s)$ to real values $s > 1$ and was able to prove Euler's formula rigorously in that case.

By rewriting the expression for $\zeta(s)$ in terms of integrals, Riemann in 1859 was able to extend its domain to the entire complex plane. He showed, in fact, that $\zeta(s)$ was finite for all values of s except $s = 1$ and also that $\zeta(-2n) = 0$ for every positive integer n. Riemann used $\zeta(s)$ in his attempts to find an analytic expression for $\pi(x)$, the number of primes less than x. Although he was not completely successful in this, he mentioned in passing that it was "very likely" that all the complex zeros of $\zeta(s)$ had their real part equal to 1/2 but that he was unable to prove this result. This statement, that all the zeros have real part equal to 1/2, has become known as the **Riemann hypothesis**. Although at first sight, it appears that the Riemann hypothesis relates only to a particular complex function, it has turned out that the truth of the Riemann hypothesis implies many other results in number theory and other areas of mathematics. Although many mathematicians have attempted to prove it since Riemann's time, and recent computer calculations have shown that the 1,500,000,000 complex zeros closest to the real line all have their real part equal to 1/2, no proof of the hypothesis has yet been found, nor, for that matter, any counterexample. Its proof is still a major mathematical challenge, one whose solution would bring fame and even fortune (see Section 25.6.4) to the solver.

22.4   VECTOR ANALYSIS

Riemann stated Green's theorem in 1851 in terms of the equality of a double integral with an integral along a curve taken with respect to the curve element ds. It was the use in physics of integrals over curves to represent work done along the curves that seems to have inspired a change in notation that occurred in the 1850s, in which the curve integral was replaced by a line integral, an integral of the form $\int p\, dx + q\, dy$. Although this notation had been used in complex integration, the physicists converted it into an expression involving vectors. Other physical concepts involving vectors led to other important integral theorems during the nineteenth century.

22.4.1   Line Integrals and Multiple Connectivity

Clerk Maxwell noted in 1855 that if α, β, and γ are the components of the "intensity of electric action" ϵ parallel to the x, y, and z axes, respectively, and if ℓ, m, n are the corresponding direction cosines of the tangent to the curve (the cosines of the angles the tangent makes with the three coordinate axes), then ϵ (considered as operating along the curve) can be written in the form $\ell\alpha + m\beta + n\gamma$. Because $\ell\, ds = dx$, $m\, ds = dy$, and $n\, ds = dz$, Maxwell wrote $\int \epsilon\, ds = \int \alpha\, dx + \beta\, dy + \gamma\, dz$. The following year this notation appeared in a physics text by Charles Delaunay (1816–1872). Delaunay was somewhat more explicit than Maxwell, noting that if F is a force and F_1 its tangential component along a curve, then the work done by the force acting along the curve could be represented by $\int F_1\, ds$. Again, if the rectangular components of F are X, Y, Z, this latter integral can be rewritten as $\int X\, dx + Y\, dy + Z\, dz$.

The line integral notation quickly became standard in physics and was adopted by Riemann in a paper of 1857 in which he studied the Riemann surfaces R on which were described the curves over which these line integrals were taken. Riemann began by observing that the integral of an exact differential $X\, dx + Y\, dy$ is zero when taken over the perimeter of a region in this surface:

> Hence, the integral $\int (X\, dx + Y\, dy)$ has the same value when taken between two fixed points along two different paths, provided the two paths together form the entire boundary of a region of R. Thus, if every closed curve in the interior of R bounds a region of R, then the integral always has the same value when taken from a fixed initial point to one and the same endpoint, and is a continuous function of the position of the endpoint which is independent of the path of integration. This gives rise to a distinction among surfaces: simply connected ones, in which every closed curve bounds a region of the surface—as, for example, a disk—and multiply connected ones, for which this does not happen—as, for example, an annulus bounded by two concentric circles.[54]

Riemann proceeded to refine the notion of multiple connectedness: "A surface F is said to be $(n+1)$-ply connected when n closed curves $A_1, \ldots, A_n$ can be drawn on it which neither individually nor in combination bound a region of F, while if augmented by any other closed curve A_{n+1}, the set bounds some region of F."[55] Riemann noted further that an $(n+1)$-ply connected surface can be changed into an n-ply connected one by means of a cut, a curve going from one boundary point through the interior to another boundary point. For example, an annulus, which is doubly connected, can be reduced to a simply connected region by any cut q that does not disconnect it. A double annulus needs two cuts to be reduced to a simply connected region.

Using the idea of cuts, Riemann was able to describe what happens when one integrates an exact differential on an $(n + 1)$-ply connected surface R. If one removes n cuts from this surface, there remains a simply connected surface R'. Integration of the exact differential $X\,dx + Y\,dy$ from a fixed starting point over any curve in R' then determines, as before, a single-valued continuous function Z of position on this surface. However, whenever the path of integration crosses a cut, the value jumps by a fixed number, dependent on the cut. There are n such numbers, one for each cut. This notion of multiple connectedness turned out to be important in physics, particularly in fluid dynamics and electromagnetism, and so it was extended to regions of three-dimensional space by such physicists as Hermann von Helmholtz (1821–1894), William Thomson (1824–1907), and Maxwell.

22.4.2   Surface Integrals and the Divergence Theorem

Physicists were interested not only in line integrals but also in surface integrals, integrals of functions and vector fields over two-dimensional regions. Recall that as early as 1760 Lagrange had given an explicit expression for the element of surface dS in the process of calculating surface areas. It was not until 1811, however, in the second edition of his *Mécanique analytique*, that Lagrange introduced the general notion of a surface integral. He noted that if the tangent plane at dS makes an angle γ with the xy plane, then simple trigonometry allows one to rewrite $dx\,dy$ as $\cos\gamma\,dS$. It followed that if A is a function of three variables, then $\int A\,dx\,dy = \int A\cos\gamma\,dS$, the second integral being taken over a region in the surface, the first over the projection of that region in the plane. Similarly, if β is the angle the tangent plane makes with the xz plane and α that with the yz plane, then $dx\,dz = \cos\beta\,dS$ and $dy\,dz = \cos\alpha\,dS$. Lagrange noted that α, β, and γ could also be considered as the angles that a normal to the surface element makes with the x, y, and z axes, respectively.

Lagrange used surface integrals in dealing with fluid dynamics. In 1813 Gauss used the same concept in considering the gravitational attraction of an elliptical spheroid. But Gauss went further than Lagrange in showing how to calculate an integral with respect to dS in the case where the surface S is given parametrically by three functions $x = x(p, q)$, $y = y(p, q)$, $z = z(p, q)$. Using a geometrical argument, he demonstrated that

$$dS = \left[\left(\frac{\partial(y, z)}{\partial(p, q)}\right)^2 + \left(\frac{\partial(z, x)}{\partial(p, q)}\right)^2 + \left(\frac{\partial(x, y)}{\partial(p, q)}\right)^2\right]^{1/2} dp\,dq$$

and hence that any integral with respect to dS can be reduced to an integral of the form $\int f\,dp\,dq$, where f is either explicitly or implicitly a function of the two variables p, q.

Gauss used his study of integrals over surfaces to prove certain special cases of what is today known as the **divergence theorem**. The general case of this theorem was, however, first stated and proved in 1826 by Mikhail Ostrogradsky (1801–1861), a Russian mathematician who was studying in Paris in the 1820s.[56] In his paper entitled "Proof of a Theorem in Integral Calculus," which came out of his study of the theory of heat, Ostrogradsky considered a surface with surface element ϵ bounding a solid region with volume element ω. With p, q, and r being three differentiable functions of x, y, z and with the angles α, β, and γ as defined above, Ostrogradsky stated the divergence theorem in the form

BIOGRAPHY

Mikhail Ostrogradsky (1801–1861)

Mikhail Ostrogradsky found his way into mathematics through a desire to be an army officer. Since he had been born into a family of modest means in the Ukraine, he could not manage the expensive lifestyle of an officer without an independent income. To support his future career, he enrolled in the University of Kharkov in 1816. He became interested in mathematics and physics and passed the exam for the degree in 1820. He did not actually receive the degree because the minister of religious affairs and national education decided to punish Ostrogradsky's teacher, T. F. Osipovsky, the rector of the university, for failing to instill the proper religious and pro-Czarist attitudes in his students.

Ostrogradsky left Russia to study in Paris for several years, where he produced some of his most important mathematical work. In 1828 he returned to St. Petersburg and was elected a member of the Academy of Sciences. He connected with his original military ambition by teaching mathematics at military academies. In 1847 he became responsible for all mathematics education in these schools and later wrote several important texts for use there (Fig. 22.9).

FIGURE 22.9

Ostrogradsky on a Russian stamp

$$\int \left(\frac{\partial p}{\partial x} + \frac{\partial q}{\partial y} + \frac{\partial r}{\partial z} \right) \omega = \int (p \cos \alpha + q \cos \beta + r \cos \gamma)\epsilon,$$

where the left-hand integral is taken over the solid V and the right-hand one over the boundary surface S. Today the theorem is generally written, by use of Lagrange's idea, in the form

$$\iiint_V \left(\frac{\partial p}{\partial x} + \frac{\partial q}{\partial y} + \frac{\partial r}{\partial z} \right) dx\, dy\, dz = \iint_S p\, dy\, dz + q\, dz\, dx + r\, dx\, dy.$$

This result, like Green's theorem, is a generalization of the fundamental theorem of calculus, so Ostrogradsky's proof used that theorem. To integrate $(\partial p / \partial x)\omega$ over a "narrow cylinder" going through the solid in the x direction with cross-sectional area $\bar{\omega}$, he used the fundamental theorem to express the integral as $\int (p_1 - p_0)\bar{\omega}$, where p_0 and p_1 are the values of p on the pieces of surface where the cylinder intersects the solid (Fig. 22.10). Because $\bar{\omega} = \epsilon_1 \cos \alpha_1$ on one section of the surface and $\bar{\omega} = -\epsilon_0 \cos \alpha_0$ on the other, where α_1 and α_0 are the angles made by the normal at the surface elements ϵ_1, ϵ_0, respectively, Ostrogradsky had demonstrated that

$$\frac{\partial p}{\partial x} \omega = \int p_1 \epsilon_1 \cos \alpha_1 + \int p_0 \epsilon_0 \cos \alpha_0 = \int (p \cos \alpha)\epsilon,$$

where the left integral is over the cylinder and the right ones over the two pieces of surface. Adding up the integrals over all such cylinders gives one-third of the desired result, the other two-thirds being done similarly. Interestingly enough, Ostrogradsky generalized his result to n dimensions in 1836, thus giving one of the earliest statements of a result in dimension greater than three.

FIGURE 22.10

Ostrogradsky's proof of the
divergence theorem

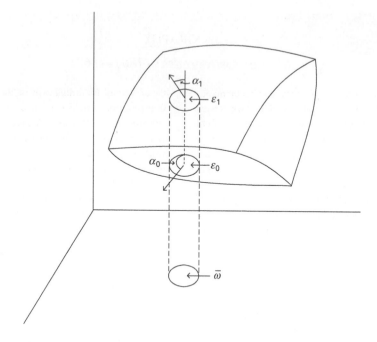

22.4.3   Stokes's Theorem

The divergence theorem relates an integral over a solid to one over the bounding surface, while Green's theorem relates an integral over a region in the plane to one over the boundary curve. A similar result comparing an integral over a surface in three dimensions to one around the boundary curve, a result now known as **Stokes's theorem**, first appeared in print in 1854. George Stokes (1819–1903) had for several years been setting the Smith's Prize Exam at Cambridge University and, in the February 1854 examination, posed the following

PROBLEM 8   *If X, Y, Z be functions of the rectangular coordinates x, y, z; dS an element of any limited surface; ℓ,m, n the cosines of the inclinations of the normal at dS to the axes; ds an element of the boundary line, shew that*

$$\iint \left[\ell \left(\frac{\partial Z}{\partial y} - \frac{\partial Y}{\partial z} \right) + m \left(\frac{\partial X}{\partial z} - \frac{\partial Z}{\partial x} \right) + n \left(\frac{\partial Y}{\partial x} - \frac{\partial X}{\partial y} \right) \right] dS$$
$$= \int \left(X \frac{dx}{ds} + Y \frac{dy}{ds} + Z \frac{dz}{ds} \right) ds$$

. . . the single integral being taken all around the perimeter of the surface.[57]

It is not known whether any of the students proved the theorem, although Maxwell did sit for the exam. However, the theorem had already appeared in a letter of William Thomson to Stokes on July 2, 1850, and the integrand on the left side had appeared in two earlier works of

BIOGRAPHY

George Stokes (1819–1903)

Although his three brothers followed his father in an ecclesiastical career in their native Ireland, Stokes was
drawn to mathematics through the influence of a teacher. In
1837 he entered Pembroke College of Cambridge University
where most of his mathematics education came from his tutor,
William Hopkins. Stokes graduated in 1841 as senior wrangler and eight years later was appointed to the Lucasian chair
of mathematics, a position he held until his death. His theoretical and experimental studies during his career spanned much
of natural philosophy, including such areas as hydrodynamics,
elasticity, and the diffraction of light. His various excursions
into pure mathematics were caused by his need to develop
methods to solve particular physical problems or to justify the
validity of the mathematical techniques he was already using. Stokes served the scientific community in various official
posts. In particular, he was the secretary of the Royal Society
from 1854 to 1885, its president from 1885 to 1890, and the
representative of the University of Cambridge in Parliament
from 1887 to 1891.

Stokes where it represented the angular velocity of a certain fluid. The first published proof
of the result appeared in a monograph of Hermann Hankel (1839–1873) in 1861, at least for
the case where the surface is given explicitly as a function $z = z(x, y)$. Hankel substituted
the value of z and $dz = (\partial z/\partial x)\, dx + (\partial z/\partial y)\, dy$ into the right-hand integral, thus reducing
it to an integral in two variables, then used Green's theorem to convert it to a double integral
easily seen to be equal to the surface integral on the left.

Stokes himself proved a related result. It was clear that

$$\frac{\partial}{\partial x}\left(\frac{\partial Z}{\partial y} - \frac{\partial Y}{\partial z}\right) + \frac{\partial}{\partial y}\left(\frac{\partial X}{\partial z} - \frac{\partial Z}{\partial x}\right) + \frac{\partial}{\partial z}\left(\frac{\partial Y}{\partial x} - \frac{\partial X}{\partial y}\right) = 0.$$

In 1849, Stokes proved what amounted to the converse, namely, if A, B, C are functions
satisfying

$$\frac{\partial A}{\partial x} + \frac{\partial B}{\partial y} + \frac{\partial C}{\partial z} = 0,$$

then there exist functions X, Y, Z such that

$$A = \frac{\partial Z}{\partial y} - \frac{\partial Y}{\partial z} \qquad B = \frac{\partial X}{\partial z} - \frac{\partial Z}{\partial x} \qquad C = \frac{\partial Y}{\partial x} - \frac{\partial X}{\partial y}.$$

This result, like Clairaut's result in two dimensions that an exact differential is the differential of a function, is only valid in certain simple domains. Neither Stokes nor Thomson,
who gave a different proof in 1851, dealt with that limitation. Their proofs required that certain differential equations be solvable, and they simply assumed that the solutions could
be found without worrying about the specific conditions that would assure this. In any
case, their result, combined with Stokes's theorem itself, shows that under the conditions
stated on A, B, and C, the surface integral $\iint (\ell A + mB + nC)\, dS$ (or, in more modern

notation, $\iint A \, dy \, dz + B \, dz \, dx + C \, dx \, dy$) does not depend on the surface but only on the boundary curve.

Both Stokes's theorem and the divergence theorem appeared in the opening chapter of Maxwell's *Treatise on Electricity and Magnetism* and were used often in the remainder of the work. Because Maxwell was an advocate of quaternion notation in physics, he wrote out these theorems in quaternion form, using the fact that if the vector operator $\nabla = (\partial/\partial x)i + (\partial/\partial y)j + (\partial/\partial z)k$ is applied to the vector $\sigma = Xi + Yj + Zk$, the resulting quaternion can be written

$$\nabla\sigma = -\left(\frac{\partial X}{\partial x} + \frac{\partial Y}{\partial y} + \frac{\partial Z}{\partial z}\right) + \left(\frac{\partial Z}{\partial y} - \frac{\partial Y}{\partial z}\right)i + \left(\frac{\partial X}{\partial z} - \frac{\partial Z}{\partial x}\right)j + \left(\frac{\partial Y}{\partial x} - \frac{\partial X}{\partial y}\right)k.$$

Maxwell named the scalar and vector parts of $\nabla\sigma$ the **convergence** of σ and the **curl** of σ, respectively, because of his interpretation of their physical meaning. Maxwell's convergence is the negative of what is today called the **divergence** of σ.

The pure vector form of these theorems finally appeared in the work of Gibbs near the end of the century. Setting $dV = dx \, dy \, dz$ to be the element of volume, $d\mathbf{a} = dy \, dz \, i + dz \, dx \, j + dx \, dy \, k$ the element of surface area, and $d\mathbf{r} = i \, dx + j \, dy + k \, dz$, Gibbs wrote the divergence theorem in the form

$$\iiint \nabla \cdot \sigma \, dV = \iint \sigma \cdot d\mathbf{a}$$

and Stokes's theorem as

$$\iint (\nabla \times \sigma) \cdot d\mathbf{a} = \int \sigma \cdot d\mathbf{r}.$$

Note that the left-hand integrands in the two theorems can also be written as **div** $\sigma \, dV$ and **curl** $\sigma \cdot d\mathbf{a}$, respectively.

Even in vector form it is not obvious that Green's theorem, Stokes's theorem, and the divergence theorem can be united into a single result. But Vito Volterra (1860–1940) in 1889 was able to unite them in the course of a study of hypersurfaces in n-dimensional space. (The study of n-dimensional space was new in 1836, but 50 years later, it was already commonplace.) Not only did Volterra state a result using a plethora of indices, of which the three theorems mentioned were all low-dimensional special cases, but he, along with Henri Poincaré (1854–1912), also generalized to higher dimensions the result of Stokes and Thomson on what Poincaré called the integrability conditions. These were the conditions on line integrals, surface integrals, and their higher-dimensional analogues that ensured the integrals did not depend on the curve, surface, or hypersurface over which they were being integrated, but only on the boundary of that geometric object. It was this generalization, now known as Poincaré's lemma, that helped provide the tools for Poincaré's study of the relationship of such multiple integrals to the topology of the domains of integration, a study that Riemann had already begun. Poincaré, in a series of papers near the turn of the twentieth century, developed this study into the beginnings of the subjects now called algebraic and differential topology, some aspects of which will be considered in Chapter 25.

EXERCISES

1. Prove the theorem of Cauchy:

   If $\lim_{x \to \infty} f(x+1) - f(x) = \infty$, then $\lim_{x \to \infty} f(x)/x = \infty$.

2. Use the theorem of Exercise 1 and the theorem on p. 768 to show that

   $$\lim_{x \to \infty} \frac{a^x}{x} = \infty \quad \text{and} \quad \lim_{x \to \infty} \frac{\log x}{x} = 0.$$

3. Show that

   $$\frac{2}{\pi} \int_0^\infty \frac{x^2 \, dt}{t^2 + x^2} = |x|.$$

4. Use the modern definition of continuity and Cauchy's trigonometric identity

   $$\sin(x + \alpha) - \sin x = 2 \sin \frac{1}{2}\alpha \, \cos\left(x + \frac{1}{2}\alpha\right)$$

   to show that $\sin x$ is continuous at any value of x.

5. Prove the following theorem of Cauchy: If $f(x)$ is positive for sufficiently large values of x and if the ratio $f(x+1)/f(x)$ converges to k as x increases indefinitely, then $[f(x)]^{1/x}$ also converges to k as x increases indefinitely.

6. Use the theorem of Exercise 5 to show that

   $$\lim_{x \to \infty} x^{1/x} = 1.$$

7. Use the Cauchy criterion to show that the series

   $$1 + \frac{1}{1!} + \frac{1}{2!} + \frac{1}{3!} + \cdots$$

   converges.

8. Show that if the sequence $\{a_i\}$ converges to a and if f is continuous, then the sequence $\{f(a_i)\}$ converges to $f(a)$.

9. Show that the series $\{u_k(x)\}$, where $u_1(x) = x$ and $u_k(x) = x^k - x^{k-1}$ for $k > 1$, satisfies the hypotheses of Cauchy's theorem 6-1-1 in a neighborhood of $x = 1$ but not the conclusion. Analyze Cauchy's proof for this case to see where it fails.

10. Use the trigonometric formula of Exercise 4 to prove that the derivative of the sine function is the cosine.

11. By putting $a^i = 1 + \beta$, use Cauchy's definition of derivative to show that the derivative of $y = a^x$ is $y' = a^x / \log_a(e)$.

12. Prove the algebraic result used by Cauchy in his main theorem about derivatives:

   If $A < \dfrac{a_i}{b_i} < B$ for $i = 1, 2, \ldots, n$, then $A < \dfrac{\sum_{i=1}^n a_i}{\sum_{i=1}^n b_i} < B$.

13. Show that if $f(x)$ is continuous on $[a, b]$ and if $a = x_0 < x_1 < \cdots < x_n = b$ is a partition of $[a, b]$ into subintervals, then the sum

   $$f(x_0)(x_1 - x_0) + f(x_1)(x_2 - x_1) + \cdots$$
   $$+ f(x_{n-1})(x_n - x_{n-1})$$

   is equal to $(b - a)f(x_0 + \theta(b - a))$ for some θ between 0 and 1.

14. Let $f(x) = x^2 + 3x$ on $[1, 3]$. Partition $[1, 3]$ into eight subintervals and determine the θ that satisfies the property of Exercise 13.

15. Complete Bolzano's proof of the least upper bound criterion by showing that the value U to which the constructed sequence converges is the least upper bound of all numbers having the property M.

16. Let A be the set of numbers in $(3/5, 2/3)$ that have decimal expansions containing only finitely many zeros and sixes after the decimal point and no other integer. Find the least upper bound of A.

17. Suppose M is the property that $x < 0$ or that $x^3 < 3$. Since this property does not belong to all x but does belong to all x less than 1, this property satisfies the conditions of Bolzano's least upper bound theorem. Beginning with the quantity $V = 1 + 1$, for which M is not valid for all x smaller than it, use Bolzano's proof method to construct an approximation to $\sqrt[3]{3}$ accurate to three decimal places.

18. Show that $\phi(x) = \alpha e^{mx}$, $\psi(y) = \beta \cos ny$, with $m^2 = n^2 = \frac{1}{A}$, are solutions to

   $$\frac{\phi(x)}{\phi''(x)} = -\frac{\psi(y)}{\psi''(y)} = A.$$

   Conclude that $v = ae^{-nx} \cos ny$ is a solution to

   $$\frac{\partial^2 v}{\partial x^2} + \frac{\partial^2 v}{\partial y^2} = 0.$$

19. Show that

   $$\int_0^\pi \sin mx \sin nx \, dx = \begin{cases} 0, & \text{if } m \neq n; \\ \frac{\pi}{2}, & \text{if } m = n. \end{cases}$$

20. Calculate the coefficients b_i in the Fourier cosine series of a function $\phi(x)$. That is, determine b_i if

   $$\frac{1}{2}\pi\phi(x) = b_0 + b_1 \cos x + b_2 \cos 2x + b_3 \cos 3x + \cdots.$$

21. Use Fourier's method of integration to calculate the Fourier series for $\phi(x) = \frac{1}{2}x$ quoted in the text. Check the correctness of this result for $x = \frac{\pi}{2}$ by using other known series sums.

22. Consider the Riemann function

$$f(x) = \sum_{n=1}^{\infty} \frac{\phi(nx)}{n^2}$$

defined on $[0, 1]$, where $\phi(x)$ is equal to x minus the nearest integer, or, if there are two equally near integers, equal to 0. Show that f is continuous except at the infinitely many points $x = p/2n$ with p and n relatively prime.

23. Prove Cauchy's theorem on the continuity of the sum of a series of continuous functions under the additional assumption that the series converges uniformly.

24. Let

$$u_k = \frac{1}{k(k+1)}$$

and

$$v_k(h) = u_k + \frac{2h}{((k-1)h + 1)(kh + 1)},$$

where h is a positive real variable. Show that $\lim_{h \to 0} v_k(h) = u_k$, that $\sum u_k$ converges, and that $\sum v_k(h)$ converges for sufficiently small h. Show also that

$$\lim_{h \to 0} \sum v_k(h) \neq \sum u_k.$$

25. Let $\mathcal{R}$ be the set of all real numbers as defined by Dedekind via his cuts. Show that this set possesses the basic attribute of continuity. Namely, show that if $\mathcal{R}$ is split into two classes $\mathcal{A}_1, \mathcal{A}_2$ such that every real number in $\mathcal{A}_1$ is less than every real number in $\mathcal{A}_2$, then there exists exactly one real number α that is either the greatest number in $\mathcal{A}_1$ or the smallest number in $\mathcal{A}_2$.

26. Define a natural ordering $<$ on Dedekind's set of real numbers $\mathcal{R}$ defined by the notion of cuts. That is, given two cuts $\alpha = (\mathcal{A}_1, \mathcal{A}_2)$, and $\beta = (\mathcal{B}_1, \mathcal{B}_2)$, define $\alpha < \beta$. Show that this ordering $<$ satisfies the same basic properties on $\mathcal{R}$ as it satisfies on the set of rational numbers.

27. Define an addition on Dedekind's cuts. Show that $\alpha + \beta = \beta + \alpha$ for any two cuts α and β.

28. Prove the theorem that every bounded increasing sequence of real numbers has a limit number, using Dedekind's cuts and also using Cantor's fundamental sequences. Which proof is easier?

29. Show that if $\{a_i\}$ and $\{b_i\}$ are fundamental sequences with $\{b_i\}$ not defining the limit 0, then $\{a_i/b_i\}$ is also a fundamental sequence.

30. Define the product AB of two fundamental sequences $A = \{a_i\}$ and $B = \{b_i\}$ as the sequence consisting of the products $\{a_ib_i\}$. Show that this definition makes sense and that if $AB = C$, then $B = C/A$, where division is defined as in Exercise 29.

31. Determine explicitly a point set P whose first and second derived sets, P', and P'', are different from P and from each other.

32. In 1890 Cantor gave a second proof that the real numbers of the interval $(0, 1)$ could not be placed in one-to-one correspondence with the natural numbers. Suppose that these numbers were in one-to-one correspondence with the natural numbers. Then there is a listing $r_1, r_2, r_3, \ldots$ of the real numbers in the interval. Write each such number in its infinite decimal form:

$$r_1 = 0.a_{11}a_{12}a_{13} \ldots$$
$$r_2 = 0.a_{21}a_{22}a_{23} \ldots$$
$$r_3 = 0.a_{31}a_{32}a_{33} \ldots .$$

Now define a number b by choosing $b = 0.b_1b_2b_3 \ldots$, where $b_1 \neq a_{11}, b_2 \neq a_{22}, b_3 \neq a_{33}, \ldots$. Show that b cannot be in the original list and thus that such a listing cannot exist.

33. Fill in all the details of the calculation using residues that shows that

$$\int_{-\infty}^{\infty} \frac{\cos x}{1 + x^2} dx = \frac{\pi}{e}.$$

34. Show using residues that

$$\int_{0}^{\infty} \frac{dx}{1 + x^6} = \frac{2\pi}{3}.$$

35. Let the complex function $w(z)$ be given as the sum $u(x, y) + iv(x, y)$. Suppose that the Cauchy-Riemann equations are satisfied, that is, that $\partial u/\partial x = \partial v/\partial y$ and $\partial v/\partial x = -\partial u/\partial y$. Show that the derivative dw/dz is equal to $\partial u/\partial x + i\partial v/\partial x$.

36. Suppose a surface S is defined by three parametric equations $x = x(p, q), y = y(p, q), z = z(p, q)$. Show geometrically that the element of surface dS can be written in the form

$$dS = \left[\left(\frac{\partial(y, z)}{\partial(p, q)} \right)^2 + \left(\frac{\partial(z, y)}{\partial(p, q)} \right)^2 + \left(\frac{\partial(x, y)}{\partial(p, q)} \right)^2 \right]^{1/2} dp \, dq.$$

37. Show that if $\sigma = Ai + Bj + Ck$ is a vector field with **div** $\sigma = 0$, then $\sigma = $ **curl** τ for some vector field τ.

38. Use Stokes's theorem to show that if curl $\sigma = 0$, then $\int_C \sigma \cdot d\mathbf{r}$ is independent of the curve C but depends only on its endpoints. Similarly, show that if div $\sigma = 0$, then $\int_S \sigma \cdot d\mathbf{a}$ is independent of the surface S but depends only on its boundary curve.

39. What does Cauchy mean by his statement that an irrational number is the limit of the various fractions that approach it? What does Cauchy understand by the term "irrational number" or, even, by the term "number"?

40. Explain the differences between Cauchy's definition of continuity on an interval and the usual modern definition of continuity at a point. Does a function that satisfies Cauchy's definition satisfy the modern one for every point in the interval? Does a function that satisfies the modern definition for every point in an interval satisfy Cauchy's definition?

41. Develop a lesson plan for teaching the concept of uniform convergence by beginning with Cauchy's incorrect theorem and proof.

42. Compare the Eulerian and the Riemannian derivations of the Cauchy-Riemann equations. Which makes a better introduction for a course in complex analysis?

REFERENCES AND NOTES

Among the best recent general works on the history of analysis in the nineteenth century are Ivor Grattan-Guinness, ed., *From the Calculus to Set Theory* (London: Duckworth, 1980), a collection of essays by experts on the various topics covered; Umberto Bottazzini, *The Higher Calculus: A History of Real and Complex Analysis from Euler to Weierstrass* (New York: Springer, 1986); and Ivor Grattan-Guinness, *The Development of the Foundations of Mathematical Analysis from Euler to Riemann* (Cambridge: MIT Press, 1970). The best treatment of Cauchy's work on calculus is Judith V. Grabiner, *The Origins of Cauchy's Rigorous Calculus* (Cambridge: MIT Press, 1981), while the best study of Cauchy's work in complex functions is Frank Smithies, *Cauchy and the Creation of Complex Function Theory* (Cambridge: Cambridge University Press, 1997).

Cauchy's *Cours d'analyse de l'École Royale Polytechnique* is reprinted in Cauchy, *Oeuvres complète d'Augustin Cauchy* (Paris: Gauthier-Villars, 1882), series 2, volume 3, while the *Résumé des leçons données á l'École Royale Polytechnique sur le calcul infinitesimal* is in series 2, volume 4. Bolzano's paper on the intermediate value theorem is available in S. B. Russ, "A Translation of Bolzano's Paper on the Intermediate Value Theorem," *Historia Mathematica* 7 (1980), 156–185, but it has also been published in Steve Russ, *The Mathematical Works of Bernard Bolzano* (Oxford: Oxford University Press, 2004), pp. 251–278. A part of Abel's paper in which he gave a counterexample to Cauchy's theorem on convergence is translated in Garrett Birkhoff, *A Source Book in Classical Analysis* (Cambridge: Harvard University Press, 1973), pp. 68–70. Parts of Riemann's "Foundations for a General Theory of Functions of One Complex Variable" and his "Theory of Abelian Functions" are also in the Birkhoff *Source Book*, pp. 48–50 and 50–55, re-

spectively. Fourier's *Analytical Theory of Heat* is available in an English translation (New York: Dover, 1955). Dedekind's work on Dedekind cuts is in Richard Dedekind, "Continuity and Irrational Numbers," translated by Wooster Beman, in Dedekind, *Essays on the Theory of Numbers* (La Salle, IL: Open Court, 1948). The same volume also contains a translation of "Was sind und was sollen die Zahlen" as "The Nature and Meaning of Numbers." Cantor's work on set theory is available as Georg Cantor, *Contributions to the Founding of the Theory of Transfinite Numbers*, translated by P. E. B. Jourdain (Chicago: Open Court, 1915). Wessel's "On the Analytical Representation of Direction" is translated in David Smith, *A Source Book in Mathematics* (New York: Dover, 1959), pp. 55–66.

1. Abel, "Investigation of the Series $1 + \frac{m}{1}x + \frac{m(m-1)}{1 \cdot 2}x^2 + \cdots$," *Crelle's Journal* 1 (1826), 311–339, p. 316. A part of this paper, including this quotation, is translated in Birkhoff, *Source Book*, pp. 68–70.

2. Sylvestre Lacroix, *An Elementary Treatise on the Differential and Integral Calculus*, translated by Charles Babbage, George Peacock, and John Herschel (Cambridge: J. Deighton, 1816), p. 2.

3. Ibid., p. 5.

4. Oystein Ore, *Niels Henrik Abel: Mathematician Extraordinary* (New York: Chelsea, 1974), preface.

5. Ibid., p. 147.

6. Cauchy, *Cours d'analyse de l'École Royale Polytechnique*, reprinted in Cauchy, *Oeuvres complète* (2), 3, p. 19. (All further page references to the *Cours d'analyse* will be from this edition. All other references to Cauchy will also be to the *Oeuvres*.) Much of the detail in the first section of this

chapter is adapted from Grabiner, *Origins*. For a brief overview of the work of Cauchy and others on the notion of continuity, see Judith V. Grabiner, "Who Gave You the Epsilon? Cauchy and the Origins of Rigorous Calculus," *American Mathematical Monthly* 90 (1983), 185–194. For more details on the development of the concept of the derivative, see Judith V. Grabiner, "The Changing Concept of Change: The Derivative from Fermat to Weierstrass," *Mathematics Magazine* 56 (1983), 195–203.

7. Cauchy, *Cours d'analyse*, p. 54.

8. Lagrange, *Théorie des fonctions analytique*, in *Oeuvres de Lagrange* (Paris: Gauthier-Villars, 1867–1892), vol. 9, p. 28. Quoted in Grabiner, *Cauchy's Rigorous Calculus*, p. 95.

9. Cauchy, *Cours d'analyse*, p. 43.

10. S. B. Russ, "Translation," p. 159. For more discussion of the work of Bolzano, see I. Grattan-Guinness, "Bolzano, Cauchy and the 'New Analysis' of the Early Nineteenth Century," *Archive for History of Exact Sciences* 6 (1970), 372–400. Grattan-Guinness claims that Cauchy took the central ideas of his definitions of continuity and convergence from Bolzano. But see also H. Freudenthal, "Did Cauchy Plagiarize Bolzano?" *Archive for History of Exact Sciences* 7 (1971), 375–392, for a contrary view.

11. Ibid., p. 162.

12. Cauchy, *Cours d'analyse*, p. 114.

13. Ibid., p. 116.

14. Quoted in A. J. Franco de Oliveira, "Anastácio da Cunha and the Concept of Convergent Series," *Archive for History of Exact Sciences* 39 (1988), 1–12, p. 4. For more on da Cunha, see João Filipe Queiró, "José Anastácio da Cunha: A Forgotten Forerunner," *The Mathematical Intelligencer*, 10 (1988), 38–43, and A. P. Youschkevitch, "J. A. da Cunha et les fondements de l'analyse infinitésimale," *Revue d'Histoire des Sciences* 26 (1973), 3–22.

15. Russ, "Translation," p. 171.

16. This treatment of Cauchy's erroneous proof is adapted from that found in unpublished notes of V. Frederick Rickey, *Using History in Teaching Calculus*.

17. Cauchy, *Resumé des leçons données á l'École Polytechnique sur le calcul infinitésimal*, in *Oeuvres* (2), 4, p. 44.

18. Lacroix, *Elementary Treatise*, p. 179.

19. Cauchy, "Mémoire sur l'intégration des équations lineares aux différentielles partielles et coefficients constantes," in *Oeuvres*, (2), 1, 275–357, p. 354.

20. Ibid., p. 334.

21. Cauchy, *Resumé*, pp. 122–123.

22. Ibid., p. 125.

23. Quoted in Grattan-Guinness, *From the Calculus to Set Theory*, p. 158.

24. Ibid., p. 153.

25. Quoted in Grattan-Guinness, *Foundations of Mathematical Analysis*, p. 104.

26. Quoted in Bottazzini, *Higher Calculus*, p. 244.

27. E. Heine, "Die Elemente der Functionenlehre," *Journal für die Reine und Angewandte Mathematik* 74 (1872), 172–188, p. 184.

28. Quoted in Ann Hibner Koblitz, *Sofia Kovalevskaia: Scientist, Writer, Revolutionary* (Boston: Birkhäuser, 1983), p. 197. This book provides a detailed nontechnical biography of Kovalevskaya as well as a sketch of her mathematics. Her mathematical work is treated in more detail in Roger Cooke, *The Mathematics of Sonya Kovalevskaya* (New York: Springer-Verlag, 1984).

29. Dedekind, "Continuity and Irrational Numbers," pp. 1–2.

30. Ibid., p. 9.

31. Ibid., p. 11.

32. Ibid., pp. 12–13.

33. Ibid., p. 15.

34. Ibid., p. 22.

35. Quoted in Bottazzini, *Higher Calculus*, p. 277.

36. Ibid., p. 278.

37. Quoted in Joseph Dauben, *Georg Cantor: His Mathematics and Philosophy of the Infinite* (Princeton, NJ: Princeton University Press, 1979), p. 49. Dauben's biography provides a detailed study of Cantor's work and how it relates to the mathematics and philosophy of his day.

38. Cantor, *Contributions to the Founding of the Theory of Transfinite Numbers*, p. 85.

39. Ibid., p. 86.

40. Quoted in Gregory H. Moore, "Towards a History of Cantor's Continuum Problem," in David Rowe and John McCleary, eds., *The History of Modern Mathematics* (San Diego, CA: Academic Press, 1989), vol. 1, 79–121, p. 82. The two-volume work of Rowe and McCleary presents the proceedings of a conference on nineteenth-century mathematics held in 1988. The papers are well worth reading.

41. Dedekind, "The Nature and Meaning of Numbers," p. 45.

42. Ibid., p. 50.

43. Ibid., p. 53.

44. Ibid., p. 63.

45. Ibid., p. 64.

46. Wessel, "On the Analytical Representation of Direction," p. 58.

47. Quoted in Bottazzini, *Higher Calculus*, p. 156, from the letter published in Gauss, *Werke* (Göttingen, 1866), vol. 8, pp. 90–92.

48. Cauchy, *Cours d'analyse*, p. 154.

49. Cauchy, *Mémoire sur les intégrales définies prises entre des limites imaginaires* (Paris, 1825), pp. 42–43, translated in Birkhoff, *Source Book*, p. 33.

50. Ibid., p. 44; p. 34.

51. Riemann, "Grundlagen für eine allgemeine Theorie der Functionen einer veränderlichen complexen Grösse," from Birkhoff, *Source Book*, p. 48.

52. Ibid., p. 49.

53. Riemann, "Theorie der Abel'sche Funktionen," *Crelle's Journal* 54 (1857), from Birkhoff, *Source Book*, p. 51.

54. Ibid., pp. 52–53.

55. Ibid.

56. More information on the divergence theorem, as well as Green's theorem and Stokes's theorem, can be found in Victor J. Katz, "The History of Stokes' Theorem," *Mathematics Magazine* 52 (1979), 146–156.

57. Stokes, *Mathematical and Physical Papers*, (Cambridge: Cambridge University Press, 1905), vol. 5, p. 320.

Probability and Statistics in the Nineteenth Century

Observations and statistics agree in being quantities grouped about a Mean; they differ, in that the Mean of observations is real, of statistics is fictitious. The mean of observations is a cause, as it were the source from which diverging errors emanate. The mean of statistics is a description, a representative quantity put for a whole group, the best representative of the group, that quantity which . . . minimizes the error unavoidably attending such practice.

—Francis Edgeworth, in "Observations and Statistics: An Essay on the Theory of Errors of Observation and the First Principles of Statistics," 1885[1]

"He rendered important services in the past; he proved that even apparently random events in social life possess an intrinsic necessity due to their periodic recurrence and their periodic mean numbers. However, he was never able to interpret this necessity. He made no progress; he just extended his observations and calculations."[2] So wrote Karl Marx in an 1869 letter to a friend about the work of Adolphe Quetelet.

The nineteenth century saw the beginning of the application of statistical methods in various fields, particularly agriculture and the social sciences. It was in fact these applications that led to the development of various standard statistical techniques in the nineteenth and twentieth centuries. In this chapter, we begin with one of the earliest statistical methods, that of least squares; then, after a brief look at Laplace's survey of the entire field of probability and statistics and Chebyshev's contributions, we consider the new interpretation of the normal curve in the middle of the nineteenth century, look at some of the developments in statistical procedures in the final decades of that century, and conclude with a brief glance at some types of statistical graphs.

 23.1

THE METHOD OF LEAST SQUARES AND PROBABILITY DISTRIBUTIONS

Perhaps the most important statistical method of the nineteenth century was that of least squares. This method was developed to give a procedure for combining observations that proved more effective than the eighteenth-century methods discussed earlier. Legendre was the first to publish this method, in 1805, but Gauss gave a much better justification of it a few years later.

23.1.1 The Work of Legendre

Although it is not known what influenced Legendre to develop the method of least squares, he discussed it in 1805 in an appendix to a work on the determination of cometary orbits. He began his discussion by giving a reason for introducing the method: "In the majority of investigations in which the problem is to get from measures given by observation the most exact results which they can furnish, there almost always arises a system of equations of the form $E = a + bx + cy + fz + \cdots$ in which $a, b, c, f, \ldots$ are the known coefficients which vary from one equation to another, and $x, y, z, \ldots$ are the unknowns which must be determined in accordance with the condition that the value of E shall for each equation reduce to a quantity which is either zero or very small."[3] In more modern terminology, one has a system of m equations $V_j(\{x_i\}) = a_{j0} + a_{j1}x_1 + a_{j2}x_2 + \cdots + a_{jn}x_n = 0$ $(j = 1, 2, \ldots, m)$ in n unknowns $(m > n)$ for which one wants to find the "best" approximate solutions $\bar{x}_1, \bar{x}_2, \ldots, \bar{x}_n$. For each equation, the value $V_j(\{\bar{x}_i\}) = E_j$ is the error associated with that solution. Legendre's aim, like those of his predecessors, was to make all the E_i small: "Of all the principles which can be proposed for that purpose, I think there is none more general, more exact, and more easy of application, than that of which we have made use in the preceding researches, and which consists of rendering the sum of the squares of the errors a minimum. By this means there is established among the errors a sort of equilibrium which, preventing the extremes from exerting an undue influence, is very well fitted to reveal that state of the system which most nearly approaches the truth."[4]

To determine the minimum of the squares of the errors, Legendre applied the tools of calculus. Namely, for the sum of the squares $E_1^2 + E_2^2 + \cdots + E_m^2$ to have a minimum when

x_1 varies, its partial derivative with respect to x_1 must be zero:

$$\sum_{j=1}^{m} a_{j1}a_{j0} + x_1 \sum_{j=1}^{m} a_{j1}^2 + x_2 \sum_{j=1}^{m} a_{j1}a_{j2} + \cdots + x_n \sum_{j=1}^{m} a_{j1}a_{jn} = 0.$$

Because there are analogous equations for $i = 2, 3, \ldots, n$, Legendre noted that he now had n equations in the n unknowns x_i and therefore that the system could be solved by "established methods." Although he offered no derivation of the method from first principles, he did observe that his method was a generalization of the method of finding the ordinary mean of a set of observations of a single quantity. For in that case (the special case where $n = 1$ and $a_{j1} = -1$ for each j), if we set $b_j = a_{j0}$, then the sum of the squares of the errors is $(b_1 - x)^2 + (b_2 - x)^2 + \cdots + (b_m - x)^2$. The equation for making that sum a minimum is $(b_1 - x) + (b_2 - x) + \cdots + (b_m - x) = 0$, so that the solution

$$x = \frac{b_1 + b_2 + \cdots + b_m}{m}$$

is just the ordinary mean of the m observations.

23.1.2   Gauss and the Derivation of the Method of Least Squares

Within 10 years of Legendre's publication, the method of least squares was a standard method in solving astronomical and geodetical problems. It appeared in 1808 in a paper by the American mathematician Robert Adrain (1775–1843) in connection with land surveying, and 10 years later Adrain used the principle in the determination of the earth's shape from the results of various measurements of the lengths of meridian arcs. But the principle also appeared in 1809 in Gauss's *Theoria motus corporum celestium* (*Theory of Motion of the Heavenly Bodies*). Gauss, like Adrain, did not quote Legendre. But Gauss also claimed that he had been using the principle himself since 1795. Gauss's statement led to a pained reaction from Legendre, who noted that priority in scientific discoveries could only be established by publication. And, in fact, whether or not Gauss used the method privately, there is no evidence that he discussed it with anyone else before Legendre's own publication.[5]

The priority dispute notwithstanding, Gauss went further with the method than Legendre. First, he realized that it was not enough to say that one can use "established methods" to solve the system of n equations in n unknowns that the method of least squares produces. In real applications, there are often many equations, and the coefficients are not integers but real numbers calculated to several decimal places. Cramer's method in these cases would require enormous amounts of calculation. Gauss therefore devised a systematic method of elimination for systems of equations, a method of multiplying the equations by appropriately chosen values and then adding these new equations together. The procedure, now known as the **method of Gaussian elimination** and virtually identical to the method used in Han-dynasty China 1800 years earlier, results in a triangular system of equations, that is, a system in which the first equation involves but one unknown, the second two, and so on. Thus, the first equation can be easily solved for its only unknown, the solution substituted in the second to get the value for the second unknown, and so on until the system is completely solved. Gauss's procedure was improved somewhat later in the century by the German geodesist Wilhelm

Jordan (1842–1899), who used the method of least squares to deal with surveying problems. Jordan devised a method of substitution, once the triangular system had been found, to further reduce the system to a diagonal one in which each equation only involved one unknown. This **Gauss-Jordan method** is the one typically taught in modern linear algebra courses as the standard method for solving systems of linear equations.[6]

Second, Gauss developed a much better justification for the method of least squares than the somewhat vague "general principle" of Legendre. He derived the method from his prior discovery of a suitable function $\phi(x)$ describing the probability of an error of magnitude x in the determination of an observable quantity, a function different from the ones worked out in the previous century. Gauss's criteria were the same as Laplace's earlier: that $\phi(x)$ should be symmetric about zero, that the curve must be asymptotic to the real axis in both directions, and that the total area under $\phi(x)$ should be 1. Gauss joined these criteria to the original problem of Legendre of determining the values of m linear functions $V_1, V_2, \ldots, V_m$ of n unknowns $x_1, x_2, \ldots, x_n$. Supposing that the observed values of these were $M_1, M_2, \ldots, M_m$, with corresponding errors $\Delta_1, \Delta_2, \ldots, \Delta_m$, Gauss noted that because the various observations were all independent, the probability of all these errors occurring was $\Omega = \phi(\Delta_1)\phi(\Delta_2) \ldots \phi(\Delta_m)$. To find the most probable set of values meant maximizing Ω, but to do this required a better knowledge of ϕ. Thus, Gauss made the further assumption that "if any quantity has been determined by several direct observations, made under the same circumstances and with equal care, the arithmetical mean of the observed values affords the most probable value."[7] Taking each V_i as the simplest linear function of one variable, namely $V_i = x_1$, he determined ϕ by supposing that $x_1 = (1/m)(M_1 + M_2 + \cdots M_m)$, the mean of the observations, gives the maximum value of Ω. That maximum occurs when $\partial\Omega/\partial x_i = 0$ for all i.

Because Ω is a product, Gauss replaced these last equations with $\frac{\partial}{\partial x_i}(\log \Omega) = 0$, or

$$\frac{\frac{\partial}{\partial x_i}(\phi(\Delta_1))}{\phi(\Delta_1)} + \frac{\frac{\partial}{\partial x_i}(\phi(\Delta_2))}{\phi(\Delta_2)} + \cdots + \frac{\frac{\partial}{\partial x_i}(\phi(\Delta_m))}{\phi(\Delta_m)} = 0. \qquad (23.1)$$

For each j, $\frac{\partial}{\partial x_i}(\phi(\Delta_j)) = \frac{\partial\phi}{\partial\Delta_j}\frac{\partial\Delta_j}{\partial x_i}$. Because $\frac{\partial\Delta_j}{\partial x_i} = 0$ for $i > 1$ and $\frac{\partial\Delta_j}{\partial x_1} = -1$ for all j, the n equations (23.1) all reduce to the single equation

$$\frac{\phi'(\Delta_1)}{\phi(\Delta_1)} + \frac{\phi'(\Delta_2)}{\phi(\Delta_2)} + \cdots + \frac{\phi'(\Delta_m)}{\phi(\Delta_m)} = 0, \qquad (23.2)$$

where we have written $\phi'(\Delta_j)$ in place of $\frac{\partial\phi}{\partial\Delta_j}$. To simplify further, Gauss supposed that each of the observations $M_2, M_3, \ldots, M_m$ were equal to $M_1 - mN$ for some value N. It followed that $x_1 = M_1 - (m-1)N$, that $\Delta_1 = M_1 - x_1 = (m-1)N$, and that $\Delta_i = -N$ for $i > 1$. Substituting these values in Equation 23.2 gave Gauss the relation

$$\frac{\phi'[(m-1)N]}{\phi[(m-1)N]} = (1-m)\frac{\phi'(-N)}{\phi(-N)}.$$

Because this is true for every positive integer m, Gauss concluded that

$$\frac{\phi'(\Delta)}{\Delta\phi(\Delta)} = k$$

for some constant k and therefore that $\log(\phi(\Delta)) = \frac{1}{2}k\Delta^2 + C$ or that $\phi(\Delta) = Ae^{(1/2)k\Delta^2}$. The Laplacian conditions on ϕ enabled Gauss to conclude that k is negative, say, $k = -h^2$, and then finally that

$$\phi(\Delta) = \frac{h}{\sqrt{\pi}}e^{-h^2\Delta^2}.$$

That this was the "correct" error function followed for Gauss because he was able easily to derive from it the method of least squares. After all, given this function ϕ, the product Ω in the general case was given by

$$\Omega = h^m\pi^{-(1/2)m}e^{-h^2(\Delta_1^2+\Delta_2^2+\cdots+\Delta_m^2)}.$$

To maximize Ω, therefore, it is necessary to minimize the $\sum \Delta_i^2$, that is, to minimize the sum of the squares of the errors, the very procedure that Legendre had developed. Ironically, Adrain had given a similar derivation of this "normal" law of errors in 1808, using the same basic idea as Gauss. But because this appeared in an American journal that lasted but one year, his result was not known in Europe at all.

That the distribution of errors is normal—that is, determined by Gauss's function—gained even more credence because it was soon supported by much empirical evidence. In particular, Friedrich Bessel made three sets of measurements of star positions for several hundred stars and compared the theoretical number of errors between given limits, according to the normal law, with the actual values. The comparison showed very close agreement.

23.1.3 The Work of Laplace

Laplace gave a new theoretical derivation of the normal law in a paper of 1810. His result was based on what is today called the **central limit theorem**, to the effect that any mean, not just the total number of successes in m trials, will, if the number of terms becomes large, be approximately normally distributed. This was a generalization of De Moivre's calculations of the previous century involving the terms of the binomial theorem. In fact, Laplace showed, under the assumption that the error arising from each observation is equally likely to be $-n$, $-n+1, -n+2, \ldots, -1, 0, 1, \ldots, n-2, n-1, n$, that for large s the probability that the sum of the errors arising from s independent observations is between

$$-2T\sqrt{\frac{n(n+1)s}{6}} \quad \text{and} \quad 2T\sqrt{\frac{n(n+1)s}{6}}$$

is $(2/\pi)\int_0^T e^{-x^2}\,dx$. Laplace derived a similar result in more general cases of error probabilities as well. In any case, the work of Laplace soon established the function $y = Ae^{-kx^2}$ as that representing error distributions and, in general, probability distributions, in a wide variety of situations.

In his book, the *Théorie analytique des probabilités* (*Analytic Theory of Probability*), published in 1812, Laplace collected all the material so far developed in probability theory, beginning with the definition of probability of an event as "the ratio of the number of cases favorable to it, to the number of possible cases, when there is nothing to make us believe that one case should occur rather than any other."[8] He thus included the statement and proof of the central limit theorem and its application to the question of the inclinations of the orbits of comets, a problem he had considered for many years. Furthermore, he dealt with the applications of the theory of probability to such topics as insurance, demographics, decision theory, and the credibility of witnesses. In fact, it was Laplace's view that, through the theory of probability, mathematics could be brought to bear on the social sciences, just as calculus was the major tool in mathematizing the physical sciences. Laplace's prediction began to come true well before the end of the century.

23.1.4 The Work of Chebyshev

FIGURE 23.1

Pafnuty Chebyshev on a Russian stamp

For Laplace, probability theory began with its definition as a ratio. But gradually, during the remainder of the nineteenth century, the thrust of the theory began to change. In particular, Pafnuty Chebyshev (1821–1894) (Fig. 23.1) began to think of probability as having to do with functions. Thus, he considered the concept of a **random variable,** that is, a function from a particular sample space that takes on a set of arbitrary values $a_1, a_2, \ldots, a_n$ with corresponding probabilities. For example, we could consider the random variable that takes the set 1, 2, 3, 4, 5, 6 of outcomes of the roll of a die to itself (by the identity), with each value occurring with probability 1/6. In this case, the probability distribution is discrete, but recall that in Bayes's example of rolling a ball on a table, the probability distribution was continuous. Chebyshev's goal, which was ultimately accomplished in the twentieth century, mostly by Russian mathematicians, was to place probability theory firmly within the general theory of functions of a real variable.

Among Chebyshev's specific contributions to probability theory is what is now called **Chebyshev's inequality,** first published in 1867, which states that in any probability distribution, nearly all the values are "close" to the mean value. (The mean value in a discrete distribution is found by a weighted average, while in a continuous distribution, one needs to calculate an appropriate integral.) More precisely, we can state the theorem in the following form:

CHEBYSHEV'S INEQUALITY   *If X is a random variable with standard deviation σ, then the probability that the outcome of X is no less than $a\sigma$ from its mean μ is no more than $1/a^2$, or*

$$P(|X - \mu(X)| \geq a\sigma) \leq \frac{1}{a^2}.$$

Chebyshev used his inequality to prove the **weak law of large numbers** for repeated independent trials. This law states that if X_n is the average value of the random variable over n repetitions, if μ is the mean (or expected) value, and if $\epsilon > 0$, then

$$\lim_{n \to \infty} P(|X_n - \mu| < \epsilon) = 1.$$

For example, in the case of rolling a die, the expected or mean value of the rolls is 3.5. The law of large numbers then states that as one rolls the die more and more times, the average value of the rolls approaches more and more closely to 3.5.

STATISTICS AND THE SOCIAL SCIENCES

The normal curve $y = Ae^{-kx^2}$ was first developed by De Moivre from computations of probabilities in a binomial experiment and then turned out to be important in minimizing errors of measurement. In its formulation by Gauss and then Laplace, it in fact represented a distribution of errors. By the middle of the nineteenth century, however, the normal curve was being applied in the social sciences.

23.2.1   Quetelet and the Average Man

FIGURE 23.2

Quetelet on a Belgian stamp

Adolphe Quetelet (1796–1874) (Fig. 23.2), a Belgian mathematician, astronomer, meteorologist, sociologist, and statistician, seized on the normal curve as the key to developing his concept of the "average man." By compiling vast numbers of statistics covering not only physical characteristics such as height and weight, but also "moral characteristics" such as the propensities for individuals to commit crimes or to become drunk, he proposed to be able to develop the idea of the representative individual in a given society at a given time. The purpose of the concept of the average man—and there were of course different average men (and, perhaps, women) for various ages and classes in each country—was to be a device for smoothing the manifold variations among people and somehow revealing the regular laws of society, a "social physics."

Quetelet noticed that many of the characteristics he gathered could be plotted in terms of a normal curve. That is, there was a mean value and "errors" from the mean that were distributed in the same way as errors of measurement. In 1846, he wrote a letter to the Grand Duke of Saxe Coburg expressing his belief in the use of the normal curve for social analysis. Suppose, he said, that one wanted to make a thousand copies of a particular statue. These copies would naturally be subject to a wide variety of errors, but, in fact, the errors would combine in a very simple fashion. In fact, he wrote, "the experiment has been already made. Yes, surely, more than a thousand copies have been measured of a statue . . . which in all cases differs little from it. These copies were even living ones."[9] Quetelet's "statues" were Scottish soldiers; he had compiled the chest circumferences of 5732 of them and noted that they were distributed normally around a mean of about 40 inches. His conclusion was that because the measurements were distributed as they would be if nature were aiming for an ideal type, this must be the case. Thus, his distribution showed that there was an "average" Scottish soldier, and the deviations from the average were simply due to a combination of accidental causes.

In his use of the normal curve in this situation and others, Quetelet's unit of deviation from the mean was the "probable error," rather than the now common "standard deviation." A data point is **one probable error** from the mean if it has a percentile rank of 25 or 75. That is, in a normal distribution, a particular value is as likely to be within one probable error of the mean as not. This measure of deviation had been introduced early in the century in connection with the theory of errors.

23.2.2   The Work of the English Statisticians

Certainly, many disagreed with Quetelet's program of looking for normal distributions everywhere, and others felt, as the quotation in the chapter opening shows, that this process would not lead to anything positive in, for example, changing the conditions leading to criminal behavior. Nevertheless, the idea of a normal distribution became central in many arguments involving statistics. Francis Galton (1822–1911), an English statistician, used Quetelet's ideas in biology, trying to mathematize Charles Darwin's theory of evolution by looking at the inheritance of variation. Galton was curious as to why the same normal curve persisted generation after generation and why the variability did not increase over the years. One experiment that he conducted in 1875 was on the size of a particular type of sweet pea. He took an equal number of pea seeds of seven sizes and studied their offspring. It turned out that the sizes in each set of progeny were normally distributed and that the variability in each group—that is, the spread of the data—was essentially the same.

But what Galton also found was that the sizes of all the seeds in the generation of offspring, looked at in its totality, were normally distributed. To explain why all the small normal distributions added together gave a "large" one, Galton created what he called the "quincunx." The quincunx is a device with a glass face and a funnel at the top. Very small steel balls were poured through the funnel and moved through an array of pins, each ball striking one pin at each level and, in principle, falling to the left or right with equal probability. The balls were then collected in compartments at the bottom. The resulting distribution was binomial, so it resembled a normal curve. But then Galton modified the device by intercepting the balls at an intermediate level (Fig. 23.3). The result at that level was again approximately normal. If the balls were released from any one compartment at the intermediate level, the distribution of balls at the bottom was approximately normal, with the mean directly under the compartment that was opened up. Of course, releasing a compartment nearer the center produced a higher curve. Releasing all the compartments at the intermediate level produced a mixture of these curves of varying sizes, but since this gave the same result as when the balls were not intercepted, the resulting curve would again be normal. Thus, this quincunx enabled Galton to show that a normal mixture of normal distributions was again normal.

FIGURE 23.3

Galton's quincunx

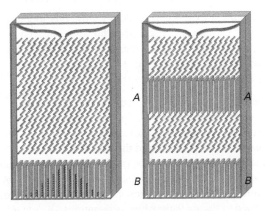

In his experiments with peas, Galton also observed that although the distribution of the seed sizes of the offspring from each parent was normal, the mean of each group was not the same as that of the parental seed sizes. In fact, the means were linearly related to those of the parent seeds, but with the slope of this line being one-third. In other words, the second generation "regressed" to the overall mean. Thus, Galton originated the statistical study of

regression (or reversion, as he first called it). Interestingly, to calculate the slope of this regression line, Galton simply plotted the points and estimated the slope of the straight line that "best" connected these points.

Galton carried out several other studies involving inheritance, including an extensive one on the heights of children as related to the heights of their parents (Table 23.1). In this case, he noted that both the heights of the children of parents of a given height and the heights of the parents of children of a given height were normally distributed. He then found that if one scaled the two sets of data by units of probable error, the slopes of the regression lines were the same. This slope became the coefficient of "co-relation" of the two variables, later to become the **correlation coefficient**. This value, which Galton could generally only roughly approximate, could be used to measure the strength of the relationship of the two variables. Galton realized, of course, that a strong correlation does not necessarily imply a causal relationship.

Other English statisticians tried to clarify and extend the work of Galton and come to grips with the basic philosophy behind looking at normal curves representing various types of distributions. Francis Edgeworth (1845–1926), quoted at the beginning of the chapter, commented on the difference between *observations*, as used in astronomy, and *statistics*, as collected in the social sciences. Thus, Edgeworth realized that the use of statistics in the social sciences would not have the "objective" character that the use of the theory of errors had in measurements in astronomy. Nevertheless, he worked to develop this new tool as best he could.

In particular, in 1885 he developed a basic significance test. Given two "means," he first estimated what he called their "fluctuations," or, in modern terms, twice the variance. Thus, if $\bar{x}$ was a sample mean, then the fluctuation was $2 \sum (x_i - \bar{x})^2 / n^2$. If c_1^2 and c_2^2 were estimates of the fluctuations of the two means, then $\sqrt{c_1^2 + c_2^2}$ estimated what he called the "modulus" of the difference of the means. (In modern terms, this modulus is $\sqrt{2}$ times the standard deviation.) He was extremely conservative, however, in coming to conclusions. He would only call the difference between two means significant if it exceeded two moduli. This is equivalent to using a two-sided test of significance at a level of 0.005, a much stronger requirement than the levels of 0.05 or 0.01 often used today.

In the last decade of the nineteenth century, two other English statisticians, Karl Pearson (1857–1936) and his student George Udny Yule (1871–1951), did further work in showing how to use statistics to come to definite conclusions about the relationships between several quantities. Pearson not only introduced the "standard deviation" in 1893 but also developed the chi-square statistic as one way of measuring the relationship between two quantities. And Yule showed how to calculate the regression equation by using, in essence, Gauss's least squares technique to find the line of best fit. But all of the procedures of these statisticians were only designed to show relationships in quantities already tabulated. A significant use of statistics today is in the design and analysis of experimental procedures that will enable, for example, farmers to determine the effectiveness of different types of fertilizer on the yield of crops and doctors to determine the efficacy of differing treatments for a particular disease. Furthering the study of statistics so that it could be applied to such problems was the work of the twentieth century.

TABLE 23.1 *Galton's data on heights.*

Height of Mid-Parent (in inches)[a]	Height of Adult Child														Total No. of Adult Children	Total No. of Mid-parents	Median
	<61.7	62.2	63.2	64.2	65.2	66.2	67.2	68.2	69.2	70.2	71.2	72.2	73.2	>73.7			
>73.0	—	—	—	—	—	—	—	—	—	—	—	1	3	—	4	5	—
72.5	—	—	—	—	—	—	—	1	2	1	2	7	2	4	19	6	72.2
71.5	—	—	—	—	1	3	4	3	5	10	4	9	2	2	43	11	69.9
70.5	1	—	1	—	1	1	3	12	18	14	7	4	3	3	68	22	69.5
69.5	—	—	1	16	4	17	27	20	33	25	20	11	4	5	183	41	68.9
68.5	1	—	7	11	16	25	31	34	48	21	18	4	3	—	219	49	68.2
67.5	—	3	5	14	15	36	38	28	38	19	11	4	—	—	211	33	67.6
66.5	—	3	3	5	2	17	17	14	13	4	—	—	—	—	78	20	67.2
65.5	1	—	9	5	7	11	11	7	7	5	2	1	—	—	66	12	66.7
64.5	1	1	4	4	1	5	5	—	2	—	—	—	—	—	23	5	65.8
<64.0	1	—	2	4	1	2	2	1	1	—	—	—	—	—	14	1	—
Totals	5	7	32	59	48	117	138	120	167	99	64	41	17	14	928	205	—
Medians	—	—	66.3	67.8	67.9	67.7	67.9	68.3	68.5	69.0	69.0	70.0	—	—	—	—	—

a. Note that "height of the mid-parent" means the average height of the parents, with mother's height scaled up by 1.08.

23.3   STATISTICAL GRAPHS

One major innovation of the nineteenth century was the use of graphs to represent data. Many types of graphs were developed, some by several different people. We give here some examples of the more important ones.

It was William Playfair (1759–1823) who developed a number of graphical designs so that he could replace tables of numbers with a visual representation. As he wrote, "a man who has carefully investigated a printed table, finds, when done, that he has only a very faint and partial idea of what he has read; and that like a figure imprinted on sand, is soon totally erased and defaced."[10] Thus, Playfair drew charts of all sorts. One example, both a line graph and a bar chart, shows the relationship between wages and the price of wheat over a 250-year period (Fig. 23.4). Another example, which uses areas of circles to represent quantity as well as a pie chart, shows the relationship between population and revenue of many of the countries of Europe (Fig. 23.5). The vertical line at the left of each circle represents population; the

FIGURE 23.4

Playfair's graph of wages and the price of wheat

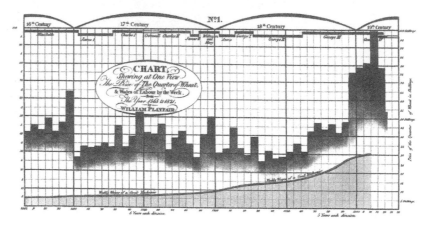

FIGURE 23.5

Playfair's graph of population and revenue, in which areas of circles represent the areas of the countries

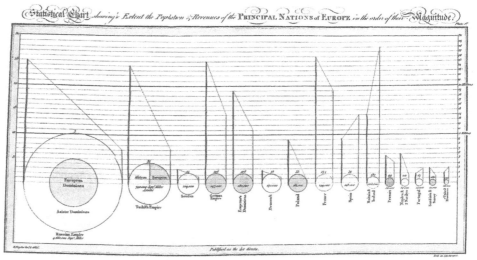

one on the right represents revenue. The slope of the lines connecting these shows whether in proportion to its population a country is burdened with heavy taxes.

Florence Nightingale (1820–1910) (Fig. 23.6), who was a nurse during the Crimean War, drew several very effective pie charts giving the monthly death rates during the war (Fig. 23.7). Each wedge is divided into three sections representing different causes of death. The innermost section shows death due to wounds, the middle section represents "other causes," while the large outer section shows death due to preventable disease. Her graphs, when published in England, led to a great outcry and caused the War Department to improve the sanitary conditions at field hospitals.

Histograms, bar charts with the horizontal axis denoting a continuous variable, were first used by A. M. Guerry in France in 1833. But they were named by Karl Pearson, who used them in an 1895 article on the mathematical theory of evolution. Figure 23.8 is an example from that article, from which it is easy to see that there are fewer examples of flowers with a greater number of petals.[11]

One final graph is the **ogive**, a graph of a cumulative frequency distribution, invented by Galton around 1875. In this example (Fig. 23.9), Galton was dealing with the strength of pull in pounds of 519 males ages 23 to 26. He plotted the strength along the vertical axis and the percentile rank along the horizontal and drew the curve representing strength as a function of percentile rank. For example, those at the 70th percentile can pull a weight of about 80 pounds. Since the population is normally distributed in this situation, Galton's function is now called the inverse normal cumulative distribution. Today, however, on such a graph, the horizontal and vertical axes are generally interchanged to determine for a given weight the percentage of the population that can pull no more than that.

FIGURE 23.6

Florence Nightingale on a stamp from the British Virgin Islands

FIGURE 23.7

Nightingale's pie chart of death rates during the Crimean War

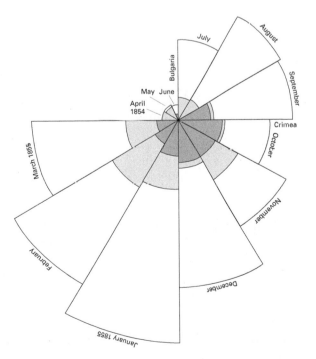

FIGURE 23.8

A histogram from Pearson's
article on evolution

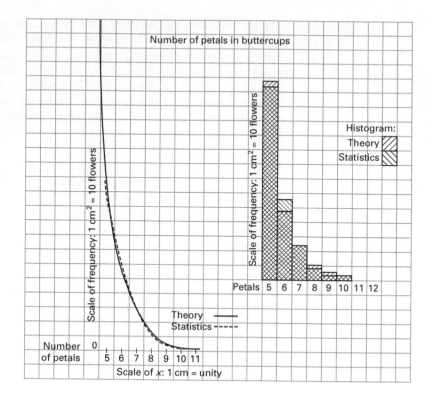

FIGURE 23.9

Galton's ogive of male
strength of pull

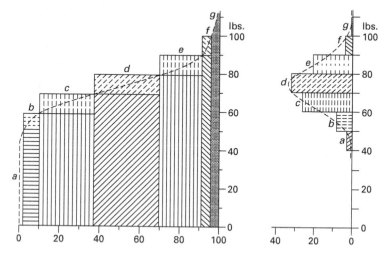

EXERCISES

1. Given four measured values for the independent variable x as $x_1 = 2.0$, $x_2 = 4.0$, $x_3 = 5.0$, $x_4 = 6.0$ and four corresponding measured values for the dependent variable y as $y_1 = 2.5$, $y_2 = 4.5$, $y_3 = 7.0$, $y_4 = 8.5$, use the method of least squares to determine the constants a and b that give the best linear function $y = ax + b$ that represents this measured relationship.

2. Given that the x value determining one standard deviation in the normal curve $y = \frac{1}{c\sqrt{\pi}} e^{-(x^2/c^2)}$ occurs at an inflection point of the curve, show that this value is given by $x = c/\sqrt{2}$.

3. Show that in a normal distribution with standard deviation σ, a measurement is one probable error from the mean if it is approximately 0.675σ from the mean and is one modulus from the mean (the value c from Exercise 2) if it is $\sqrt{2}\sigma$ from the mean. (*Hint:* Use a table of the normal curve to solve this problem.)

4. For this exercise and the next, use Galton's table (see Table 23.1). Use a histogram to graph the heights of adult children of parents of height 68.5 inches. Then graph similarly the heights of parents of children of height 69.2 inches. Show that these graphs are of an approximately normal distribution. Find the medians in each case.

5. Use histograms to graph the heights of all adult children from Table 23.1 (use the totals column) and similarly the heights of all parents. Show that these distributions are approximately normal. Find the medians in each case and calculate the standard deviations. (Use statistical software if necessary.)

6. Using Figure 23.4, show how Playfair concluded that "never at any former period [i.e., before 1820] was wheat so cheap, in proportion to mechanical labour, as it is at the present time [1821]."

7. Using Figure 23.5, show how Playfair concluded that the people of Great Britain (the sixth circle from the right) were excessively taxed in relation to the people of other European countries. (*Hint:* Note that the scales of population on the left and revenue on the right are not particularly related to one another.)

8. Recall from Figure 23.5 that the areas of the circles are proportional to the areas of the countries. The second circle from the left represents the Turkish empire. Note that it is divided into three sections representing the African, European, and Asian domains of that empire. Thus, that circle is in the form of a pie chart. Estimate the relative sizes of those three sections of the Turkish empire from the pie chart.

9. In Pearson's histogram (see Figure 23.8), 10 flowers are represented by one unit in the vertical direction. The histogram represents the number of buttercups found with a given number of petals. How many buttercups had seven petals? How many had five petals?

10. Redraw Galton's ogive (see Figure 23.9) to represent the distribution of males being able to pull a given number of pounds. Show that this distribution is approximately normal. Then redraw this as a cumulative frequency distribution, keeping the strength of pull on the horizontal axis and the numbers on the vertical axis.

11. Describe why the publication of Nightingale's pie charts would have led to an outcry in Britain. Find out what happened in the hospitals after these were published and write a brief report.

REFERENCES AND NOTES

Among the best recent general works on the history of statistics are Stephen M. Stigler, *The History of Statistics* (Cambridge: Harvard University Press, 1986), and Theodore M. Porter, *The Rise of Statistical Thinking* (Princeton: Princeton University Press, 1986). Another work with more mathematical details is Anders Hald, *A History of Mathematical Statistics from 1750 to 1930* (New York: Wiley, 1998). Probability theory in the nineteenth century is discussed in A. N. Kolmogorov and A. P. Yushkevich, eds., *Mathematics of the 19th Century* (Basel: Birkhäuser, 1992), translated from the Russian original of 1978. E. S. Pearson and M. G. Kendall, eds., *Studies in the History of Statistics and Probability*, vol. 1 (Darien, CT: Hafner, 1970), and M. G. Kendall and R. L. Plackett, eds., *Studies in the History of Statistics and Probability*, vol. 2 (New York: Macmillan, 1977), provide a valuable collection of essays on the history of probability and statistics, most of which originally appeared in *Biometrika*. Graphs of various sorts, including some of the early ones illustrated in the text, are discussed in Edward R. Tufte,

The Visual Display of Quantitative Information, second edition (Cheshire, CT: Graphics Press, 2001).

Legendre's early work on the method of least squares is available in David Smith, *A Source Book in Mathematics* (New York: Dover, 1959), pp. 576–579. Gauss's initial discussion of the method of least squares is found in translation by C. H. Davis in *Theory of Motion of the Heavenly Bodies Moving about the Sun in Conic Sections* (Boston: Little, Brown, 1857). Adolphe Quetelet's *Letters Addressed to H.R.H. the Grand Duke of Saxe Coburg and Gotha, on the Theory of Probabilities, as Applied to the Moral and Political Sciences* was translated by O. G. Downes and published in London by Charles and Edwin Layton in 1849. Galton's initial work on regression is in "Regression towards Mediocrity in Hereditary Stature," *Journal of the Anthropological Institute* 15 (1886), 246–263. Many of his conclusions about statistical methods are included in *Natural Inheritance* (London: Macmillan, 1889). Edgeworth's essay, in which occurs the quotation opening the chapter, is "Observations and Statistics: An Essay on the Theory of Errors of Observation and the First Principles of Statistics," *Transactions of the Cambridge Philosophical Society* 14 (1885), 13–169. William Playfair's *The Commercial and Political Atlas and Statistical Breviary* has recently been republished (Cambridge: Cambridge University Press, 2005), with an introduction by Howard Wainer and Ian Spence.

1. Quoted in Stigler, *History of Statistics*, p. 309, from Edgeworth's "Observations and Statistics," p. 139.

2. Quoted in Kolmogorov and Yushkevich, eds., *Mathematics of the 19th Century*, p. 245.

3. Legendre, "Sur la méthode des moindres quarrés," in Legendre, *Nouvelles méthodes pour la détermination des orbites des cometes* (Paris, 1805), translated in Smith, *Source Book*, 576–579, p. 576.

4. Ibid., p. 577.

5. For more details, see R. L. Plackett, "The Discovery of the Method of Least Squares," *Biometrika* 59 (1972), 239–251. This paper is reprinted in M. G. Kendall and R. L. Plackett, eds., *Studies in the History of Statistics and Probability*, vol. 2, pp. 279–291. For a discussion of the work of Robert Adrain, see Jacques Dutka, "Robert Adrain and the Method of Least Squares," *Archive for History of Exact Sciences* 41 (1991), 171–184.

6. For more details, see Steven C. Althoen and Renate McLaughlin, "Gauss-Jordan Reduction: A Brief History," *The American Mathematical Monthly* 94 (1987), 130–142, and Victor J. Katz, "Who Is the Jordan of Gauss-Jordan?" *Mathematics Magazine* 61 (1988), 99–100.

7. Gauss, *Theoria motus corporum celestium* (Hamburg: Pertheset Besser, 1809), translated by C. H. Davis as *Theory of Motion of the Heavenly Bodies Moving about the Sun in Conic Sections* (Boston: Little, Brown, 1857), p. 258. For more information on Gauss and the method of least squares, see O. B. Sheynin, "C. F. Gauss and the Theory of Errors," *Archive for History of Exact Sciences* 20 (1979), 21–72, and William C. Waterhouse, "Gauss's First Argument for Least Squares," *Archive for History of Exact Sciences* 41 (1991), 41–52.

8. Laplace, *Théorie analytique des probabilités* (Paris: Courcier, 1812), p. 181.

9. Adolphe Quetelet, *Letters Addressed to H.R.H. the Grand Duke of Saxe Coburg and Gotha, on the Theory of Probabilities, as Applied to the Moral and Political Sciences*, trans. by O. G. Downes (London: Charles and Edwin Layton, 1849), p. 136.

10. Playfair, *Commercial and Political Atlas*, p. xiv.

11. Karl Pearson, *Karl Pearson's Early Statistical Papers* (Cambridge: Cambridge University Press, 1948), p. 112.

24

Geometry in the Nineteenth Century

I am ever more convinced that the necessity of our geometry cannot be proved—at least not by human reason for human reason. It is possible that in another lifetime we will arrive at other conclusions on the nature of space that we now have no access to. In the meantime we must not put geometry on a par with arithmetic that exists purely a priori but rather with mechanics.

—Gauss, in a letter to Heinrich Olbers (1758–1840), 1817[1]

Every candidate to become a mathematics lecturer at the University of Göttingen had to present an inaugural lecture to the members of the philosophical faculty, a lecture designed to show that he could apply his mathematical research to more general intellectual issues. Bernhard Riemann submitted three possible topics for this *Habilitationsschrift*, the first two being closely related to some already completed research on complex functions and trigonometric series. Gauss, however, acting on behalf of the faculty, picked Riemann's third topic, "On the Hypotheses Which Lie at the Foundation of Geometry." On June 10, 1854, Riemann made his presentation. The lecture had few mathematical details but was packed with so many ideas about what geometry should be about that mathematicians have been studying it for well over a century.

The importance of analysis in the late eighteenth century, spurred to a large extent by the work of Euler and continued in the early nineteenth century by Cauchy, among others, tended to lessen the importance of pure geometry during that period. But the applications of analysis to geometry led to various important new geometrical ideas.

Although Gauss early in his career had considered aspects of what is today called differential geometry, it was only during his work on a detailed geodetic survey of the kingdom of Hanover, required of him as the Director of the Göttingen Astronomical Observatory, that he finally clarified his ideas on the subject of the theory of surfaces. He published these ideas in 1827 in a brief but densely packed paper in which he carried forward the introductory work of Euler on the theory of surfaces, applying the techniques of calculus to show that some of the basic notions of surface theory, including the notion of curvature, were intrinsic to the surface and did not depend on how the surface was situated in three-dimensional space.

In his work on surfaces, Gauss established a relationship between curvature and the sum of the angles of a triangle on the surface. This relationship turned out to be closely connected to the old question of the parallel postulate, whose truth implied that the sum of the angles of a plane triangle was equal to two right angles. Toward the end of his life, Gauss noted that he had long been convinced that the parallel postulate could not be proved and that the acceptance of alternatives could well lead to new and interesting geometries, the "truth" of which for the physical world could only be established by experiment. Nevertheless, Gauss never published any of his ideas on the subject. It was therefore left for Nikolai Lobachevsky and János Bolyai in the 1820s to publish, independently of each other, the first full treatments of a non-Euclidean geometry.

It took nearly 40 years, however, for the ideas of non-Euclidean geometry to make an impression in the mathematical community. It was only with the work of Riemann in 1854 and Hermann von Helmholtz in 1868 on the general notion of a geometrical manifold of arbitrary dimension that the meaning of these new ideas for the study of geometry took hold. Shortly thereafter, various models of non-Euclidean geometries in Euclidean space were introduced, thus convincing the mathematical public that the non-Euclidean geometries were as valid as the Euclidean one from a logical standpoint and that the question of the "truth" of Euclidean geometry for the world in which we live no longer had an obvious answer.

There were also advances in the subject of projective geometry over the early work of Pascal and Desargues, accomplished by such mathematicians as Jean-Victor Poncelet, Michel Chasles, and Julius Plücker. In 1871, Felix Klein showed a connection between projective and non-Euclidean geometry via the idea of a metric. The following year he gave a new definition of a geometry in terms of transformations, a definition that demonstrated the relationship of projective to Euclidean geometry and also the connection of geometry with the emerging theory of groups.

Graph theory, which had its beginnings in Euler's work on the problem of the bridges of Königsberg, received new emphasis with the posing of the four-color problem. Many mathematicians attempted to solve this problem of showing that four colors were sufficient to color any map. This problem was, however, not solved in the nineteenth century, but continued to attract attention through the twentieth century.

Geometry as studied to the middle of the nineteenth century dealt with objects of dimension no greater than three. But with the increasing use of analytical and algebraic methods, it became clear that for many geometric ideas there was no particular reason to limit the number of dimensions to just the number that could be physically realized. Thus, various mathematicians generalized their formulas and theorems to n dimensions, where n could be any positive integer. It was Hermann Grassmann, however, who, beginning in 1844, first attempted a detailed study of n-dimensional vector spaces from a geometric point of view. Unfortunately, Grassmann's work, like that of Bolyai and Lobachevsky, was not appreciated until the end of the century. At that time Giuseppe Peano gave a set of axioms for a finite-dimensional vector space to provide a basis for the study of higher-dimensional geometry, and Elie Cartan applied Grassmann's work to the study of differential forms.

With the creation of the various new geometries, many mathematicians toward the end of the century felt that it was time to redo the foundations of the entire subject, just as was being done in analysis. Flaws had been discovered in Euclid's reasoning, and these flaws led to certain problems in developing the non-Euclidean geometries. Thus, David Hilbert brought out a new set of axioms for Euclidean geometry that helped to remove the various flaws and clarify exactly what had to be assumed in order to develop both the old and the new geometries.

 24.1

DIFFERENTIAL GEOMETRY

Having led the survey of Hanover from 1820 to 1825, and having introduced various new methods establishing geodesy as a recognized science, Gauss was finally able by 1827 to put on paper the results of his thoughts of over a quarter century on the subject of curved surfaces. Gauss noted in the abstract of his work *Disquisitiones generales circa superficies curvas* (*General Investigations of Curved Surfaces*) that

> although geometers have given much attention to general investigations of curved surfaces and their results cover a significant portion of the domain of higher geometry, this subject is still so far from being exhausted, that it can well be said that, up to this time, but a small portion of an exceedingly fruitful field has been cultivated. Through the solution of the problem, to find all representations of a given surface upon another in which the smallest elements remain unchanged, the author sought some years ago to give a new phase to this study. The purpose of the present discussion is further to open up other new points of view and to develop some of the new truths which thus become accessible.[2]

Gauss had already solved his problem of establishing the conditions for mapping one surface conformally onto another (so that "the smallest elements remain unchanged") by 1822 in a challenge question that he had suggested the Copenhagen Scientific Society pose. The answer was that the function had to be representable as a complex analytic one in the parameters representing the two surfaces. (Gauss did not, however, use complex function theory at the time.) But in the course of developing this answer, Gauss realized that a central idea involved in the study of surfaces was that of curvature, and, in particular, he realized how curvature could be calculated in terms of an analytic description of the surface in question.

24.1.1   The Definition of Curvature

Gauss began his *Disquisitiones generales* with the notion of the curvature of a curved surface. He decided that he would only deal with surfaces, or parts of surfaces, with "continuous curvature," that is, surfaces (or parts) that possess tangent planes at all points. Thus, the vertex of a cone would not be considered, since there is no tangent plane, and no curvature, at that point. Because the sphere was Gauss's "model" surface, a surface with constant curvature analogous to the constant curvature of the circle in the plane, Gauss decided to define the curvature at a point on a surface by comparing a region around that point to a corresponding region around a point on the unit sphere. Curvature is a local property on a surface S. However it is to be defined, it is clear that the curvature may vary from point to point. On the unit sphere, however, the curvature at every point is set at 1. To make his comparison, therefore, Gauss created a mapping n (today called the Gauss normal map) from S to the unit sphere defined so that the vector from the center to $q = n(p)$ is parallel to the normal vector to S at p (that is, to the normal vector to the tangent plane to S at p) (Fig. 24.1). This mapping takes a bounded region A of the surface S to a bounded region $n(A)$ of the sphere. Gauss then defined the **total curvature** of A to be the area of $n(A)$, while he defined the more important concept of the **measure of curvature** at a point as the "quotient when the [total] curvature of the surface element about a point is divided by the area of the element itself; and hence it denotes the ratio of the infinitely small areas which correspond to one another on the curved surface and on the sphere."[3] In more modern terminology, Gauss defined the curvature at a point p to be

$$k(p) = \lim_{A \to p} \frac{\text{area of } n(A)}{\text{area of } A},$$

where the limit is taken as the region A around p shrinks to the point p itself. It follows that the total curvature of a region A is equal to $\int k\, d\sigma$, where $d\sigma$ is the element of area on the surface and the integral is taken over the region A.

FIGURE 24.1

Gauss's normal map

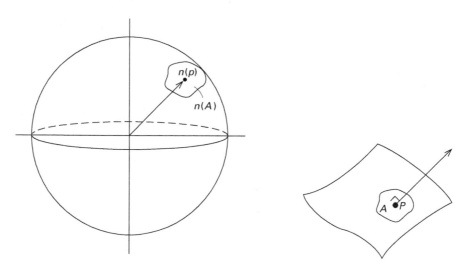

A modern reader might note at least two problems with Gauss's definition of the measure of curvature. First, how is the area on an arbitrary curved surface to be defined and, second, assuming this is done, how do we know that the limit, if it exists, is independent of the way the shrinking of the region is done? Gauss did not address these problems. In fact, he did not really define the curvature as a limit but merely as the ratio of infinitesimals. He then used his geometric intuition to assure himself that the definition made sense. For example, if the surface S is a sphere of radius r, then the area of $n(A)$ is equal to $1/r^2$ times the area of A (for any region A) and therefore the curvature at every point is equal to $1/r^2$. Similarly, if S is a plane, then $n(A)$ is equal to a point for any region A. Its area is therefore 0, as is the curvature. A somewhat more surprising result, but one that convinced Gauss his definition was correct, occurs when S is a circular cylinder. In this case, the image $n(A)$ of a region A is simply a curve on the sphere and thus again has area 0. It follows that the curvature of the cylinder, like that of the plane, is also 0. Why this result is reasonable is discussed shortly.

24.1.2   Curvature and the *Theorema Egregium*

Gauss was able to use his definition to calculate the curvature in terms of the equation of the given surface. Because the tangent plane to S at p is parallel to the tangent plane to the unit sphere at $n(p)$, the ratio of the area of $n(A)$ to that of A is equal to the ratio of the area of the projection of $n(A)$ onto the xy plane to that of the projection of A on that same plane. Considering therefore a triangular region A whose projection has the three vertices (x, y), $(x + dx, y + dy)$, $(x + \delta x, y + \delta y)$, Gauss noted that the area of the triangle is $\frac{1}{2}(dx\,\delta y - dy\,\delta x)$. Similarly, if the functions $X(x, y)$, $Y(x, y)$ represent the composition of the projection with the normal function n, then the corresponding triangle for $n(A)$ has area $\frac{1}{2}(dX\,\delta Y - dy\,\delta X)$. It follows that

$$k = \frac{dX\,\delta Y - dY\,\delta X}{dx\,\delta y - dy\,\delta x}.$$

It now remained for Gauss to determine the value of this fraction if the surface is defined either by an equation $z = z(x, y)$, by an equation $W(x, y, z) = 0$, or by the parametric equations $x = x(p, q)$, $y = y(p, q)$, $z = z(p, q)$. For the first method of representation, since

$$dX = \frac{\partial X}{\partial x}\,dx + \frac{\partial X}{\partial y}\,dy, \qquad dY = \frac{\partial Y}{\partial x}\,dx + \frac{\partial Y}{\partial y}\,dy,$$

$$\delta X = \frac{\partial X}{\partial x}\,\delta x + \frac{\partial X}{\partial y}\,\delta y, \qquad \delta Y = \frac{\partial Y}{\partial x}\,\delta x + \frac{\partial Y}{\partial y}\,\delta y,$$

Gauss determined that

$$k = \frac{\partial X}{\partial x}\frac{\partial Y}{\partial y} - \frac{\partial X}{\partial y}\frac{\partial Y}{\partial x}.$$

It is then a straightforward, although messy, calculation to rewrite this expression in terms of the partial derivatives of z to get

$$k = \frac{z_{xx}z_{yy} - z_{xy}^2}{(1 + z_x^2 + z_y^2)^2}.$$

Gauss calculated similar expressions for k when the surface is represented by an equation in three variables and by parametric equations. He was then able to derive a series of beautiful theorems.

First, he showed that the measure of curvature at a point p was expressible in terms of the curvatures of two specific sections of the surface through p. By use of a suitable choice of axes, he rewrote the equation $z = z(x, y)$ in a form where $p = (0, 0, 0)$ and $z_x(0, 0) = z_y(0, 0) = z_{xy}(0, 0) = 0$. It then followed that the maximum and minimum curvatures of all curves formed by normal sections through p are $z_{xx}(0, 0)$ and $z_{yy}(0, 0)$. Thus, the measure of curvature $k(p)$ at p is $z_{xx}z_{yy}$, the product of the two extreme curvatures of the normal sections.

Second, Gauss proved his *theorema egregium* to the effect that the measure of curvature was an isometric invariant of the surface, that is, it did not change if the surface was transformed by a distance-preserving transformation. To accomplish this, he took the parametric form of representation of the surface, $x = x(u, v)$, $y = y(u, v)$, $z = z(u, v)$, set $E = x_u^2 + y_u^2 + z_u^2$, $F = x_u x_v + y_u y_v + z_u z_v$, and $G = x_v^2 + y_v^2 + z_v^2$, and derived a formula for the curvature expressed solely in terms of E, F, G and their partial derivatives of first and second order with respect to u and v. It is not difficult to show that the distance element ds itself can also be expressed in terms of these quantities: $ds^2 = dx^2 + dy^2 + dz^2 = E\,du^2 + 2F\,du\,dv + G\,dv^2$. Thus, the curvature is determined by the distance element. Gauss then could state his "remarkable" theorem that, if one surface is "developed" onto another, that is, if there is a one-to-one function from one surface to another that preserves the element of length, then the measures of curvature at corresponding points of the two surfaces are always equal. For example, because the plane can be developed onto a cylinder, the curvature of the cylinder equals that of the plane, namely, 0. Gauss emphasized that his result was only the beginning of a new and important method of studying a surface,

> not as the boundary of a solid, but as a flexible, though not extensible, solid, one dimension of which is supposed to vanish. [In this way] the properties of the surface depend in part upon the form to which we can suppose it reduced, and in part are absolute and remain invariable, whatever may be the form into which the surface is bent. To these latter properties, the study of which opens to geometry a new and fertile field, belong the measure of curvature and the integral curvature. . . . From this point of view, a plane surface and a surface developable on a plane, e.g., cylindrical surfaces, conical surfaces, etc., are to be regarded as essentially identical.[4]

Finally, Gauss demonstrated the important relationship between the total curvature of a triangle formed by geodesic arcs on the surface (arcs of shortest length) and the sum of the angles in that triangle. In fact, he calculated that the total curvature $\int k\,d\sigma$ over a geodesic triangle was equal to $A + B + C - \pi$, where A, B, C are the measures of the three angles of that triangle. For example, on a surface of constant positive curvature every geodesic triangle has angle sum greater than π, while on a surface of constant negative curvature every such triangle has angle sum less than π. Gauss's result was a generalization of the well-known result that on a unit sphere, where the total curvature of a region equals its area, the angle sum of a triangle composed of great circle arcs (geodesics) was greater than π by a value equal to the area of the triangle.

Gauss's treatise on the differential geometry of surfaces was not only significant in and of itself but also had great consequences for future work. In particular, it turned out that the relationship of angle sum of a triangle to the intrinsic geometry on the surface helped lead to the solution of the question of the validity of Euclid's parallel postulate. Furthermore, Gauss's characterization of a surface in terms of its length element, which is in turn expressible in terms of the quantities E, F, G, proved to be the beginning of the general theory of n-dimensional manifolds, many important aspects of which were developed in the work of Riemann some 30 years later.

24.2   NON-EUCLIDEAN GEOMETRY

Recall that in the eighteenth century both Saccheri and Lambert attempted to prove Euclid's parallel postulate by assuming that it was false and trying to derive a contradiction. Saccheri believed that he had succeeded in this endeavor, but Lambert realized that his attempt was a failure. Both attacked the problem through synthetic means, trying to use the methodology of Euclid to show that he had assumed an unnecessary postulate. The nineteenth century, however, with its increasing use of analysis to solve all sorts of problems, provided a new approach to this one as well. And interestingly enough, it was the hyperbolic functions of Lambert that were called into service to make the connection between analysis and a new geometry, a connection that Lambert himself had missed.

24.2.1   Taurinus and Log-Spherical Geometry

Lambert had noted that the hypothesis of the acute angle would seem to hold on the surface of a sphere of imaginary radius, but it was Franz Taurinus (1794–1874), a man of independent means who pursued mathematics as a hobby, who actually made this connection explicit in a work of 1826. Taurinus began with a formula of spherical trigonometry connecting the sides and an angle of an arbitrary spherical triangle on a sphere of radius K, a formula we have already seen in the work of al-Battānī:

$$\cos \frac{a}{K} = \cos \frac{b}{K} \cos \frac{c}{K} + \sin \frac{b}{K} \sin \frac{c}{K} \cos A,$$

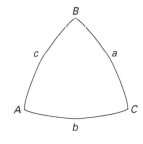

FIGURE 24.2

Spherical triangle on sphere of radius K

where the triangle has sides a, b, c and opposite angles A, B, C (Fig. 24.2). Replacing K by iK—that is, making the radius of the sphere imaginary (whatever that means)—and recalling that $\cos ix = \cosh x$ and $\sin ix = i \sinh x$, Taurinus derived the new formula

$$\cosh \frac{a}{K} = \cosh \frac{b}{K} \cosh \frac{c}{K} - \sinh \frac{b}{K} \sinh \frac{c}{K} \cos A. \qquad (24.1)$$

Taurinus called the geometry defined by this formula "log-spherical geometry," but he realized that this geometry was not possible in the plane. Exploring the consequences of the formula gives some idea of the properties of this geometry, however. For example, if the triangle is equilateral ($a = b = c$), the formula becomes

$$\cosh \frac{a}{K} = \cosh^2 \frac{a}{K} - \sinh^2 \frac{a}{K} \cos A$$

or

$$\cos A = \frac{\cosh^2 \frac{a}{K} - \cosh \frac{a}{K}}{\sinh^2 \frac{a}{K}} = \frac{\cosh \frac{a}{K} \left(\cosh \frac{a}{K} - 1 \right)}{\cosh^2 \frac{a}{K} - 1} = \frac{\cosh \frac{a}{K}}{\cosh \frac{a}{K} + 1}.$$

Because $\cosh \frac{a}{K} > 1$, it follows that $\cos A > 1/2$ and therefore that $A < 60°$. In other words, the sum of the angles of an equilateral triangle in this geometry is less than 180°. On the other hand, it is easy to see that as either the sides get smaller or the radius K gets larger, the angle A approaches 60° and the geometry approaches Euclid's geometry. In fact, one can also show (by using appropriate power series expansions) that in the limit as K

approaches ∞, Taurinus's formula (Equation 24.1) reduces to the Euclidean law of cosines $a^2 = b^2 + c^2 - 2bc \cos A$.

A second important formula of spherical trigonometry connecting the angles and a side of a spherical triangle is

$$\cos A = -\cos B \cos C + \sin B \sin C \cos \frac{a}{K}.$$

On replacing K by iK, this formula becomes a formula of log-spherical geometry:

$$\cos A = -\cos B \cos C + \sin B \sin C \cosh \frac{a}{K}. \tag{24.2}$$

For the special case where $A = 0°$ and $C = 90°$, Formula 24.2 reduces to

$$\cosh \frac{a}{K} = \frac{1}{\sin B}.$$

Naturally, a triangle with right angle at C and an angle of zero degrees at A does not exist in Euclid's geometry. Recall, however, that Saccheri had realized that the hypothesis of the acute angle led to the concept of asymptotic straight lines. Thus, Taurinus's triangle must be thought of as one in which two sides are asymptotic (Fig. 24.3). The angle B and the length a of the third side are then related through the formula $\sin B = \operatorname{sech} \frac{a}{K}$. One can rewrite this formula in the form

$$\tan \frac{B}{2} = e^{-a/K},$$

a formula that was to become fundamental in the work of Lobachevsky.

FIGURE 24.3

Triangle in which $\angle C = 90°$ and $\angle A = 0°$

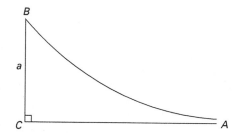

Formula 24.2 also shows that if one constructs an isosceles right triangle and splits it into two right triangles by drawing the altitude a, then the relationship between a and the base angle A of the original triangle is given by $\cosh \frac{a}{K} = \sqrt{2} \cos A$ (Fig. 24.4). It follows that the maximum possible altitude h of an isosceles right triangle occurs when $A = 0°$, that is, when the two legs of the triangle are both asymptotic to the hypotenuse. In that case $\cosh \frac{h}{K} = \sqrt{2}$, or

$$K = \frac{h}{\ln(1 + \sqrt{2})}.$$

Taurinus noted further that the area of a triangle is proportional to its defect (as Lambert had already discovered), the length of the circumference of a circle of radius r is $2\pi K \sinh \frac{r}{K}$,

FIGURE 24.4

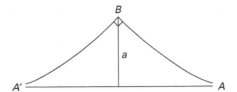

and the area of a circle of radius r is $2\pi K^2(\cosh\frac{r}{K} - 1)$. It is noteworthy that these latter results, as well as much of the work of Lobachevsky and Bolyai to be discussed shortly, had all been worked out by Gauss somewhat earlier in his private papers. Gauss, however, perhaps because he did not feel that he had proved all the various results to his own high standards, never published anything on the subject directly. On the other hand, his work relating curvature of a surface to the defect or excess of a triangle must have somehow been stimulated by his thoughts on this new geometry.

24.2.2 The Non-Euclidean Geometry of Lobachevsky and Bolyai

Despite his analytic results, Taurinus was not convinced that his geometry on an imaginary sphere applied in any "real" situation. The formulas were simply a collection of pretty results with no real content. But because neither Saccheri nor Lambert had succeeded in their attempts to refute the hypothesis of the acute angle, other mathematicians began to believe that a plane geometry in which that hypothesis was valid could exist. And that geometry would have as its analytic basis the formulas of Taurinus. The two mathematicians who first had confidence enough in their new ideas to publish them were the Russian Nikolai Ivanovich Lobachevsky (1792–1856) and the Hungarian János Bolyai (1802–1860). Both began work on the problem of parallels, determined to find a correct refutation of the hypothesis of the acute angle. And both gradually changed their minds.

In 1826, Lobachevsky gave a lecture at Kazan University in which he outlined a geometry having more than one parallel to a given line through a given point. Three years later, he published an extended version of his lecture in his Russian text *On the Principles of Geometry*. Over the next decade, he published several other versions of his new geometrical researches, including a detailed summary in 1840 in German entitled "Geometrische Untersuchungen zur Theorie der Parallellinien" ("Geometrical Investigations on the Theory of Parallel Lines"). Bolyai, also doing most of his creative work in the 1820s, published his material (in Latin) in 1831 as an appendix to a geometric work of his father, Farkas Bolyai, called "Appendix Exhibiting the Absolutely True Science of Space, Independent of the Truth or Falsity of Axiom XI [the Parallel Postulate] of Euclid (that can never be decided a priori)." Because the ideas of Lobachevsky and Bolyai turned out to be remarkably similar, we concentrate on the work of the former, as detailed in his "Geometrical Investigations."

Lobachevsky began with a summary of certain geometrical results that were true, independent of the parallel postulate. He then stated clearly his new definition of parallels: "All straight lines which in a plane go out from a point can, with reference to a given straight line in the same plane, be divided into two classes—*cutting* and *not-cutting*. The *boundary lines* of the one and other class of those lines will be called **parallel** to the given line."[5] Thus, if BC is a line, A a point not on the line, and AD the perpendicular from A to BC, one can first draw

BIOGRAPHY

Nikolai Ivanovich Lobachevsky (1792–1856)

Lobachevsky was born in Nizhni Novgorod (now Gor'kiy), Russia (a city about 250 miles east of Moscow), to parents of Polish origin. At age 14, he won a scholarship to enter the University of Kazan, the conditions being that he would teach for at least six years after receiving his degree. In fact, he spent virtually his entire adult life at the university, becoming an associate professor in 1816 and a full professor in 1822 at the age of 30. During his undergraduate years, he was influenced to turn to mathematics by J. Martin Bartels, a friend of Gauss, who had been invited to Kazan to provide mathematics instruction at the new university. Not only did Lobachevsky become an able teacher at Kazan, but he also occupied various administrative positions, including that of chief librarian and of rector. He was even able to some extent to hold the faculty together during the troubles posed by the influence of the French Revolution and the corresponding efforts of the Russian government to prevent the spread of these heretical ideas. Unfortunately, he was ultimately stripped of his positions at the university in 1846, without explanation, and lived out his last 10 years in meager circumstances (Fig. 24.5).

FIGURE 24.5

Lobachevsky on a Russian stamp

a line AE perpendicular to AD (Fig. 24.6). The line AE does not meet BC. Lobachevsky then assumed that there may be other lines through A, such as AG, that also do not meet BC, however far they are prolonged. In passing from a cutting line, such as AF, to a not-cutting line, such as AG, there must be a line AH that is the boundary between these two sets. It is AH that is the parallel to BC. The angle HAD between AH and the perpendicular AD, an angle dependent on the length p of AD, is what Lobachevsky called the **angle of parallelism**, written $\Pi(p)$. If $\Pi(p) = 90°$, then there is only one line through A parallel to BC and the situation is the Euclidean one. If, however, $\Pi(p) < 90°$, then there will be a corresponding line AK on the other side of AD from AH that also makes the same angle $\Pi(p)$ with AD. It is thus always necessary in this non-Euclidean situation to distinguish two different sides in parallelism. In any case, on each side of AD, under the non-Euclidean assumption, there are infinitely many lines through A that do not meet BC.

FIGURE 24.6

Lobachevsky's angle of parallelism

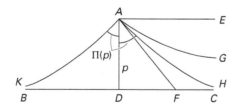

From the non-Euclidean assumption, Lobachevsky derived many results, some of which were in essence known to Saccheri and/or Lambert. For example, he showed that the property that $\Pi(p) < 90°$ for any p is equivalent to the property that the angle sum of every triangle is less than 180°. In that case, not only is the equation $\Pi(p) = \alpha$ solvable for any α less than a right angle, but also parallel lines are asymptotic to one another. To define the nature of parallel lines more precisely, Lobachevsky defined a new curve: "We call the **boundary line**

(or **horocycle**) that curve lying in a plane for which all perpendiculars erected at the mid-points of chords are parallel to each other."[6] In other words, given a line AB with A to be on the horocycle, any other point C is on the horocycle if AC makes an angle $\Pi(AC/2)$ with the line AB, for in that case the perpendicular DE to AC at its midpoint will be parallel to AB (Fig. 24.7). In fact, it turns out that all perpendiculars to the horocycle will be parallel. One can think of this curve as a circle of infinitely great radius that, under the Euclidean assumption, would be a straight line. Letting A, B be two points on a horocycle separated by a distance s and drawing two perpendiculars AA', BB' to the horocycle such that $AA' = BB' = x$, Lobachevsky constructed a new horocycle through A', B' with the distance $A'B'$ set equal to s' (Fig. 24.8). He then showed that s'/s depends only on the distance x, that is, $s'/s = f(x)$. If a new horocycle $A''B''$ is constructed at a distance x from $A'B'$ in the same manner with $A''B'' = s''$, then it turns out also that $s''/s' = f(x)$ and therefore that $f(2x) = s''/s = f(x)^2$.

FIGURE 24.7

Lobachevsky's horocycle

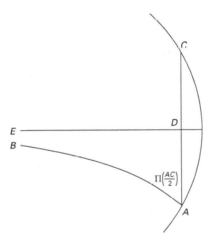

FIGURE 24.8

Perpendiculars to horocycles

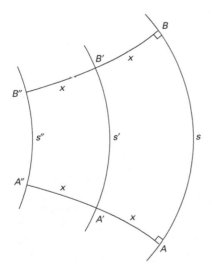

Similarly, $f(nx) = f(x)^n$, and Lobachevsky could conclude that $s' = sa^{-x}$ for some constant a. Because the units of measure are arbitrary, he took a to be e. The distance between the parallel lines AA', BB' then is given by the function $s' = se^{-x}$, where x is measured from A and/or B away from the horocycle. It follows that the parallel lines are indeed asymptotic.

Lobachevsky's most interesting results, and ones that were unknown to Saccheri and Lambert, involved the trigonometry of the non-Euclidean plane. Through a complex argument involving spherical triangles and triangles in the non-Euclidean plane, he was able to evaluate explicitly the function $\Pi(x)$ in the form

$$\tan \frac{1}{2}\Pi(x) = e^{-x},$$

essentially the same result obtained by Taurinus. It followed that $\Pi(0) = \frac{\pi}{2}$ (or that for small values of x, the geometry is close to the Euclidean one) and that $\lim_{x \to \infty} \Pi(x) = 0$. Lobachevsky could then derive new relationships among the sides a, b, c and the opposite angles A, B, C of an arbitrary non-Euclidean triangle:

$$\sin A \cot \Pi(b) = \sin B \cot \Pi(a), \tag{24.3}$$

$$\cos A \cos \Pi(b) \cos \Pi(c) + \frac{\sin \Pi(b) \sin \Pi(c)}{\sin \Pi(a)} = 1, \tag{24.4}$$

$$\cot A \sin C \sin \Pi(b) + \cos C = \frac{\cos \Pi(b)}{\cos \Pi(a)}, \tag{24.5}$$

$$\cos A + \cos B \cos C = \frac{\sin B \sin C}{\sin \Pi(a)}. \tag{24.6}$$

Lobachevsky's formulas imply standard Euclidean formulas when the sides of the triangle are small. The explicit formula for $\Pi(x)$ implies that

$$\cot \Pi(x) = \sinh x \qquad \cos \Pi(x) = \tanh x \qquad \sin \Pi(x) = \frac{1}{\cosh x}.$$

One then approximates the values of the hyperbolic functions by the terms of their power series up to degree 2 to get: $\cot \Pi(x) = x$, $\cos \Pi(x) = x$, and $\sin \Pi(x) = 1 - \frac{1}{2}x^2$. Substituting these approximations into the four formulas and neglecting terms of degree higher than 2 gives the results

$$b \sin A = a \sin B,$$

$$a^2 = b^2 + c^2 - 2bc \cos A,$$

$$a \sin(A + C) = b \sin A,$$

$$\cos A + \cos(B + C) = 0.$$

The first two results are the familiar laws of sines and cosines, respectively, while the last two, when combined with the first two, are equivalent to the result that $A + B + C = \pi$. It also follows that if one replaces the sides a, b, c of the triangle by ia, ib, ic, respectively, Lobachevsky's results transform into standard results of spherical trigonometry. Thus, Lobachevsky's geometry is essentially the same as Taurinus's log-spherical geometry on the sphere of imaginary radius.

Lobachevsky probably never read Taurinus's work. So what he saw in his trigonometric formulas was simply "a sufficient foundation for considering the assumption of [non-Euclidean] geometry as possible. Hence," he concluded, "there is no means, other than astronomical observations, to use for judging of the exactitude which pertains to the calculations of the ordinary geometry."[7] As noted in the chapter opening, Gauss, too, realized that the creation of a new and apparently valid geometry in which Euclid's parallel postulate did not hold showed that there was no "necessity" to Euclid's geometry and that one could not automatically conclude that Euclid's geometry held in the world in which we live. It was necessary to experiment to decide whether the geometry of the physical universe was Euclidean or not. Lobachevsky in fact attempted such an experiment, using data on star positions, but the results were inconclusive.

Bolyai also commented that it is not decided whether Euclidean geometry or a non-Euclidean geometry represents "reality." In fact, he claimed, "it remains to demonstrate the impossibility (apart from any supposition) of deciding a priori whether [Euclidean geometry] or some [non-Euclidean] geometry (and which one) exist. This, however, is reserved for a more suitable occasion."[8] Although the "more suitable occasion" never occurred, Bolyai did derive most of the same mathematical results as Lobachevsky. He was, however, more explicit in dealing with **absolute geometry**, the collection of theorems that were true independent of the parallel postulate. For example, he proved that "in any rectilinear triangle, the [circumferences of the] circles with radii equal to its sides are as the sines of the opposite angles."[9] In Euclidean geometry, where the circumference of a circle of radius r is $2\pi r$, this result is simply that $a : b : c = \sin A : \sin B : \sin C$, the law of sines. In non-Euclidean geometry, where the corresponding circumference is $2\pi K \sinh \frac{r}{K}$, for some constant K (each such constant determining for Bolyai a different geometry), the theorem translates into

$$\sinh \frac{a}{K} : \sinh \frac{b}{K} : \sinh \frac{c}{K} = \sin A : \sin B : \sin C,$$

a result equivalent to Lobachevsky's Equation 24.3. Bolyai was also able to prove the interesting result that in non-Euclidean geometry it is possible to construct a square of area equal to that of a circle of radius 1 by using essentially Euclidean tools.

The work of Bolyai and Lobachevsky, although responding to an age-old question about the parallel postulate, drew very little response from the mathematical community before the 1860s. There are several reasons for this, including the fact that some (though not all) of their articles were published in somewhat obscure sources and not in the major languages of the day, as well as the general difficulties of the acceptance of an entirely new idea into mathematics. But it would appear that the most important reason that the discoveries of the Hungarian and the Russian did not immediately become part of the mainstream of mathematics was that few could really understand what a non-Euclidean plane really was. Although the arguments of the founders were correct and logically coherent, and although they displayed what appeared to be reasonable mathematical formulas involving known functions, the "reality" of this new geometry simply was not accepted. Until non-Euclidean geometry could be seen as part of a more general system of geometry and be connected via this system to Euclidean geometry, it was not to be anything more than a curious sidelight.

BIOGRAPHY

János Bolyai (1802–1860)

Born in Kolozsvar, Hungary (now Cluj, Romania), Bolyai (Fig. 24.9) received his early education in Maros-Vásárhely (now Tirgu Mures), where his father, Farkas Bolyai, a friend of Gauss, was a professor. At the age of 16, he entered the imperial military academy in Vienna and became a military officer, serving in that capacity in Temesvar (Timisoara), Arad, and L'vov (Lemberg). He had to retire from the service in 1833, however, due to his physical condition. Meanwhile, his father, having been interested in the question of parallels, had corresponded with Gauss on the question over several years, without any resolution. He ultimately wearied of the matter and warned his son too about attacking this subject. Nevertheless, János persisted and, in 1823, told his father that he had made "wonderful discoveries" in the theory of parallels. He finally published the discoveries in 1831. Although János continued to develop his theory of space, his great disappointment at Gauss's response that he (Gauss) had already discovered the basic ideas of non-Euclidean geometry caused him to give up any further thought of publishing his own ideas (Fig. 24.10).

FIGURE 24.9

Bolyai on a Hungarian stamp

FIGURE 24.10

Bolyai's work on a Hungarian stamp

24.2.3   Riemann's Hypotheses on Geometry

The first mathematician to create a new general system of geometry was Riemann. As noted in the chapter opening, he presented his ideas to the professors of the philosophical faculty of Göttingen in his inaugural lecture entitled "Über die Hypothesen welche der Geometrie zu Grunde liegen" ("On the Hypotheses Which Lie at the Foundation of Geometry"). In the lecture, Riemann gave few mathematical details but devoted much time to explaining what geometry ought to be about. Thus, he began with "the task of constructing the concept of a multiply-extended quantity from general notions of quantity. It will be shown that a multiply-extended quantity is susceptible of various metric relations, so that space constitutes only a special case of a triply-extended quantity. From this, however, it is a necessary consequence that the theorems of geometry cannot be deduced from general notions of quantity, but that those properties which distinguish space from other conceivable triply-extended quantities can only be deduced from experience."[10] In other words, for Riemann the most general geometrical notion was what is today called a manifold. On a manifold one can establish various metric relations, ways of determining distance. The usual "space" of (three-dimensional) Euclidean geometry, which had generally been assumed to be the physical space in which we live, is then a special case of a three-dimensional manifold, having attached to it the **Euclidean metric**, expressed infinitesimally as $ds^2 = dx^2 + dy^2 + dz^2$. Agreeing with Gauss, he also argued that the precise nature of physical space could not be determined a priori but only by "experience."

In part one of his lecture, Riemann considered the idea of a manifold of n dimensions. He constructed such a manifold inductively, beginning with the idea of a one-dimensional manifold, or curve, "whose essential characteristic is, that from any point in it a continuous movement is possible in only two directions, forwards and backwards."[11] A two-dimensional manifold is created by having one one-dimensional manifold pass continuously into a second one; the two-dimensional manifold then consists of all the points formed by this passage. Similarly, a three-dimensional manifold is formed by the continuous passage of one two-dimensional manifold into a second and so on for ever higher dimensions. The central idea for

Riemann appears to be that the introduction of each higher dimension results in the addition of one further direction in which one can go at a point or, in more modern terminology, the addition of a new dimension to the tangent space to the manifold at a point. Riemann noted further that one can reverse the procedure in some sense and define an $(n-1)$-dimensional manifold as the zeros of a function defined on an n-dimensional one. One can take these functions to be what are today called **coordinate functions**. It then follows that each point on the manifold is determined by n numerical quantities, that is, n coordinates.

Part two of Riemann's lecture was devoted to the idea of a metric relation on the manifold, a way of determining the length of a curve on the manifold independent of its position. This is the only section of the talk that involved any mathematical formulas, but even here Riemann just presented them without any derivations. Riemann's basic assumption, based on Gauss's earlier work, was that a **metric**, the expression for the length ds of an infinitesimal element of a curve, is the square root of a homogeneous positive definite quadratic function of the dx_i, that is,

$$ds^2 = \sum_{i=1}^{n} \sum_{j=1}^{n} g_{ij}\, dx_i\, dx_j,$$

where $g_{ij} = g_{ji}$ and all of the g_{ij} are continuous functions on the manifold. Ordinary (Euclidean) space has the simplest case of this metric, namely, $ds^2 = \sum dx_i^2$. Riemann showed that it is not in general possible to transform a given metric into another. Therefore, the manifolds having this simplest metric form a special class, a class Riemann named "flat."

To deal with curved (or nonflat) manifolds, Riemann constructed a special set of coordinates, today called **Riemannian normal coordinates**. Through the use of these coordinates, Riemann defined a notion of curvature generalizing the idea of Gauss and showed that it too is intrinsic to the manifold and depends only on the coefficients g_{ij}. All the properties of the geometry of the manifold can then be described in terms of the metric and the coordinate grid. For example, one could use asymptotic parallel lines and their associated horocycles as the coordinate grid on a two-dimensional surface and, using an appropriate metric, develop all the properties of Lobachevsky's non-Euclidean geometry without making any assumption about the existence of parallels to a given line. Riemann noted that manifolds with constant curvature have the important property that "figures can be moved in them without stretching." In fact, "since the metric properties of the manifold are completely determined by the curvature, they are therefore exactly the same in all the directions around any one point as in the directions around any other. . . . Consequently, in the manifolds with constant curvature figures may be given any arbitrary position."[12]

Riemann dealt in the final section of his lecture with the relationship of his ideas to our usual concept of three-dimensional (Euclidean) space. He presented three sets of conditions, each of which he claimed is sufficient to determine whether a three-dimensional manifold is flat. One of these sets of conditions is that bodies be free to move and turn and that all triangles have the same angle sum. Riemann noted that it is difficult to determine whether these conditions hold because of the necessity of extending our observations to the immeasurably large and immeasurably small. He did state, however, that one can assume that physical space forms a three-dimensional unbounded manifold. The unboundedness, however, does not imply that space is infinite, because if the curvature is constant but positive, space would necessarily be finite. As to the immeasurably small, Riemann concluded that the metric relations do not necessarily follow from those in the large and that, in fact, the curvature may vary from point to point as long as the total curvature of every measurable portion of space is close to zero. The accurate determination of the curvature and the associated metric, however, is a matter for physics and not mathematics.

24.2.4 The Systems of Geometry of Helmholtz and Clifford

FIGURE 24.11

Helmholtz on a German stamp

Riemann made no effort to publish his lecture, perhaps because he had not initially desired to give this talk on geometry and was therefore working on several other projects at the time. Thus, although Gauss was quite impressed with it, its new ideas had very little effect elsewhere until it was published in 1868 after Riemann's untimely death. Once this occurred, however, Riemann's work met with widespread acclaim. In particular, Hermann von Helmholtz in Germany (Fig. 24.11) and William Clifford (1845–1879) in England were both influenced by Riemann's work and published their own interpretations and extensions, which helped bring it to the attention of the wider community.

Helmholtz, in a paper appearing shortly after the publication of Riemann's lecture and with a title remarkably similar, "Über die Thatsachen die der Geometrie zu Grunde liegen" ("On the Facts Which Lie at the Foundations of Geometry"), attempted to list a set of hypotheses that would provide the basis for any reasonable study of geometry. First, like Riemann, he assumed that a space of n dimensions is a manifold. His definition was somewhat more explicit than Riemann's, however, in that he assumed the existence of n independent coordinates near a point, at least one of which varies continuously as the point moves. It seems clear from his examples that he did not require that the same set of coordinates always applied in the entire manifold. Helmholtz's second axiom was that rigid bodies exist. This assumption permits one to equate two different spatial objects by superposition. Third, Helmholtz asserted that rigid bodies can move freely. In other words, any point in such a body can be moved to any other point in space; other points in the body will be carried by this motion to other points whose coordinates are related to the first by a particular set of equations. With the further assumption that $n = 3$, these hypotheses led Helmholtz to his own concept of the physical space in which we live, namely, that of a three-dimensional manifold of constant curvature. It follows that there are three possibilities for physical space: its curvature could be positive, negative, or zero. The third option leads to Euclidean geometry. Contrariwise, "if the measure of curvature is positive we have **spherical** space, in which straightest lines return upon themselves and there are no parallels. Such a space would, like the surface of a sphere, be unlimited but not infinitely great. A constant negative measure of curvature on the other hand gives **pseudospherical** space, in which straightest lines run out to infinity, and a pencil of straightest lines may be drawn, in any flattest surface, through any point, which does not intersect another given straightest line in that surface."[13] If we call a geometry Euclidean when both the parallel postulate holds and lines are unlimited (recall *Elements* I–16), then both Lobachevsky's geometry and spherical geometry are non-Euclidean. Thus, Helmholtz had succeeded in placing both of these non-Euclidean geometries into the context of Riemann's work. Furthermore, both of them could lead to possible geometries of our physical space.

In a series of lectures in England in the early 1870s, William Clifford also attempted to determine the postulates of physical space. He noted, more specifically than Helmholtz, that one way to distinguish between Euclidean and non-Euclidean spaces was by a postulate of similarity, "that any figure may be magnified or diminished in any degree without altering its shape."[14] This postulate turns out to be equivalent to the assumption of zero curvature and is therefore not true for Lobachevsky's geometry or spherical geometry. Clifford was particularly impressed, however, by the revolution wrought by Lobachevsky's ideas, comparing the effect of these with respect to Euclid to that of Copernicus's ideas with respect to Ptolemy. In both cases, humanity's view of the universe was fundamentally altered. In particular, with non-Euclidean geometry of either positive or negative curvature a possibility for physical space, it turned out that humanity's knowledge of that space, particularly in its far reaches, was limited to the distances its powers of observation could reach.

Clifford also noted in a brief paper of 1876 that although finite portions of space do have curvature zero to within the limits of our experimental accuracy, we do not really know whether all the axioms of space apply for very small portions of space. In fact, Clifford provided some new speculations that contradicted Helmholtz's concept that our space had constant curvature. Explaining these ideas, which in more recent times have come to the forefront of research in cosmology, Clifford wrote as follows:

> I hold in fact
>
> 1. That small portions of space *are* in fact of a nature analogous to little hills on a surface which is on the average flat; namely, that the ordinary laws of geometry are not valid in them.
> 2. That this property of being curved or distorted is continually being passed on from one portion of space to another after the manner of a wave.
> 3. That this variation of the curvature of space is what really happens in that phenomenon which we call the *motion of matter*, whether ponderable or etherial.
> 4. That in the physical world nothing else takes place but this variation, subject (possibly) to the law of continuity.[15]

Clifford's speculations about the physical world thus made Riemann's ideas on the theory of manifolds into an important research tool in physics. In fact, they turned out to be central in the revolutionary developments in physics related to the theory of relativity that occurred early in the twentieth century.

24.2.5 Models of Non-Euclidean Geometry

Because Lobachevsky's geometry appeared to be valid on a surface of constant negative curvature, Eugenio Beltrami (1835–1900), an Italian mathematician who held chairs in mathematics in Bologna, Pisa, Pavia, and finally Rome, attempted to construct such a surface, the so-called **pseudosphere**. It turned out that one could construct only a portion of the surface in Euclidean three-dimensional space. Nevertheless, Beltrami succeeded in determining the appropriate metric on this surface and showing the connection between this metric and Lobachevsky's trigonometric laws for non-Euclidean space. In an article in 1868, he began by parametrizing the sphere of radius k (and curvature $1/k^2$) situated in Euclidean three-dimensional space by

$$x = \frac{uk}{\sqrt{a^2 + u^2 + v^2}}, \qquad y = \frac{vk}{\sqrt{a^2 + u^2 + v^2}}, \qquad z = \frac{ak}{\sqrt{a^2 + u^2 + v^2}}$$

for some value a. It is then straightforward to calculate the metric form ds^2 on the sphere by substitution into the Euclidean form $ds^2 = dx^2 + dy^2 + dz^2$:

$$ds^2 = k^2 \frac{(a^2 + v^2)\, du^2 - 2uv\, du\, dv + (a^2 + u^2)\, dv^2}{(a^2 + u^2 + v^2)^2}.$$

To transform this result into one on a pseudosphere of curvature $-1/k^2$, Beltrami simply replaced u by iu and v by iv. The resulting metric,

$$ds^2 = k^2 \frac{(a^2 - v^2)\, du^2 + 2uv\, du\, dv + (a^2 - u^2)\, dv^2}{(a^2 - u^2 - v^2)^2},$$

turned out to have the required properties.

On the pseudosphere, the curves $u = c$ and $v = c$ are geodesics orthogonal to $v = 0$ and $u = 0$, respectively, for any constant $c < a$. Thus, Beltrami could consider a right triangle with one vertex at the origin, one leg along the curve $v = 0$, one leg along a curve $u = c$, and the hypotenuse along a geodesic through the origin that makes an angle θ with $v = 0$ (Fig. 24.12). He calculated the lengths of these three sides by integration of the appropriate metric forms. For the hypotenuse, set $u = r \cos \theta$ and $v = r \sin \theta$. It follows that

$$ds = \frac{ka\, dr}{a^2 - r^2}.$$

This element of arc is easily integrated from 0 to r to get the length ρ of the hypotenuse:

$$\rho = \frac{1}{2}k \ln \frac{a + r}{a - r}.$$

Similarly, along the curve $v = 0$,

$$ds = \frac{a\, du}{a^2 - u^2},$$

and the length s of this leg of the triangle up to a given u is

$$s = \frac{1}{2}k \ln \frac{a + u}{a - u} = \frac{1}{2}k \ln \frac{a + r \cos \theta}{a - r \cos \theta}.$$

Finally, the metric along $u = c$ is given by

$$ds = \frac{k\sqrt{a^2 - u^2}}{a^2 - u^2 - v^2}\, dv,$$

a differential whose integral up to a particular v is

$$t = \frac{1}{2}k \ln \frac{\sqrt{a^2 - u^2} + v}{\sqrt{a^2 - u^2} - v}.$$

With a bit of algebraic manipulation on the values for ρ, s, and t, Beltrami showed that

$$\frac{r}{a} = \tanh \frac{\rho}{k}, \qquad \frac{r}{a} \cos \theta = \tanh \frac{s}{k}, \qquad \frac{v}{\sqrt{a^2 - u^2}} = \tanh \frac{t}{k}.$$

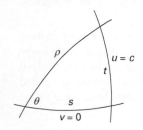

FIGURE 24.12

Beltrami's calculations on the pseudosphere:

$\rho = \frac{1}{2}k \ln \frac{a+r}{a-r}$

$s = \frac{1}{2}k \ln \frac{a+r\cos\theta}{a-r\cos\theta}$

$t = \frac{1}{2}k \ln \frac{\sqrt{a^2-u^2}+v}{\sqrt{a^2-u^2}-v}$

It then follows that

$$\cosh \frac{s}{k} \cosh \frac{t}{k} = \cosh \frac{\rho}{k}.$$

This result, identical with Taurinus's Equation 24.1 and Lobachevsky's Equation 24.4 for the case of a right triangle, shows that Beltrami's surface with its associated metric gives the same geometry as Lobachevsky's non-Euclidean plane. In other words, Beltrami's calculations showed that the apparently mysterious use of a sphere of imaginary radius by Taurinus was equivalent to the introduction of a new metric on an appropriate two-dimensional manifold.

A different way of looking at Lobachevsky's geometry is simply to consider the imaginary sphere to be projected onto the interior of the circle $u^2 + v^2 = a^2$, where u and v are the parameters given above. It then turns out that straight lines in the Lobachevskian plane are represented by chords in the circle (Fig. 24.13). Parallel straight lines are those whose intersection is at the circumference of the circle, with the circumference itself representing points at "infinity." Chords that do not intersect inside the circle represent lines that do not intersect at all in the Lobachevskian plane. Beltrami did not explicitly calculate distances between points in this model, but this gap was filled in 1872 by Felix Klein (1849–1925), who used some concepts from projective geometry to be discussed in Section 24.3.3. A similar model of Lobachevskian geometry in the interior of a circle was developed by Henri Poincaré (1854–1912) in 1882. In this model, straight lines are represented by arcs of circles that are orthogonal to the boundary circle. Parallel lines are then represented by circular arcs that intersect at the boundary. This model has the advantage that angles between circles are measured in the Euclidean way. Figure 24.13 then shows why the angle sum of a triangle is less than π.

FIGURE 24.13

(a) Parallel straight lines in Klein's model of the Lobachevskian plane
(b) Parallel straight lines in Poincaré's model of the Lobachevskian plane

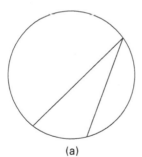

(a)

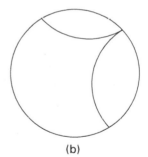

(b)

It was the use of models of Lobachevskian geometry as subsets of the ordinary Euclidean plane that helped to convince mathematicians by the end of the century that non-Euclidean geometry was as valid as Euclid's. Any contradiction in the former geometry would, by translation to the model, lead to a contradiction in the latter. Saccheri's attempts to "vindicate" Euclid had failed. With the work of Lobachevsky, Bolyai, Beltrami, Klein, and Poincaré, it was now clear that Euclid was truly vindicated. He had been completely correct in his decision 2200 years earlier to take the parallel postulate as a postulate. Because the Lobachevskian alternative to Euclid's parallel postulate led to a geometry as valid as Euclid's, it was impossible to prove that postulate as a theorem.

24.3

PROJECTIVE GEOMETRY

The work of Monge on descriptive geometry late in the eighteenth century, particularly his work on representing three-dimensional objects in two dimensions by various types of projection, led in the early nineteenth century to a renewed interest in the formal study of projective geometry, that is, the study of projective invariants of geometric figures. Certain aspects of projective geometry had been studied by artists in the Renaissance as part of their effort to master the theory of perspective, and Desargues and Pascal in the mid-seventeenth century had worked out the beginnings of the theory of the subject. But it was only in the nineteenth century that mathematicians expanded the scope of this study.

24.3.1 Poncelet and Duality

It was a student of Monge, Jean-Victor Poncelet (1788–1867), who composed the first text in synthetic projective geometry in 1822, *Traité des propriétés projectives des figures* (*Projective Properties of Figures*). Poncelet started with the theory of polars in conic sections. Given a conic section C, one can associate to any point p its **polar** π, the straight line joining the points of contact of the tangents drawn from p to C. Similarly, to any line π' crossing the conic, one can associate a point p', its **pole**, the intersection point of the tangent lines to the conic at the points where π' meets C. Poncelet saw that these concepts were reciprocal, that is, that if p' lies on π, then π', the polar of p', goes through p, the pole of π (Fig. 24.14).

FIGURE 24.14

Reciprocal relationship of poles and polars

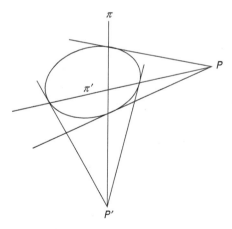

Out of this duality of pole and polar, Poncelet developed a more general notion of duality between points and lines. He saw that, in general, a true proposition about "points" and "lines" remains true if we interchange the two words. For example, the statement, "Two distinct points determine exactly one line on which they both lie" becomes "Two distinct lines determine exactly one point through which they both pass." (Note that the latter statement is not true in ordinary Euclidean geometry, but becomes true when points at infinity are added to the plane to provide intersections of parallel lines. This idea is discussed below.) As a more complicated example, recall Pascal's theorem: "If the six vertices of a hexagon lie on a conic, then the points of intersection of the three pairs of opposite sides lie on a line." The dual theorem is "If the six sides of a hexagon are tangent to a conic, then the lines joining the three pairs of opposite vertices intersect in a point." Although Poncelet did not establish the principle of duality as a theorem, he did use it as a valuable tool of discovery.

The results of Poncelet's discoveries, and the primary objective of his text, however, were the properties of central projection. Given a figure F in a plane π and a point P outside the plane, a central projection of F onto another plane π' is the figure consisting of the points of intersection with π' of all lines from P through the points of F (Fig. 24.15). For example, the projection of a square in π is a quadrilateral in π', not necessarily a square, while the projection of a circle is a conic section. Poncelet's aim was to determine which properties of figures are invariant under such a projection. Clearly, the length of a line segment is not a projective invariant, but the property that a straight line can intersect a circle in at most two points is such an invariant. Poncelet noted that because a projection may transform parallel lines into intersecting lines, and because projections preserve intersections, it was necessary to introduce the points at infinity to be used as the intersection points of ordinary parallel lines. It is then useful to assume that all the points at infinity in a given plane together make up a line at infinity. Thus, Poncelet's ideas led to the consideration of a new object, the projective plane, consisting of the ordinary points on the plane as well as the points at infinity. In the geometry of this projective plane, however, there is no distinction between an ordinary (Euclidean) point and a point at infinity, because there is always a central projection that takes any ordinary point into a point at infinity and conversely.

FIGURE 24.15

Central projection of F onto F'

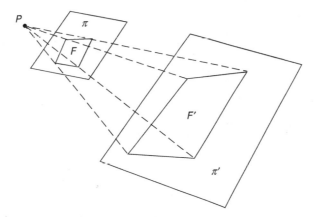

To deal with points at infinity, it proved necessary to develop a coordinate system for the projective plane. This was accomplished by Julius Plücker (1801–1868) with his introduction in 1831 of **homogeneous coordinates**. A point P in the plane with rectangular coordinates (X, Y) has the homogeneous coordinates (x, y, t), if $x = Xt$ and $y = Yt$. With this definition, a point does not have a unique set of coordinates; any two sets differ by a constant multiple. Nevertheless, the use of these coordinates means that any polynomial equation $f(X, Y) = 0$ (in rectangular coordinates) is rewritten in the form $g(x, y, t) = 0$, where all terms of g have the same degree (thus the name "homogeneous"). In addition, the points at infinity of the projective plane have homogeneous coordinates $(x, y, 0)$. Plücker noted that any straight line in the projective plane has the equation $ax + by + ct = 0$ in these coordinates. Namely, given constants (a, b, c), the set of points $\{(x, y, t)\}$ satisfying the equation all lie in a particular line. But, surprisingly, if one takes (x, y, t) as the constants, this equation also characterizes the set of lines $\{(a, b, c)\}$ that all pass through the given point (x, y, t). Thus, Poncelet's interchange of "point" and "line" in the principle of duality is justified algebraically by the interchange of "constant" and "variable" in the equation $ax + by + ct = 0$.

24.3.2   The Cross Ratio

Poncelet did not discover the most important projective invariant, the cross ratio of four points on a line. This concept was, however, thoroughly investigated by his younger countryman Michel Chasles (1793–1880) under the name of the anharmonic ratio. Recall that if a segment AB on a line p is projected onto a segment $A'B'$ on a line p' by a central projection from a point S, then the length $A'B'$ in general differs from the length AB. Similarly, if C is a point on segment AB and C' the corresponding point on $A'B'$, the ratio $AC : CB$ differs from the ratio $A'C' : C'B'$. But if C and D are two points on the segment AB whose corresponding points on segment $A'B'$ are C', D', respectively, then the **cross ratio**

$$\frac{AC}{CB} : \frac{AD}{DB}$$

is preserved by the projection; that is,

$$\frac{AC}{CB} : \frac{AD}{DB} = \frac{A'C'}{C'B'} : \frac{A'D'}{D'B'}.$$

The proof of this result is not difficult. Draw segments A_1B and AB_2 parallel to the line p' and determine the projections of C, D onto those two lines (Fig. 24.16). From the similarity of triangles ACC_2 and BCC_1 and of triangles ADD_2 and BDD_1, one gets that

$$\frac{AC}{CB} = \frac{AC_2}{C_1B} \quad \text{and} \quad \frac{AD}{DB} = \frac{AD_2}{D_1B}.$$

But $D_1B/C_1B = D_2B_2/C_2B_2$. It follows that

$$\frac{AC}{CB} : \frac{AD}{DB} = \frac{AC_2}{C_2B_2} : \frac{AD_2}{D_2B_2},$$

or that the cross ratio on line p equals the cross ratio on the line determined by segment AB_2. That the cross ratio of the points on the latter line segment is equal to the cross ratio on line p' follows from basic principles of similarity.

The standard notation for the cross ratio in the previous paragraph is (AB, CD). By permuting the four letters, one can calculate 24 cross ratios among these four points. (In this context, one considers a segment such as AB positive if A is to the left of B and negative in the contrary case.) Chasles noted that among the 24 apparent cross ratios there are really only 6 different ones, and even these are closely related. Thus, for example, $(AB, CD) = 1/(AB, DC)$ and $(AB, CD) = 1 - (AC, BD)$.

FIGURE 24.16

Cross ratios are preserved under central projection

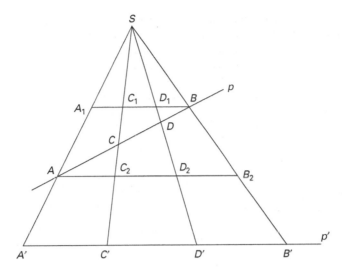

Interestingly, although projective geometry aimed to study properties of figures not dependent on such concepts as that of length, the very basis of the definition of a cross ratio was in fact the length of a line segment. It was Christian von Staudt (1798–1867) who was able to correct this problem in 1847 by outlining an axiomatic system for projective geometry, based on the notion of a projective mapping as one that preserved harmonic tetrads. A **harmonic tetrad** is a set of four points A, B, C, D such that $(AB, CD) = -1$. Although this definition again seems to require lengths, von Staudt showed in fact that given three collinear points A, B, C, one could find the "fourth harmonic" D, that is, the point such that $(AB, CD) = -1$, by a simple projective construction. Von Staudt's work in projective geometry was central in making the subject into a clearly defined area of study and set the stage for the idea of defining a notion of distance in a nonmetric geometry.

24.3.3 Projective Metrics and Non-Euclidean Geometry

It was Cayley who in 1859 first provided a definition for a metric in the projective plane. Given a conic section C, he gave a rather complex definition of a function $d_C(P_1, P_2)$, depending on C, that satisfied the basic property of a distance, namely, that if P_1, P_2, P_3 lie on the same line, then $d(P_1, P_2) + d(P_2, P_3) = d(P_1, P_3)$. Twelve years later, Klein noticed that if Cayley's conic section is a circle, then the part of the projective plane inside that circle can be considered as a model for Lobachevskian geometry and Cayley's metric can be transformed into a distance function for that geometry.

Klein defined his modified metric for the non-Euclidean plane in terms of the cross ratio. Consider the Lobachevskian plane as the interior of the circle $u^2 + v^2 = 1$. Given two points P, Q in the interior of this circle, join them by a straight line that intersects the circle at the

BIOGRAPHY

Felix Klein (1849–1925)

Felix Klein was primarily responsible for creating the mathematical institute at the University of Göttingen, an institute that transformed that small university into the mathematical center of the world for the first third of the twentieth century. After a brilliant early career, in which he contributed many beautiful ideas to the study of geometry, he suffered a nervous breakdown in the mid-1880s, following which he devoted his professional life to teaching, writing, and organizational activities. He became the editor of *Mathematische Annalen*, one of the leading mathematics journals of the time, and wrote several books consisting of edited versions of his lectures on various subjects. Many of the books were specifically aimed at mathematics teachers and dealt with the central ideas behind the mathematics taught in secondary schools. He also wrote a major work on the history of nineteenth-century mathematics and directed the publication of the *Enzyklopädie der Mathematischen Wissenschaften*, an encyclopedia whose aim, ultimately impossible to fulfill, was to collect all the results and methods in mathematics obtained up to that time.

FIGURE 24.17

Klein's Lobachevskian distance:
$d(P, Q) = c \ln\left(\frac{QR}{RP} : \frac{QS}{SP}\right)$

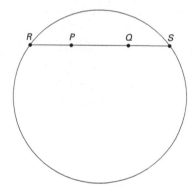

points R, S (Fig. 24.17). The (directed) distance from P to Q is then

$$d(P, Q) = c \ln(QP, RS) = c \ln\left(\frac{QR}{RP} : \frac{QS}{SP}\right) = c \ln\left(\frac{QR \cdot SP}{RP \cdot QS}\right)$$

for some constant c that determines the unit of length. It is straightforward to show that if P, Q, Q' are three points on the line, then $d(PQ) + d(QQ') = d(PQ')$, so that the function d satisfies the major property of a distance function. By putting $P = R$, for example, one can also show that the length of the entire chord is infinite, that is, that the circle itself represents the points at infinity. Klein's metric on the Lobachevskian plane is equivalent to the one derived by Beltrami on the pseudosphere. To see this, take P to be the origin and Q to be a point at Euclidean distance $r < 1$ to the right of P. Then

$$d(P, Q) = c \ln\left(\frac{-(1+r) \cdot (-1)}{1 \cdot (1-r)}\right) = c \ln\left(\frac{1+r}{1-r}\right),$$

which, for appropriate choice of c, is the same distance value Beltrami had calculated.

Klein also used the cross ratio to give a definition of the angle between two lines in the plane. This definition is, however, easier to understand in a slightly different form. Suppose that the two lines are given in projective coordinates by the triples (a_1, b_1, c_1), (a_2, b_2, c_2), that is, by the equations $a_1x + b_1y + c_1t = 0$ and $a_2x + b_2y + c_2t = 0$. These lines intersect in the point $x_0 = b_1c_2 - c_1b_2$, $y_0 = c_1a_2 - a_1c_2$, $t_0 = a_1b_2 - b_1a_2$. Their angle of intersection α is then given by

$$\alpha = \arcsin \frac{\sqrt{t_0^2 - x_0^2 - y_0^2}}{\sqrt{(a_1^2 + b_1^2 - c_1^2)(a_2^2 + b_2^2 - c_2^2)}}.$$

Notice that this formula implies that the angle of intersection is 0 if the point of intersection of the lines is on the boundary circle $x^2 + y^2 = t^2$, that is, $(x/t)^2 + (y/t)^2 = 1$ or $u^2 + v^2 = 1$.

Klein was further able to show that, with somewhat different choices of boundary curve, an analogous definition of distance would lead to either Euclidean geometry or the non-Euclidean geometry of the sphere. (Klein named the geometry of the sphere **elliptic geometry**; Euclidean geometry, **parabolic geometry**; and Lobachevsky's geometry, **hyperbolic geometry**.) In fact, Klein modified the non-Euclidean geometry of the sphere, because as it stood, two points did not always determine a unique straight line. The modification lay in identifying diametrically opposite points on the sphere. The new "half-sphere" could no longer exist in ordinary three-dimensional space (because opposite points along the equator were identified), but a distance function could be defined on it by multiplying the logarithm of a particular cross ratio by an imaginary constant. It then turned out that the total length of a geodesic is πR (where R is the radius of the sphere). Klein's unification of Euclidean geometry with the two non-Euclidean versions by use of the same type of distance function was an additional factor in convincing mathematicians that non-Euclidean geometry was as consistent as the Euclidean version.

24.3.4 Klein's *Erlanger Programm*

In 1872, Klein made another important contribution to the study of geometry in his *Erlanger Programm*. This proposal explored the notion that the various geometrical studies of the nineteenth century could all be unified and classified by viewing geometry in general as the study of those properties of figures that remained invariant under the action of a particular group of transformations on the underlying space (or manifold). In fact, Klein's demonstration of the relationship of geometry to groups of transformations helped to provide impetus for the development of the abstract notion of a group by the end of the century.

Klein's starting point may have been his realization that any projective transformation of the projective plane into itself that preserved the boundary circle of his model of the Lobachevskian plane left the cross ratio unaltered and hence preserved both distance and angle. It is these transformations that were the rigid motions on his model of the Lobachevskian plane. Having studied some of the early works in the theory of groups, Klein realized further that because the composition of any two transformations in this set belonged to the set and the inverse of any transformation in the set again belonged to it, the set of all the projective transformations preserving the boundary curve formed a group of transformations. Moreover, because the basic properties of figures in this plane are invariant under this group, one could consider the geometry of this non-Euclidean plane to be precisely the study of those invariant properties. Thus, Klein defined in general what a "geometry" was to mean: "Given a manifold and a group of transformations of the same; to investigate the configurations belonging to the manifold with regard to such properties as are not altered by the transformations of the group."[16]

Klein provided several examples of geometries and their associated groups. Ordinary Euclidean geometry in two dimensions corresponded to what Klein called the **principal group**, the group composed of all rigid motions of the plane along with similarity transformations and reflections. It is the invariants under these transformations that form the object of study in classical Euclidean geometry. Projective geometry consists of the study of those figures left unchanged by projections or, what is the same thing, by collineations, those transformations that take lines into lines. These transformations can be expressed analytically in the form

$$x' = \frac{a_{11}x + a_{12}y + a_{13}}{a_{31}x + a_{32}y + a_{33}}, \qquad y' = \frac{a_{21}x + a_{22}y + a_{23}}{a_{31}x + a_{32}y + a_{33}},$$

where $\det(a_{ij}) \neq 0$. Because the principal group can be expressed analytically as the set of transformations of the form

$$x' = ax - by + c, \qquad y' = bex + aey + d,$$

with $a^2 + b^2 \neq 0$ and $e = \pm 1$, it is clear that it is a subgroup of the projective group. It then follows that there are fewer invariants of the latter than of the former and so any theorem of projective geometry remains a theorem in Euclidean geometry, but not conversely.

Klein's original publication of his *Erlanger Programm* contained very little detailed analysis; it was intended as the basis of a research program for the study of geometry. The publication was, however, little noticed until it was translated into Italian, French, and English in the early 1890s. After that, Klein's idea that the invariants of a group of transformations were the important object of study in any field of geometry became a central facet in geometrical research well into the twentieth century.

24.4   GRAPH THEORY AND THE FOUR-COLOR PROBLEM

Recall that Euler solved the problem of the seven bridges of Königsberg algebraically. On the other hand, it is easy enough to think of the regions of Königsberg with their associated bridges as simply a set of points with line segments connecting them. Another interesting problem using points and connecting line segments was found by William Rowan Hamilton in 1856. In fact, Hamilton turned the problem into a game that was marketed in 1859. This *Icosian* game consisted of a diagram with 20 vertices on which pieces were to be placed in accordance with various conditions, the overriding consideration being that a piece was always placed at the second vertex of an edge on which the previous piece had been placed (Fig. 24.18). The first set of extra conditions Hamilton proposed was, given pieces placed on five initial points, to cover the board with the remaining pieces in succession such that the last piece placed is adjacent to the first. Hamilton gave several examples of ways in which

FIGURE 24.18

The Icosian game of William
Rowan Hamilton

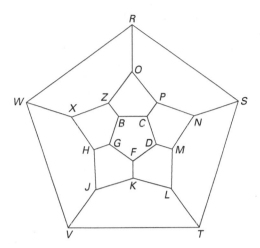

this could be accomplished but gave no general method for determining in cases other than
his special game whether or not this goal could be accomplished.

24.4.1 Precursors to Graph Theory

The ideas of Euler, Hamilton, and various other mathematicians eventually developed into the
modern subject of graph theory. One of the first to produce a definition of what is today called
a **graph** was Camille Jordan in 1869: A graph (although Jordan called it an "assemblage
of lines") consists of a nonempty set $V = v_1, v_2, \ldots, v_n$, of points (**vertices**), and a set
$E = e_1, e_2, \ldots, e_m$ of line segments (**edges**), each connecting two of the points. A **path**
in a graph is a sequence $v_a, e_b, v_c, e_d, \ldots v_k$ of vertices and edges, in which each edge joins
the vertices on either side. A **cyclic path** is a path in which $v_a = v_k$. A **connected graph** is a
graph that, for each pair v, w of vertices, contains a path beginning at v and ending at w. Thus,
in the graph-theoretic interpretation of Euler's Königsberg bridge problem, the (impossible)
goal was to produce a path that contained each edge exactly once. And the goal in Hamilton's
game was to produce a cyclic path passing containing each vertex exactly once.

The earliest purely mathematical consideration of a special class of graphs appeared in an
article of Arthur Cayley in 1857. Cayley, inspired by a consideration of possible combinations
of differential operators, defined and analyzed the general notion of a **tree**, a connected graph
with no cyclic paths and therefore in which the number of edges is one fewer than the number
of vertices. In particular, Cayley dealt with the notion of a **rooted tree**, a tree in which one
particular vertex is designated as the root. He exhibited the possible rooted trees with two,
three, and four vertices (which Cayley called **knots**), or, equivalently, with one, two, or three
edges (**branches**, to continue the botanical analogy). By a clever combinatorial argument,
Cayley then developed a recursive formula for determining the number A_r of different trees
with r branches (where "different" is appropriately defined) (Fig. 24.19). Thus, he showed,
for example, that $A_1 = 1$, $A_2 = 2$, $A_3 = 4$, $A_4 = 9$, and $A_5 = 20$. In 1874, Cayley applied

FIGURE 24.19

Different trees with r
branches, for $r = 1, 2, 3$

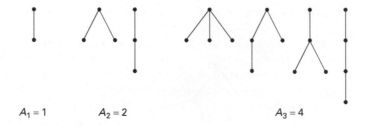

$A_1 = 1$ $A_2 = 2$ $A_3 = 4$

his results to the study of chemical isomers and a few years later succeeded in developing a
formula for counting the number of unrooted trees with a given number of vertices.

24.4.2   The Four-Color Problem

The problem that has played the greatest role in the development of modern-day graph theory,
a problem that was attacked by many mathematicians from its first formulation in 1852, is the
four-color problem. The problem was described in a letter written by Augustus De Morgan
(1806–1871) to Hamilton on October 23 of that year: "A student of mine [Frederick Guthrie]
asked me today to give him a reason for a fact which I did not know was a fact—and do not
yet. He says that if a figure be anyhow divided and the compartments differently coloured so
that figures with any portion of common boundary line are differently coloured—four colours
may be wanted, but not more. . . . My pupil says he guessed it in colouring a map of England.
The more I think of it, the more evident it seems."[17] De Morgan was not able to think of a
case of a map where five colors were required but, although he thought the sufficiency of
four colors was "evident," could give no proof of that either. Hamilton was not interested in
the problem, but in the following two decades Cayley and others spent much time in a futile
search for a proof.

There were a few basic results worked out in this period, however, that would be essential
to any proof of the **four-color theorem**, the result that Guthrie had asserted. First, we recall
Euler's formula relating the number of vertices, edges, and faces on a convex polyhedron:
$F - E + V = 2$. If we imbed the polyhedron in a sphere, then project it onto a plane by lines
through the North Pole, we get a map in the plane (Fig. 24.20). In fact, Cauchy was able to
prove Euler's formula by reducing it to a consideration of these plane maps. In any case, the

FIGURE 24.20

Map of a convex polyhedron
projected onto a plane

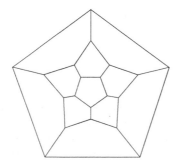

resulting plane map satisfies the relation $F - E + V = 1$, because we have "lost" an exterior region (or face) in the projection.

We can now show, using Euler's formula, that every map has at least one country with five or fewer neighbors. Let us assume the map has F countries, E boundary lines, and V vertices (or meeting points). We can assume that at least three boundary lines meet at each vertex. Thus, it appears that there are at least $3V$ boundary lines in total. However, since each boundary line has two ends, we can conclude that there are at least $\frac{3}{2}V$ boundary lines. That is, $E \geq \frac{3}{2}V$, or $V \leq \frac{2}{3}E$. Now let us assume that there is no country with five or fewer neighbors. That is, assume that every country has at least six neighbors. It follows that there are at least $6F$ boundary lines, except that again each boundary line is counted twice because there is a country on each side. So we get $E \geq 3F$, or $F \leq \frac{1}{3}E$. But then $F - E + V \leq \frac{1}{3}E - E + \frac{2}{3}E = 0$, contradicting Euler's formula. Hence, the result follows.

Now let us suppose that there are maps that require at least five colors. Pick such a map with the minimal number of countries. We note that such a map cannot contain a two-sided country (a digon) or a three-sided country (a triangle). In the second case, if the map contained a triangle, we could shrink the triangle to a point, thus removing one country, color the new map with four colors (which implies that around that point there are only three colors), then reinstate the triangle and use the fourth color to color it (Fig. 24.21). The proof in the digon case is similar. Unfortunately, this argument does not extend to minimal maps that contain a square or a pentagon. So a different argument would be needed there to complete the proof of the four-color theorem.

FIGURE 24.21

Coloring a minimal map containing a triangle

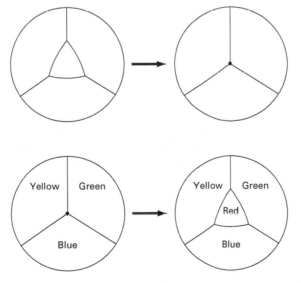

In 1879, Alfred Kempe (1849–1922) published a proof of the theorem, which seemed to solve the problem for the square and the pentagon. We outline his procedure for coloring any map. First, find a country with five or fewer neighbors. Cover this country with a blank piece of paper (a patch) of the same shape but slightly larger, and extend all the boundaries that meet this patch into the center. This amounts to shrinking the country to a point, thus reducing the number of countries by 1. Repeat this procedure until there is just one country left. This country can then be colored with one of the four colors. Then reverse the process, stripping off each patch in turn. At each stage, color the remaining country with any available color until the entire map is colored. The tricky part of Kempe's argument is, of course, the last, in which he claims there will always be one of the four colors available when a country is restored. If the restored country is a digon or triangle, there is no problem. In the cases where it is a square or pentagon, Kempe gave an involved argument dealing with chains of countries colored with two colors, beginning with two that border the restored country, an argument accepted as correct when Kempe's paper was published. However, 10 years later Percy Heawood (1861–1955) discovered a flaw in Kempe's argument for the pentagon, which could not easily be remedied. Still, enough of the argument was true so that Heawood could prove that five colors are always sufficient for coloring a map. Nevertheless, it was necessary for some new ideas to emerge before the four-color problem could eventually be solved.

 24.5

GEOMETRY IN N DIMENSIONS

The Greek limitations of geometry to three dimensions had been breached by various mathematicians in the early nineteenth century. For example, Ostrogradsky in the 1830s had generalized his divergence theorem to n dimensions almost casually (by simply adding three dots to the end of various formulas), while Cauchy even earlier had dealt with geometric objects of arbitrary dimensions in explaining his version of the diagonalization of a symmetric matrix. The actual term "geometry of n dimensions" seems to have first appeared in 1843 in the title of a paper by Cayley. The article itself, however, was purely algebraic and only touched on geometry in passing.

24.5.1 Grassmann and the *Ausdehnungslehre*

The first mathematician to present a detailed theory of spaces of dimension greater than three was Hermann Grassmann (1809–1877), a German mathematician and philologist whose brilliant work was unfortunately not recognized during his lifetime. Grassmann's aim in his *Die lineale Ausdehnungslehre (The Science of Linear Extension)* of 1844 and in its reworking in 1862 was to develop a systematic method of expressing geometrical ideas symbolically, beginning with the notion of geometrical multiplication.

Four years before he wrote his first detailed work discussing n-dimensional spaces, Grassmann was already able to deal with the multiplication of vectors in two- and three-dimensional spaces in a paper devoted to a new explanation of the theory of the tides. He defined the **geometrical product of two vectors** to be "the surface content of the parallelogram determined by these vectors" and the **geometrical product of three vectors** to be the "solid (a parallelepiped) formed from them."[18] Defining in an appropriate way the sign of such products, he was able to show that the geometrical product of two vectors is distributive and anticommutative and that the geometrical product of three vectors all lying in the same plane is zero. Because the area of the parallelogram that is the geometrical product of two vectors is equal to the product of the lengths of the two vectors and the sine of the angle between them, this product is identical in numerical value to the length

BIOGRAPHY

Hermann Grassmann (1809–1877)

Grassmann was born and lived for most of his life in Stettin in Pomerania, now Szczecin, Poland. Although at the University of Berlin he mostly studied philology and theology, after leaving the university he returned to Stettin to pursue work in mathematics and physics to prepare himself to pass the state examinations for teachers in those subjects. He subsequently taught briefly at a Berlin technical school and, after 1836, at various schools in his hometown. His great ambition in life was to qualify for a university position, but although he developed the ideas of the theory of vector spaces, few people read his efforts or recognized his great originality. Grassmann sent copies of his book to several influential mathematicians, but the only one who commented favorably on it was Hermann Hankel, a student of Riemann's, who planned to include some of Grassmann's material in his own book on complex variables. In the 1860s Grassmann turned his attention to the subject of languages and made some important scholarly contributions to the study of Sanskrit. His later mathematics works, however, were of lesser quality and he never attained his goal of a university professorship.

of the modern cross product. The difference, of course, is in the geometrical nature of the object produced by multiplying. Rather than the product being a new vector, it is a two-dimensional object. But because there is a one-to-one correspondence between Grassmann's parallelograms (considering two as equal if they have the same area and lie in the same or parallel planes) and modern vectors in three-dimensional space, determined by associating to each parallelogram the normal vector whose length equals the parallelogram's area, the two multiplications are essentially identical. The advantage of Grassmann's, however, is that it, unlike the cross product, is generalizable to higher dimensions. It is that generalization that is the basis of Grassmann's major works of 1844 and 1862.

Grassmann began the discussion in his text, particularly in the clearer 1862 version, with the notion of a vector as a straight line segment with fixed length and direction. Two vectors are to be added in the standard way by joining the beginning point of the second vector to the endpoint of the first. Subtraction is simply the addition of the negative, that is, the vector of the same length and opposite direction. Vectors are the simplest examples of what Grassmann calls an **extensive** quantity. In general, such a quantity is defined abstractly as "any expression that is derived from a system of units (none of which need be the absolute unit) by numbers."[19] Grassmann meant here that one begins with a set $\epsilon_1, \epsilon_2, \ldots, \epsilon_n$ of linearly independent quantities and then takes as an extensive quantity any linear combination of these "units." Addition of extensive quantities is by the obvious method: $\sum \alpha_i \epsilon_i + \sum \beta_i \epsilon_i = \sum (\alpha_i + \beta_i)\epsilon_i$. Similarly, one can multiply an extensive quantity by a scalar. Grassmann noted that the basic laws of algebra hold for his extensive quantities and defined the **space** of the quantities $\{\epsilon_i\}$ to be the set of all linear combinations of them.

Grassmann next defined multiplication of extensive quantities by use of the distributive law: $(\sum \alpha_i \epsilon_i)(\sum \beta_j \epsilon_j) = \sum \alpha_i \beta_j [\epsilon_i \epsilon_j]$, where each quantity $[\epsilon_i \epsilon_j]$ is called a quantity of the second order. Because this new sum must be an extensive quantity, it too must be expressible as a linear combination of units. Thus, Grassmann needed to define second-order units. Assuming that the multiplication rules defined on units extend to the same rules on any extensive quantities, he demonstrated that there were only four basic possibilities for defining second-order units. First, all of the quantities $[\epsilon_i \epsilon_j]$ could be independent. Second, one could have $[\epsilon_i \epsilon_j] = [\epsilon_j \epsilon_i]$. (This multiplication satisfies all of the ordinary algebraic multiplication rules.) Third, one could have $[\epsilon_i \epsilon_j] = -[\epsilon_j \epsilon_i]$. (This implies that $[\epsilon_i \epsilon_i] = 0$ for all i.) Finally, one could have all products $[\epsilon_i \epsilon_j] = 0$. It is the third form of multiplication, called **combinatory** multiplication, that Grassmann considered in detail in the remainder of his work. According to his condition, then, for any first-order extensive quantities A and B, the multiplication rule $[AB] = -[BA]$ holds.

With the combinatory product of two first-order units defined, it was straightforward for Grassmann to define products of three or more first-order units using the same basic rules. For example, if there are three first-order units ϵ_1, ϵ_2, ϵ_3, then there are three second-order units $[\epsilon_1 \epsilon_2]$, $[\epsilon_2 \epsilon_3]$, $[\epsilon_3 \epsilon_1]$ and one third-order unit $[\epsilon_1 \epsilon_2 \epsilon_3]$. (Any other product of three first-order units would have two factors in common and would therefore be equal to 0.) If there are four first-order units, then there are six second-order units, four third-order units, and one fourth-order unit. Grassmann noted further that the product of n linear combinations of n first-order units, $(\sum \alpha_{1i} \epsilon_i)(\sum \alpha_{2i} \epsilon_i) \cdots (\sum \alpha_{ni} \epsilon_i)$, is equal to $\det(\alpha_{ij})[\epsilon_1 \epsilon_2 \cdots \epsilon_n]$, where the bracketed expression is the single unit of nth order.

Grassmann's combinatory product of his *Ausdehnungslehre* determines, in modern terminology, the exterior algebra of a vector space. Its ideas came from Grassmann's desire to express various geometrical concepts symbolically. So, in particular, he thought of his second-order quantities as parallelograms and his third-order quantities as parallelepipeds. But even though there was no specifically geometric interpretation of higher-order quantities, Grassmann saw that the symbolic manipulations did not require the limitation to any particular number of dimensions. Not only did he construct the exterior algebra, however, but he also developed many of the important ideas relating to vector spaces, that is, to the space of all linear combinations of n units. As early as 1840 he had developed the notion of the inner product of two vectors as the "algebraic product of one vector multiplied by the perpendicular projection of the second onto it"[20] and showed that in coordinate form, the inner product was given by $(\sum \alpha_i \epsilon_i)(\sum \beta_j \epsilon_j) = \sum \alpha_i \beta_i$. And in the *Ausdehnungslehre*, he developed the notion of linear independence and basis, showed that any vector can be uniquely expressed as a linear combination of the elements in a basis, and proved that in an n-dimensional space, any vector in a basis can be replaced by another vector independent of the remaining $n - 1$ vectors. He demonstrated that an orthogonal system of quantities is linearly independent (where two vectors are orthogonal if their inner product is zero) and proved the well-known result that for two subspaces U, W of a space V,

$$\dim(U + W) = \dim U + \dim W - \dim(U \cap W).$$

Grassmann's work was not appreciated during his lifetime. But his ideas in linear algebra and the exterior algebra were rediscovered late in the nineteenth century and applied to many new areas of mathematics, including the theory of vector spaces and the theory of differential forms.

24.5.2   Vector Spaces

The basic notions of linear algebra, including those of linear independence and linear combinations, were used in many parts of mathematics during the nineteenth century, but it was not until the end of the century that an abstract definition of a vector space was formulated. The first mathematician to give such a definition was Giuseppe Peano in his *Calcolo geometrico* of 1888. Peano's aim in the book, as the title indicates, was the same as Grassmann's, namely, to develop a geometric calculus. Thus, much of the book consists of various calculations dealing with points, lines, planes, and solid figures. But in chapter IX, Peano gave a definition of what he called a **linear system**. Such a system consists of quantities provided with operations of addition and scalar multiplication. The addition must satisfy the commutative and associative laws (although these laws were not cited as such by Peano), while the scalar multiplication satisfies two distributive laws, an associative law, and the law that $1v = v$ for every quantity v. In addition, Peano included as part of his axiom system the existence of a zero quantity satisfying $v + 0 = v$ for any v as well as $v + (-1)v = 0$. Peano also defined the **dimension** of a linear system as the maximum number of linearly independent quantities in the system. In connection with this idea, Peano noted that the set of polynomial functions in one variable forms a linear system, but that there is no such maximum number of linearly independent quantities and therefore the dimension of this system must be infinite.

Peano's work, like that of Grassmann, had no immediate effect on the mathematical world. His definition was forgotten, although mathematicians continued to use the basic concepts involved. For example, Dedekind in 1893, as part of his work on algebraic number fields, defined a space Ω as the set of all linear combinations of an independent set of n algebraic numbers with coefficients in a field. He noted that the numbers of this space satisfy the basic properties we attribute to a vector space, without referring to any such definition elsewhere. And he proved, using induction, the important result that any $n + 1$ numbers in Ω are dependent. Although he did not state explicitly that no smaller set of generators would determine the space, his definition essentially assured this and thus he had shown that the dimension of a (finite-dimensional) vector space is well defined.[21]

Aspects of vector space theory continued to appear in the mathematical literature, but it was not until the twentieth century that a fully axiomatic treatment of the subject entered the mathematical mainstream.

24.5.3 Differential Forms

Grassmann's exterior multiplication found one of its most important applications in the development of the theory of differential forms by Elie Cartan (1869–1951). Naturally, differential forms, the "things under the integral sign," had been extensively used throughout the nineteenth century, particularly in line integrals, surface integrals, and volume integrals. But there was no attempt to define the forms themselves, only the integrals. Cartan, having read Grassmann's work, decided in the late 1890s that one could define differential forms in an n-dimensional space by taking for the system of units the differentials $dx_1, dx_2, dx_3, \ldots$. The multiplication of these units would be Grassmann's combinatory multiplication, while the coefficients of the units would be differentiable functions in the space. Thus, a one-form in two dimensions was an expression of the form $A(x, y)\, dx + B(x, y)\, dy$, a two-form in three dimensions would have the form $A(x, y, z)\, dx\, dy + B(x, y, z)\, dy\, dz + C(x, y, z)\, dz\, dx$, and multiplication would follow the rule $dx_i\, dx_j = -dx_j\, dx_i$ and therefore $dx_i\, dx_i = 0$.

Cartan realized, of course, that this combinatory multiplication would answer Euler's problem of finding a formal way of determining the change-of-variable formula. For if $u = u(x, y)$ and $v = v(x, y)$ are functions defining the change from variables x, y to u, v, then $du = \frac{\partial u}{\partial x}dx + \frac{\partial u}{\partial y}dy$, $dv = \frac{\partial v}{\partial x}dx + \frac{\partial v}{\partial y}dy$ and the product $du\, dv$ is given by

$$
\begin{aligned}
du\, dv &= \left(\frac{\partial u}{\partial x}dx + \frac{\partial u}{\partial y}dy\right)\left(\frac{\partial v}{\partial x}dx + \frac{\partial v}{\partial y}dy\right) \\
&= \frac{\partial u}{\partial x}\frac{\partial v}{\partial x}dx\, dx + \frac{\partial u}{\partial x}\frac{\partial v}{\partial y}dx\, dy + \frac{\partial u}{\partial y}\frac{\partial v}{\partial x}dy\, dx + \frac{\partial u}{\partial y}\frac{\partial v}{\partial y}dy\, dy \\
&= \left(\frac{\partial u}{\partial x}\frac{\partial v}{\partial y} - \frac{\partial u}{\partial y}\frac{\partial v}{\partial x}\right)dx\, dy = \frac{\partial(u,v)}{\partial(x,y)}dx, dy
\end{aligned}
$$

as desired.

Besides developing the algebra of differential forms, Cartan also developed their calculus. Namely, in 1899 he defined the derived expression (now called the **exterior derivative**) of a one-form $\omega = \sum A_i\, dx_i$ to be the two-form $d\omega = \sum dA_i\, dx_i$. For example, the derived expression of the form $A\, dx + B\, dy$ is the form

$$
d\omega = \left(\frac{\partial A}{\partial x}dx + \frac{\partial A}{\partial y}dy\right)dx + \left(\frac{\partial B}{\partial x}dx + \frac{\partial B}{\partial y}dy\right)dy = \left(\frac{\partial B}{\partial x} - \frac{\partial A}{\partial y}\right)dx\, dy.
$$

Note that this derived expression appears in the statement of Green's theorem, while the exterior derivative of the one-form $A\, dx + B\, dy + C\, dz$ appears in the statement of Stokes's theorem. In 1901, Cartan generalized his definition of the exterior derivative to forms of any degree. Namely, if $\omega = \sum a_{ij\ldots k}dx_i\, dx_j\cdots dx_k$, then the exterior derivative $d\omega$ is defined to be $\sum da_{ij\ldots k}dx_i\, dx_j\cdots dx_k$. It is straightforward to show then that the exterior derivative of the two-form $A\, dy\, dz + B\, dz\, dx + C\, dx\, dy$ is the three-form

$$
\left(\frac{\partial A}{\partial x} + \frac{\partial B}{\partial y} + \frac{\partial C}{\partial z}\right)dx\, dy\, dz,
$$

which is the expression we find in the divergence theorem.

Although Cartan realized that these three theorems of vector calculus could be easily stated using differential forms, it was Edouard Goursat (1858–1936) in 1917 who first noted that Volterra's generalization of these theorems, today called the generalized Stokes theorem, could be written in the simple form

$$
\int_S \omega = \int_T d\omega,
$$

where ω is a p-form in n-space and S is the p-dimensional boundary of the $(p+1)$-dimensional region T. Goursat also used differential forms to state and prove the Poincaré lemma and its converse, namely, that if ω is a p-form, then $d\omega = 0$ if and only if there is a $(p-1)$-form η with $\omega = d\eta$. Goursat did not notice, however, that the "only if" part of the result depends on the domain of ω and is not true in general. Cartan himself in 1922 gave a counterexample, which provided one of the impulses in the next decade for the development of the differential cohomology of differentiable manifolds.[22]

THE FOUNDATIONS OF GEOMETRY

The late nineteenth century saw the appearance of axiom systems for various types of mathematical structures. The notions of group and field were axiomatized, as was the notion of a vector space. Similarly, axioms were developed for the set of positive integers, and a great deal of effort went into the precise definition of the idea of a real number. Of course, the oldest axiom system in existence was that of Euclid for the study of geometry. In fact, that system provided the model for the creation of the various axiom systems in this period. There were, however, several flaws in Euclid's system. In particular, various mathematicians through the ages noticed that Euclid had made assumptions in some of his proofs that were not explicitly mentioned in his list of axioms and postulates. With the new developments in non-Euclidean geometry causing mathematicians to reexamine the nature of the various axioms, it is not surprising that a concerted effort was made by several to rectify the situation with regard to Euclid's work and thus put Euclid's geometry on as strong a foundation as possible.

24.6.1 Hilbert's Axioms

Of the attempts to set up a complete set of axioms from which Euclidean geometry could be derived, the most successful was by probably the premier mathematician of the late nineteenth and early twentieth centuries, David Hilbert (1862–1943). In 1899, Hilbert published his *Grundlagen der Geometrie* (*Foundations of Geometry*), essentially a record of his lectures on Euclidean geometry presented at the University of Göttingen in the winter semester of 1898–99. His aim in this work was "to choose for geometry a simple and complete set of independent axioms and to deduce from these the most important geometrical theorems in such a manner as to bring out as clearly as possible the significance of the different groups of axioms and the scope of the conclusions to be derived from the individual axioms."[23]

Hilbert's idea was to begin with three undefined terms, point, straight line, and plane, and to define their mutual relations by means of the axioms. As Hilbert noted, it is only the axioms that define the relationships. One should not have to use any geometrical intuition in proving results. In fact, one could easily replace the three notions by other ones—Hilbert suggested chair, table, and beer mug—as long as these satisfy the axioms. One sees, therefore, that Hilbert's idea of an axiom system was somewhat different from those of Euclid and Aristotle. The Greek thinkers had attempted merely to state certain "obvious" facts about concepts they already understood intuitively. Hilbert, on the other hand, like those who stated the group axioms, was determined to abstract the desired properties away from any concrete interpretation. Thus, any object could be a "point" or a "line" as long as the "points" and "lines" satisfied the axioms of the geometry, just as any set of objects could be a group as long as there was a law of "multiplication" of these objects that satisfied the axioms of a group.

Hilbert divided his axioms into five sets: the axioms of connection, of order, of parallels, of congruence, and of continuity and completeness. The first group of seven axioms established the connections among his three fundamental concepts of point, line, and plane. Thus, not only do two points determine a line (axiom I, 1), but they determine precisely one line (axiom I, 2). Similarly, three points not on the same line determine one (I, 3) and only one (I, 4) plane.

BIOGRAPHY

David Hilbert (1862–1943)

One of the last of the universal mathematicians, who contributed greatly to many areas of mathematics, Hilbert (Fig. 24.22) spent the first 33 years of his life in and around Königsberg, then capital of East Prussia, now Kaliningrad in Russia. He attended the university there and, after receiving his doctorate, joined the faculty in 1885. He only rose to prominence, however, only after he was called by Felix Klein to Göttingen, where he soon became one of the major reasons for that university's surpassing Berlin as the preeminent university for mathematics in Germany, and probably the world, through the first third of the twentieth century. Hilbert began his career with the study of algebraic forms, then turned to algebraic number theory, the foundations of geometry, integral equations, theoretical physics, and finally the foundations of mathematics. He is probably most famous for his lecture at the International Congress of Mathematicians in Paris in 1900, where he presented a list of 23 problems he felt would be of central importance for mathematics in the twentieth century. Hilbert firmly believed that it was problems that drove mathematical progress and was always confident that "*wir mussen wissen, wir werden wissen* (we must know, we will know)." After the Nazi seizure of power, Hilbert was forced to witness the demise of the Göttingen he knew and loved and died a lonely man during the Second World War.

FIGURE 24.22

David Hilbert on a stamp from the Democratic Republic of the Congo

The fifth axiom asserts that if two points of a straight line lie in a given plane, then so does the entire line. The sixth axiom says that any two planes that have a common point have at least a second common point. Finally, the seventh axiom of the first group asserts the existence of at least two points on every straight line, three noncollinear points on each plane, and four non-coplanar points in space. After listing these axioms, Hilbert noted as a consequence that two straight lines in a plane have either one point or no points in common and that two planes have either no point or a straight line in common.

Hilbert's second group of axioms enabled him to define the idea of a line segment AB as the set of points lying between the two points A and B. Euclid himself had assumed the properties of "betweenness" implicitly, probably because of the "obviousness" of the diagrams in which they occurred. Hilbert made Euclid's assumptions explicit by axiomatizing this idea of "between." For example, axiom II, 3 asserts that of any three points on a straight line, there is always one and only one that lies between the other two, while axiom II, 5 asserts that any line passing through a point of one side of a triangle and not passing through any of the vertices must pass through a point of one of the other two sides. With these axioms, Hilbert was able to deduce the important theorem that any simple polygon divides the plane into two disjoint regions, an interior and an exterior, and that the line joining any point in the interior with any point in the exterior must have a point in common with the polygon.

The third group of axioms consists solely of Hilbert's version of the parallel axiom: "In a plane α there can be drawn through any point A, lying outside of a straight line a, one and only one straight line which does not intersect the line a."[24] The fourth group asserts the basic ideas of congruence. Recall that Euclid proved his first triangle congruence theorem by "placing" one triangle on the second. Many questioned the validity of this method of superposition, and it is for this reason that Hilbert listed six axioms concerning the undefined term congruence. For example, axiom IV, 1 asserts that given a segment AB and a point A', there is always a point B' such that segment AB is congruent to segment $A'B'$, while IV, 2 states in essence that congruence is an equivalence relation. Similarly, after defining an angle as a system consisting of two distinct half-rays emanating from a single point, Hilbert asserted that given one angle and a given ray, one can determine a second angle congruent to the first. The final axiom of this group almost asserts *Elements* I, 4: If two triangles have two sides and the included angle congruent, then the remaining angles will also be congruent. Hilbert did not assert the congruence of the third sides, but proved it by an argument by contradiction. With the aid of these axioms, he also proved the other two triangle congruence theorems, Euclid's postulate that all right angles are equal to one another, the alternate-interior angle theorem, and the theorem that the sum of the angles of a triangle is equal to two right angles.

Hilbert's final set of axioms contains two concerning the basic idea of continuity. First, there is the axiom of Archimedes: Suppose A, B, C, D are four distinct points. Then on the ray AB there is a finite set of distinct points, $A_1, A_2, \ldots, A_n$ such that each segment $A_i A_{i+1}$ is congruent to the segment CD and such that B is between A and A_n. In other words, given any line segment and any measure, there is an integer n such that n units of measure yield a line segment greater than the given line segment. Among the consequences of this axiom, when used in conjunction with the earlier axioms, is that there is no limit to the length of a line. Thus, this tacit assumption of Euclid, important in Saccheri's and Lambert's rejection of the hypothesis of the obtuse angle, is now made explicit. Hilbert's final axiom states in essence that the points on the line are in one-to-one correspondence with the real numbers. In other words, there are no "holes" in the line. This axiom answers the objection made to Euclid's construction of an equilateral triangle in *Elements* I, 1 that there is no guarantee that the two constructed circles actually intersect. According to Hilbert's axiom, no points can be added to the two circles, and therefore they cannot simply "pass through" one another.

24.6.2 Consistency, Independence, and Completeness

Having stated the axioms, Hilbert proceeded to show that they were **consistent**, that is, that one could not deduce any contradiction from them, at least under the assumption that arithmetic had no contradictions. His idea, similar to that of Klein and others in showing that non-Euclidean geometry had no contradictions, was to construct a geometry, using only arithmetic operations, that satisfied the axioms. For example, beginning with a certain set Ω of algebraic numbers, Hilbert defined a point p to be an ordered pair (a, b) of numbers in Ω and a line L to be a ratio $(u : v : w)$ of three numbers in Ω, where u, v are not both 0. Then p lies on L if $ua + vb + w = 0$. With every geometric concept interpreted arithmetically in this fashion and with all the axioms satisfied in the interpretation, Hilbert had created an arithmetic model of his axioms for geometry. If the axioms led to a contradiction in geometry, there would be an analogous contradiction in arithmetic. Therefore, assuming the axioms of arithmetic are consistent, so are the axioms of geometry.

Another important characteristic of an axiom system is **independence**, that is, that no axiom can be deduced from the remaining ones. Although Hilbert did not demonstrate independence completely, he did show that the various groups of axioms were independent by constructing several interesting models in which one group was satisfied but not another. For example, he constructed a system in which all the axioms were satisfied except the axiom of Archimedes. Hilbert did not deal with a further characteristic of an axiom system, that of **completeness**, that any statement that can be formulated within the system can be shown to be either true or false. It is virtually certain, however, that Hilbert believed that his system was complete. In fact, several mathematicians soon showed that all of the theorems of Euclidean geometry could be proved using Hilbert's axioms.

The importance of Hilbert's work lay not so much in his answering the various objections to parts of Euclid's deductive scheme, but in reinforcing the notion that any mathematical field must begin with undefined terms and axioms specifying the relationships among the terms. As we have discussed, there were many axiom schemes developed in the late nineteenth century to clarify various areas of mathematics. Hilbert's work can be considered the culmination of this process, because he was able to take the oldest such scheme and show that, with a bit of tinkering, it had stood the test of time. Thus, the mathematical ideas of Euclid and Aristotle were reconfirmed at the end of the nineteenth century as still the model for pure mathematics. A century later, these ideas continue to prevail.

EXERCISES

1. Show that the area of an (infinitesimal) triangle with vertices (x, y), $(x + dx, y + dy)$, $(x + \delta x, y + \delta y)$ is equal to $\frac{1}{2}(dx\, \delta y - dy\, \delta x)$.

2. Show that if a surface is given in the form $z = z(x, y)$, then the measure of curvature k can be expressed as

$$k = \frac{z_{xx} z_{yy} - z_{xy}^2}{(1 + z_x^2 + z_y^2)^2}.$$

Hint: Show first that if X, Y, Z are coordinates on the unit sphere corresponding to the point $(x, y, z(x, y))$ on the given surface, then

$$X = \frac{-z_x}{\sqrt{1 + z_x^2 + z_y^2}}, \qquad Y = \frac{-z_y}{\sqrt{1 + z_x^2 + z_y^2}},$$

$$Z = \frac{1}{\sqrt{1 + z_x^2 + z_y^2}}.$$

3. Calculate the curvature function k of the paraboloid $z = x^2 + y^2$.

4. If $x = x(u, v)$, $y = y(u, v)$, $z = z(u, v)$ are the parametric equations of a surface and if $E = x_u^2 + y_u^2 + z_u^2$, $F = x_u x_v + y_u y_v + z_u z_v$, and $G = x_v^2 + y_v^2 + z_v^2$, show that

$$dx^2 + dy^2 + dz^2 = E\, du^2 + 2F\, du\, dv + G\, dv^2.$$

5. Calculate E, F, G on the unit sphere parametrized by $x = \cos u \cos v$, $y = \cos u \sin v$, $z = \sin u$, and show that $ds^2 = du^2 + \cos^2 u\, dv^2$.

6. Derive the formula $\cos a = \cos b \cos c + \sin b \sin c \cos A$ for an arbitrary spherical triangle with sides a, b, c and op-

posite angles A, B, C on a sphere of radius 1 by dividing the triangle into two right triangles and applying the formulas of Chapter 5.

7. Show that the formula in Exercise 6 changes to

$$\cos \frac{a}{K} = \cos \frac{b}{K} \cos \frac{c}{K} + \sin \frac{b}{K} \sin \frac{c}{K} \cos A$$

if the sphere has radius K, where $a, b,$ and c are expressed in a linear measure.

8. By using power series, show that Taurinus's "log-spherical" formula

$$\cosh \frac{a}{K} = \cosh \frac{b}{K} \cosh \frac{c}{K} - \sinh \frac{b}{K} \sinh \frac{c}{K} \cos A$$

reduces to the law of cosines as $K \to \infty$.

9. Show that Taurinus's formula for an asymptotic right triangle on a sphere of imaginary radius i, namely, $\sin B = 1/\cosh x$, is equivalent to Lobachevsky's formula for the angle of parallelism, $\tan \frac{B}{2} = e^{-x}$.

10. Show that the circumference of a circle of radius r on the sphere of imaginary radius iK is $2\pi K \sinh \frac{r}{K}$. Show that this value approaches $2\pi r$ as $K \to \infty$. (*Hint:* First determine the circumference of a circle on an ordinary sphere of radius K.)

11. Given that $\tan \frac{1}{2} \Pi(x) = e^{-x}$, where $\Pi(x)$ is Lobachevsky's angle of parallelism, derive the formulas

$$\sin \Pi(x) = \frac{1}{\cosh x} \quad \text{and} \quad \cos \Pi(x) = \tanh x$$

and show that their power series expansions up to degree 2 are $\sin \Pi(x) = 1 - \frac{1}{2}x^2$ and $\cos \Pi(x) = x$, respectively.

12. Substitute the results of Exercise 11 into Lobachevsky's formulas

$$\sin A \tan \Pi(a) = \sin B \tan \Pi(b)$$

$$\cos A \cos \Pi(b) \cos \Pi(c) + \frac{\sin \Pi(b) \sin \Pi(c)}{\sin \Pi(a)} = 1$$

to derive the laws of sines and cosines when the sides a, b, c of the non-Euclidean triangle are "small."

13. Show that if ABC is an arbitrary triangle with sides a, b, c, then the formulas $a \sin(A + C) = b \sin A$ and $\cos A + \cos(B + C) = 0$, along with the law of sines, imply that $A + B + C = \pi$.

14. Show that Lobachevsky's basic triangle formulas (Equations 24.3, 24.4, 24.5, and 24.6) transform into standard formulas of spherical trigonometry if one replaces the sides a, b, c of the triangle by ia, ib, ic, respectively. (For simplicity, assume that angle C is a right angle.)

15. Describe geometrically Beltrami's parametrization of the sphere of radius k given by

$$x = \frac{uk}{\sqrt{a^2 + u^2 + v^2}}, \qquad y = \frac{vk}{\sqrt{a^2 + u^2 + v^2}},$$

$$z = \frac{ak}{\sqrt{a^2 + u^2 + v^2}}.$$

16. Show that replacing u, v by iu, iv, respectively, transforms the sphere of Exercise 15 with curvature $1/k^2$ to a pseudosphere with curvature $-1/k^2$.

17. Show that Beltrami's formulas for the lengths ρ, s, t of the sides of a right triangle on his pseudosphere transform into

$$\frac{r}{a} = \tanh \frac{\rho}{k}, \qquad \frac{r}{a} \cos \theta = \tanh \frac{s}{k}, \qquad \text{and}$$

$$\frac{v}{\sqrt{a^2 - u^2}} = \tanh \frac{t}{k},$$

and then show that

$$\cosh \frac{s}{k} \cosh \frac{t}{k} = \cosh \frac{\rho}{k}.$$

18. Demonstrate how a central projection can transform parallel lines into intersecting lines.

19. Demonstrate the following relationships for the cross ratio:

$$(AB, CD) = 1 - (AC, BD) \qquad (AB, CD) = \frac{1}{(AB, DC)}.$$

20. Denote the cross ratios (AB, CD), (AC, DB), (AD, BC) by λ, μ, ν, respectively. Show that

$$\lambda + \frac{1}{\mu} = \mu + \frac{1}{\nu} = \nu + \frac{1}{\lambda} = -\lambda\mu\nu = 1.$$

21. Show that if the point p' lies on the polar π of a point p with respect to a conic C, then π', the polar of p', goes through p. (*Hint:* Assume first that C is a circle.)

22. Determine homogeneous coordinates of the points $(3, 4)$ and $(-1, 7)$.

23. Write the homogeneous coordinates of the point at infinity on the line $2x - y = 0$.

24. Determine rectangular coordinates of the points $(3, 1, 1)$ and $(4, -2, 2)$ given in homogeneous coordinates.

25. Determine the equation (in rectangular coordinates) of a line that passes through the point at infinity $(2, 1, 0)$.

26. Show that every circle in the plane passes through the two points at infinity $(1, i, 0)$ and $(1, -i, 0)$.

27. Given three collinear points A, B, P, show that the point Q determined by the construction in Figure 24.23 makes A, B, P, Q into a harmonic tetrad, that is, makes the cross ratio (AB, PQ) equal to -1.

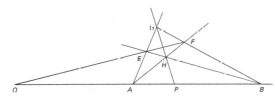

FIGURE 24.23

To determine Q so that $(AB, PQ) = -1$. Draw any two lines through A. Draw a line through P intersecting those lines at G and H. Connect G and H to B. The line through the intersection points F and E (of BG with AF and of BE with AG, respectively) intersects line APB at the desired point Q.

28. Using Klein's definition of distance d in the interior of the circle representing the Lobachevskian plane, show that if P, Q, Q' are three points on a line, then $d(P, Q) + d(Q, Q') = d(P, Q)$.

29. Find a path in Hamilton's Icosian game (see Fig. 24.18) that passes through each vertex exactly once and returns to the starting point.

30. Get an outline map of the continental United States and color it with four colors.

31. Show that a map with the minimal number of countries that requires at least five colors cannot contain a digon (a country with only two boundary edges).

32. Letting i, j, k be first-order units in three-dimensional space, determine the combinatory product of $2i + 3j - 4k$, $3i - j + k$, $i + 2j - k$.

33. Show that in Grassmann's combinatory multiplication,

$$\left(\sum \alpha_{1i}\epsilon_i\right)\left(\sum \alpha_{2i}\epsilon_i\right) \cdots \left(\sum \alpha_{ni}\epsilon_i\right)$$
$$= \det(\alpha_{ij})[\epsilon_1\epsilon_2 \cdots \epsilon_n],$$

where each linear combination is of a given set of n first-order units and where $[\epsilon_1\epsilon_2 \cdots \epsilon_n]$ is the single unit of nth order.

34. If ω is the differential one-form in three dimensions given by $\omega = A\,dx + B\,dy + C\,dz$, show that

$$d\omega = \left(\frac{\partial C}{\partial y} - \frac{\partial B}{\partial z}\right) dy\,dz + \left(\frac{\partial A}{\partial z} - \frac{\partial C}{\partial x}\right) dz\,dx$$
$$+ \left(\frac{\partial B}{\partial x} - \frac{\partial A}{\partial y}\right) dx\,dy.$$

35. Show that the exterior derivative of $\omega = A\,dy\,dz + B\,dz\,dx + C\,dx\,dy$ is the three-form

$$\left(\frac{\partial A}{\partial x} + \frac{\partial B}{\partial y} + \frac{\partial C}{\partial z}\right) dx\,dy\,dz.$$

36. Show that $d(d\omega) = 0$, where ω is a differential one-form or two-form in three-dimensional space.

37. Let ω be the two-form in $R^3 - \{0\}$ given by

$$\omega = \frac{x\,dy\,dz + y\,dz\,dx + z\,dx\,dy}{(x^2 + y^2 + z^2)^{3/2}}.$$

Show that $d\omega = 0$ but that there is no one-form η such that $d\eta = \omega$. (*Hint:* If there were such a one-form, then by Stokes's theorem, with T being the unit sphere, we would have $\int_T \omega = \int_T d\eta = \int_S \eta = 0$, because the boundary of T is empty. Then calculate $\int_T \omega$ directly.)

38. Study several new high school geometry texts. Do they follow Euclid's axioms or Hilbert's axioms or some combination? Comment on the usefulness of using Hilbert's reformulation in teaching a high school geometry class.

39. Is the analytic form of non-Euclidean geometry as presented by Taurinus, Lobachevsky, and Beltrami a better way of presenting the subject than the synthetic form? How can one make sense of a sphere of imaginary radius?

40. Read a complete version of Riemann's lecture "On the Hypotheses Which Lie at the Foundation of Geometry" (see note 10 at the end of the chapter for a reference). Describe Riemann's major new ideas and comment on how they have been followed up in the twentieth century. In particular, comment on the oft-repeated statement that Riemann's work was a precursor of Einstein's general theory of relativity.

REFERENCES AND NOTES

Some of the important works on various aspects of the history of geometry in the nineteenth century include Jeremy Gray, *Ideas of Space: Euclidean, Non-Euclidean and Relativistic* (Oxford: Clarendon Press, 1989); B. A. Rosenfeld, *A History of Non-Euclidean Geometry* (New York: Springer Verlag, 1988); Roberto Bonola, *Non-Euclidean Geometry: A Critical and Historical Study of Its Development* (New York: Dover, 1955); Julian Lowell Coolidge, *A History of Geometrical Methods* (New York: Dover, 1963); and Michael J. Crowe, *A History of Vector Analysis* (New York: Dover, 1985). A history of graph theory through original works on the subject is Norman Biggs, E. Keith Lloyd, and Robin J. Wilson, *Graph Theory, 1736–1936* (Oxford: Clarendon Press, 1986). This book contains many of the original papers on aspects of graph theory, including work on the four-color problem. The history of the four-color problem itself is well covered in Robin Wilson, *Four Colors Suffice* (Princeton, NJ: Princeton University Press, 2002).

Gauss's *Disquisitiones generales circa superficies curvas* is available in English as Karl Gauss, *General Investigations of Curved Surfaces*, Peter Pesic, ed., (Mineola, NY: Dover Publications, 2005). The book by Bonola contains translations of Lobachevsky's "Geometrical Investigations on the Theory of Parallel Lines" and Bolyai's "Appendix Exhibiting the Absolutely True Science of Space." Bolyai's work is also available in Jeremy Gray, *János Bolyai: Non-Euclidean Geometry and the Nature of Space* (Cambridge: MIT Press, 2004). Helmholtz's "On the Origin and Significance of Geometrical Axioms," is in James Newman, ed., *The World of Mathematics* (New York: Simon and Schuster, 1956), vol. 1, pp. 646–668. Beltrami's papers in which he deals with the metric for the pseudosphere and its relation to non-Euclidean geometry have been translated into English and appear in John Stillwell, *Sources of Hyperbolic Geometry* (Providence, RI: American Mathematical Society, 1996), along with fundamental papers of Klein and Poincaré. The 1844 version of *Die lineale Ausdehnungslehre* has been translated into English and appears, with other works of Grassmann, as Hermann Grassmann, *A New Branch of Mathematics*, Lloyd C. Kannenberg, trans. (Chicago: Open Court, 1995). The 1862 version has also been translated by Kannenberg and is available as Hermann Grassmann, *Extension Theory* (Providence, RI: American Mathematical Society, 2000). Kempe's "proof" of the four-color theorem appeared as "On the Geographical Problem of the Four Colours," *American Journal of Mathematics* 2 (part 3) (1879), 193–200. It is reprinted in Biggs, Lloyd, and Wilson, *Graph Theory*, pp. 96–102. Hilbert's work on geometry is David Hilbert, *The Foundations of Geometry* (La Salle, IL: Open Court, 1902).

1. Gauss, letter to Olbers, quoted in Rosenfeld, *History of Non-Euclidean Geometry*, p. 215.

2. Gauss, *General Investigations*, p. 45.

3. Ibid., p. 10. For those with experience in differential geometry, it is worth attempting to translate Gauss's ideas into modern language. A discussion of this matter is found in Michael Spivak, *A Comprehensive Introduction to Differential Geometry* (Waltham, MA: Brandeis University, 1970), vol. II. Gauss's work is also discussed in D. J. Struik, "Outline of a History of Differential Geometry," *Isis* 19 (1933), 92–120, and 20 (1933), 161–191.

4. Ibid., p. 21.

5. Lobachevsky, "Geometrical Investigations," p. 13. For more on Lobachevsky, see A. Vucinich, "Nikolai Ivanovich Lobachevski, the Man behind the First Non-Euclidean Geometry," *Isis* 53 (1962), 465–481, and V. Kagan, *N. Lobachevski and His Contribution to Science* (Moscow: Foreign Language Publishing House, 1957).

6. Ibid., p. 30.

7. Ibid., pp. 44–45.

8. Bolyai, "Appendix," p. 48.

9. Ibid., p. 20.

10. Riemann, "On the Hypotheses Which Lie at the Foundation of Geometry," translated by Spivak in his *A Comprehensive Introduction to Differential Geometry*, vol. II, pp. 4A-4–4A-20. In addition to translating Riemann's lecture, Spivak also discusses the work in detail.

11. Ibid., p. 4A-7.

12. Ibid., p. 4A-15.

13. Helmholtz, "On the Origin and Significance," in Newman, ed., *The World of Mathematics*, vol. 1, p. 657.

14. Clifford, "The Postulates of the Science of Space," in Newman, *World of Mathematics*, vol. 1, p. 564.

15. Clifford, "On the Space Theory of Matter," in Newman, *World of Mathematics*, vol. 1, pp. 568–569.

16. The original proposal was published first as Felix Klein, "Vergleichende Betrachtungen über neuere geometrische Forschungen" (Erlangen: Deichert, 1872). It was republished in *Mathematische Annalen* 43 (1893), 63–100, and then translated by M. W. Haskell and published in English as "A Comparative Review of Recent Researches in Geometry," *Bulletin of the New York Mathematical Society*, 2 (1893), 215–249. The quotation is on p. 218. There have been several recent articles on Klein's seminal work, including Thomas Hawkins, "The Erlanger Programm of Felix Klein: Reflections on Its Place in the History of Mathematics," *Historia Mathematica* 11 (1984), 442–470, and David Rowe, "A Forgotten Chapter in the History of Felix Klein's Erlanger Programm," *Historia Mathematica* 10 (1983), 448–454.

17. This letter is quoted in Biggs, Lloyd, and Wilson, *Graph Theory, 1736–1936*, pp. 90–91.

18. Grassmann, "Theorie der Ebbe und Flut," quoted in Crowe, *History of Vector Analysis*, p. 61. Crowe's work provides a detailed study of various aspects of the history of vector analysis, in particular the conflict between the adherents of quaternions and those of vectors in the last half of the nineteenth century.

19. Grassmann, *Extension Theory*, p. 4. The *Ausdehnungslehre* is discussed in detail in J. V. Collins, "An Elementary Exposition of Grassmann's *Ausdehnungslehre*, or Theory of Extension," *American Mathematical Monthly* 6 (1899), 193–198, 261–266, 297–301, and 7 (1900), 31–35, 163–166, 181–187, 207–214, 253–258, and more recently in Desmond Fearnley-Sander, "Hermann Grassmann and the Creation of Linear Algebra," *American Mathematical Monthly* 86 (1979), 809–817.

20. Grassmann, "Theorie der Ebbe und Flut," quoted in Crowe, *History of Vector Analysis*, p. 63.

21. For more details on the origins of the theory of vector spaces and abstract linear algebra, see Jean-Luc Dorier, "A General Outline of the Genesis of Vector Space Theory," *Historia Mathematica* 22 (1995), 227–261, and Gregory H. Moore, "The Axiomatization of Linear Algebra: 1875–1940," *Historia Mathematica* 22 (1995), 262–303.

22. For more details on the development of differential forms, see Victor J. Katz, "Differential Forms—Cartan to De Rham," *Archive for History of Exact Sciences* 33 (1985), 321–336.

23. Hilbert, *The Foundations of Geometry*, p. 1. See also Michael Toepell, "Origins of David Hilbert's *Grundlagen der Geometrie*," *Archive for History of Exact Sciences* 35 (1986), 329–344.

24. Ibid., p. 12.

CHAPTER 25

Aspects of the Twentieth Century and Beyond

[Emmy Noether] had great stimulating power, and many of her suggestions took final shape only in the works of her pupils or co-workers. . . . Hasse acknowledges that he owed the suggestion for his beautiful papers on the connection between hypercomplex quantities and the theory of class fields to casual remarks by Emmy Noether. She could just utter a far-seeing remark like this, 'Norm rest symbol is nothing else than cyclic algebra,' in her prophetic lapidary manner, out of her mighty imagination that hit the mark most of the time and gained in strength in the course of years; and such a remark could then become a signpost to point the way for difficult future work. . . . She originated above all a new and epoch-making style of thinking in algebra.

—From an address in memory of Emmy Noether by Hermann Weyl, 1935[1]

Resolving to complete their proof of the four-color theorem before the summer meeting of the American Mathematical Society, Kenneth Appel and Wolfgang Haken finally succeeded on July 22, 1976 with the help of a computer by showing that the last of the 1936 unavoidable configurations was reducible. This theorem, that four colors suffice to color any map, was first proposed in 1852. On July 26, the two men submitted a report of their work to the American Mathematical Society, a report that appeared in the September issue of the *Bulletin of the AMS*. Interestingly enough, in mid-June, Haken had submitted an abstract to the AMS of a paper he planned to present at the summer meeting in August. The paper was entitled "Why Is the Four Color Problem Difficult?"

A quick glance at any research library in mathematics shows that the mathematical output of the twentieth century far exceeds that of all previous centuries put together. There are shelves upon shelves of journals, most of which were started in that century. And the twentieth-century segments of even the strongest of the journals of the nineteenth century, such as *Crelle's* or *Liouville's*, dominate their designated spaces. For better or for worse, however, most of the mathematics taught to undergraduates dates from the nineteenth century or earlier. So in a text designed to be read by undergraduates, one can barely scratch the surface of the mathematics of the twentieth century. We will therefore concentrate in this chapter on only five selected areas of twentieth-century mathematics, areas that are to some extent covered in a typical undergraduate curriculum. For those who want to explore more extensively the mathematics of the twentieth century, the library shelves are open and there are many resources available as a guide.

We begin with a consideration of the problems in the foundations of mathematics at the beginning of the century. Cantor's work in the theory of infinite sets, which caused immediate controversy on its inception late in the nineteenth century, continued to cause problems early in the twentieth. For example, Cantor's attempt to prove the trichotomy law for infinite cardinals, the law that for any two cardinal numbers A, B, exactly one of the properties $A = B$, $A < B$, or $A > B$ held, ran into unexpected trouble, as did his attempt to prove that the real numbers could be well ordered. The key to solving these problems turned out to be a new axiom for set theory, the axiom of choice, an axiom that was in fact used implicitly for many years until it was explicitly stated by Ernst Zermelo in 1904. Zermelo's statement of this new axiom, however, caused new controversy, in answer to which Zermelo produced an axiomatization of set theory. Zermelo hoped that not only would this axiomatization help to resolve various paradoxes in set theory that had arisen, but also that the theory of arithmetic could be based on it and mathematics in general given a secure foundation. Things turned out differently from what most mathematicians expected, however, when in 1931 Kurt Gödel established his Incompleteness Theorems, to the effect that any theory in which the arithmetic of natural numbers could be expressed had true results that could not be proved from the axioms of that theory. Gödel's result thus closed, in some sense, the axiomatic phase of mathematics, the attempt to give complete and consistent sets of axioms on which to base the various parts of the subject.

A second aspect of twentieth-century mathematics to be discussed is the growth of topology, both point-set topology and combinatorial topology, subjects barely begun in the previous century but destined to become one of the "growth" areas in mathematics in the twentieth century. The roots of point-set topology lie in the work of Cantor on the theory of sets of real numbers, extended by numerous mathematicians into the consideration of other kinds of sets. The roots of combinatorial topology are in Riemann's attempts to integrate complex functions in regions with "holes." The beginnings of a theory of such regions, however, was in the work of Henri Poincaré in the 1890s, particularly with his definition of homology. In the 1920s the connections of combinatorial topology with algebra were recognized at Göttingen by the group of mathematicians led by Emmy Noether.

The subsequent algebraization of topology is part of the third aspect of twentieth-century mathematics to be dealt with, the growth of algebraic techniques in all areas of mathematics. This growth included new ideas in the theory of fields with the work of Kurt Hensel and Ernst Steinitz, the axiomatization of the idea of a vector space, particularly in the work of Stefan

Banach, and the intensive study of a new structure, that of a ring, centered again in the work of Noether. The growth of abstraction in algebra continued and perhaps culminated in the theory of categories and functors, introduced by Samuel Eilenberg and Saunders Mac Lane in 1945.

Statistics also exploded in importance during the twentieth century, especially with the development of techniques for designing experiments and testing hypotheses. But the use of these techniques in numerous situations only became possible with the development of electronic computers in the second half of the century. Thus, we will also look at various aspects of mathematics important in this development as well, including the early attempts at developing a computer in the work of Charles Babbage and Ada Byron, the theoretical basis for the programmable computer in the work of Alan Turing, and some aspects of the construction of the Institute for Advanced Study computer under John von Neumann.

We will then conclude with a look at four important results that were finally achieved in the late twentieth and early twenty-first centuries, results whose solutions were long sought and will have important effects in the future. These results are Fermat's Last Theorem, the classification of finite groups, the four-color theorem, and the Poincaré conjecture.

  ## 25.1   SET THEORY: PROBLEMS AND PARADOXES

Georg Cantor raised many questions about the theory of infinite sets in his work of the late nineteenth century, questions he attempted over the years to answer, in many cases unsuccessfully. Other mathematicians also attacked these questions around the turn of the twentieth century.

25.1.1 Trichotomy and Well-Ordering

Cantor had shown that for two sets M and N, with cardinality $\overline{\overline{M}}$, $\overline{\overline{N}}$, respectively, no more than one of the relations $\overline{\overline{M}} = \overline{\overline{N}}$, $\overline{\overline{M}} < \overline{\overline{N}}$, or $\overline{\overline{N}} < \overline{\overline{M}}$ can occur. It seemed obvious to him as far back as 1878, furthermore, that exactly one of these relations should hold, namely, that if the two sets did not have the same cardinality, then one set must have the same cardinality as a subset of the other. It was only later that he realized that it was not a trivial matter to deny the existence of two sets, neither one of which was equivalent to a subset of the other. In fact, in his *Beiträge* of 1895, he made explicit mention of this **trichotomy principle** and stressed that not only did he not have a proof of it but that its proof must surely be difficult. He therefore carefully avoided using this principle in other proofs in his theory.

Cantor also realized that this question of trichotomy was closely related to another principle, that every set can be **well ordered**. This principle states that for any set A, there exists an order relation $<$ such that each nonempty subset B of A contains a least element, an element c such that $c < b$ for every other b in B. The natural numbers are well ordered under their natural ordering. The real numbers, on the other hand, are not well ordered under their natural ordering, but Cantor in 1883 thought it nearly self-evident that a well-ordering existed. By the mid-1890s, however, he began to realize that this result too needed a proof, a proof he believed he found in 1897 but soon realized was incomplete.

SIDEBAR 25.1 *Hilbert's 1900 Address to the International Congress of Mathematicians*

In the winter of 1899–1900, Hilbert was invited to make one of the major addresses at the second International Congress of Mathematicians, to be held in Paris in August 1900. Hilbert took many months to decide on the topic of his address, finally deciding by July that, because a new century was just beginning, he would discuss some central open problems of mathematics, problems that he believed would set the tone for mathematics in the twentieth century. Because Hilbert decided on his topic so late, he could not give the address at the opening session. Instead, it was delivered at a joint session of the History and Teaching sections of the congress, sections regarded as less prestigious than the pure mathematical sections of arithmetic, algebra, geometry, and analysis.

Hilbert began his talk with a discussion of the criteria for a significant problem: Its statement must be clear and easily comprehended; it should be difficult but not completely inaccessible; and its solution should have significant consequences. The 23 problems Hilbert presented to his audience encompassed virtually all of the branches of mathematics. For example, from the foundations of mathematics, Hilbert asked for a proof of the continuum hypothesis as well as an investigation of the consistency of the axioms of arithmetic. From number theory came the question as to whether α^β, where α is algebraic and β is irrational, always represents a transcendental, or even just an irrational, number, and whether it can always be determined if a diophantine equation is solvable. From analysis came the question of whether all complex zeros of the Riemann zeta function have real part $1/2$ (the Riemann hypothesis), and whether one can always solve boundary value problems in the theory of partial differential equations.

Hilbert's problems have in fact proved to be central in twentieth-century mathematics. Many have been solved, and significant progress has been achieved in the remainder. And as of the beginning of the new millennium, the search for solutions to other problems continues to drive progress in mathematics.

The well-ordering principle was thought to be so important by David Hilbert that in his address to the International Congress of Mathematicians in Paris in 1900, he presented the question of whether the set of real numbers can be well ordered as part of the first of the 23 important problems he suggested for mathematicians to consider in the twentieth century (see Sidebar 25.1). As Hilbert put it, "It appears to me most desirable to obtain a direct proof of this remarkable statement of Cantor's, perhaps by actually giving an arrangement of numbers such that in every partial system a first number can be pointed out."[2] In other words, Hilbert wanted someone actually to construct an explicit well-ordering of the natural numbers.

Another troubling aspect of set theory at the beginning of the century was the appearance of a number of seeming paradoxes. One of the earliest is today called Russell's paradox, because it was published by Bertrand Russell (1872–1970) in 1903 (Fig. 25.1). Recall that Dedekind and Cantor believed that virtually any description of "objects of our thought" would define a set. However, Russell, and Ernst Zermelo (1871–1953) two years earlier, determined that defining sets that contain themselves as elements would lead to contradictions. For suppose such a set M containing each of its subsets m, m', . . . as elements exists. Then consider those subsets m that do not contain themselves as elements. These constitute a set M_0. We can prove of M_0 (1) that it does not contain itself as an element and (2) that it contains itself as an element. First, M_0, being a subset of M, is itself an element of M, but not an element of M_0. For otherwise M_0 would contain as an element a subset of M (namely, M_0 itself) that contains itself as an element, contradicting the notion of M_0. Second, it follows that M_0 itself is a subset of M that does not contain itself as an element. Thus, it must be an element of M_0.

FIGURE 25.1

Bertrand Russell on an Indian stamp

BIOGRAPHY

Ernst Zermelo (1871–1953)

Zermelo grew up in Berlin as the son of a professor. He attended the universities in Berlin, Halle, and Freiburg, finally receiving his doctorate from the University of Berlin with a dissertation on the calculus of variations. He taught for several years at Göttingen, but in 1910 moved to Zürich. Ill health forced him to resign in 1916, but David Hilbert arranged for him to receive funds in recognition of his important work on set theory so that he was able to move to the Black Forest for a long period of recuperation. He returned to teaching in 1926 at Freiburg, but resigned in 1935 because he could not accept the new Nazi policies at the university. After World War II, he returned to Freiburg, where he finished his career.

Russell himself published several other versions of this paradox, the simplest being the barber paradox: A barber in a certain town has stated that he will cut the hair of all those persons and only those persons in the town who do not cut their own hair. Does the barber cut his own hair?

By early in the twentieth century it was therefore clear that Cantor's approach to set theory, although fruitful in the development of many new concepts, had flaws needing correction. Not only did some seemingly obvious results resist any proof, but also some of his intuitive ideas apparently led to contradictions. Interestingly, although many other fields of mathematics were being axiomatized during this time period, Cantor himself did not attempt to base his set theory on any collection of axioms. His definitions and arguments grew out of his intuition, and it was his extremely broad definition of set that was at the heart of the above paradoxes. In addition, the lack of appropriate axioms prevented a solution to the problems of trichotomy and well-ordering.

25.1.2   The Axiom of Choice

The twin objections to Cantor's set theory were cleared away by Zermelo in the period from 1904 to 1908. In 1904, Zermelo published a proof of the well-ordering theorem, basing it on a principle that had appeared implicitly in various arguments for many years, the principle known today as the axiom of choice, which Zermelo first made explicit: "Imagine that with every subset M' [of an arbitrary nonempty set M] there is associated an arbitrary element m', that occurs in M' itself; let m' be called the distinguished element of M'."[3] Thus, Zermelo asserted as an axiom that there always existed a "choice" function, a function $\gamma : S \to M$ (where S is the set of all subsets of M) such that $\gamma(M') \in M'$ for every M' in S. In other words, one can always somehow "choose" an element from each subset of a given set.

Zermelo realized that the axiom he had introduced was an important principle. Not only did it allow him to prove the well-ordering theorem, but it also gave him a proof of the trichotomy principle. In addition, he noted, "this logical principle cannot . . . be reduced to a still simpler one, but it is applied without hesitation everywhere in mathematical deduction. For example, the validity of the proposition that the number of parts into which a set decomposes is less than or equal to the number of all of its elements cannot be proved except by associating with each of the parts in question one of its elements."[4] (Here, "number" means "cardinality.") As

Zermelo knew, this result had been employed by Cantor in the 1880s. In fact, the axiom of choice had been used, although not stated, even earlier. Strictly speaking, for choices in which a rule for the choice is specified, the axiom is not necessary. But as early as 1871, both Cantor and Heine, in their work publicizing the unpublished research of Weierstrass, used the axiom of choice implicitly in the case where no choice rule can be specified to prove the result that a real function sequentially continuous (see Section 25.2.2) at a point p is also continuous at p. (A sequence of numbers had to be chosen, each in a specified interval.) Dedekind too had implicitly used the axiom in picking representatives of certain equivalence classes with respect to an ideal. Notwithstanding the fact that the axiom had been used for the past 30 years, its publication by Zermelo soon raised a storm of controversy.

The essence of the controversy was whether the use of infinitely many arbitrary choices was a legitimate procedure in mathematics. This question soon became part of the broader question as to what methods were permissible in mathematics at all and if all methods must be constructive. And then of course the questions arose as to what constituted a construction or what it meant to say that a mathematical object existed. Mathematicians had rarely debated such points before, but the use of the seemingly innocuous principle of making choices now led to the proof of a result, the well-ordering theorem, of which many mathematicians were skeptical. There was therefore a wide diversity of opinion in the mathematical community as to the validity of Zermelo's results. Some accepted them fully and went on to use the axiom of choice explicitly in their own research, while others denied the validity of both the axiom and Zermelo's proof. One group of mathematicians, of which Russell was a prominent member, steered a middle ground by making a careful attempt to determine what results already accepted in set theory had proofs dependent on the axiom of choice, so as to know explicitly what a rejection of that axiom would mean. Unfortunately, even Russell was at the time unable to prove that a particular result required the axiom of choice for its proof.

25.1.3   The Axiomatization of Set Theory

One of the problems with this debate within the mathematical community was that there was no accepted collection of axioms for set theory by which one could decide what methods were acceptable. There were many principles in print, especially after Zermelo's proof, which some mathematicians accepted while others denied. Hence, Zermelo decided that to solidify his proof and to clarify the terms of the argument surrounding it, an axiomatization of the theory of sets would be appropriate in which his proof would be embedded. This axiomatization would include not only the axiom of choice but also an axiom designed to eliminate some of the paradoxes caused by Cantor's overly broad definition of a set. Zermelo's method of axiomatization, influenced by Hilbert's axioms for geometry, began with a collection of unspecified objects and a relation among them that was defined by the axioms. In other words, Zermelo started with a domain $\mathcal{B}$ of objects and a relation $\in$ of membership between some pairs of these objects. An object is called a **set** if it contains another object (except as specified by axiom 2). To say that $A \subseteq B$ meant that if $a \in A$, then also $a \in B$. Zermelo's seven axioms (with the names he gave them) were as follows:

1. (Axiom of Extensionality) If, for the sets S and T, $S \subseteq T$ and $T \subseteq S$, then $S = T$.
2. (Axiom of Elementary Sets) There is a set with no elements, called the empty set, and for any objects a and b in $\mathcal{B}$, there exist sets $\{a\}$ and $\{a, b\}$.

3. (Axiom of Separation) If a propositional function $P(x)$ is definite (see below) for a set S, then there is a set T containing precisely those elements x of S for which $P(x)$ is true.

4. (Power Set Axiom) If S is a set, then the power set $\mathcal{P}(S)$ of S is a set. (The power set of S is the set of all subsets of S.)

5. (Axiom of Union) If S is a set, then the union of S is a set. (The union of S is the set of all elements of the elements of S.)

6. (Axiom of Choice) If S is a disjoint set of nonempty sets, then there is a subset T of the union of S that has exactly one element in common with each member of S.

7. (Axiom of Infinity) There is a set Z containing the empty set and such that for any object a, if $a \in Z$, then $\{a\} \in Z$.[5]

Zermelo never discussed exactly why he chose the particular axioms he did. But one can surmise the reasons for most of them. The first axiom merely asserts that a set is determined by its members, while the second was probably motivated by Zermelo's desire to have the empty set as a legitimate set and also to distinguish between an element and the set consisting solely of that element. Similarly, the power set axiom and the axiom of union were designed to make clear the existence of certain types of sets constructed from others, types that were used in many arguments. The axiom of separation is Zermelo's method of correcting Cantor's definition of a set as defined by any property, thereby eliminating Russell's paradoxes. Namely, by this axiom, there must be first a given set S to which the function describing the property applies, and second a definite propositional function, a function defined in such a way that the membership relation on $\mathcal{B}$ and the laws of logic always determine whether $P(x)$ holds for any particular x in S. Finally, the axiom of infinity was designed by Zermelo to clarify Dedekind's argument as to the existence of infinite sets. That argument had been met with disapproval by many mathematicians, partly because it seemed to be a psychological rather than a mathematical argument. Zermelo thus proposed his own axiom asserting that an infinite set can be constructed.

The reaction to Zermelo's axiomatization was mixed. First, Zermelo was criticized for not proving his axioms to be consistent. After all, Hilbert had done so for his geometrical axioms by relying on the consistency of the real numbers. Zermelo admitted that he could not prove consistency, but felt that it could be done eventually. He was convinced, however, that his system was complete in the sense that from it all of Cantorian set theory could be derived. Second, Zermelo was criticized for the specific axioms he included and those he left out. There was certainly no consensus for the correct basis of an axiomatic system, and because Zermelo could not show that his system had no flaws, it was difficult to convince the community at large that this set of axioms would accomplish the desired goal.

To gain any consensus, two changes had to take place in Zermelo's system. First, the axioms themselves needed to be somewhat modified. On the suggestion of several mathematicians, Zermelo himself in 1930 introduced a new system, now called Zermelo-Fraenkel set theory (after Abraham Fraenkel (1891–1965)). The major change from Zermelo's original system was the introduction of a new axiom, the axiom of replacement, intended to ensure that the set $\{\mathbf{N}, \mathcal{P}(\mathbf{N}), \mathcal{P}(\mathcal{P}(\mathbf{N})), \ldots\}$ exists in the Zermelo theory, where $\mathbf{N}$ is the set of natural numbers. Fraenkel's original formulation of this axiom was "If M is a set and if M' is obtained by replacing each member of M with some object of the domain $[\mathcal{B}]$, then M' is also a set."[6] As a second change, the nature of a "definite" propositional function had to be clarified, because this was essential to the axiom of separation. It turned out that this clarification had more to do with logic than with set theory, and ultimately it became the accepted view that axiomatic set theory needed to be embedded in the field of logic. For various reasons, which we will not discuss here, there are even today certain schools of mathematicians who do not accept one or more of Zermelo's axioms. But it is fair to say that the successes achieved within mathematics on the basis of these axioms have convinced the great majority of working mathematicians that they form a workable basis for the theory of sets.

The axiom of choice itself, although probably the most controversial of Zermelo's axioms, turned out to have numerous applications. It was applied not only in analysis but also in algebra. For example, it was used to prove that every vector space has a basis and that in a commutative ring every proper ideal can be extended to a maximal ideal. It was used repeatedly in the new discipline of topology, where, among many other results, it provided the basis for the proof that the product of any family of compact spaces was compact. It also proved essential in the study of mathematical logic.

One of the most important mathematical tools derived from the axiom of choice was a maximal principle, usually known as Zorn's lemma, that was ultimately proved equivalent to the axiom. Although there were many precursors to the maximal principle, we state it in the form given by Max Zorn (1906–1993) in 1935: If $\mathcal{A}$ is a family of sets that contains the union of every chain $\mathcal{B}$ contained in it, then there is a set A^* in $\mathcal{A}$ that is not a proper subset of any other $A \subset \mathcal{A}$. (By a chain $\mathcal{B}$ is meant a set of sets such that for every two sets B_1, B_2 in $\mathcal{B}$, either $B_1 \subseteq B_2$ or $B_2 \subseteq B_1$.) Zorn's aim in stating this axiom was, in fact, to replace the well-ordering theorem in various proofs in algebra. He claimed that the latter, although equivalent to his own axiom, did not belong in algebraic proofs because it was somehow a transcendental principle. In any case, Zorn's lemma soon became an essential part of the mathematician's toolbox and was used extensively both in algebra and topology.

Even though the axiom of choice proved useful, some of its consequences were unsettling and totally unexpected. Among the most surprising of these results was the Banach-Tarski paradox first noted by Stefan Banach (1892–1945) (Fig. 25.2) and Alfred Tarski (1901–1983) in 1924. They proved, using the axiom, that any two spheres of different radii are equivalent under finite decomposition. One can take a sphere A of radius one inch and a sphere B the size of the earth and partition each into the same number of pieces A_1, A_2, $\ldots A_m$ and B_1, B_2, $\ldots B_m$, respectively, such that A_i is congruent to B_i for each i. With results such as this provable using the axiom of choice, there was great interest in clarifying its exact status with regard to the other axioms of set theory. It was certainly not clear that the axiom could not lead to a contradiction.

Zermelo realized that a proof of the consistency of his axioms would be extremely difficult. Although he and others worked on this problem through the 1920s, it was not until 1931 that Kurt Gödel (1906–1978), an Austrian mathematician who spent most of his life at the Institute for Advanced Study at Princeton, showed in essence that there could be no such proof. In fact, he showed that in any system containing the axioms for the natural numbers—Dedekind's axioms, for example, could be proved in Zermelo-Fraenkel set theory—it was impossible to prove the consistency of the axioms within that system. Gödel also showed that this system was incomplete, that is, that there are propositions expressible in the system such that neither they nor their negations are provable. Given these results, the only hope for dealing with the axiom of choice was to prove that it was relatively consistent, that is, that its addition to the set of axioms did not lead to any contradictions that would not already have been implied

FIGURE 25.2

Stefan Banach on a Polish stamp

without it. Gödel was able to give such a proof by the fall of 1935. Within the next three years, he also succeeded in showing that the continuum hypothesis was relatively consistent within Zermelo-Fraenkel set theory.

A final result in the determination of the relationship of the axiom of choice to Zermelo-Fraenkel set theory was completed by Paul Cohen (1934–2007) in 1963. Cohen, using entirely new methods, was able to show that both the axiom of choice and the continuum hypothesis are independent of Zermelo-Fraenkel set theory (without the axiom of choice). In other words, it is not possible to prove or to disprove either of those axioms within set theory, and, furthermore, one is free to assume the negation of either one without fear of introducing any new contradictions to the theory. With these and other more recent results, it seems to be the case in set theory, as had already been shown in geometry, that there is not one version but many different possible versions, depending on one's choice of axioms. Whether this will be good or bad for the progress of mathematics is a matter for history to decide.

 ## 25.2   TOPOLOGY

Topology, that part of geometry concerned with the properties of figures invariant under transformations that are continuous and have continuous inverses, grew from various roots in the nineteenth century to become a full-fledged division of mathematics by early in the twentieth. The two branches of the subject to be considered here are point-set topology, concerned with the properties of sets of points of some abstract "space," and combinatorial topology, concerned with how geometrical objects are built up out of certain well-defined "building blocks."

Point-set topology grew out of the studies of Cantor of sets of real numbers. As such, its central aim is to provide an appropriate context for generalizing such properties of the real numbers as the Bolzano-Weierstrass property and the Heine-Borel property. Both of these are closely related to the important modern notion of compactness, a notion central in the generalization of the theorem on the existence of a maximum of a continuous function on a closed interval. The **Bolzano-Weierstrass property** is the property that every bounded infinite set of real numbers contains at least one **point of accumulation**, a point in every open interval of which there is another point of the set. In other words, this property asserts the "completeness" of the real numbers. We considered Bolzano's proof of this result in Section 22.1.3.

The **Heine-Borel property**, formulated by Emile Borel (1871–1956) in 1894, states that if an infinite set A of intervals covers a finite closed interval B of real numbers, in the sense that every number of B lies in the interior of at least one element of A, then there is a finite subset of A with the same property. (Although Heine had used this result implicitly in the 1870s, it was Borel who first stated and proved this theorem with respect to a countable set A. Henri Lebesgue (1875–1941) generalized it to arbitrary infinite collections A in 1904.) Borel's proof is by contradiction. Let $A = \{A_1, A_2, \ldots, A_m, \ldots\}$. If the conclusion is not true, then for every n there is a point $b_n \in B$ such that b_n is not in A_i for every $i \leq n$. If one bisects the interval B, the same statement is true for at least one of the halves. If one continues this bisection process, one obtains a decreasing nested sequence of closed intervals B_i, each

BIOGRAPHY

Grace Chisholm Young (1868–1944)

Born Grace Chisholm at Haslemere, near London, she was educated at home and then entered Girton College, Cambridge, the first English institution where women could receive a university education. Having attained a superior score on the Cambridge Tripos exam in 1892, she decided to go to Göttingen to continue her studies because there was no possibility of advanced study in England. Felix Klein was willing to accept women students, but only after he had assured himself through a personal interview that they would succeed. (There were other members of the faculty who objected to admitting women under any conditions.) In any case, Grace Chisholm earned her PhD in 1895, being the first woman to receive a German doctorate in mathematics through the regular procedure. In 1896, she married William Young, an English mathematician who had been her tutor at Girton.

The Youngs spent the next 44 years in a partnership fruitful both in mathematics and in children (six). Although most of the more than 200 mathematical papers and books that ensued were in William Young's name, Grace had a major role in their production. As William noted in an article of 1914, he had discussed the major idea of the work with his wife, and Grace had elaborated the argument and put the paper into publishable form. Their daughter wrote that her father could only generate ideas when he was stimulated by a sympathetic audience. Not only did Grace provide this audience, but she also had the initiative and stamina to complete the various undertakings her husband proposed. William died at their home in Switzerland after the outbreak of the Second World War left Grace in England. She died in 1944, just before she was to receive an honorary degree from the Fellows of Girton College. Two of their sons as well as one of their granddaughters also became mathematicians.

of which has the same property as B itself. But $\cap B_l$ contains a point p. By hypothesis, p is in the interior of A_k for some k, so A_k must contain one of the intervals B_i, a contradiction.

The key to Borel's proof is what is often called the nested interval property, that the intersection of a nested family of closed intervals contains a point. It was this result that was later to be abstracted into the earliest definition of compactness. It was also this result that was among the earliest presented in the first systematic exposition of set theory as a whole, the 1906 *The Theory of Sets of Points* by William Young (1863–1942) and Grace Chisholm Young (1868–1944).

25.2.1 The Youngs and *The Theory of Sets of Points*

The text of the Youngs deals with sets of points on the real line or in the real plane and gives explicit definitions of many fundamental concepts that were later to be generalized. For example, a point x belonging to an interval "unclosed" at both ends is said to be an **internal point** of the corresponding closed interval. A point L is said to be a **limit point** of a given set of real numbers if inside every interval containing L as an internal point there is a point of the set other than L. A set that contains all its limit points is said to be **closed**, while one that does not is said to be **unclosed** or **open**. (Note that this is not the definition of "open" in use today.) The Youngs then reformulated the Bolzano-Weierstrass and Heine-Borel theorems in terms of these definitions and presented proofs.

The Youngs next generalized the notion of "interval" in the line to that of "region" in the plane by considering the latter to be generated by a set of triangles contained in it and then generalized the notion of limit point by replacing "interval" with "region." They noted further that, in analogy with the properties of an interval, a region divides the plane into three disjoint sets: internal points (those internal to at least one generating triangle), boundary points (those other than internal points that are limit points of internal points), and external points (those that are neither internal points nor boundary points). They then easily stated and proved the generalizations of the Bolzano-Weierstrass and Heine-Borel theorems to the plane.

Another fundamental notion in modern topology, the idea of connectedness, stems from some considerations of Cantor. Cantor, as part of his work in dealing with the "continuum," the entire set of real numbers, tried to characterize this set. As part of this attempt, he found it necessary to define what it meant for the set to be in one piece. Because he saw this idea in terms of the minimal distance between points of the set being 0, he defined "connected" in terms of distance: A set T is **connected** if given any two points p and q in it and any positive number ϵ, there is a finite number of points $t_1, t_2, \ldots, t_n$ in T such that the distances pt_1, $t_1 t_2, \ldots, t_n q$ are all less than ϵ. The Youngs, among others, realized that it would be better to give a definition not using distance, but purely in set-theoretic terms, so they translated this notion as follows: "A set of points such that, describing a region in any manner round each point and each limiting point of the set as internal point, these regions always generate a single region, is said to be a **connected set** provided it contains more than one point."[7] Using this definition, the authors then proved that a set is connected if and only if it cannot be divided into two closed components without common points.

25.2.2   Fréchet and Function Spaces

In the same year as the Youngs' text appeared, Maurice Fréchet (1878–1973) began the process of generalizing results on points in the plane to more general contexts. In his dissertation dealing with the theory of functionals, functions operating on certain sets of functions, he had to be able to decide when two functions were "close" to each other and therefore decided "to establish systematically certain fundamental principles of the Functional Calculus, and then to apply them to certain concrete examples." By doing so, "one often gains thereby from seeing more clearly that which was essential in the demonstration, . . . from the simplification, and in the freeing [of the proofs] from that which only depends on the particular nature of the elements considered."[8] In other words, Fréchet decided to reconsider the basic notions of the topology of the real line in terms of arbitrary sets and then apply these notions to his particular sets of interest. He realized that he could prove numerous results of topology in a general context once and for all and then apply them in concrete instances rather than prove the same result over and over again. He especially wanted to answer questions about convergence of functionals to limits and determine the circumstances under which the limit functional has the same properties of the functionals of which it is the limit.

Fréchet thus began, not by defining "limit," but by characterizing the notion by means of axioms. That is, a set E belongs to $\mathcal{L}$, the class for which limits are defined, if given any infinite subset $\{a_i\}$ of E it is possible to determine whether or not a unique element a exists having appropriate properties. This element is to be the limit of $\{a_i\}$. The limit element a is subject to the conditions that if each $a_i = a$, then the limit is a, and if a is the limit of $\{a_i\}$, then it is also the limit of any subsequence of $\{a_i\}$. With this abstract notion of a limit,

Fréchet stated various definitions, some of which had already been considered by Cantor: The **derived set** E' of E is the set of its limit elements. The set E is **closed** if $E' \subseteq E$ and **perfect** if $E' = E$. The point a is an **interior point** of E if a is not the limit of any sequence not in E.

As part of his study of functionals, Fréchet wanted to generalize the Heine-Weierstrass result on the existence of a maximum and minimum of a continuous function on a closed interval so that it could apply to function spaces. To accomplish this generalization, he took the central idea of the proof of the Heine-Borel theorem as a definition in a brief note of 1904: "We call a **compact** set every set E such that there always exists at least one element common to every infinite sequence of subsets $E_1, E_2, \ldots, E_n, \ldots$ of E, if these (possessing at least one element each) are closed and each is contained in the preceding one."[9] Fréchet then proved that a necessary and sufficient condition for a set E to be compact was that every infinite subset F had at least one limit element in E, an element that is the limit of some sequence of distinct points of F. Noting that compact sets by his definition had properties analogous to those of closed and bounded sets in space, Fréchet was able to generalize the Weierstrass result, using for his definition of continuity the property today generally known as sequential continuity: A function f is **continuous** at a in the closed set E if $\lim_{n \to \infty} f(a_n) = f(a)$ for all sequences $\{a_n\}$ in E that converge to a. Interestingly, in his dissertation of 1906, Fréchet took for his definition of "compact" the necessary and sufficient condition above rather than the intersection property, adding the statement that any finite set will also be considered compact.

Having shown that the concept of a distance was not necessary to develop various familiar notions, Fréchet proceeded to reintroduce that notion in a more general setting. Namely, he considered a subclass $\mathcal{E}$ of the class $\mathcal{L}$ consisting of members E in which can be defined a **metric**, a real-valued function (a, b) satisfying (1) $(a, b) = (b, a) \geq 0$; (2) $(a, b) = 0$ if and only if $a = b$; and (3) $(a, b) \leq (a, c) + (c, b)$ for any three elements a, b, c of E. Using the metric, Fréchet defined a **Cauchy sequence** to be any sequence $\{a_n\}$ such that for every $\epsilon > 0$ there is an m so that for all $p > 0$, we have $(a_m, a_{m+p}) < \epsilon$. Limit elements and the concepts associated with them are then defined in terms of the metric and Cauchy sequences. In particular, Fréchet considered the subclass of **normal** sets consisting of sets that are perfect, separable (contain a countable dense subset), and in which every Cauchy sequence has a limit (complete, in modern terminology). It is for sets of this type that he could prove a generalization of the Heine-Borel theorem: A normal set E is compact if and only if for every collection $\mathcal{G}$ of sets $\{I\}$ such that every element of E is in the interior of at least one member of that collection, there exists a finite subset of $\mathcal{G}$ with the same property.

Fréchet gave a number of examples of spaces with metrics for which his general theorems held, including examples of sets of functions. One such example was the set of real functions continuous on a given closed interval. Here the metric was the **maximum norm**

$$(f, g) = \max_{x \in I} |f(x) - g(x)|,$$

under which Fréchet proved that the set was normal. A second example was the set of all sequences of real numbers $x = \{x_1, x_2, \ldots, \}$ with the metric

$$(x, y) = \sum_{p=1}^{\infty} \frac{1}{p!} \frac{|x_p - y_p|}{1 + |x_p - y_p|}.$$

Fréchet noted that this metric is better than a more standard one, $(x, y) = \max_p |x_p - y_p|$, in the sense that the former always gives a finite distance while the latter does not. Again, Fréchet showed that this set is normal and, furthermore, that the same is true for the set of real functions defined on it with an appropriate metric.

25.2.3   Hausdorff and Topological Spaces

It was Felix Hausdorff (1868–1942) who was able to give a full axiomatization of the notion of a topological space derived from standard properties of sets of real numbers. He described this axiomatization in his 1914 text, *Grundzüge der Mengenlehre* (*Foundations of Set Theory*). Although his axioms and definitions have been modified since 1914 and numerous subsidiary definitions and concepts have been introduced, the basics of point-set topology as currently taught are all to be found in this fundamental work.

Hausdorff noted that there are three basic concepts by means of which one can base a general theory of topological spaces: the notions of distance, neighborhoods, and limits. Fréchet had in effect used both the first and the third. Hausdorff noted that if one begins with the concept of distance, one can derive the other two, while if one begins with that of neighborhood, one can define the notion of limits. In general, however, one cannot reverse these procedures. Although the method one chooses, according to Hausdorff, is a matter of taste, he decided to begin with the concept of neighborhood to define the notion of a **topological space**. Such a space, today known as a **Hausdorff space**, is a set E to each element x of which is associated a collection of subsets $\{U_x\}$ of E, called neighborhoods, which satisfy the following axioms:

1. Each point x belongs to at least one neighborhood U_x, and every neighborhood U_x contains x.
2. If U_x, V_x are two neighborhoods of the same point x, then their intersection also contains a neighborhood of that point.
3. If $y \in U_x$, then there is a neighborhood U_y such that $U_y \subseteq U_x$.
4. For two different points x, y there are two neighborhoods U_x, U_y whose intersection is empty.[10]

After showing that a metric space, with the neighborhoods defined by $U_x = \{y | (y, x) < \rho\}$ for every real number ρ, satisfies the axioms given, Hausdorff developed the same basic theory as had Fréchet. The major change was that the central concept was now the domain or, in more modern terms, the open set. For Hausdorff, a **domain** A is a subset of E containing only interior points, where the latter are defined as points x that have a neighborhood U_x contained in A. That the entire set E and each of the neighborhoods U_x are domains follows from axioms 1 and 3. Hausdorff also showed that the union of arbitrarily many domains and the intersection of finitely many are domains. A **closed** set, on the other hand, is one that contains all of its accumulation points. Hausdorff then showed that the closed sets are precisely the complements of the domains.

Using Fréchet's limit definition for compact sets, Hausdorff proved the intersection property of nested, closed, compact sets and then a generalization of the Heine-Borel theorem, modifying the original proof only slightly. He also gave a new definition of limit point and convergence general enough to apply in any topological space: The point x is a **limit** of the infinite set A, if every neighborhood U_x of x contains all but a finite number of points of A.

Further, because any set A either has a single limit x or none at all, one writes in the first case that $x = \lim A$ or that A **converges** to x.[11] Among the other new definitions was one for connectedness: "A set A differing from the empty set is **connected** if it cannot be divided into two disjoint, non-null, sets both closed relative to A [that is, such that each set is the intersection with A of a closed set in the ambient space E]."[12] Hausdorff noted that because closed sets are the complements of domains, he could equally well have specified in the definition that the components needed to be domains.

Because the idea of a neighborhood was the point of departure for Hausdorff's development, it is not surprising that he defined continuity using that concept. Namely, he took the standard $\epsilon - \delta$ definition of continuity for functions of a real variable, noted that this definition used neighborhoods on the real line, and then translated it into a general definition for topological spaces: "The function $y = f(x)$ is called **continuous** at the point a, if for each neighborhood V_b of the point $b = f(a)$ there exists a neighborhood U_a of the point a, whose image lies in V_b; i.e., $f(U_a) \subseteq V_b$."[13] Hausdorff then generated an equivalent definition: The function $f : A \to B$ is continuous at a if and only if, for each subset Q of B for which $b = f(a)$ is an interior point, the inverse image $f^{-1}(Q)$ also contains the point a as an interior point. Using this definition, it is easy to prove that a continuous function preserves connectedness and compactness. Hausdorff naturally noted that the first of these facts implies the intermediate value theorem while the second implies the existence of a maximum and minimum value for a continuous function on a closed interval. In other words, Hausdorff showed that a topological space is the natural setting for these classical results about functions of a real variable.

25.2.4 Combinatorial Topology

Combinatorial topology grew out of the study of the idea of multiple connectivity of a surface in space, an idea developed by Riemann in his work on the integration of differentials. This idea was further refined by the physicists of the mid-nineteenth century because it turned out to be important in fields such as fluid dynamics and electromagnetism. It was Enrico Betti (1823–1892), however, who in 1871 generalized the idea of multiple connectedness to n-dimensional spaces by using hypersurfaces without boundaries as the analogues of Riemann's closed curves. Betti furthermore applied this idea to study the integration of differential forms over spaces of dimension n. The differences in connectivity of two surfaces were a way of telling that the surfaces were essentially "different," that there could be no continuous invertible function from one to the other. To deal with this method of distinguishing surfaces, however, Poincaré developed the idea of homology.

In his fundamental paper, *Analysis Situs*, of 1895 and four years later in a supplementary work, Poincaré made the following definition: A homology relation exists among p-dimensional subvarieties $v_1, v_2, \ldots, v_r$ of an n-dimensional variety V, written

$$v_1 + v_2 + \cdots + v_r \sim 0,$$

if for some integer k, the set consisting of k copies of each of the v_i constitutes the complete boundary of a $(p + 1)$-dimensional subvariety W. (A "variety" for Poincaré—now generally called a manifold—was the generalization to higher dimensions of a curve in one dimension or a surface in two dimensions and was usually thought of as being defined, at least locally, either as the set of zeros of an appropriate system of functions or parametrically as the image of a certain set of such functions.) Poincaré introduced "negatives" of varieties by considering orientation. That is, $-v$ is the same variety as v but with the opposite orientation. As an example of a homology relation, let v_1 and v_2 be the outer and inner boundaries, respectively, of the ring in Figure 25.3 with the indicated orientations. Then v_1 and v_2 together form the complete boundary of the ring, and, since $-v_2$ has the opposite orientation as v_2, the relation $v_1 - v_2 \sim 0$ holds. Poincaré further observed that the varieties in homology relations can be added, subtracted, and multiplied by integers and therefore was able to call a set of varieties **linearly independent** if there were no homology relation among them with integer coefficients.

FIGURE 25.3

Homology relations

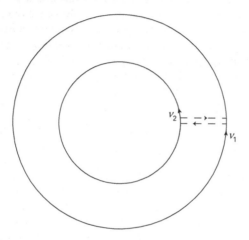

To clarify the notion of multiple connectedness, Poincaré went on to define the p-dimensional **Betti number** B_p of a variety V to be one more than the maximum number of linearly independent, closed, p-dimensional subvarieties, where a closed variety is one without a boundary. Thus, according to Poincaré the 1-dimensional Betti number of the ring in Figure 25.3 is 2, while that of the double ring is 3. On the other hand, the 1-dimensional Betti number of the disc is 1. Poincaré applied his notion of homology to the study of integration over varieties of various dimensions and also attempted to prove a duality theorem, that for compact, connected, orientable manifolds of dimension n, the relationship $B_p = B_{n-p}$ held for $1 \le p \le n - 1$. Poincaré's proof, however, and even his definition, have flaws in the eyes of modern readers. It took about 20 years to construct a homology theory containing Poincaré's basic ideas, which was rigorous by today's standards.

The modern theory of homology was developed by several mathematicians in the early twentieth century. One major simplification of Poincaré's ideas was the consideration of p-dimensional submanifolds not as solutions to systems of equations but as being formed from certain simple p-dimensional manifolds, each of which was the continuous image of a p-dimensional "triangle." The appropriate definitions were completely worked out by James W. Alexander (1888–1971) by 1926 when he defined a ***p*-simplex** to be the p-dimensional analogue of a triangle and a **complex** to be a finite set of simplexes such that no two had an interior point in common and such that every face of a simplex of the set was also a simplex

BIOGRAPHY

Henri Poincaré (1854–1912)

Poincaré, like Hilbert, was a universal mathematician, one who contributed to virtually every area of mathematics, including physics and theoretical astronomy. He was born into an upper-middle-class family, many members of whom performed various services to the French government. Poincaré displayed a strong interest in mathematics from an early age and won a first prize in the mathematics competition for students in all of the French lycées. In 1873, he entered the École Polytechnique and, after receiving his doctorate in 1879,

entered a university career, first at the University of Caen and then, in 1881, at the University of Paris. Toward the end of his life, he turned to popularization and wrote several books emphasizing the importance of science and mathematics, including such works as *Science and Hypothesis*, *The Value of Science*, and *Science and Method*. It was in the latter work that Poincaré described the psychology of discovery in mathematics, stressing the subconscious as a central factor in mathematical creativity (Fig. 25.4).

FIGURE 25.4

Poincaré on a French stamp

of the set. An elementary i-chain of a complex was defined to be an expression of the form $\pm V_0 V_1 \ldots V_i$, where the V's are vertices of an i-simplex. The expression changes sign upon any transposition of the V's, thus giving each chain an orientation. An elementary i-chain was then an i-dimensional "face" of a p-simplex, while an arbitrary i-chain was a linear combination of elementary i-chains with integer coefficients. As an example, the tetrahedron with vertices V_0, V_1, V_2, V_3 is a 3-simplex, while it together with its four faces (each a 2-simplex), its four edges (each a 1-simplex), and its four vertices (each a 0-simplex), form a complex. The face $V_0 V_1 V_2$ is then an elementary 2-chain of the 3-simplex. Alexander next defined the boundary of the elementary i-chain $K = V_0 V_i \ldots V_i$ to be the $(i-1)$-chain $K' = \sum (-1)^s V_0 \ldots \hat{V}_s \ldots V_i$ and extended this to arbitrary i-chains by linearity. Thus, the boundary of $V_0 V_1 V_2$ is $V_1 V_2 - V_0 V_2 + V_0 V_1$ (Fig. 25.5). An easy calculation with this

FIGURE 25.5

The boundary of a tetrahedron

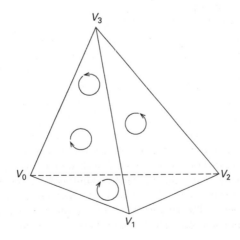

example shows that the boundary of the boundary is zero, and one can show that this result is true in general.

Alexander applied his definition of homology to **closed chains** (cycles), chains whose boundary is zero. Namely, a closed chain K is homologous to zero, $K \sim 0$, if it is the boundary of a chain L. Two chains K and K^* are homologous, $K \sim K^*$, if $K - K^*$ is homologous to zero. The pth Betti number of a complex is then the maximum number of closed p-chains that are linearly independent with respect to boundary, that is, such that no linear combination is homologous to zero. (Note that this number is one less than the number according to Poincaré's original definition.)

With a commutative operation ("addition") having an inverse being considered on the set of closed chains, it should be clear to modern readers that there is a group hiding among Alexander's definitions. Mathematicians of the 1920s saw this as well. But before discussing the applications of group theory to topology, we need first to consider the growth of algebra in general in the early part of the twentieth century.

 ## 25.3   NEW IDEAS IN ALGEBRA

Hilbert had closed out the nineteenth century by developing a new set of axioms for geometry and had shown their consistency, at least in relation to the consistency of arithmetic. As the twentieth century began, mathematicians attempted to develop sets of axioms for numerous algebraic constructs. Since Eliakim H. Moore (1862–1932) showed in 1902 that Hilbert's axioms were not independent, that is, that one of them could in fact be deduced from the others, a strong effort was made to develop sets of independent axioms in algebra. For example, in 1903 Leonard Eugene Dickson (1874–1954) developed a new set of axioms for a field, a set he believed was an improvement on the work of Weber some 10 years earlier. Dickson defined a field to be a set with two rules of combination, designated by $+$ and $\times$, satisfying the following nine axioms:

1. If a and b belong to the set, then so does $a + b$.
2. $a + b = b + a$.
3. $(a + b) + c = a + (b + c)$.
4. For any two elements a and b of the set, there exists an element x in the set such that $(a + x) + b = b$.
5. If a and b belong to the set, then so does $a \times b$.
6. $a \times b = b \times a$.
7. $(a \times b) \times c = a \times (b \times c)$.
8. For any two elements a and b of the set, such that $c \times a \neq a$ for at least one element c of the set, there exists an element x in the set such that $(a \times x) \times b = b$.
9. $a \times (b + c) = (a \times b) + (a \times c)$.

Naturally, among the axioms are the familiar closure, commutative, and associative laws of addition and multiplication, as well as the distributive law. Dickson's innovation was to replace the identity and inverse laws for both addition and multiplication by the new axioms 4 and 8. He then proceeded to prove independence by creating, for each axiom, a system with two rules of combination that did not satisfy that axiom but did satisfy each of the other eight axioms. For example, a system satisfying every axiom except for the second one consists

BIOGRAPHY

Leonard Eugene Dickson (1874–1954)

Dickson, born in Independence, Iowa, was the first recipient of a doctorate in mathematics at the University of Chicago. After further study in Leipzig and Paris and a year teaching in Texas, he returned to Chicago in 1900 for the remainder of his career. Dickson was a prolific mathematician, writing hundreds of articles and some 18 books. Among the most important of the latter was his monumental three-volume

History of the Theory of Numbers, a work in which he traced the evolution of every important concept in that field. Dickson served as editor of the *Transactions of the American Mathematical Society* from 1911 to 1916 and then as president of the A.M.S. for the following two years. He was elected to the National Academy of Sciences in 1913.

of all positive rational numbers with ordinary multiplication, but with the new addition rule, $a + b = b$. Similarly, the set of all rational numbers with ordinary multiplication but with addition given by $a + b = -a - b$ satisfies all axioms but the third. In this case, axiom 4 is satisfied by taking $x = 2b - a$.

25.3.1   The *p*-adic Numbers

As noted in Chapter 21, abstraction in algebra was increasing, and, as in the notions of topology discussed earlier, there were growing efforts to look at the structure of each of the constructs of algebra: groups, fields, ideas, and so on. Dickson's analysis of the postulates of a field was one attempt in this direction, but Dickson himself had no new fields to consider under his definition other than the ones already known by Weber. Before a complete structural analysis of the concept of fields could take place, it proved necessary for new fields to make their appearance, in particular the fields of *p*-adic numbers created by Kurt Hensel (1861–1941) around the turn of the century and discussed in detail in two books he wrote in 1908 and 1913.

Hensel started by noting that given any prime p, a positive integer A can be written uniquely in the form

$$A = a_0 + a_1 p + a_2 p^2 + \cdots + a_r p^r,$$

where each a_i satisfies $0 \le a_i \le p - 1$. In this new representation, two numbers are to be thought of as "close" relative to p if they are congruent modulo high powers of p. To develop his theory conveniently, Hensel used the analogy between this representation and the ordinary decimal representation of a fraction to write A in the form $A = a_0.a_1a_2 \cdots a_r$, where the coefficient of p^k is in the kth place after the period. Then the two numbers $3 + 2 \cdot 5$ and $3 + 2 \cdot 5 + 4 \cdot 5^{10}$, represented respectively by the numbers 3.2 and 3.2000000004, which are close if thought of as decimal fractions, were also to be thought of as "close" relative to the prime 5 because they were congruent modulo each power of 5 up through the 10th. More specifically, Hensel called the numbers $A_k = a_0.a_1a_2 \cdots a_k$ for $k < r$ the approximate values of A relative to p, since each A_k was congruent to A modulo a higher power of p than A_{k-1}.

To turn the positive integers in this new representation into a field, one needed to be able to perform the usual operations on them. Guided by the analogy with finite (and infinite) decimal fractions, Hensel showed how one must do this. Addition and multiplication are performed in nearly the standard way, by use of the "decimal" analogy, except that one works from left to right and carries forward an appropriate multiple of p when any sum or product is greater than or equal to p. For example, again using $p = 5$, one has the sum

$$
\begin{array}{r}
2.3042134 \\
+ \ 3.2413123 \\
\hline
0.10113031
\end{array}
$$

As would be expected, it is the operations of subtraction and division that force Hensel to add new elements to this set of positive integers, elements represented in a surprising way. Consider the subtraction

$$
\begin{array}{r}
3.131312 \\
- \ 4.424322 \\
\hline
\end{array}
$$

Beginning at the left and borrowing as necessary from the right, one gets 4.10243 for the first six digits of the answer. In other words, $3.131312 - 4.424322 \equiv 4.10243 \pmod{5^j}$ for $j \leq 5$. Can one, however, get an exact answer rather than just an approximation? After all, if one translates these representations back into the ordinary decimal representation, the answer is a negative integer. To introduce these "negatives," Hensel simply permitted the number of places to the right of the period to increase without limit. In this example, one can see that by placing indefinitely many 0's to the right in both the minuend and the subtrahend, the answer becomes $4.102434444 \cdots$ in the sense that if one cuts off this number at the nth place, the result will be congruent to the actual difference modulo 5^n. Hensel showed further that the use of indefinitely many digits after the period was also necessary to be able to perform divisions A/B in the special case where $B \not\equiv 0 \pmod{p}$ and that in both subtraction and this case of division, the infinite p-adic expansion was periodic.

Having rewritten the elements of a subfield of the field of rational numbers (those quotients A/B such that $B \not\equiv 0 \pmod{p}$) in a periodic p-adic expansion, Hensel extended the analogy with ordinary decimals by introducing a new set, the set of all power series of the form

$$
A = a_0 + a_1 p + a_2 p^2 + \cdots + a_n p^n + \cdots = a_0.a_1 a_2 \ldots a_n \ldots,
$$

where each coefficient a_i satisfies $0 \leq a_i < p$. As before, the finite series $A_k = a_0.a_1 a_2 \ldots a_k$, $k = 1, 2, \ldots$, are each approximations to A, because A_k is congruent to A modulo p^{k+1}. But now, because there are infinitely many such approximations, each better than the previous one, Hensel turned to analysis and wrote

$$
A = \lim_{k \to \infty} A_k.
$$

The set of power series so defined is not, however, a field. Although one can perform addition, subtraction, and multiplication by using the earlier rules on the various approximands and taking limits, the attempt to divide arbitrary power series forces one to generalize these series to those that include a finite number of terms with negative exponents. (As a very simple case,

$6 \div 5$ is written as $1 \cdot 5^{-1} + 1$.) Thus, Hensel incorporated such terms and then showed easily that the set of all series of the form

$$A = a_m p^m + a_{m+1} p^{m+1} + \cdots \quad (m \text{ any integer})$$

is indeed a field, the field today called the field $\mathbf{Q}_p$ of p-adic numbers.

Having defined a new set of fields, one for each prime p, Hensel was able to apply various concepts from the general theory of fields to these particular ones. Thus, the integers of the field are those whose smallest power of p is nonnegative, while the units are those whose smallest power of p is the zero power. (The units are precisely those integers whose multiplicative inverse is again an integer.) Hensel dealt with polynomials whose coefficients are in this field and applied the usual constructions to them, including the adjunction of roots of such polynomials to form extension fields to which Galois theory can be applied. But because the field of p-adic numbers has a natural topology, with neighborhoods of a point defined in terms of congruence to a given power of p, one can apply analytic concepts as well as the algebraic ones. For example, the notion of a Cauchy sequence of p-adic numbers can be defined as well as the notion of a limit, and it can be shown that every such Cauchy sequence has a limit.

25.3.2 The Classification of Fields

It was this creation of new fields by Hensel that influenced Ernst Steinitz (1871–1928) to undertake a new detailed structural investigation of the entire subject of fields, a study that appeared in 1910 under the title "Algebraische Theorie der Körper" ("Algebraic Theory of Fields"). Steinitz explicitly differentiated his work from that of Weber: "Whereas Weber's aim was a general treatment of Galois Theory, independent of the numerical meaning of the elements, for us it is the concept of field that represents the focus of interest." His goal, then, was "to advance an overview of all the possible types of fields and to establish the basic elements of their interrelations."[14]

The first distinction among fields involves the concept of the prime field and the characteristic. The **prime field** of any field K is its smallest subfield, that is, the intersection of all of its subfields. If ϵ is the multiplicative unit of this prime field, then there are two possibilities for the set I of all integral multiples $m\epsilon$ of ϵ. First, they could all be distinct. In this case, I is isomorphic to the set of positive integers, and the prime field is isomorphic to the field of rational numbers. K is then said to be of **characteristic 0**. The second possibility for I is that there is a smallest natural number p, of necessity prime, such that $p\epsilon = 0$. In this case I is isomorphic to the set of residue classes modulo p. Thus, I is itself a field, the finite field of p elements, and K is said to be of **characteristic p**.

Once Steinitz had analyzed the simplest possible fields, he continued toward his goal by studying various types of field extensions and determining what properties of fields are preserved by moving from simple fields to their extensions. Thus, he defined an extension L of a field K to be **algebraic** if each element of L is a root of a polynomial equation with coefficients in K, while it is a **transcendental** extension otherwise. An algebraic extension L of K is said to be **finite** and of dimension n if there are n elements of L linearly independent over K, while any set of more than n elements is linearly dependent. The extension is **infinite** if there is no maximal number of linearly independent elements. Steinitz then divided the finite algebraic extensions of a field into those to which Galois theory applies and those to which it does not apply. (Galois theory, in the fullest sense, applies only to those fields in which irreducible polynomials have distinct roots.) Fields to which Galois theory applies, including all fields of characteristic 0, are called **separable**; the others are called **inseparable**. Infinite algebraic extensions include, for example, the **algebraic closure** of a prime field K, an extension field in which every polynomial over K factors into linear factors. Steinitz showed how to construct such an algebraic closure for the prime fields of characteristic p, but noted that for the field of rational numbers, this construction involved the use of the axiom of choice. Not surprisingly, then, his general proof of the existence of an algebraic closure for an arbitrary field also involved that axiom. Finally, Steinitz discussed transcendental extensions. Defining a purely transcendental extension of a field K to be one formed by the adjunction of finitely or infinitely many unknowns, Steinitz was able, using the well-ordering theorem, to demonstrate that every extension of a field K can be formed by taking a purely transcendental extension and then an algebraic extension.

Besides giving this classification of fields, Steinitz's major innovation lay in showing that, in contrast to Dedekind and even to Weber, it was not necessary to begin the study of fields with systems of numbers with their known properties, but instead one could begin with a completely abstract idea—a set endowed with two operations satisfying certain axioms. Over the next two decades, Steinitz's ideas exploded into all areas of algebra and beyond.

25.3.3   The Axiomatization of Vector Spaces

Steinitz certainly made use of linear algebra notions in his definition of a finite algebraic extension of a field. And he went on to prove certain results about finite-dimensional vector spaces in the context of field extensions. In particular, he proved explicitly what Dedekind some years earlier had assumed implicitly, that a generating set for a finite-dimensional vector space cannot have fewer elements than the dimension of the space. His statement and proof of this result were, of course, in the context of the coordinate system determined by the generators. For suppose the dimension of L over K is n. Thus, there are n linearly independent generators $\{\alpha_i\}$ of L over K. If, in addition, there were a set of $n - 1$ generators β_j, $i = 1, 2, \ldots, n - 1$, then each α_i could be written in the form $\alpha_i = c_{i1}\beta_1 + c_{i2}\beta_2 + \cdots + c_{in-1}\beta_{n-1}$. But then the equation $d_1\alpha_1 + d_2\alpha_2 + \cdots + d_n\alpha_n = 0$ would have a nontrivial solution $\{d_i\}$. For this equation is equivalent to the system

$$
\begin{array}{ccccccc}
c_{11}d_1 & + & c_{21}d_2 & + & \cdots & + & c_{n1}d_n & = & 0 \\
c_{12}d_1 & + & c_{22}d_2 & + & \cdots & + & c_{n2}d_n & = & 0 \\
\vdots & & \vdots & & \ddots & & \vdots & & \vdots \\
c_{1n-1}d_1 & + & c_{2n-1}d_2 & + & \cdots & + & c_{nn-1}d_n & = & 0,
\end{array}
$$

and any homogeneous system with more unknowns than equations has a nonzero solution. Thus, the α_i would be dependent, contradicting the original assumption.

Although Steinitz worked out many of the basic results of the theory of finite-dimensional vector spaces, they were in the context of algebraic extensions of fields. He did not attempt to give a system of axioms for a vector space in general. And as we noted previously, the system of Peano of more than 20 years earlier had been ignored. It was Hermann Weyl (1885–1955) in his book *Raum—Zeit—Materie* (*Space—Time—Matter*) of 1918 who made a new attempt to give an axiomatic treatment of the subject as a basis to his development of the theory of relativity from basic principles. Although there is no indication that he was familiar with the work of Peano, his axiom system was virtually the same. The only difference was that, unlike his Italian predecessor, Weyl insisted that vector spaces be finite-dimensional. Thus, his final axiom stated that "there are n linearly independent vectors, but every $n + 1$ vectors are linearly dependent."[15] Unfortunately, Weyl's work was even less influential than that of Peano. The notion of a vector space thus needed a third discovery, this time in the context of analysis.

Several mathematicians in the 1920s became interested in a notion that generalized both the algebraic and the topological properties of the systems they were working with. They were aware of Fréchet's notion of a metric space and were certainly familiar with the idea of a vector space. To put these two notions together, Stefan Banach, in his dissertation of 1920, introduced the notion of what is now called a **Banach space**, a vector space possessing a norm (a distance function) under which all Cauchy sequences converged. As he wrote in the published version of 1922, "the aim of the present work is to establish certain theorems valid in different functional domains. . . . Nevertheless, in order not to have to prove them for each particular domain, which would be painful, I have chosen to take a different route; that is, I will consider in a general sense the sets of elements of which I will postulate certain properties. I will deduce from them certain theorems and then I will prove for each specific functional domain that the chosen postulates are true."[16] Thus, Banach began with a set of 13 axioms characterizing the notion of a vector space over the real numbers. And since Banach was interested in spaces of functions, his vector spaces were not limited to a finite dimension. Although Banach's axiom system contained more axioms than necessary, his paper had great influence. And by the time of the publication 10 years later of his *Théorie des opérateurs linéaires*, in which the axioms were repeated, the abstract notion of a vector space had become part of the mathematical vocabulary.

25.3.4 Algebras

The vector spaces of Steinitz and of Dedekind earlier were also fields and thus possessed a reasonable multiplication operation in addition to the additive one. A similar type of structure was that of a "linear associative algebra," first studied by the American mathematician Benjamin Peirce (1809–1880) around 1870. By a **linear associative algebra**, today just called an **algebra**, Peirce meant a finite-dimensional vector space over a field F (which Peirce limited to the field of real numbers) that possesses an associative multiplication operation that is distributive over addition. Peirce's chief aim in his work was to describe all possible algebras of dimensions 1 through 5 and some of dimension 6, by considering the possible multiplication tables for the basis elements. In the course of this work, however, which turned out to be incomplete, he introduced two important definitions: A nonzero element a of an algebra is **nilpotent** if some power a^n is zero, while it is **idempotent** if $a^2 = a$. Peirce was then able to demonstrate the

THEOREM *In every algebra there is at least one idempotent or at least one nilpotent element.*

The proof is not difficult. Because the algebra is finite-dimensional, any nonzero element A of the algebra must satisfy an equation of the form

$$\sum_{i=1}^{n} a_i A^i = 0$$

for some n. This equation can be rewritten in the form $BA + a_1 A = 0$ or $(B + a_1)A = 0$, where B is a linear combination of powers of A. It follows that $(B + a_1)A^k = 0$ for every $k > 0$ and therefore that $(B + a_1)B = 0$ or $B^2 + a_1 B = 0$. It is immediate from the last equation that if $a_1 \neq 0$, then

$$\left(-\frac{B}{a_1}\right)^2 = -\frac{B}{a_1}$$

and $-B/a_1$ is an idempotent. If $a_1 = 0$, then $B^2 = 0$ and B is a nilpotent element.

Several other mathematicians in the last quarter of the nineteenth century studied special algebras, in particular **simple** ones, those having no nontrivial two-sided ideals. (A **two-sided ideal** in an algebra is a subset I such that if α and β belong to I so does $\alpha + \beta$, $r\alpha$ and αr for any r in R. This notion generalizes the definition of Dedekind for ideals in sets of algebraic integers. Naturally, a two-sided ideal can be thought of as a subalgebra.)

Elie Cartan was able to show that every simple algebra over the complex numbers (with a multiplicative identity) was a matrix algebra, that is, was isomorphic to the algebra of $n \times n$ matrices with complex coefficients for some n. His work was generalized by Joseph Henry Maclagan Wedderburn (1882–1948) in a paper of 1907 entitled "On Hypercomplex Numbers," which contained a detailed study of the structure of algebras over any field. Among many other results, Wedderburn proved that any simple algebra is a matrix algebra, not necessarily over a field, but over a division algebra. (A **division algebra** is an algebra with multiplicative identity such that every nonzero element has a multiplicative inverse.)

To further classify algebras, it was necessary to classify division algebras. It had already been proved by Frobenius that over the field of real numbers there were only three division algebras: the real numbers, the complex numbers, and the quaternions. Wedderburn himself proved in 1909 that the only finite division algebras were the finite fields themselves, and these were well known. Division algebras over the field of p-adic numbers were classified by Helmut Hasse (1898–1979) in 1931 and over any algebraic number field by Hasse, Richard Brauer (1901–1977), and Emmy Noether (1882–1935) in 1932. Because these classifications involve many advanced concepts from algebraic number theory, including class field theory, they will not be discussed here.

25.3.5   Abstract Ring Theory

By the second decade of the twentieth century, many algebraic objects having two operations had been studied and classified. But it took time for a general definition of what we today call a ring to emerge. Hilbert in 1897 had defined a ring in a field of algebraic numbers to be the system of all polynomials in a set θ, μ, . . . of algebraic numbers with integer coefficients. Hensel, in 1913, defined a ring to be a domain satisfying all the axioms of a field except the axiom of "unrestricted and uniquely determined division." Fraenkel, a year later, defined a ring to be a set with two operations, addition and multiplication. Under addition, it was to satisfy the axioms of a commutative group, while under multiplication, it was to be associative, distributive with respect to addition, and contain an identity. This definition certainly permitted divisors of zero, so Fraenkel called those elements **regular** that were not divisors of zero. He then added two further axioms: Every regular element must have an inverse with respect to multiplication and for any two elements a, b in the ring, there were regular elements α and β such that $ab = \alpha ba$ and $ab = ba\beta$. Note that with these axioms, the set of ordinary integers does not fit the definition of a ring. Probably, in making his definition, Fraenkel had in mind the p-adic numbers, in which the units do form a multiplicative group.

Fraenkel used his definition of a ring to study factorization as well as ring extensions, in the manner of Steinitz. But it turned out that his definition, as well as the previous ones, were not adequate to generalize the notions of ideal and the process of factorization of ideals begun by Dedekind. Thus, it was Emmy Noether, in her groundbreaking paper of 1921, "Idealtheorie in Ringbereichen," who fully established the modern definition of both a ring and an ideal in a ring and used these "to translate the factorization theorems of the rational integer numbers and of the ideal in algebraic number fields into ideals of arbitrary integral domains and domains of general rings."[17] In particular, Noether gave the definition of a **ring** in use today: a set R with two operations, addition and multiplication, such that R is a commutative group under addition, such that multiplication is associative, and such that the distributive laws hold. In the 1921 paper, she focused on commutative rings with identity (rings whose multiplication is commutative and that possess a multiplicative identity) that also satisfy the condition that every ideal has a finite basis. In her honor, such rings are now called **Noetherian rings**. Noether showed that these rings satisfy the **ascending chain condition**, that every chain $I_1, I_2, \ldots, I_k, \ldots$ of ideals in the ring such that $I_k \subset I_{k+1}$ breaks off after a finite number of terms. She was then able to develop a decomposition theory for ideals analogous to Dedekind's prime factorization but applicable to rings more general than the rings of integers in algebraic number fields of her predecessor. The results were, however, somewhat weaker than unique prime factorization of ideals.

In a paper of 1926, Noether was also able to characterize those rings R for which the entire Dedekind theory of prime factorization of ideals holds by a set of axioms:

1. R satisfies the ascending chain condition.
2. The residue class ring R/A satisfies the descending chain condition for every non-zero ideal A. (The descending chain condition is the same as the ascending chain condition with the $\subset$ replaced by $\supset$.)
3. R has a multiplicative identity element.
4. R is an integral domain, that is, it has no zero divisors.
5. R is integrally closed in its quotient field. In other words, every element of the field that satisfies a polynomial equation over the ring is itself in the ring.

She then showed that if R satisfies these five axioms, its integral closure in a finite separable extension of its quotient field also satisfies them. In particular, because any **principal ideal domain** (integral domain in which every ideal is principal) satisfies these axioms, her result shows not only that all domains of integers in finite algebraic number fields possess unique factorization of ideals into primes, but also that the integrally closed subrings in finite algebraic extensions of the fields of algebraic functions in one variable have this property.

BIOGRAPHY

Emmy Noether (1882–1935)

Emmy Noether received the normal upbringing of an upper-middle-class German-Jewish girl, attending finishing school, studying the piano, and taking dance lessons. It was not until 1900, after further study of French and English, that she passed the Bavarian state examinations to qualify to teach in the schools. But about this same time her interest shifted from languages to mathematics, and she spent the next three years auditing mathematics courses at the University of Erlangen, where her father was a professor of mathematics. In fact, in her first semester she was one of only two women allowed even to audit courses. When in 1904 the University officially permitted women to register, she became a regular student and, four years later, received her doctorate with a dissertation on invariants of ternary biquadratic forms.

Noether remained at Erlangen for several more years, until in 1915 David Hilbert called her to Göttingen to assist him in his study of general relativity. Although as a woman she was not permitted to teach officially, or to receive a salary, Hilbert arranged for her to teach courses that were given under his name. He in fact argued unsuccessfully in the University Senate on her behalf: "I do not see that the sex of the candidate is an argument against her admission as *Privatdozent*. After all, the Senate is not a bathhouse."[18] It was not until after the changes in Germany at the end of World War I that she was able to receive an official position at the University and, after 1922, even a modest salary. During the next 10 years at Göttingen, her influence was felt the most, both in Germany and, through her visit in 1928–1929 to Moscow, in the Soviet Union as well. In 1932, she was the only woman mathematician invited to give a plenary lecture at the International Congress of Mathematicians in Zürich.

Her world, as well as the world of many of Germany's mathematicians, changed suddenly in early 1933 when the Nazis came to power. As a Jew, she lost her teaching position at Göttingen, and, along with many of her colleagues, was forced to take refuge abroad. A position was found for her at Bryn Mawr College, near Philadelphia, beginning in the fall of 1933, a location close enough to Princeton for her to participate regularly in activities at the Institute for Advanced Study. She died suddenly in April 1935 after seemingly successful surgery for removal of a tumor.

25.3.6   Algebraic Topology

As indicated in the opening of this chapter, Noether had great influence on her coworkers, especially in emphasizing the structural rather than the computational aspects of algebra (see Sidebar 25.2). One of Noether's suggestions, which was to create a whole new field of study, was inspired by lectures in topology given by Pavel Sergeevich Aleksandrov (1896–1982) in Göttingen in 1926 and 1927. As Aleksandrov put it in his address given in Noether's memory, "When in the course of our lectures she first became acquainted with a systematic construction of combinatorial topology, she immediately observed that it would be worthwhile to study directly the groups of algebraic complexes and cycles of a given polyhedron and the subgroup of the cycle group consisting of cycles homologous to zero; instead of the usual definition of Betti numbers and torsion coefficients, she suggested immediately defining the Betti group as the [quotient] group of the group of all cycles by the subgroup of cycles homologous to zero."[19] With Noether's remarks and the subsequent publications of Leopold Vietoris (1891–2002(!)) and Heinz Hopf (1894–1971), the subject of algebraic topology began in earnest. Vietoris in 1927 defined the **homology group** $H(A)$ of a complex A to be the quotient group of cycles modulo boundaries, as Noether recommended. About the same time, Hopf defined

SIDEBAR 25.2 *Women in Mathematics*

The attentive reader will have noticed that there are very few female mathematicians discussed in this text. The reason, of course, is that up until recently, very few women have participated in the discipline of mathematics. There were probably women whose names have been lost to history who made contributions to mathematics in ancient times, but in recorded history, both in Western and non-Western cultures, women have in general not been permitted an education that would allow them to achieve success in mathematics. A consideration of the biographies of those women who are included in this book shows that, for the most part, they had a close family member who was willing to teach them mathematics or who, at least, encouraged them to study the subject. Without such a supportive background, evidently, women could not enter the field. And even those who managed to achieve a reasonable knowledge of mathematics were often not able to participate in the mathematical community. Women simply were not supposed to engage in such intense intellectual activities.

Over the past several decades, however, the picture has been changing. Even though there are still significant obstacles for a woman to overcome, particularly the attitudes of those teaching them in school, it is now possible for women who want to be mathematicians to achieve that aim, even without a family member as a role model. In fact, in recent years approximately 20% of the new doctorates in mathematics granted in the United States have been granted to women. And women are gradually entering positions of influence in the mathematical community. The American Mathematical Society had its first female president, Julia Robinson, in the 1980s, and the Mathematical Association of America has had several recent female presidents as well as two female executive directors. The Association for Women in Mathematics has actively sought to increase opportunities for women. It has, for example, worked to find financial support for female graduate students and new doctorates and has sponsored lectures by prominent female mathematicians at major mathematics meetings. On the international level, there has been an increase in the number of female speakers at recent International Congresses, although that number is still absurdly low. In any case, it does appear that progress has been made in providing opportunities for women to enter the mathematics profession. When a history of mathematics text is written at the end of the twenty-first century, there will be far more female mathematicians discussed than in the current text.

several other Abelian groups, namely, the groups L^p, Z^p, R^p, and $\bar{R}^p$ generated by the p-simplexes, the p-cycles, the p-boundaries (those chains that were the boundary of some chain), and the p-boundary divisors (those chains for which a multiple was a boundary), respectively. Then for Hopf, the factor group $B_p = Z^p / \bar{R}^p$ was a free group (a group none of whose elements had a multiple equal to 0) whose rank (the number of basis elements) turned out to be the pth Betti number of the complex.

Matters progressed so quickly in this new field that just a year later Walther Mayer (1887–1948) published an axiom system for defining homology groups. Namely, Mayer was no longer concerned with the topological complexes themselves, but solely with the algebraic operations defined on them. Thus, a complex ring Σ was a collection of elements (complexes) $K^{(p)}$, to each of which was attached a dimension p. The p-dimensional elements formed a finitely generated free Abelian group K^p. For each p, a homomorphism $R_p : K^p \to K^{p-1}$ is defined such that $R_{p-1}(R_p(K^p)) = 0$. (R_p is called the pth boundary operator. Often, one just uses R, without subscripts, and then writes the last equation in the form $R^2 = 0$.) Given these axioms, Mayer defined the group of p-cycles C^p to be those elements K of K^p for which $R(K) = 0$ and the group of p boundaries to be $R(K^{p+1})$. Modifying Hopf's definition slightly, he defined the pth homology group of Σ to be the factor group $H_p(\Sigma) = C^p / R(K^{p+1})$.

The attachment of a group to certain topological concepts soon led to a similar group-theoretic study of other types of objects. For example, the set of differential forms defined on a manifold A with an appropriate addition can be considered as a complex of Abelian groups similar to the one defined by Mayer. By introducing an operator d (the exterior derivative), which takes k-dimensional forms to $(k + 1)$-dimensional forms, an operator that has the same property as the boundary operator, namely, $d^2 = 0$, it is possible to define in an analogous way both cycles and boundaries and then to define the **cohomology groups** of a manifold, generally written $H^k(A)$.

In the cases of both the homology groups and the cohomology groups of a manifold, the assignment of the groups to the space carries over to the assignment of functions between spaces and the corresponding groups. That is, if $f : A \rightarrow B$ is a continuous function between two manifolds A and B, considered as simplicial complexes, and if $H_k(A)$, $H_k(B)$ are the kth homology groups of A and B, respectively, then there is a well-defined group homomorphism $H_k(f) : H_k(A) \rightarrow H_k(B)$. In fact, $H_k(f)$ is defined on a k-chain by $H_k(f)(V_0 V_1 \ldots V_k) = f(V_0)f(V_1) \ldots f(V_k)$. One can prove that this assignment makes sense in terms of the homology groups, that is, that cycles are taken into cycles and boundaries into boundaries, and therefore that it defines a homomorphism of the appropriate quotient groups. A similar assignment can be made in the case of the cohomology groups when there is a differentiable function $f : A \rightarrow B$, although in that case the group homomorphism $H^k(f)$ is a mapping from $H^k(B)$ to $H^k(A)$.

25.3.7   The Structural Approach to Algebra

This study of functions between algebraic objects that preserve certain properties is a significant part of what is called the structural approach to algebra, the approach that Emmy Noether made central to algebra. This approach was clearly reflected in probably the most important algebra text of the first half of the twentieth century, *Modern Algebra*, by B. L. van der Waerden (1903–1996), first published in 1930. Recall that in Weber's algebra text of 1895, the central goal was the solving of equations. Van der Waerden's goal was entirely different— "to define diverse algebraic domains and to attempt to elucidate fully their structure."[20] Thus, van der Waerden began the book with a brief treatment of set theory and then, in subsequent chapters, discussed the ideas of groups, rings, fields, vector spaces, ideals, algebras, and so on, beginning in each case with a list of postulates characterizing the particular construct. For each type of algebraic domain, he studied such concepts as subdomains, homomorphisms between two domains, isomorphisms, direct products of domains, and residue classes. He also considered similar questions in each domain. Thus, one concept to which he returns repeatedly is the relationship between a given element of a domain and its "prime elements," the relationship we express in the case of integers as unique factorization into prime numbers. The role of "prime elements" is played, for example, by simple groups in group theory, prime ideals in rings, irreducible polynomials in rings of polynomials, and prime fields in the theory of fields. Another recurring concept is the relationship between a given domain and its subdomains or its extension domains. A third idea is whether a given domain is determined by the properties of a certain limited subset of the domain.

Although van der Waerden's text exemplified the structural approach to algebra, he never attempted to give a definition of this approach. The first American text to follow van der Waerden's approach, *A Survey of Modern Algebra*, by Garrett Birkhoff (1911–

1996) and Saunders Mac Lane (1909–2005), originally published in 1941, also did not try to give a definition of "structural approach," even though two years earlier, Mac Lane had characterized algebra as that branch of mathematics that investigates "the explicit structure of postulationally defined systems closed with respect to one or more rational operations."[21] On the other hand, Mac Lane, along with Samuel Eilenberg (1913–1998), did create one version of a definition of "mathematical structure" in their notion of a category in a paper of 1945, a notion that applied not only to algebraic structures but to structures elsewhere in mathematics. Generalizing in some sense the ideas of Klein's *Erlanger Programm*, they realized that it was always appropriate whenever a new collection of mathematical objects was defined to give a definition of mappings between these objects. Thus, a **category** $\mathcal{C}$ was defined to be a dual collection $\{A, \alpha\}$ consisting of "an aggregate of abstract elements A (for example, groups), called the **objects** of the category, and abstract elements α (for example, homomorphisms), called **mappings** of the category."[22] These mappings were subject to certain axioms, including the existence of an appropriate product mapping that satisfies the associativity property and of an identity mapping corresponding to each object A. Examples of categories besides that of groups and homomorphisms include topological spaces and continuous maps, sets and functions, and vector spaces and linear transformations.

Following their own dictum, Eilenberg and Mac Lane further introduced the concept of a functor, a function between categories. Namely, if $\mathcal{C} = \{A, \alpha\}$ and $\mathcal{D} = \{B, \beta\}$ are two categories, a (covariant) **functor** T from $\mathcal{C}$ to $\mathcal{D}$ is a pair of functions (both designated by the same letter T), an object function and a mapping function. The object function assigns to each A in $\mathcal{C}$ an object $T(A)$ in $\mathcal{D}$, while the mapping function assigns to each mapping $\alpha : A \to A'$ in $\mathcal{C}$ a mapping $T(\alpha) : T(A) \to T(A')$ in $\mathcal{D}$. This pair must further take identity mappings into identity mappings and must satisfy the condition $T(\alpha\alpha') = T(\alpha)T(\alpha')$ whenever the product $\alpha\alpha'$ exists in $\mathcal{C}$. (For a contravariant functor, the mapping function is reversed, that is, $T(\alpha) : T(A') \to T(A)$ and $T(\alpha\alpha') = T(\alpha')T(\alpha)$.) For example, homology is a covariant functor from the category of manifolds and continuous transformations to the category of Abelian groups and homomorphisms. And the assignment to each finite-dimensional vector space V of the vector space $T(V)$ of all real-valued linear functions on V induces a contravariant functor from the category of vector spaces and linear transformations to itself.

The study of categories and functors proved to be important in various recent developments in algebra as well as differential and algebraic geometry. But by 1963, even Mac Lane was no longer convinced that algebra was just the study of "structures," whether by the use of category theory or otherwise. As he wrote, "Portions of abstract algebra can, indeed, be construed as investigations of such structure theorems. But . . . the development of the subject is much more varied. Older questions, such as those about finite group theory, return to the center of interest with the development of new ideas and techniques; they cannot always simply be categorized as 'structure theory.' New types of algebraic systems arise from the application of algebra in geometry, topology, and analysis."[23] Thus, algebra, like other parts of mathematics, continues to develop and change.

25.3.8 Linear Programming

One area of algebra that is certainly not the study of structure was one that grew out of some real-world problems shortly before and during the Second World War. Linear programming deals with the problem of maximizing or minimizing a linear function $a_1x_1 + a_2x_2 + \cdots a_nx_n$ subject to constraints that are linear inequalities in the variables x_i. Strangely enough, although methods for solving systems of linear equations have been studied for over 2000 years, prior to the war virtually no attention had been paid to the study of systems of linear inequalities and even less to that of solutions that maximized a linear function, except for some initial steps by Fourier and a few others in the nineteenth century (Section 21.4.6).

The modern work on linear programming stems from two major sources, military and economic. Among the mathematicians who dealt with economic questions was the Russian Leonid V. Kantorovich (1912–1986). In 1938, Kantorovich, a professor of mathematics in Leningrad, was acting as a consultant for the Laboratory of the Plywood Trust for a very special maximization problem, a problem of distributing raw materials in order to maximize equipment productivity under certain restrictions. He noted that, mathematically, this was basically a problem of maximizing a linear function on a convex polyhedron. But the general methods of using calculus to compare the values of the function at the vertices failed here because the number of vertices, even in this rather simple problem, was enormous. Kantorovich then found that very many other economic problems had the same mathematical form. These included material cutting, the use of complex resources, distribution of orders to various suppliers, and the choosing of appropriate transport mechanisms for distribution, among others. Using his theoretical knowledge of functional analysis, he was, however, able to find an efficient method of solving such problems numerically. He then published the formulation of the basic mathematical form of the economic problem, a sketch of the solution method, and the first discussion of the economic sense of the solutions in 1939 in a book entitled *Mathematical Methods in the Organization and Planning of Production*. For this work and related work later on, Kantorovich won the 1975 Nobel Prize in Economics. Unfortunately, ideological concerns in 1939 and 1940 and then wartime conditions immediately afterward prevented Kantorovich's work from being known in the West or immediately followed up in the Soviet Union. Thus, the first widely known mathematical solutions to problems similar to those he explored were published in the United States.

It was the requirements of the Air Force staff during the Second World War that led to the consideration of linear programming problems in the United States. The specific problems they dealt with concerned such matters as deployment of particular units to particular theaters of war, scheduling of training for technical personnel, and supply and maintenance of equipment. It soon became clear that the efficient coordination of the various aspects of these problems required new mathematical techniques, which were only developed around 1947. The Air Force set up at that time a working group called Project SCOOP (Scientific Computation of Optimum Programs), among the principal members of which was George Dantzig (1914–2005). Dantzig recognized in these problems the same maximization question that Kantorovich had discovered. But he devised a new method of solution, the **simplex method**, for solving these linear programming problems.

The first step is to determine the feasible set of solutions, namely, the convex polyhedron in the appropriate dimensional space that contains all solutions to the set of linear inequalities. The next step, to move along the edges of this polyhedron from one vertex to another to maximize the linear function, was first rejected by Dantzig as inefficient, but was ultimately seen in fact to provide the most efficient way of determining the desired solution, which is always achieved at one of the vertices. Although simple linear programming problems could be solved by hand, the problems of most interest for applications all involved large numbers of variables and equations and thus required some sort of machine computation. Thus, the first test of the simplex method on a major problem was accomplished in the fall of 1947 on an early computer. Over the next several years, various computational techniques were worked out that enabled the newly developed computers to be employed in solving linear programming problems with several hundred variables and equations. In fact, the applications of linear programming have grown rapidly over the past decades in parallel with the increasing speed and computational power of the modern computer.

 25.4 ## THE STATISTICAL REVOLUTION

In general, all of the statistical procedures worked on in the nineteenth century were designed to show relationships in quantities already collected and tabulated. Today, statisticians routinely design experiments to test certain hypotheses. The methods for doing this were developed in the early twentieth century.

25.4.1 Ronald Fisher and Hypothesis Testing

Ronald Fisher (1890–1962), the chief statistician at a British agricultural experiment station, considered the question of how we prove or disprove the claim of the lady tasting tea who says she can taste the difference between tea made by first putting in the milk and tea made by adding the milk afterward. Or, to take a more serious example, he also considered how you tell whether a certain type of fertilizer will produce bigger crop yields. After all, we know from experience that the crop yields vary normally due to circumstances that we are often not entirely aware of. So what kinds of experiments can we devise?

To give an example of what Fisher did, we consider his methodology in answering the question about fertilizer. First, we need to design a reasonable experiment. So we take a field, divide it into two strips, and divide each strip into, say, 10 blocks. That is, we have 10 pairs of blocks, each at essentially the same place in the field, in order to try to remove as many other variables from the problem as possible. Then we randomly treat one block in each pair with fertilizer and leave the adjacent block untreated. After the growing season, we measure the yields. We thus have 10 pairs of numbers. Let us subtract the yield without fertilizer from the yield with fertilizer (so we probably get positive numbers). We then have a sample of 10 numbers. The question is whether this set is significantly different from a set we would get by taking a random sample from a set of numbers normally distributed with mean 0. The analysis then proceeds by calculating the so-called Student's t-statistic. This is so named because it was essentially developed by William Gosset (1876–1937) of the Guinness Brewery in Dublin, who wrote under the pseudonym "Student" in his publications on the analysis of experimental data. This statistic enabled reasonable results to be estimated with the use of only a small sample, the types of samples Gosset had to deal with at the brewery.

With the t-statistic now known in this case, we consult a table calculated from the curves of the t-statistic, which are nearly (but not quite) normal curves. (The larger the sample size, the more closely the t-curves approach the normal curve.) We want to know if our calculated t-statistic is sufficiently improbable under the assumption that it comes from a normally distributed set with mean 0. Sufficiently improbable is something we have to decide on in advance, but today that usually means with probability less than 0.05 or 0.01. If our calculated value meets this standard, we say that we reject the null hypothesis that there is no significant difference in the yields of the two types of plots. We therefore conclude that the fertilizer is effective.

The case of the lady tasting tea is a bit simpler. Fisher proposed the following test. The lady will be given eight cups of tea in a random order, four of which are made in one way and four in the other. The lady's task is to divide the eight cups into the correct two sets of four. Fisher noted that there are $\binom{8}{4} = 70$ ways of selecting a set of four out of a set of eight. A person who could not discriminate between the two processes of making tea would have only a 1 in 70 (approximately 0.014) chance of picking the correct set. Thus, if the null hypothesis is that the lady cannot discriminate, then if she does pick the correct set, it is sufficiently improbable (probability less than 0.05) that she has done this by accident that we reject the null hypothesis. Note that if the lady picks three cups correctly, the chances of this happening without the lady having the claimed discrimination are 16 out of 70, or approximately 0.23. That number is sufficiently large that we would not reject the null hypothesis in that case. Fisher noted that there are other ways of conducting this test, particularly if the lady claims not that she can always distinguish, but that she can do it more often than not. For example, if there are twelve cups, then the probability is again less than 0.05 that, without any discriminating ability, the lady could pick either five or six cups correctly.

Fisher laid out, in his two major books, *Statistical Methods for Research Workers* (1925) and *The Design of Experiments* (1935), careful instructions on how to deal with all the major parts of the design and analysis of a statistical experiment. As we noted in the two examples, there are three major parts of this analysis. First, we need a null hypothesis, basically a statement that there is no difference in two situations, say. This hypothesis must allow us to specify a unique distribution function for the test statistic. Then we have to work out the observations and order them somehow to show their relative deviation from the null hypothesis. How we do this must be determined by experience. In fact, the entire choice of a null hypothesis and observations made to determine whether to accept it is an art and cannot be reduced to a mechanical process. As Fisher wrote, "It is, I believe, nothing but an illusion to think that this process can ever be reduced to a self-contained mathematical theory of tests of significance. Constructive imagination, together with much knowledge based on experience of data of the same kind, must be exercised before deciding on what hypotheses are worth testing, and in what respects. Only when this fundamental thinking has been accomplished can the problem be given a mathematical form."[24] Finally, we need a measure of how far these observations differ from what is expected under the null hypothesis. This measure is usually expressed as a probability that this particular observation could occur, given the null hypothesis. In general, we reject the null hypothesis when this probability is less than a certain predetermined value.

Interestingly, Fisher himself realized that rejecting the null hypothesis does not necessarily prove the efficacy of the particular cause in question. One generally cannot do that on the basis of one experiment, no matter how well designed. But it does give indications that need to be confirmed by repeated experiments. As he wrote, "No isolated experiment, however significant in itself, can suffice for the experimental demonstration of any natural phenomenon. In relation to the test of significance, we may say that a phenomenon is experimentally demonstrable when we know how to conduct an experiment which will rarely fail to give us a statistically significant result."[25]

25.4.2 Alternatives to Fisher's Methods

An alternative to Fisher's methods was worked out by Egon Pearson (1895–1980) and Jerzy Neyman (1894–1981) during the late 1920s and early 1930s. Egon Pearson was the son of Karl Pearson and worked at his father's laboratory at University College, London. Neyman originally worked at the University of Warsaw before coming to London and later moving to the United States. What they decided was that looking at a single null hypothesis could not give a significant answer. They wanted a statistical test to provide a choice between alternatives. And to give a reasonable answer, they wanted a test that rarely led to error.

As they understood Fisher, there was only one kind of statistical error—rejecting the null hypothesis when it is in fact true. This they called an error of the first kind. But they also wanted to consider an error of the second kind, when one accepts a hypothesis that is false. In their basic use of their method, they considered two hypotheses, say, H_1 and H_2. The assumption is that one of these hypotheses is true. Then one decides on a so-called critical region R and conducts observations to determine whether the statistic calculated from these falls in R or not. That is, if the statistic is in R, one rejects H_1 and accepts H_2. If the statistic lies outside R, one accepts H_1 and rejects H_2. Then the probability $P(R|H_1)$ is the probability that one will reject H_1 when it is true. That is the probability of an error of the first kind, and the idea is to make this value small, say, less than 0.05 or 0.01. The probability $1 - P(R|H_2)$ is then the probability of rejecting H_2—and therefore accepting H_1—when H_1 is false. This is an error of the second kind. The value $P(R|H_2)$ is called the power of the test, and we want that value as close to 1 as possible. The relevant questions then are how to choose the alternative hypothesis H_2 and how to choose a critical region. These often involve the choice of a sample size.

For example, in the case of the lady tasting tea, Fisher's null hypothesis, which we will call H_1, is that the probability p of the lady being able to tell the type of a given cup of tea is equal to $\frac{1}{2}$. This is equivalent to the statement that the lady cannot discriminate. Neyman and Pearson would then assert an alternative hypothesis H_2, that the lady can discriminate, or that $p > \frac{1}{2}$. In fact, they would also conduct the test differently. In their analysis of the situation, they would give the lady n pairs of cups of tea, with one cup in each pair prepared with tea first and the other with milk first. We will then agree with the lady's claim if the number of pairs correctly identified is at least as great as a given value X_0 specified in advance. Suppose we say that we will agree with the lady's claim if she makes no more than two errors in identifying 10 pairs of cups of tea. That is, the statistic (the number X of correct pairs) will be in our critical region R if it is equal to 8, 9, or 10. If the statistic is in R, then we reject

H_1 and accept H_2. We first calculate the probability of an error of the first kind, that is, of rejecting H_1 when it is true. This value is calculated to be

$$P\left(X = 8 \mid p = \frac{1}{2}\right) + P\left(X = 9 \mid p = \frac{1}{2}\right) + P\left(X = 10 \mid p = \frac{1}{2}\right)$$

$$= \binom{10}{8}\left(\frac{1}{2}\right)^{10} + \binom{10}{9}\left(\frac{1}{2}\right)^{10} + \binom{10}{10}\left(\frac{1}{2}\right)^{10}$$

$$= (45 + 10 + 1)\frac{1}{2^{10}} = 0.054688.$$

This probability is not less than 0.05, but it is close. In this case, we may want to accept the result anyway, because of our calculation of an error of the second kind.

To calculate the error of the second kind, we calculate the power $P(R \mid H_2)$ as a function of the actual probability p of the lady correctly identifying the tea preparation. In the case of the given critical region, this function is

$$\binom{10}{8}p^8(1 - p)^2 + \binom{10}{9}p^9(1 - p) + p^{10}.$$

To then determine the power for a given p, it is best to graph this function (Fig. 25.6). We see, for example, that if in fact $p = 0.9$, the power is approximately 0.96. In other words, the probability of rejecting H_2 if it is true is about 0.04. That is, if the lady does not identify at least 8 pairs correctly, and we therefore accept H_1, the probability of the lady actually having the stated discriminatory ability is only 0.04. Since, in this situation, the probabilities of both errors are relatively small, Neyman and Pearson would accept the validity of this hypothesis test. Of course, it is important to note that the calculation of the power depends on p and thus we must be very clear as to exactly what we are testing. There is rarely a straightforward way to design an experiment with small errors of both kinds.

FIGURE 25.6

Graph of $\binom{10}{8} p^8(1 - p)^2 + \binom{10}{9} p^9(1 - p) + p^{10}$

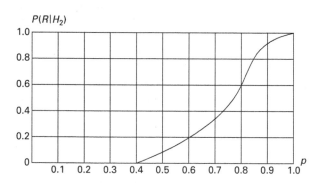

What happened in the 1930s was that neither Fisher nor Neyman really understood the other. There were polemics in which each accused the other of reaching false conclusions using their methods. Part of the problem seems to have been that the differing approaches were applicable to different types of statistical questions. So sometimes one method appeared to work better than the other, but the two sides were reluctant to recognize that. But what happened later is that the textbooks over the past 50 years or so merged the two approaches and did not even try to discuss the differences. In fact, most current textbooks ask that a researcher (1) specify the level of significance before conducting the experiment, which was what Neyman and Pearson required; (2) must not draw conclusions from a nonsignificant result, which follows Fisher; and (3) does include the errors of the first and second kinds. But if one does meet the specified significance level, an experimenter will make a definite conclusion. And Fisher himself warned that one experiment does not allow one to make a conclusion. He even suggested that nonsignificant results should be published as well as significant ones, so that the literature fairly reflects what has been happening. But this has turned out to be impossible, in particular because of the decisions of journal editors.

A major impetus for the development of better statistical procedures sprang from the requirements of the Second World War. Statisticians were brought in to help make predictions, thus providing a basis for action, in such areas as quality control, personnel selection, gunnery and bombing, and weather forecasting. Among the new techniques developed was the idea of sequential analysis. Instead of choosing n first to decide between two alternatives, as Neyman and Pearson suggested, the statistician can test sequentially by deciding at various junctures to make more observations. Thus, the sample size is not fixed in advance. The major developer of this theory, couched in terms of statistical decision functions, was Abraham Wald (1902–1950), who had escaped from Austria in 1938. Among others who worked in this area was David Blackwell (1919–), the first African American elected to the National Academy of Sciences. In fact, he systematized the entire field of decision theory in his textbook, *Theory of Games and Statistical Decisions*, of 1954.

By the 1950s, it was becoming clear that statistical computations were increasingly to be performed on computers. And once this process was well under way, it became possible to analyze larger and larger data sets. Fisher had worked on agricultural problems and developed his methods based on the relatively small data sets that could be dealt with by hand. But with computerization, the opportunity presented itself to apply statistical methods to the huge data sets being generated in biological and biomedical research. One example of such a data set was in the genome project of the late 1990s, which succeeded in mapping the human genome. In the future, it may well be possible to use the results of this analysis to develop personalized cures for various diseases in individuals.

But even though the use of computers in statistics requires the use of algorithms to perform the analyses, it is still true that, as both Fisher and Neyman suggested, there is always a need for judgment. One cannot write down mechanical rules and then come to definite conclusions by following them. To infer the truth of a statement via statistics is something that can only be done through a combination of experiments and by ones repeated as well as possible over a long period of time. Nevertheless, the theoretical advances made by the developers of the statistical methods have allowed us, using our best judgment, to come to conclusions that we have every reason to believe are correct.

 25.5 COMPUTERS AND APPLICATIONS

When a nontechnically educated person thinks of mathematics in the early twenty-first century, the most obvious aspect of the subject that comes to mind is the use of computers.

Mathematicians themselves are only gradually accepting the entrance of machine computation into their subject. For many mathematicians, pencil and paper are still the most important tools. Yet the rapid advances in computing power since the 1950s have brought the computer into the mainstream of mathematics, and a growing number of mathematicians now make use of it not only in generating examples but also in constructing proofs. In fact, as we have seen, the growth in statistics and the development of linear programming are heavily dependent on the computer. In addition, many aspects of theoretical mathematics that had lain dormant for years have received renewed attention because of their applications to the general field of computer science. Although there is not space here to give a detailed history of the development of the computer, we conclude this chapter with a sketch of the most important aspects of that history.

25.5.1   The Prehistory of Computers

The dream of mechanical calculation must have occurred as far back as the early years of our era when Ptolemy was probably forced to use a large number of human "computers" to generate the various tables that appear in his *Almagest*. Some Islamic scientists in the middle ages in fact did use certain instruments to help in their own calculations, particularly in those related to astronomy. The calculation of astronomical tables was important in Europe as well by the early seventeenth century, and logarithms were invented in part to help in this regard. In short order two Englishmen, Richard Delamain (first half of the seventeenth century) and William Oughtred, independently created a physical version of a logarithm table in the form of a slide rule, a circular (later rectilinear) arrangement of movable numerical scales that enabled multiplications and divisions, as well as computations involving trigonometric functions, to be easily performed.

About the same time, Wilhelm Schickard (1592–1635), a professor of astronomy and mathematics at the University of Tübingen, designed and built a machine that performed addition and subtraction automatically, as well as multiplication and division semiautomatically (Fig. 25.7). Schickard described the machine in letters to Kepler in 1623 and 1624, but the machine he intended to build for Kepler's own use was destroyed in a fire before it was completed. The remaining copies of the machine, as well as the designer, perished in the Thirty Years War, so Schickard's device had no influence on later work. Some 20 years later, Pascal constructed a mechanical adding and subtracting machine (Fig. 25.8), while in 1671 Leibniz constructed a machine that also did multiplication and division. Leibniz was quite sure that his machine would be of great practical use:

FIGURE 25.7

Schickard's computing machine

> We may say that it will be desirable to all who are engaged in computations which, it is well known, are the managers of financial affairs, the administrators of others' estates, merchants, surveyors, geographers, navigators, astronomers, and [those connected with] any of the crafts that use mathematics. But limiting ourselves to scientific uses, the old geometric and astronomic tables could be corrected and new ones constructed by the help of which we could measure all kinds of curves and figures. . . . Furthermore, . . . it will be easy for anyone to construct tables for himself so that he may conduct his investigations with little toil and with great accuracy. . . . Also, the astronomers surely will not have to continue to exercise the patience which is required for computation. . . . For it is unworthy of excellent men to lose hours like slaves in the labor of calculation, which could be safely relegated to anyone else if the machine were used.[26]

FIGURE 25.8

A replica of Pascal's mechanical calculating device (*Source:* Neuhart Donges Neuhart Designers, Inc.)

FIGURE 25.9

Babbage and his "computer" honored on a British stamp

25.5.2 Babbage's Difference Engine and Analytical Engine

Unfortunately, neither Leibniz's machine nor the various improved models built by others during the following century and a half were actually used to any extent in the way Leibniz envisaged. The mathematical practitioners themselves continued to do calculations by hand, probably because the machines, operated manually, provided little advantage in speed. For complicated calculations, naturally, tables were used, particularly of logarithms and trigonometric functions, even though these tables, originally calculated by hand, frequently contained errors. It was not until the industrial revolution was in full swing in England and the steam engine had been invented that another brilliant mind, that of Charles Babbage, conceived around 1821 the idea of using this new technology to drive a machine that would increase the speed as well as the accuracy of numerical computation (Fig. 25.9).

Babbage realized that the calculation of the values of a polynomial function of degree n could be effected by using the fact that the nth-order differences were always constant. To take a simple example, consider the following short table for the function $f(x) = x^2$:

x	$f(x)$	First Difference	Second Difference
1	1		
2	4	3	
3	9	5	2
4	16	7	2
5	25	9	2
6	36	11	2

Note that in this case the second differences—that is, the differences of the first differences of the values—are all 2. Thus, to calculate the values of $f(x)$, it is only necessary to perform additions, working backward from the second difference column to the first difference column to the desired tabular values. (Naturally, one must begin with certain given values, say, $2^2 = 4$ and the initial first difference, 3.) This idea was the principle behind Babbage's original machine, his Difference Engine (Fig. 25.10). The plans for the machine called for seven axes, representing the tabular values and the first six differences, each axis containing wheels that could be set to represent numbers of up to 20-decimal digits. The axes would be interconnected so that the constant set up in one of the difference axes would add to the number set up in the next lower difference axis and so on, until the tabular axis was reached. By repeating the process continually, the desired tabular values for polynomial functions of degree up to six could be calculated for as many values of the variable as desired. Babbage realized too that any continuous function could be approximated in an appropriate interval by a polynomial and therefore that the machine could be used to calculate tables for virtually any function of interest to scientists of the day. His aim, in fact, was to attach the machine to a device for making printing plates so that the tables could be printed without any new source of error being introduced. Unfortunately, although Babbage succeeded in convincing the British government to provide him with a grant to help in the building of the Difference Engine, a complete model was never constructed because of various difficulties in developing machine parts of sufficient accuracy, because ultimately the government lost interest in the project, and because Babbage himself became interested in a new project, the development of a general-purpose calculating machine, his Analytical Engine.

FIGURE 25.10

A modern model of Babbage's Difference Engine (*Source*: Neuhart Donges Neuharta Designers, Inc.)

Babbage began his new project in 1833 and had elaborated the basic design by 1838. His new machine contained many of the features of today's computers. Constructed again of numerous toothed wheels on axes as hardware, it was to consist of two basic parts, the **store** and the **mill**. The store was the section in which numerical variables were kept until they were to be processed and where the results of the operations were held, while the mill was the section in which the various operations were performed. To control the operations, Babbage took an idea of Joseph Jacquard (1752–1834), who had automated the weaving industry in France through the introduction of punched cards describing the desired pattern for the loom (Fig. 25.11). Babbage thus devised his own system of punched cards that were to contain both the numerical values and the instructions for the machine. Although Babbage never wrote out a complete description of his Analytical Engine and, in fact, never had the

FIGURE 25.11

Joseph Jacquard on a French stamp

financial resources to actually construct it, he did leave for posterity some 300 sheets of engineering drawings, each about 2 by 3 feet, and many thousands of pages of detailed notes on his ideas. Modern scholars have concluded by examining these papers that the technology of the time was probably sufficient to construct the engine, but because there was insufficient interest in the British government to finance such a massive project, the engine remained only a theoretical construct.[27]

In 1840, Babbage gave a series of seminars on the workings of the Analytical Engine to a group of Italian scientists assembled in Turin, one of whom summarized the seminars in a published article. The 17-page article was translated into English in 1843 and supplemented by an additional 40 pages of notes by Ada Byron King, Countess of Lovelace (1815–1852). In her notes, Lovelace not only expanded on various parts of the article about the detailed functioning of the engine but also gave explicit descriptions of how it would solve specific problems. Thus, she described, for the first time in print, what would today be called a computer program, in her case a program for computing the Bernoulli numbers. She began from a description of the Bernoulli numbers as the coefficients B_i in the expansion

$$\frac{x}{e^x - 1} = 1 - \frac{x}{2} + B_2 \frac{x^2}{2} + B_4 \frac{x^4}{4!} + B_6 \frac{x^6}{6!} + \cdots .$$

(See Chapter 18, Exercise 3.) By a bit of algebraic manipulation using the power series expansion for e^x, Lovelace derived the equations

$$0 = -\frac{1}{2} \frac{2n-1}{2n+1} + B_2 \left(\frac{2n}{2!} \right) + B_4 \left(\frac{2n(2n-1)(2n-2)}{4!} \right)$$
$$+ B_6 \left(\frac{2n(2n-1) \cdots (2n-4)}{6!} \right) + \cdots + B_{2n},$$

BIOGRAPHY

Ada Byron King Lovelace (1815–1852)

Augusta Ada Byron King Lovelace was the child of George Gordon, the sixth Lord Byron, who left England five weeks after his daughter's birth and never saw her again. She was raised by her mother, Anna Isabella Millbanke, a student of mathematics herself, so she received considerably more mathematics education than was usual for girls of her time. Although she never attended any university, she was tutored privately by, and was able to consult with, well-known mathematicians, including William Frend and Augustus De Morgan. In 1833 she met Charles Babbage and soon became interested in his Difference Engine. Her husband, the Earl of Lovelace, was made a Fellow of the Royal Society in 1840 and through this connection, Ada was able to gain access to the books and papers she needed to continue her mathematical studies. Her major mathematical work, discussed in the text, is a heavily annotated translation of a paper by the Italian mathematician L. F. Menabrea dealing with Babbage's Analytical Engine. Interestingly, the paper was published using only her initials, A.A.L. It was evidently not considered proper in mid-nineteenth-century England for a woman of her class to publish a mathematical work.

from which the various B_i can be calculated recursively. Thus, to calculate B_{2n}, one needs three numerical values, 1, 2, n, as well as the values B_i for $i < 2n$, values that presumably have already been calculated. Instruction cards then are needed to multiply n by 2, subtract 1 from that result, add 1 to that result, divide the two last results, multiply the result by $-\frac{1}{2}$, divide $2n$ by 2, multiply the result by B_2, and so on. The results of certain of these calculations, such as $2n - 1$, are used several times during the calculations and therefore need to be moved to various registers where the calculations will take place. At certain stages in the calculation, the machine is instructed to subtract an integer from $2n$ and then decide on the next step, depending on whether the result is positive or 0. If it is 0, the equation for B_{2n} is complete and the machine easily solves it; if it is positive, the machine repeats many of the preceding steps. It is not difficult to see that some of the basic concepts of modern-day programming, including loops and decision steps, are included in Lovelace's description. Furthermore, she had printed with her notes a detailed diagram of the above program, perhaps the first "flowchart" ever constructed (Fig. 25.12).

FIGURE 25.12

Ada Lovelace's flowchart for calculating Bernoulli numbers

Number of Operation	Nature of Operation	Variables Acted Upon	Variables Receiving Results	Indication of change in the Value on any Variable	Statement of Results	Data: $^1V_1$ 0001 [1]	Data: $^1V_2$ 0002 [2]	Data: $^1V_3$ 0004 [n]	Working Variables: $^0V_4$ 0000	$^0V_5$ 0000	$^0V_6$ 0000	$^0V_7$ 0000	$^0V_8$ 0000	$^0V_9$ 0000	$^0V_{10}$ 0000	$^0V_{11}$ 0000	$^0V_{12}$ 0000	$^0V_{13}$ 0000	Result Variables: $^1V_{21}$ B_1 in a decimal fraction [B_1]	$^1V_{22}$ B_3 in a decimal fraction [B_3]	$^1V_{23}$ B_5 in a decimal fraction [B_5]	$^0V_{24}$ 0000 [B_7]
1	×	$^1V_2 \times {}^1V_3$	$^1V_4, {}^1V_5, {}^1V_6$	$\left\{\begin{matrix}^1V_2 = {}^1V_2 \\ ^1V_3 = {}^1V_3\end{matrix}\right\}$	$= 2n$	—	2	n	$2n$	$2n$	$2n$											
2	−	$^1V_4 - {}^1V_1$	$^2V_4$	$\left\{\begin{matrix}^1V_4 = {}^2V_4 \\ ^1V_1 = {}^1V_1\end{matrix}\right\}$	$= 2n - 1$	1	—	—	$2n - 1$													
3	+	$^1V_5 + {}^1V_1$	$^2V_5$	$\left\{\begin{matrix}^1V_5 = {}^2V_5 \\ ^1V_1 = {}^1V_1\end{matrix}\right\}$	$= 2n + 1$	1	—	—	—	$2n + 1$												
4	÷	$^2V_5 \div {}^2V_4$	$^1V_{11}$	$\left\{\begin{matrix}^2V_5 = {}^0V_5 \\ ^2V_4 = {}^0V_4\end{matrix}\right\}$	$= \frac{2n-1}{2n+1}$	—	—	—	0	0	—	—	—	—	—	$\frac{2n-1}{2n+1}$						
5	÷	$^1V_{11} \div {}^1V_2$	$^2V_{11}$	$\left\{\begin{matrix}^1V_{11} = {}^2V_{11} \\ ^1V_2 = {}^1V_2\end{matrix}\right\}$	$= \frac{1}{2} \cdot \frac{2n-1}{2n+1}$	—	2	—	—	—	—	—	—	—	—	$\frac{1}{2} \cdot \frac{2n-1}{2n+1}$						
6	−	$^0V_{13} - {}^2V_{11}$	$^1V_{13}$	$\left\{\begin{matrix}^2V_{11} = {}^0V_{11} \\ ^0V_{13} = {}^1V_{13}\end{matrix}\right\}$	$= -\frac{1}{2} \cdot \frac{2n-1}{2n+1} = A_0$	—	—	—	—	—	—	—	—	—	—	0	—	$-\frac{1}{2} \cdot \frac{2n-1}{2n+1} = A_0$				
7	−	$^1V_3 - {}^1V_1$	$^1V_{10}$	$\left\{\begin{matrix}^1V_3 = {}^1V_3 \\ ^1V_1 = {}^1V_1\end{matrix}\right\}$	$= n - 1 (= 3)$	1	—	n	—	—	—	—	—	—	$n - 1$							
8	+	$^1V_2 + {}^0V_7$	$^1V_7$	$\left\{\begin{matrix}^1V_2 = {}^1V_2 \\ ^0V_7 = {}^1V_7\end{matrix}\right\}$	$= 2 + 0 = 2$	—	2	—	—	—	—	2										
9	÷	$^1V_6 \div {}^1V_7$	$^3V_{11}$	$\left\{\begin{matrix}^1V_6 = {}^1V_6 \\ ^0V_{11} = {}^3V_{11}\end{matrix}\right\}$	$= \frac{2n}{2} = A_1$	—	—	—	—	—	$2n$	2	—	—	—	$\frac{2n}{2} = A_1$						
10	×	$^1V_{21} \times {}^3V_{11}$	$^1V_{12}$	$\left\{\begin{matrix}^1V_{21} = {}^1V_{21} \\ ^3V_{11} = {}^3V_{11}\end{matrix}\right\}$	$= B_1 \cdot \frac{2n}{2} = B_1A_1$	—	—	—	—	—	—	—	—	—	—	$\frac{2n}{2} = A_1$	$B_1 \cdot \frac{2n}{2} = B_1A_1$	—	B_1			
11	+	$^1V_{12} + {}^1V_{13}$	$^2V_{13}$	$\left\{\begin{matrix}^1V_{12} = {}^0V_{12} \\ ^1V_{13} = {}^2V_{13}\end{matrix}\right\}$	$= -\frac{1}{2} \cdot \frac{2n-1}{2n+1} + B_1 \cdot \frac{2n}{2}$	—	—	—	—	—	—	—	—	—	—	—	0	$\left(-\frac{1}{2} \cdot \frac{2n-1}{2n+1} + B_1 \cdot \frac{2n}{2}\right)$				
12	−	$^1V_{10} - {}^1V_1$	$^2V_{10}$	$\left\{\begin{matrix}^1V_{10} = {}^2V_{10} \\ ^1V_1 = {}^1V_1\end{matrix}\right\}$	$= n - 2 (= 2)$	1	—	—	—	—	—	—	—	—	$n - 2$							
13	−	$^1V_6 - {}^1V_1$	$^2V_6$	$\left\{\begin{matrix}^1V_6 = {}^2V_6 \\ ^1V_1 = {}^1V_1\end{matrix}\right\}$	$= 2n - 1$	1	—	—	—	—	$2n - 1$											
14	+	$^1V_1 + {}^1V_7$	$^2V_7$	$\left\{\begin{matrix}^1V_1 = {}^1V_1 \\ ^1V_7 = {}^2V_7\end{matrix}\right\}$	$= 2 + 1 = 3$	1	—	—	—	—	—	3										
15	÷	$^2V_6 \div {}^2V_7$	$^1V_8$	$\left\{\begin{matrix}^2V_6 = {}^2V_6 \\ ^2V_7 = {}^2V_7\end{matrix}\right\}$	$= \frac{2n-1}{3}$	—	—	—	—	—	$2n - 1$	3	$\frac{2n-1}{3}$									
16	×	$^1V_8 \times {}^3V_{11}$	$^4V_{11}$	$\left\{\begin{matrix}^1V_8 = {}^0V_8 \\ ^3V_{11} = {}^4V_{11}\end{matrix}\right\}$	$= \frac{2n}{2} \cdot \frac{2n-1}{3}$	—	—	—	—	—	—	—	0	—	—	$\frac{2n}{2} \cdot \frac{2n-1}{3}$						
17	−	$^2V_6 - {}^1V_1$	$^3V_6$	$\left\{\begin{matrix}^2V_6 = {}^3V_6 \\ ^1V_1 = {}^1V_1\end{matrix}\right\}$	$= 2n - 2$	1	—	—	—	—	$2n - 2$											
18	+	$^1V_1 + {}^2V_7$	$^3V_7$	$\left\{\begin{matrix}^2V_7 = {}^3V_7 \\ ^1V_1 = {}^1V_1\end{matrix}\right\}$	$= 3 + 1 = 4$	1	—	—	—	—	—	4										
19	÷	$^3V_6 \div {}^3V_7$	$^1V_9$	$\left\{\begin{matrix}^3V_6 = {}^3V_6 \\ ^3V_7 = {}^3V_7\end{matrix}\right\}$	$= \frac{2n-2}{4}$	—	—	—	—	—	$2n - 2$	4	—	$\frac{2n-2}{4}$	—							
20	×	$^1V_9 \times {}^4V_{11}$	$^5V_{11}$	$\left\{\begin{matrix}^1V_9 = {}^0V_9 \\ ^4V_{11} = {}^5V_{11}\end{matrix}\right\}$	$= \frac{2n}{2} \cdot \frac{2n-1}{3} \cdot \frac{2n-2}{4} = A_3$	—	—	—	—	—	—	—	—	0		$\left\{\frac{2n}{2} \cdot \frac{2n-1}{3} \cdot \frac{2n-2}{4} = A_3\right\}$						
21	×	$^1V_{22} \times {}^5V_{11}$	$^2V_{12}$	$\left\{\begin{matrix}^1V_{22} = {}^1V_{22} \\ ^0V_{12} = {}^2V_{12}\end{matrix}\right\}$	$= B_3 \cdot \frac{2n}{2} \cdot \frac{2n-1}{3} \cdot \frac{2n-2}{4} = B_3A_3$	—	—	—	—	—	—	—	—	—	—	0	B_3A_3	—	—	B_3		
22	+	$^2V_{12} + {}^2V_{13}$	$^3V_{13}$	$\left\{\begin{matrix}^2V_{12} = {}^0V_{12} \\ ^2V_{13} = {}^3V_{13}\end{matrix}\right\}$	$= A_0 + B_1A_1 + B_3A_3$	—	—	—	—	—	—	—	—	—	—	—	0	$\{A_0 + B_1A_1 + B_3A_3\}$				
23	−	$^2V_{10} - {}^1V_1$	$^3V_{10}$	$\left\{\begin{matrix}^2V_{10} = {}^3V_{10} \\ ^1V_1 = {}^1V_1\end{matrix}\right\}$	$= n - 3 (= 1)$	1	—	—	—	—	—	—	—	—	$n - 3$							
					Here follows a repetition of Operations thirteen to twenty-three																	
24	+	$^4V_{13} + {}^0V_{24}$	$^1V_{24}$	$\left\{\begin{matrix}^4V_{13} = {}^0V_{13} \\ ^0V_{24} = {}^1V_{24}\end{matrix}\right\}$	$= B_7$	—	—	—	—	—	—	—	—	—	—	—	—	—	—	—	—	B_7
25	+	$^1V_1 + {}^1V_3$	$^1V_3$	$\left\{\begin{matrix}^1V_1 = {}^1V_1 \\ ^1V_3 = {}^1V_3 \\ ^5V_6 = {}^0V_6 \\ ^5V_7 = {}^0V_7\end{matrix}\right\}$	$= n + 1 = 4 + 1 = 5$; by a Variable-card; by a Variable-card	1	—	$n + 1$	—	—	0	0										

Besides discussing the basic functioning of the analytical engine, Lovelace described what kinds of jobs it could do and noted explicitly that it could perform symbolic algebraic operations as well as arithmetic ones. But, she noted,

> the Analytical Engine has no pretensions whatever to *originate* anything. It can do whatever we *know how to order it* to perform. It can *follow* analysis; but it has no power of *anticipating* any analytical relations or truths. Its province is to assist us in making *available* what we are already acquainted with. . . . But it is likely to exert an *indirect* and reciprocal influence on science itself in another manner. For, in so distributing and combining the truths and the formulas of analysis, that they may become most easily and rapidly amenable to the mechanical combinations of the engine, the relations and the nature of many subjects in that science are necessarily thrown into new lights, and more profoundly investigated. . . . It is however pretty evident, on general principles, that in devising for mathematical truths a new form in which to record and throw themselves out for actual use, views are likely to be induced, which should again react on the more theoretical phase of the subject.[28]

A better description of the computer's limitations and its implications for the development of mathematics could hardly be written today.

25.5.3   Turing and Computability

One of the reasons that Babbage's ideas were not brought to fruition with an actual Analytical Engine was that even in mid-nineteenth-century England, there was no perceived societal need for it that made it worth the enormous resources that would have been necessary for its construction. And although various computational devices and analog computers were devised in the century after Babbage's design work—devices generally adapted to solving specific mathematical problems that otherwise would require enormous amounts of manual computation—it was military necessity during the two world wars, especially the second, that led to the actual construction of the first electronic computers, in many essentials based on Babbage's ideas. Still, there were other theoretical ideas worked out in the years immediately before the Second World War that were to be fundamental in the development of these computers. One of these was the idea of computability in the work of Alan Turing (1912–1954).

Turing was interested in determining a reasonable but precise answer to the questions of what a computation is and whether a given computation can in fact be carried out. To answer these questions, Turing extracted from the ordinary process of computation the essential parts and formulated these in terms of a theoretical machine, now known as a **Turing machine**. Furthermore, he showed that there is a "universal" Turing machine, a machine that can calculate any number or function that can be calculated by any special machine, provided it is given the appropriate instructions.

Turing's machine, presented in a major paper of 1936, was formed from three basic concepts: a finite set of states, or configurations, $\{q_1, q_2, \ldots, q_k\}$; a finite set of symbols, $\{a_0, a_1, a_2, \ldots, a_n\}$, which are to be read and/or written by the machine (where a_0 is taken as a blank symbol); and a process of changing both the states and the symbols to be read. To accomplish its job, the machine is supplied with instructions by means of an (infinite) tape running through it, divided into squares, with a finite number of these squares bearing a nonblank symbol. At any given time, there is just one square, say, the rth, in the machine, bearing a symbol S_r. To simplify matters further, the possible instructions given by a symbol are limited to replacing the symbol on the square by a new one, moving the tape one square to

BIOGRAPHY

Alan Turing (1912–1954)

Alan Turing's father, an officer in the British administration in India, decided that his son would be raised in England (Fig. 25.13). Thus, Alan's parents saw him only rarely during his formative years. Turing entered King's College, Cambridge, in 1931 to study mathematics and received his MA degree four years later with a dissertation dealing with Gaussian error functions. Shortly thereafter, however, he began to work in earnest on a major new problem, Hilbert's decision problem, and resolved it in the paper in which he invented the concept of the Turing machine. At about the same time, however, Alonzo Church in Princeton published another solution to the problem, so Turing decided to go to Princeton to work with Church. Returning to England and King's College in 1938, he was called, on the outbreak of World War II, to serve at the Government Code and Cypher School in Bletchley Park in Buckinghamshire. It was there, during the next few years, that Turing led the successful effort to crack the German "Enigma" code, an effort that turned out to be central to the defeat of Nazi Germany.

After the war, Turing continued his interest in automatic computing machines and so joined the National Physical Laboratory to work on the design of a computer. He continued this work at the University of Manchester after 1948. Turing's promising career, however, came to a grinding halt when he was arrested in 1952 for "gross indecency." He was, in fact, a homosexual, and, at the time, overt homosexual acts were against the law in England. The penalty for this crime was submission to psychoanalysis and to hormone treatments designed to "cure" this "disease." Unfortunately, the cure proved worse than the disease, and, in a fit of depression, Turing committed suicide in June of 1954 by eating a cyanide-poisoned apple.

FIGURE 25.13

Alan Turing on a stamp from St. Vincent and the Grenadines

either the right or the left, and changing the state of the machine. Thus, at any given moment, the pair (q_i, S_r) will determine the behavior of the machine according to the particular functional relationship that defines the behavior. If the function is not defined on a particular pair (q_i, S_r), the machine simply halts. It is the symbols printed by the machine, or at least a determined subset of them, that represent the number to be computed. Turing's contention, backed up by many arguments in his paper, is that the operations above are all those necessary actually to compute a number.

As an example, Turing constructed a machine to compute the sequence $010101\cdots$. This machine is to have four states, q_1, q_2, q_3, q_4, and is capable of printing two symbols, 0 and 1. The tape for this machine is entirely blank initially and the machine begins in state q_1. The instructions that the machine uses are as follows:

1. If it is in state q_1 and reads a blank square, it prints a 0, moves one square to the right, and changes to state q_2.
2. If it is in state q_2 and reads a blank square, it moves one square to the right and changes to state q_3.
3. If it is in state q_3 and reads a blank square, it prints a 1, moves one square to the right, and changes to state q_4.
4. If it is in state q_4 and reads a blank square, it moves one square to the right and changes to state q_1.

It is easy enough to see that this machine does accomplish what is desired, although Turing for technical reasons arranged the printing so that there are figures only on alternate squares. And although this example does not demonstrate it, the reason for the motion of the tape in either direction is to give the machine a memory. Thus, the machine can reread a particular square and act in different ways depending on its particular state at the time. In this way the machine can "remember" numbers written earlier and use them in subsequent computations.

It is the possibility of memory that leads to perhaps the most surprising part of Turing's paper, his proof of the existence of a single machine that can compute any computable number. Turing's idea for this was to take the set of instructions for any given machine, like those written above, and turn that set systematically into a series of symbols, called the **standard description** of the machine. The universal machine is then supplied with a tape containing this standard description, followed by the symbols on the input originally supplied to that machine. Turing was in fact able to give a rather explicit description of the behavior of this universal machine in terms of a functional relationship as described earlier. The main idea is that the machine acts in cycles, each cycle representing, first, a look at the standard description of the particular machine, second, a look at one square of that machine's input, and, finally, a corresponding action. Although Turing did not at the time attempt the physical construction of a machine with the capabilities he proved could exist, it was his idea that led directly to the concept of an all-purpose computer that could be programmed to do any desired computation. Naturally, there are physical limits to the size of a machine and the length of a program, limits that did not exist in Turing's theoretical model with its infinite tape. Modern technology, however, seems to extend these physical limits so often that today's computers are better approximations to Turing's universal machine with each passing year.

25.5.4 Shannon and the Algebra of Switching Circuits

A more direct application of mathematical ideas to the construction of a computer was developed by Claude Shannon (1916–2001) in 1938 as part of his master's thesis at M.I.T. In this work, Shannon applied the algebra of logic developed by Boole a century earlier to the construction of switching circuits, which would have desired properties. It is these circuits that form the basis for the internal construction of computing machines. Shannon in fact realized that any circuit can be represented by a set of equations and that the calculus necessary for manipulating these equations is precisely the Boolean algebra of logic. Thus, given the desired characteristics of a circuit that one wanted to construct, one used this calculus to manipulate the equations into the simplest possible form, from which the construction of the circuit was then immediate. One could also perform analysis of circuits by this method, by applying the calculus to the equations of a complex circuit and thereby reducing it to a simpler form.

Shannon began his work by dealing simply with switches that may be open or closed. The open ones were represented by 1, the closed ones by 0. Placing two switches in series was represented by the Boolean operation $+$, while placing them in parallel was represented by $\cdot$ (Fig. 25.14).

FIGURE 25.14

Switches in series and in parallel

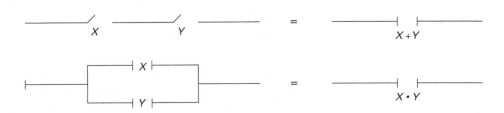

Shannon noted the following postulates, which these two operations satisfied, with their corresponding circuit interpretations, the truth of which made the analogy of switching circuits with Boolean algebra possible.

1. $0 \cdot 0 = 0; 1 + 1 = 1$. A closed circuit in parallel with a closed circuit is closed while an open circuit in series with an open circuit is open.
2. $1 + 0 = 0 + 1 = 1; 0 \cdot 1 = 1 \cdot 0 = 0$. An open circuit in series with a closed circuit is open while a closed circuit in parallel with an open circuit is closed.
3. $0 + 0 = 0; 1 \cdot 1 = 1$. A closed circuit in series with a closed circuit is closed while an open circuit in parallel with an open circuit is open.

Representing a switch in a circuit by X, Shannon noted that X could take on just the two values 0 and 1. It followed that the laws of Boolean algebra, including the two commutative laws, the two associative laws, and the distributive laws both of multiplication over addition and of addition over multiplication, could be proved by simply checking each possible case. He also introduced the negation of a variable X, written X', to be that variable that is 1 when X is 0 and 0 when X is 1, and demonstrated some additional laws. For example, $X + X' = 1$, $X \cdot X' = 0$, $(X + Y)' = X' \cdot Y'$, and $(X \cdot Y)' = X' + Y'$. Shannon then recalled Boole's expansion of functions and dualized it by interchanging multiplication and addition. Thus, for example,

$$f(X, Y, Z, \ldots) = [f(0, Y, Z, \ldots) + X][f(1, Y, Z, \ldots) + X'].$$

A very useful rule can then be established by adding X to both sides of this equation (bearing in mind the distributive laws):

$$X + f(X, Y, Z, \ldots) = X + f(0, Y, Z, \ldots).$$

With the various laws of Boolean algebra now established for switching circuits, Shannon was able both to analyze and synthesize circuits.

For example, Shannon presented the circuit (Fig. 25.15) whose algebraic representation was

$$W + W'(X + Y) + (X + Z)(S + W' + Z)(Z' + Y + S'V).$$

Using various laws of Boolean algebra, including the special law of the previous paragraph applied three times, he reduced this formula first to

$$W + X + Y + (X + Z)(S + 1 + Z)(Z' + Y + S'V),$$

then to

$$W + X + Y + Z(Z' + S'V),$$

and finally to

$$W + X + Y + ZS'V.$$

This latter formula had a much simpler circuit representation than the original.

As an example of the synthesis of a circuit having given characteristics, Shannon showed how to construct one that would add two numbers given in binary representation. If the

FIGURE 25.15

Simplifying a circuit using
Boolean algebra

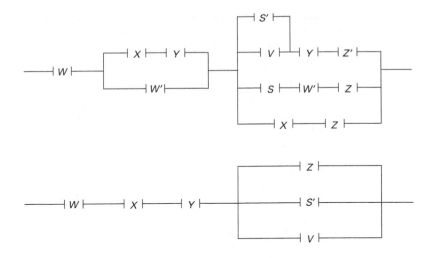

two numbers are represented by $a_n a_{n-1} \ldots a_1 a_0$ and $b_n b_{n-1} \ldots b_1 b_0$, and their sum by $s_{n+1} s_n \ldots s_1 s_0$, then s_0 is equal to 1 if $a_0 = 1$ and $b_0 = 0$ or if $a_0 = 0$ and $b_0 = 1$ and equal to 0 otherwise. There is also a carry digit c_1, which is equal to 1 if both a_0 and b_0 are 1 and 0 otherwise. Thus, s_0 is represented by the equation $s_0 = a_0 b_0' + a_0' b_0$ and c_1 is represented by $c_1 = a_0 b_0$. Each s_j for $j \geq 1$ requires the addition not only of a_j and b_j but also the carry digit c_j. Thus, the formula for s_j is $s_j = (a_j b_j' + a_j' b_j) c_j' + (a_j b_j' + a_j' b_j)' c_j$, while that for the next carry digit c_{j+1} is $c_{j+1} = a_j b_j + c_j (a_j b_j' + a_j' b_j)$. The circuit construction of an adder determined by these equations is the basis for the design of addition methods in modern-day calculators and computers.

25.5.5 Von Neumann's Computer

The work of Turing and Shannon were only two facets of the many theoretical and applied problems that had to be solved before the modern computer could be constructed, and there were numerous people who worked on these problems, particularly during the 1940s. But the man most responsible for the shape of the ultimate result was probably John von Neumann (1903–1957), who immediately after the Second World War gathered a brilliant group of scientists and engineers at the Institute for Advanced Study in Princeton. Their task was to take the experience developed during the war years in the development of two early computers, the ENIAC and the EDVAC, and combine it with recently developed theoretical knowledge to develop what one of its backers called "the most complex research instrument now in existence. . . . Scholars have already expressed great interest in the possibilities of such an instrument and its construction would make possible solutions of which man at the present time can only dream."[29]

The group under von Neumann decided to organize the computer under four main sections: an arithmetic unit, a memory, a control, and an input-output device, the first two being quite analogous to Babbage's mill and store, respectively. The arithmetic unit, now generally called the **central processing unit**, is the place where the machine performs the elementary operations, those operations that should not be reduced any further. These elementary

BIOGRAPHY

John von Neumann (1903–1957)

Born in Budapest into the family of a well-to-do Jewish banker, von Neumann received his doctorate from the University of Budapest. He taught in Berlin and Hamburg before being invited to Princeton in 1930. Three years later he was chosen as one of the charter members of the Institute for Advanced Study, a position he held for the remainder of his life. Von Neumann was one of the last mathematicians equally at home in pure and applied work. Over the years he produced a steady stream of papers in both areas. In pure mathematics, he was especially proficient in analysis and combinatorics. He had a great ability to see into complex situations and pull out appropriate axioms that would enable the subject to be treated mathematically. His talents in applied mathematics were in particular demand during the Second World War and shortly afterward, when he led the effort to develop the modern computer. He was a member of the Atomic Energy Commission from 1954 until his untimely death from cancer in 1957 (Fig. 25.16).

FIGURE 25.16

John von Neumann on a Hungarian stamp

operations are essentially wired into the machine, like, for example, the addition described above, while any other operation is built out of the elementary ones by a set of instructions. Recall that the number system of Babbage's Analytical Engine was decimal. But with the advent of electronic, rather than mechanical, devices for representing numbers, it turned out that it was simpler to represent numbers in binary, so that any particular device holding a digit would only need to have two states, on and off, to represent the two possibilities of 1 and 0. Von Neumann was in fact instrumental in designing efficient sets of decimal-binary and binary-decimal conversion instructions so that the operator could enter numbers in the normal decimal mode and receive answers in that mode as well, without compromising the speed and ease of construction of the machine.

The memory unit of the machine needed to be able to take care of two different tasks, storing the numbers that were to be used in the calculations and storing the instructions by which the calculations were to be made. But because instructions themselves can be stored in appropriate numerical code, the machine only needed to be able to distinguish between the actual numbers and the coded instructions. Moreover, in order to compromise between the "infinite" memory desired by the user of the machine and the finite memory constructible by the engineer, it was decided to organize the memory in hierarchies, such that some limited amount of memory was immediately accessible while a much larger amount could be accessed at a somewhat slower rate. It was also decided that in order to achieve a sufficiently large memory in a reasonable physical space, the units that stored an individual digit needed to be microscopic parts of some large piece.

The control unit was the section where the instructions to the machine resided, the orders that the machine could actually obey. Again, compromises had to be worked out between the desire for simplicity of the equipment and the usefulness for the sake of speed of a large number of different types of orders. In any case, one of the more important aspects of the control procedure, an aspect of which even Lady Lovelace was aware, was the ability of the machine to use a given sequence of instructions repeatedly. But because the machine must be made aware of when the repetition should end, it was also necessary to design a type of order to let the machine decide when a particular iteration was complete. Furthermore, the control unit needed to be have a set of instructions that integrated the input and output devices into the machine. Von Neumann was particularly interested, in fact, in assuring that the latter devices would allow for both printed and graphical outputs, because he realized that some of the more important results of a particular computation may best be explored graphically.

The computer eventually constructed at the Institute for Advanced Study, based on von Neumann's design and finished in 1951, proved to be the model for the more advanced computers built in succeeding years. The technological achievements in regard to computers since that time have both increased the capacity and decreased the size by factors probably undreamed of by members of the working group of the late 1940s. Computers have now become so much a part of everyday life that one can scarcely imagine how we would accomplish many common tasks without them.

OLD QUESTIONS ANSWERED

The twentieth century saw the answer to several major questions first raised much earlier. In particular, mathematicians were able to prove Fermat's Last Theorem, classify completely all simple groups, and, with the help of computers, also prove the four-color theorem. And in the opening years of the twenty-first century, another old problem, the Poincaré conjecture was also apparently solved.

25.6.1 The Proof of Fermat's Last Theorem

As we have seen, Kummer's idea enabled Fermat's Last Theorem to be proved for many prime exponents p. And during the twentieth century, various other techniques were used to show that the theorem was true for all primes less than 125,000 and also, in the special case where p does not divide x, y, or z, for all primes less than 3,000,000,000. But there is a big difference between proving the theorem for finitely many primes and proving it for *all* primes. It was becoming clear that an entirely new approach was necessary.

The new approach involved the idea of an **elliptic curve**. This is a curve of the form $y^2 = ax^3 + bx^2 + cx + d$, where a, b, c, and d are rational and where the cubic polynomial in x has distinct roots. Recall that Diophantus had determined rational solutions to at least one equation of this form, while Euler had studied the situation more generally. By late in the nineteenth century, it was known that there was a definition of "addition" on the rational points on an elliptic curve (including the "point at infinity"), that is, on the set of pairs of rational numbers that solved the equation, which turned this set into an Abelian group $E(\mathbf{Q})$. It was proved by Louis Mordell (1888–1972) in the 1920s that this group was a finitely generated Abelian group (that is, that all its elements could be written as sums of multiples of finitely many elements) and later by Carl Siegel (1896–1981) that the set of integral points on such a curve was finite.

Recall now the so-called modular group of linear fractional transformations $f(z) = (az + b)/(cz + d)$, where $ad - bc = 1$. We considered these transformations earlier as transformations on projective space over a finite field. But more generally, the name "modular group" is given to the group of these transformations with integral coefficients acting on the upper half of the complex plane $\{z = x + iy | y > 0\}$. (It is not difficult to check that in this situation $f(z)$ is also in the upper half of the complex plane.) These modular groups and their subgroups had been studied extensively beginning in the late nineteenth century, but it was the Japanese mathematicians Goro Shimura (1930–) and Yutaka Taniyama (1927–1958) who first saw a connection between the modular group and elliptic curves. Although the exact connection is beyond the scope of this book, Shimura and Taniyama conjectured that every elliptic curve comes in a very definite manner from a modular form, a function from the upper half-plane to the complex numbers that is as invariant as possible under certain subgroups of the modular group. The Taniyama-Shimura conjecture, which was made somewhat more precise by André Weil (1906–1998) in the 1960s, is frequently stated simply as "every elliptic curve is modular."

During the late 1960s, Robert Langlands (1936–) at the Institute for Advanced Study in Princeton began to believe that the unification of modular forms and elliptic curves implied by the Taniyama-Shimura conjecture was only one part of a much greater scheme of unification of aspects of number theory and analysis, many other elements of which he began to conjecture. In fact, Langlands proposed what is now known as the Langlands program, a concerted effort to prove these unifying conjectures one by one, leading ultimately to a great unification of mathematics. Although the Langlands program is still in its beginning stages, its first great triumph would be the proof of the Taniyama-Shimura conjecture.

In the early 1980s, Gerhard Frey (1944–) noted the close relationship between the Taniyama-Shimura conjecture and Fermat's Last Theorem. Suppose a solution $a^p + b^p = c^p$ of the Fermat equation existed for $p > 3$, where we may as well assume that b is even and $a \equiv -1 \pmod 4$. Frey then considered the elliptic curve $y^2 = x(x - a^p)(x + b^p)$ and, by studying various functions defined on the curve, believed that it was impossible to exist. In particular, it seemed to him that the existence of the curve would contradict the Taniyama-Shimura conjecture. The exact links between this curve and the conjecture were clarified in 1986 by Jean-Pierre Serre (1926–) and Kenneth Ribet (1947–). Basically, the three mathematicians established that the Frey elliptic curve was not modular. Thus, the truth of the Taniyama-Shimura conjecture would establish that the Frey curve could not exist, or that Fermat's Last Theorem was true.

Over the next seven years, the Princeton mathematician Andrew Wiles (1953–), who had been fascinated by Fermat's Last Theorem since reading about it while growing up in England, worked in secret in his attic study to try to establish the Taniyama-Shimura conjecture. Finally, by May of 1993, having developed numerous new techniques in number theory, Wiles believed he had proved the conjecture, at least for a certain class of elliptic curves to which the Frey curve belonged. He therefore arranged to give a series of three lectures at a number theory conference in Cambridge in June of that year. Although he did not state the goal of his lectures at the beginning, the mathematicians in attendance soon understood that a major result would be announced by the end. Thus, when Wiles concluded the third lecture on June 23 by writing the statement of Fermat's Last Theorem on the board and saying "I think I'll stop here," the audience burst into sustained applause.

The dramatic nature of Wiles's announcement notwithstanding, it turned out that there was a flaw in the proof that was only discovered during the review process of his manuscript. Wiles labored mightily over the next year to correct the flaw, finally enlisting the help of Richard Taylor (1962–), one of his former graduate students. And then on September 19, 1994, Wiles had a brilliant new insight, and all the pieces of the puzzle came together. Two new manuscripts were soon prepared, one coauthored with Taylor; the normal review process took place, and the May 1995 issue of the *Annals of Mathematics* contained the complete proof of Fermat's Last Theorem.

Wiles's work, besides leading to a proof of a very old conjecture, opened new doors to many topics in number theory, and since 1995 other mathematicians have used his ideas to push ahead. In fact, in 1999 Taylor and others published a proof of the complete Taniyama-Shimura conjecture. Although it is still a mystery as to what proof of the theorem Fermat had in mind, Wiles's proof is certainly a twentieth-century proof, one whose ideas will have ramifications into the twenty-first century and beyond.

25.6.2 The Classification of the Finite Simple Groups

Ever since Camille Jordan showed that the alternating group A_n for $n \geq 5$ and certain groups of matrices with coefficients in the field with p elements were simple groups, mathematicians attempted to find other families of such groups. By the end of the nineteenth century, Leonard Dickson and others had generalized some of Jordan's results. For example, Dickson showed that the projective special linear group $PSL(n, p^k)$ over the field with p^k elements was simple for $n > 1$ except in a few trivial cases. Similarly, Jordan had studied other subgroups of the general linear group over a field of p elements, subgroups defined by their leaving invariant certain bilinear forms. He was able to show that quotients of these groups by their centers, the subgroups that commuted with every element of the group, were also simple. Dickson generalized these results to analogous groups defined over finite fields of order p^k.

The simple groups that Dickson studied, now called the projective symplectic, orthogonal, and unitary groups, had analogues, the so-called classical groups, when the coefficients of the matrices were allowed to be complex numbers. But there were other simple groups of matrices over the complex numbers as well. Dickson was able to show that one of these families of exceptional groups had a finite analogue. It turned out that the other ones had finite analogues as well, although these were only discovered by Claude Chevalley (1909–1984) in the 1950s. Meanwhile, near the turn of the twentieth century, Frank N. Cole (1861–1927) and George A. Miller (1863–1951) showed that five groups first discovered by Emile Mathieu (1835–1890) around 1860 were simple groups that were not part of any of the known families. (Such groups are now called **sporadic** groups.) Mathieu, naturally, defined these as permutation groups. For example, one of these groups, now referred to as M_{12}, is the group generated by the following three permutations on a set of 12 elements: $A = (1, 2, 3, 4, 5, 6, 7, 8, 9, 10, 11)$, $B = (5, 6, 4, 10)(11, 8, 3, 7)$, and $C = (1, 12)(2, 11)(3, 6)(4, 8)(5, 9)(7, 10)$. It turns out that this group has order 95,040.

No new simple groups were discovered over the next 50 years, but in 1963, Walter Feit (1930–2004) and John Thompson (1932–) made a major advance in group theory by showing, in a massive paper in the *Pacific Journal of Mathematics*, that every group of odd order was solvable, thus that there were no simple groups of odd order besides the cyclic groups of prime order. Using some of the new techniques developed by Feit and Thompson in their paper, group theorists began a major attack on the problem of simple groups. Between 1965 and 1974, 21 new so-called sporadic simple groups were discovered, the largest being of order approximately 10^{54}. Mathematicians began to wonder, in fact, whether more would continue to be discovered. By 1972, however, Daniel Gorenstein (1923–1992) began to believe the opposite, namely, that there were only finitely many sporadic groups and that a complete classification of the finite simple groups was possible. In fact, he laid out a program for accomplishing this goal. And it was only nine years later, in February of 1981, that he was able to announce that the classification of the finite simple groups was complete. This result, which involved the combined efforts of several hundred mathematicians and whose proof probably covered some 10,000 pages, stated that there were four basic classes of these groups: the cyclic groups of order p, the alternating groups on $n \geq 5$ letters, 16 infinite families of "classical groups," and 26 sporadic groups.

The theorem asserting the classification of the finite simple groups is different from a normal mathematical theorem, in that its proof is impossible for any one person to absorb and verify. Nevertheless, since the individual pieces have been checked, and since the arguments made in the various papers comprising the proof often overlap, most mathematicians now believe in the truth of the result. Still, such a proof stretches our notion of what a proof should be. Further stretching is involved in the proof of the four-color theorem.

25.6.3   The Proof of the Four-Color Theorem

Once the initial disappointment of the failure of Kempe's attempted proof had worn off, mathematicians attacked the four-color problem with increasing vigor in the twentieth century. In fact, it is said that nearly every mathematician in the first half of the century tried his hand at this problem. Among the many ideas that soon emerged as crucial to the solution were those of an unavoidable set of regions and a reducible configuration. An **unavoidable set** was a set of regions, at least one of which must always appear in any map. For example, we have seen that the set consisting of a digon, triangle, square, and pentagon is unavoidable. Every map must contain one of these. Another unavoidable set consists of a digon, triangle, square, two adjacent pentagons, or a pentagon adjacent to a hexagon. A **reducible configuration** is any arrangement of regions that cannot occur in a minimal map requiring at least five colors. For if a map contains such a configuration, then any coloring of the remainder of the map with four colors can be extended to a coloring of the entire map. We have seen that a digon, a triangle, and a square are all reducible configurations and that the failure of Kempe's argument lay in his inability to prove that a pentagon was reducible.

One of the many mathematicians who sought reducible configurations was George David Birkhoff (1884–1944), who spent most of his career at Harvard. Among many other results on the problem, he showed that the configuration of Figure 25.17a was reducible. Actually, by his time, mathematicians had generally converted the map coloring problem to a problem

FIGURE 25.17

(a) Reducible configuration of the four-color theorem. (b) Graph derived from the map in part (a).

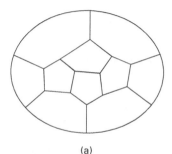

(a)

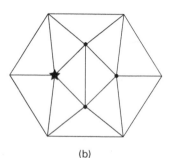

(b)

in graph theory. In other words, a map was converted to a graph whose vertices were a set in one-to-one correspondence with the regions of the map and such that two vertices were joined by an edge if and only if the two corresponding regions had a common boundary arc. (Compare Figs. 25.17a and 25.17b.) If we define a coloring of a graph to be an assignment of colors to the vertices so that no two vertices that lie on a common edge have the same color, then the four-color theorem for maps is equivalent to a four-color theorem for graphs. It was thus possible to apply many of the results of the emerging study of graphs to this old problem. In fact, Birkhoff actually stated his reducibility result as follows: A five-vertex with three consecutive five-neighbors is reducible. (The five-vertex is the vertex in Fig. 25.17b marked by a star, while the three five-neighbors are marked by circles.)

Given the two notions of unavoidable sets and reducible configurations, the goal toward which Birkhoff and others were working was to find an unavoidable set of reducible configurations, for such a set would provide a proof of the theorem. Every map would have one member of the set and each of these members would be reducible, so could not exist in a minimal map requiring five or more colors. During the first half of the century, much of the work toward this goal was piecemeal, as attempts to find unavoidable sets and reducible configurations were mostly independent. The first person who advocated a systematic search for an unavoidable set of reducible configurations was Heinrich Heesch (1906–1995). Heesch invented a method for showing that a particular configuration was unavoidable, but initially he feared that an unavoidable set of reducible configurations would contain about 10,000 configurations, each of which would have to be checked. Nevertheless, he began work on the problem and by 1948, when Wolfgang Haken (1928–) heard Heesch lecture on the subject, he had checked 500.

Nearly 20 years later, Heesch was still working on the problem, having checked thousands more configurations. Haken then suggested that computers might be useful in working through all the detailed calculations necessary. Unfortunately, Haken was not a programming expert, and it soon was clear that without efficient programs, computers in the late 1960s were not fast enough to make substantial progress on the problem. But in 1972, after Haken gave a lecture in which he stated he would quit working on the problem, Kenneth Appel (1932–), who was a programming expert, suggested that a computer attack on the problem was in fact feasible at the time. Over the next four years, the pair worked through the problem, first proving that there was an unavoidable set of reducible configurations, each of which would be amenable to a computer check, and then actually finding and checking the set. It helped that early in 1976 Haken's university, the University of Illinois, had acquired a faster computer and allowed the pair all the computer time they needed. When Appel and Haken announced that they had completed the proof in July of 1976, their unavoidable set contained 1936 configurations, but by the time they published the following year, they had succeeded in reducing the number to 1482.

A proof by an essential use of the computer is an entirely new phenomenon in mathematics. Computers, since their introduction, have been used to help mathematicians make conjectures, but the proof of the four-color theorem was the first in which the computer was actually used for the detailed work of constructing a formal proof. As was to be expected, this proof has generated much controversy since its appearance. Many mathematicians still do not accept the proof as valid, because the general standard of acceptance of a proof has always been the checking of it by many members of the mathematical community. And although the computer program itself can be checked, there is no way for mathematicians to check the various details of the work actually performed by the computer. On the other hand, in 1994 Neil Robertson, Daniel Sanders, Paul Seymour, and Robin Thomas came up with a shorter proof than the Appel-Haken one that required only 633 configurations, although it still made use of a computer. In fact, they were convinced that the computer check of reducibility was more reliable than a check by hand, because the process was so long and complicated. How this debate over the use of the computer will be resolved is not at all clear. It is possible, though it seems unlikely in the foreseeable future, that the four-color theorem will one day be proved in the traditional way, but in any case, Appel and Haken's proof initiated a new debate over what constitutes a mathematical proof.

25.6.4   The Poincaré Conjecture

Although no one mathematician presented a list of problems for mathematicians to consider in the twenty-first century, like Hilbert did at the turn of twentieth century, the Clay Mathematics Institute, a mathematics research institute among whose goals are "to increase and disseminate mathematical knowledge . . . and to recognize extraordinary achievements and advances in mathematical research,"[30] decided to produce a list of seven important unsolved problems and provide prizes of one million dollars each for solutions to these problems. The Institute believed that these prizes would "recognize achievement in mathematics of historical dimension," while at the same time "elevate in the consciousness of the general public the fact that, in mathematics, the frontier is still open and abounds in important unsolved problems." We have already mentioned one of these problems, the Riemann hypothesis (see Section 22.3.5), the only problem from Hilbert's list that is on the Clay list. The only other problem we will consider here is the Poincaré conjecture, the first of the Clay Institute's problems to be solved.

The Poincaré conjecture has its roots in the topological work of Poincaré at the beginning of the twentieth century. Recall that Poincaré defined the idea of homology on a manifold. For closed two-dimensional manifolds, it was true that the Betti numbers (or, more precisely, the homology groups defined later) characterize the manifold, in the sense that any two such manifolds with the same homology groups are homeomorphic—that is, there exists a continuous one-to-one mapping from one to the other. But when Poincaré looked at higher-dimensional manifolds, he realized the situation was more difficult. In a paper of 1900, he asserted a limited result, that any n-dimensional manifold with the same homology groups as the n-dimensional sphere (the set of points in $n + 1$–space satisfying $x_1^2 + x_2^2 + \cdots + x_{n+1}^2 = 1$) was homeomorphic to the sphere. But by 1904, he realized that he was in error and in fact produced a counterexample in the three-dimensional case.

Poincaré, however, still wanted to be able to characterize the three-dimensional sphere algebraically, so he looked at another group associated with it, or, more generally, with any manifold. This **fundamental group** arises by considering closed curves that start and end at one point on the manifold. Any two curves are defined to be equivalent if one can be deformed into the other by a continuous motion. We compose two curves by forming a new curve by traversing the first curve and then the second. The inverse of a curve is the same curve traversed in the opposite direction. It then follows that the equivalence classes of curves form a group.

For example, on a circular ring (annulus), we can consider the class of all closed curves that do not enclose the "hole." This class is the identity class. The other classes consist of curves that go around the hole once in the counterclockwise direction, twice, and so on, as well as those that go around once in the clockwise direction, twice, and so on. Thus, the fundamental group of the circular ring is an infinite cyclic group. Similarly, the fundamental group of a circular disk consists only of the identity, as does the fundamental group of a sphere of any dimension.

Given that the homology groups themselves did not characterize the three-dimensional sphere, Poincaré now asked the question as to whether the fundamental group would give a characterization. In other words, the **Poincaré conjecture** is that any closed three-dimensional manifold with trivial fundamental group (or, what amounts to the same thing, in which every closed curve can be shrunk to a point) is homeomorphic to the three-dimensional sphere. Although Poincaré only conjectured this result for three-dimensional manifolds, the analogous conjecture for higher-dimensional manifolds was proved by Stephen Smale (1930–) for dimensions five and greater in 1960 and by Michael Freedman (1951–) for dimension four in 1982. Yet Poincaré's original three-dimensional case resisted all attacks during the twentieth century.

The Clay problems were announced in May of 2000. But beginning in November of 2002, Grigory Perelman (1966–) of the Steklov Institue of Mathematics in St. Petersburg posted a proof of the conjecture in three installments on the Internet. Perhaps because of his unorthodox method of publishing, it took more than three years for the mathematical community to accept the proof. His proof included many new ideas but started with some basic work by William Thurston (1946–) in the 1970s and Richard Hamilton (1943–) in the 1980s. Thurston had shown that there were only eight possible geometric structures of a three-dimensional manifold and conjectured that in some sense every such manifold can be divided into pieces associated with one or the other of these eight (the geometrization conjecture). Only one of these structures, essentially the three-dimensional sphere, has a trivial fundamental group. Hamilton proposed a way of proving Thurston's conjecture, but the problem up until the beginning of this century was that neither Hamilton nor others working on the problem had figured out how to eliminate the possibility of some singularities when one splits up the manifold. Although Perelman did not prove Thurston's geometrization conjecture in its entirety, he was able to prove enough of it to get the Poincaré conjecture.

By the middle of 2006, even though Perelman did not publish his proof via normal channels, enough experts had studied his work and validated it so that the International Mathematical Union, meeting that year in Madrid, awarded Perelman a Fields Medal, the mathematical community's version of the Nobel Prize. Surprisingly, despite numerous entreaties from the IMU's president and others, Perelman declined to accept the prize, declaring that it was "irrelevant," and that "everybody understood that if the proof is correct then no other recognition is needed."[31] In fact, Perelman decided to drop out of mathematics entirely, and he is now living quietly in St. Petersburg with his mother. At the end of 2006, the journal *Science* honored Perelman's proof as the scientific "breakthrough of the year." But whether the Clay Institute will award him the million-dollar prize for solving the conjecture or whether he would accept it is a matter for the future to decide.

EXERCISES

1. The following is the **Richard paradox** (named after its originator, Jules Richard (1862–1956)): Arrange all two-letter combinations in alphabetical order, then all three-letter combinations and so on, and then eliminate all combinations that do not define a real number. (For example, "six" defines a real number, while "sx" does not.) Then the set of real numbers definable in a finite number of letters forms a denumerable, well-ordered set $E = \{p_1, p_2, \ldots\}$. Now define the real number $s = .a_1a_2 \ldots$ between 0 and 1 by requiring a_n to be one more than the nth decimal of p_n if this decimal is not 8 or 9 and equal to 1 otherwise. Although s is defined by a finite number of letters, it is not in E, a contradiction. How can one resolve this paradox? How is this paradox related to the barber paradox or to Zermelo's original paradox?

2. Show that the trichotomy law follows from Zermelo's well-ordering theorem.

3. Show that Zermelo's axiom of separation resolves Russell's barber paradox as well as the Richard paradox (Exercise 1) in the sense that certain "sets" are now excluded from discussion.

4. Formulate and prove the Heine-Borel theorem in the plane.

5. Prove that a connected set A, in the sense of the Youngs, cannot be expressed as $A = B \cup C$, where B and C are closed and $B \cap C = \Phi$.

6. Show that the set of rational numbers in [0, 1] is not connected.

7. Assume that a set E is compact according to Fréchet's definition in terms of nested sets. Show that every infinite subset E_1 of E has at least one limit element in E.

8. Show by use of examples that the nested set property on the real line depends on the subsets being both closed and bounded.

9. Show that a real function sequentially continuous on a closed and compact set E (in Fréchet's definition) is bounded there and attains its upper bound at least once.

10. Show that if E is closed and compact according to Fréchet's definition, and if $\{E_n\}$ is a nested sequence of closed subsets E, then the intersection $\cap_n E_n$ is not empty.

11. Prove that the space of real functions continuous on $[a, b]$ under the maximum norm metric is "normal" in the sense of Fréchet.

12. Show that the space of all infinite sequences of real numbers $\{x = \{x_1, x_2, \ldots\}\}$ with the metric

$$(x, y) = \sum_{p=1}^{\infty} \frac{1}{p!} \frac{|x_p - y_p|}{1 + |x_p - y_p|}$$

is normal.

13. Show that in the metric space defined in Exercise 12, there is a number α such that $(x, y) < \alpha$ for all x, y in the space.

14. Show that if E is a topological space and A a closed subset, that is, one containing all its accumulation points, then $E - A$ is a domain (open set) in the notion of Hausdorff. Conversely, if the subset B of E is an open set, show that $E - B$ is closed, using Hausdorff's definition.

15. Show that under Hausdorff's definition of a limit point, a given infinite set A can have no more than one.

16. Show that Hausdorff's two definitions of continuity at a point are equivalent.

17. Use Hausdorff's neighborhood definition of continuity to show that a continuous function preserves connectedness and compactness.

18. How many linearly independent closed one-dimensional subvarieties are there on the sphere? On the torus?

19. Determine the boundary of the tetrahedron $V_0 V_1 V_2 V_3$ indicated in Figure 25.5. Show that the boundary of the boundary is 0.

20. Given the face $V_1 V_0 V_3$ of the tetrahedron of the previous exercise, calculate its boundary and show that the boundary of the boundary is 0.

21. Using Dickson's axioms for a field, derive the following theorems:
    a. For any two elements a, b of the set, there exists in the set one element y such that $a + y = b$.
    b. If $a + z = a$ for a particular element a, then $b + z = b$ for every element b.
    c. If $a + b = a + b'$, then $b = b'$.

22. Show that the set $\{0, 1, -1\}$ with ordinary addition and multiplication satisfies each of Dickson's axioms except the first.

23. Show that the set $S = \{r\sqrt{2} | r \text{ rational}\}$ with ordinary addition and multiplication satisfies each of Dickson's axioms except axiom 5. Find the x in this situation that satisfies $(a \times x) \times b = b$.

24. Show that the product 4.324×3.403 is equal to 2.2312242 in Hensel's multiplication relative to $p = 5$.

25. Show that the quotient $3.12 \div 4.21$ is equal in Hensel's arithmetic relative to $p = 5$ to the periodic "decimal" $2.42204220\ldots$ by actually performing the long division.

26. Show that $3.12 \div 0.2 = 4 \cdot 5^{-1} + 0 + 1 \cdot 5$ in Hensel's field of 5-adic numbers.

27. Show for Hensel's p-adic numbers that the multiplicative inverse of a unit (a number whose smallest power of p is the zero power) is again a unit.

28. Let x be a p-adic number. Define the r-neighborhood of x, $U_r(x)$, where r is an integer, to be $U_r(x) = \{y | y \equiv x \pmod{p^r}\}$. Show that this choice of neighborhoods of x makes the field $\mathbf{Q}_p$ into a topological space in the sense of Hausdorff.

29. Any number x in $\mathbf{Q}_p$ may be written uniquely as $x = p^\alpha e$ where e is a unit. The integer α is called the order of x relative to p, written $v_p(x)$. For any x, y in $\mathbf{Q}_p$, define (x, y) to be $(1/p)^{v_p(x-y)}$. Show that (x, y) defines a metric on $\mathbf{Q}_p$ in the sense of Fréchet.

30. Using the metric of Exercise 29, define the notion of the Cauchy sequence in $\mathbf{Q}_p$ as usual. Show that every Cauchy sequence in $\mathbf{Q}_p$ converges to a limit.

31. Show that each of the following multiplication tables of two basis elements i, j determines an associative algebra of degree 2 over the real numbers. Are there any other ones?

	i	j
i	i	j
j	j	0

	i	j
i	i	j
j	0	0

	i	j
i	j	0
j	0	0

32. Find a nilpotent element in the algebra of 2×2 matrices over the rational numbers.

33. Find an idempotent element (which is not a diagonal matrix) in the algebra of 2×2 matrices over the rational numbers.

34. Give several examples of categories different from those mentioned in the text. Give several examples of functors using the categories you listed as well as the categories in the text.

35. Suppose we modify Fisher's method for verifying the claim of the lady tasting tea so that she has to taste 12 cups and pick out the set of 6 of one type. Show that the probability is less than 0.05 that she could pick 5 or 6 cups correctly if she were totally without any discriminating ability.

36. In the Neyman-Pearson version of the test for the lady tasting tea, show that $P(X = 8 | p = 1/2) = \binom{10}{8}(1/2)^{10}$.

37. In the Neyman-Pearson version of the test for the lady tasting tea, verify that the power $P(R | H_2)$ is given as a function of the probability p as $\binom{10}{8}p^8(1 - p)^2 + \binom{10}{9}p^9(1 - p) + p^{10}$.

38. Show that for a polynomial of degree n, the nth-order differences are always constant.

39. Construct a difference table that would enable Babbage's Difference Engine to calculate the pyramidal numbers, the numbers that are the sums of the triangular numbers and can be considered as representing, say, the number of cannonballs contained in triangular pyramids of a given height.

40. Show that the defining equation for the Bernoulli numbers in Exercise 3 of Chapter 18 can be transformed into the equation used by Ada Lovelace to write a program for calculating these numbers. (*Hint:* Use the power series for e^x.)

41. Use Lovelace's equation to calculate B_2, B_4, and B_6.

42. Consider a Turing machine with two states q_1, q_2 capable of printing two symbols 0 and 1. Suppose it is defined by the following instructions:

    a. If the machine is in state q_1 and reads a 1, it prints 1, moves one square to the right, and remains in state q_1.
    b. If the machine is in state q_1 and reads a blank, it prints 1, moves one square to the right, and changes to state q_2.

    (Note that there is no instruction for the machine when it is in state q_2.) Suppose the machine begins in state q_1 with a tape whose first square is blank, whose next two squares to the right have 1's, and all of the rest of whose squares to the right are blank, and suppose further that the leftmost 1 is the initial square to be read. Show that the final configuration of the tape will be the same as the initial one except that it will have three 1's instead of two. In general, interpreting a tape with n 1's as representing the number $n + 1$, show that this Turing machine will calculate the function $f(n) = n + 1$ for n any nonnegative integer.

43. Determine a Turing machine that computes the function $f(n) = 2n$.

44. Prove the following expansion theorem of Boole: $f(x, y) = [f(0, y) + x][f(1, y) + x']$, where f is a Boolean function of the two Boolean variables x, y.

45. Use the distributive law of addition over multiplication, $a + (bc) = (a + b)(a + c)$, and the theorem of Exercise 44 to establish the following law for Boolean functions: $x + f(x, y) = x + f(0, y)$.

46. Prove the Boolean expansion theorem $f(x, y) = xf(1, y) + x'f(0, y)$ and the theorem $xf(x, y) = xf(1, y)$. Note that these are the duals of the results in Exercises 44 and 45.

47. Construct a circuit representing the addition of binary numbers as outlined in the text.

48. Let $f(z) = (ax + b)/(cz + d)$, where a, b, c, d are integers and $ad - bc = 1$. Show that if $z = x + iy$ is in the upper half of the complex plane, then so is $f(z)$.

49. Consider the elliptic curve given by the equation $y^2 = x^3 + 17$. We turn the set of rational points into an Abelian group as follows: If P_1 and P_2 are rational points, first construct the line connecting them. Next, determine the point P_3' where the line intersects the curve again. Finally, let the sum $P_1 + P_2$ be the point P_3, which is the reflection of P_3' in the x axis. (If $P_1 = P_2$, then take the tangent line at that point to begin the process.) The additive identity for this group will be the point P_0 at infinity. Using this addition, show that the sum of $P_1 = (2, 5)$ and $P_2 = (4, 9)$ is $(-2, 3)$. Find the point that is double $(-2, 3)$.

50. Consider the elliptic curve given by the equation $y^2 = x^3 - 43x + 166$. Using the description of addition on elliptic curves given in Exercise 49, calculate all multiples of the rational point $(3, 8)$. Determine the order of this point.

51. Determine the order of the group $PSL(2, 8)$. (*Hint:* Recall that $PSL(n, p^k)$ is the quotient group of $SL(n, p^k)$ by its subgroup consisting of multiples mI of the identity matrix, where $m^n \equiv 1 \pmod{p^k}$.)

52. Determine the order of the group $PSL(3, 4)$. Show that this group has the same order as A_8, the alternating group on eight letters. (It turns out that the two groups are not isomorphic.)

53. In the Mathieu group M_{12} described in the text, calculate AB, BA, AC, and CA.

54. Show that the fundamental group of the torus is $\mathbf{Z} \times \mathbf{Z}$, that is, the direct product of two infinite cyclic groups.

55. Look up the Banach-Tarski paradox in a text on set theory and discuss its meaning and implications. Do you believe the result? How does this relate to your belief in the truth of the axiom of choice?

56. Could one use Hausdorff's text of 1914 (see Note 10) in a topology course today? Get a copy and compare it with a current text.

57. Find a copy of an early edition of Van der Waerden's *Modern Algebra* text such as the 1949 English translation of the second German edition (New York: Frederick Ungar, 1949). Compare it with your current algebra text.

58. Discuss the implications of a computer proof of the four-color theorem. Do you believe that a proof using computers is rigorous?

REFERENCES AND NOTES

Although there is no easily accessible general history of twentieth century mathematics, each of the topics covered in this chapter is dealt with by one or more excellent works. Gregory H. Moore, *Zermelo's Axiom of Choice: Its Origins, Development, and Influence* (New York: Springer, 1982) deals with the development of set theory early in the century. A general history of point-set topology is Jerome Manheim, *The Genesis of Point Set Topology* (New York: Macmillan, 1964). The history of algebraic topology is discussed in great detail in Jean Dieudonné, *A History of Algebraic and Differential Topology, 1900–1960* (Boston: Birkhäuser, 1989). Histories of two important concepts of point-set topology are Raymond Wilder, "Evolution of the Topological Concept of 'Connected,'" *American Mathematical Monthly* 85 (1978), 720–726, and J.-P. Pier, "Historique de la notion de compacité," *Historia Mathematica* 7 (1980), 425–443. A good part of twentieth-century algebra, including particularly the theory of algebras, is covered in part three of B. L. Van der Waerden, *A History of Algebra from al-Khwarizmi to Emmy Noether* (New York: Springer, 1985). Leo Corry, *Modern Algebra and the Rise of*

Mathematical Structure (Basel: Birkhäuser, 1996) deals with the development of structure in algebra. The history of the theory of vector spaces can be found in Gregory H. Moore, "The Axiomatization of Linear Algebra: 1875–1940," *Historia Mathematica* 22 (1995), 262–303. Two of the best studies of twentieth-century statistics are Gerd Gigerenzer et al., *The Empire of Chance: How Probability Changed Science and Everyday Life* (Cambridge: Cambridge University Press, 1989) (which also contains some work on earlier centuries), and David Salsburg, *The Lady Tasting Tea* (New York: Freeman, 2001). See George B. Dantzig, *Linear Programming and Extensions* (Princeton, NJ: Princeton University Press, 1963) for more information on linear programming and its historical development. That work also contains numerous references to original papers in the field.

Among the books that provide histories of computing in one facet or another are B. V. Bowden, ed., *Faster Than Thought* (London: Pitman, 1953), which includes a copy of Lady Lovelace's translation and notes as well as material on British computers of the immediate postwar period; Herman Goldstine, *The*

Computer from Pascal to von Neumann (Princeton, NJ: Princeton University Press, 1972), which deals primarily with work in which the author was involved in the 1940s and 1950s; and N. Metropolis et al., eds., *A History of Computing in the Twentieth Century* (New York: Academic Press, 1980), which contains the proceedings of a 1976 conference in which presentations were made by many of the pioneers of computing.

Two histories of Fermat's Last Theorem are Simon Singh, *Fermat's Enigma: The Epic Quest to Solve the World's Greatest Mathematical Problem* (New York: Walker and Company, 1997), and Amir Aczel, *Fermat's Last Theorem: Unlocking the Secret of an Ancient Mathematical Problem* (New York: Four Walls Eight Windows, 1996). Since neither of these books goes into many technical details, a reader interested in those should consult David Cox, "Introduction to Fermat's Last Theorem," *American Mathematical Monthly* 101 (1994), 3–14, and Fernando Gouvêa, "A Marvelous Proof," *American Mathematical Monthly* 101 (1994), 203–222, as well as the references in both of these articles. There are numerous surveys of the work leading up to the theorem on the classification of finite groups. Among these are Joseph Gallian, "The Search for Finite Simple Groups," *Mathematics Magazine* 49 (1976), 163–179, and Ron Solomon, "On Finite Simple Groups and Their Classification," *Notices of the American Mathematical Society* 42 (1995), 231–239. The history of the four-color theorem can be found in Robin Wilson, *Four Colors Suffice: How the Map Problem Was Solved* (Princeton, NJ: Princeton University Press, 2002), and, with more mathematical details, in Thomas L. Saaty and Paul C. Kainen, *The Four-Color Problem: Assaults and Conquest* (New York: Dover, 1986). For a discussion of the philosophical implications of the computer-assisted proof of the four-color theorem, see two articles by Thomas Tymoczko: "Computers, Proofs and Mathematicians: A Philosophical Investigation of the Four-Color Proof," *Mathematics Magazine* 53 (1980), 131–138, and "The Four-Color Problem and Its Philosophical Significance," *Journal of Philosophy* 76 (1979), 57–83. A popular discussion of the solution of the Poincaré conjecture is Donal O'Shea, *The Poincaré Conjecture: In Search of the Shape of the Universe* (New York: Walker and Company, 2007). A more technical exposition can be found in John Morgan and Gang Tian, *Ricci Flow and the Poincaré Conjecture* (Providence, RI: American Mathematical Society, 2007).

Hermann Weyl's text, which contains his definition of a vector space, is *Space—Time—Matter*, Henry Brose, trans. (New York: Dover Publications, 1952). The original texts of many of the articles on set theory discussed in this chapter can be found in Jean van Heijenoort, ed., *From Frege to Gödel: A Source Book in Mathematical Logic, 1879–1931* (Cambridge: Harvard University Press, 1967). Many of the original papers on aspects of graph theory, including those important in the four-color problem, are available in Norman Biggs, E. Keith Lloyd, and Robin J. Wilson, *Graph Theory, 1736–1936* (Oxford: Clarendon, 1976). The two major works of Ronald Fisher mentioned in the text are *Statistical Methods for Research Workers* (Edinburgh: Oliver and Boyd, 1925) and *The Design of Experiments* (Edinburgh: Oliver and Boyd, 1935). Both books went through numerous editions. Jerzy Neyman's approach to statistics is set forth in his own text *Probability and Statistics* (New York: Holt, 1950). Ada Lovelace's "Note G" to her translation of L. F. Menabrea, "Sketch of the Analytical Engine Invented by Charles Babbage," is in Philip and Emily Morrison, eds., *Charles Babbage: On the Principles and Development of the Calculator and Other Seminal Writings by Charles Babbage and Others* (New York: Dover, 1961), pp. 225–297.

1. Auguste Dick, *Emmy Noether, 1882–1935* (Boston: Birkhäuser, 1981), p. 130.

2. Felix E. Browder, ed., *Mathematical Developments Arising from Hilbert's Problems* (Providence, RI: American Mathematical Society, 1976), p. 9. This work contains Hilbert's address as well as essays discussing the progress on each of Hilbert's problems from 1900 to the time of writing.

3. Zermelo, "Beweis, dass jede Menge wohlgeordnet werden kann," *Mathematische Annalen*, 59 (1904), 514–516, translated in van Heijenoort, *From Frege to Gödel*, pp. 139–140.

4. Ibid., p. 141.

5. Zermelo, "Untersuchungen über die Grundlagen der Mengenlehre I," *Mathematische Annalen* 65 (1908), 261–281, translated in van Heijenoort, *From Frege to Gödel*, pp. 201–204.

6. Quoted in Moore, *Zermelo's Axiom of Choice*, p. 263.

7. Young and Young, *The Theory of Sets of Points* (New York: Chelsea, 1972), p. 204. This is a recent reprint of the original work.

8. Quoted in Michael Bernkopf, "The Development of Function Spaces with Particular Reference to Their Origins in Integral Equation Theory," *Archive for History of Exact Sciences* 3 (1966/67), 1–96, p. 37.

9. Fréchet, "Generalisation d'un théoreme de Weierstrass," *Comptes Rendus* 139 (1904), 848–850.

10. Hausdorff, *Grundzüge der Mengenlehre* (New York: Chelsea, 1949), p. 213. This Chelsea reprint gives the full text of the original work.

11. Ibid., p. 232.

12. Ibid., p. 244.

13. Ibid., p. 359.

14. Steinitz, "Algebraische Theorie der Körper," *Journal für die Reine und Angewandte Mathematik* 137 (1910), 167–310, p. 167.

15. Weyl, *Space—Time—Matter*, p. 19.

16. Stefan Banach, "Sur les opérations dans les ensembles abstraites et leur application aux équations intégrales," *Fundamenta Mathematicae* 3 (1922), 133–181, p. 134. This is also quoted in Gregory H. Moore, "The Axiomatization of Linear Algebra: 1875–1940."

17. Quoted in Corry, *Modern Algebra*, p. 227.

18. Quoted in Constance Reid, *Hilbert* (New York: Springer-Verlag, 1970), p. 143. More information on Noether's mathematics and the developments it inspired is found in James W. Brewer and Martha K. Smith, eds., *Emmy Noether: A Tribute to Her Life and Work* (New York: Marcel Dekker, 1981).

19. Dick, *Emmy Noether*, pp. 173–174. Aleksandrov's entire talk, as well a memorial talk by van der Waerden, are found in this volume.

20. Corry, *Modern Algebra*, p. 46.

21. Quoted in Corry, *Modern Algebra*, p. 258.

22. Eilenberg and Mac Lane, "General Theory of Natural Equivalences," *Transactions of the American Mathematical Society* 58 (1945), 231–294, p. 237.

23. Quoted in Corry, *Modern Algebra*, p. 391.

24. Quoted in Gigerenzer et al., *The Empire of Chance*, p. 95.

25. Ibid., p. 96.

26. Leibniz, in David Eugene Smith, *A Source Book in Mathematics*, pp. 180–181.

27. See Allan G. Bromley, "Charles Babbage's Analytical Engine, 1838," *Annals of the History of Computing* 4 (1982), 196–217, for more details.

28. Ada Lovelace, "Note G" to her translation of L. F. Menabrea, "Sketch," in Philip and Emily Morrison, eds., *Charles Babbage*, p. 284.

29. Quoted in Goldstine, *The Computer*, pp. 243–244.

30. J. Carlson, A. Jaffe, and A. Wiles, eds., *The Millennium Prize Problems* (Providence, RI: American Mathematical Society, 2006), p. vii.

31. Sylvia Nasar and David Gruber, "Manifold Destiny," *The New Yorker*, August 28, 2006.

Appendix

Using This Textbook in Teaching Mathematics

One of the primary goals of learning the history of mathematics is to employ it in teaching the subject, whether at the elementary, secondary, or college level. This textbook covers the history of most mathematical topics taught at these levels, so in this appendix we present tools for incorporating that history in teaching. First, for each of the standard mathematics courses taught in secondary school or in the undergraduate years, we list the topics generally covered along with the section of the text addressing its history. Of course, in any given school situation, the curriculum will vary, or topics listed in one course may be taught in a different one. Nevertheless, it should be simple enough to match your courses with what is listed here. This listing should also be useful in designing a history of mathematics course by theme. Next we present some specific ideas on how to use the history of mathematics to teach some particular mathematical ideas. Finally, there is a time line of the history of mathematics correlated to major world events to help teachers and their students relate the history of mathematics to world history in general.

 A.1   COURSES AND TOPICS

Pre-Algebra

Number systems, place value, decimals	1.1.1, 1.2.1, 2.1.1, 7.2.1, 8.2.1, 9.2, 10.4.1, 11.2.1, 11.2.2, 12.5, 13.4.1
Arithmetic algorithms	1.1.1, 1.2.1, 1.2.3, 5.2.1, 5.3.2, 7.2.1, 7.2.2, 8.2.2, 10.4.1, 12.1.1, 12.2.1, 19.1.2, 19.1.3
Lengths, areas, volumes	1.1.3, 1.2.2, 3.8, 5.3.1, 5.3.2, 7.3.1
Pythagorean theorem	1.2.3, 7.3.2

Elementary Algebra

Algebraic symbolism	6.2, 10.4.3, 12.1.1, 12.2.1, 12.2.2, 12.2.3, 12.2.4
Linear equations, proportions	1.1.2, 1.2.4, 3.3, 3.5, 10.4.1, 10.5, 19.1.1
Systems of linear equations	1.2.4, 7.4.1, 12.2.1
Algebraic manipulations, laws of exponents	3.3, 6.2, 9.3.3, 10.5.1, 12.1.1, 12.2.1, 12.2.2
Quadratic equations	1.2.4, 3.3, 3.9, 9.3.1, 12.1.2, 12.2.2, 12.2.4

Geometry

Figurate numbers, Pythagorean triples	1.2.3, 2.1.3
Logical arguments and proof	2.1.2, 1.1.3, 2.1.4, 2.3.1, 2.3.2, 2.3.3, 3.2, 6.3, 9.5.1
Axioms for geometry	3.2, 20.1.1, 24.6.1, 24.6.2
Constructions	2.1.4, 3.2, 3.3, 3.4, 4.5.4, 8.3, 9.5.2, 14.2.2, 14.2.3, 21.2.1
Theorems on triangles and parallelograms	3.2, 3.3, 8.3, 20.1.1
Theorems on circles	3.4, 8.3, 10.2.1
Ratio and proportion, similarity	3.5, 5.3.1, 7.3.2
Areas and volumes	3.8, 4.2, 4.3.1, 5.3.2, 7.3.1, 9.5.1, 10.2.1, 10.2.2, 10.2.3, 10.1.1
Solid geometry	3.8, 5.1

Intermediate Algebra

Algebraic symbolism	6.2, 10.4.3, 12.3.2, 12.4.1, 12.4.2, 14.1.1
Quadratic equations and systems	1.2.4, 3.3, 3.9, 6.2.3, 8.4, 9.3.1, 9.3.2, 10.2.1, 10.4.1, 10.4.3, 12.3.1, 12.4.1
Polynomial algebra	3.3, 6.2, 9.3.3
Solving polynomial equations	4.3.3, 7.4.2, 7.4.3, 9.3.5, 9.3.6, 12.1.2, 12.3, 12.4.2, 14.1.1, 19.1.3
Systems of linear equations	7.4.1, 10.4.3
Irrational numbers	3.7, 4.2, 9.3.2, 9.5.4, 10.5.1, 12.5
Elementary number theory	3.5, 3.6, 6.1, 9.4.2, 10.4.2
Complex numbers	12.3.2, 14.1.1, 14.1.2, 19.1.3, 21.3.3, 22.3.1
Elementary combinatorics	8.6, 9.4.1, 9.4.3, 10.3.1, 10.3.2

Trigonometry

Trigonometric functions	5.1.3, 5.2.1, 7.3.2, 8.7.1, 8.7.2, 9.6.1, 9.6.2, 10.2.4, 13.3.1
Solving plane triangles	5.2.2, 9.6.4, 10.2.4, 13.3.1, 13.4.2
Solving spherical triangles	5.2.3, 5.3.3, 9.6.3, 9.6.4, 13.3.1
Problem solving using trigonometry	5.2.3, 5.3.3, 9.6.3, 10.2.2, 10.2.3, 12.4.2, 13.2.2, 13.3.4

Elementary Probability and Statistics

Basic probability laws and calculations	14.3.1, 14.3.2, 14.3.3, 15.1.1, 18.1.2, 18.3.2, 18.3.3
Binomial distribution and normal curves	18.1.2, 18.3.1, 23.2.1, 23.2.2

Mean and standard deviation	23.2.2
Statistical graphs	23.3
Statistical inference and hypothesis testing	18.2.1, 18.2.2, 18.3.1, 18.3.2, 23.1.3, 23.1.4, 25.4.1, 25.4.2
Least squares, regression	23.1.1, 23.1.2

Precalculus

Functions	5.2.3, 17.3.4
Logarithms and exponentials	13.4, 15.2.4, 15.2.7, 17.1.4
Trigonometric functions	15.2.5, 15.2.6, 17.1.3, 17.3.4
Proof by induction	9.3.4, 10.3.2, 14.3.2
Pascal triangle, sequences and series	4.3.2, 7.4, 8.4, 8.6, 9.3.4, 10.3.2, 10.4.3, 10.5.2, 12.2.2, 14.3.2, 16.1.2, 16.2.1
Conic sections	4.4, 4.5, 9.3.5, 13.1.2, 13.1.3, 13.3.4, 13.5.2, 14.2.1, 14.2.2, 14.2.6, 14.5, 16.1.7
Analytic geometry	10.5.2, 14.2, 20.3.1, 20.3.4
Elementary mathematical physics	4.1.1, 4.1.2, 5.1, 5.3.2, 10.5, 10.5.2, 13.3.2, 13.3.3, 13.5
Elementary Diophantine equations	6.2.1, 6.2.2, 6.2.3, 7.5.1, 8.4, 10.4.2, 19.1.3
Theory of equations	2.4.2, 14.1, 14.2.3, 19.2.1, 21.3.1, 21.3.2

Calculus

Areas and volumes	3.8, 4.2, 4.3.1, 4.3.2, 7.3.1, 9.5.5, 13.3.4, 15.2, 16.1.5
Tangents, maximums, minimums	3.4, 4.5.2, 9.3.5, 15.1.1, 15.1.2, 15.1.3, 16.1.4, 17.3.2
Limits	3.7, 3.8, 4.2, 4.3.1, 7.3.1, 8.7.3, 9.2, 15.1.1, 15.2.4, 16.1.6, 16.2.4, 17.4.1, 17.4.2, 17.4.3, 22.1.1
Derivatives	8.7.3, 15.1.1, 15.1.3, 15.3.3, 16.1.3, 16.1.4, 16.1.6, 16.2.1, 16.2.2, 16.2.3, 16.3.1, 16.3.2, 17.3.1, 17.3.5, 22.1.4
Integrals	9.5.5, 15.2.1, 15.2.4, 15.2.6, 15.2.7, 16.1.2, 16.1.5, 16.2.1, 16.2.2, 17.3.3, 17.3.6, 22.1.5
Fundamental theorem of calculus	15.3.2, 15.3.3, 16.1.5, 16.2.1, 16.2.4, 17.3.2, 22.1.5
Arc length	15.3.1, 15.3.2
Power series	8.7.3, 15.3.2, 16.1.1, 16.1.2, 17.3.2, 17.3.4, 17.4.4
l'Hospital's rule	16.3.1
Celestial mechanics	13.3.2, 13.3.3, 13.3.4, 16.1.6, 16.1.7
Differential calculus in two variables	17.2.1, 17.2.2, 17.3.1, 17.3.5
Multiple integration	17.2.3, 17.2.4

Analytic geometry in three dimensions	20.3
Vector calculus	22.3.3, 22.4.1, 22.4.2, 22.4.3

Linear Algebra

Systems of linear equations	7.4.1, 19.1.2, 19.2.2, 21.4.5
Systems of linear inequalities	21.4.6, 25.3.8
Matrices	7.4.1, 21.4.1, 21.4.2
Spectral theory	21.4.3, 21.4.4
Vector spaces	21.3.3, 21.3.4, 24.5.1, 24.5.2, 25.3.3

Differential Equations

Solutions of ordinary differential equations	16.2.4, 17.1.1, 17.1.2, 17.1.3, 17.3.6, 21.4.3
Applications to physics	17.1.1, 17.1.2
Partial differential equations	17.2.5
Fourier series	17.2.5, 22.1.6
Calculus of variations	17.1.5

Abstract Algebra

Congruences	7.5.1, 7.5.2, 8.5.1, 8.5.2, 19.3.2, 21.1.1
Number theory	14.4, 19.3.2, 19.3.3, 21.1.1, 21.5.1
Groups	11.2.5, 19.2.3, 19.3.2, 21.2.2, 21.2.4, 21.2.5, 21.5.2, 21.5.3, 21.5.4, 24.3.4, 25.3.7, 25.6.2
Fermat's Last Theorem	14.4, 19.3.1, 21.1.2, 25.6.1
Galois theory	19.2.3, 21.2.1, 21.2.3, 21.2.4
Algebraic numbers and fields	21.1.2, 21.1.3, 21.1.4, 21.5.5, 25.3.1, 25.3.2
Algebras, rings, ideals	21.1.4, 21.3.3, 24.5.1, 25.3.4, 25.3.5
Boolean algebra	11.2.4, 21.3.5, 25.5.4

Modern Geometry

Attempts to prove parallel postulate	9.5.3, 20.2.1, 20.2.2
Non-Euclidean geometry	24.2.1, 24.2.2, 24.2.5, 24.3.3
Graph theory	11.2.4, 20.4, 24.4.1, 24.4.2, 25.6.3
Projective geometry	5.3.3, 13.1, 13.2.2, 14.5, 20.3.4, 24.3.1, 24.3.2, 24.3.3, 24.3.4
Differential geometry	20.3.2, 20.3.3, 20.3.4, 24.1, 24.2.3
Axioms for geometry	24.6.1, 24.6.2

Advanced Calculus

Set theory	4.3.1, 13.5.1, 15.2.1, 22.2.3, 22.2.4, 25.5.3, 25.1

Development of integers and real numbers	3.5, 3.7, 22.1.3, 22.2.1, 22.2.2, 22.2.5
Functions and series	5.2.3, 22.1.6
Limits and continuity	22.1.1, 22.1.2, 22.2.2
Convergence	22.1.3, 22.1.8
Derivatives and integrals	22.1.4, 22.1.5, 22.1.7, 24.5.3

Complex Analysis

Complex functions	22.3.2, 22.3.4
Cauchy-Riemann equation	22.3.2, 22.3.4
Complex integration	22.3.2, 22.3.3, 22.3.4
Riemann zeta function	17.3.4, 22.3.5

Topology

Point set topology	22.4.1, 25.2.1, 25.2.2, 25.2.3
Manifolds	22.4.3, 24.2.3, 24.2.4, 25.6.4
Algebraic topology	25.2.4, 25.3.6, 25.6.4

 A.2

SAMPLE LESSON IDEAS TO INCORPORATE HISTORY

There are numerous ways to incorporate history into the teaching of mathematics. Four general methods have been described by Man-Keung Siu in "The ABCD of Using History of Mathematics in the (Undergraduate) Classroom" (in Victor Katz, ed., *Using History to Teach Mathematics: An International Perspective,* Washington, DC: MAA, 2000, pp. 3–9). And although Siu writes about the "undergraduate" classroom, his ideas are certainly worthwhile for the secondary classroom as well. His four methods are A (Anecdotes), B (Broad Outline), C (Content), and D (Development of Mathematical Ideas). Here are some examples by way of explanation.

Anecdotes

There are numerous anecdotes about mathematicians to add spice and a little entertainment to a class; they may introduce a human element, forge links with cultural history, or underline a particular concept. Some are described in the text, but many anecdotes are available in online sources or in, for example, the *Mathematical Circles* books of Howard Eves. In particular, there is the story of Galois's duel, during which he recorded much of what became Galois theory in a letter to a friend the night before he was killed. There is the story of the mathematical contests in fifteenth-century Italy during the period of discovery of the solution method for cubic equations. There is Archimedes jumping out of his bath and running through the streets of Syracuse shouting "Eureka" upon his discovery of the laws of hydrostatics. And there is Wiles and the seven-year-long quest in his attic to find the proof of Fermat's Last Theorem.

Broad Outline

History can be used to sketch a broad outline of a subject before you plunge into the teaching of it. For example, before teaching trigonometry, you might describe its origins in Greece in the search to solve spherical triangles so as to predict heavenly phenomena. You might look at how trigonometry traveled to India where new trigonometric functions were developed, and on to the Islamic world where all of the six functions were tabulated and improvements were devised to solve plane and spherical triangles. Finally, you can note that trigonometry was reintroduced to Europe in the work of several mathematicians, all of whom learned significant ideas from their Islamic predecessors.

As another example, before tackling quadratic equations, you can discuss the geometric algebra of the Babylonians and the Greeks as they learned to solve problems that we consider "quadratic." You can then move on to Islam and al-Khwārizmī's first text in the subject. You could also consider how later Islamic writers applied al-Khwārizmī's algorithms to new kinds of numbers. Finally, you can consider Descartes' treatment of the geometric meaning of the solution of quadratic equations, as well as Galileo's use of quadratic methods in his study of the paths of projectiles.

Before beginning the study of calculus, you might describe some of the methods developed in Greece, India, and Islam to solve problems that were later clarified and extended by Europeans in the seventeenth century. Students should become aware of the numerous problems that were solved earlier, as well as why the algorithms of calculus rendered the often ad hoc methods of the past obsolete.

Content

The history of a particular mathematical topic can help students understand some subtle mathematical ideas. For example, Augustin-Louis Cauchy's incorrect proof of the theorem that the limit of a sequence of continuous functions is continuous provides an ideal opportunity for students in advanced calculus to learn that even great mathematicians can be wrong. If you ask students to read that proof, and then work through an example, such as a Fourier series, showing that the result cannot be true in general, they will gain a greater appreciation of the necessity for more careful definitions. They may even discover a way to repair Cauchy's argument on their own.

Euler's discussion of Cramer's paradox provides an ideal entry into the notion of the "rank" of a system of equations. Students can try to figure out how the paradox can be explained and what this means for the solution of certain systems. They can then work through the theorem of Dodgson and some relevant examples to ascertain how to tell the nature of the solution set of a particular system. With enough experience in solving systems, the notions of rank, linear dependence, and independence will appear much more natural.

On a more elementary level, the development of the quadratic formula is an ideal place to use history. Many students have greater success with understanding the formula once they realize that a quadratic equation is a statement about real squares and rectangles. Thus, if under your guidance they work their way in detail through a Babylonian or Islamic argument for solving a quadratic equation by manipulating geometric objects, they may be able to construct at least a version of the quadratic formula on their own.

As a final example, the introduction of complex numbers is often difficult for students to grasp; one year they are told that negative numbers do not have square roots, but the next year they are told the opposite. It may seem to them that mathematical rules are arbitrary. But if you develop the cubic formula and show students how complex numbers appear naturally in the solutions of certain equations with real coefficients and real solutions, they may begin to appreciate their necessity. If you then further develop the geometric notion of a complex number (as Wessel did originally) students may start to believe (as did nineteenth-century mathematicians) that complex numbers are perfectly "real" objects with their own geometric rules of manipulation, rules that can be turned into arithmetic rules as well.

Development of Mathematical Ideas

For some subjects, it can prove worthwhile to organize a course, or a portion thereof, bearing in mind the historical evolution of the subject (assuming that we can find out—at least in outline—what that is). To do this, we need to examine the history of the subject carefully, noting the high points and the discoveries that seemed necessarily to precede other developments, taking special note of those points at which great difficulty arose in understanding a crucial concept, so that we can order the topics of the course to agree as closely as possible with that history.

As one example, we might consider trigonometry. In a standard trigonometry course, one begins by defining the sine and cosine, calculating these values for $30°$, $45°$, and $60°$ via some elementary geometry, then telling the students to use their calculators if they want to calculate the sine of $37°$. Thus, when the sum and difference formulas or the half-angle formula are discussed later, they hold no meaning for the students. It would seem more reasonable to try to develop the subject based on the development by Ptolemy almost 1850 years ago.

To emulate Ptolemy's approach, we begin with the basic trigonometry definitions (noting that here we do differ from Ptolemy, who used chords rather than sines, cosines, and tangents). But then we calculate the sine values by use of geometry. We do not calculate values for just $30°$, $45°$, and $60°$ but develop them for $72°$, $18°$, $36°$, and $54°$, by use of similarity principles and the quadratic formula. We next develop, again using geometry, the half-angle formula and the sum and difference formulas. We then use these to work out trigonometry tables. Now, given that today's students do have calculators, we can allow their use, but just to do arithmetic, including square roots. And it is useful to ask them to carry out all the calculations to 8 or 10 decimal places, since that is what their calculators do anyway. It becomes clear to students that, first, the basic formulas are developed to help them calculate trigonometric values and, second, that they cannot find the sine of $1°$ via this method. Since it would be handy to be able to find such a value, we use approximation, beginning with the observation from their previous calculations that the sine is nearly linear for small values. Students can then easily get a 6-place approximation for the sine of $1°$ and then, in theory, can complete a trigonometric table.

The next stage of trigonometry involves solving plane triangles. This is done as in any trigonometry text, first dealing with right triangles and then developing the laws of sines and cosines to do other triangles. Various examples of the use of solving triangles can be given. But in keeping with the idea of historical development, one then includes a section on spherical trigonometry, where, depending on the class, one can either derive the basic formulas for right triangles or just state them. These can then be applied to two types of problems: the astronomy problems of determining the time of sunrise and sunset and the geography problems of determining distances and angles on the surface of the earth. For this latter problem, it is necessary to be able to solve non–right spherical triangles. So, students can break up the triangles into right triangles and use the earlier formulas or develop new ones. In particular, students should understand that in the spherical case, knowing the three angles determines the triangle. That is, there is no concept of "similarity" for spherical triangles. With this development, students have a firm idea of the ideas of trigonometry and of their importance.

If one wants to go further, one can also develop some analytic trigonometry, as it was done in the seventeenth and eighteenth centuries. It is not particularly difficult to develop even the power series for the sine and the cosine, assuming a knowledge of the binomial theorem and a willingness to use infinitely large and infinitely small numbers.

To take one other example, at the undergraduate level, we consider the case of abstract algebra. Most such courses start with the definition of a group, even though historically this notion was developed only after many decades of examples had been studied. Thus, a course that focuses on the historical evolution of the subject should begin with examples. One place to start is with the idea of the complex numbers as developed through the cubic formula of Cardano. Thus, as a first example of a group, one can have students consider the groups of complex roots of unity. Many of the basic theorems on Abelian groups can be proved initially by using those particular groups. One can next move on to modular arithmetic, as developed by Gauss and others. Students can study the basic properties of the fields with a prime number of elements. Then, as Galois did, they can look at polynomials over such fields. The "imaginary" roots of these polynomials enable students to construct the finite fields of prime power order. Again, these provide substantial examples of various concepts of group theory.

Since the examples so far are all commutative, it is necessary to study other examples. Accordingly, we can work our way through Lagrange's analysis of the solution methods for cubic and quartic polynomial equations, and develop the notion of a permutation of the roots of an equation. We can then have students study permutations in general, and again prove important results, many of which will be essentially the same as results proved in the earlier commutative examples. One other useful object to study at this point is the set of transformations of a geometric object, such as a square, a tetrahedron, or even a dodecahedron. The entire set of such transformations can be thought of as a "sub-object" of the set of permutations of the vertices. But these are nice, concrete geometric examples that students can hold in their hands.

It is only after a thorough study of the various examples that one should introduce the abstract notion of a (finite) group. By this point, students will be familiar enough with the basic theorems in several cases to understand how these results can be applied to the general case of an abstract finite group. At this point, one can proceed as in standard abstract algebra courses by considering, for example, the Sylow theorems and demonstrating how these determine the possible groups of various orders. Although what has been outlined here is, perhaps, only a one-semester course, there are many routes one can take to follow the historical development of other abstract algebra concepts in a second semester. (Note that

Saul Stahl's text *Introductory Modern Algebra: A Historical Approach* (New York: Wiley, 1996) embodies this approach.)

I hope that the examples given will enable teachers and prospective teachers to work out many other specific ways of using the history of mathematics in teaching mathematics. There are many other suggestions to think about in the discussion questions in the text.

 A.3 ## TIME LINE

The time line in Table A.1 highlights significant world political and cultural events that occurred in the centuries from 3000 BCE to 2000 CE, along with the major mathematical ideas that developed concurrently.

TABLE A.1 *Timeline of Mathematics*

Date	Political Events			Cultural Events	Mathematical Ideas	
	Asia	Africa	Europe		Africa/Asia	Europe
3000 BCE	Sumerian kingdoms emerge in what is now Iraq; Harappan civilization in India	Upper and Lower Egypt united		Pyramids built at Giza; writing develops in Mesopotamia and Egypt	Base-60 place value system develops in Mesopotamia; Egyptians develop base-10 grouping system for numbers	
2000	Aryan tribes move into India from the northwest; Shang Dynasty established in China	Egyptian Empire expands into western Asia	Major civilization develops on Crete, later destroyed by earthquakes and invasions	Hammurapi establishes law code in Babylon (c. 1790 BCE)	*Egypt: Rhind* and *Moscow papyri*; unit-fraction calculations; areas and volumes; *Babylonia:* Pythagorean triples; quadratic equations	
1000	King David establishes capital in Jerusalem; Persians conquer and rule southwest Asia and Egypt; Zhou Dynasty in China		Greece settled by invaders from the north; Trojan War; Roman Republic founded	Confucius flourishes (c. 600 BCE)	*India*: Pythagorean theorem; circle measurements *China*: Pythagorean theorem	Thales and mathematical proof; Pythagoras and "All is number"
500			Greek city-states flourish	Vedas codified in India; Classical Age of Greece		Hippocrates and quadrature of lunes, duplication of cube; discovery of incommensurability
400		Carthage a major power in North Africa	Alexander the Great (356–323 BCE) conquers Egypt and southwest Asia	Foundation of Roman legal system		Plato, Theaetetus, Eudoxus, Aristotle
300	Seleucids rule Babylonia; Qin unifies China; Ashoka popularizes Buddhism across India	Ptolemies rule in Egypt	Punic wars begin	Library and museum founded in Alexandria	Euclid: Apollonius	Archimedes

	Political Events			Cultural Events	Mathematical Ideas	
Date	Asia	Africa	Europe		Africa/Asia	Europe
200	Han Dynasty begins in China		Romans extend rule throughout Mediterranean		Hipparchus and beginning of trigonometry; *Suan shu shu* written in China	
100		Cleopatra defeated (30 BCE); Egypt becomes Roman province	Roman Empire established	Birth of Jesus; paper made in China	*Nine Chapters* compiled in China	
0	Fall of second temple in Jerusalem (70)		Roman Empire reaches its greatest extent	Paul's missions transform Christianity into widespread religion	Heron and applied mathematics	
100 CE				New Testament canonized	Ptolemy and the *Almagest*	
200			Barbarian invasions begin		Diophantus and the *Arithmetica*; Liu Hui and surveying	
300	Gupta Dynasty in India		Christianity becomes state religion in Roman Empire; division of Roman Empire into West and East	Council of Nicaea	Pappus	
400			Fall of Rome		Death of Hypatia; Aryabhata and sine tables	
500				Birth of Muhammad	Volume of sphere in China	
600	Muslims conquer Middle East	Muslims conquer North Africa			Brahmagupta and indeterminate equations	
700	Bagdad founded as capital of Islamic Empire		Muslims conquer Spain; Battle of Tours		*India:* development of decimal place-value system	

	Political Events			Cultural Events	Mathematical Ideas	
Date	Asia	Africa	Europe		Africa/Asia	Europe
800			Charlemagne crowned Holy Roman Emperor (800)	House of Wisdom established in Baghdad	Al-Khwārizmī and algebra	
900			Beginnings of English and French kingdoms	Russians converted to Christianity	Abū Kāmil and algebra; Abū al-Wafā and spherical trigonometry	
1000	First Crusade establishes Christian rule in Jerusalem (1095–1099)		William the Conqueror conquers England		*Egypt:* Ibn al-Haytham, sums of powers, and volumes of paraboloids *Persia:* Omar Khayyam and solution of cubics *China:* Pascal triangle developed	
1100	Northern India conquered by Muslims; Christian kingdoms established in Middle East	Islam expands into sub-Saharan Africa		Gothic art and architecture flourish	Bhaskara and Pell equation; al-Samaw'al and development of decimals and polynomial algebra; Sharaf al-Dīn al-Tūsī and cubic equations	Translations of mathematical works from Arabic into Latin; Abraham ibn Ezra and combinatorics
1200	Constantinople sacked by Crusaders (1202–1204); Genghis Khan conquers much of Asia; Marco Polo reaches China; Muslim sultanate founded in India; Muslims reestablish rule in Middle East		Christian kings defeat Muslims in Spain	Magna Carta establishes foundation of English constitutional rights; paper produced in Italy; universities founded in Paris, Cambridge, and Oxford	*Persia:* Nasīr al-Dīn al-Tūsī and trigonometry *China:* Chinese remainder theorem; solution of polynomial equations *Morocco:* development of combinatorics	Leonardo of Pisa
1300	Ming Dynasty founded in China	Timbuktu flourishes as intellectual center in Mali	Hundred Year's War wages between England and France	Great Plague ravages Europe		Levi ben Gerson and combinatorics; Mertonian school and kinematics

	Political Events			Cultural Events	Mathematical Ideas	
Date	Asia	Africa	Europe		Africa/Asia	Europe
1400	Ottomans overthrow Byzantine Empire; Chinese voyage to India and Africa	Bartholomew Dias rounds Cape of Good Hope	Renaissance begins in Italy; Columbus sets sail	Leonardo da Vinci; Gutenberg prints Bible	Power series for sine and cosine in India; al-Kāshi and decimal calculations	Luca Pacioli and the *Summa*
1500	Portuguese establish trading posts in India		Spanish colonize the Americas; Spanish Armada is defeated	Luther and the Protestant Reformation; Portuguese explorers reach China; William Shakespeare		Solution of cubic equations; Copernicus and heliocentrism; numerous algebra texts written; Viète and algebraic symbolism
1600			English colonies founded in North America; English civil war and Glorious Revolution; Royal Society and Académie des Sciences founded in London and Paris, respectively			Galileo and physics; Kepler and his laws; Newton and the *Principia*; Napier and logarithms; development of algebraic symbolism; analytic geometry and calculus; early work in probability
1700			American and French revolutions	Founding of École Polytechnique		Euler; ordinary and partial differential equations; calculus of several variables; Bayes, Laplace, and statistical inference
1800	British rule India	Africa carved up into European colonies	Napoleon conquers Europe, later meets defeat at Waterloo; Congress of Vienna; revolutions of 1848; German empire founded	Steam engine invented; railroads established; steamships sail the oceans; telephone and telegraph		Rigor in analysis; groups and fields; Galois theory; non-Euclidean geometry; growth of statistics

	Political Events		Cultural Events		Mathematical Ideas
Date	Asia	Africa	Europe	Africa/Asia	Europe
1900	Chinese overthrow emperor; Ottoman Empire collapses; Europeans carve up Middle East; India gains independence; Communist rule in China begins; Korean and Vietnamese wars; Israeli-Arab conflict begins and continues	Colonies gain independence	First World War; German, Austrian, Russian empires collapse; Bolshevik Revolution in Russia; Nazi period in Germany; Second World War; collapse of Soviet Union	Airplane invented; artificial satellites; travel to moon		Abstraction in algebra; statistical methodology; computer revolution; development of topology

General References in the History of Mathematics

Each chapter of this text provides references to works useful in providing further information on the material of that chapter. In general, however, if one wants to learn the history of a specific topic in mathematics, it pays to begin one's search in one of the following works:

- Ivor Grattan-Guinness, ed., *Companion Encyclopedia of the History and Philosophy of the Mathematical Sciences* (London: Routledge, 1994). This two-volume encyclopedia includes brief (sometimes too brief) articles on some 180 topics in the history and philosophy of mathematics, each written by an expert in the field. The encyclopedia is particularly strong on what are generally thought of as applied mathematical topics, such as mechanics, physics, engineering, and the social sciences, but is relatively weak on some of the more standard topics in the history of mathematics. Nevertheless, it is an excellent first source on the history of mathematics.
- Joseph W. Dauben, ed., *The History of Mathematics from Antiquity to the Present: A Selective Annotated Bibliography* (Providence, RI: American Mathematical Society, 2000). This CD-ROM contains approximately 4800 entries chosen by 38 experts, each of whom has attempted to pick the "best" works in their respective fields. Although the CD format is not as easy to use as a printed book, this is probably the best place to start on a quest for historical articles on a particular topic.
- Morris Kline, *Mathematical Thought from Ancient to Modern Times* (New York: Oxford University Press, 1972). This book is the most comprehensive of the recent works in the history of mathematics and pays particular attention to the nineteenth and twentieth centuries. It provides chapter bibliographies for further help. It is, however, lacking completely in information about Chinese mathematics and is very sketchy in information about the mathematics of India and the Islamic world.
- Charles C. Gillispie, ed., *Dictionary of Scientific Biography* (New York: Scribners, 1970–1990). This 18-volume encyclopedia (including two recent supplementary volumes) is in essence a comprehensive history of science organized biographically. There are articles about virtually every mathematician mentioned in this text and, naturally, articles about many who are not mentioned. There are also special essays on topics in Egyptian, Babylonian, Indian, Japanese, and Mayan mathematics and astronomy. An extensive index allows one to begin with a mathematical topic and find references to all the mathematicians who considered it.

In addition to these basic references, new books on various topics in the history of mathematics continue to appear. Two good places to look for these are in MAA Reviews, available online at www.maa.org, and in the Book List, published monthly in the *Notices of the American Mathematical Society*.

For original sources, there are several collections of material from important mathematical works, all translated into English. These include Henrietta O. Midonick, ed., *The Treasury of Mathematics* (New York: Philosophical Library, 1965); Ronald Calinger, ed., *Classics of Mathematics* (Englewood Cliffs, NJ: Prentice Hall, 1995); D. J. Struik, ed., *A Source Book in Mathematics, 1200–1800* (Cambridge: Harvard University Press, 1969); Garrett Birkhoff, ed., *A Source Book in Classical Analysis* (Cambridge: Harvard University Press, 1973); David Eugene Smith, ed., *A Source Book in Mathematics* (New York: Dover, 1959); John Fauvel and Jeremy Gray, eds., *The History of Mathematics: A Reader* (London: Macmillan, 1987); and Victor J. Katz, ed., *The Mathematics of Egypt, Mesopotamia, China, India, and Islam: A Sourcebook* (Princeton: Princeton University Press, 2007).

Naturally, more research continues to be done in the history of mathematics, and there are many journals that publish articles on the subject. The most important such journals, which can be found in most major university libraries, are *Historia Mathematica* and *Archive for History of Exact Sciences*. The former publishes in each issue a list of abstracts of recent articles in the history of mathematics. To keep up fully with current literature, however, it is best to consult *Mathematical Reviews*, published monthly by the American Mathematical Society, or the *Isis Current Bibliography of the History of Science and its Cultural Influences*, published as its fifth issue every year by *Isis*, the journal of the History of Science Society. This latter volume contains an extensive listing, by subject, of articles published during the previous twelve months on the history of science, including, of course, the history of mathematics. These sources are today available online at many research libraries.

Finally, the Internet contains numerous sources on the history of mathematics, some of which are excellent and some of which should best be ignored. Among the best sites with which to start are the following:

- David Joyce's History of Mathematics homepage (http://aleph0.clarku.edu/~djoyce/mathhist/). This is the starting point to a wealth of resources provided by David Joyce of Clark University, Worcester, MA. There are pages on regional mathematics, specific topics, books, journals, bibliographies, history of mathematics texts, and so forth, as well as an excellent list of Web Resources clearly categorized, an extensive chronology, and time lines. In addition, Joyce has the entire text of Euclid's *Elements*, with many proofs having interactive diagrams. This section also shows the relationships among the various theorems in the *Elements*.

- The Math Forum Internet Resource Collection (http://mathforum.org/library/topics/history/). This site provides an extensive list of annotated links to other sites. The sites are ordered alphabetically and the collection can be viewed in outline or annotated form. There is a well-designed search engine that allows for a variety of searches.

- St. Andrews MacTutor History of Mathematics (http://www-history.mcs.st-and.ac.uk/history/). A collection of biographies of mathematicians and a variety of resources on the developments of various branches of mathematics. An extremely rich and extensive site with some excellent pages, although the quality is not always consistent. In particular, the biographies should be viewed with care.

- Trinity College, Dublin, History of Mathematics archive (http://www.maths.tcd.ie/pub/HistMath/HistMath.html). This site, created and maintained by David Wilkins, includes biographies of some seventeenth- and eighteenth-century mathematicians, material on Berkeley, Newton, Hamilton, Boole, Riemann, and Cantor, as well as an extensive directory of history of mathematics websites.

- The British Society for the History of Mathematics (http://www.dcs.warwick.ac.uk/bshm/resources.html). This site, maintained by June Barrow-Green for the Society, contains an extensive list of history sites, carefully annotated and categorized. Anyone seriously interested in the history of mathematics should consider joining the BSHM; the Society has a wonderful newsletter and conducts numerous meetings on various topics in the history of mathematics.

- *Convergence* (http://convergence.mathdl.org/) is the Mathematical Association of America's online magazine on the history of mathematics and its use in teaching. It contains articles, book reviews, translations of original sources, quotations on mathematics, a "What happened this day in history?" feature, a collection of portraits of mathematicians, and much more.

Answers to Selected Exercises

CHAPTER 1

1. 375: $‖∩∩∩∩♉♉♉$ $\mathsf{Y Y Y} \prec \mathsf{Y Y Y}$

4856: $‖∩∩∩\,♉♉♉♉\,♉♉♉♉$ $\mathsf{Y} \lll \mathsf{Y Y Y}$

3.

1	$\overline{2}$	$\overline{14}$
$\overline{2}$	$\overline{4}$	$\overline{28}$
$\overline{4}$	$\overline{8}$	$\overline{56}$

1

5. $10 \times \overline{\overline{3}}\,\overline{30} = 7$

7. $99\,\overline{2}\,\overline{4} = 99\frac{3}{4}$

9. $16\,\overline{2}\,\overline{8} = 16\frac{5}{8}$

11. $1\,\overline{4}\,\overline{76} = 1\frac{5}{19}$

17. 1;24 0;52 0;27,30 0;39,36

19. $18 \Longleftrightarrow 3, 20$; $32 \Longleftrightarrow 1, 52, 30$; $54 \Longleftrightarrow 1, 6, 40$; $1, 04 \Longleftrightarrow 56, 15$. An integer n is regular if and only if the only prime divisors of n are 2, 3, or 5.

25. An approximate reciprocal of 1;45 is 0;34,17,09. An approximation to $\sqrt{3}$ is 1;43,55,42.

27. 67319, 72000, 98569

29. $\ell = 32; w = 24$

31. $1\ mina\ 15\frac{5}{6}\ gin$

35. $s = \frac{1}{2}$

37. $AD = 0;36$; $DE = 0;28,48$; $EF = 0;23,02,24$; $BD = 0;27$; $DF = 0;17,16,48$; $FC = 0;30,43,12$

39. $\ell = 4; w = 3$

CHAPTER 2

1. $125 = \rho\kappa\epsilon$; $62 = \xi\beta$; $4821 = {'}\delta\omega\kappa\alpha$; $23{,}855 = M^{\beta}\,{'}\gamma\omega\nu\epsilon$

3. $22\frac{2}{9} = \kappa\beta\ \varsigma'\iota'\beta'$

7. 480 feet

21. $t = 51.02$ seconds; $d = 510.2$ yards

CHAPTER 3

15. Use the intersection of the perpendicular bisectors of the three sides as the center of the circle.

17. Inscribe both the side of an equilateral triangle and the side of an equilateral pentagon in the circle, where the two sides have a common endpoint on the circle. Then the arc connecting the other two endpoints is two-fifteenths of the total circumference. Therefore, the line segment from one of the endpoints of that arc to its midpoint will be one side of a regular 15-gon.

19. gcd(963, 657) = 9; gcd(2689, 4001) = 1.

21. $33 = 2 \cdot 12 + 9$; $12 = 1 \cdot 9 + 3$; $9 = 3 \cdot 3$. $11 = 2 \cdot 4 + 3$; $4 = 1 \cdot 3 + 1$; $3 = 3 \cdot 1$. So both pairs are represented by (2, 1, 3).

25. Because $a_1 b_j = a_j b_1$ for all j, it follows that

$$a_1(b_1 + b_2 + \cdots + b_n) = a_1 b_1 + a_1 b_2 + \cdots + a_1 b_n$$
$$= b_1 a_1 + b_1 a_2 + \cdots + b_1 a_n$$
$$= b_1(a_1 + a_2 + \cdots + a_n),$$

and the proposition follows immediately.

31. ab is the mean proportional between a^2 and b^2.

37. Since BC is the side of a decagon, triangle EBC is a 36-72-72 triangle. Thus, $\angle ECD = 108°$. Since CD, the side of a hexagon, is equal to the radius CE, it follows that triangle ECD is an isosceles triangle with base angles equal to

36°. Thus, triangle EBD is a 36-72-72 triangle and is similar to triangle EBC. Therefore, $BD : EC = EC : BC$ or $BD : CD = CD : BC$ and the point C divides the line segment BD in extreme and mean ratio.

CHAPTER 4

1. $4\frac{1}{6}$ m from the 14-kg end.

5. Set $r = 1$, t_i and u_i as in the text, and P_i the perimeter of the ith circumscribed polygon. Then the first 10 iterations of the algorithm give the following:

$t_1 = .577350269$	$u_1 = 1.154700538$	$P_1 = 3.464101615$
$t_2 = .267949192$	$u_2 = 1.03527618$	$P_2 = 3.21539031$
$t_3 = .131652497$	$u_3 = 1.008628961$	$P_3 = 3.159659943$
$t_4 = .065543462$	$u_4 = 1.002145671$	$P_4 = 3.146086215$
$t_5 = .03273661$	$u_5 = 1.0005357$	$P_5 = 3.1427146$
$t_6 = .016363922$	$u_6 = 1.00013388$	$P_6 = 3.141873049$
$t_7 = .0081814134$	$u_7 = 1.000033467$	$P_7 = 3.141662746$
$t_8 = .004090638249$	$u_8 = 1.000008367$	$P_8 = 3.141610175$
$t_9 = .002045310568$	$u_9 = 1.000002092$	$P_9 = 3.141597032$
$t_{10} = .001022654214$	$u_{10} = 1.000000523$	$P_{10} = 3.141593746$

7. Let the equation of the parabola be $y = -x^2 + 1$. Then the tangent line at $C = (1, 0)$ has the equation $y = -2x + 2$. Let the point O have coordinates $(-a, 0)$. Then $MO = 2a + 2$, $OP = -a^2 + 1$, $CA = 2$, $AO = -a + 1$. So $MO : OP = (2a + 2) : (1 - a^2) = 2 : (1 - a) = CA : AO$.

9. Since $BOAPC$ is a parabola, we have $DA : AS = BD^2 : OS^2$, or $HA : AS = MS^2 : OS^2$. Thus, $HA : AS =$ (circle in cylinder) : (circle in paraboloid). Therefore, the circle in the cylinder, placed where it is, balances the circle in the paraboloid placed with its center of gravity at H. Since the same is true whatever cross section line MN is taken, Archimedes could conclude that the cylinder, placed where it is, balances the paraboloid, placed with its center of gravity at H. If we let K be the midpoint of AD, then K is the center of gravity of the cylinder. Thus, $HA : AK =$ cylinder : paraboloid. But $HA = 2AK$. So the cylinder is double the paraboloid. But the cylinder is also triple the volume of the cone ABC. Therefore, the volume of the paraboloid is 3/2 the volume of the cone ABC, which has the same base and same height.

13. Let r be the radius of the sphere. Then we know from calculus that the volume of the sphere is $V_S = \frac{4}{3}\pi r^3$

and the surface area of the sphere is $A_S = 4\pi r^2$. The volume of the cylinder whose base is a great circle in the sphere and whose height equals the diameter is $V_C = \pi r^2 (2r) = 2\pi r^3$, while the total surface area of the cylinder is $A_C = (2\pi r)(2r) + 2\pi r^2 = 6\pi r^2$. Therefore, $V_C = \frac{3}{2} V_S$ and $A_C = \frac{3}{2} A_S$, as desired.

15. Suppose the cylinder P has diameter d and height h, and suppose the cylinder Q is constructed with the same volume but with its height and diameter both equal to f. It follows that $d^2 : f^2 = f : h$, or that $f^3 = d^2 h$. It follows that one needs to construct the cube root of the quantity $d^2 h$, and this can be done by finding two mean proportionals between 1 and $d^2 h$, or, alternatively, two mean proportionals between d and h (where the first one will be the desired diameter f).

17. The focus of $y^2 = px$ is at $(\frac{p}{4}, 0)$. The length of the latus rectum is $2\sqrt{p\frac{p}{4}} = p$.

19. Let the parabola be $y^2 = px$ and the point $C = (x_0, y_0)$. Then the tangent line at C has slope $\frac{p}{2y_0}$, and the equation of the tangent line is $y = \frac{p}{2y_0}(x - x_0) + y_0$. If we set $y = 0$, we can solve this equation for x to get $x = -x_0$.

25. By *Conics* II–8, if we pass a secant line through the hyperbola $xy = 1$, which goes through points M and N on that curve and points T and U on the y axis and x axis, respectively (the asymptotes), then the segments TM and TN are equal. Thus, if we let M approach N, then the secant line approaches the tangent line at N and therefore the two line segments TN, NU between N and the asymptotes are equal. Therefore, the triangles TSN and NRU are congruent. If the coordinates of N are $(x_0, \frac{1}{x_0})$, then $TS = NR = \frac{1}{x_0}$ and $NS = x_0$. So the slope of the tangent line TNU is

$$\frac{TS}{SN} = -\frac{1/x_0}{x_0} = -\frac{1}{x_0^2}.$$

33. Apply a rectangle equal to one-fourth of the rectangle on the parameter N and the axis AB of the hyperbola to the axis (on each side) that exceeds by a square figure. This application results in two points F and G on the axis "produced by the application." These are the foci of the hyperbola. The analogue to III–48 is that lines from the two foci to any point on the hyperbola make equal angles with the tangent to the hyperbola at that point. The analogue to III–52 is the

result that the difference of the lengths of these two lines is equal to the axis AB of the hyperbola.

35. If the two parallel lines are $x = 0$ and $x = k$ and the perpendicular line is the x axis, then the equation of the curve satisfying the problem is $y^2 = px(k - x)$ or $y^2 = kpx - px^2$.

CHAPTER 5

1. crd $30° = 31;03,20$ crd $15° = 15;39,47$ crd $7\frac{1}{2}° = 7;50,54$

5. crd $12° = 12;32,36$ crd $6° = 6;16,49$ crd $3° = 3;08,29$ crd $1\frac{1}{2}° = 1;34,15$ crd $\frac{3}{4}° = 0;47,07$

11. When $\lambda = 90°$, then $\delta = 23°51'$ and $\alpha = 90°$. When $\lambda = 45°$, then $\delta = 16°37'$ and $\alpha = 42°27'$. The values at $270°$ and $315°$ are the negatives of the values at $90°$ and $45°$, respectively.

13. The length of daylight is 14 hrs, 6 min. Sunrise is at 4:57 a.m. and sunset is at 7:03 p.m., local time.

17. Approximately May 20 and July 21.

19. About April 30.

29. Since the ratio of a degree at latitude α to a degree at the equator is as $\cos\alpha$, we just need to check approximations to $\cos\alpha$: $\cos 23\frac{5}{6}° = 0.9147$, while $4\frac{7}{12} : 5 = 0.9167$; $\cos 16\frac{5}{12}° = 0.9592$, while $4\frac{5}{6} : 5 = 0.9667$.

CHAPTER 6

1. The nth pentagonal number is $\frac{3n^2 - n}{2}$. The nth hexagonal number is $2n^2 - n$.

3. Since $c : a = (c - b) : (b - a)$, we have $ac - ab = bc - ac$ or $b(a + c) = 2ac$, or the sum of the extremes multiplied by the mean is twice the product of the extremes.

5. 4, 10, 12 is a subcontrary proportion but $10 \cdot 12 \neq 2 \cdot 10 \cdot 4$

7. 84

9. $72\frac{1}{4} = (17/2)^2$ and $132\frac{1}{4} = (23/2)^2$

11. $x = 121/16$

13. $x = 12$, $y = 8$

15. $x = 5/7$, $y = 267/343$

17. One such triangle has sides 96, 28, and 100, with angle bisector 35.

19. Assume that the theorem is true, where C cuts the line AB in extreme and mean ratio. Then $AB^2 + BC^2 = 3AC^2$. But since $AB = AC + BC$, we have $(AC + BC)^2 + BC^2 = 3AC^2$. This reduces to $AC^2 + 2AC \cdot BC + 2BC^2 = 3AC^2$ or $AC \cdot BC + BC^2 = AC^2$. This in turn implies that $BC(AC + BC) = AC^2$ or that $AB \cdot BC = AC^2$. But this is precisely the statement that AB is cut in extreme and mean ratio at C.

21. Suppose the hexagon has perimeter $6d$. Then it is composed of six equilateral triangles of side d. Since such a triangle has area $\frac{\sqrt{3}}{4}d^2$, it follows that the hexagon has area $\frac{3\sqrt{3}}{2}d^2$. The square with perimeter $6d$ has side equal to $\frac{3}{2}d$ and therefore area equal to $\frac{9}{4}d^2$, which is less than the area of the hexagon.

23. 336

25. A has $15\frac{5}{7}$ coins, while B has $18\frac{4}{7}$ coins.

CHAPTER 7

1. We write, in order, the Chinese form of 56, 554, 63, and 3282:

3. 234

5. $51\frac{41}{109}$, $32\frac{12}{109}$, $16\frac{56}{109}$

7. $10\frac{15}{16}$ pounds

13. 50.5 ch'ih

15. The proof of Exercise 14 shows that

$$a - \frac{ab}{a+b+c} + b - \frac{ab}{a+b+c} = c.$$

It follows that $a + b - c = \frac{2ab}{a+b+c}$, or that the diameter D of the circle equals $a - (c - b)$. Then note that $D^2 = a^2 - 2a(c - b) + (c - b)^2 = a^2 - 2ac + 2ab + c^2 - 2bc + b^2 = 2(c^2 - bc - ac + ab) = 2(c - a)(c - b)$, so $D = \sqrt{2(c - a)(c - b)}$.

17. 57.5 feet

19. $\frac{9}{25}$, $\frac{7}{25}$, $\frac{4}{25}$

21. 6.35

23. 23

25. 9

CHAPTER 8

1. 237

5. If we calculate the sum and difference of the given fractions, we get 0.878681752. If the square of this side is equal to the area of a circle of diameter 1, then $(0.878681752)^2 = \frac{\pi}{4}$, or $\pi = 4(0.878681752)^2 = 3.088326491$.

9. $8\frac{3}{4}$ days

11. $\frac{1}{14}$ of a day; $\frac{2}{14}, \frac{3}{14}, \frac{4}{14}, \frac{5}{14}$

13. 24

15. $x = 2, y = 1000, N = 3000$

17. $x = 2, y = 731; x = 20, y = 7310$

19. 59

21. $m = 12, n = 53$

23. $x = 9, y = 82$

25. $x = 180, y = 649$

29. $\sin 15° = 890; \sin 18°45' = 1105; \sin 22°30' = 1315$

CHAPTER 9

1. 37,210,674

3. a. 3; b. 6, 4

7. a. $x = (1 + \sqrt{2} + \sqrt{13 + \sqrt{8}})^2$; b. $x = 4\frac{1}{2} - \sqrt{8}$

9. $x = \sqrt[4]{\sqrt{1250} - 50}, y = 10/x, z = 100/x^2$

15. Since triangle EGI is right, $EH : HG = HG : HI$ or $HG^2 = EH \cdot HI$. Also, $HG^2 = DH \cdot HB$. Therefore, $EH \cdot HI = DH \cdot HD$ or $DH : EH = HI : HB$. Then $(DH - EH) : EH = (HI - HB) : HB$ or $EC : EH = BI : HB$. Since $EH : HB = AE : GH$, we have $HB : GH = EH : AE = EH : EC$. So $BI : HB = EC : EH = GH : HB$ and $BI = GH$. But $EG = EB$. So $EI = EB + BI = EG + GH$.

17. Let $f(x) = x^3 - bx^2 + cx$. For three positive solutions to exist, we first must have $b^2 - 3c \geq 0$. Then, setting $x_1 = \frac{b}{3} - \frac{\sqrt{b^2 - 3c}}{3}$, we must also have $f(x_1) > d$. In the given case, $b^2 - 3c < 0$, so there is only one positive solution.

21. There are 2 solutions if $4c^3 > 27d^2$, 1 solution if $4c^3 = 27d^2$, and no solutions if $4c^3 < 27d^2$.

23. The sum of the proper divisors of 1184 is $1 + 2 + 4 + 8 + 16 + 32 + 37 + 74 + 148 + 296 + 592 = 1210$. The sum of the proper divisors of 1210 is $1 + 2 + 5 + 10 + 11 + 22 + 55 + 110 + 121 + 242 + 605 = 1184$. But these numbers are not an instance of the theorem.

29. 124°32'

31. The distance along the latitude circle is 5745 miles. The distance along a great circle is 5319 miles.

33. 3460 miles

35. 3460 miles

39. If $AB = 60°$, $AC = 75°$, and $BC = 31°$, then angle A is 29°32', angle B is 112°25', and angle C is 55°59'.

CHAPTER 10

1. 3600, 2400, 1200

3. 375 paces

5. 3.848

7. 10

9. 50 feet

11. $12\frac{1}{2}$

21. 60 *dinars* will buy 30 *litras* of the first drug, 20 *litras* of the second drug, 5 *litras* of the third drug, and 3 *litras* of the fourth drug.

25. $159\frac{11}{24}$ Pisan *denarii*

27. $10\frac{58}{77}$ pounds

29. The first has $1\frac{4}{9}$ *denarii*, while the second has $3\frac{4}{9}$ *denarii*.

33. $(\frac{25}{24})^2$

35. $x = 6, y = 4$

37. $(27 : 16)^{1/3}$

41. The total distance traveled is given by

$$D = 1 \cdot \frac{1}{2} + \left(\frac{1+2}{2}\right) \cdot \frac{1}{4} + 2 \cdot \frac{1}{8} + \left(\frac{2+4}{2}\right) \cdot \frac{1}{16}$$

$$+ 4 \cdot \frac{1}{32} + \left(\frac{4+8}{2}\right) \cdot \frac{1}{64} + \cdots$$

$$= \frac{1}{2} + \frac{3}{8} + \frac{1}{4} + \frac{3}{16} + \frac{1}{8} + \frac{3}{32} + \cdots$$

$$= \left(\frac{1}{2} + \frac{1}{4} + \frac{1}{8} + \cdots\right) + \frac{3}{4}\left(\frac{1}{2} + \frac{1}{4} + \frac{1}{8} + \cdots\right)$$

$$= 1 + \frac{3}{4}$$

$$= \frac{7}{4}$$

CHAPTER 11

3. The minimum number of days between (t_0, v_0, y_0) and (t_1, v_1, y_1) is $n \equiv 14{,}600(t_1 - t_0) - 14{,}235(v_1 - v_0) - 364(y_1 - y_0) \pmod{18{,}980}$.

5. A woman in section e has a mother in section m, a father in section f, a husband in section mf, and children in section m^2.

7. $C_{13} \oplus C_{16} = C_{14} \oplus C_1 = C_{11} \oplus C_2 = (2, 1, 2, 2)$.
   In general,

$$C_{13} \oplus C_{16} = (C_9 \oplus C_{10}) \oplus (C_{15} \oplus C_1)$$
$$= C_8 \oplus C_7 \oplus C_6 \oplus C_5 \oplus C_3 \oplus C_{14} \oplus C_1$$
$$= C_8 \oplus C_7 \oplus C_6 \oplus C_5 \oplus C_9 \oplus C_{10} \oplus C_{14} \oplus C_1$$
$$= 2C_8 \oplus 2C_7 \oplus 2C_6 \oplus 2C_5 \oplus C_{14} \oplus C_1 = C_{14} \oplus C_1$$
$$= C_{11} \oplus C_{12} \oplus C_1 = C_4 \oplus C_3 \oplus C_2 \oplus C_1 \oplus C_1$$
$$= C_4 \oplus C_3 \oplus C_2 = C_{11} \oplus C_2$$

CHAPTER 12

1. $\frac{11}{90}$ of a florin

3. 5.93 pounds

5. Third partner invested $103\frac{8}{43}$; profit of second partner was 129; profit of third partner was 153.

7. $3\frac{1}{3}$ hours

9. $5\frac{359}{389}, 4\frac{30}{389}$, or 4,6

11. $\sqrt{43} + 5 - \sqrt{18}, 5 + \sqrt{18} - \sqrt{43}$

15. $\frac{2}{7}\sqrt[3]{1225}, \frac{2}{5}\sqrt[3]{1225}$

17. $\frac{4}{3}$ hours

19. $5 + \sqrt{2}$

21. $5 - 2\sqrt{3 - 2\sqrt{2}}, 5 + 2\sqrt{3 - 2\sqrt{2}}$

23. The fourth root of 10,556,001 is 57.

25. 3, 5

31. $\sqrt[3]{4} + \sqrt[3]{2}$

33. $5, 2 \pm \sqrt{3}$

35. Francis: -48; dowry: 52

37. $4 \pm \frac{4}{3}\sqrt{3}$

39. $4 + \sqrt{-1}$

43. 13⓪3①9②5③; 22⓪8①6②4③2④;
    product is 306⓪2①6②5③9④5⑤9⑥0⑦;
    quotient is 1⓪7①.

CHAPTER 13

5. Distance from equator to 10°: 10.05 cm; distance from 10° to 20°: 10.37 cm; distance from 20° to 30°: 11.05 cm; distance from equator to 5°: 5.01 cm; distance from 10° to 15°: 5.12 cm; distance from 20° to 25°: 5.41 cm.

7. $AB = 8.46$; $AG = 14.10$

9. $a = BC = 15.78$; $c = AB = 33.06$; $b = AC = 30.06$

11. $a = 46°50'$, $b = 63°29'$, $c = 72°13'$

13. 121.41

17. 1.88 years = 687 days

19. $\text{Nlog}(xy) = \text{Nlog}\, x + \text{Nlog}\, y - \text{Nlog}\, 1$; $\text{Nlog}(x/y) = \text{Nlog}\, x - \text{Nlog}\, y + \text{Nlog}\, 1$

CHAPTER 14

1. $(b + a)(c - a)(df - aa) = 0$ becomes $bcdf + (-bdf + cdf)a + (-bc - df)aa + (b - c)aaa + aaaa = 0$. The roots are $a = -b$, $a = c$, and $a = \pm\sqrt{df}$.

3. If one substitutes $a = e + r$ into the equation $aaa - 3raa = 2xxx$, one gets, in modern notation, $(e + r)^3 - 3r(e + r)^2 = 2x^3$. If one expands this, it reduces to $eee - 3rre = 2xxx + 2rrr$. Applying this to $aaa - 6aa = 400$ gives us $eee - 12e = 416$. A root of the latter equation is $e = 8$. So a root of the former equation is $a = 10$.

5. $x = -9 \pm \sqrt{57}$

9. $xy + c = rx + sy$ is the same as $(x - s)(y - r) = rs - c$. The asymptotes are $y = r$ and $x = s$.

13. $(x - \frac{t}{4})^2 + (y - \frac{u}{4})^2 = \frac{m}{4} + \frac{1}{16}t^2 + \frac{1}{16}u^2 - \frac{1}{4}v^2 - \frac{1}{4}w^2$, where $t = x_1 + x_2 + x_3 + x_4$, $u = y_1 + y_2 + y_3 + y_4$, $v = \sqrt{x_1^2 + x_2^2 + x_3^2 + x_4^2}$, $w = \sqrt{y_1^2 + y_2^2 + y_3^2 + y_4^2}$, and m is the area.

17. $(e - cg)y^2 + (de + fgc - bcg)xy + bcfgx^2 + (dek - fg\ell c)y - bcfg\ell x = 0$

23. The equation with integral coefficients is $z^3 - 9z^2 + 26z - 24 = 0$. The roots of the original equation are $\frac{2\sqrt{3}}{9}, \frac{\sqrt{3}}{3}, \frac{4\sqrt{3}}{9}$.

29. The probability of no six in four throws is $(\frac{5}{6})^4 = \frac{625}{1296}$. Thus, the odds in favor are $(1296 - 625) : 624 = 671 : 625$.

31. $42 : 22$

35. $31 : 30$

37. $A : 1000$ vs. $B : 8139$

CHAPTER 15

1. $s = \frac{20\sqrt{3}}{3}$, $V = \frac{8000\sqrt{3}}{9} \approx 1539.6$

7. $t = \frac{pxy - 3y^3}{3x^2 - py}$

9. $t = -\frac{a^2 y^2}{b^2 x} = -\frac{a^2 - x^2}{x}$

11. The slope of the normal line is $\frac{-y_0}{v - x_0} = -\frac{\sqrt{x_0}}{1/2} = -2\sqrt{x_0}$. Therefore, the slope of the tangent line is $\frac{1}{2\sqrt{x_0}}$.

13. The maximum occurs when $x = -\sqrt{\frac{2b^2 a/3c}{9a - 3b}}$ and the minimum occurs when x is the positive square root of that expression.

15. The subtangent t is $t = \frac{2y^2}{b - 2x}$. The slope of the tangent line is given by $\frac{y}{t} = \frac{by - 2xy}{2y^2} = \frac{b - 2x}{2y}$.

23. $a_{3,\frac{1}{2}} = \frac{35}{16}, a_{3,\frac{3}{2}} = \frac{105}{16}, a_{3,\frac{5}{2}} = \frac{231}{16}$

25. $L = \frac{1024}{625}\left[\sqrt{a}\left(\frac{a^2}{5} - \frac{a}{3}\right) + \frac{2}{15}\right]$, where $a = 1 + \frac{25}{16}\sqrt{b}$.

CHAPTER 16

1. $\sqrt{1 + x} = 1 + \frac{x}{2} - \frac{x^2}{8} + \frac{x^3}{16} - \frac{5x^4}{128} + \cdots$

9. $\frac{\dot{y}}{x} = \frac{a^2 - 2x^2}{2y\sqrt{a^2 - x^2}}$

11. The curvature is $\frac{4}{(16y^2 + x^2)^{3/2}}$.

13. $\int y\,dx = \frac{a}{n}\int \frac{du}{e + fu} = \frac{a}{fn}\ln(e + fu)$

17. The ratio of the fluxions is as 1 to $-\frac{nx^{n-1}}{x^{2n}}$, or as 1 to $-\frac{n}{x^{n+1}}$.

21. The element in row n, column k is equal to $\frac{1}{n} \div \binom{n-1}{k-1} = \frac{(k-1)!(n-k)!}{n!}$.

29. $(x^x \log x + x^x + cx^{c-1} - x^{y-1}y)dx = (x^y \log x + 1)dy$

33. $\dot{y} = (1 + \log x)x^x\dot{x}$

CHAPTER 17

5. $y = x^2 + \frac{k}{x}$

11. A conic section can be written in polar coordinates as $r = de/(1 \pm e\cos\theta)$, where e is the eccentricity. If we set $x = r\cos\theta$, this equation can be rewritten as $r = de/(1 \pm ex/r) = rde/(r \pm ex)$. This is equivalent to $r \pm ex = de$ or to $\sqrt{x^2 + y^2} = \pm ex + de$, or to $\sqrt{x^2 + y^2} = \alpha x + \beta$. Note that since $|\alpha| = e$, we know that if $|\alpha| = 1$, the conic is a parabola; if $|\alpha| > 1$, the conic is a hyperbola; and if $|\alpha| < 1$, the conic is an ellipse. The value β determines the location of the center of the conic section.

15. Suppose $y = ue^{\alpha x}$ satisfies the differential equation $a^2\,d^2y + a\,dy\,dx + y\,dx^2 = 0$. We calculate that $dy = e^{\alpha x}\,du + \alpha ue^{\alpha x}\,dx$ and that $d^2y = e^{\alpha x}\,d^2u + \alpha e^{\alpha x}\,dx\,du + \alpha e^{\alpha x}\,du\,dx + \alpha^2 ue^{\alpha x}\,dx^2$. Therefore, $a^2\,d^2y + ady\,dx + y\,dx^2 = a^2 e^{\alpha x}\,d^2u + (2\alpha a^2 e^{\alpha x} + ae^{\alpha x})\,du\,dx + (a^2\alpha^2 ue^{\alpha x} + a\alpha ue^{\alpha x} + ue^{\alpha x})dx^2$. To eliminate the term in $du\,dx$, we must have $2\alpha a^2 + a = 0$, or $\alpha = -\frac{1}{2a}$. Then the coefficient of the dx^2 term is $e^{\alpha x}(a^2\frac{1}{4a^2} - a\frac{1}{2a} + 1)u$ or $\frac{3}{4}ue^{\alpha x}$. Dividing through by $e^{\alpha x}$, we find that u must be a solution to $a^2\,d^2u + \frac{3}{4}u\,dx^2 = 0$.

17. The three cube roots of 1 are 1, $\frac{-1+\sqrt{-3}}{2}$, and $\frac{-1-\sqrt{-3}}{2}$. Then, $\ln(1) = \pm 2n\pi i$, $\ln\left(\frac{-1+\sqrt{-3}}{2}\right) = \left(\frac{2}{3} \pm 2n\right)\pi i$, and $\ln\left(\frac{-1-\sqrt{-3}}{2}\right) = \left(-\frac{2}{3} \pm 2n\right)\pi i$. For the primitive fifth roots of 1, it is easier to note that they are given by $\cos\phi + i\sin\phi$, with $\phi = \pm\frac{2}{5}\pi$ and $\phi = \pm\frac{4}{5}\pi$. It follows that the logarithms are $\left(\pm\frac{2}{5} \pm 2n\right)\pi i$ and $\left(\pm\frac{4}{5} \pm 2n\right)\pi i$, respectively.

19. $xy = k$

21. $x^2 y^3 + 2x^3 y^2 + 4x^2 + 3y = k$

27. $r = \sqrt[3]{\frac{3\sqrt{2}V}{2\pi}}, h = \sqrt{2}r$

31. If the vertices of the rectangle are $(0, 0)$, $(a, 0)$, $(0, b)$, and (a, b), and if the vertex through which the line passes is $(0, 0)$, then the desired line also passes through $(a, -\sqrt[3]{a^2b})$ and $(-\sqrt[3]{ab^2}, b)$.

33. Approximately $1/16$

37. There is a maximum at $(1, \frac{3}{2})$, while there is neither a maximum nor a minimum at $(\frac{1}{2}, \frac{3}{4})$.

CHAPTER 18

1. $B_8 = -\frac{1}{30}$, $B_{10} = \frac{5}{66}$, $B_{12} = -\frac{691}{2730}$

5. Since there are $\binom{n}{n-r}$ ways of getting r successes in n trials, because each set of r successes corresponds to $n - r$ failures, and since the probability of any one of these ways of r successes and $n - r$ failures is $a^r b^{n-r}$, it follows that the total probability for r successes in n trials is $\binom{n}{n-r}a^r b^{n-r}$.

7. We calculate

$$\frac{\log c(s-1)}{\log(r+1)-\log r} - \frac{\log 19,000}{\log 31 - \log 30} = 300.465$$

and

$$\frac{\log c(r-1)}{\log(s+1)-\log s} = \frac{\log 29,000}{\log 21 - \log 20} = 210.597.$$

Thus, we take $m = 301$ and $n = 211$. Then

$$mt + \frac{st(m-1)}{r+1} = 301 \cdot 50 + \frac{20 \cdot 50 \cdot 300}{31} = 24,727.419$$

and

$$nt + \frac{rt(n-1)}{s+1} = 211 \cdot 50 + \frac{30 \cdot 50 \cdot 210}{21} = 25,550.$$

Since the latter number is the larger, we have $N(1000) = 25,550$.

9. $\frac{5}{16}$

11. $x = 6.6$, while the approximation is 6.3. With 7 trials, the odds are better than even, while with 6 they are less than even.

13. 28

15. col. $4 = \frac{2}{7m^7} \sum i^7$

$$= \frac{2}{7m^7}\left(\frac{1}{8}s^8 + \frac{1}{2}s^7 + \frac{7}{12}s^6 - \frac{7}{24}s^4 + \frac{1}{12}s^2\right)$$

$$= \frac{1}{7m^7}\left(\frac{1}{4}s^8 + s^7 + \frac{7}{6}s^6 - \frac{7}{12}s^4 + \frac{1}{6}s^2\right)$$

23. 21.4822 pounds

25. On a bet of one number, you have five chances of winning in a given drawing. The probability of any one number occurring is $\frac{1}{90}$; therefore, the probability of a win is $\frac{5}{90} = \frac{1}{18}$. Thus, the odds against winning are $17 : 1$. On a bet on a pair of numbers, there are $\binom{5}{2} = 10$ chances of winning on a single draw. The number of possible pairs is $\binom{90}{2} = 45 \cdot 89$. Thus, the probability of a win in this situation is $\frac{10}{45 \cdot 89} = \frac{1}{400.5}$, and the odds against winning are $399.5 : 1$. Finally, on a bet on a triple of numbers, there are $\binom{5}{3} = 10$ chances of winning on a single draw. The number of possible triples is $\binom{90}{3} = 15 \cdot 89 \cdot 88$. Thus, the probability of a win in this case is $\frac{10}{15 \cdot 89 \cdot 88} = \frac{1}{11,748}$, and the odds against winning are $11,747 : 1$.

27. The probability of matching three numbers is $\frac{t(t-1)(t-2)}{n(n-1)(n-2)}$; the probability of matching two numbers is $\frac{3t(t-1)(n-t)}{n(n-1)(n-2)}$; the probability of matching one number is $\frac{3t(n-t)(n-t-1)}{n(n-1)(n-2)}$.

29. The advantage to the player is

$$\frac{a}{6 \cdot 89 \cdot 22} + \frac{85b}{6 \cdot 89 \cdot 22} + \frac{7 \cdot 85c}{2 \cdot 89 \cdot 22}.$$

If α, β, γ are three positive numbers such that $\alpha + \beta + \gamma = 1$, then the three prizes are $a = 6 \cdot 89 \cdot 22\alpha = 11,748\alpha$, $b = 6 \cdot 89 \cdot 22\beta/85 = 138.21\beta$, and $c = 2 \cdot 89 \cdot 22\gamma/7 \cdot 85 = 6.58\gamma$.

CHAPTER 19

1. $\frac{8}{9}$ of a week

3. $\frac{a^2-4b^2}{2a}$

9. 70 crowns

11. $x^4 + a^4 = (x^2 + a\sqrt{2}x + a^2)(x^2 - a\sqrt{2}x + a^2)$

15. One example is $y = x^3 - 3x$ and $x = y^2 - 2$.

19. The three roots of the equation are $x_1 = 3$, $x_2 = \omega + 2\omega^2$, and $x_3 = \omega^2 + 2\omega$. The sixth-degree equation for y is $y^6 - 9y^3 + 8 = 0$. The six solutions of this equation are $y_1 = 1$, $y_2 = \omega$, $y_3 = \omega^2$, $y_4 = 2$, $y_5 = 2\omega$, and $y_6 = 2\omega^2$.

These six values of y can then be expressed as follows:

$$\frac{1}{3}(3 + \omega(\omega + 2\omega^2) + \omega^2(\omega^2 + 2\omega)) = \frac{1}{3}(7 + \omega + \omega^2) = 2$$

$$\frac{1}{3}(\omega + 2\omega^2 + \omega(\omega^2 + 2\omega) + \omega^2 \cdot 3) = \frac{1}{3}(7\omega^2 + \omega + 1) = 2\omega^2$$

$$\frac{1}{3}(\omega^2 + 2\omega + \omega \cdot 3 + \omega^2(\omega + 2\omega^2)) = \frac{1}{3}(7\omega + \omega^2 + 1) = 2\omega$$

$$\frac{1}{3}(3 + \omega(\omega^2 + 2\omega) + \omega^2(\omega + 2\omega^2)) = \frac{1}{3}(5 + 2\omega + 2\omega^2) = 1$$

$$\frac{1}{3}(\omega^2 + 2\omega + \omega(\omega + 2\omega^2) + \omega^2 \cdot 3) = \frac{1}{3}(5\omega^2 + 2\omega + 2) = \omega^2$$

$$\frac{1}{3}(\omega + 2\omega^2 + \omega \cdot 3 + \omega^2(\omega^2 + 2\omega)) = \frac{1}{3}(5\omega + 2\omega^2 + 2) = \omega$$

25. The residue of 1 is 1; of 5 is 5; of 5^2 is 12; and of 5^3 is 8. If we multiply each of these residues by 2, we get the coset $\{2, 10, 11, 3\}$. If we multiply each of the residues by 4, we get the coset $\{4, 7, 9, 6\}$.

29. $\frac{28}{5}, \frac{68}{5}, \frac{12}{5}, \frac{192}{5}$

CHAPTER 20

3. Since $DE = EF$, we have $AE = \frac{1}{2}DE = \frac{1}{2}EF = BE$. Therefore, $\triangle AEB$ is isosceles. Similarly, triangles DAC and CBF are isosceles. Also, since $AE = AD$, $BE = DC$, and $\angle E = \angle D$, triangles AEB and DAC are congruent. Similarly, both of these are congruent to triangle CBF. It follows that $AB = AC = BC$, so $\triangle ABC$ is equilateral. Next, $3\alpha + \gamma = \delta(CBF) + \delta(DAC) + \delta(AEB) + \delta(ABC) = 180 - \angle F - \angle BCF - \angle FBC + 180 - \angle D - \angle DCA - \angle CAD + 180 - \angle E - \angle EBA - \angle EAB + 180 - \angle BAC - \angle ABC - \angle BCA = 720 - \angle E - \angle D - \angle F - (\angle BAC + \angle CAD + \angle EAB) - (\angle ABC + \angle FBC + \angle EBA) - (\angle BCA + \angle BCF + \angle DCA) = 180 - \angle E - \angle D - \angle F = \delta(DEF) = \beta$. Finally, if the four smaller triangles were all congruent, then angles BCF, BCA, and ACD would be all equal. Because their sum is 180, each angle would be 60. The same would be true of all the other angles in the four triangles, so the defects would be 0, contrary to our assumption.

5. $\cosh x = \cos ix$, $\sinh x = -i \sin ix$

7. Subtangent is $\frac{x^2 - 2a^2}{x}\sqrt{\frac{2x^2 - a^2}{x^2 - a^2}}$.
Tangent is $(x^2 - 2a^2)\sqrt{\frac{1}{x^2} + \frac{1}{x^2 - a^2} + \frac{1}{x^2 - 2a^2}}$.

9. The normal vector to the plane $\alpha x + \beta y + \gamma z = a$ is $\mathbf{v} = (\alpha, \beta, \gamma)$. The angle θ between this plane and the xy plane is the same as the angle between the normal vectors, and the normal vector $\mathbf{w}$ to the xy plane is $(0, 0, 1)$. Thus, $\cos\theta = \frac{\mathbf{v}\cdot\mathbf{w}}{\|\mathbf{v}\|\|\mathbf{w}\|} = \frac{\gamma}{\sqrt{\alpha^2 + \beta^2 + \gamma^2}}$. The cosine of the angle that the plane makes with the xz plane is then $\frac{\beta}{\sqrt{\alpha^2 + \beta^2 + \gamma^2}}$, and the cosine of the angle that the plane makes with the yz plane is $\frac{\alpha}{\sqrt{\alpha^2 + \beta^2 + \gamma^2}}$.

11. The normal line to the surface $z = f(x, y)$ is a line in the direction of the gradient, namely, $(\frac{\partial z}{\partial x}, \frac{\partial z}{\partial y}, -1)$. The normal vector to the plane $z = \alpha y - \beta x + \gamma$ is $(\beta, -\alpha, 1)$. Thus, the plane will contain the normal line if these two vectors are perpendicular, that is, if their dot product is zero. This amounts to the condition $\beta\frac{\partial z}{\partial x} - \alpha\frac{\partial z}{\partial y} - 1 = 0$, as stated.

13. Monge's form of the equations of the normal line are $x - x' + (z - z')\frac{\partial z}{\partial x} = 0$ and $y - y' + (z - z')\frac{\partial z}{\partial y} = 0$. If we set $t = \frac{x - x'}{\partial z/\partial x}$, then we have $x = \frac{\partial z}{\partial x}t + x'$ and, by substituting in the first equation of Monge, $t + z - z' = 0$, or $z = -t + z'$. Then the second equation of Monge becomes $y - y' - t\frac{\partial z}{\partial y} = 0$, or $y = \frac{\partial z}{\partial x}t + y'$. These three equations form the modern vector equation of the normal line.

15. One possible Euler path in the first diagram is $EADCBAEC$. An Euler path in the second diagram is $CBDBADACAC$. In each of these, every crossing between the same two regions is on a different bridge.

CHAPTER 21

3. $2, 6, 7, 11$

7. $3 + 5i = (1 - 4i)(-1 + i)$

11. Let $\frac{z}{m} = s + ti$, where s and t are rational numbers. Choose integers q_1 and q_2 as close as possible to s and t, respectively. Let $q = q_1 + q_2 i$ and $r = z - mq$. We need to show that $N(r) < N(m)$. If $r = 0$, we are done. Otherwise, we note that $|s - q_1| \leq \frac{1}{2}$ and $|t - q_2| \leq \frac{1}{2}$. Thus, $N(\frac{z}{m} - q) = N((s + ti) - (q_1 + q_2 i)) = N((s - q_1) + (t - q_2)i) \leq (\frac{1}{2})^2 + (\frac{1}{2})^2 = \frac{1}{2}$. Therefore, $N(r) = N(z - mq) = N(m(\frac{z}{m} - q)) = N(m)N(\frac{z}{m} - q) \leq \frac{1}{2}N(m) < N(m)$, as desired.

15. If the group is $\{1, \alpha, \alpha^2, \ldots, \alpha^{17}\}$, then the cyclic subgroup of order 6 is $\{1, \alpha^3, \alpha^6, \alpha^9, \alpha^{12}, \alpha^{15}\}$ and the other cosets are $\{\alpha, \alpha^4, \alpha^7, \alpha^{10}, \alpha^{13}, \alpha^{16}\}$ and $\{\alpha^2, \alpha^5, \alpha^8, \alpha^{11}, \alpha^{14}, \alpha^{17}\}$.

19. From the equation $x^2 - \beta_1 x + 1 = 0$, where $\beta_1 = r + r^{18}$, we have

$$x = \frac{1}{2}\left(\beta_1 \pm \sqrt{\beta_1^2 - 4}\right)$$

$$= \frac{1}{2}\left(r + r^{18} \pm \sqrt{r^2 - 2 + r^{17}}\right)$$

$$= \frac{1}{2}\left(r + r^{18} \pm \sqrt{(r - r^{18})^2}\right)$$

$$= \frac{1}{2}\left((r + r^{18}) \pm (r - r^{18})\right).$$

Thus, the two roots of this equation are r and r^{18}.

23. One example is $x^5 - 3x + 2$. It can be checked that this polynomial has three real roots and two complex roots, and therefore does not meet Galois' criterion for a solvable fifth-degree polynomial.

27. $\alpha\beta = 12 - 9i + 18j + 24k$; $\frac{\alpha}{\beta} = \frac{23}{15}i + \frac{2}{3}j - \frac{4}{3}k$

29. The general form of the expansion of $f(x, y, z)$ is

$$f(1, 1, 1)xyz + f(1, 1, 0)xy\bar{z} + f(1, 0, 1)x\bar{y}z$$
$$+ f(1, 0, 0)x\bar{y}\bar{z} + f(0, 1, 1)\bar{x}yz$$
$$+ f(0, 1, 0)\bar{x}y\bar{z} + f(0, 0, 1)\bar{x}\bar{y}z$$
$$+ f(0, 0, 0)\bar{x}\bar{y}\bar{z}.$$

Then, if we evaluate $V = x - yz$ on every possible triple consisting of 0s and 1s, we find that $x - yz = 0xyz + xy\bar{z} + x\bar{y}z + x\bar{y}\bar{z} - \bar{x}yz + 0\bar{x}y\bar{z} + 0\bar{x}\bar{y}z + 0\bar{x}\bar{y}\bar{z} = xy\bar{z} + x\bar{y}z + x\bar{y}\bar{z} - \bar{x}yz.$

33. The characteristic equation of A may be written as $\lambda^2 - (a + d)\lambda + ad - bc = 0$. Thus, we need to show that $A^2 - (a + d)A + (ad - bc)I = 0$. We calculate:

$$\begin{pmatrix} a & b \\ c & d \end{pmatrix}^2 - (a + d)\begin{pmatrix} a & b \\ c & d \end{pmatrix} + \begin{pmatrix} ad - bc & 0 \\ 0 & ad - bc \end{pmatrix}$$

$$= \begin{pmatrix} a^2 + bc & ab + bd \\ ac + cd & bc + d^2 \end{pmatrix} - \begin{pmatrix} a^2 + ad & ab + bd \\ ac + cd & ad + d^2 \end{pmatrix}$$

$$+ \begin{pmatrix} ad - bc & 0 \\ 0 & ad - bc \end{pmatrix} = \begin{pmatrix} 0 & 0 \\ 0 & 0 \end{pmatrix}.$$

37. $x = \frac{2}{\sqrt{5}}u + \frac{1}{\sqrt{5}}v$; $y = -\frac{1}{\sqrt{5}}u + \frac{2}{\sqrt{5}}v$

39. Rank = 2; one basis for the set of solutions is $(-10, 18, 1, 0, 0)$, $(0, -1, 0, 1, 0)$, $(-15, 27, 0, 0, 1)$.

45. $x^3 + x + 1$ is irreducible modulo 5. Therefore, $\{a_0 + a_1\alpha + a_2\alpha^2\}$ is a field of order 5^3, where α satisfies $\alpha^3 = -\alpha - 1$ and $0 \le a_j < 5$, for $j = 0, 1, 2$.

CHAPTER 22

1. Given that $\lim_{x \to \infty} f(x + 1) - f(x) = \infty$, we know that given any positive number M, there exists a positive number N such that if $x \ge N$, then $f(x + 1) - f(x) > M$. Therefore, for $i = 1, 2, \ldots, n$, we have $f(N + i) - f(N + i - 1) > M$. The arithmetic mean of these n expressions also satisfies the same inequality. Thus,

$$\frac{f(N + n) - f(N)}{n} > M$$

or

$$\frac{f(N + n) - f(N)}{n} = M + \alpha,$$

where $\alpha > 0$. Let $x = N + n$. The equation then becomes

$$\frac{f(x) - f(N)}{x - N} = M + \alpha$$

or

$$f(x) = f(N) + (x - N)(M + \alpha).$$

Therefore,

$$\frac{f(x)}{x} = \frac{f(N)}{x} + \left(1 - \frac{N}{x}\right)(M + \alpha).$$

Since N is fixed, we see that as $x \to \infty$, $\frac{f(x)}{x}$ approaches $M + \alpha$. Thus, eventually $\frac{f(x)}{x} > M$ and therefore

$$\lim_{x \to \infty} \frac{f(x)}{x} = \infty.$$

7. Given $\epsilon > 0$, we can find n such that $\frac{1}{2^{n-2}} < \epsilon$. But $\frac{1}{n!} < \frac{1}{2^{n-1}}$. So

$$\frac{1}{n!} + \frac{1}{(n + 1)!} + \frac{1}{(n + 2)!} + \cdots$$

$$< \frac{1}{2^{n-1}} + \frac{1}{2^n} + \frac{1}{2^{n+1}} + \cdots$$

$$< \frac{1}{2^{n-2}} < \epsilon.$$

Therefore, any finite sum of the reciprocals of the factorials where $n! > 2^{n-1}$ is also less than ϵ, and by the Cauchy criterion, the series converges.

11. If $y = a^x$, then

$$y' = \lim_{i \to 0} \frac{a^{x+i} - a^x}{i} = \lim_{i \to 0} \frac{a^x(a^i - 1)}{i} = a^x \lim_{i \to 0} \frac{a^i - 1}{i}.$$

To calculate the last limit, we set $a^i = 1 + \beta$, or $i = \log_a(1 + \beta)$, where β approaches 0 with i. We need to determine now $\lim_{\beta \to 0} \frac{\beta}{\log_a(1+\beta)}$, but it is easier to work with the reciprocal. We have $\frac{\log_a(1+\beta)}{\beta} = \log_a(1 + \beta)^{1/\beta}$. As β approaches 0, the argument of this logarithm function approaches e. Therefore,

$$\lim_{i \to 0} \frac{a^i - 1}{i} = \frac{1}{\log_a e} \quad \text{and} \quad y' = \frac{a^x}{\log_a e}.$$

17. Since $(1 + 1)^3 > 3$, while $1^3 < 3$, we begin Bolzano's process by trying $1 + \frac{1}{2}$. We check that $(1 + \frac{1}{2})^3 > 3$, so we next try $1 + \frac{1}{4}$. In this case, $(1 + \frac{1}{4})^3 < 3$, so the first two terms of the desired sequence are 1, $1 + \frac{1}{4}$. We next note that $(1 + \frac{1}{4} + \frac{1}{8})^3 < 3$, and also $(1 + \frac{1}{4} + \frac{1}{8} + \frac{1}{16})^3 < 3$. So $1 + \frac{1}{4} + \frac{1}{8}$ and $1 + \frac{1}{4} + \frac{1}{8} + \frac{1}{16}$ are the next two in the sequence. After that, we try adding, in turn, $\frac{1}{32}$, $\frac{1}{64}$, and $\frac{1}{128}$, each of which gives a value too large. But $(1 + \frac{1}{4} + \frac{1}{8} + \frac{1}{16} + \frac{1}{256})^3 < 3$, so $1 + \frac{1}{4} + \frac{1}{8} + \frac{1}{16} + \frac{1}{256}$ is the next number in the sequence. Again, if we add $\frac{1}{512}$ or $\frac{1}{1024}$, the value is too large. But $(1 + \frac{1}{4} + \frac{1}{8} + \frac{1}{16} + \frac{1}{256} + \frac{1}{2048})^3 < 3$. In decimals, $1 + \frac{1}{4} + \frac{1}{8} + \frac{1}{16} + \frac{1}{256} + \frac{1}{2048}$ is equal to 1.4418945, which to three decimal places is equal to $\sqrt[3]{3}$.

25. Let A_1 be the subset of the rational numbers consisting of all rational cuts in $\mathcal{A}_1$ and A_2 the subset of the rational numbers consisting of all rational cuts in $\mathcal{A}_2$. Then the set of rational numbers has been split into two subsets A_1 and A_2 such that every element of A_1 is less than every element of A_2. Let α be the cut (or real number) determined by A_1 and A_2. Then α belongs to $\mathcal{A}_1$ or $\mathcal{A}_2$. Now suppose $\beta < \alpha$. Then there is a rational number c between β and α. By the definition of A_1, we know that $c \in A_1$. Therefore, $c \in \mathcal{A}_1$. It follows that $\beta \in \mathcal{A}_1$. Similarly, if $\beta > \alpha$, then $\beta \in \mathcal{A}_2$. Therefore, if $\alpha \in \mathcal{A}_1$, then α is the greatest number in $\mathcal{A}_1$. Similarly, if $\alpha \in \mathcal{A}_2$, then it is the smallest element in that set.

31. Example: $P = \{\frac{1}{m} - \frac{1}{m(n+1)}\}$, where m, n are positive integers, $P' = \{1, \frac{1}{2}, \frac{1}{3}, \frac{1}{4}, \ldots\}$, $P'' = \{0\}$

35. We have

$$\frac{dw}{dz} = \frac{du + i\,dv}{dx + i\,dy} = \frac{\frac{\partial u}{\partial x}\,dx + \frac{\partial u}{\partial y}\,dy + i\left(\frac{\partial v}{\partial x}\,dx + \frac{\partial v}{\partial y}\,dy\right)}{dx + i\,dy}$$

$$= \frac{\left(\frac{\partial u}{\partial x} + i\frac{\partial v}{\partial x}\right)dx + \left(\frac{\partial v}{\partial y} - i\frac{\partial u}{\partial y}\right)i\,dy}{dx + i\,dy}$$

$$= \frac{\left(\frac{\partial u}{\partial x} + i\frac{\partial v}{\partial x}\right)dx + \left(\frac{\partial u}{\partial x} + i\frac{\partial v}{\partial x}\right)i\,dy}{dx + i\,dy}$$

$$= \frac{\left(\frac{\partial u}{\partial x} + i\frac{\partial v}{\partial x}\right)(dx + i\,dy)}{dx + i\,dy} = \frac{\partial u}{\partial x} + i\frac{\partial v}{\partial x}.$$

CHAPTER 23

1. $a = 1.53$, $b = -0.87$

3. Using a table of the normal curve, we find that a percentile rank of 75 corresponds to a z score of 0.675. Therefore, one probable error from the mean, which corresponds to a percentile rank of 75, is at a distance approximately 0.675σ from the mean, where σ is the standard deviation. By the result of Exercise 2, a distance c of one modulus from the mean corresponds to $\sqrt{2}\sigma$.

9. Approximately 24 (for seven petals) and 133 (for five petals)

CHAPTER 24

1.
$$\frac{1}{2}\begin{vmatrix} dx & dy \\ \delta x & \delta y \end{vmatrix} = \frac{1}{2}(dx\,\delta y - dy\,\delta x)$$

3. $k = \frac{4}{(1 + 4x^2 + 4y^2)^2}$

5. If $x = \cos u \cos v$, $y = \cos u \sin v$, $z = \sin u$, then $x_u = -\sin u \cos v$, $x_v = -\cos u \sin v$, $y_u = -\sin u \sin v$, $y_v = \cos u \cos v$, $z_u = \cos u$, and $z_v = 0$. Therefore, we calculate $E = x_u^2 + y_u^2 + z_u^2 = \sin^2 u \cos^2 v + \sin^2 u \sin^2 v + \cos^2 u = \sin^2 u + \cos^2 u = 1$, $F = x_u x_v + y_u y_v + z_u z_v = \sin u \cos v \cos u \sin v - \sin u \sin v \cos u \cos v = 0$, and $G = x_v^2 + y_v^2 + z_v^2 = \cos^2 u \sin^2 v + \cos^2 u \cos^2 v + 0 = \cos^2 u$. So $ds^2 = E\,du^2 + 2F\,du\,dv + G\,dv^2 = du^2 + \cos^2 u\,dv^2$.

9. Since $\sin B = \frac{1}{\cosh x}$, we have $\sin B = \frac{2}{e^x + e^{-x}}$. Then $\cos B = \sqrt{1 - \sin^2 B} = \frac{e^x - e^{-x}}{e^x + e^{-x}}$. So

$$\tan\frac{B}{2} = \sqrt{\frac{1-\cos B}{1+\cos B}} = \sqrt{\frac{1-\frac{e^x-e^{-x}}{e^x+e^{-x}}}{1+\frac{e^x-e^{-x}}{e^x+e^{-x}}}}$$

$$= \sqrt{\frac{2e^{-x}}{2e^x}} = \sqrt{e^{-2x}} = e^{-x}$$

as desired. This argument works in reverse as well, so the two results are in fact equivalent.

13. From $a\sin(A+C) = b\sin A$ and the law of sines, we have $\frac{\sin B}{b} = \frac{\sin A}{a} = \frac{\sin(A+C)}{b}$. Therefore, $\sin(A+C) = \sin B$, or, interchanging A and B, $\sin(B+C) = \sin A$. From $\cos A + \cos(B+C) = 0$, we have $\cos(B+C) = -\cos A$. The sine result implies that either $A = B + C$ or $A = \pi - (B+C)$. The cosine result shows that the second equation is the correct one. Thus, $A + B + C = \pi$.

15. Consider the plane $z = a$ $(a \geq k)$ with coordinates u, v, sitting above the hemisphere $x^2 + y^2 + z^2 = k^2$. If we take an arbitrary point on the plane, say, (u, v, a), and connect it by a straight line to the origin, then we can calculate the point (x, y, z), where the line intersects the hemisphere. This line has direction vector (ut, vt, at). Its intersection point (x, y, z) with the hemisphere is found by setting the length of that vector equal to k^2. We get $u^2t^2 + v^2t^2 + a^2t^2 = k^2$, so $t^2 = \frac{k^2}{u^2+v^2+a^2}$, and $t = \frac{k}{\sqrt{a^2+u^2+v^2}}$. Thus, we have

$$x = ut = \frac{uk}{\sqrt{a^2+u^2+v^2}}, \qquad y = vt = \frac{vk}{\sqrt{a^2+u^2+v^2}}$$

and

$$z = at = \frac{ak}{\sqrt{a^2+u^2+v^2}}.$$

19. We have $(AB, CD) = \frac{AC}{CB} : \frac{AD}{DB} = \frac{AC \cdot DB}{CB \cdot AD}$. Also, $1 - (AC, BD) = 1 - (\frac{AB}{BC} : \frac{AD}{DC}) = 1 - \frac{AB \cdot DC}{BC \cdot AD} = \frac{BC \cdot AD - AB \cdot DC}{BC \cdot AD}$. But $BC = BD + DC$ and $AB = AD + DB$. Therefore,

$$1 - (AC, BD) = \frac{(BD + DC) \cdot AD - (AD + DB) \cdot DC}{BC \cdot AD}$$

$$= \frac{BD \cdot AD - DB \cdot DC}{BC \cdot AD}$$

$$= \frac{-DB \cdot (AD + DC)}{BC \cdot AD}$$

$$= \frac{-DB \cdot AC}{BC \cdot AD}$$

$$= \frac{AC \cdot DB}{CB \cdot AD} = (AB, CD).$$

Also,

$$\frac{1}{(AB, DC)} = \frac{1}{\frac{AD}{DB} : \frac{AC}{CB}} = \frac{DB \cdot AC}{AD \cdot CB} = (AB, CD).$$

23. $(1, 2, 0)$

25. $-x + 2y + 1 = 0$

29. $WRSTVJHGBCDFKLMNPOZXW$

35. If $\omega = A\,dy\,dz + B\,dz\,dx + C\,dx\,dy$, then

$$d\omega = dA\,dy\,dz + dB\,dz\,dx + dC\,dx\,dy$$

$$= \frac{\partial A}{\partial x}dx\,dy\,dz + \frac{\partial B}{\partial y}dy\,dz\,dx + \frac{\partial C}{\partial z}dz\,dx\,dy$$

$$= \left(\frac{\partial A}{\partial x} + \frac{\partial B}{\partial y} + \frac{\partial C}{\partial z}\right)dx\,dy\,dz.$$

CHAPTER 25

5. Suppose a connected set A can be expressed as $A = B \cup C$, where B and C are closed and $B \cap C$ is empty. Let $b \in B$. Then there is a region U_b containing b with $U_b \subset B$, because b cannot be a limit point of C. Similarly, around every point $c \in C$, there is a region V_c containing c with $V_c \subset C$. The union of all the U_b and V_c do not, however, generate a single region since the intersection of any U_b with any V_c is empty. This contradicts the connectedness of A.

13. The quotient $\frac{|x_p - y_p|}{1 + |x_p - y_p|}$ is always less than 1. Therefore,

$$(x, y) = \sum_{p=1}^{\infty} \frac{1}{p!}\frac{|x_p - y_p|}{1 + |x_p - y_p|} < \sum_{p=1}^{\infty} \frac{1}{p!} < e.$$

Thus, $(x, y) < e$ for all x, y in the metric space.

19. Boundary: $V_1V_2V_3 - V_0V_2V_3 + V_0V_1V_3 - V_0V_1V_2$; boundary of the boundary: $V_2V_3 - V_1V_3 + V_1V_2 - V_2V_3 + V_0V_3 - V_0V_2 + V_1V_3 - V_0V_3 + V_0V_1 - V_1V_2 + V_0V_2 - V_0V_1 = 0$.

23. If $a = r\sqrt{2}$, then $x = \frac{1}{2r}\sqrt{2}$.

25. We do the long division as follows:

$$
\begin{array}{r}
2.42204220 \\
4.21\,\overline{)\,3.12000000} \\
3.03 \\
\hline
.14444444 \\
.1111 \\
\hline
.03334444 \\
.0303 \\
\hline
.00304444 \\
.00303 \\
\hline
.00001444
\end{array}
$$

Since the .144444 from the first remainder reappears at this point, the quotient repeats in the 4-digit pattern: 4220.

27. Let $\alpha = a_0 + a_1p + a_2p^2 + \cdots$ be a unit. Thus, $a_0 \not\equiv 0$ (mod p). Further, suppose $\alpha\beta = 1$, where $\beta = b_m p^m + b_{m+1}p^{m+1} + \cdots$, with $b_m \not\equiv 0$ (mod p). Then $1 = (a_0 + a_1p + a_2p^2 + \cdots)(b_m p^m + b_{m+1}p^{m+1} + \cdots) = a_0 b_m p^m + (a_0 b_{m+1} + a_1 b_m)p^{m+1} + \cdots$. It follows that $m = 0$ and, since $a_0 b_0 \not\equiv 0$ (mod p), we have $b_0 \not\equiv 0$ (mod p). Thus, $\beta = b_0 + b_1p + b_2p^2 + \cdots$ and β is a unit.

31. One other example is the complex numbers: if the basis is taken to be i, j, then the four multiplications are $i \cdot i = i, i \cdot j = j \cdot i = j$, and $j \cdot j = -i$.

33. One example is $\begin{pmatrix} 1 & 1 \\ 0 & 0 \end{pmatrix}$.

35. Since $\binom{12}{6} = 924$, if the woman has no discriminating ability, the probability of picking 6 cups correctly is 1/924 while that of picking 5 cups correctly is 36/924. The total probability then is $37/924 = 0.04$.

39. The pyramidal numbers are numbers of the form $\binom{n}{3}$, $(n \geq 3)$, and can therefore be expressed in the form of a cubic polynomial: $\frac{1}{6}n^3 - \frac{1}{2}n^2 + \frac{1}{3}n$. These numbers are the numbers 1, 4, 10, 20, 35, ... Their first differences are the triangular numbers 3, 6, 10, 15, ... Their second differences are the integers 3, 4, 5, . . . , and their third differences are all constantly 1. Thus, to calculate the pyramidal numbers, one starts with the third differences, notes that the initial second difference is 3, the initial first difference is 3, and the initial pyramidal number is 1. We can then use the Difference Engine to calculate by finding in turn the integers, the triangular numbers, and the pyramidal numbers by repeated addition.

45. If we add x to each side of the identity from Exercise 44, we have $x + f(x, y) = x + [f(0, y) + x][f(1, y) + x'] = [x + f(0, y) + x][x + f(1, y) + x'] = [x + f(0, y)][1 + f(1, y)] = [x + f(0, y)][1] = x + f(0, y)$.

49. $2 \times (-2, 3) = (8, -23)$

51. 504

53. $AB = (1, 2, 3, 8, 4, 11, 9, 10, 6, 5, 7)$,
$BA = (1, 2, 7, 3, 10, 8, 9, 5, 4, 6, 11)$,
$AC = (1, 12, 2)(3, 7, 11)(4, 9, 6)(5, 10, 8)$,
$CA = (1, 11, 12)(2, 6, 10)(3, 8, 5)(4, 9, 7)$

Index and Pronunciation Guide

To help the student pronounce the names of the various mathematicians discussed in the book, the phonetic pronunciation of many of them is included in parentheses after the name. Naturally, since many foreign languages have sounds that are not found in English, this pronunciation guide is only approximate. To get the exact pronunciation, the best idea is to consult a native speaker of the appropriate language.

a	act, bat	j	just, fudge	ou	out, cow
ā	cape, way	k	keep, token	œ	as in German schön or in French feu
â	dare, Mary	KH	as in Scottish loch or in German ich	ʀ	rolled r as in French rouge or in German rot
ä	alms, calm	N	as in French bon or un	sh	shoe, fish
ch	child, beach	o	ox, wasp	th	thin, path
e	set, merry	ō	over, no	u	up, love
ē	equal, bee	ŏŏ	book, poor	û	urge, burn
ə	like a in alone or e in system	ōō	ooze, fool	y	yes, onion
g	give, beg	ô	ought, raw	z	zeal, lazy
i	if, big	oi	oil, joy	zh	treasure, mirage
ī	ice, bite				